Methods in

Biomedical Magnetic

Resonance

Imaging *and* Spectroscopy

ENCYCLOPEDIA OF NUCLEAR MAGNETIC RESONANCE

Methods in

Biomedical Magnetic

Resonance

Imaging *and* Spectroscopy

Edited by

Ian R. Young

Hammersmith Hospital, UK

VOLUME TWO

JOHN WILEY & SONS, LTD

Chichester • New York • Weinheim • Brisbane • Singapore • Toronto

Other Wiley Editorial Offices

John Wiley & Sons Inc., 605 Third Avenue,
New York, NY 10158-0012, USA

WILEY-VCH Verlag GmbH, Pappelallee 3,
D-69469 Weinheim, Germany

Jacaranda Wiley Ltd, 33 Park Road, Milton,
Queensland 4064, Australia

John Wiley & Sons (Asia) Pte Ltd, 2 Clementi Loop #02-01,
Jin Xing Distripark, Singapore 129809

John Wiley & Sons (Canada) Ltd, 22 Worcester Road,
Rexdale, Ontario M9W 1L1, Canada

Library of Congress Cataloging-in-Publication Data

Methods in biomedical magnetic resonance imaging and spectroscopy / edited by Ian R. Young.
    p. cm. — (Encyclopedia of nuclear magnetic resonance)
  Includes bibliographical references and index.
  ISBN 0-471-98804-9 (alk. paper)
  1. Magnetic resonance imaging. 2. Nuclear magnetic resonance spectroscopy. I.
Young, Ian R., Ph. D. II. Series.

RC78.7.N83 M48 2000
616.07'548—dc21   00-042265

British Library Cataloguing in Publication Data

A catalogue record for this book is available from the British Library.

ISBN: 0 471 98804 9

Typeset in 9/11 pt Times by Polestar Digital Data Ltd, Glasgow.
Printed and bound in Great Britain by Polestar Wheatons Ltd, Exeter.
This book is printed on acid-free paper responsibly manufactured from sustainable forestry,
in which at least two trees are planted for each one used for paper production.

Contents

VOLUME ONE

VOLUME TWO

Introduction

The *Encyclopedia of Nuclear Magnetic Resonance* covered a huge topic area, with many highly specific and esoteric aspects of great interest to those working in them but with little relevance to those operating in other fields. As a result, some of those with a more limited potential interest in the topic have been deterred from purchasing and using the eight volumes of which it is comprised. In an attempt to redress this problem, the editors felt it would be appropriate to republish sections of the Encyclopedia as independent volumes, with the intention that they would appeal very directly to the specific groups of scientists and clinicians at whom they were directed.

Because their priorities lie in other directions, with the delivery of patient care as their prime target, it was noted that one of the groups with a distinct focus for which much of the content of the Encyclopedia, while of intellectual interest, held little practical relevance, was the biomedical community. These volumes, therefore, are an abstraction from the Encyclopedia of the articles in it concerned with magnetic resonance imaging and spectroscopy. It is concerned with imaging in all its forms, and with spectroscopy in as far as it relates to *in vivo* studies, and clinical applications involving *in vitro* investigations of tissue.

The new work is substantially comprised of articles published in the Encyclopedia, some as they were originally presented, where there is little new information, and some more or less substantially revised in the light of what has been happening since the articles were first written. A number of new articles have been added where topics have either developed from fragmentary discussions in the early 1990s, or have been created *ab initio* since the first publication. The book is unique in its coverage, with a balance between imaging physics, spectroscopy and clinical studies. In many ways, it reflects the scope covered by the major international *in vivo* NMR Societies, with a conscious effort to allow the reader to understand all the elements that make up modern clinical magnetic resonance. Equally, the topic is so huge, and still evolving so fast in detail, rather than concept, that it can act as no more than an Introduction, though at quite a demanding level. Someone who peruses this book will appreciate the extent, nature and dynamics of human and animal magnetic resonance, and will have the route map to allow them to find any further information they may need. After reading this work, it should be possible for someone to design and build and operate a whole body magnetic resonance system, or even an NMR microscope. The result is unlikely to be as good as the best operators can achieve, but, then, they have had many years of practice!

This work, unlike its predecessor Encyclopedia which listed them alphabetically, has organised the various articles which comprise it into topic-based sections.

The work begins with an overview section, which includes discussion of the role of MR in practice, and covers general topics such as signal-to-noise ratio, and safety. This is followed by sections covering machine hardware, and the means by which image data is generated. Thereafter, there is a group of sections describing the methodology and techniques of *in vivo* MR, including aspects of very high resolution imaging (microscopy) and including articles on a variety of imaging topics, on flow (at all levels of molecular motion), and about observation of brain function. This part of the work concludes with sections about the very important subject (because it is responsible for so much of the contrast between tissues which is such a feature of MRI) of relaxometry, the various contrast agents now being used in MRI, and the techniques of MRS.

Next there is a section on the animal models used, principally in MRS, before the work concludes with a series of sections covering the clinical uses of MRI and MRS. These sections are grouped by body segment (head and spine; thorax, abdomen and pelvis, and the musclo-skeletal system) and mix articles on the two topics by tissue type rather than differentiating them by methodology. This is surely right, as MRS is essentially a companion tool for MRI, and is used to gain additional functional information from a region the morphology of which is already known.

Because of the way in which the work was put together, and the time period over which it was assembled, there is inevitably some unevenness of coverage in it. Its clinical coverage is less extensive than would be found in one of the Radiological textbooks; but it does, uniquely, provide substantial coverage of all aspects of both MRI and MRS as they are practised at this time.

I am very conscious of the support of the many people who have made it feasible for me to edit this work, and I would like to thank them very gratefully for it. Without contributors the work would be a thin one indeed, and I am most grateful to the many busy people who have supplied the articles in it. I am grateful to the Editors-in-Chief of the *Encyclopedia of Nuclear Magnetic Resonance* for their encouragement in attempting this further task, and, for her huge assistance, to my editor at Wiley, Ms. Jenny Cossham. The work would not have happened without her. I must also mention my secretary, Mrs. Mary Crisp, who has had much to put up with in its preparation, and who has been a tireless and most effective supporter, and Geoff Reynolds, also at Wiley, who has supervised the book's production.

Finally, I have to thank my wife, who has had to live with the gestation of the book with all its attendant traumas. She has had more to cope with than is right for anyone!

Ian R. Young

Contributors

Joseph J. H. Ackerman *Washington University, St Louis, MO, USA*

David C. Ailion *University of Utah, Salt Lake City, UT, USA*

Eben Alexander III *Brigham and Women's Hospital, Boston, MA, USA*

E. Raymond Andrew *University of Florida, Gainesville, FL, USA*

Yoshimi Anzai *University of Michigan, MI, USA*

Mitsuaki Arakawa *UCSF-RIL, San Francisco, CA, USA*

Douglas L. Arnold *Montreal Neurological Institute, McGill University, Montreal, Quebec, Canada*

Leon Axel *University of Pennsylvania, Philadelphia, PA, USA*

Peter Bachert *Forschungsschwerpunkt Radiologische Diagnostik und Therapie, Deutsches Krebsforschungszentrum (DKFZ), D-69120 Heidelberg, Germany*

Goran Bačíc *Dartmouth Medical School, Hanover, NH, USA*

Isabella Baeli *University of Rome, 'La Sapienza', Italy*

Robert S. Balaban *National Heart, Lung and Blood Institute, Bethesda, MD, USA*

James A. Balschi *Harvard Medical School, Boston, MA, USA*

Peter A. Bandettini *Medical College of Wisconsin, Milwaukee, WI, USA*

Maria L. Barnard *Royal Postgraduate Medical School, Hammersmith Hospital, London, UK*

Christine J. Baudouin *Freeman Hospital, Newcastle upon Tyne, UK*

Ronald de Beer *Delft University of Technology, Delft, The Netherlands*

Jimmy D. Bell *Imperial College School of Medicine, Hammersmith Hospital, London, UK*

Monique Bernard *CNRS, Marseille, France*

Tedros Bezabeh *National Research Council of Canada, Winnipeg, Canada*

Denis Le Bihan *Service Hospitalier Fréderic Joliot, CEA, Orsay, France*

Jeffrey R. Binder *Medical College of Wisconsin, Milwaukee, WI, USA*

Keith L. Black *University of California at Los Angeles, CA, USA*

Peter M. Black *Brigham and Women's Hospital, Boston, MA, USA*

Johan L. Bloem *Leiden University Hospital, The Netherlands*

Stefan Bluml *Huntington Medical Research Institutes, Pasadena, CA, USA*

Johannes C. Böck *Freie Universität Berlin, Germany*

Chris Boesch *University of Bern, Switzerland*

Coleen S. Bosch *Washington University, St Louis, MO, USA*

Paul A. Bottomley *Johns Hopkins University, Baltimore, MD, USA*

Wim M. M. J. Bovée *Delft University of Technology, Delft, The Netherlands*

Jerrold L. Boxerman *Massachusetts General Hospital, Charlestown, MA, USA*

William G. Bradley Jr. *Long Beach Memorial Medical Center, CA, USA*

Thomas J. Brady *Massachusetts General Hospital and Harvard Medical School, Charlestown, MA, USA*

Michael N. Brant-Zawadzki *Hoag Memorial Hospital Presbyterian, Newport Beach, CA, USA*

Rodney D. Brown III *Field Cycling Systems, Inc., River Edge, NY, USA*

Truman R. Brown *Fox Chase Cancer Center, Philadelphia, PA, USA*

David J. Bryant *GEC Hirst Research Centre, Borehamwood, Herts, UK*

Robert G. Bryant *University of Virginia, Charlottesville, VA, USA*

Thomas F. Budinger *University of California, Berkeley, CA, USA*

Michael Burl *The Robert Steiner MRI Unit, Hammersmith Hospital, London, UK*

Graeme M. Bydder *University of London, London, UK*

Ernest B. Cady *University College London Hospitals, UK*

Paul T. Callaghan *Massey University, Palmerston North, New Zealand*

Joseph Carlson *UCSF-RIL, San Francisco, CA, USA*

Carlo Catalano *University of Rome, 'La Sapienza', Italy*

Wei Chen *Center for Magnetic Resonance Research, University of Minnesota, MN, USA*

Graham R. Cherryman *University of Leicester, Leicester, UK*

M. H. Cho *Korea Advanced Institute of Science, Cheongyangni, Seoul, Korea*

Z. H. Cho *University of California, Irvine, CA, USA*

Kieran Clarke *University of Oxford, UK*

Fergus V. Coakley *University of California, San Francisco, CA, USA*

Sylviane Confort-Gouny *Centre de Résonance Magnétique Biologique et Médicale, Faculté de Médicine, Marseille, France*

Alan Connelly *Institute of Child Health and Great Ormond Street Hospital for Children, NHS Trust, London, UK*

Glyn A. Coutts *Marconi Medical Systems, UK*

Isobel Jane Cox *Imperial College School of Medicine, London, UK*

Patrick J. Cozzone *Centre de Résonance Magnétique Biologique et Médicale, Faculté de Médicine, Marseille, France*

Alex de Crespigny *Stanford University, CA, USA*

Lawrence E. Crooks *Toshiba America MRI, South San Francisco, CA, USA*

John V. Crues III *RadNet Management, Inc., Los Angeles, CA, USA*

Ashley Davidoff *University of Massachusetts, MA, USA*

Frank Davies *Oxford Magnet Technology Ltd., Witney, Oxfordshire, UK*

Nandita M. deSouza *Imperial College School of Medicine, London, UK*

Gordon DeMeester *Picker International, Highland Heights, OH, USA*

Antonio A. F. De Salles *University of California at Los Angeles, CA, USA*

Edgard E. DeYoe *Medical College of Wisconsin, Milwaukee, WI, USA*

Rosalind B. Dietrich *University of California, Irvine, Orange, CA, USA*

Ruth M. Dixon *Department of Biochemistry, University of Oxford, Oxford, UK*

W. Thomas Dixon *Emory University, Atlanta, GA, USA*

Bastiaan Driehuys *Magnetic Imaging Technologies Inc., Durham, NC, USA*

Charles L. Dumoulin *General Electric Research and Development Center, Schenectady, NY, USA*

William A. Edelstein *General Electric Corporate Research and Development, Schenectady, NY, USA*

Richard H. T. Edwards *Magnetic Resonance Research Centre and Muscle Research Centre, The University of Liverpool, Liverpool, UK*

Richard L. Ehman *Mayo Clinic, Rochester, MN, USA*

Gösta Ehnholm *Picker Nordstar Inc., Helsinki, Finland*

Richard Farb *University of Toronto, Toronto, ON, Canada*

Neil A. Farrow *California Institute of Technology, Pasadena, CA, USA*

Peter J. Feenan *Magnex Scientific Ltd, Abingdon, Oxon, UK*

Roland Felix *Freie Universität Berlin, Germany*

David N. Firmin *National Heart and Lung Institute, University of London, UK*

Scott D. Flamm *University of California at San Francisco, San Francisco, CA, USA*

James L. Fleckenstein *University of Texas Southwestern Medical Center, Dallas, TX, USA*

Margaret A. Foster *University of Aberdeen, Aberdeen, UK*

Jens Frahm *Biomedizinische NMR Forschungs GmbH am Max-Planck-Institut für Biophysikalische Chemie, Göttingen, Germany*

Francesco Fraioli *University of Rome, 'La Sapienza', Italy*

Karl J. Friston *The Wellcome Department of Cognitive Neurology, London, UK*

Peggy Fritzsche *Loma Linda University, Loma Linda, CA, USA*

David G. Gadian *Royal College of Surgeons Unit of Biophysics, Institute of Child Health, London, UK*

Allen N. Garroway *Naval Research Laboratory, Washington, DC, USA*

Henry Gibson *Magnetic Resonance Research Centre and Muscle Research Centre, The University of Liverpool, Liverpool, UK*

David J. Gilderdale *Engineering Consultant, UK*

A. Gillams *The Middlesex Hospital, UCLH, London, UK*

Jerry D. Glickson *The Johns Hopkins University School of Medicine, Baltimore, MD, USA*

Gary H. Glover *Stanford University, Stanford, CA, USA*

John R. Griffiths *St. George's Hospital Medical School, London, UK*

Robert I. Grossman *University of Pennsylvania Medical Center, Philadelphia, PA, USA*

E. Mark Haacke *Washington University, St. Louis, MO, USA*

Axel Haase *Physikalisches Institut, Universität Würzburg, Würzburg, Germany*

Donald M. Hadley *Institute of Neurological Sciences, Glasgow, UK*

Laurie D. Hall *Cambridge University, UK*

Margaret A. Hall-Craggs *The Middlesex Hospital, UCLH, London, UK*

Morley R. Halse *University of Kent, Canterbury, Kent, UK*

Jeffrey W. Hand *Hammersmith Hospital, London, UK*

Wolfgang Hänicke *Biomedizinische NMR Forschungs GmbH am Max-Planck-Institut für Biophysikalische Chemie, Göttingen, Germany*

Liliane A. Harika *Massachusetts General Hospital and Harvard Medical School, Boston, MA, USA*

Steven E. Harms *Baylor University Medical Center, Dallas, TX, USA*

David N. F. Harris *Royal Postgraduate Medical School, London, UK*

Anton N. Hasso *University of California at Irvine, CA, USA*

Jane M. Hawnaur   *Department of Diagnostic Radiology, University of Manchester, UK*

Cecil E. Hayes   *University of Washington, Seattle, WA, USA*

Joseph A. Helpern   *Nathan Kline Institute, Orangeburg, NY, USA*

R. Mark Henkelman   *Sunnybrook Health Science Center and University of Toronto, Toronto, ON, Canada*

Jürgen K. Hennig   *Universität Frieburg, Germany*

Ole Henriksen   *Danish Research Center of Magnetic Resonance, Hvidovre University Hospital, Denmark*

Robert J. Herfkens   *Stanford University School of Medicine, CA, USA*

John R. Hesselink   *UCSD Medical Center, San Diego, CA, USA*

Hoby P. Hetherington   *Brookhaven National Laboratory, Upton, NY, USA*

Sylvia H. Heywang-Köbrunner   *University of Halle, Germany*

Charles B. Higgins   *University of California at San Francisco, San Francisco, CA, USA*

G. Neil Holland   *Otsuka Electronics, Fort Collins, CO, USA*

Jan A. den Hollander   *University of Alabama, Birmingham, AL, USA*

David I. Hoult   *Institute for Biodiagnostics, National Research Council Canada, Winnipeg, MB, Canada*

Hedvig Hricak   *Sloan-Kettering Cancer Center, New York, NY, USA*

Paul S. Hsieh   *Kaiser Permanente Medical Center, San Diego, CA, USA*

Xiaoping Hu   *Center for Magnetic Resonance Research, University of Minnesota, MN, USA*

James W. Hugg   *University of Alabama, Birmingham, AL, USA*

Gregory C. Hurst   *MetroHealth Medical Center and Case Western Reserve University, Cleveland, OH, USA*

James M. S. Hutchison   *University of Aberdeen, Aberdeen, UK*

Jong-Hee Hwang   *Huntington Medical Research Institutes, Pasadena, CA, USA*

James S. Hyde   *Medical College of Wisconsin, Milwaukee, WI, USA*

Joanne S. Ingwall   *Harvard Medical School, Boston, MA, USA*

Ian Isherwood   *Department of Diagnostic Radiology, University of Manchester, UK*

Graeme D. Jackson   *Brain Imaging Research Institute, Austin and Repatriation Medical Centre, Heidelberg, West Australia, and Howard Florey Institute, University of Melbourne, Australia*

J. Randy Jinkins   *University of Texas Health Science Center, San Antonio, TX, USA*

G. Allan Johnson   *Duke University Medical Center, Durham, NC, USA*

Ferenc A. Jolesz   *Brigham and Women's Hospital and Harvard Medical School, Boston, MA, USA*

Peter M. Joseph   *Hospital of the University of Pennsylvania, Philadelphia, PA, USA*

Spyros K. Karampekios   *University Hospital of Crete, Heraklion, Greece*

Leon Kaufman   *UCSF-RIL, San Francisco, CA, USA*

Andrew P. Kelly   *Hoag Memorial Hospital Presbyterian, Newport Beach, CA, USA*

Seong-Gi Kim   *Center for Magnetic Resonance Research, University of Minnesota, MN, USA*

Seymour H. Koenig   *Relaxometry, Inc., Mahopac, NY, USA*

David Kramer   *UCSF-RIL, San Francisco, CA, USA*

Roland Kreis   *University of Bern, Switzerland*

Walter Kucharczyk   *University of Toronto, Toronto, ON, Canada*

Andrea Laghi   *University of Rome, 'La Sapienza', Italy*

Rolf M. J. N. Lamerichs   *Philips Medical Systems, Best, The Netherlands*

David J. Larkman   *Marconi Medical Systems, UK*

Gerhard Laub   *Siemens AG. Erlangen, Germany*

Alexia J. Lawrence   *Mayo Clinic, Rochester, MN, USA*

S. C. Lee   *Korea Advanced Institute of Science, Cheongyangni, Seoul, South Korea*

Laurent P. Lemaire   *Laboratoire de Biophysique Médicale Faculté de Médecine, Angers, France*

Robert E. Lenkinski   *University of Pennsylvania, Philadelphia, PA, USA*

Jonathan S. Lewin   *University Hospitals of Cleveland and Case Western Reserve University, Cleveland, OH, USA*

Frank J. Lexa   *Leonard Davis Institute of Health Economics, University of Pennsylvania, Philadelphia, PA, USA*

Debiao Li   *Washington University, St. Louis, MO, USA*

Alexander Lin   *Huntington Medical Research Institutes, Pasadena, CA, USA*

Weili Lin   *Washington University, St. Louis, MO, USA*

Martin J. Lipton   *The University of Chicago, IL, USA*

Haiying Liu   *Picker International, Highland Heights, OH, USA*

Donald B. Longmore   *Royal Brompton Hospital, London, UK*

Dolores López-Villegas   *University of Pennsylvania, Philadelphia, PA, USA*

Francesco de Luca   *Università di Roma, 'La Sapienza', Rome, Italy*

Robert B. Lufkin   *University of California, Los Angeles, CA, USA*

Nicola Lugeri   *Università di Roma, 'La Sapienza', Rome, Italy*

David J. Lurie   *University of Aberdeen, Aberdeen, UK*

Peter R. Luyten   *Philips Medical Systems, Best, The Netherlands*

Mark F. Lythgoe   *Royal College of Surgeons Unit of Biophysics, Institute of Child Health, London, UK*

James R. MacFall *Duke University Medical Center, Durham, NC, USA*

Janet S. MacFall *Duke University Medical Center, Durham, NC, USA*

Albert Macovski *Stanford University, CA, USA*

Mahmood F. Mafee *University of Illinois at Chicago, Chicago, IL, USA*

Cathy Maldjian *Mount Sinai Medical Center, New York, NY, USA*

Peter Mansfield *University of Nottingham, UK*

Bruno Maraviglia *Università di Roma, 'La Sapienza', Rome, Italy*

Claude D. Marcus *Hôpital Robert Debré, Reims, France*

Paul M. Margosian *Picker International, Highland Heights, OH, USA*

Alexander R. Margulis *University of California, San Francisco, CA, USA*

Claudia H. Martin *Brigham and Women's Hospital, Boston, MA, USA*

Peter A. Martin *Magnetic Resonance Research Centre and Muscle Research Centre, The University of Liverpool, Liverpool, UK*

Thomas J. Masaryk *Cleveland Clinic Foundation, Cleveland, OH, USA*

Gerald B. Matson *University of California San Francisco and VA Medical Center, San Francisco, CA, USA*

Paul M. Matthews *Montreal Neurological Institute, McGill University, Montreal, Quebec, Canada*

Andrew A. Maudsley *University of California San Francisco and VA Medical Center, San Francisco, CA, USA*

Shirley McCarthy *Yale University School of Medicine, New Haven, CT, USA*

Cheryl L. McCoy *St. George's Hospital Medical School, London, UK*

Paul M. J. McSheehy *St. George's Hospital Medical School, London, UK*

David K. Menon *University of Cambridge, UK*

Charles E. Metz *The University of Chicago, IL, USA*

Craig H. Meyer *Stanford University, CA, USA*

George J. Misic *Medrad, Inc., MRI Products, Indianola, PA, USA*

Michael T. Modic *Cleveland Clinic Foundation, OH, USA*

Raad H. Mohiaddin *Royal Brompton Hospital, London, UK*

Chrit T. W. Moonen *Centre Nationale de la Recherche Scientifique, University of Bordeaux 2, Bordeaux, France*

Peter G. Morris *Nottingham University, UK*

Michael E. Moseley *Stanford University, CA, USA*

Andreas Mühler *Research Laboratories Schering AG, Berlin, Germany*

Robert N. Muller *University of Mons Hainaut, Mons, Belgium*

Joseph Murphy-Boesch *USA Instruments Inc, Aurora, OH, USA*

Raja Muthupillai *Mayo Clinic, Rochester, MN, USA*

Arya Nabavi *Brigham and Women's Hospital, Boston, MA, USA*

Timothy J. Norwood *Cambridge University, UK*

Hans Oellinger *Universitäts-Klinikum Rudolf-Virchow, Freie Universität Berlin, Germany*

Seiji Ogawa *Bell Laboratories, Lucent Technologies, Murray Hill, NJ, USA*

William Okuno *University of California, San Francisco, CA, USA*

Roger J. Ordidge *University College, London, UK*

Dirk van Ormondt *Delft University of Technology, Delft, The Netherlands*

J. Stewart Orr *Hammersmith Hospital, London, UK*

Leif Østergaard *Århus University Hospital, Århus, Denmark*

Valeria Panebianco *University of Rome, 'La Sapienza', Italy*

Lawrence P. Panych *Brigham and Women's Hospital and Harvard Medical School, Boston, MA, USA*

Roberto Passariello *University of Rome, 'La Sapienza', Italy*

Pradip M. Pattany *University of Miami, Miami, FL, USA*

Paolo Pavone *University of Rome, 'La Sapienza', Italy*

John M. Pauly *Stanford University, Stanford, CA, USA*

Jacqueline M. Pennock *Royal Postgraduate Medical School, Hammersmith Hospital, London, UK*

Richard S. Pergolizzi Jr *Brigham and Women's Hospital, Boston, MA, USA*

Franca Podo *Istituto Superiore di Sanità, Rome, Italy*

Gerald M. Pohost *University of Alabama at Birmingham, AL, USA*

Brigitte Ponceleti *Massachusetts General Hospital, MA, USA*

James W. Prichard *Yale University, New Haven, CT, USA*

Basant K. Puri *MRC Clinical Sciences Centre, Imperial College School of Medicine, London, UK*

George K. Radda *University of Oxford and The John Radcliffe Hospital, Oxford, UK*

Bernd Radüchel *Research Laboratories Schering AG, Berlin, Germany*

Larry A. Ranahan *The University of Chicago, IL, USA*

Roger A. Rauch *University of Texas Health Science Center, San Antonio, TX, USA*

David L. Rayner *Magnex Scientific Ltd, Abingdon, Oxon, UK*

Stephen J. Reiderer *Mayo Clinic, Rochester, MN, USA*

Peter Reimer *Institute for Clinical Radiology, Munster, Germany*

E. Osmund R. Reynolds *University College London Medical School, UK*

John H. Richards *California Institute of Technology, Pasadena, CA, USA*

Peter A. Rinck *University of Mons-Hainaut Medical Faculty, NMR Laboratory, Mons, Belgium*

Simon P. Robinson *St. George's Hospital Medical School, London, UK*

Peter B. Roemer *Advanced NMR, Wilmington, MA, USA*

Bruce R. Rosen *Massachusetts General Hospital, Charlestown, MA, USA*

Brian D. Ross *Huntingdon Medical Research Institutes, Pasadena, CA, USA*

Gerald M. Roth *University of California, Irvine, Orange, CA, USA*

Paul M. Ruggieri *Cleveland Clinic Foundation, Cleveland, OH, USA*

Val M. Runge *University of Kentucky, Lexington, KY, USA*

Peter J. Sadler *University of Edinburgh, UK*

Mitchell D. Schnall *University of Pennsylvania Medical Center, Philadelphia, PA, USA*

Richard B. Schwartz *Brigham and Women's Hospital, Boston, MA, USA*

Leslie M. Scoutt *Yale University School of Medicine, New Haven, CT, USA*

Wolfhard Semmler *Institut für Diagnostikforschung (IDF) an der Freien Universität, D-14050 Berlin, Germany*

Raimo E. Sepponen *Helsinki University of Technology, Helsinki, Finland*

Kay J. Seymour *Huntington Medical Research Institutes, Pasadena, CA, USA*

Bruna C. de Simone *Università di Roma, 'La Sapienza', Rome, Italy*

S. Smart *The Middlesex Hospital, UCLH, London, UK*

Colin M. Smith *University of Leicester, Leicester, UK*

Ian C. P. Smith *National Research Council of Canada, Winnipeg, Canada*

Robert C. Smith *Yale University School of Medicine, New Haven, CT, USA*

Daniel K. Sodickson *Beth Israel Deaconess Medical Center and Harvard Medical School, Boston, MA, USA*

David Stark *University of Massachusetts, MA, USA*

Andrew D. Stevens *University of Leicester, Leicester, UK*

Laura C. Stewart *National Research Council, Ottawa, Ontario, Canada*

John H. Strange *University of Kent, Canterbury, Kent, UK*

Peter Styles *MRC Biochemical and Clinical Magnetic Resonance Unit, Oxford, UK*

Harold M. Swartz *Dartmouth Medical School, Hanover, NH, USA*

Jeannie Tan *Huntington Medical Research Institutes, Pasadena, USA*

Simon D. Taylor-Robinson *Hammersmith Hospital, London, UK*

E. Louise Thomas *Imperial College School of Medicine, Hammersmith Hospital, London, UK*

Jean A. Tkach *Cleveland Clinic Foundation, Cleveland, OH, USA*

Volker Tronnier *University of Heidelburg, Germany*

Supoch Tunlayadechanont *University of Pennsylvania, Philadelphia, PA, USA*

Robert Turner *Institute of Neurology, London, UK*

Patrick A. Turski *University of Wisconsin, Madison, WI, USA*

Toshihiro Ueda *The University of Iowa College of Medicine, Iowa City, IA, USA*

Kamil Urgurbil *Center for Magnetic Resonance Research, University of Minnesota, MN, USA*

Michael J. Varanelli *Yale University School of Medicine, New Haven, CT, USA*

Jean Vion-Dury *Centre de Résonance Magnétique Biologique et Médicale, Faculté de Médicine, Marseille, France*

Van J. Wedeen *Harvard Medical School, USA*

Rory J. Warner *Magnex Scientific Ltd, Abingdon, Oxon, UK*

Felix W. Wehrli *University of Pennsylvania Medical School, Philadelphia, PA, USA*

Michael W. Weiner *University of California, San Francisco, CA, USA*

Hanns-Joachim Weinmann *Research Laboratories Schering AG, Berlin, Germany*

Robert M. Weisskoff *Massachusetts General Hospital, Charlestown, MA, USA*

Ralph Weissleder *Massachusetts General Hospital and Harvard Medical School, Boston, MA, USA*

Eric C. Wong *Medical College of Wisconsin, Milwaukee, WI, USA*

Terence Z. Wong *Duke University, Durham, NC, USA*

Michael L. Wood *University of Toronto, Toronto, Ontario, Canada*

Brian S. Worthington *University of Nottingham, Nottingham, UK*

Ian R. Young *The Robert Steiner MRI Unit, Hammersmith Hospital, London, UK*

William T.C. Yuh *The University of Iowa College of Medicine, Iowa City, IA, USA*

Xiao-Hung Zhu *Center for Magnetic Resonance Research, University of Minnesota, MN, USA*

Peter C. M. van Zijl *Johns Hopkins University Medical School, Baltimore, MD, USA*

Imaging and Medical Glossary

Pulse Sequence Notation

The sequence notation is based on the recommendations of the American College of Radiology (American College of Radiology Glossary of NMR Terms, ACR, Chicago, 1983). The notation has the form SE 1500/50, where the first two letters describe the sequence (PS = partial saturation, SE = spin echo, etc). The first number is the repetition time (TR) in milliseconds; the second is the time (TE) to echo formation; the third (if present) is the inversion time in an inversion recovery (IR) sequence. Additional time intervals can be defined by third and other numbers, depending on the sequence.

The chemical shift between aqueous and lipid components in tissue is generally taken as 3.4–3.5 ppm.

Image Orientation

Transverse (transaxial) images are normally oriented in publications as if the slice was being viewed by an observer standing beyond the patient's feet. In a few non-North American instances, with little consistency even inside any one country, body images are oriented as if viewed from the feet, while the head and neck are viewed from above the top of the head.

Sagittal images are conventionally displayed as if the patient is lying on their back, with the observer looking down on the subject. The images are oriented as if viewed from the observer's left hand side (i.e. they are looking at the right hand wall of the slice).

Coronal images are oriented as if the patient were lying on their back with the observer looking down at the subject. The right side of the patient then appears at the left of the image.

Glossary terms

AA	Ascending Aorta
Acquisitions	See NSA
ADC	Apparent Diffusion Coefficient; experimentally observed value of D, in the presence of likely confusing factors such as tissue structural anisotropy; cf. D* [not to be confused with ADC, meaning analog-to-digital converter]
Adhesion	Fibrous band or structure linking two tissues not normally joined
ADP	Adenosine DiPhosphate
AE	Arterial Enhancement
Aerobic	Metabolism (usually), etc., operating using oxygen
Agenesis	Absence of an organ (from development failure onward)
AIDS	Acquired Immune Deficiency Syndrome
AMP	Adenosine MonoPhosphate
Anaerobic	Metabolism (usually), etc., operating without oxygen
Aneurysm	Region of a blood vessel in which a blockage has resulted in a swollen, weakened, and life threatening segment of the vessel
Angioplasty	Reconstruction or restructuring of a blood vessel by operation or interventional technique such as balloon dilator or laser
Anterior	Frontal; towards the front (notation used particularly in sagittal imaging)
Anteromedial	Towards the front on the center line of the body
Anteroposterior (movement)	Back to front movement towards and away from the spinal column (back/front motion); syn. AP
AP	See AnteroPosterior
Arrhythmia	Irregularity in a fixed rhythm; usually associated with heart beat irregularities (cardiac arrythmia)
Arteriole	Minute arterial branch; distribution level before, in circulatory terms, a capillary
Ataxia	Failure of muscular control and/or coordination; description of disease in other sites leading to muscular misfixation
ATP	Adenosine TriPhosphate
Atrophy	Shrinkage, loss of bulk, of an organ (e.g. atrophy of the brain)
Avascular	Without conventional blood supply; description applied to many lesions
Averages	See NSA
AVM	ArterioVenous Malformation; lesion innovating distortion and disruption of its normal vascular system

BBB	Blood Brain Barrier; largely impermeable vessel wall membrane in the brain, allowing passage of small molecules such as O_2 and CO_2; disruption of this membrane is frequently symptomatic of disease
BFAST	See Saturation
Biopsy	Procedure in which a small piece of tissue is extracted from a suspect region for histological examination
Birdcage coil (or birdcage)	Very common multiconductor cylindrical coil used in whole body MR (see **Birdcage Resonators: Highly Homogeneous Radiofrequency Coils for Magnetic Resonance** by C. E. Hayes)
'Black blood'	Process of acquiring and reconstructing angiographic data which leaves vascular structure dark against a light background; cf. Bright or White blood
BOLD	Blood Oxygen Level Dependent contrast; mechanism used in describing changes detected in functional MRI (fMRI)
BPH	Benign Prostatic Hyperplasia; common older male nonmalignant disease; symptoms are blockage of urethra and difficulty in urinating
Breath-hold	Imaging method in which a patient is asked to hold his/her breath for the duration of data acquisition (usually a few seconds)
'Bright blood'	Process of acquiring and reconstructuring data such that the vasculature appears bright against a dark ground; syn. 'White blood'; cf. Black blood
BSA	Bovine Serum Albumen; extract used in MRI, particularly for relaxation behavior studies
Ca	Carcinoma; abbreviation used with localizing description of tumor
CABS	Coronary Artery Bypass Surgery
CAD	Coronary Artery Disease (see syn. CHD)
Calcification	Large or small ('microcalcification') solid deposits visible in MRI only through lack of signal; observed in a variety of lesions, and particularly associated with some tumors
Catabolite	Product of a destructive metabolic process
Caudal	Towards the foot; associated with direction of motion, flow; cf. Cephalad
CBF	Cerebral Blood Flow
CBV	Cerebral Blood Volume
CCT	Cranial Computed Tomography
CE	Contrast Enhanced; frequently used in the context 'CE images', meaning images acquired after a contrast agent has been given to the patient
CE-FAST	T_2 weighted rapid sequence, with analogues with FISP (see below) and direct connection with STEAM (see below)
Cephalad	Towards the head; associated with direction of motion, flow; cf. Caudal
CHD	Coronary Heart Disease (see syn. CAD)
Chemsat	See FATSAT
Cho	Choline; metabolite primarily observed in proton spectroscopy but with associated metabolites also visible elsewhere
CK	Creatine Kinase; enzyme controlling major step in Krebs' metabolic cycle
CNR	Contrast-to-Noise Ratio; in practice, the signal difference between two tissues (or other entitites) being considered divided by the RMS noise
CNS	Central Nervous System; brain, brain stem, and spinal cord
Coarctation	Restriction of structure (or contraction of component)
Collateral	Used in imaging in the sense of incidental, attendant (usually unwanted) effects of a process or procedure; thus 'collateral damage' is associated with unwanted damage to normal tissue resulting from the application of therapy
Contrast	Difference in magnitude between two signals; sometimes contrast is stated as a fractional difference, at other times as an absolute difference
COPE	See ROPE
Coronal	Slice along long axis of the body parallel to its back (or front)
Corpus callosum	White matter structure in the brain, towards the centre, which provides the link between the two hemispheres
Cortex (adj. cortical)	Outer layer of brain or kidney; in the case of the former, constitutes much of the gray matter of brain
CP angle	Cerebello Pontine angle; feature of the cerebellum (lower rear portion of the brain at the level of the brain stem) [CP not to be confused with cross polarization]
Cr	Creatine; metabolite seen, as such, in the proton spectrum; in combination, also detected in other forms of spectroscopy
Cranial	To head direction
Craniocaudal (movement)	Movement up and down the body, parallel to the spine

CRF	Chronic Renal Failure
CSF	Cerebro-Spinal Fluid; water-like fluid surrounding the brain, spinal cord, and filling ventricles in the brain
CSI (CS Imaging)	Chemical Shift Imaging; class of acquisitions in which spectral and positional data are encoded together and acquired at the same time
CT	(X-ray) Computerized Tomography (often abbreviated to CT X-ray, the original computer-based imaging method)
Cytology	Study of cells, including origin, function, structure, and pathology
Cytotoxicity	Ability to produce a specific toxic action in cells in various organs
D*	See ADC
DA	Descending Aorta [not to be confused with DA meaning data acquisition]
DAC	DiAcylglycerol; metabolite seen in spectroscopy [not to be confused with DAC meaning digital-to-analog converter]
Data sets	See NSA
Diastole	Period of cardiac cycle during which the heart expands (cf. Systole, shorter period of contraction)
Diffuse (disease)	Distributed disease affecting a significant fraction of an organ without specific localizing boundaries (cf. Focal)
Dilation	Swollen or expanded (of tissue)
Distal	Remote from, far away from, any reference point (cf. Proximal, adjacent to)
Dorsal	Pertaining to the back, or rear, of anything referred to
Double oblique	Slice, the plane of which is parallel to no major axis of the machine
DPDE	DiPhospho DiEster
DRESS	Slice-selective spectroscopic method used for phosphorus spatial localization
DTPA	DiethyleneTriaminePentaAcetic acid; chelate used to minimize toxicity of otherwise desirable agent (see Gd-DTPA)
DWI	Diffusion Weighted Imaging
Dysplasia	Abnormality of development; in pathology, alteration in size, shape, and organization of cells
EC	ExtraCellular
ECG	ElectroCardioGram (loosely used to describe both the equipment used, and its output)
Ectopic (beat)	Abnormal pattern or location of organ or structure
Edema	Additional extra and extracellular fluid associated with tissue which is damaged in some way
EDTA	EthyleneDiamineTetraAcetic acid; another common chelating agent (cf. DTPA)
Effusion	Undesirable escape of fluid into tissue or structure
End-diastole	Part of cardiac cycle at end of heart expansion
Endogenous	Developing, or growing, from within a tissue or structure
EPI	Echo Planar Imaging; single-shot fast acquisition introduced by Mansfield (J. Phys. E., 1976, 9, 271)
Epidural	Structured upon, or outside, the dura mater (the outermost layer covering the brain and spinal canal); an epidural injection, for example, is one delivered, typically, into the spinal cord, outside the cord
EPRI	Electron Paramagnetic Resonance Imaging
Erythrocytes	Components of blood, known also as red blood cells, or corpuscles
ESP	Echo SPacing in multiple acquisitions such as RARE (see below) (stated as a time interval)
Etiology	Study of factors that cause disease, and their introduction to the host
ETL	Echo Train Length; number of echoes in RARE-type sequences (see RARE)
EVI	Echo Volumnar Imaging; three-dimensional version of echo planar imaging
Excitations	See NSA
Exogenous	Produced or otherwise originating outside the organism or region
Extracapsular	Situated or occurring outside a capsule (such as that, for example, surrounding a cyst)
Extradural	Situated or occurring outside the dura mater (the outer lining of the brain and spinal cord)
Extramural	Situated or occurring outside the wall of an organ
Extravasate	Escape from the vascular system
FAISE	See RARE
FATSAT	Sequence designed for the spectrally selective saturation of signals in MRI (normally used to saturate lipid components, as in breast imaging); there are a variety of methods employed for this purpose; syn. Chemsat

Fat saturation	Process of destroying the signal from lipid components in MRI to avoid chemical shift artifacts and improve lesion conspicuity in regions such as the breast
FE	Field Echo; see also GRE
Fibrillation	Uncoordinated quivering of the heart without normal blood-pumping beat
Fibrosis	Formation of fibrous tissue, or fibrous degeneration
FISP	Fast Imaging with Steady Precession; essentially a steady state derivative of FLASH (see below) in which magnetization is refocused after data acquisition, rather than being destroyed
Fistula	Abnormal passage or channel between two internal organs, or between an internal location and the body surface
FLAIR	Fluid Attenuated Inversion Recovery; inversion recovery sequence times so that signals from CSF (see above) are suppressed
FLASH	Fast Low Angle SHot; rapidly repeated (short TR) sequence with reduced precession angle and gradient recalled echo data acquisition
fMRI	Functional Magnetic Resonance Imaging; generic description of procedures in which a subject undergoes sensory stimulation, while responses are sought from brain imaging
FMRI	See fMRI
FNA	Fine Needle Aspiration
Focal (lesion)	A localized lesion with defined boundaries
FOV	Field Of View
FREEZE	See ROPE
Frequency encoding	Gradient applied to disperse spin frequencies during acquisition of data
FSE	Fast Spin Echo; see RARE
FSW	Fourier Series Windowing
fwhm	Full width half maximum; definition of linewidth, or, in context of point spread function, of resolution in imaging
$f(x, y, z)$	Spin density at point x, y, z
Gadopentate	See Gd-DTPA
Gag	An instrument for forcing open or holding open the mouth of the unconscious patient, particularly when under general anesthetic
Gating	Cardiac, respiratory; process by which machine operation is synchronized with the patient's behavior
Gd-DTPA	Gadolinium DiethyleneTriaminePentaAcetic acid; gadolinium chelate used as an in vivo contrast agent (also known as Gadopentate)
Gd-DTPA-BMA	Non-ionic variant of Gd-DTPA (last three letters represent 'BisMethylAmide')
GE	Gradient Echo; see GRE
Ghost	Artifact of motion (generally); normally an unwanted single or multiple repetition of image (or other) data in the phase encode direction(s) of a 2(3)D data set
Gibb's ringing	Artifact of the Fourier Transform process which is obvious where the resolution of the data is too coarse; analogous to effects when the data are truncated in some way; appears in images as ripples of reducing signal amplitude, adjacent to edges between sharp intensity changes
Gliosis	Excess of astroglia in damaged areas of the central nervous system
GLX	Glutamine plus glutamate; convenient code for components which are as yet unresolved in in vivo proton spectroscopy
GM	Gray Matter
GMC	Gradient Moment Compensation; see MAST
GMN	Gradient Moment Nulling; see MAST
GMR	Gradient Moment Rephasing; see MAST
Gradient echo	See GRE
GRASE	Combination of RARE and Gradient Recalled Acquisitions in a single multi-echo sequence (see RARE, EPI)
GRASS	Gradient Recalled Echo in Steady State; see GRE
GRE	Gradient Recalled Echo; commonly used acronym for 'spin warp' data acquisition (see **Spin Warp Data Acquisition** by J. M. S. Hutchinson); following a selective or nonselective pulse of significantly less than 180°, the magnetization is coded and an echo formed using gradients only; see also Spin warp
Great vessels	Large vessels entering the heart, including the aorta, pulmonary arteries and veins, and the venae cavae
Hc_t	Hematocrit (see below)
HE	Hepatic Encephalopathy; disease secondary (usually) to advanced liver disease

Hematocrit	Volume percentage of erythrocytes in whole blood
Hemorrhage	Deposit of blood products arising from a leak in the walls of, or break or damage to, a blood vessel
Heterozygote	Individual with different alleles with respect to a given characteristic
HIV	Human Immunodeficiency Virus
HMPAO-SPECT	HexaMethylPropyleneAmine-Oxide-SPECT (single photon emission computed tomography); lipid-soluble nuclear agent widely used in cardiac diagnosis
Homeostasis	Tendency to stability in the normal body states of an organism
Homozygote	Individual possessing a pair of identical alleles at a given location
Hyperintense	The component being described is brighter on the image than the region with which it is being compared (cf. Hypointense)
Hyperplasia	Abnormal multiplication of normal cells in an otherwise normal tissue structure
Hyperthermia	Class of methods in which tissue temperature is raised, particularly as a treatment for cancer; various thermal sources (rf, microwave, laser, ultrasound) are used, but the name is perhaps most closely associated with rf as the source of energy
Hypertrophy	Increase (swelling), particularly of an organ (cf. Hypotrophy)
Hypointense	The component being described is less bright on the image than that with which it is being compared (cf. Hyperintense)
Hypoplasia	Incomplete development or underdevelopment of an organ or tissue
Hypotrophy	Shrinkage, loss of size, of an organ (cf. Hypertrophy)
Hypoxia	Deficient in oxygen (in the sense of tissue being hypoxic)
IADSA	Intra Arterial Digital Subtraction Angiography; method of X-ray angiography
IC	IntraCellular
Ictus	Event, occurrence, usually in sense of a traumatic event clinically leading to consequential affects
Idiopathic	Self-originated; of unknown origin
Infarction	Result in tissue of loss of blood supply which has led to cell death; considered irreversible
Infusion	Therapeutic introduction of a fluid other than blood into a vein
Interleaving	Process (in spiral scanning, EPI, etc.) in which part of k-space is traversed in each of a series of acquisitions where the lines acquired at any one time are not contiguous
Intracranial	Inside the skull
Intramural	Within the wall of an organ
'Intrinsic' SNR	Signal-to-noise ratio calculated ignoring signal strength variations arising from the method of data acquisition (and, so, avoiding considerations such as the dependency of the relaxation time constants on field strength), i.e. signal-to-noise ratio of a fully recovered spin population with the signal measured immediately post excitation
IOP	Iron Oxide Particle; particularly in the context of superparamagnetic Fe_2O_3 particles used as contrast agent material (these have a typical diameter of 20–30 nm)
Ischemia	Process in which a region of tissue receives an inadequate blood supply for its needs; usually considered reversible
ISIS	Image Selected In vivo Spectroscopy (see **Single Voxel Whole Body Phosphorus MRS** by R. J. Ordidge)
Isointense	There exists no contrast between two tissues which are of equal signal intensity
IV	IntraVenous; particularly used to qualify an injection, for example, of contrast agents
IVC	Inferior Vena Cava
k-space	Spatial frequency space, and so that space in which the time resolved data are recovered
k_x, k_y, k_z	Coordinates of k-space along major axes
La	Lactate
LGC	Local Gradient Coil; typically a small, temporarily inserted, coil structure
Lip	In positional sense, marginal part, or section
Locular	Pertaining to a small space or cavity
Lumen	Cavity within a tube or organ
LV	Left Ventricle
Lysis	Dissolution, loosening, destruction (e.g. of cells by a specific agency); also, gradual abatement of symptoms of a disease
M_0	Fully recovered magnetization, usually taken to be the observable magnetization in whole body experiments, thus allowing for the invisibility to normal MRI and MRS methods of some nuclei and reduced contribution from others
Macrophage	Any of the large phagocytic cells occurring in walls of blood vessels and loose connective tissue

Magnevist	Trade name for the widely used MRI contrast agent Gd-DTPA (see above)
MARF	Magic Angle in the Rotating Frame (see **Imaging of Wide-band System by Line-narrowing Methods** by B. Maraviglia)
Mass effect	Shift or distortion of the normal anatomical tissue pattern due to a lesion which is occupying space otherwise available for it
MAST	Motion Artifact Suppression Technique; method in which multiple gradient lobes are used to control motion artifacts (see Pattany et al. *J. Comput. Assist. Tomogr.*, 1987, 11, 369)
Mediolateral	At midline level towards the side (as could be a location in a coronal image)
Meninges	Three membranes enveloping the brain and spinal cord
MESA	Multiple Echo Spectroscopy Acquisition; method in which the first 90° pulse is chemical shift selective, which is followed by three slice-selective 180° pulses to define a volume from which the signal is obtained
Metabolite	Any chemical component present in the tissue which may or may not take part in biochemical reactions whether or not they are observable
MI	*Myo*-Inositol; metabolic component seen in proton spectroscopy; assume this usage in spectroscopy articles
MI	Myocardial Infarction; assume this usage in clinical articles
Microvasculature	Portion of circulatory system comprising the finer vessels (typically those with an internal diameter of less than 100 μm)
MIP	Maximum Intensity Projection; method in which the highest intensity signal along a line through an image data set is taken to be the value along that line; the method is frequently used in formatting angiography data
MnDPDP	Experimental, near clinical, contrast agent, with potential liver applications, using manganese as the paramagnetic moiety (actually Manganese DiPyridoxyl DiPhosphonate)
Morbidity	Death or other grossly unfavorable outcome
Morphology	Anatomy or anatomical structure
Motion refocusing	See MAST
MPR	MultiPlanar Reconstruction
MRA	Magnetic Resonance Angiography
MRI	Magnetic Resonance Imaging
MRM	Magnetic Resonance Microscopy
MRS	Magnetic Resonance Spectroscopy
MRSI	Magnetic Resonance Spectroscopic Imaging
MRV	Magnetic Resonance Venography
MS	Multiple Sclerosis
MT	Magnetization Transfer (see magnetization transfer contrast, MTC)
MTC	Magnetization Transfer Contrast
MTF	Modulation Transfer Function
Multi-coil	Coil system comprising more than one coil assembly, each with its own preamplifiers, matching network, etc.
Multi-echo	Method in which a series of spin-echo images are acquired following a single first excitation; economical of time when different image contrasts are required quickly; cf. RARE, as the most highly developed form of the process in routine imaging, or the CPMG (Carr–Purcell–Meiboom–Gill) sequence as a more extreme case
Multi-slice	Method in which interleaved parallel slices are acquired at the same time, so employing the recovery periods (TR) in sequences profitably to reduce the time needed to cover a region of the body
Myocardial	Of, or pertaining to, the myocardium (the thickest muscle layer of the heart)
Myopathy	Any disease of a muscle
NAA	*N*-AcetylAspartate; brain metabolite dominating normal adult white matter proton brain spectrum; function is unclear, but it is considered to be associated with the presence of neurons
NAD	Nicotinamide-Adenine Dinucleotide (NADP = nicotinamide-adenine dinucleotide phosphate)
NAD$^+$	Oxidized form of NAD (NADP$^+$ = oxidized form of NADP)
NADH	Reduced form of NAD (NAD(P)H also called NADPH = reduced form of NADP)
NCO	Numerically Controlled Oscillator (or DDS = direct digital synthesizer)
NDP	Nucleotide DiPhosphate (i.e. any diphosphate)
Necrosis	See necrotic

Necrotic	Dead (noun necrosis); description of a region where cells present have died and tissue has degenerated
Neoplasia	Formation of a neoplasm (tumor)
NEX	See NSA
N(H)	Symbol for proton density; see PD
Nidus	Point of origin, or focus, of a morbid process
NPH	Normal Pressure Hydrocephalus
NPV	Negative Predictive Value
NSA	Number of Signal Averages; number of repetitions of each encoding step through k-space which are integrated; syn. acquisitions, data sets, averages, NEX, excitations, repetitions, projections, views, and profiles
NTP	Nucleotide TriPhosphate (i.e. any triphosphate)
NV	Number of Views (see NSA)
Oblique	Slice, the plane of which is parallel to one major axis only of the body
Oblique-sagittal	Slice orientation but visualized as the notation of a sagittal slice about its vertical axis; oblique-coronal, by analogy, describes a coronal slice rotated about its transverse axis (i.e. that normal to the height of the patient)
Occlusion (occluded)	Blockage in a blood vessel (partially occluded = partially blocked)
Omniscate	Trade name for the contrast agent GdDTPA-BMA
OPE	See ROPE
OR	Operating Room (theater)
PA	Pulmonary Artery
Parenchyma	Normal tissue, excluding the vasculature and other fluid channels
Paresis	Slight or incomplete paralysis
Pathogen	Any agent which is inimical to the well being of tissue
Pb EDTA	Lead-EDTA (lead chelate; see EDTA)
PC	PhosphorylCholine, if in spectroscopy article
PC	Phase Contrast, if in imaging, particularly flow-related, article
PCa	Carcinoma of the prostate; this is a common type of abbreviation needing individual interpretation in context; see Ca
PCA extract	PerChloric Acid extract
PCr	PhosphoCreatine
PD	Proton Density; frequently used as in 'PD-weighted' in clinical articles
PDCE	Proton Detected Carbon Edited; technique for editing proton spectra dependent on the presence of carbon metabolites (used to detect latter)
PDE	PhosphoDiEsters; various phosphorus metabolites with a common double ester structure; normally seen in the phosphorus spectrum as a single broad peak, resolved in extracts, and, to some degree, by proton decoupling
PDW	Proton Density Weighted; frequently as 'pD or PD weighted'; see Proton density weighted
PEACH	Paramagnetic Enhancement Accentuated by CHemical shift; method in which a contrast agent is used with fat suppression using chemical shift selective methods to visualize (particularly) breast cancer
PEAR	See ROPE
PE	Parenchymal Enhancement
PEDRI	Proton–Electron Double Resonance Imaging (see **Overhauser Effect Imaging of Free Radicals** by D. N. Lurie)
Peristalsis	Peristaltic movement; random, uncontrollable, motion of the gut
Periventricular	Round, and adjacent to, the ventricles
PEt	PhosphorylEthanolamine; phosphorus metabolite which is part of the phosphorus mono-ester complex seen in ^{31}P spectroscopy
PET	Positron Emission Tomography
Phase encoding	Process by which, using relatively short gradient pulses, a signal pattern is imposed through k-space which is present throughout the subsequent data acquisition; it is normally changed before the next data acquisition (except where multiple averages of the same data are required)
Phase image	Image in which the gray scale is a saw tooth plot of the signal for the phase angle θ in the range $-\pi < \theta \leq \pi$; if unconstrained it may vary through a number of consecutive cycles, with an appropriate pattern of increasing change from dark to light, then sudden reversion to dark; if the real and imaginary components of the signal of a voxel (i,j) are $x_{i,j}$ and $y_{i,j}$, then $\theta = \tan^{-1} y_{ij}/x_{ij}$

PhCh	PhosphatidylCholine
Phe	Phenylalanine
pH_i	Intracellular pH
Pi	Inorganic phosphate
Pixel	Two-dimensional cell which is the basic unit of an image
PM	Post Mortem
PME	PhosphoMonoEster; in phosphorus spectroscopy, large peak of in vivo spectrum contributed to by a number of metabolites which are frequently not resolved; proton decoupling can be used to assist in line resolution
pO_2	Tissue oxygen partial pressure
POPE	See ROPE
PPV	Positive Predictive Value
PR	Projection Reconstruction; method of acquisition directly analogous to the operation of the original Hounsfield translate–rotate CT X-ray scanner; has potential virtue of allowing very rapid start to data acquisition
PRE	Proton Relaxation Enhancement; typically, the result of a reduction of T_1 in situations where sequence repetition times are of the same order or less than T_1
PRESAT	See Saturation
Presaturation	Saturation of all or part of an imaging volume prior to mean sequence operation; often used to describe the situation in which a region upstream (in a blood flow sense) of a region to be studied is saturated so as to avoid artifacts due to flow effects in the region, but also used to eliminate risk of aliasing when very small fields of view, which are less than the body's natural size, are in use
Prescan	Operation of scanning sequence prior to start of data acquisition to stabilize magnetization signal
Present	In the sense 'the patient presented with such and such a symptom'; this means they were first seen, and investigated, when the symptoms noted were recorded
PRESS	Spectroscopic spatial localization method (particularly for protons) in which a selective $90°$ pulse is followed by two selective $180°$ pulses and data acquisition; each selective excitation defines one plane, and the target region is the volume common to all three
PRI	Projection Reconstruction Imaging
Profiles	See NSA
Prohance	Proprietary name for the contrast agent Gd-DOTA
Projections	See NSA
Prospective cardiac gating	Method of gating in which the position of the patient's abdomen (as it is particularly relevant to abdominal imaging) is recorded at the time of the acquisition of each line of k-space, and, by using multiple acquisitions, or convolution, attempts are made to modify the data so as to reorder it
Proton density weighted	Sequence resulting in data where the resultant contrast is mainly a function of proton density; typical examples are long TR (TR > 2000 ms), short TE (TE < 30 ms) brain imaging sequences; this can also be known as 'balanced weighting'
Proximal	Adjacent to, beside, same point of reference
PS	Partial Saturation; another name for a 90-TR-90 sequence with slice-selective $90°$ pulses and following spatial encoding and data acquisition
PSF	Point Spread Function
PSIF	See CE-FAST
PW	Proton Weighted; see Proton density weighted
QCT	Quantitative Computerized Tomography
QRS complex	Series of three consecutive peaks and troughs (including the large 'R' one) in an EKG wave form
r_1, r_2	Relaxivities of moieties (units s^{-1} $mmol^{-1}$)
r_t	Transmitter coil radius (specifically of a surface coil)
RARE	Rapid Acquisition with Relaxation Enhancement
RBC	Red Blood Cell
Refocused FLASH Methods	Derivation of FLASH (see above) in which magnetization is refocused, and the signal is pseudo steady state (acronyms, inter alia, FISP, GRASS, FAST, CE-FAST)
Relaxometry (field cycling relaxometry)	Method developed by Koenig (*Magn. Reson. Am.*, NY, 1987, p. 263) in which the relaxation times of components are measured at a series of different field levels
Repetitions	See NSA
RESCOMP	See ROPE
REST	See Saturation

RFZ	Rotating Frame Zeugmatography
ROC (curve)	Receiver Operating Characteristics (curve); method used to measure the performance of those interpreting radiological images and other similar data
RODEO	Method pioneered by the Dallas group specifically for the suppression of lipid signals in breast imaging (cf. FATSAT)
ROI	Region Of Interest
ROPE	Respiratory Ordered Phase Encoding; method of correcting for patient (particularly) abdominal motion (see Bailes et al. *J. Comput. Assist. Tomogr.*, 1985, 9, 835); though there are slight variations, the following synonyms use essentially the same concept: COPE, POPE, RESCOMP, RSPF, PEAR, phase reordering, FREEZE
ROS	Region Of Sensitivity
R–R interval (R-wave)	Interval between successive R wave peaks in an EKG waveform
RSPF	See ROPE
RT	Radiation Therapy
R-wave	Large peak in electrocardiogram measurement of heart operation
SaO_2	Oxygen saturation
Sagittal	Slice along the long axis of the body, with its plane lying from front to back
SAR	Specific Absorption Rate (in the context, usually, of rf energy deposition in tissue)
SAT	See Saturation
Saturation	Applied to saturation of magnetization of regions outside a region of interest, either to remove inflowing material (e.g. blood) to avoid artifacts or to avoid aliasing when a small field of view is used (many acronyms are used for this); note that the name is also used for many other processes (see also Presaturation)
SCHE	Sub-Clinical Hepatic Encephalopathy
SE	Spin Echo; echo formed as the result of an inverting rf pulse (selective or otherwise) and in the presence or absence of spatial encoding gradients
Section	Alternative word for slice
Sensitivity	In the clinical statistical sense, the sensitivity of a test is the ratio of the number of patients diagnosed with a disease to the total of those who have it (i.e. those correctly giving a positive result plus the false negative tests)
Shunt	Diversion or by-pass (usually channel)
SI	Signal Intensity [not to be confused with SI referring to the international system of units]
Slab	Block of tissue subdivided into slices (usually by phase encoding parallel to the slab selection direction), but falling well short of a volume acquisition; multi-slab (cf. multi-slice), series of parallel slabs acquired in an interleaved manner
SLIT DRESS	Version of DRESS in which multiple slices are interleaved
SMA	Superior Mesenteric Artery (also known as Arteria Mesenterica Superior)
SMV	Superior Mesenteric Vein (also known as Vena Mesenterica Superior)
SPAMM	SPAtial Modulation of Magnetization; technique for visualizing tissue motion
Specificity	In the clinical statistical sense, the fraction of normal subjects who give a correct negative result in a test to the total of normal subjects evaluated (i.e. the total of those correctly defined as normal plus the false positive results)
SPECT	Single Photon Emission Computed Tomography
Spin warp	Original Aberdeen-developed sequence using a gradient followed by its reversal to form an echo, through the development of which data are acquired; the method used a set of incremented pulses in a second direction to traverse *k*-space; see also GRE
SPIR	See FATSAT
Spoiled FFE	See spoiled FLASH
Spoiled FLASH	FLASH sequence with additional gradients or rf pulses to destroy residual coherent magnetization
Spoiled gradient echo	Variant of spoiled FLASH
SSFP	Steady State Free Precession; successful early method of image data acquisition; little used internationally nowadays
SSI	Solid State Imaging
ST	Saturation Transfer
ST	An abbreviation of the name of the Stejskal and Tanner pulsed gradient spin echo sequence (see PGSE) (has to be distinguished from previous entry by context)
STE	STimulated Echo (signal); see STEAM
STEAM (steam sequence)	STimulated Echo Acquisition Method of spatial localization
Stenosis	Narrowing or stricture of a duct or canal
STIR	Short T_1 Inversion Recovery; method primarily used to eliminate fat signals from images

Stroma	Supporting tissue or matrix of an organ (as distinguished from its functional component)
Subarachnoid	Located or occurring between the arachnoid and the pia motor (i.e. inner and mid-layer of meninges)
Subdural	Situated between the dura mater and the arachnoid (i.e. outer and central of the three layers of the meninges)
Supratentorial	Above the tentorium of the cerebellum
Susceptibility sensitive	In a pulse sequence, typically refers to gradient recalled echo sequences
SVC	Superior Vena Cava (large thoracic vein returning abdominal and lower blood to heart)
Synapse	(noun) Location of functional transfer between neurons; (verb) refers to the making of a connection
Systole	The part of the cardiac cycle during the contraction of the heart
$t_{1/2}$	The half-life of a process
T1W = T_1 weighted	Sequence resulting in data where the contrast is mainly a function of T_1; typical examples are short TR (300–500 ms), short TE (say 10–30 ms) brain imaging sequences
T2W = T_2 weighted	Sequence resulting in data where the contrast is mainly a function of T_2; typical examples are long TR (2000–3000 ms), long TE (TE > 60 ms) brain imaging sequences
T_2^* weighting	Conventionally, the contrast dependency found in gradient recalled acquisition sequences (or spin warp)
TCA-cycle	TriCarboxylic Acid (cycle); Krebs cycle
TE	Time to echo; time from the center of the rf pulse exciting transverse magnetization to the time at which the central point in image space is acquired (at some stage of the image acquisition process); if multiple lines are being acquired with individual excitations, TE is the time at which the central point of k-space is recovered in the data recovery without phase encoding; note that this is a more formal statement than that normally given, which simply defines TE as the time to the peak of the echo which is found at the center of the data acquisition (this version is not precise enough, however, in a number of situations)
TG	Gradient echo time; see TE
Thrombus (thrombosis)	Aggregate of blood components, primarily platelets and fibrin, frequently causing vascular obstruction
TI	Inversion time in an inversion recovery experiment; time between the center of the 180° inverting pulse and the pulse rotating magnetization into the x/y plane
Tip angle	Flip angle
TL	Locking time (in $T_{1\rho}$ experiments)
T_m	Interfacial interaction lifetime
TM	STEAM sequence time interval, between second and third rf pulses
TMA	TriMethylAmine (related to choline resonances)
TMJ	Temporo-Mandibular Joint
TMR	Topical Magnetic Resonance; early method of spatial location in spectroscopy using a carefully shaped inhomogeneous B_0 field
TOF	Time Of Flight; particularly used to describe weighting interval in some angiographic procedures where blood outside the region is preconditioned prior to the delay to allow it to reach its expected location
TOMROP	T One (T_1) by Multiple Read Out Pulses
TONE	Variable flip angle method for time-of-flight angiography
TR	Repetition time of a sequence (interval between the same stages of successive applications of a sequence to a single set of data)
Transaxial	See Transverse
Transmural	Extending through the wall of an organ
Transverse	Slice normal to the long axis of the body; syn. Transaxial
TRCF	Tilted Rotating Coordinate Frame
TSE	See RARE (means Turbo SE)
Turbo methods	Any very rapidly repeated sequence (where TR << T_1); there are a series of equivalent acronyms, e.g. snapshot FLASH = Turbo FLASH = MPRAGE = FSPGR = TFE, which do have variations in magnetization preparation but a common concept
Twave	Section of EKG waveform at end of main complex
US	UltraSound, UltraSonography
Valvular	Affecting a valve
VEC	Velocity Encoded Cone; flow imaging property
Vegetation	In a clinical sense, plant-like fungoid neoplasm, or luxuriant fungus-like growth of pathologic tissues
Ventral	Pertaining to the belly (abdomen) or towards the abdomen relative to a reference

Venule	Any small vessel collecting blood from the capillary bed, and uniting to form veins
Views	See NSA
V_{nec}	Aliasing velocity; velocity of flow where the encoding range in use means that the phase of the flowing material exceeds the $\pm\pi$ range, resulting in ambiguity
VOI	Volume Of Interest
Voxel	Unit of space into which the body is subdivided by the image formation process
VSE	Volume Selective Excitation
Wash-in effects	Enhanced signal from a relatively long T_1 component such as blood which has been allowed full recovery flowing into a volume of interest
Wash-out effects	Cf. Wash-in effects; effects arising from removal of wholly or partly saturated material moving out of a volume of interest
White blood	See Bright Blood
Window	Period of data recovery or, alternatively, visible gray scale range on a display
WM	White Matter
2-DFT imaging	Two-dimensional Fourier Transform imaging in which there is one phase encoding direction and a second (frequency encoding) direction
3-DFT imaging	Three-dimensional Fourier transform imaging; image acquisition method without slice selection having two phase encoding directions, with data acquisition along the third (frequency encoding) direction

Journal Abbreviations used in Whole Body MR Sections not commonly used otherwise in NMR

A.J.N.R	American Journal of Neuroroentgenography
AJR	American Journal of Roentgenography
Am. J. Med.	American Journal of Medicine
Am. J. Physiol.	American Journal of Physiology
Am. J. Sports Med.	American Journal of Sports Medicine
Ann. Neurol.	Annals of Neurology
Am. J. Pathol.	American Journal of Pathology
Br. J. Radiol.	British Journal of Radiology
Bull. Magn. Reson.	Bulletin of Magnetic Resonance
Clin. Chem.	Clinical Chemistry
Eur. Radiol.	European Radiology
Invest. Radiol.	Investigative Radiology
J. Anat.	Journal of Anatomy
J. Appl. Physiol.	Journal of Applied Physiology
J. Biomechanics	Journal of Biomechanics
J. Ortho Res.	Journal of Orthopedic Research
J. Comput. Assist. Tomogr.	Journal of Computer Assisted Tomography
J. Magn. Reson. Imaging	Journal of Magnetic Resonance Imaging
J. Neurosurg.	Journal of Neurosurgery
Med. Physics	Medical Physics
Magn. Reson. Med.	Magnetic Resonance in Medicine
Magn. Reson. Imaging	Magnetic Resonance Imaging
Magn. Reson. Quart.	Magnetic Resonance Quarterly
NMR Biomed.	NMR in Biomedicine
Pediatric Radiol.	Pediatric Radiology
Phys. Med. Biol.	Physics in Medicine and Biology
Proc. xth Ann. Mtg. Soc. Magn. Reson. Med.	Proceedings of the xth Annual Meeting of the Society of Magnetic Resonance in Medicine
Radiol. Clin. N. Am.	Radiology Clinics of North America
Top. Magn. Reson. Imag.	Topics in Magnetic Resonance Imaging

Part Ten
Techniques for MRS

Single Voxel Whole Body Phosphorus MRS

Roger J. Ordidge
University College, London, UK

Joseph A. Helpern
Nathan Kline Institute, Orangeburg, NY, USA

James W. Hugg
University of Alabama, Birmingham, AL, USA

and

Gerald B. Matson
University of California, San Francisco and VA Medical Center, San Francisco, CA, USA

1   INTRODUCTION

In vivo ^{31}P MRS has become an established research tool for studying metabolism in both animals and humans. Up to seven peaks can be identified in tissues (more if proton decoupling is employed). For muscle and brain tissue these peaks correspond to: phosphomonoesters (PME), inorganic phosphate (Pi), phosphodiesters (PDE), phosphocreatine (PCr), and the γ-, α-, and β-phosphates of adenosine 5′-triphosphate (ATP). The contribution of each peak to the total spectrum is expressed by calculating the area under each of the peaks and expressing the result as a percentage (mol%) of the total phosphate signal. The energetic state of brain tissue may be characterized by the PCr/Pi ratio, which is an approximate measure of the phosphorylation potential when the creatine kinase reaction is at equilibrium, and which decreases in situations of insufficient energy production. In vivo ^{31}P MRS also provides a noninvasive measure of pH by utilizing the dependence of the chemical shift of Pi on pH.[1] In addition, the relative chemical shifts of Pi, PCr, γ-, α-, and β-ATP can be used to estimate the concentration of free Mg^{2+} ions,[2] which play a role in membrane stability and are involved in phosphoryl transfer reactions during glycolysis. Finally, the presence of ATP is ubiquitous in all living tissues, and its depletion presages serious cell injury or, eventually, cell death. Therefore, in vivo ^{31}P MRS provides a wealth of information that is unobtainable by other noninvasive techniques.

The primary technological problem for in vivo ^{31}P MRS is a lack of sensitivity caused by the low metabolic concentrations of NMR-visible phosphorus metabolites of biological interest. The experimental complexity and the logistics of applying long NMR experimental procedures to study human disease have limited the widespread application of in vivo NMR spectroscopy in clinical diagnostic studies. It is, therefore, of utmost importance that the NMR spectrum is acquired with optimum sensitivity in order to minimize the experimental duration.

Interpretation of spatially localized in vivo ^{31}P NMR spectra depends upon a knowledge of the location and size of the tissue volume investigated in the experiment. Tissue specificity is essential to eliminate signals emanating from tissues of different type or metabolic status. For instance, contamination of the ^{31}P spectrum from brain tissue can arise from the scalp and neck muscles which are rich in high-energy phosphates, especially PCr, thereby giving a disproportionately intense contribution to the spectrum. It is the goal of any localization method to facilitate the acquisition of in vivo NMR spectra by minimizing contamination from undesired tissue volumes, and maximizing the S/N ratio to minimize experimental duration. In addition, many of the ^{31}P resonances of interest have short T_2 values, thus requiring an excitation and localization scheme which minimizes signal losses due to T_2 relaxation.

2   TECHNIQUES

2.1   Surface Coil and Related Methods

The first spatially localized in vivo ^{31}P MRS experiments were performed using surface coils,[3] which consist of a loop of wire placed in close proximity to the subject, and positioned so that a major component of the B_1 field of the coil is perpendicular to the B_0 field. The coils can be used for both transmission of the rf pulses and signal reception. Most of the measured signal originates from an approximately hemispherical volume of tissue of equal radius to the coil. However, the spatial sensitivity of surface coils is inherently inhomogeneous and decreases rapidly with distance from the coil. The spatial weighting of the signal in an experiment is a function of the rf pulse length and repetition time, so the spatial response pattern is both irregularly shaped in space and of variable sensitivity as a function of position. For these reasons, surface coils are often used in combination with other methods.

Although no longer being used, Topical Magnetic Resonance (TMR) provided the first three-dimensional spectroscopic localization method by generating a uniform B_0 field over a limited spatial region, with a very inhomogeneous field in surrounding regions.[4] Moreover, the TMR method could be used with surface coils for improved localization. However, the spatial localization achieved by the combined method is not sharply defined, having edges which are quite diffuse. Another major disadvantage is that the region of interest is centrally located within the magnet bore assembly, having been generated by the combination of powerful Z^2 and Z^4 shim coils, and cannot be easily moved.

'Depth' pulse methods provide another approach to improved localization over that afforded by a single surface coil.[5] In brief, depth pulse methods rely on phase cycling with multiple acquisitions to obtain signals from regions experiencing a particular value of B_1 field strength. Other, analogous, surface coil methods which retain signals from a region within a limited range of B_1 field have also been demonstrated.[6–9] However, full, three-dimensional localization requires the use of two surface coils of differing sizes.[10]

In the middle 1980s, the availability of pulsed magnetic field gradients enabled new classes of localization experiments to be developed; namely, methods relying on the application of frequency selective rf pulses in the presence of magnetic field

For References see p. 735

gradients. The first of these methods, entitled Depth Resolved Surface coil Spectroscopy, provided for observation of a plane of material parallel to the plane of the coil.[11] Another variation of the method enabled multiple parallel slices to be simultaneously investigated.[12] A disadvantage of the method is that the selected volume is a flat disk which is not conveniently shaped for the investigation of many organs. Also, T_2 relaxation occurs during the slice selective refocussing period following the 90° rf pulse, leading to some degree of spectral distortion.

All surface coil techniques benefit from the inherent high sensitivity of signal reception from material close to the coil. However, all of the above methods suffer the disadvantage that the localized region is fixed in space, and the subject must be positioned appropriately with respect to the coil(s).

2.2   Three-Dimensional Selective Excitation Methods

The combination of three selective pulses in conjunction with pulsed magnetic field gradients applied along orthogonal axes, can be used to excite a volume of material which is defined at the intersection of all three planes. Many techniques use this principle; however, the exact nature of the spatial definition process falls into three general categories:

1. Outer volume suppression methods which selectively nutate spins in unwanted regions of the sample into the transverse plane, where the signal is dephased by applied gradients. The desired signal, which has been preserved along the longitudinal axis, is then read out using a 90° rf pulse. Examples of this principle are: Volume Selective Excitation,[13] Spatially Resolved Spectroscopy,[14] localization of unaffected spins in NMR imaging and spectroscopy (LOCUS),[15] 'SPACE',[16] and 'DIGGER'.[17]
2. Direct excitation methods in which the MR signal of interest spends some portion of the experiment evolving in the transverse plane prior to forming an echo in the signal acquisition period. Examples are Pixel-Resolved Spectroscopy (PRESS),[18,19] which uses a 90°–180°–180° selective pulse sequence and acquires a spin echo, and Stimulated Echo Acquisition Method (STEAM)[20] which uses a 90°–90°–90° sequence and acquires a stimulated echo.
3. Cancelation methods in which the entire sample volume is excited, but the signal has been previously prepared using 180° selective pulses so that spatial information is encoded in the longitudinal magnetization prior to signal acquisition, and unwanted signal is canceled following combination of the signal from a series of experiments. An example of this principle is Image Selected In vivo Spectroscopy (ISIS).[21]

The advantages of all these methods is that because of the use of selective pulses in combination with field gradients, the size and position of the selected volume can be directly related to a previously obtained MR image. The selected volume is regularly shaped (usually a cube or rhomboid), and the size and position can be easily varied with reference to an MR image by modification of the selective pulse carrier frequency and applied gradients.

A disadvantage shared by all localization schemes which employ selective excitation is the presence of a chemical shift related spatial localization error.[21] When nuclei with different chemical shifts are examined, the associated differences in

NMR resonant frequency cause an erroneous displacement of the selected region for each chemical shift value. The displacement error, ΔX is given by the equation:

$$\Delta X = \frac{\sigma B_0}{G_x} \qquad (1)$$

where σ is the chemical shift value relative to a suitable reference peak, B_0 is the main magnetic field strength, and G_x is the gradient strength applied in the x direction during the spatial localization procedure. This problem can be minimized by using large magnetic field gradients for spatial localization since ΔX is inversely proportional to G_x. At a field strength of 2 T, a localized ^{31}P spectrum (σ range ± 12.5 ppm) obtained with a field gradient of 1 G cm^{-1} has associated spatial localization errors of ±2.5 mm irrespective of the size of the selected region. The disadvantage of using higher gradient strengths is an increase in the rf power required to select a region of fixed size, since the bandwidth of the rf selective excitation pulse must be increased by reducing the pulse length. The rf power is ultimately limited by the rf power amplifier.

Techniques in category (1) are difficult to implement primarily because they are susceptible to signal error arising from volumes of tissue where the NMR signal has not been effectively destroyed. This can be caused by rf field inhomogeneity, insufficient rf power, or even spin–lattice relaxation during the experiment. Without exception all the techniques in category (1) benefit from phase cycling in a series of experiments to cancel the unwanted signal. Phase cycling can sometimes be achieved in two experiments; however, the most general and accurate cancelation is only achieved in eight experiments (two experiments per spatial axis).

For in vivo ^{31}P MRS, techniques in category (2) suffer from the disadvantage of signal loss due to T_2 decay. Thus, these methods have found limited application in ^{31}P studies,[22] although they are widely used for in vivo ^{1}H MRS studies.

The disadvantage of category (3) methods and specifically ISIS, is that large NMR signals often have to be canceled in order to determine the residual signal from a small selected volume surrounded by a large mass of tissue. Such cancelation is subject to errors caused by system instability, or by subject movement. The solution is to minimize the large unwanted signal components prior to application of the ISIS method, using methods which are similar in principle to the techniques in category (1).[23] Since ISIS is, at present, the most widely used method for ^{31}P MRS, it is now described in greater detail.

The ISIS technique uses selective excitation in combination with the differencing of NMR signals from a sequence of eight experiments to select a rectangular-shaped volume of tissue for spectroscopic analysis. Upon combination of the free induction decays (FIDs) from the eight ISIS experiments, signal from the desired volume adds, whereas signals from all external regions cancel. The method relies on the principle of selective inversion of the spin population prior to data acquisition. The selective inversion pulses (180° pulses) cause inversion of longitudinal magnetization in selected slices of the sample using frequency selective rf excitation pulses in combination with applied field gradients. The selective pulses are applied in sequence along the X, Y, and Z axes. The experimental sequence is illustrated in Figure 1, and the rf selective inversion pulses either applied, or withheld, in a sequence of eight experiments are shown in the accompanying table. A delay

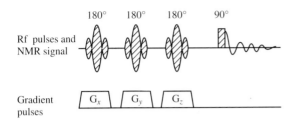

| Experiment | Selective pulse | | | Contribution to |
number	X	Y	Z	total spectrum
1	OFF	OFF	OFF	+
2	ON	OFF	OFF	−
3	OFF	ON	OFF	−
4	ON	ON	OFF	+
5	OFF	OFF	ON	−
6	ON	OFF	ON	+
7	OFF	ON	ON	+
8	ON	ON	ON	−

Figure 1 Pulse sequence used for the ISIS technique, and table indicating the order of RF pulse application during the eight experiment sequence. The addition or subtraction of relevant signals is indicated by a + or − sign

period follows application of each sequence to allow for spin–lattice relaxation.

The size of the selected region can be varied along any dimension by altering the excitation profile width of the respective rf pulse or the gradient associated with this rf pulse, thereby defining a rectangular volume. The position of the volume can be altered by adjustment of the individual frequency offset of each of the rf selective pulses. The size and position of the volume may thus be selected by reference to (and visualization with respect to) standard ^{1}H MR images of the subject.

For simplicity, the effect of a two-dimensional ISIS sequence on a two-dimensional spatial distribution of NMR spins is depicted in Figure 2. Four experiments are applied and the square array of nine elements represent three spatial regions along each axis. The hatched regions represent volumes which are inverted by ISIS prepulses in each of the four experiments. A double inversion (360° spin flip) is formed at the intersection of the inversion planes in experiment B2 (cross hatched region), and plain boxes do not receive any spin preparation prepulse. Selection of the central volume is achieved by combining the signals from the four experiments in the sequence of A1 − A2 − B1 + B2 whereupon all outer volumes correctly cancel. Three-dimensional selection (not depicted) involves extension of the scheme, and requires eight experiments, with the application of up to three selective inversion pulses, as indicated by the table which accompanies Figure 1.

The main advantage of ISIS is that signal readout immediately follows a nonselective 90° excitation pulse. This makes the method particularly useful in the study of MR signals with a short T_2 value, and specifically in the measurement of ^{31}P metabolites. Cancelation of signal outside the selected volume is accurate irrespective of the spin nutation angle experienced during each selective preparation pulse. However, selective 180° rf pulses are desirable since they provide the optimum sensitivity from spins within the ISIS volume. It is also preferable to use adiabatic, selective rf pulses such as the hyperbolic-

secant selective inversion pulse proposed by Silver et al.[24] This ensures that material within the ISIS volume contributes fully in each of the eight experiments, and the optimum S/N ratio performance is achieved virtually independent of both the rf field inhomogeneity of the coil and the rf power level applied during each selective pulse. When used with adiabatic inversion pulses, the ISIS technique can be executed very effectively with surface coils to obtain improved signal sensitivity from regions close to the surface of the subject. The experiment can be made completely independent of rf power setting by replacing the 90° readout pulse with a nonselective adiabatic 90° rf pulse. Adiabatic rf pulses are commonly used in the ISIS sequence, even when the rf coil is nominally homogeneous, because they remove the need to calibrate accurately rf nutation angles.

An additional advantage of the ISIS technique is that the rectangular volume may be tilted along any axis by replacing standard Cartesian gradient pulses by mixtures of gradients, e.g. selection along a 45° axis can be achieved by simultaneous application of equal X and Y gradient pulses during the appropriate rf selection period. The ISIS volumes may also be a parallelepiped in shape if the selection gradients are applied along nonorthogonal axes.[25] The modified ISIS volume may thus be shaped to fit better within a particular region of interest, thereby increasing S/N performance.

There are several variations in the order in which the ISIS experiments may be applied. These have been devised to (i) allow optimal signal cancelation from undesired regions when there is rf inhomogeneity present and the experimental repetition delay is insufficient for complete spin–lattice (T_1) relaxation, and (ii) reduce signal cancelation errors caused by

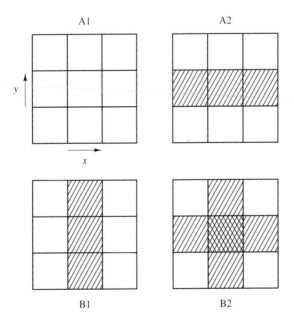

Figure 2 Diagrammatic representation of two-dimensional selection using an ISIS sequence of four experiments. The central square represents the region of interest, and is bordered along both x and y spatial dimensions by material that provides undesired signal. Inversion of a slice along x and y dimensions in experiments A2, B1, and B2 is indicated by hatched regions, and the cross hatched region (volume of interest) represents spins which have been inverted twice (e.g. 360° nutation)

For References see p. 735

subject motion. Matson et al.[26] include a 90° rf pulse following signal acquisition to ensure that T_1 relaxation commences from zero magnetization during the experimental delay period, thereby ensuring that longitudinal magnetization reaches the same state prior to each new acquisition. The experiments can then be applied in complete sequences of eight and the whole ISIS experiment is repeated successively in order to perform signal averaging. This makes the sequence less prone to movement errors and may have benefits when studying the torso. For head studies, where patient motion is not of great concern, the standard ISIS experiments may be averaged at each of the eight steps, since errors associated with relaxation will cancel upon final subtraction/coaddition of the data.[21] The absence of a 90° rf saturation pulse at the end of the signal acquisition period allows spin–lattice relaxation to proceed from the excitation pulse onwards, thus providing improved signal collection efficiency.

The method can be extended to investigate simultaneously two localized volumes centered on a common x axis but separated along the y axis. Three-dimensional ISIS with two selected cubes requires a minimum of 16 different sequences and up to four selective inversion pulses. This variation of the method is of particular use in biomedical studies since it allows simultaneous measurement of an equivalent contralateral region, thus providing an internal reference for biochemical changes in focal disease states. An example is presented in Figure 3, which shows in vivo ^{31}P brain spectra obtained from a stroke patient in the stroke-affected region and in the contralateral region for comparison.

The ISIS principle may also be extended to the simultaneous acquisition of multiple (2^n) volumes.[27] Since signal averaging is a necessity for in vivo ^{31}P spectroscopy, multiple volume spectra can be acquired with no increase in experimental duration compared with a single volume experiment. An extended sequence of selective spin inversion pulses is required in both the Hadamard transform approach[28] and the equivalent multicube ISIS method.

3   MEASUREMENT OF METABOLIC CONCENTRATIONS

Although most MRS data have been reported as spectral peak area ratios, quantitative methods have been developed for estimating molar concentrations of metabolites using in vivo MRS with single volume localization.[29,30] The principle requirements for quantitation are (a) to obtain accurate spectral integrals for the metabolite peaks of interest, (b) to determine the spatial sensitivity (B_1 field distribution) of the rf transmitter/receiver coil, (c) to compare with a calibration sample of known concentration, and (d) to determine saturation and relaxation effects (T_1 and T_2) for rapidly repeated rf excitation.

The procedure typically involves measuring a ^{31}P MRS spectrum during each localization experiment from a small sample fixed to the rf coil assembly, or to the subject (e.g. containing hexamethylphosphorus triamide with a peak chemically shifted about 110 ppm downfield from PCr). This signal is used to determine rf pulse lengths, and as a reference to account for variations in coil sensitivity (Q factor) caused by variable loading. Spectral peak areas are then compared with the signal from a calibration phantom containing a known concentration of inorganic phosphate, which is measured periodically using the same spatial localization procedure. The formula used to obtain the absolute molar concentrations of metabolites might also take account of differences in the aqueous fraction between samples, (e.g. taking account of cerebrospinal fluid volume in head studies) which requires additional MRI data. The T_1 values for all metabolite resonances must also be known to obtain a saturation correction when the experimental repetition rate is high. A more complete explanation is provided in other articles (see Related Articles).

4   APPLICATIONS

In this section we present a very brief overview of some applications of single voxel whole body ^{31}P MRS in humans, with the intent of providing examples of the kinds of metabolic information available from such ^{31}P MRS studies. Many of the organs and diseases are reported on in much more detail in other chapters (see Related Articles).

4.1   Human Brain and Brain Diseases

The contralateral ^{31}P spectrum in Figure 3 is representative of normal adult brain at 3.0 T, and illustrates prominent ^{31}P resonances as identified in the figure legend. The infant brain is characterized by higher PME levels and somewhat reduced PCr. The broad hump commonly attributed to lipid resonances is less prominent at 3.0 T than at 1.5 T as a consequence of the chemical shift anisotropy relaxation mechanism. With proton decoupling, the PDE resonance splits into separate glycerophosphocholine and other PDE resonances.[31]

Phosphorus-31 MRS single voxel studies have been performed in AIDS or HIV infection, Alzheimer's disease, deep white matter lesions, epilepsy, multiple sclerosis (MS), stroke, and brain tumors. Representative trends of the ^{31}P metabolite concentration for these diseases are indicated in Table 1. Other diseases and conditions studied by ^{31}P MRS include alcoholism (and other drug use), migraines, schizophrenia and other mental disorders, and various pediatric diseases.

4.2   Heart

Although some early single volume ^{31}P MRS studies in human heart have been performed,[26,32,33] more recent studies have used more elaborate localization methods, including multiple volume methods, to combat the need to obtain very small volumes (particularly for studies of ischemia or infarction) from a moving organ.

4.3   Liver and Kidney

A ^{31}P spectrum of human liver obtained at 1.5 T is shown in Figure 4. The most notable feature, as compared with a brain spectrum, is the lack of PCr. In fact, the absence of PCr is often taken as an indication of lack of contamination from surrounding muscle tissues which contain significant concentrations of PCr. Disease conditions studied in human liver include cancers, alcoholic hepatitis and cirrhosis, and fructose intolerence. Disease typically produces a decrease in ^{31}P

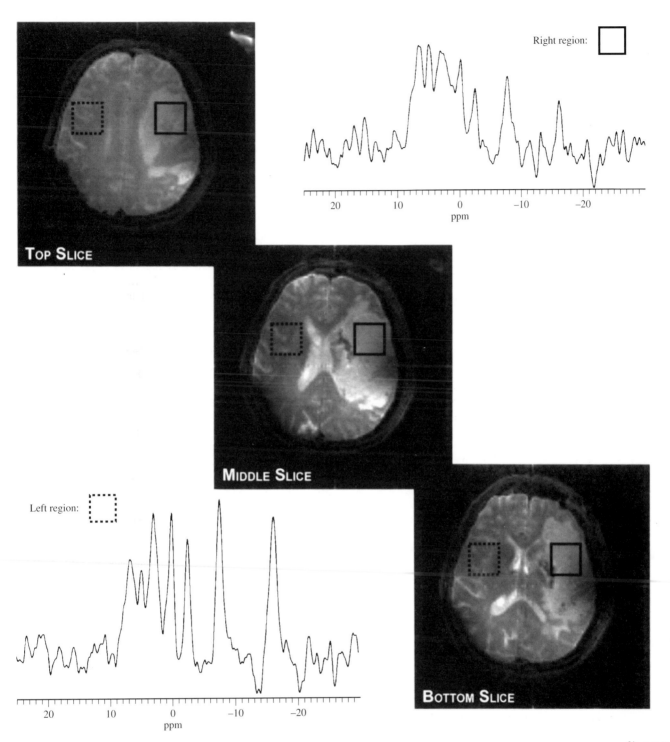

Figure 3 MR images and ³¹P spectra obtained from a stroke patient at a field strength of 3 T using ISIS. The two regions studied using ³¹P MRS were investigated simultaneously (*TR* = 1.5 s, number of acquisitions = 768, MRS study completed in 20 min), and incorporate tissue extending from the regions shown in the top slice through to those in the bottom slice (volume of each cube = 3 × 3 × 3 cm³). Peaks correspond to (from left to right) phosphomonoesters (PME), inorganic phosphate (Pi), phosphodiesters (PDE), phosphocreatine (PCr) and the γ, α-, and β-phosphates of ATP. The spectrum from stroke-affected tissue (right) should be compared with that from the unaffected contralateral region (left)

metabolites, while fructose intolerance is marked by reduction in ATP following ingestion or administration of fructose.

Few ³¹P MRS studies of normal kidney have been reported due to the depth of this organ in the body. However, transplanted kidney is typically much closer to the surface and easier to study. Several groups have made preliminary studies of transplanted kidneys and ³¹P MRS has the potential to assess transplanted kidney metabolic health, and possibly to assist in analyzing tissue status or pathology following transplantation.

For References see p. 735

Table 1 Representative Trends of ^{31}P Metabolite Concentration for Various Diseases

Diseases	Observations
Alzheimer's	Ranges from no observable changes to reported increases in PDE and possibly PME.
Deep white matter lesions	Reductions in PME, PDE, and ATP
Epilepsy	Reduced PME, increased Pi, and decreased pH (interictal studies of focal temporal lobe epilepsy)
HIV	Reduced ATP/Pi in AIDS dementia complex
MS	Ranges from no observable changes to decreases in PDE, and reduced PME and Pi (chronic lesions)
Stroke	Reductions in all metabolites; decreased pH in acute phase; increased pH in some chronic infarcts
Tumors	Variable metabolite reductions in PDE and PCr, elevated PME, depending upon tumor type

4.4 Human Muscle

By far the largest number of ^{31}P MRS studies in humans have been performed on muscle, ranging from the study of normal muscle metabolism under various conditions of (and recovery from) exercise, to studies of the alterations of muscle metabolism in various disease states. Under light to moderate exercise, normal muscle displays a fall of PCr and a fall of pH (after an initial slight rise), with maintenance of ATP levels. Alterations of metabolic concentration with work, or under recovery conditions, may be further changed by a variety of diseases, and ^{31}P MRS has proven to be clinically useful in a number of muscle disorders. However, the localization has traditionally been accomplished by the sole use of surface coils.

4.5 Body Tumors

Tumor spectra typically display a high PME, and often reduced PCr compared with the surrounding or normal tissue. Although early studies indicated that ^{31}P MRS might have considerable potential in the assessment of tumor therapy, the investigation of tumors in the body has proved difficult. While some early studies seemed to indicate that reductions in PME

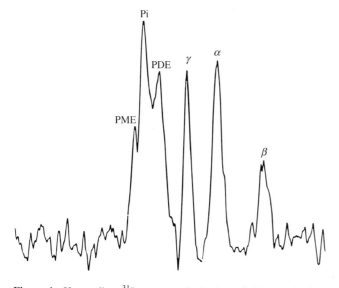

Figure 4 Human liver ^{31}P spectrum obtained at a field strength of 1.5 T using a 14 cm surface coil (number of acquisitions = 640, *TR* = 2 s). The 96 ml volume of tissue studied was localized using ISIS, and the ^{31}P spectrum confirms the absence of PCr in liver tissue. (Reproduced by permission of Academic Press from M. D. Boska et al.[34])

heralded effective therapy, more recent studies have shown that no such simple interpretations generally prevail.

5 CONCLUSION

Single voxel whole body ^{31}P spectroscopy may be performed within acceptable experimental durations in high-field magnet systems. The most commonly used method, ISIS, while not devoid of problems, offers a reasonably robust method of localization. Two cubes may be investigated simultaneously to allow an internal comparison between different regions of biological tissue.

Application of all these single voxel localization methods requires the location of the volume to be chosen from an MR image, which assumes that the operator can either observe a region of abnormality on the image, or can identify the region from clinical symptoms. If the region of metabolic dysfunction is not known, and the contrast provided by MRI is insufficient to identify the region, chemical shift imaging (CSI) offers the advantage that all regions of the sample are investigated to produce a map of each metabolite. The region of interest may then be defined on the basis of metabolic changes following data acquisition. However, CSI methods are generally more time consuming than ISIS, and rapid experimental repetition rates result in distorted spectra which must be corrected for T_1 relaxation for the purposes of quantitation. Both approaches offer advantages and disadvantages in their application to specific studies, and appropriate choice of methodology usually involves a compromise.

6 RELATED ARTICLES

Brain MRS of Human Subjects; Brain Neoplasms in Humans Studied by Phosphorus-31 NMR Spectroscopy; Chemical Shift Imaging; In Vivo Hepatic MRS of Humans; Localization and Registration Issues Important for Serial MRS Studies of Focal Brain Lesions; NMR Spectroscopy of the Human Heart; Peripheral Muscle Metabolism Studied by MRS; Proton Decoupling During In Vivo Whole Body Phosphorus MRS; Proton Decoupling in Whole Body Carbon-13 MRS; Quantitation in In Vivo MRS; Rotating Frame Methods for Spectroscopic Localization; Surface Coil NMR: Detection with Inhomogeneous Radiofrequency Field Antennas; Systemically Induced Encephalopathies: Newer Clinical Applications of MRS; Whole Body Studies: Impact of MRS.

For list of General Abbreviations see end-papers

7   REFERENCES

1. R. B. Moon and J. H. Richards, *J. Biol. Chem.*, 1973, **248**, 7276.
2. H. Halvorson, A. M. Q. Vande Linde, J. A. Helpern, and K. M. A. Welch, *NMR Biomed.*, 1992, **5**, 53.
3. J. J. H. Ackerman, T. H. Grove, G. G. Wong, D. G. Gadian, and G. K. Radda, *Nature (London)*, 1980, **283**, 167.
4. R. E. Gordon, P. E. Hanley, D. Shaw, D. G. Gadian, G. K. Radda, P. Styles, P. J. Bore, and L. Chan, *Nature (London)*, 1980, **287**, 736.
5. M. R. Bendall, and R. E. Gordon, *J. Magn. Reson.*, 1984, **53**, 365.
6. D. I. Hoult, *J. Magn. Reson.*, 1979, **33**, 183.
7. R. Tycko and A. Pines, *J. Magn. Reson.*, 1984, **60**, 156.
8. A. J. Shaka, J. Keeler, M. B. Smith, and R. Freeman, *J. Magn. Reson.*, 1985, **61**, 175.
9. A. J. Shaka and R. Freeman, *J. Magn. Reson.*, 1985, **62**, 340.
10. M. R. Bendall, J. M. McKendry, I. D. Cresshull, and R. J. Ordidge, *J. Magn. Reson.*, 1984, **60**, 473.
11. P. A. Bottomley, T. B. Foster, and R. D. Darrow, *J. Magn. Reson.*, 1984, **59**, 338.
12. P. A. Bottomley, L. S. Smith, W. M. Leue, and C. Charles, *J. Magn. Reson.*, 1985, **64**, 347.
13. W. P. Aue, S. Muller, T. A. Cross, and J. Seelig, *J. Magn. Reson.*, 1984, **56**, 350.
14. P. R. Luyten, A. J. H. Marien, B. Sijtsma, and J. A. den Hollander, *J. Magn. Reson.*, 1986, **67**, 148.
15. A. Haase, *Magn. Reson. Med.*, 1986, **3**, 963.
16. D. M. Doddrell, W. M. Brooks, J. M. Bulsing, J. Field, M. G. Irving, and H. Baddeley, *J. Magn. Reson.*, 1986, **68**, 367.
17. D. M. Doddrell, J. M. Bulsing, G. J. Galloway, W. M. Brooks, J. Field, M. Irving, and H. Baddeley, *J. Magn. Reson.*, 1986, **70**, 319.
18. P. A. Bottomley, *Ann. NY Acad. Sci.*, 1987, **508**, 333.
19. R. J. Ordidge, M. R. Bendall, R. E. Gordon, and A. Connelly, *Magn. Reson. Biol. Med.*, 1985, 387.
20. J. Frahm, K. D. Merbolt, and W. Hanicke, *J. Magn. Reson.*, 1987, **72**, 502.
21. R. J. Ordidge, A. Connelly, and J. A. B. Lohman, *J. Magn. Reson.*, 1986, **66**, 283.
22. K. D. Merboldt, D. Chien, W. Hanicke, M. L. Gyngell, H. Bruhn, and J. Frahm, *J. Magn. Reson.*, 1990, **89**, 343.
23. R. J. Ordidge, *Magn. Res. Med.*, 1987, **5**, 93.
24. M. S. Silver, R. I. Joseph, and D. I. Hoult, *J. Magn. Reson.*, 1984, **59**, 347.
25. J. C. Sharp and M. O. Leach, *Magn. Reson. Med.*, 1989, **11**, 376.
26. G. B. Matson, D. B. Twieg, G. S. Karczmar, T. J. Lawry, J. R. Gober, M. V. Valenza, M. D. Boska, and M. W. Weiner, *Radiology*, 1988, **169**, 541.
27. R. J. Ordidge, R. M. Bowley, and G. McHale, *Magn. Reson. Med.*, 1988, **8**, 323.
28. L. Bolinger and J. S. Leigh, *J. Magn. Reson.*, 1988, **80**, 162.
29. S. Wray and P. S. Tofts, *Biochim. Biophys. Acta.*, 1986, **886**, 399.
30. K. Roth, B. Hubesch, D. J. Meyerhoff, S. Naruse, J. R. Gober, T. J. Lawry, M. D. Boska, G. B. Matson, and M. W. Weiner, *J. Magn. Reson.*, 1989, **81**, 299.
31. P. R. Luyten, G. Bruntink, F. M. Sloff, J. W. A. H. Vermeulen, J. I. van der Heijden, J. A. den Hollander, and A. Heerschap, *NMR Biomed.*, 1989, **1**, 177.
32. P. R. Luyten, J. P. Groen, J. W. A. H. Vermeulen, and J. den Hollander, *Magn. Reson. Med.*, 1989, **11**, 1.
33. S. Schaefer, J. Gober, M. Valenza, G. Karczmar, G. Matson, S. A. Camacho, E. Botvinick, B. Massie, and M. W. Weiner, *J. Am. Coll. Cardiol.*, 1988, **12**, 1449.
34. M. D. Boska, B. Hubesch, D. J. Meyerhoff, B. B. Twieg, G. S. Karczmer, G. B. Matson, and M. W. Weiner, *Magn. Reson. Med.*, 1990, **13**, 228.

Biographical Sketches

Roger J. Ordidge. *b* 1956. B.S., 1977, Ph.D., 1981, physics, University of Nothingham, UK, (supervisor Sir Peter Mansfield). Bruker Instruments, 1982–1986. Lecturer, Nottingham University, 1986–1989. Professor, Oakland University, Michigan, and Researcher, Henry Ford Hospital, Detroit, MI, 1989–1993. Professor, University College of London, 1993–present. Approximately 80 publications. Research specialties, localized spectroscopy, diffusion weighted MRI, and high speed MRI.

Joseph A. Helpern. *b* 1955. B.A., Chemistry, 1977, Case Western Reserve University, Cleveland, OH. M.A., chemistry, 1979, University of North Carolina, Chapel Hill, NC. Henry Ford Hospital, Detroit, MI, 1981, helped establish NMR Research Program. Ph.D., Medical Physics, Oakland University, Rochester, MI, 1988, supervisor, Dr. Michael Chopp. Director of NMR Research, Department of Neurology, Henry Ford Hospital, 1988–present. Over 55 publications. Research interests: cerebral metabolism, blood flow measurement, and NMR techniques for assessment of brain disorder.

James W. Hugg. *b* 1952. B.S., 1973, M.S., 1974, CalTech; M.S., 1976, Ph.D., 1978, nuclear physics, Stanford. Introduced to NMR by Felix Bloch and to Medical Imaging by Al Macovski. Petroleum industry. 1978–1988. Postdoctoral work in human brain MR spectroscopy at University of California, San Francisco, 1989–92. Rabi Young Investigator Award, 1992. Senior Staff Investigator, Neurology, Henry Ford Health Science Center (Case Western Reserve University), Detroit, MI 1992–present. Over 60 publications. Research specialties: spectroscopic imaging and diffusion imaging in human brain ischemia, epilepsy, and multiple sclerosis.

Gerald B. Matson. *b* 1938, B.A. chemistry, 1960, University of California, Ph.D., 1967, physical chemistry, University of Wisconsin. Since 1986 Facilities Manager, MR Unit, Department of Veterans Affairs Medical Center, San Francisco, Since 1989 Adjunct Professor, Department of Pharmaceutical Chemistry, University of California, San Francisco. Interests include development of MR spectroscopic imaging methods for in vivo MRS of human brain and brain diseases, MRS selective excitation and localization methods, and NMR probe design.

Single Voxel Localized Proton NMR Spectroscopy of Human Brain In Vivo

Jens Frahm and Wolfgang Hänicke

Biomedizinische NMR Forschungs GmbH am Max-Planck-Institut für Biophysikalische Chemie, Göttingen, Germany

1   INTRODUCTION

The unique potential of in vivo NMR for the life sciences is based on true noninvasiveness and tremendous methodologic flexibility. A continous flow of novel techniques allows the

For References see p. 750

addressing of fundamental questions related to structural, functional, and metabolic aspects of life processes within the intact human body. In particular, in vivo spectroscopy yields detailed insights into cellular metabolism of organ systems by combining the analytical strength of high-resolution NMR with image-controlled spatial localization within a three-dimensional object. Typically, localized proton NMR spectroscopy of human brain provides access to tissue metabolites with millimolar concentrations from a few milliliter volumes and within measuring times of less than 10 min.

Early studies employed surface radiofrequency (rf) coils[1] and homogeneity profiling of the static magnetic field[2] to assess bioenergetics via phosphate metabolites.[3-5] Although the simple surface coil technique was applied to other organs and nuclei including proton NMR of human brain,[6] optimum placement of a local rf antenna suffered severe limitations and often required surgical interventions as demonstrated by animal work.[7,8] With the availability of well-controllable magnetic field gradients through advances in NMR imaging, improved localization methods took advantage of the principles of spatial encoding by magnetic field gradient pulses and slice selection by frequency-selective rf excitation in the presence of a magnetic field gradient. Two complementary approaches were developed that are able to accomplish three-dimensional spatial discrimination of the NMR signal:

- chemical shift imaging (CSI) techniques that cover a column, plane, or volume in a two- (2D), three- (3D), or four-dimensional spectroscopic image where spatial information is phase encoded onto the chemical shift information of the NMR frequencies,
- single voxel spectroscopy techniques that acquire spectroscopic time-domain data directly from a restricted volume of interest (VOI) selected prior to or during rf excitation.

The aim of this contribution is to discuss the mainstream of in vivo proton NMR spectroscopy techniques based on single voxel localization. The capability of performing human studies on whole body systems is ensured by emphasizing the compliance of pertinent sequences with safety requirements and instrumental limitations, e.g. in terms of rf power. Methodologic descriptions include an outline of specific merits and pitfalls, while selected studies of human brain metabolism illustrate particularly advantageous applications.

2   LOCALIZATION TECHNIQUES

Beyond structure elucidation of biomacromolecules, NMR studies of biologic systems comprise body fluids, cell cultures, excised tissues, perfused organs, and intact living subjects (*Body Fluids*, *Cells and Cell Systems MRS*, and *Tissue NMR Ex Vivo*). Respective applications benefit from a widespread range of instruments and sequences and cover a large diversity of basic science and clinical questions. Even when restricting this complexity to single voxel proton NMR there are multiple factors that affect the design and outcome of localization techniques. This particularly holds true for differences in (i) hardware of small-bore high-field spectrometers and whole body instruments (see Section 2.5), (ii) relaxation times and available measuring times under in vitro and in vivo con-

ditions, (iii) the extent of motion in nonliving materials, anesthetized animals, and human subjects, and (iv) cooperativity of healthy subjects and patients.

Thus, optimum strategies for localization sequences applicable to human subjects not only have to address relaxation losses and suppression factors, but also critically rely on practical simplicity and stability as crucial elements for ease of implementation and use as well as for the reliability of the results. Accordingly, the following sections discuss surface coil techniques and *i*mage *s*elected *i*n vivo *s*pectroscopy (ISIS) for historic reasons, while special emphasis is paid to *p*oint-*res*olved *s*pectroscopy (PRESS) and *st*imulated *e*cho *a*cquisition *m*ode (STEAM) spectroscopy as the major localization techniques for single voxel proton NMR on whole body units.

2.1   Surface Coils and Beyond

The popularity of local rf antennae or 'surface' coils is based on the simplicity of the concept and the excellent signal-to-noise ratio (S/N; see *Surface Coil NMR: Detection with Inhomogeneous Radiofrequency Field Antennas*). However, from a technical point of view the latter benefit is not only due to a good filling factor but is also due to a sensitive volume extending far beyond the coil radius.[9] In fact, the contributing volume is not even well-defined, since it strongly depends on the experimental parameters of the rf pulse sequence (e.g., repetition time, flip angle used) and the tissue T_1 relaxation times involved.[10]

Figure 1 depicts flip angle distributions of a single-turn surface coil for B_1 field strengths yielding flip angles of either 90° (top) or 180° (bottom) at a distance $r/2$ from the center of the coil. While pertinent flip angle distributions result in inhomogeneous *excitation* of transverse magnetizations, corresponding sensitivity distributions also affect the *acquisition* of the NMR signal by surface coil reception. Thus, transmit–receive use of an inhomogeneous coil leads to a quadratic dependence on B_1 which may be reduced to a linear dependence by combining surface coil detection with homogeneous excitation using a second circumscribing coil.

Assuming a transmit–receive situation, Figure 2 describes the sensitivity σ of a single-turn surface coil as a function of TR/T_1 and rf flip angle

$$\sigma = \int \int \int B_1(x, y, z)S(x, y, z)\mathrm{d}x\,\mathrm{d}y\,\mathrm{d}z \qquad (1)$$

with $B_1(x,y,z)$ the rf field distribution as demonstrated in Figure 1. $S(x,y,z)$ denotes the local strength of the FID of a repetitive single pulse experiment with equilibrium magnetization M_0, flip angle α, repetition time TR, assuming the absence of transverse components (i.e., $TR > 3\ T_2^*$)[11]

$$S(x, y, z) = M_0 \sin \alpha(x, y, z) \frac{1 - \exp(-TR/T_1)}{1 - \exp(-TR/T_1)\cos \alpha(x, y, z)} \qquad (2)$$

The spatial flip angle distribution

$$\alpha(x, y, z) = \gamma B_1(x, y, z)\tau_\mathrm{p}\mu I/(2\pi) \qquad (3)$$

has to be calculated according to Biot–Savart's law with γ the magnetogyric ratio, τ_p the pulse duration, I the current, and μ the permeability of the specimen.

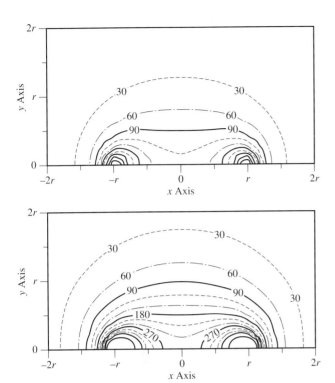

Figure 1 Flip angle distribution of a single-turn flat surface coil (radius r) in the central xy plane perpendicular to the coil for two different B_1 field strengths adjusted such that the flip angle at a distance $r/2$ from the center of the coil is either 90° (top) or 180° (bottom). Iso-B_1 curves are given for selected flip angles in steps of 30°. The coil is oriented parallel to the static B_0 field (z axis). (Reproduced by permission of Academic Press from Haase et al.[10])

Obviously, the strength of the detected NMR signal predominantly reflects the size of the excited volume. Regardless of the actual measuring parameters, the nonlocalized signal from the 'total' volume ($\sigma/TR \approx 75$, upper left in Figure 2) is considerably larger than that of a near (≈ 15, lower left) or distant (≈ 8, lower right) slice of thickness $r/5$ or that of a sphere of radius $r/2$ (≈ 4, upper right) at a distance r from the coil center. Assuming $TR/T_1 = 1$ the maximum *local* signal strength is in a 90° flip angle region close to the coil. However, although outer signal strengths are much smaller, their contributions to the received signal may easily overcompensate 'high-flux' regions close to the coil by simple volume considerations. Moreover, low flip angle excitations in outer regions effectively reduce the extent of signal saturation for acquisitions with $TR < 3$ T_1 and therefore further increase the relative contribution of unwanted signals distant from the coil. Thus, despite some gross anatomic localization of the NMR signal as given by the size and shape of the coil, it is not possible to focus properly on a desired VOI by a simple single pulse surface coil NMR experiment.

Of course, there are important biochemical questions that do not need exquisite spatial discrimination of tissue signals. Even nonlocalized experiments may provide the answer if relative changes are monitored during the course of a dynamic stimulus-driven study. Nevertheless, a number of refinements have

been proposed to improve spatial selectivity of the basic surface coil experiment:

- slice-selective excitation and refocusing of the applied magnetic field gradient to select a plane parallel to the coil, *d*epth *r*esolved *s*urface coil *s*pectroscopy (DRESS);[12]
- slice-selective presaturation, *f*ast *r*otating gradient *s*pectroscopy (FROGS),[13] and homogeneity spoiling by surface gradients[14] parallel to the coil to eliminate signal contributions from superficial tissues (fat, muscle) between the coil and the target VOI;
- phase-cycled rf excitation and composite rf pulses ('depth' pulses[15]) to reshape the sensitive volume along the B_1 gradient of the coil;
- Fourier series windowed variants of the 'rotating frame imaging'[16] experiment to profile spatially the acquired NMR signal along the B_1 gradient by multiple excitations with predefined flip angles (***Rotating Frame Methods for Spectroscopic Localization***).[17–19]

In summary, localization by placement of a surface coil is simple, easy to use, and provides good S/N. With single pulse rf excitation respective FID signals do not suffer T_2 relaxation losses. Potential pitfalls are (i) the uncontrolled extent of the sensitive volume as it depends on experimental parameters (TR, flip angle, tissue T_1), (ii) problems with absolute quantitation and comparability of studies, and (iii) limited or no access to inner organs. Most of these arguments also hold true for the refined versions mentioned above as they deliver one-dimensional (1D) solutions to a 3D problem.

Surface coils have mainly been applied in conjunction with phosphorus NMR studying alterations of cellular energy metabolism in skeletal muscle, liver, or heart (***Peripheral Muscle Metabolism Studied by MRS*** and ***Complex Radiofrequency Pulses***). In contrast, proton NMR studies of focal metabolism in brain require more sophisticated gradient-based localization techniques. While homogeneous rf coils are beneficial with respect to metabolite quantitation and study comparability, surface coils may still be used for sensitivity enhancement in a receive-only configuration. In fact, gradient-localized single voxel NMR techniques such as PRESS or STEAM are even possible in a transmit–receive mode. If the dimension of the VOI remains smaller than the surface coil radius, then localized optimization of the rf power ensures homogeneous rf excitation and signal reception.

2.2 Image-Selected In Vivo Spectroscopy (ISIS)

Prior to nonselective excitation by a 90° rf pulse (or any other rf scheme), the ISIS sequence[20] prepares the equilibrium magnetization of a 3D object by up to three slice-selective 180° inversion pulses in the presence of orthogonal magnetic field gradients. As outlined in Figure 3 the method requires a minimum of eight successive experiments to achieve spatial localization by adding/subtracting FID signals of properly prepared longitudinal magnetizations. The individual acquisitions correspond to 2^3 combinations (on/off) of the three inversion pulses. For example, subtraction of acquisitions #1 (no 180° pulse applied, excitation of the entire object) and #4 (magnetization inversion of a plane along the first gradient axis) results

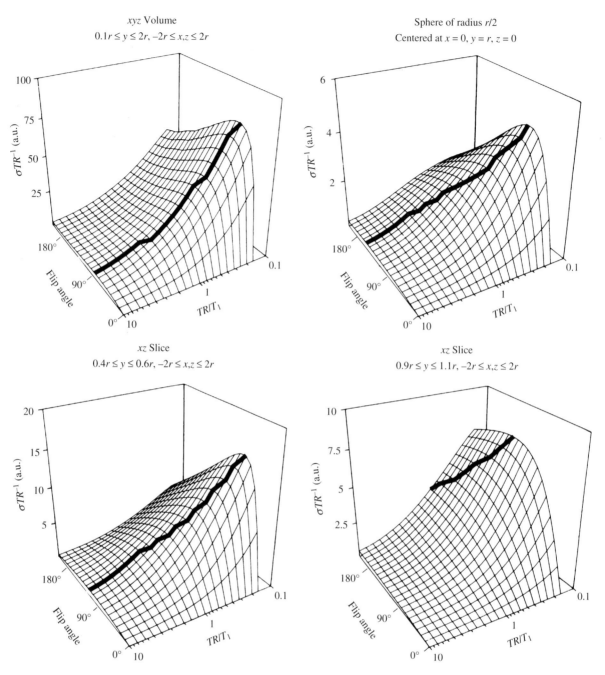

Figure 2 Sensitivity per unit time (σ/TR) as a function of TR/T_1 (logarithmic scale) and rf flip angle α for a transmit–receive single-turn flat surface coil. The NMR signal, as given in equation (1), is integrated over four different volumes: (upper left) 'total' volume covering linear dimensions of $2r$ and $4r$ perpendicular and parallel to the coil (radius r), respectively; (upper right) spherical volume of radius $r/2$ centered at a distance r from the coil center; (lower left) slice of thickness $r/5$ at a distance $r/2$ parallel to the coil; (lower right) similar slice at a distance r. The solid lines indicate maximum sensitivity. (Reproduced by permission of Academic Press from Haase et al.[10])

in twice the signal from a plane perpendicular to the first gradient axis, whereas all signals outside the inverted area have the same phase and thus cancel.

In its original form the ISIS method is only valid for fully relaxed acquisitions with $TR > 3\,T_1$. This is due to the fact that incomplete recovery of the longitudinal magnetization prior to a subsequent preparation period affects the resulting longitudinal magnetization. Since preparation of the longitudinal magnetization is different for the eight experiments, the cance-

lation of unwanted signal remains incomplete and also depends on the order in which the eight acquisitions are performed. This problem may be solved by a postacquisition saturation or 'magnetization purging' pulse[21] eliminating any residual longitudinal magnetization in between ISIS acquisitions. Further improvements are due to the use of frequency-modulated 'adiabatic' inversion pulses[21,22] that sharpen the spatial profiles of the inverted planes. In addition, signal contamination from outside the VOI has been attenuated by means of outer volume

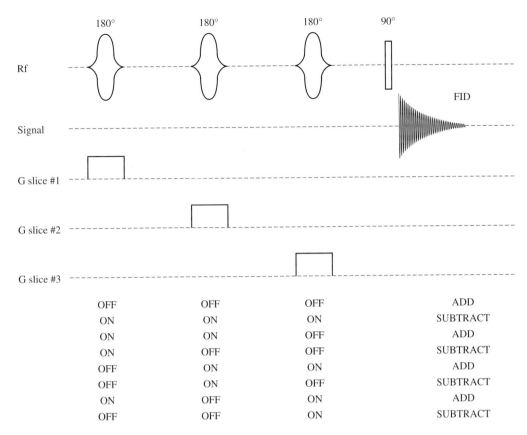

OFF	OFF	OFF	ADD
ON	ON	ON	SUBTRACT
ON	ON	OFF	ADD
ON	OFF	OFF	SUBTRACT
OFF	ON	ON	ADD
OFF	ON	OFF	SUBTRACT
ON	OFF	ON	ADD
OFF	OFF	ON	SUBTRACT

Figure 3 ISIS localization sequence comprising three slice-selective 180° inversion rf pulses along orthogonal gradient axes. A Gaussian shape has been chosen for schematic drawing only, while frequency-modulated pulses of the hyperbolic-secant type are most frequently employed. The measured NMR signal is excited by a final 90° read pulse. For full three-dimensional localization the ISIS sequence needs to be repeated at least eight times using inversion pulses on/off in all three gradient axes (2^3 combinations) with proper setting of the receiver phase (add/subtract)

suppression schemes, in particular by using noise-modulated rf excitation in combination with dephasing gradients.[23]

The main advantage of the ISIS approach is the use of the FID since it avoids signal losses for metabolite resonances with short T_2 relaxation times as frequently observed in vivo. Only very small T_1 relaxation losses occur during the initial preparation period. Effective decoupling of localization from excitation in the ISIS sequence provides potential for combinations with more complex multipulse excitation sequences including spectral editing techniques. Major disadvantages of ISIS originate from its multishot character and the fact that localization of a small VOI signal is accomplished by subtraction of large signals from almost the entire object. Thus, the technique may lead to substantial signal contamination and shows considerable sensitivity to motion-induced localization errors. Signal contamination may also arise from spins within inverted planes but outside the VOI through T_1 relaxation between inversion and excitation. Moreover, the excited though unwanted signal outside the VOI increases with decreasing size of the VOI. Finally, localized optimization of magnetic field homogeneity and water suppression as well as online monitoring of dynamic acquisitions are precluded as they require single-shot capabilities.

Despite some useful technical improvements, applications of single voxel ISIS spectroscopy have been mainly directed toward phosphorus NMR with only rare applications to proton NMR of human brain.[24–28] Figure 4 shows ISIS proton NMR

spectra of a patient with chronic infarction[27] where the nonselective 90° excitation pulse has been replaced by a semiselective spin echo sequence with different echo times TE. The excited frequency range was centered between the methyl resonances of N-acetylaspartate (NAA) and lactate to improve water suppression and to eliminate homonuclear J modulation of the lactate doublet at 1.33 ppm weakly coupled to the nonexcited methine quartet at 4.12 ppm.

2.3 Point-Resolved Spectroscopy (PRESS)

In vivo proton NMR spectroscopy of human brain is more demanding in terms of localization acuity and lack of outer volume contamination than usually can be fulfilled by surface coil and ISIS localization strategies. On the other hand, potential problems due to T_2 attenuation are considerably reduced since proton resonances of cerebral metabolites exhibit much longer T_2 relaxation times than in skeletal muscle, liver, and heart or are found in phosphorus NMR. PRESS and STEAM sequences therefore achieve localization by integrating slice selection and spectral excitation. Both methods directly excite the NMR signal from the desired VOI in a single 'shot' rather than attempt to saturate or subtract the much larger signal from outer volumes. In addition, PRESS and STEAM lend themselves to a user-friendly and partially automated spectroscopy

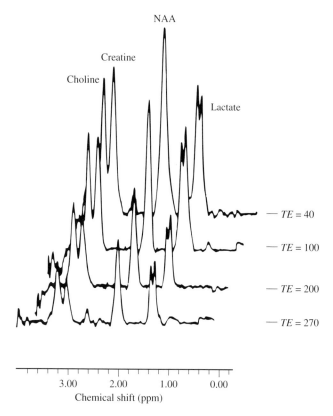

Figure 4 Localized proton NMR spectra of a patient with a chronic infarction at different echo times (*TE*) obtained using ISIS with outer volume suppression pulses and semiselective spin echo excitation. Other parameters: 2.0 T, VOI 40 × 25 × 30 mm^3, *TR* = 3000 ms, 256 transients. Frequency-selective excitation was centered between the methyl resonances of *N*-acetylaspartate (NAA) and lactate thereby eliminating *J* modulation of the lactate doublet at 1.33 ppm. Water suppression was achieved by an initial DANTE presaturation sequence. The patient spectra demonstrate a reduced NAA/creatine ratio and a strong contribution from lactate. (Reproduced by permission of Academic Press from Sappey-Marinier et al.[27])

examination ranging from image selection and VOI definition to spectral evaluation and quantitation.

Figure 5 shows a basic rf pulse and gradient scheme for a PRESS localization sequence[29,30] comprising slice-selective 90° rf excitation, dual slice-selective 180° rf refocusing, and acquisition of the resulting secondary spin echo (SE). The VOI is defined by the intersection of three orthogonal planes since only spins that experience all three rf pulses refocus as a SE signal at the double echo time *TE* = *TE*1 + *TE*2. To avoid multiple quantum interferences caused by imperfect slice-selective refocusing pulses with non-180° flip angles, the echo time *TE*1 of the first SE should be chosen as short as possible (see Section 3.2). Similar arguments hold true for a reduction of motion sensitivity (see Section 3.3). In most cases only the second half of the SE is acquired. This is due to the fact that proton signals of a well-shimmed VOI in brain exhibit rather long T_2^* relaxation times, so that even for long echo times such as *TE* = 270 ms the leading part of the SE signal becomes truncated by the localization sequence itself.

For proton NMR spectroscopy of metabolites the PRESS sequence has to be combined with water suppression. While a variety of techniques including differential relaxation, binomial excitation, and frequency-selective saturation by long pulses or pulse trains have been employed on spectrometers without shaped rf pulse capabilities, the most convenient method on modern NMR systems is the use of chemical shift selective (CHESS) water suppression.[31] The method is based on frequency-selective excitation of the resonance to be suppressed by an amplitude-modulated rf pulse followed by dephasing of the created transverse magnetization by spoiler gradients. For maximum water suppression several CHESS cycles may be applied immediately prior to the localization sequence.

The PRESS proton NMR spectra of parietal white matter of a normal human subject are shown in Figure 6 for a short (top) and long (bottom) echo time. Resonances have been assigned to *N*-acetylaspartate (NAA), creatine and phosphocreatine (Cr), choline-containing compounds (Cho), *myo*-inositol (*myo*-Ins), and glutamate/glutamine (Glx). The resonance at about 4.7 ppm represents the residual water proton signal which is further attenuated by T_2 relaxation at the echo time of *TE* = 270 ms.

The PRESS sequence represents a single-shot method that is easy to use on commercial whole body NMR systems including image selection and *localized* optimization of magnetic field homogeneity and water suppression. Since PRESS excites only those magnetizations that contain the desired VOI as part of an excited slice, most of the outer volume equilibrium magnetization remains unaffected. Gradient spoiling is applied to FID signals from outside the VOI that are due to the first 90° rf pulse as well as caused by imperfect 180° rf pulses. In contrast to ISIS, this problem decreases with decreasing size of the VOI as determined by three slice thicknesses. Of course, PRESS suffers from T_2 relaxation losses the extent of which depends on the chosen echo time, while coupled spins may be subject to substantial *J* modulation.

2.4 Stimulated Echo Acquisition Mode Spectroscopy (STEAM)

In close analogy to the PRESS technique, STEAM sequences define the VOI by the intersection of three planes excited by sequential application of slice-selective 90° rf pulses.[32–34] The schematic diagram shown in Figure 7 refers to a short echo time version corresponding to *TE* ≈ 10–30 ms. Retransformation of transverse magnetization excited by the first pulse into longitudinal magnetization components by the second pulse retains the phase information acquired during *TE*/2, but subjects the signal to T_1 relaxation during the middle interval *TM*. The FID associated with the second pulse is dephased by application of spoiler gradients during *TM* that also eliminate unwanted spin echoes. Application of the third pulse leads to two types of transverse magnetization. Firstly, components that carry phase information from the first *TE*/2 interval refocus after a total echo time (*TE*) as a stimulated echo (STE) attenuated by T_2 (and T_1) relaxation. Secondly, the FID from the equilibrium magnetization of the third slice outside the VOI as well as from T_1-relaxed longitudinal magnetizations during *TM* is dephased by the gradients following the third pulse (compare Figure 7).

It is important to note that proper functioning of the STEAM sequence requires complete dephasing of the transverse magnetization prior to application of the second rf pulse. Only randomization of transverse components ensures an equal

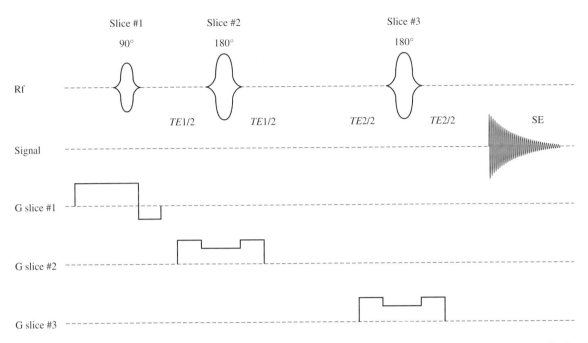

Figure 5   PRESS localization sequence comprising a slice-selective 90° rf excitation pulse followed by two slice-selective 180° rf refocusing pulses along orthogonal gradient axes. In most cases only the second half of the secondary spin echo (SE) occurring at the echo time $TE1 + TE2$ is acquired

distribution of spin moments that subsequently form the SE signal from the first two rf pulses (dephased during TM) and the STE signal following all three pulses. In the original work on stimulated echoes[35] dephasing was fulfilled by extremely poor magnet homogeneity, but has not been mentioned as a 'necessary condition' for generation of a STE signal without modulations due to resonance offsets. In fact, in the case of complete spin rephasing during $TE/2$ and on-resonance excitation of a single resonance such as water, the first two pulses of the STEAM sequence would effectively add to yield a 180° inversion. Application of the third pulse then results in a FID of the actually recovered magnetization in very much the same way as in an inversion–recovery T_1 experiment. Of course, no STE signal would be observable.

It is for the above reason that the slice-selective gradients for the second and third pulse in Figure 7 are refocused after the third rf pulse and prefocused before the second rf pulse, respectively. Since pertinent gradient pairs encode motion by inducing phase differences, it is advisable to keep TM as short as possible (also helpful to avoid T_1 losses) and as long as necessary to eliminate unwanted SE and FID signals by gradient spoiling. Typical values of $TM \approx$ 10–30 ms depend on system hardware (gradient strength) and eddy current performance.

For STEAM acquisitions at extremely short echo times, e.g. $TE = 3$ ms, refocusing of the initial transverse magnetization may be accomplished after the third rf pulse as demonstrated for the other two directions in Figure 7. Of course, a substantial reduction of TE also involves shortening of rf pulse durations which often is at the expense of slice profile (VOI) quality. Pertinent applications may be advantageous for proton NMR in the presence of unavoidable motions, e.g. in abdominal organs, or for nuclei with short T_2 relaxation times such as

phosphorus.[36] At long echo times care should be taken to keep the motion sensitivity to a minimum. Independence of TE may be accomplished by arranging respective slice-selective gradient pairs closely around the TM interval.

Figure 8 shows STEAM proton NMR spectra of parietal white (top) and gray matter (bottom) of a young healthy adult. Water suppression has been accomplished by three CHESS pulses preceding the STEAM localization sequence[34] in agreement with a detailed analysis of various suppression schemes.[37] In fact, CHESS pulses may also be incorporated into the TM interval, where longitudinal magnetizations are already restricted to a column.[38] Although such additional water suppression pulses slightly prolong the TM interval, they may be advantageous when using inhomogeneous rf coils.

Major resonances in proton NMR spectra of human brain are due to NAA, *N*-acetylaspartylglutamate (NAAG), glutamate (Glu), Cr, Cho, glucose (Glc), *scyllo*-inositol (*scyllo*-Ins), and *myo*-inositol (*myo*-Ins). Resonance assignments involved high-resolution high-field NMR spectra of tissue extracts from experimental animals and human biopsies as well as investigations of model solutions under the same conditions as used in vivo.[7,39–42] Further work is required to assign unambiguously additional resonances in the aromatic part of the spectrum as well as aliphatic resonances from mobile lipids and/or cytosolic proteins, while resonances from metabolites such as lactate, glutamine, and alanine become detectable when elevated in disease states of the brain. Further examples of STEAM spectra are given in Sections 3 and 4 discussing effects of strong spin–spin coupling and advantageous applications of single voxel proton NMR as compared with CSI, respectively.

In summary, the basic advantages of STEAM localization are identical to those of PRESS sequences (see Section 2.3). A

For References see p. 750

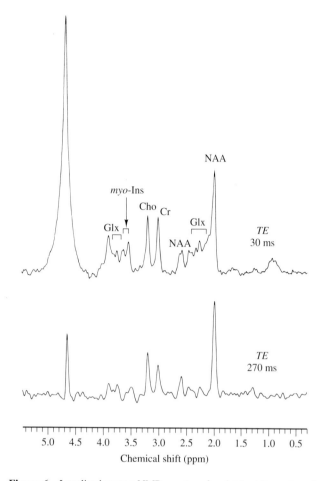

Figure 6 Localized proton NMR spectra of parietal white matter of a normal subject obtained using a PRESS sequence at $TE1 + TE2 = 30$ ms (top) and $TE1 + TE2 = 270$ ms (bottom). Other parameters: 2.0 T, VOI $25 \times 25 \times 25$ mm^3, $TR = 1500$ ms, 256 transients. Resonances are due to: N-acetylaspartate (NAA), creatine and phosphocreatine (Cr), choline-containing compounds (Cho), myo-inositol (myo-Ins), glutamate/glutamine (Glx), and residual water at 4.7 ppm. (Courtesy of T. Ernst and J. Hennig)

special merit stems from the excellent definition of the STEAM VOI via slice-selective 90° rather than 180° rf pulses. This feature translates into even less practical problems with signal contamination, e.g. when selecting a VOI position close to the lipid-containing skull. Moreover, STEAM sequences allow shorter echo times than PRESS when using the same hardware (gradient strength, pulse durations, etc.). This is a benefit from the *TM* interval in STEAM as opposed to the two echo intervals *TE*1 and *TE*2 in PRESS. Another advantage relative to PRESS is due to the fact that STEAM sequences result in only half the *J* modulation of PRESS sequences at the same echo time (see Section 3.1), which considerably improves the detectability of homonuclear spin-coupled resonances in short echo time proton NMR spectra of tissue metabolites. This particularly applies to strongly coupled multiplets suffering from incomplete refocusing and signal loss at longer echo times. A typical example at 2.0 T is given by the Glu methylene resonances in the 2.0–2.5 ppm range which are much more reliably resolved in the STEAM *TE* = 20 ms spectrum (Figure 8) than in the PRESS *TE* = 30 ms spectrum (Figure 6), where *J* modu-

lation corresponds to that of a STEAM sequence at $TE = 60$ ms.

On the negative side, STEAM results in only half the signal strength of PRESS at the same echo time. Of course, this theoretical value only holds true for exactly the same VOI: improved spatial discrimination by slice-selective 90° rf pulses in STEAM should not be mistaken as loss of signal. The STEAM technique also exhibits very small T_1 relaxation losses of the order of $[1-\exp(-1/50)]$ or about 2%. Thus, PRESS sequences are recommended for acquisitions at long echo times (e.g. $TE = 135$–270 ms), while STEAM sequences demonstrate superior performance at short echo times (e.g. $TE = 20$ ms). With respect to ISIS it is worth noting that spatial discrimination of the VOI by direct excitation with PRESS and STEAM is independent of the power of the rf excitation pulses, so that deviations from 90° or 180° flip angles reduce the resulting VOI signal but do not compromise localization acuity.

2.5 Miscellaneous Methods

This section gives a brief overview of miscellaneous localization techniques that have been described in the literature. Although of theoretical interest, many sequences suffer from practical limitations and therefore have only been applied to phantom or animal studies performed on small-bore NMR systems. In particular, a whole class of experiments relies on the use of 'hard' nonselective rf pulses that require strong rf power to be of sufficiently short duration/large bandwidth for equivalent manipulation of unwanted spins outside the desired section. Pertinent localization techniques comprise a threefold application of a sequence module [A] in order consecutively to restore longitudinal magnetization within selected sections along orthogonal gradient axes. Since localization sequences of the type

$$[A]_x - [A]_y - [A]_z - \text{excitation} \qquad (4)$$

decouple spatial selection and signal excitation, the final rf excitation sequence may be as simple as a 90°(nonsel)-FID sequence or comprise water suppression by frequency-selective excitation. While ISIS resembles a sequence of type (4) with use of slanted 180° inversion pulses, a parent method with *nonselective* rf pulses is volume selective excitation (VSE).[43] The sequence employs a pulse package [A] according to[44]

$$45°_{x'}(\text{sel}) - 90°_{-x'}(\text{nonsel}) - 45°_{x'}(\text{sel}) \qquad (5)$$

where the two selective 45° rf pulses represent the first and second half of a symmetrically selective 90° rf pulse, e.g. an amplitude-modulated pulse of Gaussian shape. Further examples are *spa*tially *r*esolved *s*pectroscopy (SPARS),[45]

$$90°_{x'}(\text{nonsel}) - 180°_{y'}(\text{nonsel}) - 90°_{-x'}(\text{sel}) \qquad (6)$$

*spa*tial and *c*hemical-shift-*e*ncoded *e*xcitation (SPACE),[46]

$$90°_{x'}(\text{sel}) - 180°_{y'}(\text{nonsel}) - 90°_{-x'}(\text{nonsel}) \qquad (7)$$

and *s*ymmetric *p*ulses for *a*ccurate *l*ocalization (SPALL),[47]

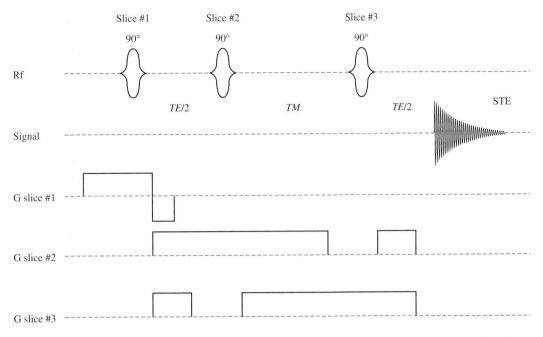

Figure 7 Short echo time STEAM localization sequence comprising three slice-selective 90° rf pulses along orthogonal gradient axes. Only the second half of the stimulated echo (STE) occurring at the echo time (*TE*) is acquired. The transverse magnetization excited by the first rf pulse is retransformed into longitudinal components by application of the second pulse, so that the intensity of the STE is subject to T_1 attenuation during the middle interval (*TM*) and to T_2 attenuation during *TE*

$$90^\circ_{x'}(\text{nonsel}) - 180^\circ_{y'}(\text{sel}) - 90^\circ_{-x'}(\text{nonsel}) \qquad (8)$$

'DIGGER'[48] is an analog of VSE using frequency-modulated sin–sinc rf pulses with reduced rf power requirements.

A further class of experiment takes advantage of outer volume suppression such as *lo*calization of *u*naffected *s*pins (LOCUS),[49] where series of slice-selective 90° rf pulses with overlapping excitation profiles are required in each direction to saturate all spins outside the desired VOI. Similar ideas apply to 'ROISTER'[50] using rotating magnetic field gradients in conjunction with noise-modulated rf pulses. In general, flat saturation is very difficult to achieve.[51] This particularly holds true for water-suppressed proton NMR studies where even minor signal contributions from outside the VOI may lead to substantial spectral contamination when integrated over the entire volume. Spatial presaturation therefore seems to be restricted to applications in NMR imaging, e.g. to avoid aliasing by outer volume/plane saturation or to manipulate directional blood flow in angiography or cardiac applications.

Another category of method is formed by localization sequences combining two or more basic techniques such as DIGGER with SPACE,[52] VSE (or SPARS) with spin echo rf excitation,[53] 1D-ISIS with PRESS,[54] and spin echo rf excitation with STEAM.[55] In addition, new rf pulse designs and foreseeable improvements in magnetic field gradient performance with regard to strength, speed, and accessible waveforms promise multidimensionally selective excitation by just one rf pulse. Preliminary results have been reported for 2D selective[56,57] and even 3D selective rf pulses[58] as well as for special hardware developments such as radial gradient coils.[59] In spoiling unwanted magnetization from planar excitations by rotating gradients, the latter approaches bear similarities to spatially selective saturation techniques. Again, current results underline some potential in NMR imaging, whereas applications for localized proton NMR spectroscopy still suffer contamination problems.

Finally, mixed forms of NMR spectroscopy and imaging have been proposed that achieve spectral localization with use of optimized point spread functions reflecting the shape of a desired VOI. For example, the *s*pectral *lo*calization with *o*ptimal *p*ointspread function (SLOOP)[60] as a generalization of the *s*pectral *lo*calization by *im*aging (SLIM)[61] results in a single voxel spectrum from an arbitrarily shaped VOI by means of a CSI acquisition with modulated phase-encoding gradients, i.e. by matched numbers of transients per gradient step. The approach is analogous to Fourier series windowing techniques[17–19] that allow focusing of a spectroscopic image in the rotating frame.

3 CHARACTERISTICS OF PRESS AND STEAM

Since both PRESS and STEAM sequences excite rf refocused echoes with use of three rf pulses, respective signals are sensitive to modulation by homonuclear spin–spin couplings or lead to the evolution of multiple quantum coherences. These effects are not seen with single pulse FID acquisitions unless surface coil or ISIS techniques are extended by appropriate excitation sequences.[62] In addition, gradient-localized sequences are subject to motion-induced amplitude and phase errors that either ought to be minimized or may be exploited for motion editing and chemically specific diffusion measurements. Some of these characteristics are discussed in the following sections.

Common properties are also due to the fact that single voxel proton NMR techniques primarily focus on studies of the central nervous system. Such applications are motivated by the

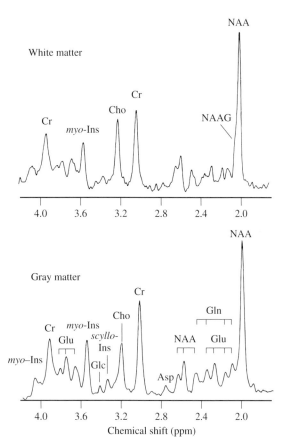

Figure 8 Localized proton NMR spectra of parietal white (top) and gray matter (bottom) of a normal subject obtained using a short echo time STEAM sequence under fully relaxed conditions. Parameters: 2.0 T, VOI 20 × 30 × 20 mm^3 (white matter) and 20 × 30 × 30 mm^3 (gray matter), *TR/TE/TM* = 6000/20/30 ms, 64 transients. Resonances are due to *N*-acetylaspartate (NAA), *N*-acetylaspartylglutamate (NAAG), glutamate (Glu), glutamine (Gln), aspartate (Asp), phosphocreatine and creatine (Cr), choline-containing compounds (Cho), glucose (Glc), *scyllo*-inositol (*scyllo*-Ins) and *myo*-inositol (*myo*-Ins). Chemical shifts are given in parts per million (ppm) and referenced to 2.01 ppm for the CH$_3$ group of NAA

importance and uniqueness of a noninvasive access to cerebral metabolism and its sensitivity to pathologic alterations in disease states of the brain ranging from neoplasms and stroke to psychiatric diseases and inborn errors of metabolism. In addition, proton NMR of the living brain is facilitated by advantageous tissue properties not found in other organ systems such as small variability of magnetic susceptibilities, metabolites with long T_2 relaxation times, and minimum motions.

For human applications the minimum accessible VOI for any of the localization techniques discussed is of the order of 1 mL (e.g. 10 × 10 × 10 mm^3). This estimate is based on the following assumptions: (i) a noninvasive approach with use of a circumscribing head coil, (ii) a reasonable measuring time of only a few minutes, and (iii) adequate spectral S/N. Obviously, VOI dimensions of several millimeters result in spectra comprising signals from both neural and glial structures as well as from vascular compartments and/or cerebrospinal fluid. Considerable improvements in spatial resolution will only be

possible by interventional procedures using miniaturized rf coils attached to catheters.

3.1 Spin–Spin Coupling

Figure 9 shows the effect of *J* modulation on spin echo proton NMR spectra of ethanol as a function of echo time *TE*. A spin–spin coupling constant of *J* = 7 Hz and a chemical shift difference of δ = 245 Hz (2.45 ppm) result in weak coupling of the methylene quartet and methyl triplet even at the low field strength of 2.35 T. Refocusing of the triplet occurs at *TE* = 1/*J* or 143 ms (inversion of quartet), while both the triplet and the quartet refocus at *TE* = 2/*J* or 286 ms. The central triplet resonance is not modulated due to the absence of a differential precession frequency and the well-known rephasing of resonance offsets (chemical shifts) by rf echoes.

Doublet resonances refocus at *TE* = 1/*J*, as frequently observed in vivo for the lactate methyl doublet in PRESS spectra of patients with cerebral disorders. This finding is in contrast to corresponding *J* modulation effects in STEAM spectra since stimulated echoes exhibit modulation frequencies of only half the value of those in spin echo sequences. Accordingly, lactate doublet resonances only refocus at *TE* = 2/*J* and integer multiples thereof. Applications of echo spectroscopy to coupled spin systems have been discussed by several authors for PRESS,[63] STEAM,[64] and both localization sequences.[65,66] In principle, it is even possible to extend these sequences to localized 2D NMR spectroscopy as has been demonstrated for a 2D *J*-resolved version of STEAM.[67]

With the prominent exceptions of lactate, alanine, and ethanol, most homonuclear proton interactions lead to strong coupling ($J/\delta \geqslant 0.1$) at low field strengths and therefore give rise to marked multiplet distortions. Such resonance patterns also deviate severely from the modulation behavior of weakly coupled spin systems. In particular, the intensities of strongly coupled multiplets tend to fade out with increasing echo time rather than to rephase at selected echo times. In practical terms it therefore turns out to be best to minimize *J* modulation by using short echo times and STEAM sequences. A typical example is shown in Figure 10 for the case of *myo*-Ins. The complex multiplet pattern at 7.0 T [Figure 10(a)] is simplified to two major resonances at 3.56 ppm and 4.06 ppm [Figure 10(b)], whereas the true singlet resonance from the six protons of *scyllo*-Ins at 3.35 ppm remains unaffected. High levels of *myo*-Ins, *scyllo*-Ins, and Cho in white matter of a patient with a mitochondrial enzyme deficiency indicate enhanced membrane turnover and/or gliosis [Figure 10(c)]. The concomitant reduction of NAA and glutamate provides evidence for a loss of vital neuroaxonal tissue.

3.2 Multiple Quantum Effects

In addition to *J* modulation, stimulated echoes may be affected by the intermediate evolution of multiple quantum coherences during *TM* (***Multiple Quantum Coherence Imaging***). For example, when using an echo time of *TE* = 1/*J* such as to provide antiphase magnetization for a doublet at the end of the first *TE*/2 interval, i.e. *TE*/2 = 1/(2*J*), then application of the second 90° rf pulse creates zero-quantum and double-quantum coherences that evolve during the middle interval *TM*. The final read pulse converts multiple quantum coherences into

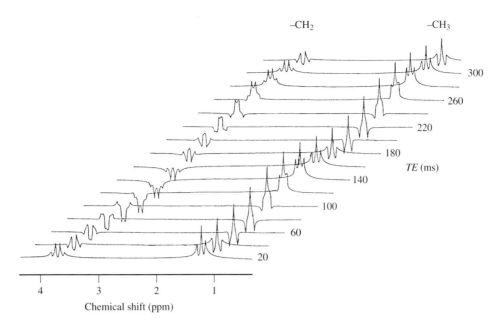

Figure 9 Proton NMR spectra of ethanol obtained using a spin echo sequence without spatial localization as a function of echo time ranging from $TE = 20–320$ ms (2.35 T). Modulation of the weakly coupled ($J/\delta = 0.03 \ll 1$) methyl triplet (1.20 ppm) and methylene quartet (3.70 ppm) occur with a period of $1/J = 143$ ms (outer triplet resonances) and $2/J = 286$ ms (central quartet resonances). The homonuclear spin–spin coupling constant is $J = 7.0$ Hz (Courtesy of T. Michaelis)

detectable transverse magnetization and causes an amplitude modulation of the resulting STE signal in correspondence to the phase of the coherence acquired during *TM*. Assuming gradient switches as for the STEAM sequence shown in Figure 7 where gradients are balanced to refocus single quantum magnetizations of static spins, double quantum coherences become automatically dephased as their precession frequencies correspond to the *sum* of chemical shifts of coupled resonances, i.e. about twice the precession frequencies of transverse magnetizations. Of course, on the one hand, adjustment of gradient refocusing according to multiple quantum order provides a simple means for order selection.[68] On the other hand, zero quantum coherences survive any gradient spoiling since their precession frequencies represent the chemical shift *difference* of coupled spins and therefore are completely independent of resonance offsets and magnetic field inhomogeneities.

The effect of zero quantum modulation as a function of *TM* is demonstrated in Figure 11 for localized STEAM spectra of an aqueous solution of lactate. The *TM* interval was incremented in steps of 0.5 ms while holding the echo time constant at $TE/2 = 1/(2J) = 68$ ms. The observed zero quantum modulation of the methyl doublet (1.33 ppm) weakly coupled to the methine quartet (4.12 ppm, $J = 7.1$ Hz) refers to the chemical shift difference of 180 Hz at 1.5 T. It is apparent that subtraction of inverted spectra, e.g. at $TM = 27$ ms and 30 ms, would result in twice the lactate methyl doublet while eliminating all noncoupled spins as well as coupled spins with other chemical shift differences, i.e. zero quantum modulation frequencies. The approach clearly indicates potential for in vivo lactate editing[69] in the presence of strong overlap with lipid signals as in skeletal muscle and kidney. Refinements of the basic technique include 2D zero-quantum and double-quantum spectroscopy,[70,71] considerations of strong coupling,[72] and most

importantly the design of single-shot multiple quantum filters for STEAM localization sequences to overcome the motion sensitivity of subtraction techniques or phase cycling schemes.[73,74]

In PRESS sequences 180° rf pulses with imperfect slice profiles subject a part of the magnetization to a stimulated echo sequence. Thus, depending on the actual echo times *TE*1 and *TE*2 and the homonuclear spin–spin couplings involved, PRESS sequences may also lead to multiple quantum modulation. Excitation of respective coherences is best avoided by not creating antiphase magnetization at the time of the second rf pulse, which requires the first half-echo interval *TE*1/2 in PRESS and *TE*/2 in STEAM to be as short as possible. This allows arbitrary echo times *TE*2 in PRESS and middle intervals *TM* in STEAM. A theoretical and experimental comparison of STEAM and PRESS sequences has addressed a number of properties including homonuclear coupling, minimum attainable *TE*, slice profiles, and motion/diffusion.[75]

3.3 Flow and Motion

In close analogy to NMR imaging, gradient-localized echo spectroscopy suffers from chemical shift artifacts in localization acuity, i.e. metabolite-specific positional displacements, and signal loss due to amplitude and phase changes associated with macroscopic and microscopic motions (*Whole Body Magnetic Resonance Angiography*; *Pulsatility Artifacts Due to Blood Flow and Tissue Motion and Their Control*; *Time-of-Flight Method of MRA*; *Phase Contrast MRA*). For proton NMR at field strengths of 1.5–2.0 T chemical shift displacements can be overcome by strong magnetic field gradients that reduce the spatial equivalents of chemically induced frequency differences

For References see p. 750

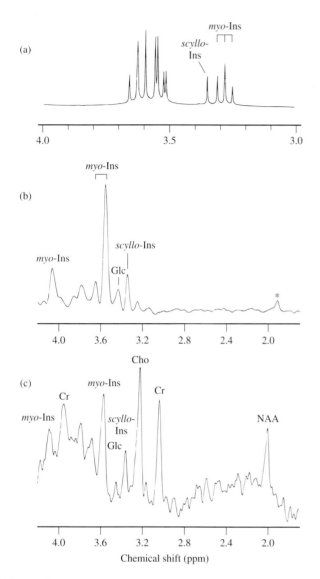

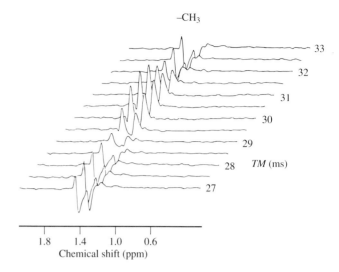

Figure 11 Localized proton NMR spectra of an aqueous solution of 100 mM lactate obtained using a STEAM sequence with a middle interval *TM* ranging from 27 to 33 ms. Other parameters: 1.5 T, VOI $40 \times 40 \times 40$ mm^3, *TR/TE* = 3000/135 ms, 4 transients. Antiphase magnetization components at *TE*/2 = 1/(2*J*) lead to multiple quantum coherences that evolve during *TM*. Elimination of all higher orders by magnetic field gradients as in Figure 7 results in exclusive zero quantum modulation of the methyl doublet (1.33 ppm) weakly coupled to the methine quartet (4.12 ppm, *J* = 7.1 Hz). The modulation period reflects the chemical shift difference of 180 Hz at 1.5 T

Figure 10 Proton NMR spectra of inositols (Ins) at different field strengths. (a) FID spectrum (7.0 T, *TR* = 10 000 ms) of a 1:18 mixture of *scyllo*-Ins (singlet at 3.35 ppm) and *myo*-Ins exhibiting a distorted triplet centered at 3.28 ppm, strongly coupled multiplets at 3.5–3.7 ppm, and a distorted triplet at 4.06 ppm (not shown). (b) Localized STEAM spectrum of a 1:12:2 mixture of *scyllo*-Ins, *myo*-Ins, and glucose (Glc) line broadened to mimic in vivo conditions (2.0 T, VOI $20 \times 20 \times 20$ mm^3, *TR/TE/TM* = 6000/20/30 ms, 32 transients, acetate reference). (c) Localized STEAM spectrum of left frontoparietal white matter of a patient with Leigh's disease (2.0 T, VOI $16 \times 16 \times 16$ mm^3, *TR/TE/TM* = 3000/20/30 ms, 128 transients). (Reproduced by permission of Wiley from Michaelis et al.[42])

to a submillimeter scale. If residual eddy currents associated with gradient switching lead to spectral distortions, then sufficient decline of frequency fluctuations prior to data acquisition is only achieved by prolonged echo times. In general, phase distortions of spectroscopic time-domain data may be corrected with use of a reference scan.[76]

Flow effects manifest themselves in two different ways: (i) inflow/outflow of spins carrying longitudinal magnetization, and (ii) motion-induced phase errors that reflect the sensitivity of transverse magnetizations to positional displacements in

magnetic field gradients with a component along the flow direction. In localized NMR spectroscopy phase changes within the VOI result in partial cancelation of individual magnetization components and thus incomplete refocusing of SE or STE signals. This particularly holds true for large VOI sizes and at long echo times. In addition, overall phase differences from scan to scan may lead to considerable signal attenuation rather than to signal accumulation during data averaging. This observation may be easily verified since the desired S/N improvement will be less than the expected factor of $\sqrt{n}$, with *n* the number of acquisitions.

Figure 12 demonstrates both the amplitude and the phase effect of flowing water on long echo time localized proton NMR spectra obtained with and without velocity compensated gradient waveforms.[77] Without rephasing, even very small flow velocities of the order of 1 mm s^{-1} cause substantial signal loss (bottom left). Velocity compensation not only eliminates such phase errors and associated signal losses, but also benefits from the inflow effect of unsaturated spins from outside the VOI when using short *TR* values. A corresponding signal increase is seen at a velocity of 1.3 mm s^{-1} as spectra were acquired under conditions of partial saturation (bottom right). Irreversible signal loss due to turbulent flow becomes dominant at even higher flow velocities (here at $\geqslant$ 2 mm s^{-1}).

In theory, PRESS sequences are less sensitive to motion than STEAM sequences. This is due to the gradient symmetry normally employed for the two slice-selective 180° rf pulses and thus a benefit from the rephasing properties of the secondary SE ('even echo rephasing'). However, for both sequences the best strategy emerges from avoiding flow/motion-induced phase errors with use of short echo times. In special circum-

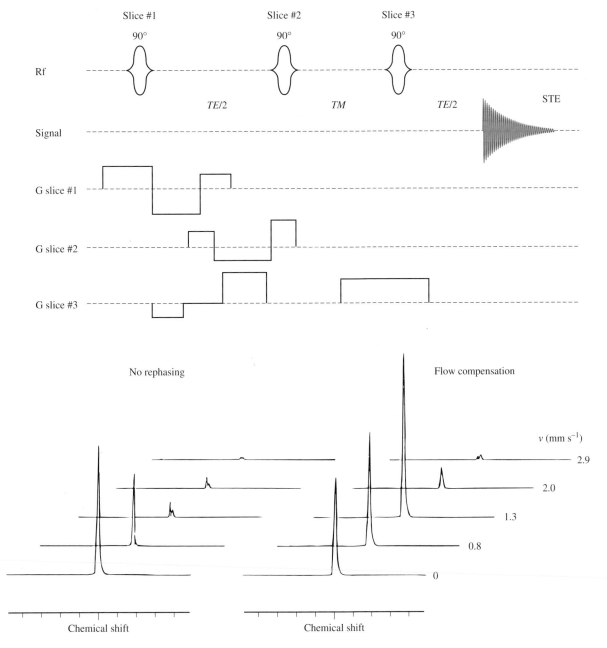

Figure 12 (Top) Long echo time STEAM localization sequence with first-order motion-rephasing gradients in all three directions. (Bottom) Localized proton NMR spectra of tap water flowing in a glass tube of 5 cm diameter at different velocities $v = 0$, 0.8, 1.3, 2.0, and 2.9 mm s^{-1}. The spectra were obtained using STEAM sequences in the absence (left) and presence (right) of a velocity-rephasing gradient waveform along the principal flow direction. Other parameters: 2.35 T, VOI $13 \times 13 \times 13$ mm^3, *TR/TE/TM* = 2000/270/100 ms, 32 transients. (Reproduced by permission of Academic Press from Gyngell et al.[77])

stances, flowing spins may be exploited for motion editing. Discrimination of stationary metabolite signals from flowing metabolites is a common problem in phosphorus spectroscopy of the heart where phosphate resonances from 2,3-diphosphoglycerate in blood within the cardiac chambers contribute to phosphorus spectra of myocardial tissue.[78] Further applications stem from elimination of resonance signals from flowing liquids used for temperature control in high-resolution NMR. This also holds true for studies of cell suspensions where contamination with substrates from the perfusate may be completely avoided by motion-encoding gradient pairs.[79]

3.4 Diffusion

NMR measurements of molecular self-diffusion coefficients in vivo provide insight into cellular dynamics and give hints as to intra- and extracellular structural organization (*Anisotropically Restricted Diffusion in MRI*; *Methods and Applications of Diffusion MRI*). While diffusion-weighted NMR imaging of water protons is unable to discriminate between multiple compartments, diffusion-weighted water-suppressed spectroscopy in most cases reflects intracellular metabolite mobility. Such studies may be even more specific since metabolites such as NAA

For References see p. 750

and *myo*-Ins have been found to be exclusively located in neuronal and glial cells, respectively.

The most suitable localization sequence for in vivo diffusion weighting is STEAM. It allows for long diffusion times Δ without detrimental effects due to strong T_2 attenuation in systems with $T_2 \ll T_1$. Thus, variable diffusion times (e.g. different *TM* intervals) may be used to study the influence of restricted diffusion when dephasing and rephasing parts of the diffusion gradient pair are separated into the first and second *TE*/2 intervals. On the other hand, to determine the free diffusion coefficient in vivo and to minimize encoding of competitive motions such as flow, diffusion weighting is best accomplished by the incorporation of a bipolar self-balanced gradient of duration δ and effective separation Δ into one of the *TE*/2 intervals of the STEAM sequence. While this strategy compensates for motion-induced phase errors as long as pertinent motions exhibit constant velocity, a further reduction of the flow/motion sensitivity of diffusion-weighted sequences is achieved by respiratory triggering and/or scan-to-scan phase alignment using postacquisition correction schemes.

Figure 13 shows diffusion-weighted localized proton NMR spectra of rat brain obtained using bipolar diffusion gradients during both *TE*/2 intervals to yield short diffusion times of $\Delta = \delta = 25$ ms at an echo time of $TE = 120$ ms.[80] The acquisition of spectra with metabolite signal intensities $S(b)$ as a function of diffusion weighting factor b allows the calculation of molecular self-diffusion coefficients D from the signal attenuation according to

$$\ln(S/S_0) = -Db \qquad (9)$$

with

$$b = \gamma^2 (\Delta - \delta/3)\delta^2 G^2 \qquad (10)$$

and G the diffusion gradient strength. Since metabolite diffusion coefficients in vivo can only be measured in the presence of residual unavoidable/uncorrectable motions, values are expressed as apparent diffusion coefficients (ADC).

The ADC value of 0.27×10^{-3} mm^2 s^{-1} found for NAA, Cr, and Cho in rat brain is about a factor of three smaller than the water proton ADC of 0.75×10^{-3} mm^2 s^{-1}. In human subjects an ADC value of 0.65×10^{-3} mm^2 s^{-1} was obtained for water protons, while values of 0.18, 0.15, and 0.13×10^{-3} mm^2 s^{-1} were reported for NAA, Cr, and Cho, respectively.[81] The finding of smaller metabolite ADC values in human brain compared with rat brain is most likely to be due to the use of a long diffusion time of $\Delta > 200$ ms in the human study that reduces the ADC by the influence of restricted diffusion.

Despite current limitations with regard to gradient strength and eddy current performance, diffusion spectroscopy is an active area of research. This is largely due to expectations that cell-specific information will contribute to the understanding of pathophysiologic events associated with stroke. In contrast to water diffusion NMR imaging, studies of intracellular metabolites may clarify the nature of the reduced water ADC observed shortly after ischemia as well as its subsequent rise to larger than normal values within 5–7 days after stroke.

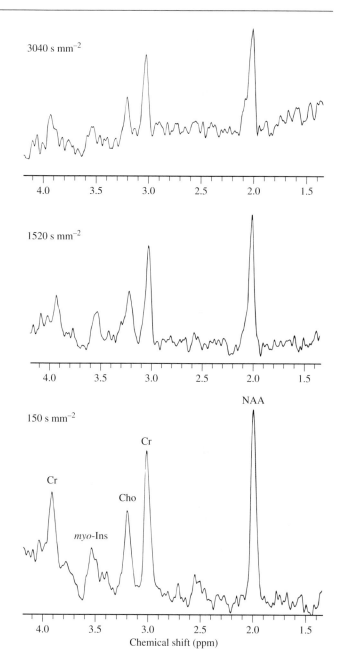

Figure 13 Diffusion-weighted localized proton NMR spectra of rat brain in vivo obtained using a STEAM sequence with bipolar diffusion gradients G in each *TE*/2 interval ($\Delta = \delta = 25$ ms). Other parameters: 2.35 T, VOI $5 \times 5 \times 5$ mm^3, *TR*/*TE*/*TM* = 3000/120/30 ms, 128 transients triggered to respiration. Spectra with increasing diffusion weighting factor b according to equation (10) are shown from bottom to top (150, 1520, and 3040 s mm^{-2}). The orientation of the diffusion gradient was along the long axis of the animal. (Reproduced by permission of Williams and Wilkins from Merboldt et al.[80])

4 SINGLE VOXEL LOCALIZATION OR CHEMICAL SHIFT IMAGING?

Biological NMR covers a wide range of diverse methods and applications. Thus, the absence of a 'single best' technique for a complex field such as in vivo NMR spectroscopy is not surprising. Instead, different problems will require different solutions and specific applications should ask for problem-oriented decisions as to which strategy to employ. For example,

Table 1 Absolute concentrations (mM) of major metabolites in human parieto-occipital white and parietal gray matter in vivo as obtained from fully relaxed (TR = 6000 ms) short echo time (TE = 20 ms) localized proton NMR spectra (STEAM) of young healthy adults (age 27 ± 4 years).[82] The values have been determined without user interference using a fully automated spectral evaluation program[83] and a library of localized model metabolite spectra calibrated with known concentrations.[84] VOI concentrations are given without corrections for partial volume effects and T_2 relaxation differences between in vitro and in vivo conditions.

	White matter (n = 61)	Gray matter (n = 45)
NAA + NAAG	8.2 ± 0.7	9.1 ± 0.7
NAA	6.1 ± 0.7	8.9 ± 0.8
NAAG	2.1 ± 0.7	0.3 ± 0.4
Cr + PCr	4.9 ± 0.6	6.5 ± 0.5
Cho	1.6 ± 0.2	1.2 ± 0.1
myo-Ins	4.2 ± 0.7	4.4 ± 0.6
Glu	5.8 ± 1.2	8.8 ± 1.1
Gln	1.8 ± 1.2	4.1 ± 1.3

single voxel localization and CSI techniques are complementary with regard to temporal resolution and volume coverage. While CSI provides simultaneous access to the spatial distribution of metabolites in one, two, or three dimensions, the strength of a single voxel spectrum arises from a more detailed and more easily quantifiable metabolic picture, as well as from its speed. These factors are often beneficial to studies of cerebral metabolism by proton NMR spectroscopy, whereas the poor sensitivity and long measuring time of localized phosphorus NMR spectroscopy are more efficiently compensated for by CSI techniques with improved S/N per unit measuring time.

Accurate definition of the spatial origin of a spectrum is a prerequisite for its biochemical (and medical) interpretation. Single voxel techniques achieve this goal by direct slice-selective excitation of the VOI, whereas the spatial characteristics of a voxel spectrum from CSI reflect the point-spread function associated with discrete 1D, 2D, or 3D Fourier transformation. Since only small numbers of 8–32 phase-encoding steps per dimension are acquired due to time constraints of in vivo studies, the resulting sidelobes may cause significant signal contamination. Typically, a 2D CSI experiment with a 32 × 32 matrix and a repetition time TR = 2000 ms already leads to a measuring time of 34 min which is often beyond patient tolerance and hampered by motion artifacts.

Single-shot localization techniques such as PRESS and STEAM allow direct 'shimming' of the VOI. Good homogeneity improves the spectral quality in two respects: firstly, in terms of chemical shift resolution, and secondly, in terms of contrast-to-noise since the relevant parameter for detectability of a resonance is its peak-to-noise ratio rather than the signal strength as given by the resonance area. The residual linewidth of in vivo proton NMR spectra of human brain does not reflect spin–spin relaxation, but is determined by microscopic susceptibility differences resulting in inhomogeneous broadening. Another benefit from homogeneity and local parameter optimization is the excellent water suppression obtainable even for short echo times.

A most attractive aspect of single voxel proton NMR is the short measuring time of only a few minutes that may be exploited for several purposes. For example, a reliable quanti-

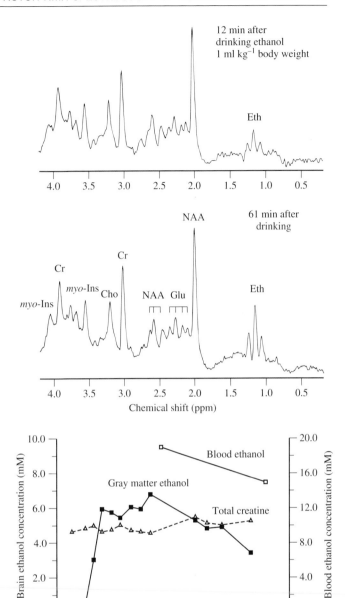

Figure 14 Uptake of ethanol (Eth) in parietal gray matter of a young healthy subject after drinking 1 mL of ethanol/kg body weight within a period of 5 min. (Upper traces) Localized proton NMR spectra reveal the presence of the methylene-coupled methyl triplet of brain ethanol at 1.20 ppm, 12 min and 61 min after drinking, respectively. Parameters: 2.0 T, VOI 16 × 30 × 30 mm^3, STEAM, $TR/TE/TM$ = 3000/20/30 ms, 128 transients. (Bottom graph) Time course of ethanol and creatine levels in gray matter as evaluated by LCModel[83] in comparison with blood plasma ethanol values (right scale). Tissue concentrations are not corrected for T_1 saturation, partial volume effects, and differential T_2 attenuation between in vivo and in vitro conditions. (Reproduced by permission of Plenum Press from Frahm[85])

tation of absolute metabolite concentrations without the need for saturation corrections may be accomplished by using long repetition times for fully relaxed acquisitions. Thus, typical recordings of localized proton NMR spectra of human brain (headcoil, 1–20 mL VOI) result in a measuring time of about 6

For References see p. 750

min assuming 64 transients with a repetition time $TR = 6000$ ms. The corresponding measuring time for the aforementioned 32×32 2D CSI experiment yields 102 min. Table 1 summarizes VOI concentrations of major metabolites in human brain in vivo from fully relaxed proton NMR spectra of white and gray matter.[82] The analysis was based on fully automated spectral evaluation[83] that takes advantage of a library of localized spectra from calibrated model metabolite solutions.[84]

Single voxel spectroscopy studies also provide immediate access to results without extensive postexamination processing as required for CSI. Control of spectra during the session improves the quality of patient studies by interactive decisions and avoids repeat studies. Most importantly, however, a temporal resolution of minutes allows monitoring of the time course of a physiological alteration or metabolic turnover via sequential acquisitions. An illustrative example is the uptake of ethanol after oral consumption of alcohol as shown in Figure 14 where proton NMR spectra from parietal gray matter reveal the methyl triplet signal only 12 min (top trace) after drinking.[85] The bottom part of Figure 14 shows the time course of the ethanol concentration in gray matter. Subsequent to a rapid increase, maximal tissue concentrations (central trace) were reached after 40–75 min (different subjects) followed by a slow decrease. Further examples of dynamic NMR spectroscopy deal with localized studies of glucose uptake and homeostasis,[28,85] glucose and lactate levels during physiologic stimulation,[25,86] and functional alterations of water proton signal strength and linewidth during brain activation.[87]

In conclusion, the selection of a localization technique should only depend on the question to be answered. In this respect, theoretical and experimental comparisons may be helpful in evaluating basic physical properties, but are of limited usefulness for a generalized judgement of relative merits. The ultimate criterion is of course given by the condition that the *results* of a localized NMR spectroscopy examination must be independent of the *method* chosen. Otherwise, in vivo NMR will not justify its potential as a unique noninvasive tool for both metabolic research and medical decision-making.

5 RELATED ARTICLES

Brain Infection and Degenerative Disease Studied by Proton MRS; Brain MRS of Human Subjects; Brain MRS of Infants and Children; Chemical Shift Imaging; In Vivo Hepatic MRS of Humans; Localization and Registration Issues Important for Serial MRS Studies of Focal Brain Lesions; Quantitation in In Vivo MRS; Single Voxel Whole Body Phosphorus MRS; Water Suppression in Proton MRS of Humans and Animals; Whole Body Studies: Impact of MRS.

6 REFERENCES

1. J. J. H. Ackerman, T. H. Grove, G. G. Wong, D. G. Gadian, and G. K. Radda, *Nature (London)*, 1980, **283**, 167.
2. R. E. Gordon, P. E. Hanley, D. Shaw, D. G. Gadian, G. K. Radda, P. Styles, P. J. Bore, and L. Chan, *Nature (London)*, 1980, **287**, 736.
3. B. Chance, Y. Nakase, M. Bond, J. S. Leigh, Jr., and G. McDonald, *Proc. Natl. Acad. Sci. USA*, 1978, **75**, 4925.
4. B. Chance, S. Eleff, J. S. Leigh, Jr., D. Sokolow, and A. Sapega, *Proc. Natl. Acad. Sci.*, 1981, **78**, 6714.
5. B. D. Ross, G. K. Radda, D. G. Gadian, G. Rocker, M. Esiri, and J. Falconer-Smith, *N. Engl. J. Med.*, 1981, **304**, 1338.
6. P. A. Bottomley, W. A. Edelstein, T. H. Foster, and W. A. Adams, *Proc. Natl. Acad. Sci.*, 1985, **82**, 2148.
7. K. L. Behar, J. A. den Hollander, M. E. Stromski, T. Ogino, R. G. Shulman, O. A. C. Petroff, and J. W. Prichard, *Proc. Natl. Acad. Sci.*, 1983, **80**, 4945.
8. K. L. Behar, D. L. Rothman, R. G. Shulman, O. A. C. Petroff, and J. W. Prichard, *Proc. Natl. Acad. Sci. U.S.A.*, 1984, **81**, 2517.
9. A. Haase. W. Hänicke, J. Frahm, and D. Matthaei, *Lancet*, 1983, **ii**, 1082.
10. A. Haase, W. Hänicke, and J. Frahm, *J. Magn. Reson.*, 1984, **56**, 401.
11. R. R. Ernst and W. A. Anderson, *Rev. Sci. Instrum.*, 1966, **37**, 93.
12. P. A. Bottomley, T. H. Foster, and R. D. Darrow, *J. Magn. Reson.*, 1984, **59**, 338.
13. R. Sauter, S. Müller, and H. Weber, *J. Magn. Reson.*, 1987, **71**, 167.
14. J. Hennig, C. Boesch, R. Gruetter, and E. Martin, *J. Magn. Reson.*, 1987, **75**, 179.
15. M. R. Bendall and R. E. Gordon, *J. Magn. Reson.*, 1983, **53**, 365.
16. D. I. Hoult, *J. Magn. Reson.*, 1979, **33**, 183.
17. J. Pekar, J. S. Leigh, Jr., and B. Chance, *J. Magn. Reson.*, 1985, **64**, 115.
18. K. R. Metz and R. W. Briggs, *J. Magn. Reson.*, 1985, **64**, 172.
19. M. Garwood, T. Schleich, B. D. Ross, G. B. Matson, and W. D. Winters, *J. Magn. Reson.*, 1985, **65**, 239.
20. R. J. Ordidge, A. Connelly, and J. A. B. Lohman, *J. Magn. Reson.*, 1986, **66**, 283.
21. T. J. Lawry, G. S. Karczmar, M. W. Weiner, and G. B. Matson, *Magn. Reson. Med.*, 1989, **9**, 299.
22. P. R. Luyten, J. P. Groen, J. W. A. H. Vermeulen, and J. A. den Hollander, *Magn. Reson. Med.*, 1989, **11**, 1.
23. A. Connelly, C. Counsell, J. A. B. Lohman, and R. J. Ordidge, *J. Magn. Reson.*, 1988, **78**, 519.
24. C. C. Hanstock, D. L. Rothman, J. W. Prichard, T. Jue, and R. G. Shulman, *Proc. Natl. Acad. Sci. U.S.A.*, 1988, **85**, 1821.
25. J. Prichard, D. Rothman, E. Novotny, O. Petroff, T. Kuwaborc, M. Avison, A. Howseman, C. Hanstock, and R. Shulman, *Proc. Natl. Acad. Sci. U.S.A.*, 1991, **88**, 5829.
26. D. L. Rothman, C. C. Hanstock, O. A. C. Petroff, E. J. Novotny, J. W. Prichard, and R. G. Shulman, *Magn. Reson. Med.*, 1992, **25**, 94.
27. D. Sappey-Marinier, G. Calabrese, H. P. Hetherington, S. N. G. Fisher, R. Deicken, C. van Dyke, G. Fein, and M. W. Weiner, *Magn. Reson. Med.*, 1992, **26**, 313.
28. R. Gruetter, D. L. Rothman, E. J. Novotny, G. I. Shulman, J. W. Prichard, and R. G. Shulman, *Magn. Reson. Med.*, 1992, **27**, 183.
29. P. A. Bottomley, *Ann. N.Y. Acad. Sci.*, 1987, **508**, 333; US Patent 4 480 228, 1984.
30. R. J. Ordidge, M. R. Bendall, R. E. Gordon, and A. Connelly, in 'Magnetic Resonance in Biology and Medicine', eds. G. Govil, C. L. Khetrapal, and A. Saran, Tata McGraw-Hill, New Delhi, 1985, p. 387.
31. A. Haase, J. Frahm, W. Hänicke, and D. Matthaei, *Phys. Med. Biol.*, 1985, **30**, 341.
32. J. Frahm, A. Haase, W. Hänicke, K. D. Merboldt, and D. Matthaei, German Patent DE 3 445 689, 1984.
33. J. Frahm, K. D. Merboldt, and W. Hänicke, *J. Magn. Reson.*, 1987, **72**, 502.
34. J. Frahm, T. Michaelis, K. D. Merboldt, H. Bruhn, M. L. Gyngell, and W. Hänicke, *J. Magn. Reson.*, 1990, **90**, 464.
35. E. L. Hahn, *Phys. Rev.*, 1950, **80**, 580.
36. K. D. Merboldt, D. Chien, W. Hänicke, M. L. Gyngell, H. Bruhn, and J. Frahm, *J. Magn. Reson.*, 1990, **89**, 343.

37. C. T. W. Moonen and P. C. M. van Zijl, *J. Magn. Reson.*, 1990, **88**, 28.
38. J. Frahm, K. D. Merboldt, W. Hänicke, and A. Villringer, *Proc. 6th Ann Mtg. Soc. Magn. Reson. Med.*, New York, 1987, p. 137.
39. C. Arús, Y. Chang, and M. Bárány, *Physiol. Chem. Phys. Med. NMR*, 1985, **17**, 23.
40. T. Michaelis, K. D. Merboldt, W. Hänicke, M. L. Gyngell, H. Bruhn, and J. Frahm, *NMR Biomed.*, 1991, **4**, 90.
41. J. Frahm, T. Michaelis, K. D. Merboldt, W. Hänicke, M. L. Gyngell, and H. Bruhn, *NMR Biomed.*, 1991, **4**, 201.
42. T. Michaelis, G. Helms, K. D. Merboldt, W. Hänicke, H. Bruhn, and J. Frahm, *NMR Biomed.*, 1993, **6**, 105.
43. W. P. Aue, S. Müller, T. A. Cross, and J. Seelig, *J. Magn. Reson.*, 1984, **56**, 350.
44. H. Post, D. Ratzel, and P. Brunner, German Patent P3 209 263.6, 1982.
45. P. R. Luyten, A. J. H. Mariën, B. Sijtsma, and J. A. den Hollander, *J. Magn. Reson.*, 1986, **67**, 148.
46. D. M. Doddrell, W. M. Brooks, J. M. Bulsing, J. Field, M. G. Irving, and H. Baddeley, *J. Magn. Reson.*, 1986, **68**, 367.
47. M. von Kienlin, C. Remy, A. L. Benabid, and M. Decorps, *J. Magn. Reson.*, 1988, **79**, 382.
48. D. M. Doddrell, J. M. Bulsing, G. J. Galloway, W. M. Brooks, J. Field, M. Irving, and H. Baddeley, *J. Magn. Reson.*, 1986, **70**, 319.
49. A. Haase, *Magn. Reson. Med.*, 1986, **3**, 963.
50. A. J. S. de Crespigny, T. A. Carpenter, and L. D. Hall, *J. Magn. Reson.*, 1989, **85**, 595.
51. P. A. Bottomley, C. J. Hardy, P. B. Roemer, and R. G. Weiss, *NMR Biomed.*, 1989, **2**, 284.
52. D. M. Doddrell, J. Field, I. M. Brereton, G. J. Galloway, W. M. Brooks, and M. G. Irving, *J. Magn. Reson.*, 1987, **73**, 159.
53. J. Briand and L. D. Hall, *J. Magn. Reson.*, 1988, **80**, 559.
54. J. Granot, *J. Magn. Reson.*, 1988, **78**, 302.
55. J. J. van Vaals, A. H. Bergman, J. H. den Boef, H. J. van den Boogert, and P. H. J. van Gerwen, *Magn. Reson. Med.*, 1991, **19**, 136.
56. C. J. Hardy, P. A. Bottomley, M. O'Donnell, and P. Roemer, *J. Magn. Reson.*, 1988, **77**, 233.
57. P. G. Morris, D. J. O. McIntyre, D. E. Rourke, and J. T. Ngo, *NMR Biomed.*, 1989, **2**, 257.
58. P. A. Bottomley and C. J. Hardy, *J. Magn. Reson.*, 1987, **74**, 550.
59. C. Y. Rim, J. B. Ra, and Z. H. Cho, *Magn. Reson. Med.*, 1992, **24**, 100.
60. M. von Kienlin and R. Mejia, *J. Magn. Reson.*, 1991, **94**, 268.
61. X. Hu, D. N. Levin, P. C. Lauterbur, and T. Spraggins, *Magn. Reson. Med.*, 1988, **8**, 314.
62. H. P. Hetherington, M. J. Avison, and R. G. Shulman, *Proc. Natl. Acad. Sci. U.S.A.*, 1985, **82**, 3115.
63. P. C. M. van Zijl, C. T. W. Moonen, and M. von Kienlin, *J. Magn. Reson.*, 1990, **89**, 28.
64. R. Kimmich, E. Rommel, and A. Knüttel, *J. Magn. Reson.*, 1989, **81**, 333.
65. T. Ernst and J. Hennig, *Magn. Reson. Med.*, 1991, **21**, 82.
66. A. H. Wilman and P. S. Allen, *J. Magn. Reson. B*, 1993, **101**, 102.
67. F. Desmoulin and J. Seelig, *Magn. Reson. Med.*, 1990, **14**, 160.
68. A. Bax, P. G. De Jong, A. F. Mehlkopf, and J. Smidt, *Chem. Phys. Lett.*, 1980, **69**, 567.
69. C. H. Sotak and D. M. Freeman, *J. Magn. Reson.*, 1988, **77**, 382.
70. C. H. Sotak, *Magn. Reson. Med.*, 1988, **7**, 364.
71. C. H. Sotak, D. M. Freeman, and R. E. Hurd, *J. Magn. Reson.*, 1988, **78**, 355.
72. N. Chandrakumar and F. Chandrasekaran, *J. Magn. Reson.*, 1992, **96**, 657.
73. A. Knüttel and R. Kimmich, *Magn. Reson. Med.*, 1989, **9**, 254.
74. A. Knüttel and R. Kimmich, *J. Magn. Reson.*, 1990, **86**, 253.
75. C. T. W. Moonen, M. von Kienlin, P. C. M. van Zijl, J. Cohen, J. Gillen, P. Daly, and G. Wolf, *NMR Biomed.*, 1989, **2**, 201.
76. U. Klose, *Magn. Reson. Med.*, 1990, **14**, 26.
77. M. L. Gyngell, J. Frahm, K. D. Merboldt, W. Hänicke, and H. Bruhn, *J. Magn. Reson.*, 1988, **77**, 596.
78. R. Zahler, S. Majumdar, B. Frederick, M. Laughlin, E. Barrett, and J. C. Gore, *Magn. Reson. Med.*, 1991, **17**, 368.
79. P. C. M. van Zijl, C. T. W. Moonen, P. Faustino, J. Pekar, O. Kaplan, and J. S. Cohen, *Proc. Natl. Acad. Sci.*, 1991, **88**, 3228.
80. K. D. Merboldt, D. Hörstermann, W. Hänicke, H. Bruhn, and J. Frahm, *Magn. Reson. Med.*, 1993, **29**, 125.
81. S. Posse, C. A. Cuenod, and D. Le Bihan, *Radiology*, 1993, **188**, 719.
82. P. J. W. Pouwels, K. Brockmann, B. Kruse, B. Wilkin, M. Wick, F. A. Hanefield, and J. Frahm, *Pediatr. Res.*, 1999, in press.
83. S. W. Provencher, *Magn. Reson. Med.*, 1993, **30**, 672.
84. T. Michaelis, K. D. Merboldt, H. Bruhn, W. Hänicke, and J. Frahm, *Radiology*, 1993, **187**, 219.
85. J. Frahm, *Adv. Exp. Med. Biol.*, 1993, **333**, 257.
86. K. D. Merboldt, H. Bruhn, W. Hänicke, T. Michaelis, and J. Frahm, *Magn. Reson. Med.*, 1992, **25**, 187.
87. J. Hennig, T. Ernst, O. Speck, G. Deuschl, and E. Feifel, *Magn. Reson. Med.*, 1994, **31**, 85.

Biographical Sketches

Jens Frahm, b 1951. Dipl.Phys., 1974, Ph.D., 1977, physical chemistry, University of Göttingen, Germany. Introduced to NMR by Hans Strehlow, research assistant at the Max-Planck-Institut für Biophysikalische Chemie in Göttingen, 1977-1992, Head of Biomedical NMR Group, 1982-1992, Head of Biomedizinische NMR Forschungs GmbH, 1993–present. Approx. 205 publications. Research specialty: NMR studies of living systems; anatomic, metabolic, and functional aspects of intact organ systems; NMR imaging, localized NMR spectroscopy. Current interests: neuroimaging.

Wolfgang Hänicke. b 1955. Dipl.Math., 1981, University of Göttingen, Germany. Member of the Department Molekularer Systemaufbau at the Max-Planck-Institut für Biophysikalische Chemie in Göttingen, 1978–1983; Member of the Biomedical NMR Unit, 1983–present. Approx. 120 publications. Current interests: localized in vivo proton spectroscopy, functional MRI, image processing.

Chemical Shift Imaging

Truman R. Brown

Fox Chase Cancer Center, Philadelphia, PA, USA

1 INTRODUCTION

This article describes chemical shift imaging (CSI), illustrating how useful it is as a tool for obtaining well-localized in vivo spectra. CSI was originally described in 1982 by Brown

For References see p. 762

et al.[1,2] A similar procedure was suggested by Maudsley et al. in 1983.[3] A brief discussion of alternative localization techniques is included for comparison.

The need for a well-defined procedure for spectral localization for in vivo studies cannot be overemphasized. Unlike high-resolution NMR studies in chemistry, where the sample is placed entirely within a uniform rf field, studies in vivo are necessarily examining an inhomogeneous tissue, often with a coil which produces a spatially variable rf field. These circumstances make it very difficult to obtain a clear understanding of the relationship between observed spectra and physiology without a precise and controlled spectral localization technique: with such a technique, relationships which exist between spectral and physiological variables may be determined.

Because the wavelengths of the frequencies involved are in the 3–30 m range, no localization can be provided by the electromagnetic field directly, as in a light microscope. Instead, all localization techniques (both for imaging and spectroscopy) depend upon controlled spatial variations in either the static magnetic field or the rf magnetic field. Thus, the basic principles of localization apply equally to imaging as well as spectroscopy. The major difference is that imaging is usually concerned with localizing only a single component, generally water. In fact, the presence of two components at different frequencies, e.g. lipid and water, causes artifacts.

Spatially variable rf fields, typically arising from a surface coil,[4] will not be discussed here although they have been used extensively. They are difficult to quantify, in spite of the fact that they make it easy to acquire spectra from relatively large volumes. With carefully controlled gradients in the rf field more precise localization can be carried out,[5] although referencing them to the imaging gradients is difficult. We shall concern ourselves below only with static field gradients, used primarily as short phase-encoding pulses, which 'label' the spins at a particular position by giving them a well-defined rf phase, different at different locations. By collecting multiple FIDs each with a different gradient, mathematical reconstruction of the spatial distribution of the spins can be done by appropriately combining the FIDs. As shown below, this is generally accomplished by a Fourier transform (FT) to yield an estimate for the spin distribution. This is the same procedure used in MR imaging to reconstruct the phase-encode directions. The significant difference is the absence of a gradient during acquisition which allows a high-resolution spectrum to be acquired and then localized to particular regions or voxels.

In addition to localizing the positions of the spins with a short gradient pulse by phase encoding, gradients in the static magnetic field can be used with a selective rf pulse to excite only a limited region of the sample.[6–8] In this case, localization in three dimensions is achieved by the application of multiple gradients in three orthogonal directions. Generally such frequency-selective techniques only obtain spectra from one, or at most a few, localized regions simultaneously,[9] although complex pulse schemes do exist for multiple regions.[10] Commonly used frequency-selective techniques are ISIS[6] and STEAM,[7] the former primarily for phosphorus and the later primarily for proton spectroscopy. These techniques function well in localizing a single species of spin to a particular region. Because ISIS involves the addition and subtraction of different acquisitions there are some problems with motion sensitivity. However, the sizes of the localized region in phosphorus spectroscopy are generally large enough so that small in vivo motions are not a serious problem. As STEAM localizes the signal in a single acquisition, it does not suffer from this type of problem.

There is one artifact, however, from which all selective excitation techniques suffer. Nuclei which resonate at different frequencies are localized to different regions in space. A frequency-selective pulse will cause nuclei of different frequency to be localized to two different regions. This happens because a shift in resonant frequency of δf is the same as a shift in position of $\delta f/\gamma G$ where γ is the gyromagnetic ratio of the spins in Hz T^{-1} and G is the strength of the applied localizing gradient in mT m^{-1}. This artifact is not found in phase-encoding techniques because they only use the gradient for a short time to label the location of a spin in space with a phase factor. No frequency-selective rf is used, so no interaction with the gradient occurs and thus no frequency shift. For example, at a field of 1.5 T two nuclei, in the same or different compound separated by 20 ppm (a typical ^{31}P chemical shift), will be localized to regions shifted 1 cm with respect to each other in a gradient of 30 mT m^{-1} (twice the typical gradient available on clinical imagers before the introduction of echo planar imaging). The consequences of this can be very significant. For a cube (voxel) of side 5 cm, the two regions to which the two nuclei are localized are shifted 20% (1 cm out of 5 cm) with respect to one another in each dimension. This 20% linear shift causes nearly 50% of the two regions to be different [only $(4/5)^3 = 64/125$ of the volumes are in common]. For smaller voxel sizes the problem is even worse. For example, for a 27 cm^3 voxel $(3\times3\times3)$, less than 1/3 (8/27) of the volumes would be in common. This means, for example, that much of the single voxel ^{31}P studies using these techniques need to be analyzed carefully if comparisons are being made between β-adenosine triphosphate (β-ATP) and inorganic phosphate (P$_i$) which are approximately 20 ppm apart.

2 THEORY OF CSI

In order to appreciate the theoretical development to follow, we need to introduce the concept of $\mathbf{k}$ space. This is the space of FT variables associated with the real space variables x, y, and z. This relationship is the same as that between the complementary FT variables, time and frequency. The associated variables are k_x, k_y, and k_z. The theorems of FT show that knowledge of a function in one domain is equivalent to knowledge of a function in the other domain.[11] Thus, if we know $\bar{\rho}(\mathbf{k}, t)$, the spin distribution function in $\mathbf{k}$ space and time, we can obtain $\rho(\mathbf{r}, \sigma)$ the spin distribution in space and frequency. They are related by the following:

$$\bar{\rho}(\mathbf{k}, t) = \int \rho(\mathbf{r}, \sigma) e^{i\mathbf{k}\cdot\mathbf{r}} e^{i\sigma} \, d\mathbf{r} \, d\sigma \qquad (1)$$

where σ is the chemical shift (frequency) variable and a 'bar' represents the transformed function.

As shown below, the fundamental theorem of NMR localization states that sampling a free induction decay is equivalent to sampling the distribution of the excited spins along a particular trajectory in $(\mathbf{k}, t)$ space. It was originally proved some time ago.[1] This theorem underlines all forms of imaging with NMR.

3 FUNDAMENTAL THEOREM OF NMR LOCALIZATION

The fundamental theorem we seek to prove is that the FID of the spins in a sample in a gradient is the FT of the detectable spatial spin distribution in the sample, $\bar{\rho}_D$, evaluated along a particular trajectory in the transform space, i.e., $S(t) = \bar{\rho}_D(k(t), t)$. As shown below, this trajectory in k space is given by the gyromagnetic ratio, γ, times the integral of the gradient in time from the beginning of the FID. That is,

$$k(t) = \gamma \int_0^t G(\tau) \, d\tau \tag{2}$$

Here $S(t)$ is the FID at time t, $\rho_D(r, \sigma)$ is the detectable concentration at position r of a spin with chemical shift σ and $G(\tau)$ is the gradient at time τ.

To prove this, consider the evolution of a single species of spins with chemical shift σ in a sample following excitation. The FID as a function of time, $S(t)$, will be the sum of the magnetization at each point r, corrected by a sensitivity factor $D(r)$ and a phase factor $\phi(r, t)$ which measures how much the magnetization at r contributes. Mathematically,

$$S(t) = \int \rho(r) D(r) e^{i\phi(r,t)} \, dr = \int \rho_D(r) e^{i\phi(r,t)} \, dr \tag{3}$$

where $D(r)$ is simply the coil detection profile while $\phi(r, t)$ is the rf phase of the spins at point r and time t. To calculate ϕ we need the expression for the frequency of the spins, ω, at r. This is given by the instantaneous magnetic field at r, $H(r, t)$:

$$\frac{d\phi}{dt} = \omega(r, t) = \gamma H(r, t) \tag{4}$$

Integrating, we obtain

$$\phi(r, t) = \int_0^t \omega(r, \tau) \, d\tau = \int_0^t \gamma H(r, \tau) \, d\tau \tag{5}$$

$H(r, \tau)$ is made up of two terms, the static magnetic field H_0 and any gradient, $G(t)$, which may be present. Substituting $H(r, \tau) = H_0 + G(\tau) \cdot r$, we find

$$\phi(r, t) = \gamma H_0 t + \gamma \int_0^t G(\tau) \cdot r \, d\tau \tag{6}$$

since H_0 is constant. Substituting this into equation (3) yields

$$S(t) = \int \rho_D(r) \exp \left\{ i \left(\gamma H_0 t + \gamma \int_0^t G(\tau) \cdot r \, d\tau \right) \right\} dr \tag{7}$$

$$= \exp\{i(\gamma H_0 t)\} \int \rho_D(r) \exp \left(i\gamma \int_0^t G(\tau) \cdot r \, d\tau \right) dr \tag{8}$$

The first part of this expression shows the phase factor of the Larmor frequency of the spins while the second can be written, using equation (2), as

$$\int \rho_D(r) e^{ik(t) \cdot r} \, dr = \bar{\rho}_D(k(t)) \tag{9}$$

Putting the time dependence back and allowing for multiple species, we have the fundamental theorem of NMR imaging, $S(t) = \bar{\rho}_D(k(t), t)$, where $\bar{\rho}_D$ is the FT in both space and frequency of the detectable spin distributions and $k(t)$ is the integral of the gradient.

Thus a FID in the presence of a constant gradient G would sample the spatial FT of the distribution of detectable spins along the line $k = \gamma G t$. Hence, by sampling uniformly in time, one also samples uniformly in k space. In fact this is what is done by the readout gradient in ordinary MRI. On the other hand, immediately following a short rectangular gradient pulse of strength G and length T_0, the FID would sample the spatial FT of the detectable spin distribution at a fixed point in k space, corresponding to $\gamma G T_0$. A combination of these two approaches, namely a constant gradient in one direction and a pulse in the other, is used in standard 2D imaging to acquire a single line in k space. This is then repeated multiple times to acquire enough different lines in k space to calculate an image of the desired resolution and field of view (FOV). A CSI dataset is acquired analogously, by repeatedly acquiring multiple FIDs with gradient pulses of different strengths and directions. In CSI, unlike standard MRI, no readout gradient is employed, as the chemical shift information must be acquired in the absence of a gradient. Each FID is acquired at a single point in the k space domain. Therefore, for two spatial dimensions, one requires phase encoding in each dimension, whereas for 3D, three separate dimensions of phase encoding are needed. In this way, a complete sampling of k space can be acquired. Since the FID is acquired in the absence of a gradient except for the very early points in the FID, multiple compounds can be detected as in ordinary high-resolution pulse-acquire NMR experiments. An estimate of the original spatial and frequency distributions is then obtained by discrete Fourier transformations (DFT) in both time and space of this dataset.

Since the gradient pulse immediately follows the rf excitation, the very early time points in the FID are unusable. This can degrade broad signals if echoes are not used. On modern clinical machines the loss of the early time points is less than 1 ms, corresponding to linewidths greater than 300 Hz. This is usually broader than most peaks of interest, although certain phospholipid peaks can be observed at such widths. In such cases care must be taken in quantitation as these spectral lines can change shape as well as amplitude.

There are several properties of the DFT which are important here. First, in order to distinguish regions separated by a distance R, one needs to sample k space on the scale of $1/R$. This sets both the FOV, the largest distance to be distinguished needing the closest sampling in k space, and the spatial resolution, the smallest distance to be resolved needing the largest spread in k space, of the resultant distribution. Another important property is how the reconstructed distribution is related to the true one. This is well known from the standard theory of DFT[11] but is worth reviewing as it is so fundamental to understanding the nature of the localized spectra in CSI experiments (as well as standard MRI).

A finite sampling of k space produces effects on the reconstructed spatial distribution of the spins which is represented by a point spread function (psf) which mathematically describes how the reconstructed spectrum at a particular lo-

For References see p. 762

cation, $\rho_{rec}(\boldsymbol{r})$, is related to the actual distribution, $\rho(\boldsymbol{r})$. This relationship takes the form of a convolution so that

$$\rho_{rec}(\boldsymbol{r}) = \int \rho(\boldsymbol{s})\mathrm{psf}(\boldsymbol{r}-\boldsymbol{s})\,\mathrm{d}\boldsymbol{s} \tag{10}$$

We now derive the form of the psf. First, from the formula of a DFT we know

$$\rho_{rec}(\boldsymbol{r}) = \sum_{n=-N/2}^{N/2-1} \bar{\rho}(\boldsymbol{k}_n)\mathrm{e}^{-\mathrm{i}\boldsymbol{k}_n \cdot \boldsymbol{r}} \tag{11}$$

Here, $\bar{\rho}(\boldsymbol{k}_n)$ is the FT of the true distribution at $\boldsymbol{k}_n$, assumed to range from $-(N/2)\,\Delta k$ to $(N/2-1)\Delta k$, where $\Delta k = 2\pi/L$, L being the FOV. Substituting for $\bar{\rho}(\boldsymbol{k}_n)$ in equation (11), we obtain

$$\begin{aligned} \rho_{rec}(\boldsymbol{r}) &= \sum_{n=-N/2}^{N/2-1} \left\{ \int \rho(\boldsymbol{r}')\mathrm{e}^{\mathrm{i}\boldsymbol{k}_n \cdot \boldsymbol{r}'}\,\mathrm{d}\boldsymbol{r}' \right\}\mathrm{e}^{-\mathrm{i}\boldsymbol{k}_n \cdot \boldsymbol{r}} \\ &= \int \rho(\boldsymbol{r}')\left\{ \sum_{n=-N/2}^{N/2-1} \mathrm{e}^{\mathrm{i}\boldsymbol{k}_n \cdot (\boldsymbol{r}'-\boldsymbol{s})}\,\mathrm{d}\boldsymbol{r}' \right\} \end{aligned} \tag{12}$$

Comparing equation (12) to equation (10), we see the psf is

$$\mathrm{psf}(\boldsymbol{r}-\boldsymbol{s}) = \sum_{n=-N/2}^{N/2-1} \mathrm{e}^{\mathrm{i}\boldsymbol{k}_n \cdot (\boldsymbol{r}-\boldsymbol{s})} \tag{13}$$

In just 1D we find

$$\begin{aligned} \mathrm{psf}(x-s) &= \sum_{n=-N/2}^{N/2-1} \mathrm{e}^{\mathrm{i}k_n(x-s)} \\ &= \sum_{n=-N/2}^{N/-1} \mathrm{e}^{\mathrm{i}n\Delta k(x-s)} \\ &= \mathrm{e}^{-\mathrm{i}(N/2)\Delta k(x-s)} \sum_{j=0}^{N-1} \mathrm{e}^{\mathrm{i}j\Delta k(x-s)} \end{aligned} \tag{15}$$

Using the identity

$$\sum_{j=0}^{N-1} x^j = \frac{1-x^N}{1-x}$$

with $x = \mathrm{e}^{\mathrm{i}\Delta k(x-s)}$, equation (15) can be evaluated to give

$$\begin{aligned} \mathrm{psf}(x-s) &= \mathrm{e}^{-\mathrm{i}(N/2)\Delta k(x-s)}\frac{1-\mathrm{e}^{\mathrm{i}N\Delta k(x-s)}}{1-\mathrm{e}^{\mathrm{i}\Delta k(x-s)}} \\ &= \mathrm{e}^{-\mathrm{i}(1/2)\Delta k(x-s)}\frac{\sin\frac{1}{2}N\Delta k(x-s)}{\sin\frac{1}{2}\Delta k(x-s)} \end{aligned} \tag{16}$$

where $\mathrm{e}^{\mathrm{i}x} - \mathrm{e}^{-\mathrm{i}x} = 2\mathrm{i}\sin x$ has been used to simplify the expression. Putting $\Delta k = 2\pi/L$ we find

$$\mathrm{psf}(x-s) = \mathrm{e}^{-\mathrm{i}\pi(x-s)/L}\frac{\sin \pi N(x-s)/L}{\sin \pi(x-s)/L} \tag{17}$$

The real and imaginary parts of this function are plotted as a function of $\Delta = x - s$ in Figure 1 for $N = 16$.

The real part has a large peak near $\Delta = 0$ with smaller oscillations extending to the edge of the FOV, L. These oscil-

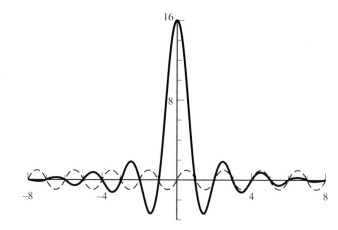

Figure 1 Point spread function for a discrete Fourier transform with $N = 16$. Solid curve is real part; dotted curve is imaginary part

lations cause the well-known 'sinc wiggles' found in DFT. Depending on the original distribution $\rho(\boldsymbol{r})$, the reconstructed distribution $\rho_{rec}(\boldsymbol{r})$ will have varying degrees of mixing of signal from outside the region. Since the function is oscillatory, these are generally small but need to be kept in mind for precise work. The imaginary part arises from the asymmetric sampling around the origin in $\boldsymbol{k}$ space ($-N/2$ to $N/2-1$).

The full 3D psf for cubic sampling is simply the product of the 1D functions:

$$\mathrm{psf}(\boldsymbol{r}-\boldsymbol{s}) = \mathrm{psf}(r_x-s_x)\mathrm{psf}(r_y-s_y)\mathrm{psf}(r_z-s_z) \tag{18}$$

A general psf can be calculated if the k-space sampling is not uniform. However it does not generally have such a simple mathematical form. In the general case, the psf is given by

$$\mathrm{psf}(\boldsymbol{r}-\boldsymbol{s}) = \sum_{\boldsymbol{k}\text{-sampled}} \mathrm{e}^{\mathrm{i}\boldsymbol{k}\cdot(\boldsymbol{r}-\boldsymbol{s})} \tag{19}$$

The psf calculated above represents a cubic sampling in $\boldsymbol{k}$ space. Numerous authors have suggested spherical or high order polyhedral as sampling schemes which can take significantly less time to carry out the acquisition.[12–14] For example, the volume of a sphere inscribed inside a cube is approximately 50% that of the cube. Thus, only one-half the time is required to sample it at equal resolution. Since the primary determinant of the resolution of the psf is the maximum separation between $\boldsymbol{k}$-space samples, such as spherical sampling will have approximately the same resolution as a cubic sampling. This has been implemented by Maudsley et al.[14]

Another potential advantage of CSI acquisition is the ability to vary the number of acquisitions at each point in $\boldsymbol{k}$ space. It has been shown by Brown and Nelson that a Weiner optimization, assuming constant noise at each point in $\boldsymbol{k}$ space, obtains the reasonable result that the $\boldsymbol{k}$ space should be sampled at a constant rate in regions where the signal-to-noise ratio (S/N) is high going smoothly to zero in regions where the S/N is low.[15] Depending upon the precise signal-to-noise ratio in a particular experimental case, a further 10% to 20% increase in sensitivity can be gathered using this procedure.

For both acquisition time considerations and spatial resolution, one generally samples only relatively small $\boldsymbol{k}$-space values in the CSI case compared with the MRI case. This is because the metabolites of interest are 1000- to 10 000-times lower in concentration than water protons. To achieve a reasonable S/N, resolutions must be on a 1 cm rather than a 1 mm scale length. Further, the three-dimensional sampling in k space required can take considerable time. For example, an $8 \times 8 \times 8$ three-dimensional grid requires 512 separate acquisitions; at a repetition time of 1 s, this is approximately 8 min.

The sampling also causes artifacts due to its finite extent if it is desired to reconstruct the distribution from the chemical shift information at a resolution more comparable to the standard MRI images. The typical way that this is accomplished is to zero-fill the datasets in k space, and then carry out a standard FT reconstruction. Note that the psf formalism [equation (10)] still describes how the reconstructed spectrum is related to the actual distribution since r can be evaluated at any point. It is clear, however, that unless two values of r are separated by a distance of the order of the size of the central peak in Figure 1 similar spectra will be obtained.

The freedom to choose any value of r for reconstruction demonstrates one of the great advantages of CSI. By obtaining information in k space the reconstructed distribution can be localized anywhere. That is to say that the psf can be centered about any point in the real space. This procedure is called voxel shifting and is accomplished by multiplying the k-space distribution by a phase factor $\exp(-i\mathbf{k}\cdot\boldsymbol{\delta})$ if we wish to shift the center of the reconstructed voxels by $\boldsymbol{\delta}$. This can be easily proven by substituting $\bar{\rho}(k_n)\exp(-i\mathbf{k}\cdot\boldsymbol{\delta})$ for $\bar{\rho}(k_n)$ in the derivation of the psf given above.

This, as shown in the examples below, is a particularly convenient way to reconstruct even a single localized region because no prior localization is required and if a mistake is made it is easy to recalculate a distribution with the psf centered at a different location.

There is no easy way to overcome the difficulties of reconstruction brought about by finite sampling while using standard FT reconstruction methodology. However, there are techniques which have been proposed which work in the reverse direction. Assume a particular distribution, for example, constrained to be within the head for a head examination, calculate the forward transform from real space into k space, adjust the distribution so that the FT agrees with what has been measured at the lower k-space regions while not forcing the higher k-space values to be zero. The primary difficulty with methods of this type is the number of degrees of freedom in the distribution are much higher than the number of acquired data points, so that extra constraints must be imposed upon the distribution to produce a unique answer. A variety of methods can be used, e.g., correlation with the proton image[16] or the use of maximum entropy or other regularizing functions to restrict the distribution. Thus far these techniques are still in their infancy, but they have great promise for removing unwanted psf effects. They are also computationally quite expensive; as computation speeds increase, this will be less of a problem.

Finally, the sensitivity of this procedure needs to be analyzed. To do this consider two extreme cases: all the spins concentrated in a very small region and a completely uniform distribution, both observed with N gradient steps. In the first case, the only effect of the different gradient pulses will be to cause the magnetization to have a different phase in each FID. The phase factor will be $\exp(ik_n\cdot r_0)$ for something located at r_0 when the k-space point k_n is observed. Thus all the DFT

does is cancel these phase factors, enabling the individual FIDs to add at the location r_0 as though N acquisitions had been carried out without gradients, resulting in an increase in S/N compared with a single acquisition of $N^{1/2}$. In the other case, only one FID will have any substantial signal. This is the one with zero gradient, since, with a large uniform sample, the signal from different parts will cancel even for the smallest gradient, corresponding to 1/FOV. In this case, when we carry out the DFT the only signal is from the $k = 0$ point and thus is constant everywhere. The noise, on the other hand, is in every acquisition and reduces the S/N, again compared with a single acquisition, by $N^{1/2}$. Another way of analyzing the uniform case is to realize that each reconstructed voxel corresponds to a volume of L^3/N and so as N increases, the spins contributing to the signal from each voxel decrease proportionally as N. Each of the N acquisitions contributes signal to the spins coming from the desired voxel, while the noise, being independent, adds incoherently. Thus, the final sensitivity is proportional to $N^{1/2}/N$ or $N^{-1/2}$, compared with a single acquisition.

As these examples make clear, the precise sensitivity factor depends upon details of the shape of the object and how its FT behaves in k space. If the resolution is increased by reducing the FOV while keeping the number of phase-encoding steps constant, the S/N will be reduced proportionately with the volume of the voxel if we are dealing with a uniform distribution. If, on the other hand, we are dealing with an unresolved bright spot, so that as the voxel gets smaller the number of spins within it stays the same, then so does the final sensitivity of that particular voxel. Alternatively, if we increase the resolution by keeping the same FOV and increasing N, then we will take more phase-encoding steps which will partially compensate for an increased resolution, of course at the expense of increased observation time.

Thus, as soon as the resolution is sufficient to begin to resolve objects into homogeneous regions, further increasing the resolution will result in substantial losses in S/N because the number of spins per resolution element drops as the cube of the voxel dimension. An equally valid analysis can be done in k space, once it is realized that the size of the FT of an object of dimension L rapidly falls to zero at k-space values greater than $1/L$. As the resolution of the sampling is increased to provide a subdivision of the uniform object, points in k space are taken with very low signal and, thus, primarily add noise to the final result.

4 DATA ANALYSIS

One of the greatest impediments to the full application of CSI to two and three dimensions is the large size of the datasets and the hundreds of spectra which need to be analyzed. There are simply too many to be done manually; an automatic procedure is required. This problem is complicated in many cases because of the low S/N found in many in vivo CSI spectra. What is needed is a robust technique for peak identification and extraction from a noisy spectrum with variable baseline, the latter coming from the absence of the early time points in the FID during the period of the gradient pulse. Such a procedure was developed by Nelson and Brown and dubbed PIQABLE (Peak, Identification, Quantification, and Automatic Base Line Estimation).[17] It is fully described in the original

For References see p. 762

paper. Briefly it relies on a statistical estimation of the probabilities of deviation of a particular spectral region under examination from a running average background. Estimates of the standard deviation are calculated from the spectrum itself. The program eliminates regions identified as peaks from noise estimation and iterates examinations of the spectrum until no further peaks are identified. It has proven to be very successful with in vivo CSI datasets and is able to deal with varying backgrounds and broad noisy peaks in a robust way. However, it is time-consuming to carry out this processing on each spectrum in every CSI dataset, thus a variety of further developments have concentrated on peak detection in specified regions.[18] This is carried out by either using a fixed frequency region or some type of localized peak search algorithm. Area estimates can then be made for each spectrum.

Several hundred numbers are still not easy to comprehend, so when one finishes analyzing a large number of spectra the problem as to how to represent these data is still present. Generally for two- and three-dimensional CSI data, image representations are an essential part of the analysis.[19] Care should be taken to examine the actual spectra as well as these images as no programs at present can be viewed as 100% reliable in converting spectra to images. Nevertheless, as a method of obtaining an overview of the entire dataset, they are excellent. Typically, different compounds, when present, are represented in different colors; this leads to a loss of gray scale sensitivity in the eye so that in cases of high sensitivity [e.g., phosphocreatine (PCr) in a skeletal muscle] a standard black and white image should be used.

Another advantage of CSI localization is that the gradients used provide the same localizing information for both the MRI proton images as well as the CSI. Thus, the MRI image and the CSI dataset are naturally in registration; no complex transformations are necessary to map from one to the other. A direct overlay of the metabolic information onto a proton image can be done as shown in several of the figures in the next section. In addition to the local metabolite concentrations, of course, one also has information regarding the specific chemical shift frequencies. By shifting compounds which are known to be independent of intracellular conditions, such as PCr, glycerolphosphocholine, etc., different regions can be aligned in frequency with the result that extremely sharp spectra can be observed from an entire organ by resuming the shifted spectra. Spatial maps of pH distribution, magnesium concentration, and other physiological variables which can shift the metabolite frequencies can also be constructed.

5   ACQUISITION

The details of acquisition of CSI sequences are generally straightforward on any standard clinical instrument. The results presented below have been acquired using a Siemens 1.5 T Magnetom. The primary criterion is the ability to generate three cycles of phase encoding which is possible on virtually all modern commercial clinical instruments. Other than that, all that is required is an excitation sequence followed immediately by a gradient pulse of as short duration as possible.

The strength of the gradient pulses can be calculated from FT considerations discussed in the previous sections. For example, a resolution of 3 cm requires the maximum distance between k-space points to be sampled, Δk_{max}, to be $2\pi \times 1/3$ cm^{-1}. From equation (10) this must be $\gamma G_{max} T_0$ for a constant gradient which turns on and off instantly. Taking T_0 to be 1 ms and realizing G_{max} can be positive or negative, we find $G_{max} = \frac{1}{2}\Delta k_{max}/\gamma T_0 = 1/2 \times 2\pi \times 1/3 \times 1/10^{-3} \times 1/\gamma$; for ^{31}P this would be 6.07 mT m^{-1} ($\gamma = 17.24$ MHz T^{-1}), while for protons it would be 2.47 mT m^{-1} ($\gamma = 42.38$ MHz T^{-1}). Depending on the FOV this maximum would be divided into eight or 16 steps, corresponding to a FOV of 24 or 48 cm, respectively. Both positive and negative values in k space need to be acquired. This is required because a spatially dependent chemical shift image is a complex function, and thus both positive and negative k-space information is needed to reconstruct it properly. This is not the case in proton imaging where, with a single frequency, there is no true complex information present in the image, and thus half Fourier space sampling can reduce acquisition times by a factor of two. This is not possible in a general CSI acquisition as different frequencies at different places beat with one another and thus appear as complex functions.

Note that long times are required to cycle through the k-space points if three dimensions are desired. For example, a 16^3 acquisition requires 4096 acquisitions, generally requiring too long a time to be practical. It is also important to have a well-shimmed sample even using localization techniques since, for example in proton-decoupled ^{31}P spectra, the resolution required is less than 0.1 ppm in a single voxel.

One point that should be mentioned is the precise positioning of the positive and negative k-space samples. Because of the finite number of points required, the sampling is often done from $-N/2$ to $N/2 - 1$ or asymmetrically around the zero point in k space. For example, if N is 8 then $-4, -3, -2, -1, 0, 1, 2$, and 3 will be sampled on a uniform grid. In general, this is the way imaging sampling is carried out for N values of 64–512. In this case, the difference between $N/2$ and $N/2-1$ is quite small. However, as pointed out in the theory section, the psf in this asymmetric case has a small imaginary component due to the one extra negative k-space sample. Thus, in general, it is better to shift the sampling points by half of the k-space grid and not sample at the precise $k = 0$ point. This produces a completely real psf with no imaginary contamination. This can be significant for broad lines since the imaginary part of the line is much wider in frequency and can thus lead to voxel contamination of a complex sort.

In addition to acquiring a single nucleus with CSI techniques, the techniques can be modified to acquire two techniques simultaneously. As shown by Gonen et al.,[20] at a slight cost in early dead time both phosphorus and proton acquisitions can be carried out in a simultaneous manner. This is accomplished by doing an early phosphorus excitation followed by part of the phase-encoding pulse, bringing the gradient to zero, exciting the protons, and finishing with another gradient pulse. The relative strengths of the gradient pulses are set so that the second pulse gives the proper proton FOV, while the combined pulses give a proper phosphorus FOV.

6   APPLICATIONS

The simplest illustration of CSI is one dimension. In this case a series of spectra is sufficient to demonstrate the localiz-

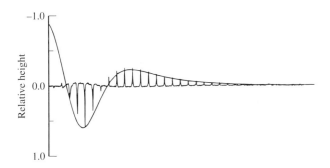

Figure 2 1D CSI experiment of a column phantom coaxial with a circular surface coil with 32 phase-encode steps. The solid curve is the theoretically expected signal. The individual peaks are the experimentally observed signal at the indicated distance from the plane of the coil

ation and no complex processing or image presentation is needed to appreciate the result. Figure 2 shows the result from 32 phase-encode steps using a single turn circular surface coil with a cylinder phantom along the coil axis. A square, rectangular rf pulse, equivalent to 540° at the center of the coil, excited the spins. The FIDs were Fourier-transformed in both time and k space, and the resultant spectra were all phased the same. The gradient amplitudes and widths corresponded to a spacing of 5 mm for a total FOV of 16 cm. The experiment was carried out on a 1.5 T Siemens Magnetom at 25.6 MHz with inorganic phosphate in the phantom. The peak region surrounding the phosphate resonance was selected from each spectrum and displaced along the axis in Figure 2 a distance corresponding to where the spectrum was localized with respect to the coil. The signals were compared with those expected theoretically, $B_1(y) \sin \gamma B_1(y) T_0$, where $B_1(y)$ is the rf field strength at the distance y along the axis, γ is the ^{31}P gyromagnetic ratio, and T_0 is the length of the rf pulse. Since the pulse was 540° at the coil center, for the circular loop $\gamma B_1(y) T_0 = 540° a^3/(y^2 + a^2)^{3/2}$, where a is the radius of the coil. This function was calculated as a function of position (5 mm per spectrum) along the axis, scaled in the vertical direction to fit the peak heights and is plotted as the solid line in Figure 2.

As can be seen the agreement is excellent. The only adjustable parameter here was the vertical scaling of the theoretical curve. These results demonstrate the ability of CSI phase-encoding measurements to provide excellent localization in space, as would be expected from their common use in the imaging field.

In two and three spatial dimensions it becomes possible to consider alternative ways of representing this spectral information, i.e. more than simply arrays of spectra. Since we have continuous localized spectra the obvious possibility is to use an image format to represent the spatial distribution. The primary difficulty with such a procedure is the determination of the peak areas or other variables of interest from the spectra in an automatic fashion. Traditional methods of measuring area often rely on hand integration, which is unsuitable in this case since we are dealing with several hundred spectra. As discussed above, to overcome this we have developed a program, PIQ-ABLE, which is capable of nonparametrically dividing a noisy spectrum into peak regions, slowly varying background and high-frequency noise.[17] This procedure can be applied auto-

matically to the CSI spectra. The results of this analysis can then be presented as images. Because there are usually only 8 × 8 or 16 × 16 pixels at the original resolution in the data, we have found these data can be much better visualized if they have been zerofilled and spatially filtered so as to provide spatial smoothing, and then presented at 64 × 64 or higher resolution.

A good illustration of these techniques in two dimensions is the observation of exercise-induced changes in the high energy phosphates in human muscle. Numerous ^{31}P studies of muscle metabolism have been carried out using surface coils in order to understand the relationship between biochemical energy demand as measured by the force developed and biochemical energy supply as measured by changes in the 'high-energy' phosphates, PCr, adenosine triphosphates (ATP), and Pi which can readily be observed in a ^{31}P spectrum of muscle. The key assumption here is that one is observing changes in metabolites in the muscle which is producing the force. This is not easy to determine when only a surface coil is used for localizing the spectrum. To investigate this more carefully, CSI was used to study single finger excitation in the human forearm. A single-turn 8 cm diameter coil was placed around the forearm; the wrist was rotated to align all the major muscle groups longitudinally so that only two-dimensional localization was needed. Steady-state studies of each finger being exercised at various levels were carried out.

The full details of the study are presented elsewhere.[21] The 2D CSI data obtained from one volunteer in this study are shown in Figures 3, 4, and 5. Figure 3 shows a cross-sectional MRI of the forearm of a volunteer with a highlighted region approximately corresponding to the *flexor digitorum profundis*. Figure 4 shows three 4 × 6 arrays of spectra observed at rest (left), during exercise of the index finger (center), and of the

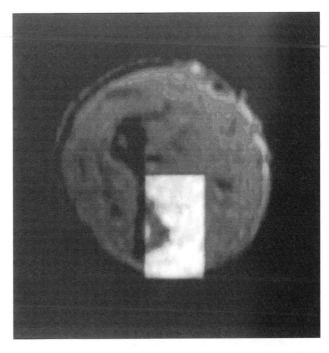

Figure 3 Axial MRI of forearm. Highlighted region indicates approximate position of *flexor digitorium profundus*

For References see p. 762

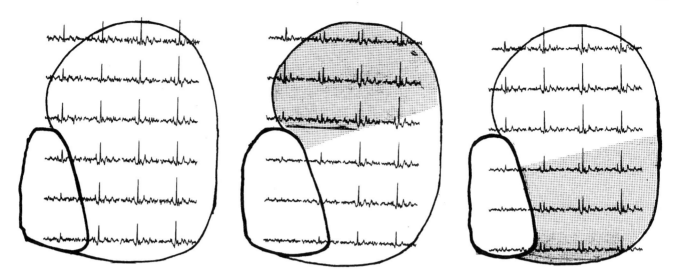

Figure 4 ^{31}P spectra from the highlighted region in Figure 3 at rest (left), during exercise of the index finger (center), and of the ring and little fingers (right)

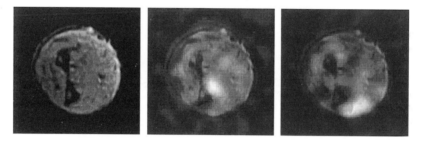

Figure 5 The image from Figure 3 (left) overlaid with a 2D CSI image calculated from the inorganic phosphate levels following excitation of only the index finger (center) and the ring and little fingers (right)

ring and little fingers (right). Careful examination of the spectra show a very different pattern of changes in PCr and Pi in the two cases. There were no changes near the surface but substantial conversion of PCr to Pi in the deeper shaded region when the index finger was exercised while the reverse was true for the ring and little fingers.

To illustrate this, the amplitudes of the Pi peaks in the two datasets have been converted into images as discussed above. These are shown in Figure 5 overlaid on the proton image in Figure 3 for anatomical reference. As can be seen, the metabolic activation of the muscle is localized to the deep interior of the arm during index finger exercise, while the superficial regions of the arm are activated if the ring and little finger are used. These observations suggest that many surface coil studies of the forearm need reevaluation since most surface coils used in human studies are 1–2 cm in diameter and only able to observe just beneath the surface of the arm. With such a coil, a bulb squeezing exercise without special care to use only the ring and little fingers, and not the index and middle fingers, would cause a confused picture when one tries to relate changes in high-energy phosphates with changes in physiological function, for example, Pi/PCr with workload.

Significantly different results are obtained if care is taken only to measure the work produced by the ring and little finger.[22] Without the localization provided by CSI there would be

no way to separate the changes at the deeper level produced by the index finger from the changes at the superficial level produced by the ring and little fingers.

These techniques can be applied in three dimensions as well as two. The first example of this was the work of Cox et al.[23] who demonstrated the feasibility of the 3D technique in human livers using a surface coil. The technique is particularly useful in studying the brain where high-quality coils can be made to encompass the entire head. Using such coils, 3D ^{31}P CSI spectra from the entire human head can be obtained with voxel sizes as small as 8 cm^3 with observation times on the order of 100 min. At lower spatial resolution, high-quality ^{31}P spectra of brain tissue can be obtained in 20–30 min.[19] This enables regional measurements of pH and pMg (from the positions of the Pi and β-ATP peaks, respectively) as well as the levels of ATP, PCr, and other phosphorylated metabolites. The Mg^{2+} levels are determined by measuring the β-ATP peak position, which is well known to titrate with Mg^{2+}.[24] Several studies have reported the intracellular brain magnesium concentration, in both normal volunteers and patients.[25,26] Interestingly, free Mg^{2+} levels in the brain appear to be substantially lower than in many other tissues.[27,28] A histogram of the position of the β-ATP peak with respect to PCr (PCr = -2.52 ppm) from normal volunteers is shown in Figure 6.

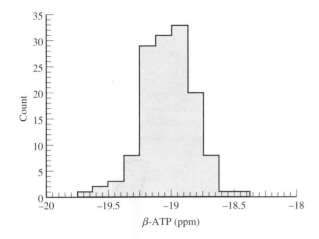

Figure 6 A histogram of the observed chemical shift of β-ATP in human brain. The data are from seven individuals and consist of measurements from 138 different voxels. The mean of the distribution is -19.03, $SD = 0.20$, corresponding to an Mg^{2+} concentration of 128 μM

The mean is approximately -19 ppm which corresponds to a free Mg^{2+} level of about 125 μM assuming a standard titration curve for brain tissue.[29] Interestingly this value is observed to change both in brain tumors as well as cachexic patients.[30] It is too early to assign significance to these changes. Although these studies can be carried out with single voxel techniques, CSI provides both a more precise localization and a distribution of Mg^{2+} values for each individual.

Recent applications of CSI also include its use in proton spectroscopy where typically 16×16 2D CSI scans through the brain have achieved excellent sensitivity with voxels sizes of the order of 1 cm^3.[31-34] The applications of these techniques to proton spectroscopy are now beginning to find their way into clinical use, as they have been implemented into the latest generation of the commercial MR imagers.

The effect of proton irradiation on ^{31}P spectra from the human brain is shown in Figure 7 which compares the ^{31}P spectra from a single 42 cm^3 voxel acquired in 35 min with and without proton irradiation.[35]

It is obvious that the phosphomonoester (PME) and phosphodiester (PDE) regions of the spectrum are considerably sharpened due to the elimination of the residual proton J coupling. In both cases individual peaks can be observed corresponding to phosphorylethanolamine (PE) and phosphorylcholine (PC) in the PME region, and glycerophosphorylethanolamine (GPE) and glycerophosphorylcholine (GPC) in the PDE region. In addition there is an NOE enhancement for some peaks (e.g., approximately 50% for PCr). The additional sensitivity that proton decoupling and NOE enhancement[36] provide has enabled metabolic images of the human brain at a resolution less than 2 cm to be constructed. Figures 8 and 9 show four coronal images of a normal volunteer and the corresponding coronal slices of the PCr intensity taken from a dataset acquired in 115 min with a voxel size of 6 cm^3. At this resolution, the basal ganglia are observed as distinct anatomical objects. Whether this is due to changes in PCr concentrations in gray versus white water or PCr relaxation times is unknown at present.

CSI techniques have also been used to advantage in obtaining spectra from tumors observed with surface coils. Even though only a few localized spectra are needed in these cases, the ability to voxel-shift the final result to correspond as precisely as possible to the tumor location outweighs the added complexity of a 3D CSI acquisition. This is particularly true as time is always short in clinical studies and no extra time is needed in CSI to obtain the correct localization. Figure 10 illustrates this, showing a proton axial MRI of the groin of a patient with a lymphoma with two localized grids; the dotted one is the original unshifted one and the solid one after voxel-shifting to optimize the spectral localization to the tumor.

Figure 11 compares the initial and final spectra most closely centered on the tumor in each case. From the absence of PCr in the spectrum (and its presence in the surrounding muscle) the spectrum is well localized to the tumor.

The ^{1}H decoupling discussed above allows clear resolution of the PME region into two components, PE and PC, with PE dominating. Without the CSI to localize the spectrum to the tumor, such detailed information is unattainable as shown by Figure 12 which is the spectrum acquired using only the surface coil for localization in the case of the patient shown in Figure 11.

An important example of the use of CSI in spectral acquisition is its use in an autoshimming procedure. This is based upon CSI localization of the water proton signals to provide a field map; through a least-squares minimization procedure, this information is used to adjust the shim currents to produce the optimum magnetic field homogeneity. The procedure produces excellent field homogeneity, as illustrated in Figure 13 which shows ^{1}H and ^{31}P spectra following the autoshim procedure to a volunteer's head.[37]

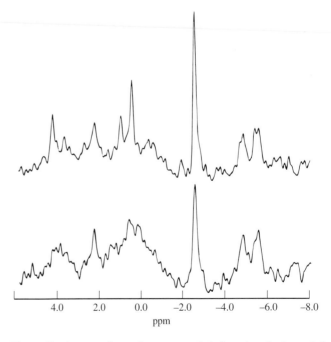

Figure 7 A comparison of proton-coupled (lower) and -decoupled (upper) ^{31}P spectra from 42 cm^3 of human brain acquired in 35 min

For References see p. 762

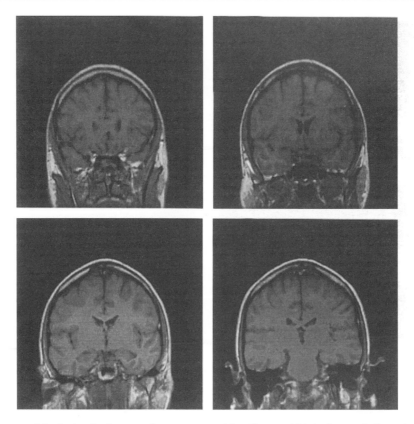

Figure 8 MRI coronal images of the brain of a human volunteer separated by 18 mm which is the voxel dimension in Figure 9

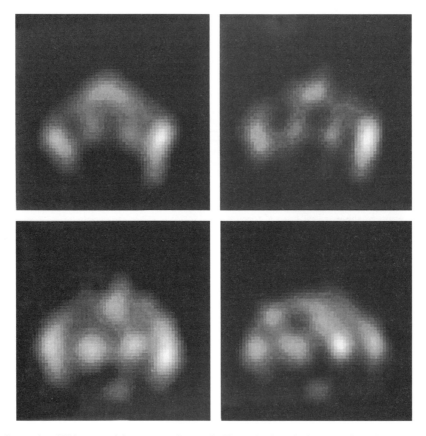

Figure 9 Adjacent phosphocreatine CSI images of the same sections as in Figure 8. Note the increased intensity in the regions of the basal ganglia as well as the superficial muscles outside the skull. These data were acquired in 115 min at a resolution of 18 mm

For list of General Abbreviations see end-papers

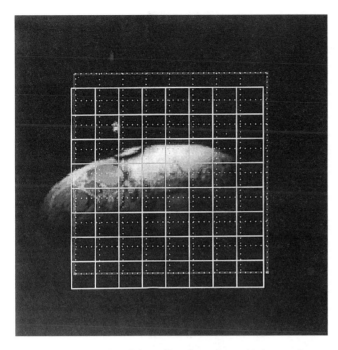

Figure 10 Axial MRI of the groin of a patient with low-grade non-Hodgkin's lymphoma with two CSI localizing grids. The dotted grid is the initial one while the solid grid is voxel shifted to align a voxel over the tumor

7 DISCUSSION

As can be seen from the previous examples, CSI localization techniques have been used in a variety of circumstances, in general with great success. Summarizing their advantages, they produce signals from the entire tissue under study with a

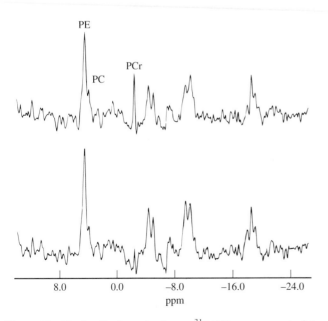

Figure 11 The localized spectra from a ^{31}P CSI measurement of the patient shown in Figure 10. The PCr in the spectrum from the initial voxel position (upper) is absent from the voxel shifted one (lower)

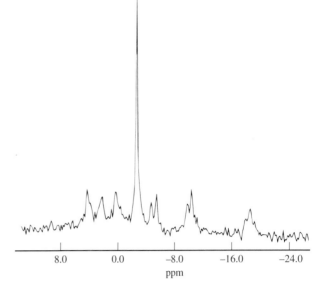

Figure 12 The nonlocalized FID from the patient shown in Figures 10 and 11

localization function that can be shifted over the sample at will, so as to optimize for the most appropriate signal from the region under investigation. Because of the well-defined properties of the FT, the difficulties due to the absence of high k-space samples can be dealt with mathematically, albeit with some loss of spatial resolution. The sensitivity of these techniques is high, as high as generally possible to obtain with a particular coil and spin concentration. Nevertheless the techniques are not without disadvantages, the primary one being the acquisition time required since each point in k space acquired requires another acquisition. As pointed out above, three-dimensional acquisitions can be quite time-consuming, particularly if high-sensitivity protons are under study. In these cases, generally two-dimensional acquisitions are carried out. A further disadvantage is the loss of early time points during the gradient pulse. This can be a serious problem for low-strength gradients. However, with most modern commercial clinical instruments, the gradients are sufficiently strong that this gradient pulse is seldom more than 1 ms for resolutions of 1 cm or so, and obtaining CSI voxels significantly smaller than this is inefficient because of signal-to-noise reasons. At this short a time, only peaks with linewidths larger than 300 Hz are seriously affected, although the rolling baseline caused by even a short pulse can affect precise spectral integration procedures.

And finally, of course, is the complexity of analysis. However, with available computers the transforms can be performed in only a few minutes; the problem then becomes developing automatic analysis procedures which are able to deal with these complex data structures. This has been done and it now seems clear that it is worth the effort of dealing with these more complicated sets of spatial information. What extra information do we obtain from these observations? There is little doubt that, in many circumstances, extra biochemical or physiological information can provide a much needed insight into both physiological responses, as well as clinical questions. This is becoming clearer as more commercial instruments make it practical to carry out these types of CSI studies, and thus it is reasonable to believe that with further development, and a

For References see p. 762

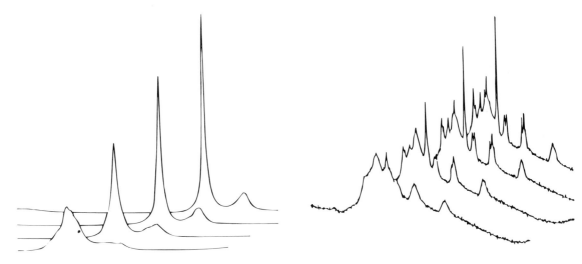

Figure 13 The ^{1}H (left) and ^{31}P (right) spectra from the whole head after several iterations of a CSI automatic shimming procedure

strong push from the user community, the instrument manufacturers will provide an integrated spectroscopy examination which is capable of easy use in a clinical setting.

This chapter has presented a brief overview of CSI and its potential. The techniques discussed here should allow a major advance in observations of metabolism in different pathological conditions. Considerable effort has been expended over the last several years to follow metabolic variations of tumors during therapeutic regimens, mostly with surface coils. Stroke and myocardial infarcts have been observed with single-voxel localization techniques which are not capable of providing as broad a picture as the multi-voxel CSI techniques discussed here. With the advent of clinical instrumentation capable of implementing these techniques, a much wider set of applications should ensue over the next few years. The ability of CSI to pinpoint metabolic changes associated with pathological states or physiological variations should provide new insights into both normal and disease processes.

8   RELATED ARTICLES

Brain MRS of Human Subjects; Brain Neoplasms in Humans Studied by Phosphorus-31 NMR Spectroscopy; In Vivo Hepatic MRS of Humans; Male Pelvis Studies Using MRI; Proton Decoupling During In Vivo Whole Body Phosphorus MRS; Single Voxel Localized Proton NMR Spectroscopy of Human Brain In Vivo; Spatial Localization Techniques for Human MRS.

9   REFERENCES

1. T. R. Brown, B. M. Kincaid, and K. Ugurbil, *Proc. Natl. Acad. Sci. USA*, 1982, **79**, 3523.
2. T. R. Brown, US Pat. 4 319 190 (9 March 1982).
3. A. A. Maudsley, S. K. Hilal, W. H. Perman, and H. E. Simon, *JMRI*, 1983, **51**, 147.
4. J. J. H. Ackerman, T. H. Grove, G. G. Wong, D. G. Gadian, and G. K. Radda, *Nature (London)*, 1980, **283**, 167.
5. R. M. Dixon and P. Styles, *Magn. Reson. Med.*, 1993, **29**, 110.
6. R. J. Ordidge, A. Connelly, and J. A. B. Lohman, *JMRI*, 1986, **66**, 283.
7. J. Frahm, H. Bruhn, M. L. Gyngell, K. D. Merboldt, W. Hanicke, and R. Sauter, *Magn. Reson. Med.*, 1989, **9**, 79.
8. R. J. Ordidge, J. A. Helpern, R. A. Knight, and K. M. A. Welch, *Proc. XIth Ann Mtg. Soc. Magn. Reson. Med.*, Berlin, 1992, p. 1925.
9. L. Bolinger and J. S. Leigh, *JMRI*, 1988, **80**, 162.
10. G. Goelman, G. Walter, and J. S. Leigh, *Magn. Reson. Med.*, 1992, **25**, 349.
11. R. N. Bracewell, 'The Fourier Transform and its Applications', McGraw-Hill, New York, 1978.
12. S. L. Ponder and D. B. Twieg, *J. Magn. Reson., Ser. B.*, 1994, **104**, 85.
13. J. C. Ehrhardt, *IEEE Trans. Med. Imag.*, 1990, **9**, 305.
14. A. A. Maudsley, G. B. Matson, J. W. Hugg, and M. W. Weiner, *Magn. Reson. Med.*, 1994, **31**, 645.
15. T. R. Brown and S. J. Nelson, *IEEE Eng. Med. Biol. Sci.*, 1990, **12**, 59.
16. S. Plevritis and A. Macovski, *Proc. XIth Ann Mtg. Soc. Magn. Reson. Med.*, Berlin, 1992, p. 3820.
17. S. J. Nelson and T. R. Brown, *JMRI*, 1987, **75**, 229.
18. S. J. Nelson and T. R. Brown, *JMRI*, 1989, **84**, 95.
19. D. B. Vigneron, S. J. Nelson, J. Murphy-Boesch, D. A. C. Kelley, H. B. Kessler, T. R. Brown, and J. S. Taylor, *Radiology*, 1990, **177**, 643.
20. O. Gonen, J. Murphy-Boesch, R. Srinivasan, J. Hu, J. Jiang, R. Stoyanova, and T. R. Brown, *J. Magn. Reson., Ser. B.*, 1994, **104**, 26.
21. J. A. L. Jeneson, S. J. Nelson, D. B. Vigneron, J. S. Taylor, J. Murphy-Boesch, and T. R. Brown, *Am. J. Physiol.*, 1992, **263C**, 357.
22. J. A. L. Jeneson, J. O. van Dobbenburgh, C. J. A. van Echteld, C. Lekkerkerk, W. J. M. Janssen, L. Dorland, R. Berger, and T. R. Brown, *Magn. Reson. Med.*, 1993, **30**, 634.
23. I. J. Cox, J. Sargentoni, J. Calam, D. J. Bryant, and R. A. Iles, *NMR Biomed.*, 1988, **1**, 56.
24. M. Cohn and T. R. Hughes, *J. Biol. Chem.*, 1962, **237**, 176.
25. N. M. Ramadan, H. Halvorsen, A. Vande-Linde, S. R. Levine, J. A. Helpern, and K. M. Welch, *Headache*, 1989, **29**, 590.
26. J. S. Taylor, D. B. Vigneron, J. Murphy-Boesch, S. J. Nelson, H. S. Kessler, L. Coia, W. Curran, and T. R. Brown, *Proc. Natl. Acad. Sci. USA*, 1991, **88**, 6810.
27. R. A. Meyer, T. R. Brown, and M. J. Kushmerick, *Am. J. Physiol.*, 1985, **248**, C279.
28. M. J. Kushmerick, P. F. Dillon, R. A. Meyer, T. R. Brown, J. M. Krisanda, and H. L. Sweeney, *J. Biol. Chem.*, 1986, **261**, 4420.

For list of General Abbreviations see end-papers

29. T. J. Mosher, G. D. Williams, C. Doumen, K. F. LaNoue, and M. B. Smith, *Magn. Reson. Med.*, 1992, **24**, 163.

30. F. Arias-Mendoza, T. Greenberg, R. Stoyanova, F. Ottery, and T. R. Brown, *Proc. XIIth Ann Mtg. Soc. Magn. Reson. Med.*, New York, 1993, p. 73.

31. C. M. Segebarth, D. F. Baleriaux, P. R. Luyten, and J. A. den Hollander, *Magn. Reson. Med.*, 1990, **13**, 62.

32. P. C. M. van Zijl, C. T. W. Moonen, J. Gillen, P. F. Daly, L. S. Miketic, J. A. Frank, T. F. DeLaney, O. Kaplan, and J. S. Cohen, *NMR Biomed.*, 1990, **3**, 227.

33. P. R. Luyten, A. J. H. Marien, W. Heindel, P. H. J. van Gerwen, K. Herholz, J. A. den Hollander, G. Friedman, and W. D. Heiss, *Radiology*, 1990, **176**, 791.

34. M. J. Fulham, A. Bizzi, M. J. Dietz, H. H. L. Shih, R. Raman, G. S. Sobering, J. A. Frank, A. J. Dwyer, J. R. Alger, and G. Di Chiro, *Radiology*, 1992, **185**, 675.

35. J. Murphy-Boesch, R. Srinivasan, L. Carvajal, and T. R. Brown, *J. Magn. Reson., Ser. B.*, 1994, **103**, 103.

36. J. H. Noggle and R. E. Shirmer, 'The Nuclear Overhauser Effect: Chemical Applications', Academic Press, New York, 1971.

37. J. Hu, T. Javaid, F. Arias-Mendoza, Z. Liu, R. McNamara, and T. R. Brown, *J. Magn. Reson.*, 1995, **108**, 213.

Acknowledgements

This work was supported by NIH Grants PO1-CA41708, RO1-CA54339, and Siemens Medical Systems. I would like to thank Fernando Arias-Mendoza, Oded Gonen, Tamela Greenberg, Jiani Hu, Tariq Javaid, Ronald McNamara, Joseph Murphy-Boesch, Sarah Nelson, William Negendank, Kristin Padavic-Shaller, Radka Stoyanova, Ravi Srinivasan, June Taylor, and Daniel Vigneron for their help and involvement in the development of the CSI techniques described here. Without the help of Doris Spindell, Jean Fink, and Lisa Siuda the chapter would never have been completed.

Localization and Registration Issues Important for Serial MRS Studies of Focal Brain Lesions

Douglas L. Arnold and Paul M. Matthews

Montreal Neurological Institute, McGill University, Montreal, Quebec, Canada

1 INTRODUCTION

Many applications have been reported illustrating the potential clinical utility of magnetic resonance spectroscopy (MRS) in the study of focal brain lesions. Examples of studies include those of brain tumors (diagnosis and monitoring response to therapy) using ^{1}H MRS[1-7] and ^{31}P MRS,[4,8-12] defining the evolution of hypoxic/ischemic brain damage in infants[13-19] and adults,[20-28] characterization of the chemical pathology and progression of focal demyelinating disease,[29-42] and localization of lesions associated with focal epilepsy, primarily using proton,[43-47] but also phosphorus,[48,49] MRS. Practical exploitation of these approaches demands: (i) the demonstration of sufficient diagnostic specificity and sensitivity; (ii) proof of the cost-effectiveness relative to current clinical test batteries for the different types of pathology; and (iii) establishment of procedures to allow studies to be reproduced with identical results at different times and with different instruments and observers. Absolute quantitation of results is desirable in the long term, but is not essential in all applications.

In this article we outline important general considerations involved in making interpretable and reproducible MRS brain studies of focal lesions. The brain is a highly heterogeneous tissue including a variety of cell types, considerable anatomic and functional specialization, and (not surprisingly, therefore) regional variations in NMR-visible metabolite resonance intensities.[32,50] Focal brain pathology is usually also anatomically, physiologically, and metabolically heterogeneous. Clinical exploitation of MRS to study focal brain lesions therefore requires that careful attention be paid to a number of fundamental problems including: (i) localization of where the magnetic resonance (MR) signals come from; (ii) registration of examinations; (iii) quantitative measurement of resonance intensities; and (iv) appropriate statistical analytical methods.

2 FUNDAMENTAL ISSUES FOR CLINICAL MRS OF FOCAL BRAIN LESIONS

2.1 Localization

The first localization problem is reliably to obtain signal only from the brain. As the fatty diploe of the skull and muscle surrounding the brain contain substantial amounts of the same compounds that are of interest in normal or pathological brain parenchyma, the ability to obtain signals from anatomically defined regions of brain without contamination from outside the volume of interest (VOI) is important. Even small proportions of signal from these regions could lead to large inaccuracies in quantitative assessment of metabolite concentrations in brain parenchyma. For example, contamination of proton brain spectra with as little as 1% of signal from extracerebral fat in the skull and scalp would lead to pathological amounts of lipid being observed in a normal brain spectrum.

Anatomical localization within the brain must also be achieved. Careful attention must be paid to potential contamination of spectra with signal from adjacent volumes having very different metabolite concentrations. Consider the problem of obtaining metabolite measurements from periventricular white matter using proton MRS, for example: periventricular white matter has a much lower lactate concentration then does cerebrospinal fluid (CSF), but concentrations of choline,

For References see p. 769

creatine, and N-acetylaspartate are very much higher in brain parenchyma than CSF.

There are significant limitations to real spatial resolution that can be achieved using current volume-selection protocols. Understanding the degree to which selectivity is achieved in any study is critical for proper interpretation of spectroscopic data.

2.1.1 Single Voxel Localization

There are many techniques that allow spectra to be obtained from more or less well localized volumes within the brain. Here we consider only those methods most commonly used with clinical instruments. These can be divided into two types: (i) those that observe free induction decays (FIDs) rather than spin echoes and rely on cancellation of unwanted signals from regions outside the chosen VOI; and (ii) those that exploit spin echo techniques to excite only the VOI. Both utilize selective excitation in the presence of magnetic field gradients and so are subject to chemical shift artifacts. These artifacts appear because the frequency selective excitation leads to slightly different excited volumes for different metabolites because of their different resonance frequencies.

As the T_2 of brain phosphates is relatively short, single voxel phosphorus MRS studies in vivo have generally relied on techniques in which FIDs are acquired, e.g. ISIS.[51,52] This avoids the signal loss due to T_2 relaxation that accompanies use of spin echo localization sequences. However, these methods depend on accurate subtractive cancellation of the relatively large amount of signal originating from outside the VOI. In practice, this is achieved only imperfectly. If one considers, for example, the problem of selecting signal from a volume of the order of 10 cm^3 out of a whole brain volume of >1000 cm^3, it is obvious that proportionally small inaccuracies in cancellation will result in spectra that are 'contaminated' by significant amounts of signal from outside the VOI.[53,54] More recent implementations of this approach minimize this problem by employing additional methods for suppressing signal from outside the VOI.[55,56]

The class of localization sequences which rely on spin echo sequences provides more reliable localization, as they excite only the VOI. Recent improvements in hardware have enabled such sequences to be employed with sufficiently short echo times to make them practical for phosphorus MRS studies without resulting in excessive signal loss or phase problems. Most proton studies of brain employ one of these sequences, most commonly either STEAM or PRESS. In proton MRS, measurement of the signal intensities of the most abundant compounds [e.g. choline (Cho), creatine (Cr), and N-acetylaspartate (NAA)] are less sensitive to echo time, as the T_2 values for these molecules are relatively large and they do not have complicated spin systems that cause phase modulation of their echoes. An additional advantage of these techniques in proton MRS is that it is easier effectively to suppress the large signals from water if one does not excite water from outside the VOI.

In proton studies spin echo sequences can be employed that use relatively long echo times to take advantage of the relatively long T_2 of some of the small soluble brain metabolites relative to fat of skin and skull in order to achieve improved localization and avoid problems from eddy currents generated by gradient coils. However, shorter echo times avoid the problem of phase modulation and allow the observation of additional metabolites with complicated spin systems and short T_2 times, e.g. amino acids and inositol.[27,57,58]

2.1.2 Spectroscopic or Chemical Shift Imaging

Single voxel techniques are inherently inefficient for applications in which chemical information from different regions is to be acquired at the same time (Figure 1). More recently developed spectroscopic imaging techniques approach an effective solution to this problem (Figure 2). Spectroscopic imaging sequences achieve localization with the use of phase-encoding pulse sequences, often combined with selective excitation so as to further minimize contamination from signal from outside the VOI. Phase encoding within the plane of the spectroscopic image is not associated with the chemical shift artifacts that arise from selective excitation. Spectroscopic imaging pulse sequences obtain signal from many relatively small voxels simultaneously. However, they tend to take a longer time to acquire than a single spectrum that represents a weighted average of resonance intensities from the same total volume. This is due to the number of averages required to obtain an adequate signal-to-noise ratio in the smaller voxels and the need to sample k-space adequately using sequences that incorporate a long sampling time in the absence of a field gradient to acquire the spectroscopic information. For proton examination in humans, current approaches allow nominal localization to volumes of about 0.5–1 cm^3 with the use of 16 × 16 or 32 × 32 phase-encoding steps (Figure 2).

Unfortunately, the relative insensitivity of phosphorus and the fact that compounds of major interest [e.g. adenosine triphosphate (ATP)] have short T_2 times, effectively limits individual voxel sizes in phosphorus MRS imaging to more than 10 cm^3, at least at 1.5 T. The large voxel size is usually combined with a low number (< 16 × 16) of phase-encoding profiles, which again tends to lead to contamination of spectra by signals adjacent to voxels.

When interpreting data from focal brain lesions it is important to appreciate that several factors cause the voxel size to be larger and less sharply defined than the nominal voxel size. It is the true shape of the excited volume that determines where signal actually comes from and, in consequence, the nature of partial volume effects. The nominal VOIs displayed on MRI images used to plan spectroscopy acquisitions may be defined very differently by different manufacturers, e.g. by the full width at half maximum amplitude of the excitation pulse or to include 90% of the excited signal. The definition of the selected volume needs to be clearly understood in terms of the percentage of signal coming from that volume versus the percentage of signal coming from outside the volume. In the case of phase-encoded spectroscopic images, the point spread function and effects of image processing also alter the voxel contours.

Defining the relationship between the volume from which the MRS signal is originating and any anatomically defined lesion on MRI is of great importance. Large focal brain lesions such as tumors, infarcts, or demyelinating plaques show wide pathological heterogeneity. Interpretation of the data therefore poses the analytical problem of inferring (at least qualitatively) the chemical pathology of a region within the selected volume element, taking into account partial volume effects. The effects of small differences in linear dimensions on partial volume effects are easily underestimated. For example, in a single

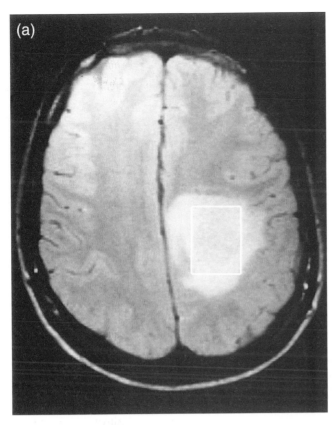

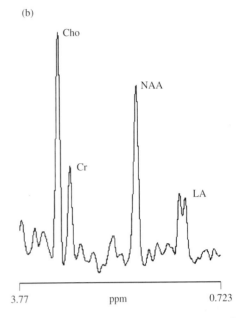

Figure 1   MRI and single voxel MRS examination of the same lesion a few days after the examination in Figure 2. The lesion appears similar on MRI (*TR* 2100, *TE* 30, slice thickness 8 mm). A single VOI for spectroscopic examination (*TR* 2000, *TE* 272, dimensions AP (anterior/posterior) 30 mm × CC (cranio/caudal) 22 mm × LR (left/right) 22 mm) is shown (in plane) in the box on the MRI. The VOI was chosen to sample as much of the lesion as possible but still accumulate essentially all of the signal selectively from within the lesion. The apparent metabolic changes represent an average over the entire volume and are therefore intermediate between those in voxels 1 and 2 in Figure 2. Quantitative comparison of the metabolic abnormalities apparent in the single voxel and MRS imaging examinations is difficult. The MRS image can be resampled in plane to correspond to the dimensions of the single voxel VOI, but the true shape of the volume from which the signal comes remains different because different techniques were used to define the voxels (i.e. phase-encoding in the MRS imaging, and selective excitation in the single voxel examination)

voxel study a spherical lesion filling an operator-defined 'cube' will contribute only 50% of the signal from the cube. If the 'cube' displayed on the monitor has margins that are defined by the full width at half maximum of the selection profile, then less than 25% of the signal will actually come from the lesion. Under these circumstances, given a normal variance of measurements of the order of 10%, only the most gross changes in metabolite concentrations could ever be detected as abnormal if the metabolic abnormality is spatially coextensive with the lesion seen on the MR image. In fact, for many focal lesions metabolic abnormalities are much more extensive than the associated imaging abnormalities,[29] and so significant results are obtained despite an apparently unfavorable partial volume of the lesion in the VOI.

2.2 Registration

2.2.1 *Definition*

The process of spatial alignment of images is referred to as 'registration'. A target image or space is defined and images to be registered with that image or space are positioned and may also be 'deformed' as necessary to match the target as closely as possible. A discussion of the methods available for registration is outside the scope of this article, but can be found in the review by Evans.[59]

2.2.2 *The Importance of Registration*

Registration of MRS examinations is important for at least three reasons. The first is that metabolic changes must be referred to anatomical pathology in the study of focal brain lesions in order to interpret the pathological significance of observed signal intensity changes. This type of registration is implicit in planning MRS volume acquisition parameters with respect to anatomical landmarks on MRI examinations (Figures 1–3). Common anatomical landmarks can be used to allow matching of spectroscopic VOIs in independent examinations of the same individual on different occasions (see Figures 2 and 3). Registration also can greatly facilitate appropriate comparisons of data collected from different individuals who may show regional variations in anatomy, by allowing matching of

For References see p. 769

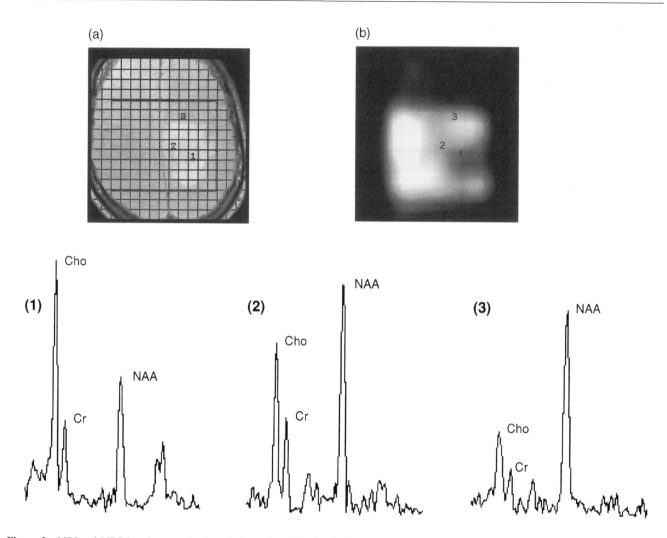

Figure 2 MRI and MRS imaging examination of a large demyelinating lesion. (a) A MRI slice through the center of the lesion (TR 2100, TE 30, slice thickness 8 mm). The phase-encoding grid for the MRS image is shown superimposed on the MRI. As long as the patient's head remains still during the entire time required for acquisition of the MR and MRS images, registration of the MRS image data is maintained. The dark and dotted lines define the VOI that was selectively excited for the MRS imaging examination. (b) The MRS image through the lesion (TR 2000, TE 272, slice thickness 20 mm). Individual spectra from voxels labeled 1, 2 and 3 are shown along the bottom of the figure. The lesion is associated with low *N*-acetylaspartate and high choline and LA. Note that the metabolic changes are different in different parts of the lesion and that abnormalities extend outside the lesion as defined on the MR image

volumes from different individuals in a standard anatomical space.[59–61] Finally, averaging of anatomically correlated metabolic information increases the intrinsically low signal-to-noise ratio of single examinations, thus facilitating comparisons of groups of image data sets.[50]

2.2.3 Technical Considerations

Registration may be accomplished before or after acquisition. Preplanned registration is performed by defining volumes for spectroscopic data acquisition with respect to anatomical landmarks before the data are acquired. Even when this is done for planning purposes within the context of a single MR session, errors in registration can occur due to patient movement between or during acquisitions, as well as to chemical shift artifacts when selective excitation is used. Pre-

planned registration of spectroscopic examinations obtained at different times requires meticulous care and has limitations in that only translations and rotations can be matched. Thus, the method is more suitable for comparisons between studies of one individual rather than between different individuals, since it cannot compensate for different head sizes and shapes. Even minor errors in registration can have serious consequences for data quantification. For example, in a three-dimensional MRS imaging examination, shifting of 1 cm^3 voxels by 2 mm in the three axes would alter the contribution of the desired voxel to the observed metabolite signals by as much as 50%.

Registration after acquisition can correct for some errors in preplanned registration and can correct for differences in head size and shape.[59] However, effective application requires three-dimensional high resolution images. These are not usually generated by conventional MRI as the slices usually have

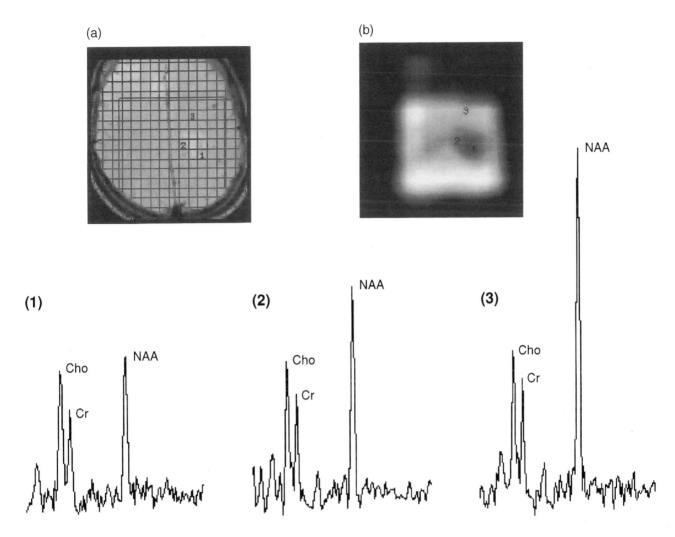

Figure 3 Follow-up MRI and MRS imaging of the same lesion as in Figure 2 approximately 1 year later, using identical acquisition techniques. The size, position, and angulation of the spectroscopic images were also matched as closely as possible to those of the first examination (Figure 2), but such a match is difficult to perform. An additional complication to registering the images is that the lesion is much smaller and the swelling which previously enlarged the hemisphere and shifted the midline towards the other side has resolved. All these factors conspire to make it unclear what the ideal volume for a single voxel spectroscopy follow-up should be. Spectroscopic images have significant advantages in this regard because of the availability of multiple small voxels. The three spectra shown are from the three voxels that most closely correspond to the three voxels chosen in the first examination (Figure 1)

relatively poor out-of-plane resolution, and significant image degradation occurs during resampling for the three-dimensional reconstruction.

For low resolution data like the MR single voxel or spectroscopic images, preplanned registration is essential. With single voxel data, changing or moving the VOI after data acquisition is not possible. Spectroscopic images permit manipulation of voxels after acquisition, but there are significant limitations. The severity of these limitations depends on whether data sets acquired are two or three dimensional. In the case of two-dimensional spectroscopic images (or multislice two-dimensional images), data must be acquired in the same plane in the studies being registered because the plane of the voxels cannot be shifted after acquisition. However, movement of voxels for registration within the plane of the slice is possible. In the case of three-dimensional spectroscopic images, translation of voxels

in all three axes is feasible. Rotation of voxels is theoretically possible, but degrades an already low-resolution image.

2.3 Measurement of Resonance Intensities

To allow chemical changes measured by MRS to be useful as indices of pathology, methods for quantitative measurement of signal intensities are necessary. In contrast to usual clinical radiology, which is based on qualitative assessments of patterns of changes in signal across images, work to date suggests that few clinical applications of brain spectroscopy will involve significant qualitative chemical changes specific to particular pathologies.[62]

There are independent (but related) problems in quantitation of signal intensities. The first major problem is to identify methods that allow accurate and reproducible measurements of

For References see p. 769

signal intensities from metabolites with varying relaxation properties and concentrations in a sample volume within which there is magnetic field inhomogeneity and, therefore, complex lineshapes. The problem is complicated in phosphorus spectra and proton spectra obtained at short echo times by the presence of overlapping and very broad peaks, which make unique fitting of peaks and the baseline impossible. Quantitative measurements with proton spectra may also be complicated by signal dynamic range problems associated with the presence of large residual signals from water. Finally, there is the problem of standardization. How can a reliable means be developed for deriving true tissue concentrations from resonance signal intensities so that data from different machines and different sites can be compared?

2.3.1 Spectral Estimation

Several methods are available for modeling either the FID or the Fourier transformed data.[63–65] The methods are generally constrained to models with limited numbers of components and generally to simple relaxation behavior as expected in a homogeneous field. These assumptions are not always well founded. Even assuming that pathological states can be modeled in the same way as normal tissue may be problematical. The nature of the constraints used in models can, in principle, significantly bias results. Differences in constraints or starting values between different clinical MRS centers could therefore be expected to limit the comparability of results. Assessment of precision of the methods is straightforward, but accuracy is difficult to evaluate because phantom systems for testing are inherently less complex physical structures than the brain.

The signal-to-noise ratio is relatively low in clinical spectra, further complicating all approaches to quantitation. The apparent signal-to-noise ratio is routinely improved by a variety of processing procedures. Signal processing can also be used to flatten baselines distorted in various ways. Effects of this processing on quantification of spectra must be taken into account.

Another problem is automation of quantitation methods. Automation facilitates evaluation by making it possible to deal with large numbers of spectra acquired in spectroscopic imaging studies. Use of prior knowledge and models facilitates automatic fitting, but introduces biases. However, these are at least consistent biases that are inherently more independent of operator bias than manually supervised methods, although they may differ between different sites if different models and approaches are used.

2.3.2 Relative and Absolute Quantification

Metabolite signal intensities from pathological tissue will not, in general, be useful alone. Many factors that vary from instrument to instrument, between individuals, and even during a prolonged examination determine observed signal intensities. Among them, coil quality, coupling of the sample to the coil, field (B_1) inhomogeneity, and relaxation times are the most important general factors. Most approaches to controlling for variability in coil responses and coupling involve use of an external (relative to the patient) standard. To be done rigorously, the process is quite complicated. Fortunately, modern instrumentation is generally stable enough for practical methods to work. Appreciating such biases is essential in comparisons of data between different sites.

More popular has been the use of internal standards. One approach is to present data as ratios of resonance intensities relative to a signal that is assumed to be relatively invariant, e.g. use of the creatine signal as an internal reference in proton MRS, or ATP in phosphorus MRS.[31,66] This method assumes that relative relaxation rates are unchanged in the pathological volumes and that concentrations of the chosen reference signal do not vary between normal-appearing and pathological tissue (note the results in Figure 3). When comparing metabolite signal intensity ratios from different regions of the brain it is necessary to be aware of the regional variation in relative signal intensities in the normal brain or to compare only identically localized volumes (Figure 2).[32]

The validity of assumptions of invariance of internal reference signals can be tested by comparisons of absolute signal intensities with anatomically homologous volumes in the (unaffected) contralateral hemisphere. Care should be taken even with this control, however, as pathology in one hemisphere may have secondary physiological, biochemical, or anatomic effects in the contralateral hemisphere because of the extensive connectivity of the two hemispheres. With this concern in mind, absolute signal intensities can also be compared directly between lesions and normal-appearing contralateral hemispheres. While this is probably valid in some situations, it will not be in others. For example, it may be true for creatine in chronic multiple sclerosis,[30,31,42] but certainly does not hold true for brain tumors.[3,12]

2.4 Statistical Analysis of Spectroscopic Data

Single voxel methods produce a single spectrum, which simplifies the problem of data quantitation and comparisons. The data are biased by subjectivity in the volume selection, but are potentially statistically powerful.

Spectroscopic image data sets are much more difficult to analyze. Comparisons based on the subjective choice of only certain voxels within the MRS image introduces bias, as in the case of single voxel examinations. Use of more than one voxel to characterize a focal lesion is complicated because the voxels are not statistically independent measurements. An appropriate correction for their spatial correlations must be made.[67] Finally, comparisons with normal data that also consist of image data sets is complex. Objective comparisons of groups of spectroscopic images and statistical analysis of the significance of differences requires the development of a new statistical methodology that avoids sampling bias, is efficient despite a low signal-to-noise ratio, and takes into account the spatial variation and correlations in the images. However, such procedures have been developed for other applications, such as the analysis of satellite photographs.[68]

3 CONCLUSIONS

Preliminary results have demonstrated that brain MRS may be able to provide potentially important information concerning localized changes in brain chemistry associated with focal pathologies. Because focal pathologies are often the object of active treatments (e.g. for abscesses or tumors) or efforts to develop effective treatments (e.g. for stroke or multiple sclerosis), this type of pathology may provide a major justification for integration of MRS into clinical neurological and neurosurgical practice. However, at the time of writing, several fundamental problems still limit the practical availability of reproducible, quantitative spectroscopic data.

The intrinsic limitations on the signal-to-noise ratio in MRS and relaxation properties of the nuclei of interest limit spectroscopic resolution relative to water-based proton images. Despite the desirability of further increases in resolution, a practical balance is necessary between resolution and the time required for acquisition in clinical studies.

Comparisons of spectroscopic image data between individuals, individuals and groups, and between serial examinations of an individual are fundamental to clinical applications. However, they are difficult to perform rigorously. Accurate preplanned registration is likely to remain essential for spectroscopic data acquisition. Techniques that allow registration to a standard anatomical space can provide averaged normative data with a much improved signal-to-noise ratio relative to individual studies for more accurate quantitative assessment of data from pathology in an individual subject.

Application of spatial statistics[68] should facilitate efficient, unbiased assessment of spectroscopic image data. Statistical approaches such as this potentially allow meaningful comparison of data obtained from a variety of modalities. Of particular interest also is the potential for simultaneous assessment of structural and biochemical MR data.

4 RELATED ARTICLES

Brain Infection and Degenerative Disease Studied by Proton MRS; Brain MRS of Human Subjects; Brain MRS of Infants and Children; Brain Neoplasms in Humans Studied by Phosphorus-31 NMR Spectroscopy; Chemical Shift Imaging; Selective Excitation Methods: Artifacts; Selective Excitation in MRI and MR Spectroscopy; Single Voxel Localized Proton NMR Spectroscopy of Human Brain In Vivo; Surface Coil NMR: Detection with Inhomogeneous Radiofrequency Field Antennas.

5 REFERENCES

1. D. L. Arnold, E. A. Shoubridge, W. Feindel, and J. G. Villemure, *Can. J. Neurol. Sci.*, 1987, **14**, 570.
2. W. Semmler, G. Gademann, P. Bachert Baumann, H. J. Zabel, W. J. Lorenz, and G. van Kaick, *Radiology*, 1988, **166**, 533.
3. H. Bruhn, J. Frahm, M. L. Gyngell, K. D. Merboldt, H. Hanicke, R. Sauta, and C. Hamburger, *Radiology*, 1989, **172**, 541.
4. M. C. Preul, Z. Caramanos, R. Leblanc, J. G. Villemure, and D. L. Arnold, *NMR Biomed.*, 1998, **11**, 192.
5. H. Kugel, W. Heindel, R. I. Ernestus, J. Bunke, R. du Mesnil, and G. Friedmann, *Radiology*, 1992, **183**, 701.
6. C. M. Segebarth, D. F. Baleriaux, P. R. Luyten, and J. A. den Hollander, *Magn. Reson. Med.*, 1990, **13**, 62.
7. M. Preul, D. Collins, R. Ethier, W. Feindel, and D. L. Arnold, *Proc. XIIth Ann Mtg. Soc. Magn. Reson. Med.*, New York, 1993, **1**, 64 (abstract).
8. J. A. den Hollander, P. R. Luyten, A. J. Marien, C. M. Segebarth, D. F. Baleriaux, R. de Beer, and D. Van Ormondt, *Magn. Reson. Q.*, 1989, **5**, 152.
9. B. Hubesch, D. Sappey-Marinier, K. Roth, D. J. Meyerhoff, G. B. Matson, and M. W. Weiner, *Radiology*, 1990, **174**, 401.
10. D. L. Arnold, J. F. Emrich, E. A. Shoubridge, J. G. Villemure, and W. Feindel, *J. Neurosurg.*, 1991, **74**, 447.
11. W. D. Heiss, W. Heindel, K. Herholz, J. Rudolph, J. Bunke, J. Jeoke, and G. Friedmann, *J. Nucl. Med.*, 1990, **31**, 302.
12. D. L. Arnold, E. A. Shoubridge, J. Emrich, W. Feindel, and J. G. Villemure, *Invest. Radiol.*, 1989, **24**, 958.
13. D. P. Younkin, M. Delivoria-Papadopoulos, J. C. Leonard, V. H. Subramanian, S. Eleff, J. S. Leigh, Jr., and B. Chance, *Ann. Neurol.*, 1984, **16**, 581.
14. P. A. Hamilton, P. L. Hope, E. B. Cady, D. T. Delpy, J. S. Wyatt, and E. O. Reynolds, *Lancet*, 1986, **i**, 1242.
15. D. T. Delpy, M. C. Cope, E. B. Cady, J. S. Wyatt, P. A. Hamilton, P. L. Hope, S. Wray, and E. O. Reynolds, *Scand. J. Clin. Lab. Invest. Suppl.*, 1987, **188**, 9.
16. D. Azzopardi, J. S. Wyatt, E. B. Cady, D. T. Delpy, J. Baudin, A. L. Stewart, P. L. Hope, P. A. Hamilton, and E. O. Reynolds, *Pediatr. Res.*, 1989, **25**, 445.
17. T. Kato, A. Tokumaru, T. O'uchi, I. Mikami, M. Umeda, K. Nose, T. Egushi, M. Haregawa, and K. Okuyama, *Radiology*, 1991, **179**, 95.
18. E. O. Reynolds, D. C. McCormick, S. C. Roth, A. D. Edwards, and J. S. Wyatt, *Ann. Med.*, 1991, **23**, 681.
19. C. J. Peden, F. M. Cowan, D. J. Bryant, J. Sargentoni, I. J. Cox, D. K. Menon, D. G. Gadian, J. D. Bell, and L. M. Dubowitz, *J. Comput. Assist. Tomogr.*, 1990, **14**, 886.
20. K. M. Welch, S. R. Levine, G. Martin, R. Ordidge, A. M. Vande Linde, and J. A. Helpern, *Neurol. Clin.*, 1992, **10**, 1.
21. S. R. Levine, K. M. Welch, J. A. Helpern, M. Chopp, R. Bruce, J. Selwa, and M. B. Smith, *Ann. Neurol.*, 1988, **23**, 416.
22. K. M. Welch, S. R. Levine, and J. A. Helpern, *Funct. Neurol.*, 1990, **5**, 21.
23. S. R. Levine, J. A. Helpern, K. M. Welch, A. M. Vande-Linde, K. L. Sowaya, E. E. Brown, N. M. Ramadan, R. K. Devashwar, and R. J. Ordidge, *Radiology*, 1992, **185**, 537.
24. J. H. Duijn, G. B. Matson, A. A. Maudsley, J. W. Hugg, and M. W. Weiner, *Radiology*, 1992, **183**, 711.
25. M. J. Fenstermacher and P. A. Narayana, *Invest. Radiol.*, 1990, **25**, 1034.
26. J. W. Berkelbach van der Sprenkel, P. R. Luyten, P. C. van Rijen, C. A. Tulleken, and J. A. den Hollander, *Stroke*, 1988, **19**, 1556.
27. G. D. Graham, A. M. Blamire, A. M. Howseman, D. L. Rothman, P. B. Fayad, L. M. Brass, O. A. Petroff, R. G. Shulman, and J. W. Prichard, *Stroke*, 1992, **23**, 333.
28. D. L. Rothman, A. M. Howseman, G. D. Graham, O. A. Petroff, G. Lantos, P. B. Fayad, L. M. Brass, G. I. Shulman, and J. W. Prichard, *Magn. Reson. Med.*, 1991, **21**, 302.
29. D. L. Arnold, P. M. Matthews, G. S. Francis, J. O'Connor, and J. P. Antel, *Ann. Neurol.*, 1992, **31**, 235.
30. P. M. Matthews, G. Francis, J. Antel, and D. L. Arnold, *Neurology*, 1991, **41**, 1251.
31. D. L. Arnold, P. M. Matthews, G. Francis, and J. Antel, *Magn. Reson. Med.*, 1990, **14**, 154.
32. J. Frahm, H. Bruhm, M. L. Gyngell, K. D. Merboldt, W. Hanicke, and R. Sauter, *Magn. Reson. Med.*, 1989, **11**, 47.
33. H. B. Larsson, P. Christiansen, M. Jensen, J. Frederikren, A. Heltberg, J. Olesen, and O. Henriksen, *Magn. Reson. Med.*, 1991, **22**, 23.
34. T. L. Richards, *Am. J. Roentgenol.*, 1991, **157**, 1073.
35. D. H. Miller, S. J. Austin, A. Connelly, B. D. Youl, D. G. Gadian, and W. I. McDonald, *Lancet*, 1991, **337**, 58.
36. R. I. Grossman, R. E. Lenkinski, K. N. Ramer, F. Gonzalez-Scarano, and J. A. Cohen, *Am. J. Neuroradiol.*, 1992, **13**, 1535.

For References see p. 769

37. L. Fu, P. M. Matthews, N. de Stefano, K. J. Worsley, S. Narayanan, G. S. Francis, J. P. Antel, C. Wolfson, and D. L. Arnold, *Brain*, 1998, **121**, 103.

38. C. A. Davie, C. P. Hawkins, G. J. Barker, A. Brennan, P. S. Tofts, D. H. Meller, and W. J. McDonald, *Proc. XIIth Ann Mtg. Soc. Magn. Reson. Med.*, New York, 1993, **1**, 133.

39. P. A. Narayana, J. S. Wolinsky, E. F. Jackson, and M. McCarthy, *JMRI*, 1992, **2**, 263.

40. J. S. Wolinsky, P. A. Narayana, and M. J. Fenstermacher, *Neurology*, 1990, **40**, 1764.

41. C. Husted, D. S. Goodin, A. A. Maudsley, J. S. Tsuruda, and M. W. Weiner, *Neurology*, 1993, **43**, A182 (abstract).

42. D. L. Arnold, G. T. Riess, P. M. Matthews, G. S. Frances, D. L. Collins, C. Wolfson, and J. P. Antel, *Ann. Neurol.*, 1994, **36**, 76.

43. P. M. Matthews, F. Andermann, and D. L. Arnold, *Neurology*, 1990, **40**, 985.

44. A. Connely, D. G. Gadian, G. D. Jackson, J. H. Gross, M. D. King, J. S. Duncan, and F. J. Kirkham, *Proc. XIth Ann Mtg. Soc. Magn. Reson. Med.*, Berlin, 1992, **1**, 234.

45. F. Cendes, F. Andermann, M. C. Preul, and D. L. Arnold, *Ann. Neurol.*, 1994, **35**, 211.

46. J. W. Hugg, K. D. Laxer, G. B. Matson, A. A. Maudsley, and M. W. Weiner, *Ann. Neurol.*, 1993, **34**, 788.

47. F. Cendes, Z. Caramanos, F. Andermann, F. Dubeau, and D. L. Arnold, *Ann. Neurol.*, 1997, **42**, 737

48. L. Fu, C. Wolfson, K. J. Worsley, N. de Stefano, D. L. Collins, S. Narayanan, and D. L. Arnold, *NMR Biomed.*, 1996, **9**, 339.

49. J. W. Hugg, K. D. Laxer, G. B. Matson, A. A. Maudsley, C. A. Husted, and M. W. Weiner, *Neurology*, 1992, **42**, 2011.

50. D. L. Collins, L. Fu, E. Pioro, A. C. Evans, and D. L. Arnold, *Magn. Reson. Med.*, 1993, **3**, 1600.

51. R. J. Ordidge, P. Mansfield, J. A. Lohman, and S. B. Prime, *Ann. NY Acad. Sci.*, 1987, **508**, 376.

52. R. J. Ordidge, R. M. Bowley, and G. McHale, *Magn. Reson. Med.*, 1988, **8**, 323.

53. G. B. Matson, D. J. Meyerhoff, T. J. Lawry, R. S. Lara, J. Duijn, R. F. Deeken, and M. W. Weiner, *NMR Biomed.*, 1993, **6**, 215.

54. S. F. Keevil, D. A. Porter, and M. A. Smith, *NMR Biomed.*, 1992, **5**, 200.

55. S. Posse, B. Schuknecht, M. E. Smith, P. C. van Zijl, N. Herschkowitz, and C. T. Moonen, *J. Comput. Assist. Tomogr.*, 1993, **17**, 1.

56. R. J. Ordidge, *Magn. Reson. Med.*, 1987, **5**, 93.

57. B. D. Ross, *NMR Biomed.*, 1991, **4**, 59.

58. R. Kreis, N. Farrow, and B. D. Ross, *NMR Biomed.*, 1991, **4**, 109.

59. A. C. Evans, in 'Principles of Nuclear Medicine', ed. H. N. Wagner, Jr., W. B. Saunders, Philadelphia, 1995.

60. P. T. Fox, J. S. Perlmutter, and M. E. Raichle, *J. Comput. Assist. Tomogr.*, 1985, **9**, 141.

61. A. C. Evans, T. S. Marrett, D. L. Collins, and T. M. Peters, *SPIE*, 1989, Medical Imaging III, 264 (abstract).

62. Arnold D. L. and P. M. Matthews, in 'MR Spectroscopy: Clinical Applications and Techniques', ed. I. R. Young and C. Charles, Martin Dunitz, London, 1995.

63. P. R. Luyten, A. J. Marien, and J. A. den Hollander, *NMR Biomed.*, 1991, **4**, 64.

64. A. Knijn, R. de Beer, and D. van Ormondt, *J. Magn. Reson.*, 1992, **97**, 444.

65. R. de Beer, A. Van den Boogaart, D. van Ormondt, W. W. Pijnappel, J. A. den Hollander, A. J. Morien, and P. R. Luyten, *NMR Biomed.*, 1992, **5**, 171.

66. D. L. Arnold, P. M. Matthews, and G. K. Radda, *Magn. Reson. Med.*, 1984, **1**, 307.

67. K. J. Worsley, A. C. Evans, S. Marrett, and P. Neelin, *J. Cereb. Blood Flow Metab.*, 1992, **12**, 900.

68. N. A. Cressie, 'Statistics for Spatial Data', Wiley, Toronto, 1991.

For list of General Abbreviations see end-papers

Biographical Sketches

Douglas L. Arnold. *b* 1951. B.Sc. (Physiology), McGill University, M.D., Cornell University Medical College. Trained in medicine and neurology at McGill before doing a postdoctoral fellowship in NMR under the supervision of Professor G. K. Radda, Department of Biochemistry, University of Oxford, UK. Currently Professor of Neurology and Neurosurgery at McGill University. Research interests: magnetic resonance spectroscopy of disorders of brain and muscle. Approx. 100 publications, including, with P. M. Matthews, the book *Diagnostic Tests in Neurology*.

Paul M. Matthews. *b* 1956. B.A. (Chemistry), D.Phil. (Biochemistry) (supervisor G. K. Radda and D. G. Gadian), University of Oxford. General medical training at Oxford and Stanford. Trained in neurology at the Montreal Neurological Institute. Currently Professor in the Department of Neurology, University of Oxford, and Director of the Oxford Centre for Functional Magnetic Resonance Imaging of the Brain. Research interests: the chemical pathology of multiple sclerosis and mitochondrial disorders.

pH Measurement In Vivo in Whole Body Systems

Neil A. Farrow and John H. Richards
California Institute of Technology, Pasadena, CA, USA

and

Brian D. Ross
Huntington Medical Research Institutes, Pasadena, CA, USA

1 INTRODUCTION

In 1973 Moon and Richards[1] described a noninvasive approach for the determination of conditions within cells, such as pH, by observing the magnetic resonance signal of ^{31}P from intracellular phosphates whose chemical shifts vary significantly with pH (see Figure 1). In this study of the pH within erythrocytes they monitored the chemical shift of the two ^{31}P signals of intracellular 2,3-diphosphoglycerate and found that the intraerythrocytic pH of whole blood was the same as that of packed cells, as well as that of hemolysates of packed red blood cells. This and analogous approaches have since been used to study a wide variety of biological systems including ion transport in plants,[2] the metabolism of flight in insects,[3] as well as human physiology and pathology, the particular focus of this discussion.

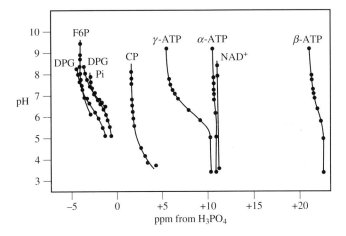

Figure 1 The pH-dependent chemical shift behavior of a variety of biological organic phosphates. Chemical shifts are reported relative to external 1.0 M phosphoric acid. F6P, fructose 6-phosphate; DPG, 2,3-diphosphoglycerate; CP, carbamyl phosphate; Pi, inorganic phosphate; ATP, adenosine triphosphate. (Reproduced by permission from R. B. Moon and J. H. Richards[1])

To determine the pH of biological samples by NMR spectroscopy the presence is required of a molecule containing a nucleus with appropriate magnetic properties. The chemical shift of the nucleus in the molecule should vary significantly with pH and the molecule should have a pK_a close to that of the environment whose pH is being determined. The ^{31}P nucleus satisfies these conditions. Its spin quantum number of $\frac{1}{2}$ provides high-resolution spectra; its 100% natural abundance and inherent sensitivity give an observational sensitivity of 6.6% relative to ^{1}H; molecules that contain ^{31}P are often present in biological samples at millimolar (mM) concentrations. These factors allow the collection of one-dimensional spectra within short periods of time to determine ^{31}P chemical shifts. Inorganic phosphate can serve as a pH-sensitive indicator as the pK_a for dihydrogen phosphate is near 7 and the two differently protonated forms are in rapid exchange ($k \sim 10^9–10^{10}$ s^{-1}).

$$H_2PO_4^- \rightleftharpoons HPO_4^{2-} + H^+ \qquad pK_a = 7.21$$

The chemical shifts of monohydrogen phosphate and dihydrogen phosphate differ by about 2.35 ppm. On a 500 MHz instrument (where ^{31}P nuclei resonate around 202 MHz) this corresponds to a frequency difference between the two differently protonated forms of 475 Hz, which is well below the chemical exchange rate of $10^9–10^{10}$ s^{-1}. In such situations one observes a single resonance for the two forms; the chemical shift represents an average weighted by the relative concentrations of the two forms. Since most biological systems normally have pH values near 7, the changing relative concentrations of these two forms with varying pH leads to a ^{31}P signal that sensitively reflects the pH of the environment.

Since features of the environment, other than pH, can also influence the chemical shifts of ^{31}P nuclei, one requires a reference signal whose chemical shift should ideally be independent of pH. For many systems phosphocreatine serves this function; it is completely ionized near physiological pH:

$$\overset{\displaystyle NH}{\overset{\|}{^-HO_3PNHC(CH_2)_3CHCO_2^-}} \rightleftharpoons \overset{\displaystyle NH}{\overset{\|}{^{2-}O_3PNHC(CH_2)_3CHCO_2^-}}$$
$$\underset{^+NH_3}{} \qquad\qquad \underset{^+NH_3}{}$$

$$pK_a \text{ of phosphate} = 4.6$$

and has a chemical shift 3.5–5.5 ppm away from that of inorganic phosphate in the range of physiological pH. Thus by constructing reference titration curves under conditions that simulate those of the biological sample (as shown in Figure 2) one can determine the pH of the biological system itself. In situations where no phosphocreatine signal is present, as in liver, the α-phosphate of ATP can serve as a reference, though one must recognize that its chemical shift changes with pH as shown in Figure 1. Other internal references, such as cell water[4] or an external reference, such as a vial of methylene diphosphonate, can also be used.

Nuclei other than ^{31}P can also serve as probes. For example esterified F-quene-1,[5] can enter a cell where the ester groups are then hydrolyzed. The resulting F-quene-1 cannot cross the cell membrane and becomes trapped intracellularly. The ^{19}F signal of F-quene-1 has been used to monitor the intracellular pH of rat liver, for example. The ^{1}H signal from C-4 of histidine in the hemoglobin of erythrocytes[6] and of anserine in muscle,[7] as well as the ^{13}C signal from C-3 of sn-glycerol 3-phosphate, have also been used as probes to determine pH.

The introduction of wide-bore magnets and whole body spectrometers that operate at fields from 1.5 to 7 T, permit observations of perfused organs, whole animals, and humans. Surface coils[8] or localization protocols such as image-selected in vivo spectroscopy[9] and chemical shift imaging[9,10] pulse sequences have become popular methods for acquiring a ^{31}P

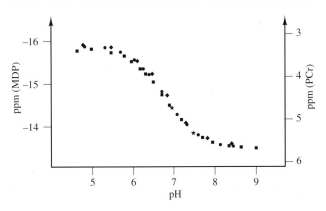

Figure 2 Variation of the chemical shift of inorganic phosphate with solution pH. The chemical shift is expressed relative to two different standards: methylene diphosphonate (MDP) and phosphocreatine (PCr) at pH 7. Solutions containing 10 mM inorganic phosphate and 10 mM phosphocreatine were adjusted to various ionic strengths by addition of KCl or NaCl and titrated at 37 °C by addition of HCl and KOH or NaOH. Symbols: ◆ 120 mM KCl; ● 160 mM KCl; ■ 200 mM KCl; ★ 160 mM NaCl. (Reproduced by permission of Clarendon Press, from D. G. Gadian, 'Nuclear Magnetic Resonance and its Applications to Living Systems', 1982, p. 32)

For References see p. 773

signal from a specific volume element of the subject. Such instruments and techniques can give important information from liver, kidneys, muscle, and brain, as well as cell cultures and ex vivo tissue samples.

2 RELIABILITY

The inherent precision of MRS measurements of pH in favorable circumstances is around 0.06 pH units.[11] In real biological samples, additional sources of error arise due to partial volume effects caused by the signal from the region of interest being compromised by signal from other regions of the sample. For example, the ^{31}P spectra of in vivo liver often contain a phosphocreatine peak whose source is the surrounding muscle. Even for a signal from a homogenous tissue or cell population, the same substance in various compartments can complicate the interpretation by giving rise to more than one signal. For example, studies on various tumors[12] reveal that the signal for inorganic phosphate originates from about 65% intracellular phosphate and 35% phosphate in the extracellular space. Accordingly, an observation of inorganic phosphate in tumors reports a pH that is somewhat weighted toward the intracellular pH despite the fact that the extracellular volume of a tumor sample exceeds that of normal tissue.

Mitochondrial preparations show two signals for inorganic phosphate, one within the mitochondria, the other from the medium.[11] The different pH values arise because of pH gradients across the membranes that separate the compartments; in the cases discussed, the magnitude of the pH differences accord with Mitchell's chemosmotic hypothesis.[13] Signals can also arise from different intracellular compartments as in aerobic preparations of rat liver; one signal represents cytoplasmic phosphate, the other comes from phosphate within the mitochondria that are so abundant in such samples.

3 CANCER

One of the earlier and significant contributions of MRS pH measurements concerned the intracellular pH of tumors, which was found to be from neutral to alkaline,[12,14–16] in direct contrast with the standard belief of the day that tumors were acidic.[17,18] Tumors produce excess lactic acid (because of appreciable anaerobic metabolism) suggesting an acidic intracellular pH, a surmise which studies with microelectrodes supported.[18] Thus MRS results gave a pH in the 6.9–7.4 range, whereas microelectrodes reported pH values in the 5.6–7.6 range. In fact, tumor cells do contain unusually high amounts of anionic lactate[19] (not lactic acid as was once assumed) because they probably export the protons which then accumulate in the extracellular regions in unusually large amounts because of the incomplete equilibration that results from the compromised blood circulation in many solid tumors. In these circumstances microelectrodes report more the extracellular acidic pH than the neutral intracellular pH that MRS allows one to observe noninvasively. This result has proved significant in rational design of anticancer drugs.

For list of General Abbreviations see end-papers

4 MUSCLE

Moderate exercise of muscle leads to intracellular acidosis. Indeed the pH can drop from 7.1 to values as low as 6.0.[20] After stopping exercise, the pH recovers to normal values within 1 min. Some protocols for moderate exercise show little or no decrease in intracellular pH.[21] As one would expect, more vigorous exercise, or exercise under ischemic conditions, causes a more profound intramuscular acidosis,[22] as shown in Figure 3. Interestingly, and reassuringly for athletes and fitness fans, similar levels of exercise after training cause significantly less acidification than that for untrained individuals and this benefit extends even to those muscles not specifically trained.[20] Exercise protocols can also be used to study the pathology of

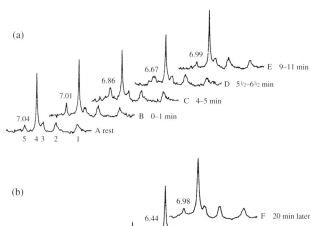

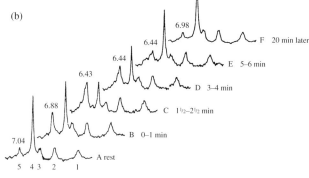

Figure 3 Phosphorus-31 NMR spectra in a control subject showing the effects of (a) normal aerobic exercise and (b) ischemic exercise. A 1.9 T, 20 cm bore magnet and a 4.5 cm surface coil were used to acquire signal from about 25 mL of the subject's flexor compartment of the forearm. Phosphorus-31 spectra were recorded at 32.5 MHz with an interpulse delay of 2 s, sweep width 4 kHz, and a line broadening of 6 Hz used before Fourier transformation. The exercise protocol involved flexion of fingers at 2 s intervals. The signals are assigned as follows: 1, 2, and 3, the β, α, and γ phosphates of ATP; 4, phosphocreatine; and 5, inorganic phosphate. The pH value given above each inorganic phosphate signal was determined from the frequency separation of the inorganic phosphate and phosphocreatine signals. (a) Zero minutes corresponds to the start of exercise and was maintained for 5 min. Spectra A and E were accumulated with 64 scans; the other spectra were accumulated with 32 scans. The pH value is an average in spectra where the inorganic phosphate signal is split into two components. (b) Zero minutes corresponds to the start of exercise and was maintained for 1.5 min. Arterial occlusion was maintained from zero minutes to 3 min with arterial flow being restored after this period. (Reproduced by permission from B. D. Ross, G. K. Radda, D. G. Gadian, G. Rocker, M. Esiri, and J. Falconer-Smith, *N. Engl. J. Med.*, 1981, **304**, 1338)

various muscle diseases. For example, patients with myotonic dystrophy show a smaller decrease in intramuscular pH with exercise, and their muscles also return more rapidly to a normal value than do those of controls.[23] Furthermore, patients who lack glycogen phosphorylase (McArdles syndrome) actually manifest the expected increase in intramuscular pH due to the unmasking of the alkalinizing hydrolysis of phosphocreatine to phosphate and creatine.

5   HEART

To monitor transient ischemia, the effects on heart muscle of occlusion of the coronary artery followed by reperfusion have been monitored by ^{31}P MRS. In a perfused heart ligation, occlusion of the left anterior descending artery[24,25] caused a decrease in pH from 7 to 6.7. In a globally ischemic heart, the pH falls to 6.2.[26,27] In dogs, occlusion of the left descending coronary artery caused the pH to decrease to 5.9; in some cases the pH recovered spontaneously to preocclusion values presumably because of the development of collateral flow to the affected myocardium.[28] Other hearts showed more permanent changes reflecting a decreased pH that was consistent with the ischemia. These pH changes occurred within minutes of the occlusion.

One can even observe the results of exercise-induced ischemia in the human heart in vivo, despite the large ^{31}P signal from the 2,3-diphosphoglycerate in the red cells of the blood within the ventricles. Heart muscle cells have a relatively high density of mitochondria (30%) such that one might actually be able to observe the pH gradient between the mitochondria and the cytosol. Such differences have been demonstrated in other tissues, for example lineshape analysis of ^{31}P spectra leads to a pH gradient of 0.5 in skeletal muscle[29,30] and kidney.[31] In the perfused rat heart, deoxyglucose, which is unable to enter mitochondria, and upon phosphorylation becomes trapped in the cytoplasm, gives the same pH value as that determined by using inorganic phosphate.[32] This study suggested that the pH observed in heart muscle reflects that in the cytoplasm, possibly because intramitochondrial inorganic phosphate cannot be observed with present-day technology due to its low concentration and/or due to a broadened signal. Alternatively, the pH gradient may be small in heart muscle.

Calculations have shown that a pH difference of 0.3–0.4 is necessary to resolve the peaks for inorganic phosphate within or outside a cell when 65% of the observed MRS signal represents intracellular phosphate.[33] Nonetheless, separate MRS signals for Walker's tumors were unobservable even though the ΔpH was 0.6, as calculated from the distribution of lactate across the cell membrane; this failure was attributed to the poor signal-to-noise characteristic of this type of tumor.

6   BRAIN

Studies in the brain of animal models also show a similar decrease in pH due to ischemia. Unfortunately, correlation of these findings with other parameters such as preischemic glucose levels or the exogenous introduction of base to alleviate

trauma, have provided inconclusive, and even in some cases contradictory, results.[34,35]

7   FUTURE

Difficulties occasioned by low signal-to-noise or poor resolution hamper many studies now being undertaken with in vivo MRS studies of human pathologies. The introduction of whole body spectrometers with higher fields, the use of alternative nuclei such as ^{15}N (spectra of which can now[36,37] be acquired in vivo), the ability to obtain high-resolution data from smaller, well-defined volume elements, the use of other endogenous metabolites or the introduction of exogenous reporting substances, provide hopeful new avenues to achieve noninvasively more reliable information about the metabolic status of organs, tissues, and other biological samples. The resulting insights will greatly broaden our understanding of the complex integrated metabolism of intact living systems. For humans, these advances can significantly enhance the power of diagnosis.

8   RELATED ARTICLES

Brain MRS of Human Subjects; Brain MRS of Infants and Children; Brain Neoplasms in Humans Studied by Phosphorus-31 NMR Spectroscopy; In Vivo Hepatic MRS of Humans; Kidney, Prostate, Testicle, and Uterus of Subjects Studied by MRS; Peripheral Muscle Metabolism Studied by MRS; Tissue Behavior Measurements Using Phosphorus-31 NMR; Whole Body Studies: Impact of MRS.

9   REFERENCES

1. R. B. Moon and J. H. Richards, *J. Biol. Chem.*, 1973, **248**, 7276.
2. C. M. Spickett, N. Smirnoff, and R. G. Ratcliffe, *Plant Physiol.*, 1993, **102**, 629.
3. G. Wegener, N. M. Bolas, and A. A. G. Thomas, *J. Comp. Physiol. B*, 1991, **161**, 247.
4. A. Madden, M. O. Leach, D. J. Collins, and G. S. Payne, *Magn. Reson. Med.*, 1991, **19**, 416.
5. J. S. Beech and R. A. Iles, *Magn. Reson. Med.*, 1991, **19**, 386.
6. F. F. Brown, I. D. Campbell, P. W. Kuchel, and D. L. Rabenstein, *FEBS Lett.*, 1977, **82**, 12.
7. S. R. Williams, D. G. Gadian, E. Proctor, D. B. Sprague, D. F. Talbot, F. F. Brown, and I. R. Young, *Biochem. Soc. Trans.*, 1985, **13**, 839.
8. J. J. H. Ackerman, T. H. Grove, G. G. Wong, D. G. Gadian, and G. K. Radda, *Nature (London)*, 1980, **283**, 167.
9. R. J. Ordidge, A. Connelly, and J. A. B. Lohman, *J. Magn. Reson.*, 1986, **66**, 283.
10. T. R. Brown, B. M. Kincaid, and K. Ugurbil, *Proc. Natl. Acad. Sci. USA*, 1982, **79**, 3523.
11. S. R. Smith, R. D. Griffiths, P. A. Martin, and R. H. Edwards, *Radiology*, 1989, **173**, 572.
12. M. Stubbs, Z. M. Bhujwalla, G. M. Tozer, L. M. Rodrigues, R. J. Maxwell, R. Morgan, F. A. Howe, and J. R. Griffths, *NMR Biomed.*, 1992, **5**, 351.
13. P. Mitchell, *Biochem. Soc. Trans.*, 1976, **4**, 399.
14. S. Ogawa, H. Rottenberg, T. R. Brown, R. G. Schulman, C. L. Castillo, and P. Glynn, *Proc. Natl. Acad. Sci. USA*, 1978, **75**, 1796.

For References see p. 773

15. J. R. Griffiths, A. N. Stevens, R. A. Iles, R. E. Gordon, and D. Shaw, *Biosci. Rep.*, 1981, **1**, 319.

16. R. A. Iles, A. N. Stevens, and J. R. Griffiths, *Prog. Nucl. Magn. Reson. Spectrosc.*, 1982, **15**, 49.

17. J. R. Griffiths, E. Cady, R. H. T. Edwards, V. R. McCready, D. R. Wilkie, and E. Wiltshaw, *Lancet*, 1983, **1**, 1435.

18. J. R. Griffiths, *Br. J. Cancer*, 1991, **64**, 425.

19. J. L. Wike-Hooley, J. Haveman, and H. S. Reinhold, *Radiother. Oncol.*, 1984, **2**, 343.

20. O. Warburg, 'The Metabolism of Tumors', Translated into English by F. Dickens, Constable, London, 1930.

21. J. R. Griffiths, Z. M. Bhujwalla, R. C. Coombes, R. J. Maxwell, C. J. Midwood, R. J. Morgan, A. H. Nras, P. Perry, M. Prior, and A. Prysor Jones, *Ann. N.Y. Acad. Sci.*, 1987, **508**, 193.

22. B. D. Ross, N. Farrow, and D. M. Freeman, '31-P MRS: A Tool for Studying Muscle Metabolism', Presented at the 8th Congress of Sports Medicine of the A.Z. ST, Brugge, Belgium, 1989.

23. G. D. Marsh, D. H. Paterson, J. J. Potwarka, and R. T. Thompson, *J. Appl. Physiol.*, 1993, **75**, 648.

24. D. J. Taylor, G. J. Kemp, C. G. Woods, J. H. Edwards, and G. K. Radda, *J. Neurolog. Sci.*, 1993, **116**, 193.

25. H. Mayr, D. M. Freeman, and B. D. Ross, *J. Appl. Cardiol.*, 1989, **4**, 153.

26. J. C. Chatham, D. M. Freeman, P. B. Barker, E. A. Moress, and B. D. Ross, *J. Appl. Cardiol.*, 1989, **4**, 117.

27. W. E. Jacobus, and M. L. Weisfeldt, *Biophys. J.*, 1981, **33**, 33.

28. P. B. Garlick, G. K. Radda, and P. J. Seeley, *Biochem. J.*, 1979, **184**, 547.

29. P. A. Bottomley, L. S. Smith, S. Brazzamano, L. W. Hedlund, R. W. Redington, and R. J. Herfkens, *Magn. Reson. Med.*, 1987, **5**, 129.

30. S. J. W. Busby, D. G. Gadian, G. K. Radda, R. E. Richards, and P. J. Seeley, *Biochem. J.*, 1978, **170**, 103.

31. D. I. Hoult, S. J. W. Busby, D. G. Gadian, G. K. Radda, R. E. Richards, and P. J. Seeley, *Nature (London)*, 1974, **252**, 285.

32. I. A. Bailey, S. R. Williams, G. K. Radda, and D. G. Gadian, *Biochem. J.*, 1981, **196**, 171.

33. A. A. Maudsley, S. K. Hilal, W. H. Perman, and H. E. Simon, *J. Magn. Reson.*, 1983, **51**, 147.

34. R. J. T. Corbett, A. R. Laptook, R. L. Nunally, A. Hassan, and J. Jackson, *J. Neurochem.*, 1988, **51**, 1501.

35. P. L. Hope, E. B. Cady, D. T. Delpy, N. K. Ives, R. M. Gardiner, and E. O. R. Reynolds, *J. Neurochem.*, 1988, **50**, 1394.

36. K. Kanamori, F. Parivar, and B. D. Ross, *NMR Biomed.*, 1993, **6**, 21.

37. K. Kanamori and B. D. Ross, *J. Neurochem.*, 1997, **68**, 1209.

Biographical Sketches

Neil A. Farrow. *b* 1966. B.Sc., 1988, Portsmouth (UK). Currently graduate student Caltech; expected graduation date 1996. One publication. Research interests include structure/function relationships of proteins with DNA, NMR spectroscopy of proteins/DNA, in vivo MRS, and electron transfer in proteins.

John H. Richards. *b* 1930. B.S., 1951, U.C. Berkeley; B.Sc., 1953, Oxford (Rhodes Scholar); Ph.D., 1955, U.C. Berkeley; Harvard, 1955–1957; Caltech 1957–present. Approx. 150 publications. Research interests include structure–function relationships in biological molecules and biophysical studies of protein function.

Brian D. Ross. *b* 1938. B.Sc., 1958, University College, London. D.Phil., 1966, University of Oxford. M.B., 1961, University College Hospital, London. F.R.C.S., 1973, Royal College of Surgeons, London. M.R.C.Path., 1976, Royal College of Pathologists, London. 1989, F.R.C.Path. University of Oxford lecturer, Metabolic Medicine. Director, Renal Metabolism Unit and Consultant Chemical Pathologist, Radcliffe Infirmary, Oxford, 1976–84. Director of Clinical Spectroscopy Programs at Radcliffe Infirmary, Oxford, 1981–84, Hammersmith Hospital, London, 1986–88, and Huntington Medical Research Institutes, Pasadena, CA, 1986–present. Visiting Associate, California Institute of Technology, 1986–present.

For list of General Abbreviations see end-papers

Proton Decoupling During In Vivo Whole Body Phosphorus MRS

Rolf M. J. N. Lamerichs and Peter R. Luyten

Philips Medical Systems, Best, The Netherlands

1 INTRODUCTION

In vivo MRS in humans started with the measurement of phosphorus spectra in muscle tissue (see also **Whole Body Studies: Impact of MRS**). These first measurements were directed toward the assessment of muscle metabolism by using surface coils to localize the spectral information to a confined region of interest. With the advent of localization techniques, in vivo whole body phosphorus spectroscopy could also be used to study metabolism in other regions of the human body. Localized brain spectra have been used to study the metabolic changes in intracranial tumors, stroke, epilepsy, and white matter disorders. Liver and heart phosphorus spectroscopy have provided metabolic information on stress and ischemic conditions.

In spite of mature localization techniques for spatially selective phosphorus spectroscopy, the clinical efficacy of these techniques is hampered by a number of physical limitations. First, phosphorus spectroscopy is an eternal struggle for a better signal-to-noise ratio. Volumes of interest smaller than 10 cm^3 will usually result in excessive measurement times, excluding those techniques for relatively focal pathologies. Higher field strengths may overcome this fundamental problem to some extent. However, the linewidth of in vivo NMR signals is predominantly determined by tissue susceptibility, which results in broader lines at higher field strengths. This effect limits the advantages of increasing the field strengths. Another reason for poor ^{31}P signal detection is the low intensity for some ^{31}P resonances due to complex *J* coupling patterns with protons. Further complications may result from the overlap of resonance frequencies between different components.

One technique which may alleviate these limitations to some extent is the use of proton decoupling in combination with whole body ^{31}P spectroscopy.[1] Proton decoupling may simplify the structure of the spectrum by suppressing the heteronuclear *J* modulation patterns due to coupling with the spin–spin protons. Moreover, excitation or saturation of the surrounding protons in the tissue may contribute to the signal strength of the individual phosphorus signals by virtue of the nuclear Overhauser effect (NOE).[2]

The use of double resonance techniques in whole body ^{31}P MRS results in considerable signal enhancement which improves the clinical utility of ^{31}P MRS.

2   DECOUPLING OF ^{31}P SIGNALS

The limited chemical shift dispersion of in vivo ^{31}P NMR spectra obtained at the relatively low field strengths used for human applications may be the cause of poor spectral resolution. Excellent shimming over the region of interest is an absolute necessity for good in vivo spectral resolution. Depending on the relevant tissue susceptibility and volume of interest size, a homogeneity of 0.05–0.3 ppm (full width at half maximum) can be achieved. Methods to accomplish this homogeneity either employ the spatially resolved water signal in an iterative shim process or generate B_0 maps obtained from phase-sensitive imaging methods. After shimming, the linewidth will be 1.5–7 Hz at 1.5 T for in vivo ^{31}P NMR resonances. These linewidths are sufficient to resolve the homonuclear [31]P J couplings of ± 17 Hz between the three nuclear ^{31}P spins of adenosine triphosphate (ATP). Since the three-bond isotropic heteronuclear coupling between protons and phosphorus is in the range of 6–10 Hz, this coupling may become the dominant factor in the residual resonance linewidth of ^{31}P nuclei coupled to one or more protons, such as in 2,3-diphosphoglycerate (2,3-DPG), nicotinamide adenine dinucleotide (NAD$^+$), phosphoethanolamine (PE), phosphocholine (PC), glycerophosphoethanolamine (GPE), glycerophosphocholine (GPC), and α-ATP. Therefore, proton decoupling at a field strength of 1.5 T combined with an excellent local homogeneity can improve spectral resolution substantially, resulting in more resolved resonances and more reliable, quantitative information. At higher field strengths proton decoupling of ^{31}P MR signals will be less effective due to the increased linewidths, which may become larger than the actual J couplings.

3   THE NUCLEAR OVERHAUSER EFFECT AND ^{31}P SIGNALS

Irradiation of protons during in vivo ^{31}P spectroscopy results in signal increase due to collapse of multiplet structures into singlets. Furthermore, this irradiation may introduce NOE which also results in significant signal increase. This signal enhancement affects not only ^{31}P nuclei that are coupled to protons within the same molecule, but also noncoupled resonances, e.g. the phosphocreatine (PCr) signal, may show a substantial increase in signal intensity during saturation of the surrounding water protons. Due to the heteronuclear dipolar relaxation mechanism and the long dipolar cross-relaxation times between the ^{31}P and ^{1}H water nuclei, the NOE may be quite effective. After application in animal studies,[3] different irradiation schemes have been suggested for whole body in vivo applications that result in enhancement of the ^{31}P signal. These methods range from low-power single-frequency irradiation at the H_2O resonance, which proved to be quite effective in gen-

erating NOE in ^{31}P spectra of calf muscle,[4] to truncated driven or transient NOE techniques.[5] These methods are relatively simple to implement on a whole body MR scanner equipped with a second rf channel. Like decoupling, the NOE may result in relatively larger signal enhancements at lower field strengths. This is related to the inherent relaxation mechanisms for the ^{31}P nuclear spins. The two most important relaxation mechanisms for in vivo ^{31}P signals originate either from dipolar interactions or from the shielding anisotropy. Because this latter relaxation mechanism is proportional to B_0^2, it becomes the predominant relaxation mechanism at higher field strengths, resulting in a decreased NOE.

4   EXPERIMENTAL IMPLEMENTATIONS

A variety of effective pulse sequences have been developed for decoupling purposes, especially for high-resolution NMR applications. A well-defined decoupling sequence requires a homogeneous decoupling field B_1, whose magnitude exceeds the strength of the J couplings involved and covers the entire spectral range of the proton spectrum that has to be decoupled. Broadband decoupling techniques generally rely on repeated inversion of the proton spins. Therefore, they require inversion pulses with a repetition time that is short compared with $1/J$. For proton decoupling of in vivo ^{31}P NMR signals, these conditions are easily fulfilled with the current 1.5 T whole body technology.[1] First of all, the J couplings involved are very small, only of the order of 6–10 Hz. This allows for relatively long pulses, and concomitantly for low rf power levels to decouple sufficiently. Furthermore, the relevant in vivo proton resonances coupled to phosphorus nuclei resonate in a small chemical shift range of ± 100 Hz around the H_2O resonance. This again allows for low rf power without affecting the broadband decoupling performance.

Sufficient decoupling power can be generated utilizing a WALTZ-4 sequence using a basic 90° pulse length of 900 μs, corresponding to an expected decoupling bandwidth of ± 280 Hz around the water resonance. These values allow for the use of large resonators such as standard whole body ^{1}H transmission coils employed on clinical MR scanners. For NOE the conditions are even more favorable. As most NOE contributions are expected to originate from the tissue water only, a low-power single-frequency irradiation of the water protons can be sufficient to generate large NOE enhancements.

5   APPLICATIONS

As the availability of second rf channels on clinical 1.5 T scanners is limited, the number of in vivo applications is quite restricted. However, for all applications of ^{31}P MRS, especially for studies on focal pathologies, proton decoupling showed an important improvement in the sensitivity of the technique. An illustration of the effect of broadband proton decoupling is given Figure 1. The lower part shows a coupled ^{31}P spectrum of the human calf. The upper part shows the effect of proton decoupling, which clearly demonstrates how the multiplets of GPC and NAD resolve into two singlets. Another example is given in Figure 2, showing the resolved PE, GPC, and GPE resonances in the localized ^{31}P spectrum of the brain of a 6-

For References see p. 777

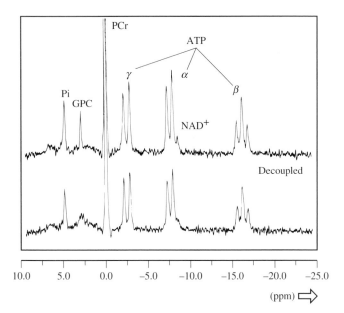

Figure 1 1.5 T ^{31}P NMR spectra of the human calf. The spectra were obtained for 64 transients and a repetition time of 3 s using a 14 cm diameter surface coil without spatial localization. (Reproduced by permission of Wiley from Luyten et al.[1])

year-old girl. These particular resonances are important indicators for membrane metabolism. These mono- and diester signals are used for the assessment of tumor metabolism and therapy response. Increased resolution will greatly enhance the

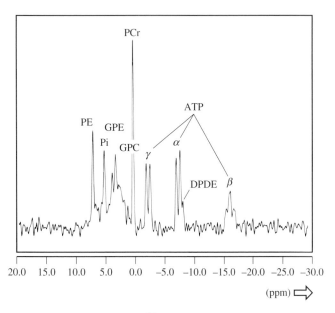

Figure 2 Proton decoupled ^{31}P NMR spectrum of the brain of a 6-year-old girl who has a history of chemotherapy for leukemia. The spectrum was acquired using a small ^{31}P NMR head coil, and a regular ^{1}H imaging head coil for decoupling. WALTZ-4 decoupling was applied; the spectrum was obtained for 256 transients, with a 3.5 s repetition time. Proton decoupling has helped to resolve resonances from GPC, GPE, PE, and NAD$^+$. DPDE = diphosphodiester. (Reproduced by permission of Wiley from Luyten et al.[1])

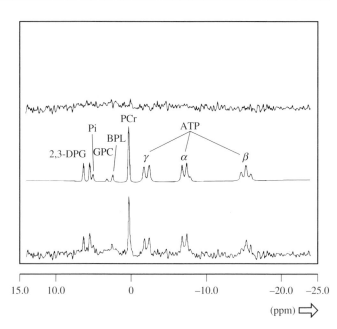

Figure 3 Proton decoupled ^{31}P spectrum of the left ventricular myocardium of a healthy subject. The acquired spectrum (bottom), simulated spectrum (middle), and the difference between the acquired and simulated spectra (top) are illustrated. Indicated are the signals from 2,3-DPG, inorganic phosphate (Pi), GPC, phospholipid signal from the blood (BPL), PCr, and ATP. Three-dimensional localization was applied by combining one-dimensional phase encoding and two-dimensional image-selected in vivo spectroscopy column selection. (Reproduced by permission of the Society of Magnetic Resonance Imaging from de Roos et al.[6])

potential of localized ^{31}P MRS in several important applications.

Figure 3 shows the effect of proton decoupling on the localized ^{31}P spectrum in the normal human myocardium. The lower trace shows the spectrum without any data processing, whereas the middle spectrum is the result of a time domain fitting routine (see also **Quantitation in In Vivo MRS**). The upper trace represents the residual signal between raw and fitted data. The spectrum shows a resolved inorganic phosphate (Pi) signal. Without any proton decoupling this signal would have been obscured by the overlapping 2,3-DPG signal from blood. This resolution is absolutely required to determine the intracellular pH, which is 7.15 for this particular spectrum.

Figure 4 summarizes signal enhancements that can be expected from NOE and decoupling separately. Spectrum A was obtained without the use of any proton irradiation. Broadband decoupling during ^{31}P signal acquisition was applied to obtain spectrum B, whereas spectrum C was obtained by broadband decoupling combined with an inversion pulse on the water resonance to create a transient NOE. Continuous irradiation during the entire experiment results in the largest enhancement (spectrum D): continuous low-power (5 W) single-frequency irradiation of the water resonance during all relaxation delays and broadband proton decoupling during ^{31}P acquisition were used. The NOE factors for PCr in spectra B, C, and D are 0.35, 0.50, and 0.64, respectively. For purely ^{31}P–^{1}H dipolar relaxation and under extreme narrowing conditions the maximum enhancement factor is given by η =

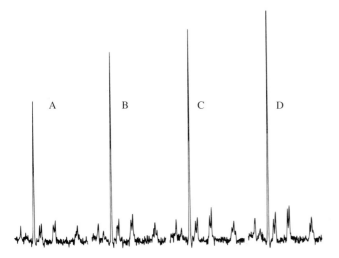

Figure 4 Proton decoupling and NOE applied to the human calf. The spectra were obtained for eight transients with a repetition time of 3 s. Spectrum A was acquired without any ^{1}H irradiation, spectrum B with proton decoupling (WALTZ-4) during acquisition of the ^{31}P signal only, spectrum C with decoupling and an inversion pulse on the ^{1}H signal 1000 ms before ^{31}P data acquisition (transient NOE), and spectrum D by combining WALTZ-4 decoupling during ^{31}P data acquisition with continuous single-frequency low-power irradiation of the water resonance during the relaxation delay

$\gamma(^1\text{H})/2\gamma(^{31}\text{P}) = 1.24$. This implies that combined broadband decoupling during acquisition and single-frequency low-power irradiation at the water resonance frequency during all relaxation delays results in a substantial signal increase. This effect may have an important impact on the quantification of ^{31}P MR signals that are obtained using double resonance techniques.

6   SAFETY CONSIDERATIONS

Continuous irradiation or broadband decoupling during relatively long acquisition intervals may result in excessive sample heating. This phenomenon is well-known in high-resolution NMR, where sample cooling during the experiment is common practice. For in vivo experiments this requires extremely careful precautions to prevent tissue damage. Fortunately, proton decoupling of in vivo ^{31}P signals requires power levels which are on the safe side of the current US Food and Drug Administration safety limits for rf irradiation. The total power deposition required to obtain the result depicted in Figure 4, spectrum D, results in a continuous single-frequency power deposition of 5 W to obtain NOE enhancement, whereas the broadband decoupling consumed 150 W during the 512 ms acquisition interval. Using a *TR* of 3 s, this resulted in a worst-case average power absorption (under the assumption that all power is dissipated in the tissue, i.e. zero reflection and no cable losses) of 30 W. As these values relate to transmission with the body coil, total power absorption of this experiment is well within the safety guidelines. However, extreme precautions including a dedicated power delimiter for a second rf channel are still an absolute necessity for performing these experiments, especially when decoupling is performed with smaller (surface) coils, which may result in excessive tissue heating close to the conductors of the coil.

7   RELATED ARTICLES

Proton Decoupling in Whole Body Carbon-13 MRS; Quantitation in In Vivo MRS; Whole Body Studies: Impact of MRS.

8   REFERENCES

1. P. R. Luyten, G. Bruntink, F. M. Sloff, J. W. A. H. Vermeulen, J. I. van der Heijden, J. A. den Hollander, and A. Heerschap, *NMR Biomed.*, 1989, **1**, 177.
2. J. H. Noggle and R. E. Schirmer, 'The Nuclear Overhauser Effect', Academic Press, New York, 1971.
3. R. L. Nunnally and P. A. Bottomley, *Science*, 1981, **211**, 177.
4. P. A. Bottomley and C. J. Hardy, *Magn. Reson. Med.*, 1992, **24**, 384.
5. P. Bachert and M. E. Belleman, *J. Magn. Reson.*, 1992, **100**, 146.
6. A. de Roos, J. Doornbos, P. R. Luyten, L. J. M. P. Oosterwaal, E. E. van der Wall, and J. A. den Hollander, *J. Magn. Reson. Imaging*, 1992, **2**, 711.

Biographical Sketches

Rolf M. J. N. Lamerichs. *b* 1959. Ph.D., 1989, chemistry, University of Utrecht (thesis on two-dimensional NMR studies on protein and DNA structures). Department of MR Systems and Methods, Philips Medical Systems, 1990–present.

Peter R. Luyten. *b* 1954. Ph.D., 1984, physical chemistry, Free University of Amsterdam (dissertation on NMR relaxation studies on small solutes in liquid crystals). Visiting scholarship (1 year): University of California, San Diego, 1982; Department of MR Predevelopment, Philips Medical Systems, 1984. Currently Director of Clinical Science, Philips Medical Systems.

Proton Decoupling in Whole Body Carbon-13 MRS

Glyn A. Coutts
Marconi Medical Systems, UK

and

David J. Bryant
GEC Hirst Research Centre, Borehamwood, Herts, UK

1   INTRODUCTION

High-resolution ^{13}C NMR spectroscopy has utilized the full 200 ppm range of the ^{13}C spectrum for exploring metabolism in cells.[1] The multiplet structure of the spectrum arising from heteronuclear *J* coupling to protons has been exploited for the assignment of resonances. The spin-$\frac{1}{2}$ ^{13}C isotope has only

For References see p. 780

1.1% natural abundance, but by using ^{13}C-enriched labeled substrates, metabolic pathways can be followed dynamically. In vivo human spectroscopy has so far mainly been limited to studies of triglycerides in adipose tissue and glycogen storage in liver and muscle because of the very low sensitivity of the ^{13}C signal compared with that from protons or ^{31}P. In order to increase the range of metabolites accessible to ^{13}C spectroscopy, or to improve the localization within the human body, every effort has to be made to improve the S/N. One route for improvement of the S/N is to exploit the heteronuclear coupling, either through manipulation of the population of the ^{13}C energy levels through the nuclear Overhauser effect (NOE) or polarization transfer, or by collapsing the multiplet lines through direct decoupling from the protons. The gated decoupling experiment imposes stringent technical demands. The proton and carbon channels need to be well isolated, and coil systems need to be designed with a large degree of decoupling of their mutual inductances, which are adapted to the relevant anatomy and capable of delivering the decoupling energies at the required volume of tissue. However the most stringent limit in experiment design arises from the near continuous rf irradiation at proton frequencies with the requirement that the local deposition of rf energy does not exceed the various national guidelines, and presents no risk to the human subjects.

2   THE EFFECT OF PROTON IRRADIATION ON THE ^{13}C SPECTRUM

Irradiation at proton frequencies can affect the ^{13}C spectrum in two ways. The first of these is enhancement of the ^{13}C sensitivity by using the heteronuclear coupling to alter the population of ^{13}C energy levels. This further divides into two techniques: NOE[2,3] and polarization transfer. The NOE uses a continuous secondary rf (B_2) irradiation to saturate the proton levels of the coupled system, whilst cross relaxation transfers the high proton polarization to the ^{13}C levels leading to a new steady state determined by the various relaxation processes. For purely dipolar relaxation the enhancement is $1 + \gamma_H/2\gamma_C$ or a factor of 3. Polarization transfer uses rf pulses applied to the protons to create population changes which are then transferred to the carbon levels. The SINEPT (SINE-dependent polarization transfer) sequence,[4] for example, consists of two 90° pulses separated by a time determined by the coupling constant of the particular resonance to be enhanced and the proton transverse relaxation time. Although the enhancement is optimized for those multiplets exhibiting a particular J coupling, the enhancement available is γ_H/γ_C, or a factor of 4.

The second effect of proton irradiation on the ^{13}C spectrum is through coherent effects causing collapse of the multiplet lines. The idea is that if nuclear spins X, in our case the protons, of an AX coupled system are irradiated at a frequency centered on the chemical shift of X and at an intensity B_2 such that $\gamma B_2 \approx 2\pi J_{AX}$ then in the rotating frame the evolution of the A, or ^{13}C, magnetization vectors due to coupling is continually reversed and the multiplet line coalesces to a single peak. This is known as continuous wave (CW) decoupling.[5] The effectiveness of the decoupling diminishes with offset of the proton frequency from the X resonance and B_2 intensity. Broadband decoupling attempts to apply decoupling equally for

all metabolites in the ^{13}C spectrum. One method of doing this is to modulate the B_2 irradiation with a noise source.[6] A second method uses composite pulses[7,8] which are designed to give inversion over a wide range of frequencies. The most commonly used of the composite pulses is the WALTZ pulse[9] consisting of the triplet 90°(+X)180°(−X)270°(+X). Phase cycling of these pulses produces WALTZ-4, WALTZ-8, or WALTZ-16 with progressively less sensitivity to errors in rf phase shifts. For full decoupling the pulses need to be applied at a frequency greater than $2J$, where J is the largest coupling constant for the metabolites under consideration, and the average B_2 energy utilization is much greater than for CW decoupling.

Switching of the B_2 irradiation, or gated decoupling, depending on whether data are being acquired, can give any combination of sensitivity enhancement or multiplet collapsing since the timescales of the former are those of the relaxation processes whilst the coherent effects are instantaneous. Suitable gating of the B_2 irradiation throughout the duty cycle is therefore useful for minimizing the overall rf power deposition.

3   HARDWARE, COIL DESIGN, AND RF DEPOSITION

For a gated decoupling experiment a second transmit channel with frequency synthesizer, phase modulator, and power amplifier is required. Most workers have also specifically designed the system of coils needed for proton transmit and carbon transmit and receive. The coil design most widely used for human in vivo ^{13}C studies consists of a planar, circular, ^{13}C transmit-receive surface coil with a butterfly or figure-of-eight proton transmit coil, the crossover point being coaxial with the carbon coil center. This arrangement minimizes the inductive coupling of the two coils, and can be easily sited over the anatomy of interest with the long axis of the proton coil parallel to B_0. Distributed capacitances are of course required to prevent electrical breakdown. In addition the proton transmit coil is generally spaced off from the subject to avoid regions of large B_2 field near to the coil. Further filtering is generally required to give the high level of magnetic isolation of the two frequencies so that the carbon receive signal is not degraded by the simultaneous proton frequency irradiation. For example Bottomley et al.,[10] working at 1.5 T, using a 3-turn spiral 6.5 cm diameter ^{13}C coil and 8×13 cm air-cooled proton coil, achieved a 90 dB isolation of the 64 MHz signal at 16 MHz.

Rf power deposition in human NMR studies through the induction of eddy currents and Joule heating has long been a cause of concern and has led to guidelines limiting the specific absorption ratio (SAR) experienced by human subjects. In the USA the Food and Drug Administration (FDA) guidelines[11] are a body average SAR of 0.4 W kg^{-1} and a peak SAR of 8 W kg^{-1} in any 1 g of tissue, whilst in the UK the National Radiological Protection Board guidelines[12] are a body average SAR of 0.4 W kg^{-1} and a peak SAR of 4 W kg^{-1} in any 1 g of tissue. There is also a limit on the body temperature increase of 1 °C. Bottomley and Roemer[13] have used a homogeneous tissue model to investigate the power deposition for the coil arrangement described above and have compared their model with experimental observations of power deposition on the basis of the coil Q value using $P_{absorbed} = P_{total}(1 - Q_{loaded}/Q_{unloaded})$. They concluded that, although the theoretical homogeneous models probably overestimated the local power deposition in adipose tissue close to the coils, nevertheless, without care, the guidelines for peak SAR could easily be exceeded. Bottomley et al.[10] in their studies on human limbs and chest using a noise-modulated decoupling source limited the decoupling power to below 8 W, which according to the above model limited the power deposited in tissue to 3 W. Using one-dimensional phase encoding methods in a phantom they estimated that some level of decoupling was sustained at depths of 6 cm whilst significant NOE was realized to depths of 4 cm.

Working at 4 T, Bomsdorf et al.[14] used an 11 cm diameter carbon coil and a 7×14 cm butterfly proton coil, achieving 40 dB of isolation. It was concluded that broadband decoupling would not be possible at this field strength without exceeding the SAR limits, but studies on the calves of human volunteers were performed with gated CW decoupling. Decoupling power levels of 6 W were used for subcutaneous fat at a depth of 1 cm, and 24 W for muscle at a depth of 2.5 cm. Heating sensations for one volunteer at the higher power levels were eliminated by reducing the duty cycle, i.e. increasing the recovery period between data acquisitions. In addition, these workers, by careful positioning of the carbon surface coil, were able to use the body coil for proton transmit in polarization transfer experiments, where there is no proton irradiation during data collection. In this case for a repetition time of 200 ms the SAR was estimated to be ~1 W kg^{-1} and hence well within guidelines.

Saner et al.[15] working at 1.5 T were able to use a 7 cm diameter carbon coil in combination with a standard head coil for CW decoupling, and were able to conduct polarization transfer studies without hardware modification. Applying the gated CW decoupling during a 64 ms data acquisition these authors estimated that the average SAR was 3.9 W kg^{-1} whilst the peak SAR was twice this, and thus just within FDA guidelines, but were able to demonstrate decoupling within calf muscle of a human leg.

Systems of concentric and coplanar circular coils have also been successfully used for decoupling. Beckmann et al.[16] were able to use broadband decoupling at 1.5 T with circular coils of 8 cm (carbon) and 13 cm (proton) diameter. Using a WALTZ-8 sequence intercalated throughout the 70 ms data acquisition, and interspersing rest periods within the total data acquisition time to reduce the duty cycle, they estimated that the average SAR did not exceed 2.8 W kg^{-1} and also demonstrated decoupling within the calf muscle of a human volunteer. Finally, with this coil arrangement, gated CW decoupling through a 20 ms data acquisition and a 10% duty cycle, Avison et al.[17] were able to demonstrate decoupling on a human calf muscle at 4.7 T.

4 HUMAN ^{13}C STUDIES

In whole body ^{13}C spectroscopy adipose tissue is the most accessible for study, especially if proton decoupling is required. Moonen et al.[18] showed that all the resonances in the ^{13}C spectrum are due to triglycerides, and using the resonances of the olefinic signal from unsaturated carbons (C*=C*) at 129 ppm

were able to determine the relative concentrations of poly- and monounsaturated fatty acids. This noninvasive technique has the additional advantage that adipose tissue from specific sites, e.g. waist or thigh, can be studied. This study was also performed on patients with cystic fibrosis.[19] Beckmann et al.[20] by using long repetition times, were able to achieve sufficient S/N to quantify the relative concentrations of poly- and monounsaturated fatty acids in relation to dietary intake of fat without decoupling, but noted the poor resolution of the resonances at 130 ppm arising from long range couplings compared with the high resolution of the decoupled spectrum. Bryant et al.,[21] using both coupled and decoupled spectra in a study of the adipose tissue of vegans and control subjects, demonstrated by this noninvasive technique a significant increase in unsaturated fatty acid content in the vegan subjects compared with controls. Ende and Bachert,[22] using broadband decoupled ^{13}C spectroscopy of human calf and breast, sought to optimize the NOE enhancement, achieving maximum enhancements of 80%. With acquisition times of ~15 min they detected resonances other than those from triglycerides in breast tissue.

Glycogen in many ways has provided a test case for the implementation of human in vivo ^{13}C spectroscopy. It plays a central role in carbohydrate storage and metabolism in both muscle and liver, at depths of a few centimeters as far as NMR detection is concerned. Although fully NMR visible,[23] concentrations of around 84 mmol L^{-1} in muscle and up to 300 mmol L^{-1} in liver test the sensitivity of ^{13}C studies. The C-1 resonance of glycogen appears as a doublet at 95 and 106 ppm, but can be decoupled by CW decoupling, irradiating the proton spectrum at 5 ppm. In addition broadband decoupling is expected to give an S/N improvement of slightly more than 2 through the suppression of long range couplings. Short T_1 and T_2 values mean that the irradiation duty cycle times can be as short as 25 ms. Finally labeling studies can be easily and safely performed by the ingestion of ^{13}C-enriched glucose.

Decoupling and detection of the C-1 resonance of glycogen in human muscle and liver at 2.1 T was demonstrated by Jue et al. in 1987.[24] Using WALTZ-4 broadband decoupling at 1.5 T, Heerschap et al.[25] were able to follow the depletion and recovery of glycogen in the gastrocnemius muscle of a fit volunteer following a 23 km run. They note that the broadband proton decoupling aids quantification both through narrowing the region covered by the high triglyceride signals and by allowing referencing against other metabolite signals, for example creatine, at 157 ppm. Depletion and repletion studies of glycogen in skeletal muscle following 2 h each of cycling and walking were recently performed[26] at 2.0 T without decoupling, with acquisition times to achieve quantifiable S/N of ~20 min.

A patient with glycogen storage disease has been studied by Beckmann et al.[16] Muscle and liver spectra were obtained using WALTZ-8 proton decoupling. In calf muscle not only was an elevated C-1 glycogen resonance observed, but also resonances in the region 73–77 ppm, tentatively assigned to the C-4 and C-2,3,5 resonances of glycogen. For the spectra obtained from the abdominal region the authors note the problems of localization to the liver given the intense triglyceride signals for the overlying adipose tissue. The fact that the fatty acid content of the livers of normal western subjects tends to be high, together with the general broadening of liver signals, suggests that quantification will be difficult without decoupling.

For References see p. 780

They were able to demonstrate the increased C-1 glycogen signal in the patient compared with controls, as well as resonances in the region 73–77 ppm. These authors also presented the first proton-decoupled ^{13}C spectra from the human head, with a tentative assignment of a resonance at 54 ppm to creatine.

It should be noted that Saner et al.[15] have compared the enhancement of the glycogen signal in human leg by CW decoupling with that of the much less rf intensive polarization transfer using the SINEPT technique, and concluded that, in vivo, both methods gave a gain of approximately 2.5. Work has also begun in gaining improved S/N from ^{13}C signals by using labeling in conjunction with decoupling. Jue et al.[27] gave fasted volunteers a ^{13}C-enriched glucose drink and obtained CW decoupled spectra using a surface coil placed over the right lobe of the liver. With a 15 min acquisition time they were able to obtain a baseline decoupled C-1 glycogen signal, and were then able to follow the increase in this resonance following ingestion of the glucose. They were also able to observe the α- and β-glucose resonances at 93 and 96 ppm.

Beckmann et al.[28] were able to compare glycogen formation in the human liver following ingestion or intravenous infusion of nonlabeled glucose as well as [1-^{13}C]glucose with a low level of enrichment. Using WALTZ-8 decoupling the resonances of glycogen and α- and β-glucose were followed over time, showing greater glucose uptake by the oral route. Other labeling studies have also been performed. In a recent study[29] at 2.0 T, glutamine and glutamate in the human brain were observed in a ^{13}C spectrum following infusion of [1-^{13}C]glucose. Sensitivity and resolution were enhanced by the use of NOE and WALTZ-16 broadband decoupling, achieving linewidths of less than 2 Hz.

Other human in vivo ^{13}C studies have been considered. For example, in 1988 Sillerud et al.[30] working at 1.5 T were able to observe the central resonance of the C-2,5 citrate triplet at 47 ppm in the human prostate, but were unable to resolve the outer pair of signals at 54 and 38 ppm. They noted that with narrowband decoupling and NOE enhancement a factor of 9 gain in S/N would theoretically be achieved, but to date no decoupling studies have been attempted in this region. The trend has been rather to use improved technique and experience to move away from the technically demanding requirements of the gated decoupling experiment.

5 RELATED ARTICLES

Chemical Shift Imaging; In Vivo Hepatic MRS of Humans; Multifrequency Coils for Whole Body Studies; Peripheral Muscle Metabolism Studied by MRS; Proton Decoupling During In Vivo Whole Body Phosphorus MRS; Quantitation in In Vivo MRS; Radiofrequency Systems and Coils for MRI and MRS.

6 REFERENCES

1. J. B. Stothers, 'Carbon-13 NMR Spectroscopy', Academic Press, New York, 1982.
2. A. W. Overhauser, *Phys. Rev.*, 1955, **92**, 411.
3. I. Solomon, *Phys. Rev.*, 1955, **99**, 559.

4. H. S. Jakobsen, O. W. Sorensen, and H. Bilboe, *J. Magn. Reson.*, 1983, **51**, 157.
5. R. Freeman and W. A. Nelson, *J. Chem. Phys.*, 1962, **37**, 85.
6. R. R. Ernst, *J. Chem. Phys.*, 1966, **45**, 3845.
7. M. H. Levitt, *Prog. NMR. Spectrosc.*, 1986, **18**, 61.
8. J. S. Waugh, *J. Magn. Reson.*, 1982, **49**, 517.
9. A. J. Shaka, J. Keeler, and R. Freeman, *J. Magn. Reson.*, 1983, **53**, 313.
10. P. A. Bottomley, C. J. Hardy, P. B. Roemer, and O. M. Mueller, *Magn. Reson. Med.*, 1989, **12**, 348.
11. United States Food and Drug Administration (FDA), 'Guidance for Content and Review of a Magnetic Resonance Diagnostic Device 510(k) Application: Communication to Manufacturers', FDA, Rockville, MD, 1988.
12. NRPB (National Radiological Protection Board) ad hoc Advisory Group on Nuclear Magnetic Resonance Clinical Imaging, *Br. J. Radiol.*, 1983, **56**, 974.
13. P. A. Bottomley and P. B. Roemer, *Ann. N. Y. Acad. Sci.*, 1992, **649**, 144.
14. H. Bomsdorf, P. Roschmann, and J. Wieland, *Magn. Reson. Med.*, 1991, **22**, 10.
15. M. Saner, G. McKinnon, and P. Boesiger, *Magn. Reson. Med.*, 1992, **28**, 65.
16. N. Beckmann, J. Seelig, and H. Wick, *Magn. Reson. Med.*, 1990, **16**, 150.
17. M. J. Avison, D. L. Rothman, E. Nadel, and R. G. Shulman, *Proc. Natl. Acad. Sci. USA*, 1988, **85**, 1634.
18. C. T.W. Moonen, R. J. Dimand, and K. L. Cox, *Magn. Reson. Med.*, 1988, **6**, 140.
19. R. J. Dimand, C. T. W. Moonen, S. C. Chu, E. M. Bradbury, G. Kurland, and K. L. Cox, *Pediatr. Res.*, 1988, **24**, 243.
20. N. Beckmann, J.-J. Brocard, U. Keller, and J. Seelig, *Magn. Reson. Med.*, 1992, **27**, 97.
21. D. J. Bryant, J. D. Bell, E. L. Thomas, S. D. Taylor-Robinson, J. Simbrunner, J. Sargentoni, M. Burl, G. A. Coutts, G. Frost, M. L. Barnard, S. Cunnane, and R. A. Iles, *Proc. XIIth Ann Mtg. Soc. Magn. Reson. Med.*, New York, 1993, p. 1048.
22. G. Ende, and P. Bachert, *Proc. XIIth Ann Mtg. Soc. Magn. Reson. Med.*, New York, 1993, p. 84.
23. S. O. Sillerud and R. G. Shulman, *Biochemistry*, 1983, **22**, 1087.
24. T. Jue, J. A. B. Lohman, R. Ordridge, and R. G. Shulman, *Magn. Reson. Med.*, 1987, **5**, 377.
25. A. Heerschap, P. R. Luyten, J. I. van der Heyden, L. J. M. P. Oosterwaal, and J. A. den Hollander, *NMR Biomed.*, 1989, **2**, 124.
26. L. A. Domico, K. F. Kendrick, R. Reddy, B. Chance, and J. S. Leigh, in *Proc. XIIth Ann Mtg. Soc. Magn. Reson. Med.*, New York, 1993, p. 1145.
27. T. Jue, D. L. Rothman, B. A. Tavitian, R. DeFronzo, G. I. Shulman, and R. G. Shulman, *Proc. VIIth Ann Mtg. Soc. Magn. Reson. Med.*, San Francisco, 1988, p. 357.
28. N. Beckmann, R. Fried, I. Turkalj, J. Seelig, U. Keeler, and G. Stalder, *Magn. Reson. Med.*, 1993, **29**, 583.
29. R. Gruetter, E. J. Novotny, S. D. Boulware, D. L. Rothman, W. V. Tamborlane, and R. G. Shulman, *Proc. XIth Ann Mtg. Soc. Magn. Reson. Med.*, Berlin, 1992, p. 1921.
30. L. O. Sillerud, K. R. Halliday, R. H. Griffey, C. Fenoglio-Preiser, and S. Sheppard, *Magn. Reson. Med.*, 1988, **8**, 224.

Biographical Sketches

Glyn A. Coutts. *b* 1959. B.A. (Physics) University of Oxford, UK, 1980, D.Phil. University of Oxford 1985. Senior Research Associate at Picker Research UK, 1987–99. Marconi Medical Systems, UK, 1999–present. Approx. 30 publications in the field of MRI and MRS. Current interest: interventional MRI.

D. J. Bryant. *b* 1956. B.Sc., 1977, Ph.D., 1981, University of Nottingham, supervisor E. Raymond Andrew. Research fellow with Raymond

Andrew, 1980–82. GEC Hirst Research Centre working in laboratory headed by Ian R. Young, 1982–present. Approx. 50 publications. Research interests: MRI and MRS on whole-body systems in the 0.15–1.5 T range.

Rotating Frame Methods for Spectroscopic Localization

Peter Styles

MRC Biochemical and Clinical Magnetic Resonance Unit, Oxford, UK

1 INTRODUCTION

Rotating frame spectroscopic localization techniques enable high-resolution spectra to be obtained from spatially resolved slices within a sample. In contrast to most localization schemes where space is encoded by means of switched gradients in the B_0 field, the rotating frame method uses a gradient rf transmitter field (B_1) to effect the spatial discrimination. Most applications have been in ^{31}P spectroscopy of metabolism of animals and humans. Being a one-dimensional (spatial) technique, the method is most appropriately employed in situations where the anatomy consists of approximately planar tissues.

In the late 1970s, there was great interest in developing methodologies for NMR imaging, and the concept of using a gradient B_1 field to encode position was proposed by Hoult.[1] He demonstrated that a two-dimensional proton image could be produced by defining one dimension by a static B_0 gradient, and the second dimension by a gradient transmitter field. A straightforward modification of this 'rotating frame zeugmatography' was to apply the method to spectroscopic localization. The B_0 read gradient was omitted, preserving the chemical shift information at the expense of one dimension of spatial resolution.[2] Spatial encoding is achieved by applying, in separate data acquisitions, an incremented transmitter pulse, and performing a two-dimensional Fourier transformation to obtain a plot of chemical shift versus B_1 strength. Thus, rotating frame spectroscopic imaging is a one-dimensional chemical shift imaging (CSI) technique. A second form of implementation enables spectra to be obtained from just a limited number of regions by combining data which together form a Fourier series defining the required voxel shape and position.[3]

A review article[4] provides a more comprehensive reference list than can be accommodated here.

2 PRINCIPLES OF ROTATING FRAME SPECTROSCOPIC LOCALIZATION

2.1 Rotating Frame Spectroscopic Imaging

When a nuclear spin is irradiated by a transmitter field at the Larmor frequency, then, in the rotating frame of reference, the spin precesses around the B_1 field at a nutational frequency Ω given by

$$\Omega = \gamma B_1 \tag{1}$$

In a homogeneous B_1 field, detection of this frequency is trivial, and is undertaken every time one determines a $\pi/2$ pulse length. Signal amplitude is measured as a function of pulse duration, and a plot of amplitude versus pulse length produces a sine wave of frequency Ω.

If, instead of using a conventional probe, we employ a transmitter coil which produces a gradient in, say, the x direction [$B_1 = f(x)$], then the nutational frequency is directly related to the position of the sample along the x axis:

$$\Omega = \gamma f(x) \tag{2}$$

Detection of Ω is again a simple matter of determining signal amplitude as a function of pulse length, but we now have a range of frequencies reflecting the distribution of the sample in the gradient field. Therefore, a Fourier transform is required to interpret the variation of amplitude with pulse length. The result of this transform is a plot of amplitude versus Ω, and so represents the signal strength as a function of B_1 and hence position in the x direction.

The experimental sequence may be represented as

$$r\theta_x - \text{Acq.} \ (r = 1, \ldots, n) \tag{3}$$

where the pulse θ_x is increased by incremental steps in each of n data accumulations. In the general case, where the sample consists of various chemical species, a two-dimensional Fourier transform of these n FIDs produces a data set with chemical shift on one axis and position in the x direction on the other. The protocol is illustrated in Figure 1, and was the first to be used for obtaining chemical shift images from a human.[5]

2.2 Phase-Modulated Rotating Frame Imaging

The experiment described by sequence (3) produces signals which are amplitude modulated as a function of Ω, and this incurs an inevitable loss in sensitivity. The maximum signal is only recovered when $r\theta_x$ is an odd multiple of $\pi/2$, and passes through a null for even multiples of $\pi/2$. This disadvantage can be avoided by converting the amplitude modulation into phase modulation by using the following pulse sequence:

$$r\theta_x - (\pi/2)_y - \text{Acq.} \ (r = 1, \ldots, n) \tag{4}$$

The second pulse, which is not incremented, rotates the magnetization from the zy plane into the xy plane, and thus the maximum signal is obtained irrespective of the tip angle generated by the $r\theta_x$ pulse. This 'phase-modulated rotating frame' protocol[1,6] is illustrated in Figure 2.

For References see p. 785

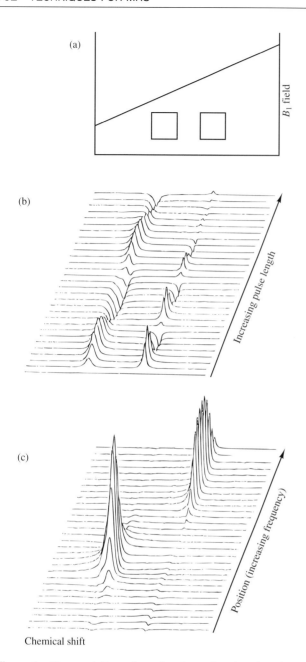

Figure 1 The acquisition of rotating frame imaging data from a phantom. (a) Schematic representation of a two-compartment phantom placed in a gradient B_1 transmitter field. (b) A set of 'pulse and collect' spectra obtained from the phantom; the pulse length is incremented in successive data collections. (c) The data generated by performing a second Fourier transformation on the set of spectra shown in (b). The vertical axis represents the frequency with which the signal modulates as a function of the increasing pulse length, and therefore displays the position of the phantom in the gradient B_1 field. (Reproduced by permission of Springer-Verlag from P. Diel, E. Fluck, H. Gunther, R. Kosfeld, and J. Seelig (eds), 'In Vivo Magnetic Resonance Spectroscopy. II: Localization and Spectral Editing', Berlin, 1992, p. 45)

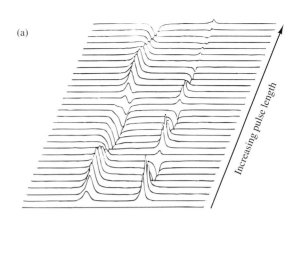

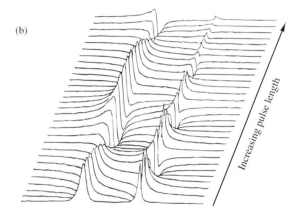

Figure 2 A comparison of spectra collected by (a) amplitude and (b) phase-modulated protocols. The data were obtained from the same phantom as in Figure 1, and are sets of individual spectra prior to the second Fourier transformation. Note that the full signal strength is obtained at all pulse lengths in the phase-modulated experiment. (Reproduced by permission of Springer-Verlag from P. Diel, E. Fluck, H. Gunther, R. Kosfeld and J. Seelig (eds), 'In Vivo Magnetic Resonance Spectroscopy. II: Localization and Spectral Editing', Berlin, 1992, p. 45)

pulses altered by π. To process the data, each set of FIDs is separately transformed, one transform is then reversed in the spatial dimension and the two sets added. The dispersion components now subtract to obtain pure absorption spectra.

2.3 Fourier Series Methods

In certain instances, it is sufficient to obtain signals from a limited number of slices, and then the rotating frame method can be implemented in a simpler way. Instead of collecting a full two-dimensional data set, transmitter pulses may be chosen such that the region of interest experiences tip angles which are odd multiples of $\pi/2$. Appropriate combination of the several FIDs (addition or subtraction, depending on the sign of the signal) will result in the maximum signal from the region of interest, but destructive interference of signals from elsewhere. The shape and extent of the sensitive region is determined by the number of pulse angles used, and the weighting function applied to successive data acquisitions. In essence, the spatial

One difficulty of the phase-encoded sequence is that the data generate phase twisted lines with dispersion components obscuring both spectral and spatial information. The solution is to repeat the complete sequence with the phase of one of the

response of each collection forms a Fourier component of the final selected region, and so the technique is usually described as the 'Fourier series window' method.[3] Provided that a reasonable number of FIDs have been collected, it is possible to combine them in different ways, thus defining more than one window from a single data set. Developments of this approach have included the use of adiabatic pulses to reduce artifacts, and a combination of B_0 and B_1 techniques to improve slice definition.[7]

2.4   Relaxation Time Measurements and Magnetization Transfer Sequences

The basic rotating frame sequences can be combined with saturation, inversion or spin echo pulses, thus facilitating the determination of spatially localized relaxation times, or providing a means of spectral editing. One of the benefits of B_1 localization over B_0 methods is that no time is lost waiting for gradient recovery, and therefore nuclear species with short relaxation times may be detected and quantitated.[8]

Magnetization transfer experiments may also be implemented,[9] giving localized measurement of certain chemical exchange processes in vivo.

3   ROTATING FRAME LOCALIZATION IN PRACTICE

3.1   Probes for Rotating Frame Spectroscopy

The generation of a linear gradient in the B_1 field is less straightforward than in the B_0 field. Hoult[1] used a saddle-shaped transmitter coil with three turns on one side and one turn, wound in the opposite sense, on the other. To perform the phase-encoded protocol, a second orthogonal transmitter coil is needed to generate the $(\pi/2)_y$ pulse. This arrangement has never been used for in vivo spectroscopy due to the excessive amounts of rf power which would be required. Instead, surface coils have been used for production of the B_1 gradient. This approach is, at best, a pragmatic solution on several counts. First, a surface coil does indeed provide a gradient field, but it is only approximately linear over a restricted region of space. Secondly, although the isofield contours are relatively flat close to the axis of the coil, they curve towards the plane of the coil away from this central region. Finally, the phase encoding experiment ideally needs a homogeneous pulse in addition to the gradient pulse for spatial encoding. These problems constitute substantial drawbacks, but the particular suitability of the surface coil for in vivo spectroscopy has encouraged workers to devise methods which maintain the integrity of the experiment within acceptable bounds. Three approaches have been adopted:

1. Single surface coil transmit/receive probes can be used, provided that the sample is restricted to a region close to the coil axis. This method is not of general application, but has been used, for example, to obtain information from the excised bovine globe.
2. A single surface coil probe may be used if some other localization strategy is incorporated into the protocol to restrict the localized volume to a column bounded within the surface coil perimeter. B_0 methods such as ISIS[10] can be used to define the lateral extent of the detected signal, and rotating frame localization subdivides this region into planar slices.[6]
3. Separate surface coils for transmit and receive overcome many of the difficulties encountered.[5] A surface coil transmitter produces a nearly linear field gradient over a region between about $0.3r_t$ and r_t (r_t is the radius of the transmitter coil), and close to the coil axis. If a smaller receiver coil is positioned coaxially with the transmitter, but offset by the appropriate amount, then the detected signals will come from this linear transmitter region. The two coils must, of course, incorporate suitable circuitry to avoid the electrical coupling caused by their mutual inductance.

3.2   Phase-Modulated Rotating Frame Implementation

The most efficient rotating frame protocol is the phase-modulated version, but this experiment requires a homogeneous $\pi/2$ phase-encoding pulse. In practice, the provision of yet another coil to produce this irradiation has been avoided. Instead, the phase-encoding pulse is usually applied by the gradient coil. Clearly, this will only produce an accurate $\pi/2$ rotation in a single plane. However, provided that the region of interest does not extend over too large a range of B_1 field, the experiment still gives good results. Away from the optimum plane, the main consequence is a small loss of sensitivity ($\leqslant 15\%$) where the phase-encoding pulse is within the range $\pi/4$ to $3\pi/4$.

An alternative solution to this problem is to employ an adiabatic pulse in place of the hard $\pi/2$ pulse.[11] B_1 insensitive plane rotation pulses are now available, their main drawback being that they are longer than hard pulses, and so introduce relaxation dependent losses into the sequence.

4   APPLICATIONS OF ROTATING FRAME METHODS

Rotating frame methods[4] are most usefully applied in situations where a surface coil provides the optimum method of data collection, and the anatomy is suited to planar localization. In humans, the heart and liver have both been studied extensively. In each case, the tissue of interest needs to be distinguished from the overlying intercostal muscles. The liver is of particular note because, unlike muscle, it contains no creatine phosphate. Thus, the efficacy of the localization can be easily tested using the liver as an 'in vivo phantom' (Figure 3). In other studies of the human, spectra have been obtained from the brain, breast, leg muscle, and several types of tumor. A variety of disorders have been investigated, including heart failure, hepatic diseases, and trauma of the brain.

Many experiments have also been performed on animal models, perhaps the most striking being the detection of metabolite concentration differences across the ventricular wall of the exposed beating heart.[12,13] Although most studies have used the phosphorus nucleus, carbon, fluorine, and sodium spectra have also been obtained.

The absence of time-consuming switched field gradients is important for the detection and measurement of short T_2

For References see p. 785

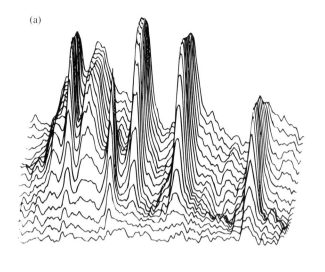

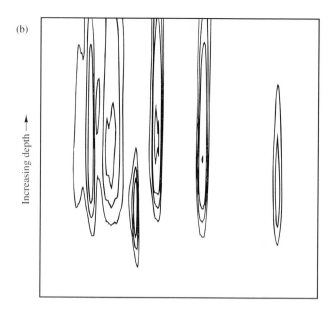

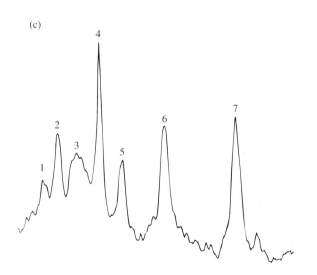

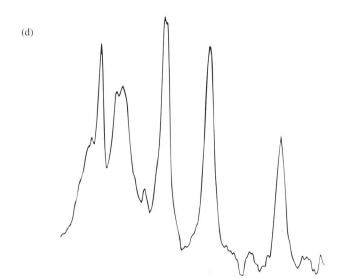

Figure 3   A phase-modulated rotating frame spectroscopic image obtained from the chest of a human subject in the region where the intercostal muscles overlie the liver. The data are presented as: (a) a stacked plot of the spectra; (b) a contour plot; (c) a single spectrum from the muscle region; (d) a single spectrum from the liver region. Peak assignment: 1, phosphomonoesters; 2, inorganic phosphate; 3, phosphodiesters; 4, creatine phosphate; 5–7, mainly adenosine trisphosphate. Note the localization of the creatine phosphate to the superficial region corresponding to the intercostal muscles

species such as the broad phosphodiester resonances in brain.[8] In the liver, proton spin echo rotating frame spectroscopy has been used to quantitate iron overload,[14] which manifests itself by a dramatic decrease in the T_2 of water.

5   THE ROLE OF ROTATING FRAME TECHNIQUES

Techniques for localized spectroscopy have undergone considerable development over the last 10 years, and it is

instructive to consider both the merits and disadvantages of rotating frame methods in comparison to other localization schemes.

There is no doubt that the rotating frame approach lacks the versatility of phase-encoded Fourier imaging using switched B_0 field gradients. B_1 methods give localization in only one dimension, suffer from off-resonance effects, and often require specialized hardware in the form of multicoil probes. Also, there is no direct correlation between the localized spectroscopic data and anatomical information obtained from an MRI image taken during the same investigation.

On the other hand, the absence of switched gradients imparts several benefits. The signal is collected immediately after the transmitter pulse, thus avoiding phasing problems and loss of signal due to T_2 effects. The experiment is silent, which can be particularly important when a subject is nervous or young. Finally, the localization is defined relative to the probe (as distinct to the magnet), and so moving objects such as the chest wall or exposed beating heart can be scanned by attaching the probe directly to the tissue.

Therefore, rotating frame localization is a technique which does not match the generality of B_0 Fourier methods, but can offer very significant advantages in specific applications such as those outlined above.

6 RELATED ARTICLES

Chemical Shift Imaging; Spatial Localization Techniques for Human MRS; Whole Body Studies: Impact of MRS.

7 REFERENCES

1. D. I. Hoult, *J. Magn. Reson.*, 1979, **33**, 183.
2. S. J. Cox and P. Styles, *J. Magn. Reson.*, 1980, **40**, 209.
3. The Fourier series experiment was described simultaneously by several workers, including: K. R. Metz and R. W. Briggs, *J. Magn. Reson.*, 1985, **64**, 172; J. Pekar, J. S. Leigh, and B. Chance, *J. Magn. Reson.*, 1985, **64**, 115; M. Garwood, T. Schleich, B. D. Ross, G. B. Matson, and W. D. Winters, *J. Magn. Reson.*, 1985, **65**, 239.
4. P. Styles, in 'In Vivo Magnetic Resonance Spectroscopy. II: Localization and Spectral Editing', ed. P. Diehl, E. Fluck, H. Günther, R. Kosfeld, and J. Seelig, Springer, Berlin, 1992, p. 45.
5. P. Styles, C. A. Scott, and G. K. Radda, *Magn. Reson. Med.*, 1985, **2**, 402.
6. M. J. Blackledge, B. Rajagopalan, R. D. Oberhaensli, N. M. Bolas, P. Styles, and G. K. Radda. *Proc. Natl. Acad. Sci. USA*, 1987, **84**, 4283.
7. P.-M. Robitaille, H. Merkle, E. Sublett, K. Hendrich, B. Lew, G. Path, A. H. L. From, R. J. Bache, M. Garwood, and K. Ugurbil. *Magn. Reson. Med.*, 1989, **10**, 14.
8. P. M. Kilby, J. L. Allis, and G. K. Radda, *FEBS Lett.*, 1990 **272**, 163.
9. T. A. D. Cadoux-Hudson, M. J. Blackledge, and G. K. Radda, *FASEB J.*, 1989, **3**, 2660.
10. R. J. Ordidge, A. Connelly, and J. A. B. Lohman, *J. Magn. Reson.*, 1986, **66**, 283.
11. K. Hendrich, H. Merkle, S. Weisdorf, W. Vine, M. Garwood, and K. Ugurbil, *J. Magn. Reson.*, 1991, **92**, 258.
12. P.-M. Robitaille, H. Merkle, B. Lew, G. Path, K. Hendrich, P. Lindstrom, A. H. L. From, M. Garwood, R. J. Bache, and K. Ugurbil, *Magn. Reson. Med.*, 1990, **16**, 91.
13. B. Rajagopalan, J. D. Bristow, and G. K. Radda. *Cardiovasc. Res.*, 1989, **23**, 1015.
14. R. M. Dixon and P. Styles, *Magn. Reson. Med.*, 1993, **29**, 110.

Biographical Sketch

Peter Styles. *b* 1946. B.Sc., 1968, University of Sussex, D. Phil., 1984. University of Oxford, Introduced to NMR by Rex Richards and David Hoult, then worked with George Radda. Initial employment as a designer of a high field spectrometer, later as an in vivo spectroscopist. Currently acting director, Medical Research Council Biochemical and Clinical Magnetic Resonance Unit, University of Oxford. Approx. 90 publications. Research interests include: instrumental and experimental aspects of in vivo spectroscopy, metabolic control in skeletal muscle, and application of MRI and MRS in clinical research.

Spatial Localization Techniques for Human MRS

David J. Bryant and Glyn A. Coutts
GEC Hirst Research Centre, Borehamwood, Herts, UK

1 INTRODUCTION

The surface receiver coil has received widespread use in the acquisition of in vivo magnetic resonance (MR) spectra.[1-3] Its characteristic of high magnetic field (B_1) flux density in close proximity to the coil provides spectra of good signal-to-noise ratio (S/N) and provides a level of spatial localization. Applications involving deep-seated organs and pathology give inferior results through lack of sensitivity and contamination from unwanted surface signals.

A number of techniques have been developed over the years in order to provide flexibility in positioning the regions from which spectroscopic data can be acquired. In this article a general description of the localization methods investigated to date are described. The predominant techniques and their applications now in use are described in the related articles listed at the end. The wide range of techniques involve manipulation of the applied magnetic field B_0, either by static profiling or by the use of B_0 field gradients, or the use of radiofrequency B_1 field gradients. The choice of localization technique will often be as a result of the availability of specific hardware components. Radiofrequency (rf) techniques are particularly tied to the use of the surface transmit coil. The flexibility of location and size or field-of-view offered by the B_0 field gradient methods places a greater premium on the control of eddy currents, high homogeneity rf coils, and rf pulse design. These additional features are also included in the related articles listed at the end of this article.

2 STATIC FIELD PROFILING

The higher order spatial harmonics of the magnetic field B_0 are designed to limit the size of the homogeneous volume from which high-quality spectra can be obtained. Signals outside and on the periphery of this volume are inhomogeneously broa-

For References see p. 790

dened and are either absent from the signal or can be removed by postprocessing. Topical magnetic resonance (TMR) has been demonstrated,[4,5] in conjunction with surface transmit and receive coils, to give high-quality spectra. The homogeneous volume from which the signal is derived is not particularly well-defined and there is very little control over its position and dimensions. Realignment of the patient is required in order to achieve this level of control.

A related approach has been suggested[6] which employs a surface B_0 field gradient coil that degrades the quality of field in its close proximity. The nonlinear gradient is pulsed during the first τ interval of a spin echo sequence to ensure that only signals remote from the surface gradient coil contribute to the echo. Irreversible dephasing of surface signals has also been achieved by the use of immobilized ferrite particles with a simple pulse and collect acquisition of the data.[7] Both of these approaches give a robust method of removing the intense unwanted signals near to an rf surface receive coil.

3 RADIOFREQUENCY (B_1) METHODS

This family of techniques relies upon the use of the surface coil as a transmitter system. The spatial variation in the transverse components of B_1 perpendicular to the static field B_0 can be manipulated to produce spatially constrained spectra. It would not be uncommon for the transmitter and receiver of the rf system to be physically different coils such that localization attributes can be optimized with the transmit coil while the reception qualities are not prejudiced.

The axial components of the B_1 field along the axis of a single current carrying loop are given by:

$$B_1(r) = \frac{\mu_0 i}{2} \cdot \frac{R^2}{(R^2 + r^2)^{3/2}} \tag{1}$$

where μ_0 is the magnetic permeability, i the current in the loop of radius R, and r the position along the axis of the coil. Assuming a square excitation pulse of duration t_p, the transverse magnetization M_{xy} at each point along the axis is proportional to $\sin \theta(r)$, where $\theta(r)$ is the spatially-dependent pulse angle given by $\gamma B_1(r)t_p$ (radians). The sinusoidal dependence of M_{xy} has been exploited in a number of ways to project the region of interest away from the proximity of the coil.

3.1 Depth and Composite Pulses

Analysis of equation (1) demonstrates that a 90° excitation is experienced at $0.77R$ when $B_1(0)$, the field at the center of the coil, is set to 180°. Peak signal is generated at $0.58R$ for a transmit/receive surface coil arrangement where the signal is proportional to $\theta(r) \sin \theta(r)$, where $\theta(r)$ is the spatially-dependent pulse angle. Contour plots of the B_1 field show that the excited volume is poorly defined.[3] However it can be considerably improved by combining several pulses together to form a composite 'depth' pulse.[8] The profile of selected M_{xy} has been demonstrated to be $\sin^n \theta(r)$, instead of $\sin \theta(r)$, where the value of n will depend upon the number of pulses that compose the depth pulse and upon the extent of phase cycling employed within the pulse. Depth pulses are often based upon the spin echo $(\theta(r); 2\theta(r)(\pm x, \pm y))$, where the use of $\theta(r), 2\theta(r)$ empha-

sizes the spatial variability of the B_1 field, ';' denotes a small time delay, and $\pm x, \pm y$ refer to the quadrature phases of the rf pulses. Analysis of the depth pulse approach involving the inversion–recovery sequence $(2\theta(r); \theta(r)(x); [2\theta(r)(\pm x, \pm y)_m];)$ has also been demonstrated allowing volume localized relaxation measurements to be carried out.[9]

Composite pulses, whose initial usage was to compensate for pulse imperfections,[10] have also been applied to the problem of spatial localization and complement the depth pulse approach.[11,12] These composite pulses excite M_{xy} and subsequent 180° pulses ensure that only signals experiencing a very narrow range of B_1 amplitudes actually form a spin echo. Contour plots of the inversion efficiency of these composite pulses demonstrate the potential of the method.[13,14]

The duration and extent of phase cycling for both depth and composite pulse approaches can become large as the degree of localization is improved. Comparisons of the two techniques showed that depth pulses had improved performance with respect to frequency offset although they did require more phase cycling.[15–17]

3.2 Fourier Methods

Analysis of the depth pulse described above demonstrated that the spatial profile of M_{xy} generated by the surface coil transmitter could depend on $\sin^n \theta(r)$, where n is an odd integer and $\theta(r)$ the spatially-dependent pulse angle. Pekar et al. showed that any required region at depth from the transmitter coil can be constructed from the differing spatial harmonics generated with differing pulse power in the transmitter.[18] Each harmonic is suitably weighted before accumulating a final signal response prior to Fourier transformation to obtain the localized spectrum. This harmonically analyzed sensitivity profile (HASP) technique has also been called Fourier series analysis,[19] or Fourier series windowing (FSW).[20,21] Unless each transient is accorded equal weight there is a loss of sensitivity; this loss is particularly exaggerated when the localization is required in the minimum number of transients. If unequal numbers of transients are accumulated for each of the generated harmonics, then the correct spatial profile can be obtained with no loss in sensitivity, but at the expense of increased scan acquisition time. There is the possibility of saving time by recognizing that some harmonics will not contribute greatly to the required spatial profile and need not be acquired.

The rotating frame zeugmatographic method of Hoult has been extended to the acquisition of spatially localized, high-resolution spectra.[22,23] Incrementing the rf power level through a transmitter system generating an rf gradient produces a series of spatially-encoded transients. Each transient has to be recorded and stored separately rather than weighted and coadded as in the Fourier series method described previously. The resultant two-dimensional data set can be Fourier transformed in two directions to generate planes of conventional spectra.[23–25] In this simplified method there is a loss of sensitivity of $\sqrt{2}$ since the transients are amplitude modulated by the exciting pulses and the longitudinal component of magnetization M_z remains undetected. However this loss of sensitivity can be recovered by an orthogonal 90° B_1 pulse which precesses all magnetization into the transverse plane, converting amplitude spatial encoding to phase modulation.[22,26] This additional pulse should not be dependent upon the B_1 field and will usually be either of a composite or adiabatic nature. It should also be noted that the same approach can be applied to the Fourier series window technique described in the preceding paragraph. Strictly, the B_1 gradient should be spatially linear and Helmholtz transmitter coils have been proposed for this purpose.[22] It has been noted that the surface coil response [equation (1)] is approximately linear along its axis in the range $0.3R$ to $0.8R$ and that this linearity is insignificantly distorted for small off-axis distances ($<0.5R$).[25] More complex rf systems are required to introduce a second spatial dimension by providing an additional orthogonal linear B_1 field gradient. Applications employing these Fourier rf techniques are described in ***Rotating Frame Methods for Spectroscopic Localization***.

4 B_0 FIELD GRADIENT TECHNIQUES

The wider application of MRI hardware in the form of magnetic field (B_0) gradient coils has allowed the development of a number of methods that provide much of the required flexibility for the acquisition of in vivo MRS data. These techniques may involve oscillatory gradients, slice selective methods, or Fourier techniques.

4.1 Sensitive Point Method

The magnetic field can be dynamically profiled by the application of sinusoidally varying B_0 field gradients.[27] Non-time-dependent signals are derived only from the isocenter of a field gradient system and these can be extracted by low-pass filtering. The isocenter can be controlled by adjustment of the balance of the current between the two halves of the gradient coil set. Localization to a volume in space has been achieved by the use of three orthogonal gradients with differing time-dependences. Signal is usually sampled under the steady-state regime ($TR \leqslant T_2, T_1$) and then the rf pulse repetition rate must exceed the gradient frequency. If this is not achieved, then the gradient oscillations are superimposed upon the time domain signal and artifactual sidebands can occur in the resultant spectra. By sampling the echo rather than the FID of the steady-state signal, it has been demonstrated that these artifacts can be reduced.[28]

4.2 Slice-Selective Techniques

The degree of spatial localization offered by slice-selective pulses has led to their widespread use for in vivo MRS. Control of voxel dimension by gradient strength and rf pulse bandwidth and voxel position with rf carrier offset provide flexible positioning of regions from which signals are derived.

Common problems concern the generation of eddy currents by the pulsed gradients employed with these techniques, chemical shift artifact, and loss of initial portions of the signal. The absence of eddy currents remains an underlying strength of the B_1 gradient methods described earlier. The problem has been alleviated by a number of compensation (see ***Eddy Currents and Their Control***) and correction schemes (see ***Quantitation in In Vivo MRS***) and by the development of self-shielded gradient systems (see ***Gradient Coil Systems***).

Chemical shift dispersion amongst individual resonances of a spectrum leads to a misplacement in position when pulsed B_0 field gradients are employed in slice-selective procedures. The extent of this chemical shift artifact or mis-registration is given by $\delta/\gamma G$ mm, where δ is the dispersion expressed in hertz (Hz) and γG is the field gradient (Hz mm^{-1}). The positional difference becomes more pronounced as the chemical shift increases, at higher field strengths, or in ^{13}C rather than ^{1}H MRS, but can be alleviated by use of the largest gradient possible. The use of high gradient values places a requirement for wider bandwidth, higher power rf pulses, and a greater penalty in terms of increased eddy currents. In reality, following these measures, the misplacement for in vivo ^{1}H MRS may be less than 1 mm. More detail is given in ***Selective Excitation Methods: Artifacts***.

The loss of initial data points of the digitized signal can arise through the finite time required to switch the gradient from one value to the next. Each data point lost represents a 180° phase roll across the resultant spectrum, which in poorly-resolved in vivo MRS, has proved intractable in the most rigorous sense. A number of baseline correction schemes have been suggested or, as an alternative, fitting the time domain data directly rather than introducing the artifact with the use of the Fourier transform (see ***Quantitation in In Vivo MRS***). The extent of signal loss can be reduced with greater gradient strength and switch rates, eddy current control, and sequence development. The latter provide the material for the discussion below.

Slice-selective pulses are employed in three distinct ways; the signal from the required volume may be prepared along the z axis or in the xy plane, or alternatively the magnetization from unwanted regions may be presaturated by a series of slice-selective pulses.

4.2.1 Longitudinal Spatial Preparation

A number of techniques have been described that prepare signal from the required region of interest parallel to the static field B_0. During the volume-selective procedure, unwanted M_{xy} is dephased by field gradients or alternatively by a phase cycling scheme during the acquisition of the data. The required signal, which is maintained along the z axis, is observed with a single broadband pulse. Loss of signal and time delay to the start of data acquisition are kept to a minimum by this approach which can be extended to include spin echo techniques as required.

Image selected in vivo spectroscopy (ISIS) consists of three orthogonal slice-selective 180° pulses in combination with a phase cycling scheme.[29] The phase cycling and data accumulation required introduce a minimum of eight transients before the volume localized spectrum can be observed. This represents a serious inconvenience when shimming is undertaken prior to the accumulation of the data. The technique has been extended to include the multiple acquisition of several voxels simultaneously,[30] and conformation of the voxel shape to match required pathological regions.[31] ISIS, these developments, and its applications to ^{31}P MRS are described in ***Single Voxel Whole Body Phosphorus MRS***.

Volume selective excitation (VSE) is a single shot localization technique based upon a composite selective procedure (45°, 90°, 45°) where both the 45° pulses are slice selective and the 90° is broadband.[32] The required 'in-slice' material is inverted longitudinally while the unwanted magnetization is left in the transverse plane. Care must be taken to avoid the

For References see p. 790

generation of stimulated echoes. Three orthogonal composite slice-selection pulses are required to define a single voxel and these are performed sequentially. VSE requires high rf power if the gradient is *not* switched during the broadband 90° pulse. VSE has been applied to the acquisition of ^{13}C and ^{31}P spectra.[33,34] A comparable technique (DIGGER)[35] has improved the off-resonance capability and does not rely upon phase cycling by paying attention to the qualities of the individual rf excitation pulses.

The solvent suppressed spatially resolved spectroscopy (SPARS) technique alleviates the high rf requirement of VSE by introducing gradient switching and the spin echo into the localization to reduce the effects of the duration of the procedure.[36] Each of the required spatially selective pulses is a 90°, 180°, 90° sandwich, where *only* the final 90° pulse is slice selective. Unwanted signal remains in the transverse plane to be dephased and its cancellation can be improved by phase cycling. A similar technique (SPACE)[37] has also been proposed that employs the same rf pulse composite as SPARS. However, in this case, the first 90° is spatially selective and subsequent pulses are phase cycled to improve the suppression of unwanted signals.

4.2.2   Transverse Spatial Preparation

These techniques prepare the required signal in the transverse plane which can make them particularly vulnerable to loss of signal through T_2^x decay unless spin echo methods are introduced. Significant initial data loss can also occur, with subsequent increase in baseline roll in the resultant spectra, without the use of the spin echo. Despite these shortcomings, these methods have been widely employed to acquire ^{31}P and ^{1}H spectra.

Depth resolved surface coil spectroscopy (DRESS) represents the simplest of localization techniques, a single slice selective 90° followed by the acquisition of the data.[38] The lateral extent of the selected plane is controlled by the use of the surface receive coil and the selected region is usually assumed to be a disk. A slice-interleaved version of DRESS (SLIT-DRESS)[39] allows more spectroscopic data to be acquired by using offset slices during the inevitable relaxation delays. DRESS methods and applications are described in *NMR Spectroscopy of the Human Heart*.

Spin echo techniques can be introduced to alleviate time delay problems. Point resolved surface spectroscopy (PRESS) employs three selective pulses (90°, 180°, 180°) to leave M_{xy} from the required voxel to contribute to the final echo.[40] PRESS is particularly amenable to the acquisition of ^{1}H spectra where the increased echo time can be integrated into the scan as an additional suppression of unwanted fat and lipid signals. In this case the localization is preceded by a chemical shift selective (CHESS) pulse in order to presaturate the large water resonance (see *Selective Excitation in MRI and MR Spectroscopy*). A related technique that is applicable to ^{1}H MRS, the multiple echo spectroscopic acquisition (MESA) method,[41–43] employs three mutually orthogonal slice selective pulses to define to a given region of interest. Initial transverse magnetization is generated by a binomial excitation where the large water resonance remains along the B_0 axis and undetected (see *Water Suppression in Proton MRS of Humans and Animals*). The MESA sequence (1331, 180°, 180°, 180°), like many multiecho techniques, places emphasis upon the quality of the

refocusing 180° rf pulses, and unwanted signals can be substantially reduced by phase cycling.

Stimulated echo pulse sequences have been proposed to perform spectroscopic localization.[44–46] The stimulated echo acquisition mode (STEAM) sequence consists of three selective 90° pulses (90°, 90°, 90°), one in each of the three spatial orientations. During the period after the second 90° pulse, the *TM* period, the required signal is orientated along the z axis and so there is no loss in signal due to transverse relaxation. At all other points of the sequence, the required signal is in the transverse plane and as such the STEAM sequence is included in this section. As with the multiecho PRESS sequence, ^{1}H spectra can be acquired by preceding the STEAM localization with a series of narrow band, CHESS water suppression pulses. Stimulated echoes are intrinsically half the intensity of their spin echo counterparts which would appear to be a considerable disadvantage. However voxel definition is superior, as a result of employing selective 90° rather than 180° pulses, and STEAM consistently provides shorter echo time sequences than its spin echo counterparts PRESS and MESA. In practice these features account for the widespread use of STEAM in the acquisition of ^{1}H spectra (see *Single Voxel Localized Proton NMR Spectroscopy of Human Brain In Vivo*).

4.2.3   Spatial Presaturation

Maintaining the coherence of the signal from the required voxel can be alleviated by the use of spatial presaturation. Slice-selective pulses are used preferentially to excite the unwanted regions into the transverse plane to be dephased prior to the use of a single broadband pulse to observe the required signal. Time delays to the start of data acquisition can be reduced to a minimum suitable for shorter T_2 species.[47] This approach has been integrated into other localization schemes and has been called outer volume suppression (OVS).[48–50] This approach has been particularly useful in the acquisition of ^{1}H spectra where the spatial presaturation of high scalp fat and lipid signals is required.

4.3   Fourier Methods

In contrast to the B_1 Fourier methods described earlier, where amplitude spatial-encoding has to be converted to phase modulation in a two-step process,[22] B_0 field gradients act directly upon the transverse components of magnetization and generate spatial phase modulation directly into the signal. In most other respects the B_0 methods introduced here mirror those B_1 techniques described earlier.

The harmonically analyzed approach to spectral localization has been described by the application of a field gradient pulse within a spin echo pulse sequence.[51] A number of transients are recorded with differing combinations of gradients in the x, y, and z directions to achieve full volume localization. Each transient can be Fourier-transformed to give spatially phase distorted spectra. Suitable weighting of each of these spectra prior to accumulation gives the desired volumetric region. The weighting factors employed are simply the Fourier coefficients of the desired region of interest. Weighting is best achieved by accumulating different numbers of transients for the various gradient permutations in order to maintain sensitivity.[52] It has also been noted that the required region may be repositioned with respect to the gradient isocenters by multiplying each time domain transient in the acquired matrix by a phase factor $\exp(-i\gamma G_r r_0 t)$, where r and r_0 may refer to either x, y, or z components of the voxel's position.

Rather than weight and accumulate signals to achieve the required level of localization, it is possible to submit the data to a multidimensional Fourier transform.[53–56] The resultant chemical shift image is a matrix of voxels, each containing a high-resolution spectrum. Chemical shift imaging or phase-encoded spectroscopy has been widely applied to the acquisition of ^{1}H, ^{19}F, ^{31}P, and ^{13}C in vivo spectra and is described in detail in *Chemical Shift Imaging*. In situations of low S/N, which is inevitable for human MRS at low field with a poor filling factor for the remote regions of interest, this Fourier approach represents the most efficient of data acquisition procedures. In addition, since spatial resolution is expected to be low because of low concentration, the gradient usage is considerably less than the slice-selective techniques described in preceding sections with commensurate decrease in eddy current problems.

Consideration of the sensitivity, spatial response shape, and spatial location with respect to both harmonically analyzed and full Fourier approaches is given by Mareci and Brooker.[57]

4.4 Nonlinear Gradients

A method involving surface gradients has been described earlier which involves the use of nonlinear, pulsed B_0 field gradients.[6] It has also been noted that slice-selective procedures in nonlinear gradients can generate multidimensional spatial localization.[58] However the development of two- and three-dimensional spatially-selective pulses and spectro-spatial-selective pulses, both with the convenience of linear gradients, may overshadow this work (see *Complex Radiofrequency Pulses*).

5 PRACTICAL APPROACHES AND HYBRID TECHNIQUES

It is very difficult to make statements concerning the best technique for spatially-resolved in vivo MRS given the wide range of methods available. It is not surprising, given this comment, that the practical approach to spectral localization is to employ some hybrid of the techniques described beforehand. In a number of cases this hybrid approach may naturally occur through the availability of the relevant components of hardware.

The lateral spatial extent of signal, when employing a surface coil, may be controlled with any of the slice-selective techniques described earlier. This may require the use of adiabatic, hyperbolic secant[59] or frequency modulated rf pulses to compensate for the inhomogeneous B_1 field if a surface transmitter coil is employed. ISIS has been employed in this manner, in combination with depth pulses, to acquire ^{1}H spectra.[60] A similar approach, employing a two-dimensional ISIS method in sequence with DRESS,[40,61,62] has been used to acquire examples of good quality ^{31}P muscle spectra. Two-dimensional ISIS, which defines a column of tissue after a four transient addition/subtraction cycle, has been employed with the B_1 Fourier series window technique.[63] In this case nine Fourier harmonics were acquired over 96 transients to obtain canine myocardia spectra in conjunction with a 29-mm surface transmit/receive coil. Phase modulation was implemented to

improve sensitivity with this ISIS/FSW approach,[26,64] where additional flexibility was provided by voxel shifting the acquired data matrix to overlay the required region of interest.

Proton MRS in many ways presents the largest technical problems in spectral localization due to the intense resonances of water and lipid. This is compounded by the narrow range of ^{1}H chemical shift dispersion available at the relatively low fields offered by whole body in vivo MR systems. Robust practical approaches involve spatial presaturation (OVS), in addition to the volume selected procedure offered by STEAM, PRESS, and MESA,[48–50] in order to control the large signals from peripheral fat layers. Water suppression may be either by narrow-band CHESS or binomial pulses. Fourier encoding has also been employed to localize further the ^{1}H signals within the defined region of interest. In a number of examinations the extent of this localization may not be fully required or warranted, but will provide a level of robustness across a number of eventualities.

The slice-selection techniques are usually introduced as single voxel methods that generate a single high-resolution spectrum from a predetermined position located by MRI. These offer the possibility of obtaining good quality spectra from regions where B_0 homogeneity may not be at its best. Local shimming of these selected volumes can compensate, to an extent, for the local difficulties caused by tissue interfaces or hemorrhagic pathology for example.[65,66] The Fourier, or multi-voxel, methods require a much larger volume of high-quality field since data are acquired from all voxels simultaneously with the possibility of studying a wider field of view. When S/N is low and extensive data accumulation is necessary, and local deviations in field homogeneity are small, it makes sense to employ Fourier techniques by running the required spatial encoding in parallel with the required averaging. Since signal and noise are derived with equal weight and from exactly the same tissue volume, these methods make for efficient data acquisition. However the limited number of spatial harmonics that may be acquired, particularly when localization is extended into three dimensions, means that the degree of localization is intrinsically poor. Large changes in spectral appearance, when they occur very locally, are not well perceived due to voxel bleeding.[67]

A number of comparisons have been made in relation to the variety of slice-selection methods. The use of selective 90° pulses for STEAM localization has been demonstrated to give better voxel definition than the inversion pulses of PRESS.[68] ISIS, PRESS, and STEAM have also been compared with respect to selection efficiency, selection contamination, and water suppression.[69] The minimum echo time TE for ISIS is much less than either STEAM or PRESS, and this tends to dominate such comparisons. The addition/subtraction requirement of ISIS was identified as a source of baseline distortion that is not readily apparent in the single-shot PRESS and STEAM techniques. Further comparisons between STEAM and PRESS have been made with respect to spin displacement (flow and diffusion effects),[68] minimum TE,[68] effect of TE upon ^{1}H spectra quality and appearance,[70] water suppression,[68] effects of rf pulse miscalibration,[68] and homonuclear coupling effects.[68,71,72]

An important point concerning rf power deposition is highlighted in a comparison between SPARS and STEAM,[73] where the broadband pulses of the SPARS method put it at a serious disadvantage. It was also noted in this work that the intrinsic

For References see p. 790

insensitivity of the stimulated echo is not usually an overbearing issue since SPARS includes periods of irreversible dephasing during its localization period.

All the techniques presented previously have the attribute of acquiring conventional MR absorption spectra. Chemically shift selective pulses can be employed to excite a single resonance *within* a spectrum. Conventional and fast MRI techniques can then be employed to perform the degree of spatial localization required with greater efficiency.

6 BEYOND LOCALIZATION

A number of techniques discussed earlier, and in particular those applied to the acquisition of ^{1}H spectra, have moved beyond simple localization requirements and include additional methods for reducing the magnitude of strong, unwanted signals. In addition, a number of techniques have been investigated for the incidental or artifactual effects of flow and perfusion,[68] and homonuclear coupling phenomena.[71,72] It is likely that in vivo MRS will parallel earlier developments in high-resolution NMR by employing such dependences to improve spectral sensitivity. There have been reports of both diffusion-weighted MRS,[74,75] and editing,[76] polarization transfer,[77] and double quantum filtering.[78,79]

McKinnon and Boesinger[76] have defined the three consecutive slice-selective pulses, as employed in the MESA technique, as a volume selective refocusing (VSR) sequence that may be subsequently integrated into a wide range of existing spectral editing techniques. More recent developments in spectro-spatial pulses give potential to allow the spatial localization and editing processes to occur simultaneously (see also *Complex Radiofrequency Pulses*) and this will further extend the quality of spectral information derived from the relatively low fields currently offered by whole body MR scanners.

7 RELATED ARTICLES

Birdcage Resonators: Highly Homogeneous Radiofrequency Coils for Magnetic Resonance; Chemical Shift Imaging; Complex Radiofrequency Pulses; Eddy Currents and Their Control; Gradient Coil Systems; Multifrequency Coils for Whole Body Studies; Multiple Quantum Coherence Imaging; NMR Spectroscopy of the Human Heart; Quantitation in In Vivo MRS; Radiofrequency Systems and Coils for MRI and MRS; Rotating Frame Methods for Spectroscopic Localization; Selective Excitation Methods: Artifacts; Selective Excitation in MRI and MR Spectroscopy; Single Voxel Localized Proton NMR Spectroscopy of Human Brain In Vivo; Single Voxel Whole Body Phosphorus MRS; Surface Coil NMR: Detection with Inhomogeneous Radiofrequency Field Antennas; Surface and Other Local Coils for In Vivo Studies; Water Suppression in Proton MRS of Humans and Animals.

8 REFERENCES

1. J. J. H. Ackerman, T. H. Grove, G. G. Wong, D. G. Gadian, and G. K. Radda, *Nature (London)*, 1980, **283**, 167.
2. R. L. Nunnally and P. A. Bottomley, *Science*, 1981, **211**, 177.
3. A. Haase, W. Hanicke, and J. Frahm, *J. Magn. Reson.*, 1984, **56**, 401.
4. R. E. Gordon, P. E. Hanley, D. Shaw, D. G. Gadian, G. K. Radda, P. J. Bore, and L. Chan, *Nature (London)*, 1980, **287**, 736.
5. R. D. Oberhaensli, G. J. Galloway, D. J. Taylor, P. J. Bore, and G. K. Radda, *Br. J. Radiol.*, 1986, **59**, 695.
6. M. G. Crowley and J. J. H. Ackerman, *J. Magn. Reson.*, 1985, **65**, 522.
7. Y. Geoffrion, M. Rydzy, K. W. Butler, I. C. P. Smith, and H. C. Jarrell, *NMR Biomed.*, 1988, **1**, 107.
8. M. R. Bendall and R. E. Gordon, *J. Magn. Reson.*, 1983, **53**, 365.
9. T. C. Ng, J. D. Glickson, and M. R. Bendall, *Magn. Reson. Med.*, 1984, **1**, 450.
10. M. H. Levitt and R. Freeman, *J. Magn. Reson.*, 1979, **33**, 473.
11. A. J. Shaka and R. Freeman, *J. Magn. Reson.*, 1980, **63**, 596.
12. R. Tycko and A. Pines, *J. Magn. Reson.*, 1984, **60**, 156.
13. A. J. Shaka and R. Freeman, *J. Magn. Reson.*, 1984, **59**, 169.
14. A. J. Shaka, J. Keeler, M. B. Smith, and R. Freeman, *J. Magn. Reson.*, 1985, **61**, 175.
15. M. R. Bendall and D. T. Pegg, *J. Magn. Reson.*, 1985, **63**, 494.
16. M. R. Bendall and D. T. Pegg, *J. Magn. Reson.*, 1986, **68**, 252.
17. M. R. Bendall, D. Foxall, B. G. Nichols, and J. R. Schmidt, *J. Magn. Reson.*, 1986, **70**, 181.
18. J. Pekar, J. S. Leigh, and B. Chance, *J. Magn. Reson.*, 1985, **64**, 115.
19. K. R. Metz and R. Briggs, *J. Magn. Reson.*, 1985, **64**, 172.
20. M. Garwood, T. Schleich, B. D. Ross, G. B. Matson, and W. D. Winters, *J. Magn. Reson.*, 1985, **65**, 239.
21. M. Garwood, T. Schleich, and M. R. Bendall, *J. Magn. Reson.*, 1985, **65**, 510.
22. D. I. Hoult, *J. Magn. Reson.*, 1979, **33**, 183.
23. S. J. Cox and P. Styles, *J. Magn. Reson.*, 1980, **40**, 209.
24. A. Haase, C. Malloy, and G. K. Radda, *J. Magn. Reson.*, 1983, **55**, 164.
25. P. Styles, C. A. Scott, and G. K. Radda, *Magn. Reson. Med.*, 1985, **2**, 402.
26. M. J. Blackledge, B. Rajagopalan, R. D. Oberhaensli, N. M. Bolas, P. Styles, and G. K. Radda, *Proc. Natl. Acad. Sci. USA*, 1987, **84**, 4283.
27. P. A. Bottomley, *J. Magn. Reson.*, 1982, **50**, 335.
28. K. N. Scott, H. R. Brooker, J. R. Fitzsimmons, H. F. Bennett, and R. C. Mick, *J. Magn. Reson.*, 1982, **50**, 339.
29. R. J. Ordidge, A. Connelly, and J. A. B. Lohman, *J. Magn. Reson.*, 1986, **66**, 283.
30. R. J. Ordidge, R. M. Bowley, and G. McHale, *Magn. Reson. Med.*, 1988, **8**, 323.
31. J. C. Sharp and M. O. Leach, *Proc. VIIth Ann Mtg. Soc. Magn. Reson. Med.*, San Francisco, 1988, p. 705.
32. W. P. Aue, S. Muller, T. A. Cross, and J. Seelig, *J. Magn. Reson.*, 1984, **56**, 350.
33. W. P. Aue, S. Muller, and J. Seelig, *J. Magn. Reson.*, 1985, **61**, 392.
34. S. Muller, W. P. Aue, and J. Seelig, *J. Magn. Reson.*, 1985, **63**, 530.
35. D. M. Doddrell, J. M. Bulsing, G. J. Galloway, W. M. Brooks, J. Field, M. G. Irving, and H. Baddeley, *J. Magn. Reson.*, 1986, **70**, 319.
36. P. R. Luyten, A. J. H. Marien, B. Sijtsma, and J. A. Den Hollander, *J. Magn. Reson.*, 1986, **67**, 148.
37. D. M. Doddrell, W. M. Brooks, J. M. Bulsing, J. Field, M. G. Irving, and H. Baddeley, *J. Magn. Reson.*, 1986, **68**, 367.
38. P. A. Bottomley, T. B. Foster, and R. D. Darrow, *J. Magn. Reson.*, 1984, **59**, 338.
39. P. A. Bottomley, L. S. Smith, W. M. Leue, and C. Charles, *J. Magn. Reson.*, 1985, **64**, 347.
40. P. A. Bottomley, *Ann. NY. Acad. Sci.*, 1987, **508**, 333.

41. D. Lampman, G. Hurst, J. McNally, and M. Paley, *Magn. Reson. Imag.*, 1986, **4**, 115.

42. D. K. Menon, C. Baudouin, D. Tomlinson, and C. Hoyle, *J. Comp. Assist. Tomogr.*, 1990, **14**, 882.

43. C. J. Peden, F. M. Cowan, D. J. Bryant, J. Sargentoni, D. K. Menon, D. G. Gadian, J. D. Bell, and L. M. S. Dubowitz, *J. Comp. Assist. Tomogr.*, 1990, **14**, 886.

44. J. Granot, *J. Magn. Reson.*, 1986, **70**, 488.

45. J. Frahm, K. D. Merboldt, and W. Hanicke, *J. Magn. Reson.*, 1987, **72**, 502.

46. J. Frahm, H. Bruhn, M. L. Gyngell, K. D. Merboldt, W. Hanicke, and R. Sauter, *Magn. Reson. Med.*, 1989, **9**, 79.

47. J. C. Sharp, M. O. Leach, A. Hind, V. R. McCready, H. Weber, and R. Sauter, *Proc. VIth Ann Mtg. Soc. Magn. Reson. Med.*, New York, 1987, p. 600.

48. J. H. Dujin, G. B. Matson, A. A. Maudsley, and M. W. Weiner, *Magn. Reson. Imag.*, 1992, **10**, 315.

49. S. Posse, B. Schuknecht, M. E. Smith, P. C. van Zijl, N. K. Herschkowitz, and C. T. W. Moonen, *J. Comput. Assist. Tomogr.*, 1993, **17**, 1.

50. D. C. Shungu and J. D. Glickson, *Magn. Reson. Med.*, 1993, **30**, 661.

51. T. H. Mareci and H. R. Brooker, *J. Magn. Reson.*, 1984, **57**, 157.

52. H. R. Brooker, T. H. Mareci, and J. Mao, *Magn. Reson. Med.*, 1987, **5**, 417.

53. T. R. Brown, B. M. Kincaid, and K. Ugurbil, *Proc. Natl. Acad. Sci. USA*, 1982, **79**, 3523.

54. A. A. Maudsley, S. K. Hilal, W. H. Perman, and H. E. Simon, *J. Magn. Reson.*, 1983, **51**, 147.

55. D. R. Bailes, D. J. Bryant, G. M Bydder, H. A. Case, A. G. Collins, I. J. Cox, P. R. Evans, R. R. Harman, A. S. Hall, S. Khenia, P. McArthur, A. Oliver, M. R. Rose, B. D. Ross, and I. R. Young, *J. Magn. Reson.*, 1987, **74**, 158.

56. D. R. Bailes, D. J. Bryant, H. A. Case, A. G. Collins, I. J. Cox, A. S. Hall, R. R. Harman, S. Khenia, P. McArthur, B. D. Ross, and I. R. Young, *J. Magn. Reson.*, 1988, **77**, 460.

57. T. H. Mareci and H. R. Brooker, *J. Magn. Reson.*, 1991, **92**, 229.

58. K. J. Jung, S. K. Hilal, and Z. H. Cho, *Proc. XIth Ann Mtg. Soc. Magn. Reson. Med.*, Berlin, 1992, p. 3838.

59. M. S. Silver, R. I. Joseph, and D. I. Hoult, *J. Magn. Reson.*, 1984, **59**, 347.

60. C. C. Hanstock, D. L. Rothman, T. Jue, and R. G. Shulman, *J. Magn. Reson.*, 1988, **77**, 583.

61. W. I. Jung and O. Lutz, *Z. Naturforsch.*, 1988, **43a**, 909.

62. W. I. Jung, O. Lutz, K. Muller, M. Pfeiffer, and K. Kuper, *Proc. VIIIth Ann Mtg. Soc. Magn. Reson. Med.*, Amsterdam, 1989, p. 637.

63. K. Hendrich, Y. Xu, S.-G. Kim, and K. Ugurbil, *Magn. Reson. Med.*, 1994, **31**, 541.

64. K. Hendrich, H. Liu, H. Merkle, J. Zhang, and K. Ugurbil, *J. Magn. Reson.*, 1992, **97**, 486.

65. G. A. Coutts, D. J. Bryant, A. G. Collins, I. J. Cox, J. Sargentoni, and D. G. Gadian, *NMR Biomed.*, 1989, **1**, 190.

66. I. R. Young, I. J. Cox, G. A. Coutts, and G. M. Bydder, *NMR Biomed.*, 1989, **2**, 329.

67. T. J. Lawry, D. B. Twieg, A. A. Maudsley, M. W. Weiner, and G. B. Matson, *Proc. VIIth Ann Mtg. Soc. Magn. Reson. Med.*, San Francisco, 1988, p. 944.

68. C. T. W. Moonen, M. von Kienlin, P. C. M. van Zijl, J. Cohen, J. Gillen, P. Daly, and G. Wolf, *NMR Biomed.*, 1989, **2**, 201.

69. N. M. Yongbi, G. S. Payne, and M. O. Leach, *Proc. XIth Ann Mtg. Soc. Magn. Reson. Med.*, Berlin, 1992, p. 3840.

70. R. E. Lenkinski, D. P. Flamig, and J. Listerud, *Proc. IXth Ann Mtg. Soc. Magn. Reson. Med.*, New York, 1990, p. 1073.

71. A. H. Wilman and P. S. Allen, *Proc. XIth Ann Mtg. Soc. Magn. Reson. Med.*, Berlin, 1992, p. 3842.

72. Th. Ernst and J. Hennig, *Magn. Reson. Med.*, 1991, **21**, 82.

73. P. A. Narayana, E. F. Jakson, J. D. Hazle, L. K. Fotdar, M. V. Kulkarni, and D. P. Flamig, *Magn. Reson. Med.*, 1988, **8**, 151.

74. C. T. W. Moonen, P. van Gelderen, P. C. M. van Zijl, D. DesPres, and A. Olson, *Proc. Xth Ann Mtg. Soc. Magn. Reson. Med.*, San Francisco, 1991, p. 141.

75. S. Posse, C. A Cuenod, and D. Le Bihan, *Radiology*, 1993, **188**, 719.

76. G. C. McKinnon and P. Boesiger, *Magn. Reson. Med.*, 1988, **8**, 355.

77. A. Dolle, *J. Magn. Reson.*, 1991, **94**, 596.

78. W. I. Jung and O. Lutz, *J. Magn. Reson.*, 1991, **94**, 587.

79. J. F. Shen and P. S. Allen, *J. Magn. Reson.*, 1991, **92**, 398.

Biographical Sketches

David J. Bryant. *b* 1956. B.Sc., 1977, Ph.D., 1981, University of Nottingham, supervisor E. Raymond Andrew. Research fellow with Raymond Andrew, 1980–82. GEC Hirst Research Centre working in laboratory headed by Ian R. Young, 1982–present. Approx. 50 publications. Research interests: MRI and MRS on whole body systems in the 0.15–1.5 T range.

Glyn A. Coutts. *b* 1959. B.A., 1980, Physics, D.Phil., 1985, University of Oxford, UK. Research Associate at GEC Hirst Research Centre in laboratory headed by Ian R. Young, 1987–present. Approx. 30 publications. Research interests: MRI and MRS, interventional MRI.

Water Suppression in Proton MRS of Humans and Animals

Chrit T. W. Moonen

Centre National de la Recherche Scientifique, University of Bordeaux 2, Bordeaux, France

and

Peter C. M. van Zijl

Johns Hopkins University Medical School, Baltimore, MD, USA

1 INTRODUCTION

In vivo NMR spectroscopy offers a noninvasive window on metabolism and physiology in healthy and diseased humans and animals. Most initial efforts in NMR spectroscopy of living systems were devoted to the phosphorus and carbon nuclei. Because proton NMR is more sensitive, and allows access to a range of different metabolites, attention to proton spectroscopy has been increasing. However, without specific efforts to suppress water and fat resonances, the latter compounds

For References see p. 801

completely dominate the in vivo proton NMR spectrum. For example, the effective concentration of water hydrogen atoms may be as high as $110\,\text{M}$, whereas we wish to study metabolites in the millimolar range. Pioneering work on in vivo proton spectroscopy was performed at the laboratory of Shulman.[1,2] Recent advances are now permitting in vivo proton spectroscopy and spectroscopic imaging of the human brain in a routine examination. The topic of this review is to outline the possible strategies to suppress the water resonance, and other unwanted resonances such as those of fat (triglycerides).

The suppression of solvent resonances is an area of research that has long been in the domain of high-resolution (HR) NMR. Many techniques were well established before in vivo proton NMR was even contemplated. Recent developments such as shielded field gradient technology, and shaped rf excitation pulses have played an important role in NMR imaging. Now, these advanced tools are among the most powerful assets for water-suppressed proton spectroscopy in vivo, as well as in vitro.

When compared with HR NMR, in vivo spectroscopy has to overcome several typical problems related to B_0 and B_1 inhomogeneity, and motion, regardless of the nucleus being studied. These problems increase dramatically in proton spectroscopy because of the additional problem of water suppression and the narrow proton chemical shift range. For example, the range of water resonance frequencies over the human head amounts to several ppm (especially outside the brain area), comparable to the chemical shift range of most observable compounds. Therefore, chemical-shift-based water suppression cannot work unless a very high degree of localization is achieved. The use of surface coil receivers may help to avoid water signals from outside areas with a different resonance frequency. However, with the increasing attention being paid to spectroscopic imaging of large brain areas simultaneously, this solution is only of limited use. Motion effects further aggravate these problems.

Owing to limitations of space, in this article we emphasize general principles only, and highlight some important new developments. Because of the technical nature of this review, we do not restrict ourselves to human spectroscopy but also include important results in proton spectroscopy on animals. Detailed general reviews of water suppression in HR NMR and in vivo NMR have been published before.[3–6]

2 OVERVIEW OF SUPPRESSION METHODS BASED ON NMR AND OTHER BIOPHYSICAL PROPERTIES

2.1 Chemical Shift

Most suppression techniques are based on the frequency of the solvent resonance. Within this class of suppression strategies, we discriminate between methods that

1. accomplish a selective saturation of a frequency band before excitation of the spins of interest is started;[7–19] or
2. avoid altogether the excitation of a frequency band;[3,5,20,21] or
3. filter out a frequency band after signal reception.

In general, these methods have the disadvantage that all resonances of interest within the suppressed frequency band are also eliminated.

2.1.1 Selective Saturation

The general pulse sequence is

$$P_1-d_1(G_1)-P_2 \quad \text{Acq} \tag{1}$$

where P_1 is the soft rf pulse, d_1 is a short delay, preferably containing a gradient spoiling pulse G_1, and P_2 is the excitation pulse followed by acquisition time or the rest of a more complicated pulse sequence. This sequence, with a constant amplitude P_1 and a duration of the order of the solvent T_1 (the 'presaturation' method[7]) is one of the most common water suppression methods in HR NMR. It is now well established that it is more effective to employ a selective (shaped) P_1 pulse using a flip angle slightly larger than $90°$ to take T_1 relaxation during d_1 into account, thus nulling solvent magnetization at the time of rf pulse P_2.[8–14] The latter technique, including a spoiling gradient,[8,9] is often referred to as a CHESS sequence (chemical shift suppression). The method does not truly reflect a classical 'saturation' experiment. However, it leads to the condition of zero M_z magnetization and dephased transverse magnetization. In addition, selective magnetization inversion can be used followed by a waiting period to allow for zero M_z magnetization.[15]

The saturation procedure does not have to take place prior to the rest of the sequence. For example, a CHESS sequence can also be employed in the middle (TM) period of the stimulated echo sequence in in vivo NMR, when the magnetization due the resonances of interest is longitudinal. The basic sequence can be repeated for increased efficiency. However, care should be taken that different gradient directions and powers are used to avoid unwanted echoes of the solvent resonance. A detailed analysis has been presented previously.[12] Examples are given in Figure 1.

A B_1 gradient may also be helpful in suppressing a frequency band.[16–19] A combination of hard pulses followed by spin locking procedures has been proposed for in vivo NMR.[18,19] The basic idea is that all off-resonance magnetization will dephase in the inhomogeneous spin locking field. In addition, a fast pulse train of hard pulses (DANTE[20]) given with an inhomogeneous B_1 field has also been shown to lead to rapid effective nulling of longitudinal magnetization of solvent.[16,17] These techniques are particularly useful for surface coils.

2.1.2 Excitation Profiles with a Gap at the Water Frequency

The difference in evolution frequency of transverse magnetization between solvent and solute can be employed to create a gap in the effective excitation profile at the solvent frequency.[3,5,21] The best-known and widely used is the so-called jump–return sequence of Plateau and Guéron.[21] The sequence starts with a $\frac{1}{2}\pi$ rf pulse with the transmitter on the solvent resonance. Following an evolution period t, a second $\frac{1}{2}\pi$ rf pulse is given with opposite phase. The magnetization of the solvent resonance is thus put back along the z axis. However, for a resonance with a relative frequency of $(4t)^{-1}$ with respect to the offset, the transverse magnetization will have rotated during t exactly $90°$, and thus will remain unchanged by the second rf

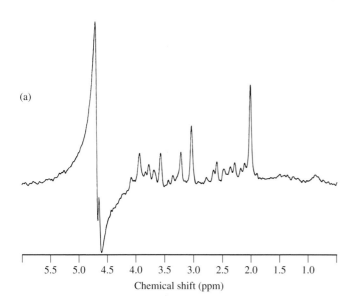

(a)

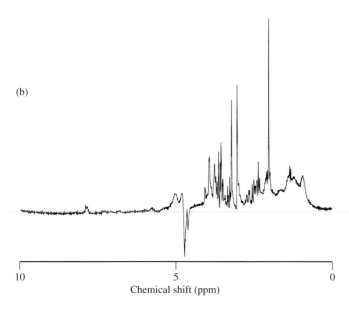

(b)

Figure 1 Examples of high quality water suppression in in vivo proton spectroscopy of (a) human and (b) animal brain. (a) In vivo proton NMR spectrum of an 18 ml volume obtained from normal gray matter of a healthy volunteer. The spectrum is obtained with STEAM localization (*TR/TE/TM* = 6000/20/30 ms), and three preceding CHESS water suppression procedures. The spectrum was obtained in approx. 6 min. A 2.0 T instrument was used, with a conventional quadrature head coil. Only exponential multiplication was used prior to FT, and no further data manipulation was applied. Courtesy of Dr. Jens Frahm, Biomedizinische NMR Forschungs GmbH, Max Planck Institut für Biophysikalische Chemie, Göttingen, Germany. (b) In vivo proton NMR spectrum from a 0.7 mL volume located in the superior part of the cat brain using STEAM localization (*TR/TE/TM* = 1000/18/80 ms). A 4.7 T instrument was used, equipped with 20 mT m^{-1} shielded gradients. A surface coil (outer diameter 48 mm) was used for transmission and reception. Three preceding CHESS water suppression sequences, and three CHESS water suppression sequences in the *TM* period were used. Lorentzian-to-Gaussian transformation was used prior to FT. For more details, see Moonen and van Zijl.[12]

pulse. The frequency profile of such a sequence is approximately a sine function. Instead of two rf pulses, an increasing number of rf pulses can be used to improve the frequency profile. A disadvantage is the sometimes complicated phase roll over the spectrum. Shaped, self-refocusing rf pulses can improve the phase response dramatically.[22] Guéron et al.[5] have given a detailed review of this class of suppression techniques, along with further improvements.

Continuous wave (CW) and rapid scan techniques are based on a rapid sweep of the frequency band of interest and avoidance of the solvent resonances. However, these methods are now rarely used because of their low S/N per unit time.

2.1.3 Suppression of Water Resonance Following the Complete Pulse Sequence

When radiation damping is not a problem, the solvent resonance can be suppressed following the complete pulse sequence. The disadvantage is that even a very large dynamic range analog-to-digital converter ADC can often not handle the large voltage differences due to resonances of interest and solvent[3,6] (Section 3.3).

2.2 Relaxation

Differences in relaxation times T_1 and T_2 between solutes and solvent can be exploited to achieve a relative suppression of the water resonance. T_1 and T_2 values of some brain compounds have been given by Frahm et al.[23] If the solvent suppression mechanism is entirely based on relaxation differences, signal loss for the resonances of interest can be significant, because differences in relaxation times are not dramatic. The advantage of relaxation-based water suppression is that its mechanism is not frequency-selective, and therefore the method works regardless of the actual resonance frequency of water.

2.2.1 T_1-Based Methods

The general pulse sequence is as above. When P_1 in sequence 1 is a π pulse, and d_1 is a delay adjusted for zero longitudinal magnetization of water,[15,24,25] the sequence is often called a WEFT (water-eliminated Fourier transform). To avoid T_1 weighting of the resonances of interest, a selective π pulse can be used (see also Section 2.1.2). T_1-based methods require more time than CHESS-type methods to null M_z solvent magnetization (because of T_1), and are optimal only for a single T_1. In the case of more than two solvent populations with different T_1s, more inversion pulses can be used, with adjustable delays between π pulses.[25]

2.2.2 T_2-Based Methods

A general spin echo (or stimulated echo) sequence can be used, such as

$$P_1 - \tau - P_2 - \tau \quad \text{Acq} \qquad (2)$$

where P_1 is the $\frac{1}{2}\pi$ excitation pulse, 2τ is the echo time, and P_2 is the refocusing (π) pulse. Acquisition is started at the top of the echo, or immediately following the P_2 pulse. Apart from water suppression, the spin echo sequence has the added advantage of

For References see p. 801

the removal of broad resonances (owing to short T_2). The reduced overlap is often helpful for quantifying resonance intensities, for example in proton spectroscopic imaging. A disadvantage is the phase modulation due to J coupling. However, when several refocusing pulses are used, specific J modulation patterns of molecules of interest can be turned into an advantage, for example for selecting lactate and avoiding uncoupled resonances.[26]

2.3 Scalar Coupling

Many modern methods use nuclear coupling characteristics to select only coupled resonances and thus suppress the (singlet) solvent resonance automatically.[27–49] These methods are identical to many of the so-called 'editing' techniques. Some techniques can be understood in sufficient detail using a classical vector presentation, whereas some newer methods can best be treated with the modern coherence pathway formalism. We shall first explain some basic methods for heteronuclear applications (e.g. proton detection of ^{13}C-labeled compounds). For the sake of simplicity, we shall limit the discussion to weakly coupled AX systems.

2.3.1 Indirect Detection by Spin Echo Methods

The heteronuclear spin echo difference method (POCE, proton-detected carbon editing) is a two-scan experiment in which the heteronuclear π pulse is turned on/off in alternating experiments:[43,44,46,47]

$$^1\text{H} : \tfrac{1}{2}\pi\text{--}\tau\text{--}\pi\text{--}\tau \quad \text{Acq}$$
$$^{13}\text{C} : \qquad \pi \quad (\text{on/off}) \tag{3}$$

The optimum delay τ is $(2J)^{-1}$. The evolution of the multiplet can be easily followed in the rotating frame. If the heteronuclear π pulse is off, the coupled spins change the direction of their evolution at the time of the π pulse. As a result, the components of the multiplet rephase at the top of the echo. If both π pulses are on, the chemical shift is rephased at the top of the echo, but evolution due to the coupling has continued for a period J^{-1}. As a consequence, the multiplet has opposite phase. Subtraction thus gives resonances only for coupled spins and eliminates those for all noncoupled spins.

2.3.2 Indirect Detection by Polarization Transfer and Multiple Quantum Methods.

So far, we have discussed spin systems in terms of longitudinal and transverse magnetization. A more general treatment involves the analysis of coherence orders during the pulse sequence, and is particularly useful for the understanding of polarization transfer and multiple quantum pulse sequences. Coherence orders are denoted by p, where $p = 0, \pm1, \pm2$ for zero, single, and double quantum order, respectively. For an introduction to the coherence formalism, the reader is referred to the literature.[27–29] We shall review two fundamental heteronuclear (^{1}H–^{13}C) sequences here. Proton spins will be indicated by I, carbon spins by S. The single quantum coherences, $\hat{I}_x$ and $\hat{I}_y$, refer to the classical transverse (observable) proton magnetization, $\hat{S}_x$ and $\hat{S}_y$ to single quantum carbon magnetization. $2\hat{I}_x\hat{S}_y$ consists of double quantum coherence and zero quantum coherence, and is not directly observable. $2\hat{I}_x\hat{S}_z$

and $2\hat{I}_z\hat{S}_x$ are so-called antiphase single quantum coherences. Radiofrequency pulses may generate transitions between coherence orders, but gradient pulses cannot. The complete pulse sequence can be seen as a coherence pathway vector $\boldsymbol{p}$ with as many elements as rf pulses in which the coherence order in period i is indicated with order p_i.

As an example, we shall follow an HMQC (heteronuclear multiple quantum coherence)[38,39,48,49] experiment (for the pulse sequence see Figure 2a). Following excitation of the protons ($\hat{I}_z \rightarrow -\hat{I}_y$) by the $\tfrac{1}{2}\pi$ rf pulse along the x axis in the rotating frame and neglecting the chemical shift, evolution of $\hat{I}_y$ will then lead to

$$-\hat{I}_y \xrightarrow{\text{evolution}} -\hat{I}_y \cos(\pi J_{IS} t) + 2\hat{I}_x\hat{S}_z \sin(\pi J_{IS} t) \tag{4}$$

The carbon pulse after an evolution period of $(2J)^{-1}$ will transfer the antiphase single quantum coherence into a multiple quantum coherence:

$$2\hat{I}_x\hat{S}_z \xrightarrow{\tfrac{1}{2}\pi^S_x} -2\hat{I}_x\hat{S}_y \tag{5}$$

The π proton pulse converts the zero-quantum into a double-quantum coherence, and the double-quantum into a zero-quantum coherence. The final $\pi/2$ carbon pulse transfers the $2\hat{I}_x\hat{S}_y$ coherences into observable antiphase proton magnetization ($2\hat{I}_x\hat{S}_z$ coherence), which is first allowed to refocus and then detected.

Figure 2(b) shows an HSQC heteronuclear polarization transfer experiment (heteronuclear single quantum coherence[50]). Following evolution as in the HMQC experiment [equation (4)] and refocusing of the chemical shift, the essential step is the simultaneous proton and carbon $\tfrac{1}{2}\pi$ pulses, which create antiphase carbon magnetization from antiphase proton magnetization ($2\hat{I}_x\hat{S}_z \rightarrow 2\hat{I}_z\hat{S}_y$). This is an effective polarization transfer from the proton spins to the coupled carbon spins. Now, the carbon nuclei can be detected with an enhanced polarization (the enhancement is γ_H/γ_C). The basic experiment is called the INEPT experiment. For maximum enhancement, a second (inverse) polarization transfer is carried out, and the sensitive hydrogen nucleus can be detected for optimum efficiency. Note the coherence pathway diagram in Figure 2(b).

When selecting a unique coherence pathway of Figure 2, noncoupled resonances are automatically eliminated. Therefore, coupled resonances under the water line can, in principle, be measured. However, note that the coherence pathway selection is generally a multiscan experiment (see Section 3.1 for important exceptions). The reason is that the rf pulses also lead to undesired coherences. This may be caused by the inhomogeneous $\boldsymbol{B}_1$ field, or by unwanted coherences that are generated even when ideal $\tfrac{1}{2}\pi$ pulses can be used (for example, owing to relaxation effects). In order to avoid the detection of unwanted coherences, the phase of the rf pulses and the receiver are varied using a phase cycling scheme.[27–29] For the purpose of water suppression, it is important to note that in a multiscan experiment, the ADC resolution may still be a limiting factor.

Limitations of space preclude a thorough analysis here of all hetero- and homonuclear editing pulse sequences with automatic water suppression. Generally, similar principles hold for homo- and heteronuclear coherence pathways.[27,28] However, note that single quantum evolution frequencies are similar in homonuclear sequences, whereas they are dependent on the

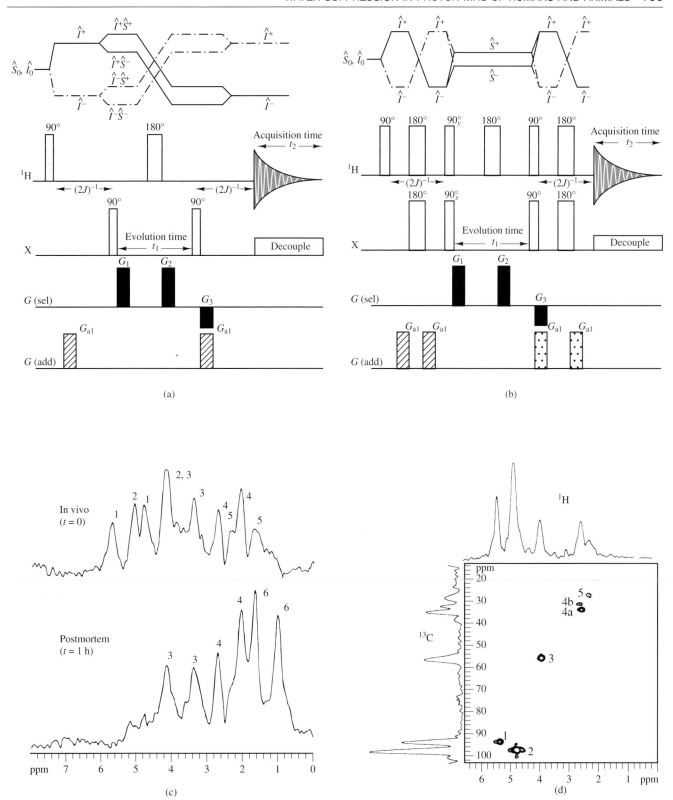

Figure 2 Coherence pathways and pulse sequences for (a) heteronuclear multiple quantum coherence (HMQC) and (b) heteronuclear single quantum coherence (HSQC) experiments. The phase of the rf pulses is irrelevant for the sequence when gradients G_{sel} and G_{add} are used for coherence selection. The phase of the rf pulses, and the multiscan phase cycle, is important when no gradients are used. Reproduced, with permission, from Ruiz-Cabello et al.[48] (c) $^{13}C-^{1}H$ HMQC spectra obtained from cat brain in vivo (upper spectrum), and 45 min following sacrifice (lower spectrum). Both spectra were obtained in 3 min 19 s from an 8 mm slice through the cat brain following infusion of ^{13}C-labeled glucose (final blood glucose level was 16.6 mM). Spectra are not ^{13}C-decoupled, and show ^{13}C multiplets. Note the complete absence of the water resonance. Assignments are: (1) α-[1-^{13}C]glucose; (2) β-[1-^{13}C]glucose; (3) [2-^{13}C]glutamate and glutamine; (4) [4-^{13}C]glutamate and glutamine; (5) [3-^{13}C]glutamate and glutamine; (6) [3-^{13}C]lactate. (d) Two-dimensional $^{13}C-^{1}H$ HMQC correlation spectrum obtained in a fully decoupled mode by incrementing the multiple quantum period. Total acquisition time was 30 min. The projections along the ^{1}H and ^{13}C axes are shown as well. Assignments are as in (c). Note the separation in peak 4 corresponding to a separation of the glutamate (peak a) and glutamine (peak b)

gyromagnetic ratio of nucleus I or S in the heteronuclear case. Thus, a homonuclear editing sequence should involve either at least one rf pulse that avoids water excitation, or a multiple quantum order in part of the sequence in order to inherently eliminate the singlet solvent resonance.

In short, multipulse editing sequences, based on polarization transfer and multiple quantum coherences, can be used for 'automatic' water suppression. Multiscan phase cycling schemes are necessary for pure coherence pathway selection. In Section 3.1.3, an important exception to this rule will be reviewed, namely, the use of field gradients for single scan coherence pathway selection.

2.4 Diffusion

The diffusion constant of water is considerably higher than that of small metabolites and much higher than that of macromolecules. Pulse sequences can be sensitized to diffusion by the use of $\boldsymbol{B}_0$ magnetic field gradients without any effect on stationary molecules.[51–54] For two gradient pulses of strength G and duration δ, and with duration Δ between their starts, the attenuated signal S resulting from free diffusion can be expressed as

$$S/S_0 = \exp[-\gamma^2 G^2 \delta^2 (\Delta - \tfrac{1}{3}\delta)D] \qquad (6)$$

where S_0 is the starting signal and D is the diffusion coefficient. For long diffusion times, restrictions by cell membranes and binding to macromolecules have to be taken into account. An additional advantage for water suppression is that the permeability of cell membranes to water is generally much higher than to most other compounds. The main disadvantage is that diffusion-sensitized pulse sequences are also very sensitive to motion. Therefore, such methods appear to have more potential for HR NMR than for in vivo NMR.[52,54]

2.5 Exchange

Exchange of protons between a dissolved compound and solvent water can result in different relaxation properties of the water protons. These exchange properties can be used to attenuate the solvent resonance.[55] For example, the saturation of exchangeable spins, whether by rf irradiation, T_2 relaxation, or dephasing in a magnetic field, can be transferred to the water resonance. This exchange mechanism is the basis of the so-called 'magnetization transfer contrast' in MRI,[56] but has not been used for the specific purpose of solvent suppression in in vivo NMR.

3 ADVANCED HARDWARE AND SOFTWARE TOOLS

3.1 Pulsed Field Gradients

The development of high-quality pulsed field gradients is probably the single most important tool that has made the routine applications of proton spectroscopy possible in humans. Historically, pulsed field gradients were used in HR NMR to dephase (spoil) the transverse magnetization—hence the often used name of a homospoil pulse.[57] Although the fundamental

advantages of field gradients were known,[57–60] the limited quality with respect to amplitude, ramp times, current stability, and particularly residual gradients and vibration effects, prevented their general use. The extreme demands in MRI—in particular, with respect to the very fast echo planar imaging method[61]—have dramatically accelerated hardware developments. The advent of self-shielded gradient coils with minimal residual gradients should thus be viewed as a milestone.[62,63] An outer gradient coil is used that cancels the outside field of the inner coil, thus eliminating eddy currents in the cryostat. Pulsed field gradient technology is now also becoming a routine tool in HR NMR. Field gradients are used for the following purposes:

1. rapid dephasing of transverse magnetization (homospoil pulse);
2. spatial encoding and spatially selective excitation (in conjunction with shaped rf pulses);
3. single shot coherence pathway selection;
4. creating diffusion sensitivity (Section 2.1.4).

3.1.1 Rapid Dephasing of Transverse Magnetization

Following a gradient pulse of duration δ with magnitude G_α in direction α, the phase ϕ_α at location r_α as a function of coherence order p is

$$\phi_\alpha(p) = p\gamma r_\alpha \delta G_\alpha \qquad (7)$$

Therefore, a phase dispersion results that attenuates the transverse magnetization by a factor f:

$$\frac{1}{f} = \left| \frac{1}{r_2 - r_1} \int_{r_1}^{r_2} e^{-i\varphi_\alpha} \, dr_\alpha \right| \qquad (8)$$

The limits r_2 and r_1 are the voxel dimensions in the gradient direction α. Equation (8) assumes equal spin density, and a homogeneous field. Note that the attenuation works as a result of the integration of the phase over the dimension of the volume.[12] In the case of spectroscopic imaging, the integration is over the voxel dimensions, not over the full object. The attenuation is higher with a higher gyromagnetic ratio, and with a higher coherence order. Homonuclear zero quantum order will thus not be dephased by the gradient pulse. Note also that the efficiency of the gradient crusher depends on the local static field homogeneity. For a detailed account, see Moonen et al.[64] Shimming is therefore important in the full region where high efficiency of gradient crushers is necessary, not only in the region of interest.

3.1.2 Spatial Encoding and Spatially Selective Excitation

The essence of field gradients in imaging is the encoding of spatial information,[65] whether used for phase encoding, frequency encoding, or spatially selective excitation. In proton spectroscopy of humans,[66–82] field gradients are often employed for the same purpose. For spectroscopic imaging,[83–89] phase encoding is generally used for spatial encoding in two or three directions. In conjunction with shaped rf pulses (leading to a desired frequency profile), field gradient pulses lead to spatially selective excitation or refocusing (see Section 4.2 for a more detailed account). The importance of spatial selection for water suppression purposes is further explained in Section 4.1.

Spatial and spectral selection can be combined in a single rf pulse together with a special gradient waveform.[89–91]

3.1.3 Single Shot Coherence Pathway Selection

Conventionally (Section 2.3), a coherence pathway is selected using a multiscan approach using specific phase cycles of rf pulses and receiver. However, this ideal result is often not reached in vivo as a result of motion or of instrumental instabilities. The selection can also be achieved using pulsed field gradients, often completely eliminating the need for phase cycling. For example, a π refocusing pulse at the end of a pulse sequence may generate transverse magnetization as a result of an inhomogeneous B_1 field. Using two scans with opposite phase of the π pulse, transverse magnetization is canceled. A pair of identical gradient pulses around the π pulse will lead to cancelation of the unwanted transverse magnetization in a single scan.

We can generalize equation (7) with respect to the phase of transverse magnetization at location r as a function of coherence order p during all periods i of the pulse sequence:

$$\phi_i(r) = \left(\sum_j p_j \gamma_j \right) r G_i t_i \tag{9}$$

where p_j is the coherence order of the individual nuclei (e.g. ^{13}C, ^{1}H) in period i and all other symbols have the same meaning as in equation (7). Thus, the phase evolves as a function of coherence order and gyromagnetic ratio, and equation (9) can be used to keep track of the phase evolution during the entire pulse sequence. A certain coherence order can thus be selectively rephased if the following condition is satisfied:

$$\sum_i \phi_i(r) = 0 \tag{10}$$

In the ideal case, spins of interest are refocused after the last rf pulse, and a gradient scheme is employed that avoids any rephasing of undesired coherences.[61,62] The required gradient power depends on the signal strength of the desired versus undesired coherences.

Localization, achieved by one or more spatially selective rf pulses, can be conveniently treated as a type of coherence path-

way selection.[12,64] When more rf pulses are employed during a pulse sequence, it becomes harder to ensure sufficient dephasing of all undesired coherences. However, the use of gradients in all three principal axes with different amplitudes, and using different combinations at different time points in the sequence, offers great flexibility to optimize the single scan coherence selection.

Using equation (9), it can be easily seen that a coherence pathway selection involving a double quantum evolution ($p = \pm 2$) leads to dephasing of the solvent resonance without recourse to any frequency-dependent water suppression. The example shown in Figure 3 demonstrates that resonances under

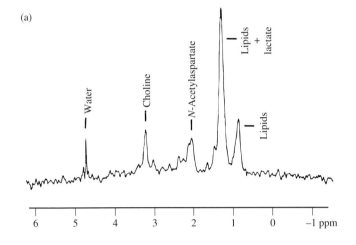

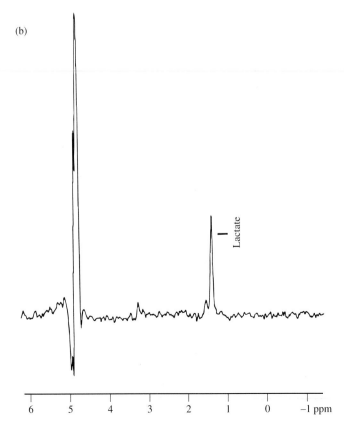

Figure 3 Unedited (a) and lactate-edited (b) localized proton spectra (volume $200\,\mu l$) of an intracerebral glioma in the rat obtained at $4.7\,T$ with a surface coil for excitation in conjunction with adiabatic pulses (*TE/TR* = 272/2000 ms; 4 min acquisition time for each spectrum). The basic pulse sequence is a spin echo method with jump–return (Section 2.1.2) excitation and refocusing pulses for water suppression, combined [in (b)] with a frequency-selective inversion pulse at the resonance frequency of the C-α proton (4.1 ppm). The inversion pulse is alternated on/off, and the resulting signals are subtracted for the editing procedure.[47] All rf pulses have been replaced with adiabatic pulses. Instead of hard jump–return pulses, solvent-suppressive adiabatic pulses were used. The inversion pulse was an adiabatic DANTE pulse. To maintain better phase coherence in the editing procedure, a second (2π) adiabatic DANTE was inserted in the second $\frac{1}{2}TE$ period. Phase cycling using the EXORCYCLE procedure was performed. Localization was achieved with a modified 3D ISIS method using BIR-4 inversion pulses. Further details are given by Schapp et al.[92] and de Graaf et al.[93] Courtesy of Michael Garwood, Center for Magnetic Resonance Research, University of Minnesota, Minneapolis, MN

For References see p. 801

the water line may be observed, such as the resonances of the biologically important glucose molecule. Heteronuclear polarization transfer can also be achieved in a single scan with complete water suppression. However, homonuclear polarization transfer or any homonuclear sequence with $p = 0, \pm 1$, without chemical-shift-selective pulses, cannot achieve this, and additional water suppression methods must be employed. However, even in such cases, the use of field gradient pulses is beneficial because of the elimination of phase cycling (less motion sensitivity), and, in conjunction with other solvent suppression techniques, decreased water signal per scan and thus fewer demands on the receiver system.[94,95]

The use of field gradients for single shot coherence selection also has some disadvantages. Most important is the loss of a factor of two in the signal (except in the case of gradient pulses around a π refocusing pulse, and in special sequences using selective refocusing pulses[37] or sensitivity-enhanced gradient experiments[96]). The reason is that only one coherence pathway is selected, whereas phase cycling allows pathways of opposite signs to be selected simultaneously. For correlated multidimensional NMR, a second disadvantage (and a consequence of the first) is that phase-sensitive 2D spectra cannot be obtained using a single scan per t_1 increment. When using a second scan with selection of coherence path with opposite sign of coherence order in the t_1 period, phased resonances can be obtained. Of course, if these disadvantages are serious, pulsed field gradients may still be used only around refocusing pulses, and thus still eliminate part of the phase cycle.

3.2 Shaped rf Pulses

Shaped rf pulses play an important role in MRI, especially in slice-selective excitation and refocusing.[90-92,97-107] To a first approximation, the frequency response resulting from a particular rf pulse shape is given by the Fourier transform of the waveform. This is known as the linear response. For flip angles close to, and above 90°, linear response theory is no longer completely accurate, and the Bloch equations have to be used to determine the response.[5,108] Corrections to the basic sinc waveform are now commonly used in NMR imaging. In localized proton spectroscopy, shaped rf pulses are used for the same purposes. In addition, a frequency-selective excitation is often used to select a narrow frequency band at the water resonance. In contrast to slice-selective pulses, phase-coherent response is not desired for frequency-selective suppression. In other applications, it is useful to combine spatial and spectral selection in a single pulse. The different specifications in different pulse sequences clearly indicate the opportunities for a specific design of rf pulses. In modern applications, simple Fourier transform of a desired frequency profile is often used as a starting point for a shaped rf pulse. Then, an optimization routine is used to arrive at improved shapes using the complete Bloch equations. Recent advances use complex polynomials for mapping the rf pulse and then solving them analytically.[91,107]

The inhomogeneity in the B_1 field (e.g., of a surface coil) has advantages and disadvantages in this respect. One can use the B_1 gradient to advantage in order to arrive at a desired phase dispersal of the water magnetization following excitation (see Section 2.1.1). On the other hand, inhomogeneous B_1 fields lead to loss of signal due to a spread in flip angle and to

a spatially dependent population of desired and undesired coherence pathways, often neccessitating the use of large gradient crushers for coherence path selection. The latter disadvantage can be overcome by using adiabatic pulses.

3.2.1 Adiabatic rf Pulses

Most water suppression techniques demand precise flip angles. Despite high B_1 homogeneity in the modern volume coils that are now routinely available in whole body MR instruments, the range of the B_1 field may still exceed the requirements—in particular, for applications of water-suppressed proton spectroscopic imaging. Accurate flip angles, independent of the local B_1 field, can be achieved with adiabatic rf pulses.[92,97-106] The first such pulses accomplished excitation and spin inversion, and employed a continuous frequency ramp of the B_1 field while maintaining a constant amplitude. The effect of this pulse can be visualized in the rotating frame by analyzing the direction of the effective field (a function of the B_1 field and the rf frequency relative to the Larmor frequency). During the pulse, the effective field B_e rotates from the positive z axis to the negative z axis (in the case of inversion) or to the (x,y) plane (in the case of excitation). So long as the B_1 field exceeds a threshold value (the adiabatic condition), the magnetization continues to rotate effectively with a small precession angle around the field B_e during the entire pulse. In other words, the magnetization M remains collinear with B_e above a threshold B_1. Therefore, if we let B_e rotate by π then M will also rotate by π, and spin inversion is accomplished. The flip angle is independent of the B_1 field if the rate of change of the angle of the field B_e with the z axis remains much smaller than the rotation frequency of magnetization M around the effective field B_e (i.e., the adiabatic criterion). The frequency sweep does not need to be accomplished with a linear ramp. In fact, a whole range of frequency (or phase) modulation functions have been described. The effect of the different modulation functions affect the B_1 and off-resonance range where the pulse remains adiabatic. The disadvantages of these original adiabatic pulses are that

1. they do not result in good slice profiles when used for slice selection;
2. they require increased rf power;
3. plane rotation is not possible.

The latter disadvantage, in particular, has prevented many applications, especially in areas related to coherence pathway selection and water suppression. However, recent advances have made it possible to perform adiabatic plane rotations and thus refocusing. In addition, slice-selective refocusing with adiabatic pulses has been made possible by using two consecutive adiabatic inversion pulses.[109]

The problem of plane rotation using the above adiabatic pulses is that, unlike magnetization that is initially collinear with the effective field B_e, magnetization perpendicular to B_e at location r will precess through an angle β depending on the actual magnitude of B_e at location r. This angle therefore varies with location, and, as a result, the possibility of B_1-independent rotation is apparently lost. The central idea behind the solution of Garwood, Ugurbil and colleagues[92,102-104] is that if the effective field B_e is suddenly inverted, and if B_e is properly rotated, leading to a precession of $-\beta$ at location r, then the magnetization M will undergo an effective plane rotation. One of the most versatile of this new class of adiabatic plane rotation pulses is the so-called BIR-4 pulse.[105] The pulse consists of three sections: an adiabatic half-passage in reverse, adiabatic inversion, and adiabatic half-passage. Two inversions of B_e occur, for example, between periods one and two, and between two and three, accompanied by two discontinuous phase jumps, $\Delta\phi_1$ and $\Delta\phi_2$. The BIR-4 pulse can accomplish uniform plane rotations through any angle, the magnitude being determined by the phase jumps $\Delta\phi_1$ and $\Delta\phi_2$.

The feature of BIR-4 that is uniquely suitable for water suppression purposes or coherence pathway selection purposes is the fact that the phase jumps can be accomplished not only using the phase of the B_1 field, but also by the Larmor precession frequency or by J coupling.[104,106] The first possibility is achieved simply by inserting a delay at the time of the phase jumps, leading to frequency-selective adiabatic pulses. The second possibility leads to adiabatic editing based on spin–spin coupling and adiabatic polarization transfer.[106] These elegant features can be accomplished with a high insensitivity of more than 10-fold variations in the B_1 field. Figure 3 shows an example.

3.3 Postacquisition Frequency Filters

Even when the pulse sequence has resulted in water suppression of such high quality that the preamplifier has not been overloaded and the dynamic range of the ADC is sufficient with respect to the S/N of the experiment, it may be advantageous further to reduce the water line using data processing algorithms. The reason lies in some remaining disadvantages of the (suppressed but still dominating) water resonance. First, the base of the water line extends far into other regions of the spectrum. Second, insufficient apodization and truncation of the time domain can cause 'ringing' over a large portion of the spectrum. The latter problem is especially relevant if the acquisition time is short, such as in fast spectroscopic imaging using multiple spin echoes.

Several data processing methods can be used to extract the information from the raw time domain signals. Baseline corrections and convolution difference methods have been used extensively.[3–6] Much progress has been made in fitting in the time and frequency domains.[110–112] The latter methods are preferred whenever additional a priori information is available about the spectrum. A simple approach—the frequency filter—is demonstrated in Figure 4. The method is robust, fast, and automatic.[113–115] First, with water on resonance, a moving average filter over n data points is applied to generate a time domain signal from the raw data, which thus contain only data around zero frequency. Generally, a Gaussian weighting function is used, together with some extrapolation to determine the first and last $\frac{1}{2}n$ points. Second, the filter output is subtracted from the raw

Figure 4 The use of a simple postprocessing frequency filter to limit detrimental effects of the suppressed but still dominant water resonance on spectral quality. The example is taken from a voxel of a water-suppressed spectroscopic image (for pulse sequence details, see Moonen et al.).[64] The raw time domain signals, and magnitude mode presentation following FT are shown in (a) and (b), respectively. The effect of the time domain frequency filter is shown on the filtered time domain (c) and the corresponding frequency spectrum (d) in magnitude mode presentation. Courtesy of Geoffrey Sobering, In Vivo NMR Research Center, NCRR, NIH, Bethesda, MD

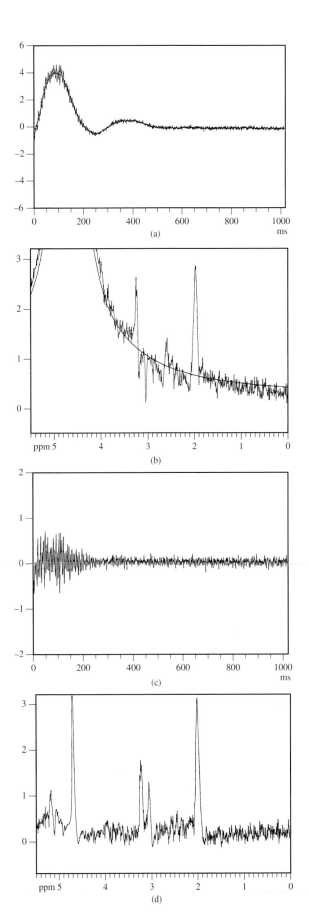

data. It has been shown that hardware frequency filters can also be implemented before the preamplifier stage.[116]

4   EVALUATION OF EXPERIMENTAL PROBLEMS

4.1   Spatial Selection

Most human proton spectroscopy is performed with the standard quadrature resonator. Therefore, signal detection is about equally efficient over the whole head, including the neck region. Owing to susceptibility differences over the complete sensitive volume, and also in part to nonideal shimming or magnet imperfections, water in some regions will have similar resonance frequencies to the metabolites of interest in the region to be examined. Two important conclusions can be drawn. First, frequency-based water suppression may work in a well-shimmed area, but not in the complete sensitive coil volume. Second, frequency-based water suppression only makes sense in conjunction with high-quality localization.

Localization may be performed using a surface coil for reception and transmission purposes.[117] The relatively small coil limits the sensitive area and thus the source of artifactual signals. In addition, B_1 gradients generated by the surface coil can be used for additional localization.[118] Far more common is the use of B_0 gradients, which are, of course, routinely available on every clinical MRI instrument. One of the first methods used a field profiling method to spoil resonances originating outside the region of interest.[119] However, most methods employ switched B_0 gradients for localization—either for phase encoding or for slice selection purposes.[66–82] Slice selection in one direction is achieved similarly to common methods in MRI. Volume selection is achieved using a combination of three frequency-selective rf pulses in the presence of three mutually orthogonal gradients. This can be done in a multiscan[66–70,73–76] or single scan[10,70,71,78–81] approach. Selection in two or three different directions can also be achieved simultaneously.[90,120] Outer volume suppression may help the localization process. Because of the very high degree of localization, and the presence of motion, single shot methods are preferred. Note the important additional role of the B_0 gradients in single shot localization (e.g., crushers used for coherence pathway selection without rf phase cycling). Phase encoding employing B_0 gradients can be used for additional localization in spectroscopic imaging with Fourier transform[84] or alternative data processing methods using a priori information.[121,122] The quality of localization using phase encoding can be analyzed with the point spread function.

4.2 B_0 Inhomogeneity

Perhaps the most significant problem in in vivo proton spectroscopy is the attainable homogeneity in the static magnetic field B_0. Air–tissue and fat–tissue interfaces may lead to susceptibility gradients of the same magnitude as the generally used frequency range in in vivo proton spectroscopy. High-quality field homogeneity can be obtained in the brain, but is rather difficult in almost all other organs and tissues. This is the reason why proton spectroscopy is most commonly used in brain. Limited B_0 homogeneity is not only detrimental to spec-

tral appearance, but has also other disadvantages. Among these are

1. limitations in frequency-based water suppression methods;
2. errors in localization based on slice selection principles;
3. limitation in the efficiency of B_0 gradients for coherence selection and for rapid dephasing;
4. echo shifting when the effect of an incompletely compensated B_0 gradient is countered by the field inhomogeneity.

There are no complete solutions to the problem of limited B_0 homogeneity. Partial remedies include optimal (possibly local) shimming, selecting small voxels, and using self-shielded gradients. Methods have been presented that result in sharp lines despite B_0 inhomogeneities.[123] However, this result comes at the price of a severe S/N penalty.

4.3 B_1 Inhomogeneity

The disadvantages of limited B_1 homogeneity have been discussed in Section 2.1. In the context of water (or fat) suppression using the popular CHESS methods, the flip angle must be defined precisely. Recent advances in adiabatic pulse design may provide elegant improvements.

4.4   Motion

Motion is a common problem in MRI. Despite the much larger voxels in proton spectroscopy, it can be an even bigger problem in the latter technique—in particular, because of the extreme demands on water suppression and localization. In this regard, we may distinguish not only macroscopic motion of uncooperative patients, but also motion of arterial and venous blood, of CSF, and of brain tissue. The preferred method to minimize the detrimental effects of motion is to perform water suppression and localization in a single scan.[12] Multiscan phase cycling methods lead to incomplete suppression of unwanted signals in the presence of motion. However, single scan methods employing B_0 gradients are not without problems. For example, motion during the scan may result in a phase shift of the echo. Tracking the phase using an additional echo[123] may help if the phase shift is identical for the complete voxel, but is of limited use if it is spatially dependent. In addition, if signal averaging is used, the motion sensitivity may be decreased when magnitude spectra are averaged, instead of pure phase spectra. In addition, shifts in the echo position may occur as a result of an imbalanced gradient in the presence of field inhomogeneities resulting from the motion.

4.5   Spectroscopic Imaging

The typical problems and solutions for solvent suppression in single voxels are similar to those in spectroscopic imaging methods. However, since modern spectroscopic imaging[86–91] is performed in a single session covering a large part of the brain, problems with B_0 and B_1 inhomogeneities are more severe. The problem of B_0 homogeneity leads to the need for a larger frequency band affected by the water suppression procedure. Postprocessing water suppression (Section 3.3) is commonly used to minimize the effects of spatially dependent water suppression. Adiabatic pulses—now already routinely used for spin inversion purposes in spectroscopic imaging—may bring further improvements. Motion during spectroscopic imaging may lead to severe artifacts.

4.6 Fat Suppression

This article has dealt with suppression of the water resonance. Resonances of fat, or indeed, the family of triglycerides, can also dominate an in vivo proton NMR spectrum. Triglycerides lead to many resonances covering a large part of the useful proton spectrum. Therefore, typical frequency-selective suppression is not satisfactory. Three features of fat resonances are used for suppression purposes.

1. T_1 values are short (about 280 ms at 1.5 T), and can be used conveniently when preceding a pulse sequence with an inversion–recovery sequence.
2. In healthy brain tissue, no significant contributions originate from brain triglycerides.[2] The fat resonances arise predominantly from the skull tissue and can thus be suppressed by
(a) outer volume suppression[67] (e.g., by spatial excitation of the entire skull area followed by rapid dephasing); or
(b) avoiding excitation of skull area using accurate localization procedures (see Section 4.1).
3. Coupling patterns in trygliceride resonances, as distinct from those in the lactate doublet, are often used to separate the two compounds. Since the spatial origin of fat resonances is in the skull, phase encoding is helpful in defining the spatial origin of the fat signals, and thus suppressing the fat resonances in brain spectra.[124]

4.7 Safety Issues

The potential safety issues in water-suppressed proton spectroscopy are the same as for routine MRI. There are no specific safety issues with regard to water suppression, except in some cases heat deposition by rf irradiation when employing (semi)-continuous rf irradiation or B_1 gradients to achieve scrambling of the water magnetization. However, most methods described in this article use less rf power per unit time than routinely used in multiple spin echo MRI methods.

5 SUMMARY

Solvent suppression has been, and remains, a rich area of research. Many advanced methods, developed over at least 30 years of high-resolution NMR, have been adapted for use in in vivo proton NMR spectroscopy. Water-suppressed human proton spectroscopy has now become almost a routine tool for brain NMR examination at 1.5 T. Although numerous developments have contributed to this extraordinary achievement, two fundamental technologies have been crucial:

1. the development of high-quality, fast switching, B_0 gradient coils, in particular, the invention of the self-shielded gradient coil;
2. the achievement of excellent rf phase and amplitude stability and control.

The latter technology is especially relevant for adiabatic and other shaped pulses. Together with excellent progress in the field of single shot localization, these developments have made high-quality, water-suppressed proton spectroscopy of human and animal brain a reality. Solvent suppression in spectroscopic imaging is similar, but not identical, to the problem of single voxel solvent suppression. However, recent progress indicates that automated single slice and multislice spectroscopic imaging of the human brain will soon be available on standard 1.5 T instruments. The technological challenge that still remains largely unsolved is the routine use of proton spectroscopy of organs other than the brain.

6 RELATED ARTICLE

Single Voxel Localized Proton NMR Spectroscopy of Human Brain In Vivo.

7 REFERENCES

1. K. L. Behar, J. A. Den Hollander, M. E. Stromski, Y. Ogino, R. G. Shulman, O. A. C. Petroff, and J. W. Pritchard, *Proc. Natl. Acad. Sci. USA*, 1983, **80**, 4945.
2. C. C. Hanstock, D. L. Rothman, T. J. Jue, and R. G. Shulman, *Proc. Natl. Acad. Sci. USA.*, 1988, **85**, 1821.
3. P. J. Hore, *Methods Enzymol.*, 1989, **176**, 64.
4. J. E. Meier and A. G. Marshall, in 'Biological Magnetic Resonance', eds. L. J. Berliner and J. Reuben, Plenum Press, New York, 1990, Vol. 9, p. 199.
5. M. Guéron, P. Plateau, and D. Decorps, in 'Progress in Nuclear Magnetic Resonance Spectroscopy', eds. J. W. Emsley, J. Feeney and L. H. Sutcliffe, Pergamon Press, Oxford, 1991, Vol. 23, p. 135.
6. P. C. M. van Zijl, and C. T. W. Moonen, in 'NMR Basic Principles and Progress', eds. P. Diehl, E. Fluck, and R. Kosfeld, Springer-Verlag, Berlin, 1992, Vol. 26, p. 67.
7. D. I. Hoult, *J. Magn. Reson.*, 1976, **21**, 337.
8. A. Haase, J. Frahm, W. Hänicke, and D. Matthaei, *Phys. Med. Biol.*, 1985, **30**, 341.
9. D. M. Doddrell, G. J. Galloway, W. M. Brooks, J. Field, J. M. Bulsing, M. G. Irving, and H. Baddeley, *J. Magn. Reson.*, 1986, **70**, 176.
10. J. Frahm, K. D. Merboldt, and W. Hänicke, *J. Magn. Reson.*, 1987, **72**, 502.
11. I. M. Brereton, G. J. Galloway, J. Field, M. F. Marshman, and D. M. Doddrell, *J. Magn. Reson.*, 1989, **81**, 411.
12. C. T. W. Moonen and P. C. M. van Zijl, *J. Magn. Reson.*, 1990, **88**, 28.
13. R. H. Griffey and D. P. Flamig, *J. Magn. Reson.*, 1990, **88**, 161.
14. R. J. Ogg, P. B. Kingsley, and J. S. Taylor, *J. Magn. Reson.*, 1994, **104B**, 1.
15. C. A. G. Haasnoot, *J. Magn. Reson.*, 1983, **52**, 153.
16. D. Canet, J. Brondeau, E. Mischler, and F. Humbert, *J. Magn. Reson.*, 1993, **105A**, 239.
17. W. E. Maas and D. G. Gory, *J. Magn. Reson., Ser. A*, 1994, **106A**, 256.
18. P. Blondet, M. Decorps, and J. P. Albrand, *J. Magn. Reson.*, 1986, **69**, 403.
19. D. Bourgeois and P. Kozlowski, *Magn. Reson. Med.*, 1993, **29**, 402.
20. G. A. Morris and R. Freeman, *J. Magn. Reson.*, 1978, **29**, 433.
21. P. Plateau and M. Guéron, *J. Am. Chem. Soc.*, 1982, **104**, 7310.
22. H. Liu, K. Weisz, and T. L. James, *J. Magn. Reson., Ser. A*, 1993, **105**, 184.

For References see p. 801

23. J. Frahm, H. Bruhn, M. L. Gyngell, K.-D. Merboldt, W. Hänicke, and R. Sauter, *Magn. Reson. Med.*, 1989, **11**, 47.

24. T. Inubushi and E. D. Becker, *J. Magn. Reson.*, 1983, **51**, 128.

25. J. H. Duijn, G. B. Matson, A. A. Maudsley, J. W. Hugg, and M. W. Weiner, *Radiology*, 1992, **183**, 711.

26. J. H. Duijn, J. A. Frank, and C. T. W. Moonen, *Magn. Reson. Med.* 1995, **33**, 101.

27. R. R. Ernst, G. Bodenhausen, and A. Wokaun, 'Principles of Nuclear Magnetic Resonance in One and Two Dimensions', Clarendon Press, Oxford, 1987.

28. N. Chandrakumar and S. Subramianan, 'Modern Techniques in High-Resolution FT-NMR', Springer-Verlag, New York, 1987.

29. O. W. Sørensen, in 'Progress in Nuclear Magnetic Resonance Spectroscopy', eds. J. W. Emsley, J. Feeney, and L. H. Sutcliffe, Pergamon Press, Oxford, 1989, Vol. 21, p. 503.

30. C. L. Dumoulin and D. Vatis, *Magn. Reson. Med.*, 1986, **3**, 282.

31. C. H. Sotak, D. M. Freeman, and R. E. Hurd, *J. Magn. Reson.*, 1988, **78**, 355.

32. R. E. Hurd and D. M. Freeman, *Proc. Natl. Acad. Sci. USA*, 1989, **86**, 4402.

33. D. M. Freeman, C. H. Sotak, H. H. Muller, S. W. Young, and R. E. Hurd, *Magn. Reson. Med.*, 1990, **14**, 321.

34. A. Knuttel and R. Kimmich, *Magn. Reson. Med.*, 1989, **10**, 404.

35. D. M. Doddrell, I. M. Brereton, L. N. Moxon, and G. J. Galloway, *Magn. Reson. Med.*, 1989, **9**, 132.

36. C. H. Sotak, *J. Magn. Reson.*, 1990, **90**, 198.

37. L. A. Trimble, J. F. Shen, A. H. Wilman, and P. S. Allen, *J. Magn. Reson.*, 1990, **86**, 191.

38. A. Knüttel, R. Kimmich, and K.-H. Spohn, *J. Magn. Reson.*, 1990, **86**, 526.

39. J. M. Bulsing and D. M. Doddrell, *J. Magn. Reson.*, 1986, **68**, 52.

40. C. J. Hardy and C. L. Dumoulin, *J. Magn. Reson.*, 1987, **5**, 75.

41. M. von Kienlin, J. P. Albrand, B. Authier, P. Blondet, S. Lotito, and M. Décorps, *J. Magn. Reson.*, 1987, **75**, 371.

42. A. Knüttel and R. Kimmich, *Magn. Reson. Med.*, 1989, **9**, 254.

43. D. L. Rothman, K. L. Behar, H. P. Hetherington, and R. G. Shulman, *Proc. Natl. Acad. Sci. USA*, 1984, **81**, 6330.

44. A. A. de Graaf, P. R. Luyten, J. A. den Hollander, W. Heindel, and W. M. M. J. Bovée, *Magn. Reson. Med.*, 1993, **30**, 231.

45. J. E. van Dijk, A. F. Mehlkopf, and W. M. M. J. Bovée, *NMR Biomed.*, 1992, **5**, 75.

46. D. L. Rothman, K. L. Behar, H. P. Hetherington, J. A. den Hollander, M. R. Bendall, O. A. C. Petroff, and R. G. Shulman, *Proc. Natl. Acad. Sci. USA*, 1985, **82**, 1633.

47. T. Jue, *J. Magn. Reson.*, 1987, **73**, 524.

48. J. Ruiz-Cabello, G. W. Vuister, C. T. W. Moonen, P. van Gelderen, J. S. Cohen, and P. C. M. van Zijl, *J. Magn. Reson.*, 1992, **100**, 282.

49. P. C. M. van Zijl, A. S. Chesnick, D. DesPres, C. T. W. Moonen, J. Ruiz-Cabello, and P. van Gelderen, *Magn. Reson. Med.*, 1993, **30**, 544.

50. G. Bodenhausen and D. J. Rubin, *Chem. Phys. Lett.*, 1980, **69**, 185.

51. E. O. Stejskal and J. E. Tanner, *J. Chem. Phys.*, 1965, **42**, 288.

52. P. C. M. van Zijl and C. T. W. Moonen, *J. Magn. Reson.*, 1990, **87**, 18.

53. J. Kärger, H. Pfeifer, and W. Heink, in 'Advances in Magnetic Resonance', ed. J. S. Waugh, Academic Press, New York, 1988, Vol. 12, p. 1.

54. P. C. M. van Zijl, C. T. W. Moonen, P. Faustino, J. Pekar, O. Kaplan, and J. S. Cohen, *Proc. Natl. Acad. Sci. USA*, 1991, **88**, 3228.

55. D. L. Rabenstein, S. Fan, and T. T. Nakashima, *J. Magn. Reson.*, 1985, **64**, 541.

56. S. D. Wolff and R. Balaban, *Magn. Reson. Med.*, 1989, **10**, 135.

57. R. L. Vold, J. S. Waugh, M. P. Klein, and D. E. Phelps, *J. Chem. Phys.*, 1968, **48**, 3831.

58. A. A. Maudsley, A. Wokaun, and R. R. Ernst, *Chem. Phys. Lett.*, 1978, **55**, 9.

59. A. Bax, P. G. de Jong, A. F. Mehlkopf, and J. Smidt, *Chem. Phys. Lett.*, 1980, **69**, 567.

60. P. Barker and R. Freeman, *J. Magn. Reson.*, 1985, **64**, 334.

61. P. Mansfield, *J. Phys. C*, 1977, **10**, L55.

62. P. Mansfield and B. Chapman, *J. Phys. E*, 1986, **19**, 540.

63. P. B. Roemer, W. A. Edelstein, and J. S. Hickey, *Proc. Vth Ann Mtg. Soc. Magn. Reson. Med.*, Montreal, 1986, p. 1067.

64. C. T. W. Moonen, G. S. Sobering, P. C. M. van Zijl, J. Gillen, M. von Kienlin, and A. Bizzi, *J. Magn. Reson.*, 1992, **98**, 556.

65. P. C. Lauterbur, *Nature (London)*, 1973, **242**, 190.

66. W. P. Aue, *Rev. Magn. Reson. Med.*, 1986, **1**, 21.

67. R. Sauter, S. Müller, and H. Weber, *J. Magn. Reson.*, 1987, **71**, 167.

68. P. R. Luyten, A. J. H. Mariën, B. Sijtsma, and J. A. den Hollander, *J. Magn. Reson.*, 1986, **67**, 148.

69. P. A. Bottomley, *Ann. NY Acad. Sci.*, 1987, **508**, 333.

70. R. J. Ordidge, P. Mansfield, J. A. B. Lohman, and S. B. Prime, *Ann. NY Acad. Sci.*, 1987, **508**, 376.

71. P. C. M. van Zijl, C. T. W. Moonen, J. R. Alger, J. S. Cohen, and A. S. Chesnick, *Magn. Reson. Med.*, 1989, **10**, 256.

72. C. T. W. Moonen, M. von Kienlin, P. C. M. van Zijl, J. Gillen, P. Daly, J. S. Cohen, and G. Wolf, *NMR Biomed.*, 1989, **2**, 201.

73. W. P. Aue, S. Müller, T. A. Cross, and J. Seelig, *J. Magn. Reson.*, 1984, **56**, 350.

74. P. A. Bottomley, T. H. Foster, and R. D. Darrow, *J. Magn. Reson.*, 1984, **59**, 338.

75. R. J. Ordidge, A. Connelly, and J. A. B. Lohman, *J. Magn. Reson.*, 1986, **66**, 283.

76. T. Mareci and H. R. Brooker, *J. Magn. Reson.*, 1985, **57**, 157.

77. A. Connelly, C. Counsell, J. A. B. Lohman, and R. J. Ordidge, *J. Magn. Reson.*, 1988, **78**, 519.

78. R. J. Ordidge, M. R. Bendall, R. E. Gordon, and A. Connelly, in 'Magnetic Resonance in Biology and Medicine', eds. Govil, Khetrapal, and Sran. McGraw-Hill, New Delhi, 1985, p. 387.

79. J. Granot, *J. Magn. Reson.*, 1986, **70**, 488.

80. G. McKinnon, *Proc. 5th Ann. Mtg. Soc. Magn. Reson. Med.*, Montreal, 1986, p. 168.

81. R. Kimmich and D. Hoepfel, *J. Magn. Reson.*, 1987, **72**, 379.

82. J. Frahm, H. Bruhn, M. L. Gyngell, K.-D. Merboldt, W. Hänicke, and R. Sauter, *Magn. Reson. Med.*, 1989, **9**, 79.

83. T. R. Brown, B. M. Kincaid, and K. Ugurbil, *Proc. Natl. Acad. Sci. USA*, 1982, **79**, 3523.

84. A. A. Maudsley, S. K. Hilal, W. H. Perman, and H. E. Simon, *J. Magn. Reson.*, 1983, **51**, 147.

85. P. R. Luyten, A. J. H. Mariën, W. Heindel, P. H. J. van Gerwen, K. Herholz, J. A. den Hollander, G. Friedmann, and W. D. Heiss, *Radiology*, 1990, **176**, 791.

86. J. H. Duijn, G. B. Matson, A. A. Maudsley, and M. W. Weiner, *Magn. Reson. Med.*, 1992, **25**, 107.

87. J. H. Duijn, J. Gillen, G. Sobering, P. C. M. van Zijl and C. T. W. Moonen, *Radiology*, 1993, **188**, 277.

88. J. H. Duijn and C. T. W. Moonen, *Magn. Reson. Med.*, 1993, **30**, 409.

89. D. M. Spielman, J. M. Pauly, A. Macovski, G. Glover, and D. R. Enzmann, *J. Magn. Reson. Imaging*, 1992, **2**, 253.

For list of General Abbreviations see end-papers

90. P. G. Morris, in 'NMR Basic Principles and Progress', eds. P. Diehl, E. Fluck, and R. Kosfeld, Springer-Verlag, Berlin, 1992, Vol. 26, p. 149.

91. J. Pauly, P. Le Roux, D. Nishimura, and A. Macovski, *IEEE Trans. Med. Imaging*, 1991, **10**, 53.

92. D. G. Schapp, H. Merkle, J. M. Ellermann, Y. Ke, and M. Garwood, *Magn. Reson. Med.*, 1993, **30**, 1.

93. R. A. de Graaf, Y. Luo, M. Terpstra, H. Merkle, and M. Garwood, *J. Magn. Reson., Ser. B*, 1995, **109**, 184.

94. R. E. Hurd and B. K. John, *J. Magn. Reson.*, 1991, **91**, 648.

95. M. von Kienlin, C. T. W. Moonen, A. van der Toorn, and P. C. M. van Zijl, *J. Magn. Reson.*, 1991, **93**, 423.

96. J. Cavanagh, A. G. Palmer III, P. E. Wright, and M. J. Rance, *J. Magn. Reson.*, 1991, **429**, 91.

97. J. Baum, R. Tycko, and A. Pines, *J. Chem. Phys.*, 1983, **79**, 4643.

98. M. S. Silver, R. I. Joseph, and D. I. Hoult, *J. Magn. Reson.*, 1994, **59**, 347.

99. P. G. Morris, D. J. O. McIntyre, D. E. Rourke, and J. T. Ngo, *Magn. Reson. Med. Biol.*, 1989, **11**, 167.

100. S. Conolly, D. Nishimura, and A. Macovski, *J. Magn. Reson.*, 1989, **83**, 324.

101. C. J. Hardy, W. A. Edelstein, and D. Vatis, *J. Magn. Reson.*, 1986, **66**, 470.

102. K. Ugurbil, M. Garwood, and M. R. Bendall, *J. Magn. Reson.*, 1986, **72**, 177.

103. M. Garwood and K. Ugurbil, in 'NMR Basic Principles and Progress', eds. P. Diehl, E. Fluck, and R. Kosfeld, Springer-Verlag, Berlin, 1992, Vol. 26, p. 110.

104. B. D. Ross, H. Merkle, K. Hendrich, R. S. Staewen, and M. Garwood, *Magn. Reson. Med.*, 1992, **23**, 96.

105. M. Garwood and Y. Ke, *J. Magn. Reson.*, 1991, **94**, 511.

106. M. Garwood and H. Merkle, *J. Magn. Reson.*, 1990, **94**, 180.

107. M. Shinnar, L. Bolinger, and J. S. Leigh, *Magn. Reson. Med.*, 1989, **12**, 88.

108. M. Goldman, 'Quantum Description of High-Resolution in Liquids', Clarendon Press, Oxford, 1988, Chap. 1.

109. S. Connolly, G. Glover, D. Nishimura, and A. Macovski, *Magn. Reson. Med.*, 1991, **18**, 28.

110. R. de Beer and D. van Ormondt, in 'NMR Basic Principles and Progress', eds. P. Diehl, E. Fluck, and R. Kosfeld, Springer-Verlag, Berlin, 1992, Vol. 26, p. 201.

111. A. A. de Graaf, J. E. van Dijk, and W. M. M. J. Bovée, *Magn. Reson. Med.*, 1989, **13**, 343.

112. J. S. Nelson and T. R. Brown, *J. Magn. Reson.*, 1989, **84**, 95.

113. H. Barkhuijsen, R. de Beer, W. M. M. J. Bovée, and D. van Ormondt, *J. Magn. Reson.*, 1985, **61**, 465.

114. D. S. Stephenson, in 'Progress in Nuclear Magnetic Resonance Spectroscopy', eds. J. W. Emsley, J. Feeney, and L. H. Sutcliffe, Pergamon Press, Oxford, 1988, Vol. 20, p. 515.

115. D. Marion, M. Ikura, and A. Bax, *J. Magn. Reson.*, 1989, **84**, 425.

116. O. Gonen and G. Johnson, *J. Magn. Reson., Ser. B*, 1993, **102**, 98.

117. J. J. H. Ackerman, T. H. Grove, G. G. Wong, D. G. Gadian, and G. K. Radda, *Nature (London)*, 1980, **283**, 167.

118. M. R. Bendall and R. E. Gordon, *J. Magn. Reson.*, 1983, **53**, 365.

119. R. E. Gordon, P. E. Hanley, D. Shaw, D. G. Gadian, G. K. Radda, P. Styles, P. J. Bore, and L. Chan, *Nature (London)*, 1980, **287**, 367.

120. C. J. Hardy, P. A. Bottomley, M. O'Donnell, and P. Roemer, *J. Magn. Reson.*, 1988, **77**, 233.

121. X. Hu, D. N. Levin, P. C. Lauterbur, and T. Spraggins, *Magn. Reson. Med.*, 1988, **8**, 314.

122. M. von Kienlin and R. Mejia, *J. Magn. Reson.*, 1991, **94**, 268.

123. L. D. Hall and T. J. Norwood, *J. Magn. Reson.*, 1986, **67**, 382.

124. S. Posse, B. Schuknecht, M. E. Smith, P. C. M. van Zijl, N. Herscovitch, and C. T. W. Moonen, *J. Comput. Assist. Tomogr.*, 1993, **17**, 1.

Biographical Sketches

Chrit Moonen. *b* 1955. M.S., 1980, Ph.D., 1983, Agricultural University, Wageningen, The Netherlands. Introduced to protein NMR by Kurt Wüthrich, Zürich, 1979. Faculty member, Department of Biochemistry at Wageningen 1984–86; sabbatical with George Radda, Department of Biochemistry, Oxford; one year with Morton Bradbury at University of California at Davis, 1987. Head of the NIH In Vivo NMR Research Center 1987–1996. Director of Research at the Centre National de la Recherche Scientifique, University of Bordeaux 2 1997–present. Approximately 100 publications. Research interests: mapping of brain function, physiological MR, diffusion and perfusion imaging, interventional MRI, MRI guided focused ultrasound.

Peter C. M. van Zijl. *b* 1956. M.S., 1980, Ph.D., 1985, Free University, Amsterdam, The Netherlands. Postdoctoral fellow at Carnegie Mellon University, Department of Chemistry, 1985–87; visiting associate at National Institutes of Health, National Cancer Institute, 1987–90. Research assistant professor at Georgetown University, Department of Pharmacology, 1990–92; Professor at Johns Hopkins Medical School, Department of Radiology, 1992–present. Approx. 75 publications. Research interests: in vivo spectroscopy and imaging of brain function, including diffusion imaging and imaging of metabolite levels and active metabolism of magnetically labelled precursors. Mechanism of ischemia. Design of NMR technology for HR NMR.

Quantitation in In Vivo MRS

Wim M. M. J. Bovée, Ronald de Beer and Dirk van Ormondt

Delft University of Technology, Delft, The Netherlands

Peter R. Luyten

Philips Medical Systems, Best, The Netherlands

and

Jan A. den Hollander

University of Alabama, Birmingham, AL, USA

1 INTRODUCTION

The development of in vivo magnetic resonance spectroscopy (MRS) has provided a noninvasive means of

For References see p. 809

chemically analyzing tissue volumes of a few cc's in the human body.[1–3] Ideally, diagnostic statements in the clinic should be based on numerical criteria derived from the measurement of absolute metabolite concentrations in normal and pathological tissue. Both the MRS measurement technique and the data analysis method may introduce substantial errors in obtaining absolute, or even relative, quantitative MRS data. For the introduction of MRS in the clinic as a standard method of diagnosis and of assessment of response to therapy, the development of a reliable user-unbiased quantitation methodology is therefore essential.

In this article, the quantitation of MRS data, rather than their acquisition, is discussed. The first step is to determine signal intensities. Ideally, the amplitude of an MRS signal of a metabolite, be it FID or echo, is proportional to the concentration of that metabolite. The method of choice for estimating the amplitude is to fit an appropriate model function to the data. This can be executed directly in the time (measurement) domain, or in the frequency domain after Fourier transformation (FT). Problem areas are, among others, low signal-to-noise ratio (S/N), coincidence of spectral frequencies, and magnetic field inhomogeneity. In the frequency domain, an additional concern may be estimation of the baseline. If no appropriate model function is available, one may either decompose the signal into a limited number of known well-behaved functions or apply integration in the frequency domain. Often the quality of quantitation can be increased dramatically by exploiting all available prior knowledge about experiment and signal. In addition, reference should be made to the fundamental limits of precision at the given S/N (Cramér–Rao bounds).

As a second step, the signal intensity ratios obtained may need corrections for the effects of relaxation, scalar coupling, several instrumental and pulse-sequence-dependent signal intensity distortions, and imperfect localization. Calibration of instrumentation and measurement methodology permits suitable correction factors to be derived, and to obtain the prior knowledge mentioned above. Relative metabolite concentrations can be obtained in this way, and in principle also absolute ones, but at the cost of much greater effort.

There are a number of reviews of quantitative NMR.[1–11]

2   DATA ANALYSIS

2.1   Model Functions

The basic signal types encountered in in vivo MRS are FIDs and echoes. Since an echo can be looked upon as two FIDs back to back, most modeling aspects pertain to both types of signals.

A FID consists of a sum of damped sinusoids. Well-known functions for modeling the damping are $\exp(\alpha t)$, $\exp(\beta t^2)$, and $\exp(\alpha t + \beta t^2)$, in which α and β are negative. The frequency-domain counterparts of these functions have Lorentzian, Gaussian, and Voigt lineshapes, respectively. In practice, more complicated damping functions may be required—for instance when modeling an unresolved multiplet by a single sinusoid. In addition, the frequency of a sinusoid may be ill defined. The latter occurs when anisotropic interactions are not properly

averaged out, or when the magnetic field is inhomogeneous.[12] For brevity, we consider only two cases:

1. the exponentially damped sinusoid and its frequency domain counterpart;
2. unknown functions and shapes.

The basic exponential function suitable for describing FIDs as well as echoes is

$$\hat{x}_n = \sum_{k=1}^{K} c_k e^{(\alpha_k^\pm + i\omega_k)t_n + i\varphi_k}, \quad n = 0, 1, \ldots, N-1 \qquad (1)$$

where c_k is the real-valued amplitude, $\alpha_k^\pm$ the damping factor, ω_k the angular frequency, and φ_k the phase of sinusoid k. All φ_k may be equal (an example of prior knowledge). The signal is sampled at times $t = t_n$.

If the signal is a FID, the time origin is set at the actual beginning of the precession. t_0 is the time of the first usable sample, relative to the origin just defined. In principle, t_0 can be inferred from experimental knowledge. In the absence of the latter, t_0 may be estimated along with the other unknown parameters.[10]

If the signal is an echo, the time origin is chosen such that at the echo maximum (t_{em}), $t = t_{em} = 0$. As for the damping factors, we have $\alpha_k^\pm \equiv \alpha_k^+ > 0$ for $t < 0$, whereas $\alpha_k^\pm \equiv \alpha_k^- < 0$ for $t > 0$. Note that $|\alpha_k^+|$ and $|\alpha_k^-|$ need not be equal. $t_{em} - t_0$ may be known from the settings of the spectrometer. Alternatively, $t_{em} - t_0$ may be estimated along with the other unknown model parameters,[13] as mentioned in the previous paragraph.

2.2   Estimation of Model Parameters

Quantitation amounts to estimation of the model parameters appearing in equation (1). Good results are usually obtained by nonlinear least-squares (NLLS) fitting, which includes the variable projection method.[14] If the noise is Gaussian, NLLS is identical to maximum likelihood estimation.[10] An advantage of NLLS is that available prior knowledge about model parameters can be exploited, which yields significant reduction of the errors, especially at low S/N. A disadvantage is that good starting values of the parameters have to be provided in order to ensure correct convergence. Problems related to starting values and convergence can be avoided by resorting to simulated annealing[15] or genetic algorithms,[16] but these techniques require rather more computer time.

An alternative to maximum likelihood estimation is Bayesian estimation (BE),[17] which requires prior probability distributions rather than starting values. One advantage of BE is that unwanted parameters (nuisance parameters) can be 'integrated out', and subsequently ignored. This may simplify the estimation process considerably. In addition, BE was successfully used for rendering maximum entropy (MEM) suitable for quantitation.[18]

Both NLLS and BE can be executed either directly in the time domain, or in the frequency domain following Fourier transformation (FT).[17–24] In the time domain, the sample times t_n need not be uniformly distributed, which may be useful in multidimensional experiments, for example, for time saving.[10] In addition, data truncation at either end does not affect the fit procedure. Frequency domain estimation is equally insensitive to data truncation if one obtains the model function from the discrete FT (DFT), rather than the usual continuous FT, of equation (1).[8,21] When using the continuous FT, the resulting model function does not reflect data truncation, which in turn may necessitate judicious processing of the experimental baseline.[25,26] Finally, we mention that in both domains, one can restrict the fit to only a small subset of the peaks (sinusoids), provided the frequencies in the subset differ sufficiently from those in the remainder.[4,17,27] This property can yield substantial reduction of processing time. A new, very different, method capable of the same is that of *wavelets*.[28]

If the residue of a properly converged fit clearly exhibits features above the noise, the wielded model function is evidently inadequate. In a number of cases, replacement of the Lorentzian model by the Gaussian or Voigt model may significantly improve the fit. However, notorious problem cases are the signals of, for example, water and fat, whose model functions are usually nondescript. The following two paragraphs indicate how one can still estimate the intensity of a nondescript signal, in both domains.

First, we consider the frequency domain, which is traditionally chosen for the task. If the baseline has been correctly estimated and the peak in question does not overlap with other peaks, the area of that peak is proportional to the amplitude of the related FID at the time origin. The FID amplitude, in turn, is proportional to the wanted number of molecules of interest. The gist of the method is that the peak area can be estimated by integration, without knowledge of the peak shape. Statistical and systematic errors attendant to the procedure have been treated by McLeod and Comisarov[29] and Nadjari and Grivet.[30] As with model fitting, one can restrict integration to just the peak(s) of interest. In cases where the model function is known, comparison of model fitting and integration has been done.[31] Simulations showed that statistical errors of integration can be about equal to those of model fitting. Systematic integration errors, on the other hand, can be substantial.[31]

Recently, time domain quantitation of nondescript signals has been shown to be equally feasible.[32] This is based on the fact that bell-shaped peaks of the kind encountered in spectroscopy can be represented in the time domain by a relatively small number of exponentially damped sinusoids. In agreement with intuition, the frequencies of the largest sinusoids are concentrated in the most intense region of the peak. Algorithms based on singular value decomposition (SVD) are available for the task.[33,34] In NMR, such algorithms are known as LPSVD[5-10] (linear prediction SVD), HSVD[5-10] (Hankel SVD), HLSVD[32] (Hankel–Lanczos SVD), HD[35] (Hankel diagonalization), EPLPSVD[36] (enhanced procedure LPSVD), and MV-HTLS[37] (minimum variance Hankel total least squares). The methodology is illustrated in Figure 1. Note that if the signal at hand is exponentially damped, SVD-based methods are viable alternatives to NLLS. An important advantage of these methods is that automation is easy. However, prior knowledge other than the number of sinusoids cannot be accommodated, and the S/N below which performance degradation begins (the threshold S/N) is higher than that of NLLS. However, a substantial reduction of the threshold S/N of SVD-based methods has recently been

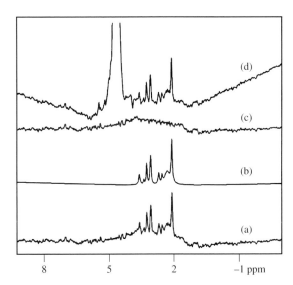

Figure 1 ^{1}H MRS STEAM signals, echo time 20 ms, from the right parietal lobe of a human brain (experiment by P. Gilligan, J. MacEnri and J. T. Ennis, The Institute of Radiological Sciences, Dublin). The signal processing comprised (1) truncation of the data record at the beginning from 1024 to 1022 points and subsequently at the end from 1022 to 512 points; (2) removal of the residual water signal by subtracting exponentially decaying sinusoids, fitted by the HLSVD method;[32] (3) quantitation in the time domain by the variable projection method,[14] thereby multiplying initial data and model function by the same weighting function in order to account for a broad background feature.[27] The figure shows the cosine DFT of (a) the experimental signal after removal of the residual water signal; (b) the fitted model function; (c) the residue (a)−(b); (d) the original experimental signal

achieved.[36,37] In addition, the considerable computational load of SVD is being lowered continually.[32,38]

2.3 Precision of the Model Parameters and Prior Knowledge

The standards against which one measures the precision of parameter estimates are the so-called Cramér–Rao bounds (CRBs).[39,40] These are the smallest possible standard deviations under the given conditions. No estimator is capable of dodging the CRBs. The CRBs are derived from

1. the model function (including the 'true' values of the model parameters);
2. the noise level;
3. the sample times;
4. prior knowledge.

Prior knowledge may pertain to, for example, relations between model parameters of overlapping members of multiplets.[13,41–43] Numerical studies of the CRBs reveal that prior knowledge is very beneficial for the precision of the remaining free model parameters. Other advantages of imposing prior knowledge are stabilization of the convergence process, lowering of the threshold S/N, and reduction of the computation time. Naturally, prior knowledge should be trustworthy.

Table 1 is a summary of data analysis techniques.

For References see p. 809

Table 1 Summary of Data Analysis Methods

	NLLS / BE	SVD-based	Integration
Model function	User-supplied	(see text)	—
Domain	Time/frequency	Time	Frequency
Prior knowledge	Accommodated	—	—
Starting values / distribution	User-supplied	—	—
Sampling interval	Arbitrary (time)	Uniform	Uniform
	Uniform (frequency)		
Truncation	Accommodated (time)	Accommodated	Baseline correction
	Baseline correction (frequency)		

Additional aspects such as robustness, computing efficiency, automation, and threshold S/N are difficult to quantitate, because they depend strongly on the particular implementation of the algorithm

3 METABOLITE CONCENTRATIONS FROM SIGNAL INTENSITIES

3.1 Correction of Signal Intensities

Converting signal intensities to metabolite concentrations may need corrections for the effects of relaxation, scalar coupling, the applied pulse sequence, and instrumental parameters. The correction factors may be different for the separate lines.

3.1.1 Relaxation and Off-Resonance Effects

For repetition times T_r smaller than about four times the longitudinal relaxation time T_1, the signal intensity of a line j in a FID-like experiment is affected by partial saturation, and depends on T_r, T_{1j}, the total dephasing time T_{2j}^*, the flip angle θ_j, and the offset frequency ω_j, in combination with the rf amplitude.[44] Corrections can be made because T_r, T_{1j}, T_{2j}^*, θ_j, and ω_j are known or can be determined. Partial saturation can be avoided, at the cost of the signal-to-noise ratio per unit time, by using large T_r values or small flip angles, if the in vivo experiment allows this. With inhomogeneous rf fields, the flip angle depends on the position relative to the coil. In multi-pulse sequences, there is an effective flip angle depending on all applied pulses, and the signal intensity also has to be corrected for transverse relaxation during the delays. Experimental determination of T_2 values is easy for uncoupled but very difficult for strongly coupled spin systems. Metabolite T_{1j} and T_{2j} values may be age-dependent[45] and disease-dependent,[46] may show regional variations,[47,48] and may change during the development of pathologies. If a reference compound is used to calibrate the instrument (see below), the T_1 and T_2 values of this compound should also be taken into account.

3.1.2 Radiofrequency Field Inhomogeneity

This causes flip angles and signal strengths to vary over the VOI, and degrades slice profiles. Surface coils are notorious in this respect. This spatially varying sensitivity can be corrected.[11,49,50] The effects on the excitation of the spins can be minimized by using adiabatic rf pulses,[51] but remain for the detection. Correction is (especially for surface coils) more essential for magnetic resonance spectroscopic imaging (MRSI) than for single voxel MRS, because the dimensions of the field of view for MRSI are in general larger than those of the VOI for MRS. Methods used to correct effects of B_0 inhomogeneity and eddy currents (see below) can easily be adapted to correct

also for rf field inhomogeneity (see also *Surface Coil NMR: Detection with Inhomogeneous Radiofrequency Field Antennas*).

3.1.3 Static Field Inhomogeneity and Eddy Currents

These effects distort the resonance frequencies, intensities, phases, widths, and shapes of the lines over the VOI, complicating the data analysis and the calculation of spectroscopic images from the line intensities in each MRSI voxel. These distortions can be derived from a reference signal and used for corrections. An example is given in Figure 2. Based on this, several methods have been proposed in the literature[12,52–54] and references cited therein.

3.1.4 Localization Effects

3.1.4.1 Radiofrequency field and B_0 inhomogeneity and eddy currents. For these, see Sections 3.1.2 and 3.1.3.

3.1.4.2 Slice and VOI Profiles. In single voxel MRS, slice and VOI profiles are not perfect, and depend on the rf inhomogeneity and pulse type, and the localization method. The signal from the intended VOI may become increasingly contaminated with signals from outside when the VOI becomes relatively smaller, especially in add–subtraction techniques such as ISIS,[55] where all spins within the sensitive region of the coil are excited. Single shot techniques for localization are therefore to be preferred. Moreover, the exact volume of the VOI is unknown, hampering the calculation of absolute concentrations. This requires exact calibration of the actual volume of the selected regions of interest.[56]

In MRSI signal leaks from each voxel to the other voxels due to a discrete truncated sampling in k-space.[57] This results in severe contamination, especially when the sample contains discontinuities or sharp edges. In this way, very intense signals (e.g., of lipids) may obscure the small metabolite signals by leaking over several voxels. Appropriate weighting in k-space diminishes these effects, but decreases the spatial resolution somewhat.

3.1.4.3 Chemical shift. Slice selection with frequency-selective pulses and gradients shifts the slices of spins with different chemical shifts over a distance proportional to the ratio of their chemical shift and the gradient strength. For protons at 1.5 T, with a relevant chemical shift range of about 5 ppm, and a gradient strength of $10\,mT\,m^{-1}$, this displacement is about 1 mm. This is insignificant with respect to the VOI size of about 1 cm, but at higher field strengths, or for nuclei

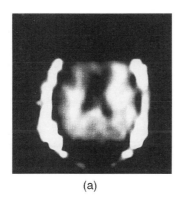
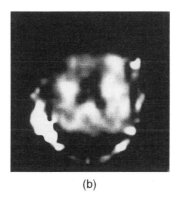

(a)   (b)

Figure 2 NAA spectroscopic image of a human brain, reconstructed from a 3D in vivo ^{1}H MRSI data set (experiment by R. M. N. J. Lamerichs, P. R. Luyten, and H. van Vroonhoven, Philips Medical Systems, Best). Important steps in the signal processing comprised (1) suppression of residual water signals and correction for phase and frequency variations over the pixels by postprocessing techniques; (2) quantitation by frequency domain integration or time domain NLLS fitting, in the latter case combined with exploiting prior knowledge of spectral parameters; (3) reconstruction of spectroscopic images from frequency domain peak areas (a) or time domain amplitudes (b). Note that lipid contamination is significantly reduced by the time-domain quantitation approach

with larger chemical shift ranges (^{31}P and ^{13}C), this is no longer the case.

When the rf pulse bandwidth is not large compared with the chemical shift difference of coupled spins, multiplet components in different parts of the VOI are differently refocused in echo experiments, resulting in considerable signal loss.[58] Consider, for instance, lactate in the 3D PRESS experiment, with an rf pulse bandwidth of 1 kHz, and an echo time of 138 ms; at 1.5 T, 9%, and at 4 T, 24% of the CH_3 signal is lost.

It is obvious that localization errors will be more important when the sample is more heterogeneous. To minimize these errors, phantoms and spectroscopic or imaging techniques can be used to determine the size, shape and position of the selected VOI, and contamination and signal loss upon localization.[56]

3.1.5 Motions

These degrade the localization performance in single voxel single shot MRS; in multishot add–subtract experiments like ISIS[55] this effect is further increased. In MRSI,[57] motions produce large artifacts and contamination in the spectra of each voxel, and ghosting in the spectroscopic images. The effects of motions can be minimized by using single shot experiments, short echo times, and methods that trigger from the motion, as used in MRI.

3.1.6 J Modulation Effects

Transverse multiplet components of scalar-coupled spins dephase under the influence of the coupling. This dephasing is not refocused, partly refocused, or totally refocused in multipulse experiments, depending on pulse distances,[59] the offset frequencies of the spins, and the bandwidth of the refocusing pulses.[58] J modulation effects may therefore lead to complex modulation patterns in multipulse localization experiments; the several spectral components are phase and amplitude-modulated in different ways—the modulation depending on coupling constants, chemical shifts (in case of strong coupling), pulse delays and pulse angles. Moreover, if the spectral components are not all resolved then (partial) cancellation of signals with

different phases may result.[43] The resulting intensity distortions therefore also depend on the B_0 homogeneity; see Figure 3.

The intensity distortions can be determined experimentally on model solutions under well-defined conditions,[43] and used to correct the observed signal strengths, or as prior information in the NLLS fitting techniques to improve the accuracy. The linewidth has to be carefully controlled to keep the (partial) cancellation of signals with phase abnormalities comparable.[12,43]

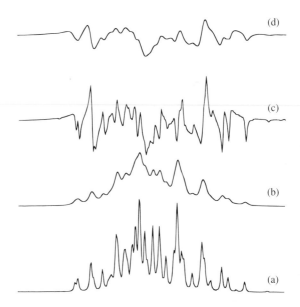

(d)

(c)

(b)

(a)

Figure 3 Real parts of the spectra obtained after FT of the 1.5 T γ and β simulated proton signals of glutamate: (a) and (b) were obtained after a single 90° pulse applied to the spin system in its equilibrium state; (c) and (d) were obtained from the echoes after a PRESS sequence with an echo time of 55 ms. In (a) and (c), the Lorentzian linewidth is 1 Hz; in (b) and (d), it is 3 Hz. The integrals of the spectra (a) and (b) are 16.4 and 16.3 (arbitrary units); the integrals of the absolute value mode signals corresponding to (c) and (d) are 13.6 and 7.8 (arbitrary units). Note the considerable signal loss in (d) as a result of phase anomalies due to J modulation combined with line overlap due to line broadening

For References see p. 809

3.1.7 Frequency-dependent Intensity Distortions by the Applied Pulse Sequence

Well-known examples are several water-suppressing pulse sequences. The effect can either be calculated or determined experimentally, allowing a correction.

3.1.8 Decoupling, Static, and Dynamic Nuclear Overhauser Effects; Exchange

Proton decoupling may result in completely different spectral patterns and intensity ratios in ^{31}P and ^{13}C spectra due to the decoupling and the heteronuclear Overhauser effect. Water suppression by selective inversion of the water resonance, followed by a relaxation delay to null the water signal, may result in distorted lineshapes and metabolite dependent intensity changes due to a dynamic nuclear Overhauser effect and/or exchange of protons. These effects can, in principle, be corrected for by determining them experimentally. See also *Magnetization Transfer and Cross Relaxation in Tissue*; *Proton Decoupling in Whole Body Carbon-13 MRS*; *Proton Decoupling During In Vivo Whole Body Phosphorus MRS*.

3.1.9 Partial NMR Visibility

Metabolites with short T_2 values are visible as broad lines in the spectrum, or are totally invisible. For instance, the lipid signals in the normal brain in ^{1}H MRS are invisible, bone from the skull shows up as a broad peak in ^{31}P MRS, and glutamate is only about 65% NMR-visible in the ^{1}H spectrum of rat brain.[60] Corrections for (partial) NMR visibility are difficult.

3.2 Relative and Absolute Metabolite Concentrations

In this section, it is assumed that the corrections for distortions as described in Section 3.1 have been performed as far as possible. Relative metabolite concentrations are then obtained as signal intensity ratios, and often reported as such. This is a simple procedure, but global concentration changes are obscured. Absolute metabolite concentrations are of more clinical importance when tissue destruction or cellular damage is expected,[11] and allow better comparison of data from different studies. By acquiring, besides the metabolite signals, the signal of an internal or external reference compound with known concentration and volume, the instrument-dependent relation between signal intensity and concentration can be calibrated, and absolute concentrations obtained. Before discussing this further, some general remarks will be made.

1. To compare concentrations, the volumes occupied by the reference compound and the metabolites both have to be taken into account. These volumes can be determined by imaging. MRS localization errors may be different for the two volumes.
2. The detected NMR signal is an average over the VOI; biological tissue is heterogeneous, and the distribution over the VOI may be different for the reference and the several metabolites.
3. Signal intensity corrections as described in the previous section for both the reference and the metabolites have to be performed.

4. Loading modifies the rf field strength and coil sensitivity. By measuring the loaded and unloaded quality of the coil, or by mimicking coil loading by phantoms with conducting solutions,[11,61] a correction is possible.

3.2.1 Internal Reference

The use of an endogenous reference with a known concentration is the simplest way to obtain absolute metabolite concentrations. The PCr/Cr peak, NAA, or tissue water have been used for ^{1}H NMR, ATP, and also tissue water (as a heteronuclear reference) for ^{31}P MRS.[11,61] An internal reference has the advantages that the same experimental conditions and volume hold for reference and metabolites, making corrections for localization, rf inhomogeneity, and loading superfluous. A heteronuclear reference needs correction for the different resonance frequencies.[11] Disadvantages are that concentration and relaxation times of the reference may change with pathology and physiology. The reference peak may also contain signals from other compounds, an example being a contribution of *N*-acetylaspartylglutamate to the NAA peak.[3] The spatial distributions of reference and metabolites can be different.

3.2.2 External Reference

Several procedures have been described.[3,11] One possibility is to acquire together with the in vivo signal, a signal from a reference with known concentration and volume in a capillary close to the coil.[62] In a separate experiment, the patient is replaced by a phantom containing a known metabolite concentration. The same VOI is localized as in the in vivo experiment, acquiring signal from capillary and phantom. Relating the four signals gives the absolute concentrations. Alternative methods and corrections for partial volume effects when using water as reference can be found in the literature.[63–65]

3.2.3 Some Absolute Concentrations

The above considerations might give the impression that the determination of absolute concentrations by MRS is hardly possible. It is possible within the limitations mentioned, but it requires a lot of effort. In several studies, absolute concentrations have been reported,[3,11] and satisfactory agreement with other analysis methods was obtained. Some literature values of metabolite concentrations for normal adult human brain obtained from ^{1}H MRS are[45] (in mmol kg^{-1}) NAA 8.9 PCr + Cr 7.5, and Cho 1.3. For ^{31}P, adult human brain concentrations found were[11] (in mmol kg^{-1}) PME $\approx$ 3.5, Pi $\approx$ 2.0, PDE $\approx$ 11.0, PCR $\approx$ 4.5, and NTP (nucleotide triphosphates) $\approx$ 3.0; the corresponding values for human skeletal muscle are 2.2, 5.2, 4.2, 34, and 6.8 mmol kg^{-1}. For more data, see the review papers.[3,11,45]

4   RELATED ARTICLES

Chemical Shift Imaging; Proton Decoupling in Whole Body Carbon-13 MRS; Single Voxel Localized Proton NMR Spectroscopy of Human Brain In Vivo; Single Voxel Whole Body Phosphorus MRS; Surface Coil NMR: Detection with Inhomogeneous Radiofrequency Field Antennas.

5 REFERENCES

1. P. Diehl, E. Fluck, H. Günther, R. Kosfeld, and J. Seelig, eds. 'NMR Basic Principles and Progress' Springer-Verlag, Berlin, 1992, Vols. 26–28.
2. J. D. de Certaines, W. M. M. J. Bovée, and F. Podo eds., 'Magnetic Resonance Spectroscopy in Biology and Medicine', Pergamon Press, Oxford, 1992.
3. F. A. Howe, R. J. Maxwell, D. E. Saunders, M. M. Brown, and J. R. Griffiths, *Magn. Reson. Quart.*, 1993, **9**, 31.
4. J. C. Lindon and A. G. Ferrige, in 'Progress in Nuclear Magnetic Resonance Spectroscopy', eds. J. W. Emsley, J. Feeney, and L. H. Sutcliffe, Pergamon Press, Oxford, 1980, Vol. 14, p. 27.
5. J. C. Haselgrove, V. H. Subramanian, R. Christen, and J. S. Leigh, *Rev. Magn. Reson. Med.* 1987, **2**, 167.
6. D. S. Stephenson, 'Progress in Nuclear Magnetic Resonance Spectroscopy', 1988, Vol. 20, p. 515.
7. J. C. Hoch, *Meth. Enzymol.*, 1989, **176**, 216.
8. H. Gesmar, J. J. Led, and F. Abildgaard, 'Progress in Nuclear Magnetic Resonance Spectroscopy', eds. J. W. Emsley, J. Feeney, and L. H. Sutcliffe, Pergamon Press, Oxford, 1990, Vol. 22, p. 255.
9. R. E. Hoffman and G. C. Levy, 'Progress in Nuclear Magnetic Resonance Spectroscopy', eds. J. W. Emsley, J. Feeney, and L. H. Sutcliffe, Pergamon Press, Oxford, 1991, Vol. 23, p. 211.
10. R. de Beer and D. van Ormondt, in 'NMR Basic Principles and Progress', eds. P. Diehl, E. Fluck, A. Günther, R. Kosfeld, and J. Seelig, Springer-Verlag, Berlin, 1992, Vol. 26, p. 201.
11. E. B. Cady, in 'NMR Basic Principles and Progress', eds. P. Diehl, E. Fluck, A. Günther, R. Kosfeld, and J. Seeling, Springer-Verlag, Berlin, 1992, Vol. 26, p. 249.
12. A. A De Graaf, J. E. van Dijk, and W. M. M. J. Bovée, *Magn. Reson. Med.*, 1989, **13**, 343.
13. J. E. van Dijk, A. F. Mehlkopf, D. van Ormondt, and W. M. M. J. Bovée, *Magn. Reson. Med.*, 1992, **27**, 76.
14. G. H. Golub and V. Pereyra, *SIAM J. Numer. Anal.*, 1973, **10**, 413.
15. F. S. Digennaro, and D. Cowburn, *J. Magn. Reson.*, 1992, **96**, 582.
16. C. B. Lucasius and G. Kateman, *Chemometrics and Intelligent Laboratory Systems*, 1993, **19**, 1.
17. G. L. Bretthorst, *J. Magn. Reson.*, 1992, **98**, 501.
18. J. Skilling, *American Laboratory*, 1992, October, 32J.
19. J. S. Nelson and T. R. Brown, *J. Magn. Reson.*, 1989, **84**, 95.
20. F. Montigny, J. Brondeau, and D. Canet, *Chem. Phys. Lett.*, 1990, **170**, 175.
21. M. Joliot, B. M. Mazoyer, and R. H. Huesman, *Magn. Reson. Med.*, 1991, **18**, 358.
22. C. Burger, G. McKinnon, R. Büchli, and P. Boesiger, *Magn. Reson. Med.*, 1991, **21**, 216.
23. E. K.-Y Ho, R. E. Snyder, and P. S. Allen, *J. Magn. Reson.*, 1992, **99**, 590.
24. A. van den Boogaart, M. Ala-Korpela, J. Jokisaari, and J. R. Griffiths, *Magn. Reson. Med.*, 1994, **31**, 347.
25. A. R. Mazzeo and G. C. Levy, *Magn. Reson. Med.*, 1991, **17**, 483.
26. S. A. Rashid, W. R. Adam, D. J. Craik, B. P. Shenan, and R. M. Wellard, *Magn. Reson. Med.*, 1991, **17**, 213.
27. A. Knijn, R. de Beer, and D. van Ormondt, *J. Magn. Reson.*, 1992, **97**, 444.
28. N. Delprat, B. Escudié, P. Guillemin, R. Kronland-Martinet, P. Tchamitchian, and B. Torrésani, *IEEE Trans. Information Theory*, 1991, **38**, 644.
29. K. McLeod and M. B. Comisarov, *J. Magn. Reson.*, 1989, **84**, 490.
30. R. Nadjari and J.-P. H. Grivet, *J. Magn. Reson.*, 1991, **91**, 353.
31. R. de Beer, P. Bachert-Baumann, W. M. M. J. Bovée, E. Cady, J. Chambron, R. Domisse, C. J. A. Echteld, R. Mathur de Vré, and S. Williams, *Magn. Reson. Imaging*, 1995, **13**, 169.

32. W. W. F. Pijnappel, A. van den Boogaart, R. de Beer, and D. van Ormondt, *J. Magn. Reson.*, 1992, **97**, 122.
33. D. V. Bhaskar Rao and K. S. Arun, *Proc. IEEE*, 1992, **80**, 283.
34. A. J. van der Veen, E. F. Deprettere, and A. L. Swindlehurst, *Proc. IEEE*, 1993, **81**, 1277.
35. P. Mutzenhardt, F. Montigny, and D. Canet, *J. Magn. Reson., Ser A*, 1993, **101**, 202.
36. A. Diop, A. Briguet, and D. Graveron-Demilly, *Magn. Reson. Med.*, 1992, **27**, 318.
37. H. Chen, S. Van Huffel, C. Decanniere, and P. Van Hecke, *J. Magn. Reson., Ser. A.*, 1994, **109**, 46.
38. H. Park, S. van Huffel, and L. Eldén, in 'Proceeding of International Conference on Acoustics, Speech, and Signal Processing, Adelaide, 1994', Vol. 4, pp. 25–28.
39. J. P. Norton, 'An Introduction to Identification', Academic Press, London, 1986.
40. A. van den Bos, in 'Handbook of Measurement Science', John Wiley & Sons, Chichester, 1982, Vol. 1, p. 331.
41. A. de Roos, J. Doornbos, P. R. Luyten, L. J. M. P. Oosterwaal, E. E. van der Wall, and J. A. den Hollander, *J. Magn. Reson. Imaging*, 1992, **2**, 711.
42. J. W. C. van der Veen, R. de Beer, P. R. Luyten, and D. van Ormondt, *Magn. Reson. Med.*, 1988, **6**, 92.
43. A. A. de Graaf and W. M. M. J. Bovée, *Magn. Reson. Med.*, 1990, **15**, 305.
44. R. R. Ernst, in 'Advances in Magnetic Resonance', ed. J. S. Waugh, Academic Press, New York, 1966, Vol. 2, p. 1.
45. R. Kreis, T. Ernst, and B. D. Ross, *Magn. Reson. Med.*, 1993, **30**, 424.
46. D. Sappey-Marinier, G. Calabrese, H. P. Hetherington, S. N. G. Fisher, R. Deicken, C. van Dyke, G. Fein, and M. W. Weiner, *Magn. Reson. Med.*, 1992, **26**, 313.
47. J. Frahm, H. Bruhn, M. L. Gyngell, K. D. Merboldt, W. Hänicke, and R. Sauter, *Magn. Reson. Med.*, 1989, **11**, 47.
48. P. A. Narayana, D. Johnston, and D. P. Flamig, *Magn. Reson. Imaging*, 1991, **9**, 303.
49. G. B. Matson, D. J. Meyerhoff. T. J. Lawry, R. S. Lara, J. Duijn, R. F. Deicken, and M. W. Weiner, *NMR Biomed.*, 1993, **6**, 215.
50. K. Roth, K. Hubesch, D. J. Meyerhoff, S. Naruse, J. R. Gober, T. J. Lawry, M. D. Boska, G. B. Matson, and M. W. Weiner, *J. Magn. Reson.*, 1989, **81**, 299.
51. M. Garwood and K. Ugurbil, in 'NMR Basic Principles and Progress', eds. P. Diehl, E. Fluck, H. Günther, R. Kosfeld, and J. Seelig, Springer-Verlag, Berlin, 1992, Vol. 26, p. 109.
52. G. Johnson, K. J. Jung, E. X. Hu, and S. K. Hilal, *Magn. Reson. Med.*, 1993, **30**, 255.
53. J. R. Roebuck, D. O. Hearshen, M. O'Donnell, and T. Raidy, *Magn. Reson. Med.*, 1993, **30**, 277.
54. P. Webb, D. Spielman, and A. Macovski, *Magn. Reson. Med.*, 1992, **23**, 1.
55. R. J. Ordidge, A. Connelly, and J. A. B. Lohman, *J. Magn. Reson.*, 1986, **66**, 283.
56. W. M. M. J. Bovée, in 'Magnetic Resonance Spectroscopy in Biology and Medicine', eds. J. D. de Certaines, W. M. M. J. Bovée, and F. Foder, Pergamon Press, Oxford, 1992, p. 209.
57. M. Decorps and D. Bourgeois, in 'NMR Basic Principles and Progress', eds. P. Diehl, E. Fluck, H. Günther, R. Kosfeld, and J. Seelig, Springer-Verlag, Berlin, 1992, Vol. 27, p. 119.
58. J. Slotboom, A. F. Mehlkopf, and W. M. M. J. Bovée, *J. Magn. Reson., Ser A*, 1994, **108**, 38.
59. A. Abragam, 'The Principles of Nuclear Magnetism', Clarendon Press, Oxford, 1961, pp. 480–510.
60. A. A. de Graaf, N. E. P. Deutz, D. K. Bosman, R. A. F. M. Chamuleau, J. G. De Haan, and W. M. M. J. Bovée, *NMR Biomed.*, 1991, **4**, 31.
61. R. Büchli and P. Boesiger, *Magn. Reson. Med.*, 1993, **30**, 552.

For References see p. 809

62. P. R. Luyten, J. P. Groen, J. W. Vermeulen, and J. A. den Hollander, *Magn. Reson. Med.*, 1989, **11**, 21.
63. T. Ernst, R. Kreis, and B. D. Ross, *J. Magn. Reson., Ser. B*, 1993, **102**, 1.
64. R. Kreis, T. Ernst, and B. D. Ross, *J. Magn. Reson., Ser. B*, 1993, **102**, 9.
65. P. Christiansen, O. Henriksen, M. Stubgaard, P. Gideon, and H. B. W. Larsson, *Magn. Reson. Imaging*, 1993, **11**, 107.

Biographical Sketches

Wim M. M. J. Bovée. *b* 1938. M.Sc., 1969, chemistry, Ph.D., 1975, physics, Delft University. Billiton Research Laboratories, Arnhem, 1969–70. Delft University 1970–present. Research specialties: MR relaxation, in vivo MRS and MRI.

Ronald de Beer. *b* 1944. M.Sc., 1967, Ph.D., 1971, physics, Delft University. Delft University 1967–present. Research specialty: signal processing for MR.

Dirk van Ormondt. *b* 1936. M.Sc., 1959, Ph.D., 1968, physics, Delft University, 1969–70, postdoc., University of Calgary with H. A. Buckmaster, 1970–71, postdoc., Clarendon Laboratory, Oxford with J. M. Baker. Delft University 1971–present. Research specialty: signal processing for MR.

Peter R. Luyten. *b* 1954. M.Sc., 1980, Ph.D., 1984, Free University, Amsterdam. Philips Medical Systems, Best, 1984–present. Research specialty: in vivo MRS.

Jan A. den Hollander. *b* 1947. M.Sc., Ph.D., physical chemistry, University of Leiden. Exchange visitor at the Biophysics Research Department, Bell Laboratories, Murray Hill, New Jersey, 1977–79. Research associate, Department of Molecular Biophysics and Biochemistry, Yale University, New Haven, Connecticut, 1979–83. Staff scientist, Max-Planck Institut für Systemphysiologie, Dortmund, Germany, 1983–84. Philips Medical Systems, Best, The Netherlands, 1984–1991. Professor of Medicine, University of Alabama at Birmingham, 1991–present.

Sodium-23 Magnetic Resonance of Human Subjects

Peter M. Joseph

Hospital of the University of Pennsylvania, Philadelphia, PA, USA

1   INTRODUCTION

Sodium-23 is a nucleus with spin quantum number $\frac{3}{2}$ whose gyromagnetic ratio, γ, corresponds to a frequency of 11.262 MHz tesla^{-1}, which is approximately one-quarter that of pro-

tons. It is widely but inhomogeneously distributed in body tissues. It is present in highest concentrations in extracellular water in a concentration of about 150 mM. Hence in vivo sodium NMR signals are less than one-thousandth those of hydrogen so that sodium MR images can never hope to compete with proton images in quality. Only little preliminary research effort has been directed toward the clinical utility of sodium MRI.

However, Na MRI may be able to contribute unique medical/physiological information that is not available in proton MRI. Intracellular Na is typically at the level of about 10–20 mM in most healthy cells. The enormous difference between intracellular (IC) and extracellular (EC) concentrations is maintained by the Na–K pump which actively transports the two kinds of ions across the cell membrane. Certain pathologies alter the function of that pump resulting in a significant rise in IC Na. There is evidence from in vitro experiments that this occurs in certain disease states, especially ischemia, infarction, and cancer. Thus one aim of Na MRI is to demonstrate a rise in the IC sodium concentrations. While there is still debate as to whether that goal has been achieved with in vivo MRI, it is one of the factors driving research interest in this field. There are various subtle aspects of the Na NMR signal that are thought to differ in the IC and EC environments, and much research has been directed toward clarifying this question.

2   BIOPHYSICS OF ^{23}Na NMR

The major difference between Na and ^{1}H NMR is that Na has spin quantum number $I = \frac{3}{2}$. This means there will be four energy levels, corresponding to the quantum numbers $m = I_z = -\frac{3}{2}, -\frac{1}{2}, \frac{1}{2},$ and $\frac{3}{2}$. Since normal NMR excitation and radiation permits only transitions whose m values differ by 1, there will in general be three possible Larmor frequencies; these are classified as 'inner', meaning $\frac{1}{2}$ to $-\frac{1}{2}$, and 'outer', meaning either $\frac{3}{2}$ to $\frac{1}{2}$ or $-\frac{1}{2}$ to $-\frac{3}{2}$. If there are no other interactions in the system, the energy levels are equally spaced and the three Larmor frequencies are equal.

However, the Na nucleus also has an electric quadrupole moment, denoted by eQ.[1] (e = electron charge.) This implies that the energy of the nucleus will be influenced by the presence of an electric field gradient, which is described by a second-rank tensor of strength eq. In this case there is another energy term in the Hamiltonian given by

$$\hat{\mathcal{H}}_Q = \frac{e^2 qQ[3m^2 - I(I+1)][3\cos^2(\theta) - 1]}{8I(2I - 1)} \quad (1)$$

where θ is the angle between the major axis of the quadrupolar tensor and the z axis. From this Hamiltonian comes a quadrupolar energy E_Q and frequency $\omega_Q = 2\pi E_Q/h$ depending on the quantum state.

Equation (1) has important consequences for the energy levels of the nuclei, the relaxation rates, and signal strength. First, we see that $\hat{\mathcal{H}}_Q$ depends quadratically on m; this changes the energy level differences of the outer levels but not the inner. Hence the quadrupolar interaction does not change the Larmor frequency corresponding to the inner transition. Both the magnitude and the sign of the energy level shift of the external transitions depend on the angle θ. If the local environ-

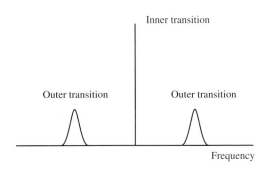

Figure 1 Schematic diagram of the spectrum from ^{23}Na in an ordered environment with electric quadrupolar interaction in a first-order approximation. The width of the outer transition lines depends on the strength and correlation time of the quadrupolar interaction. The inner transition carries 40% of the total signal

ment is isotropic, then averaging over angles gives zero average energy displacement. However, in situations of exceptional order (such as liquid crystals) the orientations are not random and a net shifting of the energy levels can occur. The resulting spectrum will show three lines as illustrated in Figure 1. Note that whereas the central line, corresponding to the inner transition, is narrow, the external spectral lines may be broadened by an amount dependent on the details of the microscopic electrostatic environment.

In this situation, quantum theory predicts that only 40% of the total radiated energy will lie in the unshifted inner line, while the remaining 60% may be distributed over the broad outer lines.[2] Depending on the experimental conditions, it is possible that the outer transitions will be so broadened as to be virtually undetectable with most NMR spectrometers. This leads to the widely discussed phenomenon of 'NMR invisibility' of sodium; i.e. several experiments have suggested that the observed Na NMR signal strength is less than what should be observed based on measured Na concentration levels. For example, Shporer and Civan proposed that this would explain previously reported discrepancies between the strength of the Na NMR signal observed and that expected from tissue measurements.[3] In particular, they demonstrated this effect for the relatively simple system of sodium linoleate in water. Joseph and Summers[4] noted that if the loss of Na NMR visibility were in fact due to a nonoverlapping displacement of the energy levels, then quantum theory predicts that the rate of nutation of the inner (unshifted) spectral component is *twice* that predicted by the usual classical vector model of the NMR excitation. In other words, if an NMR spectrometer is adjusted on a homogeneous phantom (such as aqueous NaCl) so that a particular radiofrequency pulse gives 90° of spin flip, then the same pulse applied to the split Na system will rotate the unsplit component by 180°. They demonstrated the vanishing of the sodium linoleate signal under conditions in which a NaCl phantom gave optimal signal.[4] This suggests that this 'flip angle effect' could be used to test for the presence of quadrupolar splitting in vivo without the need actually to detect the invisible Na spectral lines. To date, however, no one has demonstrated this effect using realistic in vivo biological material.

The more common interpretation of the quadrupolar interaction is that the orientation angle is isotropically distributed

and is rapidly changing with a correlation time, τ_c, on the order of microseconds or less. In this situation the effect of $\hat{\mathcal{H}}_Q$ is not to induce energy level splitting but to broaden the lines, i.e., to induce relaxation. The physics of this situation is rather complex.[2,5,6] The result depends both on the whether the product $\omega_0 \tau_c$ is large or small compared with unity as well as the product $\omega_Q \tau_c$. If $\omega_0 \tau_c \ll 1$, the signal will show only one spectral line. If $\omega_0 \tau_c > 1$ (but with $\omega_Q \tau_c \ll 1$) this line may have two distinct decay rates (i.e., two rates for T_1 and two for T_2). The inner and outer transitions give the same spectral frequency but differ in their relaxation rates. The outer transitions produce 60% of the signal and the inner 40%. This phenomenon of biexponential relaxation (BR) is widely observed in nature. For example, most biological tissue shows a short T_2 on the order of a few milliseconds and a longer one on the order of 20–30 ms. These results are interpreted in terms of a correlation time on the order of microseconds. In contrast, the relaxation times of NaCl–water are usually 50 ms or less for both T_1 and T_2 (at 2 tesla); this is due to a very short τ_c on the order of 10^{-11} s.

The phenomenon of BR is related to the question of Na visibility, i.e., if $T_{2f} \ll TE$ then the fast component will have died out before the signal is detected in a spin echo experiment. Obviously, this definition of 'invisibility' is highly dependent on the NMR equipment, and especially the shortest TE used. The use of a pulse spectrometer with very short TE is equivalent to being able to detect a very broad spectral component with a CW spectrometer. Most recent experiments have reported high visibility in biological materials. Joseph and Summers[4] found essentially 100% visibility in excised cat brain and porcine skeletal muscle. Kohler et al.[7] found only 80% visibility for Na in the vitreous humor of enucleated eyes, but it increased to 100% after digestion with collagenase.

The actual environment of biological Na ions is considerably more complex than that of a simple solution. Even in the EC space there is always a multitude of proteins, fatty acids, sugars, and other hydrophilic molecules with which the Na ions can interact. One expects short-term 'bonding' and exchange of the Na ions among the various bound and free pools. Such exchange processes are known also to influence the relaxation rates.[6,8] It is often assumed that the degree of IC Na bonding is much greater than the EC, so that IC Na will have a shorter T_2 than EC. This conclusion is subject to considerable doubt however, since even homogeneous solutions can produce BR very similar to that seen in vivo. For example, the relatively simple system consisting of NaCl ions in 4% agarose gel shows T_2 values of 5.5 and 23 ms.[4]

Since one motivation for doing Na MRI is the hope of imaging IC versus EC Na, we will review what is currently known about the scientific basis for this. All of this work relies on the use of a hyperfine shift reagent (SR),[9] such as dysprosium tripolyphosphate (Dy-TPP), which shifts the Larmor frequency of the EC Na only; this means that IC and EC Na have separate spectral lines. Such experiments show that IC Na often displays BR. Shinar and Navon[10] observed BR in red blood cells (RBC) with T_2 values of about 6.6 and 23.6 ms. However, the precision with which one can measure two T_2 components is limited and somewhat dependent on the assumptions made in modeling the decay. For example, Shinar and Navon *assumed* a 40/60 distribution between the slow and fast components (T_{2s}/T_{2f}) as predicted by theory for a single population of spins

For References see p. 815

with BR. From their data one can probably not rule out a somewhat shorter T_{2f} with a different short/fast distribution ratio. Later work by Shinar and Navon[11] in the nuclei of chicken RBCs showed T_{2f} values in the 2–8 ms range, where the decay rate had a strong dependence on the IC concentration of DNA.

The BR of the IC Na is only one-half of the question, however. One wants to know if the BR phenomenon is found *only* in the IC Na. In 1986, Shinar and Navon[12] found that none of blood plasma, serum nor various solutions of proteins in water gave BR. This would suggest that the BR phenomenon is limited to IC Na. However, from a detailed analysis of the T_1 and T_2 decay rates they concluded that there was a significant contribution to quadrupolar relaxation from a pool of bound Na even in this noncellular environment. That is, one assumed that the physics of BR was operative but the effect was too weak to be resolvable into two distinct T_2 values. This experiment was done without the use of shift reagents.

Since blood is a relatively small fraction of the fluid in most tissues, one wonders whether the results on blood plasma can be applied to interstitial Na in living tissues. A very careful study of this question was conducted by Foy and Burstein[13] using Na spectroscopy of perfused hearts in both frog and rat models. They eliminated the contribution of blood by a selective inversion–recovery technique. In part of their experiment they used a shift reagent to distinguish IC from EC Na, with the result that 85% of the signal was EC in this model. They analyzed the transverse decay of the parenchymal signal (i.e., both IC and EC but excluding the blood) without SR and found BR with $T_{2s} \approx 25$ ms and $T_{2f} \approx 2$ ms in both models. Since the fast component was about 40% of the total signal it could not have come solely from the IC compartment. This is clearly a counter example to the hypothesis that short T_2 components are found *only* in the IC compartment of living tissues. However, it is still possible that the *average* IC T_2 could be less than that in the EC space.

Pekar et al. proposed using double quantum (DQ) filtration to distinguish IC from EC sodium.[14] The main point is that DQ, and in general multiple quantum (MQ), signals can emerge only if the physical parameters (strength of the quadrupolar constant and correlation time) are such that BR occurs. Pekar et al. used a SR and found the DQ sodium signal to be limited to the IC space. This encouraged the hope that IC could be distinguished from EC solely by the physical characteristics of the NMR signal. However, subsequent work by Jelicks and Gupta[15] demonstrated that the presence of the SR in the EC would itself quench an MQ NMR signal. This means that the absence of EC MQ signal as observed by Pekar et al. did not contradict the finding of Foy and Burstein that BR does occur in the EC space. However, by detecting an MQ signal in the presence of SR it does provide an alternative method for obtaining a pure IC signal.[16,17]

3 IMAGING TECHNIQUES

The principles of imaging sodium do not differ from those used to image protons; basically, in addition to exciting the Na spins and reading out the signal, some means must be provided for spatial encoding.[18] The main differences are due to the characteristics of the Na signal already discussed, namely, the

low S/N and the short values of T_1 and T_2. Short T_1, on the order of 50 ms in tissue, is an advantage since it implies rapid recovery of longitudinal magnetization after excitation so that short TR values can be used with little loss of signal. The low S/N forces one to employ techniques with relatively poor spatial resolution and long imaging times; typical values are 3 × 3 mm in-plane resolution and 10 mm slice thickness. Imaging times are usually a large fraction of 1 h.

The most important aspect of Na imaging is the need to achieve short echo times. One wants quantitation of the short versus long T_2 components, and the goal of imaging is to demonstrate these differences anatomically. A common technique is to use gradient echo (GE), which is equivalent to imaging the free induction decay (FID). The fact that GE is more sensitive than spin echo (SE) to background field gradients is rarely of importance for sodium, partly because the low γ implies less sensitivity of phase shift to the background fields and partly because the TE values achieved are usually less than 3 ms.

There are several ways in which spatial encoding can be applied. The single slice two-dimensional Fourier transform (2DFT) technique widely used in proton images can be used. More commonly, however, for multislice imaging, three-dimensional Fourier transform (3DFT) is used (Figure 2). The initial excitation may be a hard, nonselective pulse or a 'slab' which consists of a set of slices to be imaged. The slice definition is obtained by phase encoding the z axis. To reduce aliasing one can saturate spins outside of the desired region.[19] As shown in Figure 2, it is possible to obtain both a FID and SE signals. Some workers have employed a prephasing G_x pulse during

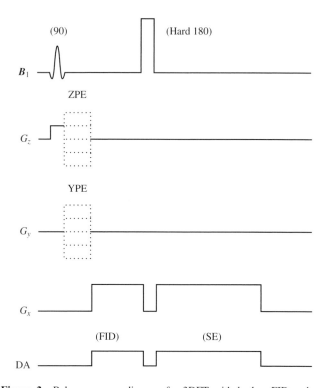

Figure 2 Pulse sequence diagram for 3DFT with both a FID and a spin echo (SE) detected. The 90° pulse is meant to excite a slab of slices in the z direction. The dotted lines on the G_z and G_y pulses indicate phase encoding which varies from one excitation to another (ZPE, z phase encoding; YPE, y phase encoding)

the phase-encoding interval so that the $k_x = 0$ echo is delayed; this is called a GE. The advantage of the GE is that both positive and negative k_x values are obtained in the same data acquisition (DA), which leads to fewer imaging artifacts due to background gradients, eddy currents, etc. However, the resulting delay in the minimum echo time is disadvantageous for detecting the short T_2 component.

Both of the FT techniques require the use of phase-encoding gradient pulses applied after excitation and before signal read out. This is a major disadvantage because they delay the time at which signal acquisition can begin and will therefore severely attenuate the short T_2 component. A technique that avoids that delay is projection–reconstruction imaging (PRI). PRI works by reconstructing an object from measurements of its projections along many lines or planes over a wide range of angles. The two dimensional (2D) form of PRI is the basis of all clinical X-ray computerized tomography (CT) scanners, and was also the first technique used for MRI.[20] Shepp[21] showed how the 2D PRI method could be generalized to three dimensions. In this method, whose pulse sequence is illustrated in Figure 3, each excitation is completely nonselective and is followed immediately by readout of signal in the presence of all three gradients. The signal $S(t)$ will be the Fourier transform of the spin density along the direction defined by the gradient vector, g:

$$S(t) = \iiint M(x,\,y,\,z) \exp(\mathrm{i}\gamma\,\boldsymbol{g}\cdot\boldsymbol{r}t)\,\mathrm{d}x\,\mathrm{d}y\,\mathrm{d}z \qquad (2)$$

where $\boldsymbol{r} = (x,\,y,\,z)$ is the position vector and $M(x,\,y,\,z)$ is the spin density at point $\boldsymbol{r}$, weighted by the relevant relaxation factor. If we can neglect T_2 decay during the readout period,

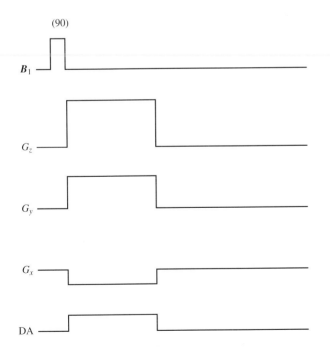

Figure 3 Pulse sequence diagram for 3D PRI. The 90° pulse is hard. The relative magnitudes of the G_x, G_y, and G_z pulses are varied from pulse to pulse such that the gradient vector sweeps over a spherical surface in $\boldsymbol{k}$-space

Fourier transform of the signal yields a frequency distribution $S_\mathrm{f}(\omega)$ given by

$$S_\mathrm{f}(\omega) = \int S(t)\exp(-\mathrm{i}*\omega t)\,\mathrm{d}t = \iiint M(x,\,y,\,z)\,\delta(\omega - \gamma\boldsymbol{g}\cdot\boldsymbol{r})\,\mathrm{d}x\,\mathrm{d}y\,\mathrm{d}z$$

$$(3)$$

where $\delta()$ is the Dirac delta function. Equation (3) indicates that the signal corresponding to any given frequency ω comes from a *plane* perpendicular to the $\boldsymbol{g}$ vector and displaced by distance $\omega/(\gamma g)$ from the origin. To reconstruct the density $M(x,\,y,\,z)$ it is necessary to collect a set of signals in which the gradient vector ranges, more or less uniformly, over the full range of the 4π solid angle of a sphere.[22] In the language of $\boldsymbol{k}$-space, each gradient vector defines a vector in three-dimensional (3D) $\boldsymbol{k}$-space, which varies with time t as

$$\boldsymbol{k}(t) = \int \gamma\boldsymbol{g}\,\mathrm{d}t = \gamma t\boldsymbol{g} \qquad (\text{for } \boldsymbol{g} = \text{const}) \qquad (4)$$

Hence, the 3D PRI method essentially scans $\boldsymbol{k}$-space along a series of spokes starting from the origin and extending out to some maximum value. To reconstruct the function $M(x,\,y,\,z)$ the data in $\boldsymbol{k}$-space are first multiplied by the factor k^2 then Fourier transformed back to the image domain to yield a set of modified planar data.[23] These planar ('filtered') data are then directly projected onto the 3D image matrix. The method is easily generalized to include multiple spin echoes.[24]

The major advantage of 3D PRI for Na MRI is that data acquisition can begin immediately after excitation, without the need to wait for gradient pulses to play out. This is especially advantageous when eddy currents and gradient amplifier limitations add ≈ 1 ms to the nominal time duration of gradient pulses. The method is intrinsically three dimensional, so one is committed to a rather large number of pulses to sample $\boldsymbol{k}$-space adequately. However, like 3DFT, the method is very efficient for signal to noise ratio (S/N) because each excitation includes the entire volume imaged. However, the method is sensitive to frequency offset and $\boldsymbol{B}_0$ inhomogeneity errors, but for Na MRI these are rarely a significant problem.

This method has no dephasing gradient prior to signal readout; this means that on any excitation only one-half of the k region is covered; i.e. $k > 0$ on one excitation and $k < 0$ on another. This technique is called 'half echo', and results in the earliest possible time for the $k = 0$ part of phase space. However, if $T_{2\mathrm{f}}$ is much less than the DA time window, high k values will be attenuated and the image will suffer loss of spatial resolution. This is a consequence of imaging objects with very short T_2 using frequency encoding.

An interesting variant of this technique combines Fourier encoding in the slice direction with PRI in the transverse plane; a Hahn SE time of 3.6 ms has been obtained on a clinical MRI machine.[25]

Another approach to the problem of achieving rapid data acquisition after excitation is to use gradients in the rf $\boldsymbol{B}_1$ field. This method has been applied to Na MRI, but the image quality obtained was less than that obtained by workers using the more conventional $\boldsymbol{B}_0$ gradient methods.[26]

When shift reagents are used, it is necessary somehow to image separately the two spectral lines produced. This can be done using various methods well known from proton chemical shift imaging.[27] For example, Kohler et al.[28] used a 2D phase-

For References see p. 815

encoding technique equivalent to that originally proposed by Brown et al.[29] This technique uses a standard spatially selective excitation produced by the application of the z gradient, followed by phase encoding in both the x and y planes. The signal is read out in the absence of a gradient, so that Fourier transform will yield the spectrum. This sequence commits the experimenter to a relatively large number of excitations ($Nx \cdot Ny$). However, for Na imaging this is not a problem since one is usually forced to use a long scan time to achieve acceptable S/N. This technique can be generalized to use multiple echoes (on the long T_2 component) and a matched filter can be used to optimize S/N.[30]

The previous techniques are classed as single quantum (SQ). There is interest in obtaining MQ signals and using them for imaging. Wimperis and co-workers have investigated this, and the details are discussed elsewhere.[31] Triple quantum (TQ) is a better choice than DQ for this purpose. The minimum pulse sequence to generate MQ coherence consists of three rf pulses: α_1-t_1-α_2-t_2-α_3-t_3-DA, where the optimal values for the flip angles α_i depend on the type of experiment, and the optimal interpulse time intervals t_i depend on the relaxation properties of the Na. This pulse sequence gives a mixture of coherence orders, and it is necessary to 'filter' the signal to obtain the desired coherence. This is done by cycling the phase (ϕ in Figure 4) of the last rf pulse; the choice of flip angle = 109° maximizes the TQ signal component for this sequence.[32]

The simplest technique for MQ MRI is simply to append one of the previous SQ imaging pulse sequences to the previous MQ preparation sequence. However, a more ingenious imaging pulse sequence has been suggested by Wimperis and Wood[31] which uses the unique properties of the TQ coherence to select the TQ signal without phase cycling. The pulse sequence is shown in Figure 4. The first 90-180-90 set of pulses refocuses dephasing due to $\boldsymbol{B}_0$ inhomogeneity and/or chemical shift effects. During time interval t_2 there is TQ coherence. This implies that the phases of the relevant matrix elements evolve at three times the SQ Larmor frequency. Thus, the gradient pulses applied during that time need only be one-third as strong (or long) as in SQ techniques. This is an advan-

tage in systems whose gradients are designed for proton imaging and in which it is difficult to achieve the factor of four increase needed to obtain high-resolution Na images. Furthermore, the TQ signal decays during that time at a rate comparable to T_{2s}, so there is less need to minimize the time duration.[31] Application of the frequency-encoding prephasing pulse (on G_x) during this time sets up a typical echo during the readout time as shown. However, in this case the area under the prephase pulse needs to be only one-third that of the readout gradient pulse; this unusual condition acts to select only the TQ desired coherence component so that phase cycling is in principle not necessary. (However, in practice the SQ signals are so strongly excited that phase cycling is recommended.) The strength of the TQ signal developed, S, depends on both t_1 and t_3 according to the function

$$S(t_1, t_3) \propto f(t_1)\, f(t_3) \tag{5}$$

where

$$f(t) = \exp(-t/T_{2s}) - \exp(-t/T_{2f}) \tag{6}$$

When $T_{2s} \gg T_{2f}$, $f(t)$ starts at zero, builds to a maximum in time $\approx T_{2f}$, and decays with time constant T_{2s}. Hence it is not necessary to achieve ultrashort echo or DA times with this sequence.

The drawback to MQ imaging is the reduction in signal amplitude relative to SQ.[15] In particular, the use of gradients to filter the MQ coherence reduces signal by a factor of two.[32] Wimperis et al.[33] have modified the technique in an attempt to increase S/N. However, they demonstrated only modest image quality in a phantom of 200 mM Na, which is 10 times the amount present IC. Keltner et al.[34] have used a technique similar to Figure 4 to image a phantom containing 12 mM Na, obtaining an excellent S/N in voxels of $1.5 \times 1.5 \times 1.5$ cm size. To date, no one has published in vivo MQ images of Na with an image quality acceptable for biological application.

4   BIOLOGICAL APPLICATIONS

To date, most applications of Na MRI to biomedical problems have been experimental. These applications can be roughly categorized as concerning edema, infarction, and neoplasia.

4.1   Imaging Studies of Edema

Edema means the pathological increase in water content of tissues. When the edema is due to leaking blood vessels, the excess water is EC and its Na content should be about 150 mM, i.e., much greater than the IC level. This will result in increased Na signal. Furthermore, since the IC [Na] is so low, the amount of signal increase will largely reflect the increase in the size of the EC space. However, these same factors lead to increased signal in conventional T_2-weighted proton images, so Na MRI has not found widespread use as a clinical tool for this purpose.

One application of this principle was described by Lancaster et al.[35] who imaged experimental lung edema in rats. Proton MRI is especially problematic in this case because of severe suppression of signal due to the inhomogeneous magnetic sus-

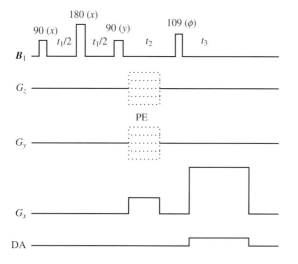

Figure 4   Pulse sequence diagram for triple quantum Na imaging. The dotted lines for the G_z and G_y pulses indicate phase encoding (PE). The x axis is frequency encoded during the data acquisition period (DA)

ceptibility of normal lung tissue. Sodium MRI is less sensitive to this effect because of its reduced γ. They demonstrated much more dramatic increases in Na signal than proton signal in their model of permeability edema. They also attempted to isolate the vascular component of Na by using an iron oxide superparamagnetic particle suspension to suppress the vascular Na.

4.2 Imaging of Stroke and Infarction

The rationale for this application is that if a cell dies, its Na–K pump will shut down and the IC [Na] will increase. However, to demonstrate the effect experimentally sometimes involves reperfusing the organ following vascular occlusion in order to provide a source of the excess Na. Working with such a dog heart model, Cannon et al.[36] demonstrated increased Na signals in regions of ischemic damage. Some early work by Hilal et al.[37] demonstrated increased Na signal in MRI of the brain in one patient; both an old and a recent (2 h) stroke were shown. However, visualization of old strokes is easily accomplished with T_2-weighted proton MRI. The demonstration of early stroke would be very valuable clinically if it could be accomplished in a rapid and routine fashion. There has been little subsequent work in this area, and current research on MRI of stroke is more focused on the possible utility of diffusion-weighted proton MRI.

4.3 Imaging Studies of Cancer

There is much work in the area of distinguishing IC Na levels in normal and malignant cells. For example, Liebling and Gupta[38] found that IC Na of a benign uterine leiomyoma was relatively depressed at only 5.1 mM, while a malignant leiomyosarcoma had the elevated value of 34.6 mM. There have been many in vitro nonimaging studies of tumors in animal models which will not be reviewed here. Imaging is useful when it is desirable to see the spatial distribution of Na in more detail. Summers et al.[39] used Na MRI to study the development of necrosis in an animal model of human neuroblastoma; they found that the total Na signal (not distinguishing IC from EC) tended to rise with increasing tumor necrosis. Lin et al.[40] used Na MRI to distinguish two different grades of tumor in a rat prostate tumor model; they found the two differed in the magnitude of T_{2s} in a way that could not be seen using proton MRI.

Several groups have demonstrated increased Na signal in human brain cancer. However, since most tumors are visualized using proton MRI, the interest is whether Na MRI can make more specific diagnoses. In some cases, the technique uses a relatively short gradient echo as well as several spin echoes with longer TE. In the normal brain, if TE > 10 ms the only structures visible are those containing cerebrospinal fluid and the vitreous humor of the eye. Looking only at echo times greater than 13 ms, Turski et al.[41] found that a brain stem glioma showed less enhancement than an astrocytoma in the pons. Hashimoto et al.[42] used a short TE (1.9 ms) gradient echo as well as longer spin echoes. However, they reported that the most useful differences between the Na and proton images of a meningioma were derived from the long TE images; the short TE images were obviously blurred. Schuierer et al.[43] looked at a variety of supratentor-

ial lesions with Na MRI at 4.0 tesla using only a TE = 11 ms echo; they found that Na MRI did not contribute any information that was not provided by proton MRI or X-ray CT.

Hilal and co-workers[44] have energetically applied Na MRI to brain cancer. Skirting the recognized difficulties in differentiating IC from EC Na, they derive an image of 'putative intracellular' Na from a gradient echo starting at TE = 0.2 ms and two spin echoes (TE = 12 and 24 ms). The long T_2 signal, assumed to be purely EC, is used to estimate the EC volume fraction by comparison with the vitreous humor. The short T_2 component is computed by subtracting the value of the EC signal extrapolated to TE = 0 (equivalent to what is called 'curve stripping'). From these numbers an estimate of the putative IC concentration {[Na]/(IC volume)} is obtained for each pixel in the image. From their preliminary sample of 23 patients, they found a correlation between the IC [Na] and the histological grade of gliomas and astrocytomas. Like Hashimoto et al., they also found that Na MRI often reveals a larger tumor than proton MRI.

Thus, there is ample evidence that Na levels are increased in various types of tumor. However, whether this is due to increases in the IC or EC space is a complex question, and the answer probably will depend on the type of tumor and which organ it is in. In particular, many tumors have a loosely organized cellular system and so have a relatively large EC space compared with normal tissues. In these cases, the increase in EC Na would probably correlate with the signal increase seen in proton MRI.

5 RELATED ARTICLES

Chemical Shift Imaging; Image Formation Methods; Projection–Reconstruction in MRI.

6 REFERENCES

1. C. P. Slichter, 'Principles of Magnetic Resonance', 3rd edn., Springer-Verlag, New York, 1990, pp. 494–500.
2. W. D. Rooney and C. S. Springer, *NMR Biomed.*, 1991, **4**, 209.
3. M. Shporer and M. M. Civan, *Biophys. J.*, 1972, **12**, 114.
4. P. M. Joseph and R. M. Summers, *Magn. Reson. Med.*, 1987, **4**, 67.
5. T. E. Bull, *J. Magn. Reson.*, 1972, **8**, 344.
6. S. Forsen and B. Lindman, *Methods Biochem. Anal.*, 1981, **27**, 289.
7. S. J. Kohler, N. H. Kolodny, D. J. D'Amico, S. Balasubramanian, P. Mainardi, and E. Gragoudas, *J. Magn. Reson.*, 1989, **82**, 505.
8. U. Eliav and G. Navon, *J. Magn. Reson.*, 1990, **88**, 223.
9. M. S. Albert, W. Huang, J. H. Lee, J. A. Balschi, and C. S. Springer, *NMR Biomed.*, 1993, **6**, 7.
10. H. Shinar and G. Navon, *Biophys. Chem.*, 1984, **20**, 275.
11. H. Shinar and G. Navon, *Biophys. J.*, 1991, **59**, 203.
12. H. Shinar and G. Navon, *Magn. Reson. Med.*, 1986, **3**, 927.
13. B. D. Foy and D. Burstein, *Biophys. J.*, 1990, **58**, 127.
14. J. Pekar, P. F. Renshaw, and J. S. Leigh, *J. Magn. Reson.*, 1987, **72**, 159.
15. L. A. Jelicks and R. K. Gupta, *J. Magn. Reson.*, 1989, **83**, 146.
16. J. L. Allis, A. M. L. Seymour, and G. K. Radda, *J. Magn. Reson.*, 1991, **93**, 71.

17. H. Shinar, T. Knubovets, U. Eliav, and G. Navon, *Biophys. J.*, 1993, **64**, 1273.
18. S. J. Kohler and N. H. Kolodny, *Prog. NMR Spectrosc.*, 1992, **24**, 411.
19. W. H. Perman, D. M. Thomasson, M. A. Bernstein, and P. A. Turski, *Magn. Reson. Med.*, 1989, **9**, 153.
20. P. C. Lauterbur, *Nature (London)*, 1973, **242**, 190.
21. L. A. Shepp, *J. Comput. Assist. Tomogr.*, 1980, **4**, 94.
22. A. K. Louis, *J. Comput. Assist. Tomogr.*, 1982, **6**, 334.
23. O. Nacioglu and Z. H. Cho, *IEEE Trans. Nucl. Sci.*, 1984, **31**, 553.
24. J. B. Ra, S. K. Hilal, and C. H. Oh, *J. Comput. Assist. Tomogr.*, 1989, **13**, 302.
25. J. B. Ra, S. K. Hilal, and Z. H. Cho, *Magn. Reson. Med.*, 1986, **3**, 296.
26. J. P. Boehmer, K. R. Metz, J. Mao, and R. W. Briggs, *Magn. Reson. Med.*, 1990, **16**, 335.
27. D. Burstein and M. Mattingly, *J. Magn. Reson.*, 1989, **83**, 197.
28. S. J. Kohler, E. K. Smith, and N. H. Kolodny, *J. Magn. Reson.*, 1989, **83**, 423.
29. T. R. Brown, B. M. Kincaid, and K. Ugurbil, *Proc. Natl. Acad. Sci. U.S.A.*, 1982, **79**, 3523.
30. D. Lu and P. M. Joseph, *Magn. Reson. Imaging*, 1995, **13**, in press.
31. S. Wimperis and B. Wood, *J. Magn. Reson.*, 1991, **95**, 428.
32. J. W. C. VanderVeen, S. Slegt, J. H. N. Creyghton, A. F. Mehlkopf, and W. M. M. J. Bovee, *J. Magn. Reson., Ser. B*, 1993, **101**, 87.
33. S. Wimperis, P. Cole, and P. Styles, *J. Magn. Reson.*, 1992, **98**, 628.
34. J. R. Keltner, S. T. S. Wong, and M. S. Roos, *J. Magn. Reson., Ser. B*, 1994, **104**, 219.
35. L. Lancaster, A. R. Bogdan, H. L. Kundel, and B. McAffee, *Magn. Reson. Med.*, 1991, **19**, 96.
36. P. J. Cannon, A. A. Maudsley, S. K. Hilal, H. E. Simon, and F. Cassidy, *J. Am. Coll. Cardiol.*, 1986, **7**, 573.
37. S. K. Hilal, A. A. Maudsley, J. B. Ra, H. E. Simon, P. Roschmann, S. Wittekoek, Z. H. Cho, and S. K. Mun, *J. Comput. Assist. Tomogr.*, 1985, **9**, 1.
38. M. S. Liebling, R. K. Gupta, and H. L. Kundel, *Ann. N.Y. Acad. Sci.*, 1987, **508**, 149.
39. R. M. Summers, P. M. Joseph, and H. L. Kundel, *Invest. Radiol.*, 1991, **26**, 233.
40. R. Lin, A. R. Bogdan, L. Lancaster, K. Meyer, H. L. Kundel, E. Kassab, V. Liuolsi, M. Salscheider, and P. M. Joseph, *Proc. IXth Ann Mtg. Soc. Magn. Reson. Med.*, New York, 1990, Vol. 1, p. 722.
41. P. A. Turski, L. W. Houston, W. H. Perman, J. K. Hald, D. Turski, C. M. Strother, and J. F. Sackett, *Radiology*, 1987, **163**, 245.
42. T. Hashimoto, H. Ikehira, H. Fukuda, A. Yamaura, O. Watanabe, Y. Tateno, R. Tanaka, and H. E. Simon, *Am. J. Physiol. Imaging*, 1991, **6**, 74.
43. G. Schuierer, R. Ladebeck, H. Barfuss, D. Hentschel, and W. J. Huk, *Magn. Reson. Med.*, 1991, **22**, 1.
44. S. K. Hilal, C. H. Oh, I. K. Mun, and A. J. Silver, in 'Magnetic Resonance Imaging' eds. D. Stark and W. Bradley, Mosby Year Book, St. Louis, MO, 1992, pp. 1091–1112.

Biographical Sketch

Peter M. Joseph. *b* 1939. B.S., 1959, Ph.D., 1967, physics, Harvard University, USA; assistant professor of high energy physics, Carnegie Mellon University, 1970–72. NIH postdoctoral fellow in medical physics, Memorial Sloan Kettering Inst., New York, 1972–73. Instructor and assistant professor of clinical radiology, Columbia-Presbyterian Medical Center, New York, 1973–80. Associate Professor of Diagnostic Imaging Physics, University of Maryland, Baltimore, 1980–82. Associate and full professor, University of Pennsylvania, 1983. Approx. 120 publications in high energy physics, X-ray and computerized tomography scanning, and MRI. Current research specialty: electron beam tomography, sodium and fluorine MRI.

Phosphorus-31 Magnetization Transfer Studies In Vivo

Ruth M. Dixon

Department of Biochemistry, University of Oxford, Oxford, UK

1   BACKGROUND AND THEORY

The rates of certain chemical reactions can be studied by magnetization transfer techniques, the theory of which was first developed by Forsén and Hoffman.[1] The theory and uses of these techniques for measuring enzyme-catalyzed reaction rates are reviewed extensively elsewhere.[2–6] This article will therefore give only a brief outline of the theoretical background, concentrating on the applications of the methods in living animals, and will include the small number of studies in human subjects.

1.1   Techniques

Perturbation of the magnetization of a species in chemical exchange with others causes a change in the magnetization of the other species that depends in general not only on the exchange rate constants $k_{forward}$ and $k_{reverse}$, but also on the spin–lattice relaxation times (T_1) of the exchanging species. For a reaction involving two species, A$\leftrightarrow$B, the unidirectional fluxes through the reaction are $k_{forward}$[A] for the forward reaction and $k_{reverse}$[B] for the reverse reaction. The measured rate constants (k) are always first-order, or pseudo first order, that is, for reactions involving several substrates, the apparent rate constant is a function of the concentrations of the other metabolites. Two main types of experiment may be performed to measure the forward and reverse rate constants. In the first, a continuous radiofrequency irradiation selectively saturates one of the exchanging species and the changes in the other species are observed, either as a function of the time of irradiation, or at steady state. This experiment is known as saturation transfer (ST). In the second type of experiment [inversion transfer (IT)], a selective inversion pulse is applied to one of the exchanging resonances, and the temporal changes of all of the resonances are observed. In both experiments the various recovery curves of the exchanging species are fitted to suitable solutions of the Bloch equations modified for chemical exchange.[1–6] For instance, in a steady-state ST experiment where species B is irradiated and species A is measured, the rate constant is given by:

$$k_{A \rightarrow B} = \frac{1}{T_1} \left[\frac{M_0}{M_s} - 1 \right]$$

where T_1 is the longitudinal relaxation time of A in the absence of chemical exchange, M_s is the signal from A in the presence of a saturating pulse on B, and M_0 is the signal from A in the presence of a saturating pulse an equal distance on the opposite side of A. In this type of experiment, a separate determination of T_1 (with irradiation of the exchanging species) is required. (See also *Selective Excitation in MRI and MR Spectroscopy*).

Two-dimensional (2D) techniques (analogous to NOESY experiments) give cross peaks between resonances that are in chemical exchange. The rate constants can be derived from the dependence of the cross-peak volumes on the mixing time. Selective pulses are not required, but a separate 2D experiment is required for each delay time. Two-dimensional experiments have been implemented in vivo (reviewed in Kuchel[6]) but there have been few applications in whole organisms because of time constraints.

The range of exchange rates that can be measured by magnetization transfer techniques is limited by the chemical shift difference of the exchanging species (in Hz), and by the T_1 times. For exchange rates that are fast compared with the chemical shift difference, only one resonance at an average chemical shift is observed, while for very slow reactions, where $k \ll 1/T_1$, relaxation will occur before significant exchange has taken place. Therefore rate constants of the same order of magnitude as $1/T_1$ are optimal for measurement by magnetization transfer.

1.2 Reactions Studied In Vivo

The reaction most widely studied in vivo is the reaction catalyzed by creatine kinase (CK):

$$PCr + ADP + H^+ \rightleftharpoons Cr + ATP$$

The phosphate of phosphocreatine (PCr) exchanges with the γ-phosphate of adenosine 5'-triphosphate (ATP γ-P). The role of CK is not fully understood, but various possibilities have been discussed (reviewed in Wallimann et al.[7] and by Saks et al.[8]). It may simply buffer the free ATP and adenosine diphosphate (ADP) levels during periods of energy utilization, or PCr may act particularly in muscle as a transporter for 'high-energy' phosphate groups, by diffusing from the mitochondria to the myofibrils. Since PCr and Cr have higher diffusion coefficients in vivo than ATP and ADP, and Cr concentration is much higher than that of ADP, they may facilitate the exchange between the energy-producing and energy-utilizing sites in the cell. The calculation of free ADP concentration from the NMR-measured PCr/ATP ratio, the proton concentration, and the CK equilibrium constant, assumes that the reaction is at equilibrium in each tissue of interest. Magnetization transfer

measurements have been used to confirm whether this is indeed the case, using the principle that if the flux through CK is much faster than the net utilization (or production) of ATP by the cell, the enzyme may be considered to be at equilibrium. The interpretation of these studies is, however, not always straightforward, and it is suggested that not all of the Cr flux is visible to NMR as a consequence of compartmentalization and immobilization of enzymes and substrates.[9]

Other enzymatic reactions that have been studied by magnetization transfer methods in vivo are those involving ATP exchange, for instance, inorganic phosphate (Pi) $\rightarrow$ ATP γ-P exchange, catalyzed by the mitochondrial $F_1 F_0$ ATPase, glyceraldehyde-3-phosphate dehydrogenase, and phosphoglycerate kinase, and ATP γ-P $\rightarrow$ Pi exchange, catalyzed by various ATPases. The exchange between ADP and ATP β-phosphates is also catalyzed by many enzymes, and so an apparent exchange is often seen between ATP γ-P and ATP β-P, which is in fact due to the saturation of ADP β-P, at less than 1 ppm upfield of ATP γ-P. As yet unexplained is the apparent exchange between ATP γ-P and ATP α-P that has been seen in some systems.

A rather different set of studies, that may also be termed magnetization transfer, involves the irradiation of the broad resonance that underlies the spectra from brain, liver, and some other tissues. These are discussed in Section 4.

2 WHOLE ANIMAL STUDIES

Most ^{31}P magnetization transfer studies in vivo have involved the CK reaction in skeletal muscle, heart, and brain. Extensive investigations of isolated systems such as perfused organs, immobilized cells, and tissue slices, have also been undertaken,[2–5,10] but these are beyond the scope of this article (see also *Cells and Cell Systems MRS*).

2.1 Skeletal Muscle

Creatine kinase is found in the skeletal muscle of all mammalian species. During exercise, ATP is hydrolyzed to ADP, and this is rephosphorylated to ATP by CK. This process is so rapid that the level of ATP generally remains constant even during severe exercise, until the PCr level falls to zero. In vitro, the rate of ATP production from PCr increases as the ADP concentration is increased, suggesting that the reaction is substrate-controlled. In studies of intact rat skeletal muscle in vivo, however, no such increase with workload and calculated [ADP] was observed.[11] Indeed in one study, the flux from PCr to ATP was significantly lowered as the workload increased, despite a calculated ninefold increase in [ADP], which would be expected to result in a 70% increase in flux.[12] The CK reaction does not appear to be obligatory for muscle metabolism, in that rats fed with an analog of creatine, β-guanidinopropionic acid (β-GPA), have a 12-fold lower flux through CK (PCr $\rightarrow$ ATP) and 10% of the normal PCr concentration, but normal physiological performance under electrical stimulation.[13] It was suggested that the results supported a role for CK in preventing sudden large changes in [ADP] (which may regulate oxidative ATP synthesis), but a role for PCr in facilitating diffusion of high-energy phosphates was not ruled out, since adaptive changes in muscle fiber type or blood flow could have occurred over the 6–10 weeks of β-GPA feeding. A more direct manipulation of CK activity was achieved by Brosnan et al.,[14] who expressed the B isozyme of CK (normally found only in brain and heart) in the skeletal muscle of transgenic mice. The normal levels of the M isozyme were also expressed, resulting in a 49% increase in total CK activity in vitro. No changes in metabolite levels were found (confirming that the reaction is at equilibrium in normal and transgenic mice) but the ATP → PCr flux (measured by saturation transfer) doubled. It is possible that this reflects the lower Michaelis constant (K_m) for ADP of the B isozyme, although earlier studies did not show a simple relationship between [ADP] and k.[11,12] Transgenic mice lacking the M isozyme of CK were produced by van Deursen et al.[15] These mice had only mitochondrial CK activity in their muscles, and normal levels of PCr and ATP, but no detectable flux through CK, measured by IT. Surprisingly, the PCr level fell during exercise and recovered with a similar timecourse to the controls, but this could have been catalyzed by the small amount of the mitochondrial isozyme, without the flux being detectable in the IT experiment. Studies of CK kinetics in transgenic mice have been reviewed by Nicolay et al.[16]

During steady state exercise and at rest, the ratio of the forward to reverse fluxes of CK should be equal, since the metabolite concentrations are constant. This was found to be the case in a study of resting rat muscle, obtained by saturation transfer.[17] In a study of rabbit muscle[18] the forward and reverse fluxes were equal when measured by fitting the initial recovery rates of an inversion transfer experiment, and the forward flux (PCr → ATP) determined by saturation transfer agreed well with the inversion recovery rates, but the reverse flux determined by saturation transfer was significantly lower. Although competing ATPase reactions would be able to account for this discrepancy, no ATP γ-P to Pi exchange was observed. It has been suggested that compartmentation of intracellular ATP may account for this type of discrepancy.[19] A 2D study of rat muscle found no significant differences between the forward and reverse fluxes, and no exchange between ATP γ-P and Pi, ATP α-P, or ATP β-P,[20] in contrast to the saturation transfer studies.[11,12] The differences arise from the fact that the 2D technique, like the IT experiment, follows the transfer of a transient pulse of magnetization, and would not detect small pools of exchanging metabolites that would be detected in the ST experiment, in which the pool is continuously labeled. A saturation transfer study of tetanically stimulated rat muscle showed that ATP turnover (Pi → ATP flux) showed an approximately linear dependence on calculated free [ADP], up to approximately an ADP concentration of about 90 μM.[21] There are difficulties with measuring the ATP β-P to ADP β-P flux, as shown by Le Rumeur et al.[22] They found that the overall rate of β-phosphoryl conversion increased with exercise in rat muscle but did not reach the expected rate, which should equal or exceed that of the reverse CK reaction, ATP→PCr (since this reaction includes β-phosphoryl conversion). The authors suggest that the flux is underestimated either because of compartmentation of ATP or because of imperfect saturation of ATP β-P (if its signal is broadened by binding to intracellular proteins).

Since PCr takes part only in the reaction catalyzed by CK, the forward flux (PCr → ATP) can be measured accurately by saturation of ATP γ-P. An image of the intensity ratio of PCr with and without irradiation of ATP γ-P therefore reflects localized CK activity. Such an image was produced by Hsieh and Balaban,[23] who showed that CK activity did not vary across the resting hind limb muscles of the rabbit.

2.2 Heart

The CK system has been investigated extensively in the heart, both in the isolated perfused heart and in vivo, in order to understand the metabolic coupling between oxidative phosphorylation and energy demand. Studies of perfused hearts led to contradictory results concerning the relationship between the flux through CK and oxidative metabolism, and questions remain about the regulation in vivo.[24] Ingwall and co-workers showed that the pseudo-first-order rate constant (k) for myocardial PCr → ATP exchange in the open-chested rat correlated closely with the rate pressure product (a measure of cardiac function) whether this was depressed by hypoxia,[25] or increased by norepinephrine infusion.[26] The change in k was accompanied by significant changes in [PCr], [ATP], and calculated free [ADP]. It was concluded that the flux through CK is controlled by the availability of ATP during hypoxia, and by ADP levels during inotropic stimulation. In contrast, in the pig heart in vivo, a similar infusion of norepinephrine produced no significant increase in rate constant over a 2.5-fold increase in rate pressure product, nor any changes in [PCr] or [ATP] (and hence in calculated [ADP]).[27] Pacing the canine heart in vivo so as to double oxygen consumption, led to no changes in CK flux. The flux was estimated to be at least 10 times the flux through oxidative phosphorylation, confirming that the CK reaction is at equilibrium in this tissue.[28] The transmural variation of the CK reaction was studied recently in the open-chested dog heart by combining one-dimensional (1D) localization with saturation transfer measurements. A 61% increase in the PCr → ATP flux was observed from the endocardium to the epicardium, in the presence of constant ATP/PCr ratios. This difference could not be explained by differences in CK levels, oxygen consumption, or substrate levels, and led the authors to invoke possible differences in the rate constant of the reaction.[29] Discrepancies between the forward and reverse fluxes were found in the rat heart in situ, which could not be attributed to ATP→ADP or ATP→Pi exchange, but it was suggested that they could be due to a small pool of exchanging ATP in or near the mitochondria.[3,30]

ATP synthesis rates were measured in the intact dog heart by measuring Pi → ATP flux before and after a period of ischemia, and an estimate of the contribution of the glycolytic enzymes glyceraldehyde 3-phosphate dehydrogenase and phosphoglycerate kinase to this exchange was made.[31]

An interesting use of saturation transfer (ST) to measure myocardial pH was demonstrated in sheep, dog, and cat hearts in vivo.[32] The NMR signal from intracellular Pi in the heart is obscured by resonances from 2,3-diphosphoglycerate and Pi in the blood. ATP is in exchange with Pi in the heart, so difference spectra with and without saturation of ATP γ-P contained a resonance from myocardial Pi (a similar exchange in blood could not be demonstrated). The pH could be obtained from the chemical shift of Pi and was found to be about pH 7.0, that is, considerably lower than previous estimates (***pH Measurement In Vivo in Whole Body Systems***).

2.3 Brain

The brain also contains high levels of CK. The first study in vivo showed that the reaction is at equilibrium in the anesthetized rat brain, since the forward flux through CK (PCr→ATP) was about five times the steady state rate of ATP synthesis.[33] A lower reverse flux through CK was observed in this ST study, while a 2D study showed equal forward and reverse rates.[20] This problem was addressed by Degani et al.[34] in a study of the rabbit brain, in which IT data were compared with ST data from the same animals, and by Mora et al.[35] in the monkey brain. The forward and reverse fluxes were equal, within experimental error, when IT data were analyzed, but the reverse flux was undetectably low in an ST experiment when the PCr resonance was irradiated. These results agree with other observations that ST experiments underestimate the ATP→PCr flux when inhomogeneous tissue is studied.[4] Sauter and Rudin found that the CK forward flux (and $k_{forward}$) correlated linearly with the electroencephalogram intensity in rat brain when brain function was depressed or stimulated pharmacologically.[36]

Mora et al.[37] also produced a map of CK activity in the monkey brain by combining a Hahn spin echo imaging sequence with selective saturation of ATP γ-*P*. The difference between this image and one obtained with control irradiation showed that CK activity was approximately uniform across the monkey brain in the sagittal plane, but since no slice selection was used, each voxel contained a mixture of white and grey matter. In the developing piglet brain, the CK rate constant and the forward flux were measured by ST localized by spectroscopic imaging and were found to be higher in white than in grey matter.

2.4 Other Organs

Magnetization transfer techniques have found few applications in other organs in the whole animal. Koretsky et al.[39] studied the rat kidney in vivo with chronically implanted radio-frequency coils. The Pi→ATP rate constant was estimated to be 0.12 ± 0.03 s^{-1}, somewhat less than that obtained in the isolated perfused rat kidney (0.35 ± 0.05 s^{-1}). The CK reaction has also been studied in an implanted tumor in the hind limb of a rat.[6]

3 HUMAN STUDIES

Few studies have involved human subjects, and none, so far, have involved pathological conditions in patients. In skeletal muscle at 1.9 T, Rees et al.[40] found that the PCr→ATP flux was significantly reduced with exercise to 64% of the resting value of 8.5 ± 0.7 mM·s^{-1} in 12 normal subjects. This decrease was similar to that found in rat muscle,[11,12] but in contrast to the expected dependence of the flux on [ADP] and pH. More recently, Goudemant et al. found that if the exercise intensity was varied, the decrease in flux was found only at the highest workload.[41] These authors argue that the results are not consistent with a 'shuttle' role for PCr in skeletal muscle, since this would lead to a functional coupling between energy pro-

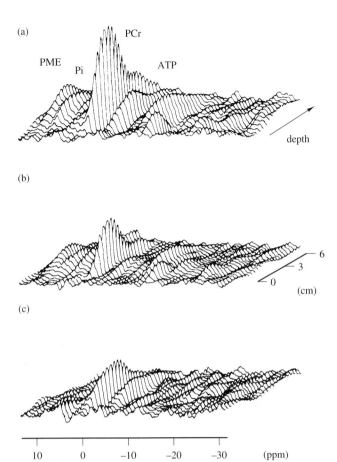

Figure 1 Stack plots of ^{31}P phase-modulated rotating frame images of human brain with (a) control irradiation; (b) irradiation of ATP γ-*P*; (c) the subtraction of (b) from (a). Repetition delay = 15 s, total number of transients for each data set = 80. (Reproduced by permission of FASEB from Cadoux-Hudson et al.[43])

duction and the CK reaction, which was not, in fact, found. Instead, the role of the CK reaction is likely to be energy buffering. They suggest that the results could be explained by the formation at high workloads of a dead-end enzyme–Cr–ADP complex, stabilized by anions such as chloride and bicarbonate, as suggested by McFarland et al. from studies of cat muscle ex vivo.[42] Walliman, however, argued that NMR may not accurately determine the total CK flux in muscle, and that these results are caused by varying degrees of compartmentalization during exercise.[9]

A 1D map of CK activity in brain was obtained by Cadoux-Hudson et al.[43] using phase-modulated rotating frame imaging combined with ST (see also ***Rotating Frame Methods for Spectroscopic Localization***). Figure 1 shows a typical data set, with control and on-resonance irradiation. The CK activity was significantly greater in regions of predominantly gray matter ($k = 0.30 \pm 0.04$ s^{-1}, flux = 1.56 ± 0.30 mmol^{-1} L^{-1} s^{-1}) than in white matter ($k = 0.16 \pm 0.02$ s^{-1}, flux = 0.72 ± 0.12 mmol^{-1} L^{-1} s^{-1}) in six normal subjects. In contrast, Bottomley and Hardy[44] found no significant variation of CK activity with brain region in an ST 2D chemical shift imaging (CSI) experiment in one subject at 4 T ($k = 0.42$ s^{-1}, flux = 2.1 mmol^{-1} kg wet weight^{-1} s^{-1}). One-dimensional CSI with ST in the human heart resulted in a forward rate constant of 0.51 s^{-1}, and a flux

For References see p. 820

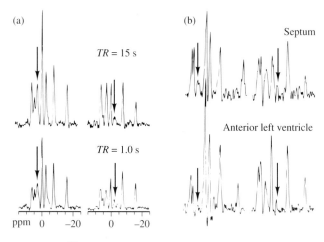

Figure 2 (a) ^{31}P NMR spectra of human brain (entire axial section) with saturating irradiation applied at ±2.7 ppm (arrows). Repetition delay $(TR) = 1$ s or 15 s. (b) Localized ^{31}P NMR spectra from 1-cm-thick sections through the anterior wall (lower) and septum (upper) of the heart in a resting volunteer with saturating irradiation applied at ±2.7 ppm. (Reproduced by permission of Academic Press, from Bottomley and Hardy[44])

of 5.6 mmol^{-1} kg wet weight^{-1} s^{-1}.[44] Figure 2 shows spectra from this study, with control and on-resonance irradiation of ATP γ-P in heart and brain. The decrease in PCr with irradiation of ATP γ-P can clearly be seen in both tissues.

The technique of functional MRI based changes dependent on blood oxygenation was recently used together with localized ^{31}P ST to measure the activation of the visual cortex during visual stimulation.[45] This study showed that the forward rate constant for PCr→ATP γ-P increased significantly within the visual cortex from 0.56 ± 0.19 s^{-1} (control) to 0.76 ± 0.29 s^{-1} during visual stimulation, in the absence of any detectable change in PCr/ATP or PCr/Pi ratios.

4 BASELINE FLATTENING AND PHOSPHOLIPID BILAYERS

The ^{31}P NMR spectra of brain, liver, kidney, and breast all show a broad signal in the phosphodiester region that cannot wholly be accounted for by the water-soluble phosphodiesters glycerophosphorylcholine and glycerophosphorylethanolamine. A study of liver and brain slices in vitro, and of liver in vivo, showed that this resonance was characteristic of phospholipid bilayers broadened by shielding anisotropy.[46] Irradiation of the wings of the broad resonance, 28 ppm from its maximum, completely eliminated the broad resonance from rat liver spectra in vivo, allowing the remaining resonances to be quantified more accurately.

5 RELATED ARTICLES

Cells and Cell Systems MRS; pH Measurement In Vivo in Whole Body Systems; Rotating Frame Methods for Spectroscopic Localization; Selective Excitation in MRI and MR Spectroscopy.

For list of General Abbreviations see end-papers

6 REFERENCES

1. S. Forsén and R. A. Hoffman, *J. Chem. Phys.*, 1963, **39**, 2892.
2. J. R. Alger and R. G. Shulman, *Q. Rev. Biophys.*, 1984, **17**, 83.
3. A. P. Koretsky and M. W. Weiner, in 'Biomedical Magnetic Resonance', eds. T. L. James and A. R. Margulis, Radiology Research and Education Foundation, San Franscisco, 1984, p. 209.
4. K. M. Brindle, *Prog. NMR Spectrosc.*, 1988, **20**, 257.
5. M. Rudin and A. Sauter, in 'NMR, Basic Principles and Progress', eds. P. Diehl, E. Fluck, H. Günther, R. Kosfeld, and J. Seelig, Springer, Berlin, 1992, Vol. 27, p. 257.
6. P. W. Kuchel, *NMR Biomed.*, 1990, **3**, 102.
7. T. Wallimann, M. Wyss, D. Brdiczka, K. Nicolay, and H. M. Eppenberger, *Biochem. J.*, 1992, **281**, 21.
8. V. A. Saks, Z. A. Khuchua, E. V. Vasilyeva, O. Y. Belikova, and A. V. Kuznetsov, *Mol. Cell. Biochem.*, 1994, **133/134**, 155.
9. T. Wallimann, *J. Muscle Cell Res. Cell Motil.*, 1996, **17**, 177.
10. P. G. Morris, *Annu. Rep. NMR Spectrosc.*, 1988, **20**, 1.
11. E. A. Shoubridge, J. L. Bland, and G. K. Radda, *Biochim. Biophys. Acta*, 1984, **805**, 72.
12. E. Le Rumeur, L. Le Moyec, and J. D. de Certaines, *Magn. Reson. Med.*, 1992, **24**, 335.
13. E. A. Shoubridge and G. K. Radda, *Biochim. Biophys. Acta*, 1984, **805**, 79.
14. M. J. Brosnan, S. P. Raman, L. Chen, and A. P. Koretsky, *Am. J. Physiol.*, 1993, **264**, C151.
15. J. van Deursen, A. Heerschap, F. Oerlemans, W. Ruitenbeek, P. Jap, H. ter Laak, and B. Wieringa, *Cell*, 1993, **74**, 621.
16. K. Nicolay, F. A. van Dorsten, T. Reese, M. J. Kruiskamp, J. F. Gellerich, and C. J. A. van Echteld, *Mol. Cell Biochem.*, 1998, **184**, 195.
17. J. A. Bittl, J. DeLayre, and J. S. Ingwall, *Biochemistry*, 1987, **26**, 6083.
18. P. S. Hsieh and R. S. Balaban, *Magn. Reson. Med.*, 1988, **7**, 56.
19. A. P. Koretsky, V. J. Basus, T. L. James, M. P. Klein, and M. W. Weiner, *Magn. Reson. Med.*, 1985, **2**, 586.
20. R. S. Balaban, H. L. Kantor, and J. A. Ferretti, *J. Biol. Chem.*, 1983, **258**, 12787.
21. K. M. Brindle, M. J. Blackledge, R. A. J. Challis, and G. K. Radda, *Biochemistry*, 1989, **28**, 4887.
22. E. Le Rumeur, N. Le Tallec, F. Kernec, and J. D. de Certaines, *NMR Biomed.*, 1997, **10**, 67.
23. P. S. Hsieh and R. S. Balaban, *J. Magn. Reson.*, 1987, **74**, 574.
24. F. Joubert and J. A. Hoerter, *Cell. Mol. Biol.*, 1997, **43**, 763.
25. J. A. Bittl, J. A. Balschi, and J. S. Ingwall, *Circulation Res.*, 1987, **60**, 871.
26. J. A. Bittl, J. A. Balschi, and J. S. Ingwall, *J. Clin. Invest.*, 1987, **79**, 1852.
27. J. F. Martin, B. D. Guth, R. H. Griffey, and D. E. Hoekenga, *Magn. Reson. Med.*, 1989, **11**, 64.
28. L. A. Katz, J. A. Swain, M. A. Portman, and R. S. Balaban, *Am. J. Physiol.*, 1989, **256**, H265.
29. P. M. Robitaille, A. Abduljalil, D. Rath, H. Zhang, and R. L. Hamlin, *Magn. Reson. Med.*, 1993, **30**, 4.
30. A. P. Koretsky, S. Wang, M. P. Klein, T. L. James, and M. W. Weiner, *Biochemistry*, 1986, **25**, 77.
31. P.-M. Robitaille, H. Merkle, E. Sako, G. Lang, R. M. Clack, R. Bianco, A. H. L. From, J. Foker, and K. Ugurbil, *Magn. Reson. Med.*, 1990, **15**, 8.
32. K. M. Brindle, B. Rajagopalan, D. S. Williams, J. A. Detre, E. Simplaceanu, C. Ho, and G. K. Radda, *Biochem. Biophys. Res. Commun.*, 1988, **151**, 70.
33. E. A. Shoubridge, R. W. Briggs, and G. K. Radda, *FEBS Lett.*, 1982, **140**, 288.
34. H. Degani, J. R. Alger, R. G. Shulman, O. A. C. Petroff, and J. W. Pritchard, *Magn. Reson. Med.*, 1987, **5**, 1.

35. B. N. Mora, P. T. Narasimhan, and B. D. Ross, *Magn. Reson. Med.*, 1992, **26**, 100.

36. A. Sauter and M. Rudin, *J. Biol. Chem.*, 1993, **268**, 13166.

37. B. Mora, P. T. Narasimhan, B. D. Ross, J. Allman, and P. B. Barker, *Proc. Natl. Acad. Sci., USA*, 1991, **88**, 8372.

38. D. Holtzman, R. Mulkern, M. Tsuji, C. Cook, and R. Meyers, *Dev. Neurosci.*, 1996, **18**, 535.

39. A. P. Koretsky, S. Wang, J. Murphy-Boesch, M. P. Klein, T. L. James, and M. W. Weiner, *Proc. Natl. Acad. Sci., USA*, 1983, **80**, 7491.

40. D. Rees, M. B. Smith, J. Harley, and G. K. Radda, *Magn. Reson. Med.*, 1989, **9**, 39.

41. J. F. Goudemant, M. Francaux, I. Mottet, R. Demeure, M. Sibomana, and X. Sturbois, *Magn. Reson. Med.*, 1997, **37**, 744.

42. E. W. McFarland, M. J. Kushmerick, and T. S. Moerland, *Biophys. J.*, 1994, **67**, 1912.

43. T. A. Cadoux-Hudson, M. J. Blackledge, and G. K. Radda, *FASEB J.*, 1989, **3**, 2660.

44. P. A. Bottomley and C. J. Hardy, *J. Magn. Reson.*, 1992, **99**, 443.

45. W. Chen, X. H. Zhu, G. Adriany, and K. Ugurbil, *Magn. Reson. Med.*, 1998, **38**, 551.

46. E. J. Murphy, B. Rajagopalan, K. M. Brindle, and G. K. Radda, *Magn. Reson. Med.*, 1989, **12**, 282.

Biographical Sketch

Ruth M. Dixon. *b* 1959. B.A. Churchill College, Cambridge, 1981; Ph.D. University of Warwick, 1985; Postdoctoral research fellow with Professor G. Lowe, Dyson Perrins Laboratory, University of Oxford, 1986; MRC nonclinical scientist at the MRC Biochemical and Clinical Magnetic Resonance Unit, Oxford 1986–present. Alexander von Humboldt research fellow, working with Dr. J. Frahm at the Max-Planck-Institute, Göttingen, 1992–1993. Research interests include the study of liver and brain metabolism by NMR in vivo, phospholipid metabolism in tumors and rapidly dividing tissues, and the development of techniques for NMR spectroscopy in vivo.

Applications of ¹⁹F-NMR to Oncology

Paul M. J. McSheehy

St George's Hospital Medical School, London, UK

Laurent P. Lemaire

Laboratoire de Biophysique Médicale Faculté de Médecine, Angers, France

and

John R. Griffiths

St George's Hospital Medical School, London, UK

1 INTRODUCTION

The first reported ¹⁹F MRS experiments that applied directly to chemotherapy date from 1984, in which the metabolism of an anticancer fluoropyrimidine, 5-fluorouracil (5FU), was studied in mouse liver and a subcutaneous mouse tumor.[1] Subsequently, most preclinical ¹⁹F MRS studies have indeed been concerned with 5FU and its metabolism in liver, and a variety of tumors in different animal models. This work, and other applications of ¹⁹F MRS, have been extensively reviewed since the late 1980s.[2–5] The most recent review[6] demonstrates that ¹⁹F MRS is of increasing use in the clinic and could, in the future, be used to individualize and optimize chemotherapy.

This review will briefly describe the rationale for the development of fluoropyrimidines as anticancer drugs, the key ¹⁹F MRS experiments that have led to the exciting developments in the clinic, and the future mechanistic research required to realize fully the potential of this technique. Reference will also be made to other types of fluorinated compound which have been studied by ¹⁹F MRS to investigate tumor hypoxia, blood flow and pH, and drug pharmacokinetics in animal models. These other studies show how ¹⁹F MRS can be used at an earlier stage in drug development as a research tool.

2 5-FLUOROPYRIMIDINES: ANIMAL MODELS AND CLINICAL APPLICATIONS

The 5-fluoropyrimidine (FP) 5FU was synthesized in 1957, and is an example of rational synthesis in which the substitution of a fluorine atom for hydrogen in position 5 of the pyrimidine uracil was intended to lead to misincorporation of the drug into RNA, and inhibition of DNA synthesis.[7] Anticancer activity is now known to be mediated via conversion to cytotoxic 5-fluoronucleotides (Fnuct) (see Figure 1). Specifically, FdUMP inhibits the key DNA synthesis enzyme thymidylate synthase (TS), FdUTP becomes misincorporated into DNA, and misincorporation of FUTP interferes with RNA maturation.[8] Effective prodrugs of 5FU have also been synthesized which first require conversion to 5FU in the liver, or tumor, and can have a better therapeutic ratio. These prodrugs include capecitabine and the FPs, Flox (5-fluorouridine), Tegafur (tetrahydro-2-furanyl-5-fluorouracil), and Doxfluoridine (5-deoxy-5-fluorouridine)[9]; however, 5FU remains the FP of choice, predominantly for the treatment of breast, head and neck, and gastrointestinal tumors. Despite problems with toxicity, 5FU remains the only drug with significant activity against colorectal adenocarcinoma, which is a tumor frequently metastasizing to the liver. At least 50% of 5FU is catabolized in the liver, i.e. deactivated, with probably small contributions from some types of tumors (Figure 1).[10,11] The metabolism of 5FU can be modulated to increase cytotoxicity, e.g. by increasing Fnuct formation using methotrexate, interferon, or *N*-(phosphonacetyl)-L-aspartate (*N*-PALA), or by increasing binding of FdUMP to TS using leucovorin, or decreasing catabolism by coadministration of thymidine. Because of the relatively wide chemical shift range of ¹⁹F MRS, many of the metabolites of 5FU can be resolved, especially in vitro, where pH-dependent differences in chemical shift can be exploited. It is this property, along with a relatively high sensitivity (~80% of ¹H), and the absence of a background signal, that makes ¹⁹F MRS ideally suited to monitoring the pharmacokinetics of 5FU and other fluorine-containing drugs. That said, in models in vivo, and especially in the clinic, where fields of only 1.5 T are

For References see p. 824

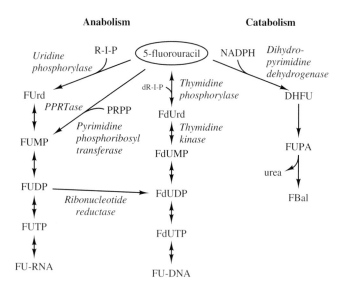

Figure 1 Anabolic and catabolic pathways of 5FU. DHFU, dihydrofluorouracil; FUPA, fluoro-β-ureidopropionic acid; FBal, α-fluoro-β-alanine; FUrd, 5-fluorouridine; FUMP, 5-fluorouridine monophosphate; FUDP, 5-fluorouridine diphosphate; FUTP, 5-fluorouridine triphosphate; FdUrd, 5-fluorodeoxyuridine; FdUMP, 5-fluorodeoxyuridine monophosphate; FdUDP, 5-fluorodeoxyuridine diphosphate; FdUTP, 5-fluorodeoxyuridine triphosphate; pyrimidine phosphoribosyl transferase

employed, the 5-fluoronucleosides (Fnucs) and Fnuct, and the fluorocatabolites (Fcat) of fluoro-β-ureidoproprionic acid (FUPA) and α-fluoro-β-alanine(FBal) are not easily distinguishable. Nevertheless, 5FU, Fcat, and often Fnuct, can be detected in both liver and tumor in vivo at doses near or equivalent to those in clinical use, and this has allowed real-time pharmacokinetics to be used, which in some cases can predict tumor response. Figure 2 shows these signals, in this case those observed in a rat mammary tumor following treatment with 5FU.

After the demonstration that ^{19}F MRS could detect 5FU and metabolites in vivo,[1] research was initiated to relate the amount of signal to tumor response. Decreasing doses of 5FU administered as an intravenous bolus (120, 50, and 25 mg kg^{-1}) to rats bearing the Walker carcinosarcoma led to decreased 5FU signal and Fnuct formed in the tumors. Tumors showed a significant response to the 50 mg kg^{-1} dose, but not to the 25 mg kg^{-1} dose.[12] Other studies[13,14] showed a significant correlation between shrinkage of mouse (RIF-1) tumors with the amount of Fnuct formed in the tumor. Furthermore, there was a negative correlation between Fcat in the mouse liver and Fnuct formed in the tumor, indicating liver deactivation and tumor activation of 5FU were competing processes. This latter observation was important for clinical studies, where the tumor signal-to-noise ratio (S/N) may be very low, but analysis of liver signals may still be of value for prognosis. However, in these animal studies the correlation between 5FU retention in the tumor and response to treatment was weaker or not significant.[13–15] Detection of increased Fnuct formation by ^{19}F NMS, signifying increased cytotoxicity, has also been recorded in ascites tumor cells.[16]

Modulation of 5FU metabolism by a range of substances, including allopurinol,[12] methotrexate,[17–20] interferon[20,21], N-

PALA and leucovorin,[20] and thymidine,[21,22] has been observed in animal tumor models. In some cases this has been shown to lead to a change in the halflife ($t_{1/2}$) for 5FU clearance,[12,13,18,21,22] and, more significantly, to an increase in the rate[18,19] and final amount of Fnuct formed;[17–19,21,22] results that have also been confirmed in extracts.[12,18–21] In general, the biochemical basis for these modulations is fairly well established. Fluorine-19 MRS demonstrated this could be observed in situ, noninvasively, and in real time, providing additional data on the rate of elimination of 5FU from the tumor, which normally can only be obtained by biopsy, indirectly via blood or urine samples, or by sacrifice of many animals. Furthermore, in some cases, increased Fnuct formation was also correlated with tumor response.[19,22]

Besides biochemical modulation of 5FU metabolism, the tumor environment can be transiently altered to favor increased drug uptake, an approach with potentially important applications in the clinic where effective drug delivery to solid tumors can be a major problem. For example, breathing of carbogen gas by rodents can lead to increased tumor blood flow and oxygenation,[23,24,25] and in RIF-1 tumors this led to increased 5FU uptake and FNuct formation, with a corresponding increase in tumor growth inhibition.[14]

In the clinic, the SNR is lower, predominantly because of the lower field strengths employed: 1.5–2 T. This means that in contrast to the experimental tumors in animals, generally only the 5FU and Fcat signals are observed, and rarely the cytotoxic species Fnuct. However, results accumulating in the clinic have shown correlations between 5FU retention in the tumor and patient response, whether partial, or complete.[26–29] In three independent studies,[26–28] 5FU signals have been observed in 78–100% of responders, but in only 8–23% of nonresponders. Furthermore, there is clearly a slower clearance of 5FU from tumors ($t_{1/2}$ of 20–60 min) compared with blood (8–20 min). This phenomonon is known as 'tumor trapping', and its significance is described below. Some studies on the modulation of 5FU metabolism have also begun in the clinic, and changes have been detected using methotrexate, leucovorin, and interferon.[6] Previously undetected, or more intense and prolonged, 5FU signals were observed, and, furthermore, when using combination chemotherapy, Fnuct signals were also detected. The clinical utility of this type of information would be the detection of nonresponders during the early phases of treatment, and the optimization of scheduling for others.

A number of preclinical studies have been performed to identify the mechanism(s) for the trapping of 5FU that is observed clinically. It is already clear that the tumor energy status is likely to play a major role, since a high-energy status (high NTP/Pi ratio) would enable a more effective activation of 5FU to FNuct and would be likely to reflect a well-vascularized tumor, making 5FU access to the tumor easier. In two multinuclear MRS studies,[15,30] there were significant correlations between the pretreatment tumor Pi, PCr, or NTP, the final amount of FNuct formed, and the response of the RIF-1[15] or primary rat[30] tumors to 5FU treatment. Consistent with these observations, treatment with hydralazine (a vasodilator that de-energizes tumors[31]) favored the conversion of FNuct to 5FU.[6] In contrast, in both the mouse[15] and rat model,[30] no significant correlation was observed between the pretreatment pH measured by MRS (i.e. intracellular pH, pH$_i$) and treatment response. However, in a rat fibrosarcoma model, a 2.5-fold

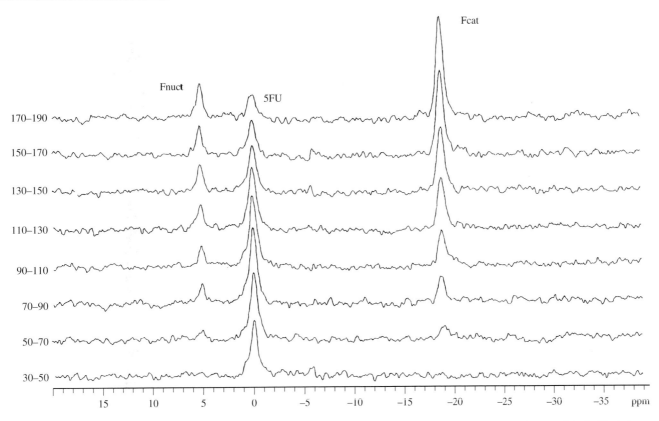

Figure 2 Fluorine-19 MRS spectra of 5FU metabolism by a chemically induced primary rat mammary tumor. Spectra (20 min blocks) were acquired at 4.7 T using a one-turn surface coil on a 5 g tumor following injection of 5FU (150 mg kg^{-1} intraperitoneally)

increase in the 5FU $t_{1/2}$ (increased 5FU trapping) was observed when the pH$_i$ was reduced by increased tumor glycolysis,[32] suggesting tumor pH could have a role in tumor retention of 5FU. What may be of significance is the low extracellular pH (pH$_e$) of solid tumors,[33] which in combination with a neutral pH$_i$ leads to a reverse, or negative, pH gradient across the tumor plasma membrane (i.e. $-\Delta$pH).[34] One study using isolated tumor cells showed that 5FU uptake was increased when the $-\Delta$pH was increased;[35] a similar relationship was observed in vivo.[21] An explanation of these apparently conflicting different studies is that in the rat fibrosarcoma model, the induced decrease in pH$_i$ probably casued a larger decrease in pH$_e$,[35] thereby increasing the $-\Delta$pH. Consequently, it may be that pH$_e$ and/or $-\Delta$pH measurements in vivo will correlate with 5FU trapping. These observations on tumor pH and energy status provide a clear direction for further studies that could lead to protocols with a potentially improved therapuetic index for 5FU.

For chemotherapy, it is also beneficial to study the pharmacology of a drug in other tissues besides tumor. With respect to the FPs, this normally involves liver studies to determine the rate of drug detoxification and/or liver damage.[11,36,37] Measurement of the rate of drug disappearance in the liver, formation or destruction of 5FU, and the appearance of catabolites, with or without rescue schedules, have been performed. In principle, these studies may also relate to tumor cytotoxicity.[13] Similar studies have been performed clinically, the first in 1987 by Wolf et al.,[38] and most recently to provide parameters such as the maximal velocity of metabolic conversion, using only the proportional peak areas of the signals from 5FU and its metabolites.[39]

The relatively high doses of 5FU used (0.6–1 g m^{-2} d^{-1} over 5 days) can also lead to a number of adverse clinical reactions including myelosuppression, diarrhea, and mucositis. The cardiotoxicity of 5FU was first reported in 1975,[40] and incidences of about 8% have been reported.[41] Various hypotheses were proposed to explain the cardiotoxicity of this drug, including ischemia secondary to coronary artery spasms, interaction of 5FU with the coagulation system, immunoallergic phenomena, and direct toxicity of 5FU on the myocardium.[42] The precise biochemical mechanism underlying this cardiotoxicity was unclear. It has been suggested that the major catabolite of 5FU, Fbal, might be transformed into fluoroacetate (FAC), a highly cardiotoxic and neurotoxic poison.[43] Fluorine-19 MRS demonstrated, using an isolated perfused rat liver model, that although Fbal is indeed metabolized into FAC, commercial solutions of 5FU also contain 'impurities'. Degradation compounds are gradually formed in the basic medium, which is indispensable to the solubilization of 5FU. Two cardiotoxic compounds were identified: fluoroacetaldehyde, and fluorosemialdehyde malonic acid, the former being metabolized into FAC, a violent cardiotoxic compound.[44,45]

3   OTHER FLUORINATED COMPOUNDS

Apart from the FP work described that has led directly to similar clinical studies, ^{19}F MRS has been applied to other

For References see p. 824

fluorinated compounds as a research tool to investigate tumor hypoxia and pH and the pharmacokinetics of drugs in development.

The useful clinical activity of 5FU as a TS inhibitor encouraged the development of other TS inhibitors, such as the antifolate CB 3717. This drug had clinical anticancer activity against breast and hepatoma, but dose-limiting nephrotoxicity. Animal studies revealed renal and hepatic toxicity that was partly due to long-term retention of the drug. A fluorinated version of the drug, CB 3988, was synthesized to determine tissue distribution and concentration by NMR. Pharmacokinetics were studied in vivo in the mouse and rat using both surface and solenoid coils for fluorine imaging and spectroscopy.[46] These studies demonstrated, noninvasively, both hepatic and renal clearance and, ultimately, localization of the drug to the abdominal cavity. The $t_{1/2}$ values determined were similar to those measured by high-performance liquid chromatography of body fluids. Recently, another fluorinated TS inhibitor, ZD9331, has been developed and is currently in phase II/III clinical trials. Attempts to study a fluorinated alkylating agent, β,β-difluorochlorambucil, were limited because of binding of the drug to proteins in blood plasma and tumors and excessive toxicity.[47] A difluorinated nucleoside analog, gemcitabine, developed in the 1980s is now showing considerable success in the clinic against a range of solid tumors. The high doses that are used (approximately 1.5 g m^{-2}) have permitted ^{19}F-imaging in rat tumors;[48] ^{19}F MRS showed increased uptake in gemcitabine-sensitive murine human xenografts compared with less sensitive xenografts.[49]

The energy state of a tumor is critical in determining chemo- and radiosensitivity, and a major regulator of this is the blood flow to the tumor. A number of methods have been applied to measure directly or indirectly tumor hypoxia, or blood supply. A number of studies have been performed using ^{19}F-labeled nitromidazoles, such as fluoromisonidazole, Ro-070741,[47] and SR-4554.[50] These are bioreductive agents that in hypoxic conditions undergo nitroreduction to reactive intermediates such as hydroxylamines, which bind to macromolecules. Selective retention was seen in tumor compared with normal tissue, and increased retention was seen in mouse tumors that were known to have a greater hypoxic fraction.[14,47,50] Perfluorocarbons (PFCs) are fluorinated organic molecules which provide a high S/N for ^{19}F MRS studies. The ^{19}F spin–lattice relaxation rates of the PFC resonances are sensitive to the oxygen tension (PO$_2$), and this has been exploited to measure tumor oxygenation. The principle has been reviewed,[4] and new methods,[51] and PFC types,[25,52] including one with 12 equivalent fluorine atoms,[53] are continually developed. One group has used a PFC to demonstrate potential chemotherapeutic applications.[24,54] A PFC emulsion was injected intraperitoneally, and 3 days later ^{19}F MRS was performed prior to and immediately following administration of the radiosensitizer nicotinamide. Highly significant increases in PO$_2$ were recorded in the mouse RIF-1 tumor with the maximum effect at 60–70 min posttreatment.[54]

Tumor pH is another important determinant of chemotherapeutic response, as described for 5FU. Although in vivo pH is generally measured by ^{31}P MRS, a number of specific fluorinated probes have been developed by Deutsch and colleagues to measure intracellular pH.[55,56] While these studies provided useful information on cell function, there is some doubt about

their applicability to studying tumors in vivo; other probes, while confirming ^{31}P measurements of pH, have suffered from poor solubility.[57] The Fnuct formed from FPs is an unequivocal intracellular pH marker,[12,15] although the S/N is generally low, and the composite nature of the peak (many Fnucs each with a slightly different pK_a) precludes absolute pH measurements. Nevertheless, dynamic 5FU-induced increases in tumor pH$_i$ were recorded using FNuct in RIF-1 tumors.[14] Recently, new pH probes have been described that are sufficiently soluble and sensitive to allow studies in vivo.[58,59] One, ZK 150471, is a difluorinated extracellular probe that has been shown to give comparable measurements of tumor pH$_e$ when compared with a ^{31}P-pH probe.[60] The other, F6POL, is a monofluorinated probe that has been shown to measure both pH$_i$ and pH$_e$ in perfused rat heart[59] and may permit measurements in situ of tumors.

Finally, the fluorinated 2-deoxyglucose analog 2-fluoro-2-deoxyglucose (FDG) has been used to study glucose uptake in ascites tumors.[61,62] Contrary to dogma, FDG metabolism does not cease with phosphorylation to FDG 6-phosphate. By ^{19}F MRS, it has been shown that FDG 6-phosphate may be epimerized to the mannose 6-phosphate derivative; this undergoes further conversions by ascites cells as well as in brain, heart, and muscle to a number of FDG metabolites.[62] Significantly, one of these metabolites, NDP-FDM, persists for 24–48 h in ascites cells at sufficiently high concentrations to permit ^{19}F imaging, allowing the possibility of using ^{19}F imaging of FDG metabolites for the detection of tumors in the body.

4   SUMMARY

The high sensitivity of fluorine, absence of a background signal, and wide chemical shift range make the fluorine atom ideal for noninvasive measurements of drug pharmacokinetics. In some cases the fluorine is already an integral part of the drug, which allows direct applications of animal models to the clinic, with the increasingly realizable potential of optimizing drug schedules. Other fluorinated compounds can be used as probes of tumor glycolysis, pH, and hypoxia, which are important determinants of chemo- and radiosensitivity.

5   RELATED ARTICLES

Brain Neoplasms in Humans Studied by Phosphorus-31 NMR Spectroscopy; In Vivo Hepatic MRS of Humans; Localization and Registration Issues Important for Serial MRS Studies of Focal Brain Lesions; Spectroscopic Studies of Animal Tumor Models.

6   REFERENCES

1. A. N. Stevens, P. G. Morris, R. A. Iles, P. W. Sheldon, and J. R. Griffiths, *Br. J. Cancer*, 1984, **50**, 113.
2. P. M. J. McSheehy and J. R. Griffiths, *NMR Biomed.*, 1989, **2**, 133.

3. M. C. Malet-Martino, R. Martino, and J. P. Armand, *Bull. Cancer Paris*, 1990, **77**, 1223.
4. M. J. W. Prior, R. J. Maxwell, and J. R. Griffiths, in 'NMR Basic Principles and Progress', ed. Springer-Verlag, Berlin, 1992, Vol. 28, Chap. 3
5. R. J. Maxwell, *Cancer Surv.*, 1993, **17**, 415.
6. M. P. N. Findlay and M. O. Leach, *Anticancer Drugs*, 1994, **5**, 260.
7. C. Heidelberger, P. V. Danenberg, and R. G. Moran, *Adv. Enzymol.*, 1983, **54**, 57.
8. H. M. Pinedo, and G. F. J. Peters, *J. Clin. Oncol.*, 1988, **6**, 1653.
9. Y. O. Rustum, A. Harstrick, S. Cao, U. Vanhoefer, M.-B. Yin, H. Wilke, and S. Seeber, *J. Clin. Oncol.*, 1997, **15**, 389.
10. F. N. M. Naguib, M. H. el Kouni, and S. Cha, *Cancer Res.*, 1985, **45**, 5405.
11. M. J. W. Prior, R. J. Maxwell, and J. R. Griffiths, *Biochem. Pharm.*, 1990, **39**, 857.
12. P. M. J. McSheehy, M. J. W. Prior, and J. R. Griffiths, *Br. J. Cancer*, 1989, **60**, 303.
13. P. E. Sijens, Y. Huang, N. J. Baldwin, and T. C. Ng, *Cancer Res.*, 1991, **51**, 1384.
14. P. M. J. McSheehy, S. P. Robinson, A. S. E. Ojugo, E. O. Aboagye, M. B. Cannell, M. O. Leach, I. R. Judson, and J. R. Griffiths, *Cancer Res.*, 1998, **58**, 1185.
15. P. E. Sijens, N. J. Baldwin, and T. C. Ng, *Magn. Reson. Med.*, 1991, **19**, 373.
16. P. M. J. McSheehy, R. J. Maxwell, and J. R. Griffiths, *NMR Biomed.*, 1991, **4**, 274.
17. J. A. Koutcher, R. C. Sawyer, A. B. Kornblith, R. L. Stolfi, D. S. Martin, M. L. Devitt, D. Cowburn, and C. W. Young, *Magn. Reson. Med.*, 1991, **19**, 113.
18. A. El-Tahtawy and W. Wolf, *Cancer Res.*, 1991, **51**, 5806.
19. P. M. J. McSheehy, M. J. W. Prior, and J. R. Griffiths, *Br. J. Cancer*, 1992, **65**, 309.
20. Y. J. L. Kamm, I. M. C. M. Rietjens, J. Vervoort, A. Heerachap, G. Rosenbusch, H. Hofs, and D. J. T. Wagner, *Cancer Res.*, 1994, **54**, 4321.
21. P. M. J. McSheehy, M. T. Seymour, A. S. E. Ojugo, L. M. Rodrigues, M. O. Leach, I. R. Judson, and J. R. Griffiths, *Eur. J. Cancer*, 1997, **33**, 2418.
22. P. E. Sijens and T. C. Ng, *Magn. Reson. Imaging*, 1992, **10**, 385.
23. S. P. Robinson, L. M. Rodrigues, A. S. E. Ojugo, P. M. J. McSheehy, F. A. Howe, and J. R. Griffiths, *Br. J. Cancer*, 1997, **75**, 1000.
24. K. G. Helmer, S. Han, and C. H. Sotak, *NMR Biomed.*, 1998, **11**, 120.
25. S. Hunjan, R. P. Mason, A. Constantinescu, P. Peschke, E. W. Hahn, and P. P. Antich, *Int. J. Radiat. Oncol. Biol. Phys.*, 1998, **41**, 161.
26. C. A. Presant, W. Wolf, M. J. Albright, K. L. Servis, R. Ring, D. Atkinson, R. L. Ong, C. Wiseman, M. King, D. Blayney, P. Kennedy, A. El-Tahtawy, M. Singh, and J. Shani, *J. Clin. Oncol.*, 1990, **8**, 1868.
27. M. P. N. Findlay, M. O. Leach, D. Cunningham, D. J. Collins, G. S. Payne, J. Glaholm, J. L. Mansi, and V. R. McCready, *Ann. Oncol.*, 1993, **4**, 597.
28. C. A. Presant, W. Wolf, V. Waluch, C. Wiseman, P. Kennedy, D. Blayney, and R. R. Brechner, *The Lancet*, 1994, **343**, 1184.
29. H.-P. Schlemmer, P. Bachert, W. Semmler, P. Hohenberger, P. Schlag, W. J. Lorenz, and G. Van Kaick, *Magn. Reson. Imaging*, 1994, **12**, 497.
30. L. P. Lemaire, P. M. J. McSheehy, and J. R. Griffiths, *Cancer Chemother. Pharmacol.*, 1998, **42**, 201.
31. G. M. Tozer, R. J. Maxwell, J. R. Griffiths, and P. Pham, *Br. J. Cancer*, 1990, **62**, 553.
32. J.-L. Guerquin-Kern, F. Leteurtre, A. Croisy, and J.-M. Lhoste, *Cancer Res.*, 1991, **51**, 5770.
33. J. R. Griffiths, *Br. J. Cancer*, 1991, **64**, 425.
34. M. Stubbs, L. M. Rodrigues, F. A. Howe, J. Wang, K. S. Jeong, R. L. Veech, and J. R. Griffiths, *Cancer Res.*, 1994, **54**, 4011.
35. A. S. E. Ojugo, P. M. J. McSheehy, M. Stubbs, G. Alder, R. J. Maxwell, C. L. Bashford, M. O. Leach, I. R. Judson, and J. R. Griffiths, *Br. J. Cancer*, 1998, **77**, 873.
36. M. Harada, K. Koga, I. Miura, and H. Nishitani, *Magn. Reson. Med.*, 1991, **22**, 499.
37. Y. Kanazawa, S. Kuragi, S. Shinahara, Y. Noda, and K. Masuda, *Chem. Pharm. Bull.*, 1994, **42**, 774.
38. W. Wolf, M. J. Albright, M. S. Silver, H. Weber, U. Reichardt, and R. Sauer, *Magn. Reson. Imaging*, 1987, **5**, 165.
39. R. E. Port, H.-P. Schlemmer, and P. Bachert, *Eur. J. Clin. Pharmacol.*, 1994, **47**, 187.
40. R. G. Dent and I. McColl, *The Lancet*, 1975, **1**, 347.
41. W. J. Gradishar and E. E. Vokes, *Ann. Oncol.*, 1990, **1**, 409.
42. N. J. Freeman and M. E. Costanza, *Cancer*, 1988, **61**, 36.
43. I. Matsubara, J. Kamiya, and S. Imai, *Jpn. J. Pharmacol.*, 1980, **30**, 871.
44. L. Lemaire, M. C. Malet-Martino, M. de Forni, R. Martino, and B. Lasserre, *Br. J. Cancer*, 1992, **66**, 119.
45. L. Lemaire, M. C. Malet-Martino, M. de Forni, R. Martino, and B. Lasserre, *Oncol. Rep.*, 1994, **1**, 173.
46. D. R. Newell, R. J. Maxwell, and B. T. Golding, *NMR Biomed.*, 1992, **5**, 273.
47. P. Workman, R. J. Maxwell, and J. R. Griffiths, *NMR Biomed.*, 1992, **5**, 270.
48. M. E. Belleman, G. Brix, U. Haberkorn, J. Blatter, L. Gerlach, F. Oberdorfer, and W. J. Lorenz, *Proc. 3rd Ann Mtg. Int. Soc. Magn. Reson. Med.*, Nice, 1995, **3**, 1685.
49. P. E. G. Kristjansen, B. Quistorff, M. Spang-Thomsen, and H. H. Hansen, *Ann. Oncol.*, 1993, **4**, 157.
50. E. O. Aboagye, R. J. Maxwell, A. D. Lewis, P. Workman, M. Tracey, and J. R. Griffiths, *Br. J. Cancer*, 1998, **77**, 65.
51. R. P. Mason, H. Shukla, and P. P. Antich, *Magn. Reson. Med.*, 1993, **29**, 296.
52. B. J. Dardzinski and C. H. Sotak, *Magn. Reson. Med.*, 1994, **32**, 88.
53. C. H. Sotak, P. S. Hees, H. N. Huang, M. H. Hung, C. G. Krespan, and S. Raynolds, *Magn. Reson. Med.*, 1993, **29**, 188.
54. P. S. Hees and C. H. Sotak, *Magn. Reson. Med.*, 1993, **29**, 303.
55. C. J. Deutsch and J. S. Taylor, *Biophys. J.*, 1989, **55**, 799.
56. M. Bental and C. J. Deutsch, *Am. J. Physiol.*, 1994, **266**, C541.
57. J. S. Beech and R. A. Iles, *Magn. Reson. Med.*, 1991, **19**, 386.
58. T. Frenzel, S. Kossler, H. Bauer, U. Niedballa, and H. J. Wienmann, *Invest. Radiol.*, 1994, **29**, S220.
59. S. Hunjan, R. P. Mason, V. D. Mehta, V. Padmakar, P. V. Kulkarni, V. Arora, and P. P. Antich, *Mag. Res. Med.*, 1998, **39**, 551.
60. A. S. E. Ojugo, P. M. J. McSheehy, D. J. O. McIntyre, C. McCoy, M. Stubbs, M. O. Leach, I. R. Judson, and J. R. Griffiths, *NMR Biomed.*, 1999, **12**, 000.
61. M. Kojima, S. Kuribayashi, Y. Kanazawa, T. Hardahira, Y. Maehara, and H. Endo, *Chem. Pharm. Bull.*, 1988, **36**, 1194.
62. Y. Kanazawa, K. Umayahara, T. Shimmura, and T. Yamashita, *NMR Biomed.*, 1997, **10**, 35.

Biographical Sketches

Paul M. J. McSheehy. *b* 1959. B.Sc.(Hons), 1980, Biochemistry, University of Sussex, Ph.D., 1984, Biochemistry, University of London, UK. Introduced to in vivo MRI by Professor J. R. Griffiths in 1986. Bone endocrinologist, currently Research Fellow at St. George's Hospital Medical School UK, studying the action mechanisms, and pharmacokinetics of chemotherapeutic drugs.

Laurent P. Lemaire. *b* 1968. B.S. 1989, Biochemistry–Biophysics, Ph.D., 1993, Pharmacology, University of Toulouse, France. Intro-

For References see p. 824

duced to NMR by P. Cozzone in 1989. Current research speciality: metabolism of anticancer drugs by NMR.

John R. Griffiths. *b* 1945. M.B. B.S., 1969, London, D. Phil., 1974, Biochemistry, Oxford, Studied ESR with George Radda, 1970–74 and (with David Gadian and Richard Isles) ^{31}P MRS of livers, 1979–80. Approx. 100 publications. Including (with E. R. Andrews et al.,) 'Clinical Magnetic Resonance: Imaging and Spectroscopy', publ. Wiley, Chichester, 1990. Since 1980 research interests have been in MRS as applied to cancer, latterly also MRI of nerves.

Fluorine-19 MRS: General Overview and Anesthesia

David K. Menon

University of Cambridge, UK

1 INTRODUCTION

Fluorine-19 has 100% natural abundance and possesses a spin of $\frac{1}{2}$ and an NMR sensitivity of 80–85% relative to protons. Consequently, fluorine-containing compounds produce NMR signals that are nearly as easily detected as proton signals. The range of chemical shifts for ^{19}F compounds is about 1000 ppm, far greater than that observed for ^{1}H spectroscopy, leading to much greater spectral dispersion. These attributes, coupled with the fact that ^{19}F magnetic resonance (MR) does not require solvent suppression, would be expected to make it an ideal nucleus for study in biological systems. Unfortunately, no naturally occurring molecules of biological consequence contain adequate concentrations of MR-visible ^{19}F (bone and teeth contain a substantial amount of ^{19}F, but this is contained in compounds that possess extremely short T_2 values). However, several pharmacologically active compounds contain fluorine, and ^{19}F magnetic resonance spectroscopy (MRS) has been used for the study of the biodistribution, pharmacokinetics, and metabolism of several drugs.[1] Other groups have also used ^{19}F-containing compounds as tracers for blood,[2] either to image tissue blood flow (perfusion) or intravascular volume (angiography). In addition, ^{19}F chemical shifts and relaxation parameters can be greatly influenced by the physical and chemical environment, and ^{19}F containing molecules in biological systems exhibit significant and variable ^{19}F{^{1}H} nuclear Overhauser effects (NOEs) which depend on their chemical environment and mobility.[3] These properties allow ^{19}F MR to be used to probe molecular interactions in biological systems.

Section 2 of this article provides an overview of the applications of in vivo ^{19}F MR to biological systems, while Section 3 covers the application of ^{19}F MR to research in anesthesia.

For list of General Abbreviations see end-papers

2 APPLICATIONS OF ^{19}F MR: AN OVERVIEW

In the following discussion illustrative examples are described, the aim being to outline the principles of applications of ^{19}F MR. The discussion will focus on experiments that have involved whole body in vivo MR either in experimental animals or in humans, but occasional reference will be made to in vitro studies of tissue preparations or cell cultures to illustrate a potential in vivo application.

2.1 Biodistribution and Pharmacokinetics

^{19}F MRS has been used to study the biodistribution of volatile anesthetic agents and antimitotic compounds, and studies addressing this area are discussed in some detail in later sections of this article. In addition, the technique has also been used to study the metabolism of other drugs, two examples of which are discussed here.

Karson et al.[4] studied subjects being treated with the antidepressant agent fluoxitine, and found brain concentrations of 1.3–5.7 μg mL^{-1} on a daily dose of 40 mg day^{-1} in volunteers. They found acquisition of signals from patients more difficult, but suggested a correlation between dose and measured brain concentrations. In a more recent paper,[5] the same group describe the use of a quadrature coil for localized ^{19}F MRS of fluoxitine in patients and reported on T_1 and T_2 values for the fluoxitine resonance, which was predominantly intracerebral in location. Ex vivo ^{19}F MRS of brain obtained from patients at autopsy showed that at least part of the in vivo resonance was due to the active metabolite norfluoxitine, and in vitro analysis of lipid extracts of brain showed far higher concentrations (12.3–18.6 μg mL^{-1}) than were observed in vivo, suggesting that some of the compound may be NMR invisible in vivo. These studies clearly demonstrate that, while ^{19}F MRS is technically suitable for clinical pharmacokinetic research, in vivo MRS needs to be underpinned in its initial phases by ex vivo and in vitro data. Relevant clinical studies might investigate the relationship of brain concentrations to drug response, and the behavior of brain concentrations in relation to changes in dosage schedules.

Lee et al.[6] used ^{19}F MRS to study the rate of absorption of the nonsteroidal antiinflammatory agent flubiprofen from a topical preparation in an animal model. This technique may be easily transferred to human studies, and may provide substantial advantages over traditional methods (which either involve excised human skin or require the use of radioisotope labeling) for studying transcutaneous drug absorption.

2.2 Metabolism

Most work reported in the literature has concentrated on the metabolism of drugs, and several examples are discussed in later sections in the context of volatile anesthetic agents and antimitotics. However, ^{19}F MRS may also be used to study the metabolism of modified physiological substrate molecules.

Some recent studies have investigated the use of fluorinated glucose analogs, such as 2-fluoro-2-deoxy-D-glucose (2-FDG), in imaging glucose utilization in intact animals.[7-9] Such studies provide a regional map of glucose utilization in a manner analogous to ^{18}FDG positron emission tomography (PET), but without the need to expose subjects to radiation.[7] In addition,

^{19}F MRS also provides the ability to interrogate individual metabolic pathways by detecting and quantifying downstream metabolites. Thus, 3-fluoro-3-deoxy-D-glucose (3-FDG) has been used to study glucose metabolism to sorbitol via the aldose reductase pathway,[9] a process that is thought to be intimately involved in the pathogenesis of diabetic cataract and peripheral neuropathy. This is of particular relevance because of the need to evaluate new aldose reductase inhibitors which may be of use in postponing or avoiding the long-term complications of diabetes mellitus. Unfortunately, while many animal studies are underway, the large doses of FDG required, coupled with the potential toxicity of the molecule, dictate that clinical studies are not likely to be possible in the immediate future.

2.3 Insights into Molecular Mechanisms of Drug Action and Metabolism

The interaction of ^{19}F-containing compounds with other molecules at their site of action may be elucidated by studying changes in NMR properties, including T_1, T_2, and NOE values. Such information may provide insight into the mechanisms of action or metabolism of drugs. For example, the demonstration of short T_2 environments for volatile anesthetics in the brain in some studies implies that the agents may be immobilized at their site of residence (see later), and that this environment is substantially different from that in adipose tissue, where these agents possess long T_2 values. This distinction may be of considerable importance in validating putative mechanisms of action of volatile anesthetic agents.

The T_1 relaxation of fluorinated compounds in biological systems is dominated by [1]H–[19]F interactions,[10] and the resulting NOEs can provide useful information regarding these interactions.[3] In small molecules, irradiation of the proton frequency often results in an enhancement of signal intensity (a positive NOE). In other systems, where the ^{19}F nucleus is bound to a large molecule (e.g. a protein), the NOE is typically negative (i.e. there is a reduction in signal intensity with irradiation of the proton frequency). In situations where the ^{19}F nucleus under study is part of a small molecule which attaches reversibly to a macromolecule, the observed NOE depends on the mole fraction of the small molecule that is bound and its rate of dissociation from the complex. In many biological systems the observed effects are dominated by macromolecular binding, even when the bound fraction is small and the rate of dissociation slow. Jacobson et al.[3] have used these effects to investigate the interaction of a cytochrome P450 inducer and a novel herbicide with macromolecules in the liver using in vivo ^{19}F MRS, and the technique may be applicable to many more experimental situations. Direct transfer to human research must be preceded by studies that estimate energy deposition produced by ^{1}H irradiation, but this is unlikely to be major problem, since similar issues have been addressed in the context of ^{13}C and ^{31}P MRS.

2.4 Information Regarding Physicochemical Environment

The T_1 of perfluorocarbons varies with, and may thus provide information regarding, their chemical and physical environment. The paramagnetic O_2 molecule reduces their T_1 from 1–4 to 0.3–0.5 s in linear proportion to the local partial pressure of oxygen.[11] When administered intravenously, these compounds are sequestered in tissues, where their T_1 value provides a measure of pO_2. Alternatively, the compound may be directly introduced into the site of interest (e.g. the vitreous humor). This technique has been used to estimate tissue pO_2 in experimental tumors, vitreous humor in animals and humans, and in animal myocardium.

Unfortunately, the T_1 of fluorocarbons also varies with temperature, making estimation of pO_2 difficult in experimental situations such as brain ischemia where there may be simultaneous changes in temperature. Berkowitz et al.[12] suggest the use of the compound perfluorotributylamine (FTBA) in this situation. While the T_1 value of FTBA changes with pO_2 and temperature, the chemical shift of its ^{19}F resonance is independent of pO_2, but shows a linear change with temperature. ^{19}F MRS of this compound can thus provide an independent measure of the tissue temperature from chemical shift data, and the resulting information can be used to correct T_1 data to provide independent and valid measures of pO_2.

The chemical shift of other ^{19}F compounds is sensitive to pH, and ^{19}F MRS has been used to measure pH in biological systems. For example, Beech and Iles[13] used the chemical shift of exogenously administered F-Quene 1 to estimate intracellular pH (pH_i) in rat liver in vivo. While the technique was practicable, it did not work consistently, and they found small differences when comparisons were made to estimates of pH_i from ^{31}P MRS (using the chemical shift of Pi).

Finally, when ^{19}F is covalently attached to various chelators (e.g. 5,5'-difluoro-1,2-bis(o-aminophenoxy)ethane-N,N,N',N'-tetraacetic acid (5-FBAPTA)), the ^{19}F chemical shift of the resultant compound is sensitive to binding by cations, and the local concentration of many cations (of which calcium is the most biologically important) can be estimated by measuring the relative concentration of the bound species. 5-FBAPTA and other fluorinated chelates have been used to study intracellular Ca^{2+} concentrations in cell preparations and tissue slices,[14] and more recent studies have used the techniques to quantify other cations.[15] However, no in vivo study has been reported.

2.5 Imaging of Large Vessel Flow and Tissue Perfusion

The perfluorocarbons have been used as 'contrast agents' for ^{19}F angiography, with varying degrees of success.[2,16] While their distribution in tissues provides a measure of local tissue blood flow, they do not give a true estimate of perfusion since they are retained in the intravascular compartment, and are not freely diffusible tracers. However, ^{19}F-containing gases and vapors are freely diffusible and can be used as tracers for estimating brain perfusion. Rudin and Sauter[17] used the washout of halothane to estimate cerebral blood flow (CBF) in rats. However, halothane is known to be a potent cerebral vasodilator and increases CBF in the doses used in this study. Consequently, its use to determine CBF is inappropriate. However, Pekar et al.[18] administered the inert diffusible gas trifluoromethane via the inhalational route in cats, and estimated CBF from wash-in and wash-out data. Their results suggest that the measurement of rCBF may be possible using this technique, with a spatial resolution of 0.4 mL, at least in experimental animals.

For References see p. 836

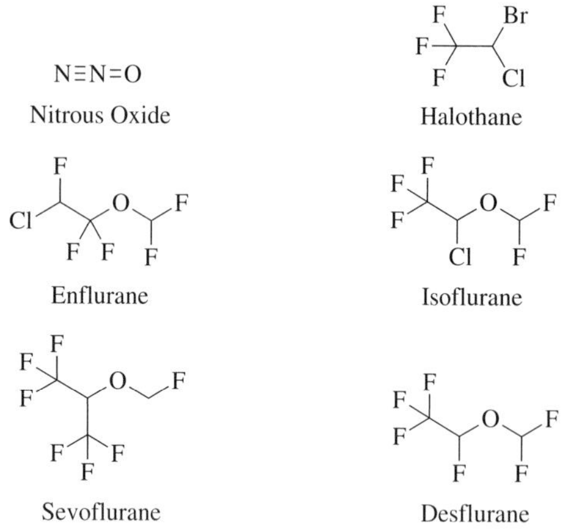

N≡N=O

Nitrous Oxide

Halothane

Enflurane

Isoflurane

Sevoflurane

Desflurane

Figure 1 Chemical structures of general anesthetic agents in common clinical use

3 ¹⁹F MRS STUDIES OF FLUORINATED ANESTHETIC AGENTS

3.1 Clinical and Pharmacological Context

Recent ^{19}F MRS studies of fluorinated anesthetics have been the source of some controversy, as results from some groups challenge generally held notions of anesthetic action. Two main issues are pertinent: (a) theories regarding the site of action of anesthetic agents; and (b) the duration of residence of modern anesthetic agents in the brain. Conventional viewpoints on these issues are briefly described below, in order to put the following discussion into some sort of context.

General anesthetic agents vary widely in chemical structure (Figure 1), but still continue to have surprisingly similar behavioral effects. However, we have little information on how and where they act in the brain. Two opposing theories claim very different sites of anesthetic action.[19] One school of thought attributes general anesthesia to action at a nonspecific hydrophobic site in the lipid bilayer of cell membranes which is 'disordered' by the entry of the anesthetic molecule.[20] The opposing school of thought suggests that these agents may act at specific stereoselective hydrophilic sites on membrane proteins.[21] Data that suggest molecular specificity of the interaction between anesthetic agents and the brain would provide support to the second theory.[21]

The fluorinated volatile agents constitute the most widely used class of general anesthetics in the world, and individual agents that are presently in use (Figure 1) include halothane, enflurane, isoflurane, sevoflurane, and desflurane.[22] These compounds are characterized, on clinical grounds, by an apparently rapid onset and offset of action. Conventionally, the rapidity of wash-in and wash-out of an anesthetic agent is thought to vary inversely with its solubility in blood and tissue.[22] On this basis, the agents mentioned above rank as follows: desflurane > sevoflurane > isoflurane > enflurane > halothane, where halothane has the highest blood gas solubility, and hence slowest wash-in and wash-out characteristics[22] (Figure 2). Nevertheless, even with halothane, the offset of clinical anesthesia is relatively rapid, with patients awakening within a few minutes of ceasing administration of the agent. However, patients are reported to

continue to have subtle psychomotor impairments for several hours to days after general anesthesia.[23]

3.2 Pharmacokinetics of Fluorinated Anesthetics: Is Anesthetic Residence Prolonged in the Brain?

Over the last 10 years there have been numerous papers from several groups that have used ^{19}F MRS to study anesthetic action in the brain. These address two main issues; the first of which is discussed in this section; the second issue is discussed in Section 3.3.

The first in vivo surface coil study of fluorinated anesthetics demonstrated the feasibility of such studies, and suggested that the compounds could be detected in the brain for substantially longer than had been expected.[24] Further studies from Wyrwicz et al.[25] conducted with 2.8 cm surface coil on the scalp, suggested that, following a 2-h exposure to 1% halothane, halothane was washed out of rabbit brain with a biexponential temporal profile (Figure 3). The initial rapid decay had a time constant of approximately 25 min, while the later slower washout phase had a time constant of 320 min. These findings were confirmed by in vitro ^{19}F MRS of extracts of excised rabbit brain at different intervals after the cessation of anesthesia (Figure 4). Although figures for washout half-times were not presented for the in vitro data, they were reported to show an 'elimination profile similar to that observed in intact animals'. These in vitro studies provided better spectral resolution than in vivo studies, and showed that, starting at 90 min after cessation of halothane anesthesia, brain halothane could be resolved into two resonances: a doublet with a 5.6 Hz proton J coupling that was attributed to the trifluoromethyl resonance of halothane, and a second peak 0.7 ppm downfield whose proportion increased with time, such that it represented 40% of the residual ^{19}F signal by 6 h. These two resonances were also differentiated by their relaxation properties. The halothane resonance showed a T_1 of 1.3 s and a T_2 of 3.8 ms, while the new singlet resonance had a T_1 of 2.8 s and a T_2 of 10.6 ms, suggesting that the two resonances were in different chemical environments. Further subcellular fractionation showed that the singlet resonance was confined to the cytosol, while the

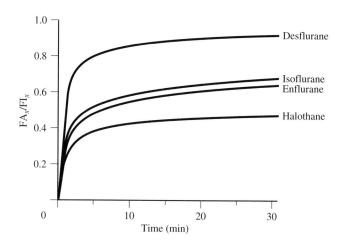

Figure 2 Increase in the alveolar fractional concentration (FA_x) toward that of inspired fractional concentration (FI_x) with time, compared for different fluorinated volatile agents. Note that the less soluble agents (isoflurane and desflurane) show a more rapid rise

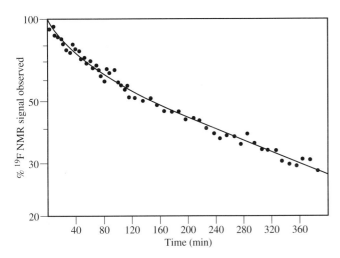

Figure 3 A representative time course of decay of the ^{19}F signal from halothane during recovery from anesthesia. Spectra were obtained during a pulse-and-collect experiment with a surface coil placed over the head of a rabbit after a 120-min exposure to 1% halothane. All signal intensities are expressed as a percentage of the initial signal obtained. (Reproduced with permission from A. M. Wyrwicz, C. B. Conboy, B. G. Nichols, K. R. Ryback, and P. Eisle, *Biochim. Biophys. Acta*, 1987, **929**, 271)

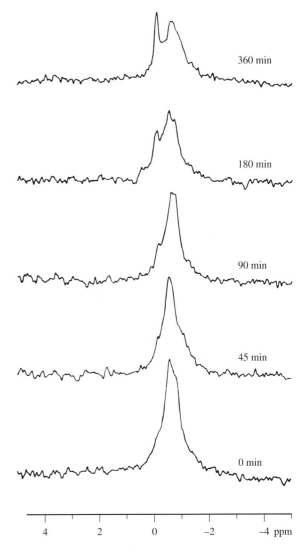

Figure 4 High-resolution ^{19}F MRS experiments on excised rabbit brain removed after varying periods of recovery (between 0 and 6 h) following halothane anesthesia. Chemical shifts are reported relative to an external 5 mM trifluoroacetic acid standard. Note the appearance of a new resonance in the 90-min spectrum, attributed to trifluoroacetic acid. (Reproduced with permission from A. M. Wyrwicz, C. B. Conboy, B. G. Nichols, K. R. Ryback, and P. Eisle, *Biochim. Biophys. Acta*, 1987, **929**, 271)

halothane trifluoromethyl resonance was widely present, and was loosely bound, being dissociated from cellular components by washing with 0.32 M sucrose. The authors concluded that the new singlet resonance represented a nonvolatile metabolite of halothane, possibly trifluoroacetate.

These findings were surprising, since they implied prolonged halothane residence in the brain, a concept that flew in the face of conventional perfusion-limited models of anesthetic elimination.[26] The possibility that a nonvolatile metabolite might be responsible for the prolonged ^{19}F signal provides some explanation for these findings, but did not account for the fact that even 6 h after anesthesia, 60% of the observed ^{19}F signal in excised brain appeared to arise from the trifluoromethyl resonance of unmetabolized halothane.[25] Furthermore, this theory could not explain the fact that in a similar experiment Wyrwicz et al.[27] observed similar two-compartment elimination for isoflurane (Figure 5), a compound that is only minimally (<1%) metabolized.[28] The initial decay phase for both halothane and isoflurane were similar ($t_{1/2}$ of 25 and 26 min, respectively), but the later elimination of halothane was significantly slower than isoflurane ($t_{1/2}$ of 320 and 174 min, respectively). The authors suggested that the slow, late elimination of halothane could be attributed to its greater tissue solubility and the presence of nonvolatile metabolites.

While the two pharmacokinetic compartments might represent brain tissue with varying perfusion (e.g. gray and white matter), the $t_{1/2}$ values were at odds with data obtained from non-NMR techniques. Cohen et al.[29] administered labeled halothane intravenously, and found minimal (<10%) retention of radiolabel after 20 min. Similarly, Wolff[30] found that 97% of a subanesthetic dose of halothane was eliminated within 3 h. However, it has been argued by Topham and Longshaw[31] that the pharmacokinetics of halothane may be significantly dependent on the route of administration and the dose administered. Consequently, the preceeding two experiments may be irrele-

vant to the process of clinical anesthesia. Indeed, Divakaran et al.[32] used gas chromatography to measure halothane levels in excised brain, and found elimination rates that correlate well with the figures obtained by Wyrwicz et al. However, Strum et al.[33] used gas chromatography to measure residual concentrations of isoflurane in rabbit tissue following 90 min of 1.3% isoflurane administration, and showed that brain concentrations of isoflurane were reduced to about 10% by 90 min.

In an effort to address these discrepancies, the research group at the University of California studied elimination of halothane and isoflurane in animal models. They hypothesized that much of the delayed elimination in published in vivo studies arose from contaminating signals in extracranial tissue, where volatile anesthetics tend to be retained, owning to lower perfusion. Accordingly, they used a spatially selective depth

For References see p. 836

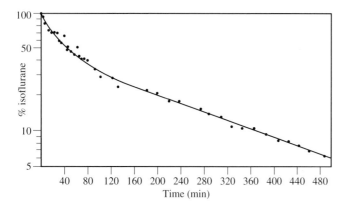

Figure 5 A representative time course of decay of the ^{19}F signal from isoflurane during recovery from anesthesia. Spectra were obtained during a pulse-and-collect experiment with a surface coil placed over the head of a rabbit after a 90-min exposure to 1.5% isoflurane. All signal intensities are expressed as a percentage of the initial signal obtained. (Reproduced with permission from A. M. Wyrwicz, C. B. Conboy, K. R. Ryback, B. G. Nichols, and P. Eisele, *Biochim. Biophys. Acta*, 1987, **927**, 86)

pulse in an effort to obtain a selective signal from deeper brain tissue during ^{19}F MRS studies of halothane elimination in rats.[34] They found that the ^{19}F NMR signal from halothane decreased to 40% of its initial value by 34 min (in contrast to about 240 min as reported by Wyrwicz et al.[25]). Several factors may have explained this discrepancy. First, halothane elimination may be significantly faster in rats when compared with rabbits. Second, the depth pulse imposed an acquisition delay of about 0.5 ms, and may have resulted in significant loss of signal from the short T_2 component. In a separate paper,[35] the same group reported data on isoflurane elimination in rabbits

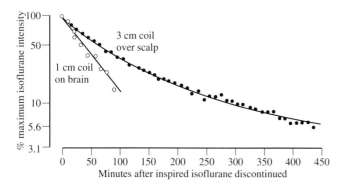

Figure 6 Time course of decay of the ^{19}F signal from isoflurane during recovery from anesthesia. Spectra were obtained during a pulse-and-collect experiment with a surface coil placed over the head of a rabbit after a 90-min exposure to 1.5% isoflurane. All signal intensities are expressed, on a log scale, as a percentage of the initial signal obtained. (○) Signal obtained from a 1.0 cm diameter surface coil placed directly over the dura in craniectomized animals, fitted to a single exponential decay. (●) Signal obtained from a similar experiment, acquired using a 3.0 cm diameter surface coil placed over the intact scalp of the rabbit; the line represents the optimum least-squares fit for a biexponential decay. (Reproduced with permission from P. Mills, D. I. Sessler, M. Moseley, W. Chew, B. Pereora, T. L. James, and L. Litt, *Anesthesiology*, 1987, **67**, 169)

with two protocols, both obtained with pulse-and-collect sequences (Figure 6). In the first set of experiments they used a 3 cm surface coil over the scalp, similar to that employed by Wyrwicz et al.,[27] and found similar kinetics for isoflurane elimination. In another experiment they placed a 1 cm surface coil directly over the exposed dura in rabbits. This study revealed a monoexponential decay of isoflurane concentrations in the brain, with a reduction to 15% of initial values by 90 min. These data agreed well with findings of Strum et al.,[33] who used non-NMR methods and showed a reduction to 10% of initial values by 90 min. The University of California group concluded that the later prolonged elimination kinetics for isoflurane observed by Wyrwicz et al. arose from signal contamination by isoflurane in extracerebral tissues.

The papers discussed above show that the issue of the spatial localization of the ^{19}F signal from anesthetic agents is an important one. Attempts have been made to clarify this point using ^{19}F MRI studies. While both spin echo[36] and gradient echo[37] sequences have been used to image halothane and enflurane in anesthetized animals, the echo times for these imaging studies have been 9 and 10 ms, respectively. As the T_2 of volatile anesthetics in the brain is short (typically 3–4 ms),[25] it is not surprising that these studies were unable to demonstrate any anesthetic in the brain, despite being conducted while the animal was still anesthetized. On the other hand, they clearly demonstrate fairly high concentrations of fluorinated agent in extracerebral tissues. When these distributions were compared with either ^{1}H lipid imaging[36] or ^{13}C imaging,[37] it appeared that the anesthetics localized in lipid rich areas in adipose tissue stores, glands, and muscle. Localized spectroscopy of volatile anesthetic agents therefore demands techniques that do not involve appreciable delays in data acquisition.

Wyrwicz and Conboy[38] used rotating frame zeumatography (RFZ) to examine the distribution of halothane in the rat head. Localized spectra were obtained from animals that were allowed to recover for varying periods after 90 min of anesthesia with 1% halothane (Figure 7). The localization of signals was necessarily imperfect since RFZ localizes signals along a single axis only. However, the ^{19}F spectra obtained immediately after anesthesia demonstrated very little variation in chemical shift across most of the rat head, but a heterogeneous distribution of ^{19}F signal across the animal's head with apparently lower peak heights in the 'brain' slices [Figures 7(a) and (b)]. The authors did not report on variations in T_2 or linewidths across the blocks of spectra, so estimates of signal intensity are impossible to obtain. RFZ experiments after 4.5 and 6.5 h of recovery from anesthesia showed signals from both halothane and halothane metabolite in all areas of the head [Figures 7(c) and (d)]. With time, the proportion of ^{19}F signal arising from halothane decreased, but this reduction followed a heterogeneous pattern across the head with more rapid reductions being seen in the 'brain' slices. Nevertheless, spectra obtained at 6.5 h (390 min) continued to show clearly detectable peaks from halothane in 'brain' slices. As the metabolite signal contribution increased, its distribution also changed, with a larger proportion of the ^{19}F signal arising from the brain. By 10 h after anesthesia, virtually all the ^{19}F signal arose from the metabolite resonance, and was mainly confined to 'brain' slices. No information regarding relative changes in signal intensity over time in different regions was provided, so even crude estimates of regional elimination halflives are impossible.

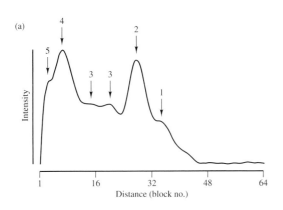

(a)

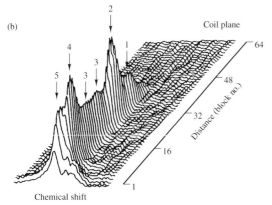

(b)

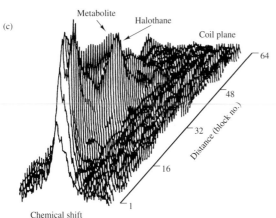

(c)

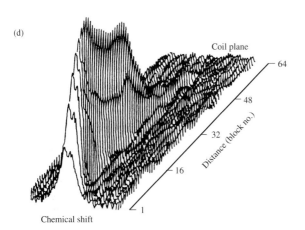

(d)

In a more recent study, Lockhart et al.[39] measured the cerebral uptake and elimination of halothane, isoflurane, and desflurane from rabbit brain, using a 1 cm surface coil positioned over the exposed dura. They found that rates of change of cerebral concentrations of anesthetic paralleled alveolar levels, and both cerebral uptake and elimination of desflurane were 3 times as fast as for halothane and 1.7 times as fast as for isoflurane (Figure 8). These findings would fit with standard clinical and pharmacokinetic premises, since desflurane has a rapid onset and offset of action owing to its lower blood and tissue solubility. The data for the elimination of isoflurane were similar to those obtained from the previous University of California study, with a monoexponential decay and only 10% of initial levels retained in the brain 90 min after discontinuation of isoflurane. The data also clearly demonstrate a hysteresis between alveolar and cerebral levels of all three volatile agents; alveolar concentrations were higher during the uptake phase, while cerebral concentrations were higher during the elimination phase (Figure 9). This is consistent with conventional pharmacokinetic modeling. However, it was found that the degree of hysteresis was identical for all three agents, suggesting that the rates of transfer between brain and blood are governed by factors other than blood and tissue solubility.

Chen et al.[40] studied elimination of isoflurane from the rat brain as a function of age using ^{19}F MRS with a surface coil. They found a biphasic elimination of isoflurane in both age groups with $t_{1/2}$ values of 7–9 and 100–115 min for the fast and slow components, respectively. The $t_{1/2}$ estimates for the fast compartment are consistent with data for elimination of isoflurane from the brain obtained using other techniques. Consequently, the authors attributed the slow component to washout of isoflurane from intracranial fatty tissue. They also found that older rats showed a slower elimination of isoflurane than younger rats, a difference that the authors attributed to impaired cardiovascular function.

What inferences can we draw from this review of ^{19}F MRS studies of the cerebral pharmacokinetics of anesthetic agents? Clearly, ^{19}F MRS provides a convenient technique for investigating these questions, but several issues need to be addressed. These include efficient spatial localization without acquisition delays, which will result in the loss of signal from short T_2 components. While many of the

Figure 7 Results of localized ^{19}F MRS using rotating frame zeumatography performed following administration of 1.5% halothane for 90 min. (a) One-dimensional projection across the rabbit head (from above downwards) showing the location of different blocks of spectra: 1, soft tissue (skin and muscle); 2, brain, muscle, and calvarium; 3, brain; 4, brain, muscle, and calvarium; 5, tissue below the cranium. (b) Stacked, spatially resolved spectra obtained immediately after discontinuing anesthesia, blocks are numbered as in (a). (c) Stacked, spatially resolved spectra obtained 4.5 h after cessation of anesthesia. Note the appearance of a new resonance to the left of the halothane resonance, which is attributed to trifluoroacetic acid. (d) Stacked, spatially resolved spectra obtained 6.5 h after cessation of anesthesia. Note the decrease in intensity of the halothane resonance, and the increase in the intensity of the metabolite resonance. (Reproduced with permission from A. M. Wyrwicz and C. B. Conboy, *Magn. Reson. Med.*, 1989, **9**, 219)

For References see p. 836

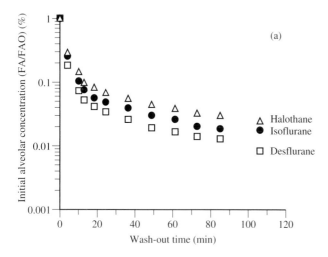

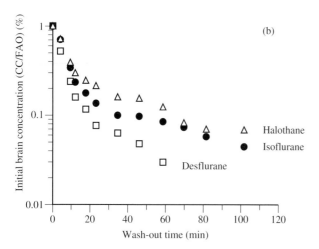

Figure 8 (a) Alveolar washout obtained from gas chromatography, and (b) cerebral washout obtained from in vivo ^{19}F MRS of anesthetic agent following anesthesia with desflurane (□), isoflurane (●), and halothane (△). (Reproduced with permission from S. M. Lockhart, Y. Cohen, N. Yasuda, B. Freire, S. Taheri, L. Litt, and E. I. Eger II, *Anesthesiology*, 1991, **74**, 575)

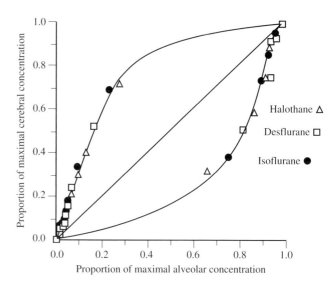

Figure 9 Plot of the relationship between cerebral and alveolar concentrations of volatile anesthetics during induction of and recovery from anesthesia (lower and upper halves of ellipsoid, respectively). Both parts of the curve deviate from the line of identity (straight line), suggesting a significant hysteresis between alveolar and brain levels of anesthesia. Note the unexpected result showing that all three anesthetic agents studied seem to fit to the same curve. (Reproduced with permission from S. H. Lockhart, Y. Cohen, N. Yasuda, B. Freire, S. Taheri, L. Litt, and E. I. Eger II, *Anesthesiology*, 1991, **74**, 575)

anesthetic retention may provide clues regarding the mechanisms of anesthetic action.

3.3 Sites of Anesthetic Residence in the Brain: Physical Environment and its Significance, Saturability

Many authors have reported that volatile anesthetics in the brain exhibit shorter T_2 values than when they are dissolved in lipid solvents or adipose tissue. These findings have been interpreted as implying immobilization of the fluorinated anesthetic molecules, and have led to speculation that the short T_2 environment may be intimately related to the process of anesthesia. These latter speculations were further fuelled by the results of a study by Evers et al.[41] in which T_2 values of four fluorinated anesthetic agents (methoxyflurane, enflurane, isoflurane, and fluoroxene) and one nonanesthetic fluorinated hydrocarbon (hexafluoroethane) in the brain were compared. The T_2 values of the four anesthetic agents were significantly shorter than in adipose tissue (0.5–4.5 versus 200–400 ms, respectively) and correlated linearly with the anesthetic potency of these agents (Figure 10). The nonanesthetic fluorocarbon exhibited a substantially higher T_2 (18.5 ms) in brain tissue. These data were interpreted as supporting the premise that the short T_2 environment represented a molecular site of anesthetic action.

In a later study, Evers et al.[42a] studied spontaneously breathing rats, anesthetized with halothane, using high resolution in vitro ^{19}F MRS of excised brain at different anesthetic concentrations and in vivo MRS of whole animals at 4.7 T. These experiments showed increasing cerebral levels of anesthetic with increasing inspired concentrations of halothane up to 2.5%. However, little or no further increases in cerebral halothane concentrations were observed when inspired

studies that reported prolonged anesthetic residence in the brain may have been flawed due to contamination by signal from extracerebral tissues, three points cannot be ignored. First, many MR studies tend to suggest slower anesthetic elimination from the brain when compared to non-NMR techniques. However, two more recent studies[39,40] suggest brain isoflurane kinetics that are similar to those obtained from non-NMR techniques. Second, the RFZ studies strongly suggest the presence of residual unmetabolized anesthetic in the brain as long as 6.5 h after anesthesia.[38] Finally, prolonged residence of at least some proportion of halothane is supported by the in vitro studies of brain extracts performed by Wyrwicz et al.[25] However, as several of the later studies have shown, this may be an extremely small proportion of the initial concentrations.

These small quantities of residual cerebral anesthetic are important for two reasons. First, they may provide an explanation for the minor but prolonged psychomotor deficits that are recognized to persist for up to days after general anesthesia.[23] Second, elucidation of the mechanisms that result in prolonged

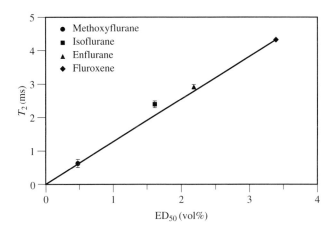

Figure 10 Relationship of anesthetic potency (measured as ED_{50}) to ^{19}F spin–spin relaxation times of fluorinated volatile anesthetics in the brain. (Reproduced with permission from A. S. Evers, J. C. Haycock, and D. A. d'Avignon, *Biochem. Biophys. Res. Commun.*, 1988, **151**, 1039)

halothane concentrations were increased from 2.5% to 4% (Figure 11). These data were interpreted as evidence for abundant saturable binding sites for halothane, a concept that did not support a nonspecific hydrophobic site of action of volatile anesthetic agents. Evers et al. also measured the T_2 relaxation parameters for the brain halothane resonance, and found that there were at least two components, with one T_2 value of about 3.6 ms (short T_2) and a second one of about 43 ms (long T_2).

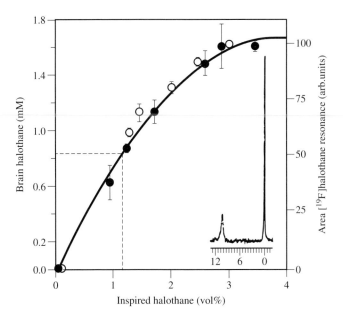

Figure 11 Brain halothane concentrations in rats as a function of inspired halothane concentrations, measured using either in vivo ^{19}F MRS (○) or ex vivo ^{19}F MRS (●) of excised brain. The horizontal dashed line indicates the half-maximal concentration of brain halothane concentration achieved in this experiment. This value (about 1.2%) corresponds closely to the ED_{50} for halothane in rats. The inset shows a representative ^{19}F MR spectrum from the ex vivo experiments. The resonance at 0 ppm arises from methoxyflurane in an external standard, while the peak at 10.2 ppm corresponds to halothane in the brain. (Reproduced with permission from A. S. Evers, B. A. Berkowitz, and D. A. d'Avignon, *Nature*, 1987, **328**, 157)

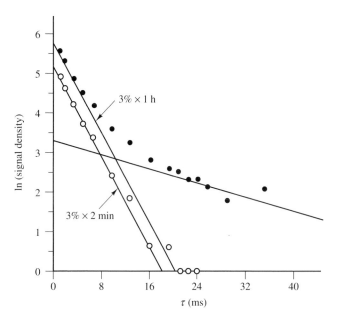

Figure 12 Decay of the ^{19}F MR signal from brain halothane in rats as a function of echo time (expressed as τ) in a Carr–Purcell–Meiboom–Gill sequence from rats exposed to 3% halothane for 2 min (○) or 1 h (●). Note the biexponential T_2 decay of the signal from rats exposed to halothane for 1 h, and the minimal increase in the amount of signal with short T_2 characteristics with longer exposure. (Reproduced with permission from A. S. Evers, B. A. Berkowitz, and D. A. d'Avignon, *Nature*, 1987, **328**, 157)

They also showed that the short T_2 site represented 85% of the halothane molecules present in the brain, and was preferentially occupied during the initial period of induction of anesthesia (Figure 12). Continuing anesthetic administration beyond this point seemed to occupy the long T_2 site. These findings were thought to provide further proof for the importance of the short T_2 site in the process of anesthesia. Unfortunately, these conclusions had to be revised owing to further experimental results.

First, both Evers et al.[42b] and Litt et al.[43] reported that the apparent saturability of brain halothane concentrations was a physiological artifact, being the consequence of respiratory depression at high inspired concentrations of halothane in spontaneously breathing animals. When the studies were repeated with artificially ventilated animals, the cerebral anesthetic concentrations scaled linearly with inspired concentrations of halothane (Figure 13). Furthermore, the short T_2 environment was not specific to brain,[44] but was also found in other tissues, including red blood cells, liver, and kidney, while the long T_2 behavior was exhibited by halothane in serum.

These more recent results have prompted a thorough review of the original paper by Evers et al.,[41] and have led to a reappraisal of the pharmacological relevance of the T_2 behavior of volatile anesthetic agents in the brain. The area clearly requires further study.

3.4 Metabolism of Fluorinated Agents

Many of the previous ^{19}F MRS studies of halothane provided some data on the production of metabolites of the agent. Other groups have used the technique to study directly the

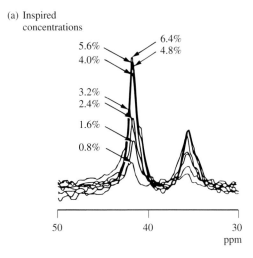

(a) Inspired concentrations

5.6%
4.0%
6.4%
4.8%
3.2%
2.4%
1.6%
0.8%

50   40   30
                      ppm

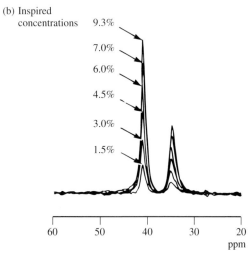

(b) Inspired concentrations   9.3%

7.0%
6.0%
4.5%
3.0%
1.5%

60   50   40   30   20
                            ppm

Figure 13 Superimposed ^{19}F MR spectra of brain isoflurane obtained at 4.7 T at different inspired concentrations of the agent, in spontaneously breathing (a) and artificially ventilated (b) animals. The apparent saturability of brain isoflurane at high inspired concentrations seen in (a) is a reflection of respiratory depression, since the spectra in (b) show a linear increase in brain halothane levels with inspired concentration. (Reproduced with permission from S. H. Lockhart, Y. Cohen, N. Yasuda, F. Kim, L. Litt, E. I. Eger II, L.-H. Chang, and T. James, *Anesthesiology*, 1990, **73**, 455)

metabolism of volatile anesthetics in the liver.[46–48] This issue is of considerable importance, since reactive metabolites of halothane and other agents are thought to act as haptens and lead to rare but fatal immunologically mediated liver failure.[49] For example, Preece et al.[50] studied the metabolism of enflurane in the liver of rats anesthetized with thiobarbitone. They demonstrated an elimination $t_{1/2}$ of 76 min for enflurane in the rat liver, and showed that this figure could be reduced to 39 min by pretreatment with isoniazid, which is known to enhance hepatic metabolism of enflurane (Figure 14). Similar studies may be useful in delineating pharmacogenetic differences in metabolism between individuals, or in elucidating the effects of drugs that enhance or inhibit the metabolism of volatile anesthetic agents. Data obtained from such studies may be quantitatively different, as in the studies by Preece et al.,[50] or show qualitative differences in metabolic pathways in different individuals or induced by different agents. Such processes

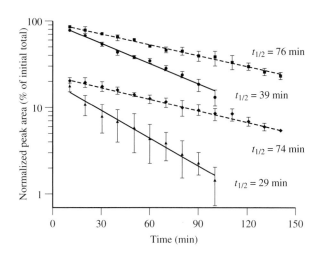

Figure 14 Decrease in the normalized signal intensities (as a percentage of the initial total) of enflurane in the liver with time. The plots show the decay of the major and minor resonances in control (■, ◆) and isoniazid pretreated rats (●, ▲). Points and bars represent the mean ± SEM of four rats; the calculated $t_{1/2}$ values for signal decay are also shown. (Reproduced with permission from N. E. Preece, J. Challands, and S. C. Williams, *NMR Biomed.*, 1992, **5**, 101)

could result in an increase or decrease in the production of toxic metabolites.

3.5 Clinical Studies

Clinical ^{19}F MRS studies of volatile anesthetics may have substantial advantages, since they enable the assessment of subtle cognitive dysfunction produced by trace residual anesthetic concentrations, and may be technically easier since the human head is larger, and the improved signal-to-noise ratio may enable more effective and specific volume selection. However, such studies also present considerable logistic and ethical problems. Menon et al.[51] reported the use of ^{19}F MRS to study

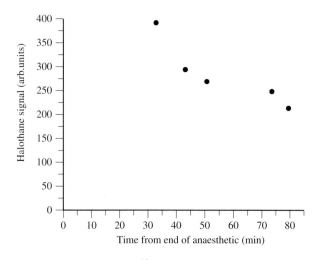

Figure 15 Decay of the ^{19}F signal intensity (arbitrary units) of halothane resonance with time after anesthesia in a patient where a single ^{19}F resonance was observed, and localized to the brain using a two-dimensional CSI sequence. (Reproduced with permission from D. K. Menon, G. G. Lockwood, C. J. Peden, I. J. Cox, J. Sargentoni, and J. D. Bell, *Magn. Reson. Med.*, 1993, **30**, 680)

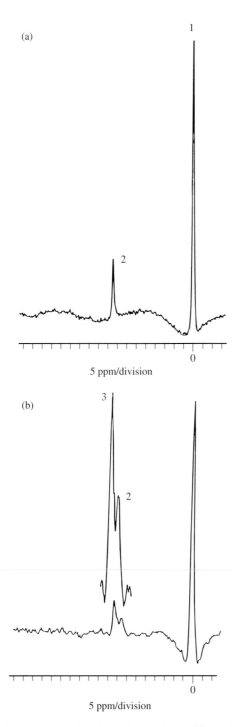

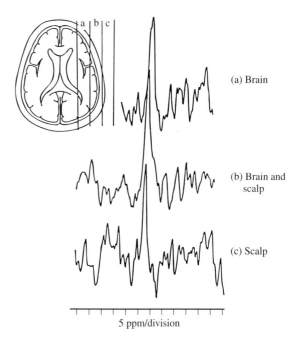

Figure 17   Spectra from three contiguous planes of the head acquired during the course of a two-dimensional CSI sequence, the location of which is shown on the line diagram of a transverse image of the head. Note that the resonances from superficial and deep slices have different chemical shifts (+38 and +41 ppm, respectively) and linewidths. (Reproduced with permission from D. K. Menon, G. G. Lockwood, C. J. Peden, I. J. Cox, J. Sergentoni, and J. D. Bell, *Magn. Reson. Med.*, 1993, **30**, 680)

Figure 16   Spectra acquired from the whole sensitive volume of a surface receiver coil in two patients recovering from halothane anesthesia. In both spectra the large resonance at 0 ppm arises from an external NaF standard. (a) The single resonance at +43 ppm is from halothane. (b) The two resonances at +38 and +41 ppm are clearly seen in the inset, where the peaks are enlarged. (Reproduced with permission from D. K. Menon, G. G. Lockwood, C. J. Peden, I. J. Cox, J. Sargentoni, and J. D. Bell, *Magn. Reson. Med.*, 1993, **30**, 680)

halothane in the brain of eight patients recovering from halothane anesthesia of short duration. A 6 cm surface receiver coil was used, and resonances attributable to halothane were observed in these patients up to 90 min after withdrawal of the

anesthetic agent (Figure 15). Localized spectroscopy with two-dimensional chemical shift imaging (CSI) showed that, although some of the signal arose from extracranial tissues, a substantial proportion arose from brain. The signal-to-noise ratio for an unlocalized spectrum was typically 20 with data collection times of 2 min.

In seven patients a single resonance was seen with a mean (± SD) chemical shift of +43.3 (± 1.8) ppm, referenced to NaF at 0 ppm [Figure 16(a)]. This chemical shift is comparable to that observed in animal studies of halothane anesthesia. This resonance exhibited a T_1 value of 0.5–1 s, and a T_2^* (estimated from the linewidth of the resonance) value of 3.5–10 ms. In one patient two resonances were observed with chemical shifts of +38 and +41 ppm [Figure 16(b)]. These two resonances may represent halothane in two different chemical environments within the brain, since phantom studies showed that the chemical shift of halothane in different environments (such as water, olive oil, methanol, and lecithin) could vary to an extent that accounted for the two resonances seen. The two resonances observed in this patient had different T_1 values and linewidths (and, by implication, different T_2^* values). Intriguingly, the two-dimensional CSI localization in this patient suggested that the broader resonance at +38 ppm was confined to the brain, while the narrow resonance at +41 ppm arose from overlying scalp (Figure 17). Although susceptibility effects could not be excluded, it is at least possible that these two resonances represent halothane in two different tissues.

While the results discussed above do not resolve the problems discussed in earlier sections, they do demonstrate the

For References see p. 836

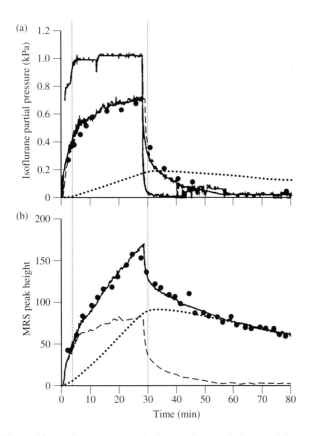

Figure 18 Isoflurane anesthesia in a volunteer during wash-in and wash-out.[53] The subject was unresponsive to verbal stimuli during the period between the two vertical dotted lines. (a) Partial pressure of isoflurane measured in inspired gas (upper continuous line), expired gas (lower continuous line) and arterial blood (solid circles). (b) Actual ^{19}F MRS data in arbitrary units (solid circles) and the summed intensities from the two-compartment model calculated for the data (continuous line). The fast (broken line) and slow (dotted line) components of the model are calculated from the MR data. In (a) the calculated fast-compartment trace is largely identical with the experimental data and is not visible. (Reproduced by kind permission of Oxford University Press from G. G. Lockwood, D. P. Dob, D. J. Bryant, J. A. Wilson, J. Sargentoni, S. M. Sapsed-Byrne, D. N. Harris and D. K. Menon, Magnetic resonance spectroscopy of isoflurane kinetics in humans, Part II: Functional localisation. *British Journal of Anaesthesia*, **79**, 586–9 (1999).)

feasibility of in vivo ^{19}F MRS studies of fluorinated volatile agents in humans, and confirm several findings observed in animal studies. In further studies, Lockwood et al.[52,53] have used ^{19}F MRS to study the pharmacokinetics of isoflurane in volunteers. They found two-compartment kinetics within the head with equilibrium half-times of 3.5 min and approximately 1 h with respect to expired isoflurane concentrations.[52] Using critical fusion flicker frequency as an objective measure of the cerebral effect of isoflurane, they found evidence to identify the fast component as in brain tissue. Responsiveness to command was lost at a brain partial pressure of 0.3% isoflurane.[52] It was concluded that these measurements exactly matched the predictions of the classical perfusion-limited model. These compartments showed decay half-times of 9.5 and 130 min, but the signal

was too weak to localize the compartments spatially. If the fast compartment is assumed to be the brain then these results match the predictions of the classical perfusion-limited pharmacokinetic model of inhalation anesthesia (Figure 18).[53]

4 RELATED ARTICLES

Brain MRS of Human Subjects; Cation Movements across Cell Walls of Intact Tissues Using MRS; pH Measurement In Vivo in Whole Body Systems; Temperature Measurement Using In Vivo NMR; Whole Body Studies: Impact of MRS.

5 REFERENCES

1. M. C. Malet-Marino and R. Martino, *Biochimie*, 1992, **74**, 785.
2. P. M. Joseph, J. E. Fishman, B. Mukherji, and H. A. Sloviter, *J. Comput. Assist. Tomogr.*, 1985, **9**, 1012.
3. A. R. Jacobsons, L. A. Sylvia, D. L. Veenstra, and J. T. Gerig, *J. Magn. Reson.*, 1992, **96**, 387.
4. C. N. Karson, J. E. Newton, P. Mohanakrishnan, J. Sprigg, and R. A. Komoroski, *Psychiatr. Res.*, 1992, **45**, 95.
5. R. A. Komoroski, J. E. O. Newton, D. Cardwell, J. Sprigg, J. Pearce, and C. N. Karson, *Magn. Reson. Med.*, 1994, **31**, 204.
6. D. J. Lee, C. T. Burt, and R. L. Koch, *J. Invest. Dermatol.*, 1992, **99**, 431.
7. T. Nakada, I. L. Kwee, B. V. Griffey, and R. H. Griffey, *Magn. Reson. Imag.*, 1988, **6**, 633.
8. T. Nakada, I. L. Kwee, B. V. Griffey, and R. H. Griffey, *Radiology*, 1988, **168**, 823.
9. T. Nakada, I. L. Kwee, P. J. Card, N. A. Matwiyoff, B. V. Griffey, and R. H. Griffey, *Magn. Reson. Med.*, 1988, **6**, 307.
10. J. T. Grieg, *Methods Enzymol.*, 1989, **177**, 1.
11. L. C. Clark Jr, J. L. Ackerman, S. R. Thomas, R. W. Millard, R. E. Hoffman, R. G. Pratt, H. RagleCole, R. A. Kinsey, and R. Janakiraman, *Adv. Exp. Med. Biol.*, 1984, **180**, 835.
12. B. A. Berkowitz, J. T. Handa, and C. A. Wilson, *NMR Biomed.*, 1992, **5**, 65.
13. J. S. Beech and R. A. Iles, *Magn. Reson. Med.*, 1991, **19**, 386.
14. H. S. Bachelard, R. S. Badar-Goffer, K. J. Brooks, S. J. Dolin, and P. G. Morris, *J. Neurochem.*, 1988, **51**, 1311.
15. R. Badar-Goffer, P. Morris, N. Thatcher, and H. Bachelard, *J. Neurochem.*, 1994, **62**, 2488.
16. P. M. Joseph, Y. Yuas, H. L. Kundel, B. Mukherji, and H. A. Sloviter, *Invest. Radiol.*, 1985, **20**, 504.
17. M. Rudin and A. Sauter, *NMR Biomed.*, 1989, **2**, 98.
18. J. Pekar, L. Ligeti, T. Sinnwell, C. T. Moonen, J. A. Frank, and A. C. McLaughlin, *J. Cereb. Blood Flow Metab.*, 1994, **14**, 656.
19. J. K. Alifimoff and K. W. Miller, 'Principles and Practice of Anesthesiology', M. C. Rogers, J. H. Tinker, B. G. Covino, and D. E. Longnecker ed., Mosby, St Louis, 1993, pp. 1035–1052.
20. K. W. Miller, *Int. Rev. Neurobiol.*, 1985, **27**, 1.
21. N. P. Franks and W. R. Lieb, *Nature*, 1994, **367**, 607.
22. D. E. Longnecker and F. L. Miller, 'Principles and Practice of Anesthesiology', ed. M. C. Rogers, J. H. Tinker, B. G. Covino, and D. E. Longnecker, Mosby, St Louis, 1993, pp. 1053–1086.
23. M. Herbert, T. E. J. Healy, J. B. Bourke, I. R. Fletcher, and J. M. Rose, *Br. Med. J.*, 1983, **286**, 1539.
24. A. Wyrwicz, M. H. Pszenny, J. C. Schofield, P. C. Tillman, R. E. Gordon, and P. A. Martin, *Science*, 1983, **222**, 428.

25. A. M. Wyrwicz, C. B. Conboy, B. G. Nichols, K. R. Ryback, and P. Eisle, *Biochim. Biophys. Acta*, 1987, **929**, 271.
26. E. I. Eger II, 'Anaesthetic Uptake and Action', Williams & Wilkins, Baltimore, 1974, p. 230.
27. A. M. Wyrwicz, C. B. Conboy, K. R. Ryback, B. G. Nichols, and P. Eisele, *Biochim. Biophys. Acta*, 1987, **927**, 86.
28. J. H. Lewis, H. J. Zimmerman, K. G. Ishak, and F. G. Mulluk, *Ann. Int. Med.*, 1983, **98**, 984.
29. E. N. Cohen, K. L. Chow, and L. M. Mathers, *Anesthesiology*, 1972, **37**, 324.
30. M. S. Wolff, *J. Toxicol. Environ. Health*, 1977, **2**, 1079.
31. J. C. Topham and S. Longshaw, *Anesthesiology*, 1972, **37**, 311.
32. P. Divakaran, F. Joiner, B. M. Rigor, and R. C. Wiggins, *J. Neurochem.*, 1980, **34**, 1543.
33. D. P. Strum, B. H. Johnson, and E. I. Eger II, *Science*, 1986, **234**, 1586.
34. L. Litt, R. Gonzalez-Mendez, T. L. James, D. I. Sessler, P. Mills, W. Chew, M. Moseley, B. Pereira, J. W. Severinghaus, and W. K. Hamilton, *Anesthesiology*, 1987, **67**, 161.
35. P. Mills, D. I. Sessler, M. Moseley, W. Chew, B. Pereora, T. L. James, and L. Litt, *Anesthesiology*, 1987, **67**, 169.
36. T. L. James, L.-H. Chang, W. Chew, R. Gonzalez-Mendez, L. Litt, P. Mills, M. Moseley, B. Pereira, D. I. Sessler, and P. R. Weinstein, *Ann. NY. Acad. Sci.*, 1987, **508**, 64.
37. T. Hashimoto, H. Ikehira, H. Fukuda, Y. Ueshima, and Y. Tateno, *Magn. Reson. Imag.*, 1991, **9**, 577.
38. A. M. Wyrwicz and C. B. Conboy, *Magn. Reson. Med.*, 1989, **9**, 219.
39. S. H. Lockhart, Y. Cohen, N. Yasuda, B. Freire, S. Taheri, L. Litt, and E. I. Eger II, *Anesthesiology*, 1991, **74**, 575.
40. M. Chen, J. I. Olsen, J. A. Stolk, M. P. Schweizer, M. Sha, and I. Ueda, *NMR Biomed.*, 1992, **5**, 121.
41. A. S. Evers, J. C. Haycock, and D. A. d'Avignon, *Biochem. Biophys. Res. Commun.*, 1988, **151**, 1039.
42. (a) A. S. Evers, B. A. Berkowitz, and D. A. d'Avignon, *Nature*, 1987, **328**, 157; (b) correction: 1989, **341**, 766.
43. S. H. Lockhart, Y. Cohen, N. Yasuda, F. Kim, L. Litt, E. I. Eger II, L.-H Chang, and T. James, *Anesthesiology*, 1990, **73**, 455.
44. H. J. C. Yeh, E. J. Moody, and P. Skolnick, *Nature*, 1990, **346**, 227.
45. H. Zimmerman, *Hepatology*, 1991, **13**, 1251.
46. D. S. Selinsky, M. Thompson, and R. E. London, *Biochem. Pharmacol.*, 1987, **36**, 413.
47. B. S. Selinsky, M. E. Perlman, and R. E. London, *Mol. Pharmacol.*, 1988, **33**, 559.
48. B. S. Selinsky, M. E. Perlman, and R. E. London, *Mol. Pharmacol.*, 1988, **33**, 567.
49. H. Zimmerman, *Hepatology*, 1991, **13**, 1251.
50. N. E. Preece, J. Challands, and S. C. R. Williams, *NMR Biomed.* 1992, **5**, 101.
51. D. K. Menon, G. G. Lockwood, C. J. Peden, I. J. Cox, J. Sargentoni, G. A. Coutts, and J. G. Whitesam, *Magn. Reson Med.*, 1993, **30**, 680.
52. G. G. Lockwood, D. P. Dob, D. J. Bryant, J. A. Wilson, J. Sargentoni, S. M. Sapsed-Byrne, D. N. Harris, and D. K. Menon, *Br. J. Anaesth.*, 1997, **79**, 581.
53. G. G. Lockwood, D. P. Dob, D. J. Bryant, J. A. Wilson, J. Sargentoni, S. M. Sapsed-Byrne, D. W. Harris, and D. K. Menon, *Br. J. Anaesth.*, 1997, **79**, 586.

Biographical Sketch

David Krishna Menon. *b* 1956. M.B.B.S. 1977, M.D. 1982, M.R.C.P. 1984, F.R.C.A. 1988, Ph.D. 1995, F. Ac. Med. Sci. UK 1998, F.R.C.P. 1999; internal medicine residency, Jawaharlal Institute, Pondicherry, India 1978–1983; Medical Registrar, Professorial Medical Unit, Leeds General Infirmary 1985–86; Registrar in Anaesthesia, Royal Free Hospital, London 1987–1988; MRC Research Fellow, NMR Unit, Hammersmith Hospital 1989–91. Presently: lecturer in anaesthesia, Cambridge University, and Director, Neurosciences Critical Care Unit, Addenbrooke's Hospital, Cambridge, UK. Current research interests: metabolic imaging of acute brain injury and brain inflammation following ischemia and trauma.

Body Fat Metabolism: Observation by MR Imaging and Spectroscopy

E. Louise Thomas and Jimmy D. Bell

Imperial College School of Medicine, Hammersmith Hospital, London, UK

1 INTRODUCTION

The importance of lipids in both health and disease is increasingly recognized. The major causes of morbidity and mortality in the western world include cancer, coronary heart disease (CHD), diabetes mellitus, and obesity. Lipids appear to have a major role in both treatment and prevention.[1,2] There is, therefore, a growing need for in vivo methods for studying human lipid metabolism and investigating lipid depots in the body. The development and use of nuclear magnetic resonance (NMR) techniques of imaging (MRI) and spectroscopy (MRS) for lipid studies is a major recent advance.

Since the late 1980s, interest has increased in the use of both in vivo and in vitro NMR for investigation of human and animal metabolism. Researchers are taking advantage of the nondestructive and noninvasive nature of the technique. These characteristics are particularly important for widespread, population-based studies, which often require serial examinations.

The initial applications of in vivo NMR to the study of lipids were principally concerned with development of the NMR methodology, rather than focusing on specific biochemical problems.[3,4] However, even at this early stage, the potential of MRI and MRS when applied to lipid metabolism could be envisaged. In this chapter, we will review the recent use of these techniques both in vivo and in vitro for investigation of human lipid metabolism and body fat composition and deposition.

2 USEFUL NUCLEI FOR NMR STUDIES OF LIPIDS

MRS with ^{1}H, ^{2}H, ^{13}C, and ^{31}P nuclei has been applied to the study of lipid metabolism in animals and humans.[3-9] The ^{1}H nucleus has the highest natural abundance in biological tis-

For References see p. 843

sues and is the highest sensitivity in NMR compared with other useful nuclei. In the past, the application of in vivo ^{1}H MRS to the study of lipids was not considered to be a useful technique. Problems arising from the intrinsically small chemical shift range resulted in severe signal overlap, which had limited its use. Recently, however, ^{1}H NMR has been shown to be an extremely powerful noninvasive method to determine intramuscular lipid content.[7,8]

MRS using ^{13}C in vivo has been successfully applied to studies of lipid metabolism and adipose tissue composition.[3,4,9] Compared with studies using ^{1}H, ^{13}C MRS is relatively insensitive because of the low gyromagnetic ratio and low natural abundance (1.1%) of the ^{13}C nucleus. Its relative sensitivity is further reduced by the ^{1}H–^{13}C coupling, which splits the ^{13}C signal, effectively reducing intensity. These problems are partially offset by the short relaxation times of most ^{13}C resonances, the high adipose tissue content of ^{13}C-containing compounds, and the use of decoupling techniques. Moreover, the very fact that ^{13}C has 1.1% natural abundance has been exploited by using ^{13}C-enriched metabolites in turnover studies, similar to classical ^{14}C-tracer studies.

MRS with ^{2}H has only had limited use in in vivo metabolic studies. The deuterium nucleus is quadrupolar ($I = 1$). It has both a low gyromagnetic ratio and natural abundance (0.015%), which leads to a very low sensitivity relative to proton spectroscopy (1.45×10^{-6}). These unfavorable properties are offset by its short relaxation times and high body content (12 mmol/l), allowing its detection by natural abundance in vivo MRS. Furthermore, the tissue content can be readily increased by use of deuterated water (D_2O), a fact that Brereton et al. have utilized to great effect in lipid turnover studies.[5]

Limited information can be obtained from in vivo ^{31}P MRS of lipids because ^{31}P NMR detects only phosphate-containing compounds. Consequently, ^{31}P MRS has principally been applied to lipid studies in vitro.[10–12] One area of research where in vivo ^{31}P MRS has been applied is to the study of membrane phospholipids. However, as the signals from the bound phospholipids tend to be very broad, the useful resonances come principally from phospholipid precursors (phosphocholine and phosphoethanolamine) or breakdown (glycerophosphocholine and glycerophosphoethanolamine) products and, therefore, will not be discussed further in this chapter.

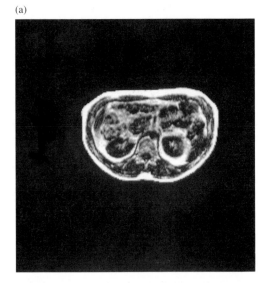

(a)

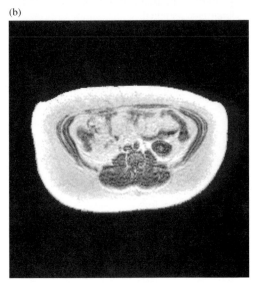

(b)

Figure 1 Typical abdominal transverse MR images showing distribution of internal and subcutaneous fat in a lean (a) and an obese (b) volunteer. Images were acquired using a rapid T_1-weighted spin echo sequence

3 MRI OF BODY FAT CONTENT AND DISTRIBUTION

The accurate determination of total body fat content and measurement of regional fat depots has become an important issue as the contribution of body fat to diseases such as non-insulin-dependent diabetes mellitus and CHD has become clearer. There are many techniques that have long been used to give estimates of total and peripheral body fat content with varying degrees of accuracy. However, until the development of techniques such as X-ray computed tomography (CT) and MRI, it was not possible to differentiate between subcutaneous and internal fat depots (Figure 1).

Internal fat depots, in particular visceral fat, may be a key factor in disease development. CT gives an accurate direct measurement of visceral fat depots, though exposure to ionizing radiation makes whole body fat measurements, especially for serial studies, impractical; consequently only single slices tend to be acquired.

Initial studies using MRI to study body fat have focused on the validation of the technique. MRI has been validated in phantoms, animals, and human cadavers and has been shown to measure both muscle and adipose tissue in vivo accurately, showing good agreement with values produced by dissection and chemical analysis.[13–17] MRI has also been compared with other techniques such as underwater weighing, anthropometry, body water dilution, impedance, and dual-energy X-ray absorptiometry (DEXA).[18–24] Generally there is good degree of correlation between the different methods, though agreement between methods for individuals can be quite variable.[24]

A wide variety of methodologies have been used in the application of MRI to the study of body fat. Different parameters and methods of data collection ranging from extrapolation of single-slice or multiple-slice acquisitions over selected regions of the body to whole body fat measurements have been used. Many studies have been published with single-slice MR data from the abdomen, generally at the level of L4/L5; however, there are significant drawbacks with this approach.

For an accurate measurement of body fat content, it is important that sufficient data be collected from the whole region of interest so that subtle changes are not missed or overinterpreted. A change or lack of change reported using single-slice CT or MRI scans from a selected region of the abdomen might not reflect the effect of the intervention on the entire adipose tissue depot. Indeed, it has previously been shown that a single CT scan obtained at the level of the umbilicus contains a substantial amount of retroperitoneal fat, which is less metabolically active than other visceral fat depots. It has been suggested that changes occurring in the entire visceral fat depot may be 'diluted out' by the presence of the less-active retroperitoneal fat in the single slice. Factors such as this can have a profound effect on the final results and their interpretation. It is, therefore, more appropriate to collect sufficient data from the entire depot to be studied. One approach is to obtain multislice data from the whole body, as shown in Figure 2.

The main applications of MRI to the measurement of body fat depots in human subjects have been following interventions such as diet and exercise.[25–27] Ross et al., using whole body MRI, have shown significant reductions in total and regional body fat content in obese subjects following a combination of diet and exercise (resistance or aerobic) and diet alone.[25,26] They found similar changes in body composition in response to diet combined with resistance exercise and diet combined with aerobic exercise.[26] Furthermore, significantly more subcutaneous fat was lost from the abdomen compared with the lower body and there was a greater loss of visceral fat from the upper than from the lower abdomen. The combination of diet and exercise resulted in a greater fat loss than occurred with diet alone.[25] Thomas et al., also using whole body MRI, have suggested that there is a preferential loss of visceral fat in lean women following moderate aerobic exercise without dietary restriction.[27] Interestingly, the change in body composition was only detected using MRI; weight and body fat content measured by impedance and anthropometry were not significantly different following exercise.

A preferential loss of visceral fat has also been reported in obese individuals following dietary restriction and treatment with dexfenfluramine.[28–31] However, these studies evaluated regional body fat distribution by measuring the area on a single MRI scan, which, as discussed above, could give misleading results.

4 IN VIVO MRS

4.1 Proton MRS

The application of in vivo proton MRS to the study of lipids has until recently been rather limited because of the small chemical shift range of ¹H resonances and the intense water signal. However, a number of researchers have shown that in vivo ¹H MRS can be used to determine noninvasively the triglyceride content within muscle cells known as intramyo-

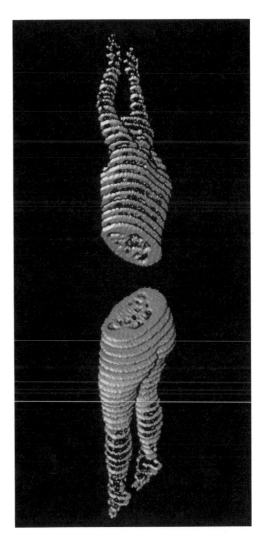

Figure 2 Transverse MR images showing distribution of internal and subcutaneous fat from a whole body MRI data set from a healthy female volunteer age 21 years, basal metabolic index 27.9 kg/m^{-2}, waist to hips girth ratio 0.81, subcutaneous fat 29.6 l, and visceral fat 2.25 l. Images were acquired using a rapid T_1-weighted spin echo sequence from the volunteer's fingertips to her toes by acquiring 10 mm thick transverse images with 30 mm gaps between the slices

cellular lipids (IMCL). This is particularly important as there is evidence to suggest that these 'muscle triglycerides' may be implicated in the pathogenesis of insulin resistance. Schick et al., using ¹H MRS to detect signals from human skeletal muscle, reported that the methyl and methylene lipid signals each consisted of two well-resolved peaks (Figure 3).[7] They suggested that the two pairs of peaks corresponded to IMCL and triacylglycerols in adipocytes between muscle fibers (extramyocellular lipids, EMCL).

Evidence in support of this interpretation has come from other groups.[32,33] Boesch et al. compared measurement of IMCL by in vivo ¹H MRS with morphometry and chemical analysis of human biopsy samples and suggested that measurement of IMCL by ¹H MRS had the best correlation with the estimation of the 'true' level of IMCL.[32] Szczepaniak et al., in a very elegant study of subjects with congenital lipodystrophy, a condition associated with almost complete absence of EMCL,

For References see p. 843

(a)

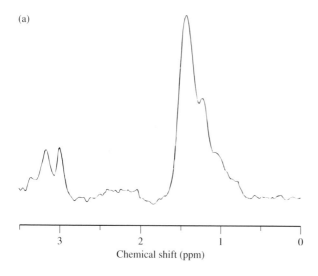

Chemical shift (ppm)

(b)

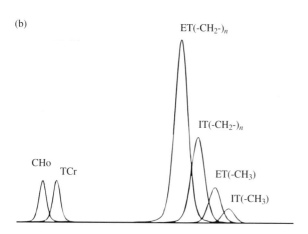

ET(-CH$_2$-)$_n$

IT(-CH$_2$-)$_n$

CHo
TCr
ET(-CH$_3$)
IT(-CH$_3$)

Figure 3 In vivo ^{1}H MR spectrum from the soleus muscle of a healthy volunteer before (a) and after (b) line fitting. Cho, choline and carnitine; TCr, creatine and phosphocreatine; ET($-$CH$_2-$)$_n$, extracellular muscle triglycerides methylene; IT($-$CH$_2-$)$_n$, intracellular muscle triglycerides methylene; ET($-$CH$_3$)$_n$, extracellular muscle triglycerides methyl; IT($-$CH$_3$)$_n$ intracellular muscle triglycerides methyl

showed that the in vivo ^{1}H NMR spectra revealed only single methylene and methyl resonances, corresponding to IMCL.[33]

Using ^{1}H MRS, Rico-Sanz et al. have shown diversity in the level of IMCL (as well as other muscle metabolites) in human skeletal muscle.[34] Levels of IMCL were significantly lower in the tibialis muscle than in the soleus and gastrocnemius muscles, possibly resulting from differences in fiber type composition and deposition of metabolites owing to adaptation of the muscles for locomotion.

To date, most studies have concentrated on looking at the effects of exercise on IMCL. Boesch et al. reported a significant decrease (about 40%) in IMCL in tibialis anterior muscle following 3 h of intensive cycling.[8] However, it appears that the nature and intensity of the exercise may also be important, as Rico-Sanz et al. showed no changes in IMCL levels in the tibialis, gastrocnemius, or soleus muscles following two different 90 min moderate exercise protocols.[35]

Interestingly, there have been several reports in abstract form demonstrating that IMCL content assessed by ^{1}H MRS is elevated in insulin-resistant individuals.[36,37] However, this relationship was not found in subjects from all ethnic groups.[37]

NMR images of the leg also tend to be acquired during the spectroscopy examination. These provide additional information from which it is possible to measure subcutaneous fat, bone marrow, bone, and levels of EMCL. The role of EMCL in muscle metabolism is not fully understood, but it is thought that EMCL and IMCL may have different roles. The combination of MRI and ^{1}H MRS will be excellent tools for increasing our knowledge of the metabolism of these two lipid depots.

4.2 Carbon-13 MRS

High-resolution ^{13}C MRS has found widespread use in the study of lipids in vitro.[38,39] In particular, the clear distinction of multiple, different fatty acid groups allows this technique to be used, often quantitatively, in studies of dietary oils.[40–43] This degree of resolution cannot be achieved in vivo, which limits the utility of this technique in noninvasive human research. However, major lipid groups can be distinguished and, within these constraints, important clinical and biochemical work has been carried out using proton-coupled ^{13}C and proton-decoupled {^{1}H} ^{13}C MRS. A typical ^{13}C NMR spectrum of human adipose tissue is shown in Figure 4.

Canioni et al. first showed that in vivo ^{13}C {^{1}H} MRS could detect differences in the lipid composition of adipose tis-

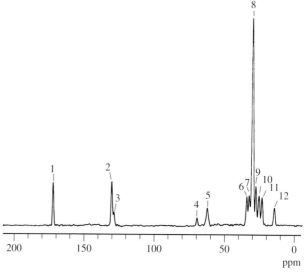

Figure 4 Natural abundance in vivo ^{13}C{^{1}H} NMR spectrum of human adipose tissue, dominated by signal from triglycerides. Peak assignment (referenced to $-$**C**H$_3$): 1. **C**=O (171.79 ppm); 2. $-$**C**H=**C**H$-$ (monounsaturated) and $-$**C**H=**C**H$-$CH$_2$$-$**C**H=**C**H$-$ (polyunsaturated) (129.83 ppm); 3. $-$CH=CH$-$**C**H$_2$$-$CH=CH$-$ (polyunsaturated) (128.13 ppm); 4. **C**2 glycerol (69.21 ppm); 5. **C**1, **C**3 glycerol (62.05 ppm); 6. $-$**C**H$_2$CH$_2$$-CO-O-$R (33.93 ppm); 7. $-$**C**H$_2$$-CH_2$$-CH_3$ (32.14 ppm); 8. $-$(**C**H$_2$)$_n$$-$ (29.69 ppm); 9. $-$**C**H$_2$$-$CH=CH$-$ (27.39 ppm); 10. $-$**C**H$_2$$-CH_2$$-CO-O-$R (25.04 ppm); 11. $-$CH$_2$$-$**C**H$_2$$-CH_3$ (22.94 ppm); 12. $-$CH$_2$$-$**C**H$_3$ (14.1 ppm)

sue and liver in rats fed a diet high in polyunsaturated fatty acids.[3] Further work by Sillerud et al. examined $^{13}C\{^1H\}$ NMR of triacylglycerols in rat adipocytes in vitro and advanced our knowledge particularly of signal assignment in biological systems.[44] Moonen et al. developed in vivo $^{13}C\{^1H\}$ NMR to characterize adipose tissue in human subjects.[4] They confirmed that linoleic acid ($C_{18:2n-6}$), usually the principal polyunsaturated fatty acid present in human adipose tissue, dominated the polyunsaturated fatty acid carbon signal observed in vivo, allowing estimation of this stored essential fatty acid.

In vivo ^{13}C MRS has been used to study the fatty acid composition of adipose tissue in rats fed diets based on significantly different fatty acid mixtures.[3,45,46] In animals fed fats with different fatty acid content (butter/lard, olive oil, sunflower oil, fish oil), clear differences were found in the in vivo spectra, consistent with a significant effect of dietary fatty acid intake on the tissue fatty acid profile. Progressively increasing total unsaturated and polyunsaturated fatty acid content in the diet was reflected in the adipose tissue composition. However, data from both groups appeared to show discrepancy in the results obtained by ^{13}C MRS compared with GLC, which worsened in adipose tissue from animals on a diet high in complex polyunsaturated fat, a finding later confirmed in human subjects.[47]

In humans, the turnover of fatty acids in adipose tissue is slow (half-life 600 days).[48] The fatty acid profile of human subcutaneous fat, therefore, provides an index of the habitual dietary fatty acid intake over the previous 2 to 3 years.[49] This information is important for long-term epidemiological surveys but has previously been limited by the need for repeat tissue biopsies for fatty acid estimation. The application of in vivo ^{13}C MRS to the noninvasive analysis of adipose tissue composition is, therefore, of particular value for human nutritional studies.

Beckmann et al. have used in vivo ^{13}C MRS to show a significant change in human adipose tissue fatty acid composition following a fat-reduced diet, with a correlation between diet and tissue monounsaturated fatty acids.[9] No difference was found in the degree of polyunsaturation, and this may reflect the limited period (6 months) of dietary change. Thomas et al. studied vegan subjects as a defined population, with an established long-term diet high in polyunsaturated fatty acids.[50] A significantly increased adipose tissue total unsaturated and mono- and polyunsaturated fatty acid carbon content was shown in the ^{13}C spectra from the vegan group compared with omnivore controls (Figure 5). Interestingly they found no significant differences in adipose tissue composition between vegetarian subjects and the omnivore controls.

MRS with ^{13}C has also been used to investigate the influence of maternal diet on infant adipose tissue composition. Thomas et al. compared the adipose tissue composition of breast-fed infants of women who maintained either an omnivore or a vegan diet.[51] The adipose tissue composition of infants directly reflects that of their mothers, with the vegan infants having 70% more polyunsaturated fatty acid carbons than the omnivore infants. Although the consequences of essential fatty acid deficiency in formula-fed infants are well documented, less is known about the effects of very high levels of long-chain polyunsaturated fatty acids in the infant diet. The long-term effects for vegan infants having such high levels of

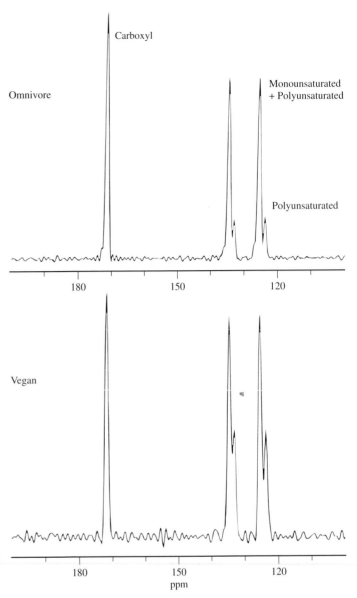

Figure 5   Alkene region from natural abundance in vivo coupled ^{13}C NMR spectra of human thigh adipose tissue in vegan and omnivore subjects. Spectra are scaled to the carboxyl signal. The vegan subjects, compared with omnivores, have an increased adipose tissue content of both total unsaturated and polyunsaturated fatty acid carbons (polyunsaturated: $3.45\pm0.87\%$ versus $2.39\pm0.55\%$ of total fatty acid carbons, $p<0.05$). This reflects at least 3 years' adherence to the vegan diet, which, in the absence of all animal products, is high in polyunsaturated fats (polyunsaturates $33.13\pm7.4\%$ versus $25.2\pm8.5\%$ total dietary fat polyunsaturated, $p<0.05$)

polyunsaturated fatty acids in their diet and adipose tissue are unknown. In the same study, Thomas et al. studied the adipose tissue composition of term and pre-term infants at birth, 6 weeks, and 6 months and compared them with their mothers. They found an increase in unsaturated fatty acids with increasing gestational age and maturity (Figure 6).

The complex interactions between lipids in plasma and those in the body tissues are not fully defined and are becom-

For References see p. 843

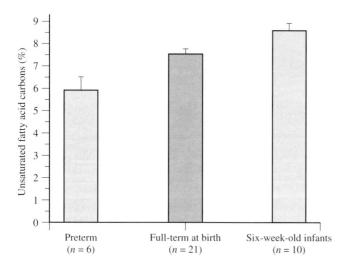

Figure 6 Fatty acid composition of adipose tissue from preterm infants and full-term infants at birth and after 6 weeks of development

ing an area for NMR-based, dynamic studies of nutrition and lipid metabolism.

In vivo ^{13}C MRS has also been used in combination with GLC to study the adipose tissue composition in malnourished patients with cirrhosis of the liver.[52] No significant differences were found in overall adipose tissue composition in the patients compared with healthy volunteers by either technique, although GLC did reveal significant differences in individual fatty acids. Following liver transplantation and subsequent recovery, the patients were re-scanned by ^{13}C MRS. The resulting increase in body fat mass was accompanied by a preferential increase in saturated fatty acids.[52] This may be a dietary effect, as high levels of saturated fatty acids in patients' diets following liver transplantation would result in the deposition of saturated fatty acids in the adipose tissue. Alternatively, this may be secondary to a general repletion of membrane polyunsaturated fatty acids or the use of essential fatty acids (polyunsaturated) for biosynthesis of eicosanoids in the postoperative period.

Dimand et al. have shown that the composition of human adipose tissue may also be a useful marker of fatty acid status in diseases such as cystic fibrosis.[53] They found levels of polyunsaturated fatty acids to be reduced and monounsaturated fatty acids elevated in patients compared with healthy controls.

Several studies have investigated the influence of exercise on adipose tissue composition.[27,54] Intensive exercise was shown to have a significant independent effect on adipose tissue composition, with a significant decrease in polyunsaturated fatty acids following 10 weeks of basic military training.[54] Interestingly, no changes were found in adipose tissue composition following more moderate exercise.[27]

The use of in vivo ^{13}C MRS in humans and animals has not solely been applied to adipose tissue. Barnard et al. used in vivo ^{13}C{^{1}H} NMR spectroscopy to examine the hepatic fatty acid profile in rats fed fats with different fatty acid patterns.[55] The liver has a pivotal role in lipid metabolism. Hepatic receptor-mediated low-density lipoprotein (LDL) clearance is a major process controlling plasma LDL concentrations. Dietary saturated fat may increase plasma LDL by suppression of hepatic LDL-receptor activity, whereas dietary substitution with

polyunsaturated fatty acids may increase hepatic LDL clearance and lower plasma LDL, possibly by increasing membrane fluidity.[56] A significant dietary effect was shown with an increasing hepatic content of unsaturated fatty acids and an increasing degree of polyunsaturation as these fatty acids increased in the diet.

Investigation of the metabolism of specific fatty acids is largely unexplored and a possible future application of ^{13}C MRS. The use of ^{13}C-labeled fatty acids to trace metabolic pathways and kinetic rates is an exciting prospect. These studies may be restricted by the limited availability of labeled fatty acids, their oxidation after administration, and the sensitivity of the system, given the pre-existing strong lipid signals. However, preliminary work has been performed by Cunnane et al. and this demonstrated the ability of in vitro ^{13}C{^{1}H} NMR to detect carbon-specific incorporation of injected [U-^{13}C]-eicosapentaenoic acid in extracted rat liver lipids.[57]

4.3 Deuterium Spectroscopy

There is considerable interest in monitoring lipid turnover in both health and disease. A novel approach to the study of dynamic in vivo tissue lipid metabolism was developed by Brereton et al.[5] They investigated the use of in vivo ^{2}H MRS in mice. Administration of D_2O (10% v/v) in the drinking water for 3–4 days allowed the detection of deuterium-enriched tissue lipid resonances in spectra acquired in less than 2 min, as shown in Figure 7a. The spectra consist of resonances from deuterated water (HOD) and the CHD-group of the tissue lipids. Removal of D_2O from the drinking water led to clear changes in the intensity of both signals (Figure 7b). The loss of ^{2}H from the water signal was significantly faster than from tissue lipids. The loss of ^{2}H from the lipid resonance was, therefore, proposed as a noninvasive measure of the rate of fat utilization. This method has since been applied to the study of fat utilization in obesity.

Obesity is one of the most common medical disorders and is characterized by excess adipose tissue and elevated plasma nonesterified fatty acids. Furthermore, obesity is known to lead to glucose intolerance and insulin resistance. Fat turnover, as reflected by the levels of plasma nonesterified fatty acids, may be implicated in this process. However, the lack of in vivo techniques for measuring fat utilization has greatly hampered progress in this area. Body fat turnover in mouse models of obesity and diabetes mellitus have been studied in vivo using the ^{2}H NMR technique.[58] Brereton et al. showed that the rates of fat utilization in obese mice were significantly lower than the rates for nonobese mice; the induction of diabetes did not affect utilization of fat as a metabolic fuel.[58] These studies clearly suggest that ^{2}H MRS will provide researchers with a powerful method for noninvasive assessment of fat turnover rates.

5 IN VITRO MRS

Plasma concentrations of LDL and high-density lipoproteins (HDL) are well-established markers for CHD risk assessment. Standard methods for quantification and compositional analysis of plasma lipoproteins require laborious and time-consuming physical separation based on particle size, density, or apolipo-

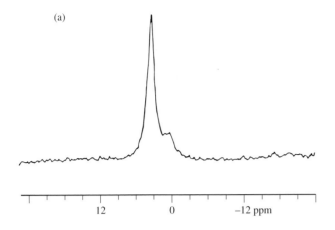

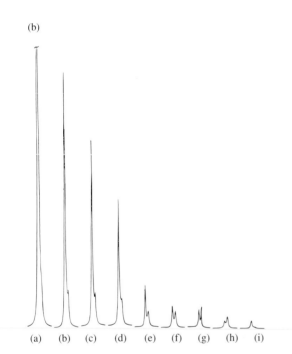

Figure 7 In vivo ^{2}H MRS in mice. (a) Natural abundance spectrum of a mouse. The resonance linked to the CHD-group of lipids is to high field at 1 ppm. Chemical shifts were reference to the deuterated water (HOD) resonance assigned to 4.8 ppm. (b) Spectra of a mouse following the removal of 10% (v/v) D_2O from its drinking water. The spectra were recorded at 1, 4, 5, 8, 11, 15, 16 and 21 days after the resumption of normal drinking water, indicated by (a) to (h), respectively. Spectrum (i) was recorded prior to the administration of D_2O to drinking water. (Adapted with permission of Heyden & Son Ltd from I. M. Brereton, D. M. Doddrell, S. M. Oakenfull, D. Moss, and M. G. Irving, *NMR Biomed.*, 1989, **2**, 55.)

protein content. Bell et al. have shown that MRS can be readily applied to distinguish lipoprotein fractions as well as to study alterations in plasma lipoproteins associated with dietary manipulation, both in intact plasma and in the isolated lipoprotein classes.[11,12] For example, sophisticated computer deconvolution methods (line-fitting analysis) are being developed to identify and quantify lipoprotein fractions in the ^{1}H NMR spectra of intact plasma samples (500 μl) within minutes.[59,60]

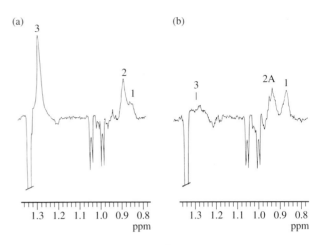

Figure 8 Expansions of spin echo ^{1}H NMR spectra of intact human plasma (a) before and (b) after 7 days of fish oil supplementation. Peaks 1, 2: terminal $-$**CH**$_3$ group from LDL/HDL and very-low-density lipoprotein (VLDL), respectively; peak 3: $-$(**CH**$_2$)$_n-$ of VLDL. Peak 2A arises from n-3/ω-3 fatty acids incorporated into plasma lipoproteins

The effects of dietary manipulation on the ^{1}H NMR profile of lipoproteins are illustrated in Figure 8. Changes in composition following fish oil supplementation are shown by the presence of a new lipid resonance (peak 2A) arising from fish oil n-3/ω-3 fatty acids incorporated into plasma lipoproteins, while the methylene ($-$(**CH**$_2$)$_n-$) signal (peak 3) is markedly reduced. Furthermore, T_2 analysis of the lipid resonances showed that these changes in lipid composition of the LDL particles are accompanied by alteration in the structural characteristics of these particles.[11]

6   SUMMARY

In this chapter, we have reviewed the current state of in vivo MRS as applied to the study of lipids. It is clear that this is a relatively new area of NMR research. However, it offers many exciting prospects for future research into human lipid metabolism and body composition.

7   RELATED ARTICLES

Brain MRS of Human Subjects; Imaging and Spectroscopy of Muscle; Whole Body Studies: Impact of MRS.

8   REFERENCES

1. J. E. Kinsella, B. Lokesh, and R. A. Stone, *Am. J. Clin. Nutr.*, 1990, **52**, 1.
2. R. A. DeFronzo and E. Ferrannini, *Diabetes Care*, 1991, **14**, 173.
3. P. Canioni, J. R. Alger, and R. G. Shulman, *Biochemistry*, 1983, **22**, 4974.
4. C. T. W. Moonen, R. J. Dimand, and K. L. Cox, *Magn. Reson. Med.*, 1988, **6**, 140.
5. I. M. Brereton, M. G. Irving. J. Field, and D. M. Doddrell, *Biochem. Biophys. Res. Commun.*, 1986, **137**, 579.

For References see p. 843

6. P. C. Dagnelie, J. D. Bell, S. C. R. Williams, I. J. Cox, D. K. Menon, J. Sargentoni, and G. A. Coutts, *NMR Biomed.*, 1993, **6**, 2.

7. F. Schick, B. Eismann, W. I. Jung, H. Bongers, M. Bunse, and O. Lutz, *Magn. Reson. Med.*, 1993, **29**, 158.

8. C. Boesch, J. Slotboom, H. Hoppeler, and R. Kreis, *Magn. Reson. Med.*, 1997, **37**, 484.

9. N. Beckmann, J. J. Brocard, U. Keller, and J. Seelig, *Magn. Reson. Med.*, 1992, **27**, 97.

10. J. D. Bell, O. Lavender, V. C. Morris, and P. J. Sadler, *Magn. Reson. Med.*, 1991, **17**, 414.

11. J. D. Bell, J. C. C. Brown, R. E. Norman, and P. J. Sadler, *NMR Biomed.*, 1988, **1**, 90.

12. J. D. Bell, M. L. Barnard, H. G. Parkes, E. L. Thomas, C. H. Brennan, S. C. Cunnane, and P. C. Dagnelie, *J. Lipid Res.*, 1996, **37**, 1664.

13. R. Ross, L. Léger, R. Guardo, J. de Guise, and B. G. Pike, *J. Appl. Physiol.*, 1991, **70**, 2164.

14. P. A. Fowler, M. F. Fuller, C. A. Glasbey, G. G. Cameron, and M. A. Foster, *Am. J. Clin. Nutr.*, 1992, **56**, 7.

15. N. Abate, A. Garg, R. Coleman, S. M. Grundy, and R. M. Peshock, *Am. J. Clin. Nutr.*, 1997, **65**, 403.

16. M. L. Barnard, J. V. Hajnal, J. E. Schwieso, E. L. Thomas, J. D. Bell, N. Saeed, G. Frost, and S. R. Bloom, *NMR Biomed.*, 1996, **9**, 156.

17. N. Mitsiopoulos, R. N. Baumgartner, S. B. Heymsfield, W. Lyons, D. Gallagher, and R. Ross, *J. Appl. Physiol.*, 1998, **85**, 115.

18. P. Tothill, T. S. Han, A. Avenell, G. McNeill, and D. M. Reid, *Eur. J. Clin. Nutr.*, 1996, **50**, 747.

19. A. Sohlström, L. O. Wahlund, and E. Forsum, *Am. J. Clin. Nutr.*, 1993, **58**, 830.

20. M. A. Staten, W. G. Totty, and W. M. Kohrt, *Invest. Radiol.*, 1989, **24**, 345.

21. R. Ross, L. Léger, D. Morris, J. de Guise, and R. Guardo, *J. Appl. Physiol.*, 1992, **72**, 787.

22. P. A. Fowler, M. F. Fuller, C. A. Glasbey, M. A. Foster, G. G. Cameron, G. McNeill, and R. J. Maughan, *Am. J. Clin. Nutr.*, 1991, **54**, 18.

23. R. Ross, D. S. Kimberley, Y. Martel, J. de Guise, and L. Avruch, *Am. J. Clin. Nutr.*, 1993, **57**, 470.

24. E. L. Thomas, N. Saeed, J. V. Hajnal, A. E. Brynes, A.P. Goldstone, G. Frost, and J. D. Bell, *J. Appl. Physiol.*, 1998, **85**, 1778.

25. R. Ross, H. Pedwell, and J. Rissanen, *Am. J. Clin. Nutr.*, 1995, **61**, 1179.

26. R. Ross and J. Rissanen, *Am. J. Clin. Nutr.*, 1994, **60**, 695.

27. E. L. Thomas, A. Byrnes, J. V. Hajnal, N. Saeed, G. Frost, and J. D. Bell, *Proc. VIth Ann Mtg. (Int) Soc. Magn. Reson. Med.*, Sydney, 1998, (Vol. 3), p. 1812.

28. D. S. Gray, K. Fujioka, P. M. Colletti, H. Kim, W. Devine, T. Cuyegkeng, and T. Pappas, *Am. J. Clin. Nutr.*, 1991, **54**, 623.

29. R. Leenen, K. van der Kooy, P. Deurenberg, J. C. Seidell, J. A. Weststrate, F. J. Schouten, and J. G. Hautvast, *Am. J. Physiol.*, 1992, **263**, E913.

30. K. van der Kooy, R. Leenen, J. C. Seidell, P. Deurenberg, and J. G. Hautvast, *Am. J. Clin. Nutr.*, 1993, **58**, 853.

31. S. J. Marks, N. R. Moore, M. L. Clark, B. J. Strauss, and T. D. Hockaday, *Obes. Res.*, 1996, **4**, 1.

32. C. Boesch, R. Kreis, H. Howald, S. Matter, R. Billeter, B. Essen-Gustavsson, and H. Hoppeler, *Proc. VIth Ann Mtg. (Int) Soc. Magn. Reson. Med.*, Sydney, 1998, (Vol. 3), p. 1785.

33. L. S. Szczepaniak, D. T. Stein, F. Schick, A. Garg, and J. D. McGarry, *Proc. Vth Ann Mtg. (Int) Soc. Magn. Reson. Med.*, Vancouver, 1997, (Vol. 2), p. 1334.

34. J. Rico-Sanz, E. L. Thomas, G. Jenkinson, S. Mierisova, R. Iles, and J. D. Bell, *J. Appl. Physiol*, 1999, **87** (in press).

35. J. Rico-Sanz, J. V. Hajnal, E. L. Thomas, S. Mierisova, M. Ala-Korpela, and J. D. Bell, *J. Physiol.*, 1998, **510**, 615.

36. D. T. Stein, R. Dobbins, L. Szczepaniak, C. Malloy, and J. D. McGarry, *Diabetes*, 1997, **46**, (Suppl 1), 23A.

37. N. G. Forouhi, G. Jenkinson, S. Mullick, E. L. Thomas, U. Bhonsle, P. M. McKeigue, and J. D. Bell, *Proc. Br. Diabetic Assoc.*, 1998, S36.

38. J. G. Batchelor, R. J. Cushley, and J. H. Prestegard, *J. Org. Chem.*, 1974, **39**, 1698.

39. J. Bus, I. Sies, and M. S. F. Lie Ken Jie, *Chem. Phys. Lipids*, 1976, **17**, 501.

40. F. D. Gunstone, M. R. Pollard, C. M. Scrimgeour, and L. Vedanayagam, *Chem. Phys. Lipids*, 1977, **18**, 115.

41. J. N. Shoolery, *Prog. NMR Spect.*, 1977, **11**, 79.

42. N. G. Soon, *Lipids*, 1985, **20**, 778.

43. F. D. Gunstone, *Chem. Phys. Lipids*, 1991, **59**, 83.

44. L. O. Sillerud, C. H. Han, M. W. Bitensky, and A. A. Francendese, *J. Biol. Chem.*, 1986, **261**, 4380.

45. M. L. Barnard, J. D. Bell, S. C. R. Williams, T. A. B. Sanders, H. G. Parkes, K. K. Changani, J. S. Beech, M. L. Jackson, and S. R. Bloom, *Proc. XIth Ann Mtg. Soc. Magn. Reson. Med.*, Berlin, 1992, p. 3339.

46. T. W. Fan, A. J. Clifford, and R. M. Higashi, *J. Lipid Res.*, 1994, **35**, 678.

47. E. L. Thomas, S. C. Cunnane, and J. D. Bell, *NMR Biomed.*, 1998, **11**, 290.

48. J. Hirsch, J. W. Farquhar, E. H. Ahrens, M. L. Peterson, and W. Stoffel, *Am. J. Clin. Nutr.*, 1960, **8**, 499.

49. A. C. Beynen, R. J. J. Hermus, and J. G. A. J. Hautvast, *Am. J. Clin. Nutr.*, 1980, **33**, 81.

50. E. L. Thomas, G. Frost, M. L. Barnard, D. J. Bryant, J. Simbrunner, S. D. Taylor-Robinson, G. A. Coutts, M. Burl, S. R. Bloom, K. D. Sales, and J. D. Bell, *Lipids*, 1996, **31**, 145.

51. E. L. Thomas, J. D. Hanrahan, M. Ala-Korpela, G. Jenkinson, D. Azzopardi, R. A. Iles, and J. D. Bell, *Lipids*, 1997, **32**, 645.

52. E. L. Thomas, S. D. Taylor-Robinson, M. L. Barnard, G. Frost, J. Sargentoni, B. R. Davidson, S. C. Cunnane, and J. D. Bell, *Hepatology*, 1997, **25**, 178.

53. R. J. Dimand, C. T. W. Moonen, S. Chu, E. M. Bradbury, G. Kurland, and K. L. Cox, *Pediatr. Res.*, 1988, **24**, 243.

54. E. L. Thomas, A. Byrnes, G. Jenkinson, M. Jubb, G. Frost, and J. D. Bell, *J. Magn. Reson. Anal.* 1999, in press.

55. M. L. Barnard, J. D. Bell, S. C. R. Williams, T. A. B. Sanders, and S. R. Bloom, *Proc. XIIth Ann Mtg. Soc. Magn. Reson. Med.*, New York, 1993, p. 92.

56. J. Loscalzo, J. Freedman, A. Rudd, I. Barsky-Vasserman, and D. E. Vaughan, *Arteriosclerosis*, 1987, **7**, 450.

57. S. C. Cunnane, R. J. McDonagh, S. Narayan, and D. J. Kyle, *Lipids*, 1993, **28**, 273.

58. I. M. Brereton, D. M. Doddrell, S. M. Okenfull, D. Moss, and M. G. Irving, *NMR Biomed.*, 1989, **2**, 55.

59. M. Ala-Korpela, Y. Hiltunen, J. Jokisaari, S. Eskelinen, K. Kiviniitty, M. Savolainen, and Y. A. Kesaniemi, *NMR Biomed.*, 1993, **6**, 225.

60. J. D. Otvos, E. J. Jeyarajah, D. W. Bennett, and R. M. Krauss, *Clin. Chem.*, 1993, **38**, 1632.

Biographical Sketches

E. Louise Thomas. *b* 1970; B.Sc. Biochemistry University of London, 1992; Ph.D. London, 1996. Ph.D. thesis on in vivo ^{13}C NMR spectroscopy under the supervision of J. D. Bell and K. D. Sales. Currently a Senior Research Fellow at Imperial College School of Medicine. Over 20 publications. Research interests include application of MRI/MRS to the study of human fat and muscle metabolism.

Jimmy D. Bell. *b* 1958; B.Sc. Biochemistry University of Warwick, 1982; Ph.D. London, 1987. Lecturer Hammersmith Hospital 1989–94. Senior lecturer Royal Postgraduate Medical School 1994–97. Senior lecturer at MRC Clinical Sciences Centre, Imperial College 1997–present. Approximately 90 publications. Research interests include application of MRI/MRS in clinical research, and MRS to the chemistry of tissue and body fluids.

EPR and In Vivo EPR: Roles for Experimental and Clinical NMR Studies

Harold M. Swartz and Goran Bačíc

Dartmouth Medical School, Hanover, NH, USA

1 INTRODUCTION

1.1 Scope and Rationale of Coverage

The use of electron paramagnetic resonance (EPR or, completely equivalently, electron spin resonance) in biomedicine is a vast and still developing subject, with many areas of direct or indirect relevance to the subject of this volume (related articles are listed at the end of the text). We will not, however, attempt to cover all of these topics but instead will restrict our discussion to four topics that are especially important to the use of MRI and MRS in vivo for experimental and clinical purposes:

1. measurement of the concentration of oxygen, especially in tissues in vivo;
2. measures of redox metabolism and/or distribution of drugs;
3. measurement of the viability of cells;
4. measurement of the concentration and distribution of contrast agents, especially in vivo.

These topics were chosen because they are areas of great importance for the use of NMR in biomedicine and are measurements for which EPR techniques can provide critically needed information that cannot be obtained as easily and/or as well by other techniques, including NMR. Our coverage of these topics will focus on overviews of the current status and the likely future developments; we will not attempt to provide extensive reviews of the rapidly expanding literature on these subjects (see[1–5] for recent reviews).

The measurement of the concentration of oxygen in cells and tissues, with an emphasis on the determination of hypoxia, seems especially important. Many of the uses of NMR, especially spectroscopy, really are aimed at making such measurements, and/or studying phenomena that are very dependent on the concentration of oxygen. This is certainly the case for most phosphorus NMR studies, which usually are aimed at measurements of the redox sensitive concentrations of phosphorus metabolites. Similarly much of proton NMR spectroscopy is aimed at measurements of intermediates such as lactic acid which reflect the concentration of oxygen in the tissues and, indeed, in many cases these measurements are used to provide indirect measures of the extent of hypoxia because more direct measurements are not available. As we will see EPR can make very sensitive and accurate measurements of the concentration of oxygen in tissues but it cannot, as yet, achieve the depth of penetration and the spatial resolution of NMR imaging. As a consequence, there is a great potential for the combined use of EPR and NMR in which the EPR techniques provide the needed determinations of the relationships between NMR spectroscopic measurements and the actual concentrations of oxygen. Such data will then make it much more feasible to use NMR techniques to obtain accurate insights into the state of oxygen dependent metabolism and therefore better address the diagnosis and therapy of diseases in which hypoxia plays a significant role (e.g. cancer, ischemia, and inflammation).

From the preceding discussion it seems evident that to the extent that EPR can provide additional measures of metabolism, beyond the direct measurement of the concentration of oxygen, EPR will further complement NMR measurements significantly. The metabolism of nitroxides appears to provide such a measure. EPR also can be used to follow the distribution and metabolism of some drugs, which can provide additional information on redox metabolism as well as pharmacokinetics. These studies are feasible through the use of metabolizable paramagnetic spin labels which are either the objects of interest themselves or attached to drugs, etc. as labels which then can be followed selectively by EPR.

The measurement of the viability of cells by EPR, while still at an early stage of development, provides an essentially unique capability that is very complementary to NMR studies. At the present time the results of this approach are available only in aggregates of cells (spheroids) and isolated cells. In the former they provide perhaps the most advanced and effective applications of EPR imaging (EPRI) in a biological system.

The use of EPR to study NMR contrast agents in many ways is very straightforward and logical, although it has not been applied as widely as would seem desirable. Virtually all NMR contrast agents are active because they have unpaired electrons and therefore they are directly observable by EPR. This is so because EPR sensitively and exclusively responds to electronic magnetic moments; usually those associated with unpaired electrons. While this approach has been very productive for the study of some NMR contrast agents such as the nitroxides, there are some significant limitations in the application of this technique in vivo to many of the metal ion based contrast agents due to unfavorably short relaxation times and/or spectral features. (The metal ion contrast agents, of course, can be studied very effectively by EPR in vitro.)

1.2 Brief Summary of the Principles of EPR Applied to the Study of In Vivo Systems

Although there are few fundamental differences between the principles of electron and nuclear magnetic resonance, differences in physical and chemical properties of the resonant

For References see p. 854

species (unpaired electrons vs. nuclei with net spin) lead to profound differences in the techniques that are used to record spectra. Perhaps the greatest differences arise because the gyromagnetic ratio of an unpaired electron is ~700 times larger than that of a proton, so the resonance frequency/magnetic field ratio for the electron is 28 GHz T^{-1}, vs. 42.5 MHz T^{-1} for the proton. Consequently, standard EPR spectrometers operate at much higher frequencies and lower fields than conventional NMR spectrometers. While EPR spectroscopy is now performed using a wide range of frequencies from rf to far-infrared, most standard, commercial EPR spectrometers operate at 9.5 and 35 GHz (X- and Q-band). At these frequencies non-resonant absorption of the electromagnetic radiation by the liquid water in biological systems presents a serious problem, thus limiting the sample size to a thickness of less than 1 mm. Larger aqueous samples or animals can be studied only by reducing the operating frequency to 1 GHz or less, but, of course, this results in reduced sensitivity. To date most of the published successful applications of in vivo EPR techniques to actual biomedical problems have used spectroscopy rather than imaging and have been done at 1 GHz (L-band). Several groups are investigating the feasibility of extending these approaches to imaging and/or lower frequencies, but sensitivity may be a limiting factor. The use of lower frequencies (e.g. 300–500 MHz) would be especially attractive because this would extend the sensitive depth from about 10 mm to about 70 mm.

The lack of sufficient amounts of naturally occurring para-magnetic materials is another potential source of problems. Both direct observations and theoretical calculations show that endogenous paramagnetic species such as paramagnetic metal ions and free radicals are present in insufficient concentrations to be detected directly by EPR in vivo.[4] (The only naturally occurring exception is the polymer, melanin.[4]) This presents an obvious problem since the paramagnetic substance has to be introduced, which can cause problems with regard to sensitivity and, potentially, toxicity. However, this has some beneficial aspects, since the only paramagnetic species that will be observed in vivo will be those introduced by the experimenter. The absence of background signals is particularly convenient for in vivo EPR spectroscopy (e.g. in contrast to proton spectroscopy where water suppression is required). With the addition of appropriate paramagnetic species the sensitivity of EPR, on a molar basis, is 700 times greater than for NMR. Paramagnetic species which are very stable and have appropriate lineshapes have been developed recently and these have been a significant factor in the successful development of in vivo EPR techniques.[4]

The short relaxation times of most paramagnetic species can be an additional potential source of problems. The timescale of electron spin relaxation of most paramagnetic species is nano-seconds (in contrast to NMR where the relaxation times typically are milliseconds); consequently, essentially all in vivo studies have been done using CW-EPR and have not been able to utilize most of the pulse techniques developed for NMR in biological systems. At the same time, the large widths of EPR resonance lines necessitate the use of high magnetic field gradient strengths for imaging that are at least one order of magnitude greater than those used in NMR imaging. Additional problems for EPRI arise from the fact that the EPR spectra of most paramagnetic species have multiple lines and multiple lineshapes; these spectral features, however, can be very valuable for EPR spectroscopy because they can provide very sensitive parameters of the environment in which the paramagnetic species are located.

As a consequence of the above factors the development of in vivo EPR techniques, especially imaging, has not occurred as rapidly as with NMR. Recently, however, there has been a significant increase of productivity in these areas, especially with in vivo spectroscopy. The emphasis on spectroscopy vs. imaging has not turned out to be a very significant limitation because the most useful applications of EPR in vivo can probably be achieved best by localized spectroscopy.

To date almost all of the in vivo EPR studies have been done with small animals and there are only two reports of human studies. It appears very likely, however, that rapid progress will occur in the next few years, because of the very promising data which have been obtained recently, especially in regard to oximetry.

1.3 Experimental Considerations

The first EPR experiment conducted by Zavoisky in 1944[6] was performed at 133 MHz. With the development of commercial EPR spectrometers the usual frequency was raised by 2–3 orders of magnitude to increase sensitivity, but this simultaneously reduced the size of nonfrozen aqueous samples that could be studied by standard cylindrical or rectangular EPR cavities to less than 1 mm, due to an increase in the nonresonant absorption of microwave by aqueous samples, thus considerably limiting applications of EPR to biological systems. In the first in vivo EPR study performed in 1975 at 9 GHz (X-band) these limitations were avoided by implanting a helix antenna into the rat liver,[7] and more recently in vivo EPR studies in small animals have been carried out at 9 GHz by inserting the tail into the cavity. The obvious constraints of such approaches eventually led to the development of low-frequency EPR instruments, thus increasing the ability to penetrate deeper into biological tissues; however, it took almost 10 years before the next studies aimed at in vivo EPR were performed.[8,9] Subsequently developments have occurred at an ever-increasing rate and have led to a number of very productive applications of in vivo EPR techniques. The key technical advances occurred in four areas: development of low-frequency EPR spectrometers suitable for use with living objects,[10–14] development of detectors suitable for in vivo studies,[15,16] recognition and development of paramagnetic species with properties suited for particular applications (especially the measurement of pO_2),[17] and improved methods for data acquisition and analysis.[18–21] Limitations of space prevent us from describing these developments in detail here, but they are illustrated in the following descriptions of applications and further details are available in the references that are cited.

While these developments have greatly expanded the use of EPR in living subjects, some fundamental limitations remain and therefore the application of EPR techniques to clinical medicine is unlikely to reach the extent of NMR techniques. Probably the key factors are a combination of depth of penetration and sensitivity. These are, of course, related because in order to achieve greater depth of penetration of the microwaves one must decrease the frequency and this inevitably leads to decreased sensitivity. Many of the principal uses of EPR in vivo are likely to occur with the use of frequencies around 1 GHz. At 1 GHz there is relatively high sensitivity but the practical limit for depth is about 10 mm. This can be partially overcome by the use of catheters, etc. to insert the detector deeper into the body. If the frequency is reduced to 300–500 MHz the practical limits on depth increase by a factor of up to 10 but there is a concomitant decrease in sensitivity. Very useful data have already been obtained in experimental animals with in vivo EPR at both ends of these frequency ranges, and feasibility studies have been successful in human subjects, so it is therefore clear that EPR will be quite useful in vivo. The extent of the usefulness of in vivo EPR seems certain to be extended by combining it with complementary approaches such as NMR.

2 MEASUREMENT OF THE CONCENTRATION OF OXYGEN, ESPECIALLY IN TISSUES IN VIVO, BY EPR OXIMETRY

One of the most effective and rapidly developing uses of EPR in biological systems is the measurement of pO_2. The importance to NMR of these measurements rests on (a) the profound role of oxygen in many physiological and pathophysiological processes (ischemic diseases, reperfusion injury, response of tumors to therapy, etc.); (b) the lack of an adequate alternative methodology to make accurate in vivo measurements under biologically pertinent conditions (these conditions include making the measurements directly in the tissues, at the relatively low concentrations of oxygen which occur in tissues in vivo); and (c) the suitability of NMR techniques, especially spectroscopy, to measure metabolites whose concentrations and importance are linked closely to pO_2.

The term EPR oximetry encompasses a number of distinct experimental techniques; however, most of these methods rely on the paramagnetic properties of molecular oxygen. The ground state of oxygen has two unpaired electrons and a very rapid relaxation rate, which via Heisenberg spin exchange between oxygen and other paramagnetic species provides an efficient relaxation mechanism for the latter. As a consequence, the presence of oxygen has an effect on both T_1 and T_2 of spectra and these effects can be followed quantitatively by a variety of spectral parameters (e.g. linewidth, resolution of superhyperfine structure, and microwave power saturation). An NMR counterpart of this type of oximetry uses the effect of oxygen on the spin–lattice relaxation time of ^{19}F in tissues preloaded with fluorine compounds.

The use of EPR to measure the concentration of oxygen in biological samples was introduced more than 20 years ago and while it has been a useful technique for many studies of model systems and cells, it is only recently that it has been applied successfully in vivo. The latter applications have been very successful because of the development of new techniques and new oxygen sensitive paramagnetic materials. Some of the latter appear to have properties that are virtually ideal for the measurement of the concentration of oxygen in tissues in vivo, including a high degree of chemical and physical stability, essentially complete inertness in tissues, an apparent lack of toxicity, high sensitivity to the concentration of oxygen, and retention of their response to oxygen for periods of years or longer.[22]

There are two principal types of oxygen sensitive materials which are now available for the direct measurement of pO_2 in biological systems: soluble, stable free radicals (especially nitroxides and their derivatives), and stable paramagnetic particles with oxygen sensitive spectra (especially fusinite coal, lithium phthalocyanine, India ink, and chars of carbohydrates). Each of these has properties that may be especially useful for particular applications and, in aggregate, they provide a versatile and effective set of approaches to measure the concentration of oxygen under most or all conditions that are likely to occur experimentally and clinically.

The use of particulate materials has already led to some very useful data on pO_2 in vivo.[22–26] These materials share several very desirable properties for such studies including:

1. relative noninvasiveness and repeatability (after an initial placement of the paramagnetic materials, measurements can be made noninvasively as frequently as desired over a period of a year or more);
2. sensitivity (measurements can, at low pO_2, resolve differences of the order of 0.1 torr);
3. accuracy (repeated measurements have a small variability and correlate closely with measurements of pO_2 by other methods);
4. provide localized measurements (the spatial resolution is the same as the size of the paramagnetic particles, which can be as small as a single particle of less than 0.2 mm in diameter. At the same time, accurate localization of the probe can be achieved using MR imaging);[24]
5. capability of making several measurements simultaneously (this is accomplished by inserting multiple discrete solid particles and applying a magnetic field gradient such that sites less than 1 mm apart can be resolved);[25]
6. little or no effects of chemical and physical conditions likely to be encountered in viable biological systems (changes in pH or the presence of oxidants, reductants, or other paramagnetic materials do not affect measurements of pO_2);
7. little or no toxicity (the paramagnetic materials are very inert in biological systems as assayed in both cell culture and in vivo);
8. capability of following changes in pO_2 with a time resolution of seconds or less;
9. response to pO_2 (in contrast to nitroxides and other oxygen sensitive materials which may respond to the concentration of oxygen or a product of concentration and the rate of diffusion).

The organs that have been studied by this approach include brain, heart (both in situ and as isolated perfused systems, especially the Langendorf preparation), lung, liver, kidney, skeletal muscle, and joints. These approaches have also been applied to the study of several important pathological and physiological conditions including tumors, ischemia, inflammation, and anesthesia.

The capabilities of these approaches are indicated in Figure 1 which shows the change in pO_2 in tumors as a function of time and therapeutic irradiation. The ability to resolve small changes in pO_2 (less than 0.2 torr) in unanesthetized animals provides a powerful tool for understanding and interpreting NMR spectroscopic studies of this type of system.

While it appears that many studies in the future will utilize the oxygen-sensitive particulate paramagnetic materials, there

For References see p. 854

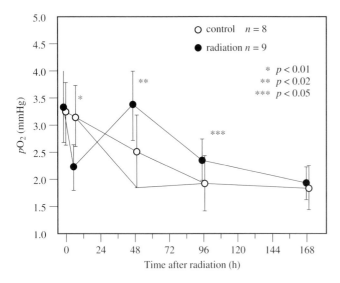

Figure 1 In vivo EPR measurements of changes of pO_2 as measured by India ink in a mouse tumor (mammary adenocarcinoma) after a single dose of radiation (20 Gy)

also are a number of very effective uses of the nitroxides in measuring pO_2 (in general the nitroxides actually measure the product of the concentration of oxygen and the rate of diffusion but this measurement can usually be converted directly to pO_2 under most experimental conditions). EPR oximetry with nitroxides can utilize both the physical and the physiological effects of oxygen, and its application to cellular systems has been extensively reviewed and will not be considered further here.

The nitroxides have also been used for oximetry in larger systems, including intact animals. The results of these studies, while very promising, have not yet been applied to the direct study of pathological processes, although they have produced some very exciting data on pO_2 in perfused hearts and some promising data on in vivo systems.[11,27–29] The advantages of nitroxides for such studies includes their versatility (nitroxides can be synthesized with virtually any chemical and physical properties) and that they are usually available as soluble materials that can diffuse throughout the system. Their potential disadvantages include a tendency to undergo bioreduction and their relatively weaker response to changes in pO_2 (it seems unlikely that they will be able to resolve changes of less than 5 torr in vivo).

Nitroxides have also been used in extended systems where EPR imaging or in vivo EPR spectroscopy were used to study the concentration, distribution, and diffusion of oxygen in macroscopic biological samples. Several experiments have been performed on small samples with higher frequency EPR (9 GHz) using one-dimensional (1D) and two-dimensional (2D) EPR. 1D EPR imaging is based on the use of conventional EPR spectroscopy in the presence of a unidirectional linear magnetic field gradient. The obvious restriction is that the sample has to possess a high degree of symmetry and/or isotropic properties; however, 1D EPR allows rapid accumulation (e.g. 2 s per spectrum) and is particularly suitable for studying fast diffusion processes such as diffusion of oxygen. Interaction of oxygen with nitroxides shortens relaxation times; hence the

diffusion of oxygen can be measured in samples preloaded with oxygen sensitive paramagnetic species. To illustrate this approach, fast 1D EPRI has been used to follow oxygen diffusion in aqueous media[30] and the results agree well with the value of the oxygen diffusion coefficient measured by classical methods.

A few 2D EPRI experiments have been reported, including studies to examine the feasibility of visualizing the distribution of oxygen in model and biological samples and the basic mechanisms for the contrast enhancement.[31,32] The effects of oxygen on the relaxation times of nitroxides (analogous to the action of paramagnetic agents on proton relaxation) were utilized for contrast enhancement. Shortening of T_2 of the nitroxides causes line broadening and a concomitant decrease of line height, while shortening of T_1 induces differences in microwave power saturation. Using proper selection of EPR imaging parameters, both effects can be utilized to produce either positively or negatively enhanced oxygen dependent images.[31] Other properties such as differential solubility of oxygen in lipophilic and hydrophilic environments also can be used to distinguish different regions of the tissue. This was demonstrated on a tissue slice containing fat and muscle regions, all of them containing the same amount of nitroxide, but they have different intensities in the image due to the higher concentration of oxygen in fat-rich areas.[32]

Although EPR imaging can give insights on the distribution of oxygen it is probably not suitable for precise and direct measurements of pO_2. Such measurements will probably require a form of spectral-spatial imaging, where the EPR spectrum can be displayed as a function of spatial localization.[18,19,33] This considerably increases the time required for the measurement. The feasibility of doing spectral spatial oximetry has been explored in model systems.[33] It is not clear whether sufficient sensitivity can be achieved to apply this technique to animals, but it does appear to be very promising for the study of detailed distribution of pO_2 in systems amenable to study at higher frequencies, especially the tumor model, spheroids, at 9 GHz.

3   MEASURES OF REDOX METABOLISM AND/OR DISTRIBUTION OF DRUGS BY EPR

There are several interrelated approaches, based on nitroxides (although in principle other metabolizable, stable free radicals could also be used), that provide data on redox metabolism and the distribution of drugs which should be very complementary for NMR studies.

The oxygen dependent metabolism of nitroxides has been advocated as an approach to the measurement of pO_2 and redox metabolism.[34] Functional biological systems can reversibly reduce the nitroxides to the nonparamagnetic hydroxylamines.[35] While this effect was originally considered only to be a drawback, subsequently it has been demonstrated to be a potentially powerful tool for following redox metabolism. In principle this effect could be followed both by direct EPR studies and by NMR (the latter by using the change in contrast from the change in local concentrations of the nitroxides). Extensive studies of the reduction of nitroxides and its reverse, the oxidation of hydroxylamines back to nitroxides by various cellular systems, have shown that these metabolic rates can be very sensitive indicators of the presence of profound hypoxia, with the principal changes in the rate of metabolism occurring at $pO_2 < 1$ torr.[36] This approach is similar to NMR measurements of phosphorus metabolites (e.g. Pi/ATP ratio) and lactic acid production, both of which appear to change especially when the pO_2 becomes very low. To date, however, data are lacking on the correlation between these changes as followed by NMR and the actual concentration of oxygen.

The nonoxygen dependent metabolism of nitroxides should also be a very useful parameter for evaluating redox metabolism in vivo. The reduction of the nitroxides to the hydroxylamines occurs primarily through reduction by the mitochondria, with the reducing equivalents coming from approximately the level of ubiquinone. Therefore the rate of reduction of nitroxides can provide a useful measure of this activity that may not be obtainable by other means in vivo. Redox processes at different sites can be followed selectively by the use of nitroxides whose physical or chemical properties (e.g. solubility, charge, reactivity of side groups) result in their selective localization.

Recent studies illustrate the potential for using the metabolism of nitroxides as a means of investigating redox metabolism in vivo. These uses may be extended to the study of other biophysical parameters which can be reflected in the EPR spectra of nitroxides. There are many regions of interest such as tumors which fall within the sensitive depth for 1 GHz EPR and there are also some techniques in which an insertable probe can be used to probe depths up to several cm with this frequency.[4] These approaches may be especially useful in studies of the skin since the depth of penetration of microwaves is sufficient to cover the region of interest using a surface probe. We describe some of these studies in skin as an indication of what might be obtained in other systems as well.

Nitroxides have been applied topically and their diffusion into skin and their metabolism deduced from temporal changes of 1D EPR spectra in the presence of a field gradient orthogonal to the skin surface. Both the rate and spatial distribution of the physiological processes involved in the metabolism of nitroxides or nitroxide-labeled compounds can be studied. Fuchs et al.[37] studied diffusion of several nitroxides into skin biopsies of hairless mice and found that penetration of lipophilic nitroxides is much faster than penetration of more hydrophilic nitroxides. They also studied percutaneous absorption and distribution of a nitroxide-labeled drug (estradiol) and a local anesthetic (procaine) finding that the former readily penetrates into deep skin layers while the latter does not, even in the presence of a penetration enhancer, such as dimethyl sulfoxide. Using spectral spatial imaging and analysis of spectral features, the same authors showed spatially resolved (surface, epidermis, dermis) partitioning of the nitroxide di-*t*-butyl nitroxide (DTNB) between polar and apolar regions (Figure 2), as well as the spatial distribution of thiol groups by studying the diffusion of proxylmaleimide, which chemically reacts with thiol groups.

The study of Gabrijelcic et al.[38] illustrates the capability of studying potential applications of liposomes as drug carriers for topical treatment of skin disorders. They found that entrapment in liposomes promotes the penetration of otherwise impenetrable, charged nitroxides into pig skin and that the size and

phospholipid composition of liposomes significantly influences their efficiency. Combining these results with the study of reduction of free nitroxides, the loss of integrity of liposomes in different layers of the skin was assessed.[39] The physiological information obtained by EPR in these studies cannot be obtained by other imaging modalities (MRI, ultrasound), with the same spatial resolution (well below 100 μm). Although these experiments were performed at 9 GHz and used tissue slices, there are no apparent obstacles for the use of these techniques in living animals (or humans) using lower frequency spectrometers.

Another area of overlapping interest for both EPR and MRI is the study of redox metabolism of ischemia/reperfusion injury of the heart. Free radicals derived from oxygen are believed to be involved in reperfusion injury of cardiac tissue, but current techniques do not permit their direct observation. However, generation of free radicals and underlying molecular mechanisms can be studied indirectly by spin-trapping, and the potential of this approach has been investigated in isolated perfused hearts[40] using nitroxides which resemble the molecular structure of the most commonly used spin-trapping molecules. Zweier & Kuppusamy[11,28] studied the kinetics of nitroxide uptake from perfusate and their subsequent reduction upon induction of ischemia. The average oxygen concentration was simultaneously determined from measurements of linewidth. By including spectral spatial imaging in the experimental protocol and synchronizing acquisition with the heart beat these experiments could provide valuable data.

There is also a wide range of biophysical parameters that can be studied by EPR since nitroxide spectra are sensitive, in addition to pO_2, to viscosity, polarity, membrane order and fluidity, pH, redox processes, and other metabolic properties. Some of these parameters are not amenable to study by other means and could be very useful for understanding changes detected by NMR techniques.

4   MEASUREMENT OF THE VIABILITY OF CELLS, ESPECIALLY BY EPR MICROSCOPY

EPR has the capability of measuring the viability of cells in an accurate, nonperturbing, and repeatable manner, under appropriate circumstances.[41] This essentially unique capability is based on the use of nitroxides in combination with the presence of significant concentrations of charged, paramagnetic metal ions. Under these circumstances the EPR signal from the nitroxides is observable only where the nitroxide is not in close proximity to the metal ion. This occurs inside viable cells; when the cells lose their viability the EPR signal from those cells is suppressed. This measure of viability, which reflects the function of the membrane, appears to correlate well with other measures of viability.[41]

This approach is most suitable at higher frequencies (e.g. 9 GHz) where there is maximum sensitivity; at such frequencies this approach can be used for microscopic imaging with resolution of the order of cell dimensions or less.

This approach has been used successfully in spheroids, which are an important and frequently used in vitro model of solid tumors.[42] The authors have demonstrated that noninvasive assessment of spatial distribution of viable and necrotic regions

For References see p. 854

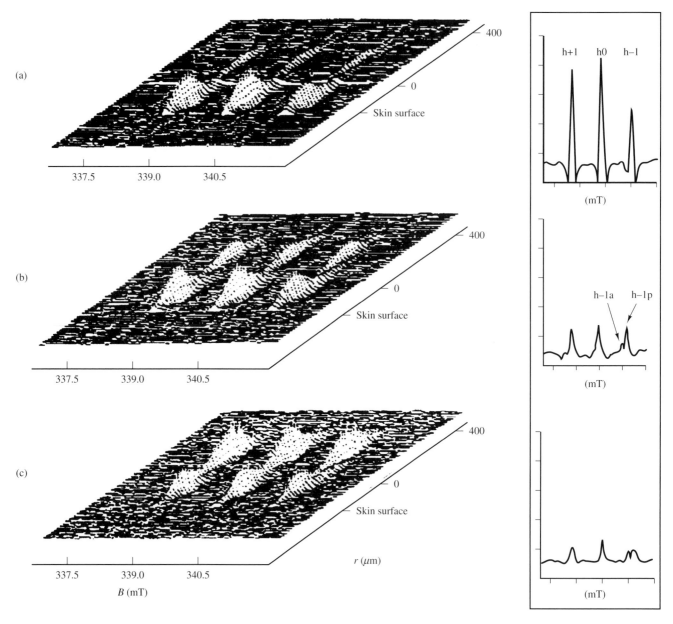

Figure 2 Penetration of DNTB into mouse skin as monitored by EPR imaging at 9 GHz. Punch biopsy samples were pretreated with *N*-ethylmaleimide prior to topical application of DNTB. Left: 2D spectral spatial images of the time dependent changes of the distribution of DNTB: (a), 5 min; (b) 15 min; (c) 45 min. Right: EPR spectra of DNTB (2nd harmonics) in different layers of the skin: (a) epidermal surface; (b) epidermis; (c) dermis. A splitting of the high-field resonance line, h-1, into two peaks indicates distribution of DNTB between apolar (h-1a) and polar (h-1p) microenvironments within the epidermis and dermis. (Reproduced by permission of Elsevier Science from J. Fuchs, N. Groth, T. Herrling, R. Milbradt, G. Zimmer, and L. Packer, *J. Invest. Dermatol.*, 1992, **98**, 713)

of spheroids can be done using 2D EPRI microscopy (Figure 3). They used a water soluble, cell permeable nitroxide (^{15}N substituted, perdeuterated Tempone) and a paramagnetic broadening agent (chromium oxalate), which is excluded from viable cells, but enters dead cells thus leaving observable only nitroxide signals from living cells. Using a magnetic field gradient of 1 T m^{-1}, a linear resolution of 10–20 μm was obtained; this is quite comparable to that obtained in a similar MRI study.[43] The method was used to resolve microscopic areas of necrosis in spheroids at different stages of their growth. The same

approach has also been used to study the cytotoxicity of an antitumor drug (Adriamycin). The advantages of this approach include that it can be applied to single spheroids and that the same spheroid can be studied repeatedly over periods of several weeks.[41]

This technique could potentially be used in vivo, although this would necessarily involve lower resolution because of the lower frequency that is needed for in vivo EPR, and a method would need to be developed for achieving locally high concentrations of a suitable paramagnetic metal ion.

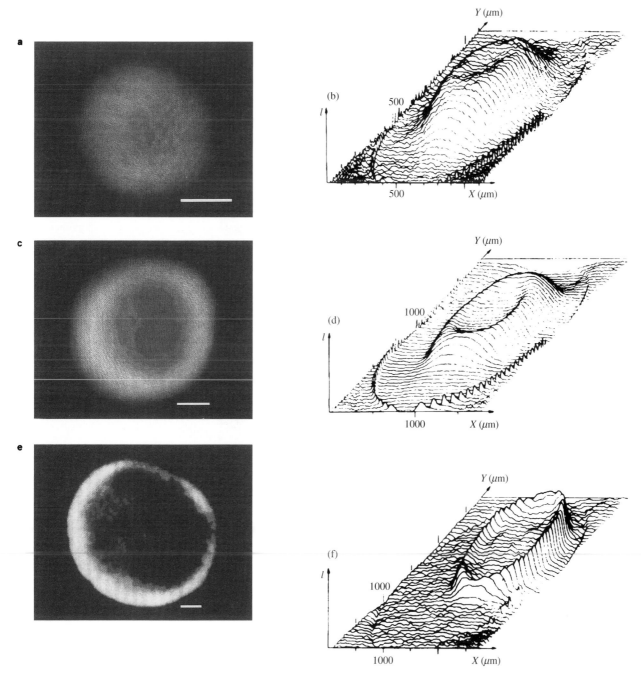

Figure 3 Two-dimensional EPRI microscopy (gray scale and surface plots) of cell viability in spheroids at different stages of growth: (a) and (b) = 12 days; (c) and (d) = 3 weeks; (e) and (f) = 5 weeks, or diameter of 0.58, 0.99 and 1.49 mm. The intensity of the image corresponds to the amount of viable cells in different regions. Note that the thickness of the viable rim remains the same regardless of the size of the spheroid (bars are approximately 0.2 mm). For each image 64 projections were collected with a magnetic field gradient of 0.9 T m^{-1}. (Reproduced by permission of Macmillan Press from J. W. Dobrucki, F. Demsar, T. Walczak, R. K. Woods, G. Bacic, and H. M. Swartz, *Br. J. Cancer*, 1990, **61**, 221)

Another potential approach to the use of EPR to measure viability in vivo was demonstrated by Berliner et al.[44] in the first reported EPR image of a living animal. Using an EPR spectrometer operating at 1.5 GHz and a loop-gap resonator they obtained images of a melanoma which was implanted into the tail of a mouse, using serial intravenous (i.v.) injections of a nitroxide (CTPO; 3-carbamoyl-2,2,5,5-tetramethyl- pyroline-1-oxyl) as the imaging agent. Although reconstructed from only four projections, the resulting 2D image of the tumor showed distinguishable necrotic and viable areas. Additional studies are needed to determine if the distribution plus metabolism of nitroxides can provide a reproducible means of distinguishing between viable and nonviable regions. If so, the resolution obtainable when imaging at 1 GHz may be sufficient to provide very useful information, especially in connection with studies of metabolism by NMR spectroscopy.

For References see p. 854

5 MEASUREMENT OF THE CONCENTRATION AND DISTRIBUTION OF CONTRAST AGENTS, ESPECIALLY IN VIVO

5.1 Nitroxides

The main interest in studying nitroxides as potential MRI contrast agents arises from the concept of metabolically responsive contrast agents.[34,45] Namely, if the administered paramagnetic contrast agent can undergo reduction or oxidation to a nonparamagnetic product, then it may be possible to obtain a set of images where the contrast reflects regions with different redox metabolism. The productive use of nitroxides as metabolically responsive contrast agents for MRI requires knowledge of at least three elements: their metabolic responsiveness; their effectiveness as relaxation enhancers; and their pharmacological properties.

5.1.1 Metabolic Responsiveness of Nitroxides

In vitro EPR studies of mammalian cells in suspensions[46] showed that virtually all nitroxides undergo metabolism to a nonparamagnetic state under appropriate conditions and that the rate of reduction is primarily affected by the concentration of oxygen. Susceptibility to reduction varies by a factor of more than 100 between different nitroxides, with the principal factors being whether the nitroxide readily enters the cells and the structure of the ring to which the nitroxide group is attached (nitroxides based on five membered rings are generally more resistant to reduction than nitroxides based on the six-membered piperidine ring). The mechanism of metabolic reduction in vivo has been studied only to a limited extent[29,47–49] but it seems that the relationship between the structure of the nitroxides and their susceptibility to bioreduction is similar to that observed in vitro systems. It has also been shown that cells can oxidize hydroxylamines back to paramagnetic nitroxides,[50] which raises the possibility of administering hydroxylamines and obtaining differential amounts of contrast in direct proportion to the tissue pO_2.

5.1.2 Relaxivity of Nitroxides

The ability of nitroxides to affect the relaxation times of protons (i.e. relaxivity) is an order of magnitude lower than that of metal ions. The nitroxides are spin-$\frac{1}{2}$ systems with a single unpaired electron, which is a disadvantage when compared with metal ions (e.g. Mn or Gd) that have several unpaired electrons. A second disadvantage is that nitroxides do not bind water tightly, in contrast to the metal ions and their complexes, so that relaxation occurs via relatively inefficient outer sphere mechanisms.[51] However, the relaxivity of nitroxides can be improved by binding them to macromolecules which changes their rotational mobility and the correlation times of the interaction of the unpaired electron and the proton. In order to optimize such an approach, two major points should be considered: the possibility of binding multiple nitroxides to macromolecules and the degree of the immobilization of the nitroxide groups. EPR plays a key role in designing such types of contrast agents since both the number of nitroxides per macromolecule and their molecular dynamics can be deter-

mined directly from EPR spectra. Several nitroxide–macromolecule (serum albumin, polysaccharide, fatty acids) complexes have been investigated by EPR, and their relaxivity and in vivo distribution have been determined by NMR.[51–55] In addition to increasing relaxivity, binding of nitroxides to certain vector molecules such as arabinogalactans, which are specifically taken up by hepatocytes, increases organ selective delivery while retaining metabolic responsiveness.[55]

5.1.3 Pharmacological Studies of Nitroxides

The nitroxide groups themselves appear to play a relatively minor role in the distribution of the nitroxides, which is usually determined by the other parts of the molecule based on the usual factors such as size, charge, and lipophilicity. The in vivo kinetics appear to be highly complex, depending upon the type of nitroxide, amount of reduction that can occur in the blood, specific accumulation in particular organs, cellular uptake and intracellular reduction rates, and clearance mechanisms. Figure 4(a) and (b) illustrate an in vivo EPR spectroscopy experiment in which patterns of distribution and reduction of three nitroxides were followed.[47] A surface detector was placed over the region of interest to localize the spectra. The different behavior of the three nitroxides [Figure 4(a)] reflects differences in the rate of membrane permeation and bioreduction and is consistent with results from in vitro studies. These experiments also demonstrated the presence of ascorbate in the bladder. Reduction by ascorbate in the bladder caused the decay of the EPR signal of Cat_1 [Figure 4(b)]. Subsequent in vivo EPR spectroscopic studies further analyzed the pharmacology of nitroxides in these and other organs such as brain and lungs.[29,48,49] Although these experiments were designed to study nitroxide metabolism and their potential use as a metabolically sensitive contrast agent for MRI or EPR imaging, the same technique, combined with the use of spin-labeled analogs of drugs, could also be used for detailed pharmacokinetic studies of a wide range of drugs taking advantage of the high specificity and sensitivity of EPR spectroscopy.

Low-frequency EPRI (operating frequencies 1.2 GHz to 280 MHz) has also been used in pharmacological studies.[56,57] While demonstrating that potentially both whole body distribution and pharmacokinetics can be simultaneously studied, these papers also outlined some of the principal problems facing in vivo EPRI. As already mentioned, the major problem arises from the fact that the imaging substance (nitroxide) has to be introduced externally. Toxicity considerations limit the maximum amount to be administered and although in vivo EPR spectrometers can in principle detect very low concentrations of nitroxides (1–10 μmol kg^{-1}), imaging requires at least 100-fold higher concentrations and even then the resulting image may have a rather poor signal/noise ratio. The inability of in vivo EPRI to utilize pulsed, slice selected imaging procedures results in a long imaging time and/or a small number of projections. The acquisition time for 2D images is 2–10 min which is acceptable, but results in superposition of different planes so that it can be difficult to interpret images on an anatomical basis. True zeugmatographic 3D images require 20–80 min, during which a concentration of nitroxide in a selected region has to be maintained as constant. Reported inplane resolution for rat head at 780 MHz was 0.5 mm with an effective

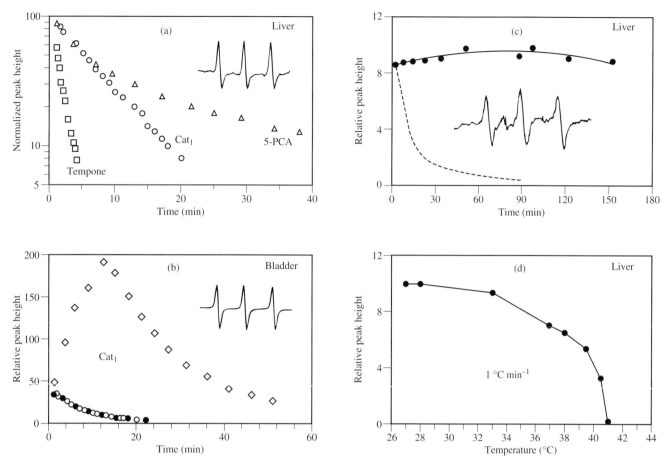

Figure 4 In vivo EPR spectroscopy. (a) and (b) Pharmacokinetics of nitroxides. (a) Decrease of the intensities of the EPR signals of nitroxides from the liver region of mice following the i.v. injection of three different nitroxides: Tempone (4-oxo-2,2,6,6-tetramethylpiperidine-1-oxyl), Cat$_1$ (4-trimethylammonium-2,2,6,6-tetramethylpiperidine-1-oxyl), and PCA (3-carboxy-2,2,5,5-tetramethylpyrroline-1-oxyl) at doses 0.10–0.15 mmol kg^{-1} body wt. The signals of all nitroxides are normalized to the signal intensity at the termination of injection. The temperature of the mice was maintained at 37 °C. The inset shows the in vivo spectrum of PCA obtained 5 min after injection. (b) The EPR intensities of Cat$_1$ from the bladder region under the same experimental conditions. The decrease of the signal from the vascular pool, measured by the placement of the surface probe over regions without localized uptake, is also shown (solid and open circles indicate points from two different experiments). (c) and (d) Studies of the status of liposomes. Changes of the intensity of the EPR signal of a nitroxide from the liver region of a mouse following the i.v. injection of Cat$_1$ encapsulated in dipalmitoyl phosphatidylcholine/DPPG liposomes. The injected dose was 0.03 mmol kg^{-1}. (c) Changes of the signal intensity vs. time after injection. The temperature of the mouse was maintained at 30 °C, i.e. below the phase transition point of liposomes. The inset is a typical spectrum obtained 30 min after injection. The broken line shows a typical blood clearance curve for these liposomes. (d) Changes of the signal intensity vs. temperature of the mouse. Heating of the mouse by flowing warm air (1 °C min^{-1}) was started 30 min after injection of the liposomes. (Reproduced by permission of Academic Press from G. Bacic, M. J. Nilges, R. L. Magin, T. Walczak, and H. M. Swartz, *Magn. Reson. Med.*, 1989, **10**, 266)

slice thickness of 1.5 mm. Resolution can be improved by using larger field gradients, but this would, besides being technically demanding, further reduce the S/N ratio of the image. In addition, all resonators currently used have a nonhomogeneous rf field distribution, particularly along the axial direction, which further complicates accurate image reconstruction. Use of larger resonators can alleviate this problem, but at the expense of a further reduction in sensitivity. A potential solution for many of the above mentioned problems is the development of pulsed EPRI techniques. The frequency ranges of in vivo EPRI machines nowadays overlap with high-field MRI machines, which should make development of pulsed EPRI easier. Until then, it seems that much more reliable images of spin-labels in animals can be obtained using standard MRI[55] or pro-

ton electron double resonance imaging (see *Overhauser Effect Imaging of Free Radicals*).

5.2 Other Contrast Agents

Metal ion (e.g. Mn, Gd) complexes or aggregates (iron particles) are paramagnetic species and it would appear that their pharmacology can be directly studied using in vivo EPR in the same manner as for nitroxides. However, at the low fields, required for in vivo EPR, their EPR spectra have some unfavorable characteristics that effectively prevent their detection in vivo. For example, the spectrum of Gd-DTPA at 9.5 GHz contains one broad line spanning 60 mT and measurements of physiological concentrations of species with such broad res-

For References see p. 854

onant lines are well below the sensitivity of *L*-band EPR spectrometers. Despite these limitations, a few studies demonstrate that properly designed EPR studies can substantially add to our understanding of the pharmacology of MRI contrast agents. For example, liposomes have been extensively studied as an alternative means for targeted delivery of contrast agents. Encapsulation of contrast agents within liposomes increases selective delivery, but restricts the accessibility of the tissue water to the relaxation center. It has been found that a liposome encapsulated contrast agent (Mn^{2+}) delivered to the liver remains MRI silent as long as it is confined within liposomes or Kupffer cells due to the limited exchange of water.[58] Consequently, measurements of the rates of liposome breakdown and subsequent redistribution of contrast agents are important steps in evaluating the usefulness of liposomes as carriers of contrast agents. MRI detects these processes indirectly via changes in water relaxation and the results are often difficult to interpret, but EPR can be used effectively for these measurements as illustrated in Figure 4(c) and (d).[47] The experiment took advantage of the rapid bioreduction of the nitroxide, Cat_1, to follow the integrity of the liposomes in the liver, because intact liposomes protect encapsulated Cat_1 from bioreduction. When liposomes were heated to near their transition temperature, their contents began to leak out and the EPR signal decreased as the Cat_1 became reduced. The clearance from the blood of the intact liposomes could also be followed by this technique.[58] Besides the localization within the tissue, the state of the contrast agent (bound/free) also has a profound effect on relaxation enhancement. Again MRI measures this indirectly, but EPR can measure free and bound fractions directly. In an experiment similar to the one in Figure 4(c) and (d) liposomes were loaded with Mn^{2+} and liver samples at different time points were analyzed by EPR and NMR. It was determined that as long as manganese is within the Kupffer cells and is predominantly an aquoion, no substantial changes in the relaxation of the protons of water in the liver were observed. Subsequent release of manganese from Kupffer cells resulted in binding of Mn^{2+}, as detected by EPR, yielding a pronounced increase in relaxation enhancement.[58]

6 CONCLUSIONS

EPR spectroscopy has capabilities which are quite complementary to those of NMR, especially for studies in vivo which involve contrast agents, and for measurements related to pO_2 and/or redox metabolism. The appropriate use of EPR in such studies can significantly enhance the value of the NMR studies. It is already quite clear that for *experimental* studies EPR provides data that are widely applicable and useful for NMR studies, especially the capability of measuring pO_2 in tissues in vivo. It is unlikely that EPR will become a widespread *clinical* tool with the very important exception that it is quite possible that in the near future it will become the method of choice for measuring pO_2 in tissues in vivo.

7 RELATED ARTICLES

Cells and Cell Systems MRS; Contrast Agents in Magnetic Resonance: Operating Mechanisms; Contrast Agents in Whole Body Magnetic Resonance: An Overview; Overhauser Effect Imaging of Free Radicals; Spectroscopic Studies of Animal Tumor Models; Surface and Other Local Coils for In Vivo Studies.

8 REFERENCES

1. G. R. Eaton, S. S. Eaton, and K. Ohno, eds. 'EPR Imaging and in vivo EPR', CRC Press, Boca Raton, FL, 1991.
2. K. Ohno and N. Tsuchihashi, *Trends Phys. Chem.*, 1991, **2**, 79.
3. S. Colacicchi, M. Ferrari, and A. Sotgiu, *Int. J. Biochem.*, 1992, **24**, 205.
4. H. M. Swartz and T. Walczak, *Phys. Med.*, 1993, **9**, 41.
5. J. H. Freed, *J. Chem. Soc., Faraday Trans.*, 1990, **86**, 3173.
6. E. Zavoisky, *J. Phys. (Moscow)*, 1945, **9**, 211.
7. A. Feldman, E. Wildman, G. Bartolini, and L. H. Piette, *Phys. Med. Biol.*, 1975, **20**, 602.
8. S. J. Lukiewicz and S. G. Lukiewicz, *Magn. Reson. Med.*, 1984, **1**, 297.
9. L. J. Berliner and H. Fujii, *Science*, 1985, **227**, 517.
10. M. J. Nilges, T. Walczak, and H. M. Swartz, *Phys. Med.*, 1989, **5**, 195.
11. J. L. Zweier and P. Kuppusamy, *Proc. Natl. Acad. Sci. U.S.A.*, 1988, **85**, 5703.
12. H. J. Halpern, D. P. Spencer, J. van Polen, M. K. Bowman, A. C. Nelson, E. M. Dowey, and B. A. Teicher, *Rev. Sci. Instrum.*, 1989, **60**, 1040.
13. M. Alecci, S. Della Penna, A. Sotgiu, L. Testa, and I. Vannucci, *Rev. Sci. Instrum.*, 1992, **63**, 4263.
14. J. A. Brivati, A. D. Stevens, and M. C. R. Symons, *J. Magn. Reson.*, 1991, **92**, 480.
15. W. Froncisz and J. S. Hyde, *J. Magn. Reson.*, 1982, **47**, 515.
16. A. Sotgiu, *J. Magn. Reson.*, 1985, **65**, 206.
17. J. F. Glockner and H. M. Swartz, in 'Oxygen Transport to Tissue', eds. W. Erdmann and D. F. Bruley, Plenum, New York, 1993, Vol. 14.
18. U. Ewert and T. Herrling, *Chem. Phys. Lett.*, 1986, **129**, 516.
19. M. M. Maltempo, S. S. Eaton, and G. R. Eaton, *J. Magn. Reson.*, 1987, **72**, 449.
20. K. Ohno, *Spectrosc. Rev.*, 1986, **22**, 1.
21. R. K. Woods, G. Bačić, P. C. Lauterbur, and H. M. Swartz, *J. Magn. Reson.*, 1989, **84**, 247.
22. N. Vahidi, R. B. Clarkson, K. J. Liu, S. W. Norby, M. Wu, and H. M. Swartz, *Magn. Reson. Med.*, 1994, **31**, 139.
23. K. J. Liu, P. Gast, M. Moussavi, S. W. Norby, N. Vahidi, T. Walczak, M. Wu, and H. M. Swartz, *Proc. Natl. Acad. Sci. U.S.A.*, 1993, **90**, 5438.
24. G. Bačić, K. J. Liu, J. A. O'Hara, R. D. Harris, K. Szybinski, F. Goda, and H. Swartz, *Magn. Reson. Med.*, 1993, **30**, 568.
25. A. I. Smirnov, S. W. Norby, R. B. Clarkson, T. Walczak and H. M. Swartz, *Magn. Reson. Med.*, 1993, **30**, 213.
26. H. M. Swartz, K. J. Liu, F. Goda, and T. Walczak, *Magn. Reson. Med.*, 1994, **31**, 229.
27. W. K. Subczynski, S. Lukiewicz, and J. S. Hyde, *Magn. Reson. Med.*, 1986, **3**, 747.
28. J. L. Zweier and P. Kuppusamy, *J. Bioenerg. Biomembr.*, 1991, **23**, 855.
29. M. Ferrari, S. Colacicchi, G. Gualtieri, M. T. Santini, and A. Sotgiu, *Biochem. Biophys. Res. Commun.*, 1990, **166**, 168.
30. F. Demsar, T. Walczak, P. D. Morse II, G. Bačić, Z. Zolnai, and H. M. Swartz, *J. Magn. Reson.*, 1988, **76**, 224.
31. G. Bačić, F. Demsar, Z. Zolnai, and H. M. Swartz, *Magn. Reson. Med. Biol.*, 1988, **1**, 55.
32. G. Bačić, T. Walczak, F. Demsar, and H. M. Swartz, *Magn. Reson. Med.*, 1988, **8**, 209.

33. R. K. Woods, W. B. Hyslop, and H. M. Swartz, *Phys. Med.*, 1989, **5**, 121.

34. H. M. Swartz, in 'Advances in Magnetic Resonance Imaging', ed. E. Feig, Ablex Publishing, Norwood, NJ, 1989, Chapt. 2.

35. K. Chen and H. M. Swartz, *Biochim. Biophys. Acta*, 1989, **992**, 131.

36. K. Chen, J. F. Glockner, P. D. Morse II, and H. M. Swartz, *Biochemistry*, 1989, **28**, 2496.

37. J. Fuchs, N. Groth, T. Herrling, R. Milbradt, G. Zimmer, and L. Packer, *J. Invest. Dermatol.*, 1992, **98**, 713.

38. V. Gabrijelcic, M. Sentjurc and J. Kristl, *Int. J. Pharm.*, 1990, **62**, 75.

39. V. Gabrijelcic, M. Sentjurc, and M. Schara, *Period. Biol.*, 1991, **93**, 245.

40. G. M. Rosen, H. J. Halpern, L. A. Brunsting, D. P. Spencer, K. E. Strauss, M. K. Bowman, and A. S. Wechsler, *Proc. Natl. Acad. Sci. U.S.A.*, 1988, **85**, 7772.

41. J. W. Dobrucki, R. M. Sutherland, and H. M. Swartz, *Magn. Reson. Med.*, 1991, **19**, 42.

42. J. W. Dobrucki, F. Demsar, T. Walczak, R. K. Woods, G. Bačíc, and H. M. Swartz, *Br. J. Cancer*, 1990, **61**, 221.

43. L. O. Sillerud, J. P. Freyer, M. Neema, and M. A. Mattingly, *Magn. Reson. Med.*, 1990, **16**, 380.

44. L. J. Berliner, H. Fujii, X. Wan, and S. J. Lukiewicz, *Magn. Reson. Med.*, 1987, **4**, 380.

45. R. C. Brasch, *Radiology*, 1983, **147**, 781.

46. H. M. Swartz, M. Sentjurc, and P. D. Morse II, *Biochim. Biophys. Acta*, 1986, **888**, 82.

47. G. Bačíc, M. J. Nilges, R. L. Magin, T. Walczak, and H. M. Swartz, *Magn. Reson. Med.*, 1989, **10**, 266.

48. F. Gomi, H. Utsumi, A. Hamada, and M. Matsuo, *Life Sci.*, 1993, **52**, 2027.

49. K. Takeshita, H. Utsumi, and A. Hamada, *Biochem. Biophys. Res. Commun.*, 1991, **177**, 874.

50. K. Chen and H. M. Swartz, *Biochim. Biophys. Acta*, 1988, **970**, 270.

51. H. F. Bennett, R. D. Brown III, S. H. Koenig, and H. M. Swartz, *Magn. Reson. Med.*, 1987, **4**, 93.

52. G. Sosnovsky, N. U. M. Rao, J. Lukszo, and R. C. Brasch, *Z. Naturforsch. Teil B.*, 1986, **41**, 1170.

53. H.-C. Chan, K. Sun, R. L. Magin, and H. M. Swartz, *Bioconjugate Chem.*, 1990, **1**, 32.

54. B. Gallez, R. Debuyst, R. Demeure, F. Dejehet, C. Grandin, B. Van Beers, H. Taper, J. Pringot, and P. Dumont, *Magn. Reson. Med.*, 1993, **30**, 592.

55. B. Gallez, V. Lacour, R. Demeure, R. Debuyst, F. Dejehet, J.-L. De Keyser, and P. Dumont, *Magn. Reson. Imag.*, 1994, **12**, 61.

56. V. Quaresima, M. Alecci, M. Ferrari, and A. Sotgiu, *Biochem. Biophys. Res. Commun.*, 1992, **183**, 829.

57. S.-I. Ishida, S. Matsumoto, H. Yokoyama, N. Mori, H. Kumashiro, N. Tsuchihashi, T. Ogata, M. Yamada, M. Ono, T. Kitajima, H. Kamada, and E. Yoshida, *Magn. Reson. Imag.*, 1992, **10**, 109.

58. G. Bačíc, M. R. Niesman, R. L. Magin, and H. M. Swartz, *Magn. Reson. Med.*, 1990, **13**, 44.

Biographical Sketches

Harold M. Swartz. *b* 1935. M.D., 1959, University of Illinois, Ph.D., 1969, Biophysics, Georgetown University. Professor of Radiology and Biochemistry, Medical College of Wisconsin, Milwaukee, WI, USA, 1970–1979; Professor of Physiology & Biophysics, Medicine, and Bioengineering, University of Illinois, Urbana-Champaign, IL, USA, 1980–1991; Professor of Radiology, Physiology, and Bioengineering Dartmouth, Hanover, NH, USA, 1991–present. Approx. 225 publications in applications of magnetic resonance to biology and medicine, measurements of pO_2 in cells and in vivo, structure and function of melanins, radiation biology, free radicals in biology and medicine, contrast agents for MRI and MRS.

Goran Bačíc. *b* 1951. B.S., 1976, M.S., 1980, Ph.D., 1985, Physical Chemistry, University of Belgrade, YU. Postdoctoral work at University of Illinois, USA (with H.M. Swartz). Successively research associate and assistant professor at University of Belgrade. Currently visiting research professor at Dartmouth College, NH, USA. Approx. 50 publications. Research specialties: in vivo EPR and EPR imaging, proton NMR relaxation, MRI contrast agents.

For References see p. 854

Part Eleven
MR of Animal and Cell Models

Body Fluids

Jimmy D. Bell

Imperial College School of Medicine, Hammersmith Hospital, London, UK

and

Peter J. Sadler

University of Edinburgh, UK

1 INTRODUCTION

1.1 Historical Perspective

The full potential for the use of high resolution NMR spectroscopy as applied to the study of body fluids has only become apparent in the last decade. However, Eric Odeblad, a Swedish physicist and gynecologist, used ^{1}H NMR to study body fluids as far back as 1957.[1,2] Although the spectra were relatively uninformative, he advocated the establishment of a 'department of medical research with NMR.' In the early 1970s, the introduction of Fourier transform spectroscopy, together with technical advances (field strength, coil designs, pulse sequences, and computing) enabled the use of high-resolution NMR for the detection and identification of different chemical species in biological systems. In 1979, Ohsaka et al.[3] reported high resolution proton NMR spectra of intact body fluids and attempted serodiagnosis of cancer. Their spectra consisted of sharp resonances from low-molecular-mass metabolites superimposed on broad resonances from macromolecules, with an elevated lactate signal as a criterion for diagnosis of cancer. This was closely followed by a study of high-resolution NMR spectra of urine samples from subjects with diabetes and renal failure.[4] Marked changes in the concentration of a number of then unidentified endogenous metabolites were observed. From these preliminary studies, it became clear that high resolution NMR could play an important role in the study of body fluids.

The first full clinical use of NMR to study body fluids was in a study of metabolic disorders by Bock and coworkers.[5,6] Elevated concentrations of D-lactate were detected in serum samples from human subjects following jejunoileal bypass. Full characterizations of different body fluids by high-resolution NMR and application to a wider range of biochemical and clinical problems came soon afterwards. Since then, the study of body fluids by high resolution NMR spectroscopy has become an increasingly expanding area of research.

1.2 The Range of Body Fluids

The biochemical composition of body fluids is known to reflect the metabolic status of the donor. Thus, detailed knowledge of the composition of body fluids is likely to provide good insight into the biochemical and clinical status of the donor. This has led workers to apply a variety of techniques to obtain metabolic profiles of body fluids. However, most of the standard techniques require extensive sample preparation together with careful selection of analytical conditions. This can clearly restrict the number of compounds that can be studied at a given time. Moreover, a number of fluids contain a range of molecules with diverse chemical and molecular masses, including amino acids, proteins and lipoproteins, all or some of which can be affected in disease. The nonselective nature of NMR spectroscopy makes it an ideal technique for obtaining metabolic profiles of all available body fluids. Furthermore, NMR spectroscopy can give useful insights into molecular interactions within intact fluids which are not readily available by standard biochemical techniques. Thus, NMR spectroscopy, although relatively insensitive, has been successfully applied to the detection and quantitation of a variety of molecules in intact body fluids, including blood, blood plasma, serum, urine, cerebrospinal fluid (CSF), bile, synovial fluid, seminal fluid, milk, saliva and sweat (Table 1).[7–60] Representative ^{1}H NMR spectra of some body fluids are shown in Figure 1.

The physicochemical composition of different body fluids varies widely. Fluids such as plasma, bile, CSF, seminal fluid and synovial fluid contain both macromolecules, low-molecular-mass compounds and inorganic ions. These compounds range from complex lipoproteins and glycoproteins to relatively simple organic acids, amino acids and metal ions. Less complicated body fluids include urine, aqueous humor, saliva, and sweat, which contain relatively low levels of macromolecules, but often a very large number of different low-molecular-mass metabolites (Table 1). The ^{1}H NMR chemical shifts of some common metabolites are listed in Table 2. NMR spectroscopy can provide qualitative and quantitative information about most of these body fluids components, often in just a few minutes, as well as information about molecular structures, molecular interactions and dynamics.

1.3 Sample Preparation and NMR Data Acquisition

NMR spectra can be readily obtained from intact body fluids, requiring little or no sample preparation. Samples (usually about 500 μL) can be examined untreated in the spectrometer, except that normally a small amount of a deuterated solution is added to provide a field/frequency lock (e.g. 5–10% v/v D_2O or saline). This can be added directly to the sample, or placed in a capillary tube, thus avoiding dilution of sample. Particular care should be taken with the storage of samples. For example, contaminating bacteria can introduce exogenous substances, and freezing and thawing can cause chemical changes such as precipitation and protein denaturation. Buffering can be a problem. For example, the main natural buffer in blood plasma is bicarbonate/CO_2 and the pH of samples can rise rapidly after storage due to the volatility of CO_2. This can be counteracted by adding sodium bicarbonate (25 mM) or a small amount of another buffer (e.g. phosphate).

Reasonably resolved ^{1}H NMR spectra of body fluids can be obtained on commercially available high-frequency spectrometers (>400 MHz). Higher frequency NMR spectrometers (600 and 750 MHz) give increased resolution and relative sensitivity,[21] and improve the detection and identification of body fluid components; however, these instruments are relatively expensive.

For References see p. 870

Table 1   Endogenous Metabolites detected in Body Fluids by High-Resolution ^{1}H NMR Spectroscopy[a]

Amniotic fluid

Acetate α-alanine, D-3-hydroxybutyrate, citrate, creatine, creatinine, formate, glucose, glutamate, glycerophosphorylcholine, isoleucine, lactate, lecithin, leucine, lipids, lysine, methionine, ornithine, 2-oxoisovalerate, phospholipids, proline, sphingomyelin, threonine, tyrosine, valine[55-57]

Aqueous humor

Acetate, alanine, ascorbate, citrate, creatine, formate, glucose, glutamine/glutamate, histidine, lactate, phenylalanine, threonine, tyrosine, valine[44,45]

Bile

Acetate, acetone, acetoacetate, α-alanine, cholesterol, cholesterol esters, choline, citrate, creatine, creatinine, ethanol, ergothioneine, formate, glucose, glutamate, glutamine, glycine, glycocholic acid, D-3-hydroxybutyrate, isoleucine, lactate, propionate, phosphatidylcholine, taurine, taurocholic acid, urea, valine[39,40]

Blood plasma and serum

Acetate, acetone, acetoacetate, acetylcarnitine, N-acetyl sugars, N-acetyl glycoproteins, acylglycerols, alanine, aspartate, betaine, carnitine, cholesterol, cholesterol esters, choline, citrate, creatine, creatinine, dimethylamine, dimethylglycine, ethanol, ergothioneine, formate, glucose, glutamate, glutamine, glycine, glycerol, histidine, D-3-hydroxybutyrate, p-hydroxyphenyllactate, isoleucine, lactate, leucine, lysine, lipoproteins (high density, low density, very low density), 1-methylhistidine, 3-methylhistidine, myoinositol, phenylalanine, phosphatidylcholine, pyruvate, succinate, threonine, trimethylamine N-oxide, tyrosine, urea, valine

Cerebrospinal fluid

Acetate, acetone, acetoacetate, α-alanine, aspartate, arginine, choline, citrate, citrulline, creatine, creatinine, dimethylamine, formate, fumarate, GABA, galactose, glucose, glutamine, glutamate, glycine, glycerol, D-3-hydroxybutyrate, histidine, indoxyl sulfate, inositol, isoleucine, lactate, leucine, lysine, mannitol, methionine, proline, pyruvate, phenylalanine, spermidine, succinate, taurine, threonine, trimethylamine N-oxide, tyrosine, valine.

Cystic fluid

Acetate, N-acetylglycoproteins, alanine, arginine, citrate, creatine, dimethylamine, ethanol, formate, glucose, glutamine, histidine, isoleucine, lactate, lysine, phenylalanine, succinate, threonine, tyrosine, urea, valine[58]

Follicular fluid

Acetate, N-acetyl sugars, N-acetyl glycoproteins, alanine, creatine, creatinine, ethanol, glucose, glycine, D-3-hydroxybutyrate, lactate, valine[49]

Gastric fluid

Histidine, tyrosine, phenylalanine[50]

Milk

Acetate, alanine, citrate, creatine, formate, glucose, lactose, lactate, lipids, taurine[51]

Saliva[41,42]

Seminal fluid

Acetate, alanine, arginine, aspartate, asparagine, D-3-hydroxybutyrate, choline, citrate, creatine, cystine, dimethylamine, formate, fumarate, fructose, glucose, glutamate, glutamine, glycine, glycerophosphorylcholine, histidine, inositol, isoleucine, lactate, leucine, lysine, methionine, phenylalanine, phosphorylcholine, proline, pyruvate, spermidine, spermine, taurine, threonine, trimethylamine N-oxide, tryptophan, tyrosine, uridine, valine[46-48]

Sweat

Acetate, alanine, D-3-hydroxybutyrate, citrate, glutamate, glutamine, glycine, histidine, lactate, tyrosine, valine[43]

Synovial fluid

Acetate, acetone, acetoacetate, N-acetyl sugars, N-acetyl glycoproteins, α-alanine, cholesterol, cholesterol esters, choline, citrate, creatine, creatinine, ethanol, formate, glucose, glutamate, glycine, histidine, D-3-hydroxybutyrate, lactate, lipoproteins, phenylalanine, threonine, tyrosine, valine[43,52-54]

Urine

Acetate, acetamide, acetone, acetoacetate, acetylcarnitine, N-acetyl sugars, α-alanine, β-alanine, allantoin, anserine, ascorbate, betaine, butanone, carnitine, choline, citrate, creatine, creatinine, dimethylamine, dimethylglycine, ethanol, formate, glucose, glutamate, glutamine, glutarate, glycine, guanine, hippurate, histidine, hydroxybenzoate, D-3-hydroxybutyrate, 2-hydroxyisocaproate, 2-hydroxyisovalerate, 2-hydroxy-3-methylvalerate, hydroxypropionate, indoxyl sulfate, inositol, isoleucine, isovalerylglycine, 2-oxoglutarate, lactate, leucine, lysine, methanol, methionine, methylamine, methylmalonate, N-acetylglycoproteins, 2-oxoisocaproate, 2-oxoisovalerate, 2-oxo-3-methylvalerate, phenylalanine, propionate, propionylcarnitine, propionylglycine, succinate, taurine, trimethylamine, trimethylamine N-oxide, threonine, tyrosine, urea, valine

Vitreous fluid

Acetate, alanine, citrate, creatine, formate, glutamine/glutamate, lactate, valine[45]

[a]References given are mostly to work not described in the text.

In general, ^{1}H NMR spectra of body fluids with good signal-to-noise ratios can be obtained within a few minutes. The acquisition of NMR spectra of body fluids requires consideration of two major factors: T_1 and T_2 relaxation, and water suppression. Fully-relaxed ^{1}H NMR spectra can usually be obtained with 35° pulses and a 3-s pulse repetition time. ^{13}C NMR spectra require up to 10 s to allow full relaxation of some lipid and protein components. For some body fluids, spin echo techniques are useful for the suppression of broad resonances from macromolecules.[12] T_2 effects, which can affect the

quantitation of metabolites, must be taken into account during the analysis of these spectra.

Water suppression is crucial for ^{1}H NMR experiments on body fluids since the concentration of water protons is about 10^5 times that of the protons in the metabolites of interest. Although freeze-drying and redissolution in D_2O is very effective, volatile metabolites may be lost and the structures of some macromolecules may be altered by this process.[23] The most commonly used method for suppression is presaturation[61] (secondary irradiation during the relaxation delay, e.g. for 1 s), although this does not allow detection of peaks very close to the water peak. For the latter purpose the WATR method[62] (water attenuation by T_2 relaxation) is very effective. This involves addition of a water relaxation agent (e.g. NH_4Cl) and use of the CPMG spin echo sequence with a short refocusing time so that normal multiplets without phase modulation are obtained. Details of other methods can be found elsewhere in this volume. Many suppression methods lead to distortions in peak intensities and this has to be checked if quantitative information is required. It seems likely that increasing use will be made of field gradient water suppression methods in future. Other advances are likely to include the use of microprobes requiring only about 50 μL of sample, and, at the other extreme, larger sample volumes (e.g. 4 mL in a 10 mM tube) which provide high sensitivity.

2 DIAGNOSIS AND MOLECULAR INTERACTIONS

2.1 Organ Dysfunction

The biochemical profile of a body fluid is known to reflect the metabolic status of the donor. Thus, the unique exploratory nature of NMR makes it an ideal technique for the study of body fluids from subjects with organ dysfunction This is clearly reflected in published work which includes analysis of plasma, CSF, urine and bile from subjects suffering from liver failure,[8,39] brain tumors,[24] organ transplant,[21] and renal failure.[16] For example, increased concentrations of aromatic amino acids have been observed in NMR spectra of plasma of subjects with hepatic failure,[8,39] while variations in plasma N-acetyl resonances correlate with acute heart rejection.[14]

An important area of clinical and biochemical research has been the application of high-resolution NMR spectroscopy to the study of the biochemical abnormalities associated with chronic renal failure (CRF).[16,21] Renal damage in both animals and humans has been shown to give rise to altered profiles for low-molecular-mass metabolites of urine and plasma, which in turn is reflected in the NMR spectral fingerprint. Bell et al.[16] used ^{1}H NMR to show that plasma and urine from subjects' CRF had increased levels of creatinine, trimethylamine oxide (TMAO) and dimethylamine. These amines had not been previously detected in body fluids from human subjects and have since become valuable markers to monitor renal transplant function. Since then several novel markers of renal dysfunction have been identified by NMR,[21] including TMAO, dimethylamine (DMA), dimethylglycine (DMG), 1-methylhistidine and 3-methylhistidine, thus giving a new perspective to the uremic condition and the effects of hemodialysis on body fluid composition. The contribution of NMR spectroscopy of body fluids to renal transplantation has been clearly highlighted by Foxall

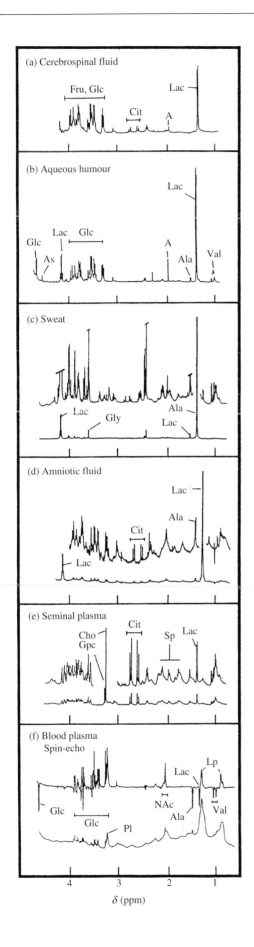

Figure 1 Representative ^{1}H NMR spectra of the aliphatic regions of some human body fluids. A, acetate; Ala, alanine; As, ascorbate; Cho, choline; Cit, citrate; Fru, fructose; Glc, glucose; Gly, glycine; Gpc, glycerophosphorylcholine; Lac, lactate; Lp, lipoprotein; NAc, N-acetyls of glycoproteins; Pl, phospholipids; Sp, spermine; Val, valine. (Adapted from J. D. Bell, J. C. C. Brown and P. J. Sadler, *Chem. Br.* 1988, **24**, 1021)

For References see p. 870

Table 2 ^{1}H NMR Chemical Shifts of some Common Metabolites in Body Fluids at pH of about 7[a]

Metabolite	δ/ppm (multiplicity)
Acetate	1.92 (s)
N-Acetylaspartate	2.03 (s), 2.50 (dd), 2.70 (dd), 4.40 (dd)
Acetoacetate	2.29 (s)
Acetone	2.24 (s)
Adenosine	6.06 (d), 8.22 (s), 8.33 (s)
(S)-Adenosylmethionine	6.11 (d), 8.27 (s), 8.29 (s)
Alanine	1.47 (d), 3.76 (q)
β-Alanine	2.56 (t), 3.18 (t)
Allantoin	5.40 (s)
p-Aminohippurate	3.93 (d), 6.87 (d), 7.68 (d)
Arginine	1.66 (m), 1.91 (m), 3.24 (t), 3.75 (dd)
Asparagine	2.88 (dd), 2.95 (dd), 4.00 (dd)
Aspartate	2.68 (dd), 2.79 (dd), 3.89 (dd)
α-Hydroxy-n-butyrate	0.90 (t), 1.70 (m), 3.99 (m)
D-3-Hydroxybutyrate	1.20 (d), 2.05 (m), 2.51 (m), 4.18 (m)
Betaine	3.27 (s), 3.90 (s)
Choline	3.21 (s), 3.56 (t), 4.05 (t)
Citrate	2.67 (d), 2.80 (d)
Citrulline	1.57 (m), 1.85 (m), 3.15 (t)
Creatine	3.04 (s), 3.93 (s)
Creatinine	3.05 (s), 4.01 (s)
Cysteine	3.04 (dd), 3.12 (dd), 4.00 (dd)
Diphosphoglycerate	4.05 (m), 4.09 (m), 4.55 (dd)
Dimethylamine	2.72 (s)
EDTA	3.29 (s), 3.67 (s)
Formate	8.46 (s)
Fumarate	6.52 (s)
γ-Aminobutyrate	1.89 (m), 2.30 (t), 3.03 (t)
Galactose	β: 4.52 (d)
α-Glucose	3.41 (t), 3.53 (dd), 3.71 (t), 3.74 (m), 3.84 (m), 5.23 (d)
β-Glucose	3.24 (dd), 3.40 (t), 3.49 (t), 3.72 (dd), 3.90 (dd), 4.64 (d)
Glutamate	2.09 (m), 2.34 (dt), 3.75 (t)
Glutamine	2.14 (m), 2.45 (m), 3.77 (t)
Glycerophosphorylcholine	3.23 (s), 3.52 (dd), 3.58 (t), 3.59 (m), 3.82 (m), 4.19 (t)
Glycerol	3.56 (dd)
Glycine	3.56 (s)
Hippurate	3.97 (d), 7.64 (t), 7.84 (d)
Histidine	3.15 (dd), 3.25 (dd), 4.00 (dd), 7.11 (s), 7.89 (s)
Hydroxyproline	2.17 (m), 2.44 (m), 3.37 (d), 3.49 (dd), 4.35 (dd)
Hypoxanthine	8.19 (s), 8.22 (s)
Indoxyl sulfate	7.20 (m), 7.28 (m), 7.54 (d), 7.73 (d)
Inosine monophosphate	6.15 (d), 8.23 (s), 8.57 (s)
Inositol	3.27 (t), 3.53 (dd), 3.61 (t), 4.05 (t)
Isoleucine	0.93 (t), 1.00 (q), 1.25 (m), 1.45 (m), 1.96 (m), 3.65 (m)
α-Ketoglutaric acid	2.47 (t), 3.01 (t)
Lactate	1.33 (d), 4.11 (q)
Leucine	0.94 (d), 0.96 (d), 1.69 (m), 1.72 (m), 3.69 (t)
Lysine	1.45 (m), 1.73 (m), 3.01 (t), 3.46 (t)
Malate	2.37 (dd), 2.67 (dd), 4.30 (dd)
Mannitol	3.70 (dd), 3.82 (dd), 3.90 (dd)
Methionine	2.14 (s), 2.16 (m), 2.64 (t), 3.84 (dd)
Nicotinamide	7.60 (q), 8.25 (d), 8.72 (d), 8.94 (s)
Ornithine	1.81 (m), 1.95 (m), 3.06 (t), 3.79 (t)
Oxaloacetate	2.38 (s)
Phenylalanine	3.13 (dd), 3.24 (dd), 7.31 (m), 7.44 (m)
Phosphocreatine	3.05 (s), 3.94 (s)
Phosphorylcholine	3.22 (s), 3.61 (t), 4.19 (t)
Phosphorylethanolamine	3.26 (m), 4.07 (dd)
Polyamines	1.78 (m), 2.12 (m), 3.06 (t), 3.14 (m)
Proline	2.00 (m), 2.11 (m), 2.31 (m), 3.33 (t), 3.40 (t), 4.14 (t)
Pyruvate	2.40 (s)
Sarcosine	2.74 (s), 3.61 (s)
Serine	3.84 (dd), 3.96 (dd)
Succinate	2.41 (s)
Taurine	3.26 (t), 3.42 (t)
Threonine	1.34 (d), 3.58 (d), 4.25 (m)
Trimethylamine	2.88 (s)
Trimethylamine N-oxide	3.27 (s)
Tyrosine	3.05 (dd), 3.15 (dd), 3.93 (dd), 6.88 (d), 7.18 (d)
Uracil	5.80 (d), 7.53 (d)
Valine	0.98 (d), 2.26 (m), 3.55 (d)

[a]Note that the shifts may vary with pH, ionic strength and temperature.

2.2 Metabolic Disorders

et al.[21] (Figure 2). They have shown that significantly higher concentrations of TMAO can be observed for urine collected from patients during episodes of graft dysfunction compared with urine from patients with good graft function or from healthy control subjects. Furthermore, they suggest that graft dysfunction is associated with damage to the renal medulla, which causes the release of TMAO into the urine, thus providing a novel urinary marker for posttransplant graft dysfunction.

The nonselective nature of NMR spectroscopy makes it the method of choice for the study of intact body fluids, especially for the study of congenital errors of metabolism. NMR spectroscopy allows rapid and accurate diagnosis of different metabolic disorders, with the added advantage that novel metabolites related to a given disease can be detected, thus improving diagnosis and possible therapy. A large number of inherited metabolic disorders have been successfully studied by NMR spectroscopy, including maple syrup urine disease, isovaleric aciduria, dicarboxylic aciduria, and tyrosinemia.[31,33,38,43] A few examples are used here to illustrate the potential of NMR spectroscopy in this area of research.

By simple NMR urinalysis, Iles and Chalmers[31] were able to monitor the progression of the disease, the fate of the therapeutic compounds and the restoration of metabolic balance in a girl with a defect in propionic acid metabolism, methylmalonic aciduria. The ^{1}H NMR spectrum of urine from this patient is shown in Figure 3. The subject was given intravenous carnitine in an attempt to resolve a severe episode of acute metabolic decompensation.

Carnitine is known to reduce the propionate pool by combining with propionyl-CoA to form propionylcarnitine. The latter is excreted in the urine. Spectra taken during this acute episode showed: (1) the range of metabolites associated with the disorder, including methylmalonate and creatine; (2) the metabolic fate of the therapeutic intervention in the form of unchanged carnitine and propionylcarnitine; and (3) the appearance of acetylcarnitine, signifying restoration of acetyl-CoA metabolism on lowering the level of intracellular propionyl-CoA. The appearance of acetylcarnitine coincided with clinical

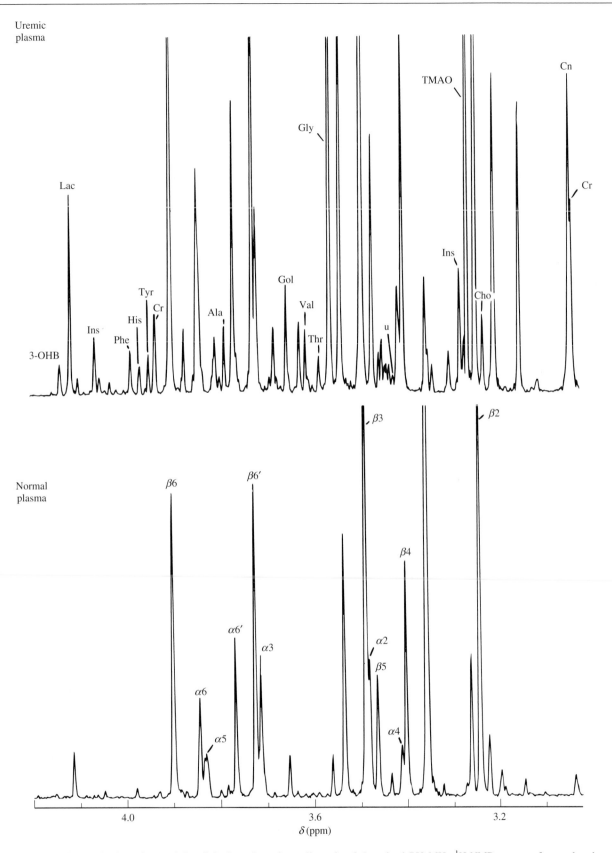

Figure 2 A 'skyline' projection of part of the aliphatic region of two-dimensional *J*-resolved 750-MHz ^{1}H NMR spectra of normal and uremic human blood plasma. α and β refer to resonances of glucose anomers. Cr, creatine; Cn, creatinine; Cho, choline; TMAO, trimethylamine *N*-oxide; Ins, myoinositol; Gly, glycine; Thr, threonine; Val, valine; Gol, glycerol; Ala, alanine; Tyr, tyrosine; His, histidine; Phe, phenylalanine; Lac, lactate; 3-OHB, D-3-hydroxybutyrate. (Adapted from P. J. D. Foxall, M. Spraul, R. D. Farrant, L. C. Lindon, G. H. Neild, and J. K. Nicholson, *J. Pharmacol. Biomed. Anal.*, 1993, **11**, 267)

For References see p. 870

improvement of the patient. Since the acquisition of an NMR spectrum from urine takes less than 3 min (at 500 MHz), this type of research opens up the possibility of 'real-time' therapeutic monitoring.

Although, ^{1}H NMR spectroscopy is the preferred method for the study of congenital errors of metabolism, dynamic metabolic studies have been carried out via ^{13}C NMR spectroscopic studies of body fluids following the infusion of ^{13}C-enriched metabolites. Tanaka et al.[63] used ^{13}C NMR urinalysis to monitor the fate of valine in methylmalonic aciduria. Following intravenous injection of DL-[1-^{13}C]- and DL-[1,2-^{13}C]valine to a boy with cobalamin-responsive methylmalonic aciduria, ^{13}C NMR spectra of urine showed that propionate was an obligatory intermediate from valine to methylmalonyl-CoA. It was previously thought that propionate was bypassed via methylmalonic semialdehyde. A similar approach was taken by Lapidot[13] in studies of glycogen storage and inherited fructose intolerance diseases in children. By means of a combination of ^{13}C NMR spectroscopy, and isotopomer analysis with gas chromatography–mass spectrometry, following nasogastrically infused [U-^{13}C]glucose, they were able to gain an insight into the biochemical altera-

tions in specific metabolic pathways which accompany these diseases.

2.3   Proteins and Lipoproteins in Intact Fluids

Blood plasma and serum are heterogeneous fluids containing lipoprotein particles (about 5–1200 nm in diameter), macromolecules (proteins), small molecules and ions. Any of the normal anticoagulants can be used for preparing plasma for NMR studies, e.g. EDTA and lithium heparin. The former is often preferred for lipoprotein work and gives rise to peaks for CaEDTA (NCH$_2$ singlet at 2.57 ppm) and MgEDTA (2.76 ppm) in spectra. No interfering peaks are usually seen for heparin.

Simplification of ^{1}H NMR spectra of plasma (which are similar to those of serum) can be achieved by the use of spin echo methods to filter out the broadest resonances.[7,9,12] No resonances remain in the aromatic region with t values greater than about 30 ms. Spin echo spectra can be used to study small molecules such as valine, acetate, alanine, lactate, glucose, and the ketone bodies acetone, acetoacetate and D-3-hydroxybutyrate in cases of insulin-dependent diabetes or fasting,[7,9] as well

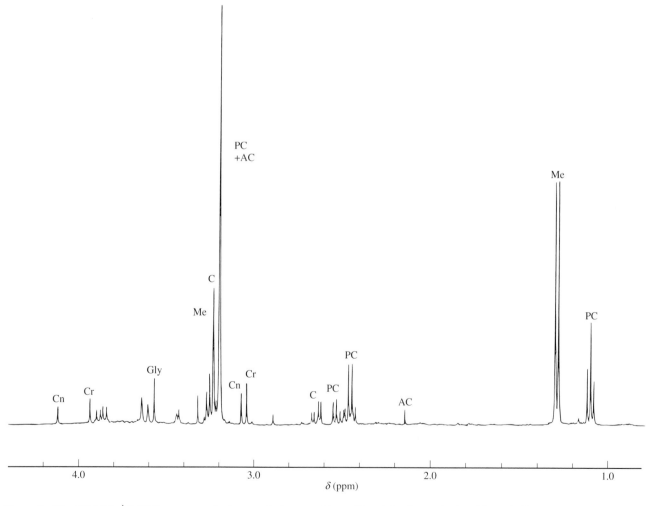

Figure 3   The 500 MHz ^{1}H NMR spectrum of urine sample from a subject with a defect in propionic acid metabolism. PC, propionylcarnitine; Me, methylmalonic acid; AC, acetylcarnitine; C, carnitine; Cr, creatine; Cr, creatinine; Gly, glycine. (Adapted from figure supplied by R. A. Iles)

For list of General Abbreviations see end-papers

as mobile segments of lipoproteins and acute-phase glycoproteins.

2.3.1 Proteins

Broad resonances at 2.04 and 2.08 ppm in ^{1}H NMR spectra of blood plasma have been assigned to the N-acetyl groups of N-acetylglucosamine and N-acetylneuramic acid residues of the glycan chains of 'acute-phase' glycoproteins such as α_1-acid glycoprotein.[64] These resonances have been used to monitor increases in maternal levels of these proteins (which are two-fold) compared with the matched neonatal cord plasma,[65] and the progress of heart transplant rejection.[66]

The application of resolution enhancement to high-field ^{1}H NMR spectra of blood plasma enables the detection of specific resonances for the major plasma protein albumin (66 kDa), including those for the N-terminal amino acids Asp1-Ala2-His3.[20] The ^{1}H NMR resonances of $H_{15}3$ of albumin can be used to monitor the state of Cys34 of albumin in blood plasma, and the formation of disulfides with, for example, the alcohol-aversive drug disulfiram and adducts with gold antiarthritic drugs.[67]

2.3.2 Lipoproteins

There is particular interest in the determination of lipoprotein concentrations in blood plasma on account of their association with coronary heart disease, hyperlipidemias, and possibly cancer and other diseases.

Broad resonances from the CH_2 and CH_3 groups of the fatty acid constituents of, for example, triacylglycerols, phospholipids and cholesterol esters in lipoproteins occur near 1.2 and 0.9 ppm, respectively. Comparisons of the ^{1}H NMR spectra of the individual isolated lipoproteins[10] enabled specific assignments to be made for each of the four major classes of lipoprotein (Figure 4), and suggested that ^{1}H NMR spectroscopy would allow a rapid determination of lipoproteins in intact plasma or serum. Both the CH_2 and CH_3 peaks shift slightly to high field as the density of the particle increases. The origin of the shifts is not yet clear, but may be related to susceptibility effects, protein interactions [greatest for high-density lipoprotein (HDL)], and the different types of lipid involved (higher proportion of phospholipids in denser particles). Lipoprotein classes can also be distinguished by their differing relaxation properties; in Hahn spin echo spectra, the $NMe_3{}^+$ peak from choline headgroups of phospholipids in low-density lipoprotein (LDL) and HDL decays more rapidly with increasing refocusing time for LDL than for HDL, and the CH_2 of LDL but not HDL contributes to the methyl and methylene region with longer TE values (60 ms).

Two approaches have subsequently been used to simulate the CH_2 and CH_3 regions of spectra of blood plasma and determine lipoprotein concentrations. In the first,[15] mean experimental 250 MHz spectra of isolated lipoprotein fractions have been used to construct plasma lipid CH_3 resonances using a linear least-squares fitting routine. Since this region of ^{1}H spectra of plasma from different donors is relatively similar for different donors, this procedure is reasonably successful. It has been refined[68] by including subspecies distributions within each lipoprotein class [very-low-density lipoprotein (VLDL), LDL, HDL], effectively broadening the peaks for each [Figure 5(a)]. For LDL the subspecies distributions derived from NMR correlated well with those from gradient-gel electrophoresis.

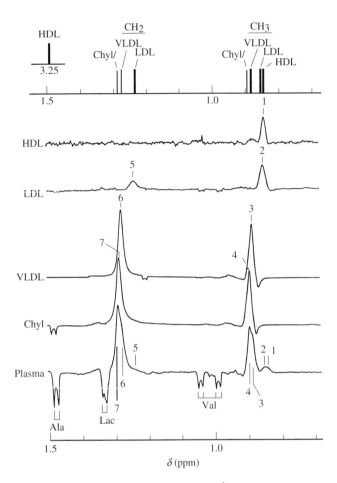

Figure 4 Assignments for the CH_3 and CH_2 ^{1}H NMR resonances of lipoproteins in Hahn spin echo spectra (TE = 60 ms). Chyl, chylomicrons; VLDL, very-low-density lipoproteins; LDL, low-density lipoproteins; HDL, high-density lipoproteins. The peak at 3.25 ppm in the insert is from the choline headgroup of HDL phospholipids. (Adapted from J. D. Bell, P. J. Sadler, P. J. Macleod, P. R. Turner, and A. LaVille, *FEBS Lett.*, 1987, **219**, 239)

These NMR measurements were carried out at 45 °C so as to avoid the problems caused by the increasing linewidth with temperature of peaks for LDL (the midpoint of the phase transition is at about 32 °C).

In a second approach the spectra of the CH_2 peaks of 400 MHz spectra of isolated individual lipoproteins at 30 °C were simulated by resolvable Lorentzian component lines:[17] two for VLDL, three for LDL, and two for HDL [Figure 5(b)]. NMR determinations of the relative amounts of each lipoprotein fraction in the plasma of normolipidemic and hyperlipidemic subjects using these model resonances gave good agreements with biochemical determinations.

Further refinements of these procedures can be envisaged involving inclusion of all the lipid peaks in the spectrum in the analysis, and consideration of modified lineshapes for lipoprotein particles with a high content of polyunsaturated fatty acids, e.g. for subjects on high fish oil content diets.[69] Halogenated anesthetics can mobilize lipids within lipoprotein particles[70] and there may also be lipophilic drugs which have such effects. ^{13}C and ^{31}P NMR analyses may be useful complements to ^{1}H work.

For References see p. 870

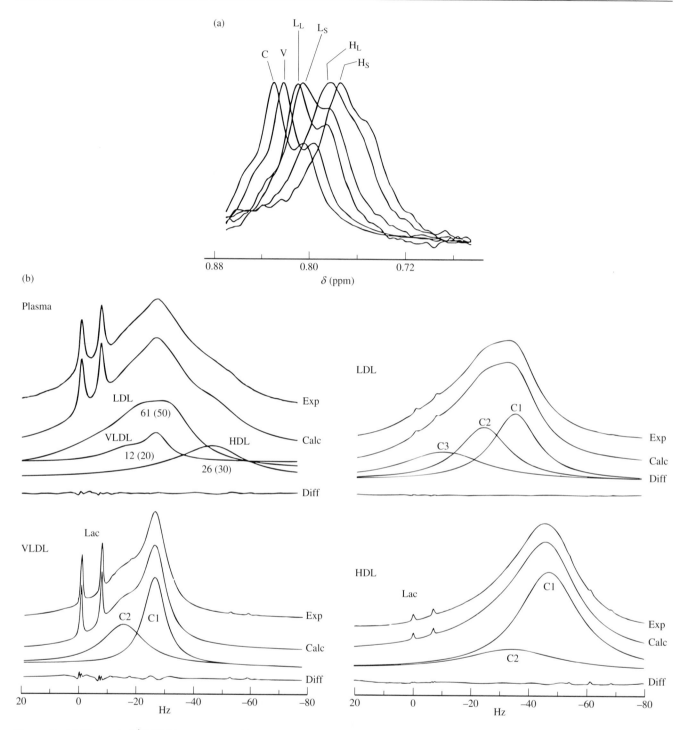

Figure 5 (a) The methyl ^{1}H NMR resonances of lipoproteins at 250 MHz, 45 °C. VLDL, very-low-density, lipoproteins; L_L and L_S, large and small LDL; H_L and H_S, large and small HDL. (Adapted from J. D. Otvos, E. J. Jeyarajah, D. W. Bennett, and R. M. Krauss, *Clin. Chem.*, 1992, **38**, 1632). (b) Experimental and calculated lipoprotein ^{1}H NMR CH$_2$ peaks for VLDL, LDL and HDL, and intact human blood plasma. (Adapted from M. Ala-Korpela, Y. Hiltunen, and J. Jokisaari, *NMR Biomed.*, 1993, **6**, 225)

Figure 6 (a) Lactate binding in blood plasma. 500 MHz single pulse (bottom) and Hahn spin echo (TE = 60 ms) ^{1}H NMR spectra (top) of human blood plasma before and after addition of ammonium chloride. The increase in intensity of the lactate methyl resonance with ammonium chloride concentration is plotted in the inset. (Adapted from J. D. Bell, J. C. C. Brown, G. Kubal, and P. J. Sadler, *FEBS Lett.*, 1988, **235**, 81.) (b) Ni^{2+} binding to the N-terminus of albumin in human blood plasma. Resolution-enhanced (combined exponential and sine-bell functions) 500 MHz ^{1}H NMR spectra of the aromatic region of human blood plasma before (lower) and after (upper) addition of Ni^{2+} (0.5 equivalents with respect to albumin). Alb-His3 refers to the third residue in albumin. Ni^{2+} binds to the N-terminal Aspl-Ala2-His3 residues forming a square-planar, diamagnetic complex. (Adapted from S. U. Patel, P. J. Sadler, A. Tucker, and J. H. Viles, *J. Am. Chem. Soc.*, 1993, **115**, 9285)

For list of General Abbreviations see end-papers

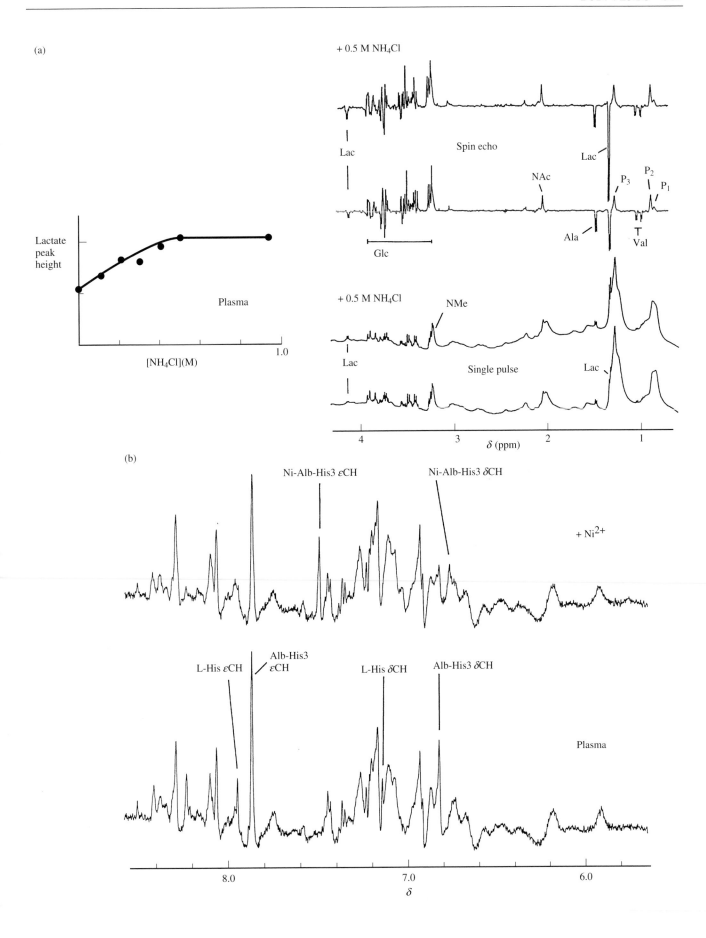

(a)

For References see p. 870

A rapid method for separation of lipoproteins from plasma for NMR work is that described by Hoffman et al.,[71] which involves the use of 1.5 mL of a gel such as Sepharose 6B in a disposable Pasteur pipet eluted with a deuterated salt solution.

It should be noted that, despite initial claims,[72] there is no useful relationship between the linewidths of lipoprotein CH_2 and CH_3 ^{1}H NMR resonances and cancer diagnosis, although they are reported to be of value for monitoring heart transplant rejection.[73] The use of 'star plots' (graphical display of peak height ratios) has been advocated for the detection of metabolic modifications brought about by cancer evolution or therapy.[74]

Kaartinen et al. have demonstrated that clinically relevant lipid classifications can be obtained from self-organizing map analysis of ^{1}H NMR spectra of blood plasma.[75] It has been suggested that the pattern of resonances at 1.17, 1.18 and 1.20 ppm can be used to measure other levels of oxidized low-density lipoprotein in human serum with possible application to cardiovascular risk factor profiles.[76] The advantages of using high frequency ^{1}H (800 MHz) and ^{13}C NMR spectroscopy for improving the resolution of lipid peaks are apparent.[77]

2.3.3 Molecular Interactions and Binding

Considerable caution has to be exercised in the interpretation of the intensities of peaks in spin echo spectra of plasma. The long T_1 values of some small molecules have to be taken into account when pulsing conditions for quantitation experiments are chosen (e.g. acetate 3.6 s, formate 6.6 s, glycine 2.4 s, valine 1.1 s: serum, pH 7, 400 MHz, 22 °C, 0.8 M NH_4Cl

added).[78] To allow the rapid quantitation of small molecules in the plasma from subjects with chronic renal failure (CRF), Grasdalen et al.[79] applied correction factors to compensate for the effects of relaxation and water irradiation on peak intensities.

Reliable quantitation can be obtained for alanine and valine with standard additions, but lactate and some other carboxylic acids bind to plasma macromolecules. Lactate has been shown to bind to both transferrin[11] and albumin[80] in slow exchange on the NMR timescale. It can be displaced from binding by addition of NH_4Cl [Figure 6(a)],[11] the antihypertensive drug Captopril,[80] and the anticonvulsant agent [Cu_2(DIPS)$_4$] (DIPS is 3,5-diisopropylsalicylate).[18] Intriguingly, there appears to be less lactate binding in plasma from subjects with CRF than in normal plasma.[16] A noncovalent interaction between Captopril and albumin has been detected in blood plasma and the *cis* isomer undergoes faster exchange between free and bound forms than does the *trans* isomer.[80] Binding of Cu^{2+} to formate and histidine in urine has been studied by NMR, and the time-course of the reaction between copper ethylenediaminetetra-acetate (CuEDTA) and blood plasma has been monitored.[18] Even at relatively low levels of Al^{3+} ions (50 μM), binding to citrate can be detected, which is reversed by the therapeutic chelating agent desferrioxamine.[19] The binding of Ni^{2+} to the N-terminal site of albumin (the antigenic determinant in cases of Ni^{2+} allergy) can be studied directly in plasma[20] [Figure 6(b)]. These kinds of experiment offer promise in the metallodrugs field, since ligand exchange and redox reactions in biological media are common, and alternative methods which involve the separation of the components can lead to changes in metal speciation.

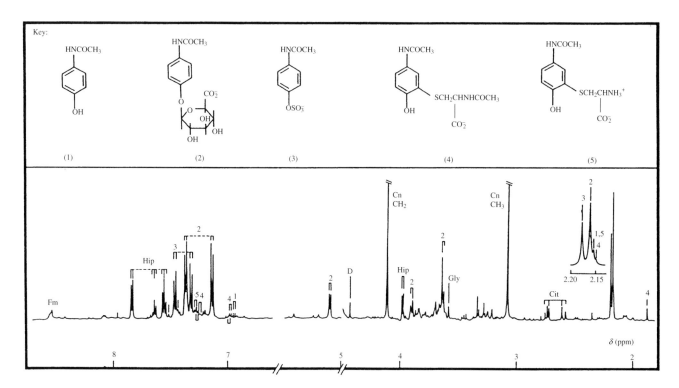

Figure 7 The 500 MHz ^{1}H NMR spectrum of human urine 6 h after a therapeutic 1 g (6.6 mmol) dose of the analgesic drug paracetamol. As well as natural metabolites (Cn, creatinine; Cit, citrate; D, dihydroxyacetone; Fm, formate; Gly, glycine; Hip, hippurate), peaks for the drug (1) and all its major metabolites (2–5) are seen. (Adapted from J. D. Bell, J. C. C. Brown, and P. J. Sadler, *Chem. Br.*, 1988, **24**, 1021)

Although a useful reference for chemical shifts and concentrations in body fluids with low levels of macromolecules, trimethylsilylpropionate (TSP) is often undetectable at <2.5 mM in ^{1}H NMR spectra of blood plasma, due to binding.[78] Similarly, resonances for phenylalanine and tyrosine are suppressed in spin echo spectra of blood plasma at normal pH compared with highly acidic pH values (<2.5). The interaction of glutamine with other components of CSF is markedly changed after freeze drying and redissolving the sample, only then are its ^{1}H NMR resonances visible.[23]

3 DRUG METABOLISM AND TOXICOLOGY

NMR spectroscopy is finding increasing uses in the study of drug metabolism and toxicology.[59] Conventional techniques require the use of radioactive labeling and extensive use of separation techniques, e.g. high-performance liquid chromatography (HPLC). Moreover, valuable information can be lost using conventional techniques since exogenous and endogenous metabolites can go undetected. NMR spectroscopy, however, allows observation of all metabolites in one experiment[27–30] (Figure 7), although it suffers from inherently low sensitivity and 'chemical noise'. Chemical noise refers to the many low-intensity resonances found in body fluids that may mask the resonances under study. In part these problems can be overcome by concentrating the samples and by the use of multidimensional/multinuclear NMR. A more recent and powerful approach is the separation of metabolites by solid-phase extraction/chromatography in conjunction with NMR (SPEC-NMR) or HPLC coupled with high-field NMR.

NMR spectroscopy has been successfully applied to the study of a variety of xenobiotics including, acetaminophen, 5-fluorouridine, ifosfamide, ibuprofen, penicillamine, flucloxacillin, ampicillin, hydrazine, N-methylformamide, and carboplatin.[59,81] ^{1}H and ^{19}F are usually the nuclei of choice in NMR studies of xenobiotics, although work using ^{31}P, ^{2}H and ^{15}N has been published. The advantages of ^{1}H NMR are clearly based on its high sensitivity and the fact that protons are ubiquitous to most drugs. However, ^{1}H NMR spectra often suffer from severe overlap of signals from exogenous and endogenous metabolites. On the other hand, ^{19}F NMR is particularly good for these studies in that it has high sensitivity, 100% natural abundance and no interference from naturally-occurring fluorinated compounds, and therefore no chemical noise. A good illustration of the use of NMR spectroscopy in drug metabolism is that of 5-fluorouracil (5FU).[32] The complete urinary excretion profile of cancer patients on a drug course of 5FU was followed by ^{19}F NMR spectroscopy with and without methotrexate pretreatment. The parent drug (5FU) and four of its metabolites were detected in the urine, including F$^-$ which is produced as a degradation product. The major metabolite found in these studies was α-fluoro-β-alanine.

A new area of research has been the development of NMR of body fluids in analytical toxicology. Nicholson and co-workers[34–36] have used NMR for the rapid multicomponent analysis of body fluids following xenobiotic insults. The effects of known nephrotoxins on the NMR profiles of rat urine were investigated. The nephrotoxins consisted of sodium chromate (pars convoluta of proximal tubule), cisplatin, hexachlorobutadiene, mercury(II) chloride (pars recta of the proximal tubule),

propylene imine, and 2-bromoethanamine (renal papilla). These compounds induced damage in specific regions of the rat kidney. Urinalysis by ^{1}H NMR spectroscopy showed clear biochemical fingerprints for site-specific nephrotoxic states (Figure 8). Aminoaciduria, glycosuria, and lactic aciduria were characteristic of exposure to all proximal tubular toxins (except for cisplatin), whereas papillary insult resulted in early elevations in urinary TMAO and DMA. TMAO and DMA were suggested as markers of site-specific renal papillary injury in the rat.

Toxicological studies in humans have been more limited. However, a number of cases have arisen from accidental or self-inflicted poisoning episodes, e.g. paracetamol (acetaminophen) overdose and accidental skin adsorption of phenol.[35] Although these studies are partly hampered by the lack of control of body fluids, useful clinical and biochemical information

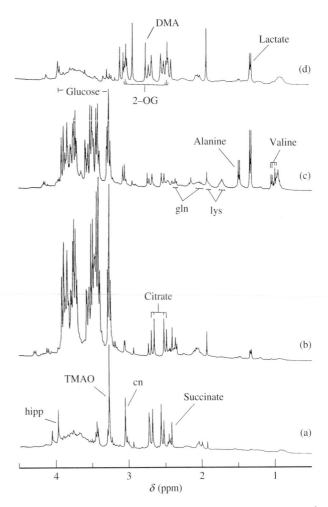

Figure 8 Effects of administration of various nephrotoxins on the ^{1}H NMR spectra (400 MHz) of rat urine. (a) Control; (b) 24–48 h after dosing with 20 mg kg^{-1} sodium chromate (pars convoluta proximal tubular toxin); (c) 24–48 h after dosing with 50 mg kg^{-1} p-aminophenol (pars recta proximal tubular toxin); (d) 24–48 h after dosing with 20 ml kg^{-1} propylene imine (produces renal medullary necrosis). DMA, dimethylamine; 2-OG, 2-oxyglutarate; hipp, hippurate; cn, creatine; gln, glutamine; lys, lysine; TMAO, trimethylamine N-oxide. (Adapted from K. P. R. Gartland, F. W. Bonner, and J. K. Nicholson, *Mol. Pharmacol.*, 1989, **35**, 242)

For References see p. 870

has been obtained. These results clearly demonstrate the strength of the NMR technique as applied to toxicological studies.

Proton NMR spectra of body fluids consist in many cases of a very large number (>100) of resonances, giving rise to a wealth of, as yet complex, information. Spectral quantitation and interpretation can thus become tedious and time-consuming. Furthermore, metabolic alterations in the body can lead to subtle, yet consistent changes in the composition profiles of some body fluids. In many cases standard quantitative and statistical analysis of the spectral data are not robust enough to detect some of these changes. Alternative computer-based pattern-recognition methods[36,59] have been applied to analyses of NMR spectra of body fluids. Preliminary results suggest that pattern-recognition methods applied to NMR data may be powerful for the characterization of metabolic abnormalities associated with diseased states and toxicological events.

4 RELATED ARTICLE

Tissue and Cell Extracts MRS.

5 REFERENCES

1. E. Odeblad and B. Westin, *Acta Radiol.*, 1958, **49**, 390.
2. E. Odeblad, *Acta Isot.*, 1961, **1**, 27.
3. A. Ohsaka, K. Yoshikawa, and T. Matuhasi, *Jpn. J. Med. Sci. Biol.*, 1979, **32**, 305.
4. K. Matsushita, K. Yoshikawa, and A. Ohsaka, *JEOL News*, 1982, **18**, 54.
5. J. L. Bock, *Clin. Chem.*, 1982, **28**, 1873.
6. M. Traube, J. L. Bock, and J. L. Boyer, *Ann. Intern. Med.*, 1983, **98**, 171.
7. J. K. Nicholson, M. J. Buckingham, and P. J. Sadler, *Biochem. J.*, 1983, **211**, 605.
8. J. R. Bales, J. D. Bell, P. J. Sadler, J. K. Nicholson, J. A. Timbrell, P. N. Bennett, R. D. Hughes, and R. Williams, *Magn. Reson. Med.*, 1988, **6**, 300.
9. J. K. Nicholson, M. P. O'Flynn, P. J. Sadler, A. F. Macleod, S. M. Juul, and P. H. Sonksen *Biochem. J.*, 1984, **217**, 365.
10. J. D. Bell, P. J. Sadler, A. F. Macleod, P. R. Turner, and A. LaVille, *FEBS Lett.*, 1987, **219**, 239.
11. J. D. Bell, J. C. C. Brown, G. Kubal, and P. J. Sadler, *FEBS Lett.*, 1988, **235**, 81.
12. D. L. Rabenstein, K. K. Millis, and E. J. Strauss, *Anal. Chem.*, 1988, **60**, 1380A.
13. A. Lapidot, *J. Inherited Metab. Dis.*, 1990, **13**, 466.
14. H. Pont, J. Vion-Dury, R. Faure, D. Maraninchi, J. R. Harle, S. Confort-Gouny, M. Sciaky, E. Fontanarava, P. Viout, and P. J. Cozzone, *Lancet*, 1991, **337**, 792.
15. J. D. Otvos, E. J. Jeyarajah, and D. W. Bennett, *Clin. Chem.*, 1991, **37**, 377.
16. J. D. Bell, J. A. Lee, H. A. Lee, P. J. Sadler, D. R. Wilkie, and R. H. Woodham, *Biochim. Biophys. Acta*, 1991, **1096**, 101.
17. M. Ala-Korpela, Y. Hiltunen, J. Jokisaari, S. Eshelman, K. Kiviniity, M. J. Savolainen and Y. A. Resaniemi, *NMR Biomed.*, 1993, **6**, 225.
18. S. W. Bligh, H. A. Boyle, A. B. McEwen, P. J. Sadler, and R. H. Woodham, *Biochem. Pharmacol.*, 1992, **43**, 137.
19. J. D. Bell, G. Kubal, S. Radulovic, P. J. Sadler, and A. Tucker, *Analyst*, 1993, **118**, 241.
20. S. U. Patel, P. J. Sadler, A. Tucker, and J. H. Viles, *J. Am. Chem. Soc.*, 1993, **115**, 9285.
21. P. J. D. Foxall, M. Spraul, R. D. Farrant, L. C. Lindon, G. H. Neild, and J. K. Nicholson, *J. Pharmacol. Biomed. Anal.*, 1993, **11**, 267.
22. O. A. C. Petroff, R. K. Yu, and T. Ogino, *J. Neurochem.*, 1986, **47**, 1270.
23. J. D. Bell, J. C. C. Brown, P. J. Sadler, A. F. Macleod, D. H. Sonksen, R. D. Hughes, and R. Williams, *Clin. Sci.*, 1987, **72**, 563.
24. F. Koschorek, H. Gremmel, J. Stelten, W. Offermann, E. Kruger, and D. Leibfritz, *AJNR*, 1989, **10**, 523.
25. B. C. Sweatman, R. D. Farrant, E. Holmes, F. Y. Ghauri, J. K. Nicholson, and J. C. Lindon, *J. Pharmacol. Biomed. Anal.*, 1993, **11**, 651.
26. F. Nicoli, J. Vion-Dury, J. M. Maloteaux, C. Delwaide, S. Confort-Gouny, M. Sciaky, and P. J. Cozzone, *Neurosci. Lett.*, 1993, **154**, 47.
27. J. R. Bales, D. P. Higham, I. Howe, J. K. Nicholson, and P. J. Sadler, *Clin. Chem.*, 1984, **30**, 426.
28. J. R. Bales, P. J. Sadler, J. K. Nicholson, and J. A. Timbrell, *Clin. Chem.*, 1984, **30**, 1631.
29. J. R. Bales, J. K. Nicholson, and P. J. Sadler, *Clin. Chem.*, 1985, **31**, 757.
30. J. Bernadou, J. P. Armand, A. Lopez, M. C. Malet-Martino, and R. Martino, *Clin. Chem.*, 1985, **31**, 846.
31. R. A. Iles and R. A. Chalmers, *Clin. Sci.*, 1988, **74**, 1.
32. W. E. Hull, R. E. Port, R. Herrmann, B. Britsch, and W. Kunz, *Cancer Res.*, 1988, **48**, 1680.
33. S. E. C. Davies, R. A. Chalmers, E. W. Randall, and R. A. Iles, *Clin. Chim. Acta*, 1988, **178**, 241.
34. K. P. R. Gartland, F. W. Bonner, and J. K. Nicholson, *Mol. Pharmacol.*, 1989, **35**, 242.
35. P. J. Foxall, M. R. Bending, K. P. R. Gartland, and J. K. Nicholson, *Hum. Toxicol.*, 1989, **8**, 491.
36. E. Holmes, F. W. Bonner, B. C. Sweatman, J. C. Lindon, C. R. Beddell, E. Rahr, and J. K. Nicholson, *Mol. Pharmacol.*, 1992, **42**, 922.
37. P. J. Foxall, G. J. Mellotte, M. R. Bending, J. C. Lindon, and J. K. Nicholson, *Kidney Int.*, 1993, **43**, 234.
38. S. E. C. Davies, D. A. Woolf, R. A. Chalmer, J. E. M. Rafter, and R. A. Iles, *J. Nutr. Biochem.*, 1993, **3**, 523.
39. K. P. R. Gartland, C. T. Eason, K. E. Wade, F. W. Bonner, and J. K. Nicholson, *J. Pharmacol. Biomed. Anal.*, 1989, **7**, 699.
40. J. P. M. Ellul, G. M. Murphy, H. G. Parkes, R. Z. Slapa, and R. H. Dowling, *FEBS Lett.*, 1992, **300**, 30.
41. H. Harada, H. Shimizu, and M. Maeiwa, *Forensic Sci. Int.*, 1987, **34**, 189.
42. A. Yamada-Nosaka, S. Fukutomi, S. Uemura, T. Hashida, M. Fujishita, Y. Kobayashi, and Y. Kyogoku, *Arch. Oral. Biol.*, 1991, **36**, 697.
43. J. D. Bell, J. C. C. Brown, and P. J. Sadler, *NMR Biomed.*, 1989, **2**, 246.
44. J. C. C. Brown, P. J. Sadler, D. J. Spalton, S. M. Juul, A. F. Macleod, and P. H. Sonksen, *Exp. Eye Res.*, 1986, **42**, 357.
45. J. V. Greiner, L. A. Chanes, and T. Glonek, *Opthalmic Res.*, 1991, **23**, 92.
46. S. Hamamah, F. Seguin, C. Barthelemy, S. Akoka, A. Le-Pape, J. Lansac, and D. Royere, *J. Reprod. Fertil.*, 1993, **97**, 51.
47. P. M. Robitaille, P. A. Robitaille, P. A. Martin, and G. G. Brown, *Comp. Biochem. Physiol. B*, 1987, **87**, 285.
48. M. J. Lynch, J. Masters, J. P. Pryor, J. C. Lindon, M. Spraul, P. J. D. Foxall, and J. K. Nicholson, *J. Pharmacol. Biomed. Anal.*, 1994, **12**, 5.
49. R. G. Gosden, I. H. Sadler, D. Reed, and R. H. Hunter, *Experientia*, 1990, **46**, 1012.

50. J. J. Powell, S. M. Greenfield, H. G. Parkes, J. K. Nicholson, and R. P. H. Thompson, *Food Chem. Toxicol.*, 1993, **31**, 449.

51. J. D. Bell and P. J. Sadler, unpublished work.

52. H. G. Parkes, M. C. Grootveld, E. B. Henderson, A. Farrel, and D. R. Blake, *J. Pharmacol. Biomed. Anal.*, 1991, **9**, 75.

53. M. Grootveld, E. B. Henderson, A. Farrell, D. R. Blake, H. G. Parkes, and P. Haycock, *Biochem. J.*, 1991, **273**, 459.

54. K. Albert, S. Michele, U. Gunther, M. Fial, H. Gall, and J. Saal, *Magn. Reson. Med.*, 1993, **30**, 236.

55. T. R. Nelson, R. J. Gillies, D. A. Powell, M. C. Schrader, D. K. Manchester, and D. H. Pretorius, *Prenatal Diagn.*, 1987, **7**, 363.

56. J. M. Pearce, M. A. Shifman, A. A. Pappas, and R. A. Komoroski, *Magn. Reson. Med.*, 1991, **21**, 107.

57. M. Dittmer, A. C. Kuesel, T. Kroft, M. Villeneuve, R. Benzeie, and I. C. P. Smith, *Proc. XIth Ann Mtg. Soc. Magn. Reson. Med.*, Berlin, 1992, Vol. 2, 3417.

58. P. J. D. Foxall, R. G. Price, J. K. Jones, G. H. Neild, F. D. Thompson, and J. K. Nicholson, *Biochim. Biophys. Acta*, 1992, **1138**, 305.

59. J. K. Nicholson and I. D. Wilson, *Progr. NMR Spectrosc.*, 1989, **21**, 449.

60. J. D. Bell, in 'Magnetic Resonance Spectroscopy in Biology and Medicine: Functional and Pathological Tissue Characterization', eds. J. D. de Certaines, W. M. M. J. Bovee, and F. Podo, Pergamon, Oxford, 1992.

61. P. J. Hore, *JMRI*, 1983, **55**, 283.

62. D. L. Rabenstein and S. Y. Fan, *Anal. Chem.*, 1986, **58**, 3178.

63. K. Tanaka, I. M. Armitage, H. S. Ramsdell, Y. E. Hsia, S. R. Lipsky, and L. E. Rosenberg, *Proc. Natl. Acad. Sci. USA*, 1975, **72**, 3692.

64. J. D. Bell, J. C. C. Brown, J. K. Nicholson, and P. J. Sadler, *FEBS Lett.*, 1988, **235**, 81.

65. J. D. Bell, J. C. C. Brown, P. J. Sadler, D. Garvie, A. F. Macleod, and C. Lowy, *NMR Biomed.*, 1989, **2**, 61.

66. J. Vion-Dury, A. Mouly-Bandini, P. Viout, M. Sciaky, S. Confort-Gouny, J. R. Monties, and P. Cozzone, *C.R. Acad. Sci. Paris*, 1992, **315**, 479.

67. J. Christodoulou, P. J. Sadler, and A. Tucker, *FEBS Lett.*, 1995, **376**, 1.

68. J. D. Otvos, E. J. Jeyarajah, D. W. Bennett, and R. M. Krauss, *Clin. Chem.*, 1992, **38**, 1632.

69. J. D. Bell and P. J. Sadler, *Chem. Br.*, 1993, **7**, 597.

70. J. D. Bell, C. Hahn, J. K. Lodge, S. U. Patel, J. Sear, and P. J. Sadler, unpublished work.

71. D. W. Hoffman, R. A. Venters, S. F. Shedd, and L. D. Spicer, *Magn. Reson. Med.*, 1990, **13**, 507.

72. E. T. Fossel, J. M. Carr, and J. McDonagh, *New Engl. J. Med.*, 1986, **315**, 1369.

73. M. Eugene, L. LeMmoyec, J. de Certaines, M. Desruennes, E. Le Reumeur, J. B. Fraysse, and C. Cabrol, *Magn. Reson. Med.*, 1991, **18**, 93.

74. J. Vion-Dury, R. Favre, M. Sciaky, M. Kriat, S. Confort-Gouny, J. R. Harle, N. Grazziani, P. Viout, F. Grisoli, and P. J. Cozzone, *NMR Biomed.*, 1993, **6**, 58.

75. J. Kaartinen, Y. Hiltunen, P. T. Koranen, and M. Ala-Korpela, *NMR Biomed.*, 1998, **11**, 168.

76. J. Jankowski, J. R. Nofer, M. Tepel, et al., *Clin. Sci.*, 1998, **95**, 489.

77. W. Wilker and D. Leibfritz, *Magn. Res. Chem.*, 1998, **36**, 579.

78. M. Kriat, S. Confort-Gouny, J. Vion-Dury, M. Sciaky, P. Viout, and P. J. Cozzone, *NMR Biomed.*, 1992, **5**, 179.

79. H. Grasdalen, P. S. Belton, J. S. Prior, and G. T Rich, *Magn. Res. Chem.*, 1987, **25**, 811.

80. D. A. Keire, S. V. Mariappan, J. Peng, and D. L. Rabenstein, *Biochem. Pharmacol.*, 1993, **46**, 1059.

81. J. D. Ranford, P. J. Sadler, K. Balmanno, and D. R. Newell, *Magn. Res. Chem.*, 1991, **29**, 125.

Biographical Sketches

J. D. Bell. *b* 1958. B.Sc. Biochemistry 1982, University of Warwick. Ph.D., (supervisor P. J. Sadler), 1987, London. Senior NMR Research Fellow (with I. Young and D. G. Gadian), Hammersmith Hospital, 1989–93. Faculty of Biochemistry, Royal Postgraduate Medical and Imperial College School of Medicine. Approx. 100 publications. Research interests include application of MRI/MRS in clinical research, and NMR spectroscopy to the chemistry of tissues and body fluids.

P. J. Sadler. *b.* 1946. B.A., 1969, M.A. D.Phil., 1972, Oxford. Introduced to NMR by R. E. Richards, H. A. O. Hill and R. J. P. Williams. MRC Research Fellow University of Cambridge and National Institute for Medical Research (NMR Unit), 1971–73. Faculty of Chemistry, Birkbeck College, University of London, 1973–96. Faculty of Chemistry, University of Edinburgh 1996–present. Approx. 280 publications. Research interests include applications of NMR spectroscopy to metallodrugs, metal transport proteins in blood plasma, and the chemistry of body fluids.

Cation Movements Across Cell Walls of Intact Tissues Using MRS

James A. Balschi

Harvard Medical School, Boston, MA, USA

Kieran Clarke

University of Oxford, UK

Laura C. Stewart

National Research Council, Ottawa, Ontario, Canada

Monique Bernard

CNRS, Marseille, France

and

Joanne S. Ingwall

Harvard Medical School, Boston, MA, USA

1 INTRODUCTION: THE IMPORTANCE OF CATION GRADIENTS FOR CELL FUNCTION

Intracellular cations play critical roles in cell division and maintaining normal cell function in both prokaryotic and eukaryotic cells, and cells actively regulate their intracellular cation levels. Large concentration gradients of the monovalent cations

For References see p. 878

Na$^+$, K$^+$, and H$^+$, and the divalent cations Ca^{2+} and Mg^{2+}, exist across the plasma membrane. The quintessential example of the role of cation movements in normal cell function is the control of excitation–contraction coupling in striated muscle. During a twitch in skeletal muscle and during each cardiac cycle, Na$^+$, Ca^{2+}, H$^+$, and K$^+$ (as well as other charged species) move across the plasma membrane to initiate a sequence of events leading to contraction: Na$^+$ and Ca^{2+} move from the interstitium to the intracellular space while K$^+$ and H$^+$ move out of the cell. Muscle relaxation occurs when the concentrations of these critical cations (most importantly Ca^{2+}) return to baseline. Recovery of resting cell intracellular cation concentrations requires energy. Furthermore, maintenance of intracellular Na$^+$ and Ca^{2+} levels (Na$_i$ and Ca$_i$), which in the quiescent cell are below their electrochemical equilibrium values, requires energy. The source of energy for these cation movements is the hydrolysis of ATP, either directly via membrane proteins that stoichiometrically couple ATP hydrolysis and ion translocation, or indirectly through the Na$^+$ gradient.

Intracellular concentrations of Na$^+$, K$^+$, H$^+$, and Ca^{2+} are determined by the integrated action of these membrane proteins: ion pumps, exchangers, and channels. Membrane proteins from organelles within the cell, namely the endoplasmic reticulum and mitochondria, are also important in the regulation of cation concentrations. A diagram summarizing the major components of this network is shown in Figure 1. Cation NMR has been used to examine the relative contributions of these specific components to net ion movements and how this may change in pathological states. The major advantage of using NMR to measure cation contents is that repetitive nondestructive measurements can be made for intact functioning tissue, usually over long periods of time. Moreover, some of these techniques allow simultaneous measurement of extra- and intracellular ion content, thereby permitting quantitative assessment of net ion movement.

In this chapter we first describe the MRS methods for measuring monovalent cation contents and movements in intact organs. To illustrate their power, we then describe their application to the heart made ischemic by an abrupt cessation in blood flow. Our focus is on Na$^+$ because it has received the most attention experimentally.

2 METHODS FOR MEASURING CATION CONTENT AND MOVEMENTS IN VIVO

Hydrogen, Na$^+$ and its biological congener Li$^+$, K$^+$ and its biological congener Rb$^+$, Mg^{2+}, and Ca^{2+} all have nuclides that are active in NMR[1] (see Table 1). The high natural abundance, relatively high concentration in biological tissues, and short values for spin–lattice relaxation times combine to make Na$^+$ and K$^+$ amenable to direct study in biological samples using NMR. The low natural abundance of the NMR active isotopes of Mg^{2+} and Ca^{2+} precludes direct NMR observation. Because Li$^+$ and Rb$^+$ are biological congeners of Na$^+$ and K$^+$, respectively, they are also useful nuclides for the study of cation movements in biological samples. Estimates of intracellular H$^+$, Mg^{2+}, and Ca^{2+} contents by MRS are made using indirect techniques and are not discussed here.

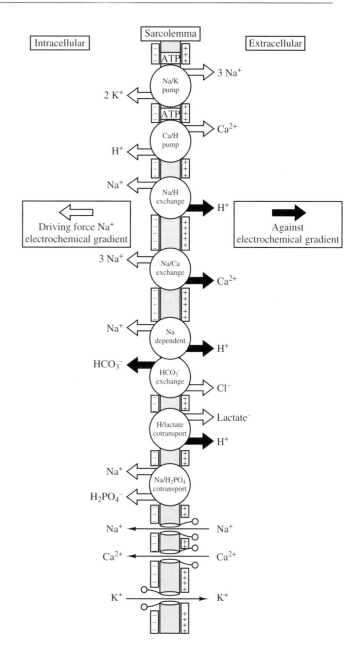

Figure 1 Diagramatic representation of the plasma membrane of a muscle cell (sarcolemma) showing the primary pathways of Na$^+$, K$^+$, and Ca^{2+} movement: the ATP-dependent Na$^+$ and Ca^{2+} pumps, electroneutral Na$^+$/H$^+$ exchange and electrogenic Na$^+$/Ca^{2+} exchange, Na$^+$-dependent HCO$_3^-$ exchange, H$^+$/lactate and Na$^+$/H$_2$PO$_4^-$ cotransport, and the voltage-dependent Na$^+$, K$^+$, and Ca^{2+} channels. Intracellular/extracellular concentrations for Na$^+$(10/144 mM), K$^+$ (85–130/5 mM), Ca^{2+} (10–4/1.25 mM), and H$^+$ (pH 7.15/pH 7.40)

2.1 MRS of ^{23}Na and ^{39}K

Because of its inherent NMR sensitivity, high natural abundance, high concentration in biological samples, and short T_1 values, ^{23}Na is the second (next to ^{1}H) most NMR-sensitive nucleus found in biological tissues (Table 1). The first ^{23}Na NMR spectrum of a biological sample was obtained in 1956; the first ^{23}Na image of tissue was obtained in 1981 for perfused heart.[1] Although it is the cation present in highest

Table 1   NMR Properties of Selected Nuclides

Nucleus	Spin quantum number	Natural abundance (%)	Effective sensitivity[a] ($\times 10^{-2}$)	T_1[b] (ms)	T_2[b] (ms)	Resonant frequency at 8.4 T (MHz)
^{23}Na	$\frac{3}{2}$	100	9.3	70	54	95.2
^{7}Li	$\frac{3}{2}$	93	29.4	25×10^3	18×10^3	139.8
^{39}K	$\frac{3}{2}$	93	0.05	54	54	16.8
^{87}Rb	$\frac{3}{2}$	28	17.7	2.5	2.5	117.7
^{25}Mg	$-\frac{5}{2}$	10	–	–	–	22.0
^{43}Ca	$-\frac{7}{2}$	0.14	–	–	–	24.2

[a]Relative to hydrogen. [b]Measured in aqueous solution.

concentration in the intracellular space, low frequency and hence low sensitivity of the ^{39}K nuclide combine to make its detection by NMR difficult. The first ^{39}K NMR signal from tissue was made in 1970;[2] the first in perfused heart in 1985.[3] The ^{23}Na and ^{39}K nuclei do not exhibit any significant chemical shifts in aqueous solutions. Cations, such as Na$^+$ or K$^+$, do not bond covalently but bind electrostatically at suitable anionic sites. Hence, the electronic environment about the nucleus is relatively constant. While such binding may give rise to a small chemical shift, it is more likely to result in line broadening or intensity changes. For the most part, Na$^+$ and K$^+$ will remain as hydrated cations in aqueous solution. The electron density around the nucleus does not change, and there are no differences in chemical shift. Therefore, the resonance frequencies of intracellular and extracellular cations are indistinguishable.

Two general approaches have been attempted to discriminate between the resonances of intracellular and extracellular cations. The first approach attempts to use characteristic intrinsic properties of the cation nuclei. The ^{23}Na (and presumably all spin-$\frac{3}{2}$ nuclides) resonance of cells and tissues is predominantly homogeneous biexponential: narrow central resonance ($\frac{1}{2} - \frac{1}{2}$ transition) superimposed on two satellite coherences ($\frac{3}{2} - \frac{1}{2}$) and ($-\frac{1}{2} - \frac{3}{2}$).[4] Since spin $\frac{3}{2}$ nuclei have four nuclear spin states and three single-quantum NMR transitions, more than one relaxation rate exists in biological samples. Several investigators have explored the possibility that multiple quantum NMR can detect differences in relaxation rates of nuclei located in the intra- versus extracellular spaces in biological samples.[5–7] It is clear that more than one rate constant is required to describe observed relaxation of cations present in both the intra- and extracellular spaces. Consequently, quantitative analysis of Na$^+$ (or K$^+$) signals distributed in either intra- or extracellular spaces using a multiple quantum approach is technically problematic. However, in practical terms, changes in multiple quantum signals may reflect primarily changes in either the intra- or extracellular Na$^+$ (or K$^+$), allowing for good estimates to be made.

The second approach to discriminate between the intra- and extracellular Na$^+$, signals is to use hyperfine frequency-shift reagents. In the early 1980s, two groups independently developed shift reagents for cations.[8,9] Dysprosium (III) tripolyphosphate (Dy(PPP)$_2$$^{7-}$)[9] and triethylenetetramine-hexaacetate-dysprosium (III) (DyTTHA^{3-})[10] were widely employed. More recently, another shift reagent, the thulium (III) complex of 1,4,7,10-tetraazacyclo-dodecane-*N,N′,N′,N‴*-tetra(methylene-

phosphonate) (TmDOTP^{5-}) was introduced.[11] The putative structures of these three shift reagents are shown in Figure 2. These shift reagents have been successfully used to discriminate between ^{23}Na and ^{39}K NMR signals arising from the intra- and extracellular spaces in a wide variety of biological samples, ranging from single cells such as yeast[12,13] and red blood cells,[14] to isolated perfused tissues such as heart[3] and to organs within the intact animal[15–17].

2.1.1   Properties of Frequency Shift Reagents

Aqueous shift reagents enable the spectroscopist to vary the resonance frequency of nuclei such as ^{23}Na, ^{39}K, ^{25}Mg, and ^{43}Ca. All shift reagents for cations consist of a paramagnetic ion, typically one of the lanthanides such as dysprosium, chelated to an anionic ligand (or ligands). The net negative charge

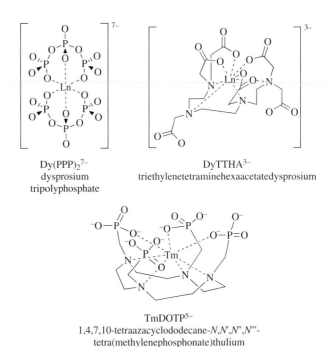

Dy(PPP)$_2$$^{7-}$
dysprosium
tripolyphosphate

DyTTHA^{3-}
triethylenetetraminehexaacetatedysprosium

TmDOTP^{5-}
1,4,7,10-tetraazacyclododecane-*N,N′,N″,N‴*-
tetra(methylenephosphonate)thulium

Figure 2   Putative structures for the three shift reagents currently in use to discriminate intra- and extracellular Na$^+$, K$^+$, and Li$^+$ NMR signals in biological systems. (Reproduced by permission of Kluwer Academic Press from J. S. Ingwall, 'Cardiovascular Magnetic Resonance Spectroscopy', 1992, Ch. 13, p. 195)

For References see p. 878

on the chelated lanthanide complex is necessary for interaction between the cation and the shift reagent. The anionic character of shift reagents also minimizes binding to biological macromolecules and to membranes, essentially precluding their entry into the cell.

In general, the change of the observed chemical shift of a cation induced by the presence of a paramagnetic compound includes a diamagnetic contribution owing to a complex formation, contact hyperfine shift, and dipolar hyperfine (pseudocontact) shift. In the case of shift reagents for cations, the cation does not bond directly to the lanthanide; consequently, the contact hyperfine shift does not contribute. The complex formation shift term is negligible. As a result, the shift induced by a shift reagent is primarily from the dipolar hyperfine shift.

A magnetic dipolar interaction between the paramagnetic metal ion of the shift reagent and the cation nucleus shifts the resonance frequency of the observed nucleus. The nature of the dipolar interaction requires a specific binding site (or sites) for the cation on the complex relatively close to the paramagnetic center. The binding of the cation is relatively weak or labile; as a result, all the cationic nuclei in the sample interact briefly with the site and the lifetime of the bound cation is very short. If the average lifetime of the cation is much less than the inverse of the frequency difference between the bound and the free cation, only one resonance frequency will be observed (rapid exchange on the NMR time scale).

The direction of the observed shift, either upfield (decreasing frequency) or downfield (increasing frequency), is a characteristic of the particular shift reagent complex and the lanthanide ion. For example, substituting the lanthanide thulium for dysprosium with the same chelate results in a reversal of the direction of the observed shift. Different chelates of the same lanthanide ion can cause shifts in opposite directions. The magnitude of the observed shift for each particular shift reagent depends on the ratio of shift reagent to cation, adding more shift reagent at constant cation concentration results in a greater shift. The magnitude also depends on the total cation content. For example, adding more K^+ to a shift reagent solution decreases the observed shift in a ^{23}Na NMR spectrum. Divalent ions, such as Ca^{2+}, typically bind more strongly to the shift reagent and are, therefore, more effective at reducing the observed shift.

The dipolar shift described above is usually the dominant mechanism by which cation resonance shifts are induced by shift reagents. However, in any NMR sample the chemical shift can be affected by bulk magnetic susceptibility (BMS). The susceptibility shifts are the result of shielding effects produced by the surrounding medium, not the local electrons. Paramagnetic compounds produce large magnetic susceptibility effects. This type of paramagnetic susceptibility shift is observed in samples that have compartments that are not spherical. BMS shifts depend on differences in the susceptibility, shape, and orientation of each compartment with respect to the magnetic field. BMS shifts are predictable in certain well-defined situations.[18] Based on experiments using parallel coaxial glass cylinders,[18] simple cell suspensions,[4] and skeletal muscle containing oriented fibers,[19] it is clear that the ^{23}Na spectra of a multicompartment system in which only some of the compartments contain shift reagent are determined by a combination of hyperfine shift and BMS shifts. The BMS shifts in most bio-

logical samples are not predictable. Since the BMS shifts are variable in direction and extent, the net effect is an increase in observed linewidths.[20] The shift of intracellular resonances induced by shift reagents in other compartments is a result of the susceptibility shift.

2.1.2 Comparison of Shift Reagents

NMR properties: Each of the shift reagents currently in use to distinguish extracellular Na^+ (Na_o) and Na_i in biological samples produces spectra with both hyperfine and BMS shifts, but their relative contributions differ. For the same concentration of shift reagent, the hyperfine shift is much larger for $Dy(PPP)_2^{7-}$ than for $TmDOTP^{5-}$, which in turn is larger than for $DyTTHA^{3-}$. The contribution of BMS shift effects is smaller for $TmDOTP^{5-}$ than for $DyTTHA^{3-}$. Importantly, at concentrations of 3–10 mM, each of these shift reagents shifts the Na_o signal at least 2–4 ppm, allowing discrimination of Na_o and Na_i signals.

Physiological requirements: Shift reagents must also be compatible with the physiological requirements of the system under study. By assessing cardiac performance (heart rate, aortic pressure, left ventricular pressures) and the contents of the high-energy phosphate compounds ATP and phosphocreatine (or some other sensitive marker) in vivo and in isolated hearts, short-term toxicity has been assessed. Subject to action by phosphatases, $Dy(PPP)_2^{7-}$ is not stable in most biological samples, yielding free lanthanide which is toxic. Precipitates form at physiological Ca^{2+} concentrations. Consequently, this reagent has limited biological usefulness. Neither $DyTTHA^{3-}$ nor $TmDOTP^{5-}$ disturbs heart function or high-energy phosphate content in isolated hearts provided that care is taken to maintain normal levels of extracellular cations, especially Ca^{2+}. Supplying $DyTTHA^{3-}$ (maximum concentration of 16 mM in blood)[21] to the living rat had no effects on hemodynamics or high-energy phosphate content of skeletal muscle.[15] Preservation of cardiac function and high-energy phosphate content also supports the conclusion that the sarcolemma is not permeable to these shift reagents, thereby allowing discrimination of Na_i and Na_o.

In summary, $TmDOTP^{5-}$ provides good shifts with little BMS-induced broadening. Its major limitation is loss of solubility with Ca^{2+} levels greater than 1 mM. Nonetheless, $TmDOTP^{5-}$ has been used successfully in the intact rat.[17,22] $DyTTHA^{3-}$ is quite stable in solutions containing high $[Ca^{2+}]$. However, it does not offer large hyperfine shifts. Moreover, the $DyTTHA^{3-}$ concentrations employed typically induce bulk magnetic susceptibility shifts, which broadens both the intra- and extracellular resonances. $DyTTHA^{3-}$ has been successfully used in the intact rat,[15] pig,[16] and dog[23,24].

2.2 Information Available from ^{23}Na NMS with Shift Reagents for Cations

2.2.1 Information Derived from Changes in Chemical Shift

Information derived from changes in chemical shift of Na_o resonance with time includes kinetics of shift reagent permeation into the tissue and its washout, the extent of perfusion, and the distribution volume of shift reagent. These have been described for skeletal muscle in vivo.[16,21] The chemical shift of Na_i signal in tissues perfused with shift reagents is small, typically 0.25 ppm; this shift results from BMS effects and does not provide information.

2.2.2 Information Derived from Changes in ^{23}Na Signal Intensities

The Na_o signal is proportional to the amount of Na in all spaces accessible to shift reagent, while the Na_i signal is proportional to the amount of free Na^+ in the compartment(s) not accessible to shift reagent. Electron probe X-ray microanalysis of heart muscle confirms the assumption that almost all cytolic Na^+ is unbound and also provides evidence that Na^+ is homogeneously distributed through the intracellular space, at least on this scale of measurement. The amounts of Na_i measured by ^{23}Na NMR with shift reagent are the volume-weighted averages over the entire cell population contained within the NMR-sensitive volume. For many tissues, there is significant heterogeneity of cell type; however, for striated muscle, myocytes comprise 80% or more of tissue volume. Therefore, changes in the Na_i signal faithfully report changes in the amount of Na_i in myocytes.

Visibility factor: The amount of Na^+ in any compartment is proportional to the product of its concentration, the volume of the compartment, and the NMR visibility factor. The NMR visibility factor is the ratio of amount of Na^+ detected by ^{23}Na NMS to the amount measured by other chemical (putatively accurate) techniques. Thus, if volume and visibility factor were known, ^{23}Na resonance areas could be converted to cation concentrations. For this reason, the history of ^{23}Na NMR is replete with attempts to establish both the NMR visibility factor of Na_i and to offer explanations for NMR 'invisibility'.

Published visibility factors for ^{23}Na in biological systems range from 0.1 to 1.0. For a system in which all the Na is the aqua ion and the ion is in a simple bulk solution, the three single quantum transitions of the spin-$\frac{3}{2}$ nuclide are degenerate and the visibility is 1.0. In contrast, for at least some biological systems, it has been thought that only the central transition is visible, yielding a visibility factor of 0.4. Others have suggested that each of the transitions were equally visible or invisible, and, therefore, signal from each transition must contribute equally to the observed signal. In an elegant theoretical and experimental analysis of all six relaxation rate constants of the Na^+ resonance obtained for a single population of isolated spins in fast exchange, Rooney and Springer demonstrated that a major determinant of the so-called 'visibility' is the delay between the hard pulse and enabling of the receiver, the 'dead time'.[4,25] The shorter the dead time, the higher the visibility factor, i.e. the more of the signal that will be captured. This dependence may explain some of the variability in literature values.

In skeletal muscle, Balschi et al. showed that the visibility factors of the interstitial and intracellular Na are essentially the same or, more precisely, that the visibility of the Na^+ which moves from the interstitium to the intracellular spaces is unchanged.[15] An analogous analysis of the Na^+ signals for the vascular compartment of an isolated perfused heart and its surrounding bath during collapse of the vasculature (caused simply by changing perfusion pressure from 100 to 0 mmHg)

showed that the visibility of Na^+ moving from vascular to the bath is not altered by changing environments. Therefore, it is likely that the visibility factors are similar for interstitial and intracellular spaces (probably <1.0) and for vascular and bath (probably 1.0) in a perfused heart. By combining approaches such as these with standard curves, carefully constructed to mimic lossy environments, it should be possible to assess at least changes in visibility factors.

Calculation of concentration: There are two ways of calculating Na_i using ^{23}Na NMR spectra, the ratio method and external standard method. The ratio method takes advantage of the simultaneous measurement of Na_o and Na_i signals in ^{23}Na spectra with shift reagent. Rooney and Springer provide a strategy to quantitate intracellular sodium area (see their Figure 8c and discussion).[4] Briefly they recommend that the total resonance area of Na_i and Na_o be integrated, curve resolution being employed to integrate Na_i. The Na_i area should then be multiplied by 2.5 (this analysis predicts that only the central resonance, the $\frac{1}{2} - -\frac{1}{2}$ transition, is accurately quantified) to provide the total Na_i area and this area subtracted from the total to provide the Na_o area.[4] The value of Na_i can be calculated using the relationship:

$$\frac{A_{Na_i}}{A_{Na_o}} = \frac{[Na^+]_i V_i v_i}{[Na^+]_o V_o v_o} \tag{1}$$

where A_{Na_o} is the shifted peak area, A_{Na_i} is the unshifted resonance area, $[Na^+]_i$ is intracellular Na^+ concentration, $[Na^+]_o$ is extracellular interstitial Na^+ concentration, V_i is the intracellular volume, V_o is the extracellular volume, v_i is the visibility of Na_i and v_o is the visibility of Na_o. Solving for $[Na^+]_i$ yields:

$$[Na^+]_i = \frac{A_{Na_i} V_o v_o [Na^+]_o}{A_{Na_o} V_i v_i} \tag{2}$$

The ratio method is appropriate for surface coil experiments where the NMR-sensitive volume has an irregular shape. In experiments where the sample is entirely contained in the NMR-sensitive volume, an external or reference standard may be used to relate area to amount. In this case, the visibility of the compartments reported by Na_o and Na_i must be known.

In both cases, the volume of each compartment must be known. For some tissues, such as resting skeletal muscle, there is good agreement in the literature. However, for isolated organs, there is little or no information about the magnitude of the vascular bed during perfusion or how it changes with time and intervention. The approach described by Clarke and colleagues using ^{31}P-NMR to measure phenylphosphonate and dimethylmethylphosphonate in the total and extracellular water spaces in vivo provides a means to solve this problem.[26,27] Another method to measure volumes is to use $TmDOTP^{5-}$ as a combined cation shift reagent and extracellular space marker, thereby allowing quantification of $[Na^+]_i$ in vivo by NMR.[28,29].

Results from both NMR and classical experiments show that $[Na^+]_i$ differs for different striated muscles for a given species: 5 mM for rat skeletal muscle by NMR[15] and 10 mM by neutron activation analysis[30] versus ~15 mM for rat heart by ion-selective electrodes.[31] Even for the same tissue, $[Na^+]_i$ differs among species: $[Na^+]_i$ in the hearts of rat, rabbit, and guinea pig are 15, 10, and 10 mM, respectively (note that these

For References see p. 878

estimates assume the visibility factor is 0.6). A more complete comparison of $[Na^+]_i$ determined by classical techniques and by ^{23}Na NMS for heart has been recently summarized.[32]

3 APPLICATION OF ^{23}Na, ^{7}Li, ^{39}K, AND ^{87}Rb MRS TO STRIATED MUSCLE: DEFINING CATION MOVEMENTS DURING ISCHEMIA IN THE RAT HEART

Cation MRS with and without shift reagents has been used to measure intra- and extracellular Na^+, Li^+, K^+, and Rb^+ contents. Many studies have examined cation changes during total global ischemia (i.e. zero flow) in isolated hearts of small mammals, typically rats. We have also examined changes in Na_i in the intact heart of the open chest pig and dog during regional ischemia. We illustrate these applications with several examples.

In the submerged isolated heart, there are five major spatial compartments that contain cations: the intracellular space, the extracellular space, the coronary vasculature, the cardiac chambers (atria and ventricles), and the volume outside the heart within the NMR tube, which we term the bath. For all calculations with this model, we assumed that the extracellular (interstitial, vascular, and chamber) ion concentration, [ion], was equal to the perfusate ion concentration during preischemia and reperfusion. For the rat heart during buffer perfusion at 37°C and 100 mmHg pressure we used the following volumes: V_i 0.48 ml g^{-1} blotted weight and V_o 1.45 ml g^{-1} blotted weight.[26]

3.1 MRS of ^{23}Na

3.1.1 Isolated Heart Studies

The basic protocol consisted of 28 min perfusion with buffer containing the shift reagent followed by 28 min of total global normothermic ischemia. The heart was bathed with ion-free buffer containing shift reagent and mannitol under control of a separate pump. The bath was used to minimize cation signal from the bath. Use of shift reagent in both the Krebs–Henseleit perfusion fluid and the mannitol bath separated the signals from Na_i and Na_o (Figure 3a). The broad resonances underlying the heart signals, ranging from 0 to 9 ppm, represent Na^+ in the mannitol-containing bath, whereas the large shifted peaks, at 1.2–5 ppm, represent buffer Na^+ in the vascular, chamber, and interstitial spaces of the heart (Na_o). The small upfield shoulder on the shifted peak at 1.2–0 ppm represents the resonance of Na_i.

During zero-flow ischemia, the large shifted peak (Na_o) moves ~2 ppm downfield because of the movement of Na^+ from the myocardial vasculature and chambers to the mannitol bath. The area of this peak decreases during the course of ischemia; concomitantly there are progressive increases in the areas of Na_i resonance and in the shifted resonances of the bath. Using the peak areas from the spectra shown in Figure 3a, we estimate that by 28 min of ischemia the Na_i signal increased to 3.6 times the preischemic values: from 9.1 ± 1.3 mM (uncorrected for visibility factor) in the preischemic rat heart to 32.1 ± 1.2 mM at 28 min ischemia. During this time,

79% of the extracellular Na^+ signal moved from the heart to the bath. Assuming that the only water movement was from the heart to the bath and that no water moved from the extracellular to the intracellular space during ischemia, the changes in $[Na^+]_i$ and V_o allow calculation of the changes in $[Na^+]_o$, which decreased from 144 mM in the preischemic heart to 108 mM after 28 min ischemia. Therefore, the ratio of $[Na^+]_o$ to $[Na^+]_i$ during preischemia was 16 (144/9.1) and after 28 min ischemia was 3.4 (108/32). Using the calculated values for $[Na^+]_o$ and $[Na^+]_i$ in the Nernst equation, these data show that $[Na^+]_i$ in the preischemic heart would be in electrochemical equilibrium with a membrane potential of +74 mV.

A number of studies have attempted to define the Na^+-transport mechanisms during global ischemia in the isolated rat heart. During both low- and zero-flow ischemia, Na_i did not increase in hearts treated with the Na^+/H^+ exchange inhibitor, 5-(N-ethyl-N-isopropyl) amiloride (EIPA).[33] During reperfusion recovery, left ventricular developed pressure was improved for zero-flow hearts treated with EIPA. These studies show that Na^+/H^+ exchange is an important mechanism for Na_i accumulation during myocardial ischemia.[33,34] After 10 min of global ischemia, lidocaine, a class IB fast Na^+ channel blocker known to protect ischemic myocardium, resulted in lower $[Na^+]_i$.[35]

Other studies have attempted to define the Na^+-transport mechanisms during reperfusion following ischemia. A study that tested the effects of limiting Na^+-K^+ ATPase activity with ouabain found that Na^+_i increased upon reperfusion in hearts treated with ouabain[36] but did not increase if EIPA was also added to inhibit Na^+/H^+ exchange. This indicates the existence of a Na^+/H^+ exchange-mediated Na^+ influx upon reperfusion, which is usually masked by Na^+-K^+ ATPase activity. Another recent study has shown that altering the $[Ca^{2+}]_o$ can modify the rate at which Na_i decreases during reperfusion. Following reperfusion, low $[Ca^{2+}]_o$ slowed the decrease in Na_i while high $[Ca^{2+}]_o$ accelerated the decrease in Na_i.[37] This implicates the role of the Na^+/Ca^{2+} exchanger in reperfusion ion movements.

3.1.2 Intact Animal Studies

Isolated cells and perfused hearts are relatively well-controlled systems. There are, however, many differences between isolated systems and the in situ heart. The technical difficulty of accurately measuring Na_i in an intact heart is perhaps the main reason that few of these values for the in situ heart exist. We tested the hypothesis that Na_i increases during regional reduced pressure ischemia in the in situ pig heart. Using ^{23}Na NMR in combination with the shift reagent DyTTHA^{3-} the Na_o and Na_i contents of the pig heart could be distinguished (Figure 3b). We estimated a $[Na^+]_i$ of 7 mM for the pig left ventricle prior to ischemia.[16]

In a model of reduced-pressure regional ischemia, Na_i did not change during the first 18 min of ischemia with Na_i set to 1 for preischemia. Subsequently, Na_i increased from 1.3 at 22 min to 1.6 at 38 min, the end of ischemia. Levels remained elevated during early reperfusion; after 10 min of reperfusion Na_i had increased further to 1.9. This increase of Na_i and its delayed recovery during postischemia did not arise from insufficient reperfusion because phosphorylated metabolites, measured by concomitant ^{31}P NMR, all rapidly recovered during early reperfusion while Na_i was increasing. This

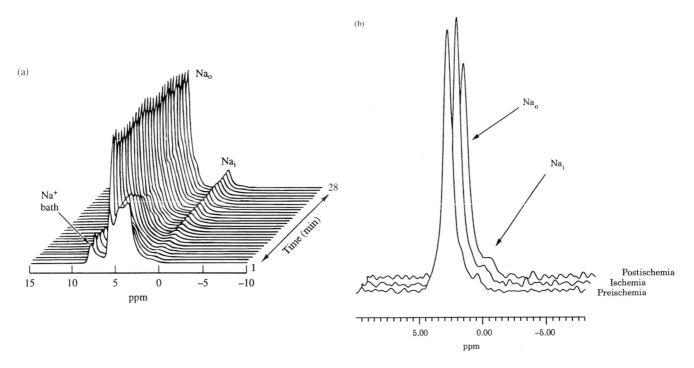

Figure 3 (a) Sodium-23 NMR spectra of an isolated buffer-perfused rat heart showing the increase in intracellular Na^+ (Na_i) during 28 min total, global ischemia. Spectra were acquired using a pulse angle of $90°$ and a recycle time of 250 ms. Each spectrum consisted of 240 FID, providing a time resolution of 1 min. The spectral width was ± 2 kHz and 1000 data points were collected. Prior to Fourier transformation, the data were multiplied by a double exponential function ($DM = 12$, GE software) for resolution enhancement. (b) An offset stacked plot of ^{23}Na NMR spectra obtained from the in situ pig heart after infusion with the shift reagent $DyTTHA^{3-}$. The spectra are displayed from the preischemia (front), ischemia, and postischemia (rear) periods. The ^{23}Na NMR resonances of extracellular Na^+ (Na_o) and of intracellular Na^+ (Na_i) are identified

remarkable response of Na_i to reperfusion may reflect enhanced Na^+ influx, possibly by Na^+/H^+ exchange, with or without reduced Na^+/K^+ ATPase efflux. Levels of Na_i returned to preischemia values at 18 min of reperfusion.[16]

3.2   MRS of ^{7}Li

Lithium-7 is 93% abundant with an NMR sensitivity that is twice that of ^{23}Na (Table 1). These NMR properties, combined with the ability of Li^+ to replace Na^+ in the fast Na^+/H^+ exchanger (see Figure 1) makes ^{7}Li MRS a useful tool to study Na^+ movements in biological samples.

Lithium-7 MRS has been used to show that the Na^+/H^+ antiporter contributes to net Na^+ influx during ischemia in isolated perfused hearts.[38] Fully relaxed ^{7}Li NMR spectra [such as shown in Figure 4(a)] were acquired throughout the protocol. The kinetics of Li^+ uptake were determined by perfusing hearts with a buffer containing 78 mM Li^+.

Movement of Li^+ into the intracellular space of the heart is monoexponential with a rate constant of 0.068 min^{-1}, a $t_{1/2}$ value of 10.3 min, and an initial rate of increase of 5.27 mM min^{-1}. The Li^+ concentration after 24 min of control perfusion was 57 ± 7 mM. ICP–AES analysis gave an intracellular Li^+ concentration of 50 ± 2 mM after 20 min perfusion showing that NMR visibility was approximately 1. Uptake in the well-perfused heart was unaffected by the Na^+/H^+ exchange inhibitor amiloride. At the end of the 28 min ischemia, [Li]$_i$ was 84 ± 10 mM. Amiloride completely prevented the uptake of Li^+

that occurred during ischemia, showing that Na^+/H^+ exchange contributes importantly to the increase in Na_i during ischemia.

3.3   MRS of ^{39}K

The ^{39}K NMR spectrum of a rat heart perfused with buffer containing shift reagent has two well-defined peaks: a large intracellular K^+ resonance and a smaller downfield extracellular resonance, the area of which is proportional to the sum of K^+ in the vascular, interstitial, and bath spaces [Figure 4(b)]. The areas of both signals remained constant throughout control perfusion. During control perfusion the K_i signal comprised 74% of the total ^{39}K signal.

During ischemia, the vasculature collapses within minutes (owing to zero perfusion pressure) and perfusate is extruded into the bath. Hence, the extracellular compartment in exchange with the intracellular compartment is primarily the interstitium. Accordingly, the interstitial K^+ signal can be calculated as the difference between the sum of vasculature and bath K^+ signals and the total extracellular K^+ signal. By 28 min of ischemia, there was no detectable change in [K^+]$_i$ but interstitial [K^+] decreased. The change corresponded to an increase in the potassium membrane potential from -88 mV to -46 mV.

3.4   MRS of ^{87}Rb

Rubidium-87 is a well-established congener for K. Relatively high natural abundance and high NMR sensitivity

For References see p. 878

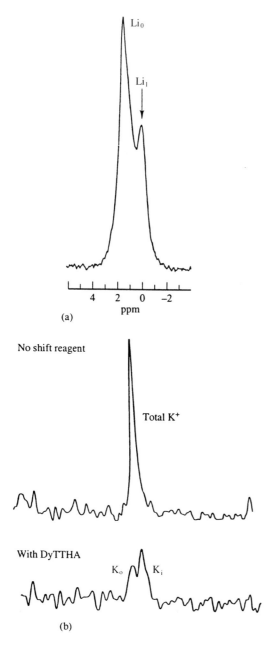

Figure 4 (a) Lithium-7 NMR spectra of a rat heart perfused with buffer containing shift reagent and 78 mM Li^+ (intracellular Li^+, Li_i; ecxtracellular Li^+, Li_o). Spectra were acquired at 8.34 T using a pulse angle of $90°$ and a recycle time of 45 s. Each spectrum consisted of 8 FID, providing a time resolution of 6 min. Prior to Fourier transformation, the data were multiplied by an exponential function giving a line broadening of 10 Hz for resolution enhancement. (b) Potassium-39 NMR spectra of a rat heart perfused with buffer containing the shift reagent $Dy(TTHA)^{3-}$ (intracellular K^+, K_i; extracellular K^+, K_o). Spectra were collected as described as in Figure 3. (Reproduced by permission of Springer-Verlag from J. S. Ingwall, 'Magnetic Resonance (MR) Imaging', 2nd ed., 1992, p. 131)

combined with low biological abundance make ^{87}Rb a good tracer for K^+ influx and efflux studies. Unlike ^{23}Na or ^{39}K NMR studies where changes in steady-state concentrations are measured with the aid of shift reagents, the rates of influx (in

the presence of extracellular Rb^+) and efflux (in the absence of extracellular Rb^+) have been measured in the absence of shift reagents.

Measured rates of influx of Rb^+ have been used to estimate Na^+/K^+ ATPase activity in living tissues using radioisotopes, atomic absorption spectroscopy, and MRS. It has been shown using ^{87}Rb MRS that, under normal physiological conditions, Rb^+ influx occurs mainly through Na^+-K^+ ATPase; the contribution of the Na^+-$K^+/2Cl^-$ cotransporter and K^+ channels to Rb^+ influx is small.[39] Therefore, ^{87}Rb NMR can be used to assess the activity of the Na^+-K^+ ATPase in the perfused heart under normal and ischemic conditions.[40]

4   SUMMARY

MRS with ^{23}Na and ^{39}K in combination with shift reagents allows measurement of intra- and extracellular $[Na^+]$ and $[K^+]$ in biological systems. The use of ^{87}Rb MRS provides a measurement of Na^+-K^+ ATPase activity. These techniques provide unique information on the transmembrane movements of Na^+ and K^+ in intact tissues.

5   RELATED ARTICLES

Animal Methods in MRS; NMR Spectroscopy of the Human Heart; Sodium-23 Magnetic Resonance of Human Subjects.

6   REFERENCES

1. J. D. DeLayre, J. S. Ingwall, C. Malloy, and E. T. Fossel, *Science,* 1981, **212**, 935.
2. F. W. Cope and R. Damadian, *Nature (London),* 1970, **228**, 76.
3. M. M. Pike, J. C. Frazer, D. F. Dedrick, J. S. Ingwall, P. D. Allen, C. S. Springer, Jr., and T. W. Smith, *Biophys. J.,* 1985, **48**, 159.
4. W. D. Rooney and C. S. Springer, *NMR Biomed.,* 1991, **4**, 209.
5. L. A. Jelicks and R. K. Gupta, *Magn. Reson. Med.,* 1993, **29**, 130.
6. J. M. Dizon, J. S. Tauskela, D. Wise, D. Burkhoff, P. J. Cannon, and J. Katz, *Magn. Reson. Med.,* 1996, **35**, 336.
7. J. S. Tauskela, J. M. Dizon, J. Whang, and J. Katz, *Magn. Reson.,* 1997, **127**, 115.
8. M. M. Pike, S. R. Simon, J. A. Balschi, and C. S. Springer, Jr., *Proc. Natl. Acad. Sci. USA,* 1982, **79**, 810.
9. R. K. Gupta and P. Gupta, *J. Magn. Reson.,* 1982, **47**, 344.
10. S. C.-K. Chu, M. M. Pike, E. T. Fossel, T. W. Smith, J. S. Balschi, and C. S. Springer, Jr., *J. Magn. Reson.,* 1984, **56**, 33.
11. D. C. Buster, M. C. A. Castro, C. F. G. C. Geraldes, C. R. Malloy, A. D. Sherry, and T. C. Siemers, *Magn. Reson. Med.,* 1990, **15**, 25.
12. J. A. Balschi, V. P. Cirillo, and C. S. Springer, Jr., *Biophys. J,* 1982, **38**, 323.
13. H. Hoefeler, D. Jensen, M. M. Pike, J. L. DeLayre, V. Cirillo, C. S. Springer, Jr., E. T. Fossel, and J. A. Balschi, *Biochemistry,* 1987, **26**, 4953.
14. M. M. Pike, E. T. Fossel, T. W. Smith, and C. S. Springer, Jr., *Am. J. Physiol.,* 1984, **246**, C528.
15. J. A. Balschi, J. A. Bittl, C. S. Springer, Jr., and J. S. Ingwall, *NMR Biomed.,* 1990 **3**, 47.

16. J. A. Balschi, *Basic Res. Cardiol.*, 1999, **94**, 60.

17. N. Bansal, M. J. Germann, I. Lazar, C. R. Malloy, and A. D. *JMRI*, 1992, **2**, 385.

18. S. C.-K. Chu, Y. Xu, J. A. Balschi, and C. S. Springer, Jr., *Magn. Reson. Med.*, 1990, **13**, 239.

19. S. J. Kohler, S. B. Perry, L. C. Stewart, D. E. Atkinson, K. Clarke, and J. S. Ingwall, *Magn. Reson. Med.*, 1991, **18**, 15.

20. J. A. Balschi, S. J. Kohler, J. A. Bittl, C. S. Springer, Jr., and J. S. Ingwall, *J. Magn. Reson.*, 1989, **83**, 138.

21. M. S. Albert, W. Huang, J. H. Lee, J. A. Balschi, and C. S Springer, Jr., *NMR Biomed.*, 1993, **6**, 7.

22. V. Seshan, A. D. Sherry, and N. Bansal, *Magn. Reson. Med.*, 1997, **38**, 821.

23. S. M. Eleff, I. J. McLennan, G. K. Hart, Y. Maruki, R. J. Traystman, and R. C. Koehler, *Magn. Reson. Med*, 1993, **30**, 11.

24. J. O. Hai and J. A. Balschi, *Circulation*, 1995, **92**, I-635.

25. W. D. Rooney and C. S. Springer, Jr., *NMR Biomed*, 1991, **4**, 227.

26. K. Clarke, J. F. Nedelec, S. M. Humphrey, L. C. Stewart, S. Neubauer, J. A. Balschi, A. G. Kleber, C. S. Springer, Jr., T. W. Smith, and J. S. Ingwall, *NMR Biomed.*, 1993, **6**, 278.

27. K. Clarke, R. E. Anderson, J. F. Nedelec, D. O. Foster, and A. Ally, *Magn. Reson. Med*, 1994, **32**, 181.

28. P. M. Winter, V. Seshan, J. D. Makos, A. D. Sherry C. R. Malloy, and N. Bansal, *J. Appl. Physiol*, 1998, **85**, 1806.

29. J. D. Makos, C. R. Malloy, and A. D. Sherry, *J. Appl. Physiol.*, 1998, **85**, 1800.

30. M. I. Linginger and G. J. F. Heigenhauser, *J. Appl. Physiol.*, 1987, **63**, 426.

31. M. J. Shattock and D. M. Bers, *Am. J. Physiol.*, 1989, **256**, C813.

32. J. S. Ingwall, 'Cardiovascular Magnetic Resonance Spectroscopy.', Kluwer Academic Press, Norwell, MA, 1992, Chap. 13, p. 195.

33. M. M. Pike, C. S. Luo, M. D. Clark, K. A. Kirk, M. Kitakaze, M. C. Madden, E. J. Cragoe, Jr., and G. M. Pohost, *Am. J. Physiol.*, 1993, **265**, H2017.

34. E. Murphy, M. Perlman, R. E. London, and C. Steenbergen, *Circ. Res.*, 1991, **68**, 1250.

35. N. B. Butwell, R. Ramasamy, I. Lazar, A. D. Sherry, and C. R. Malloy, *Am. J. Physiol.*, 1993, **264**, H1884.

36. J. G. van Emous, J. H. Schreur, T. J. Ruigrok, and C. J. van Echteld, *J. Mol. Cell. Cardiol.*, 1998, **30**, 337.

37. K. Imahashi, H. Kusuoka, K. Hashimoto, J. Yoshioka, H. Yamaguchi, and T. Nisimura, *Circ. Res.*, 1999, **84**, 1401.

38. C. A. Keon, J.-F. Nedelec, and K. Clarke, *Circ. Res.*, 1991, **84**, II-7.

39. V. V. Kupriyanov, L. C. Stewart, B. Xiang, J. Kwak, and R. Deslauriers, *Circ. Res.*, 1991, **76**, 839.

40. H. R. Cross, G. K. Radda, and K. Clarke, *Magn. Reson. Med.*, 1995, **34**, 673.

Biographical Sketches

James A. Balschi, *b.* 1950. B.A. 1973, Temple University, Ph.D., 1984, State University of New York at Stony Brook. Postdoctoral studies in cardiac metabolism and in vivo NMR at Harvard Medical School, 1983–86, Faculty in Medicine, Harvard Medical School and Brigham and Women's Hospital, 1986–88, University of Alabama at Birmingham, in Medicine, 1988–98, Harvard Medical School and Brigham and Women's Hospital, 1998–present. Approx. 30 publications. Current research interests: relationship between energy metabolism and cation transport studied using ^{31}P and ^{23}Na MRS.

Kieran Clarke. *b* 1951. B.Sc., 1973, Flinders University, Ph.D., 1986, University of Queensland. Postdoctoral studies in ^{31}P and ^{23}Na NMR at Harvard Medical School, 1987–89; National Research Council, Canada, 1989–91; Oxford University, UK, 1991–present. Approx. 50 publications. Research interests: use of NMR spectroscopy to study

myocardial cation movements as the subcellular cause of ischemic damage and control of myocardial oxidative phosphorylation.

Laura C. Stewart. *b* 1959. B.Sc., 1982, University of Toronto, Ph.D., 1987, National Research Council. Faculty in Medicine, Harvard Medical School and Brigham and Women's Hospital, 1988–90, National Research Council, 1990–93, Science Consultant for research institutes, government and industry, 1993–present. Approx. 30 publications. Research interests: ^{31}P, ^{23}Na, ^{39}K, and ^{87}Rb NMR spectroscopy to study cation movements in striated and smooth muscles.

Monique Bernard. *b* 1958. M.S., 1980, Ph.D., 1983, University of Aix-Marseille. Faculty in Medicine, Centre National de la Recherche Scientifique, 1983–present; Faculty of Medicine, Harvard Medical School and Brigham and Women's Hospital, 1989–1990. Approx. 40 publications. Research interests: NMR applications to biological systems, energy metabolism in *E. coli* cells, myocardial protection during ischemia and reperfusion, and ^{13}C NMR conformational dynamics of proteins in solution.

Joanne S. Ingwall. *b* 1941. B.S., 1963, LeMoyne College, Ph.D., 1968, Cornell University. Postdoctoral training in Cell Biology and Biochemistry, University of California, San Francisco, 1968–73. Faculty in Medicine Biochemistry, University of California, San Diego, 1973–76, Faculty in Medicine (Physiology), Harvard Medical School and Brigham and Women's Hospital, 1977–present. Approx. 160 publications. Current research interests: regulation of cardiac energy metabolism and cation transport using ^{31}P and ^{23}Na NMR spectroscopy.

Cells and Cell Systems MRS

Jerry D. Glickson

The Johns Hopkins University School of Medicine, Baltimore, MD, USA

1 INTRODUCTION

Magnetic resonance spectroscopy (MRS) of cells and cell systems (or cellular MRS) includes studies of intact living cells or assemblies of cells. Perfused organs and tissues, and cell extracts are not included in this topic (see *Tissue and Cell Extracts MRS*).

A key advantage of cellular MRS is that it can be conducted under well-defined and well-controlled conditions that generally cannot be achieved in the intact organism. The cellular model is often designed to facilitate the interpretation of in vivo MRS data. Systemic physiological effects, which often limit the interpretation of in vivo MRS, are absent in ex vivo

For References see p. 886

cellular studies or can be systematically simulated and evaluated. However, some applications of cellular MRS focus on issues unique to the cellular system, e.g. characterization of bioreactors for production of various biological products, cell preservation, artificial organs, and microbiological studies. The range of cell systems that have been examined includes isolated bacteria and other prokaryotes, yeast, plant cells, mammalian cells, and even giant single mammalian cells. In most instances the goal is to study a uniform cell population; however, the effects of heterogeneity can be systematically investigated using multicellular assemblies such as tumor spheroids or renal microtubules. Subcellular organelles such as mitochondria and chromaffin granules have also been investigated. A variety of nuclear species have been monitored, including ^1H, ^2H, ^3H, ^7Li, ^{13}C, ^{15}N, ^{19}F, ^{23}Na, ^{31}P, ^{39}K, and ^{133}Cs, the main limitation being sensitivity and cost.

A number of excellent reviews of cellular MRS have been published.[1,2] We will limit ourselves to a description of key methods, their advantages and disadvantages, and will also present a few examples of important applications. We will note some recent developments that permit noninvasive discrimination between intracellular and extracellular metabolites. Finally, we will evaluate the potential of this method, and indicate possible areas for development.

2 PROBE DESIGN

2.1 Cell Suspensions

Early cellular NMR studies were performed on unstirred high-density suspensions of cells in conventional NMR tubes. Cells were often cooled in order to preserve viability. Spectral quality was generally poor, and cells died as substrates were depleted and toxic waste products accumulated. These methods were most suitable for nonoxygen-dependent cells that can tolerate high cell densities, such as red blood cells and anaerobic bacteria. Despite these limitations, the early studies demonstrated the potential utility of in vivo measurements of cellular metabolism.

The first cellular NMR studies were ^{13}C NMR measurements by Matwiyoff's laboratory on suspensions of blood cells[3] containing ^{13}C-labeled CO_2, CO, and CN^- and on the metabolism of [1-^{13}C]glucose by yeast.[4] Moon and Richards[5] first demonstrated the ability of ^{31}P NMR to measure intracellular pH from the chemical shift of the 2,3-diphosphodiglycerate (DPG, an intracellular metabolite) peaks and from the intracellular inorganic phosphate (Pi) resonance. Shulman's group at Bell Laboratories and later at Yale University initiated ^{31}P and ^{13}C studies of energy and carbohydrate metabolism, sodium transport and pH regulation in bacteria,[6,7] yeast,[8] liver cells,[9] and cultured tumor cells.[10] These studies had a seminal influence on the development of in vivo NMR spectroscopy, and demonstrated the unique ability of this method to examine key metabolic processes noninvasively.

Bubbling of oxygen into the NMR tube was introduced in order to oxygenate and stir the suspension, but the timing of spectral accumulations had to be adjusted to avoid intervals when bubbles were in the tube. Some of the limitations of these early methods could be overcome by using larger chambers equipped with mechanical stirrers and provided

with ports for delivery of gases and sensors for monitoring temperature and pO_2; cells were examined at 37 °C for up to 2 h.[11,12] Placement of the receiver/transmitter coil inside the sample volume[12] improved sensitivity, but the apparent relaxation times were dependent on the mixing rate. Another variation of this approach is the 'airlift chamber'.[13,14] This assembly consists of a large NMR tube with a smaller-diameter concentric tube (open at both ends) inside it. A rapid stream of oxygen bubbles is delivered through a capillary located above the region of the receiver coil. This oxygenates the cells without introducing bubbles into the sampling region and causes efficient circulation of the cell suspension. Taylor and Deutsch[15] compared the airlift probe with a flow cell assembly for measuring oxygen concentrations and pH in suspensions of Paracoccus denitrificans (see Section 3.2). Since spectral characteristics, particularly relaxation times, depend on flow rate and bubble size, the flow and airlift systems have to be recalibrated for each experiment.

In summary, the key limitation of suspension systems is that they are applicable only to nonanchorage-dependent cells; they do not control the concentration of oxygen and vital substrates, and they allow toxic metabolic end-products to accumulate. Bubbling oxygen through the cell suspension, or mechanical stirring, often causes damage to fragile cells and modifies relaxation rates.

2.2 Perfusion Systems

2.2.1 General Characteristics

The solution to most of these problems is a perfused cell system, which allows long-term experiments to be performed on isolated cells, cell aggregates, or small organisms under conditions of well-defined medium composition. The system consists of an NMR probe in which the cells are retained by attachment to beads, entrapment in a gel, or containment in a chamber with a semipermeable membrane. The perfusion system is attached to a reservoir that provides medium at a fixed temperature, usually 37 °C. For reasons of economy, recirculating systems are preferred when an expensive medium containing isotopically enriched substrates or expensive drugs is used. The recirculating system is adequate provided its capacity is large enough to prevent depletion of substrates or accumulation of toxic waste products; alternatively, a single-pass delivery system can be employed. Provision is generally made for delivery of gases, usually 5% CO_2/95% O_2 or air. Sterility should be maintained throughout the experiment by sterilizing the media and perfusion system, and by using large-capacity filters or closed systems. Perfusion systems have been designed for studies of both anchorage-dependent and anchorage-independent cells. Ng et al.[16] have constructed a system suitable for γ-irradiation of cells directly in the NMR tube without interruption of perfusion.

2.2.2 Microcarrier Beads

In 1981, Ugurbil et al.[17] introduced the first perfusion system for NMR investigation of anchorage-dependent cultured mammalian cells. Normal and transformed mouse embryo fibroblasts were grown on collagen-coated dextran microcarrier beads; at confluence the cells were transferred into a cylindrical glass chamber fitted with Plexiglas end-pieces containing nylon mesh screen filters for retention of the beads. A four-turn solenoidal coil was fitted around this sample chamber. Theoretically, a solenoid provides a 2.6-fold enhancement in sensitivity over a Helmholtz (saddle) coil.[18] However, in practice at least equivalent spectral quality can be achieved with a simpler system developed by Neeman and Degani[19] which employs a conventional 10 mm NMR tube and uses a commercial high-resolution Helmholtz coil (Figure 1). The improvement in magnetic field homogeneity of this design makes up for the higher intrinsic sensitivity of the solenoidal coil. The perfusate reservoir can easily be replaced for changing the perfusion medium, and the spectrometer can be programmed to monitor various nuclei according to a predefined protocol. Figures 2 and 3 show representative [31]P (202.5 MHz) and [13]C (125.7 MHz) spectra obtained with this system.

The key advantage of the microcarrier method is that it maintains cells in a uniform well-perfused state; several days of perfusion can be readily achieved with currently available perfusion probes (see below). The cell density is relatively low, and only a small fraction of the volume of the sample chamber is occupied by the cells (~6%). This improves homogeneity of perfusion since minimal depletion of oxygen and substrates avoids creation of local nutrient gradients.

Cell replication generally occurs, at least until the plateau phase of growth. Finding the proper conditions for attachment of cells to the beads is essential. Beads manufactured from var-

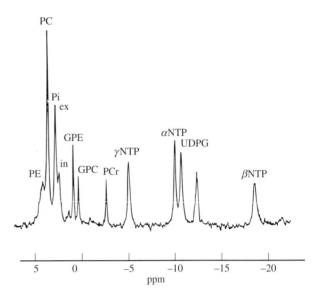

Figure 2 Fully relaxed ^{31}P spectrum of perfused T47D-clone 11 (5.2×10^7) cells treated for 5 days with 17β-estradiol (3×10^{-6} M). Each spectrum was obtained after approximately 30 min of perfusion, by accumulating 720 transients (2 h total accumulation) and processed using 10 Hz line broadening. NTP, nucleoside triphosphate; PC, phosphorylcholine; PE, phosphorylethanolamine; ex, extracellular; in, intracellular; GPE, glycerol–PE; GPC, glycerol–PC; PCr, phosphocreatine; UDPG, uridine diphosphoglucose. (Reproduced by permission of American Association for Cancer Research, Inc. from Neeman and Degani[19])

ious materials such as polyacrolein, polystyrene, glass, polyacrylamide, and agarose polyacrolein are commercially available. Coating with polylysine or collagen can be used to improve cell attachment. Gelatin beads or dextran beads covered with a layer of denatured collagen (Cytodex 3) are enzymatically degradable, which facilitates recovery of cells

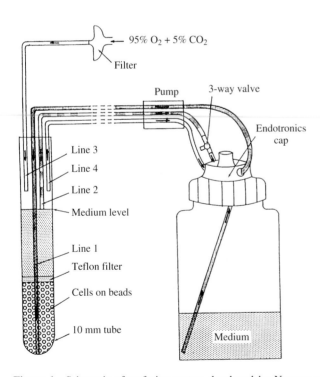

Figure 1 Schematic of perfusion system developed by Neeman and Degani[19] employing a 10 mm outer diameter NMR tube as the perfusion chamber. Line 1, inflow of perfusate; Line 2, removal of used medium mixed with 95% O_2 5% CO_2; Line 3, inflow of gas mixture; *Line 4*, additional return line. (Reproduced by permission of American Association for Cancer Research, Inc. from Neeman and Degani[19])

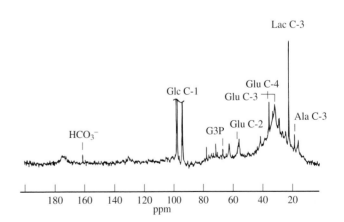

Figure 3 Carbon-13 NMR spectra of T47D-clone 11 (6.0×10^7) cells treated with estrogen as in Figure 2 after 1 h perfusion with medium containing [1-^{13}C]glucose. Spectra were Fourier transformed using 10 Hz line broadening. The designated glucose signals are due to nonenriched carbons of the glucose in the medium. (Reproduced by permission of American Association for Cancer Research, Inc. from Neeman and Degani[19])

For References see p. 886

for assay of cell number and viability. Disadvantages of the microcarrier bead system include that it has relatively low spectral sensitivity, that it is limited to anchorage-dependent cells, and that exposure to cytotoxic agents (e.g. cancer drugs) can detach cells from the beads. Smaller beads increase the total cell volume and, hence, increase spectral sensitivity.

2.2.3 Gels

Cells have been immobilized by encapsulation in a variety of gelatinous materials. Encapsulation is usually simpler than attachment to microcarrier beads, and is applicable to anchorage-independent as well as to anchorage-dependent cells. An important advantage for studies of therapeutic agents is that encapsulation of the cells prevents cell detachment as a result of exposure to cytotoxic agents such as cancer drugs or γ-irradiation. The porosity of the matrix determines the rate of diffusion of substrates and metabolic end-products. The gel may also modify cellular growth characteristics. Preparation of cells for encapsulation in the gel can involve treatment with enzymes to isolate them from the primary culture system and exposure to low temperature. These procedures may modify important cellular characteristics. Removal of cells from the gel in a viable state suitable for clonogenic assay (ex vivo measurement of cellular capacity to replicate) can in some instances be difficult.

2.2.4 Agarose Threads or Beads

This method was first introduced by Foxall and Cohen in 1983.[20] Cells are mixed with liquid agarose in phosphate-buffered saline, and placed in a bath at 37 °C for 5–7 min to form a gel. The mixture is extruded under low pressure through 0.5 mm internal diameter tubing in an iced bath into an NMR tube containing growth medium. The solid threads are concentrated without compression at the bottom of the tube by a plastic insert. Perfusion is maintained by a peristaltic pump, and the system is aerated by diffusion of air through Teflon tubes in the system. The system is stable for more than 24 h when perfused with whole growth medium.

The agarose threads are permeable to proteins such as albumin and also to various polypeptides, thus permitting investigation of their effects on cellular metabolism and growth characteristics. The system is suitable for encapsulation of yeast, bacteria, protozoa, and a variety of mammalian cells. Small anchorage-independent cells such as lymphocytes tend gradually to wash out of (0.9%) agarose gel threads.[1] Additional limitations include the requirement for cooling cells in an ice bath and reduced cell growth rates in the agarose threads.

Encapsulation of cells in agarose beads is accomplished by mixing the cell suspension with warm liquid agarose and dispersing the suspension in paraffin oil with magnetic stirring. The mixture is cooled in an ice bath after the droplets have formed, to produce small solid cell-containing capsules, which are collected by centrifugation and placed in an appropriate medium and NMR perfusion system. This method has been used for studies of microorganisms, plant cells, and algae.[1] Entrapment in agarose did not alter metabolism, and allowed cells to proliferate. Lymphocyte proliferation has been examined in 3% agarose gel beads, but beads composed of 1.2% agarose were unstable.[21] Mouse hybridoma cells encapsulated

in agarose were metabolically active and produced monoclonal antibodies.[1]

2.2.5 Matrigel

Growth rates of anchorage-dependent cells comparable with in vivo rates (of tumor cells) can be achieved by using a basement membrane matrix preparation extracted from a mouse tumor line (commercially available as Matrigel).[1] Cells are introduced into the liquid Matrigel in an ice bath; the suspension is warmed to 37 °C to form a gel, which is extruded through tubing to produce threads. Typically, threads are prepared at low cell density, and cells are allowed to grow in the threads before examination. Normal cells grown in Matrigel become morphologically and functionally similar to their in vivo counterparts; tumor cells simulate in vivo tumor cell growth, producing a high-density model tumor Figure 4. Nonuniform perfusion occurs in later stages of tumor growth, as it does in in vivo tumors and in spheroids (see below).

DAY 1
DAY 4

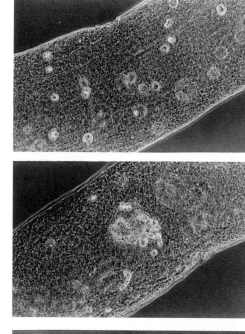

DAY 8

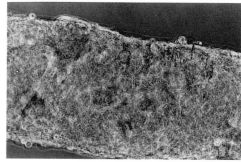

Figure 4 MDA-231 breast cancer cells growing in Matrigel threads (500 μM diameter). (Reproduced by permission of Springer-Verlag from Kaplan et al.[1])

2.2.6 Alginate Beads

Narayan et al.[22] introduced a method for NMR studies of perfused tumor cells encapsulated in calcium alginate beads. Cells isolated from monolayer culture by trypsinization are suspended in a sodium alginate solution; beads form when droplets of this suspension are introduced into a $CaCl_2$ solution. The beads are washed with fresh medium and perfused in a system similar to the one depicted in Figure 1.

Cells perfused in alginate retain stable levels of phosphate metabolites for at least 72 h of perfusion. Some cells replicate in this medium; however, RIF-1 tumor cells do not replicate at the high cell densities used for NMR studies (7×10^7 cells/mL), even though they do at lower cell densities. Furthermore, RIF-1 tumor cells recovered by dissolving the alginate beads with sodium citrate do not retain clonogenic capacity. Exposure of cells to high concentrations of calcium could be responsible for loss of proliferative capacity.

Tumor cells, plant cells, and algae have been investigated.[1] Lymphocytes, which are too small to be retained in (0.9%) agarose threads (see above), can be studied in alginate beads.

McGovern et al.[23] introduced a technique for monitoring oxygen concentration during NMR experiments by incorporating perfluorocarbons into the alginate cell suspension. The ^{19}F spin–lattice relaxation rate of the perfluorocarbon indicates the average oxygen concentration (see Section 3.2). Mathematical modeling permits estimation of the oxygen gradient within the alginate beads. This method could readily be adapted for measuring rates of oxygen consumption and can be used in any gel-based immobilization system.

2.2.7 Other Immobilization Matrixes

Aiken et al.[24] used Cultisphere G-L collagen beads (Hyclone Labs, Logan, UT, USA) for studies of RIF-1 cells. The cells grow within crevices as well as on the surface of the beads. Unlike the alginate system (see above), the collagen beads allow (RIF-1) cells to replicate at a rate equivalent to that in monolayer culture. Cells are easily recovered by treating the beads with Dispase (Collaborative Research, Bedford, MA, USA); cloning efficiency of RIF-1 is similar to that of cells trypsinized from monolayer culture. This system is suitable for studying the effects of antineoplastic agents during normal cell replication.

Gamcsik (personal communication) has recently employed a 'sponge-like' collagen gel matrix, Spongostan (Ferrosan, Denmark) in NMR studies of perfused MCF-7 tumor cells. This matrix can provide more mechanical stability than Cultisphere G-L beads, which tend to settle out or move around in the perfusion chamber.

2.2.8 Hollow Fibers

The hollow fiber methods, first introduced by Gonzalez-Mendez et al.,[25] deliver oxygen and substrate to the cells through a compartment separated from the cells by a semipermeable membrane. These bioreactor systems are described elsewhere in detail. The key advantage of these systems is that they allow very high-density growth of tumor cells, approaching that of intact tissue. This also produces their key deficiency—the development of oxygen, nutrient, and metabolite gradients. These systems are considerably more complex than the immobilization techniques described above. Nonuniform distribution of cells can occur, and it is difficult to access the cells for counting or bio-

logical assays during the course of an experiment. However, these are the best systems for very long-term investigations, and they can achieve the highest spectral sensitivity of all the perfused cell systems.

2.2.9 Spheroids

Multicellular spheroids are spherical aggregates of tumor cells, which serve as models of heterogeneously perfused in vivo tumors. Cells near the external surface are well perfused, metabolically active, and rapidly proliferating; those further into the interior of the spheroid are relatively poorly perfused and metabolically and proliferatively quiescent; in larger spheroids a central core of necrotic cells is also encountered.

Lin et al.[26] introduced a very simple perfusion system for NMR study of spheroids, in which spheroids were allowed to settle at the bottom of a 10 mm NMR tube while perfusate was slowly pumped through the system for up to 80 h. No mechanical restraints were introduced into the system. This system has a number of deficiencies, the most important one being the tendency of spheroids to aggregate, which results in substantial modification of the metabolic and proliferative properties of the tumor cell.

Ronen and Degani[27] addressed this problem by encasing small MCF-7 spheroids in agarose beads. Studies were performed in the conventional perfusion system shown in Figure 1. They achieved a very high cell density (10^8 cells/1.5 mL), which permitted measurement of fully relaxed ^{31}P NMR spectra in 7.5 min. Cell growth followed Gompertzian kinetics; decreases in levels of NTP and the rate of production of ^{13}C-labeled lactate (from metabolism of [1-^{13}C]glucose monitored by ^{13}C NMR) in the plateau phase were attributed to the development of central necrosis in the spheroids.

Freyer et al.[28] have designed an efficient perfusion system for spheroids (Figure 5). It consists of a 20 mm NMR tube equipped with a mechanical stirrer and multiple inlet lines and sensors for oxygen concentration, pH, and temperature. EMT6/

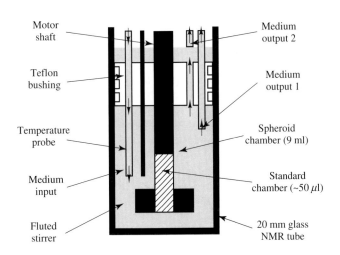

Figure 5 Diagram of the spheroid NMR perfusion chamber. The arrows indicate the direction of medium flow. The sensing volume of the NMR coil is indicated by the hatched area in the spheroid chamber. (Reproduced by permission of John Wiley and Sons from Freyer et al.[28])

For References see p. 886

Ro spheroids are grown in the NMR tube under conditions identical to those employed in batch culture, and exhibit identical growth characteristics, cytokinetic parameters, and clonogenic viability. Except for changes due to cell growth, the system is extremely stable (<5% variation over a 24 h period in the NMR tube or in culture). Transient interruption of perfusion or stirring produces significant, largely reversible, changes in ^{31}P NMR characteristics. The cell density (~5 × 10^8 cells/9.5 mL) is substantially lower than that of the Ronen and Degani system, and the maximum medium flow rate is 34 mL min^{-1} compared with 1.5 mL min^{-1} employed by Ronen and Degani.[27] Freyer's system has two additional advantages: (i) the spheroids are uniformly distributed in the perfusion system, and (ii) the spheroids can readily be recovered for cytokinetic and radiobiological assays.

3   SPECIAL METHODS

3.1   Distinction Between Intracellular and Extracellular Metabolites

Distinction between intracellular and extracellular compartments by MRS has been accomplished on the basis of differences in chemical shifts, relaxation times, and diffusion constants of metabolites in these compartments or by the introduction of reagents that are confined to one or another compartment. In most instances this has required the introduction of special reagents, such as paramagnetic contrast agents or exploitation of unique properties of the cellular target (e.g. magnetic field gradients within erythrocytes). A more general approach to this problem which does not require introduction of reagents or unique properties in the target cell is the use of pulsed field gradients. This technique, first introduced by Andrasko[29] in 1976 for studies of Li$^+$ transport into red blood cells, was recently adapted to ^{1}H NMR studies of perfused cell systems by van Zijl et al.[30] It exploits differences in the effective diffusion constants of intracellular and extracellular metabolites. These investigators employed a diffusion-sensitive stimulated echo pulse sequence to isolate signals from MCF-7 cells encapsulated in agarose threads (see above). By proper choice of gradients and delays, one can determine the ratio of extracellular to intracellular volumes and eliminate the resonances of all the extracellular metabolites, which generally have a larger diffusion constant because they are flowing and are not restricted to the intracellular compartments.

This method could be extended to other nuclei such as ^{19}F, ^{31}P, ^{13}C, and ^{7}Li, but lower γ-nuclei will require proportionately steeper gradients. These are becoming available on the newer instruments.

3.2   Indicator Compounds

Many cellular parameters cannot be directly measured by NMR. Exogenous reagents have been developed to measure parameters such as intracellular and extracellular pH, concentrations of Ca^{2+}, and Na$^+$, as well as redox state, pO_2, cell volume, and membrane potential. Szwergold[2] has reviewed these, and they will be discussed only briefly here.

3.2.1   pH

Deutsch's laboratory[31] has developed a variety of fluorinated pH indicators that are trapped inside cells. A variety of other indicators of intracellular and extracellular pH which can be monitored by ^{31}P NMR are also available.[2]

3.2.2   Calcium Ions

Intracellular calcium concentrations can be monitored by ^{19}F NMR with the fluorinated chelator 5,5'-difluoro-1,2-bis(2-aminophenoxy)ethane-N,N,N',N'-tetraacetic acid (F-BAPTA),[32] but the high calcium buffering capacity of this chelator has generated concern about its perturbing effect on cell physiology. F-BAPTA can also bind other divalent cations. The fluorinated chelator is introduced into the cell as a membrane permeable ester and is trapped inside the cell after hydrolysis by endogenous esterases. The calcium concentration is determined from the dissociation constant of the calcium complex of F-BAPTA and the ratio of intensities of resonances of the free and bound chelate.

3.2.3   Magnesium Ions

Intracellular magnesium can be measured from the chemical shift of the β-resonance of ATP, but the K_d is so low that ATP is usually saturated under physiological conditions. 4-Methyl-5-fluoroaminophenol-N,N,O-triacetic acid (MF-APTRA) and 5-fluoroaminophenol-N,N,O-triacetic acid (5F-APTRA) are fluorinated chelators with considerably greater promise because of their lower dissociation constants.[2]

3.2.4   Sodium Ions

Paramagnetic shift and relaxation reagents have been used to distinguish intracellular from extracellular sodium ions,[33,34] but these agents can be toxic and some are slowly degraded on long-term contact with cells (see **Cation Movements across Cell Walls of Intact Tissues Using MRS**). These problems can be avoided by not recycling the shift reagent (G. Elgavish, personal communication). The use of multiple quantum coherence methods to distinguish intracellular sodium has been proposed;[35] however, extracellular sodium ions that are immobilized by interaction with a gel matrix or with cell surfaces at high cell densities may also exhibit multiple quantum coherence transfer and, hence, be indistinguishable from intracellular sodium ions.

3.2.5   Redox Potential

Redox potentials can be determined from the ratio of oxidized to reduced forms of metabolites, using the Nernst equation and the standard reduction potential of the couple. This principle has been applied to a variety of ^{13}C-labeled couples, including lactate/pyruvate, β-hydroxybutyrate/acetoacetate, NAD(P)/NAD(P)H, and the reduced and oxidized forms of the radiation sensitizer WR3689.[2]

3.2.6   Oxygen

The increase in the spin–lattice relaxation rate due to paramagnetic oxygen molecules, $1/T_{1P}$, varies linearly with oxygen concentration. With proper calibration the oxygen concentration can be determined from measurements of T_{1P} of the ^{19}F resonance of perfluorocarbons.[15,23]

3.2.7 Membrane Potential

Membrane potentials are calculated from the ratio of intracellular and extracellular metabolites, which are in slow exchange on the chemical shift timescale. The membrane potential of erythrocytes has been measured by ^{31}P NMR using the hypophosphite ion as a probe; trifluoroacetate was employed in a ^{19}F NMR study.[2]

3.2.8 Cell Volume

Cell volume has been measured by ^{31}P and ^{19}F NMR, using dimethyl methylphosphonate or trifluoroacetamide as an extracellular probe, and also by ^{23}Na NMR, by utilizing the low concentration of intracellular compared to extracellular sodium.[2]

4 APPLICATIONS

Below we present a few examples of cellular NMR studies which illustrate the versatility of this method. They demonstrate its utility for studying fundamental principles of bioenergetics, for investigating mechanisms of antineoplastic agents in tumor cells, for elucidating biochemical pathways, and for evaluating mechanisms underlying therapy-induced in vivo spectral changes in tumors.

4.1 Cell Suspensions

4.1.1 The Chemiosmotic Hypothesis

As originally proposed by Mitchell,[36] certain membranes of microorganisms and organelles (such as mitochondria and chloroplasts) are impermeable to protons. Oxidation of substrates via the electron transport chain results in transport of protons across these membranes, generating an electrical potential ($\Delta\psi$) and a pH gradient that are coupled to the free energy, ΔG_{ATP}, for the ATP synthase reaction (Pi + ADP $\rightleftharpoons$ ATP) by the equation

$$\Delta G_{ATP} = -nF\,\Delta\psi + 2.3nRT\,\Delta pH \qquad (1)$$

where F is the Faraday constant, and n is the number of protons transported per ATP molecule synthesized. When the ATP synthase-mediated reaction runs in the forward direction, protons are transported back across the membrane so as to collapse the electrochemical potential.

Ugurbil et al.[6] used ^{31}P NMR simultaneously to measure changes in ΔpH and in phosphorylation potential ([ATP]/[ADP][Pi]) in *Escherichia coli* cells grown on succinate or glucose under aerobic and anaerobic conditions. Effects of ATPase inhibitors and ATPase-deficient mutants and uncouplers were investigated. These results, together with independent determinations of the membrane potential,[37] were in qualitative agreement with equation (1), providing confirmation of the chemiosmotic hypothesis.

4.2 Perfused Cells and Spheroids

4.2.1 Cyclophosphamide Metabolism

Cyclophosphamide (**1**) is a potent alkylating agent used in the treatment of a wide range of human cancers. The key steps in the metabolism of this drug are depicted in Figure 6. In the liver, microsomal oxidation of (**1**) produces 4-hydroxycyclophosphamide (**2**), which is subsequently converted to aldophosphamide (**3**) in the tumor. The latter is converted to acrolein (**5**) and phosphoramide mustard (**4**), the active species which cross-links DNA. Boyd et al.[38] measured the apparent rate of alkylation of DNA in perfused tumor cells immobilized in agar threads. They incorporated (**2**) into the perfusate, monitored its conversion into (**3**) and (**4**), and then eliminated (**2**) and (**3**) by washing the cells with drug-free perfusate. Subsequent disappearance of (**4**) followed first-order kinetics, and was attributed to formation of an NMR-invisible adduct of DNA. While drug concentrations were in the millimolar range, well above clinical levels, this experiment demonstrates the unique ability of NMR spectroscopy to monitor complex pharmacological reactions inside intact tumor cells.

4.2.2 Phospholipid Metabolism

Ronen and Degani[39] have used ^{31}P and ^{13}C NMR spectroscopy to examine the effect of spheroid size on the metabolism of phospholipids. The largest variations occurred in levels of the phospholipid precursors, phosphocholine (PC) and phosphoethanolamine (PE). Phosphatidylcholine (PhCh) is syn-

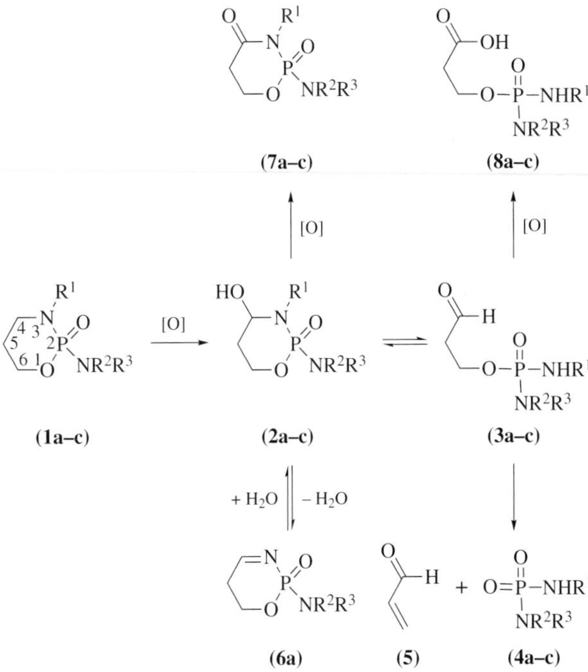

(a) $R^1 = H$; $R^2 = R^3 = CH_2CH_2Cl$
(b) $R^1 = R^2 = CH_2CH_2Cl$; $R^3 = H$
(c) $R^1 = R^2 = R^3 = CH_2CH_2Cl$

Figure 6 Generally accepted scheme for the metabolism of cyclophosphamide in the liver and tumor cells. (Reproduced by permission of American Chemical Society from Boyd et al.[38])

For References see p. 886

thesized via the Kennedy pathway by which choline is phosphorylated to PC by choline kinase, then PC is combined with cytosine triphosphate (CTP) via the action of CTP-phosphocholine transferase, forming cytosine diphosphocholine, which then combines with diacylglycerol (DAG) to form PhCh. Although phosphatidylethanolamine (PhEt) can be synthesized from PE via the Kennedy pathway, if ethanolamine is limiting, an alternative pathway, decarboxylation of phosphatidylserine, may be used. The reaction effectively converts serine (which is abundant in most media) into ethanolamine, which can then be used for synthesis of PhEt.

To monitor the kinetics of these reactions Ronen and Degani[39] grew T47D human breast spheroids in the presence of $[1,2-^{13}C]$choline, and $[1,2-^{13}C]$ethanolamine and $[3-^{13}C]$serine. The specific pathway for phospholipid metabolism could be deduced from the labeling pattern of the metabolites in cells and in lipid extracts. They found that the ratio of PhCh to PhEt was independent of spheroid size, but the rates of the reactions in the Kennedy and serine decarboxylase pathways varied with the availability of substrates. The serine decarboxylase reaction decreased with increasing ethanolamine in the medium, and increased with spheroid size; the latter effect was attributed to decreased permeability of large spheroids to exogenous ethanolamine.

4.2.3 Mechanism of Chemotherapy- and Radiation-Induced MRS Changes in In Vivo Tumors

During untreated growth, most transplantable tumor models in rodents exhibit a progressive decline in their bioenergetic status, a decrease in levels of high-energy phosphates compared with low-energy phosphates and an acid shift in pH (see *Spectroscopic Studies of Animal Tumor Models*).[16,24] When treated effectively with radiation or a number of chemotherapeutic agents, many of these tumors exhibit an anomalous reversal of this pattern which does not simply reflect a decrease in tumor size. These metabolic changes could result from direct effects on cellular metabolism or they could result from 'tissue effects' (changes in perfusion, cell density, etc.). To evaluate direct effects on cellular metabolism, Ng et al.[16] and Aiken et al.[24] examined the effects of γ-irradiation and of chemotherapy with 4-hydroperoxycyclophosphamide (an active derivative of cyclophosphamide), respectively, on RIF-1 cells perfused after immobilization on alginate beads or Cultisphere G-L beads (see above). The in vivo effects of these agents were not observed in the perfused cells. In fact, radiation induced a decline in bioenergetic status prior to cell death. The duration of these experiments corresponded to the timescale over which the bioenergetic 'improvements' were observed in the in vivo tumor. These data suggest that 'tissue effects' are responsible for the changes induced in this tumor by radiation and chemotherapy in vivo. This conclusion is consistent with reports that increases in tumor blood flow are correlated with radiation- and chemotherapy-induced improvements in tumor bioenergetic status.[16,24]

5 SUMMARY AND FUTURE DIRECTIONS

A wide variety of perfusion systems are available to support various applications of NMR spectroscopy to the study of cellular metabolism. Each method has distinct advantages and limitations. The particular system should be chosen to meet the needs of the application. The key advantage of NMR methods is that they permit investigation of metabolism in the intact cell. The key disadvantage of all NMR methods is limited spectral sensitivity. The examples we have chosen illustrate the utility of the method for investigation of key metabolic processes and for elucidation of mechanisms underlying in vivo spectral changes.

It is likely that major technical advances such as the use of gradient-based methods to discriminate intracellular and extracellular compartments described above will be employed in studies of cellular transport and metabolism. Multiple quantum editing techniques and imaging methods to measure diffusion constants and flow will probably be used in future cell studies. Isotopes such as ^{15}N and ^{17}O should receive more attention. Tagging of cells, multicellular organisms, or organelles with paramagnetically or isotopically labeled reagents may be used to follow embryonic development. Specialized NMR probes for monitoring oxygen in the concentration range relevant to radiation biology and respiration are likely to be developed (see **EPR and In Vivo EPR: Roles for Experimental and Clinical NMR Studies**). Molecular biological techniques will doubtlessly have a major impact on cellular NMR studies. Overall, the field of cellular MRS, like the other MR fields, is undergoing very rapid development and is likely to continue to do so for some time.

6 RELATED ARTICLES

Cation Movements across Cell Walls of Intact Tissues Using MRS; EPR and In Vivo EPR: Roles for Experimental and Clinical NMR Studies; Spectroscopic Studies of Animal Tumor Models; Tissue and Cell Extracts MRS.

7 REFERENCES

1. O. Kaplan, P. C. M. van Zijl, and J. S. Cohen, in 'NMR Basic Principles and Progress', Springer-Verlag, Berlin, 1992, Vol. 28, pp. 3–54.
2. B. S. Szwergold, *Annu. Rev. Physiol.*, 1992, **54**, 775.
3. N. A. Matwiyoff and T. E. Needham, *Biochem. Biophys. Res. Commun.*, 1972, **49**, 1158.
4. R. T. Eakin, L. O. Morgan, C. T. Gregg, and N. A. Matwiyoff, *FEBS Lett.*, 1972, **28**, 259.
5. R. B. Moon and J. H. Richards, *J. Biol. Chem.*, 1973, **248**, 7276.
6. K. Ugurbil, R. G. Shulman, and T. R. Brown, in 'Biological Applications of Magnetic Resonance', ed. R. G. Shulman, Academic Press, New York, 1979, pp. 537–589.
7. A. M. Castle, R. M. Macnab, and R. G. Shulman, *J. Biol. Chem.*, 1986, **261**, 7797.
8. J. K. Barton, J. A. den Hollander, J. J. Hopfield, and R. G. Shulman, *J. Bacteriol.*, 1982, **151**, 177.
9. S. M. Cohen and R. G. Shulman, *Philos. Trans. R. Soc. London Ser. B.*, 1980, **289**, 407.
10. G. Navon, S. Ogawa, R. G. Shulman, and T. Yamane, *Proc. Natl. Acad. Sci. USA*, 1977, **74**, 87.
11. R. S. Balaban, D. G. Gadian, G. K. Radda, and G. G. Wong, *Anal. Biochem.*, 1981, **116**, 450.

12. A. J. Meehan, C. J. Eskay, A. P. Koretsky, and M. M. Domach, *Biotech. Bioeng.*, 1992, **40**, 1359.
13. J. E. Jentoft and C. D. Town, *J. Cell Biol.*, 1985, **101**, 778.
14. H. Santos and D. L. Turner, *J. Magn. Reson.*, 1986, **68**, 345.
15. J. Taylor and C. Deutsch, *Biophys. J.*, 1988, **53**, 227.
16. C. E. Ng, K. A. McGovern, J. P. Wehrle, and J. D. Glickson, *Magn. Reson. Med.*, 1992, **27**, 296.
17. K. Ugurbil, D. L. Guernsey, T. R. Brown, P. Glynn, N. Tobkes, and I. S. Edelman, *Proc. Natl. Acad. Sci. USA*, 1981, **78**, 4843.
18. D. I. Hoult and P. C. Lauterbur, *J. Magn. Reson.*, 1979, **34**, 425.
19. M. Neeman and H. Degani, *Cancer Res.*, 1989, **49**, 589.
20. D. L. Foxall and J. S. Cohen, *J. Magn. Reson.*, 1983, **52**, 346.
21. M. Bental and C. Deutsch, *Magn. Reson. Med.*, 1993, **29**, 317.
22. K. S. Narayan, E. A. Moress, J. C. Chatham, and P. B. Barker, *NMR Biomed.*, 1990, **3**, 23.
23. K. A. McGovern, J. S. Schoeniger, J. P. Wehrle, C. E. Ng, and J. D. Glickson, *Magn. Reson. Med.*, 1993, **29**, 196.
24. N. R. Aiken, K. A. McGovern, C. E. Ng, J. P. Wehrle, and J. D. Glickson, *Magn. Reson. Med.*, 1994, **31**, 241.
25. R. Gonzalez-Mendez, D. Wemmer, G. Hahn, N. Wade-Jardetzky, and O. Jardetzky, *Biochim. Biophys. Acta*, 1982, **720**, 274.
26. P.-S. Lin, M. Blumenstein, R. B. Mikkelsen, R. Schmidt-Ullrich, and W. W. Bachovchin, *J. Magn. Reson.*, 1987, **73**, 399.
27. S. Ronen and H. Degani, *Magn. Reson. Med.*, 1989, **12**, 274.
28. J. P. Freyer, N. H. Fink, P. L. Schor, J. R. Coulter, M. Neeman, and L. O. Sillerud, *NMR Biomed.*, 1990, **3**, 195.
29. J. Andrasko, *J. Magn. Reson.*, 1976, **21**, 479.
30. P. C. M. van Zijl, C. T. W. Moonen, P. Faustino, J. Pekar, O. Kaplan, and J. S. Cohen, *Proc. Natl. Acad. Sci. USA*, 1991, **88**, 3228.
31. C. J. Deutch and J. S. Taylor, in 'NMR Spectroscopy of Cells and Organisms', ed. R. J. Gupta, CRC Press, Boca Raton, 1987, Vol. II, p. 55.
32. G. A. Smith, R. T. Hesketh, J. C. Metcalfe, J. Feeney, and P. G. Morris, *Proc. Natl. Acad. Sci. USA*, 1983, **80**, 7178.
33. R. K. Gupta, in 'NMR Spectroscopy of Cells and Organisms', ed. R. K. Gupta, CRC Press, Boca Raton, 1987, Vol. I, pp. 1–32.
34. C. J. Springer, *Annu. Rev. Biophys. Biophys. Chem.*, 1987, **16**, 375.
35. J. Pekar, P. F. Renshaw, and J. S. Leigh, *J. Magn. Reson.*, 1987, **72**, 159.
36. P. Mitchell, 'Chemiosmotic Coupling and Energy Transduction', Glynn Research, 1968.
37. A. M. Castle, R. M. Macnab, and R. G. Shulman, *J. Biol. Chem.*, 1986, **261**, 3288.
38. V. L. Boyd, J. D. Robbins, W. Egan, and S. M. Ludeman, *J. Med. Chem.*, 1986, **29**, 1206.
39. S. M. Ronen and H. Degani, *Magn. Reson. Med.*, 1992, **25**, 384.

Biographical Sketch

Jerry D. Glickson. *b* 1941. B.A., 1963, M. A., 1964, Ph.D., 1969, Columbia University. Introduced to NMR by Roy King at Iowa State University, USA, 1965–68. Postdoctoral fellow at DuPont Central Research with Bill Phillips and Cam McDonald, 1968–70. Faculty in Biochemistry and the Comprehensive Cancer Center, University of Alabama in Birmingham, 1970–84, Faculty in Radiology, Biological Chemistry, and Biophysics and Biophysical Chemistry, The Johns Hopkins School of Medicine, USA, 1984–present. Approx. 150 publications. Current research specialty: in vivo NMR spectroscopy of cancer.

Tissue and Cell Extracts MRS

Patrick J. Cozzone, Sylviane Confort-Gouny, and Jean Vion-Dury

Centre de Résonance Magnétique Biologique et Médicale, Faculté de Médecine, Marseille, France

1 INTRODUCTION

Biochemistry has been based for a long time on the identification of molecules involved in the complex architecture and organization of living systems. Prerequisites to the identification of these molecules are their extraction from cells, their chemical and structural analysis, and their quantitation. High-resolution magnetic resonance spectroscopy (MRS) of tissue and cell extracts has been playing a growing role over the past 15 years in the identification of key molecules participating in biochemical processes (metabolism). In the 1990s, studies by MRS of extracts have generated more than 100 publications per year and are often considered as a required complement to in vivo and ex vivo experiments.

MRS of extracts is gradually developing into a major analytical technique. It combines the most advanced magnetic resonance technology (spectrometers) and methods (pulse sequences) with the methods of biochemical extraction and separation based on chemical and physicochemical properties of molecules. In most cases, the MR spectrometer analyses complex mixtures of compounds in a way which is similar to the detector of a gas chromatograph or the chamber of a mass spectrometer. In some cases, chromatographic separation and subsequent analysis of compounds by MRS have been combined online in the same apparatus as in the coupling of a MR spectrometer with a high-performance liquid chromatograph.

There are many compelling reasons for an MR spectroscopist to analyze tissue and cell extracts. The first pertains to the assignments of resonances in spectra recorded ex vivo and in vivo on organs, animals, and humans, accounting for about 20% of publications on the subject. The second reason of interest is the description and monitoring of normal and pathological metabolism in cells and organs (respectively 34% and 21% of publications), essentially using isotope labeling. Attempts at diagnosing pathologies based on the analysis of human tissue extracts are being made (6% of publications). Finally, the field of pharmacology (bioconversion of drugs, pharmacokinetics, effect of drugs on metabolism) benefits significantly from the MRS analysis of extracts (19% of publications).

In this chapter, we are not aiming to cover exhaustively the field of MRS of extracts. Instead of presenting a comprehensive review, we will try to illustrate what is unique to MRS of tissue and cell extracts in the general context of analytical techniques and modern studies of metabolism.

For References see p. 891

2   TECHNICAL ASPECTS

The extraction of compounds to be analyzed by MRS has to meet two requirements. It must be *selective* in order to separate molecules of interest from unwanted compounds or cellular substructures (cell debris, membranes etc.). Also, the extraction procedure has *to avoid/limit the degradation and chemical transformation* of molecules. Treatment by perchloric acid is a widely used general method of extraction of water-soluble metabolites.[1] This method requires neutralization and allows removal of salts and proteins by centrifugation (pellet). Perchloric extraction usually follows freeze-clamping of tissues to stop metabolism. It is routinely conducted at a temperature below 5 °C in order to avoid/limit enzymatic or chemical degradation of extracted metabolites. In some cases, inhibitors of enzymatic activity can be added in the supernatant. Other methods are more specific of particular metabolites. For example, extraction by NaOH is used to obtain glycogen and extraction of lipids requires an organic phase such as a 2:1 mixture of chloroform and methanol.[2]

Extracts have to be stored at very low temperature (deep freezer at −80 °C or liquid nitrogen) to limit degradation. Progressive thawing is also critical since chemical modifications of metabolites may occur during this process.

It should be kept in mind that extraction procedures may give a distorted image of the biological reality. Aggressive procedures such as extraction by strong acids or organic solvents modify very significantly the physicochemical environment of metabolites, as compared with the one they experience in the integer cell. As an example, the ionic environment is modified at very low pH (perchloric acid) or when protons and water molecules are lost in organic phases.

Finally, all metabolites are not equal with respect to extraction procedures. They display wide variations in chemical stability, binding properties, conformational rearrangement etc. Hence, one should exert caution when comparing metabolic profiles obtained by in vivo MRS with results obtained on extracts by in vitro MRS. The comparison can be at best misleading if appropriate control experiments are not run since the information offered by extracts is, in essence, different from the information obtained by localized in vivo MR spectroscopy.

MRS of extracts does not require customized MR spectrometers. Standard high-field spectrometers (4.7–18 T) can be used with either narrow-bore or wide-bore magnets. No specific developments in the hardware or software are necessary and MRS of extracts can be performed at any MR laboratory with a basic knowledge of NMR techniques.

3   TISSUE EXTRACTS FOR RESONANCE ASSIGNMENTS

Specific assignment of the resonances recorded in vivo on an MR spectrum remains a critical issue, and a growing one, in consideration of the multiplicity of new applications of MRS in metabolic, pharmacological, and medical studies. Assignment is undoubtedly facilitated by the analysis of extracts using data banks (chemical shifts, coupling patterns, coupling constants), pH titration, and selective addition of authentic compounds to extracts.[3] Also, one can use the whole arsenal of pulse sequences and multinuclear approaches that the chemists have

developed in high-resolution MRS of organic compounds.[4–6] With the limitation already mentioned, in vitro MRS of extracts provides the best basis for the assignments of in vivo and ex vivo spectra. (See Figure 1.)

Extracts reflect a 'frozen situation' in cell metabolism which is often an advantage. Spectral resolution is always superior in vitro vs. in vivo and ex vivo by several orders of magnitude and facilitates the interpretation of spectra recorded on living cells, tissues, and organs. Interestingly, assignments of resonances in extracts have revealed the presence of unexpected compounds in plants, yeasts, bacterias, and human tissues,[7,8] the identification of which has required chromatographic separation and chemical structure identification by standard analytical techniques, including high-resolution MRS. In these instances, MRS of extracts has directly contributed to a better understanding of metabolic events occurring in the cell.

4   TISSUE EXTRACTS IN METABOLIC STUDIES

MRS of tissue extracts has played a major role in the study of metabolism of cells and organs. Two strategies have been used. The first strategy is 'passive' and consists of the detection and dosage of the metabolites present in the tissue extract reflecting, as a snapshot, the metabolic status of the living tissue at the time of extraction. Quantitation can be achieved with respect to wet weight of tissue, protein content, or an internal reference (e.g., an endogenous compound of known concentration). Proton and ^{13}C MR spectra bear information on the intermediary metabolism (amino acids, lipids, Krebs cycle intermediates, etc.). Phosphorus-31 MR spectra carry different information depending on the type of extraction. Aqueous extracts contain phosphorylated molecules of energy metabolism (ATP, creatine phosphate, phosphomonoesters and phosphodiesters, inorganic phosphate, etc.). (See Figure 2.) The polar head groups of many phospholipids can be assayed on lipidic extracts. In fact, ^{31}P MRS provides a viable alternative to the quantitative assay of phospholipids. (See Figure 3.)

The second strategy is 'active' and involves the perfusion of a selectively ^{13}C-enriched substrate prior to the freeze of metabolic activity and subsequent extraction. By choosing appropriately both the enriched molecule and the chemical site of the enrichment on the molecule, specific metabolic pathways can be singled out and monitored and metabolic fluxes can be quantitated under a variety of conditions, in relation to pathological state, cell differentiation,[9] cell growth, aging etc. This approach is widely used on perfused organs and cultured cells. In these experiments, extracts are prepared and analyzed at specific times to follow up the conversion and biodistribution of the ^{13}C marker. Basic metabolic events have been documented such as the phosphorylation status of liver under normal conditions[10] or in the presence of ethanol, the metabolism of acetate,[11] or the redox state of the brain,[12,13] etc.

Cellular compartmentation can also be studied from a metabolic standpoint and particularly interesting information has been obtained in brain tissue to shed additional light on the respective role of glial and neuronal cells.[14]

Particular attention has been devoted to the study of metabolic disorders in relation to cancer. The reader is referred to the abundant literature on this subject[15,16] which illustrates the large extent to which MRS of extracts has become an indispen-

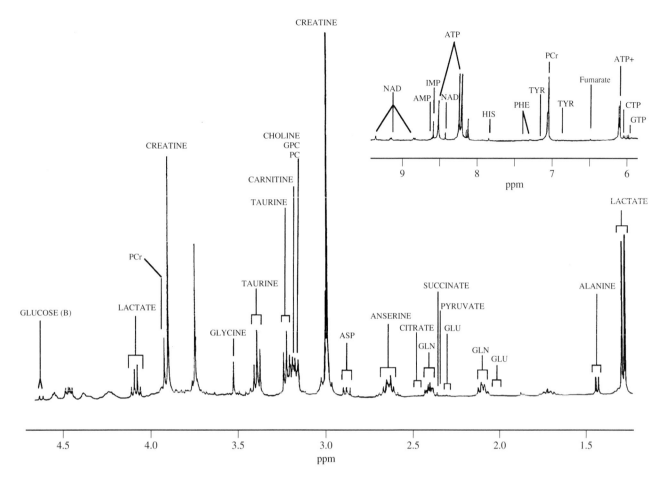

Figure 1 Proton MR spectrum (400 MHz) of a perchloric extract of rat gastrocnemius muscle obtained after freeze-clamping. The supernanant is lyophilized in order to eliminate water. Then the metabolites are dissolved in D_2O. Main Figure: aliphatic region of the spectrum (0–5 ppm). Inset: aromatic region of the spectrum (5–12 ppm). Abbreviations: PC, phosphorylcholine; GPC, glycerophosphorylcholine; PCr, phosphocreatine; ASP, aspartate, GLN, glutamine; GLU, glutamate; NAD, nicotinamide-adenine dinucleotide; AMP, adenosine monophosphate; IMP, inosine monophosphate; HIS, histidine; PHE, phenylalanine; TYR, tyrosine; CTP, cytidine 5′-triphosphate; GTP, guanosine 5′-triphosphate; ATP, adenosine triphosphate

sable analytical tool to study normal and pathological metabolism in cells and tissues.

5   APPLICATIONS TO PHARMACOLOGY

Besides the study of cell metabolism, and naturally occurring metabolites, MRS of extracts can document the metabolic fate and the metabolic impact of xenobiotics. Bioconversion of drugs can be readily studied with unsurpassed accuracy. The same MRS analysis can detect, identify (structure), and quantitate the metabolites to which a drug is converted and the kinetics of their appearance.[17]

Prototype studies have been conducted on antimitotic, antifungal, and antiviral drugs.[18,19] The analysis of drug metabolites benefits from the use of selective isotope labeling and multinuclear (^{31}P, ^{19}F, ^{13}C, ^2H) and multidimensional MRS experiments.

A second pharmacological application of MRS of extracts is the evaluation of the metabolic changes induced by the administration of a drug. As an example, the impact of the different interleukins on cell metabolism is not well known. However, obtaining additional information is critical to develop, adapt,

and optimize anticancer treatment using this new family of molecules. MRS plays a key role in evaluating the effect of these peptides on the general metabolism of the patient toward eliciting the adequate immune response to the presence of a cancer.[20] Drug resistance, another new problem of public health, is currently tackled by MRS of extracts in bacterial and eukaryotic cells, animal models, and even patients in an attempt to establish some metabolic basis to the alarming decrease in therapeutic efficacy of a number of antibiotic and antimitotic drugs.[21,22]

6   NEW DIRECTIONS: TOWARD MEDICAL DIAGNOSIS BY MRS OF TISSUE EXTRACTS

Tissue characterization in medicine is essentially based on the methods developed in pathology and cytology with a large role devoted to optical and electronic microscopy, and the use of selective dyes and immunological labeling. Biochemical studies of excised tissues is not common at the hospital, with the exception of blood and some body fluids. The analysis of tissue biopsies is a growing area of application for the MRS of extracts. Muscle and brain biopsies have been studied but most

For References see p. 891

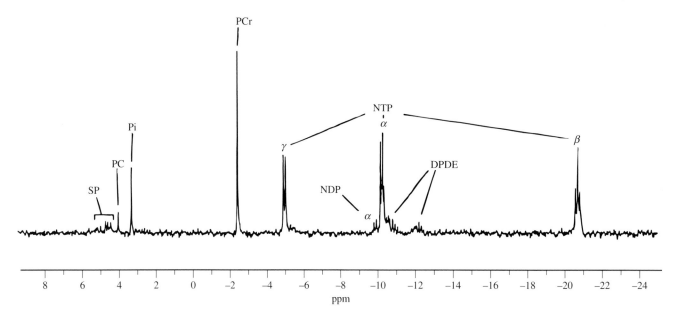

Figure 2 Phosphorus-31 MR (162 MHz) spectrum of a perchloric extract of RINm5F cells cultured in Dulbecco's Modified Eagle's Medium containing glucose. This spectrum (sum of 4800 scans, with complete proton decoupling) corresponds to the perchloric extract of 6×10^8 RINm5F cells cultured in flasks. Abbreviations: PC, phosphorylcholine; SP, sugar phosphate; PCr, phosphocreatine; DPDE, diphosphodiesters; NTP, nucleoside triphosphates; NDP, nucleoside diphosphates

of the work done to date has focused on human tissues biopsies: breast cancer,[23] uterus,[24] colon.[25] As a matter of fact, the knowledge of the metabolic features of a cancer tumor is of the utmost interest in diagnosis, prognosis and therapeutic decisions.[26] Specific metabolic profiles are observed from extracts of malignant tumors with deviations in the metabolism of amino acids and lipids/phospholipids.[24,27] (See Figure 3.)

The wealth of information that is available on the spectra (usually proton spectra) of biopsy extracts has prompted new strategies for data processing and analysis, based on pattern recognition, neural networks, and chemometrics.[28,29] This integrated and automated approach to spectral processing should facilitate very significantly the acceptance of MRS of biopsy extracts in routine medical diagnosis.

7 CONCLUSIONS

Obviously, MRS of extracts is not 'the MRS of the poor' that scientists and clinicians use when they do not have access to sophisticated methods of in vivo spectroscopy. It is making a definite contribution to the understanding of normal and pathological cell metabolism when it is submitted to a variety of perturbations. MRS of extracts is a cost-effective, highly informational, easy-to-use analytical procedure which is now entering the field of clinical biology by providing unique information that the medical profession can readily use for diagnosis, treatment, and prognosis. (See Figure 4.)

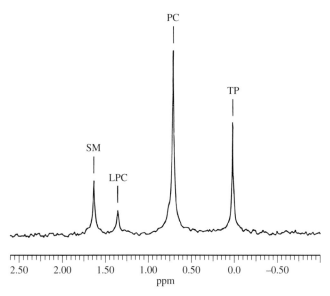

Figure 3 Phosphorus-31 MR (162 MHz) spectrum of a control plasma lipid extract. A line broadening of 4 Hz was applied. Abbreviations: SM, sphingomyelin; PC, phosphatidylcholine; LPC, lysophosphatidylcholine; TP, triethyl phosphate

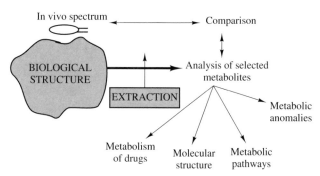

Figure 4 Synthetic scheme summarizing the position of extracts in analysis of biological activity

For list of General Abbreviations see end-papers

8 RELATED ARTICLES

Body Fluids; Cells and Cell Systems MRS; Whole Body Studies: Impact of MRS.

9 REFERENCES

1. M. Leach, L. Le Moyec, and F. Podo, in 'Magnetic Resonance Spectroscopy in Biology and Medicine', eds. J. De Certaines, W. M. M. J. Bovée, and F. Podo, Pergamon Press, Oxford, 1992, Chap. 18.
2. J. Folch, M. Lees, and G. H. S. Stanley, *J. Biol. Chem.*, 1957, **226**, 497.
3. O. Kaplan, Van Zijl, and J. S. Cohen, in 'NMR Basic Principles and Progress', eds. P. Diehl, E. Fluck, H. Günther, R. Kosfeld, and J. Seelig, Springer, Berlin, 1992, Vol. 28, Chap. 1.
4. G. Navon, T. Kushnir, N. Askenasy, and O. Kaplan. in 'NMR Basic Principles and Progress', eds. P. Diehl, E. Fluck, H. Günther, R. Kosfeld, and J. Seelig, Springer, Berlin, 1992, Vol. 27, Chap. 10.
5. C. Arus, Y. C. Chang, and M. Barany, *Physiol. Chem. Phys. Med. NMR*, 1985, **17**, 23.
6. M. Barany, C. Arus, and Y. C. Chang, *Magn. Reson. Med.*, 1985, **2**, 289.
7. N. De Tomasi, S. Piacente, F. De Simone, C. Pizza, and Z. L. Zhou, *J. Nat. Prod.*, 1993, **56**(10), 1669.
8. F. H. Kormelink, R. A. Hoffmann, H. Gruppen., A. G. Voragen, J. P. Kamerling, and J. F. Vliegentart. *Carbohydr. Res.*, 1993, **249**(2), 369.
9. J. P. Galons, J. Fantini, J. Vion-Dury, P. J. Cozzone, and P. Canioni. *Biochimie*, 1989, **71**, 949.
10. R. A. Iles, A. N. Stevens, J. R. Griffiths, and P. G. Morris. *Biochem. J.*, 1985, **229**, 141.
11. S. R. Williams, E. Proctor, K. Allen, D. G. Gadian, and H. A. Crockard. *Magn. Reson. Med.*, 1988, **7**, 425.
12. O. Ben-Yosef, S. Badar-Goffer, P. G. Morris, and H. S. Bachelard. *Biochem. J.*, 1993, **291**, 915.
13. C. Remy, C. Arus, A. Ziegler, E. Sam Lai, A. Moreno, Y. Le Fur and M. Décorps. *J. Neurochem.*, 1994, **62**(1), 166.
14. J. Urenkak, S. R. Williams, D. G. Gadian, and M. Noble. *J. Neurosci.*, 1993, **13**(3), 981.
15. C. E. Mountford, C. L. Lean, and W. B. Mackinnon. *Annu. Rep. NMR Spectrosc.*, 1993, **27**, 174.
16. M. Czuba and I. C. P. Smith. *Pharmacol. Ther.*, 1991, **50**, 147.
17. J. Vion-Dury, S. Confort-Gouny, and P. J. Cozzone, in 'Human Pharmacology', eds. H. Kuemmerle, T. Shibuya, and J. P. Tillement, Elsevier, Amsterdam, 1991, Chap. 13.
18. M. O. F. Fasoli, D. Kerridge, P. G. Morris, and A. Torosantucci. *Antimicrob. Agents Chemother.*, 1990, **34**(10), 1996.
19. R. A. Vere-Hodge, S. J. Darlinson, and S. A. Readshaw. *Chirality*, 1993, **5**(8), 577.
20. E. Prioretti., F. Belardelli., G. Carpinelli, M. Di Vito, D. Woodrow, J. Moss, P. Sestelli, W. Fiers, I. Gresser and F. Podo. *Int. J. Cancer*, 1988, **42**, 582.
21. A. Ferretti, L. L. Chen, M. Di Vito, S. Barca, M. Tombesi, M. Cianfriglia, A. Bozzi, R. Srom, and F. Podo. *Anticancer Res.*, 1993, **13**, 867.
22. G. L. May, L. C. Wright, M. Dyne, W. B. Mackinnon, R. M. Fox, and C. E. Mountford. *Int. J. Cancer*, 1988, **42**, 728.
23. T. A. D. Smith, C. Bush, C. Jameson, J. C. Titley, M. O. Leach, D. E. V. Wilman, and V. R. McCready. *NMR Biomed.*, 1993, **6**, 318.
24. C. E. Mountford, E. J. Delikatny, M. Dyne, K. T. Holmes, W. B. Mackinnon, R. Ford, J. C. Hunter, I. D. Truskett, and P. Russell. *Magn. Reson. Med.*, 1990, **13**, 324.
25. A. Moreno, M. Rey, J. M. Montane, J. Alonso, and C. Arus. *NMR Biomed.*, 1993, **6**, 111.
26. D. G. Gadian, T. E. Bates, S. R. William, J. D. Bell, S. J. Austin, and A. Connelly. *NMR Biomed.*, 1991, **4**, 85.
27. M. Kriat, J. Vion-Dury, S. Confort-Gouny, R. Favre, P. Viout, M. Sciaky, H. Sari, and P. J. Cozzone. *J. Lipid Res.*, 1993, **34**, 1009.
28. S. L. Howells, R. J. Maxwell, and J. R. Griffiths. *NMR Biomed.*, 1992, **5**, 59.
29. S. L. Howells, R. J. Maxwell, A. C. Peet, and J. R. Griffiths. *Magn. Reson. Med.*, 1992, **28**, 214.

Biographical Sketches

Patrick J. Cozzone. *b* 1945. Ph.D., 1971, University of Marseille, MBA, University of Chicago. Professor of Biochemistry, University of Marseille, 1975–90; Professor of Biophysics, Faculty of Medicine, Marseille, 1990–present; Director, Centre de Résonance Magnétique Biologique et Médicale (CRMBM), 1986–present.

Sylviane Confort-Gouny. *b* 1957. Ph.D., 1984, University of Grenoble, MRS instrumentation. Research engineer, Centre National de la Recherche Scientifique; in charge at Centre de Résonance Magnétique Biologique et Médicale, Marseille, of methodological developments and clinical transfers of MRS preclinical protocols, 1986–present.

Jean Vion-Dury. *b* 1956. M.D. 1983, School of Medicine of Marseille, Ph.D., 1988, University of Marseille. Assistant Professor of Biophysics, Faculty of Medicine, Marseille, 1994–present. NMR research at the Centre de Résonance Magnétique Biologique et Médicale, 1986–present.

Tissue NMR Ex Vivo

Ian C. P. Smith and Tedros Bezabeh
National Research Council of Canada, Winnipeg, Canada

1 INTRODUCTION

NMR of tissue specimens is now an established, powerful adjunct to histopathology for classification of human cancers. The present article is an update of earlier reports.[1,2] The initial study that provided insight into the potential of NMR for the early detection and diagnosis of cancer was reported in 1980 for mouse thymus.[3] Spectral changes preceded the cellular changes observed by histopathology, suggesting that the tissue NMR method could detect preinvasive changes prior to any morphological manifestation. This was supported by a study of a rat mammary model for adenocarcinoma metastasis.[4] Spectra from the excised tumor closely resembled those from suspensions of the same cells grown in vitro. The T_2 relaxation time

For References see p. 897

for the strong peak at 1.3 ppm correlated well with the metastatic potential of this tumor in the rat.

This early success with a rat model led to similar studies with human tissues. Human colorectal tumors yielded spectra with some parameters similar to those obtained from rat mammary adenocarcinoma. Over 90% of the human colorectal tumors gave long T_2 relaxation values for the resonance at 1.3 ppm, implying a high metastatic potential.[5] Patient follow-up showed no false positives; that is, no patients whose tumor generated only a short T_2 relaxation value for the resonance at 1.3 ppm developed secondary tumors.[6] Many of those with long T_2 relaxation values did develop secondary tumors.

From these studies it was apparent that the strong signals from fat were superimposed upon those of potentially diagnostic species present in the spectra from colorectal biopsies, a problem now known to exist also for lymph node and breast tissues. The implementation of two-dimensional (2D) NMR spectroscopy was a valuable complement, especially when 1D spectra were complex.[7]

Tissue NMR ex vivo has so far been primarily performed on ^{1}H nuclei. This is mainly because of the high sensitivity of ^{1}H NMR spectroscopy and the stability of the proton-containing compounds/metabolites in an excised tissue that are of diagnostic interest. Natural abundance ^{13}C NMR spectroscopy would take unreasonably long acquisition time and, therefore, is not ideal for a study of an excised/isolated piece of tissue. The instability (hydrolysis) of the high-energy phosphates [e.g., phosphocreatine (PCr), ATP] after the removal of the tissue from the system makes the use of ^{31}P NMR spectroscopy impractical.

Tissue NMR ex vivo offers an advantage over NMR of cell cultures and extracts in that it provides a situation similar to that in vivo. In particular, it helps in the examination of the biological properties of tissue that are dependent on architecture and morphology, which is impossible to do with extracts and isolated cells. Another advantage of tissue NMR ex vivo is its nondestructive character; the tissue can be submitted for histopathological assessment following the NMR experiment, enabling a direct correlation between the histopathological and MR spectral features of the same tissue specimen. As a result, no additional biopsy need be taken for the study since the clinical diagnosis can be made on the MRS sample, with no compromises or artifacts.

2 METHODS

All successful studies of viable tissues have relied heavily on careful sample handling prior to ^{1}H MRS experiments.[8] Tissues are placed into sterile tubes immediately after excision, immersed in liquid nitrogen, and stored at $-70\,^\circ$C. Either a glass wool platform supports the sample surrounded by buffer,[7] or the specimen is inserted into a capillary containing buffer,[8] which in turn is inserted into an NMR tube containing the same buffer and a reference. This latter method of sample positioning has several advantages including measurable volume for quantification and easier homogeneity adjustment (Figure 1).

Time constraints imposed by the limited viability of cell and tissue specimens dictate careful selection of acquisition parameters for 2D NMR spectroscopy. In order to obtain a

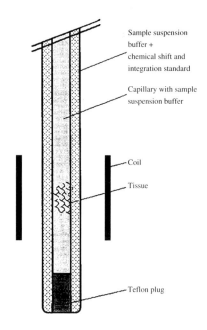

Figure 1 Schematic of the positioning of a small tissue specimen in a capillary tube. (Reproduced with permission from Kuesel et al., *Magn. Reson. Med.*, 1992, **27**, 349–355. Copyright © 1992, Wiley-Liss)

pathological assessment of the sample used for NMR, time in the instrument must be minimized. Therefore, the trade-off between signal strength, spectral resolution, and sample viability must be addressed for each tissue type. Cells and tissues contain a variety of biochemical species at a range of concentrations, with T_2 relaxation times ranging from <100 ms to >1 s and, therefore, care must be taken with signal processing. It is imperative that histological examination be undertaken on the NMR specimen itself, rather than relying upon the routine hospital report. Spectral analysis is best made by nonsubjective multivariate methods (see below).

We describe below studies on tissue from a variety of organs, each representing a solution to different technical/clinical problems: (1) the distinction between preinvasive and invasive cancers; (2) the presence of significant levels of fat, obscuring signals of interest; and (3) the detection of cellular abnormalities which are not morphologically manifest.

3 CERVIX

Cervical epithelium has preinvasive states that are well documented by histopathology. Independent clinical studies in two laboratories have shown that ^{1}H NMR can distinguish between preinvasive lesions, including carcinoma in situ (CIS), and frankly invasive cervical cancer with $p < 0.0001$ (Student t-test).[7–9] The ^{1}H NMR acquisitions are completed in 15 min, and the tissue is not subject to the vagaries of sampling since the entire punch biopsy specimen is examined. Two different approaches have been used for spectral analysis. In one, manual measurements of peak heights were performed. Typical spectra of preinvasive and invasive cervical epithelium are characterized by intense lipid

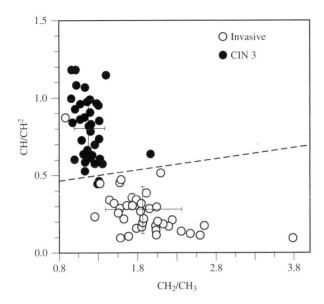

Figure 2 Peak height ratios from one-dimensional ¹H NMR spectra.[12] Severely dysplastic epithelium (CIN3) is compared with invasive carcinoma. The cross bars show the mean ± standard deviation. (Reproduced with permission from Delikatny et al., Proton NMR and human cervical neoplasia: ex vivo spectroscopy allows distinction of invasive carcinoma of the cervix from carcinoma in situ and other preinvasive lesions. *Radiology*, 1993, **188**, 791–796)

resonances at 0.9, 1.3, and 5.2 ppm. In contrast, spectra of preinvasive lesions are free of signals from high-resolution lipid.[9] A second spectral region which provides diagnostic information lies between 3.4 and 4.2 ppm. The ratio of the strongest resonance between 3.4 and 4.2 ppm and the methylene resonance at 1.3 ppm, is plotted against the ratio of the methylene to methyl resonances at 1.3 and 0.9 ppm, respectively (Figure 2). The separation between the mean ratio values for CIN3 and invasive cancer is significant. The *p* values obtained when comparing invasive cancer to the low, moderate, or severe levels of dysplasia are all <0.0001, with a specificity of 0.94 and sensitivity of 0.98.[9]

Alternatively, a mathematical treatment of the data sets may be used. A variety of multivariate analyses, such as linear discriminant analysis, neural nets, and genetic programming were applied to these data. Using a consensus approach for a series of 150 spectra of cervical tissue, spectra were allocated to characteristic groups. Groupings were then compared with the data from histopathology,[10] and yielded sensitivities and specificities greater than 0.99 for the various levels of dysplasia. In a recent study of 196 cervical punch biopsies (nondysplastic and dysplastic of varying grades), it was found that classification of the spectra into four diagnostic groups (i.e., nondysplastic, CIN1, CIN2, and CIN3) was difficult.[11] This could be because the inter-class biochemical changes that help to differentiate between these groups are very small compared with the sensitivity of NMR spectroscopy. However, multivariate classification of the spectra into two groups, i.e., nondysplastic and CIN1 versus CIN2 and CIN3, yielded a much better classification accuracy.

Chemical shift imaging (CSI) has the potential to revolutionize the detection of cancers since it can provide in-

formation on the biochemical composition and distribution, and the spatial location, of neoplasms. In CSI the image is constituted from a single specific frequency. Water-based CSI images (excitation at 4.8 ppm) of cervical punch biopsies from invasive adenocarcinoma and low-grade squamous dysplasia show that both biopsies appear with equal intensity. However, when a lipid-based CSI (excitation at 1.3 ppm) is performed on the biopsies, only the malignant specimen shows a substantial accumulation of MR-visible lipids.[12]

4 THYROID

The thyroid is the first organ where NMR has distinguished between benign adenoma and carcinoma which are morphologically identical under the light microscope.[13] Histological diagnosis of follicular thyroid cancer relies upon visualization of capsular invasion or the patients having demonstrated secondary tumors. Therefore, the morphological similarity of malignant and benign follicular cells leads to surgery solely for diagnostic purposes. One-dimensional ¹H NMR spectroscopy clearly distinguishes normal thyroid tissue from morphologically distinct invasive papillary carcinoma, based on differences in the relative intensities of resonances at 1.7 and 0.9 ppm,[14] demonstrating the potential to distinguish genuinely benign follicular adenomas from follicular carcinomas. The MR parameters separated follicular neoplasms into two categories, each of which was directly comparable with either normal thyroid or papillary carcinoma. The patients with clinically proven or histologically invasive follicular carcinoma were categorized by ¹H MR together with papillary cancers, as were a number of patients with clinically or morphological atypical neoplasms. It was subsequently established that the same diagnostic information could be obtained from fine needle aspiration biopsies, without recourse to surgery.[15] Multivariate analysis undertaken in a blind study delineated five out of six patients with follicular carcinoma (identified clinically by the presence of secondary tumors).[13] A clinical trial is underway in Australia where patients with normal NMR profiles from fine needle aspiration biopsies are given the option not to undergo surgery for diagnostic purposes.

Although the 1.7/0.9 ppm intensity ratio can distinguish thyroid cancer from normal tissue with a high sensitivity,[16] its specificity is relatively low. The probability of correctly identifying patients without cancer was significantly raised by using the 2.05/0.9 intensity ratio.[17] Furthermore, 2D NMR spectroscopy yielded an increased incidence of cross-peaks from cholesterol/cholesteryl esters and di-/triglycerides and a decreased incidence of two unassigned cross-peaks, unique to the thyroid, for cancer specimens.[17] The best accuracy was obtained via a computed consensus diagnosis (see section 9.5 below). Analysis of 107 thyroid biopsies resulted in 100% sensitivity and specificity in the training set, and 100% specificity and 98% sensitivity on samples of known malignancy in the test set.[13]

5 COLON

Colon tissue presents the technical difficulty of naturally high levels of fat,[6] necessitating 2D NMR spectroscopy,[6] or T_2

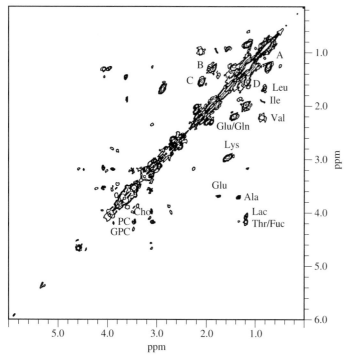

Figure 3 Proton 360 MHz NMR symmetrized COSY spectrum of colon tumor in phosphate-buffered saline/D$_2$O, 37 °C. Lac, lactic acid; Leu, leucine; Val, valine; Glu/Gln, glutamic acid/glutamate; Lys, lysine; Ile, isoleucine; Ala, alanine; Thr, threonine; Fuc, fucose; Cho, choline-containing compounds; PC, phosphatidylcholine; GPC, glycerophosphorylcholine; A, B, C, and D, triacylglycerols

relaxation filters[18] or very careful sample preparation[19] to define the diagnostic resonances. The 2D ^{1}H–^{1}H COSY spectra of colon-invasive adenocarcinoma indicate such diagnostic parameters as choline, choline-based metabolites, and altered lipid profiles (Figure 3). Tissues histopathologically classified as normal, while remaining distinct from the malignant spectral profile, were found to fit into two categories, one of which had some of the spectral characteristics of malignancy.[18] These results suggest that ^{1}H MRS identifies colorectal mucosa with abnormality that is not morphologically manifest.

The NMR studies of colorectal cancer biopsies have relied heavily on the correlation of biological and genetic information with the NMR data from cultured cell lines.[20] It was shown that 2D NMR data for cell surface fucosyl residues identified the various preinvasive states as well as a feature unique to the carcinoma in situ.[21] These pathological states, identifiable by NMR, include the distinction between malignant tumors and normal tissue; carcinoma in situ and frankly invasive adenocarcinoma; and the extents of dysplasia in preinvasive polyps.

Colon tumor and morphologically normal specimens were examined recently by ^{1}H NMR.[22] The spectra from 55 specimens were subjected to multivariate analysis: classification accuracy of 100% was obtained in both the training and the test sets. The concern with surgical specimens of colon has been the contamination of the sample with the nonmucosal layers of the colonic wall.[19,23] Of particular interest is the submucosal layer, which gives rise to intense lipid signals that have no diagnostic value but mask other potentially useful metabolites present at lower concentrations. Therefore, in such

studies an effort must be made to examine only the mucosa. Another way to circumvent this problem is to study endoscopic biopsies where the composition is dominantly mucosal. A recent study of such biopsies aimed to differentiate between ulcerative colitis and Crohn's disease, two distinct forms of inflammatory bowel disease (IBD) sharing many similar etiologic, clinical and pathological features. A strong similarity between these two diseases poses difficulty in making an accurate differential clinical diagnosis. To date, there are no definitive methods available to distinguish Crohn's disease from ulcerative colitis, especially when the endoscopy, radiology and histology are equivocal. Preliminary results showed that ^{1}H NMR spectroscopy can differentiate between these two diseases with a good classification accuracy.[24]

Another useful contribution of tissue NMR ex vivo has been in providing an insight into premalignant colon tissue. A study performed on rat colon tissue showed that the NMR spectra of aberrant crypt foci (preneoplastic lesions in the colon) have spectral features intermediate between those of normal mucosa and tumor, supporting the widely held adenoma–adenocarcinoma sequence in colorectal cancer.[25] Similar studies are also underway on human polyps in our laboratory.

CSI performed on normal human colon tissue helped to characterize the different layers of the colon wall.[26] Whereas the water-based image helped to delineate the mucosal layer, the submucosal layer was identified on a lipid-based image. The presence of high levels of neutral lipid distributed in the submucosal layer agrees with the spectral profile of specimens largely composed of submucosa (i.e., with intense lipid resonances).[19,20,27]

6 PROSTATE

Distinction between benign and malignant abnormalities of the prostate is sometimes difficult. Human prostate tissue contains many metabolites that are detectable using NMR. It is particularly rich in the citrate anion. Human prostate tissue specimens obtained from either radical prostatectomy (complete removal of the prostate) or transurethral resection of the prostate have been examined by ^{1}H NMR.[28] Comparison of the spectra from benign prostatic hyperplasia (BPH) and prostatic adenocarcinoma indicated significant differences. As shown in Figure 4, the resonance of the citrate anion (at ~2.5 ppm) has the potential to discriminate between the two classes. As indicated in the figure, citrate levels become significantly reduced (almost to below detection levels) in prostatic adenocarcinoma. Whether the citrate levels show any correlation with the grade or stage of the cancer is yet to be determined.

Other NMR-visible metabolites with diagnostic potential in the prostate include glutamic acid, taurine and choline. The use of computed multivariate methods of tissue NMR data analysis developed at our Institute has given highly accurate results with a sensitivity and specificity of 100 and 95.5%, respectively, in classifying benign from malignant prostatic tissue.[28]

Benign prostatic hyperplasia, which is an age-related enlargement of the prostatic gland, may derive from two different types of tissue: stromal and glandular. Knowing the type of BPH in a particular patient, however, is critical for the proper management of the disease. Proton NMR of tissue obtained from dominantly glandular and stromal specimens have indi-

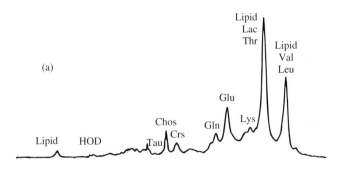

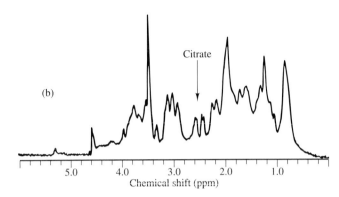

Figure 4 Proton MR spectra (360 MHz) of prostate tissue specimens. (a) Cancer (Gleason grade 3 + 3); (b) benign prostatic hyperplasia. Chos, choline-containing compounds; Crs, creatines; Gln, glutamine; Glu, glutamic acid; Leu, leucine; Lys, lysine; Tau, Taurine; Thr, threonine; Val, valine. Assignments do not imply that these are the only substances contributing to a particular resonance. (Reproduced with permission from Hahn et al., Classification of benign and malignant human prostate tissue by multivariate analysis of ^{1}H magnetic resonance spectra. *Cancer Res.*, 1997, **57**, 3398)

cated significantly higher levels of citrate anion in the glandular cases (Figure 5).[28]

Besides its value in solving diagnostic problems, tissue NMR ex vivo may also have a role to play in assessing treatment responses and post-therapy recurrences. For example, one of the problems in assessing post-radiotherapy response is distinguishing between radiation necrosis and recurring tumor. We are currently examining prostate biopsies obtained by transrectal ultrasound guidance from patients who underwent radiation therapy 2 years previously.

Other studies of fresh tissue prostate specimens have also indicated higher levels of spermine (a polyamine with NMR resonances at 1.8, 2.1 and 3.1 ppm) in both normal and benign hyperplasia compared with tumors. However, these resonances may not be present in tissue specimens that have been frozen and thawed since the positively charged spermine may bind to negatively charged macromolecules (DNA/RNA, lipids) and become invisible to NMR.[29]

Kurhanewicz et al. have been able to perform in vivo MR spectroscopic studies of the human prostate using an endorectal coil.[30–32] Although their spectra have lower resolution than that of the ex vivo spectra, they were able to see discrete resonances from some of the diagnostically useful metabolites such

as the citrate anion. By comparing MRI and MR spectroscopy plus MRI, they were able to show that the specificity and sensitivity of detecting cancer in the prostate gland, and the estimation of the extracapsular extension of the cancer, improve significantly when the combined approach is used.[31,32]

7 BREAST

The combination of physical examination, mammography, and fine-needle aspiration cytologic analysis is currently the most sensitive method for the preoperative diagnosis of breast lesions. While the sensitivity of such a diagnostic approach is satisfactory, its specificity remains inadequate. Proton NMR spectroscopy of tissue biopsies has the potential to augment/complement this triple assessment method and to increase both the sensitivity and the specificity. Experiments performed on fine-needle biopsies obtained from patients undergoing diagnostic biopsy or definitive treatment (lumpectomy, quadrantectomy, or mastectomy) for histologically proven invasive breast cancer indicate significantly higher ratios of the peak height intensities of spectral resonances at 3.25 to 3.05 ppm for infiltrating/invasive carcinoma compared with benign lesions (Figure 6). While the 3.25 ppm resonance results from choline-based metabolites, the 3.05 ppm resonance could be from creatine or lysine. Using this spectral ratio, the sensitivity and specificity obtained in differentiating between benign and malignant breast lesions were 95% and 96%, respectively.[33] The increase in this ratio may be primarily a consequence of

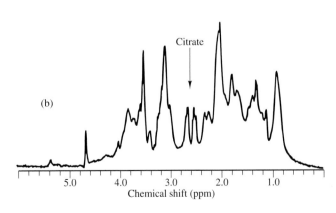

Figure 5 Proton MR spectra (360 MHz) of tissue specimens from benign prostatic hyperplasia. (a) 95% stromal, 5% glandular; (b) 90% glandular, 10% stromal. (Reproduced with permission from Hahn et al., Classification of benign and malignant human prostate tissue by multivariate analysis of ^{1}H magnetic resonance spectra. *Cancer Res.*, 1997, **57**, 3398)

For References see p. 897

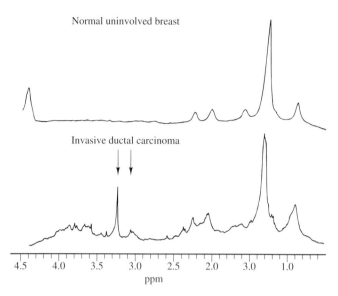

Normal uninvolved breast

Invasive ductal carcinoma

Figure 6 One-dimensional ^{1}H MR spectra in fine-needle biopsy specimens obtained in a patient who underwent mastectomy. Spectra for a normal uninvolved breast and a breast with invasive ductal carcinoma. The distinction between normal breast and invasive carcinoma is based on an increase in the N-trimethyl resonance at 3.25 ppm normalized to that of creatine at 3.05 ppm (peaks indicated by arrows). (Reproduced with permission from Mackinnon et al., Fine-needle biopsy specimens of benign breast lesions distinguished from invasive cancer ex vivo with proton MR spectroscopy. *Radiology*, 1997, **204**, 661)

the elevation of choline metabolites in malignant tissues, consistent with what has been observed in other tissue types (e.g., brain, and colon). Interestingly, all atypical or suspicious fine-needle aspiration cytologic results, in subsequently confirmed benign specimens, had NMR spectroscopic ratios indicating benign disease.

NMR spectroscopy performed before cytologic analysis of aspirates may well be a valuable adjunct because it improves the specificity of diagnosis. Therefore, NMR spectroscopy of fine-needle biopsy specimens could reduce the number of biopsies performed in benign lesions and could lead to a more conservative approach, such as continued observation or surveillance with repeat fine-needle aspiration cytologic analysis and NMR spectroscopy. Breast tissue is naturally rich in fat; suppression of the fat signal may be critical to detect other metabolites that may have diagnostic potential but are present at low concentrations.

8   OVARY

Two of the most crucial issues in ovarian cancer are the frequent late detection of the disease and the resistance of most of these tumors to current therapeutic modalities. Proton NMR spectroscopy of ovarian tissue obtained during surgery has been performed to address these issues.[34] Multivariate analysis was performed on the data to maximize the sensitivity and specificity of the classification; a sensitivity of 100% and a specificity of 95% were obtained in distinguishing ovarian cancer from normal ovarian tissue. Moreover, the multivariate

analysis was able to distinguish untreated ovarian cancer from recurrent ovarian cancer with a sensitivity of 92% and a specificity of 100%. A 2D study of ovarian tissue showed the presence of multiple cross-peaks in cancer specimens[35] attributable to cell surface fucosylation.[20] The complexity of this fucosylation correlated with tumor grade and loss of cellular differentiation. Such multiple cross-peaks were absent in both normal and benign ovarian specimens.[35]

9   METHODOLOGICAL CONSIDERATIONS

9.1   Magnetic Field Strength

Most of the work reported in this review was performed at magnetic field strength of 8.5 T (360 MHz for ^{1}H). Although higher magnetic fields would undeniably generate spectra with better detection sensitivity and resolution, sufficient diagnostic information appears to be available at this moderate field strength. Higher magnetic fields with gradient echo-editing capabilities would be preferable for 2D NMR experiments in order to increase the detection sensitivity and resonance dispersion, shorten the acquisition time, and reduce the effects of T_1 noise.

9.2   Tissue Spectral Stability

In our studies, tissue MR spectral profiles were found not to change significantly over the duration of the NMR experiment. This is particularly true for 1D experiments where the total time the tissue stays inside the magnet (for set up and data acquisition) does not exceed 1 hour. Another factor related to tissue spectral viability is the temperature at which the experiment is performed. Most of the studies reported here were performed at 37 °C, in order to mimic the in vivo physiological condition. However, lower temperatures could be used for experiments requiring longer acquisition times in order to preserve the tissue integrity/viability.

9.3   Fresh versus Frozen Tissue

Although it may be better to use fresh tissue specimens for such NMR experiments, comparison between spectra obtained from fresh and frozen tissue specimens (at −70 °C) has shown few differences. In a multivariate data analysis where the computer training was carried out with spectra from fresh specimens and tested on spectra from frozen specimens, and vice versa, classification accuracies were comparable.[36] Therefore, since performing the NMR experiments on freshly obtained tissue may be difficult for logistical/practical considerations, experiments performed on frozen tissue would be equally useful but more convenient. It is our experience that tissue specimens can be left frozen for as long as about 6 weeks without any significant biochemical alteration or degradation.

9.4   Magic Angle Spinning (Spectral Resolution)

Recently, the use of magic angle spinning was introduced to tissue NMR ex vivo by Cheng et al.[37] with the objective of improving the spectral resolution over that obtained using standard high-resolution NMR techniques. The technique, which is widely used in solid-state NMR spectroscopy, involves mechanically spinning the sample at the 'magic' angle of $54°44'$ (which meets the requirement $3\cos^2\theta - 1 = 0$, where θ is the angle between the static magnetic field and the internuclear vector). By doing so, the contribution from such interactions, as dipole–dipole couplings and chemical shift anisotropy, which produce spectral broadening, are reduced. Such studies have already been successfully applied to breast and brain tissues and yielded useful clinical information.[38,39] A potential disadvantage of this technique, however, is that it adds many resonances to an already crowded spectrum, possibly decreasing the discriminating power.

9.5 Data Analysis

Although tissue NMR spectra are rich with biodiagnostic information, advanced methods of data analysis are required to extract all available information reliably. Conventionally, peak height/area ratios of preselected resonances/metabolites have been used to analyze the data. However, such an approach does not make use of the entire information content. Multivariate methods of analysis that are robust, nonsubjective and make use of all spectral information have been successfully employed to classify normal and cancerous tissue spectra of different organs, namely, colon, prostate, brain, cervix and ovary.[10,11,22,28,34,40] Adequate data set size and proper data preprocessing/reduction are essential requirements for a successful multivariate analysis. The use of computed consensus diagnosis, which involves the cross-validated training of several classifiers (linear discriminant analysis, neural net-based methods and genetic programming) on the same data and subsequently combining their outcomes, has been successfully used in the classification of thyroid neoplasms.[13] The use of such a consensus diagnostic approach resulted in higher sensitivity and specificity than that obtained with a single classifier. It is also worth noting here that it was difficult to reach conclusive diagnoses by simply using the peak height ratio approach. In a similar study of prostate tissue specimens, comparison between the sensitivity and specificity obtained by multivariate analysis and by using the peak height ratios shows the former to be far superior.[28]

10 CONCLUSIONS

It has now been established that the tissue NMR parameters allow pathological and in some cases clinical classification of human cancers. More importantly, it has been demonstrated that ^1H NMR can discern changes to the cellular chemistry in human tissues prior to their histological manifestation. This has significant implications for identifying the extent of cellular abnormality in the preinvasive states, subjects with a predisposition to cancer, and pathologies that have to date eluded the light microscope.

Recent studies show that tissue NMR ex vivo studies using a portable bench top spectrometer could be clinically useful by providing histopathological information that is not evident to the pathologist.[41] With the help of multivariate classifiers, it would be convenient to obtain the NMR spectrum of the tissue immediately after the biopsy procedure and perform the classification/diagnosis nonsubjectively. This would be useful for rapid and inexpensive early detection of abnormalities and for making/confirming the diagnoses of borderline cases. By obviating unnecessary surgery, such a technique will help to develop a more conservative and efficient strategy for disease diagnosis and management.

The ^1H NMR spectra of biopsy specimens have demonstrated powerful diagnostic strength. The next step in the application of the technique to human medicine is in vivo spectroscopy. As will be seen in other chapters, the in vivo studies are considerably more difficult, especially since they are usually performed at much lower magnetic field values (1.5 to 3.0 T). Difficulties with radiofrequency field homogeneity, volume localization, detection of resonances with short T_2 values, and lower dispersion of resonances remain a challenge. However, the ultimate goal of accurate diagnosis in vivo is now within reach, based on the expansive data set obtained by ^1H NMR ex vivo.

11 RELATED ARTICLES

Body Fluids; Cells and Cell Systems MRS; Kidney, Prostate, Testicle, and Uterus of Subjects Studied by MRS; Tissue and Cell Extracts MRS.

12 REFERENCES

1. I. C. P. Smith and C. E. Mountford, 'Encyclopedia of NMR', 1995, Vol. 8, 4776.
2. I. C. P. Smith and D. E. Blandford, *Biochem. Cell Biol.*, 1998, **76**, 472.
3. C. E. Mountford, G. Grossman, P. A. Gatenby, and R. M. Fox, *Br. J. Cancer*, 1980, **41**, 1000.
4. C. E. Mountford, W. B. Mackinnon, M. Bloom, E. E. Burnell, and I. C. P. Smith, *J. Biochem. Biophys. Methods*, 1984, **9**, 323.
5. C. E. Mountford, G. L. May, P. G. Williams, M. H. N. Tattersall, P. Russell, J. K. Saunders, K. T. Holmes, R. M. Fox, I. R. Barr, and I. C. P. Smith, *Lancet*, 1986, **i**, 651.
6. I. C. P. Smith, E. J. Princz, and J. K. Saunders, *J. Can. Assoc. Radiol.*, 1990, **41**, 32.
7. C. E. Mountford, E. J. Delikatny, M. Dyne, K. T. Holmes, W. B. Mackinnon, R. Ford, J. C. Hunter, I. D. Truskett, and P. Russell, *Magn. Reson. Med.*, 1990, **13**, 324.
8. A. C. Kuesel, T. Kroft, J. K. Saunders, M. Préfontaine, N. Mikhael, and I. C. P. Smith, *Magn. Reson. Med.*, 1992, **27**, 349.
9. E. J. Delikatny, P. Russell, J. C. Hunter, R. Hancock, K. H. Atkinson, C. van Haaften-Day, and C. E. Mountford, *Radiology*, 1993, **188**, 791.
10. R. L. Somorjai, A. E. Nikulin, A. C. Kuesel, M. Préfontaine, N. Mikhael, and I. C. P. Smith, *Proc. XIth Ann Mtg. Soc. Magn. Reson. Med.*, Berlin, 1992, **1**, 56.
11. L. Friesen, Ph.D. Thesis, University of Manitoba, Winnipeg, Canada, 1999.
12. B. Kunnecke, E. J. Delikatny, P. Russell, J. C. Hunter, and C. E. Mountford, *J. Magn. Reson.*, 1994, **10**, 135.
13. R. L. Somorjai, A. E. Nikulin, N. Pizzi, D. Jackson, G. Scarth, B. Dolenko, H. Gordon, P. Russell, C. L. Lean, L. Delbridge, C. E. Mountford, and I. C. P. Smith, *Magn. Reson. Med.*, 1995, **33**, 257.
14. P. Russell, C. L. Lean, L. Delbridge, G. L. May, S. Dowd, and C. E. Mountford, *Am. J. Med.*, 1994, **96**, 383.

For References see p. 897

15. C. L. Lean, P. Russell, L. Delbridge, G. L. May, S. Dowd, and C. E. Mountford, *Proc. XIIth Ann Mtg. Soc. Magn. Reson. Med.*, New York, 1993, **1**, 71.

16. L. Lean, L. Delbridge, P. Russell, G. L. May, W. B. Mackinnon, S. Roman, T. J. Fahey III, S. Dowd, and C. E. Mountford, *J. Clin. Endocrinol. Metab.*, 1995, **80**, 1306.

17. W. B. Mackinnon, L. Delbridge, P. Russell, C. L. Lean, G. L. May, S. Doran, S. Dowd, and C. E. Mountford, *World J. Surg.*, 1996, **20**, 841.

18. C. L. Lean, R. C. Newland, D. A. Ende, E. L. Bokey, I. C. P. Smith, and C. E. Mountford, *Magn. Reson. Med.*, 1993, **30**, 525.

19. K. M. Briere, A. C. Kuesel, R. B. Bird, and I. C. P. Smith, *NMR Biomed.*, 1995, **8**, 33.

20. C. L. Lean, W. B. Mackinnon, E. J. Delikatny, R. H. Whitehead, and C. E. Mountford, *Biochemistry*, 1992, **31**, 11 095.

21. D. A. Ende, C. L. Lean, W. B. Mackinnon, P. Chapuis, R. Newland, P. Russell, E. L. Bokey, and C. E. Mountford, *Proc. XIIth Ann Mtg. Soc. Magn. Reson. Med.*, New York, 1993, **2**, 1033.

22. T. Bezabeh, I. C. P. Smith, E. Krupnik, R. L. Somorjai, D. G. Kitchen, C. N. Bernstein, N. M. Pettigrew, R. P. Bird, K. J. Lewin, and K. M. Briere, *Anticancer Res.*, 1996, **16**, 1553.

23. A. Moreno, M. Rey, J. M. Montane, J. Alonso, and C. Arús, *NMR Biomed.*, 1993, **6**, 11.

24. T. Bezabeh, C. N. Bernstein, R. L. Somorjai, and I. C. P. Smith, *Proc. Vth Ann Mtg. (Int.) Soc. Magn. Reson. Med.*, Vancouver, 1997, **2**, 1279.

25. E. Krupnik, K. M. Briere, R. P. Bird, C. Littman, and I. C. P. Smith, *Anticancer Res.*, 1999, **19**, 1699.

26. D. Ende, A. Rutter, P. Russell, and C. E. Mountford, *NMR Biomed.*, 1996, **9**, 179.

27. C. E. Mountford, W. B. Mackinnon, P. Russell, A. Rutter, and E. J. Delikatny, *Anticancer Res.*, 1996, **16**, 1521.

28. P. Hahn, I. C. P. Smith, L. Leboldus, C. Littman, R. L. Somorjai, and T. Bezabeh, *Cancer Res.*, 1997, **57**, 3398.

29. M. van der Graaf, R. G. Schipper, G. O. N. Oosterhof, J. A. Schalken, A. A. J. Verhofstad, and A. Heerschap, *Proc. VIth Ann Mtg. (Int.) Soc. Magn. Reson. Med.*, Sydney, 1998, **1**, 613.

30. J. Kurhanewicz, D. B. Vigneron, S. J. Nelson, H. Hricak, J. M. Macdonald, B. Konety, and P. Narayan, *Urology*, 1995, **45**, 459.

31. R. Males, D. Vigneron, S. Nelson, J. Scheider, H. Hricak, P. Carroll, and J. Kurhanewicz, *Proc. VIth Ann Mtg. (Int.) Soc. Magn. Reson. Med.*, Sydney, 1998, **1**, 487.

32. J. Kurhanewicz, R. Males, D. Sokolov, S. Nelson, J. Scheidler, K. Yu, H. Hricak, P. Carroll, and D. Vigneron, *Proc. VIth Ann Mtg. (Int.) Soc. Magn. Reson. Med.*, Sydney, 1998, **2**, 967.

33. W. B. Mackinnon, P. A. Barry, P. L. Malycha, D. J. Gillett, P. Russell, C. L. Lean, S. T. Doran, B. H. Barraclough, M. Bilous, and C. E. Mountford, *Radiology*, 1997, **204**, 661.

34. J. C. Wallace, G. P. Raaphorst, R. L. Somorjai, C. E. Ng, M. Fung Kee Fung, M. Senterman, and I. C. P. Smith, *Magn. Reson. Med.*, 1997, **38**, 569.

35. W. B. Mackinnon, P. Russell, G. L. May, and C. E. Mountford, *Int. J. Gynecol. Cancer*, 1995, **5**, 211.

36. R. L. Somorjai, D. Kitchen, B. Dolenko, A. Nikulin, G. Scarth, D. Ende, R. Newland, P. Russell, C. E. Mountford, T. Bezabeh, K. Briere, C. N. Bernstein, N. M. Pettigrew, K. J. Lewin, and I. C. P. Smith, *Proc. IIIrd Ann Mtg. (Int.) Soc. Magn. Reson. Med.*, Nice, 1995, **3**, 1938.

37. L. L. Cheng, M. J. Ma, L. Becerra, T. Ptak, I. Tracey, A. Lackner, and R. G. Gonzalez, *Proc. Natl. Acad. Sci., U.S.A.*, 1997, **94**, 6408.

38. L. L. Cheng, I. W. Chang, B. L. Smith, and R. G. Gonzalez, *J. Magn. Reson.*, 1998, **135**, 194.

39. L. L. Cheng, I. W. Chang, D. N. Louis, and R. G. Gonzalez, *Cancer Res.*, 1998, **58**, 1825.

40. R. L. Somorjai, B. Dolenko, A. K. Nikulin, N. Pizzi, G. Scarth, P. Zhilkin, W. Halliday, D. Fewer, N. Hill, I. Ross, M. West, I. C. P. Smith, S. M. Donnelly, A. C. Kuesel, and K. M. Brière, *JMRI*, 1996, **6**, 437.

41. C. E. Mountford, S. Doran, C. L. Lean, and P. Russell, *Biophys. Chem.*, 1997, **68**, 127.

Biographical Sketches

Ian C. P. Smith. *b* 1939. B.Sc. 1961, M.Sc. 1962, Manitoba; Ph.D. Cambridge, 1965 (Research Director, Alan Carrington); Fil. Dr. (H.C.) Stockholm, 1986; D.Sc. (H.C.) Winnipeg, 1990. Introduced to NMR by W. G. Schneider, NRC, Ottawa, 1960. Postdoctoral work with H. M. McConnell, Stanford, 1965–66 and R. G. Shulman, Bell Labs, 1966–67. Research Officer, Institute for Biological Sciences, NRC, 1967–87; Director-General, 1987–91; Director-General, Institute for Biodiagnostics, NRC, 1992–present. Approx. 400 publications. Research specialty: applications of modern spectroscopic and computational methods in medicine, with emphasis on magnetic resonance spectroscopy.

Tedros Bezabeh. *b* 1965. B.Sc. 1986, Asmara, Eritrea; M.A. 1989, Ph.D. 1993, Washington University in St Louis (Thesis Advisor: Joseph J. H. Ackerman). Postdoctoral work with Ian C. P. Smith, Institute for Biodiagnostics, National Research Council of Canada, 1993–98. Research Officer, Institute for Biodiagnostics, National Research Council of Canada, 1998–present. Approx. 10 publications. Research specialty: applications of NMR spectroscopy in the diagnoses of diseases, with emphasis on cancer.

Animal Methods in MRS

David G. Gadian

Royal College of Surgeons Unit of Biophysics, Institute of Child Health, London, UK

1 INTRODUCTION

Magnetic resonance spectroscopy studies of animal models have contributed in numerous ways to our understanding of the metabolic processes that take place in normal and diseased tissue. Applications to tumor metabolism are covered in accompanying articles (see *Spectroscopic Studies of Animal Tumor Models*), and are therefore not discussed here. Space limitations do not allow a comprehensive review; instead, we concentrate in this article on three major themes in this research area, namely (i) ^{31}P MRS and tissue energetics, (ii) ^{1}H MRS of brain metabolites, and (iii) ^{13}C studies of intermediary metabolism, and we conclude with some illustrative examples of the role of MRS in the investigation of stroke models.

2 PHOSPHORUS-31 MRS AND TISSUE ENERGETICS

Following earlier NMR studies of cellular and tissue metabolism in the 1970s,[1] in 1980 it was reported that the use of an unusual type of radiofrequency coil, termed a surface coil (see *Surface and Other Local Coils for In Vivo Studies*), provided a simple and effective means of probing skeletal muscle and brain metabolism noninvasively in small animals.[2] The ^{31}P spectra that were obtained in these studies (Figure 1) illustrate the type of information that is provided by ^{31}P NMR. The spectra of skeletal muscle and brain include ^{31}P signals from ATP, phosphocreatine (PCr), and inorganic phosphate (Pi), metabolites that play central roles in energy metabolism. In addition to measuring the relative concentrations of these metabolites, it is also possible to determine the intracellular pH from the chemical shift of the inorganic phosphate signal. A number of interesting points emerged from these surface coil studies. For example, the ratio of phosphocreatine to ATP was on the high side of values commonly obtained using invasive rapid freezing techniques, while the concentrations of inorganic phosphate and of free cytoplasmic ADP were lower than those measured with invasive techniques.

While the measurement of relative concentrations often provides an adequate basis for interpretation, there is increasing

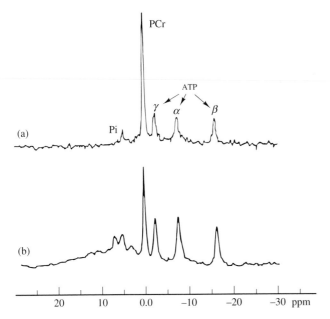

Figure 1 ^{31}P spectra obtained at 74 MHz showing signals from the α-, β-, and γ-phosphates of ATP, phosphocreatine (PCr), and inorganic phosphate (P$_i$) in (a) skeletal muscle and (b) the brain of an anesthetized rat. The brain spectrum also shows signals from phosphomonoesters (6–8 ppm) and phosphodiesters (2–4 ppm). (Reproduced with permission from *Nature*, Ackerman et al., *Nature (London)*, 1980, **283**, 167, Copyright (1980) Macmillan Magazines Limited)

emphasis on the determination of absolute concentrations. A number of studies have shown that NMR and freeze clamping measurements tend to give similar values for tissue ATP concentrations.[3] In some cases, however, the two types of measurement may differ somewhat because of the presence of NMR-invisible ATP pools. Also, comparison of the two types of measurement is complicated by the fact that nucleotides other than ATP can make contributions to the triphosphate signals; in recognition of this, these signals are sometimes labeled NTP rather than ATP. In contrast to these ATP observations, the concentrations of ADP and inorganic phosphate that have been measured in various tissues using noninvasive NMR methods are often much lower than those measured by analysis of tissue extracts.[4–6] These discrepancies in the measurements of ADP and inorganic phosphate could reflect the presence of NMR-invisible intracellular fractions, resulting from tight binding to macromolecules or perhaps from sequestration within certain intracellular compartments, including the mitochondria. Alternatively, in some cases they may reflect an unavoidable breakdown of high-energy phosphates in invasive measurements, for example in human biopsy analyses where there is an inevitable delay between removal and freezing of the sample. Regardless of the precise explanation for the differences between these NMR and chemical measurements, the low levels of free inorganic phosphate and ADP have important kinetic and thermodynamic implications; in particular, these findings strongly influence our understanding of those reactions that are subject to control by inorganic phosphate and ADP.[7]

^{31}P MRS provides an excellent experimental approach to the investigation of changes in energy metabolism associated with impaired oxygen delivery, as in ischemia or hypoxia. Metabolic changes can be monitored sequentially in a single animal, for example throughout a period of ischemia and subsequent reperfusion, offering considerable advances over invasive methods of analysis. One of the main areas of interest is the evaluation of agents or procedures that may protect against the damaging consequences of inadequate oxygen supply. This is of relevance to conditions such as stroke, to methods used for the preservation of organs prior to transplantation, and to cardioplegic techniques for use during open heart surgery. To exemplify this approach, a later section of this article describes some studies of experimental stroke.

Extensive studies have also been carried out of the metabolic changes that are associated with muscular exercise. While a number of such studies have been carried out in animals, the advantages of spectroscopy are most apparent for investigations in humans, where there are obvious benefits over biopsy procedures (see *Whole Body Studies: Impact of MRS*). The large changes in energy demand and in the concentrations of the energy metabolites make it reasonable to expect that metabolites such as ADP should play a key role in the control of energy metabolism in skeletal muscle, and numerous studies have indicated that this is so. However, it is less obvious that this should be the case for other tissues such as cardiac muscle, liver, and brain which do not have such extensive variations in energy demand. Indeed, ^{31}P NMR studies of the heart (discussed more fully below), brain, and kidney have shown little correlation between the rate of oxidative phosphorylation in the steady state and the cytoplasmic concentrations of ADP, inorganic phosphate, and ATP.[7] This does not mean that these metabolites no longer have the capacity to control oxidative

For References see p. 903

phosphorylation in these tissues. However, it does suggest that, at least under the conditions of these particular studies, there are other factors, including substrate and oxygen delivery, that presumably have a stronger influence on the regulation of oxidative phosphorylation.

It is difficult to measure free ADP directly from ^{31}P spectra, partly because its signals overlap with the much larger signals from ATP, but also because it is apparent that, even if overlap were not a problem, the free ADP is commonly present at such low concentrations that its signals would not be visible above the noise. The evidence for this is obtained by making use of the creatine kinase equilibrium, which permits the concentration of free ADP to be calculated on the basis of the known concentrations of the other creatine kinase reactants. While this method is feasible for skeletal muscle, cardiac muscle, and brain, it cannot be used for tissues that do not express significant amounts of creatine kinase, such as liver and kidney. An interesting means of circumventing this problem has been described by Brosnan et al.[8] who used the transgenic mouse technique to express high levels of creatine kinase in the liver. They were thus able to show that the free ADP in the liver was about 60 μmol g^{-1} wet weight. This is similar to the value previously determined by Veech et al.[9] on the basis of similar arguments involving enzymes that are believed to catalyze reactions that are near to equilibrium. Such transgenic models could clearly contribute significantly to our understanding of tissue metabolism and its control.

^{31}P NMR can make further contributions to our understanding of the kinetic and thermodynamics of oxidative phosphorylation, through its ability to monitor the unidirectional flux between ATP and inorganic phosphate in intact tissues. This measurement can be achieved by magnetization transfer techniques, in an analogous manner to the measurement of creatine kinase activity, and a particularly extensive series of studies has been carried out on cardiac muscle.[10]

These magnetization transfer studies of cardiac muscle were carried out on isolated perfused preparations. The study of cardiac muscle in vivo is a lot more difficult, partly because of the effects of motion, and also because blood can contribute significant amounts of signal, making measurements based on the phosphomonoester and inorganic phosphate signals particularly problematical. Some of the problems can be circumvented by the use of surgical procedures enabling the radiofrequency coil to be positioned adjacent to the region of interest. A number of such studies have been carried out on the dog heart in vivo. For example, a series of ^{31}P studies has been carried out examining the relationship between workload and phosphorus energy metabolites. It was found that the concentration of ADP and the phosphocreatine/ATP ratio in the heart did not change significantly despite an almost fourfold change in ATP hydrolysis, as estimated from the myocardial oxygen consumption.[11] It was subsequently found that there were no significant changes in ATP, ADP, phosphocreatine, or inorganic phosphate until the workload apparently outstripped the ability of the heart to produce ATP, because of limitations either in coronary blood flow or in oxidative phosphorylation capacity.[12] As discussed above, these findings are relevant to our understanding of metabolic control in the heart.

The spectral contributions from blood, which as mentioned above cause problems with observations based on the inorganic phosphate signal (including measurements of myocardial pH),

can be eliminated or at least greatly reduced by using appropriate localization techniques. For example, Robitaille et al.[13] described the use of one-dimensional spectroscopic imaging in conjunction with an implanted surface coil to detect the transmural metabolite distribution in the dog heart. Amongst their findings, they noted that their measured pH value of 7.0–7.1 was in agreement with measurements on perfused hearts, and suggested that in some in vivo studies the pH might have been overestimated due to the contributions of 2,3-diphosphoglycerate and inorganic phosphate from the blood.

3 ^{1}H MRS AND BRAIN METABOLISM

Following earlier studies of metabolism in red blood cells,[14] the first high-resolution ^{1}H spectra of intact animals were reported by Behar et al.[15] In their studies of the rat brain, they were able to observe ^{1}H signals from a number of metabolites, including N-acetylaspartate, creatine + phosphocreatine, and choline-containing compounds, and they demonstrated an increase in the lactate signal on hypoxia, with a return to normal on subsequent oxygenation (Figure 2). Numerous technical developments followed, and ^{1}H MRS is now extensively used for the noninvasive detection of human brain metabolites. The range of metabolite signals that has been unequivocally identified by ^{1}H MRS has widened considerably, and a variety of animal models of disease have shown the presence of unusually high signals from specific metabolites. For example, histidine has been detected in mice with histidinemia,[16] elevated glutamine has been seen in hyperammonemic rats,[17–19] unusually high concentrations of choline-containing compounds (and in particular betaine) are present in animal models of multiple sclerosis,[20,21] and high levels of γ-aminobutyrate (GABA) have been observed in rats treated with anticonvulsants that act as inhibitors of GABA transaminase.[22,23] Studies of this type have pointed the way forward to analogous investigations of disordered cerebral metabolism in man; they can also aid interpretation and assignment of clinical spectra. In addition, of course, they can highlight potential pitfalls in interpretation. For example, in clinical spectroscopy the assignment of signals to brain lipids may need to be treated with some caution, in view of the demonstration that certain brain proteins can give rise to peaks that show some similarities with lipid signals.[24,25]

A number of animal studies, including some of those mentioned above, have shown that there could be many conditions in which the ^{1}H spectra are more responsive than ^{31}P signals to brain disease. However, in relating animal studies to clinical investigations, it must be appreciated that many animal models are studied in the acute phase, whereas the majority of clinical cases are investigated in the subacute or chronic phase. This is just one reason why some degree of caution must be exercised in making comparisons between studies of animal models and of man. One metabolite that is of particular relevance in this context is N-acetylaspartate. Several lines of evidence suggest that almost all of the N-acetylaspartate within the brain is neuronal, and so in clinical studies a reduction in the N-acetylaspartate signal is commonly interpreted in terms of neuronal loss or damage.[26] This interpretation is consistent with the spectral changes that have been observed in a number of disorders for which neuronal loss can be expected, including stroke, glio-

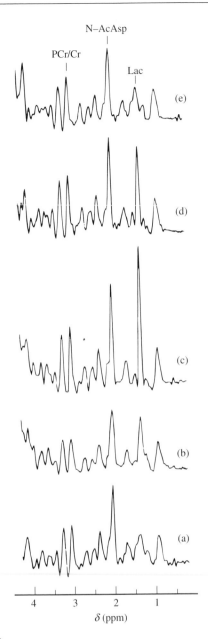

Figure 2 ^{1}H NMR spectra obtained at 360 MHz from the rat brain in vivo, showing the time course of lactate production and clearance during hypoxia and recovery. Phosphocreatine and creatine (PCr/Cr), N-acetylaspartate (NAc-Asp), and lactate (Lac) signals are labeled. Conditions were as follows: a, normoxic, 25% O_2; b, hypoxic, 5 min after administration of 4% O_2; c, after 17 min at 4% O_2; d, recovery, 14 min after administration of 25% O_2; e, after 41 min at 25% O_2. (Reproduced with permission from Behar et al., *Proc. Natl. Acad. Sci. USA*, 1983, **80**, 4945)

mas, AIDS, and epilepsy (see ***Brain Infection and Degenerative Disease Studied by Proton MRS***). However, extensive loss of N-acetylaspartate is less commonly observed in animal models, primarily because the pathological processes resulting in such loss may have a timescale of many hours or days, whereas the animal models are often studied in the acute phase. Thus in ^{1}H studies of animal stroke models, the emphasis has tended to be more on the detection of acute changes in lactate than on chronic changes in N-acetylaspartate. In more chronic

animal models, a major loss of N-acetylaspartate has indeed been observed, and in kainate-induced status epilepticus this loss has been found to correlate well with neuronal injury.[27]

4 ^{13}C MRS AND INTERMEDIARY METABOLISM

The natural abundance of ^{13}C is only 1.1%, and this has led to the emergence of two categories of investigation. In one category, the compounds of interest are present at sufficiently high concentrations to permit detection without ^{13}C labeling. These natural abundance studies are primarily of storage compounds, in particular glycogen, since these tend to be present in the appropriate concentration range. In the early 1980s, it was reported that glycogen gives a well-resolved ^{13}C NMR spectrum in which the carbons are almost completely NMR-visible.[28] This was somewhat surprising in view of the fact that large molecules with relatively little mobility tend to give very broad signals. Presumably, glycogen has a high degree of internal mobility, as a result of which its carbons give fairly narrow signals. The high concentration of glycogen in liver and muscle means that its ^{13}C signals can be detected in these tissues without any need for isotopic enrichment. In common with other developments in spectroscopy, the main applications for the detection of glycogen may well prove to be in man.

The second category of ^{13}C studies exploits the use of ^{13}C-labeling for the investigation of specific metabolic pathways. For example, Kunnecke et al.[29] have investigated the metabolism of [1,2-^{13}C$_2$]glucose and of multiply-labeled 3-hydroxybutyrate in rat brain, obtaining spectra both in vivo and in vitro. Figure 3 shows in vivo ^{13}C spectra of rat brain before infusion (a) and during infusion of each of the two substrates (b and c), while Figure 4 shows the corresponding spectra obtained from perchloric acid extracts of the brain. Analysis of the spin-coupling patterns in such spectra provides detailed information about many aspects of cerebral metabolism, including the pathways of glutamate, glutamine, and GABA synthesis, and reveals various features of metabolic compartmentation that can readily be attributed to the differing metabolic characteristics of neurons and glial cells. These studies can also contribute to our understanding of the role of N-acetylaspartate, which was referred to above in relation to ^{1}H MRS studies of the brain. In particular, it was found that while free aspartate is ^{13}C-enriched to a similar extent as glutamate and glutamine, the labeling of the aspartyl moiety of N-acetylaspartate remained at the natural abundance level. In contrast, the acetyl moiety of N-acetylaspartate was significantly enriched, lending further support to the view that this metabolite may act as an acetyl carrier or as a storage form of acetyl-CoA.

An alternative technical approach exploits the influence of the ^{13}C spin on the signals from neighboring protons. For example, the CH$_3$ protons of lactate are split into a doublet if the C3 carbon is ^{13}C-labeled, and the detection of such splitting in ^{1}H spectra has formed the basis for demonstrating turnover of brain lactate in human stroke. As an extension of this indirect approach to the detection of ^{13}C label, the ^{1}H signals can be detected with and without ^{13}C-decoupling, and this provides a means of measuring fractional enrichment of the ^{13}C label. Such proton-observe-carbon-edit methods[30] have been used to measure the rate of label entry from [1-^{13}C]glucose into glutamate. Since the glucose carbon is incorporated

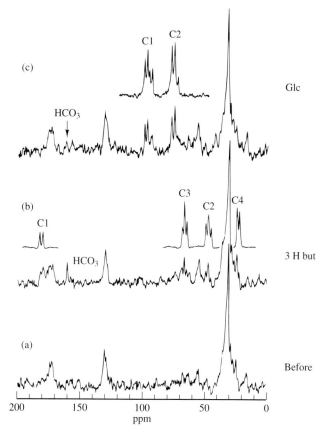

Figure 3 ^{13}C NMR spectra of rat brain obtained at 20 MHz in vivo before (a) and during the infusion of [U-$^{13}C_4$]3-hydroxybutyrate (b), or [1,2-$^{13}C_2$]glucose (c). The insets show the spectra of [U-$^{13}C_4$]3-hydroxybutyrate (b) and [1,2-$^{13}C_2$]glucose (c) in saline solution. (Reproduced with permission from Kunnecke et al., *NMR Biomed.*, 1993, **6**, 264)

into glutamate by rapid exchange with the tricarboxylic acid cycle (TCA) intermediate α-ketoglutarate, it is possible to use the rate of glutamate labeling to obtain an estimate of the TCA cycle activity. The method has now been extended from animal studies to humans. Since ^1H MRS is intrinsically more sensitive than ^{13}C MRS, it is likely that this type of approach will greatly extend the applicability of ^{13}C-labeling studies in vivo. These ^{13}C studies, and in particular their applications to humans, are discussed extensively in the article on **Brain MRS of Human Subjects**.

5 MRS AND CEREBRAL ISCHEMIA

Ischemia refers to reduced blood flow, and has been defined as 'blood flow... too low to supply enough oxygen to support cellular function'.[31,32] One question that arises is how well one can define a threshold value for cerebral blood flow, above which energy supply is sufficient but below which energy failure ensues. The precise relationship between flow and energy metabolism has been difficult to establish because of problems in making concurrent measurements of flow and metabolism in vivo in a single animal as a function of time. These problems can be overcome with the aid of MRS measurements.

The combination of ^1H and ^{31}P spectroscopy, together with simultaneous measurements of regional cerebral blood flow using the hydrogen clearance technique, has demonstrated that major changes in energy metabolites occur in the gerbil brain when the flow falls from control values of about 60 mL per 100 g per min to about 20 mL per 100 g per min or below.[33] These changes reflect the characteristic metabolic features of energy failure; that is, there is a decrease in high-energy phosphates and intracellular pH, and an increase in inorganic phosphate and lactate. The flow value of 20 mL per 100 g per min is similar to that at which, in a variety of species (including man), electrical activity ceases, and is also similar to the flow level at which water accumulates in the gerbil brain. These results therefore provide evidence suggesting that the thresholds for electrical function and edema are a direct consequence of energy failure. A model such as this could also prove to be useful when examining pharmaceuticals or other forms of therapy that may influence flow and/or metabolism. For example, one might anticipate that a treatment could be beneficial if it lowers the threshold flow value from 20 to, say, 10 mL per 100 g per min, as this would imply that the brain can tolerate lower flow values without undergoing energy failure. To illustrate this point, it was shown that the decline in energy status at reduced flow values is less severe under hypothermic (30 °C) conditions than at normal body temperature, consistent with a protective effect of hypothermia.[34]

Ischemia is also associated with failure of the transmembrane ionic pumps, and this aspect of energy failure can be investigated with ^{23}Na NMR. As discussed in the article *Cation Movements across Cell Walls of Intact Tissues Using MRS*, the ^{23}Na signals of intracellular and extracellular Na$^+$ ions differ in their relaxation properties, and this provides a means of exploring changes in the distribution of Na$^+$ ions across the cell membrane. Thus the relationship of ion homeostasis to tissue energetics can be carried out by combining ^{31}P with ^{23}Na NMR.[35–37] For example, one study showed that upon production of cerebral ischemia there is a delay of approximately 2 min before the ATP level begins to fall and the intracellular Na$^+$ begins to rise.[37] These findings, together with the flow thresholds that have been observed for energy failure, are relevant to observations that have been made in recent years using diffusion-weighted MRI, a technique that shows remarkable sensitivity to early events in cerebral ischemia.[38,39] On the basis of the relationships between the spectroscopy and diffusion-weighted imaging findings, it appears that the diffusion-weighted imaging changes observed early after the onset of ischemia are sensitive to the disruption of tissue energy metabolism or to a consequence of this disruption, including water movements associated with loss of ion homeostasis. This has raised the interesting possibility of using diffusion-weighted imaging to visualize regions of compromised energy metabolism in man, with the spatial resolution characteristic of MRI; i.e. with much superior spatial resolution to that provided by ^{31}P or ^1H MRS of energy metabolites.

6 GENERAL DISCUSSION

While the focus of attention in MRS studies is inevitably moving towards noninvasive studies of tissue metabolism in man, it is apparent that studies of animal models of disease

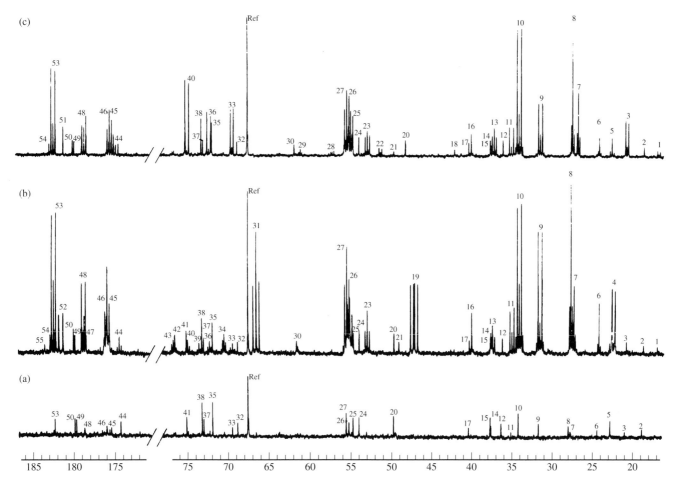

Figure 4 ^{13}C NMR spectra obtained at 100 MHz from perchloric acid extracts of rat brains after the infusion of saline solution (a), [U-$^{13}C_4$]3-hydroxybutyrate (b), and [1,2-$^{13}C_2$]glucose (c). Only the aliphatic and carboxylic regions of the spectra are shown. (Reproduced with permission from Kunnecke et al., *NMR Biomed.*, 1993, **6**, 264, where the numerous spectral assignments are given)

will continue to complement and aid interpretation of clinical observations. The above studies serve to illustrate the central role that animal studies can play in the investigation of tissue biochemistry and pathophysiology; they also show the importance of combining these MRS studies with in vitro analyses of tissue extracts, and with MRI investigations. It is this type of integrated approach that offers much of the excitement for the future.

7 RELATED ARTICLES

Brain Infection and Degenerative Disease Studied by Proton MRS; Brain MRS of Human Subjects; Cation Movements across Cell Walls of Intact Tissues Using MRS; Spectroscopic Studies of Animal Tumor Models; Surface and Other Local Coils for In Vivo Studies; Whole Body Studies: Impact of MRS.

8 REFERENCES

1. Numerous articles in *Philos. Trans. R. Soc. London*, 1980, **289B**, 379.

2. J. J. H. Ackerman, T. H. Grove, G. G. Wong, D. G. Gadian, and G. K. Radda, *Nature (London)*, 1980, **283**, 167.

3. E. B. Cady, *NMR Basic Principles and Progress*, 1992, **26**, 249.

4. D. J. Taylor, P. J. Bore, P. Styles, D. G. Gadian, and G. K. Radda, *Mol. Biol. Med.*, 1983, **1**, 77.

5. D. Freeman, S. Bartlett, G. K. Radda, and B. D. Ross, *Biochim. Biophys. Acta*, 1983, **762**, 325.

6. R. A. Iles, A. N. Stevens, J. R. Griffiths, and P. G. Morris, *Biochem. J.*, 1985, **229**, 141.

7. R. S. Balaban, *Am. J. Physiol.*, 1990, **258**, C377.

8. M. J. Brosnan, L. H. Chen, T. A. van Dyke, and A. P. Koretsky, *J. Biol. Chem.*, 1990, **265**, 20 849.

9. R. L. Veech, J. W. R. Lawson, N. W. Cornell, and H. A. Krebs, *J. Biol. Chem.*, 1979, **254**, 6538.

10. P. B. Kingsley-Hickman, E. Y. Sako, P. Mohanakrishnan, P. M. Robitaille, A. H. From, J. E. Foker, and K. Ugurbil, *Biochemistry*, 1987, **26**, 7501.

11. R. S. Balaban, H. L. Kantor, L. A. Katz, and R. W. Briggs, *Science*, 1986, **232**, 1121.

12. L. A. Katz, J. A. Swain, M. Portman, and R. S. Balaban, *Am. J. Physiol.*, 1989, **256**, H265.

13. P.-M. Robitaille, H. Merkle, E. Sublett, K. Hendrich, B. Lew, G. Path, A. H. L. From, R. J. Bache, M. Garwood, and K. Ugurbil, *Magn. Reson. Med.*, 1989, **10**, 14.

14. F. F. Brown, I. D. Campbell, P. W. Kuchel, and D. C. Rabenstein, *FEBS Lett.*, 1977, **82**, 12.

For References see p. 903

15. K. L. Behar, J. A. den Hollander, M. E. Stromski, T. Ogino, R. G. Shulman, O. A. C. Petroff, and J. W. Prichard, *Proc. Natl. Acad. Sci. USA*, 1983, **80**, 4945.

16. D. G. Gadian, E. Proctor, S. R. Williams, E. B. Cady, and R. M. Gardiner, *Magn. Reson. Med.*, 1986, **3**, 150.

17. S. M. Fitzpatrick, H. P. Hetherington, K. L. Behar, and R. G. Shulman, *J. Neurochem.*, 1989, **52**, 741.

18. T. E. Bates, S. R. Williams, R. A. Kauppinen, and D. G. Gadian, *J. Neurochem.*, 1989, **53**, 102.

19. D. K. Bosman, N. E. P. Deutz, A. A. de Graaf, R. W. M. Hulst, H. M. H. van Eijk, W. M. M. J. Bovee, M. A. W. Maas, G. G. A. Jorning, and R. A. F. M. Chamuleau, *Hepatology*, 1990, **12**, 281.

20. R. E. Brenner, P. M. G. Munro, S. C. R. Williams, J. D. Bell, G. J. Barker, C. P. Hawkins, D. N. Landon, and W. I. McDonald, *Magn. Reson. Med.*, 1993, **29**, 737.

21. N. E. Preece, D. Baker, C. Butter, D. G. Gadian, and J. Urenjak, *NMR Biomed.*, 1993, **6**, 194.

22. K. L. Behar and D. Boehm, *J. Cereb. Blood Flow Metab.*, 1991, **11**, Suppl. 2, 5783.

23. N. E. Preece, S. R. Williams, G. Jackson, J. S. Duncan, J. Houseman, and D. G. Gadian, *Proc. Xth Ann Mtg. Soc. Magn. Reson. Med.*, San Francisco, 1991, p. 1000.

24. K. L. Behar and T. Ogino, *Magn. Reson. Med.*, 1993, **30**, 38.

25. R. A. Kauppinen, T. Niskanen, J. Hakumaki, and S. R. Williams, *NMR Biomed.*, 1993, **6**, 242.

26. D. G. Gadian, 'Nuclear Magnetic Resonance and its Applications to Living Systems', Oxford University Press, Oxford, 1995.

27. T. Ebisu, W. D. Rooney, S. H. Graham, Z. Wu, M. W. Weiner, and A. A. Maudsley, *Proc. XIIth Ann Mtg. Soc. Magn. Reson. Med.*, New York, 1993, p. 513.

28. L. O. Sillerud and R. G. Shulman, *Biochem. J.*, 1983, **221**, 1087.

29. B. Kunnecke, S. Cerdan, and J. Seelig, *NMR Biomed.*, 1993, **6**, 264.

30. D. L. Rothman, K. L. Behar, H. P. Hetherington, M. R. Bendall, O. A. C. Petroff, and R. G. Shulman, *Proc. Natl. Acad. Sci., USA*, 1985, **82**, 1633.

31. G. F. Mason, D. L. Rothman, K. L. Behar, and R. G. Shulman, *J. Cereb. Blood Flow Metab.*, 1992, **12**, 434.

32. N. A. Lassen and J. Astrup, in 'Protection of the Brain from Ischemia', eds. P. R. Weinstein and A. I. Faden, Williams and Wilkins, Baltimore, MD, 1990, p. 7.

33. H. A. Crockard, D. G. Gadian, R. S. J. Frackowiak, E. Proctor, K. Allen, S. R. Williams, and R. W. Russell, *J. Cereb. Blood Flow Metab.*, 1987, **7**, 394.

34. K. L. Allen, A. L. Busza, E. Proctor, M. D. King, S. R. Williams, H. A. Crockard, and D. G. Gadian, *NMR Biomed.*, 1993, **6**, 181.

35. H. Naritomi, M. Sasaki, M. Kanashiro, M. Kitani, and T. Sawada, *J. Cereb. Blood Flow Metab.*, 1988, **8**, 16.

36. S. M. Eleff, Y. Maruki, L. H. Monsein, R. J. Traystman, R. N. Bryan, and R. C. Koehler, *Stroke*, 1991, **22**, 233.

37. J. Pekar, L. Ligeti, Z. Ruttner, T. Sinnwell, C. T. W. Moonen, and A. C. McLaughlin, *Proc. Xth Ann Mtg. Soc. Magn. Reson. Med.*, San Francisco, 1991, p. 149.

38. M. E. Moseley, Y. Cohen, J. Mintorovitch, L. Chileuitt, H. Shimizu, J. Kucharczyk, M. F. Wendland, and P. R. Weinstein, *Magn. Reson. Med.*, 1990, **14**, 330.

39. A. L. Busza, K. L. Allen, M. D. King, N. van Bruggen, S. R. Williams, and D. G. Gadian, *Stroke*, 1992, **23**, 1602.

Biographical Sketch

David G. Gadian. *b* 1950. B.A. (Physics), 1971, D. Phil., 1975, University of Oxford, UK (supervisor Rex Richards). Postdoctoral research with Rex Richards and George Radda in Oxford. Moved in 1983 to the Royal College of Surgeons in London. Currently Rank Professor of Biophysics and Head of the RCS Unit of Biophysics and of the Radiology and Physics Unit at the Institute of Child Health, London. Approx. 140 publications. Research interests: development and application of magnetic resonance techniques for noninvasive investigation of brain metabolism and physiology.

Animal Models of Stroke Studied by MRI

Mark F. Lythgoe and David G. Gadian
Royal College of Surgeons Unit of Biophysics, Institute of Child Health, London, UK

1 INTRODUCTION

This chapter describes the development and adaptation of the various animal models of stroke for use with MRI and MRS. The evolution of animal models for use with MR systems has been driven by the introduction of new MR techniques and by the increasing awareness of the role of these techniques in investigating the pathophysiology and treatment of stroke.

Early research in this area used in vitro MRS of cortical and white matter tissue to investigate cerebral edema and demonstrated the potential applicability of MR for the noninvasive detection of human stroke.[1] With the advent of imaging systems in the early 1980s and the development of animal models for in vivo research, regional changes in T_1 and T_2 were observed in animal models several hours following the initial insult.[2] During this time, in vivo MRS was used to follow changes in high-energy phosphates, inorganic phosphate, intracellular pH, lactate, and changes in Na^+ distribution in various animal models of ischemia, anoxia, and hypoxia.[3] From these early experiments it became apparent that NMR, together with other techniques such as specific gravity measurements or histology, is well suited for use in the investigation and characterization of stroke.[4] Later in the 1980s, more elaborate experiments were designed, permitting the simultaneous acquisition of cerebral blood flow (CBF) and MRS data for the determination of CBF thresholds in cerebral ischemia.[5] These experiments were carried out by producing graded ischemia during the MR investigation using a remote-controlled method for gradually reducing the CBF through a series of desired levels. This novel model was subsequently used to demonstrate the relationship between diffusion-weighted imaging, energy metabolism, and CBF (Figure 1).[6,7] Even though MR systems place constraints on methodology through the confined space of the magnet bore, simultaneous physiological monitoring of

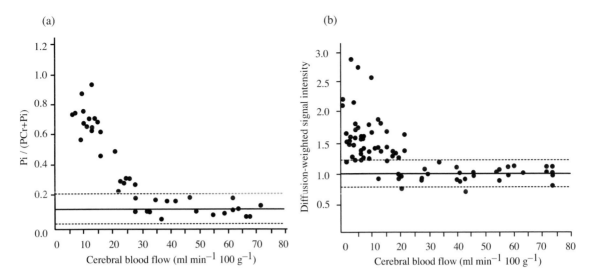

Figure 1 Remote-controlled graded ischemia produced using a snare system around both common carotid arteries in a Mongolian gerbil model. This animal model was developed to investigate the relationship between cerebral blood flow (CBF), measured with a H_2 clearance technique, diffusion-weighted MRI, and energy metabolism. CBF thresholds are similar for diffusion and for changes in Pi/(PCr+Pi). This suggests that the depletion of high-energy phosphates, leading to energy failure and an inevitable accumulation of intracellular osmotically obliged water (cytotoxic edema), results in a diffusion change through the redistribution of the water molecules. (Kindly provided by Dr Albert Busza)

electrical activity, blood gases, and blood pressure was found to be feasible.[8]

The introduction of new MRI techniques such as diffusion-weighted imaging[9] which allows the investigation of a stroke within minutes of the initial event, led to the modification of animal models to take advantage of this information about hyperacute changes. This type of animal model required remote-controlled access, that is, the ability to modulate the animal physiology from outside the MR system; this is perhaps the most challenging of adaptations (see Section 3).[10,11] Recent developments include the application of optical techniques for noninvasive cerebral metabolic monitoring and histological studies in conjunction with MR.[12] Near-infrared spectroscopy can be used to monitor hemoglobin oxygenation, the oxidation state of cytochromes, and cerebral blood volume. Laser Doppler flowmetry is used for the noninvasive measurement of CBF, and invasive bioluminescence techniques may be used to monitor metabolic disturbances. All of these methods have been interfaced with animal models of stroke and MR.[13–15]

Since the mid-1980s comprehensive reviews have described the numerous possible animals models in the investigation of cerebral ischemia.[16–20] Further, the relevance of these animal models to human disease has also been considered;[21–23] consequently, this area is not a focus for discussion in this article. The following sections contain an overview of the most widely used small animal models for MR investigations of stroke, including a discussion of the early work and subsequent modifications of these techniques for MR use.

2 FROM THE BENCH TO THE MR SYSTEM

Experimental models of cerebral ischemia may be broadly classified into global or focal models, which may, in turn, be permanent or reversible in nature. In deciding which animal model should be used to test a hypothesis, the MR researcher must take into account issues that may not be a consideration when experimenting 'on the bench'. When choosing an animal model, the investigator must consider aspects of the MR system such as the bore size of the magnet and access to the animal once it is positioned in the magnet. Another consideration is whether the magnet is horizontal or vertical, which may influence the physiology of the animal, although this is becoming less of an issue as more and more researchers use dedicated horizontal systems. Issues of temperature and its control, together with the physiological monitoring and administration of anesthetics, can also be difficult to address and must be taken into account during experimental design.

Once the choices between global and focal models, and between different animal species have been made, the issue of whether to induce the insult outside the magnet arises and whether the insult is to be treated or reversed from within the magnet. Initial experimental MR studies of stroke required that, following acquisition of the preischemia data, the animal was removed from the magnet so that the lesion could be induced on the bench, after which the animal was then replaced and the post-insult data were acquired. Improvements in design that permit remote access have allowed the investigation of the pathological consequences of the stroke immediately following the insult and have allowed direct comparison with control images without the need for image registration. This, of course, is not possible with all procedures, but the advantages are clear if remote access is achieved.

To acquire data within the MR scanner, the animal should be immobilized in a nonmagnetic probe with ear and bite bars rather like a stereotactic head holder, allowing the head to be positioned relative to the gradient coils to attain the correct imaging slice or region of interest. In MR magnets with a small bore, the use of ear and bite bars may not be possible, in which case the skull may be glued directly to the probe to reduce movement artefacts.[24] Temperature control is commonly

For References see p. 912

attained through the use of warm air blown over the animal, an electrically heated mat, or warm water jacket or mat. Animals are readily anesthetized by administration of anesthetics through a nose cone for spontaneously breathing rats, by intraperitoneal injection, or by mechanical ventilation. Physiological monitoring of electrocardiographs (ECG) and respiration rate may be performed simultaneously using implanted chest electrodes.[25] Blood pressure is conventionally monitored using an intra-arterial line or, more recently, a piezo-electric pulse transducer.[26] Intravenous lines may be inserted for blood gas measurements or administration of drugs. Nonmagnetic graphite electrodes, which reduce susceptibility effects, may be used for electroencephalography (EEG).[11]

3 FOCAL ISCHEMIA

A number of experimental models of focal ischemia have been developed, including permanent, reversible, and partial occlusions (Table 1).

3.1 Permanent Middle Cerebral Artery Occlusion

Rats with permanent occlusion of the middle cerebral artery (MCA) are frequently used for exploring the pathogenesis of infarction and for the evaluation of new therapeutic agents.[54] In 1975, Robinson et al. developed a model of MCA occlusion in the rat in which the MCA was coagulated distal to the rhinal fissure, the procedure being performed through a craniotomy.[55]

To improve the reproducibility of this model, Tamura et al. adopted a subtemporal approach in which the proximal MCA was occluded between the rhinal branch and the lateral striate arteries in Sprague-Dawley rats.[31] A less surgically demanding procedure was described by Chen et al., involving a more distal occlusion of the MCA above the rhinal fissure, coupled with permanent ipsilateral and temporary contralateral occlusion of the common carotid arteries (CCA).[56] Several modifications of the above model have been described in conjunction with occlusion of the MCA to increase reproducibility further, such as MCA occlusion following a brief period of hypotension[57] or permanent occlusion of the ipsilateral CCA.[58] The advantages of this type of model include its reproducible nature[16] and the ease of adaptation for MR use. However, as the occlusion is performed on the bench, the early stage of ischemia is lost because of the time taken to place the animal in the MR system from the bench, and the shimming and MR setup period. Permanent occlusion of the MCA may also be performed using the intraluminal suture approach, which has been used extensively with MR, as described below (Figure 2).[59]

3.2 Reversible Middle Cerebral Artery Occlusion

Permanent MCA occlusion models do not permit the investigator to monitor the effects of reperfusion following the initial insult. To address this problem, experimenters have developed a number of methods for occluding and reperfusing the MCA.

Table 1 Focal models of ischemia

Method of occlusion	Species	Remote/reversible	References	
			On bench occlusion	MR examples
Occlusion of the middle cerebral artery				
Intraluminal occlusion				
Suture	Rat	Rev	27	40
Suture	Mouse		28	41
Suture (remote)	Rat	Remote/rev	27	10
Silicon cylinder	Rat		29	42
Transorbital				
Balloon occluder	Cat	Remote/rev	30	43
Coagulation	Pig			44
Transcranial				
Coagulation	Rat		31	45
Coagulation	Mouse		32	46
Embolization				
Autologous blood clots	Rat	Remote/rev	33	47
Photothrombosis	Rat		34	48
Vasospastic occlusion with endothelin-1	Rat	Rev	35	49
Occlusion of the common carotid arteries				
Unilateral occlusion	Gerbil		36	4
Bilateral snares, graded	Gerbil	Remote/rev	36	6
Unilateral plus hypoxia	Rabbit	Rev	37	50
Unilateral plus hypoxia	Rat	Remote/rev	37	51
Unilateral plus hypoxia	Neonatal rat	Remote/rev	38	52
Both arteries, silicon cylinder	Rat		39	53

Suture, intraluminal thread (nylon monofilament), which may be coated and retractable; remote, insult may be performed from outside the magnet; rev, mechanism for insult may be reversed to control state; graded, insult may be performed in a controlled graded fashion; coagulation, electrocoagulation of vessel.

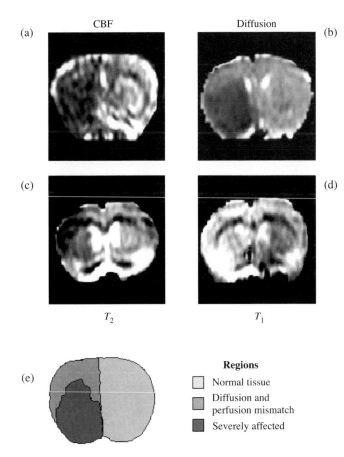

(a) CBF (b) Diffusion

(c) (d)

T_2 T_1

(e)

Regions

Normal tissue

Diffusion and perfusion mismatch

Severely affected

Figure 2 Remote-controlled permanent intraluminal suture occlusion of the middle cerebral artery allows MR images to be acquired throughout the acute period. (a) A map of cerebral blood flow (CBF) (measured using arterial spin labeling) with reduced flow in the occluded side (acquired 12 min postocclusion). (b) The apparent diffusion coefficient of water (25 min postocclusion) with a region of reduced diffusion indicating the area of cytotoxic edema. (c) A T_2 map (1 min postocclusion) with a region of reduced T_2. (d) A T_1 map (5 min postocclusion), with a region of increased T_1. (e) Schematic representation of the three main areas: 'normal tissue'; 'diffusion and perfusion mismatch area', with reduced CBF but normal diffusion; and 'severely affected area', in which both the CBF and diffusion were significantly reduced. (Kindly provided by Mr Fernando Calamante)

In 1985, Shigeno et al. placed a snare ligature around the stem of the MCA in a rat model, just distal to the lenticulostriate branches, and by pulling and releasing the thread, they occluded then reperfused the MCA.[60] This technique and many of the above are very invasive, requiring craniotomy, durotomy, and arachnoid incision, with associated drainage of cerebrospinal fluid and altered intracranial pressure. The next section describes a rat model of reversible MCA occlusion that avoids many of the above disadvantages. This type of model is now used with MR to investigate the early changes following cerebral ischemia.

In 1986, Koizumi et al. developed a new reversible focal ischemic model.[27] The MCA was occluded by a silicone rubber cylinder attached to a thread inserted through the internal caro-

tid artery in Wistar rats. Recirculation was accomplished by pulling the thread out of the artery. This model has the advantage of avoiding craniotomy and has an important role in MR investigations of ischemia. The intraluminal thread method of occlusion was later modified by Zea Longa et al. using a blunted nylon monofilament, in an attempt to increase reproducibility.[61]

In one study comparing these two techniques, it was suggested that there were differences in the levels of reduced CBF on occlusion, and that the Koizumi method was more reliable than that of Zea Longa.[62] However, other investigators have not supported these conclusions. The discussions for these variations focused on the degree of silicone coating of the thread, the filament size and length, and the body weight of animals.[63,64] It is interesting to note that, when comparing two uncoated 4-0 nylon monofilaments from different manufacturers, subtle differences in material and diameter resulted in variations in the final infarct size following MCA occlusion.[65] There is still much debate as to whether to coat the suture and what coating to use; one investigator has suggested that a simple coating of poly-l-lysine can reduce interanimal variability and increase infarct size.[66]

Two recent studies have investigated, at some length, the intraluminal suture model. Li et al., using a silicone-coated 4-0 nylon occluder adapted for remote occlusion during MRI (see below), achieved successful occlusions in 88% of animals without subarachnoid hemorrhage.[67] The failures included preocclusion damage (1/67), occluding device sliding out of the outer holding catheter (1/67), no occlusion (2/67), and arterial perforation (4/67). The other critical evaluation of the suture method, by Schmid-Elsaesser et al., indicated that a 4-0 silicone-coated suture, when compared with a 3-0 uncoated suture, produced less subarachnoid haemorrhage (8% compared with 30%).[68] However, with both these types of suture, premature reperfusion occurred in the first minutes of MCA occlusion in approximately 25% of studies, and this contributed to variability in the lesion size. Consequently, institutions using equivalent methods observe different lesion variability, which indicates the difficulty with these methods. One cannot overestimate the lengths to which a researcher must go in order to standardize the intraluminal MCA occlusion technique.

3.3 Use of Reversible Middle Cerebral Artery Occlusion with MRI

The advantage of the suture technique is that it allows remote-controlled occlusion (Figure 2) and reperfusion of the MCA from outside the magnet. Remote occlusion is essential to investigate the hyperacute MR changes postocclusion and also to allow a direct comparison of pre- and postocclusion image data.[67,69] This technique was adapted for MR use by Roussel et al. using a 2×0.25 mm silicon embolus on a nylon thread,[69] by Kohno et al. using a 3-0 monofilament coated with glue,[11] and, more recently, by Li et al. using a silicon-coated 4-0 monofilament.[67] This early work was performed in large horizontal bore imaging magnets and has been further modified for remote MCA occlusion studies in a 5 cm diameter bore vertical 8.5 T magnet.[59] This is not a trivial experiment but, as shown by Li et al., successful occlusion can be achieved in 88% of animals studied.[67]

For References see p. 912

Table 2   Global models in which cerebral blood flow is reduced

Method of occlusion	Species	Remote/reversible	References	
			Non-MR models	MR examples
Occlusion of the common carotid arteries				
Bilateral, remote	Gerbil	Remote/rev	71	77
Bilateral, graded	Gerbil	Remote/rev	71	7
Plus hypotension	Cat	Remote/rev	72	78
Plus hypoxia	Neonatal pig	Remote/rev		79
Occlusion of innominate and left subclavian, remote	Cat	Remote/rev	73	80
Occlusion of common carotid and vertebral arteries				
Bilateral carotid arteries, remote	Rat	Remote/rev	74	81
Bilateral carotid arteries, graded	Rat	Remote/rev	74	82
Cardiac arrest				
Fibrillation, remote	Cat	Remote/rev	75	83
With KCl	Cat	Remote		78
With KCl	Neonatal rat	Remote	76	84
Anoxia	Rat	Remote		10

See Table 1 for definition of terms.

3.4   Incomplete or Partial Occlusion

Models of focal oligemia are not common, although they are finding increasing use in the investigation of the pathophysiology of penumbra, in which regions of oligemic misery perfusion are present.[70] The aim of these experimental models is to produce a large focal lesion in which the CBF is moderately reduced throughout the MCA territory (see Table 4). The degree of reduction in CBF is dependent upon the method used and the degree of stenosis or partial occlusion. In 1994, Derugin et al. developed a reversible MCA stenosis.[85] More recently, a rat model was developed for MR use in which a combination of hypotension (induced by placing the rat vertically) and electrocoagulation of the distal MCA produced a focal hypoperfusion lesion.[86] However, only the first of these models allows reperfusion to take place and neither permits occlusion from outside an MRI scanner to observe the hyperacute phase following occlusion. To avoid these problems, Thomas et al. developed a model based on the intraluminal suture method that used an undersize suture to produce a partial obstruction of the MCA and a moderate reduction in CBF throughout the MCA territory (Figure 3).[87]

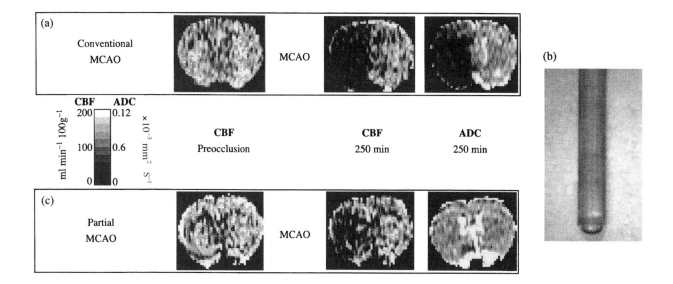

Figure 3   Intraluminal suture to produce partial and complete middle cerebral artery occlusion (MCAO). (a) Conventional MCAO in the rat using a 0.24 mm diameter nylon monofilament. Following occlusion there is a concomitant CBF and apparent diffusion coefficient (ADC) decline. (b,c) Partial occlusion of the middle cerebral artery using an undersized embolus (0.2 mm) with a bullet-shaped tip induces a moderate CBF decrease without ADC decrease

Table 3 Other models for ischemia

Method of occlusion	Species	Remote/reversible	References	
			On bench	MR examples
Secondary energy failure				
MCA ligation plus CCAs	Rat	Rev	56	90
MCA suture occlusion	Rat	Remote/rev	27	91
Unilateral CCA plus hypoxia	Rat	Remote/rev	37	51
Unilateral CCA plus hypoxia	Neonatal rat	Remote/rev	38	52
Occlusion of CCAs/hypoxia	Piglet	Remote/rev		92
Occlusion of CCAs	Gerbil	Remote/rev	36	100
Subarachnoid hemorrhage				
Perforation of the circle of Willis	Rat	Remote	89	88

MCA, middle cerebral artery; CCA, common carotid artery; for other terms see Table 1.

3.5   Focal Lesion: Other Methods

Further methods of inducing focal ischemia that avoid a craniotomy include injection of homologous blood clot fragments[47] and intraluminal occlusion with a silicon cylinder.[39,42] Another method for producing focal infarction is the injection of rose bengal dye, followed by photochemical activation to induce focal thrombosis and cerebral ischemia.[48] This model has the advantage of being relatively noninvasive, as well as providing several cerebral locations for the placement of the lesion, although it does result in microvascular injury. Focal cerebral ischemia can also be induced using a local application of endothelin-1, a vasoconstrictive agent, which produces a blood flow decrease with a half-time of 45 to 60 min.[35] An animal model of subarachnoid hemorrhage has been produced using modification of the intraluminal suture model, which allows perforation of the circle of Willis and subsequent production of the hemorrhage (Table 3). This technique may be performed remotely and therefore monitored from the hyperacute phase onwards using MR.[88,89]

4   GLOBAL ISCHEMIA

Cardiac arrest has been extensively employed for the study of the pathological consequences of global ischemia and reperfusion (see Table 2).[20] Fischer et al. have extended this work for MR and demonstrated the feasibility of resuscitation following cardiac arrest within the magnet, thus enabling them to follow cerebral events during reperfusion.[83] As well as the cardiac arrest models, there are essentially three commonly used small animal models of global cerebral ischemia for use with MR, two in the rat and one in the gerbil. The four-vessel occlusion model in the rat produces global ischemia, while the other two models induce a bilateral forebrain ischemia. Global insults may also have both a hypoxic and ischemic component. With the aim of modeling human birth asphyxia, Thornton et al. have recently described the temporal and anatomical variation of the cerebral apparent diffusion coefficient (ADC) of water, following a hypoxic/ischemic insult in a piglet model (Figure 4).[92]

4.1   Four-vessel Occlusion Model

In 1979, Pulsinelli and Brierley introduced the four-vessel occlusion model of reversible ischemia in rats.[74] Modifications by the same group increased the proportion of successful studies.[95] Briefly, on the day before the experiment, atraumatic clasps are placed loosely around both CCAs, and the vertebral arteries are electrocauterized through the foramen alar in the first cervical vertebra. Global ischemia is induced by tightening the sutures around the CCAs. A period of between 10 and 30 min of ischemia, followed by reperfusion, produces cell changes that provide evidence of selective vulnerability and delayed neuronal damage.[96] This technique has been used quite extensively with MR for both single and repeated occlusions,[81] and also to produce graded ischemia.[82]

Table 4 Oligemic models of reduced cerebral blood flow

Method of occlusion	Species	Remote/reversible	References	
			On bench	MR examples
Occlusion of the middle cerebral artery				
Suture remote	Rat	Remote/rev	27	87
Stenosis	Cat	Rev	30	85
Coagulation and hypotension	Rat		56	86
Hypotension				
Hemorrhagic hypotension	Piglet	Remote/rev	93	94

See Table 1 for definition of terms.

For References see p. 912

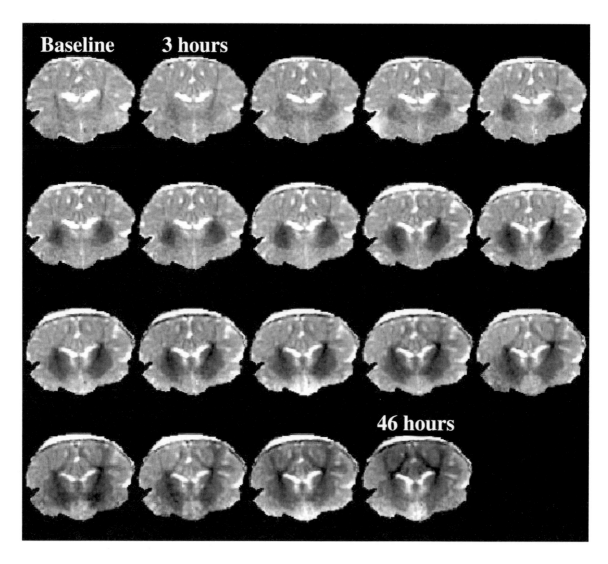

Figure 4 Coronal diffusion images (apparent diffusion coefficient maps) following a global hypoxic/ischemic insult and recovery in a piglet model of perinatal asphyxia. The initial diffusion change is limited to the basal ganglia and subsequently expands with time, apparently coincident with major white matter tracts. (Kindly provided by Dr John Thornton and Professor Roger Ordidge)

4.2   Gerbil Bilateral Common Carotid Occlusion Model

In most animals, as in humans, occlusion of the CCAs will not produce ischemia. In 1966, Levin and Payen demonstrated that in gerbils there was an anomaly in the circle of Willis such that unilateral occlusion of the CCAs resulted in ipsilateral forebrain ischemia in 40% of the animals.[36] The anomaly involved the absence of the posterior communicating arteries, which connect the anterior and posterior circulation to the brain. Subsequently, it was demonstrated that almost all gerbils exhibited neurological signs of ischemia if both CCAs were ligated. This method of occlusion has been used for both single (Figure 5)[97] and graded[6] occlusion MR experiments. Early work using this model demonstrated regions of delayed tissue damage following a brief period of bilateral CCA occlusion.[98] More recently, early subtle changes in diffusion imaging have been observed in regions known to present with delayed damage, indicating the gradual evolution of these lesions (Figure 6; Table 3).[99–101]

4.3   Rat Two-vessel Occlusion Model

This type of model is not commonly used by MR researchers, although it is technically easier to perform than the four-vessel model. Due to the intact circle of Willis in the rat, bilateral occlusion of the CCAs must be combined with systemic hypotension to produce reversible ischemia.[18] Hypotension is induced by bleeding[102] or with the administration of trimethaphan or phentramine.[103] This model is influenced by the need to monitor and correct for alterations in systemic hypotension, which may limit its applicability to MR studies.

4.4   Graded Global Ischemia

The methods for occlusion that have been mentioned so far do not provide any flexibility in the level of reduction of CBF. The degree of ischemia is primarily dependent on the vessel

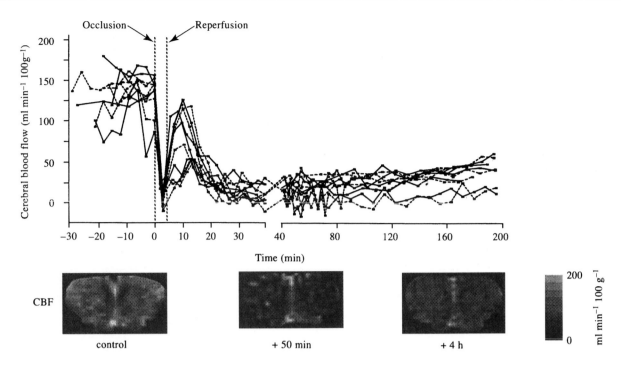

Figure 5 Forebrain ischemia induced for 5 min via remote-controlled bilateral carotid artery occlusion in a gerbil. Noninvasive MR measurement of cerebral blood flow using an arterial spin labeling technique demonstrated a prolonged and delayed hypoperfusion following reperfusion. (Kindly provided by Dr Gaby Pell)

occluded, the quality of occlusion, and the collateral supply to that region. Most types of intervention give little or no control over the level of the induced ischemia. Controllable graded ischemia, however, allows investigation of cerebral tissue at the desired level of CBF. Using a gerbil model, Allen et al.

have placed adjustable snares around the CCAs that can be tightened to decrease the internal diameter of the CCAs, thereby reducing the flow to the forebrain.[6] Forebrain blood flows ranging from 80 to 8 ml min^{-1} per 100 g tissue are attainable with this method (see Figure 1).

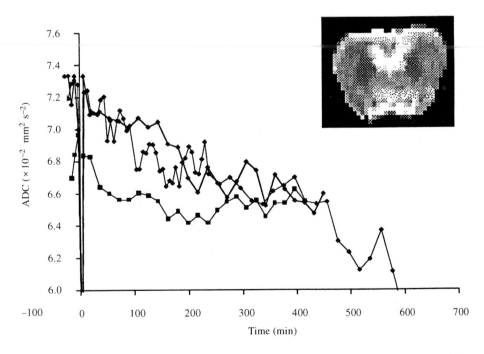

Figure 6 Diffusion (apparent diffusion coefficient, ADC) images over 10 h following 5 min of bilateral common carotid artery occlusion in three gerbils; one representative image is shown. The region of diffusion change in the image is localized to the lateral portion of the striatum, which is an area known to present with delayed tissue damage.[101] In this region a gradual diffusion decrease is observed throughout the reperfusion period

For References see p. 912

5 CONCLUSION

Adaptation of some animal models for use with MRI is a relatively simple practice, while for others that require remote access it is technically more demanding. The ability of MR to follow time courses in single animals has promoted the development of new animal models in which data can be acquired before and after the insult without removing the animal from the magnet. The development of a range of physiological MRI techniques, including diffusion, perfusion and T_2*-weighted imaging, as well as various MRS approaches, provides new opportunities for understanding the evolving pathophysiology of lesions in these models and for evaluating novel forms of therapy.

6 RELATED ARTICLES

Anisotropically Restricted Diffusion in MRI; Cerebral Perfusion Imaging by Exogenous Contrast Agents; Ischemic Stroke; Methods and Applications of Diffusion MRI.

7 REFERENCES

1. K. Gwan and H. T. Edzes, *Arch Neurol.*, 1975, **32**, 462.
2. N. van Bruggen, T. P. Roberts, and J. E. Cremer, *Cerebrovasc. Brain Metab. Rev.*, 1994, **6**, 180.
3. S. R. Williams, H. A. Crockard, and D. G. Gadian, *Cerebrovasc. Brain Metab. Rev.*, 1989, **1**, 91.
4. K. R. Thulborn, G. H. du Boulay, L. W. Duchen, and G. K. Radda, *J. Cereb. Blood Flow Metab.*, 1982, **2**, 299.
5. D. G. Gadian, R. S. Frackowiak, H. A. Crockard, E. Proctor, K. L. Allen, S. R. Williams, and R. W. Ross Russell, *J. Cereb. Blood Flow Metab.*, 1987, **7**, 199.
6. K. L. Allen, A. L. Busza, E. Proctor, M. D. King, S. R. Williams, H. A. Crockard, and D. G. Gadian, *NMR Biomed.*, 1993, **6**, 181.
7. A. L. Busza, K. L. Allen, M. D. King, N. Van Bruggen, S. R. Williams, and D. G. Gadian, *Stroke*, 1992, **23**, 1602.
8. H. Naritomi, M. Sasaki, M. Kanashiro, M. Kitani, and T. Sawada, *J. Cereb. Blood Flow Metab.*, 1988, **8**, 16.
9. M. E. Moseley, Y. Cohen, J. Mintorovitch, L. Chileuitt, H. Shimizu, J. Kucharczyk, M. F. Wendland, and P. R. Weinstein, *Magn. Reson. Med.*, 1990, **14**, 330.
10. S. A. Roussel, N. van Bruggen, M. D. King, and D. G. Gadian, *J. Cereb. Blood Flow Metab.*, 1995, **15**, 578.
11. K. Kohno, T. Back, M. Hoehn-Berlage, and K. A. Hossmann, *Magn. Reson. Imaging*, 1995, **13**, 65.
12. M. Hoehn-Berlage, *NMR Biomed.*, 1995, **8**, 345.
13. S. Punwani, R. J. Ordidge, C. E. Cooper, P. Amess, and M. Clemence, *NMR Biomed.*, 1998, **11**, 281.
14. M. Hoehn-Berlage, D. G. Norris, K. Kohno, G. Mies, D. Leibfritz, and K. A. Hossmann, *J. Cereb. Blood Flow Metab.*, 1995, **15**, 1002.
15. O. Kloiber, T. Miyazawa, M. Hoehn-Berlage, and K. A. Hossmann, *NMR Biomed.*, 1993, **6**, 144.
16. M. D. Ginsberg and R. Busto, *Stroke*, 1989, **20**, 1627.
17. K. A. Hossman, *Cerebrovasc. Dis.*, 1991, **1**, 2.
18. S. Takaizawa, A. M. Hakim, *Cerebrovasc. Dis.*, 1991, **1**, 16.
19. T. L. C. Luvisotto and G. R. Sutherland, in 'Magnetic Resonance Spectroscopy and Imaging in Neurochemistry', eds. H. Batchelard, Plenum Press, New York, 1997, p. 117.
20. K. A. Hossmann, *Cardiovasc. Res.*, 1998, **39**, 106.
21. D. O. Wiebers, H. P. Adams, and J. P. Whisnant, *Stroke*, 1990, **21**, 1.
22. J. A. Zivin and J. C. Grotta, *Stroke*, 1990, **21**, 981.
23. C. Millikan, *Stroke*, 1992, **23**, 795.
24. M. F. Lythgoe, A. L. Busza, F. Calamante, C. H. Sotak, M. D. King, A. C. Bingham, S. R. Williams, and D. G. Gadian, *Magn. Reson. Med.*, 1997, **38**, 662.
25. N. Van Bruggen, J. Syha, A. L. Busza, M. D. King, G. W. H. Stamp, S. R. Williams, and D. G. Gadian, *Magn. Reson. Med.*, 1990, **15**, 121.
26. R. E. Widdop, and X. C. Li, *Clin. Sci.*, 1997, **93**, 191.
27. J. Koizumi, Y. Yoshida, T. Nakazawa, and G. Ooneda, *Jpn J. Stroke*, 1986, **8**, 1.
28. E. S. Connolly, C. J. Winfree, D. M. Stern, R. A. Solomon, and D. J. Pinsky, *J. Neurosurg.*, 1996, **38**, 523.
29. J. H. Turner, *Stroke*, 1975, **6**, 703.
30. M. D. O'Brien and A. G. Waltz, *Stroke*, 1973, **4**, 201.
31. A. Tamura, D. I. Graham, J. McCulloch, and G. M. Teasdale, *J. Cereb. Blood Flow Metab.*, 1981, **1**, 53.
32. F. C. Barone, D. J. Knudsen, A. H. Nelson, G. Z. Feuerstein, and R. N. Willette, *J. Cereb. Blood Flow Metab.*, 1993, **13**, 683.
33. M. Kudo, A. Aoyama, S. Ichimori, and N. Fukunaga, *Stroke*, 1982, **13**, 505.
34. B. D. Watson, W. D. Dietrich, R. Busto, M. S. Wachtel, and M. D. Ginsberg, *Ann. Neurol.*, 1995, **17**, 497.
35. J. Sharkey, I. M. Ritchie, and P. A. Kelly, *J. Cereb. Blood Flow Metab.*, 1993, **13**, 865.
36. S. Levine and H. M. Payne, *Exp. Neurol.*, 1966, **16**, 255.
37. S. Levine, *Am. J. Pathol.*, 1960, **36**, 1.
38. P. Andine, M. Thordstein, and L. Kjellmer, *Thromb. Res.*, 1990, **35**, 253.
39. T. Takeda, T. Shima, Y. Okada, S. Matsumura, Y. Nishi, and T. Uozumi, *J. Cereb. Blood Flow Metab.*, 1987, **7**, S66.
40. J. Mintorovitch, M. E. Moseley, L. Chileuitt, H. Shimizu, Y. Cohen, and P. R. Weinstein, *Magn. Reson. Med.*, 1991, **18**, 39.
41. R. Hata, G. Mies, C. Wiessner, K. Fritze, D. Hesselbarth, G. Brinker, and K. A. Hossmann, *J. Cereb. Blood Flow Metab.*, 1998, **18**, 367.
42. E. Busch, C. M., Kerskens, and M. Hoehn-Berlage, *Proc. Vth Annu. Mtg. (Int.) Soc. Magn. Reson. Med.*, Vancouver, 1997, **1**, 599.
43. A. J. de Crispigny, M. F. Wendland, N. Derugin, E. Kozniewska, and M. E. Moseley, *Magn. Reson. Med.*, 1992, **27**, 391.
44. L. Rohl, L. Ostergaard, M. Sakoh, C. Z. Simonsen, P. Vestergaard-Poulsen, R. Sangil, H. Stodkilde-Jorgensen, and C. Gyldensted, *Proc. VIIth Annu. Mtg. (Int.) Soc. Magn. Reson. Med.*, Philadelphia, 1999, **2**, 453.
45. R. A. Knight, R. J. Ordidge, J. A. Helpern, M. Chopp, L. C. Rodolosi, and D. Peck, *Stroke*, 1991, **22**, 802.
46. M. D. King, J. Houseman, D. G. Gadian, and A. Connelly, *Magn. Reson. Med.*, 1997, **38**, 930.
47. Q. Jiang, R. L. Zhang, Z. G. Zhang, J. R. Ewing, G. W. Divine, and M. Chopp, *J. Cereb. Blood Flow Metab.*, 1998, **18**, 758.
48. N. van Bruggen, B. M. Cullen, M. D. King, M. Doran, S. R. Williams, D. G. Gadian, and J. E. Cremer, *Stroke*, 1992, **23**, 576.
49. K. Fuxe, B. Bjelke, B. Andbjer, H. Grahn, R. Rimondini, and L. F. Agnati, *Neuroreport*, 1997, **8**, 2623.
50. H. E. D'Arceuil, A. J. De Crispigny, J. Rother, S. Seri, M. E. Moseley, D. K. Stevenson, and W. Rhine, *JMRI*, 1998, **8**, 820.

For list of General Abbreviations see end-papers

51. R. M. Dijkhuizen, S. Knollema, H. B. van der Worp, G. J. ter Horst, D. J. de Wildt, J. W. Berkelbach van der Spenkel, C. A. F. Tullenken, and K. Nicolay, *Stroke*, 1998, **29**, 695.

52. U. I. Tuor, P. Kozlowski, D. R. Del-Bigo, B. Ramjiawan, S. Su, K. Malisza, and J. K. Saunders, *Exp. Neurol.*, 1998, **150**, 321.

53. S. C. Jones, A. D. Perez Trepichio, M. Xue, A. J. Furlan, and I. A. Awad, *Acta Neurochir.*, 1994, **60**, 207.

54. L. Edvinsson, E. T. MacKenzie, and J. McCulloch, 'Cerebral Blood Flow and Metabolism', eds. L. Edvinsson, E. T. MacKenzie, and J. McCulloch, Raven Press, New York, 1993, p. 618.

55. R. G. Robinson, W. J. Shoemaker, M. Schlumpf, T. Valk, and F. E. Bloom, *Nature*, 1975, **255**, 332.

56. S. T. Chen, C. Y. Hsu, E. L. Hogan, H. Maricq, and J. D. Balentine, *Stroke*, 1986, **17**, 738.

57. K. A. Osborne, T. Shigeno, A. M. Balarsky, I. Ford, J. McCulloch, G. M. Teasdale, and D. I. Graham, *J. Neurol. Neurosurg. Psychiatr.*, 1987, **50**, 402.

58. S. Brint, M. Jacewicz, M. Kiessling, J. Tanabe, and W. A. Pulsinelli, *J. Cereb. Blood Flow Metab.*, 1988, **8**, 474.

59. F. Calamante, M. F. Lythgoe, G. S. Pell, D. L. Thomas, M. D. King, C. H. Sotak, A. L. Busza, S. R. Williams, R. J. Ordidge, and D. G. Gadian, *Magn. Reson. Med.*, 1999, **41**, 479.

60. T. Shigeno, G. M. Teasdale, J. McCulloch, and D. I. Graham, *J. Neurosurg.*, 1985, **63**, 272.

61. E. Longa Zea, P. R. Weinstein, S. Carlson, and R. Cummins, *Stroke*, 1989, **20**, 84.

62. R. J. Laing, J. Jakubowski, and R. W. Laing, *Stroke*, 1993, **24**, 294.

63. J. H. Garcia, *Stroke*, 1993, **24**, 1423.

64. J. P. Holland, S. G. C. Sydserff, Taylor W. A. S., and A. B. Bell, *Stroke*, 1993, **24**, 1423.

65. Y. Kuge, K. Minematsu, T. Yamaguchi, and M. Yoshihiro, *Stroke*, 1995, **26**, 1655.

66. L. Belayev, A. F. Alonso, R. Busto, W. Zhao, and M. D. Ginsberg, *Stroke*, 1996, **27**, 1616.

67. F. Li, S. Han, T. Tatlisumak, R. A. D. Carano, C. Irie, C. H. Sotak, and M. Fisher, *Stroke*, 1998, **29**, 1715.

68. R. Schmid-Elsaesser, S. Zausinger, E. Hungerhuber, A. Baethmann, and H. J. Reulen, *Stroke*, 1998, **29**, 2162.

69. S. A. Roussel, N. van Bruggen, M. D. King, J. Houseman, S. R. Williams, and D. G. Gadian, *NMR Biomed.*, 1994, **7**, 21.

70. M. F. Lythgoe, S. R. Williams, A. L. Busza, L. I. Wiebe, A. J. B. McEwan, D. G. Gadian, and I. Gordon, *Magn. Reson. Med.*, 1999, **41**, 706.

71. M. Kobayashi, W. D. Lust, and J. Passonneau, *J. Neurochem.*, 1977, **29**, 53.

72. F. A. Welsh, M. J. O'Connor, and V. R. Marcy, *J. Neurochem.*, 1977, **31**, 311.

73. E. F. Gildea and S. Cobb, *Acta Neurol. Psychiatr.*, 1930, **23**, 876.

74. W. A. Pulsinelli and J. B. Brierley, *Stroke*, 1979, **10**, 267.

75. V. Hossmann and K. A. Hossmann, *Brain Res.*, 1973, **60**, 423.

76. J. Korf, C. Klein, K. Venema, and F. Postema, *J. Neurochem.*, 1988, **50**, 1087.

77. K. L. Allen, A. L. Busza, H. A. Crockard, R. S. Frackowiak, D. G. Gadian, E. Proctor, R. W. Ross Russell, and S. R. Williams, *J. Cereb. Blood Flow Metab.*, 1988, **8**, 816.

78. D. Davis, J. A. Ulatowski, S. Eleff, M. Izuta, S. Mori, D. Shungu, and P. C. M. Van Zijl, *Magn. Reson. Med.*, 1994, **31**, 454.

79. A. Lorek, Y. Takei, E. B. Cady, J. S. Wyatt, J. Penrice, A. D. Edwards, D. Peebles, M. Wylezinska, H. Owen-Reece, V. Kirkbride, C. E. Cooper, R. F. Aldridge, S. C. Roth, G. Brown, D. T. Delpy, and E. O. R. Reynolds, *Pediatr. Res.*, 1994, **36**, 699.

80. C. Pierpaoli, J. R. Alger, A. Righini, J. Mattiello, R. Dickerson, D. Des Pres, A. Barnett, and G. di Chiro, *J. Cereb. Blood Flow Metab.*, 1996, **16**, 892.

81. B. T. Andrews, P. R. Weinstein, M. Keniry, and B. Pereira, *Neurosurgery*, 1987, **21**, 699.

82. O. H. J. Grohn, M. I. Kettunen, M. Penttonen, P. C. M. Van Zijl, and R. A. Kauppinen, *Proc. VIIth Annu. Mtg. (Int.) Soc. Magn. Reson. Med.*, Philadelphia, 1999, **2**, 329.

83. M. Fischer, K. Bockhorst, M. Hoehn-Berlage, B. Schmitz, and K. A. Hossmann, *Magn. Reson. Imaging.*, 1995, **13**, 781.

84. A. van der Toorn, E. Sykova, R. M. Dijkhuizen, I. Vorisek, L. Vargova, E. Skobisova, M. van Lookeren Campagne, T. Reese, and K. Nicolay, *Magn. Reson. Med.*, 1996, **36**, 52.

85. N. Derugin and T. P. L. Roberts, *Microsurgery*, 1994, **15**, 70.

86. O. H. J. Grohn, J. A. Lukkarinen, J. M. E. Oja, P. C. van Zijl, J. A. Ulatowski, R. J. Traystman, and R. A. Kauppinen, *J. Cereb. Blood Flow Metab.*, 1998, **18**, 911.

87. D. L. Thomas, G. S. Pell, M. F. Lythgoe, F. Calamante, C. H. Sotak, A. L. Busza, S. R. Williams, M. D. King, D. G. Gadian, and R. J. Ordidge, *Proc. Vth Annu. Mtg. (Int.) Soc. Magn. Reson. Med.*, Vancouver, 1999, p. 606.

88. E. Busch, C. Beaulieu, A. J. Crispigny, and M. E. Moseley, *Stroke*, 1998, **29**, 2155.

89. J. B. Bederson, I. M. Germano, and L. Guarino, *Stroke*, 1995, **26**, 1086.

90. M. van Lookeren Campagne, G. R. Thomas, H. Thibodeaux, J. T. Palmer, S. P. Williams, D. G. Lowe, and N. Van Bruggen, *J. Cereb. Blood Flow Metab.*, 1999, **19**, 1354.

91. M. D. Silva, F. Li, M. Fisher, and C. H. Sotak, *Proc. VIIth Annu. Mtg. (Int.) Soc. Magn. Reson. Med.*, Philadelphia, 1999, p. 52.

92. J. S. Thornton, R. J. Ordidge, J. Penrice, E. B. Cady, P. N. Amess, S. Punwani, M. Clemence, and J. S. Wyatt, *Magn. Reson. Med.*, 1998, **39**, 920.

93. P. Gygax, N. Wiernsperger, W. Meier-Ruge, and T. Baumann, *Gerontology*, 1978, **24**, 14.

94. A. R. Laptook, R. J. T. Corbett, H. T. Nguyen, J. Peterson, and R. L. Nunnally, *Pediatr. Res.*, 1988, **23**, 206.

95. T. Mima 'Central Nervous System Trauma', eds. S. T. Ohnishi and T. Ohnishi, CRC Press, New York, 1995, 107.

96. W. A. Pulsinelli, J. B. Brierley, and F. Plum, *Ann. Neurol.*, 1982, **11**, 491.

97. G. S. Pell, D. L. Thomas, M. F. Lythgoe, F. Calamante, A. M. Howseman, D. G. Gadian, and R. J. Ordidge, *J. Magn. Reson. Med.*, 1999, **41**, 829.

98. T. Kirino, *Brain Res.*, 1982, **239**, 57.

99. G. S. Pell, M. F. Lythgoe, D. L. Thomas, F. Calamante, M. D. King, D. G. Gadian, and R. J. Ordidge, *Stroke*, 1999, **30**, 1263.

100. M. F. Lythgoe, G. S. Pell, D. L. Thomas, F. Calamante, M. D. King, S. R. Williams, R. J. Ordidge, and D. G. Gadian, *Proc. VIIth Annu. Mtg. (Int.) Soc. Magn. Reson. Med.*, Philadelphia, 1999, **1**, 55.

101. B. J. Crain, W. D. Westerkam, A. H. Harrison, and J. V. Nadler, *Neuroscience*, 1988, **27**, 387.

102. B. Eklof and B. K. Siesjo, *Acta Physiol. Scand.*, 1972, **86**, 155.

103. M. L. Smith, G. Bendek, N. Dahigren, I. Rosen, T. Wieloch, and B. K. Siesjo, *Acta Neurol. Scand.*, 1985, **69**, 385.

Acknowledgements

The authors thank Dr David Thomas for his constructive suggestions during the preparation of the manuscript and Dr Gaby Pell, Mr Fernando Calamante, Dr Albert Busza and Dr John Thornton for the use of images.

For References see p. 912

Biographical Sketches

Mark F. Lythgoe. b 1962. M.Sc. (Behavioural Biology), Surrey, 1993, Ph.D. (Biophysics), London, 1999. Research in pediatric nuclear medicine at the Great Ormond Street Hospital: investigation of experimental cerebral ischemia using MRI. Research interests: the use of diffusion and perfusion MRI, and novel MR contrast mechanism to examine the underlying pathophysiology of stroke.

David G. Gadian. b 1950. B.A. (Physics) Oxford, 1971, D.Phil. Oxford, 1975. Application of NMR spectroscopy, including initial ^{31}P investigation of muscle metabolism. Current research interests: development and application of MR techniques to the investigation of brain diseases and its functional consequences, with an emphasis on disorders of childhood.

Spectroscopic Studies of Animal Tumor Models

Simon P. Robinson, Cheryl L. McCoy, and John R. Griffiths

St George's Hospital Medical School, London, UK

1 INTRODUCTION

In the clinic, there are practical constraints on what can be done to study tumors in patients as well as ethical considerations, i.e. an untreated tumor cannot be monitored through growth. Animal tumor models allow studies of basic tumor biology, metabolic stress, and response to therapeutic regimes in large enough numbers to obtain statistical significance. For basic, routine cancer research a regular supply of reproducible tumors is required. Many tumor models have been developed; the majority are grown subcutaneously in standard laboratory rodents, and are well characterized and widely available, e.g. the RIF-1 fibrosarcoma grown in C_3H mice.[1] Such tumor models have been extensively studied by MRS, to monitor tumor biology and physiology, and also for the development of new pulse sequences or other technical improvements. The use of MRS to study experimental tumor models,[2] and general work on rodent tumor models[3] have been reviewed. Now that there are many clinical MR instruments, the noninvasive examination of rodent tumors by MRS is directly applicable to humans. In this article we consider the MR nuclei that have been observed in spectroscopic studies of animal tumor models.

2 ^{31}P MAGNETIC RESONANCE SPECTROSCOPY

The nucleus most commonly observed in MRS studies of tumors is ^{31}P which has several desirable spectroscopic proper-

ties. Phosphorus-31 is 100% abundant and is one of the more sensitive nuclei (see Table 1), and ^{31}P spectra are simple and easy to interpret. Also, several important phosphorus-containing compounds occur in living systems at high enough concentrations to be detectable by ^{31}P MRS (>0.2 mM), thus providing an ideal means of monitoring the energetic state of tumors.

In vivo ^{31}P MR spectra of animal tumors were first reported using simple surface coils and pulse-acquire protocols.[4–7] A typical spectrum taken from a mouse tumor is shown in Figure 1(a). Metabolites detected include three resonances from the α, β, and γ phosphates of NTP (predominantly ATP), inorganic phosphate (Pi), phosphocreatine (PCr), and both reduced and oxidized pyridine dinucleotides (NAD$^+$ and NADH) which appear as a shoulder on the high field side of α-NTP. Phospholipid metabolites observed include phosphomonoesters (PME), primarily consisting of phosphocholine (PC) and phosphoethanolamine (PE), and phosphodiesters (PDE), primarily consisting of glycerophosphocholine (GPC) and glycerophosphoethanolamine (GPE). These resonance assignments have been confirmed by ^{31}P high resolution MRS analysis of PCA extracts.[8] Intracellular tumor pH can also be derived from the chemical shift of the Pi signal.[9]

Several changes of ^{31}P spectra are consistently seen during unperturbed tumor growth. These include larger Pi resonances than observed in normal tissues, high PME resonances that increase with tumor growth, decreasing PCr levels when detectable, and a size-dependent decline in NTP relative to Pi, coupled with a decrease in tumor pH. As in nutritional deprivation in tumors, NTP decreases as Pi increases and the resultant ratio of 'high energy' to 'low energy' phosphates, i.e. NTP/Pi, is often used as a measure of metabolic state.

The decline in the bioenergetic status of the tumor is consistent with the tumor vasculature becoming incapable of providing a nutritive blood (and hence oxygen) supply to the rapidly proliferating tissue, leading to the onset and development of hypoxia.[10,11] Tumor acidification also reflects poor vascularity, as the large amounts of H$^+$ ions produced by glycolysis accumulate extracellularly.[12] The large PME and PDE resonances in ^{31}P spectra are related to cell membrane turnover in tumors. As both PC and PE are membrane synthesis substrates and GPC and GPE are membrane breakdown products, changes in the PME/PDE ratio may correspond to high rates of membrane synthesis within rapidly proliferating tissue or membrane decomposition within necrotic regions.[13]

Phosphorus-31 MR spectroscopy has also revealed alterations in the energy and phospholipid metabolism of tumors during response to treatment. The first use of ^{31}P MRS to monitor the response of animal tumors to radiotherapy showed an increase in intensity of the PCr resonance at 28 and 40 h post irradiation.[6] Subsequent studies have shown a common pattern of response to radiation therapy, with an observed increase in PCr/Pi, NTP/Pi, and intracellular pH, the extent of response being dose dependent and not a systemic response to radiation toxicity.[14] This observed improvement in energetic status is consistent with tumor reoxygenation, i.e. the reacquisition of radiosensitivity of cells that have survived irradiation because they were hypoxic at the time of exposure; it may

Table 1

Nuclide	Sensitivity	% Natural Abundance	Chemical Shift Range (ppm)	Comments
^{1}H	1.00	100	10	amino acids, lactate
^{31}P	0.066	100	30	energy metabolism, pH, phospholipid metabolites
^{13}C	0.016	1.1	200	tumor metabolism
^{19}F	0.833	100	250	drug metabolism, blood flow
^{2}H	0.010	0.015	10	blood flow, pharmacokinetic analysis

result from increases in tumor blood perfusion after irradiation.[15] Another possible explanation for the improvement in cell energy status is the reduced oxygen demand, which is a consequence of cessation of growth, as well as an improved blood supply as the tumor shrinks.[16]

The response of animal tumors to a range of chemotherapeutic agents has also been observed by ^{31}P MRS. Agents such as cyclophosphamide and 1,3-bis(2-chloroethyl)-1-nitrosourea (BCNU) increase the levels of high-energy phosphates,[16,17] the reverse of the ^{31}P trends seen during unperturbed tumor growth. This response is accompanied by a decrease in tumor volume, yet the changes in ^{31}P MR spectra are detectable early after initiation of treatment, before measurable changes in tumor size are evident.[16] Other agents such as flavone acetic acid (FAA) induce metabolic deactivation of animal tumors as observed by ^{31}P MRS, i.e. decreased NTP levels. FAA causes a reduction in perfusion by causing vascular collapse, though the exact mechanism of action, as with many of these drugs, is still unclear.[18]

Improving or decreasing tumor blood flow, and hence tumor oxygenation, could facilitate tumor therapy. Acute manipulation of tumor blood flow by a range of noncytotoxic modifiers has been studied by ^{31}P MRS. Injection of agents such as nicotinamide or administration of 100% oxygen have led to improved bioenergetic status as seen by ^{31}P MRS, thus potentially rendering the tumor more susceptible to radiotherapy or improving drug delivery.[19,20] Conversely, modifiers such as hydralazine, a vasodilator acting directly on host vascular smooth muscle, decrease tumor perfusion, resulting in a decrease in NTP/Pi and pH in transplanted tumors, as depicted in Figure 1(b). These changes occur as blood is 'stolen' from the tumor, leading to vascular collapse. Such a response, however, has not been observed in spontaneous animal tumor models.[21]

The marked changes in energy metabolism seen in animal tumors are rarely observed in patients. However, the PME peak does change markedly in tumors in patients; it rises during tumor growth and decreases after successful therapy.[22] Such changes have also been observed and studied in animal models.[23]

3 ^{2}H MAGNETIC RESONANCE SPECTROSCOPY

The use of ^{2}H in spectroscopic studies of animal tumors has centered upon the observation of the uptake or clearance of the freely diffusible, nonradioactive tracer deuterium oxide ($^{2}H_2O$) to measure tumor blood flow (TBF).[24] Tumor blood flow is an important determinant of the response of solid tumors to treatment and can also be altered by treatment.

Two distinct approaches have been used. Initially, $^{2}H_2O$ clearance after a direct intratumor bolus injection was measured to give information on local TBF.[25] This method is easy to implement and generates numerical values in mL (100 g min^{-1})

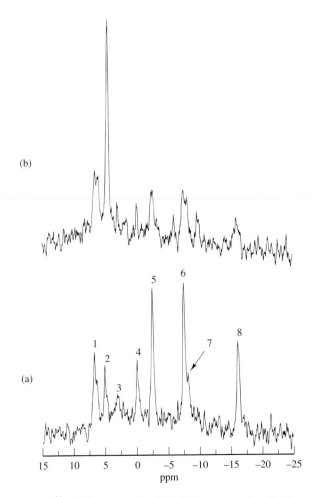

Figure 1 ^{31}P ISIS spectra of an RIF-1 fibrosarcoma in a C_3H mouse (a) before and (b) after i.v. injection of 5 mg kg^{-1} hydralazine. Peak assignments are: 1, PME; 2, Pi; 3, PDE; 4, PCr; 5, γ-NTP; 6, α-NTP; 7, NAD$^+$/NADH; and 8, β-NTP

For References see p. 918

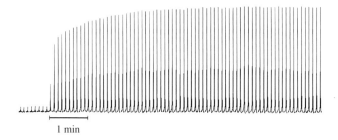

Figure 2 Uptake of ^2H_2O into a GH3 prolactinoma in a Wistar Furth rat as a function of time. Eight background FIDs collected prior to a bolus injection of 300 μl ^2H_2O

for TBF by fitting the data to an appropriate model, but it only measures TBF in the local area of injection and the injection itself may disrupt the tumor vasculature. Alternatively, ^2H_2O uptake by the tumor after i.v. injection has been observed by ^2H MRS (Figure 2); this method does not directly affect the tumor tissue and reports TBF throughout the whole tumor. This latter approach is ideal for measuring relative TBF and the response to blood flow modifiers, though quantification of TBF is more complicated and requires knowledge of the arterial input function, i.e. the ^2H_2O concentration in the arterial blood.[26]

Administration of ^2H_2O allows repeated measurements to be made on the same tumor; hence changes in TBF in response to therapy or blood flow modifiers can be easily assessed. For example, relative TBF in response to hyperthermia showed a significant difference at 45 °C compared with control tumors at 37 °C.[26] Relative TBF was also shown to decrease in response to the vasodilator hydralazine, as observed by ^2H MRS.[27]

Adaptation of these techniques to a ^2H imaging experiment has also been accomplished, and imaging the distribution of uptake of ^2H_2O provides an indication of tumor vascularity. From measurement of ^2H_2O accumulation into each pixel, a map of the spatial blood flow distribution can be created which is useful in assessing the characteristic heterogeneity of tumor blood flow.[28] Deuterium MRS provides reliable data on TBF when used with caution and provides a useful, noninvasive method of assessing an important determinant of tumor physiology.

Hydrogen-2 MRS has also been used for the direct in vivo measurement of drug concentrations. Using deuterated drugs such as ^2H-misonidazole, the concentration of the drug can be determined from a single ^2H NMR spectrum with a limit of detection of approximately 0.5 mM for a drug labeled with three equivalent deuterons.[29]

4 ^{13}C MAGNETIC RESONANCE SPECTROSCOPY

Carbon-13 MRS has been little used to study cancer. This is due in part to the relatively low natural abundance and sensitivity of the ^{13}C nucleus relative to ^1H (see Table 1) which causes two problems: (i) it is necessary to use high concentrations of ^{13}C labeled substrates, which can be expensive to synthesize; and (ii) the large amount of the compound that is administered may alter tumor metabolism. In spite of these limitations, the use and application of both enrichment studies and natural abundance ^{13}C studies to monitor tumor biochemis-

try has progressed. Most ^{13}C studies in animals have used labeled substrates injected directly into a tumor to increase the signal-to-noise ratio. Also, more efficient probe designs have been developed to overcome the limitations associated with ^{13}C MRS.[30]

The major application of in vivo ^{13}C MRS for studying cancer is to monitor glycolysis and the rate of energy utilization of tumors relative to normal tissue. Compounds such as glucose, citrate, and alanine can be detected. The presence (or absence) of several compounds can be detected, along with their concentration, mobility, and relaxation characteristics, simultaneously and noninvasively in one animal.

In vivo ^{13}C studies are routinely performed to obtain a background spectrum before injection of labeled substrates (see below). Typical studies take several minutes using ^1H decoupling (which increases the ^{13}C spectral resolution by removing the coupling interactions with neighboring protons whilst allowing NOE) to obtain spectra with adequate signal-to-noise ratios. For example, natural abundance ^{13}C MRS of Dunning rat tumors, a model for human prostate cancer, was used to examine the relationship between prostatic tumors and the presence or absence of citrate.[31]

Recently, ^{13}C MRS has been used in combination with ^{31}P MRS and traditional biochemical methods to examine glucose metabolism in RIF-1 tumors.[32,33] Natural abundance ^{13}C spectra were acquired from tumors to obtain a background signal from endogenous lipids ($-CH_2-$) which was used as an internal chemical shift marker; the mice were then injected intratumorally with ^{13}C labeled glucose and proton-decoupled ^{13}C spectra were acquired for 70 min. The levels of 3-^{13}C lactate and 3-^{13}C alanine were increased and the level of lactate approached a steady state at approximately 40 min after injection.

Carbon-13 MRS has also been used after intraperitoneal and intravenous administration of glucose to study tumor metabolism (Figure 3) and to monitor the generation of lactic acid. In addition to the study of subcutaneous (s.c.) tumors, brain tumor metabolism has also been examined with ^{13}C MRS.[34]

So far, ^{13}C MRS has been used to study the relationship between glucose availability and tumor metabolism. Although these natural abundance and enrichment studies demonstrate the feasibility of oncological applications of in vivo ^{13}C MRS, to date, there have been no reported studies in which it has been used to monitor tumor metabolism after therapy.

Another application of ^{13}C MRS is to follow the pharmacokinetics of ^{13}C-labeled chemotherapeutic agents. This labeling is attractive as most drugs contain carbon and usually there is one position that can be labeled during drug synthesis, thus removing the need for a label that might modify drug activity (e.g. ^{19}F). However, detection of the drug in vivo presents a challenge because of the poor sensitivity. One study has demonstrated the detection of a ^{13}C label on the active methyl group of the methylating agent temozolamide in RIF-1 fibrosarcomas with sufficient sensitivity to determine a halflife for elimination from the tumor.[35]

5 ^{19}F MAGNETIC RESONANCE SPECTROSCOPY

The ^{19}F nucleus is an excellent nuclide to monitor by MRS because it is 100% naturally occurring and has a sensitivity

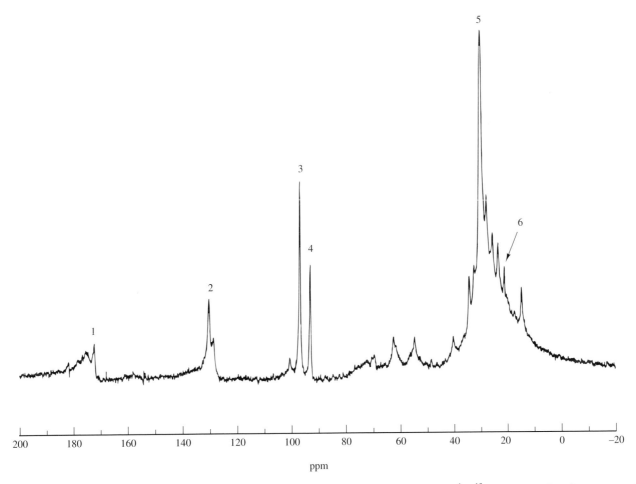

Figure 3 $^{13}C\{^1H\}$ MR spectrum of an RIF-1 tumor in a C_3H mouse after i.v. injection of 7 mg kg^{-1} 1-^{13}C-glucose. Peak assignments are: 1, carbonyl carbons (C=O); 2, alkene carbons (C=C); 3, β-glucose; 4, α-glucose; 5, methylene carbons (CH$_2$), and 6, lactate. Figure supplied by Dr R. J. Maxwell

nearly equal to that of ^{1}H (see Table 1) and there is no background signal in vivo. The noninvasive nature of ^{19}F MRS also makes it a useful tool for monitoring in vivo drug metabolism since it is not necessary to sacrifice a large number of animals and each animal can serve as its own control.

Flourine-19 MRS has been extensively used in oncology. Most of the work has used fluorine-substituted chemotherapeutic drugs such as the fluoropyrimidines. There has also been work done using perfluorocarbons (hydrocarbons whose protons have been replaced with fluorine nuclei), which are ideal candidates for both spectroscopy and imaging. Other applications of ^{19}F MRS in cancer research in murine tumors include the use of compounds containing ^{19}F as probes for tumor hypoxia,[36,37] intracellular pH,[38] and for monitoring tumor therapy.[39]

The fluoropyrimidine that has been most widely studied by ^{19}F MRS is 5-fluorouracil (5-FU), an effective anticancer agent that has been used clinically since 1957.[40] 5-FU was synthesized as an antimetabolite of uracil and is almost always metabolized similarly to uracil and its breakdown products (Figure 4). It is commonly used in combination chemotherapy with agents such as methotrexate or adriamycin.[41] The first reported study monitored 5-FU metabolism in the livers and tumors of mice via ^{19}F MRS.[42] A more recent review explores

the potential of ^{19}F MRS of fluoropyrimidines for monitoring the therapy of single chemotherapeutic agents or combinations.[43] There have also been studies using ^{19}F MRS to monitor 5-FU therapy in patients.[44]

Recently, ^{19}F MRS of perfluorocarbon emulsions has been used to examine changes in the oxygenation status of mouse RIF-1 tumors in response to treatment with nicotinamide.[45] Consistent with the hypothesis that nicotinamide improves tumor oxygenation (see above), there was a significant increase in pO_2 for treated tumors versus control tumors. The perfluorocarbon emulsions have several features that make them ideal candidates for ^{19}F MRS measurements of tumor oxygenation status: (i) they are selectively sequestered in solid tumors; (ii) oxygen binding changes the spin–lattice (T_1) relaxation time of the fluorocarbon; and (iii) there is a linear correlation between the T_1 value of the perfluorocarbon in the tumor and the tumor oxygenation.[45,46]

^{19}F MRS can also be used to observe the behavior of organofluorides such as halothane in vivo. The inhalant anaesthetic, halothane, is a hydrophobic probe that can be used for monitoring plasma membrane alterations and thus may have potential for determining the effectiveness of treatments such as hyperthermia.[47]

For References see p. 918

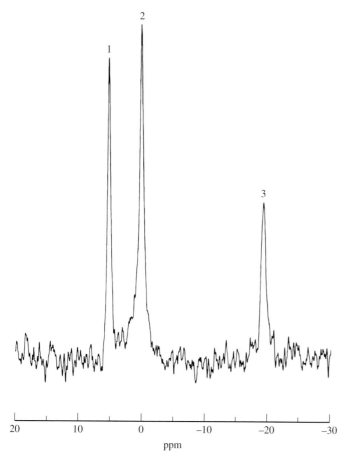

Figure 4 ^{19}F spectrum showing uptake of 5-FU into a GH3 prolactinoma in a Wistar Furth rat after i.v. injection of 60 mg kg^{-1} 5-FU. Peak assignments are: 1, fluoronucleotides; 2, 5-fluorouracil; and 3, fluoro-β-alanine. Figure supplied by Dr R. J. Maxwell

6 ^{1}H MAGNETIC RESONANCE SPECTROSCOPY

Studies of animal tumors by ^{1}H MRS are few, even though protons are naturally occurring and ^{1}H is the most sensitive nucleus (Table 1). Although ^{1}H MRS allows observation of several metabolites that are not detected by MRS of other nuclei, and nearly every compound in living tissues contains hydrogen atoms, there are several disadvantages associated with its application in vivo. The major obstacles are the high concentrations of tissue water (ca. 40 M) and lipids (ca. 1 mM) which produce intense background signals that must be suppressed to observe the metabolites of interest. The small chemical shift range of ^{1}H signals means the B_0 field must be shimmed precisely. Local hemorrhaging within the tumor, or motion, can degrade spectral resolution.[48] Because of these problems, most of the studies to date have been limited to intracranial (brain) tumor models which have very little motion and less intense lipid signals.[49,50]

The resonances that appear in a water-suppressed ^{1}H spectrum differ, depending on the echo time that is used. Typical resonances observed at 'long' echo times (e.g. 240–270 ms) may include total creatines, total cholines, and lactate. Reson-ances attributed to inositol, glucose, and mobile lipids may appear at 'short' echo times (e.g. 20–50 ms).

Recently, water- and lipid-suppressed ^{1}H spectra of subcutaneous RIF-1 and EMT-6 tumors have been published.[51] There were no changes in metabolite levels during untreated tumor growth, and tumor response to treatment with 5-FU resulted in a decrease in all metabolites. Acute TBF reduction by hydralazine (see above) caused a twofold increase in lactate. Also, ^{1}H spectroscopic images of RIF-1 tumors showed that radiotherapy significantly decreased lactate levels.[52,53] These experiments demonstrate the feasibility of ^{1}H MRS outside the brain.

Preliminary ^{1}H MRS studies in animal tumors demonstrate the wealth of information that can be exploited once the technical obstacles have been overcome. Such approaches may therefore be extrapolated into the clinical arena.

7   SUMMARY

Animal tumor models have already been extensively studied by MRS: a recent review cites more than 100 papers on ^{31}P MRS alone.[23] The ability to monitor endogenous intracellular metabolism, drug metabolism, tumor blood flow, or tumor intracellular pH repeatedly and noninvasively in a single animal, and to combine various monitoring procedures in a single experiment, has provided a great advance in experimental oncology.

8   RELATED ARTICLES

Animal Methods in MRS; Applications of ^{19}F-NMR to Oncology; Brain Neoplasms in Humans Studied by Phosphorus-31 NMR Spectroscopy; Brain Neoplasms Studied by MRI; Cells and Cell Systems MRS; Fluorine-19 MRS: General Overview and Anesthesia; MRI of Musculoskeletal Neoplasms; Single Voxel Localized Proton NMR Spectroscopy of Human Brain In Vivo; Single Voxel Whole Body Phosphorus MRS.

9   REFERENCES

1. P. R. Twentyman, J. M. Brown, J. W. Gray, A. J. Franko, M. A. Scoles, and R. F. Kallman, *J. Natl. Cancer Inst.*, 1980, **64**, 595.
2. J. Denekamp, *NMR Biomed.*, 1992, **5**, 234.
3. 'Rodent Tumor Models in Experimental Cancer Therapy', ed. R. F. Kallman, Pergamon Press, Oxford, 1987, pp. 1–310.
4. J. R. Griffiths, A. N. Stevens, R. A. Iles, R. E. Gordon, and D. Shaw, *Bioscience Rep.*, 1981, **1**, 319.
5. J. R. Griffiths and R. A. Iles, *Bioscience Rep.*, 1982, **2**, 719.
6. T. C. Ng, W. T. Evanochko, R. N. Hiramoto, V. K. Ghanta, M. B. illy, A. J. Lawson, T. H. Corbett, J. R. Durant, and J. D. Glickson, *J. Magn. Reson.*, 1982, **49**, 271.
7. W. T. Evanochko, T. C. Ng, J. D. Glickson, J. R. Durant, and T. H. Corbett, *Biochem. Biophys. Res. Commun.*, 1982, **109**, 1346.
8. W. T. Evanochko, T. T. Sakai, T. C. Ng, N. R. Krishna, H. D. Kim, R. B. Zeidler, V. K. Ghanta, R. W. Brockman, L. M. Schiffer, P. G. Braunschweiger, and J. D. Glickson, *Biochim. Biophys. Acta*, 1984, **805**, 104.
9. R. B. Moon and J. H. Richards, *J. Biol. Chem.*, 1973, **248**, 7276.

10. P. Vaupel, F. Kallinowski, and P. Okunieff, *Cancer Res.*, 1989, **49**, 6449.
11. P. Okunieff, J. A. Koutcher, L. Gerweck, E. McFarland, B. Hitzig, M. Urano, T. Brady, L. Neuringer, and H. D. Suit, *Int. J. Radiat. Oncol. Biol. Phys.*, 1986, **12**, 793.
12. J. R. Griffiths, *Br. J. Cancer*, 1991, **64**, 425.
13. P. F. Daly, R. C. Lyon, P. J. Faustino, and J. S. Cohen, *J. Biol. Chem.*, 1987, **262**, 14 875.
14. S. S. Rajan, J. P. Wehrle, S.-J. Li, R. G. Steen, and J. D. Glickson, *NMR Biomed.*, 1989, **2**, 165.
15. G. M. Tozer, Z. M. Bhujwalla, J. R. Griffiths, and R. J. Maxwell, *Int. J. Radiat. Oncol. Biol. Phys.*, 1989, **16**, 155.
16. R. G. Steen, *Cancer Res.*, 1989, **49**, 4075.
17. S.-J. Li, J. P. Wehrle, S. S. Rajan, R. G. Steen, J. D. Glickson, and J. Hilton, *Cancer Res.*, 1988, **48**, 4736.
18. J. L. Evelhoch, M.-C. Bissery, G. G. Chabot, N. E. Simpson, C. L. McCoy, L. K. Heilbrun, and T. H. Corbett, *Cancer Res.*, 1988, **48**, 4749.
19. P. J. Wood, C. J. R. Counsell, J. C. M. Bremner, M. R. Horsman, and G. E. Adams, *Int. J. Radiat. Oncol. Biol. Phys.*, 1991, **20**, 291.
20. P. Okunieff, E. McFarland, E. Rummeny, C. Willett, B. M. Hitzig, L. Neuringer, and H. D. Suit, *Am. J. Clin. Oncol.*, 1987, **10**, 475.
21. S. B. Field, S. Needham, I. A. Burney, R. J. Maxwell, J. E. Coggle, and J. R. Griffiths, *Br. J. Cancer*, 1991, **63**, 723.
22. W. G. Negendank, *NMR Biomed.*, 1992, **5**, 303.
23. J. D. de Certaines, V. A. Larsen, F. Podo, G. Carpinelli, O. Briot, and O. Henriksen, *NMR Biomed.*, 1993, **6**, 345.
24. J. J. H. Ackerman, C. S. Ewy, N. N. Becker, and R. A. Shalwitz, *Proc. Natl. Acad. Sci. USA*, 1987, **84**, 4099.
25. S.-G. Kim and J. J. H. Ackerman, *Cancer Res*, 1988, **48**, 3449.
26. J. Mattiello and J. L. Evelhoch, *Magn. Reson. Med.*, 1991, **18**, 320.
27. I. A. Burney, R. J. Maxwell, S. B. Field, C. L. McCoy, and J. R. Griffiths, *Acta Oncol.*, 1995, **34**, 367.
28. J. B. Larcombe McDouall and J. L. Evelhoch, *Cancer Res.*, 1990, **50**, 363.
29. J. L. Evelhoch, C. L. McCoy, and B. P. Giri, *Magn. Reson. Med.*, 1989, **9**, 402.
30. R. C. Lyon, R. G. Tschudin, P. F. Daly, and J. S. Cohen, *Magn. Reson. Med.*, 1988, **6**, 1.
31. L. O. Sillerud, K. R. Halliday, J. P. Freyer, R. H. Griffey, and C. Fenoglio-Preiser, in 'Magnetic Resonance in Experimental and Clinical Oncology', eds. J. L. Evelhoch, W. Negendank, F. A. Valeriote, and L. H. Baker, Kluwer Academic, Boston, 1990, p. 149.
32. I. Constantinidis, J. C. Chatham, J. P. Wehrle, and J. D. Glickson, *Magn. Reson. Med.*, 1991, **20**, 17.
33. Z. M. Bhujwalla, I. Constantinidis, J. C. Chatham, J. P. Wehrle, and J. D. Glickson, *Int. J. Radiat. Oncol. Biol. Phys.*, 1992, **22**, 95.
34. B. D. Ross, R. J. Higgins, J. E. Boggan, J. A. Willis, B. Knittel, and S. W. Unger, *NMR Biomed.*, 1988, **1**, 20.
35. D. Artemov, Z. M. Bhujwalla, R. J. Maxwell, J. R. Griffiths, I. R. Judson, M. O. Leach, and J. D. Glickson, *Magn. Reson. Med.*, 1995, **34**, 338.
36. J. A. Raleigh, A. J. Franko, L. A. Trimble, and P. S. Allen, *Chem. Mods. Cancer Treat., 6th Annu. Conf.*, Paris, 1–2, 1988.
37. R. J. Maxwell, P. Workman, and J. R. Griffiths, *Int. J. Radiat. Oncol. Biol. Phys.*, 1989, **16**, 925.
38. A. Joseph, C. Davenport, L. Kwock, C. T. Burt, and R. E. London, *Magn. Reson. Med.*, 1987, **4**, 137.
39. W. E. Hull, W. Kunz, R. E. Port, and N. Seiler, *NMR Biomed.*, 1988, **1**, 11.
40. C. Heidelberger, in 'Antineoplastic and Immunosuppressive Agents', eds. A. C. Sartorelli and D. G. Johns, Springer Verlag, New York, 1974, Vol. 38, p. 193.
41. D. Cunningham, *Br. J. Cancer*, 1988, **58**, 695.
42. A. N. Stevens, P. G. Morris, R. A. Iles, P. W. Sheldon, and J. R. Griffiths, *Br. J. Cancer*, 1984, **50**, 113.
43. P. M. J. McSheehy and J. R. Griffiths, *NMR Biomed.*, 1989, **2**, 133.
44. A. El-Tahtawy and W. Wolf, *Cancer Res.*, 1991, **51**, 5806.
45. P. S. Hees and C. H. Sotak, *Magn. Reson. Med.*, 1993, **29**, 303.
46. C. H. Sotak, P. S. Hees, H. N. Huang, M. H. Hung, C. G. Krespan, and S. Raynolds, *Magn. Reson. Med.*, 1993, **29**, 188.
47. C. T. Burt, R. R. Moore, and M. F. Roberts, *NMR Biomed.*, 1993, **6**, 289.
48. F. A. Howe, R. J. Maxwell, D. E. Saunders, M. M. Brown, and J. R. Griffiths, *Magn. Reson. Q.*, 1993, **9**, 31.
49. H. Bruhn, T. Michaelis, K. D. Merboldt, W. Hanicke, M. L. Gyngell, C. Hamburger, and J. Frahm, *NMR Biomed.*, 1992, **5**, 253.
50. C. Remy, M. von Kienlin, S. Lotito, A. Francois, A. L. Benabid, and M. Decorps, *Magn. Reson. Med.*, 1989, **9**, 395.
51. D. C. Shungu, Z. M. Bhujwalla, J. P. Wehrle, and J. D. Glickson, *NMR Biomed.*, 1992, **5**, 296.
52. Z. M. Bhujwalla, D. C. Shungu, and J. D. Glickson, *Magn. Reson. Med.*, 1996, **36**, 204.
53. Z. M. Bhujwalla and J. D. Glickson, *Int. J. Radiat. Oncol. Biol. Phys.*, 1996, **36**, 635.

Biographical Sketches

S. P. Robinson. *b* 1969. B.Sc. (Hons) 1991, City University, Ph.D., 1995, London University. Introduced to in vivo MR by Dr D. G. Reid. Research specialties: tumor oxygenation, perfusion and bioenergetics, development of more clinically relevant tumor models in animals.

C. L. McCoy. B.S., 1983, Fisk University, Ph.D., 1991, Wayne State University. Introduced to in vivo MR by Dr J. L. Evelhoch. Research specialties: tumor oxygenation, blood flow.

J. R. Griffiths. *b* 1945. M.B., B.S., 1969, London, D. Phil. (Biochemistry), 1974, Oxford. Studied ESR with George Radda 1970–74 and ^{31}P MRS of livers with David Gadian and Richard Iles, 1979–80. Approx. 140 publications. Research specialty: MRS as applied to cancer, latterly MRI of nerves.

In Vivo ESR Imaging of Animals

Graham R. Cherryman, Andrew D. Stevens, and Colin M. Smith

University of Leicester, Leicester, UK

1 INTRODUCTION

The first successful electron spin resonance (ESR) experiment was reported in 1945 by Zavoisky.[1] ESR spectroscopy

For References see p. 923

rapidly developed into an essential tool for the analytical chemist. It is only since the mid 1960s that its value to the biologist has been established. ESR is a technique that can detect and characterize molecules (or fragments of molecules) with an unpaired electron without destroying the molecule.

The existence of radical species with unpaired electrons was first proposed in the early 1900s. The importance of radicals in biology was first recognized in the 1930s by Leonor Michaelis (1875–1949) with his work on quinones and semiquinones. Today radical species are perceived to be important in understanding the pathogenesis of common and important diseases such as cancer and ischemia/reperfusion injury in stroke and myocardial infarction. Knowing the importance of radicals in initiating tissue damage also makes them targets for therapy with appropriate antioxidant therapies aimed at reducing tissue damage resulting from the overproduction of radicals. Medical scientists are now interested in the possibility of obtaining real-time in vivo ESR data on intact animal and human subjects.

When a paramagnetic substance is placed in a magnetic field, the degeneracy of the energy states for the unpaired electron is lifted. Upper and lower energy states are produced and transitions occur as a result of the absorption or emission of electromagnetic (em) energy at the resonant frequency. An ESR signal is obtained as a result of the net absorption of em energy which results from the slightly greater population of the lower energy level under nonsaturating conditions. The type of radical present determines the em frequency and magnetic field strength where this absorption occurs (g-factor). For technical reasons the absorption spectrum is usually plotted as a first derivative. The influence of magnetic nuclei within the molecule will split the ESR signal into several components (hyperfine splitting), which in turn will help characterize the radical species present. The linewidth of the ESR spectrum is also influenced by variations in the local paramagnetic environment. ESR signals are specific for the presence of unpaired electrons and can be obtained even when the majority of molecules present are nonradicals.

Biological ESR experiments are typically conducted at microwave frequencies, on tiny body fluid and tissue samples. Specimens are often frozen to 77 K to slow down the rapid bioreduction of short-lived natural radical species. Under these conditions ESR signals can be obtained from almost all tissues. The validity and usefulness of many of these observations remains to be determined.

In order to obtain adequate signal from larger aqueous samples and animals, ESR experiments must be conducted at a lower frequency than currently used in the laboratory so as to achieve sufficient tissue penetration or depth. Existing spectrometers may be modified to experiment on small animals or parts of an animal, e.g. a rat tail can be imaged at low microwave L-band frequencies, but for serious animal work and eventual human studies the ideal em frequency will lie in the radiofrequency region.

The present challenge is to develop radiofrequency ESR spectrometer design to a point where spectra and images can be obtained with suitable sensitivity on large biological specimens and to develop satisfactory methods of image creation and processing.

2 RADIOFREQUENCY ESR SPECTROMETERS AND IMAGERS

These are still at an early stage of development.[2,3] The essential components are (i) an rf bridge with transmitter and receiver circuitry, (ii) a resonator or surface coil, (iii) a magnet assembly and power supply for producing B_0, (iv) magnets and power supplies for producing G_x, G_y and G_z, and (v) a host computer(s) for control, data accumulation and image processing.

Continuous wave (CW) irradiation is normally used, in contrast to the pulse techniques used in MRI, because T_2 values are normally very short (about $10^{-9}–10^{-7}$ s) and the bandwidth is very broad (about 90 MHz for nitroxides). The frequency must be low enough to give complete tissue penetration without a prohibitive loss of sensitivity. The best compromise between these factors is difficult to predict. Models for tissue absorption have to be adapted to account for resonator geometry and the precise relationship between sensitivity and frequency is still a matter for conjecture. For human studies, 250 MHz is probably the upper limit.[4] Our 250 MHz and 300 MHz rf bridges use phase-locked crystal oscillator sources which have better spectral purity than the voltage controlled oscillators normally used.[5,6] Various systems can be incorporated into the transmitter and receiver circuitry to reduce instrumental and motional artifacts and we have designed automatic rf coupling, resonator tuning and rf phase controls for this.[5,7]

Resonators with moderately good quality factors (Q), the most advantageous B-field characteristics and the least disadvantageous E-field geometries have to be used. Loop-gaps appear to be the most suitable. We have constructed a range of multigap, single loop and multiloop, multigap resonators up to 0.4 m in diameter.[8] Figure 1 shows a section through a double

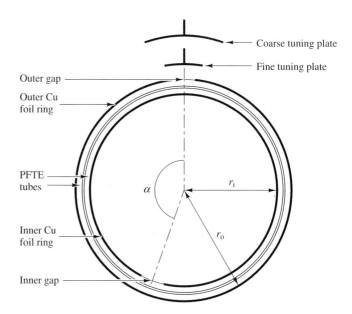

Figure 1 Schematic cross section of the 250 MHz double split-ring resonator. The inner radius (r_0) is 27.5 mm. The resonator is tuned by rotating the inner split ring (changing α) and by use of the servo driven tuning plates

split-ring type loop-gap which we use for whole body studies of rats and which can also be used as a surface coil for larger specimens.

B_0 is usually produced using air-cored Helmholtz coils, and is swept through the resonance position. A field of only 8.9 mT is required at 250 MHz for resonance at 'free spin' (g = 2.0023). The field is modulated using additional coils. Subsequent phase-sensitive detection at the modulation frequency produces a first derivative signal. Because of the large signal bandwidth, much larger field gradients than used for MRI are necessary, which can present an engineering problem with large samples. G_z is normally produced by an oil or water cooled Maxwell pair, G_x and G_y by Anderson coils or variants.[9] Figure 2 shows the coil arrangement for our 250 MHz imager, which can produce continuous gradients of up to 15

mT m^{-1}, sufficient to give a resolution of about 5 mm using nitroxides. A multipole magnet for producing field gradients of up to 150 mT m^{-1} has also been described.[10]

A spectrometer for FT ESR spectroscopy and imaging at 300 MHz has recently been constructed, but is only suitable for spectroscopy using lithium phthalocyanine crystals in vacuo, for which the FID lasts about 4 μs,[11] and imaging phantom systems comprising solvated electrons, for which the FID lasts about 9 μs. At present the signal-to-noise ratio does not appear to be better than can be obtained using CW EPR, but this is bound to improve as the technology is developed. In addition, biologically useful spin probes suited to this technique will undoubtedly be produced.

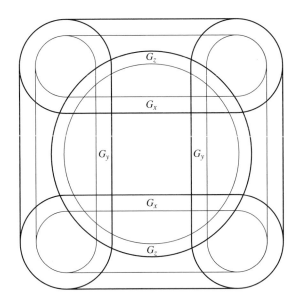

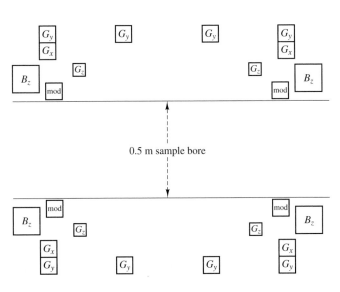

Figure 2 The arrangement of magnet coils in the 250 MHz imager, viewed from above and also from the side along the y axis. The B_0 and modulation coils (mod) are not shown in the top view for the sake of clarity

3 IMAGE RECONSTRUCTION FROM ESR SPECTRA

An ESR lineshape of a chemical species with unpaired electrons (free radical), is a function, $z(B)$, which relates ESR activity (a measure of spin density) to magnetic field and is independent of spatial position. If a constant field gradient is applied in the spectrometer, then the local magnetic field experienced in a spatial distribution of free radicals will depend upon their positions, and the measured ESR spectrum will be the mathematical convolution of the ESR lineshape of the system (the spectrum when no field gradient is applied) and the projection $p(r,\theta)$ of the spatial distribution function $F(x,y)$[12,13] (here we show results for the two-dimensional case) along the ray at angle θ defining the field gradient. In other words, the spatial extent of the free radicals causes ESR spectra to be broadened in the presence of a field gradient. If the spatial distribution is approximated on a finite grid by an array F_{ij}, then:

$$p(r, \theta) = \sum_{ij} F_{ij}\delta_{r,(x_i \cos\theta + y_j \sin\theta)} \qquad (1)$$

where $\delta_{i,j} = 1$ only if $i=j$ and 0 otherwise. If units are chosen so that the field gradient has unit magnitude and the spatial origin corresponds to the middle of the magnetic field scan range, the ESR spectrum measured for nonzero field gradient is:

$$
\begin{aligned}
d(B, \theta) &= \sum_i p(r_i, \theta)z(B - r_i) \\
&= \sum_i \left[\sum_{kl} F_{kl}\delta_{r_i(x_k \cos\theta + y_l \sin\theta)} \right] z(B - r_i) \\
&= \sum_{kl} F_{kl}z(B - x_k \cos\theta - y_l \sin\theta) \qquad (2)
\end{aligned}
$$

We actually measure the first derivative spectra and integrate to form the functions $z(B)$ and $d(B,\theta)$. The first derivative spectra may be used directly but better results are obtained using $z(B)$ and $d(B,\theta)$.

In principle, the projection of the spatial distribution may be recovered exactly by deconvolution, i.e. dividing the Fourier transform of the ESR spectrum with nonzero field gradient by that of the lineshape, entry by entry, and inverse Fourier trans-

For References see p. 923

forming the result. With a complete (infinite) set of projections, the free radical spatial distribution may be found using the convoluted back-projection formula for the inverse Radon transform and an acceptable reconstruction may be found using a large, finite set.[12] The field gradients are chosen to have the same magnitude along uniformly distributed projection angles θ between 0 and 180°. Although some workers use the Fourier reconstruction formula for the inverse of the Radon transform because it is much quicker computationally, we prefer the convoluted back-projection formula because it gives clearer reconstructions.

In practice, deconvolution is numerically unstable and a filter must be used to neglect or modify[14] the Fourier coefficients that are polluted by noise. Also, having a small number of projections introduces ripple artifacts into the reconstruction whose magnitude depends inversely on the projection number. With in vivo experiments time considerations dictate that our spectra are very noisy and sometimes we must be satisfied with only 16 projections. We have therefore developed an indirect reconstruction method that avoids the numerical instability and which does not admit ripples.[15] From a given spatial distribution of free radicals, one may calculate a set of ESR spectra. Starting from a simple guess and iterating, the spatial distribution, whose projections broaden the lineshape to the ESR spectra measured in a field gradient, can be found by minimizing the mean squared error between observed and calculated spectra. If the observed spectrum at angle θ is denoted $d(B,\theta)$ we define the mean squared error as:

$$S^2 = \frac{1}{n} \sum_{ij} \left[\sum_{kl} F_{kl} z(B_i - x_k \cos\theta_j - y_l \sin\theta_j) - d(B_i, \theta_j) \right]^2 \quad (3)$$

where n is the total number of measured data points. S^2 is minimized initially using a simple steepest descent update, which is chosen to ensure that F_{ij} are all nonnegative, using exact first derivatives and then following with a maximum entropy[16] algorithm to give a reconstruction that is closely consistent with the real data without fitting the random noise. This method is much slower than the convoluted back-projection reconstruction, but it does avoid the numerical instability and does not produce the ripple artifacts. Figure 3 shows a maximum entropy reconstructed image from 16 projections of a rat using our ESR spectrometer.

4 ESR IMAGING EXPERIMENTS

In 1971 Hutchison and Mallard reported endogenous ESR signals from a live mouse using a 100 MHz ESR spectrometer.[17] In 1987 Berliner et al.[18] demonstrated ESR signals arising from a melanoma tumor implanted into a mouse tail and imaged at L-band, following the intravenous injection of the nitroxide spin label CTPO (2,2,5,5-tetramethyl-1-oxyl-pyrrolidine-3-carboxamide). Four projections were used to construct a low resolution, filtered, back-projection image, which appeared to show central necrosis within the tumor. Subsequent serial studies showed dynamic differences in spin density between different regions of the tumor.

The same group[19] then followed the in vivo pharmacokinetics of the same intravenous nitroxide (CTPO) by serial ESR spectra obtained from a rat tail inserted into a cavity resonator

again operating at L-band. They observed that bioreduction of the nitroxide to the corresponding hydroxylamine in the plasma accounted for the bulk of signal loss rather than urinary or biliary excretion of the tracer.

A method of three-dimensional (3D) reconstruction was first proposed by Colacicchi et al. in 1988,[20] followed in 1989 by reports on spatial localization of signal and oximetry in mice.[21] In 1990 Alecci et al.[22] published a 3D reconstruction of the injected spin label 3-carbamoyl-2,2,5,5-tetramethyl-3-pyrrolidin-1-yloxy in the tail vein of a 300 g anesthetized rat. The solution (1 mL, 50×10^{-3} M) was injected into the caudal vein and spectra collected over the following 18 min. The images were calculated by Fourier interpolation from 32×8 field gradient orientations. The experiment was performed at 1.2 GHz. The same group published their experience with free

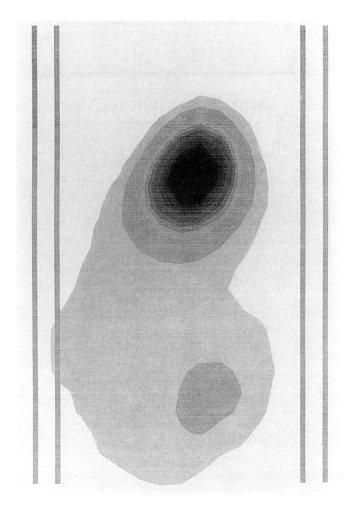

Figure 3 Sagittal maximum entropy reconstructed 250 MHz ESR image of a rat after injection of ^{15}N PCA (2,2,5,5-tetramethyl-1-oxyl-[1-^{15}N]pyrrolidine-d_{15}-3-carboxylic acid) (1 mL, 16 mM). Field gradients of 15 mT m^{-1} and a power level of 80 mW were used. Data were accumulated in 7 min and corrected for signal decay due to metabolism. The resonator split rings are shown for reference. The animal was positioned with the head toward the bottom of the resonator. A variation in spin density is seen with the highest concentration in the bladder. An area of increased spin density in the abdomen is most probably due to the concentration of the spin probe in the kidneys

radical pharmacokinetics in vivo.[23] In Japan, spin labeled images of a rat head were obtained at *L*-band.[24]

With the success of *L*-band ESR imaging the move to radiofrequency became inevitable. A new generation of ESR equipment was required. Fortunately CW ESR imagers could be constructed at relatively low cost. The Leicester (UK) group published their description of a small bore 300 MHz machine.[5]

In vivo radiofrequency ESR experiments have now been published from several groups including some from Italy[2,25] and Leicester, UK. In Leicester a much larger 250 MHz ESR imager has been built and used for in vivo imaging in the rat.[6,7,15,26,27] A surface coil has been constructed for use with larger animal and human subjects.[26]

5 THE FUTURE

There is sufficient evidence to suggest that in vivo spin label ESR spectroscopy and imaging at radiofrequency using nitroxide tracers works. We should expect to see in the next few years rapid development of the technique as a method of investigating intra and extracellular redox status.[28] Nitroxides constitute an extensive and diverse area of organic chemistry. There are already nitroxide analogs of many molecules of interest in medicine. Nitroxides with variable side chains can be used to localize within specific body compartments. When considering the development of nitroxide compounds for human use as ESR spin labels it is reassuring to note that some nitroxides, including PCA, which at one time were under development as possible paramagnetic contrast agents for MRI, showed remarkably little toxicity.[29–31]

Other ESR spin labels will also emerge. To develop ESR imaging there is also a need to develop single line spin labels with longer T_1 and T_2 relaxation times than found at present with nitroxide compounds. This will allow the development of 'pulse' ESR equipment capable of working in a similar fashion to FT MRI. In the laboratory experiments on lithium phthalocyanine and compounds based on carbon, semiquinones and the $Ph_3C\cdot$ radical have shown some of the necessary characteristics.

Swartz and colleagues have pioneered the use of spin labels to measure tissue oxygenation in vivo.[32] Promising compounds include lithium phthalocyanine and fusinite,[33] nitroxide–liposome systems,[34] carbohydrate chars[35] and India ink.[36]

Will we ever be able directly to image endogenous free radicals of medical interest? This seems unlikely due to their low concentrations and ultrashort half-lives. It is possible that suitable, safe, spin traps with identifiable spin adducts will evolve and provide some useful information about the turnover of the highly reactive radicals found in the body.[37] These same spin trap compounds may have a therapeutic effect through trapping the reactive radicals. This possibility is likely to act as a stimulus to the development of effective spin traps.

Animal experimentation is crucial to the development of human ESR imaging. It is only by demonstrating the ability of in vivo ESR imaging to answer important clinical questions that suitable, safe, spin labels and spin traps will be developed for human use.

6 RELATED ARTICLES

EPR and In Vivo EPR: Roles for Experimental and Clinical NMR Studies; Projection–Reconstruction in MRI; Relaxometry of Tissue.

7 REFERENCES

1. E. Zavoisky, *J. Phys.*, 1945, **9**, 211.
2. S. Colacicchi, M. Ferrari, and A. Sotgiu, *Int. J. Biochem.*, 1992, **24**, 205.
3. eds. G. R. Eaton, S. S. Eaton, and K. Ohno 'EPR Imaging and In-Vivo EPR', CRC Press, Boca Raton, FL, 1991.
4. H. J. Halpern and M. K. Bowman, in 'EPR Imaging and In-Vivo EPR', eds. G. R. Eaton, S. S. Eaton, and K. Ohno, CRC Press, Boca Raton, FL, 1991, pp. 45–61.
5. J. A. Brivati, A. D. Stevens, and M. C. R. Symons, *J. Magn. Reson.*, 1991, **92**, 480.
6. A. D. Stevens, *Br. J. Radiol.*, 1994, **67**, 1243.
7. A. D. Stevens and J. A. Brivati, *Meas. Sci. Technol.*, 1994, **5**, 793.
8. A. D. Stevens, J. A. Brivati, M. C. R. Symons and C. M. Smith, 'A 250 MHz EPR Spectrometer for the In Vivo Imaging of Large Specimens', *Proc. XIth Ann Mtg. Soc. Magn. Reson. Med.*, Berlin, 1992, 4111.
9. R. W. Quine, G. R. Eaton, K. Ohno, and S. S. Eaton, in 'EPR Imaging and In-Vivo EPR', eds. G. R. Eaton, S. S. Eaton, and K. Ohno, CRC Press, Boca Raton, FL, 1991, pp. 15–24.
10. M. Alecci, G. Gualtieri, A. Sotgiu, L. Testa, and A. Varoli, *Meas. Sci. Technol.*, 1991, **2**, 32.
11. J. Bourg, M. C. Krishna, J. B. Mitchell, R. G. Tschudin, T. J. Pohida, W. S. Friauf, P. D. Smith, J. Metcalfe, F. Harrington, and S. Subramanian, *J. Magn. Reson.*, 1993, **102**, 112.
12. G. T. Herman, 'Image Reconstruction from Projections. Computer Science and Applied Mathematics', Academic Press, London, 1980.
13. R. K. Woods, W. B. Hyslop, R. B. Marr, and P. C. Lauterbur, in 'ESR Imaging and In Vivo EPR', eds. G. R. Eaton, S. S. Eaton, and K. Ohno, CRC Press, Boca Raton, FL, 1991.
14. F. Momo, S. Colacicchi, and A. Sotgui, *Meas. Sci. Technol.*, 1993, **4**, 60.
15. C. M. Smith and A. D. Stevens, *Br. J. Radiol.*, 1994, **67**, 1186.
16. J. Skilling and R. K. Bryan, *Mon. Not. R. Astron. Soc.*, 1984, **211**, 111.
17. J. M. S. Hutchison and J. R. Mallard, *J. Phys. E*, 1971, **4**, 237.
18. L. J. Berliner, H. Fujii, X. M. Wan, and S. J. Lukiewicz, *Magn. Reson. Med.*, 1987, **4**, 380.
19. L. J. Berliner and X. M. Wan, *Magn. Reson. Med.*, 1989, **9**, 430.
20. S. Colacicchi, P. L. Indovina, F. Momo, and A. Sotgiu, *J. Phys. E*, 1988, **21**, 910.
21. S. Colacicchi, M. Ferrari, G. Gaultieri, M. T. Santini, and A. Sotgiu, *Phys. Med. Biol.*, 1989, **5**, 297.
22. M. Alecci, S. Colacicchi, P. L. Indovina, F. Momo, P. Pavone, and A. Sotgiu, *Magn. Reson. Imaging*, 1990, **8**, 59.
23. M. Ferrari, S. Colacicchi, G. Gaultieri, M. T. Santini, and A. Sotgiu, *Biochem. Biophys. Res. Commun.*, 1990, **166**, 168.
24. S. I. Ishida, S. Matsumoto, H. Yokoyama, N. Mori, H. Kumashiro, N. Tsuchihashi, T. Ogata, M. Yamada, M. Ono, T. Kitajima, H. Kamada, and E. Yoshida, *Magn. Reson. Imaging*, 1992, **10**, 109.
25. V. Quaresima, M. Alecci, M. Ferrari, and A. Sotgiu, *Biochem. Biophys. Res. Commun.*, 1992, **183**, 829.
26. A. D. Stevens, C. M. Smith, J. A. Brivati, A. R. Moody, G. R. Cherryman, and M. C. R. Symons 'Radiofrequency EPR Imaging Using a Surface Coil', *Proc. XIIth Ann Mtg. Soc. Magn. Reson. Med.*, New York, 1993, Vol. 2, p. 948.

27. A. D. Stevens, C. M. Smith, and G. R. Cherryman, *Br. J. Radiol.*, 1994, **67** (Congress Supplement), 106.
28. H. M. Swartz and T. Walczak, *Phys. Med. Biol.*, 1993, **9**, 41.
29. V. Afzal, R. C. Brasch, D. E. Nitecki, and S. Wolff, *Invest. Radiol.*, 1984, **19**, 549.
30. W. R. Couet, U. G. Eriksson, R. C. Brasch, and T. N. Tozer, *Pharmacol. Res.*, 1985, **2**, 69.
31. L. K. Griffeth, G. M. Rosen, E. J. Rauckman, and B. P. Drayer, *Invest. Radiol.*, 1984, **19**, 553.
32. H. M. Swartz, K. Chen, M. Pals, M Sentjung, and P. D. Morse, *Magn. Reson. Med.*, 1986, **3**, 169.
33. H. M. Swartz, S. Boyer, P. Gast, J. F. Glockner, H. Hu, K. J. Liu, M. Moussavi, S. W. Norby, N. Vahidi, T. Walczak, M. Wu, and R. B. Clarkson, *Magn. Reson. Med.*, 1991, **20**, 333.
34. J. F. Glockner, H. C. Chan, and H. M. Swartz, *Magn. Reson. Med.*, 1991, **20**, 123.
35. S. W. Norby, A. I. Smirnov, S. Boyer, H. M. Swartz, and R. B. Clarkson, 'Measurement of [O2] In Vivo Utilizing Synthetic Carbohydrate Char and EPR', *Proc. XIth Ann Mtg. Soc. Magn. Reson. Med.*, Berlin, 1992, 4107.
36. H. M. Swartz, K. J. Liu, F. Goda, and T. Walczak, *Magn. Reson. Med.*, 1994, **31**, 229.
37. H. M. Swartz, *Free Radical Res. Commun.*, 1990, **9**, 399.

Biographical Sketches

Graham R. Cherryman. *b* 1952. M.B., Ch.B., Cape Town, 1975, F.R.C.R., London, 1981. Introduced to NMR by J. R. Mallard and F. W. Smith, University of Aberdeen, 1983 and ESR by M. C. R. Symons and co-workers, University of Leicester, 1991. Approx. 60 publications on imaging and MR. Research interests include the application of radiofrequency ESRI to imaging, fast MRI of heart and brain and the assessment of new health technologies.

Andrew D. Stevens. *b* 1957. B.Sc., Chemistry, Nottingham University, 1978, Ph.D., Leicester University, 1985. Introduced to ESR in 1981 by M. C. R. Symons at Leicester and R. S. Eachus at Eastman Kodak, New York. Approx. 15 publications on ESR including five on radiofrequency ESRI. Research interests include the use of ESR to investigate the physics of the photographic process and high-temperature superconductors and the design and construction of equipment for low frequency ESRI.

Colin M. Smith. *b* 1955. B.Sc., Mathematics, Exeter University, 1977, Ph.D., Nottingham University, 1981. Spent 10 years studying applications of optimization in quantum chemistry, in England and Japan. Introduced to ESR imaging in 1992 by A. D. Stevens. Approx. 15 publications including four on imaging. Current research interests: nonlinear minimization image reconstruction.

Part Twelve
MR of Head, Neck, and Spine

Ischemic Stroke

William T. C. Yuh and Toshihiro Ueda

The University of Iowa College of Medicine, Iowa City, IA, USA

and

J. Randy Jinkins and Ronald A. Rauch

University of Texas Health Science Center, San Antonio, TX, USA

1   INTRODUCTION

Stroke affects 750 000 people annually in the USA and is the third leading cause of death.[1] However, only a small percentage of patients die as a result of stroke, making this the leading cause of disability and the third most costly disease affecting adults in this country.[2] Because many diseases can mimic ischemic stroke symptoms (epilepsy, migraine, tumor, syncope, and psychiatric disorders), neuroimaging, particularly MRI with contrast injection, becomes essential for early diagnosis and prompt treatment.

2   ETIOLOGY OF STROKE

Ischemic stroke is the result of failure to perfuse and/or oxygenate the brain. This is usually caused by arterial occlusion. However, it may also be seen after systemic hypotension and/or hypoxia. The underlying pathophysiology during acute brain ischemia can be explained by the hypothetical ischemic models proposed by Virapongse et al.: complete ischemia (no perfusion) and incomplete ischemia (some perfusion).[3] The MRI appearance and clinical outcome are usually distinctly different in these two ischemic models (Figures 1–4).[4-6] Because of the paucity of collateral circulation, complete ischemia, caused by severe compromise of the arterial supply, generally involves most of the cerebral tissue supplied by the vessel and leads to complete infarction (Figures 1 and 2). Limited or no contrast agent can reach the ischemic tissue, thus early parenchymal enhancement is not expected. Incomplete ischemia, on the other hand, caused by a transient vascular

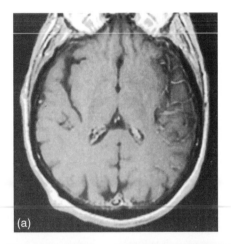

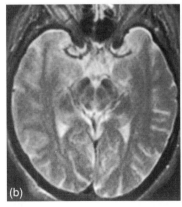
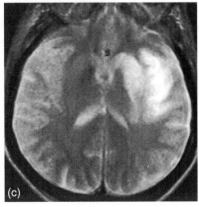

Figure 1   Complete ischemia in a 62-year-old man who presented with acute left hemispheric stroke symptoms. (a) Axial T_1-weighted contrast-enhanced image (SE700/20) obtained 3 hours after onset showing abnormal arterial enhancement (AE) in the distribution of the left middle cerebral artery. No abnormal parenchymal enhancement (PE) is noted (AE $\propto$ 1/PE). (b) The axial T_2-weighted image (SE2200/90) at the corresponding level shows no parenchymal T_2 signal changes ($T_2\Delta$ values) at this time. (c) Axial T_2-weighted image (SE2000/100) obtained 24 hours after (b) showing extensive T_2 signal changes consistent with ischemia of the left middle cerebral artery (AE $\propto$ $T_2\Delta$ $\propto$ 1/PE). (Reproduced by permission of the American Society of Neuroradiology from W. T. C. Yuh, M. R. Crain, D. J. Loes, G. M. Greene, T. J. Ryals, and Y. Sato, *Am. J. Neuroradiol.*, 1991, **12**, 565)

For References see p. 937

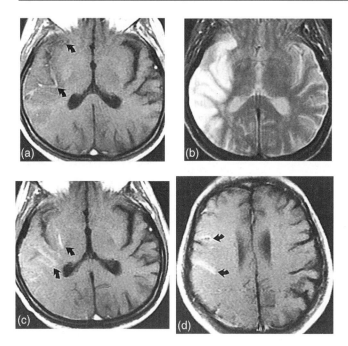

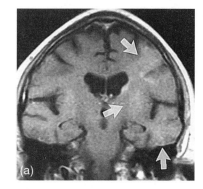

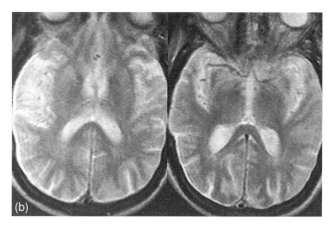

Figure 2 Complete ischemia with an inverse relationship between arterial enhancement (AE) and parenchymal enhancement (PE). (a) Axial T_1-weighted contrast-enhanced image (SE583/20) and (b) the corresponding T_2-weighted image (SE2000/100) obtained within the first 24 hours of ischemic symptoms. Arterial enhancement (arrowed) is demonstrated in the distribution of the right middle cerebral artery (a). An intense signal on T_2-weighted images ($T_2\Delta$ values) without parenchymal enhancement is present at this time (b). These findings are typical of complete ischemia expressed as $T_2\Delta \propto AE \propto 1/PE$. (c, d) Follow-up axial T_1-weighted contrast-enhanced images (SE583/20) obtained 7 days after the onset of acute ischemic symptoms when arterial enhancement (AE) has completely resolved and significant parenchymal gyriform enhancement (PE) has developed (PE $\propto 1/AE$). (Reproduced by permission of the American Society of Neuroradiology from Crain et al.[6])

occlusion (complete or partial), tends to produce a less severe ischemic insult and may be associated with a potentially reversible or minimal neurological deficit (Figures 3 and 4). Contrast material may be able to reach the ischemic tissue, and therefore early parenchymal enhancement is possible.

Watershed infarction has a somewhat different pathophysiological basis from that of complete or incomplete ischemia.[4–6] This region is located at the edges of the vascular territory between major cerebral arteries (Figure 5). A decrease in perfusion pressure from either systemic hypotension or partial occlusion may therefore affect only the regions of the brain that have the most marginal blood supply. Watershed infarction is a hybrid of complete and incomplete ischemia in that collateral circulation and/or antegrade arterial supply are intact but inadequate. Despite the severe ischemic insult, contrast material can still be delivered to the ischemic tissue as in incomplete ischemia. The clinical outcome is less favorable, however, and is more characteristic of complete ischemic infarction.

3   ACUTE CEREBRAL ISCHEMIA

The MRI appearance of acute cerebral ischemia is dependent on several physiological factors (Table 1) that are often

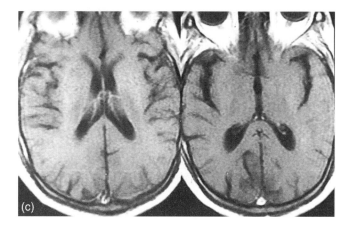

Figure 3 Incomplete ischemia in a patient with transient neurological symptoms and early parenchymal enhancement. This patient developed transient left hemispheric symptoms after 2 minutes of test balloon occlusion of the left internal carotid artery. (a) Coronal T_1-weighted contrast-enhanced image (SE538/20) obtained at 2 hours showing diffuse parenchymal enhancement (PE) (arrows) in the distribution of the left middle cerebral artery without arterial enhancement (AE) (PE $\propto 1/AE$). Also noted is enhancement of the caudate nucleus. (b) The axial T_2-weighted images (SE2000/100) show no apparent signal change at this time. (c) Axial T_1-weighted contrast-enhanced images (SE583/20) at 24 hours show no evidence of parenchymal or arterial enhancement. The resolution of parenchymal enhancement by 24 hours parallels the rapid neurological improvement in this patient. (Reproduced by permission of the American Society of Neuroradiology from Crain et al.[6])

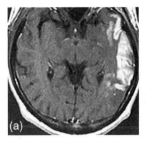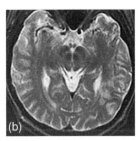

Figure 4 Incomplete ischemia with early, intense parenchymal enhancement (PE) in a patient who had a transient ischemic episode with right-sided weakness and aphasia that resolved almost completely in 2 hours. (a) Axial T_1-weighted contrast-enhanced image (SE700/20) showing diffuse, intense cortical enhancement in the distribution of the left middle cerebral artery without evidence of arterial enhancement (AE) (PE $\propto$ 1/AE). (b) The corresponding axial T_2-weighted image (SE2000/100) shows a much less extensive area ($T_2\Delta$ values) of signal abnormality (PE $\propto$ 1/AE $\propto$ 1/$T_2\Delta$). (Reproduced by permission of the American Society of Neuroradiology from Crain et al.[6])

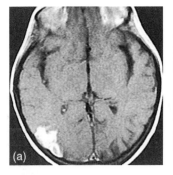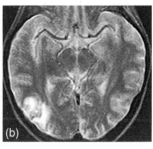

Figure 5 Watershed infarction in a patient with severe neurological sequelae and early, intense parenchymal enhancement. (a) Axial T_1-weighted contrast-enhanced image (SE516/20) showing early and intense parenchymal enhancement in the right posterior watershed zone without arterial enhancement. (b) Axial T_2-weighted image (SE2000/100) showing an area of signal abnormality approximating the area of enhancement in (a). (Reproduced by permission of the American Society of Neuroradiology from Crain et al.[6])

coexistent.[5] Four types of MRI abnormality occur soon after an acute cerebral insult: (1) alteration in arterial blood flow, (2) parenchymal swelling, (3) parenchymal signal change, and (4) abnormal parenchymal enhancement.[2,4,7–19]

Vascular flow abnormalities can be detected by MRI as an absence of flow void and/or abnormal arterial enhancement (Figures 1, 2, 6, and 7). Normal rapid flow in large arteries, such as the internal carotid or basilar arteries, typically produces a flow void on standard spin echo MRI, especially when studied with long *TE* pulse sequences. The loss of a flow void denotes thrombosis or abnormally slow flow (Figures 6 and 7).[5–9] Although angiography is necessary to discriminate slow flow from thrombosis, the loss of flow void on MRI suggests cerebral vascular disease.

In certain cases, contrast enhanced MRI can further accentuate the underlying vascular flow abnormalities. On T_1-weighted spin echo images ($TE \geqslant 20$ ms) obtained without flow compensation, arterial enhancement usually does not occur after the intravenous injection of paramagnetic contrast agents. In vessels where there is abnormally slow flow, caused either by

a high degree of stenosis or by occlusion without significant collateral supply, enhancement of arterial structures may be evident (see Figures 1 and 2).[2,5,6,10,11] Although the absence of a flow void may represent either slow flow or thrombosis, arterial enhancement has been shown to correlate well with angiographic evidence of slow flow.[12] In cortical ischemia, patients with arterial enhancement typically have more severe clinical symptoms and a poorer outcome, whereas absent or minimal symptoms are usually seen in patients with evidence of cerebral ischemia without arterial enhancement. Although arterial enhancement occurs early, it generally lasts only about 1 week and seldom persists beyond 11 days (Figure 8).[6] This contrasts with absence of flow void, which can be permanent.[5,7–9] The disappearance of arterial enhancement probably indicates the reestablishment of rapid flow in the involved vessels and coincides with the 'luxury perfusion' demonstrated on computerized tomography or scintigraphy (occurring at 7–14 days). The early disappearance of arterial enhancement within 1 or 2 days has been seen in patients with transient ischemic attacks and coincides with the rapid resolution of the

Table 1 Pathophysiological Mechanisms for MRI Findings in Ischemia

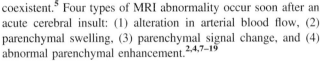

Mechanism	MRI Findings	Possible causes	Estimated time (h)[a]
Flow kinetics	Absent flow void	Slow flow; occlusion	Immediately
	Arterial enhancement	Slow flow	Immediately
Biophysiological	T_1 morphological change (swelling)	Cytotoxic edema (free water)	2–4
	T_2 signal change	BBB breakdown; vasogenic edema; macromolecular binding	8
	T_1 signal change	BBB breakdown; vasogenic edema; macromolecular binding	16–24
Combination	Progressive parenchymal enhancement[b]	Impaired delivery of significant contrast agent	>120[c]
	Early exaggerated enhancement[d]	Intact delivery of contrast agent; focal hyperemia	2–4

BBB, blood–brain barrier.
[a]Time at which findings generally could first be detected by available MRI examinations; this does not necessarily imply the exact time of onset.
[b]Typical findings in completed cortical infarctions.
[c]Usually not detected before 5–7 days.
[d]Found in cases with transient or partial occlusions and in watershed infarctions.
(Reproduced by permission of the American Society of Neuroradiology from W. T. C. Yuh, M. R. Crain, D. J. Loes, G. M. Greene, T. J. Ryals, and Y. Sato, *Am. J. Neuroradiol.*, 1991, **12**, 621)

For References see p. 937

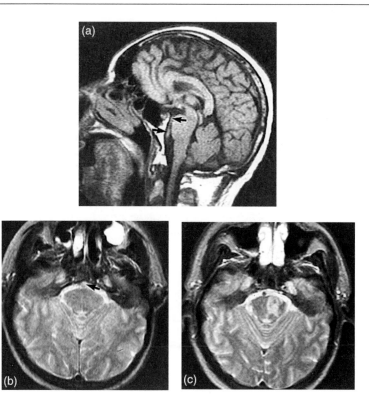

Figure 6 Arterial thrombosis in a 50-year-old man with symptoms of brain stem ischemia. (a) Parasagittal T_1-weighted image (SE350/26) obtained 4 hours after the onset showing a linear signal isointense to brain in the prepontine region (arrows) along the course of the basilar artery, suggesting an intraluminal clot. (b) Axial T_2-weighted image (SE2000/100) showing the absence of a flow void in the basilar artery (arrow). No apparent T_2 signal abnormality is detected within the pons at this time. (c) Repeat axial T_2-weighted image (SE2000/100) obtained 48 hours after (a) showing the interval development of ischemic changes in the pons. (Reproduced by permission of the American Society of Neuroradiology from W. T. C. Yuh, M. R. Crain, D. J. Loes, G. M. Greene, T. J. Ryals, and Y. Sato, *Am. J. Neuroradiol.*, 1991, **12**, 565)

underlying proximal vascular disorder.[6] Because watershed infarction usually has intact although insufficient blood flow, arterial enhancement is seen infrequently.

4 BRAIN SWELLING

Another early finding in ischemic stroke is brain swelling. Mass effect or brain swelling is probably caused by an abnormal accumulation of tissue water related to a complex combination of intra- and extracellular edema.[5] This swelling can be recognized by the distortion of normal brain anatomy, with sulcal obliteration being the first observable sign of cortical ischemia (see Figure 7). T_1-Weighted images offer the best definition of anatomy with minimal interference from cerebrospinal fluid and are generally superior for recognition of early swelling. Brain swelling often occurs as early as 2 hours before the onset of T_2 signal changes, presumably caused by cytotoxic edema (see Figures 3 and 7). This early (cytotoxic) edema primarily represents a shift of free water from the extracellular space into the intracellular space without an associated protein shift.

The swelling may then progress over several days, mostly caused by the development of vasogenic (interstitial) edema, and is associated with an abnormal signal on long TE sequences.[5,13–18] In the early phase of acute ischemia, the signal change due to vasogenic edema is generally not observed until 8 hours[5] (see Figures 1, 3, and 6), and is not fully developed until 24 hours after the infarction.[5,19] The fact that the

signal change usually occurs after tissue swelling has begun supports the hypothesis of two phases of edema (cytotoxic and vasogenic). In transient ischemic attacks, swelling can be seen transiently on T_1-weighted images without the development of T_2 hyperintensity.[20] In fact, reversible ischemia seldom is associated with major T_2-weighted parenchymal signal changes. The absence of a T_2 signal change in cytotoxic edema is probably related not only to the small amount of free water shift (estimated at 3%) but also to the absence of a major change in the interactions between the water protons and the macromolecular proteins.[5,15] Vasogenic edema, by comparison, is readily detected on T_2-weighted images and usually becomes visible 8 hours after the onset of symptoms (see Figures 1, 2, and 6). It is associated with a significant amount of water and protein shift (exudate) from the intravascular space to the extracellular space. Because signal changes usually do not occur until 2 hours after the onset of the blood–brain barrier breakdown, appreciable signal changes detected by MRI may require a gradual accumulation of a sufficient amount of water in the extracellular space (the amount of water protons) as well as an alteration in the relaxation time of water molecules (the binding state of proton molecules). The maximal signal changes noted on T_2-weighted images usually occur in 24–48 hours (see Figures 1, 5, and 6).[5,21]

Signal changes are usually best seen on standard T_2-weighted images. Abnormalities in the cortex or near the ventricle may be difficult to separate from the normal hyperintensity of adjacent cerebrospinal fluid. For this reason, the abnormal tissue signal may be easier to recognize on the

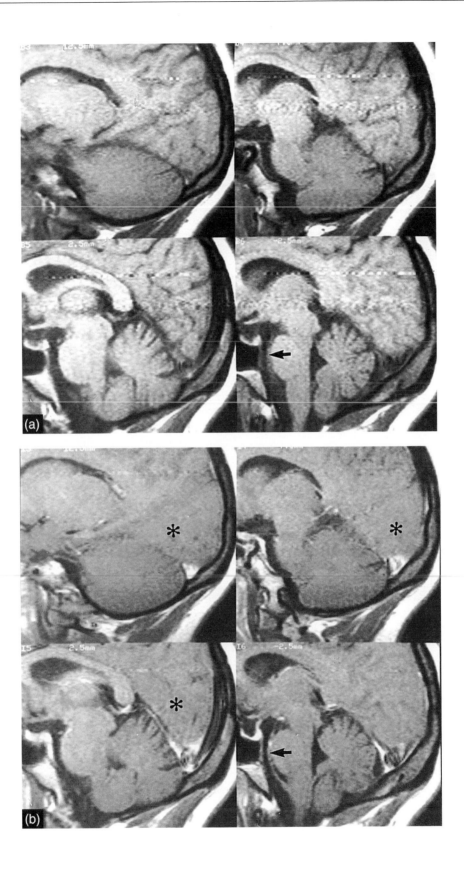

Figure 7 (a,b)

For References see p. 937

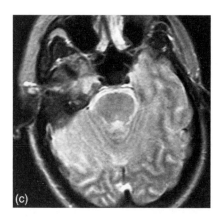

Figure 7 (c) Early development of brain swelling associated with complete ischemia in a 35-year-old woman. (a) Parasagittal T_1-weighted nonenhanced images (SE450/20) obtained 2.5 hours after the onset showing signals isointense to brain in the prepontine cistern (arrow). (b) Corresponding parasagittal T_1-weighted contrast enhanced images (SE450/20) obtained 40 minutes after (a) showing development of massive brain swelling in the occipital lobes (asterisk) and cerebellum and progression of a blood clot in the basilar artery (arrows). (Reproduced by permission of the American Society of Neuroradiology from W. T. C. Yuh, M. R. Crain, D. J. Loes, G. M. Greene, T. J. Ryals, and Y. Sato, *Am. J. Neuroradiol.,* 1991, **12**, 565)

first echo of a double-echo T_2-weighted study, also referred to as a proton density weighted image.

5 ABNORMAL ENHANCEMENT FOLLOWING ADMINISTRATION OF A CONTRAST AGENT

Abnormal enhancement of cerebral tissue with intravenously administered gadolinium after an ischemic insult is associated with a breakdown of the blood–brain barrier and/or a loss of arterial autoregulation.[4–6] Parenchymal enhancement usually occurs late in complete ischemia. During the acute phase of complete ischemia, severe arterial obstruction coupled with deficient collateral supply prevents both blood and contrast

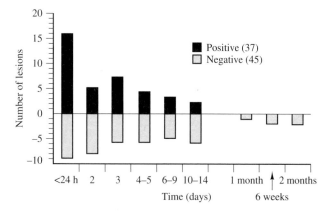

Figure 8 Graph depicting the number of lesions that demonstrated arterial enhancement at the time of the initial MRI examination. Arterial enhancement frequently was detected within the first 24 hours, rarely after 7 days, and not at all after 11 days, suggesting the reestablishment of the circulation or development of collateral circulation. (Reproduced by permission of the American Society of Neuroradiology from Crain et al.[6])

material from reaching the ischemic tissue (see Figures 1, 2, and 7). Therefore, no parenchymal enhancement is expected during the early phase of complete ischemia. By 5–7 days, the parenchymal enhancement will appear due to the reestablishment of the blood supply by recanalization of the occluded vessel and/or improved collateral circulation (see Figure 2). This reestablished blood supply will not only permit contrast material to reach the ischemic tissue, resulting in parenchymal enhancement, but also allow sufficiently rapid flow to cause the disappearance of arterial enhancement. This abnormal parenchymal enhancement will continue until the blood–brain barrier is reinstated with the maturation of the neovasculature, a process that normally takes several months after the infarction.[20]

Early enhancement can be seen in either incomplete ischemia or watershed infarction (see Figures 3–5). In incomplete ischemia, the intact delivery of contrast material may allow for the possibility of early parenchymal enhancement (see Figure 2). We have seen parenchymal enhancement within a few hours after the onset of symptoms. Such enhancement is most likely caused by a loss of autoregulation or hyperemia and typically resolves on subsequent scans performed 24 hours later. Minimal or no T_2 signal abnormality develops. Similarly, in a watershed infarction (hypoperfusion with intact but inadequate blood supply), it is uncommon to see early parenchymal enhancement (see Figure 5). Unlike incomplete ischemia, however, the early enhancement observed in a watershed infarction does not indicate a favorable clinical outcome.

The complex relationship between the abnormal MRI findings, including extent of brain swelling (EBS), T_2 signal changes ($T_2\Delta$ values), arterial enhancement (AE), and parenchymal enhancement (PE), and clinical severity (CS) in acute ischemia generally can be expressed by the following formula:

$$CS \propto T_2\Delta \propto AE \propto EBS \propto 1/PE \qquad (1)$$

This formula expresses the direct relationship between clinical severity, signal changes, extent of brain swelling, and arterial enhancement, which in turn have an inverse relationship to parenchymal enhancement.

The two exceptions that do not apply to equation (1) are the relationship between clinical severity in a watershed infarction and arterial enhancement in a small noncortical infarction. In a watershed infarction there is intact delivery of the contrast agent as in incomplete ischemia; however, it is insufficient. Therefore, despite the fact that the MRI findings suggest incomplete ischemia by equation (1), the prognosis and clinical severity are more like those of complete ischemia. The other exception is the lack of arterial enhancement in noncortical ischemia. Because the blood vessels involved in these types of infarction tend to be small, arterial enhancement usually is not seen. Nevertheless, the clinical severity of such cases may be quite grave.

6 NEW MAGNETIC RESONANCE MODALITIES

6.1 Diffusion MRI

Recent advances in MRI techniques now provide new information directly related to the underlying pathophysiology at the cellular and microscopic levels and may allow the diagnosis of acute ischemia and estimation of infarction size. Diffusion imaging reflects abnormal water movement in acute ischemic tissues caused by the failure of the high-energy Na^+–ATP pump and is sensitive to hyperacute ischemia within a few minutes after onset.[22] In both animal models and clinical cases, early ischemic changes have been reported on diffusion imaging when standard T_2-weighted imaging has shown no abnormalities.[22,23] The abnormality in diffusion imaging is shown as a high signal intensity area. This increase in signal demonstrates a region of diminished net diffusion, which is dark on the apparent diffusion coefficient (ADC) images. The ADC is obtained from an exponential fit of the signal intensities of a series of diffusion imaging as a function of b factor. Diffusion abnormality can be identified at high b values. ADC changes are a sensitive marker of ionic equilibrium because they show a function of intra–extracellular water homeostasis.

In animal studies, the signal abnormality in diffusion imaging appears a few minutes after occlusion of the middle cerebral artery and enlarges over the next 24 hours.[22,24] The abnormality in diffusion imaging is equivalent to that in T_2-weighted imaging by 24 hours after the onset of ischemia;[25] however, cytotoxic and vasogenic edema may cause an overestimation of lesion volume. Kohno et al. reported regional correspondence of hyperintensity in diffusion imaging with perfusion deficits at cerebral blood flow (CBF) thresholds of 34 ml/min per 100 g tissue after 30 minutes and 41 ml/min per 100 g tissue after 2 hours of a major coronary artery occlusion.[26] Their reports suggest that the threshold for ADC changes at a given CBF depending on the duration of ischemia.

In clinical studies, diffusion imaging has also been reported to have a high sensitivity and specificity for acute ischemia. Lövblad et al. reported that the sensitivity and specificity of diffusion imaging in acute ischemic patients within 24 hours of onset were 88% and 95%, respectively.[27] However, the pathophysiological changes in acute ischemia in humans are likely to be more heterogeneous than those found in animal models. Human subjects reportedly have two phases in the time course of ADC changes: a significant reduction for 96 hours from stroke onset and an increasing trend from reduction to pseudo-normalization to elevation at later subacute to chronic time points (> 7 days).[28]

Previous reports do not conclude whether the diffusion imaging signal changes are indicative of reversibly or irreversibly injured tissue. The diffusion abnormality in global ischemia is reversible by early reperfusion within 12 minutes.[29] The regression of the diffusion abnormality was also reported in reperfusion models. Miyabe et al. indicated that if reperfusion occurred before ADC value decreased to approximately 70% or less of control values for 10–20 minutes, the ADC changes were usually reversible.[30] Ueda et al. suggested that diffusion imaging had the highest sensitivity but was not as specific as the regional cerebral blood volume (rCBV) map in predicting acute ischemic injury and tended to overestimate infarction size in patients studied within 72 hours of stroke onset.[31] In addition, 25% (4 out of 16 lesions) of ADC abnormalities were false positives or reversible ischemia. These results may support the suggestion that the diffusion abnormality indicates early changes of both reversible and irreversible ischemia.

6.2 Perfusion MRI

Perfusion imaging provides direct information related to a reduction in blood flow that reflects the primary underlying pathophysiology of acute ischemia. Compared with the conventional spin-echo pulse sequence, echo-planar imaging is more susceptible to field inhomogeneity and is more advantageous in the evaluation of the T_2^* effect. The combination of dynamic echo-planar T_2^*-weighted imaging and intravenous bolus injection of contrast material produces hemodynamic information such as mean transit time (MTT) and CBV. Although perfusion imaging cannot produce absolute values but only semiquantitative data in estimating MTT and CBV maps, it can be performed quickly and has proven essential in the emergent management of patients with acute ischemic stroke. Furthermore, higher temporal resolution and multi-section images can be achieved by the echo-planar imaging techniques.

In an animal model, perfusion imaging demonstrated diminished perfusion within minutes of arterial occlusion.[32] Perfusion imaging shows a signal or a delay in peak signal loss in the vascular distribution when an artery is occluded. In clinical studies, perfusion imaging was reported to be superior to diffusion imaging in the assessment of hemodynamic changes of chronic cerebral hypoperfusion. Maeda et al. indicated that ischemic tissue with prolonged regional MTT (rMTT) and a marked decrease in rCBV tended to suffer irreversible damage.[33] A mild decrease in rCBV with prolonged rMTT may suggest an area of reversible ischemia. A marked increase in rCBV may show the state of luxury perfusion in subacute ischemia.

Combined diffusion and perfusion imaging can provide more important information in the management of patients with acute ischemic stroke. The volume of ischemic tissue demonstrated by both diffusion and perfusion MRI has been reported to have a high correlation with neurologic outcome as measured by the National Institutes of Health (NIH) stroke scale, and these imaging techniques, therefore, can play a useful prognostic role in acute ischemic stroke patients (Figure 9).[34]

Currently, the mismatch between diffusion and perfusion imaging in patients with acute ischemic stroke has been reported in several studies. Sorenson et al. studied patients within 10 hours of onset and showed that the abnormality in rMTT maps was larger than those in rCBV maps and diffusion imaging.[35] Rordorf et al. demonstrated that diffusion lesion volumes were smaller than the volumes of rCBV map abnormality in patients studied within 12 hours of onset, and the early CBV abnormality was slightly better than the diffusion abnormality as a predictor of final infarction size.[36] Röther et al. suggested that it was possible to differentiate between severely ischemic tissue and peri-infarct parenchyma by rCBV maps in hyperacute ischemia.[37] A recent report by Ueda et al. is similar to those reports in which the rMTT map overestimated the final infarction volume and the rCBV map provided the best estimation but differed in that their diffusion imaging underestimated final infarction volume.[31] The size of the abnormality in diffusion and perfusion imaging depends on the imaging time from the onset of symptoms. Quantitative assessment of ischemic tissue viability and/or reversibility requires further study.

For References see p. 937

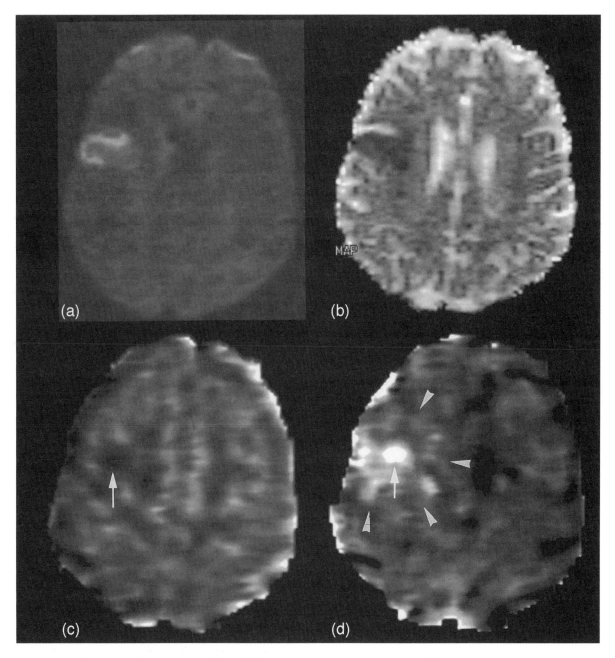

Figure 9   MRI within 12 hours in a 50-year-old male with acute onset of left hemiparesis. (a) Diffusion-weighted imaging; (b) apparent diffusion coefficient (ADC) map; (c) relative cerebral blood volume (rCBV); (d) relative mean transit time (rMTT) map. Both (a) and (b) demonstrate right hemisphere lesions in the middle cerebral artery distribution. The rMTT map (d) shows an area with hypoperfusion (arrow heads) that is much larger than that demonstrated by (a) and (b). The infarction core (arrow) may show the highest signal intensity (most severe hypoperfusion). In (c), the infarction core has a depletion of blood volume consistent with an inadequate collateral circulation (arrow)

6.3   MR Spectroscopy

Clinical application of MR spectroscopy (MRS) in the diagnosis and management of acute stroke has not been well established. There are several possible causes for the underutilization of spectroscopy in patients with strokes. (1) Most MR centers do not have sufficient scientific background and clinical expertise to facilitate such an application, and frequently avoid using such a technique. (2) The acquisition time for clinical MRS, which in the past has been long and not very reproducible, is not adequate for the management of acute stroke. (3) The diagnosis of acute stroke is usually quite straightforward by the conventional clinical examination and radiological means. Consequently, clinical MRS is usually only applied for diagnosis of problematic cases. The capability of MRS in the assessment of tissue viability/reversibility has not been well established, although it may have great potential if the acquisition times can be shortened with improved techniques.

MRS provides a means to assess the biochemical characteristics of brain disease through direct and noninvasive assay of cerebral metabolites.[38,39] Practically, phosphorus and protons are being measured in clinical applications for central nervous system disease. In ischemic brain tissue, MRS provides infor-

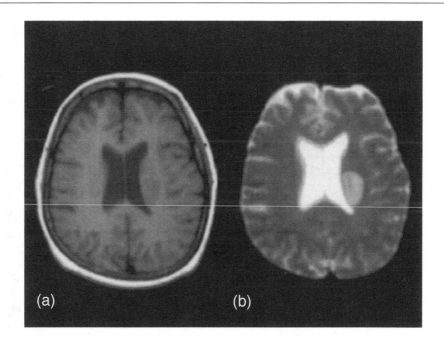

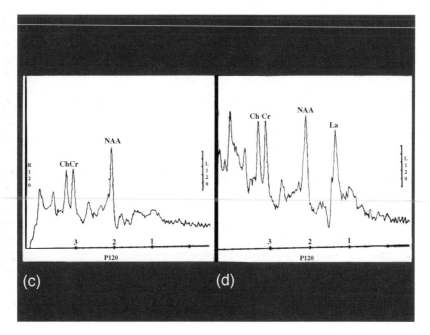

Figure 10 MRI examination of a 45-year-old patient with acute right side weakness: (a) T_1-weighted image; (b) T_2-weighted image. A rounded lesion at the left-periventricular region has hypointensity on the T_1-weighted image (a) and hyperintensity on the T_2-weighted image (b). This lesion does not demonstrate any contrast enhancement. Although the referring physician has high suspicion of acute stroke, the MR findings are not characteristic of acute stroke and are more consistent with a mass lesion. Proton spectra were obtained from normal parenchyma on the contralateral side (c) and from the lesion (d). This lesion shows a large lactate (La) peak and a diminished N-acetyl aspartate (NAA) peak at 1.3 and 2.0 ppm, respectively (d), compared with the normal tissue (c). (Ch, choline; Cr, creatine.) Follow-up MR examination obtained several months later demonstrated atrophy and gliosis at the same location, consistent with chronic infarction. In this particular patient, spectroscopy was valuable not only because it enabled surgical biopsy to be avoided, but also because it allowed a correct diagnosis for acute stroke

mation about energy status and oxidative metabolites related to the underlying pathophysiology (ischemia). Proton MRS is more effective because hydrogen atoms are more abundant than phosphorus in the brain parenchyma and proton MRS is easier to perform with the clinical unit.

There are two types of approach: localized and spectroscopic imaging (MRSI). Localized proton MRS methods can be separated into those with long and short echo times. Long *TE* (135 or 270 ms) acquisitions generally have proved easiest to use in clinical practice. Localization methods commonly

For References see p. 937

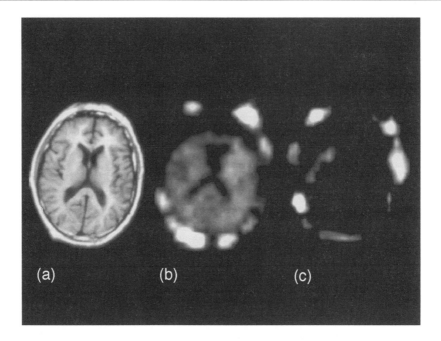

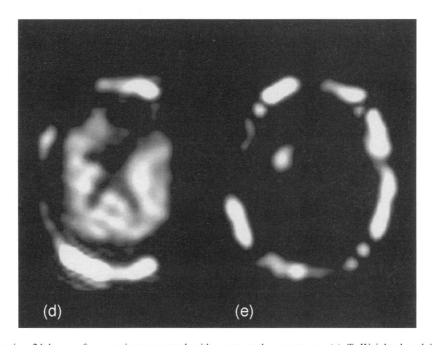

Figure 11 MRI examination 24 hours after a patient presented with acute stroke symptoms. (a) T_1-Weighted and T_2-weighted images were unremarkable. (b), Spectroscopic imaging of N-acetyl aspartate (NAA) did not demonstrate abnormality (area with decreased NAA). (c) Spectroscopy imaging of lactate showed an area of increased lactate in the right basal ganglia. (d) Follow-up spectroscopic imaging of NAA at 8 days demonstrated a focal area with decreased NAA at the right basal ganglia that was much smaller than the initial abnormalities demonstrated on the spectroscopy imaging of lactate (c) and was consistent with a focal infarction demonstrated on the T_2-weighted imaging obtained at the same day (not shown). (e) Spectroscopy imaging of lactate at the same time again showed persistent elevations of lactate in the infarcted tissue. (Courtesy of Dr Nick Bryan, Diagnostic Radiology Department, NIH)

used in clinical proton MRS include depth-resolved surface coil spectroscopy (DRESS), point-resolved surface coil spectroscopy (SPARS), and the stimulated-echo method (STEAM).[40–42] For practical purposes, STEAM allows for shorter echo times, thereby improving resolution of metabolites (e.g. myoinositol, glutamate, glutamine, and glycine).

Magnet designs have until recently favored the use of long TE sequences that are insensitive to eddy currents and can be easily implemented on the commercial scanner. Using long TE, the signal from most metabolites in the brain is lost except for that from four: choline, creatine, N-acetyl aspartate (NAA), and lactate. However, improvements in MRS techniques have allowed short *TE* sequences taking 10–15 minutes; consequently these may realistically be incorporated into a routine imaging study without a significant time penalty.

The characteristic findings of a NAA and lactate peak can reflect the underlying pathophysiology of acute ischemia through a biochemical parameter. The NAA peak is attributable to its *N*-acetylmethyl group, which resonates at 2.0 ppm (Figure 10(c,d)). This peak also contains contributions from less-important *N*-acetyl groups. NAA is considered as a neuronal marker and is not present in tumors outside the central nervous system. Its concentration decreases with many diseases of the brain.[43] Similar to NAA, glutamate and *N*-acetylaspartyl glutamate are also localized in neurons. Glutamate is an excitatory neurotransmitter that plays a role in mitochondrial metabolism.[44] Glutamine plays a role in detoxification and regulation of neurotransmitter activities. These two metabolites resonate closely together and they are commonly represented by their sum as peaks located between 2.1 and 2.5 ppm. Breakdown of *N*-acetylaspartyl glutamate releases both NAA and glutamate, and subsequent breakdown of NAA leads to aspartate. The compounds are excitatory amino acids that increase with ischemia and cause 'toxic' effects, resulting in expanded tissue damage. Therefore, the concentrations of *N*-acetylaspartyl glutamate and glutamate may serve to monitor treatments designed to protect brain tissues by blocking excitatory amino acids.

The lactate peak has a special configuration and occurs at 1.32 ppm. It consists of two distinct, resonant peaks called a 'doublet' and is caused by the magnetic field interactions between adjacent protons (J coupling). Lactate levels of the normal brain parenchyma are low. The presence of lactate generally indicates that the normal cellular oxidative respiration mechanism is no longer in effect, and that carbohydrate catabolism is taking place.[45] Confirmation that a peak at 1.32 ppm corresponds to lactate may be obtained by altering the *TE*. At a *TE* of 272 ms, lactate projects above the baseline, whereas at a *TE* of 136 ms the lactate doublet is inverted below the baseline.

In humans, proton MRS performed within the first 24 hours after a stroke shows elevation of lactate, suggesting that anaerobic glycolysis is occurring as a result of ischemia (Figures 2 and 3).[46] Decreased NAA can be seen as early as 4 days after acute ischemia, suggesting neuronal loss (infarction) (Figures 10 and 11).[47] In chronic infarctions, there is a decrease in NAA, creatine, and choline, but no evidence of lactate.[48] Experimentally, an increase in lactate may be detected after only 2 to 3 minutes of cerebral ischemia.[49] In these animals, the lactate returned to normal when the underlying ischemia was reversed. In the evaluation of human hyperacute cerebral infarction, shortening of examination time as well as the better tolerance of the motion artifact will be needed in order to make spectroscopy a realistic tool in the management of the stroke patient.

The age-related white matter changes contain normal levels of NAA and creatine[50] but do not contain lactate. The increased choline levels are suggestive of an alteration of the white matter phospholipids.

7 CONCLUSIONS

An understanding of the spectrum of MRI findings in acute ischemia may facilitate a correct diagnosis of stroke, particularly in the hyperacute stage, and allow differentiation from other etiologies. Recent advances in diffusion and perfusion MRI indicate a potential for providing important information concerning factors that determine tissue viability and/or reversibility; this will assist clinical decisions in selecting the appropriate patients for thrombolytic therapy.

8 RELATED ARTICLES

Anisotropically Restricted Diffusion in MRI; Brain MRS of Human Subjects; Brain MRS of Infants and Children; Diffusion: Clinical Utility of MRI Studies; Hemorrhage in the Brain and Neck Observed by MRI; ; Magnetic Resonance Imaging of White Matter Disease.

9 REFERENCES

1. Stroke update, US Department of Public Health, HHS, National Institutes of Health, 1989, p. 3.
2. S. Imakita, T. Nishimura, H. Naito, N. Yamada, K. Yamamoto, M. Takamiya, Y. Yamada, Y. Sakashita, J. Minamikama, and H. Kikuohi, *Neuroradiology*, 1987, **29**, 422.
3. C. Virapongse, A. Mancuso, and R. Quisling, *Radiology*, 1986, **161**, 785.
4. W. T. C. Yuh and M. R. Crain, *Neuroimaging Clin. North Am.*, 1992, **2**, 421.
5. W. T. C. Yuh, M. R. Crain, D. J. Loes, G. M. Greene, T. J. Ryals, and Y. Sato, *Am. J. Neuroradiol.*, 1991, **12**, 621.
6. M. R. Crain, W. T. C. Yuh, G. M. Greene, D. J. Loes, T. J. Ryals, Y. Sato, and M. N. Hart, *Am. J. Neuroradiol.*, 1991, **12**, 631.
7. J. Biller, W. T. C. Yuh, G. W. Mitchell, A. Bruno, and H. P. Adams Jr., *Stroke*, 1988, **19**, 297.
8. A. Uchino, T. Mori, and M. Ohno, *Clin. Imaging*, 1991, **15**, 176.
9. J. I. Lane, A. E. Flanders, H. T. Doan, and R. D. Bell, *Am. J. Neuroradiol.*, 1991, **12**, 819.
10. A. D. Elster and D. M. Moody, *Radiology*, 1990, **177**, 627.
11. S. Imakita, T. Nishimura, N. Yamada, H. Naito, M. Takamiya, Y. Yamada, J. Minamikama, H. Kikuchi, M. Nakamura, and T. Swada, *Neuroradiology*, 1988, **30**, 372.
12. D. P. Mueller, W. T. C. Yuh, D. J. Fisher, K. B. Chandran, M. R. Crain, and Y.-H. Kim, *Am. J. Neuroradiol.*, 1993, **14**, 661.
13. M. Brant-Zawadzki, B. Pereira, P. Weinstein, S. Moore, W. Kucharczyk, T. Berry, M. McNamara, and N. Derugin, *Am. J. Neuroradiol.*, 1986, **7**, 7.
14. K. A. Hossmann and F. J. Schuier, *Stroke*, 1980, **11**, 583.
15. F. J. Schuier and K. A. Hossmann, *Stroke*, 1980, **11**, 593.
16. B. A. Bell, L. Symon, and N. M. Branston, *J. Neurosurg.*, 1985, **62**, 31.
17. O. Gotoh, T. Asano, T. Koide, and K. Takakura, *Stroke*, 1985, **16**, 101.
18. M. Brant-Zawadzki, P. Weinstein, H. Bartkowski, and M. Moseley, *Am. J. Neuroradiol.*, 1987, **8**, 39.
19. R. N. Bryan, *Radiology*, 1990, **177**, 615.
20. T. M. Simonson, T. J. Ryals, W. T. C. Yuh, G. P. Farrar, K. Rezai, and H. T. Hoffman, *Am. J. Roentgenol.*, 1992, **159**, 1063.
21. T. J. Masaryk, *Curr. Opin. Radiol.*, 1992, **4**, 79.
22. M. E. Moseley, Y. Cohen, J. Mintorovitch, L. Chileuitt, H. Shimizu, J. Kucharczyk, M. F. Wendland, and P. R. Weinstein, *Magn. Reson. Med.*, 1990, **14**, 330.
23. M. P. Marks, A. de Crespigny, D. Lentz, D. R. Enzmann, G. W. Albers, and M. E. Moseley, *Radiology*, 1996, **199**, 403.
24. R. A. Knight, M. O. Dereski, J. A. Helpern, R. J. Ordidge, and M. Chopp, *Stroke*, 1994, **25**, 1252.
25. I. Loubinoux, A. Volk, and J. Borredon, *Stroke*, 1997, **28**, 419.

For References see p. 937

26. K. Kohno, M. Hoehn-Berlage, G. Mies, T. Back, and K. A. Hossmann, *Magn. Reson. Imag.*, 1995, **13**, 73.
27. K. O. Lövblad, H. J. Laubach, A. E. Baird, F. Curtin, G. Schlaug, R. R. Edelman, and S. Warach, *Am. J. Neuroradiol.*, 1998, **19**, 1061.
28. G. Schlaug, B. Siewert, A. Benfield, R. R. Edelman, and S. Warach, *Neurology*, 1997, **49**, 113.
29. D. Davis, J. Ulatowski, S. Eleff, M. Izuta, S. Mori, D. Shungu, and P. C. M. van Zijl, *Magn. Reson. Med.*, 1994, **31**, 454.
30. M. Miyabe, S. Mori, P. C. M. van Zijl, J. R. Kirsch, S. M. Eleff, R. C. Koehler, and R. J. Traystman, *J. Cereb. Blood Flow Metab.*, 1996, **16**, 881.
31. T. Ueda, W. T. C. Yuh, J. E. Maley, J. P. Quets, P. Y. Hahn, and V. A. Magnotta, *Am. J. Neuroradiol.*, 1999, **20**, 983.
32. D. A. Finelli, A. L. Hopkins, W. R. Selman, R. C. Crumrine, S. U. Bhatti, and W. D. Lust, *Magn. Reson. Med.*, 1992, **27**, 189.
33. M. Maeda, W. T. C. Yuh, T. Ueda, J. E. Maley, D. L. Crosby, M. W. Zhu, and V. A. Magnotta, *Am. J. Neuroradiol.*, 1999, **20**, 43.
34. D. C. Tong, N. A. Yenari, G. W. Albers, M. O'Brien, M. P. Marks, and M. E. Moseley, *Neurology*, 1998, **50**, 864.
35. A. G. Sorensen, F. S. Buonanno, R. G. Gonzalez, L. H. Schwamm, M. H. Lev, F. E. Huang-Hellinger, T. G. Reese, R. M. Weisskoff, T. L. Davis, N. Suwanwela, U. Can, J. A. Moreira, W. A. Copen, R. B. Look, S. P. Finklestein, B. R. Rosen, and W. J. Koroshetz, *Radiology*, 1996, **199**, 391.
36. G. Rordorf, W. J. Koroshetz, W. A. Copen, S. C. Cramer, P. W. Schaefer, R. F. Budzik, L. H. Schwamm, F. Buonanno, A. G. Sorenson, and G. Gonzalez, *Stroke*, 1998, **29**, 939.
37. J. Röther, F. Gückel, W. Neff, A. Schwartz, and M. Hennerici, *Stroke*, 1996, **27**, 1088.
38. B. Ross and T. Michaelis, *Magn. Reson. Q.*, 1994, **10**, 191.
39. M. Castillo, L. Kwock, S. K. Mukherji, *Am. J. Neuroradiol.*, 1996, **17**, 1.
40. P. A. Bottomley, T. B. Foster, and R. B. Darrow, *J. Magn. Reson.*, 1984, **59**, 338.
41. P. R. Luyten, A. J. H. Marien, B. Systma, et al., *J. Magn. Reson.*, 1989, **9**, 126.
42. J. Frahm, H. Bruhn, M. L. Gyngell, K. D. Merboldt, W. Hanicke, and R. Sauter, *Magn. Reson. Med.*, 1989, **9**, 79.
43. B. L. Miller, *NMR Biomed.*, 1991, **4**, 46.
44. M. S. van der Knapp, B. Ross, and J. Valk, in 'Magnetic Resonance Neuroimaging', ed. J. Kucherarczyk, M. Mosely, A. J. Barkovich, CRC Press, Boca Raton, FL, 1994, pp. 245–318.
45. J. A. Sanders, in 'Functional Brain Imaging', ed. W. W. Orrison, J. D. Lewine, J. A. Sanders, M. F. Harthshorne, Mosby, St Louis, MO, 1995, pp. 419–467.
46. P. B. Barker, J. H. Gillard, P. C. M. van Zijl, B. J. Seher, D. F. Hanley, A. M. Agildere, S. M. Oppenheimer, and R. N. Bryan, *Radiology*, 1994, **192**, 723.
47. H. Bruhn, J. Frahm, M. L. Gyngell, K. D. Merboldt, W. Hanick, and R. Sauter, *Magn. Reson. Med.*, 1989, **9**, 126.
48. J. H. Duijn, G. B. Matson, A. A. Maudsley, J. W. Hugg, M. W. Weiner, *Radiology*, 1992, **183**, 711.
49. K. L. Behar, J. A. den Hollander, M. E. Stromski, T. Ogino, R. G. Shulman, O. A. Petroff and J. W. Prichard, *Proc. Natl. Acad. Sci. USA*, 1983, **80**, 4945.
50. Sappey-Marinier, G. Calabrese, H. P. Hetherington, S. N. Fisher, R. Deicken, C. Van Dyke, G. Fein, and M. W. Weiner, *Magn. Reson. Med.*, 1992, **26**, 313.

Biographical Sketches

William T. C. Yuh. *b* 1947. B.S., 1971, Chiao-tung University, Taiwan. M.S.E.E., 1974, Auburn University. M.D., 1980, University of Alabama-Birmingham, USA. Internship, Lloyd Noland Hospital. Radiology residency, UCLA Medical Center. Nuclear medicine fellowship, V. A. Wadsworth-UCLA. Magnetic resonance fellowship, UCLA

Medical Center. Instructor and assistant professor, neuroradiology fellowship, associate professor, professor, Department of Radiology, The University of Iowa, 1994–present. Approx. 150 publications. Research specialties: magnetic resonance contrast agents, central nervous system ischemia.

Toshihiro Ueda. *b* 1960. M.D., 1987, Ehime University School of Medicine, Japan, Neurosurgery residency and fellowship at Ehime University School of Medicine, Ph.D., 1995, Postgraduate School of Ehime University School of Medicine. Associate (Staff Physician), Department of Neurosurgery, Ehime University School of Medicine, 1995–96. Research fellow, Department of Radiology, the University of Iowa, 1996–98. Visiting Assistant Professor, Department of Radiology, the University of Iowa, 1998–present. Approx. 40 publications. Research interests: cerebral ischemia, neurointervention, diffusion perfusion MRI, thrombolytic therapy.

J. Randy Jinkins. *b* 1949. B.A., 1971, Biology, University of Texas, Austin, USA. M.D., 1975, University of Texas, Galveston, USA. Radiology Residency, Emory University, Atlanta, USA. Neuroradiology Fellowship, Massachusetts General Hospital, Harvard Medical School, Boston, MA, USA. Currently Associate Professor of Radiology, University of Texas, San Antonio, USA. Approx. 125 publications. Current research specialty: pathophysiologic aspects of disease as they pertain to medical neuroradiologic imaging.

Ronald A. Rauch. *b* 1953. B.A. (Biochemistry), 1975, University of Kansas M.D., 1979, Baylor College of Medicine. Neurology Resident, Stanford, 1981–84. Radiology Resident, University of California, Irvine, 1985–88. Neuroradiology fellow, Long Beach Memorial, 1988–89, and UCLA, 1989–90. Currently assistant professor of radiology, University of Texas Health Science Center at San Antonio. Approx 20 publications. Current research interests: use of MRI to quantify corpus callosum morphology, MRI of white matter changes, especially those associated with dementia, and MRI of spondylolisthesis associated with spondylolysis.

Magnetic Resonance Imaging of White Matter Disease

Donald M. Hadley

Institute of Neurological Sciences, Glasgow, UK

1 INTRODUCTION

1.1 Characteristics

The white matter of the brain constitutes the core of the hemispheres, brainstem, and cerebellum. It is composed of axons, which transmit chemically mediated electrical signals, and glial supporting cells set in a mucopolysaccharide ground substance. The glial cells—oligodendrocytes, astrocytes, ependymal cells, and microglia—account for about half the brain's volume and 80–90% of its cells. The oligodendrocytes provide an insulating sheath of myelin by invagination, wrapping concentric layers of their cell membrane around the axons. Astrocytes are now known to influence and communicate through their long foot processes, which are in intimate contact with capillaries, neurones, synapses, and other astrocytes. The ependymal cells form the lining of the brain's internal cavities, while the microglia, normally relatively inconspicuous, are capable of enlarging and becoming active macrophages.

The white matter fibers are grouped into location-specific tracts, which can be divided into three main types:

(a) projection fibers that allow efferent and afferent communication between the cortex and target organ;

(b) long and short association fibers, which connect cortical regions in the same hemisphere; and

(c) commissural fibers, which connect similar cortical regions between hemispheres.

The formation and maturation of axons has been reviewed by Barkovich et al.[1] After development of the axons and their synapses, the final process of myelination occurs. This is crucial to the appearance on MRI.

1.2 Evolution and Imaging

The contrast obtained between gray and white matter is largely due to the myelination of the white matter tracts. Myelin is composed of a bilayer of lipids (phospholipids and glycolipids), cholesterol, and large proteins. In 1974, Parrish et al.[2] showed differences in the relaxation times of gray and white matter by spectroscopy before imaging was possible. These differences were later confirmed by MRI. In white matter, the myelin lipids themselves contain few mobile protons visible to routine MRI, but they are hydrophobic, and therefore, as myelination progresses, there is loss of brain water and a decrease in T_2 signal. Cholesterol tends to have a short T_1, and the increased protein also decreases the T_1 of water. This results in white matter having a reduced intensity on T_2-weighted images and increased intensity on T_1-weighted images compared with unmyelinated fibers or gray matter.

Myelination of the white matter is first noted in the cranial nerves during the fifth fetal month, and continues throughout life.[3] By birth, myelination is present in the medulla, dorsal midbrain, cerebellar peduncles, posterior limb of the internal capsule, and the ventrolateral thalamus. In general, myelination is completed from caudal to cephalad, from central to peripheral, and from dorsal to ventral. Key landmarks include the pre- and post-central gyri, which are myelinated at one month, with the motor tracts completed by three months. At this time, myelination is completed in the cerebellum, and progresses through the pons in the corticospinal tracts, cerebral peduncles, the posterior limb of the internal capsule, and up to the central portion of the centrum semiovale, to be completed by six months. The optic radiations and the anterior limb of the internal capsule are myelinated by three months. Myelination in the subcortical white matter is first noted in the occipital region at three months, and proceeds rostrally to the frontal lobes. This posterior-to-anterior maturation is noticeable in the corpus callosum, with the splenium first showing myelin at four

months, progressing to the genu, where it is complete at six months.

This normal development is best visualized by MRI with age-related heavily T_1-weighted (e.g., inversion–recovery) sequences for the first six months, by which time the appearance is close to adult; after this, T_2-weighted sequences are most helpful, with all the major tracts assuming an adult appearance by 18 months. The cause of these differences is not fully understood, but is thought to be related to the initial hydrophilicity, with its associated increase in hydrogen bonding with water. Next, the T_2 shortening may be caused by the subsequent tightening of the myelin sheath, further redistributing the free water components. It has been shown that the very earliest changes of myelination are shown better by diffusion- or fluid-attenuated T_2-weighted sequences than by conventional T_1- or T_2-weighted sequences.[4,5] It must be noted, however, that some areas around the trigones of the lateral ventricles may not fully myelinate in normal children until they are 10 years old.

1.3 Classification of Abnormalities

There are a bewildering number of white matter diseases with multiple etiologies and pathological mechanisms. Although MRI is very sensitive to any white matter abnormality, it is rarely possible for the radiologist to make a specific diagnosis.[6] It is, however, useful to divide them into three main groups:

(a) a dysmyelinating group in which there is a biochemical defect in the production or maintenance of normal myelin; some of the individual enzyme deficiencies have been identified and will be discussed below;

(b) a demyelinating group in which myelin is formed normally but is later destroyed;

(c) a vascular group in which normally myelinated white matter is destroyed by a critical reduction in blood flow to a particular region; this may also involve the adjacent gray matter or a large segmental part of the brain, depending on the extent and severity of the reduction in blood flow and the susceptibility of the cells involved.

In many of these conditions, the diagnosis is biochemical, but the radiologist has an important role in suggesting the diagnosis, and documenting progression, response to therapy, or complications.

2 DYSMYELINATION DISEASE

2.1 Leukodystrophies

These are diseases where dysmyelination occurs as a result of the production and maintenance of abnormal myelin. Becker[7] and Kendall[8] have produced excellent reviews of this subject.

2.1.1 Alexander's Disease (Fibrinoid Leukodystrophy)

This usually presents in the first few weeks of life with macrocephaly and failure to attain developmental milestones. There is progressive spastic quadriparesis and intellectual fail-

For References see p. 950

ure. Death ensues in infancy or early childhood, although cases have been reported in adolescents and adults. An enzyme defect has not yet been identified.

MRI shows increased T_1 and T_2 relaxation times, starting in the frontal lobes, and progressing to the parietal and capsular regions.[9,10] With the accumulation of Rosenthal fibers around blood vessels, there may be disruption of the blood–brain barrier, producing frontal periventricular enhancement. Frank cystic changes develop in the frontal lobes in the later stages, with atrophy of the corpus callosum.

2.1.2 Canavan's Disease (Spongiform Leukodystrophy)

This is a lethal autosomal recessive neurodegenerative disorder of Jewish infants caused by a deficiency of aspartoacylase. The disease progresses with marked hypotonia, macrocephaly, seizures, and failure to attain motor milestones in the first few months of life, although sometimes it starts as early as a few days, progressing to spasticity, intellectual failure, and optic atrophy. Death usually occurs in the second year of life. The radiological features may be seen before the full clinical picture has developed, but the diagnosis depends on the biochemical testing.

The demyelinated white matter shows increased T_1 and T_2 relaxation times, preferentially in the arcuate U fibers of the cerebral hemispheres. The occipital lobes are more involved than the frontal, parietal, and temporal lobes. Initially, it may spare the corpus callosum, deep white matter, and internal and external capsules, but, as it progresses, diffuse white matter involvement occurs, which leads to eventual cortical atrophy.[11]

2.1.3 Krabbe's Disease (Globoid Cell Leukodystrophy)

This is a rare, lethal, autosomal recessive leukodystrophy (locus now mapped to chromosome 14) due to a deficiency of the first of the two galactocerebroside β-galactosidases. This arrests the normal breakdown of cerebroside, disrupts the turnover of myelin, and results in the accumulation of galactosylsphingosine. This is toxic to oligodendrocytes, and causes a marked loss of myelin, although the minute amount of myelin remaining is normal. Changes become evident between one and six months old, although it is occasionally noted earlier, and leads to death within one to three years. The clinical diagnosis is based on an assay of β-galactosidase from leukocytes or skin fibroblasts.

Early in the disease process, MRI can be normal[12] and a spectrum of lesions then develops over several months. These are nonspecific symmetrical patchy changes in the periventricular white matter similar to many other demyelinating diseases such as multiple sclerosis, with increased T_1 and T_2 signals.[8,13] The thalami, central white matter, and cerebellar white matter may show decreased T_1 and normal or slightly decreased T_2 signals.[14] These changes are reflections of the increased attenuation sometimes seen on computerized tomography (CT), and are probably the result of paramagnetics such as crystalline calcification. In advanced disease, there is diffuse cerebral atrophy.[12]

2.1.4 Pelizaeus–Merzbacher Disease

This term has been used to cover the five subtypes of sudanophilic leukodystrophy,[15] but here it will be taken to mean the slowly progressive X-linked recessive leukodystrophy. The

dysmyelination is now thought to be due to a point mutation in the PLP gene coding for the myelin–protein proteolipid protein. It presents in infancy, and runs a very chronic course leading to death in adolescence or early adulthood.

On MRI, there is a general lack of myelination without white matter destruction. The brain has the appearance of the newborn, with high signal intensity only appearing in the internal capsule, optic radiations, and proximal corona radiata on T_1-weighted images, and practically no low signal in the supratentorial region on T_2-weighted sequences.[16] A 'tigroid' pattern consisting of normal myelinated white matter within diffuse dysmyelination can be seen later on T_2-weighted sequences. When severe, there may be a complete absence of myelin. Cortical sulcal enlargement may be seen.

2.1.5 Metachromatic Leukodystrophy

The commonest of the sphingolipidoses is due to a deficiency in the activity of arylsulfatase A. This enzyme is responsible for normal metabolism of sulfatides, which are important constituents of the myelin sheath. The disease is subdivided into:

(a) neonatal, with a rapid downhill course leading to early death;

(b) infantile, presenting between one and four years with polyneuropathy, ataxia, progressive retardation, and spastic tetraparesis;

(c) juvenile, with dementia[17] and behavioral disorders progressing to spastic tetraparesis;

(d) the rare adult type, presenting at any age with dementia and spastic paraparesis.

The imaging findings are nonspecific, with symmetrical areas of increased T_1 and T_2 relaxation times in the centrum semiovale, representing progressive dysmyelination and gliosis within areas of normal myelination. The peripheral white matter, including the arcuate U fibers, is spared until late in the disease. As there is no inflammation, enhancement is not a feature. These appearances allow differentiation from the gross lack of myelination seen in Pelizaeus–Merzbacher disease. As the disease progresses, brain atrophy becomes more prominent than the white matter signal changes. Proton MRS may have a clinical role in the diagnosis.[18]

2.1.6 Adrenoleukodystrophy (Childhood Type: X-Linked)

This is seen exclusively in males, and was thought to be due to a deficiency of acyl-CoA synthetase. Long-chain fatty acids are incorporated in cholesterol esters, replacing the normal nonesterified cholesterol. It usually presents between the ages of five and ten years, with a disturbance of gait and intellectual impairment, with fairly rapid progression and the development of hypotonia, seizures, visual impairment, and bulbar symptoms. Neurological complaints are classically preceded by adrenal insufficiency and skin pigmentation, which may be precipitated by an intercurrent infection; however, they sometimes may never appear.

Symmetrical long T_1 and T_2 signals are usually first seen in the peritrigonal regions extending into the splenium of the corpus callosum. Although typical, these may rarely be seen in other white matter diseases such as Krabbe's. These signal changes gradually extend to involve the occipital lobes and more anterior regions such as the medial and lateral geniculate bodies, thalami, and the inferior brachia. The pyramidal tracts and the occipito-temporo-parieto-pontine fibers show progressive alteration, with sparing of the fronto-pontine fibers in the medial part of the crus cerebri. The lateral lemnisci and the cerebellar white matter may also become involved. Although this is the usual pattern, atypical symmetrical or asymmetrical involvement of other lobes sometimes occurs[19] (Figure 1).

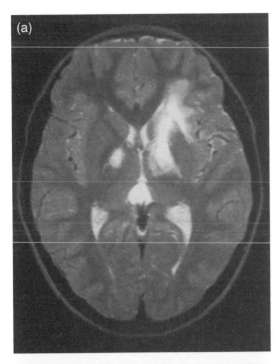

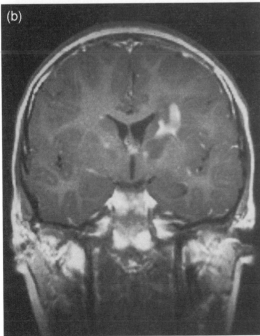

Figure 1 Adrenoleukodystrophy: (a) T_2-weighted and (b) gadolinium-enhanced T_1-weighted sections of a five-year-old boy showing bilateral focal areas of white matter abnormality with marginal enhancement (biopsy-proven)

Three zones of abnormal long T_1 and T_2 signals can be recognized:

(a) a central region of gliosis with necrosis and cavitation, next to

(b) an intermediate region of active inflammatory demyelination that shows enhancement due to blood–brain barrier breakdown, surrounded by

(c) a peripheral less marked zone of demyelination without inflammatory reaction.

As the disease progresses, atrophy becomes the more dominant feature on MRI.[8]

2.1.7   Adrenoleukodystrophy (Adrenomyeloneuropathy—Adult Type)

This often presents in the same family as the childhood type, but occurs in adult life. It is caused by a similar enzyme defect. The abnormal myelination is most marked in the corticospinal and spinocerebellar tracts, but can extend into the brainstem, involving the pyramidal tracts running into the posterior limb of the internal capsules, and the frontopontine and occipito-temporo-parieto-pontine fibers. The cerebellar white matter is usually affected, with sparing of the cerebral white matter. MR changes usually appear late in the course of the disease.

2.1.8   Adrenoleukodystrophy (Neonatal)

Several disorders have been grouped under this heading, but with active research at present underway, the classification of the enzyme defects may change. Severe progressive neurological impairment occurs, with psychomotor retardation, dysmorphic facial features, hypotonia, seizures, and defective liver function. In contradistinction to the childhood type, these abnormalities are present from birth. The enzyme defect may be confined to fatty acyl-CoA oxidase resulting in defective very long chain fatty acid oxidation.

There is diffuse degeneration of cerebral white matter, causing atrophy at a very early age. Progressive MRI changes have been described[20] in a single case followed for three years, with delayed myelination followed by symmetrical demyelination of the corona radiata, optic radiations, and pyramidal tracts.

2.1.9   Phenylketonuria

This is an autosomal recessive metabolic encephalopathy due to a defect in phenylalanine hydroxylase conversion of phenylalanine to tyrosine, resulting in hyperphenylalaninemia. Strict dietary restrictions must be maintained from early infancy to prevent profound mental retardation. Other cofactor defect variants may result in lesser or greater degrees of encephalopathy.[21] When there is a defect of dihydropteridine reductase, there are severe neurological and cognitive abnormalities in spite of adequate dietary restrictions. Severe white matter changes have been noted,[22] with cystic degeneration and loss of parenchyma.

Subtle abnormalities of white matter have been shown by MRI in older children and adults who have classical phenylketonuria despite having maintained a degree of dietary restriction.[23,24] This possibly provides some evidence for continuing the restrictive diet and phenylalanine-free protein supplements and into adulthood.[25]

For References see p. 950

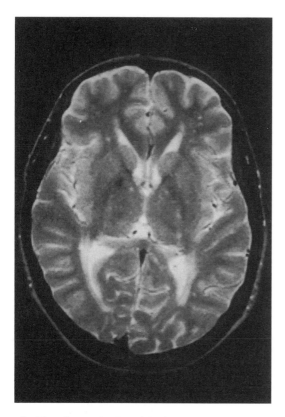

Figure 2 Phenylketonuria: T_2-weighted section showing subtle white matter hyperintensities in the optic radiations in spite of apparently adequate dietary control in a 13-year-old

Varying degrees of periventricular white matter abnormality have been shown, with focal and diffuse lengthening of T_1 and T_2 relaxation times most easily seen on the T_2-weighted images (Figure 2). In some studies, these changes were found to correlate loosely with the adequacy of the reduction and maintenance of serum phenylalanine levels,[25,26] while other workers showed no clear relationship.[24]

2.2 Miscellaneous

Over 600 individual dysmyelination disorders that may affect MRI appearances have been identified in childhood alone. Continuing research is progressively isolating the individual enzyme or gene defects, which will in time allow more specific classification. Meanwhile, the following less well defined groups of disorders will be considered.

2.2.1 Neurodegenerative Disorders

These occur in a number of devastating developmental disorders of childhood, which are either congenital or acquired. Clinical findings are usually nonspecific, and laboratory tests have to be selected carefully. Imaging demonstrates the results of abnormal cellular function on parenchymal morphology. MRI is sensitive, but specificity is limited, and it must be integrated with the other clinical findings. Proton MRS may be able to detect abnormal metabolite levels and allow an earlier and more specific determination of neurodegeneration.[27]

2.2.2 Lysosomal Disorders

Several of these have been mentioned under specific enzyme defects above, such as metachromatic leukodystrophy and Krabbe's disease. All lack activity of a specific lysosome enzyme, which is inherited in an autosomal recessive manner except Hunter II, which is X-linked. Abnormal materials build up in the lysosomes. The CNS is affected directly or secondary to the metabolic abnormality in adjacent structures.

Dysmyelination is shown as an increase in T_1 and T_2 relaxation times in the white matter, with a variable degree of involvement of the arcuate U fibers. In Fabry's disease, involvement of the small arteries may cause multifocal small infarcts visible on MRI. The gangliosidosis in addition may show focal decreases of T_2 relaxation time in the thalami, possibly reflecting the calcification seen on CT.[7,8]

2.2.3 Peroxisomal Disorders

These relate to deficiencies in the activity of respiratory enzymes and organelles of most cells. In most of these diseases, the CNS is involved. Adrenoleukodystrophy has already been discussed above, but there are multiple other rarer diseases that belong to this classification. In general, they produce dysmyelination. Several also show disturbances of neuronal migration. In some, there is an additional inflammatory response.

2.2.4 Mitochondrial Encephalopathies

These group of disorders are characterized by functionally or structurally abnormal mitochondria in the CNS or muscle. They are transmitted by non-Mendelian maternal inheritance, resulting in slowly progressive multisystem diseases with a wide range of clinical presentations, usually appearing in childhood but showing considerable variability depending on their severity.

Imaging is nonspecific.[28] There is diffuse but variable white matter atrophy and lengthened T_1 and T_2 signals in the basal ganglia. Focal infarcts may also be seen (Figure 3). Abnormal metabolites, including lactate, have been shown by MRS in the brain lesions. This is thought to be due to impaired aerobic metabolism of pyruvate.[29,30]

In Leigh's disease, there may be spongy degeneration with astroglial and microglial reaction, with vascular proliferation affecting the basal ganglia, brainstem, and spinal cord. Cerebellar and cerebral white matter undergoes demyelination, with preservation of the nerve cells and axons resulting in hyperintensity and hypointensity on T_2- and T_1-weighted images, respectively.

3 DEMYELINATION DISEASE

3.1 Idiopathic

3.1.1 Multiple Sclerosis

Multiple sclerosis (MS) is an idiopathic inflammatory and demyelinating disorder of the central nervous system (CNS). The definitive clinical and pathological features of the disease were established by Charcot[31] as long ago as 1868. Since then, the disorder's characteristics have been refined, with improvements in imaging giving the most recent insights into its pathophysiology. It is now one of the commonest reasons given for requesting MRI in the northern latitudes of the Wes-

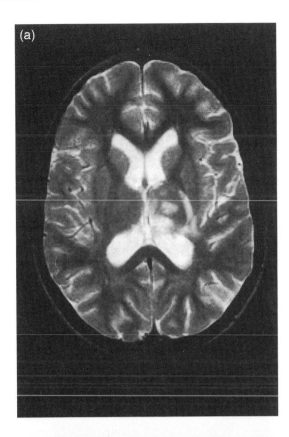

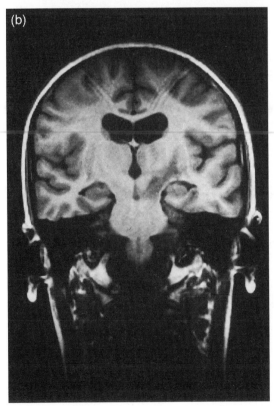

Figure 3 Mitochondrial encephalopathy: (a) T_2-weighted and (b) T_1-weighted sections in a nine-year-old boy showing increased T_1 and T_2 relaxation times representing focal infarction in the posterior limb of the left internal capsule and thalamus

tern world, and this diagnosis has huge social and economic consequences. It is therefore considered in some depth in the following paragraphs.

Clinically, MS usually follows a fluctuating course, with symptoms varying from paroxysmal and brief to slowly progressive and chronic. The lesions affect single or multiple sites simultaneously, usually involving long white matter tracts, but clinical–pathological correlation is often poor. The disorder leads to visual loss, numbness, tingling in the limbs, spastic weakness, and ataxia.[32] Diagnosis is allowed when a combination of signs and symptoms localize lesions in separate and distinct areas of the CNS disseminated in time and space.[33] Supportive laboratory and imaging data can now be defined for research studies, dividing the disease into clinically definite and probable MS, with or without laboratory support.[34,35]

Although some CT studies using high-dose iodine-enhanced delayed imaging have reported sensitivities as high as 72% when patients are in an acute relapse,[36] generally MRI has proved to have considerably greater sensitivity, and, by using intravenous gadolinium-based contrast agents, can separate acute from subacute and chronic lesions.[37]

Initially, MS lesions were shown at low field on T_1-weighted (inversion–recovery) sequences,[38] but within four years several rigorously controlled studies demonstrated the effectiveness of spin echo sequences where the T_2-dependent contrast can be organized to maximize the sensitivity between normal and abnormal tissue while minimizing partial volume effects between cerebrospinal fluid (CSF) and adjacent lesions.[39,40] Although no single sequence will detect all lesions, multifocal supratentorial white matter abnormalities have been shown on moderately T_2-weighted images in 96.5% of a group of 200 consecutive patients with clinically definite MS.[41] One or more periventricular lesions were seen in 98%, and lesions discrete from the ventricle in 92.5%, with cerebellar lesions in just over half of the group. Normal scans were found in 1.5% of patients. The majority of these lesions were clinically silent; therefore MRI can produce the extra information that helps to fulfill the criteria of dissemination in space (Figure 4). It can also exclude other causes of the patients' signs and symptoms, such as Arnold Chiari malformations and spinocerebellar degeneration. Serial studies with careful repositioning may also fulfill the criterion of dissemination in time by showing new, often asymptomatic, lesions[42,43] This high sensitivity has now been confirmed by many workers.[36,42,44]

The T_1 of apparently normal white matter in patients with MS may be increased,[42,45] and an apparent increase in the iron content has been found in the thalamus and striatum at high field.[46] Post-mortem studies have confirmed that the long T_2 lesions found correspond to MS plaques.[42] This sensitivity makes MRI the most appropriate modality for examining a patient with suspected MS.

Unfortunately, these multifocal white matter lesions may be indistinguishable from other conditions that produce demyelination, gliosis, or periventricular effusions[47] (Table 1), and between 5 and 30% of apparently normal controls older than 50 years have been shown to have white matter lesions, probably due to asymptomatic cerebrovascular disease. The patient's age and pattern of lesions can help to improve the specificity of the MRI examination.[48] Fazekas et al.[49,50] have shown that if at least three areas of increased T_2 signal intensity are present with two of the following features—abutting

For References see p. 950

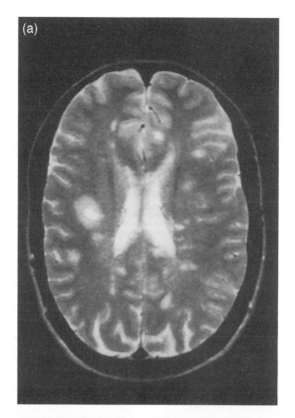

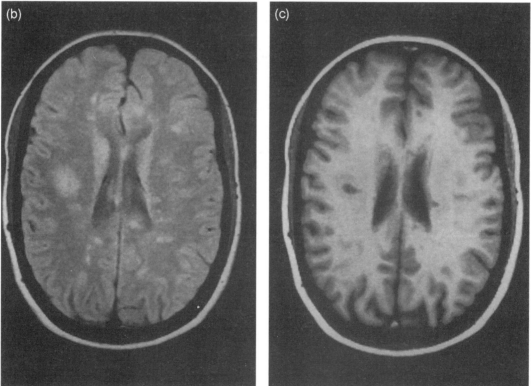

Figure 4   Multiple sclerosis: (a) T_2-weighted, (b) proton density, and (c) T_1-weighted sections showing the typical signal changes found in the multiple focal and coalescing acute and chronic plaques

the body of the lateral ventricles, infratentorial location, and size greater than 5 mm—then the sensitivity is decreased to 88%, but the specificity increases to 96% in patients with clini-

cally definite multiple sclerosis. Although the relaxation times of acute plaques have been found to be longer,[51] the range was wide, and the age of an individual lesion could not be deter-

Table 1 Differential Diagnosis of Multifocal White Matter Lesions

Multiple sclerosis
Aging, small vessel vascular disease, lacunar infarcts
Infarction
Acquired immune deficiency syndrome
Encephalitis (ADEM), (SSPE)
Progressive multifocal leukoencephalopathy
Metastases
Trauma
Radiation damage
Granulomatous disease (e.g., sarcoid)
Inherited white matter disease
Normal (in healthy elderly, especially hypertensive)
Hydrocephalus with CSF interstitial edema

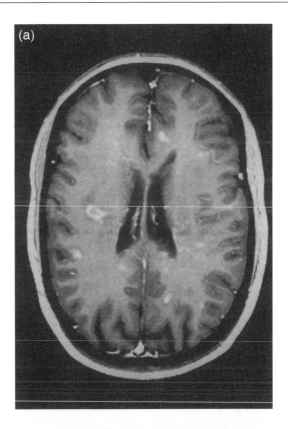

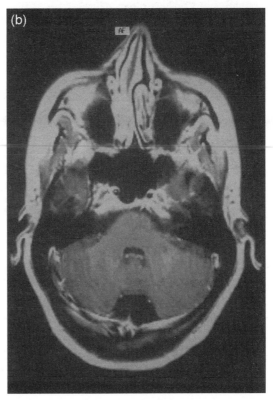

Figure 5 Multiple sclerosis: gadolinium-enhanced T_1-weighted sections showing (a) multiple enhancing acute lesions including a ring enhancing plaque and (b) a nonenhancing chronic cerebellar peduncular lesion

mined by T_1 or T_2 measurement alone. It is important to be able to define new and active lesions to differentiate between multiphasic (MS) and monophasic (ADEM) disease and to determine whether there is evidence of continuing disease progression (e.g., in clinical therapeutic trials).

The areas of perivenular inflammation and edema associated with the acute MS plaque[52] cause a transient disruption of the blood–brain barrier[53,54] and allow leakage of intravascular contrast agents. This is shown as enhancement on T_1-weighted images (Figure 5), and is safer and more effective than high-dose delayed contrast-enhanced CT.[53] Correlation with the lesions seen on T_2-weighted sections is good. Only a small number of cortical or subcortical plaques were seen solely on enhanced T_1-weighted images. Enhancement is now considered a consistent feature of recognizably new lesions or new parts of existing plaques,[54] although occasionally blood–brain barrier breakdown develops in older previously nonenhancing plaques associated with no increase in their size. The inflammatory demyelination has been shown pathologically to begin in a perivenular distribution and spread centrifugally, corresponding with the ring enhancement noted in several studies[37,54] (Figure 5). Gadolinium enhancement is particularly useful in the clear delineation of lesions in the spinal cord and optic nerves, especially if T_1-weighted fat saturation chemical shift sequences are used to reduce the high signal from surrounding periorbital fat.[55,56]

Differences in the enhancement pattern of primary and secondary progressive MS, the two major clinical patient groups, have been identified.[57] The secondary progressive group had more new lesions (18.2 lesions per patient per year), of which a larger proportion (87%) enhanced. In addition, there was enhancement at the edge of preexisting lesions. This compares with few new lesions (3.3 lesions per patient per year), of which only one enhanced in the primary progressive group at the time when there were no differences over six months in the rates of clinical deterioration between the two groups. This suggests a difference in the underlying dynamics of the inflammatory component of the disease. With improved imaging techniques, e.g., fast scanning methods, and particularly with real-time echo planar imaging,[58] the complex morphology of the initial phase of gadolinium enhancement after intravenous injection may be further elucidated and related to the lesions on the unenhanced scan. Correlation with lipid imaging may allow study of the relationship between demyelination and inflammation. Advances in the use of MRI and MRS have been reviewed by Paty et al.[59]

For References see p. 950

Now that new and biologically active lesions can be identified routinely in clinic patients by gadolinium-enhanced T_1-weighted scans, the association of blood–brain barrier leakage in some but not all of multiple plaques shown on T_2-weighted images indicates the presence of dissemination in time[60] and refines the diagnosis of MS. The method can be used to subdivide clinical groups, and will be useful in monitoring and possibly shortening the time required for therapeutic trials (e.g., with steroids[61] and interferon[62]) in MS.

3.1.2 Schilder's Disease (Myelinoclastic Diffuse Sclerosis)

This is a rare but distinctive acute demyelinating condition, which can be defined by biochemical, pathological, and electrophysiological criteria, yet remains faithful to Schilder's original description of 1912.[63] There is a severe selective inflammatory demyelination with sparing of the subcortical U fibers, and extensive attempts at remyelination.

MR white matter changes have been described,[64] with bilateral involvement of the anterior hemispheres, extensive fluctuating increased relaxation times, mass effect, and varying partial ring enhancement indicating changes in blood–brain barrier breakdown. Other leukodystrophies with a similar appearance such as adrenoleukodystrophy and Pelizaeus–Merzbacher disease must be excluded by biochemical testing.

3.2 Postinflammatory: Viral, Allergic, Immune-mediated Responses to Previous Infection

3.2.1 ADEM

Acute disseminated encephalomyelitis is a demyelinating disease that is thought to be an immune-mediated disorder secondary to a recent viral infection or more rarely to vaccination. It has an acute onset and a monophasic course, in contradistinction to MS. Most patients make a complete recovery, with no neurological sequelae. This makes it one of the most important differential diagnoses in the acute clinical situation.

Pathologically, there is a diffuse perivenous inflammatory process resulting in confluent areas of demyelination. These frequently occur at the corticomedullary junction, with gray matter much less often involved than white. Large areas of increased T_1 and T_2 relaxation time are found on MRI in both hemispheres, but the effects are usually asymmetrical.[65] Although in the acute stage, the demyelinating lesions may enhance, the blood–brain barrier quickly returns to normal. As with MS, MRI is much more sensitive than CT at demonstrating these lesions.

3.2.2 SSPE

Subacute sclerosing panencephalitis is a rare progressive demyelination resulting from reactivation of the measles virus due to a defect in immunity that allowed the virus to remain latent.[66] There is a variable rate of progression, with death between two months and several years after reactivation.

There is perivascular infiltration by inflammatory cells, cortical and subcortical gliosis, and white matter demyelination progressing from occipital to the frontal lobes and from the cerebellum to the brainstem and spinal cord. This is reflected in an increase in T_1 and T_2 relaxation times of the multifocal patchy white matter lesions.[67,68]

3.2.3 PML

Progressive multifocal leukoencephalopathy is a demyelinating disease probably caused by the papova viruses (e.g., JC and SV40-PML). These are universal childhood infections that are reactivated in the immunosuppressed patient. It is characterized by demyelination, with abnormalities of oligodendrocytes in the white matter. Initially, the lesions are widely disseminated, but later tend to become confluent, producing large lesions. It is now seen with increasing incidence in patients with AIDS[69] and those treated with immunosuppressive drugs. It most commonly involves the subcortical white matter of the posterior frontal and parietal lobes extending to the level of the trigones and occipital horns of the lateral ventricles (Figure 6). The lesions give an increased T_2 and slightly increased T_1 relaxation time, and because there is only occasional perivenular inflammation in the acute stage, they do not have mass effect and gadolinium enhancement is not usually a feature. These are useful differentiating features from lymphoma and toxoplasmosis.[70]

3.3 Posttherapy

3.3.1 Disseminated Necrotizing Leukoencephalopathy

When patients are treated with intrathecal antineoplastic agents such as methotrexate for disseminated lymphoma, leukemia, or carcinomatosis, a necrotizing leukoencephalopathy can occur despite the fact that the agent does not usually cross the blood–brain barrier.[71] Radiation therapy potentiates this neurotoxicity. There is endothelial injury, loss of oligodendroglial cells, coalescing foci of demyelination, and axonal swelling. The damaged endothelium responds by attempts at repair, resulting in hyalinization, fibrosis, and mineralization of the vessel walls. This causes a relative tissue ischemia, demyelination, and necrosis.

On MRI,[72] there is a diffuse increase in T_1 and T_2 relaxation times, with no mass effect and little or no gadolinium enhancement, reflecting only the edema and demyelination present.

The patterns of injury to the white matter from radiation therapy on its own are divided into three stages: (a) acute, (b) early, and (c) delayed. In the first days or weeks after therapy, vasogenic edema may be produced because of transient disruption of the blood–brain barrier, and some enhancement will occur. In the weeks or months following radiation, demyelination may occur. MRI may show focal or diffuse increases in T_1 and T_2 relaxation times in a patchy often periventricular distribution, which may be asymmetrical, affecting the white matter but generally sparing the corpus callosum, internal capsules, and basal ganglia.[73,74] The gray matter is only involved in severe cases. Delayed effects can occur months to years after therapeutic irradiation. These are less common, and develop later with hyperfractionated doses. There is endothelial hyperplasia, fibrinoid necrosis of perforating arterioles, and thrombosis. Cerebral necrosis supervenes, with blood–brain barrier disruption, edema, and mass effect. MRI at this stage shows mass effect, increased T_1 and T_2 relaxation times, and enhancement after intravenous gadolinium. Therefore this cannot be differentiated from recurrent tumor with MRI. There is some evidence that measures of the lesion's metabolism with

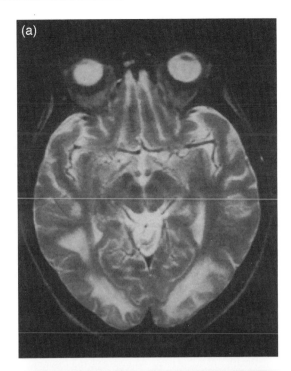

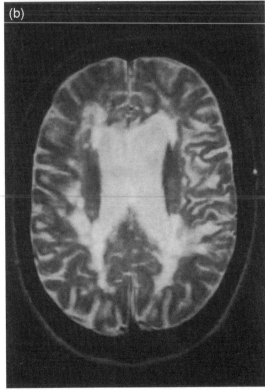

Figure 6 PML in AIDS: T_2-weighted images showing (a) extensive subcortical demyelination in the trigonal and occipital regions and, in a different patient, (b) gross loss of white matter substance with periventricular coalescing hyperintensities

^{18}F-deoxyglucose positron emission tomography, ^{201}Tl or L-3-[^{123}I]iodo-α-methyltyrosine single photon emission computerized tomography will selectively differentiate the hypometabolic radiation necrosis from the hypermetabolic malignant tumor.[75,76]

A reversible acute cerebellar and cerebral syndrome has been reported[77] after systemic high-dose cytarabine therapy used for treatment of postremission and refractory leukemia treatment. Diffuse patchy areas of increased intensity on T_2-weighted images were shown in the deep white matter of the frontal, occipital, and parietal lobes. Punctate enhancement was observed in the occipital lobes. Over a month, the symptoms and white matter abnormalities resolved. At post-mortem later, there was no evidence of white matter disease.

3.4   Toxic and Degenerative Disease

3.4.1   Central Pontine Myelinolysis

In this condition, there is loss of myelin and oligodendroglia in the central pons, which may extend to the lateral thalamus and mesencephalon, sparing the ventrolateral longitudinal fibers.[78] It is usually associated with rapidly corrected hyponatremia, often in alcoholics.[79,80] When there is only a tiny lesion or the patient is in coma due to the underlying disease process, it may be asymptomatic, but usually there is tetraparesis with a pseudobulbar palsy or a 'locked-in' syndrome. Mild cases with full recovery have now been reported.[81]

The demyelination is depicted as T_1 and T_2 prolongation with no mass effect, and although there is a single report of ring enhancement, there is usually no blood–brain barrier breakdown.[82]

3.4.2   Marchiaflava–Bignami Disease

This is characterized by a toxic demyelination of the corpus callosum in alcoholics.[83,84] A rapidly fatal form and a more chronic form have been recognized. In the acute form, extensive lesions have been reported in the centrum semiovale and corpus callosum, while at the chronic stage only corpus callosum lesions are seen, and occasionally there is a favorable outcome.[85] MRI shows these as small areas of increased T_1 and T_2 relaxation times, with no mass effect. Enhancement has not been reported.

3.4.3   Carbon Monoxide Encephalopathy

As carbon monoxide binds to the hemoglobin molecule and displaces oxygen, it induces hypoxia and vulnerable cells are destroyed. Although the gray matter structures are damaged first, the white matter is also involved, especially when there is episodic or chronic exposure.

MRI shows areas of increased T_1 and T_2 relaxation time in the thalamus, basal ganglia, hippocampus, and centrum semiovale. These areas may show enhancement with gadolinium in the acute stage. The lesions are usually bilateral and symmetrical, but can be patchy. Laminar necrosis has been reported as high signal cortical foci on T_2-weighted images.[86]

3.4.4   Substance Abuse

Inhalation of organic solvents and black market drugs such as heroin vapors (pyrolysate) and cocaine produce a wide variety of acute and chronic neurological signs and symptoms. The effects on the white matter will depend largely on the chemical constituent involved.

Xiong et al.[87] have shown that in toluene (one of the constituents of paint sprays) abuse, there is generalized cerebral,

For References see p. 950

cerebellar, and corpus callosum atrophy, with a loss of gray–white matter contrast associated with diffuse multifocal hyperintensity of the cerebral white matter on T_2-weighted sequences. Additionally, hypointensity of the thalami are also seen.

Adulterated and synthetically produced drugs can produce severe leukoencephalopathy. Tan et al.[88] reported on four patients who inhaled contaminated heroin vapor and developed extensive, symmetrical lesions of the white matter of the cerebrum, cerebellum, and midbrain. Selective involvement of the corticospinal tract and lemniscus medialis was also found.

These have to be differentiated from the effects of cocaine abuse, where there is generally neurovascular damage with vasculitis, vasospasm, and thrombosis. Eventually, cerebral atrophy can be seen.[89]

3.4.5 Hypoxic–Ischemic Encephalopathy

This generally refers to brain damage in the fetus and infant. It may be focal or diffuse. When focal, it may be the cause of territorial infarcts such as can occur in cyanotic congenital heart disease when emboli can bypass the filtering effect of the lungs. This will be discussed in Section 3.4.6. In asphyxia, there is diffuse hypoxia hypercarbia, acidosis, and loss of the brain's normal vascular autoregulation, resulting in pressure-passive flow and reduced perfusion. Capillary permeability is also altered. Sudden reperfusion of these weakened capillaries can result in rupture and intracerebral hemorrhage. The periventricular white matter is particularly susceptible, lying at the distal end of the supply zone of the long narrow centripetal arteries that run from the cerebral surface.[90]

When 100 high-risk neonates of different gestational ages were followed prospectively with MRI, CT, and ultrasound examinations,[91,92] it was found that lesions associated with hypoxic–ischemic encephalopathy such as coagulative necrosis and germinal matrix hemorrhage were best shown on MRI. In the analysis, ultrasound showed 80%, while CT only showed 40% of those lesions depicted on MRI.

This has raised interest in medico-legal circles, since the timing of the insult may be more clearly defined. The appearances of the brain damage on MRI can now give important clues as to the time and nature of the asphyxia. Barkovich and Truwit[93] found that when the asphyxia occurred before the 26th week of gestation, there was dilatation of the ventricles without any signal changes, whereas in older fetuses there was increasing periventricular gliosis. Both periventricular and more peripheral white matter gliosis with associated general atrophy were found in cases who had been partially asphyxiated, or where asphyxia had occurred near term or in postmature fetuses. Total asphyxia involves the deep gray matter nuclei and the brainstem, and presents a different pattern.

These MRI patterns may have prognostic value, with initial studies[94] reporting good correlations between imaging findings at 8 months and neurodevelopmental outcome at 18 months.

3.4.6 Trauma (Contusions–Shear Injuries)

In children, the effects of asphyxia and mechanical trauma, be they accidental or nonaccidental, may initially produce the same imaging appearance, with generalized cerebral swelling resulting in blurring of the clear distinction between the gray matter and white matter boundaries, ventricular compression, and loss of CSF from the sulci and cisterns. This is due to a combination of edema and a failure of autoregulation producing an increase in cerebral blood volume. This can result in watershed ischemia and infarction, with eventual loss of white matter producing ventricular and sulcal dilatation.

Focal cerebral contusions involve the gyral crests, and can extend into the subcortical and deeper white matter regions, depending on their severity. There is edema and petechial perivascular hemorrhage, but tissue integrity is largely preserved in small lesions. With more severe contusions, the petechial hemorrhages coalesce into focal hematomas, which have some space-occupying effect. These are well shown by MRI at all stages,[95,96] although, in practice, CT is easier to perform and gives clinically adequate information in the acute stage.

Blunt trauma resulting in sudden acceleration or deceleration of the skull, especially when this is rotational, sets up shear–strain deformation at the moment of impact in response to the inertial differences between tissues of different density and viscosity.[97] This can cause immediate and irreversible structural damage to axons, and has been termed diffuse axonal injury.[98–100]

Diffuse axonal injury is a pathological diagnosis, and imaging may only show a few apparently focal lesions in the lobar white matter. It is, however, very important to recognize these as the 'hallmark' of associated widespread microscopic axonal disruption. Although CT is the most commonly conducted examination in the acute situation, MRI is much more sensitive[95,101] and essential when the clinical state is not explained by the CT imaging appearances (Figure 7). In the acute situation, foci of edema that may or may not contain macroscopic hematomas can be seen on T_2-weighted MRI in the corpus callosum, the parasagittal frontal white matter close to the gray–white matter interface, the basal ganglial regions and the dorso-lateral quadrant of the rostral brainstem.[96] At the subacute stage, hemorrhage will be better depicted on T_1-weighted images, but both acutely and in the chronic phase T_2* gradient echo sequences are most sensitive to deoxyhemoglobin and hemosiderin respectively. MRI can rarely appear entirely normal[102] in severe diffuse axonal injury, and it is only on followup that the widespread white matter damage is reflected in atrophy with ventricular and cortical sulcal enlargement.[100,103,104]

4 VASCULAR DISEASE

4.1 Infarction

Stroke remains one of the commonest causes of hospital admission in the developed world, has a high morbidity and mortality, and consumes more healthcare resources than any other single disease. Ninety percent of cases are due to ischemia (10% to cerebral hemorrhage) from thrombosis of a nutrient artery, with only a small number due to emboli from the heart or other vessels. Despite treatment advances, the mortality remains at around 50%. It is thought that only preventative public health measures and earlier thrombolytic therapy can improve this situation.[105] This requires the accurate early identification of patients with acute infarcts and those with transient ischemia at risk of completing their infarcts.

Routine unenhanced MRI can detect abnormalities within about 8 h of the onset of symptoms (although changes on MRI

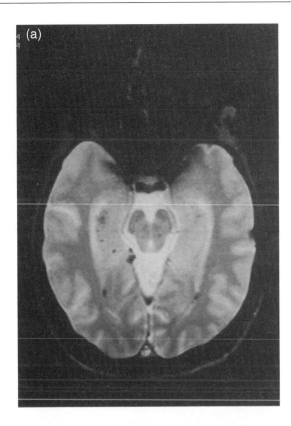

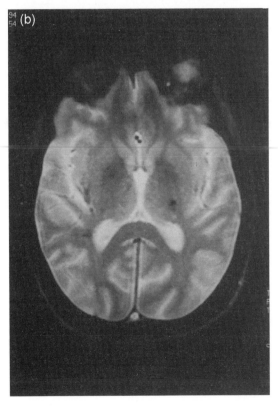

Figure 7 Head injury: T_2* gradient echo sections in an 25-year-old unconscious patient with a normal CT scan. Multiple focal hypointensities represent hemorrhage at white/gray matter interfaces—evidence of diffuse axonal injury

with a vessel occlusion stroke model were shown as early as 1–2 h without paramagnetic contrast),[106,107] whereas CT is normal for at least 14 h and if perfusion is not re-established remains 'bland' for several days. MRI initially shows subtle swelling and an increase in T_1 and T_2 signals due to failure of the 'sodium–potassium' pump and increasing intracellular water–cytotoxic edema. At this stage, function is lost but structure is maintained. It is only with continuing ischemia that blood–brain barrier breakdown occurs, structural integrity is lost, and vasogenic edema supervenes. Although there are anecdotal reports of MRI-defined cytotoxic edema being reversed on treatment of ischemia in humans,[108,109] and more rigorous demonstrations in cats using diffusion sequences,[107] it has not been established in routine clinical practice whether these MRI changes, unlike those on CT, are reversible.

Recent gadolinium-enhanced MRI studies[110–112] of the first 24–48 h after the ictus in the clinical population has shed light on this crucially important acute stage. Sato and colleagues[113] studied six patients within 8 h and a further two between 8 and 26 h. They showed areas of cerebral ischemia/infarction using gadolinium-enhanced T_1-weighted spin echo sequences. Abnormal curvilinear areas of enhancement thought to represent cortical arterial vessels with markedly slowed circulation were seen adjacent to affected brain. This tissue was shown to progress to frank infarction on follow-up CT and MRI. These features have been confirmed and extended by the Iowa group.[110,114] They demonstrated the vascular flow-related abnormalities with absence of normal flow voids and the presence of arterial enhancement detected within minutes of the onset of symptoms. Brain swelling on T_1-weighted images without signal changes on T_2-weighted images was detected within the first few hours. In contrast to the usual absence of parenchymal enhancement typically found in cortical infarctions in the first 24 h, a few lesions showed paradoxical early exaggerated enhancement. These were the transient or partial occlusions and isolated watershed infarcts. Longer term prospective observations through the first fortnight have defined the subacute appearances.[115] Three stages have been demonstrated:

(a) vascular enhancement—days 1–3, seen in 77% of cases;
(b) leptomeningeal enhancement—days 4–7, seen only in larger infarcts;
(c) brain parenchymal enhancement—days 7–14, seen in 100% of cases studied.

Enhancement is not noted after two to three months. These changing patterns of enhancement reflect the underlying pathophysiology, and may have prognostic significance; if this proves to be so then gadolinium enhancement will be crucial in the evaluation of early ischemia and its response to intervention.

Once the parenchymal long T_1 and T_2 signal changes are established, the differential diagnosis must be made in a similar way to conventional CT, following consideration of

(a) the site: vascular territory, watershed region, deep gray or white matter tracts;
(b) the shape: wedge, involving gray and white matter with subtle bowing of interfaces;
(c) the margins: sulcal effacement, blurring of the gray and white matter borders;
(d) the degree of edema;
(e) the sequence of resolution of mass effect over three to four weeks.[116]

For References see p. 950

Hypointensity on heavily T_2-weighted sequences and the use of gradient echoes or susceptibility mapping can often show petechial hemorrhage in the second week that is not seen on CT. While frank hemorrhage correlates with a worsening clinical state, fine interstitial bleeding mainly due to diapedesis relates neither to anticoagulation nor to a poorer clinical condition.[116]

The patency of the extracranial and major intracerebral arteries can be assessed on routine MRI sequences as a 'flow void' or with slower laminar flow even as echo rephasing. At present, projectional images produced by time-of-flight and phase contrast angiographic sequences are being evaluated, and may yet replace preoperative conventional cerebral angiography.[117,118]

Both diffusion imaging[119,120] and spectroscopy[121] are being used experimentally in clinical populations to try to gain an understanding of microscopic water shifts as the different types of edema develop and to give an insight into the progressive cycles of bioenergetic exchange as oxidative metabolism breaks down in the ischemic brain cells.

4.2 Ischemic White Matter Disease—Normal Aging

Focal and confluent white matter abnormalities seen on MRI do not necessarily represent actual necrosis and infarction, but can be due to a spectrum of chronic cerebrovascular insufficiency.[122] These merge with the changes found in as many as 30% of the normal aging population over 60 years of age who show no clinical cognitive deficit, but which are seen with increasing frequency in patients with hypertension, diabetes mellitus, and coronary artery disease (Table 1).[123]

Dilated perivascular spaces give a CSF signal, are usually smaller than lacunar infarcts, and occur in typical locations in the base, deep white matter, and cortex of the brain. Gliosis may become more confluent around these vessels as the vascular insufficiency progresses and produces an increased signal on proton density images in addition to the increased T_2 signal differentiating it from CSF. This has now been confirmed microscopically.

In a small post-mortem study,[124] histological examination showed that the larger lesions were characterized centrally by necrosis, axonal loss, and demyelination, and therefore represent true infarcts. Reactive astrocytes oriented along the degenerated axons were identified at distances of up to several centimeters from the central infarct. This isomorphic gliosis shows hyperintensity on T_2-weighted images, and increases the apparent size of the central lesion. Confirmation has been provided by Munoz et al.,[125] who investigated the pathological correlates of increased T_2 signal in the centrum ovale in an unselected series of 15 post-mortems. On the basis of size, greater than and less than 10 mm, two types of lesion were described, namely, extensive and punctate. The extensive areas of hyperintensity on T_2-weighted images were found to show myelin pallor that spared the subcortical U fibers. There was diffuse vacuolation and reduction in glial cell density. The punctate abnormalities were less well defined, and were found to be due to dilated Virchow–Robin spaces.

The white matter changes seen on MRI are therefore non-specific, and although seen with increased frequency in ischemic brains, there is often little or no correlation with the clinical state in the elderly patient.

5   CONCLUSIONS

Over the last 15 years, MRI has become the main diagnostic tool in the investigation of white matter disease. In some conditions, its sensitivity is the key to selecting patients for further attention, while in others it identifies more specific features that in turn lead to further laboratory investigations leading to a final diagnosis. MRI can be used to select patients for treatment and to monitor the effects of this treatment. The implementation of new sequences, faster scanning techniques, and better patient–machine ergonomics will ensure the preemptive position of MRI for the investigation of white matter diseases for the foreseeable future.

6   RELATED ARTICLES

Brain MRS of Infants and Children; Brain Neoplasms Studied by MRI; Diffusion: Clinical Utility of MRI Studies; Echo-Planar Imaging; Gadolinium Chelate Contrast Agents in MRI: Clinical Applications; Hemorrhage in the Brain and Neck Observed by MRI; Intracranial Infections; Localization and Registration Issues Important for Serial MRS Studies of Focal Brain Lesions.

7   REFERENCES

1. A. J. Barkovich, G. Lyon, and P. Evrard, *Am. J.N.R.*, 1992, **13**, 447.
2. R. G. Parrish, J. R. Kurland, W. W. Janese, and L. Bakay, *Science (Washington, DC)*, 1974, **483**, 349.
3. A. J. Barkovich and T. V. Maroldo, *Top. Magn. Reson. Imaging*, 1993, **5**, 96.
4. Y. Nomura, H. Sakuma, K. Takeda, T. Tagami, Y. Okuda, and T. Nakasasa, *Am. J.N.R.*, 1994, **15**, 231.
5. A. Oatridge, J. V. Hajnal, F. M. Cowan, C. J. Baudouin, I. R. Young, and G. R. Bydder, *Clin. Radiol.*, 1993, **47**, 82.
6. B. E. Kendall, *J. Inherited Metab. Dis.*, 1993, **16**, 771.
7. L. E. Becker, *Am. J.N.R.*, 1992, **13**, 609.
8. B. E. Kendall, *Am. J.N.R.*, 1992, **13**, 621.
9. G. B. Bobele, A. Garnica, G. B. Schaefer, J. C. Leonard, D. Wilson, W. A. Marks, R. W. Leech, and R. A. Brumback, *J. Child. Neurol.*, 1990, **5**, 253.
10. T. Ichiyama, T. Hayashi, and T. Ukita, *Brain Dev.*, 1993, **15**, 153.
11. J. Brismar, G. Brismar, G. Gascon, and P. Ozand, *Am. J.N.R.*, 1990, **11**, 805.
12. D. A. Finelli, R. W. Tarr, R. N. Sawyer, and S. J. Horwitz, *Am. J.N.R.*, 1994, **15**, 167.
13. T. J. Farley, L. M. Ketonen, J. B. Bodensteiner, and D. D. Wang, *Pediatr. Neurol.*, 1992, **8**, 455.
14. S. Choi and D. R. Enzmann, *Am. J.N.R.*, 1993, **14**, 1164.
15. J. C. Koetsveld-Baart, I. E. Glaudemans-van-Gelderen, J. Valk, and P. G. Barth, *Ned. Tijdschr. Geneeskd.*, 1993, **137**, 2494.
16. M. Ishii, J. Takanashi, K. Sugita, A. Suzuki, M. Goto, Y. Tanabe, K. Tamai, and H. Niimi, *No To Hattatsu*, 1993, **25**, 9.
17. E. G. Shapiro, L. A. Lockman, D. Knopman, and W. Krivit, *Neurology*, 1994, **44**, 662.
18. B. Kruse, F. Hanefeld, H. J. Christen, H. Bruhn, T. Michaels, W. Hanicke, J. Frahm, *J. Neurol.*, 1993, **241**, 68.
19. P. J. Close, S. J. Sinnott, and K. T. Nolan, *Pediatr. Radiol.*, 1993, **23**, 400.

20. M. S. van der Knaap and J. Valk, *Neuroradiology*, 1991, **33**, 30.
21. J. Brismar, A. Aqeel, G. Gascon, and P. Ozand, *Am. J.N.R.*, 1990, **11**, 135.
22. R. Sugita, I. Takahashi, K. Ishii, K. Matsumoto, T. Ishibashi, K. Sakamoto, and K. Narisawa, *J. Comput. Assist. Tomogr.*, 1990, **14**, 699.
23. D. W. W. Shaw, K. R. Maravilla, E. Weinberger, J. Garretson, C. M. Trahms, and C. R. Scott, *Am. J.N.R.*, 1991, **12**, 403.
24. K. D. Pearsen, A. D. Gean-Marton, H. L. Levy, and K. R. Davis, *Radiology*, 1990, **177**, 437.
25. A. J. Thompson, I. Smith, D. Brenton, B. D. Youl, G. Rylame, D. C. Davidson, B. Kordall, and A. J. Lees, *Lancet*, 1990, **336**, 602.
26. A. J. Thompson, S. Tillotson, I. Smith, B. Kendall, S. G. Moore, and D. P. Brenton, *Brain*, 1993, **116**, 811.
27. A. A. Tzika, W. S. Ball Jr., D. B. Vigneron, R. S. Dunn, and D. R. Kirks, *Am. J.N.R.*, 1993, **14**, 1267.
28. A. J. Barkovich, W. V. Good, T. K. Koch, and B. O. Berg, *Am. J.N.R.*, 1993, **14**, 1119.
29. P. M. Matthews, F. Andermann, K. Silver, G. Karpati, and D. L. Arnold, *Neurology*, 1993, **43**, 2484.
30. B. Barbiroli, P. Montagna, P. Martinelli, R. Lodi, S. Lotti, P. Cortelli, R. Fanicello, and P. Zaniol, *J. Cereb. Blood Flow Metab.*, 1993, **13**, 469.
31. J. M. Charcot, *Gaz. des Hôp. Civ. Mil., Paris*, 1868, **41**, 554.
32. W. I. McDonald and D. H. Silberberg, 'Multiple Sclerosis', Butterworths, London, 1986.
33. G. A. Schumacher, G. Beebe, R. F. Kibler, L. T. Kurland, J. F. Kurtzke, F. McDowell, B. Nagler, W. A. Sibley, W. W. Tourtellotte, and T. L. Willmon, *Ann. NY Acad. Sci.*, 1965, **122**, 552.
34. C. M. Poser, D. W. Paty, L. Scheinberg, W. I. MacDonald, F. A. Davis, G. C. Ebers, K. P. Johnson, W. A. Sibley, D. H. Silberberg, and W. H. Tourtellote, *Ann. Neurol.*, 1983, **13**, 227.
35. A. K. Asbury, R. M. Herndon, H. F. McFarland, W. I. McDonald, W. J. McIlroy, D. W. Paty, J. W. Prineas, L. C. Scheinberg, and J. S. Wolinsky, *Neuroradiology*, 1987, **29**, 119.
36. D. W. Paty and D. K. B. Li, in 'Clinical Neuroimaging', ed. W. H. Theodore, Alan R. Liss, New York, 1988, Vol. 4, Chap. 10.
37. R. I. Grossman, B. H. Braffman, J. R. Brorson, H. I. Goldberg, D. H. Silberberg, and F. Gonzalez-Scarano, *Radiology*, 1988, **169**, 117.
38. I. R. Young, A. S. Hall, C. A. Pallis, N. J. Legg, G. M. Bydder, and R. E. Steiner, *Lancet*, 1981, **ii**, 1063.
39. S. A. Lukes, L. E. Crooks, M. J. Aminoff, L. Kaufman, H. S. Panitch, C. Mills, and D. Norman, *Ann. Neurol.*, 1983, **13**, 592.
40. I. E. C. Ormerod, G. H. du Boulay, and W. I. McDonald, in 'Multiple Sclerosis', ed. W. I. McDonald and D. H. Silberberg, Butterworths, London, 1986.
41. D. H. Miller, *MRI Decis.*, 1988, **2**, 17.
42. I. E. C. Ormerod, D. H. Miller, W. I. McDonald, E. P. G. H. du Boulay, P. Rudge, B. E. Kendall, I. F. Moseley, G. Johnson, P. S. Tofts, and A. N. Halliday, *Brain*, 1987, **110**, 1579.
43. D. W. Paty, *Can. J. Neurol. Sci.*, 1988, **15**, 266.
44. D. W. Paty, J. J. F. Oger, L. F. Kastrukoff, S. A. Hashimoto, J. P. Haage, A. A. Eisen, K. A. Eisen, S. T. Purves, M. D. Low, and V. Brandejs, *Neurology*, 1988, **38**, 180.
45. D. Lacomis, M. D. Osbakken, and G. Gross, *Magn. Reson. Med.*, 1986, **3**, 194.
46. B. P. Drayer, P. Burger, B. Hurwitz, D. Dawson, and J. Cain, *Am. J.N.R.*, 1987, **8**, 413.
47. D. H. Miller, I. E. C. Ormerod, A. Gibson, E. P. G. H. du Bouley, P. Rudge, and W. I. McDonald, *Neuroradiology*, 1987, **29**, 226.

48. F. Z. Yetkin, V. M. Haughton, R. A. Papke, M. E. Fischer, and S. M. Rao, *Radiology*, 1991, **178**, 447.
49. F. Fazekas, H. Offenbacher, S. Fuchs, R. Schmidt, K. Niederkorn, S. Horners, and H. Lechner, *Neurology*, 1988, **38**, 1822.
50. H. Offenbacher, F. Fazekas, R. Schmidt, W. Freidl, E. Floch, F. Payer, and H. Lechner, *Neurology*, 1993, **43**, 905.
51. I. E. C. Ormerod, A. Bronstein, P. Rudge, G. Johnson, D. G. P. MacManus, A. M. Halliday, H. Barratt, E. P. du Boulay, B. E. Kendall, and I. F. Moseley, *J. Neurol. Neurosurg. Psychiatry*, 1986, **49**, 737.
52. J. Prineas, *Hum. Pathol.*, 1975, **6**, 531.
53. R. I. Grossman, F. Gonzalez-Scarano, S. W. Atlas, S. Galetta, and D. H. Silberberg, *Radiology*, 1986, **161**, 721.
54. H. Miller, P. Rudge, B. Johnson, B. E. Kendall, D. G. MacManus, I. F. Moseley, D. Barnes, and W. I. McDonald, *Brain*, 1988, **111**, 927.
55. E.-M. Larsson, S. Holas, and O. Nilsson, *Am. J.N.R.*, 1989, **10**, 1071.
56. S. F. Merandi, B. T. Kudryk, F. R. Murtagh, and J. A. Arrington, *Am. J.N.R.*, 1991, **12**, 923.
57. A. J. Thompson, A. J. Kermode, D. Wicks, D. G. MacManus, B. E. Kendall, D. P. Kingley, and W. I. McDonald, *Ann. Neurol.*, 1991, **29**, 53.
58. M. K. Stehling, P. Bullock, J. L. Firth, A. M. Blamire, R. J. Ordidge, B. Coxon, P. Gibbs, and P. Mansfield, *Proc. VIIIth Ann Mtg. Soc. Magn. Reson. Med.*, Amsterdam, 1989, p. 358.
59. D. W. Paty, *Curr. Opin. Neurol. Neurosurg.*, 1993, **6**, 202.
60. R. Heun, L. Kappos, S. Bittkau, D. Staedt, E. Rohrbach, and B. Schuknecht, *Lancet*, 1988, **ii**, 1202.
61. M. J. Kupersmith, D. Kaufman, D. W. Paty, G. Ebers, M. McFarland, K. Johnson, J. Reingold, and J. Whitaker, *Neurology*, 1994, **44**, 1.
62. D. W. Paty and D. K. Li, *Neurology*, 1993, **43**, 662.
63. P. Schilder, *Z. Gesamte Neurol. Psychiatr.*, 1912, **10**, 1.
64. M. F. Mehler and L. Rabinowich, *Am. J.N.R.*, 1989, **10**, 176.
65. S. W. Atlas, R. I. Grossman, H. I. Goldberg, D. B. Hackney, L. T. Bilanuck, and R. A. Zimmerman, *J. Comput. Assist. Tomogr.*, 1986, **10**, 798.
66. A. J. Barkovich, 'Pediatric Neuroimaging', Raven Press, New York, 1996, p. 597.
67. R. Murata, H. Hattori, O. Matsuoka, T. Nakajima, and H. Shintaku, *Brain. Dev.*, 1992, **14**, 391.
68. S. Yagi, Y. Miura, S. Mizuta, A. Wakunami, N. Kataoka, T. Morita, K. Morita, S. Ono, and M. Fukunaga, *Brain. Dev.*, 1993, **15**, 141.
69. A. S. Mark and S. W. Atlas, *Radiology*, 1989, **173**, 517.
70. L. Ketonen and M. Tuite, *Semin. Neurol*, 1992, **12**, 57.
71. F. Ebner, G. Ranner, I. Slavc, C. Urban, R. Kleinert, H. Roulner, R. Ernspieler, and E. Justich, *Am. J.N.R.*, 1989, **10**, 959.
72. R. Asato, Y. Akiyama, M. Ito, M. Kubota, R. Okumura, Y. Miki, J. Konishi, H. Mikaua, *Cancer*, 1992, **70**, 1997.
73. J. T. Curnes, D. W. Laster, M. R. Ball, T. D. Koubek, D. M. Moody, and R. L. Witcofski, *Am. J.N.R.*, 1986, **7**, 389.
74. W. J. Curran, C. Hecht-Leavitt, L. Schut, R. A. Zimmerman, and D. F. Nelson, *Int. J. Radiat. Oncol. Biol. Phys.*, 1987, **13**, 1093.
75. R. B. Schwartz, P. A. Carvalho, E. Alexander III, J. S. Loeffler, R. Folkerth, and B. L. Holman, *Am. J.N.R.*, 1991, **12**, 1187.
76. Karl-J. Langen, H. H. Coenen, N. Roosen, P. Kling, O. Muzik, H. Herzog, T. Kuwort, G. Stocklin, and L. E. Femendegen, *J. Nucl. Med.*, 1990, **31**, 281.
77. D. J. Vaughn, J. G. Jarvik, D. Hackney, S. Peters, and E. A. Stadtmauer, *Am. J.N.R.*, 1993, **14**, 1014.

For References see p. 950

78. Y. Korogi, M. Takahashi, J. Shinzato, Y. Sakamoto, K. Mitsuzaki, T. Hirai, and K. Yoshizumi, *Am. J.N.R.*, 1993, **14**, 651.

79. M. Mascalchi, M. Cincotta, and M. Piazzini, *Clin. Radiol.*, 1993, **47**, 137.

80. R. D. Laitt, M. Thornton, and P. Goddard, *Clin. Radiol.*, 1993, **48**, 432.

81. V. B. Ho, C. R. Fitz, C. C. Yoder, and C. A. Geyer, *Am. J.N.R.*, 1993, **14**, 163.

82. K. J. Koch and R. R. Smith, *Am. J.N.R.*, 1989, **10**, S58.

83. M. E. Charness, *Alcohol Clin. Exp. Res.*, 1993, **17**, 2.

84. P. Tomasini, D. Guillot, P. Sabbah, C. Brosset, P. Salamand, and J. F. Briant, *Ann. Radiol. (Paris)*, 1993, **36**, 319.

85. S. Canaple, A. Rosa, and J. P. Mizon, *Rev. Neurol. (Paris)*, 1992, **148**, 638.

86. A. L. Horowitz, R. Kaplan, and G. Sarpel, *Radiology*, 1987, **162**, 787.

87. L. Xiong, J. D. Matthes, J. Li, and R. Jinkins, *Am. J.N.R.*, 1993, **14**, 1195.

88. T. P. Tan, P. R. Algra, J. Valk, and E. C. Wolters, *Am. J.N.R.*, 1994, **15**, 175.

89. E. Brown, J. Prager, H. Y. Lee, and R. G. Ramsey, *Am. J. Roentgenol.*, 1992, **159**, 137.

90. M. D. Nelson, I. Gonzalez-Gomez, and F. H. Gilles, *Am. J.N.R.*, 1991, **12**, 215.

91. S. E. Keeney, E. W. Adcock, and C. B. McArdle, *Pediatrics*, 1991, **87**, 421.

92. S. E. Keeney, E. W. Adcock, and C. B. McArdle, *Pediatrics*, 1991, **87**, 431.

93. A. J. Barkovich and C. L. Truwit, *Am. J.N.R.*, 1990, **11**, 1087.

94. P. Byrne, R. Welch, M. A. Johnson, J. Darrah, and M. Piper, *J. Pediatr.*, 1990, **117**, 694.

95. E. Teasdale and D. M. Hadley, in 'Handbook of Clinical Neurology, 2nd Series: Head Injury', ed. R. Braakman, Elsevier, Amsterdam, 1990, Vol. 13, Chap. 7.

96. D. M. Hadley, *Curr. Imaging*, 1991, **3**, 64.

97. A. H. S. Holbourn, *Lancet*, 1943, **ii**, 438.

98. J. H. Adams, D. I. Graham, L. S. Murray, and G. Scott, *Ann. Neurol.*, 1982, **12**, 557.

99. T. A. Gennarelli, G. M. Spielman, T. W. Langfitt, P. L. Gildenberg, T. Harrington, J. A. Jane, L. F. Marshall, J. D. Miller, and L. H. Pitts, *J. Neurosurg.*, 1982, **56**, 26.

100. A. D. Gean, 'Imaging of Head Trauma', Raven Press, New York, 1994.

101. A. Jenkins, G. M. Teasdale, D. M. Hadley, P. Macpherson, and J. O. Rowan, *Lancet*, 1986, **ii**, 445.

102. D. M. Hadley, P. Macpherson, D. A. Lang, and G. M. Teasdale, *Neuroradiology*, 1991, **33**, 86.

103. K. D. Wiedmann, J. T. L. Wilson, D. Wyper, D. M. Hadley, G. M. Teasdale, and D. N. Brooks, *Neuropsychology*, 1990, **3**, 267.

104. J. T. L. Wilson, K. D. Wiedmann, D. M. Hadley, B. Condon, G. M. Teasdale, and J. D. N. Brooks, *J. Neurol. Neurosurg. Psychiatry*, 1988, **51**, 391.

105. C. D. Forbes, *Scot. Med. J.*, 1991, **36**, 163.

106. M. Brant-Zawadzki, B. Pereira, P. Weinstein, S. Moore, W. Kusharczyk, I. Berry, M. McNamara, and N. Derugin, *Am. J.N.R.*, 1986, **7**, 7.

107. M. E. Moseley, Y. Cohen, J. Mintorovitch, L. Chileuitt, H. Shimizer, W. Kueharczyk, M. F. Wendland, and P. R. Weinstein, *Magn. Reson. Med.*, 1990, **14**, 330.

108. A. M. Aisen, T. O. Gabrielsen, and W. J. McCune, *Am. J.N.R.*, 1985, **6**, 197.

109. W. G. Bradley, *Neurol. Res.*, 1984, **6**, 91.

110. W. T. C. Yuh, M. R. Crain, D. J. Loes, G. M. Greene, T. J. Ryals, and Y. Sato, *Am. J.N.R.*, 1991, **12**, 621.

111. S. Warach, W. Li, M. Ronthal, and R. R. Edelman, *Radiology*, 1992, **182**, 41.

112. R. N. Bryan, L. M. Levy, W. D. Whitlow, J. M. Killian, T. J. Preziosi, and J. A. Rosario, *Am. J.N.R.*, 1991, **12**, 611.

113. A. Sato, S. Takahashi, Y. Soma, K. Ishii, T. Watanabe, and K. Sakamoto, *Radiology*, 1991, **178**, 433.

114. M. R. Crain, W. T. C. Yuh, G. M. Greene, D. J. Loes, T. J. Ryals, Y. Sato, and M. N. Hart, *Am. J.N.R.*, 1991, **12**, 631.

115. A. D. Elster and D. M. Moody, *Radiology*, 1990, **177**, 627.

116. W. G. Bradley, in 'MRI Atlas of the Brain', eds. W. G. Bradley and G. Bydder, Martin Dunitz, London, 1990, Chap. 5.

117. T. J. Masaryk, G. A. Laub, M. T. Modic, J. S. Ross, and E. M. Haacke, *Magn. Reson. Med.*, 1990, **14**, 308.

118. A. W. Litt, *Am. J.N.R.*, 1991, **12**, 1141.

119. M. Doran and G. M. Bydder, *Neuroradiology*, 1990, **32**, 392.

120. R. M. Henkelman, *Am. J.N.R.*, 1990, **11**, 932.

121. M. Brant-Zawadzki, P. R. Weinstein, H. Bartkowski, and M. Moseley, *Am. J. Roentgenol.*, 1987, **148**, 579.

122. M. L. Bots, J. C. van-Swieten, M. M. Breteler, P. T. de-Jong, J. Van Gijn, A. Hofman, and D. E. Grobbee, *Lancet*, 1993, **341**, 1232.

123. T. Horikoshi, S. Yagi, and A. Fukamachi, *Neuroradiology*, 1993, **35**, 151.

124. V. G. Marshall, W. G. Bradley, C. E. Marshall, T. Bhoopat, and R. H. Rhodes, *Radiology*, 1988, **167**, 517.

125. D. G. Munoz, S. M. Hastak, B. Harper, D. Lee, and V. C. Hachinski, *Arch. Neurol.*, 1993, **50**, 492.

Biographical Sketch

Donald M. Hadley. *b* 1950. M.B.Ch.B., 1974, Ph.D., 1980, D.M.R.D., 1981, Aberdeen University, Scotland; F.R.C.R., 1983, London, UK. Introduced to NMR by Professor John Mallard and Dr Francis Smith while carrying out postdoctoral work in the Department of Bio-medical Physics, University of Aberdeen 1981. MRC research fellow, Glasgow University 1984, consultant and director of Neuroradiology 1992, Institute of Neurological Sciences, Glasgow, UK. Approx. 200 publications. Current research interests: MRI and MRS investigation of acute trauma, epilepsy, metabolic white matter diseases and stroke.

Brain Neoplasms Studied by MRI

Andrew P. Kelly and Michael N. Brant-Zawadzki

Hoag Memorial Hospital Presbyterian, Newport Beach, CA, USA

1 INTRODUCTION

Magnetic resonance imaging (MRI) has become the imaging modality of choice for the evaluation of brain neoplasms. There are two main reasons why MRI has supplanted computerized axial tomography (CT) scanning at many institutions in the USA. First and foremost is the superior sensitivity that MRI possesses in detecting alterations in brain tissue caused by neoplasms. As location of an intracranial neoplasm is a factor with important diagnostic as well as prognostic implications, the ability of MRI to image in multiple planes is the second reason. This advantage helps to determine whether a lesion is intraaxial (of parenchymal origin, such as a glioma) versus extraaxial (of dural origin, for example, such as a meningioma).

A short article on MRI of brain neoplasms leaves important areas uncovered. For more extensive discussion of the topic, the reader is referred to the excellent recent reviews on brain neoplasms in the books edited by Stark and Bradley[1] and Atlas.[2-4] Also, cross references to pertinent related subjects discussed elsewhere in this volume are given at the end of this article. The initial part of this article focuses on the general features exhibited by many brain neoplasms and why MRI is sensitive to these changes. Technical considerations behind deciding appropriate imaging pulse sequences are also discussed. The second section discusses some of the most common brain neoplasms and their characteristics, as displayed by MRI.

2 GENERAL NEOPLASM FEATURES SEEN ON MRI

MRI of the brain and its pathology takes advantage of the exponential time constants, T_1 and T_2, exhibited by normal brain tissue, and alterations in these values caused by brain neoplasms. The reader is referred to other chapters in this volume for discussions on the complex basis of signal intensity and magnetic relaxation characteristics of normal brain and brain pathology. Essentially, calculated T_1 and T_2 values of brain tumors, such as astrocytomas, are longer than normal gray and white matter, but widespread differences exist even within single histological classifications, and attempts at histologic stratification of brain neoplasms by quantitative T_1 and T_2 relaxation analysis have proven futile.[5] The presence of edema, hemorrhage, necrosis, cyst development, and even calcification can help characterize neoplasms in the brain.

Standard MRI sequences employed by most institutions include a sagittal short TR, short TE (T_1-weighted) localizing sequence, axial long TR, long TE (T_2-weighted) sequences, and axial T_1-weighted sequences, usually obtained after administration of a paramagnetic contrast agent such as gadolinium-DTPA. Gradient recalled sequences are sometimes added to help characterize lesions in regards to foci of hemorrhage or calcification, as these sequences optimize magnetic susceptibility effects. Additional planes are included when deemed necessary for further information on tumor location and extension. Because T_2-weighted spin echo (SE) MRI sequences have the disadvantage of long acquisition times resulting in degradation of image quality secondary to patient motion, many scans are now performed using fast spin echo (FSE) techniques. The usefulness of FSE sequences already has been demonstrated in the imaging of pathologic intracranial conditions.[6,7] Utilizing these techniques, a routine entire brain scan can now be completed in under 15 min.

By analyzing tumor signal intensity on different pulse sequences, some insight can be gained as to the nature of the cells making up the tumor. For example, neoplasms containing a high nuclear/cytoplasm ratio, such as lymphoma and meningioma, may display this lack of water content as low signal intensity on T_2-weighted sequences.

3 SPECIFIC TUMOR FEATURES SEEN ON MRI

3.1 Location, Mass Effect, Infiltration and Hydrocephalus

The multiplanar capability of MRI allows the determination of the location of a tumor, i.e., intra-axial versus extraaxial, and thus helps to identify the potential cell of origin. The multiplanar capability of MRI aids in defining the extent of certain tumors that can be infiltrative, such as gliomas and primary lymphomas, which can extend via white matter tracts such as the corpus callosum. By assessing ventricular size and shape as well as gray and white matter interfaces, the presence or absence of mass effect and hydrocephalus can be determined. Recently, mass effect and necrosis, as displayed by and graded on MRI, were found to be statistically significant characteristics for grading astrocytic series gliomas, as compared with biopsy findings, which sometimes can be subject to sampling error.[8]

3.2 Edema

One of the major diagnostic advantages of MRI over CT is in the detection of brain edema, which is one of the most striking features associated with tumors. Termed 'vasogenic' edema, it is probably secondary to neovascularity devoid of the usual blood–brain barrier or stretching of normal vessels secondary to mass effect.[9] Edema, by increasing bulk water, causes prolongation of T_1 and T_2, as most tumors do; thus, changes detected on T_2-weighted images actually represent tumor in addition to edema. MRI is highly sensitive to these changes, but specifically defining areas representing actual tumor versus edema is difficult if using T_2-weighted images alone.

3.3 Necrosis and Cyst Formation

In most cases, intratumoral necrosis is considered to be a sign of a more aggressive lesion, and necrosis is well demonstrated by MRI. Necrosis can be cystic or noncystic; therefore, a varied appearance can be seen on MRI. Cystic necrosis with increased bulk water content will prolong T_1 and T_2 relaxation times, while nonnecrotic cysts can show shortened T_1 and T_2 relaxation times due to hemorrhage and accumulation of proteinaceous debris, leading to high signal intensity on T_1-weighted images.[2]

Benign or malignant lesions may have cystic components. Benign lesions, such as arachnoid cysts, will exhibit the behavior of 'true' cysts, following CSF signal intensity on all pulse sequences. Other benign cysts, such as colloid cysts or craniopharyngiomas, show greater variability in cyst components which determine signal changes on MRI. Malignant lesions may contain cystic components for a number of reasons, ranging from true tumoral cysts, to hemorrhage into solid lesions with subsequent clot lysis, and other causes of necrosis. Certain tumors may present as mural nodules in the cyst wall, particularly childhood astrocytomas and juvenile pilocytic astrocytomas, the latter having a very low malignant potential (Figure 1).

For References see p. 957

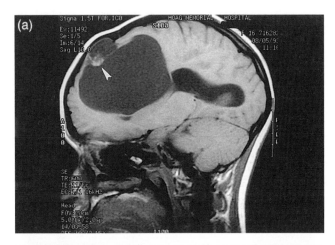

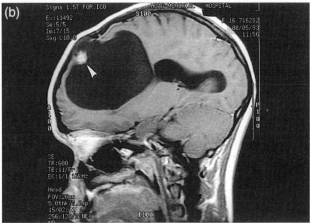

Figure 1 A 16-year-old female with a large juvenile pilocytic astrocytoma occupying much of the right frontal lobe. (a) Before gadolinium-DTPA T_1-weighted sagittal image obtained on a 1.5-T GE Signa magnet shows the large cystic tumor with a mural nodule (white arrowhead). (b) Note the dense enhancement of the mural nodule after contrast administration

3.4 Hemorrhage in Brain Neoplasms

Many primary brain neoplasms may initially be discovered secondary to symptoms related to intratumoral hemorrhage, which, when accompanied by vasogenic edema, can cause a confusing picture on MRI. The physics behind the appearance of blood on MRI is beyond the scope of this article, but the evolution of blood breakdown products from oxyhemoglobin through paramagnetic methemoglobin to hemosiderin is followed well by MRI and affords the potential for timing the event.[4] Certain tumors, especially metastases from melanoma, lung and renal cell carcinoma, have a propensity to bleed, as do a small percentage of gliomas. Hemorrhage itself should not be considered specific for tumor, since most cerebral bleeds have other causes.

3.5 Use of Paramagnetic Contrast Agents and Enhancement

A highly regulated and consistent internal milieu must be maintained for optimal brain and spinal cord function. Specialized capillaries, with endothelial tight junctions, provide this

'blood–brain barrier' in the normal state, aided by foot processes from nearby astrocytes. By administration of paramagnetic contrast agents such as gadolinium-DTPA, areas of breakdown in the blood–brain barrier caused by tumor can be detected. This is most helpful in determining to a close approximation the extent of neoplasm versus edema by comparing post-gadolinium T_1-weighted images with T_2-weighted images. By enhancing the relaxation of nearby water protons, contrast agents can decrease T_1 values in areas where a breakdown in the blood–brain barrier has occurred. The vast majority of glioblastoma multiforme tumors will enhance in a heterogeneous, thick, irregular pattern, but it is important to note that degree of enhancement does not correlate with aggressiveness of tumor.[10] Metastases also enhance in almost all instances. Contrast is most useful in detecting lesions that are isointense on T_1-weighted images, show little edema on T_2-weighted images, but which strongly enhance. This is seen in such tumors as meningiomas and acoustic neuromas. Postoperatively, contrast agents can help detect areas of suspected tumor recurrence or residual tumor, but these changes can be nonspecific, as discussed, below.

3.6 Postoperative Changes and Radiation

MRI with gadolinium is indicated following surgery, radiation or chemotherapy to follow tumor size, but recurrent tumor is not all that enhances. Local enhancement secondary to leptomeningeal scarring may persist for years after surgery, and only a size increase on sequential studies in a region of enhancement is likely to be definitive evidence of recurrent tumor.[10]

Many intracranial neoplasms now undergo radiotherapy as a mainstay for treatment, and most high-grade astrocytomas are now treated by gross resection followed by high energy local radiotherapy (radiation implants or radiosurgery). Sites of radiation necrosis can enhance and show edema, simulating tumor.[11]

3.7 Tumoral Calcification

Certain neoplasms, such as oligodendrogliomas and craniopharyngiomas, have a tendency to display areas of calcification, but identification of tumoral calcification in itself seldom causes one to favor a particular neoplasm over another. Secondary to the low resonant proton density of calcified tissue, calcium may be missed on MRI; when seen, it is generally noted to appear hypo- or isointense on T_1- and T_2-weighted spin echo sequences, and with gradient echo sequences signal loss is more profound due to the sensitivity of this method to the heterogeneous magnetic susceptibility found in calcified tissue. Recent articles have described occasional calcified brain lesion as appearing bright on T_1-weighted images due to shortening of T_1 relaxation times by a surface relaxation mechanism.[12]

4   SPECIFIC NEOPLASMS AND CHARACTERISTICS DISPLAYED ON MRI

Since location as well as histologic subtype are important features that help determine the clinical presentation and prognosis of most brain neoplasms, an attempt has been made to categorize the most common neoplasms according to site of origin, with subcategorization according to cell of origin.

4.1 Intra-axial Lesions

4.1.1 Gliomas and Tumor of Glial Cell Origin

Almost 50% of primary brain tumors are gliomas, and three major tumor types are recognized, corresponding to types of glial cell: astrocytes, oligodendrocytes and ependymal cells. Since neoplastically transformed astrocytes give rise to 75–95% of all gliomas, the discussion here will center on astrocytomas.[11]

The classification system providing the greatest prognostic validity is a three-level system where: grade I refers to low grade benign astrocytomas, such as the juvenile pilocytic astrocytoma; grade II indicates anaplastic astrocytoma with intermediate grade of malignancy; and grade III refers to the highly malignant astrocytomas and glioblastoma multiforme.

MRI features suggesting a more benign variety include absence of necrosis, well-defined margins, minimal edema and little mass effect. The hallmark of the glioblastoma multiforme, on the other hand, is necrosis, marked edema and mass effect, with prominent enhancement (Figure 2). Wide variability exists, however, since highly malignant and infiltrative tumors may not show edema or demonstrate significant enhancement.

The reader is referred to other texts on brain neoplasms for discussions of oligodendrogliomas and ependymomas.[1,2]

4.1.2 Nonglial-Cell Intra-axial Tumors

Once considered rare, the frequency of primary intracranial lymphoma is increasing due to its occurrence in patients with immunodeficiencies, particularly patients with the acquired immune deficiency syndrome (AIDS). Because of this and its fairly characteristic MRI appearance, lymphoma deserves mention. More than 50% of cases are multifocal, and can exist in supratentorial and infratentorial locations. Lymphomas can infiltrate and cross the corpus callosum, a property shared with gliomas. Dense hypercellularity causes this tumor to appear isointense to hypointense on T_2-weighted images. Most lymphomas enhance densely and homogeneously after contrast is administered, and tend to show less edema than gliomas of the same size, although wide variability in enhancement patterns and edema can exist (Figure 3).

Medulloblastoma, a common primary intracranial neoplasm in children, is one of several tumors occurring more commonly in an infratentorial location. Other tumors commonly seen in this location include cerebellar astrocytomas, juvenile pilocytic astrocytomas, hemangioblastomas, and fourth ventricular ependymomas. Because of its usually cerebellar location, multiplanar MRI is important in diagnosis. Tending to be hypercellular like lymphoma, medulloblastoma enhances diffusely and is usually isointense to hypointense on T_2-weighted images.[13]

An important use of MRI is in diagnosis of CSF dissemination of tumor. Termed leptomeningeal spread, this is a common pathway for this malignancy to metastasize or recur. Gadolinium-enhanced views of the brain and spinal cord are used to evaluate for CSF seeding.[11]

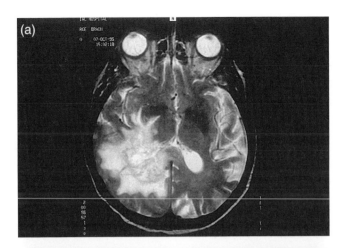

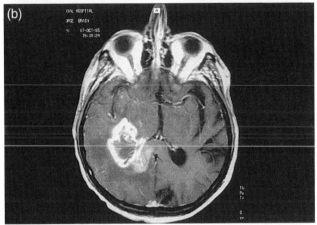

Figure 2 A 69-year-old man with a high-grade astrocytoma. (a) T_2-weighted axial image obtained on a 1.5-T Siemens Magneton magnet demonstrates a large necrotic tumor with extensive surrounding edema in the right temporal and parietal lobes. (b) Postgadolinium T_1-weighted image shows the ring-enhancing mass lesion, but the edema is not as evident on this sequence

In searching for *intracranial metastatic spread* from extracranial primary tumors such as breast, lung, and colon carcinomas and melanoma, MRI with gadolinium is the optimal screening test.[14] Metastases incite greater edema as compared with primary tumors. The presence of multiple lesions strongly favors metastic disease over a primary tumor such as glioma, although a small percentage of gliomas are multicentric.

4.2 Extra-axial Tumors

4.2.1 Meningiomas

Comprising 10–20% of intracranial tumors, meningiomas are the most common extra-axial tumor, with an autopsy prevalence of 1–2%.[3] Common sites of occurrence include the cerebral convexities, the falx, and the sphenoid wing. Meningiomas originate from the dural layer covering the brain parenchyma.

For References see p. 957

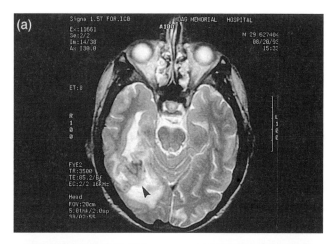

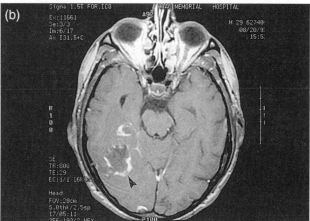

Figure 3 A 29-year-old male with AIDS and primary intracranial lymphoma. (a) Note the low signal intensity (black arrowhead) on this T_2-weighted axial sequence obtained on a 1.5-T GE Signa magnet. (b) Post-gadolinium T_1-weighted axial image shows little enhancement in this case

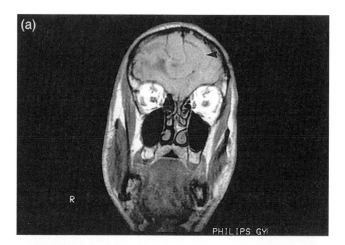

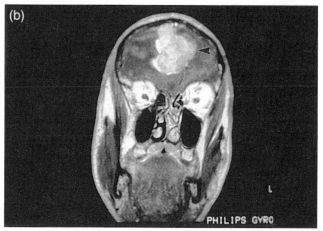

Figure 4 A 60-year-old male with a meningioma originating from the falx. (a) On this coronal T_1-weighted precontrast image, the signal intensity of the tumor is similar to the brain parenchyma (black arrowhead). The images were obtained on a 1.5-T Siemens Magneton magnet. (b) After gadolinium administration, the borders of this densely enhancing meningioma are easily defined

Because most meningiomas show only a slight increase in T_1 over white matter and a T_2 within normal range for brain, they may appear mildly hypointense on T_1-weighted images and isointense to hyperintense on T_2-weighted images. Heterogeneous signal intensity, especially on T_2-weighted images, may be seen secondary to vascular flow voids, calcification, and cystic foci. Usually little mass effect or significant edema is seen, though these tumors can be quite large. A hallmark is intense enhancement after contrast administration, and a tapered extension of enhancement along the tumor base, called a 'dural tail', may be present, (Figure 4).

4.2.2   Acoustic Neuroma: a Cerebellopontine Angle Tumor

Being a relatively common site of tumor occurrence, the cerebellopontine angle (CPA) is well imaged by MRI. The prototype tumor occurring in this location is the acoustic neuroma arising from the eighth cranial nerve. On precontrast images, enlargement of the seventh and eighth cranial nerve complex is seen, and intense enhancement is demonstrated with contrast. Extension into the internal auditory canal highly suggests this tumor type.

4.2.3   Pituitary Gland Tumors

MRI has become the primary modality for diagnosis of hormone-secreting pituitary microadenomas as well as other tumors that may occur in the sella or suprasellar locations, such as craniopharyngiomas. Coronal and sagittal pre- and post-gadolinium images are usually employed. Timing of the imaging with the administration of contrast agent is important, as enhancement of a microadenoma will be delayed compared with the normal enhancement of the rest of the gland. Larger pituitary tumors, such as macroadenomas, may displace the carotid artery or invade the cavernous sinus, both features being depicted well by MRI.

4.2.4   Other Extra-axial Tumors

Tumors may originate in the bones comprising the base of the skull, chordomas and chondrosarcoma being two such examples. A difficult area to image by CT, MRI demonstrates skull base masses well and may depict infiltrative changes in the marrow-containing portions of the skull base, as can be seen with metastatic involvement of the clivus. A number of tumors may originate in the pineal gland, and multiplanar imaging is important in identifying the pineal gland as the origin. Ependymoma, a tumor which originates in the ependymal lining cells of the ventricular system, is the most common intraventricular brain neoplasm and can cause expansion of the ventricle at the site of origin.[15] Extrusion through various ventricular foramina, such as the foramen of Magendie or Luschka, highly suggest this tumor type when it originates in the fourth ventricle.

5 SUMMARY

MRI has proven its usefulness in the diagnosis and follow-up of brain neoplasms. As discussed, its major limitations are in the diagnosis of tumor recurrence after surgical excision and in distinguishing tumor recurrence from radiation necrosis. Recent developments in magnetic resonance spectroscopy (MRS) and positron emission tomography (PET) offer further help in such cases. Experimental work with proton MRS, for example, has shown that increased levels of lactate, choline and lipids may be associated with certain malignancies.[16] In the future, the noninvasive assessment of brain neoplasm histology will probably combine the efforts of MRI, MRS and PET as the two latter technologies are further developed.

6 RELATED ARTICLES

Brain MRS of Infants and Children; Brain Neoplasms in Humans Studied by Phosphorus-31 NMR Spectroscopy; Cranial Nerves Investigated by MRI; Gadolinium Chelate Contrast Agents in MRI: Clinical Applications; Gadolinium Chelates: Chemistry, Safety, and Behavior; Hemorrhage in the Brain and Neck Observed by MRI; MRI in Clinical Medicine; Relaxation Measurements in Imaging Studies.

7 REFERENCES

1. A. N. Hasso, K. E. Kortman, and W. G. Bradley, in 'Magnetic Resonance Imaging', 2nd edn., eds. D. D. Stark and W. G. Bradley, Mosby Year Book, St. Louis, 1992, Chap. 25.
2. S. W. Atlas, in 'Magnetic Resonance Imaging of the Brain and Spine', ed. S. W. Atlas, Raven, New York, 1991, Chap. 10.
3. H. I. Goldberg, in 'Magnetic Resonance Imaging of the Brain and Spine', ed. S. W. Atlas, Raven, New York, 1991, Chap. 11.
4. K. R. Thulborn and S. W. Atlas, in 'Magnetic Resonance Imaging of the Brain and Spine', ed. S. W. Atlas, Raven, New York, 1991, Chap. 9.
5. M. Just and M. Thelen, Radiology, 1988, 169, 779.
6. S. W. Atlas, D. B. Hackney, D. M. Yousem, and J. Listerud, Radiology, 1991, 181, 165.
7. G. H. Zoarski, J. K. Maskey, Y. Anzai, W. N. Hanafee, P. S. Melki, R. V. Mulkern, F. A. Jolesz and R. B. Lufkin, Radiology, 1993 188, 323.
8. B. L. Dean, B. P. Drayer, C. R. Bird, R. A. Flom, et al., Radiology, 1990, 174, 411.
9. W. M. Kelly and M. Brant-Zawadzki, in 'Radiology, Diagnosis, Imaging, Intervention', eds. J. M. Taveras and J. T. Ferrucci, J. B. Lippincott, Philadelphia, 1989, Chap. 53.
10. W. G. Bradley, Jr., W. T. C. Yuh, and G. M. Bydder, J. Magn. Reson. Imaging, 1993, 3, 199.
11. R. B. Schwartz and M. T. Mantello, Semin. Ultrasound CT MR, 1992, 13, 449.
12. R. M. Henkelman, J. F. Watts, and W. Kucharczyk, Radiology, 1991, 179, 199.
13. S. P. Meyers, S. S. Kemp, and R. W. Tarr, Am. J. Roentgenol., 1992, 158, 859.
14. J. H. Bisese, Semin. Ultrasound CT MR, 1992, 13, 473.
15. J. Jelinek, J. G. Smirniotopoulos, J. E. Parisi, and M. Kanzer, Am. J. Roentgenol., 1990, 155, 365.
16. P. R. Luyten, J. H. Marien, and W. Heindel, P. M. van Gerwen, K. Herholz, J. A. den Hollander, G. Friedmann, and W. D. Heiss, Radiology, 1990, 176, 791.

Biographical Sketches

Andrew P. Kelly. b. 1960. B.A. Chemistry, 1983, California State University, Fullerton; M.D., 1988, University of California, Davis. Fellow in MRI, Hoag Memorial Hospital Presbyterian, Newport Beach, CA, 1993–present.

Michael Brant-Zawadzki. Approx. 150 articles, including the textbook 'Magnetic Resonance Imaging of the Central Nervous System'. A frequent lecturer on MR imaging applications and contrast agents.

Intracranial Infections

Spyros K. Karampekios
University Hospital of Crete, Heraklion, Greece

and

John R. Hesselink
UCSD Medical Center, San Diego, CA, USA

1 INTRODUCTION

Despite the development of many effective antibiotic therapies and the evolution of new neurosurgical techniques, central nervous system (CNS) infections persist. Also the increasing, devastating effect of acquired immune deficiency syndrome (AIDS) has created a whole group of life-threatening opportunistic infectious diseases. CNS involvement usually occurs as a result of an infection from another organ system, or as a manifestation of a systemic disease. The brain, particularly, is well protected from invading agents by the calvarium, the meninges and the blood–brain barrier. However, different types of pathogens including bacteria, viruses, fungi, and parasites, can reach the brain by hematogenous spread (related to septicemia or endocarditis) and less likely by direct extension (bony erosion from an adjacent infected paranasal sinus, mastoid or middle ear). Other possible pathways for intracranial infections are via

For References see p. 965

anastomotic veins from the face and scalp, along peripheral and cranial nerves (viruses) and after a penetrating head trauma.

Once in the intracranial cavity, pathogens can involve the parenchyma (cerebritis–abscess, encephalitis), the meninges (meningitis–ependymitis), and other extraaxial spaces (subdural and epidural empyemas). The brain's unique response to infection is due in part to the absence of draining lymphatics, the differentiation of vascular supply in gray and white matter, the presence of the blood–brain barrier, and the existence of perivascular cerebrospinal fluid (CSF) containing spaces (Virchow–Robin spaces). CSF assists in the dissemination of an infectious disease, acting as a perfect culture medium for microbial growth.

Magnetic resonance imaging (MRI) provides the most sensitive imaging modality to detect cerebral infection because of its optimal contrast resolution, the multiplanar imaging capability and the absence of signal from the surrounding bone. Intravenous contrast administration of gadolinium–diethylenetriaminepentaacetic acid (Gd-DTPA) increases the sensitivity of MRI and has allowed earlier detection of many infections compared to computed tomography (CT).[1] The only disadvantage of MRI is its poorer ability to detect calcifications, an important finding in some cerebral infections.

2 CEREBRITIS AND BRAIN ABSCESS

Cerebritis and abscess formation constitute a spectrum of the same process. Regardless of the pathogen, the brain tissue reacts in a predictable way to focal parenchymal infection, developing initially an area of focal cerebritis, which consists of vascular congestion, petechial hemorrhage, and brain edema. Progression from the cerebritis stage to an encapsulated abscess with a central area of liquefied, necrotic material requires 10–14 days, although the rapidity of this process depends on patient's immunocompetence and the organism.[2] In the late mature abscess stage, when the collagenous capsule is fully formed, the surrounding edema decreases, and surgical drainage is facilitated. A brain abscess is usually caused by direct extension from an adjacent infected sinus or mastoid, or by hematogenous spread from an extracerebral source. Rarely is an abscess secondary to meningitis. In up to 20% of cases the source of infection is not discovered. In children the majority of cerebral abscesses are associated with cyanotic congenital heart disease. Anaerobic bacteria are isolated most frequently in the brain abscesses, although a mixture of pathogens are often found. Overall, in otherwise immunocompetent individuals, the commonest cultured organisms are *Staphylococcus* and *Streptococcus* species.[3] Patients present usually in the late cerebritis–early abscess phase with nonspecific symptoms of headache, confusion, seizures or focal neurologic deficits. Fever and leukocytosis are common during the invasive phase of an abscess, but may resolve as it matures.

The MRI features of cerebritis–brain abscess depend on the stage of the infectious process at the time of imaging. In the early cerebritis stage MRI depicts the changes earlier than CT because of its superior sensitivity to alterations in water content. The area of cerebritis appears mildly hypointense on T_1-weighted images. On the same sequence the early, subtle

mass effect is best demonstrated. The infectious focus depicts high signal on T_2-weighted images, both centrally from inflammation and peripherally from edema. As the infection matures, it increases in size due to an increase in edema and necrotic debris accumulates centrally, while the body attempts to isolate the infection by forming a capsule. The abscess capsule is thicker along the cortical surface because the vascularity of gray matter is much greater than that of white matter, resulting in increased local cortical reaction. The thinner portion of the capsule toward the white matter accounts for the predilection of abscesses to rupture into the ventricles producing ependymitis. At this stage of abscess formation, MRI demonstrates lengthening of T_1 and T_2 relaxation times in the core of the lesion. Peripherally there is moderate amount of vasogenic edema, which is mildly hypointense on T_1- and hyperintense on T_2-weighted images. On T_1-weighted images, against the hypointense areas of the necrotic center and the surrounding edema, the abscess capsule stands out as an iso- or slightly hyperintense ring. On T_2-weighted images the ring is markedly hypointense (Figure 1(a)). There is still discussion about the causes of capsular intensity. The hyperintensity on T_1-weighted images is attributed to capsular hemorrhage by some authors. More recently the signal properties of the abscess capsule have been attributed to paramagnetic hemoglobin degradation products, or free radicals within macrophages. Macrophages are abundant in the capsule, and their activity is highest at the late cerebritis–early abscess phase, exactly when the signal intensity of the capsule on the T_2-weighted images is particularly low.[4]

Contrast administration is very helpful in the evaluation of brain abscesses. Gd-DTPA produces mottled, heterogenous areas of enhancement in the cerebritis stage, with an enhancing rim developing as the abscess matures. The ring-like enhancement reflects the damaged blood–brain barrier and is typically smooth, well defined and thin walled. The ring is thinnest at its medial margin and often points to the adjacent ventricle. If an abscess ruptures into a ventricle and secondary ependymitis develops, there is enhancement of the ventricular wall, suggesting a very poor prognosis [Figure 1(b)].

Brain abscesses produced by nonpyogenic organisms typically occur in immunocompromised patients; these will be discussed in the section of AIDS-related infections.

3 MENINGITIS

Meningitis is an acute or chronic inflammation of the piaarachnoid (leptomeninges) and the adjacent CSF. Patients with meningitis present with fever, headache, neck stiffness, photophobia, and altered consciousness. Acute meningitis (purulent) is usually caused by bacteria. The most common pathogens are *Haemophilus influenza, Neisseria meningitidis*, and *Streptococcus pneumonia*. Significant morbidity and mortality occurs with meningitis. The overall mortality rate ranges from 5% to 15% and severe, persistent neurologic deficits may be seen in 10–25%.[5] In neonates and immunosuppressed patients Gram-negative meningitis is common, caused by *Escherichia coli, Pseudomonas aeruginosa* and *Klebsiella*. Viruses can also cause an acute meningitis (lymphocytic), particularly enteroviruses and the mumps virus. Viral meningitis is usually self-

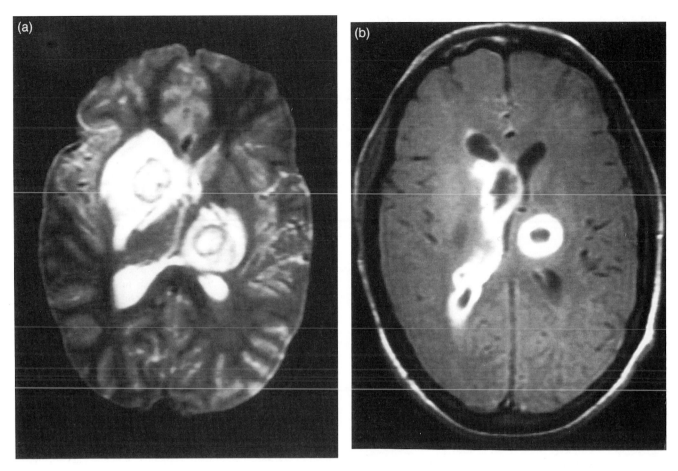

Figure 1 Streptococcal abscesses. (a) Axial, T_2-weighted (SE 2500/70) image shows the abscess capsules as hypointense rings surrounding the necrotic central core. There is also significant amount of peripheral edema. (b) Axial, postcontrast T_1-weighted (SE 800/20, Gd-DTPA) image one week later demonstrates intense ependymal enhancement of the right lateral ventricle from the resulting ependymitis secondary to rupture of an abscess into the ventricular system. (Reproduced by permission of W. B. Saunders from R. R. Edelman and J. R. Hesselink (eds), *Clinical MRI*, 1990, Chap. 19, p. 578)

limited, with less significant symptoms and complications than those of bacterial origin.

Chronic meningitis is usually of tuberculous origin and presents as a long-standing, indolent process, in which vasculitis and cerebral infarctions from basal meningeal inflammation are more prevalent. In patients with an immunologic dysfunction, meningitis can be the result of fungal infection, with the main representatives being cryptococcosis, coccidiomycosis, and blastomycosis. Cryptococcal meningitis will be discussed in the section of AIDS-related infections. Finally, sarcoidosis is a noninfectious granulomatous disease that involves the CNS in 5% of patients, producing inflammation of the leptomeninges in the basal cisterns, the optic chiasma and the infundibulum. Other noninfectious processes which cause meningeal disease and simulate infections are meningeal carcinomatosis, postoperative meningeal irritation, chemical meningitis, meningeal reaction adjacent to cerebral infarction, and subarachnoid hemorrhage.

Neuroimaging plays a limited role in the diagnosis of meningitis, which is made by history, physical examination and laboratory CSF findings. CT and MRI are mainly focused on the detection of associated complications, which include

vascular thrombosis, infarctions, cerebritis–brain abscess, ventriculitis, hydrocephalus, empyemas of epidural and subdural space and subdural effusions. In case of uncomplicated acute meningitis, unenhanced MRI scans are usually unremarkable. More severe chronic cases may disclose hyperintense CSF on T_1-weighted and proton density (PD) images in the basal cisterns secondary to obliteration from the inflammatory exudate and meningeal hyperemia. Contrast administration is very helpful in the evaluation of suspected meningeal infection because the involved meninges enhance diffusely and intensely. In the majority of cases of acute bacterial or viral meningitis, the meningeal enhancement occurs predominantly over the cerebrum, especially involving the frontal and parietal lobes, and the interhemispheric and sylvian fissures. However, in cases of tuberculous, fungal and sarcoid meningitis, the meningeal enhancement is more prominent in the basal cisterns (Figure 2). Gd-DTPA enhanced MRI appears to be much more sensitive than postcontrast CT in the detection of the meningeal enhancement, particularly when it occurs near the convexity.[6] However, according to some authors, postcontrast MRI does not completely correlate with the extent of the inflammatory cell infiltration. Using animal models they have proved that

For References see p. 965

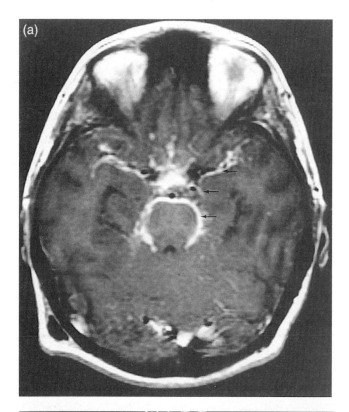

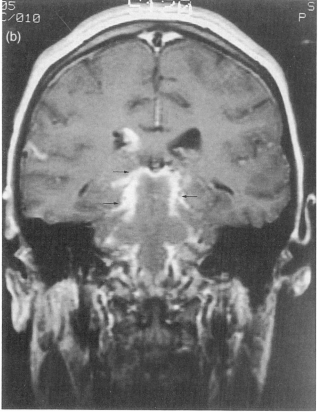

Figure 2 Coccidiomycosis meningitis. Contrast-enhanced T_1-weighted (SE 500/15, Gd-DTPA) axial (a) and coronal (b) scans demonstrate abnormal, extensive meningeal enhancement in the basal and ambient cisterns (arrows). (Reproduced by permission of the American Society of Neuroradiology from C. J. Wrobel, S. Meyer, R. H. Johnson, and J. R. Hesselink, *Am. J. Neuroradiol.*, 1992, **13**, 1243

above a threshold of inflammation, meningeal enhancement was visualized, but areas that exhibited mild meningitis histologically did not enhance.[7] While more sensitive, postcontrast MRI is no more specific, in that any process that causes meningeal irritation can also cause meningeal enhancement.

In untreated or more severe cases of meningitis significant complications may occur in the following days or weeks. The exposure of the blood vessels to the inflammatory exudate may result in vasculitis and thrombosis, a common complication in cases of tuberculous meningitis. Occlusion of small perforating arteries results in focal infarcts in the basal ganglia, while involvement of larger vessels can produce massive infarcts. Multiple hemorrhagic infarcts in the white matter are the result of cortical venous or dural sinus thrombosis.[8] Another severe complication of meningitis is cerebritis and brain abscess formation. Sometimes, rupture of a preexisting brain abscess into the CSF spaces can produce secondary meningitis. Ventriculitis and ependymitis can result from direct extension of an abscess, progression of meningitis, or from an infected intraventricular shunt. In cases of ependymitis, MRI exhibits increased periventricular signal on T_2- and PD-weighted images, that often has a nodular and irregular character to distinguish it from transependymal CSF flow associated with obstructive hydrocephalus. Also, on postcontrast scans ependymal enhancement outlines the ventricles. If purulent debris fills the ventricles, the CSF may give a higher signal than normal. It has been proved more accurate to compare the intraventricular CSF with the vitreous of the globe, since associated meningitis may prevent comparison with CSF of the basal cisterns.[9] Another frequent complication of meningitis is either obstructive, or communicating hydrocephalus, which occurs secondary to cellular debris obstructing the CSF pathways, or to arachnoid adhesions impairing extraventricular CSF flow and absorption. Ventricular dilatation may be the only abnormal finding in patients with meningitis and is adequately evaluated with both CT and MRI. Subdural collections can also complicate meningitis; these are discussed in the section on extraaxial empyemas.

4 ENCEPHALITIS

Encephalitis refers to a diffuse parenchymal inflammation of the brain caused primarily by viruses. Clinically, acute encephalitis should be suspected if the patient presents with convulsions, altered consciousness, delirium, aphasia, or ataxia. Particularly in herpetic encephalitis the symptoms reflect the propensity to involve the subfrontal and temporal lobes, with hallucinations, seizures, and personality changes. Viral encephalitis is usually acute, although it can occur from reactivation of a latent virus. The most common invading viruses are herpes simplex type 1 and type 2, herpes zoster, arbo- and enteroviruses, which produce almost the same reaction in brain tissue and appear similar on CT and MRI.[10] In patients with AIDS, viral encephalitis may be caused by the human immunodeficiency virus (HIV), the cytomegalovirus (CMV) and the papovavirus (PML); these are discussed in the section on AIDS-related infections. Acute disseminated encephalomyelitis (ADEM) represents an immune-mediated complication following an antecedent viral infection, especially measles, mumps, rubella, varicella, or a preceding vaccination. Various forms of viral encephalitis (slow virus) are described in association with Creutzfeldt–Jacob disease and subacute sclerosing panencephalitis (SSPE). Finally, neonatal encephalitis is caused frequently by the group of TORCH pathogens (from the initials of toxoplasma, rubella, cytomegalovirus, and herpes simplex type 2) and share some common features like microcephaly, brain atrophy, hydrocephalus, and cerebral calcifications.

Herpes simplex virus type 1 (HSV-1) accounts for 95% of herpetic encephalitis in adults. The mortality rate approaches 70% and most of the survivors exhibit severe, persistent neurologic impairment. The virus usually invades the brain after reactivation of a latent form, which is frequently located in the trigeminal (gasserian) ganglion. The marked predilection for temporal lobe involvement supports the proposed theory that the infection spreads intracranially from the trigeminal ganglion along the meningeal branches of the trigeminal nerve. The resulting necrotizing encephalitis rapidly disseminates in the brain, sparing the basal ganglia and producing mass effect and edema as well as small petechial hemorrhages. Early diagnosis is paramount in order to institute effective medical therapy (vidarabine or acyclovir) and avoid devastating and irreversible brain damage. Definitive diagnosis of HSV-1 encephalitis is done after isolation of the virus from brain biopsy. However, given the appropriate clinical presentation, with or without MRI confirmation, treatment should be instituted immediately. In the early course of the infection the characteristic distribution is almost pathognomonic for HSV-1 encephalitis. By MRI, the early edematous changes appear as ill-defined areas of low signal on T_1- and high signal on T_2-weighted images, usually beginning unilaterally but rapidly progressing to both hemispheres (Figure 3). Variable mass effect and gyral enhancement may also be present. Occasionally, foci of hemorrhage are visualized as areas of high intensity on both T_1- and T_2-weighted images.[11] Note that, unlike HSV-1, HSV-2 encephalitis, the most common neonatal viral encephalitis, is a panencephalitis without any predilection in the temporal lobes.

5   EXTRA-AXIAL EMPYEMAS

Accumulation of inflammatory debris may occur in the subdural or, less frequently, in the epidural space. Most of the extraaxial empyemas present acutely secondary to sinusitis or mastoiditis. Empyemas that occur secondary to an infected posttraumatic extra-axial hematoma or postcraniotomy cavity have a more prolonged and indolent course. Empyemas of the extra-axial compartments can also occur (particularly in infants) as complications of meningitis.

5.1   Subdural Empyemas

Pathogens may enter the subdural space via retrograde thrombophlebitis, via local dural erosion, or after contamination of a meningitis-induced subdural effusion. Most frequently, subdural empyemas are secondary to frontal sinusitis (40%).[12] Patients usually present with fever, seizures, focal neurologic deficit, or coma, and require urgent surgical and medical intervention. If completely or partially untreated, the empyema may be complicated with venous thrombosis, infarcts and parenchymal abscesses. The long *TR*, long *TE*

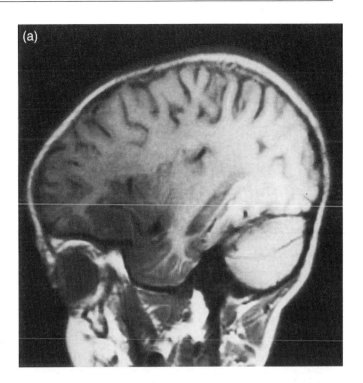

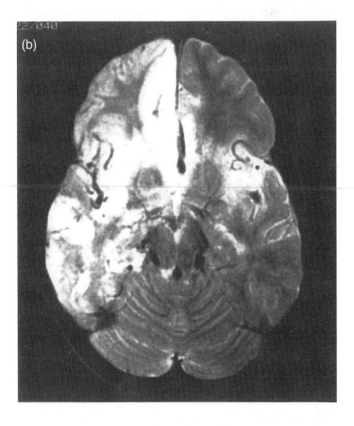

Figure 3 Herpes simplex encephalitis. (a) Sagittal, T_1-weighted (SE 800/20) image shows area of low signal intensity in the right temporal lobe and frontal operculum. (b) Axial, T_2-weighted (SE 2800/80) image demonstrates extensive hyperintense areas in the subfrontal and temporal lobes bilaterally, though more severe on the right hemisphere. (Reproduced by permission of W. B. Saunders from R. R. Edelman and J. R. Hesselink, (eds), *Clinical MRI*, 1990, Chap. 19, p. 584)

For References see p. 965

sequence is the most sensitive in detecting small, crescentic fluid collections, especially when they are located near the inner table or along the falx. MRI can also distinguish subdural empyemas from subdural effusions or hematomas, depending on signal differences. T_1-weighted images demonstrate the purulent collections as areas with more signal than pure CSF, due to the increased content of proteins and inflammatory debris. Subdural hematomas (subacute) can be easily distinguished because the extracellular methemoglobin exhibits markedly hyperintense signal on both T_1- and T_2-weighted images. MRI can also evaluate mass effect on the adjacent brain or CSF spaces, as well as the underlying parenchymal abnormalities.

5.2 Epidural Empyemas

In the case of epidural empyema, the purulent collection tends to localize outside the inelastic and firm dura, which protects the underlying brain from undesirable concomitant abnormalities. Thus, patients usually have a more silent clinical course. As with subdurals, MRI is the most sensitive method for the evaluation of epidural disease. The signal characteristics of the lentiform extraaxial collections are similar to subdural empyemas on both T_1- and T_2-weighted images. Mass effect on the adjacent brain can also be seen, usually without any evidence of parenchymal changes. On postcontrast scans there is profound enhancement of the inflamed dura, often thicker than that observed in subdurals.

6 CYSTIC LESIONS

Parasitic infections of the brain can commonly produce focal, cystic areas. We review the imaging features of the most frequent parasitic diseases with emphasis on neurocysticercosis. Toxoplasmosis is included in the section on AIDS-related infections.

6.1 Cysticercosis

This is the most common parasitic infection of the CNS in immunocompetent individuals and is caused by the pork tapework *Taenia solium*. Humans ingest the eggs with food contaminated by human or pig feces. In the stomach the eggs release embryos (oncospheres) which enter the intestinal wall and spread hematogenously, invading any human tissue and developing the larvae (cysticerci). Skeletal muscles and CNS are more frequently affected by cysticercosis. CNS infection is reported in 70–90% of cases and constitutes the commonest cause of seizures in young patients in developing countries with poor hygiene. Neurocysticercosis may involve the brain parenchyma, the ventricles, or the subarachnoid space. In parenchymal cysticercosis as the larvae initially invade the brain, they produce a mild inflammatory reaction, which on MRI can be demonstrated as focal, nonenhancing area of edema. Typically, at this very early, active stage of the disease, CT findings are unremarkable. In the next stage (3–12 months after infection) cystic lesions with thin walls and clear fluid are

formed. They are usually located at the gray-white matter junction and appear as 1–2 cm spherical cysts with signal intensity identical to CSF on all MRI sequences (Figure 4). There is no evidence of associated edema or contrast enhancement. At this stage, a mural nodule (scolex), which is pathognomonic for cysticercosis, can be seen as a small (1–2 mm) focal, mural projection, isointense to the brain parenchyma. Visualization of the scolex is much easier with magnetic resonance than CT, due to the superior contrast resolution of MRI. Later on, as the parasites degenerate and die, they induce an inflammatory response with associated thickened cystic walls, increased signal intensity from the cyst's content owing to protein accumulation, marked surrounding edema and ring-like enhancement around the cyst. Finally, at the end-stage of the disease process, the cysts are completely mineralized and only punctate foci of calcifications may be seen. Obviously CT is more effective than MRI in the detection of calcifications, although specific MR sequences (gradient echo) are relatively sensitive to calcified lesions.[13] Intraventricular cysticercosis occur in approximately 20% of cases. Cysts located near the aqueduct of sylvius or other CSF pathways can cause acute, obstructive hydrocephalus. The signal characteristics of the intraventricular cysts also follow CSF. Furthermore, on T_1- and PD-weighted images, the cystic wall and the scolex can be more apparent against the adjacent ventricular CSF, even though the cystic component is indistinguishable from CSF.[14] The unique feature in the ventricular variety of cysticercosis is that the degenerated cysticerci rarely calcify. In about 10%, cysticercosis may occur in the subarachnoid space, where the cysts have a racemose form and are larger, sterile, and usually without scoleces. Sometimes the only imaging finding is asymmetry of the CSF spaces. In the case of a profound inflammatory reaction, adjacent brain abnormalities such as edema, gliosis, or communicating hydrocephalus can be seen.

6.2 Hydatid Disease

This is a parasitic infection which results from the ingestion of contaminated dog feces, containing eggs of *Echinococcus granulosus*. After ingestion, the eggs release embryos (oncospheres) in the gastrointestinal tract that spread hematogenously via the portal circulation to other human tissues. The majority of the lesions affect the liver and the lungs, although some can reach the brain (2–5%), where the larvae develop into large, unilocular cysts. Either CT or MRI can demonstrate adequately the large, commonly solitary cerebral hydatid cyst as a spherical thin-walled structure containing clear, sterile fluid with CSF imaging characteristics. Calcification of the cystic wall may occasionally occur and is better shown by CT. Hydatid (echinococcal) cysts are not associated with edema or contrast enhancement, except in case of cystic rupture and leakage, when an extensive inflammatory reaction is created.[15]

7 AIDS-RELATED INFECTIONS

Patients with AIDS often develop neurologic complications. At least 10% of all AIDS patients present with problems re-

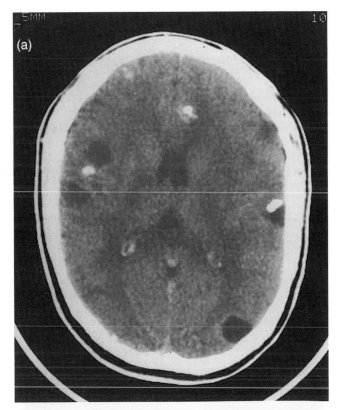

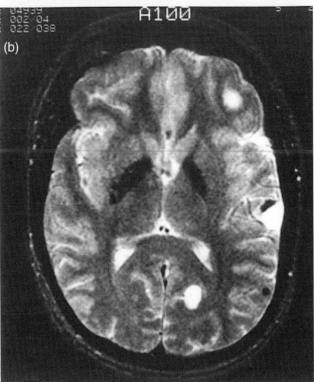

Figure 4 Parenchymal cysticercosis. (a) Noncontrast CT scan demonstrates multiple cysts in the brain parenchyma, several of which are calcified. (b) Axial, T_2-weighted (SE 2500/70) image at approximately the same level, reveals high signal intensity within the cysts. Small calcified foci are difficult to identify, although the larger calcifications exhibit markedly low signal intensity. (Reproduced by permission of W. B. Saunders from R. R. Edelman and J. R. Hesselink, (eds), *Clinical MRI*, 1990, Chap. 19, p. 580)

lated to the nervous system, and more than one-third of them will manifest a clinically apparent neurologic disorder during the course of the disease.[16] At autopsy it was revealed that 80–90% of patients with AIDS had neuropathologic abnormalities, and usually more than one disease process was present.[17] AIDS is due to the HIV, which on the one hand damages the nervous system by direct infection (HIV encephalitis), and on the other produces immune system dysfunction, especially of cell-mediated immunity. The resultant immunodeficiency leaves the person with AIDS vulnerable to many opportunistic CNS infections. The most important and the commonest of these will be further analyzed below.

7.1 Toxoplasmosis

Toxoplasmosis is the most frequent opportunistic brain infection in AIDS patients. It is caused by the parasite *Toxoplasma gondii* which is an obligate intracellular protozoan. Patients may present with clinical manifestations of focal mass effect such as seizures, focal neurologic deficits, or cranial nerve palsies, as well as more generalized symptoms like headache, confusion, lethargy, and declining mental status. MRI is the most sensitive imaging modality, demonstrating lesions in patients with normal CT scans, particularly lesions at an earlier stage or located in the posterior fossa. Postcontrast MRI scans show multiple ring or nodular enhancing lesions surrounded by vasogenic edema and an associated mass effect. Toxoplasma lesions are typically located at the corticomedullary junction and the basal ganglia. Hemorrhage is uncommon. On T_2-weighted images the lesions are either hyperintense or iso- to hypointense to brain parenchyma.[18] The neuroimaging findings with both CT and MRI are not pathognomonic for toxoplasmosis and can also be seen in various infectious or noninfectious processes, such as brain metastases, intracerebral lymphoma, Kaposi's sarcoma, cryptococcoma, and tuberculoma. If the imaging studies are suggestive of toxoplasmosis and confirmatory toxoplasma titers are present, empiric antitoxoplasma medication is begun and the response of the treatment can be monitored with follow-up CT or MRI scans. If any or all of the lesions fail to respond to therapy by imaging criteria, toxoplasma may not be the causative infective agent, or there may be another, concurrent disease process present, and brain biopsy should be considered.[19]

7.2 Cryptococcosis

Cryptococcus neoformans causes the most common CNS fungal infection in AIDS patients, which in terms of relative frequency ranks third after HIV encephalitis and toxoplasmosis. It involves the CNS most commonly as a meningitis with minimal inflammatory response and extends mainly in the distribution of the perforating brain arteries and the perivascular (Virchow–Robin) spaces. On MRI there are four patterns: focal cryptococcomas; dilated Virchow–Robin spaces in the basal ganglia and midbrain; miliary enhancing nodules in the brain parenchyma and leptomeningeal-cisternal spaces; and a mixed pattern.[20]

For References see p. 965

7.3 Other Brain Abscesses

Brain abscesses caused by nonpyogenic organisms typically occur in immunocompromised patients and share almost the same characteristics as the bacterial ones. Some of their unique features have been attributed to the diminished host response. Thus, multiplicity and extraparenchymal spread are frequently encountered. Tuberculous brain abscesses usually spread hematogenously from the lungs. They contain encapsulated pus with viable tubercle bacilli and differ from the more common tuberculomas (granulomas) which are smaller and contain caseous debris. Other associated lesions such as basal meningitis, multiple granulomas, and deep cerebral infarctions are very helpful clues to the diagnosis of tuberculosis.[21] Another cause of non-bacterial brain abscess is fungal infections, which have recently increased substantially because of the large number of AIDS patients. The fungi may involve intracranial blood vessels, meninges, or the brain parenchyma (granulomas or abscesses). Fungal abscesses are caused primarily by *Aspergillus*, *Mucor* and *Candida* species, owing to their large hyphal forms which permit only limited access to the meningeal microcirculation. *Aspergillus* spreads to the brain hematogenously from the lung, gastrointestinal tract, or directly from the nasal cavity (paranasal sinuses). It readily invades vascular structures and causes hemorrhagic infarcts. In that case the resulting brain abscesses may mimic other hemorrhagic lesions such as metastases. CNS mucormycosis occurs in patients with compromised natural defense and uncontrolled diabetes and is spread by direct extension from adjacent facial compartments. Like *Aspergillus*, *Mucor* invades blood vessels and causes abscesses, more often at the inferior temporal and frontal lobes. Candidiasis may involve the brain parenchyma, producing multiple, scattered microabscesses.[22]

7.4 HIV Encephalitis

This is the most common CNS complication in AIDS patients, resulting in a progressive subcortical dementia with associated motor and behavioral abnormalities (AIDS dementia complex).[23] The hallmark of HIV encephalitis is the isolation of multinucleated giant cells and microglial nodules in the brain. Neither CT nor MRI are sensitive enough to detect those characteristic pathologic changes, showing only a nonspecific diffuse brain atrophy with a central predominance (ventricular enlargement). HIV leukoencephalopathy, which occurs in more severe cases of the encephalitis, is frequently located in the periventricular white matter and centrum semiovale. MRI depicts the white matter disease as extensive, nearly symmetric, ill-defined areas of high intensity on T_2-weighted images, without any evidence of mass effect or contrast enhancement[24] (Figure 5).

7.5 CMV Encephalitis

CMV is a herpes virus which can reactivate in the immuno-suppressed host and produce a necrotizing encephalitis and ependymitis. CNS infection by CMV is found in more than one-third of autopsies of AIDS patients. Pathologically, there is necrosis of the periventricular parenchymal tissue, associated

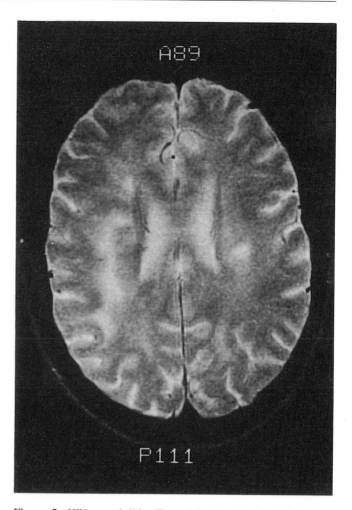

Figure 5 HIV encephalitis. The axial, T_2-weighted (SE 2500/70) image demonstrates multiple, ill-defined, hyperintense lesions in the periventricular white matter, becoming confluent in the right hemisphere. (Reproduced by permission of W. B. Saunders from R. R. Edelman and J. R. Hesselink (eds), *Clinical MRI*, 1990, Chap, 19, p. 590)

with accumulations of enlarged cells containing the typical inclusion bodies of CMV. However, the correlation between clinical diagnosis and postmortem results is very poor in cases of CMV encephalitis, for three reasons: firstly, the signs and symptoms of the encephalitis are subtle and nonspecific; secondly, there is no characteristic CSF profile; and, finally, the neuroimaging findings are usually absent or nonspecific. PD- and T_2-weighted magnetic resonance images may demonstrate a periventricular, thick, or nodular hyperintensity, which often involves the corpus callosum. CMV infection usually has a centrifugal spread from the ventricular ependyma involving diffusely the gray and white matter. On postcontrast scans there is diffuse subependymal enhancement around the lateral ventricles, representing the changes of ependymitis and making this an important differentiating point from HIV encephalitis or PML. The common simultaneous occurrence of both HIV and CMV viruses within the brain of AIDS patients supports the proposed theory that an interaction between them may play a role in the pathogenesis of encephalitis in patients with

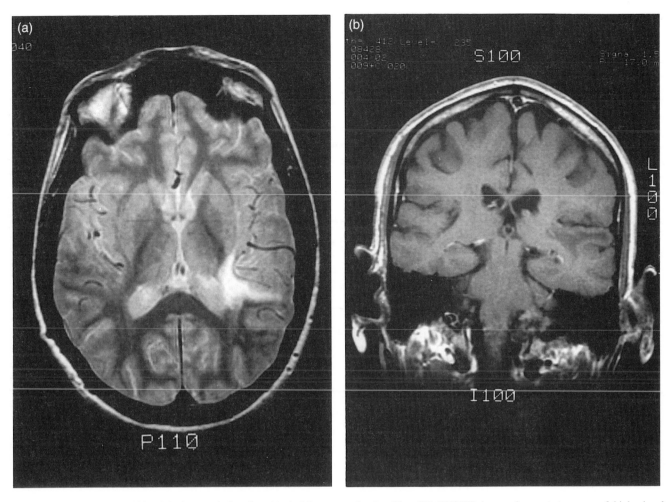

Figure 6 Progressive multifocal leukoencephalopathy. (a) Axial, proton density (Fast SE 3600/17) image demonstrates area of high signal intensity in the subcortical white matter of the left, posterior temporal lobe. (b) Postcontrast, T_1-weighted (SE 550/20, Gd-DTPA) coronal image discloses low signal intensity in the same region without any evidence of enhancement or mass effect

AIDS. There is also some evidence from recent studies that CMV by itself has immunosuppressive properties which may compromise a patient's natural defense against HIV or other opportunistic pathogens.[25,26]

7.6 Progressive Multifocal Leukoencephalopathy (PML)

PML is a progressive demyelinating disease of the brain, caused by reactivation of a latent papovavirus, and affecting 3–5% of all AIDS patients. The main target of PML is the oligo-dendrocyte which is the myelin-producing cell. The resulting demyelination develops a rapidly deteriorating clinical syndrome, with mental status changes, limb weakness, visual loss, or ataxia, and death usually ensues within 6 months. Initially, T_2-weighted images demonstrate focal, round, or oval areas of high intensity which become larger and confluent with time. These areas are predominantly located in the white matter of the parieto-occipital region. However, involvement of the cortical gray matter, as well as the basal ganglia, the cerebellum, or the brainstem, is not uncommon.[27] Generally the lesions in

PML do not enhance and there is no evidence of significant edema or mass effect (Figure 6).

8 RELATED ARTICLES

Brain Neoplasms Studied by MRI; CSF Velocity Imaging; Gadolinium Chelate Contrast Agents in MRI: Clinical Applications; Hemorrhage in the Brain and Neck Observed by MRI; Magnetic Resonance Imaging of White Matter Disease; MRI in Clinical Medicine.

9 REFERENCES

1. G. Schroth, K. Kretzschmar, J. Gawehn, and K. Voigt, *Neuroradiology*, 1987, **29**, 120.
2. D. R. Enzmann, R. H. Britt, and R. Placone, *Radiology*, 1983, **146**, 703.
3. G. Sze and S. H. Lee, in 'Cranial MRI and CT', 3rd edn, eds. S. H. Lee, K. C. V. G. Rao, and R. A. Zimmerman, McGraw-Hill, New York, 1992, Chap. 13.

For References see p. 965

4. A. B. Haimes, R. D. Zimmerman, S. Morgello, K. Weingarten, R. D. Becker, R. Jennis, and M. D. F. Deck, *Am. J. Roentgenol.*, 1989, 152, 1073.

5. Z. A. McGee and J. R. Baringer, in 'Principles and Practice of Infectious Diseases', 3rd edn., eds. G. L. Mandell, R. G. Douglas, and J. E. Bennett, Churchill Livingstone, New York, 1990, Chap. 66.

6. K. H. Chang, M. H. Han, J. K. Roh, I. O. Kim, M. C. Han, and C. W. Kim, *Am. J. Neuroradiol.*, 1990, 11, 69.

7. V. P. Mathews, M. A. Kuharik, M. K. Edwards, P. G. D'Amour, B. Azzarelli, and R. G. Dreesen, *Am. J. Neuroradiol.*, 1988, 9, 1045.

8. G. Sze and R. D. Zimmerman, *Radiol. Clin. N. Am.*, 1988, 26, 839.

9. T. J. Barloon, W. T. C. Yuh, L. E. Knepper, J. Biller, T. J. Ryals, and Y. Sato, *J. Comput. Assist. Tomogr.*, 1990, 14, 272.

10. G. Sze, in 'Magnetic Resonance Imaging', 2nd edn., eds. W. G. Bradley and D. D. Stark, Mosby, 1992, Vol. 1, Chap. 22.

11. R. D. Tien, G. J. Felsberg, and A. K. Osumi, *Am. J. Roentgenol.*, 1993, 161, 167.

12. K. Weingarten, R. D. Zimmerman, R. D. Becker, L. A. Heier, A. B. Haimes, and M. D. F. Deck, *Am. J. Roentgenol.*, 1989, 152, 615.

13. H. R. Martinez, R. Rangel-Guerra, G. Elizondo, J. Gonzalez, L. E. Todd, J. Ancer, and S. S. Prakash, *Am. J. Neuroradiol.*, 1989, 10, 1011.

14. G. P. Teitelbaum, R. J. Otto, M. Lin, A. T. Watanabe, M. A. Stull, H. J. Manz, and W. G. Bradley, *Am. J. Roentgenol.*, 1989, 153, 857.

15. K. H. Chang, S. Y. Cho, J. R. Hesselink, M. H. Han, and M. C. Han, *Neuroimag. Clin. N. Am.*, 1991, 1, 159.

16. R. M. Levy, D. E. Bredesen, and M. L. Rosenblum, *J. Neurosurg.*, 1985, 62, 475.

17. P. L. Lantos, J. E. McLaughlin, C. L. Schoitz, C. L. Berry, and J. R. Tighe, *Lancet*, 1989, i, 309.

18. B. C. Bowen and M.J.D. Post, in 'MRI of the Brain and Spine', ed. S. W. Atlas, Raven, New York, 1991, Chap. 16.

19. J. A. Cohn, A. McMeeking, W. Cohen, J. Jacobs, and R. S. Holzman, *Am. J. Med.*, 1989, 86, 521.

20. R. D. Tien, P. K. Chu, J. R. Hesselink, A. Duberg, and C. Wiley, *Am. J. Neuroradiol.*, 1991, 12, 283.

21. C. Campi de Castro and J. R. Hesselink, *Neuroimag. Clin. N. Am.*, 1991, 1, 119.

22. C. Bazan III, M. G. Rinaldi, R. R. Rauch, and J. R. Jinkins, *Neuroimag. Clin. N. Am.*, 1991, 1, 57.

23. B. A. Navia, B. D. Jordan, and R. W. Price, *Ann. Neurol.*, 1986, 19, 517.

24. H. S. Chrysikopoulos, G. A. Press, M. R. Grafe, J. R. Hesselink, and C. A. Wiley, *Radiology*, 1990, 175, 185.

25. C. A. Wiley and J. A. Nelson, *Am. J. Pathol.*, 1988, 133, 73.

26. R. L. Yarrish, in 'AIDS and other Manifestations of HIV Infection', 2nd edn., ed. G. P. Wormser, Raven, New York, 1992, Chap. 17.

27. A. S. Mark and S. W. Atlas, *Radiology*, 1989, 173, 517.

Biographical Sketches

Spyros K. Karampekios. *b* 1959. M.D., 1984, M.D. Thesis, 1989, University of Athens, Greece, 1989. Faculty in the Department of Radiology, University Hospital of Crete, Greece, 1990–present. Currently lecturer in radiology, University of Crete, School of Medicine. Postdoctoral work as a research fellow in neuroradiology, University of California, San Diego. Approx. 15 publications. Research interests include MRI of central nervous system infections and magnetic resonance evaluation of spinal cord cavities.

For list of General Abbreviations see end-papers

John R. Hesselink. *b* 1945. M.D., University of Wisconsin, 1971. Neuroradiology fellowship, Massachusetts General Hospital, 1979. Assistant Professor of Radiology, Harvard Medical School, 1979–84. Professor of Radiology and Neurosciences, University of California, San Diego, 1984–present. Approx. 150 publications and coeditor of textbook with Robert Edelman entitled *Clinical MRI*. Research interests include MRI of infections in immune compromised patients, fat suppression imaging techniques, magnetic resonance angiography and functional magnetic resonance.

Hemorrhage in the Brain and Neck Observed by MRI

Robert I. Grossman

University of Pennsylvania Medical Center, Philadelphia, PA, USA

1 HISTORY

Any new imaging modality that purported to compete with computerized tomography, in the late 1970s and early 1980s, had to be capable of detecting hemorrhage reliably. The success of MRI was intimately related to its ability for characterizing all stages of hemorrhage. Initially, at low imaging field strengths (up to 0.3 T), demonstration of acute hemorrhage (within the first 24–48 hours) was problematic at best.[1] With the introduction of higher field (0.5–1.5 T) MRI scanners in approximately 1984, it was clear that MRI could reliably image all stages of hemorrhage from the acute stage (within the first 3 days or so after the acute event) to chronic hemorrhages (those that occurred months to years following the event). Both the ability to recognize older hemorrhages and localize them precisely were major factors enabling MRI to become the primary imaging modality for the vast majority of hemorrhagic conditions affecting the central nervous system.

It is interesting to note the magnetic properties of dried blood had been first studied by Faraday almost 150 years ago.[2] Pauling and Coryell in 1936 recognized that deoxyhemoglobin was paramagnetic and oxyhemoglobin was diamagnetic.[3] The T_1 relaxation mechanisms of solutions of methemoglobin were investigated by Davidson and Gold in 1957 and further elucidated by Koenig et al. in 1981.[4,5] These studies were based on the Bloembergen et al. theory (1948) of 'outer sphere' relaxation of protons by paramagnetic centers.[6]

Thulborn et al. in 1982 observed that, at Larmor frequencies from 80 to 469 MHz, deoxyhemoglobin inside intact red blood cells (RBCs) would cause significant heterogeneity of magnetic susceptibility resulting in selective T_2 shortening.[7] They also demonstrated that this phenomenon was proportional to the square of the magnetic field. De La Paz et al. attempted to characterize the MRI features of acute intracranial hemorrhage.[8] They appreciated that the relaxation times of hemorrhagic tissue could be related to several interacting factors including red cell integrity, oxygen saturation, and hemoglobin concentration. Bradley and Schmidt in 1984 modeled subarachnoid hemorrhage on a spectrometer operating at 20 MHz and concluded that methemoglobin may be at least partially responsible for the observed high intensity on some magnetic resonance images.[9] This was confirmed experimentally by Di Chiro et al. in an animal model of intraparenchymal hemorrhage.[10] Gomori et al. unequivocally showed on a 1.5 T MRI scanner that a variety of different stages in the evolution of hemorrhage could be detected.[11] The gradient echo technique was initially utilized clinically by Bydder and Young in 1985 and then applied by Edelman et al. to increase the sensitivity to susceptibility differences, thus improving the detectability of hemorrhage at all field strengths.[12,13]

This review begins with the biophysical principles relevant to hemorrhage. These include a brief discussion of paramagnetism and magnetic susceptibility followed by structure of hemoglobin, the effect of protein on MRI signal intensity, and the importance of heterogeneity of magnetic susceptibility in the appearance of hemorrhage. The benefits and limitations of common imaging pulse sequences is then described. Lastly, we shall apply those principles to understand and interpret the diverse, yet characteristic, findings produced by blood on MRI.

2 BIOPHYSICAL PRINCIPLES

2.1 Paramagnetism and Magnetic Susceptibility

The magnetic moment of an electron is the result of its momentum and electric charge. Electrons have very large magnetic moments because of their light mass (~1/2000th of the proton's mass). The term 'magnetic dipole' is applied to the electrons because of their north and south magnetic poles which are separated by a distance. Magnetic fields are induced by moving electric charges. Diamagnetic substances have paired electrons in their atomic and molecular orbitals. Paramagnetic molecules or atoms, on the other hand, contain unpaired orbital electrons. This produces a situation in which the magnetic moment of the unpaired electron is unopposed. Paramagnetic and diamagnetic molecules do not have magnetic fields of their own. In the absence of an external magnetic field, the electron dipoles of these substances randomly align, resulting in zero net magnetization. However, when placed in the magnetic field of an MRI scanner the electron dipoles of the paramagnetic substances line up in such a manner that the majority of the electrons point in the same direction as the external field. This results in augmentation of the external magnetic field. The ratio of the additional magnetic field strength to the strength of the applied magnetic field is termed the magnetic susceptibility of the paramagnetic substance. Magnetic susceptibility is thus a measure of how easily the substance may be magnetized.

Diamagnetic substances, which possess paired electrons, may still produce a magnetic field (when placed in an external magnetic field) by virtue of the fact that they are orbiting around the nucleus. This magnetic field is oriented to oppose the main magnetic field, and is called a Lenz field. It is far weaker than the field generated by the unpaired electron spins of paramagnetic substances. Paramagnetic substances, by virtue of having unpaired electrons, have the effect of magnetic susceptibility which augments the main magnetic field and also the much smaller effect of the Lenz field opposing the main magnetic field.

In the absence of unpaired electrons, protons relax (realign with the main magnetic field) by fluctuations in their magnetic fields caused by the motion of adjacent protons. Because of their large magnetic moments, unpaired electrons create fluctuations in local magnetic fields. These enable protons to realign faster with the external magnetic field producing a shortened T_1 relaxation time. This phenomenon is termed proton–electron dipole–dipole proton relaxation enhancement (PEDD PRE). In order for this interaction to occur the water proton must come within 0.3 nm of the unpaired electron.

If a paramagnetic substance is nonuniformly distributed and the unpaired electrons are configured in a position such that water protons cannot come close (within 0.3 nm) to the unpaired electrons then the only interaction the water protons will appreciate is the locally increased magnetic field produced by the unpaired electrons (paramagnetic susceptibility). Protons will precess at a rate proportional to the local field strength which varies with the local susceptibility. The precession rate (Larmor frequency) is proportional to the local field strength. If a paramagnetic substance is heterogeneously distributed then the local field strength will vary throughout the region in which it is distributed. Protons, diffusing through these different areas of local field variation produced by susceptibility differences, will precess at different rates. Thus, protons sensing a slightly higher local field will precess more rapidly than those in a slightly lower magnetic field environment. The net effect is for the protons to develop phase differences with shortening of the apparent T_2 relaxation time (T_2 proton relaxation enhancement or T_2 PRE). The T_2 PRE occurs when there is a heterogeneous distribution of paramagnetic substances with diffusion of water protons.

In summary, paramagnetic substances enhance relaxation by two mechanisms, either PEDD PRE or T_2 PRE. The former requires the water protons to approach the unpaired electrons and produces shortening of both T_1 and T_2; however, the effect is predominantly noted as a T_1 effect. This relates to the fact that the T_1 relaxation rate ($1/T_1$) is much smaller than the T_2 relaxation rate ($1/T_2$). A paramagnetic relaxation effect added equally to either relaxation rate would contribute proportionally more to the much smaller term ($1/T_1$). Shortening of the T_1 relaxation time produces high intensity on the short TR/short TE (so-called T_1 weighted) images. The T_2 PRE produces low intensity on long TR images (so-called proton density and T_2 weighted images). Hypointensity from T_2 PRE can, at times, also be observed on T_1 weighted images. This is because there is a T_2 component to intensity even on a T_1-weighted image, and substances with a very short T_2 (heterogeneously paramagnetic) can produce hypointensity on a T_1-weighted image. The presence of susceptibility effects can be visualized by observing significant increases in hypointensity as the images become more T_2-weighted (i.e. as the TE is increased).

2.2 Structure of Hemoglobin

Hemoglobin, the primary oxygen carrier in the bloodstream, is composed of four protein subunits weighing approximately 16 000 Da apiece. Each subunit contains one heme molecule

For References see p. 971

consisting of a porphyrin ring and an iron atom which provides the binding site for oxygen.[14] Binding of oxygen to the heme molecule of an individual subunit produces a conformational change in that subunit and adjacent subunits. The iron atom sits near the center of the porphyrin ring and is bound to one of the imidazole nitrogens of histidine-92. Molecular oxygen binds to the iron atom on the opposite side of the porphyrin ring.

In oxyhemoglobin, the oxidation state of the Fe is formally in the ferrous state which has six d orbital electrons. Molecular oxygen has two unpaired electrons. Oxyhemoglobin is diamagnetic, indicating that there must be pairing of the unpaired electrons of molecular oxygen and the d electrons of the heme iron atom. Suffice to say that theories have been proposed to conceptualize the electronic state of oxyhemoglobin which are beyond the scope of this discussion.[15,16]

There are four unpaired electrons in deoxyhemoglobin making it a paramagnetic molecule. When a hemoglobin subunit loses its oxygen molecule to form deoxyhemoglobin, the heme protein undergoes a small but significant change in its tertiary structure. The histidine ligand of the heme pulls the Fe^{2+} atom of deoxyhemoglobin out of the plane of the porphyrin ring and causes the porphyrin itself to dome. As a consequence, water molecules are unable to bind to the heme iron as they do in methemoglobin (see below), effectively preventing water from approaching close enough to the paramagnetic iron (<0.3 nm) to undergo PEDD PRE. However, deoxyhemoglobin can still produce local magnetic susceptibility changes (T_2 PRE).

Deoxyhemoglobin can be oxidized by one electron to methemoglobin by a number of different mechanisms. The Fe^{3+} is closer to the plane of the porphyrin ring in methemoglobin (in contrast to the Fe^{2+} atom of deoxyhemoglobin) so that water molecules can rapidly bind to the heme iron. Water must virtually bind to the heme iron in order to experience significant PEDD PRE. As the number of heme molecules is relatively small compared with the number of water molecules, this effect would be small except that the exchange rate of the water molecules is rapid relative to *TR*, allowing many water molecules to bind to the heme during the course of imaging. Normally an enzyme system within the RBCs rapidly reduces methemoglobin back to deoxyhemoglobin, but this process requires glucose and NADPH, which most likely are in short supply in sequestered hematomas.

2.3 The Effect of Protein

Hemorrhagic lesions generally contain significant concentrations of proteins. The effect of such protein is to slow the rotation of the water molecules. The much larger protein molecule has an intrinsically slower rate of rotation arising from Brownian motion due to its much larger mass. A strongly bound hydration shell of water molecules which surrounds the protein molecule with a zone of 'structured' water possesses slower rotational rates than the 'bulk' water molecules.[17,18] In hemorrhagic situations the concomitant increased protein results in a greater number of water molecules rotating more slowly, and associated T_1 shortening. A physical correlate of the change in rotational rate that occurs with increasing the protein concentration is the increased viscosity of solutions with high protein concentrations (i.e. acute and subacute intraparenchymal hematomas). The rotational correlation time is a measurement of the time it takes for this rotation to occur. The

T_1 relaxation time is noted to decrease with increasing protein concentration. At very high concentrations, the T_1 shortening produced by the addition of protein molecules plateaus. Slowing of the rotational rate of the water molecules by the addition of protein also produces a decrease in T_2.

There are thus two major effects produced by the hemoglobin (deoxyhemoglobin or methemoglobin) molecule: (1) the paramagnetic effects secondary to the iron within the heme molecule and (2) the diamagnetic effects of the protein portion (apoprotein) of the hemoglobin molecule.

2.4 Heterogeneity of Hemorrhage

Heterogeneity of an RBC solution is proportional to Hct × (100 − Hct), where Hct is the hematocrit. Thus heterogeneity would be maximal at Hct = 50 and nil at Hct = 0 or Hct = 100. Since the Hct of clot is ~90, heterogeneity is present with a considerable susceptibility effect (36% of the maximal effect). Furthermore, retracted in vivo clot packs less efficiently and regularly than under experimental conditions. This further increases this predicted heterogeneity of magnetic susceptibility.

3 IMAGING PULSE SEQUENCES IN HEMORRHAGE

3.1 Spin Echo Imaging

The 90° rf pulse in a spin echo sequence rotates the protons into the transverse plane, and subsequently they precess at rates proportional to the field strength they sense. If there is heterogeneous distribution of paramagnetic molecules the protons will rapidly develop phase incoherence. This is because protons diffusing to different regions are experiencing varying magnetic strengths and precessing at slightly different rates. The 180° rf pulse in the spin echo sequence was designed to obviate local static magnetic inhomogeneities. The analogy is made to slow and fast runners that must reverse course half-way through a race so that all runners wind up at the finish line at precisely the same time. If, however, the protons diffuse randomly to different regions containing varying magnetic fields, then, the 180° rf pulse cannot rephase these diffusing protons to precisely the same location they had previously been at just prior to the pulse. This loss of coherence of the spins in the transverse plane results in the T_2 PRE and hypointensity on T_2-weighted images.

The longer the interecho interval (the time between 180° pulses) the shorter the T_2 in regions of heterogeneous distribution of paramagnetic molecules. This is because the protons have more time to diffuse through different regions of varying field strengths when the interecho times are increased. Protons will thus lose phase coherence faster and have a shorter T_2 under conditions of heterogeneity of magnetic susceptibility and long interecho times. The selective shortening of T_2 varies as the square of the magnitude of the applied magnetic field, and the square of the variation in the magnetic susceptibility.

3.2 Gradient Echo Imaging

Gradient refocused echo images are more sensitive to susceptibility changes because the resultant local field gradients superimpose upon the applied phasing and rephasing gradients.

The presence of local field gradients from some blood products adds linearly to the phasing and rephasing gradients. This leads to rapid dephasing of the protons and signal loss on these T_2*-weighted images. This imaging technique is most sensitive for detection of heterogeneity of magnetic susceptibility but lacks some specificity with respect to particular hemorrhagic lesions. It is also problematic for routine imaging because of susceptibility differences at the skull base imposed by the air–tissue interfaces.

3.3   Fast Spin Echo Imaging

In the case of this pulse sequence there are up to 16 pulses of 180° per TR interval. This decreases the diffusion time of water protons and hence decreases the T_2 PRE. Susceptibility differences are decreased, and those elements of hemorrhage whose detection depends upon such differences (i.e. acute and chronic hemorrhage) become slightly more difficult to detect.[19]

4   CLINICAL SITUATIONS

4.1   Acute Hemorrhage

The biophysical effects just outlined sum to produce the appearance of acute hemorrhage as hypointensity on T_2-weighted images. In a simple intraparenchymal hemorrhage, blood is extravasated into the brain parenchyma. The RBCs are isolated from the circulation and cannot be reoxygenated. The RBCs which initially contain oxyhemoglobin become desaturated. Oxyhemoglobin has no unpaired electrons and is not paramagnetic. Imaging a hematoma containing only oxyhemoglobin will be iso- or slightly hypointense on short TR images and hyperintense on long TR images. This is related to the water content in the serum of the extravasated blood. Occasional images of hyperacute hematomas (within a few hours after an ictus) have noted a rim of hypointensity on long TR/long TE images. This rim has been postulated to be the result of early accumulation of iron or deoxyhemoglobin.[20]

Over hours the RBCs become desaturated. In addition, the inability of the isolated hematoma to rid itself easily of metabolic wastes with the possibility of increased lactic acid and CO_2 will tend to shift the oxyhemoglobin dissociation curve to the right (Bohr effect) which at any pO_2 will result in less hemoglobin saturation and more deoxyhemoglobin. Such desaturation can usually be appreciated very early following an ictus (within a few hours). The high protein content of the clot, as evidenced by the hyperdensity noted on CT, shortens T_1 relative to cerebrospinal fluid causing the hematoma to be hyperintense relative to CSF and isointense to the brain parenchyma on the short TR/short TE images (T_1-weighted images). Since solvent water molecules are unable to approach close enough to the Fe^{2+} heme of deoxyhemoglobin, there is no additional T_1 shortening due to PEDD PRE interactions. On long TR/long TE images (T_2-weighted images), the protein also shortens T_2, rendering the clot hypointense relative to cerebrospinal fluid. Superimposed is a marked hypointensity secondary to susceptibility effects arising from the local field inhomogeneity produced by encapsulation of the paramagnetic deoxyhemoglobin within the RBCs. This effect is magnified on

gradient refocused images, which as stated above are more sensitive to susceptibility effects. The end-result is the profound hypointensity of the acute hematoma on long TR/long TE and gradient refocused images (Figure 1). There is usually a penumbra of high intensity on long TR images surrounding the

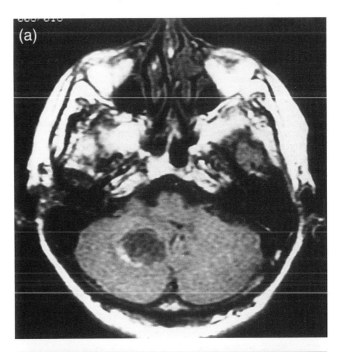

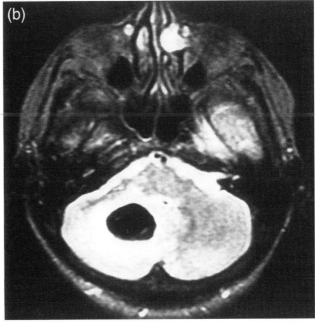

Figure 1 Acute hematoma. (a) Short TR/short TE image of right cerebellar hemorrhage. This image was performed approximately 72 hours after the ictus. Note the hypointensity of the hematoma compared with normal cerebellum. There is a slight rim of hyperintensity representing methemoglobin which is just beginning to form at the periphery of the hematoma. (b) Long TR/long TE image of (a). There is striking hypointensity of the hematoma which is the result of heterogeneity of magnetic susceptibility due to deoxyhemoglobin in intact RBCs. Surrounding the hematoma is a penumbra of high intensity from serum and/or vasogenic edema

For References see p. 971

acute hematoma. This is a consequence of clot retraction with extrusion of serum into the surrounding normal brain and, later, the development of vasogenic edema (over days from the initial ictus). The high intensity of the surrounding edema and the mass effect of the hemorrhage should disappear within 4–6 weeks. If persistent high intensity remains surrounding the hemorrhage, then one should be suspicious of an intratumoral hemorrhage.

4.2 Early Subacute Hemorrhage

Within 3–7 days there is oxidation of the deoxyhemoglobin to methemoglobin inside of the RBCs, occurring initially at the periphery of the clot. Unlike deoxyhemoglobin, water molecules are able to approach within 0.3 nm of the paramagnetic heme of methemoglobin, permitting PEDD PRE interactions which shorten T_1. It is this effect, in concert with the T_1 shortening from the high protein concentration, that gives methemoglobin its characteristic hyperintensity on T_1-weighted images. Because the paramagnetic methemoglobin remains encapsulated within the RBCs, marked hypointensity is also seen on the long TR/long TE and gradient refocused images due to the same susceptibility mechanism described above for deoxyhemoglobin. The PEDD PRE interaction also shortens T_2, but this effect is insignificant compared with the much larger susceptibility effects.

4.3 Mid Subacute Hemorrhage

The oxidation of deoxyhemoglobin proceeds centrally with time as evidenced by the development of hyperintensity within the center of the hematoma on the short TR/short TE images (Figure 2). Again the hematoma remains hypointense on the long TR, long TE and gradient refocused images due to susceptibility effects.[21]

4.4 Late Subacute Hemorrhage

Methemoglobin is less stable than deoxyhemoglobin, and the heme group can be lost spontaneously from the protein molecule. This free heme and/or other exogenous compounds (including peroxide and superoxide) can produce RBC lysis. Concomitantly there is protein breakdown and dilution of the remaining extracellular methemoglobin. Hyperintensity persists on the short TR/short TE images despite the decrease in the protein concentration due to the PEDD PRE of methemoglobin even at relatively low concentrations. The hematoma signal intensity increases on long TR/long TE images, approaching that of cerebrospinal fluid as there is loss of local field inhomogeneity upon RBC lysis as well as a decrease in the protein concentration. Also, the T_2 shortening effects of the paramagnetic heme due to PEDD PRE are rather small and unimportant except at the highest concentrations. Paralleling the breakdown of the methemoglobin is an accumulation of the iron molecules as hemosiderin and ferritin within macrophages at the periphery of the lesion which can be identified in the mid subacute stage of hemorrhage as well [Figure 2(b)].[22] The iron cores of hemosiderin and ferritin contain ~2000 iron molecules

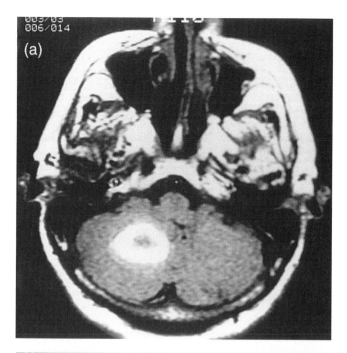

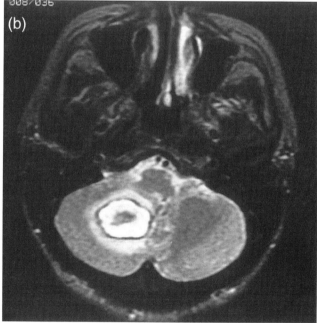

Figure 2 Subacute hematoma. Same patient as in Figure 1 except approximately 7 days after ictus. (a) Short TR/short TE image revealing ring of high intensity. This high intensity from the PEDD PRE of methemoglobin is the hallmark of the subacute stage of hemorrhage. (b) Long TR/long TE image demonstrates a rim of low intensity representing peripheral hemosiderin. Free methemoglobin is observed inside the hemosiderin rim. In the centermost portion of the hematoma is deoxyhemoglobin. There is edema surrounding the hemosiderin rim

ferromagnetically coupled to produce a 'super paramagnetic' substance which exhibits a very large susceptibility effect. The result is a black ring surrounding the lesion visible on short TR/short TE images but increasingly prominent on long TR, long TE spin echo and gradient refocused images.

For list of General Abbreviations see end-papers

4.5 Chronic Hemorrhage

After months, there is near complete breakdown and resorption of the fluid and protein within the clot. The iron atoms from the metabolized hemoglobin molecules are deposited in hemosiderin and ferritin molecules unable to exit the brain parenchyma due to an intact blood–brain barrier. The susceptibility effects of the 'super paramagnetic' iron cores of hemosiderin produce hypointensity on all spin sequences but are most prominent on the long TR/long TE and gradient refocused images. There is no edema or mass effect associated with the hemorrhage. The methemoglobin (central high intensity) is gradually resorbed over months to years while the hemosiderin remains permanently in the brain. Very old hemorrhages often appear as slit-like cavities lined with hemosiderin and ferritin (Figure 3).

5 OTHER CLINICAL CONCEPTS

MRI is the technique of choice in virtually all hemorrhagic conditions of the head and neck. With its ability to detect hemorrhages of various ages, diagnosis of recurrent hemorrhagic events is facilitated. Conditions in which these situations are important include child abuse, siderosis of the central nervous system, and amyloid angiopathy.

It should be emphasized that the pO_2 of the hemorrhagic environs is important. Tumors are generally more hypoxic than normal brain so that an acute hemorrhage into a tumor may be more hypointense on long TR images because of increased amounts of deoxyhemoglobin. Conversely, in subarachnoid hemorrhage the red blood cells are bathed in the cerebrospinal fluid which has a pO_2 of approximately 43 mmHg. At this pO_2, 72% of the blood is in the oxyhemoglobin form, and 28% in the deoxyhemoglobin state. As the amount of T_2 shortening varies as the square of the concentration of the paramagnetic $(0.28)^2 \times 100\% = 7.8\%$ of the T_2 shortening would be expected in this case when compared with that from 100% deoxyhemoglobin.[23] Thus, acute subarachnoid hemorrhage is the lone condition in which MRI is not the imaging technique of choice.

In general the vast majority of hemorrhagic lesions follow the continuum of MRI changes that have just been described. This makes MRI an elegant probe for most hemorrhagic conditions including intraparenchymal hemorrhage, intratumoral hemorrhage, traumatic hemorrhage, hemorrhagic cortical infarction, vascular dissection, venous thrombosis, and cavernous hemangioma.

6 RELATED ARTICLES

Brain Neoplasms Studied by MRI; Contrast Agents in Magnetic Resonance: Operating Mechanisms; Contrast Agents in Whole Body Magnetic Resonance: An Overview; MRI in Clinical Medicine.

7 REFERENCES

1. J. T. Sipponen, R.E. Sepponen, and A. Sivula, *J. Comput. Assist. Tomogr.*, 1983, **7**, 954.
2. L Pauling and C. Coryell, *Proc. Natl. Acad. Sci. USA*, 1936, **22**, 210.
3. L Pauling and C. Coryell, *Proc. Natl. Acad. Sci. USA*, 1936, **22**, 159.
4. S. H. Koenig, R. D. Brown III, and T. R. Lindstrom, *Biophys. J*; 1981, **34**, 397.
5. N. Davidson and R. Gold, *Biochim. Biophys. Acta*, 1957, **26**, 370.
6. N. Bloembergen, E. M. Purcell, and R. V. Pound, *Phys. Rev.*, 1948, **73**, 679.
7. K. R. Thulborn, J. C. Waterton, P. M. Matthews, and G. K. Radda, *Biochim. Biophys. Acta*, 1982, **714**, 265.
8. R. L. De La Paz, P. F. J. New, F. S. Buonanno, J. P. Kistler, R. F. Oot, B. R. Rosen, J. M. Taveras, and T. J. Brady, *J Comput. Assist. Tomogr.*, 1984, **8**, 599.
9. W. G. Bradley Jr and P. G. Schmidt, *Radiology*, 1985, **156**, 99.
10. G. Di Chiro, R. A. Brooks, M. E. Girton, T. Caporale, D. C. Wright, A. J. Dwyer, and M. K. Harne, *Am. J. Neuroradiol.*, 1986, **7**, 193.
11. J. M. Gomori, R. I. Grossman, H. I. Goldberg, R. A. Zimmerman, and L. T. Bilaniuk, *Radiology*, 1985, **157**, 87.
12. G. M. Bydder and I. R. Young, *J. Comput. Assist. Tomogr.*, 1985, **9**, 1020.
13. R. R. Edelman, K. Johnson, R. Buxton, G. Shoukimas, B. R. Rosen, K. R. Davis, and T. J. Brady, *Am. J. Neuroradiol.*, 1986, **7**, 751.
14. R. Dickerson and I. Geis, 'Hemoglobin: Structure, Function, Evolution and Pathology', Benjamin/Cummings, Menlo Park, CA, 1983.
15. J. Weiss, *Nature*, 1964, **202**, 83.
16. W. A. Goddard and B. D. Olafson, *Proc. Natl. Acad. Sci. USA*, 1975, **72**, 1335.
17. S. Koenig, 'The Dynamics of Water-Protein Interactions', in ACS Symposium Series, No. 127, ed. S. P. Rowland, American Chemical Society, 1980, p. 157.

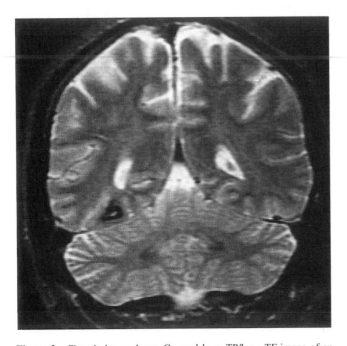

Figure 3 Chronic hemorrhage. Coronal long TR/long TE image of an old right occipital hemorrhagic cavity delineates the hypointensity of hemosiderin lining the walls of the collapsed hematoma cavity. Note that there is no mass effect or edema associated with this chronic hemorrhage

For References see p. 971

18. G. Fullerton, M. Finnie, K. Hunter, V. A. Ord, and I. L. Cameron, *Magn. Reson. Imaging*, 1987, **5**, 353.
19. K. M. Jones, R. V. Mulkern, M. T. Mantello, P. S. Melki, S. S. Ahn, P. D. Barnes, and F. A. Jolesz, *Radiology*, 1992, **182**, 53.
20. K. R. Thulborn and S. W. Atlas, in 'Magnetic Resonance Imaging of the Brain and Spine', ed. S. W. Atlas, Raven Press, New York, 1991, p. 175.
21. J. M. Gomori, R. I. Grossman, and D. B. Hackney, *Am. J. Neuroradiol.*, 1987, **8**, 1019.
22. K. R. Thulborn, A. G. Sorensen, N. W. Kowall, A. McKee, A. Lai, R. C. McKinsky, J. Moore, B. R. Rosen, and T. J. Brady, *Am. J. Neuroradiol.*, 1990, **11**, 291.
23. R. I. Grossman, S. S. Kemp, I. C. Yu, J. E. Fishman, J. M. Gomori, P. M. Joseph, and T. Asakura, *Acta Radiol.*, 1986; **369** (suppl), 56.

Biographical Sketch

Robert I. Grossman. *b* 1947. B.S., 1969, M.D., 1973 University of Pennsylvania School of Medicine, USA. Internship Beth Israel Hospital, Boston, USA, 1973–74. Neurosurgical Residency, University of Pennsylvania, USA, 1974–76. Radiology Residency, University of Pennsylvania, USA, 1976–79. Neuroradiology Fellowship, Massachusetts General Hospital, USA, 1979–81. Professor of Radiology, Neurosurgery, and Neurology, Associate Chairman of Radiology, Section Chief Neuroradiology, University of Pennsylvania Medical Center, 1981–present. Approx. 200 publications. Current research interests: white matter diseases, stroke, head trauma.

Eye, Orbit, Ear, Nose, and Throat Studies Using MRI

Mahmood F. Mafee

University of Illinois at Chicago, Chicago, IL, USA

1 MRI OF THE EYE

1.1 Introduction

The eye consists of three primary layers: (1) the sclera, or outer layer, composed primarily of collagen-elastic tissue; (2) the uvea or middle layer, a richly vascular and pigmented tissue consisting of the iris, ciliary body, and choroid; and (3) the retina or inner layer, the neurosensory stratum of the eye.[1] The Tenon's capsule (bulbar fascia) is a fibroelastic membrane that covers the sclera and envelops the eyeball. Tenon's capsule encloses the posterior four-fifths of the eyeball separating it from the orbital fat. Tenon's capsule is separated from the sclera by the potential Tenon's space.[2,3]

1.2 Anatomy of the Globe

1.2.1 Anterior and Posterior Segments of the Globe

The ciliary body and the iris divide the globe into an anterior and a posterior segment. The anterior segment which includes the anterior and posterior chambers is the space lying between the cornea and the crystalline lens. The posterior (vitreous) segment contains the vitreous chamber (body), which acts as a biologic shock absorber. The vitreous chamber is a gel-like, transparent, extracellular matrix, composed of a meshwork of 0.2% collagen fibrils interspersed with 0.2% hyaluronic acid polymers, 98–99% water, and small amounts of soluble protein.[4,5]

1.2.2 Lens

The lens is one of the least hydrated organs of the body containing only 66% water.[5] The remainder is composed of long, thin cells (lens fibers) whose main cytoplasmic content is protein.

1.2.3 Sclera

The sclera, the outer, white, leathery coat of the globe is primarily composed of cellular bundles of collagen. Its external surface lies adjacent to Tenon's capsule. Its internal surface blends with the uveal tissues.

1.2.4 Uvea (Iris, Ciliary Body and Choroid)

The uvea lies between the sclera and retina and provides vascular supply to the eye and regulates ocular temperature. The choroid is a portion of the uvea that lies between the sclera and the retinal pigment epithelium (RPE). The ciliary body is a direct anterior continuation of the choroid, and the iris is a further extension of the ciliary body itself.

1.2.5 Retina

The external surface of the retina is in contact with the choroid and the internal surface is adjacent to the vitreous body; posteriorly, the retina is continuous with the optic nerve. Grossly the retina has two layers: (1) the inner layer is the sensory retina, composed of photoreceptors, first and second order neurons (ganglion cells) and neuroglial elements and (2) the outer layer is the RPE, which consists of a single lamina of cells whose nuclei are adjacent to the basal membrane of the choroid, the so-called Bruch's membrane.

1.3 MRI Anatomy of the Globe

The globe is unique in that it contains the most (vitreous) and the least (lens) water concentration of all the tissues in the body.[4,5] The eye is an ideal organ to be evaluated by MRI because the wide variations in water content produce different water proton relaxation times of the various tissues. The eye can be imaged using head coil or surface coil. Images obtained with surface coil provide better spatial resolution. A routine MRI of the eye usually includes single echo short repetition time (*TR*), short echo time (*TE*) (*TR/TE*, 400–600/20–25 ms), sagittal and axial views and an axial view SE double echo (*TR/TE*, 2000/20–25; 70–100 ms) pulse sequences. Following the intravenous (i.v.) administration of gadolinium diethylenetria-

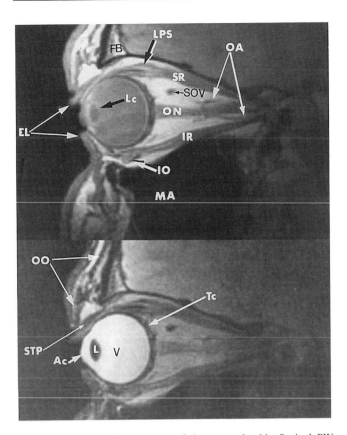

Figure 1 Normal MRI anatomy of the eye and orbit. Sagittal PW (top) and T_2W (bottom) MR images. Ac = anterior chamber, EL = eyelids, FB = frontal bone, IO = inferior oblique muscle, IR = inferior rectus muscle, L = lens, Lc = lens capsule, LPS = levator palbebra superioris, MA = maxillary antrum, OA = ophthalmic artery, ON = optic nerve, OO = fibers of the orbicularis oculi muscle, SOV = superior ophthalmic vein, SR = superior rectus muscle, STP = superior tarsal plate, Tc = Tenon's capsule

minepentaacetic acid (Gd-DTPA), single echo, short *TR*, short *TE* images in axial and other views are obtained with or without fat suppression techniques. The section thickness for MRI of the eye should be 3 mm. The normal lens has long T_1 and short T_2 relaxation properties on MRI (Figure 1). The nucleus of the lens has both lower water content and shorter T_2 than the cortex of the lens. The sclera has a low signal intensity on various pulse sequences which is probably related to the short T_2 caused by a greater proportion of bound water.[5] Differentiation of individual layers of the retina, choroid, and posterior hyaloid membrane is impossible by MRI in the normal eye; however, in pathological conditions, detachment of different layers of the eye can be visualized on MR images.

1.4 MRI of Ocular Pathology

MRI has served as an important diagnostic method in ophthalmology for the last decade.[2,6–14] Excellent contrast between various normal structures and high sensitivity in detection of pathologic states is due to the intrinsic differences in proton density and in proton relaxation times.[2,4]

1.5 Posterior Hyaloid, Retinal, and Choroidal Detachment

The posterior hyaloid space is a potential space between the posterior hyaloid (vitreous) membrane and the sensory retina. The posterior hyaloid membrane is thin and is invisible on MRI. It becomes visible when blood or other fluid fills the potential posterior hyaloid space. On MRI, a detached posterior hyaloid membrane is seen as a thin membrane within the globe, in front of the optic disk.

Retinal detachment results from separation of the sensory retina from the RPE with accumulation of fluid in the potential subretinal space.

Since the retina is very thin, it cannot be directly visualized on MRI scans. However, it may be shown when it is outlined by significant contrast differences between the subretinal effusion and the vitreous chamber. On axial MRI, the detached retina is seen as a V-shaped structure, with its apex at the optic disk and its extremities toward the ora serrata just behind the ciliary body, where the sensory retina ends. On coronal MR scans, retinal detachment appears as a characteristic folded membrane. The subretinal exudate is rich in protein, therefore has higher MR signals than the vitreous cavity on T_1- and often on T_2-weighted MR images.

Choroidal detachment results from the accumulation of fluid (serous) or blood (hemorrhagic choroidal detachment) within the potential space between the choroid and the sclera. MRI is an excellent technique in the evaluation of choroidal detachment, particularly when ultrasonography or computerized tomography (CT) in conjunction with the clinical examination have not provided sufficient information. Serous choroidal detachment appears as a semilunar or ring-shaped area of higher MR signal than vitreous on T_1- and T_2-weighted MR images.[9] The appearance of choroidal detachment on MR images may be confused with retinal detachment. However, choroidal detachment is often restricted by the anchoring effect of the vortex veins and short posterior ciliary arteries. This restriction results in a characteristic appearance in which the leaves of the detached choroid, unlike the detached retina, frequently do not extend to the region of the optic disc.

The MRI appearance of hemorrhagic posterior hyaloid, retinal and choroidal detachments varies with the age and the degree of organization of the hemorrhage. A hematoma less than 48 h old usually appears iso- to slightly hypointense relative to the normal vitreous body on T_1-weighted (T_1W) and proton-weighted (PW) MR images, but is markedly hypointense on T_2-weighted (T_2W) images. When the hematoma is approximately 5 days old, it appears relatively hyperintense on PW and T_1W images but is hypointense on T_2W images. At this stage, hematoma may be confused with an ocular melanoma.[6,8,9] The hematoma usually continues to increase in signal intensity on T_1W, PW, and T_2W images and usually becomes markedly hyperintense by day 14 on all MR pulse sequences.

1.6 Retinoblastoma

Retinoblastoma, the most common intraocular tumor of childhood, is a highly malignant retinal tumor. The tumor arises from neuroectodermal cells (nuclear layer of the retina) destined to become retinal photoreceptors.[2,10] Leukokoria, a pink–white or yellow–white pupillary reflex (cat's eye), is the

For References see p. 983

most common presenting sign of retinoblastoma. Computerized tomography is most sensitive in demonstrating the presence of the typical deoxyribonucleic acid (DNA)–calcium complexes within the tumor.[2,10] Magnetic resonance imaging is not as specific as CT scanning in the diagnosis of retinoblastoma due to its lack of sensitivity in detecting calcification. However, it is more specific than CT in differentiating retinoblastoma from other lesions that simulate retinoblastoma clinically, such as persistent hyperplastic primary vitreous (PHPV), Coats' disease, retinopathy of prematurity (ROP), toxocariasis, retinal detachment, and other so-called pseudogliomas and leukokorias.[2,10,12,13] Lesions less than 2–3 mm in thickness are not recognized as discrete areas by MR technology to date. At times lesions less than 3 mm thick may be better visualized on post Gd-DTPA contrast MR scans due to increased contrast resolution. On MRI, retinoblastomas appear slight to moderately hyperintense to vitreous on T_1W and slightly or moderately hypointense to vitreous on T_2W MR scans. Even extensive calcification noted on CT scans may be missed in all MR pulse sequences. Retinoblastomas demonstrate moderate to marked contrast enhancement following i.v. administration of Gd-DTPA contrast material. Involvement of the Tenon's capsule, optic nerve and retrobulbar space is best evaluated on post Gd-DTPA fat-suppression T_1W MR images. MRI including Gd-DTPA contrast study appears to be the study of choice for the evaluation of intracranial spread of retinoblastoma and the association of bilateral retinoblastoma, suprasellar and pineal primitive neuroectodermal tumors (tetra lateral retinoblastoma).[14]

1.7 Persistent Hyperplastic Primary Vitreous (PHPV)

Persistent hyperplastic primary vitreous is caused by failure of the embryonic hyaloid vascular system to regress normally. The ocular malformation can reflect either an isolated congenital defect or may be a manifestation of more extensive ocular or systemic involvement.[2,15,16] The MRI findings of PHPV include: (1) microphthalmos, (2) lack of calcification, (3) posterior hyaloid or retinal detachment, (4) tubular, cylindrical or other discrete intravitreal images suggestive of persistence of fetal tissue, (5) fluid level within the globe, reflecting the presence of serosanguinous fluid in either the subhyaloid or subretinal space, (6) enhancement of abnormal intravitreal tissue following i.v. administration of Gd-DTPA contrast material, (7) altered signal of lens of the involved eye compared with the other normal eye.

1.8 Coats' Disease

Coats' disease is a primary vascular anomaly of the retina, characterized by idiopathic retinal telangiectasia and exudative retinal detachment (exudative retinopathy). It is almost always unilateral and generally affects boys of 4 to 6 years of age, although it may be seen in younger children. In the early stage of the disease, MRI may yield little information. In the later stages, the exudative retinal detachment is seen as hyperintense images on all MRI pulse sequences. At times, due to lipoproteinaceous exudate, typically seen in Coats' disease, the subretinal exudate may be less hyperintense than the vitreous

body on T_2W MR scans. The leaves of the detached retina in Coats' disease may be very thickened and may show moderate to significant enhancement following i.v. administration of Gd-DTPA contrast material.

1.9 Retinopathy of Prematurity (ROP)

Retinopathy of prematurity, formerly termed retrolental fibroplasia, is believed to be related to a combination of developmental (prematurity) and environmental (supplemental oxygen therapy) factors that prevent normal retinal vasculogenesis. The MRI findings of the early stage of ROP are nonspecific. The eyes may be microphthalmic. In the more advanced stages the differential diagnoses include PHPV, retinoblastoma, endophthalmitis as well as a number of pathologic conditions in which retinal detachment is a common feature.[2,17] Retinoblastoma has rarely been reported in microphthalmic eyes with or without ROP or PHPV.

1.10 Ocular Toxocariasis

Toxocariasis results from ingestion of eggs of the nematode, *Toxocara canis*. The death of the larva results in a wide spectrum of intraocular inflammatory reactions. MRI is superior to CT for the evaluation of ocular toxocariasis. The MR findings of ocular toxocariasis consist of localized or diffuse intravitreal mass with moderate to marked contrast enhancement. The subretinal exudate has variable hyperintense signal on T_1W, PW, and T_2W MR scans.[17]

1.11 Malignant Uveal Melanoma

Malignant uveal melanoma is the most common intraocular tumor in adults. They are uncommon in blacks, the white:black ratio being about 15:1[6] Metastasis primarily involves the liver. Uveal melanomas and other intraocular lesions more than 3 mm in thickness are usually well visualized on MR scans.[6,17] Smaller lesions are better studied with ultrasonography. The MR characteristics of melanotic lesions are thought to be related to the paramagnetic proton relaxation by stable radicals in melanin.[18] The stable radicals cause a proton relaxation enhancement that shortens both T_1 and T_2 relaxation times.[11] Hence, uveal melanomas appear as areas of moderately to markedly high signal (greater signal intensity than vitreous) on T_1W and PW MR images. On T_2W images, melanomas appear as areas of moderately to markedly low signal (lesser signal intensity than vitreous). These MR characteristics are similar to those of retinoblastoma. Some of the melanomas may not be hypointense on T_2W images. Melanomas demonstrate moderate enhancement following i.v. administration of Gd-DTPA contrast media (Figure 2). Exudative retinal detachment, associated with choroidal melanoma is seen on MRI as dependent areas of moderate to high intensity on T_1W, PW and T_2W images. Organized exudative chronic retinal detachment and hemorrhagic subretinal fluids may simulate melanoma on MRI scans. Invasion of the sclera, extension of tumor into the Tenon's capsule or optic disk and extraocular invasion can be best detected by MR scans, particularly on post Gd-DTPA fat-suppression T_1W MR images.

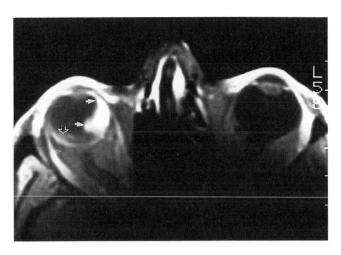

Figure 2 Malignant choroidal melanoma. Axial T_1W postcontrast MR image. A hyperintense enhancing mass (solid arrows) is demonstrated. The melanoma has caused retinal detachment (hollow arrow). The tumor can be easily differentiated from subretinal exudate

1.12 Choroidal Hemangioma

Choroidal hemangiomas are congenital vascular hamartomas. Two different forms have been described: a circumscribed or solitary type and a diffuse angiomatosis.[19] Diffuse choroidal hemangiomas are usually seen in association with Sturge–Weber disease. Choroidal hemangiomas can be diagnosed by ophthalmoscopy, fluorescein angiography, ultrasonography or CT. However, misdiagnoses are not uncommon, particularly when the angiomas are concealed by a detachment of the retina.[2] On MRI, choroidal hemangiomas appear hypointense to slightly hyperintense to vitreous on T_1W images and hyperintense on T_2W images. They show intense enhancement on post Gd-DTPA T_1W MR scans. MRI scans are, therefore, extremely useful in differentiating choroidal hemangiomas from choroidal melanomas.

1.13 Uveal Metastases

A most common source of secondary carcinoma within the eye is from breast or lung. Metastatic lesions of the uvea extend chiefly in the plane of the choroid, causing relatively little increase in its thickness. Unlike uveal melanomas, which tend to form a protuberant mass, uveal metastases usually have a mottled appearance and a diffuse outline on MRI scans. Most uveal metastases appear as areas of low signal on T_1W MR scans. On T_2W images, metastases appear as areas of high signal intensity. A mucin-producing metastatic lesion (adenocarcinoma) may simulate a uveal melanoma because the proteinaceous fluid tends to decrease T_1 and T_2 relaxation times.[2] Uveal metastatic carcinoid lesions may simulate uveal melanomas on MRI scans.[20] Uveal metastases demonstrate moderate to marked enhancement following i.v. administration of Gd-DTPA contrast material. Exudative retinal detachment associated with uveal metastases can be best differentiated from metastasis on postcontrast T_1W images.

1.13.1 Choroidal Hemorrhage and Detachment

Choroidal hemorrhage and detachment may be easily mistaken for choroidal tumors.[2,8,9] MRI is invaluable in differentiating choroidal detachments from uveal melanomas.[8]

1.13.2 Other Uveal Tumors

Neurofibroma and schwannoma of the choroid and ciliary body, adenoma, leiomyoma, and lymphoreticular proliferative disorders of the uveal tract are rare lesions involving the choroid and ciliary body which can cause diagnostic confusion with uveal melanoma. If these lesions are large enough (3 mm or more) they can be detected by MRI.[10]

2 MRI OF THE ORBIT

2.1 Introduction

The orbits are two recesses that contain the globes, the extrinsic ocular muscles, the blood vessels, lymphatic, nerves (II, III, IV, and V), adipose and connective tissues, and most of the lacrimal apparatus.[1] The orbit is bounded by the periosteum (periorbita) and separated from the globe by Tenon's capsule. Anteriorly are the orbital septum and the lids. The orbital cavity is pyramidal; its apex is directed posteriorly and medially and its base is directed anteriorly and laterally as the orbital opening. Its bony walls separate it from the anterior cranial fossa superiorly, the ethmoid and sphenoid sinuses and nasal cavity medially, the maxillary sinuses inferiorly, and the lateral surface of the face and temporal fossa laterally and posteriorly. At the apex of the orbit, the optic canal and the superior and inferior orbital fissures allow various structures to enter and leave the orbit.[1,21]

2.2 Anatomy of the Orbit

2.2.1 Orbital Septum and Eyelids

The orbital septum is a weak, membranous sheet that forms the fibrous layer of the eyelids and is attached to the margins of the bony orbit where it is continuous with the periorbita. Each eyelid or palpebra from without inwards consists of skin, subcutaneous areolar tissue, fibers of the orbicularis oculi, tarsus, and orbital septum.

2.2.2 Orbital Fatty Reticulum, Extraocular Muscles, and Optic Nerve

Within the orbit all structures are embedded in a fatty reticulum. The fibroelastic tissue that makes up the reticulum divides the fat into lobes and lobules.[21] The six striated extraocular muscles, including the four recti and the two oblique muscles, control eye movement. The rectus muscles arise from the annulus of Zinn, which is a tunnel-shaped tendinous ring that encloses the optic foramen and the medial end of the superior orbital fissure.[22]

The optic nerve is a nerve fiber tract that conveys visual sensation. It traverses the optic canal with the ophthalmic artery. The optic nerve is covered by three layers, the pia

For References see p. 983

mater, the arachnoid, and the dura, all of which sheathe the nerve.

2.2.3   Lacrimal Apparatus

The lacrimal gland lies in the superolateral angle of the orbit in the lacrimal fossa. Tears produced by the gland are drained into the lacrimal sac. The lacrimal sac lies in the lacrimal groove, located in the anterior-inferomedial angle of the orbit. From the sac, the tears drain through the nasolacrimal duct into the inferior meatus of the corresponding nasal cavity.

2.3   MRI Techniques

Excellent natural contrast between various orbital tissues (water, fat, muscles, and nerves) makes the orbit an ideal organ to study by MRI. The three-dimensional capability of MRI along with its superior contrast resolution is of great assistance in localizing orbital pathology. Using a head coil, a routine MRI of the orbit usually includes a single echo short TR, short TE (TR/TE, 500–800/20–25 ms), SE sagittal view, an axial SE double echo (TR/TE, 2000–2500/20–25, 70–100 ms) and a single echo, short TR, short TE, SE axial or coronal view. Following the intravenous i.v. administration of Gd-DTPA, single echo, short TR, short TE images are obtained with or without fat-suppression techniques. Images obtained with a surface coil provide better spatial resolution; however, the apical lesions and intracranial extension of the orbital lesions may not be fully evaluated because of signal dropout, proportional to the distance from the surface coil. We use the surface coil for those lesions situated in the anterior orbit. In general, the orbital MRI study and the use of paramagnetic contrast material should be tailored according to the suspected problem of individual patients.[22]

2.4   MRI Anatomy of the Orbit

MRI provides the best cross sectional imaging anatomy of the orbit. The SE, short TR and short TE pulse sequences provide the most spatial resolution with exquisite anatomical details.[22] The extraocular muscles, the intermuscular septae, the optic nerve, superior ophthalmic vein, ophthalmic artery, the lacrimal gland, lacrimal sac, and nasolacrimal duct are very well visualized on routine MR images (Figure 3). Smaller structures such as divisions of the ophthalmic and third cranial nerves, lacrimal vein and artery, and even ciliary vessels can be visualized on high resolution MR images, obtained with a surface coil. The orbital septum, structures of the eyelids including tarsal plates, the annulus of Zinn, and most of the tendinous portion of the extraocular muscles can be best visualized on images obtained with a surface coil.[22]

2.5   MRI of Orbital Pathology

2.5.1   Developmental Orbital Cysts

MRI is particularly helpful in the evaluation of congenital anomalies of the orbit and optic system. The most frequent

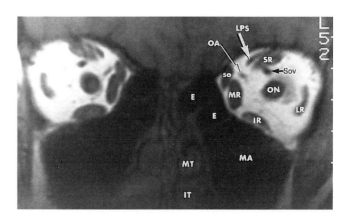

Figure 3 Normal orbital anatomy. Coronal 3 mm thick MR scan, demonstrating various anatomical structures of the orbits, nasal and paranasal sinuses. E = ethmoidal sinus, IR = inferior rectus muscle, IT = inferior turbinate, LPS = levator palpebra superioris, LR = lateral rectus muscle, MA = maxillary antrum, MR = medial rectus muscle, MT = middle turbinate, OA = ophthalmic artery, ON = optic nerve, so = superior oblique muscle, Sov = superior ophthalmic vein, SR = superior rectus muscle

developmental cysts involving the orbit and periorbital structures are epidermoid and dermoid cysts and teratomas. Epidermoid and dermoid cysts appear hypointense to brain on T_1W and hyperintense on T_2W MR images. Portions of the dermoids which contain fatty tissue appear hyperintense on T_1W and hypointense on T_2W images. Epidermoid and dermoid cysts do not show enhancement with Gd-DTPA.

2.5.2   Inflammatory Conditions

Preseptal and postseptal orbital inflammation, subperiosteal phlegmon and abscess, ophthalmic vein and cavernous sinus thrombosis can be demonstrated by MRI. The main contribution of MRI to the diagnosis of sinonaso-orbital infections is its clear demonstration of the relationship between nasal, sinus, orbital and cranial disease.[22]

2.5.3   Idiopathic Orbital Inflammations (Pseudotumors)

Pseudotumors are defined as nonspecific, idiopathic, orbital inflammatory conditions for which no local identifiable cause or systemic disease can be found.[22–24] Classically the rapid development of unilateral, painful ophthalmoplegia, proptosis and chemosis, with a rapid and lasting response to steroid therapy in an otherwise healthy patient, is highly suggestive of the diagnosis of pseudotumor. On MRI, pseudotumors usually appear as an infiltrative process which is isointense or hypointense to muscles on T_1W and hyperintense on T_2W images. On MRI, the enhancement of pseudotumors with Gd-DTPA is best demonstrated on fat-suppressed T_1W images. There are no specific imaging characteristics or clinical signs to establish the diagnosis of pseudotumor with absolute certainty. Idiopathic orbital myositis may simulate thyroid myopathy. However, typically, enlargement of the muscles in thyroid myopathy involves the muscle belly, sparing its tendinous portion.

2.6 Orbital Tumors

2.6.1 Lymphoma

Lymphomas, the most common malignant orbital tumors in adults, are solid tumors of the immune system. Most are composed of monoclonal B cells. The most common cytologic forms of malignant lymphoma involving the orbit are histiocytic and lymphocytic in various degrees of differentiation. True lymphoid tissue in the eye is found in the subconjunctival and lacrimal gland. This explains why most orbital lymphomas are commonly seen in the anterior portion of the orbit. The MRI findings of orbital lymphomas are usually nonspecific and at times are impossible to differentiate from those of orbital pseudotumors. Both pseudotumors and lymphomas may have an intermediate or hypointense signal in T_1W images (Figure 4) and appear isointense or hyperintense and even hypointense (although less common) to fat in T_2W images (Figure 4). Both pseudotumors and lymphomas demonstrate moderate contrast enhancement on postcontrast MRI scans.

2.6.2 Rhabdomyosarcoma

Rhabdomyosarcoma is the most common malignant mesenchymal tumor of childhood, as well as the most common primary malignant tumor of the orbit in children.[22,25] The tumor is notorious for producing a rapidly developing unilateral proptosis. Any rapidly developing proptosis in childhood must presumptively be diagnosed as rhabdomyosarcoma. The differential diagnosis also includes leukemia and metastatic deposits (neuroblastoma), Langerhans histiocytosis, ruptured dermoid cyst, hemorrhage in a pre-existing lymphangioma, subperiosteal infection and hematoma after trauma. Rhabdomyosarcomas are both aggressive bone-destroying lesions and bone-pushing lesions. Rhabdomyosarcoma appears hypointense to brain on T_1W and isointense to slightly hyperintense to brain on T_2W images. The use of paramagnetic contrast media results in moderate to marked enhancement.[25] Magnetic resonance imaging is the most sensitive imaging study to differentiate rhabdomyosarcoma from most simulating lesions.[22,25]

2.6.3 Hemangiomas and Lymphangiomas

Capillary hemangiomas are the most common orbital vascular tumors that occur in infants during the first year of life. The tumor often increases in size for 6 to 10 months and then gradually involutes. On MRI, capillary hemangiomas have long T_1 and long T_2 characteristics. At times, prominent vessels can be identified within the lesion. They demonstrate marked contrast enhancement. On MR angiography, the vascular blush may not be appreciated.

Cavernous hemangiomas, the most common orbital vascular tumor in adults, appear as well-defined masses, frequently occurring within the intraconal space. On MRI, they appear as well-defined, sharply marginated, homogeneous, rounded, ovoid, or lobulated masses, which appear hypointense to brain on T_1W and hyperintense on T_2W images. They demonstrate moderate to marked contrast enhancement on MRI. The MRI characteristics of orbital schwannomas and hemangiopericytomas may be similar to cavernous hemangiomas.[22,26]

Orbital lymphangiomas occur in children and young adults. Lymphangiomas may have distinct borders but are typically diffuse and not well capsulated. Spontaneous hemorrhage within the lesion is common.[22,26] On MRI scans lymphangiomas are seen as homogeneous or heterogeneous, and hypointense to relatively hyperintense on T_1W and, usually, heterogeneously very hyperintense on T_2W scans. Their MRI characteristics usually help to differentiate them from hemangiomas, pseudotumors, rhabdomyosarcoma and many other lesions.[22,26]

2.6.4 Optic Nerve Sheath Meningioma

Optic nerve sheath meningioma (ONSM) arises from the meningoendothelial cells of the arachnoid that are situated along the optic nerve sheath. Meningiomas are usually seen in middle-aged and elderly women. ONSM presents as very slowly progressive axial proptosis and loss of vision. On MRI scans, ONSM can be seen as localized or fusiform enlargement of the optic nerve. ONSM frequently appears hypointense on T_1W and hyperintense on T_2W images as compared with the brain. ONSM demonstrates moderate to marked enhancement on post Gd-DTPA MR scans.[22] Postcontrast fat-suppressed T_1W MR scans are the most sensitive images to detect ONSM. Involvement of the optic nerve by sarcoidosis may result in focal or diffuse enlargement of the optic nerve, mimicking an optic nerve meningioma on all MR pulse sequences.

2.6.5 Optic Nerve Glioma

Optic nerve glioma is a tumor arising from the neuroglia (glial cells) of the optic nerve. It is usually a tumor of childhood (2 to 8 years of age). In general, optic nerve glioma in children is considered a benign, well-differentiated, slowly growing, and noninvasive astrocytoma (pilocystic astrocytoma).

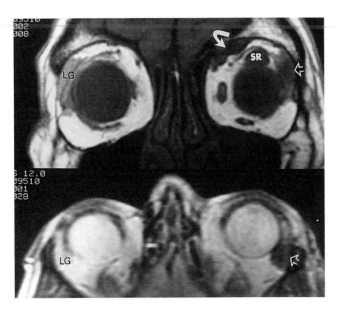

Figure 4 Small cell lymphoma of the orbit. Coronal T_1W (top) and axial T_2W (bottom) MR scans demonstrating normal right lacrimal gland (LG) and abnormal left lacrimal gland (hollow arrows) due to involvement by lymphoma. Notice lymphomatous involvement of left superior rectus muscle (SR) as well as extraconal fat (curved arrow)

For References see p. 983

However, optic nerve glioma in adults is a rare but invasive malignant astrocytoma (usually glioblastoma multiformis). Bilateral optic nerve gliomas are characteristic of neurofibromatosis type 1 (NF1). Optic nerve gliomas are seen on MRI as a well-defined, fusiform enlargement of the optic nerve. Diffuse, tortuous enlargement of the optic nerve with a characteristic kinking and buckling appearance is a characteristic feature of the childhood form of optic glioma. On T_1W MR scans, gliomas appear isointense or slightly hyperintense compared with the white matter. On T_2W images they appear hyperintense compared with the white matter.[22] The use of paramagnetic contrast media results in moderate to marked enhancement. MRI remains the study of choice for the evaluation of optic gliomas. Intracranial extension of optic nerve gliomas and associated intracranial pathological changes in patients with neurofibromatosis can be best appreciated on MRI scans.[22]

2.7 Optic Neuritis

Optic neuritis is an acute inflammatory process involving the optic nerve. The process may be idiopathic, immune-mediated, metabolic, and infective in nature.[27,28] Multiple sclerosis is the most common cause of optic neuritis;[28,29] visual loss is typically unilateral in patients with demyelinative optic neuritis. A recent study by Beck et al.[29] has shown that intravenous methylprednisolone followed by oral prednisone will speed the recovery of visual loss due to optic neuritis. On MR, the optic nerve in patients with optic neuritis may infrequently appear diffusely enlarged. On T_2W MR scans, the involved optic nerve may be slightly or moderately more hyperintense than the one on the normal side. Most patients with acute idiopathic optic neuritis and with no other neurologic disorder have the same areas of high-intensity signal in the cerebral white matter on T_2W MR scans that are found in patients with multiple sclerosis. In patients with optic neuritis, contrast enhancement on MRI is often subtle or present in a short segment of the optic nerve. It is best demonstrated by comparing pre- and post-contrast T_1W MR images or on postcontrast fat-suppressed T_1W MR scans.[22]

2.8 Lacrimal Gland Tumors

Epithelial tumors represent approximately 50% of masses involving the lacrimal gland. Half of these are pleomorphic (benign mixed) adenomas; the other half are malignant.[30] Of the malignant tumors, adenoid cystic (adenocystic) are the most common.[31] The nonepithelial masses of lacrimal gland fossa include lymphoid-inflammatory lesions, that is, benign dacryoadenitis, pseudotumor, and malignant lymphoma.[31] MRI is the study of choice for the evaluation of lacrimal gland tumors. Benign, mixed tumors appear as well-defined encapsulated masses and have long T_1 and long T_2 MR signal characteristics. The tumor may be homogeneous or heterogeneous. The use of paramagnetic contrast material results in moderate to marked enhancement. The heterogeneity within the tumor may be better appreciated on T_2W and postcontrast T_1W MR images. Malignant lacrimal gland tumors may show poor margins and an infiltrative character. Invasion of the surrounding tissues and intracranial extension can be best evaluated on MRI scans.[22,31]

3 MRI OF THE EAR

3.1 Introduction: Anatomy of the Ear

The ear is essentially a distance receptor concerned with the collection, conduction, modification, amplification and parametric analysis of the complex soundwaves which impinge on the head. The ear is subdivided into three major areas: the outer ear, the middle ear, and the inner ear. Each region has a specific function. The outer ear channels sound to the tympanic membrane (ear drum). The middle ear is an air-filled space (tympanic cavity), connected via the eustachian tube to the nasopharynx. Lying across the tympanic cavity are three small bones; the malleus, the incus and the stapes. The stapes footplate touches a membrane called the oval window, located between the middle ear and the inner ear.[32] A second membrane-covered opening called the round window is located in the cochlea below the oval window. This connects the inner ear with the middle ear and maintains a constant (perilymphatic fluid) pressure within the cochlea. The essential function of the tympanic cavity and its ossicles is the efficient transfer of energy from the air to the perilymphatic fluids which surround the cochlea.[32] The inner ear contains sensory receptors for hearing and balance. It consists of three main parts: the cochlea, the vestibule, and the semicircular canals. The inner ear is composed of the membranous labyrinth and the osseous labyrinth. The membranous labyrinth has two major subdivisions, a sensory portion called the sensory labyrinth and a nonsensory portion designated as the nonsensory labyrinth. The sensory labyrinth lies within the petrous portion of the temporal bone. It contains two intercommunicating portions: (1) the cochlear labyrinth that consists of the cochlea and is concerned with hearing, and (2) the vestibular labyrinth that contains the three semicircular canals, the saccule and utricle, (two small sacs), occupying the vestibule, all of which are concerned with equilibrium. These hollow chambers are filled with fluid, known as endolymph, that resembles intracellular fluid. The nonsensory element of the membranous labyrinth is formed by the endolymphatic duct and sac, whose main function is believed to be the degradation and absorption of the endolymph, produced in the sensory labyrinth, i.e. the cochlea and vestibule. All of the structures of the membranous labyrinth are enclosed within hollowed-out bony cavities that are considerably larger than their membranous contents. These bony cavities assume the same shape as the membranous chambers and are referred to as the osseous labyrinth. The bony cavities of the osseous labyrinth contain fluid, known as perilymph, that bathes the external surface of the membranous labyrinth.

The sound vibrations travel through the middle ear and then through the ossicular chain to the oval window. This in turn sets the perilymph in the cochlear labyrinth in motion. Stimulated by the fluid motion, sensory receptors (hair cells) in the cochlea generate nerve impulses that are transmitted by the cochlear nerve and auditory pathway to the auditory center in the temporal lobe of the brain. Impulses generated in the sen-

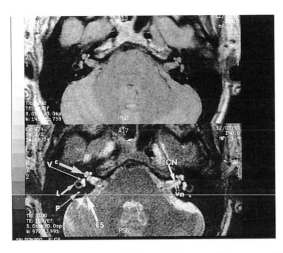

Figure 5 Normal anatomy of the ear: 3 mm axial proton-weighted (top) and T_2W (bottom) MR scans. The cochlea (c), vestibule (V), lateral semicircular canal (L), posterior semicircular canal (P), endolymphatic sac (ES), cochlear nerve (CN) and inferior vestibular nerve (vn) are shown

sory cells of the semicircular canals and vestibule are carried to the brain by the vestibular nerve.

The essential function of the vestibular labyrinth is to provide the central nervous system (CNS) with a constant flow of information concerning the static position of the head in space, or its state of linear or angular acceleration or deceleration.[32,33]

3.2   Acoustic and Facial Nerves

Acoustic nerve consists of the cochlear nerve and the vestibular nerves (superior and inferior). These are joined into a common trunk that enters the internal auditory canal with the facial nerve. The facial nerve then travels within the temporal bone for some 33 mm through a tortuous bony canal, known as the fallopian canal, before it enters the parotid gland.

3.3   MRI Anatomy of the Ear

MRI remains the study of choice for imaging evaluation of the membranous labyrinth, and the vestibulo-cochlear and facial nerves. The osseous labyrinth, cortical bones, and the air filled mastoid cells appear hypointense in all pulse sequences. The endolymphatic and perilymphatic fluid within and surrounding the membranous labyrinth provides MR signal, resulting in visualization of the membranous labyrinth on various MR pulse sequences (Figure 5). MRI remains the study of choice for the evaluation of the membranous labyrinth, and the vestibulo-cochlear and facial nerves. Cochlear and vestibular divisions of the acoustic nerve and the entire course of facial nerve including its parotid segment can be visualized with exquisite details on MR scans.

3.4   Pathology using MRI of the Ear

The main contribution of MRI to the diagnosis of ear disorders is its clear demonstration of the membranous labyrinth, cochlear, vestibular, and facial nerves. The use of paramagnetic

contrast material enhances the sensitivity and at times specificity of MRI in the evaluation of certain pathological conditions. Gd-DTPA contrast has been shown to play an important role in the detection of inflammatory (Figure 6), autoimmune, and neoplastic diseases of the membranous labyrinth as well as soft tissue tumors and inflammatory and infiltrative processes of the ear.[34,35]

3.5   Developmental Anomalies of the Ear

Computerized tomography scanning remains the study of choice for the evaluation of ear anomalies. MRI provides little information in the evaluation of patients with anomalies of the external and middle ears. In these patients MRI may be used to evaluate the facial nerve. Inner ear anomalies such as Mondini anomaly, large vestibular aqueduct, and Michel anomaly are best evaluated on CT scans. In Mondini anomaly, MRI reveals fluid-filled cystic dilatation of the membranous labyrinth. In patients with large vestibular aqueduct, MRI reveals fluid-filled cystic dilatation of the endolymphatic duct and sac.[33]

3.6   Inflammatory Conditions of the Ear

CT remains the study of choice for the evaluation of acute and chronic otomastoiditis and their complications, including acquired cholesteatomas.[34] The main contribution of MRI to the evaluation of the inflammatory conditions of the ear is its clear demonstration of the involvement of membranous labyrinth and associated intracranial complications (Figure 6). T_1-weighted MRI sequences obtained following the use of paramagnetic contrast material are most informative, as they readily demonstrate abnormal intralabyrinthine enhancement (Figure 6). In patients with vestibulocochlear or facial neuronitis, MRI

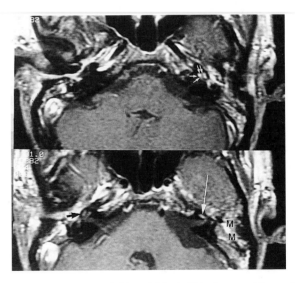

Figure 6 Acute otomastoiditis and labyrinthitis: 3 mm axial post-Gd-DTPA T_1W MR scans. There is marked enhancement of granulation tissues within the mastoid air cells (M). Notice marked enhancement of left membranous cochlea (white, short and long arrows). There is slight enhancement of the right cochlea (black arrow); patient had bilateral acute otomastoiditis, left more so than right

For References see p. 983

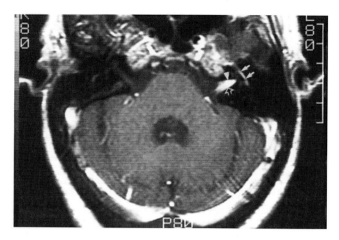

Figure 7 Ramsey hunt syndrome: 3 mm axial post-Gd-DTPA T_1W MR scan. There is marked enhancement of the meatal segment (arrow head) and tympanic segment (solid arrows) of the facial nerve. Note enhancement of the superior vestibular nerve (hollow arrow)

may demonstrate moderate to marked contrast enhancement of the involved nerve (Figure 7).

3.7 Cholesteatomas and Cholesterol Granulomas

Cholesteatomas of the ear may be congenital or acquired. Histologically the pathology of congenital cholesteatomas is similar to that of the acquired varieties, namely, a cavity, lined with keratinizing stratified squamous epithelium which encapsulates desquamated epithelial cells, keratomatous materials, and minimal cholestrin.

Cholesterol granulomas of the ear are believed to be a tissue response to hemorrhage and an irritant foreign body, i.e. cholesterol crystals and other blood byproducts. Cholesteatomas and cholesterol granulomas of the petrous apex are the most common expansile, slowly destructive, lesions of the petrous bone.[34] On CT scans they are very difficult to differentiate from each other. However, on MRI, cholesteatomas have a long T_1 and long T_2 characteristics, whereas, cholesterol granulomas have a short T_1 and long T_2 characteristic. Both lesions if not infected demonstrate no contrast enhancement. CT scanning is superior to MRI for the detection of middle ear cholesteatomas.[34] MRI however, is superior to CT scanning for the evaluation of infected cholesteatomas, petrous apex and intradural cerebellopontine (CPA) cholesteatomas (epidermoids), as well as for the evaluation of cholesteatomatous involvement of the facial nerve, membranous labyrinth, and intracranial structures.

3.8 Tumors of the Ear

The diagnostic approach to evaluating a patient suspected of having a lesion of acoustic and facial nerves or a CPA tumor has changed as the resolution of MRI has improved and as the sensitivity and specificity of MRI has been established. MRI remains the initial imaging study of choice for the evaluation of acoustic and facial neuromas, and most CPA tumors. The most common tumors of the temporal bone are Schwann-cell tumors that arise from the vestibular divisions of the acoustic nerve; as such, they are actually vestibular schwannomas. Bi-

lateral vestibular schwannomas are characteristic of neurofibromatosis type 2 (NF2).

Vestibular schwannomas are the most common CPA masses.[36] Meningiomas are the second and epidermoid cysts are the third most common CPA masses. Magnetic resonance imaging is the most useful diagnostic test to diagnose as well as to differentiate these CPA masses.[37] The use of paramagnetic contrast material significantly increases the sensitivity of MRI. Tumors as small as 2 mm in size can be detected along the course of facial and acoustic nerves. The use of paramagnetic contrast has allowed for the first time the detection of tiny intralabyrinthine schwannomas,[38] and inflammatory processes of the facial nerve which were extremely difficult to diagnose even with high-resolution CT scan.[39–41] Paragangliomas or chemodectomas (glomus tumors) are the second most common tumors of the temporal bone.[42] Paragangliomas appear as regions of hypointensity on T_1W and hyperintensity on T_2W MR scans.[42] There are frequently areas of signal void related to rapid arterial flow present in the matrix of these hypervascular tumors. Paragangliomas demonstrate significant enhancement following intravenous (i.v.) administration of paramagnetic contrast material.[42] Other benign and malignant tumors that involve the ear, including adenoma, carcinomas, adenocarcinomas, papillary adenocarcinomas, adenoid cystic carcinomas, rhabdomyosarcomas, chondrosarcomas, osteosarcomas, osteoclastomas, chordomas, and other mesenchymal tumors can be adequately evaluated by MRI.

4 MRI OF THE NASAL CAVITY AND PARANASAL SINUSES

4.1 Introduction

The nasal cavity, the first of the respiratory passages, extends from the roof of the mouth upwards to the base of the skull. It is divided by the nasal septum into two halves which open on the face through the nostrils, and communicates behind with the nasopharynx.[43] The lateral wall of the nasal cavity is irregular, owing to the presence of three bony projections, the inferior, middle, and superior nasal conchae (turbinates). The conchae project downward and slightly medially and divide the passageway into meatuses, or channels of air. The inferior meatus receives the orifice of the nasolacrimal duct. The middle meatus receives the opening of the ipsilateral anterior ethmoidal air cells, frontonasal duct of the frontal sinus and the ostium of the maxillary sinus. The superior meatus receives the opening of the ipsilateral posterior ethmoidal air cells. A narrow interval, the spheno-ethmoidal recess, separates the superior turbinate from the anterior surface of the body of the sphenoid, through which each sphenoidal sinus opens into the nasal cavity.[43] The nasal mucous membrane lines the nasal cavity. It is continuous with the mucous membrane of the nasopharynx and paranasal sinuses.

The paranasal sinuses are air-filled cavities and include the frontal, ethmoidal, sphenoidal and maxillary sinuses. They are lined with upper respiratory mucosa, continuous with that of the nasal cavity.[43] The frontal sinuses are rudimentary or absent at birth, they are usually fairly well developed by the seventh or eighth years, and they reach their full size after puberty. The ethmoidal sinuses are small, but of clinical importance, at birth; they grow rapidly between the sixth and eighth years and after puberty.[43] The sphenoidal sinuses are present at birth, as minute cavities within the body of the sphenoid bone, but their main development takes place between the third and tenth years and after puberty. The maxillary sinuses are present as minute cavities at birth, but do not, however, reach full size until after the eruption of all the permanent teeth. The natural opening of the maxillary sinus is high above its floor and communicates with the lower part of the hiatus semilunaris to be drained into the middle meatus.

4.2   MRI Anatomy of Nasal and Paranasal Sinuses

The nasal cavity and paranasal sinuses can be visualized by MRI, using head coil or surface coil. Images obtained with head coil are preferred because of the size and deep location of posterior sinonasal cavities. The cortical bone and air of the sinonasal structures do not provide MR signal, and therefore they appear hypointense in all pulse sequences. The mucosa of nasal structures, however, appear as intermediate signal on T_1W and increased signal (hyperintense) on T_2W images. The normal mucosa of the paranasal sinuses are very thin and cannot be visualized on MR scans.

4.3   MRI of Sinonasal Pathology

4.3.1   Congenital Anomalies

Magnetic resonance imaging remains the study of choice for the evaluation of congenital sinonasal anomalies such as sphenoidal, sphenoethmoidal, ethmoidal, frontonasal, and intraorbital encephaloceles. Magnetic resonance imaging is very useful for the evaluation of intranasal–extranasal dermoids and nasal gliomas.

4.3.2   Inflammatory Conditions

Standard sinus roentgenograms, CT and, more recently, MRI are currently the most used imaging methods to assess paranasal pathology.[44] Computerized tomography remains the imaging study of choice for appropriate evaluation of patients with acute sinusitis associated with complications such as subperiosteal or orbital abscesses and for patients with chronic sinusitis.[44,45] Magnetic resonance imaging is also very sensitive in the evaluation of acute sinonasal pathology. The main contribution of MRI to the diagnosis of sinonasal-orbital infections is its clear demonstration of the relationship between nasal, sinus, orbital and cranial diseases. Orbital cellulitis, subperiosteal phlegmon and abscess can be readily visualized on MR scans. Magnetic resonance imaging is superior to CT in demonstrating sinogenic complications such as thrombosis of the superior ophthalmic vein, thrombosis of the cavernous sinus, and intracranial complications such as epidural abscess, focal or diffuse encephalitis and brain abscesses.

4.3.3   Tumors

Magnetic resonance imaging provides excellent delineation of tumor from surrounding soft tissue, inflammatory tissue, and retained secretions within the sinuses.[46–52] It might be anticipated that the typical appearance of edema or retained secretions within the sinuses on MR images would be of low

intensity on T_1W and high intensity on T_2W images, reflecting the high water content associated with the excessive interstitial fluid or retained secretions. However, because of the frequent chronic nature of these benign processes in patients who have a nasal cavity or paranasal sinus tumor, especially in those patients in whom an advanced tumor has been diagnosed, sufficient time has elapsed to allow for the concentration of high water affinity mucoproteins and the absorption of free water. This leads to various degrees of shortening of both T_1 and T_2 relaxation times.[45] The areas of complete desiccation appear as low signal on T_1W, PW, and T_2W MR images.

Sinonasal carcinomas are highly cellular tumors with little free water; as such, they appear as low to intermediate signal intensity on T_1- and T_2W images (Figure 8). Inflammatory diseases have intermediate signal intensity on T_1W and high signal intensity on T_2W images [Figure 8(b)]. Benign pathological processes within the nasal cavity and paranasal sinuses, such as polyps, papillomas, low grade adenocarcinomas, (minor salivary gland tumors), and schwannomas have higher signal

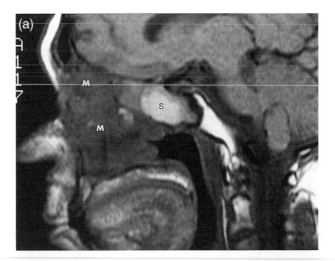

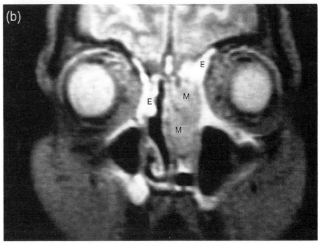

Figure 8   (a) Sinonasal squamous cell carcinoma. Sagittal T_1W MR image. A hypointense mass (M) is demonstrated, involving the nasal cavity and ethmoidal sinus. The tumor has caused obstruction of the ostium of the sphenoid sinus, resulting in hyperintense retained secretions within the adjacent sphenoid sinus (S). (b) Coronal T_2W MR images. A relatively hypointense mass (M) is demonstrated, compared with hyperintense retained secretions in the ethmoidal sinuses (E)

For References see p. 983

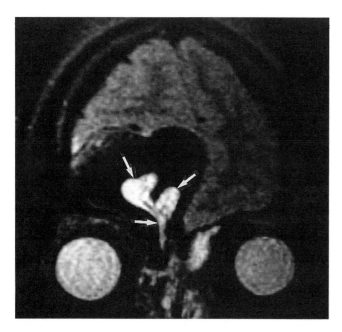

Figure 9 Frontal sinus polyp. Coronal T_2W MR image. A hyperintense polyp (arrows) is demonstrated within an otherwise aerated frontal sinus. The signal intensity of the polyp (benign process) is greater than that of the carcinoma (malignant process), seen in Figure 8(b)

of paramagnetic contrast material often offers additional information.[52] In most instances, Gd-DTPA contrast enhancement of the tumor is less than the enhancement of the normal mucosa and associated inflammatory changes. Gd-DTPA is extremely helpful in detecting intracranial extension of a sinonasal tumor.

Magnetic resonance imaging has proved valuable in the diagnosis of most other tumors of the sinonasal cavity such as rhabdomyosarcomas, lymphomas, angiomas, angiofibromas, hemangiopericytomas, chondrosarcomas, osteogenic sarcomas malignant melanomas, and metastases.[45]

5   MRI OF THE NECK

5.1   Introduction: Anatomy of the Neck

The neck is the junctional region between the head and thorax and upper limbs. Its rostral limits are the base of the skull posteriorly and the inferior border of the mandible anteriorly. Caudally it is bounded by the thoracic inlet and pectoral girdle (clavicle, manubrium of sternum, acromion and spine of the scapula), a plane which parallels the first rib. The neck contains the cervical vertebrae and muscles, the pharynx, the larynx, cervical esophagus, cervical trachea, thyroid gland, salivary glands lymph nodes, and vessels and nerves that either supply the various organs of the neck or are in transit between the head, thorax, and upper limbs.[53]

5.2   MRI Techniques

Our protocol for the upper neck (above the angle of the mandible) uses a head coil and a neck coil for the lower neck.

The nasopharynx, parapharyngeal space, infratemporal fossa, masticatory space, oropharynx, tongue, and salivary glands are ideally suited for evaluation by head coil. The larynx is best evaluated by using a surface coil. Thyroid gland and parathyroid glands and lower neck may be evaluated by both head and surface coils. A routine MRI of the neck should include at least a sagittal T_1W localizer and axial SE double echo pulse sequences. At times, additional coronal T_1W or T_2W images may be obtained. The use of Gd-DTPA contrast material, gradient echo images, and fat-suppression pulse sequences should be tailored according to the clinical information and initial evaluation of the sagittal T_1W and axial SE double echo pulse sequences.

5.3   MRI Anatomy of the Neck

Magnetic resonance imaging offers the best soft tissue contrast resolution and therefore remains the study of choice for imaging the anatomy of the soft tissue of the neck.[54] The larynx, pharynx, tongue, floor of the mouth, salivary glands, thyroid gland, parapharyngeal space, infratemporal fossa, masticatory space, and vascular structures of the neck can be visualized with exquisite anatomical detail.[54–56]

5.4   MRI Pathology of the Neck

5.4.1   Developmental Neck Cysts

Magnetic resonance imaging is extremely helpful in the evaluation of congenital anomalies of the neck.[55] The most frequent developmental conditions involving the neck are brachial cleft anomalies, cystic hygromas (lymphangiomas), dermoids and epidermoids, and teratomas. Most of these lesions have characteristic MRI findings that should help the radiologists make the diagnosis.[55]

5.4.2   Inflammatory Conditions

The main contribution of MRI to the diagnosis of neck infections is its clear demonstration of the relationship between various potential spaces within the neck and adjacent structures including base of the skull and intracranial structures. Soft tissue induration, edema, and fluid collections (pus) can be readily demonstrated on MRI (Figure 10). Magnetic resonance imaging is highly sensitive for the detection of early and late inflammatory changes of the bone marrow of the bony structures of the head and neck as well as the temporo-mandibular joint.

5.5   Neck Tumors

Magnetic resonance imaging is capable of depicting most benign and malignant tumors of the neck. On MRI, most tumors have long T_1 and T_2 characteristics.[54–57] The T_2W images are most valuable for detecting differences in signal intensity of the pathological process.[54–60] Magnetic resonance imaging remains the study of choice for imaging evaluation of nasopharyngeal, and parapharyngeal, salivary gland tumors. In most patients with laryngeal tumor, MRI appears to be more appropriate than a computerized tomography (CT) scan for evaluating the extent of laryngeal tumor.[57] Involvement of the pre-epiglottic space, paralaryngeal space, postcricoid region, as

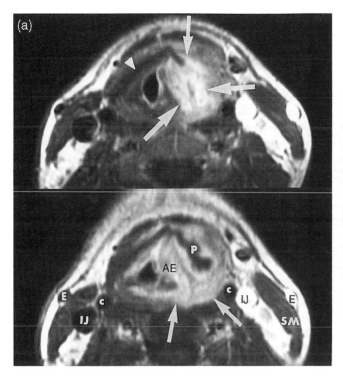

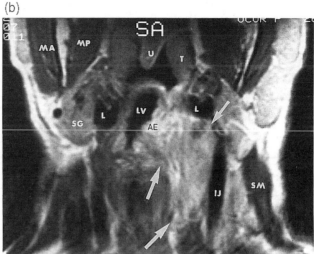

Figure 10 (a) Chronic laryngeal and perilaryngeal infection: 5 mm axial post-Gd-DTPA T_1W MR scans. A large enhancing inflammatory induration (arrows) is demonstrated, associated with marked thickening of the left aryepiglottic fold (AE), and an area of fluid loculation, proved to be pus (P), in the paralaryngeal space. Notice the normal right thyroid cartilage (arrowhead) and partial destruction on the left side. c = common carotid, E = external jugular vein, IJ = internal jugular vein, P = pus, SM = sternocleidomastoid muscle. (b) Coronal post-Gd-DTPA MR scan (4 mm). This coronal MR image demonstrates the caudal and cephalad extension of laryngeal and paralaryngeal inflammation (arrows). Notice marked thickening of the left aryepiglottic fold (AE), lateral displacement of the left internal jugular vein (IJ), and bilateral laryngoceles (L). LV = laryngeal vestibule, MA = masseter muscle, MP = medial pterygoid muscle, SG = submandibular gland, SM = sternocleidomastoid muscle, T = palatine tonsil, U = uvula

well as cartilage invasion by laryngeal tumor can be readily visualized on MR scans.[57] Magnetic resonance imaging is as sensitive as CT for demonstrating nodal (lymph) diseases of the neck.[59] Gd-DTPA contrast studies have been shown to play an important part in the detection and diagnosis of certain lesions of the neck.[61]

6 RELATED ARTICLES

Cranial Nerves Investigated by MRI; Head and Neck Investigations by MRI; Hemorrhage in the Brain and Neck Observed by MRI; Temperomandibular Joint MRI.

7 REFERENCES

1. R. Warwick and P. L. Williams, eds. 'Gray's Anatomy', 35th edn. (British), Saunders, Philadelphia, 1973.
2. M. F. Mafee, in 'Radiology of the Eye and Orbit', eds. T. H. Newton and L. T. Bilaniuk, Raven Press, New York, 1990 p. 2. 1–23.1.
3. M. F. Reeh, J. L. Wobij and J. D. Wirtschafter, 'Ophthalmic Anatomy: A Manual With Some Clinical Applications', American Academy of Ophthlmology, San Francisco, 1981, p. 11.
4. J. B. Aguayo, B. Glaser, A. Mildvan, H. M. Cheng, R. G. Gonzalez, and T. Brady, *Invest. Ophthalmol. Visual Sci.*, 1985, **26**, 692.
5. B. J. terPenning, H. M. Cheng, P. Barnett, J. Seddon, D. Sang, M. Latina, J. B. Aguayo, R. G. Gonzalez and T. Brady, *J. Comput. Assist. Tomogr.*, 1986, **10**, 551.
6. M. F. Mafee, G. A. Peyman, J. E. Grisolano, M. E. Fletcher, D. G. Spigos, F. W. Wehrli, F. Rasouli, and V. Capek, *Radiology*, 1986, **160**, 773.
7. M. F. Mafee, G. A. Peyman, J. H. Peace, S. B. Cohen, and M. W. Mitchell, *Ophthalmology*, 1987, **94**, 341.
8. M. F. Mafee, B. Linder, G. A. Peyman, B. G. Langer, K. H. Choi, and V. Capek, *Radiology*, 1988, **168**, 781.
9. M. F. Mafee and G. A. Peyman, *Radiol. Clin. North Am.*, 1987, **25**, 487.
10. M. F. Mafee, M. F. Goldberg, S. B. Cohen, E. D. Gotsis, M. Safran, L. Chekuri, and B. Raofi, *Ophthalmology*, 1989, **96**, 965.
11. J. M. Gomori, R. I. Grossman, J. A. Shields, J. J. Augsburger, P. M. Joseph, and D. DeSimeone, *Radiology*, 1986, **158**, 443.
12. R. G. Peyster, J. J. Augsburger, J. A. Shields, B. L. Hershey, R. Eagle, and M. E. Haskin, *Radiology*, 1988, **168**, 773.
13. B. G. Haik, L. Saint Louis, M. E. Smith, R. M. Ellsworth, D. H. Abramson, P. Cahill, M. Deck, and J. Coleman, *Ophthalmology*, 1985, **92**, 1143.
14. M. F. Mafee, in, 'Head and Neck Disorders (Fourth Series) Test and Syllabus', ed. P. M. Som, American College of Radiology, Reston, VA, 1992, p. 71.
15. M. F. Mafee, M. F. Goldberg, G. E. Valvassori, and V. Capek. *Radiology*, 1982, **145**, 713.
16. M. F. Mafee and M. F. Goldberg, *Radiol. Clin. North Am.*, 1987, **25**, 683.
17. M. F. Mafee, *Radiol. Clin. North Am.* 1998, **36**, 1083.

For References see p. 983

18. R. Damadian, K. Zaner, D. Hor, and T. Dimaio, *Physiol. Chem. Phys.* 1973, **5**, 381.

19. M. F. Mafee, D. J. Ainbinder, A. A. Hidayat, and S. M. Friedman, *Int. J. Neuroradiol.* 1995, **1**, 67.

20. R. G. Peyster, M. D. Shapiro, and B. G. Haik, *Radiol. Clin. North Am.* 1987, **25**, 647.

20. R. Warwick and P. L. Williams, eds. 'Gray's Anatomy', 35th edn. (British), Saunders, Philadelphia, 1973, p. 262–264.

21. M. F. Reeh, J. L. Wobig, and J. D. Wirtschafter, 'Ophthalmic Anatomy: A Manual with Some Clinical Applications', American Academy of Ophthalmology, San Francisco, 1981, p. 11.

22. M. F. Mafee, in 'Imaging of the head and neck', eds. G. E. Valvassori, M. F. Mafee, and B. L. Carter, Thieme, Stuttgart, 1995, p. 158.

23. F. C. Blodi and J. D. Gass, *Trans. Am. Acad. Opthalmol. Otolaryngol.*, 1967, **71**, 303.

24. A. E. Flanders, M. F. Mafee, and V. M. Rao, *J. Comput. Assist. Tomogr*, 1989, **13**, 40.

25. M. F. Mafee, E. Pai, and B. Philip, *Radiol. Clin. North Am.*, 1998, **36, 2**, 15.

26. M. F. Mafee, A. Putterman, G. E. Valvassori, M. Campos, and V. Capek, *Radiol. Clin. North Am.*, 1987, **25**, 529.

27. S. Lessell, *N. Engl. J. Med.*, 1992, **329**, 634.

28. G. C. Ebers, *Arch. Neurol.*, 1985, **42**, 702.

29. R. W. Beck, P. A. Cleary, M. M. Anderson, J. L. Keltner, W. T. Shults, D. I. Kaufman, E. G. Buckley, J. J. Corbett, M. J. Kupersmith, and N. R. Miller, *N. Engl. J. Med.*, 1992, **326**, 581.

30. M. F. Mafee and B. G. Haik, *Radiol. Clin. North Am.*, 1987, **25**, 767.

31. M. F. Mafee, D. P. Edward, K. Keller, and S. Darodi, *Radiol. Clin. North Am.*, 1998, **37**, 219.

32. R. Warwick and P. L. Williams, eds. 'Gray's Anatomy', 35th edn. (British), Saunders, Philadelphia, 1973, p. 1134.

33. M. F. Mafee, D. Charletta, A. Kumar, and H. Belmont, *Am. J. Neuroradiol.*, 1992, **13**, 805.

34. M. F. Mafee, *J. Otolaryngol.*, 1993, **22**, 240.

35. A. S. Mark and D. Fitzgerald, *Am. J. Neuroradiol.*, 1993, **14**, 991.

36. M. F. Mafee, *Otolaryngol. Clin. North Am.*, 1995, **28**, 407.

37. G. E. Valvassori, *Otolaryngol. Clin. North Am.*, 1988, **21**, 337.

38. M. Brogan and D. W. Chakeres, *Am. J. Neuroradiol.*, 1990, **11**, 407.

39. M. F. Mafee, G. E. Valvassori, A. Kumar, et al. *Otolaryngol. Clin. North Am.*, 1988, **21**, 349.

40. M. F. Mafee, C. S. Lachenauer, A. Kumar, P. M. Arnold, R. A. Buckingham, and G. E. Valvassori, *Radiology*, 1990, **174**, 395.

41. D. L. Daniels, L. F. Czervionke, S. J. Millen, T. J. Haberkamp, G. A. Meyer, L. E. Hendrix, L. P. Mark, A. L. Williams, and V. M. Haughton. *Radiology*, 1989, **171**, 807.

42. P. D. Phelps, *Clin. Radiol.*, 1990, **41**, 301.

43. eds. R. Warwick and P. L. Williams, 'Gray's Anatomy', 35th edn. (British), Saunders, Philadelphia, 1973, p. 1087.

44. M. F. Mafee, *J. Am. Med. Assoc.*, 1993, **269**, 2608.

45. P. M. Som and H. D. Curtin, *Radiol. Clin. North Am.*, 1993, **31**, 33.

46. J. M. Chow, J. P. Leonetti, and M. F. Mafee, *Radiol. Clin. North Am.*, 1993, **31**, 61.

47. M. F. Mafee, *Radiol. Clin. North Am.*, 1993, **31**, 75.

48. P. M. Som, M. D. Shapiro, H. F. Biller, C. Sasaki, and W. Lawson, *Radiology*, 1988, **167**, 803.

49. G. A. S. Lloyd, V. J. Lund, P. O. Phelps, and D. J. Howard, *Br. J. Radiol.*, 1987, **60**, 957.

50. M. G. M. Hunink, R. G. M. de Slegte, G. J. Gerritsen, and H. Speelman, *Neuroradiology*, 1990, **32**, 220.

51. P. van Tassell and Y. Y. Lee, *J. Comput. Assist. Tomogr.*, 1991, **15**, 387.

52. C. F. Lanzieri, M. Shah, D. Krauss, and P. Lavertu, *Radiology*, 1991, **178**, 425.

53. J. A. Gosling, P. F. Harris, and J. R. Humpherson, eds. 'Atlas of Human Anatomy with Integrated Text', Lippincott, Philadelphia, 1985.

54. M. F. Mafee, F. Rasouli, D. G. Spigos, G. E. Valvassori, H. Friedman, and M. Capek, *Otolaryngol. Clin. North Am.*, 1986, **19**, 523.

55. M. F. Mafee, in 'Head and Neck Imaging', eds. G. E. Valvassori, R. A. Buckingham, B. L. Carter, W. N. Hanafee, and M. F. Mafee, Thieme Medical, New York, 1988, p. 253.

56. R. B. Lufkin and W. N. Hanafee, *Invest. Radiol.*, 1988, **23**, 162.

57. J. A. Castelijns, R. P. Golding, C. Van Schaik, J. Valk, and G. B. Snow, *Radiology*, 1990, **174**, 669.

58. H. D. Curtin, *Radiology*, 1989, **173**, 1.

59. P. M. Som, *Am. J. Radiol.*, 1992, **158**, 961.

60. L. M. Teresi, R. B. Lufkin, D. G. Worthman, E. Abemayor, and W. N. Hanafee, *Radiology*, 1989, **163**, 405.

61. T. Vogl, S. Dresel, L. T. Bilaniuk, G. Grever, K. Kang, and J. Lissner, *Am. J. Neuroradiol.*, 1990, **154**, 585.

Biographical Sketch

Mahmood F. Mafee. *b* 1941. M.D., 1969 Tehran University, Iran; Resident in Radiology, 1973, Albert Einstein University Hospital, New York; Resident in Radiology, University of Illinois Hospital at Chicago, 1974–76, Fellowship in Neuroradiology-Head and Neck, University of Illinois Hospital and Eye and Ear Infirmary, 1976–1977. Faculty in Radiology, Illinois State University 1976–present. Approx. 170 publications. Research interests: the clinical applications of MRI in the diagnosis of head and neck disorders and neuro-otological and neuro-ophthalmological disorders.

Head and Neck Investigations by MRI

Yoshimi Anzai

University of Michigan, USA

and

Robert Lufkin

University of California, Los Angeles, CA, USA

1 INTRODUCTION

MRI has revolutionized head and neck imaging and has replaced computerized tomography (CT) as the study of choice for many lesions of the extracranial head and neck. MRI easily surpasses CT in its ability to differentiate subtle differences in soft tissue boundaries and extensions of tumors of the head and neck. The beam hardening artifacts on CT images from dental amalgam and dense cortical bone of the mandible, skull

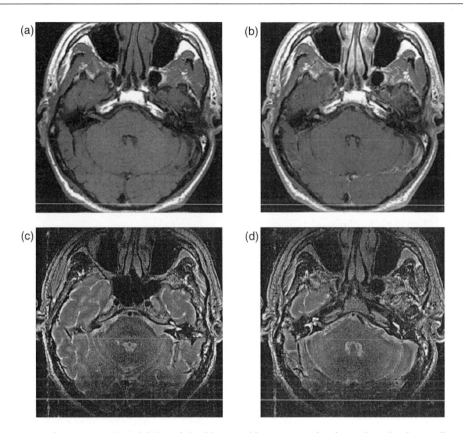

Figure 1 Intracanalicular acoustic neuroma. (a) Axial T_1-weighted image without contrast barely resolves the abnormality. (b) Following contrast administration, there is dense enhancement by the tumor. (c) High resolution (512 matrix) surface coil T_2-weighted axial image through the contralateral side shows clearcut depiction of the normal internal auditory canal contents. (d) View through the abnormal side shows the tumor without the use of a contrast agent. Some investigators are now questioning the added value of the contrast study

base, shoulders, and other areas are also not a problem with MRI. Multiplanar imaging capabilities and lack of ionizing radiation make MRI the preferred imaging study for many head and neck processes.

The exceptions where CT is still indicated are many inflammatory diseases, congenital malformations of the temporal bone, fibro-osseous lesions, and fractures. In patients with a contraindication to MRI, difficult or confusing cases where the MRI findings are inconclusive, or in situations where MRI is not available, CT is often very valuable. In the paranasal sinuses and middle ear disease where lesions are often defined by their relationship to thin cortical bone and air, CT is the study of choice. However, MRI is the preferred imaging modality for the majority of neoplastic conditions in most areas of the head and neck.

Cancers in the head and neck are often more accessible by visual inspection or palpation than similar lesions in brain, lung, abdomen, and pelvis. Therefore the role of any imaging modality in the evaluation of head and neck cancer is usually to define the deep extension of tumors, rather than to detect their presence. Mucosal changes are usually best evaluated by direct inspection during physical examination. Some submucosal tumors do not show findings other than fullness or bulging on physical examination. MRI is a powerful tool for the determination of deep disease extension and the relationship of the tumor to significant structures such as major vessels and nerves, which helps define therapeutic options.

2 TECHNIQUE

Surface coils greatly increase the MRI signal-to-noise ratio and allow higher spatial resolution of most examinations of the head and neck. While acceptable studies of the sinuses and naso- and oropharynx may be obtained with the standard head coil, the use of surface coils for examination of the larynx, hypopharynx, neck, and temporal bone is practically essential.[1,2]

The new bilateral phased array surface coil provides high resolution with a higher signal-to-noise ratio compared with standard surface coils. Each phased array coil element has its own preamplifier, receiver channel, and digitizer, allowing the collection of data and generation of images individually and subsequent combination into a composite image. This approach is now routinely used in many centers for MRI of the temporal bone and temporomandibular joint (Figure 1).

While specialized surface rf coils are important factors that make quality magnetic resonance images of the head and neck possible, perhaps the most significant factor in the superiority of MRI over CT scanning in this area is the high soft tissue contrast resolution of MRI. Unlike the central nervous system where virtually no MRI signal from fat is present, the abundance of fat–water interfaces in the extracranial head and neck, which literally defines the anatomy of this region, greatly influences the selection of pulse sequences for optimum image contrast.

For References see p. 992

Although there is considerable variation in imaging approaches to the head and neck, a standard MRI sequence would include T_1- and T_2-weighted axial images, and T_1-weighted coronal images for most lesions. T_1-weighted sagittal images are necessary for the larynx, mastoid portion of the facial nerve, and midline lesions.

Currently, T_2-weighted fast spin echo (FSE) imaging has replaced the regular T_2-weighted spin echo image for most studies of the head and neck at many centers.[3] In FSE acquisition, multiple refocused echoes are generated during a single TR (usually 8–16 echoes train), resulting in a short scan time and a better signal-to-noise ratio. The reduced scan time allows higher image matrix acquisition, such as 512 instead of 256, for improved spatial resolution. The drawbacks of FSE include fewer slices obtained for the same TR and less sensitivity to susceptibility effects.

Other parameters commonly used in head and neck MRI are an FOV (field of view) of 18–20 cm and slice thickness of 3–4 mm. A 192×256 matrix is routinely used in most of head and neck MRI, except for high resolution temporal bone MRI where a 384×512 matrix is used.

The role of gadolinium contrast agents in MRI examinations of the head and neck has been investigated.[4] Although gadolinium does enhance most head and neck tumors, it may also obscure the tumor margin in some cases because of the presence of enhancement within normal tissue resulting in reduced fat/tumor contrast. Fat suppression gadolinium enhancement MRI, therefore, was introduced to improve the lesion conspicuity following contrast administration. This technique is most valuable in regions with large amounts of fat, such as the orbit. In other regions of the head and neck the advantages of this technique are less well defined.

Although gadolinium administration adds information about whether the suspected tumor does or does not enhance, the clinical relevance of this information has been questioned in some cases. When intracranial extension is present, however, gadolinium can clearly improve the detection of the blood–brain barrier and leptomeningeal pathology. For these reasons many centers now limit their use of gadolinium to studies of the orbit and other regions close to the skull base where leptomeningeal and perineural pathology is a possibility.

3 TEMPORAL BONE

MRI is the modality of choice to evaluate sensorineural hearing loss and to a large extent to evaluate the facial nerve in its course through the temporal bone. High-resolution technique is critically important in the imaging of the temporal bone.[5] In our institution, a bilateral phased array coil is used with 3 mm interleaved slices. T_1-weighted spin echo and T_2-weighted FSE images as well as gadolinium enhanced T_1-weighted images are obtained. Gadolinium may also help in the detection of small intracanalicular acoustic tumors, although the possibility of inflammatory enhancement simulating a neoplasm must always be considered.[6]

The most common CP angle tumor is the acoustic neuroma that appears as an enhancing mass often extending into the internal acoustic canal (IAC) (Figure 2). While gadolinium enhanced MRI clearly delineates the tumor extension and improves detection of the intracanalicular acoustic neuroma,

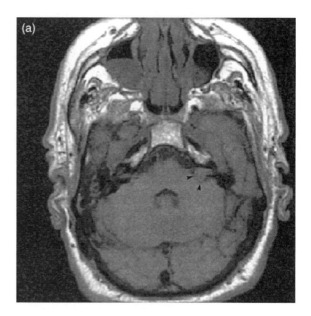

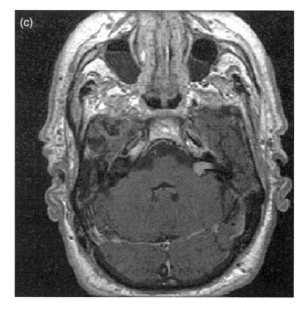

Figure 2 Acoustic neuroma. This middle-aged man presented with left sided sensorineural hearing loss. (a) Axial T_1-weighted image shows a rounded soft tissue mass (arrowed) centered over the right internal auditory canal extending to the cerebellopontine angle. (b) Axial T_2-weighted image shows similar findings. (c) Axial T_1-weighted image following contrast administration shows uniform enhancement

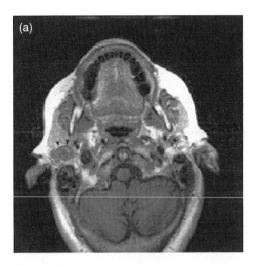

recent studies using high-resolution T_2-weighted images now question its value.[7]

Meningioma is the second most common CP angle tumor, and characteristically shows broad dural attachment. They are isointense to the cerebral cortex on T_2-weighted images in the majority of cases. Hyperostosis and dural thickening are often associated with meningioma. Following gadolinium administration, meningioma typically shows enhancing dural extension of the tumor.

Epidermoid (primary cholesteatoma) is a congenital lesion resulting from inclusion of ectoderm during early fetal life. The epidermoid contains keratin debris and solid cholesterin. The signal intensity of the tumor depends on the proportion of these two elements. When keratin debris predominates, the tumor appears of low signal on T_1- and of high signal on T_2-weighted images. The cholesterin portion gives a high signal on T_1-weighted images. The absence of gadolinium enhancement is characteristic.

CT remains the imaging study of choice at the present time for inflammatory disease and most other conditions resulting in conductive hearing loss. Important information about the status of the ossicles and bony covering of the facial nerve canal is still best obtained with CT. CT can also best visualize tumors and metabolic diseases affecting the middle ear and mastoid air cells. For the remainder of lesions of the temporal bone, CT and MRI will continue to play complementary roles in the imaging workup. Depending on the location and particular pathology, each of the imaging modalities will contribute information.

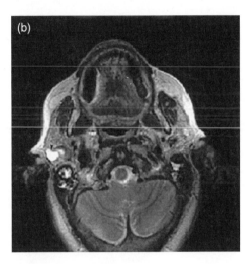

4   SALIVARY GLANDS

The preferred method of imaging the salivary glands has evolved with new developments in imaging technology as well as clinical needs. While sialography is still used to evaluate inflammatory or autoimmune diseases of the salivary glands, MRI has replaced sialo CT scanning for the imaging evaluation of the majority of masses in the major salivary glands.[8–10]

While 80% of major salivary gland tumors occur in the parotid gland, inflammatory lesions are the most common in the submandibular gland. The sublingual gland is the smallest major salivary gland. Tumors originating from the sublingual gland are relatively rare. Therefore, MRI is most frequently indicated for evaluation of parotid masses.

T_1-weighted images in the axial plane provide excellent contrast among most tumors, parotid gland, and parapharyngeal fat, allowing differentiation of an intraparotid from extraparotid gland masses (Figure 3). These views are supplemented with coronal or sagittal views when there is a suggestion of temporal bone involvement. Although poor tumor margin and low signal intensity on T_2-weighted images has been reported to predict the high grade of malignancies in parotid gland tumors to some extent this is not universally accepted, and determinations of histology by imaging studies is generally not possible. This is in part due to the unusual, and variety of, histology found in parotid tumors. Fortunately, fine needle aspiration (FNA) is a simple, lower cost and relatively noninvasive method to determine the nature of the tumor, and has been widely applied in large clinical series.

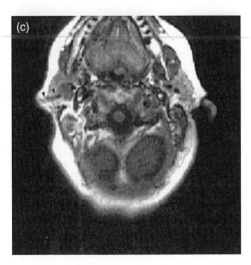

Figure 3   MRI guided aspiration cytology of a parotid mass. Several attempts at needle biopsy in the clinic without imaging guidance were uninformative. (a) Axial T_1-weighted image shows the superficial parotid mass (arrowed). (b) Axial T_2-weighted image shows heterogeneous areas of high signal which are nonspecific. This may be seen with a variety of benign and malignant conditions. (c) Axial T_1-weighted image following MRI compatible needle placement shows the linear signal defect of the needle entering the mass (arrowed). The cytology revealed adenocarcinoma

For References see p. 992

The real advantage of MRI in evaluating parotid masses is its ability to determine more accurately whether the intraparotid tumor is within the superficial or deep lobe. The superficial and deep lobes of the parotid gland are not based on an actual anatomical structure but, rather, are defined by their location relative to the main trunk of the facial nerve. Coronal images are sometimes useful to determine the distance of the parotid mass from the stylomastoid foramen, which is crucial information for a surgeon. Malignant tumors in close relation to the facial nerve may require excision of part of that structure and subsequent microsurgical repair or transposition of nerve VII. Access to larger, deeper lesions may require division and reflection of the ramus of the mandible. This valuable information is easily obtained with MRI.[11–13]

5   PARANASAL SINUSES

As stated before, MRI is primarily indicated for sinus tumors rather than inflammatory or fibro-osseous diseases. Although cortical bone and air do not return a signal on MRI, cortical bone can often be visualized as a lower signal between the layers of high signal soft tissue and mucosa covering the sinus wall. MRI allows excellent delineation of solid tissue masses surrounded by secretions within the paranasal sinuses.[14,15] Erosions of the bony walls can be demonstrated on MRI as the absence of signal void normally present in cortical bone (Figure 4). For extensions beyond the bony sinuses, MRI is the clear

study of choice because it differentiates skeletal muscle from tumor extension. In cases where there is a question of extension to the anterior or middle cranial fossa, MRI with gadolinium enhancement remains the study of choice.

MRI is also valuable for repeated follow-up studies for diseases of children such as juvenile angiofibroma where multiple studies are required and radiation dose becomes a factor. The high soft tissue contrast capabilities of MRI in distinguishing low protein fluids, high protein fluids, soft tissues, and normal musculature make MRI ideal to differentiate sinus tumor from secondary obstructed sinuses.[14]

One possible pitfall encountered in MRI of inflammatory sinus disease is signal void seen in patients with desiccated lesions such as fungus infection. Unlike nonfungus sinusitis, mycetoma is of low signal intensity on both T_1- and T_2-weighted images, simulating well pneumatized normal sinuses. This is due to the presence of paramagnetic substances such as manganese, iron, and calcium as well as desiccated debris within mycetoma which do not contain any free water. The fungus infection can be clearly demonstrated on CT, which is another reason that the mainstay for imaging inflammatory sinus disease will continue to be CT.

6   NASOPHARYNX

The lack of motion and abundant facial planes of the nasopharynx result in high quality MRI.[16] Retropharyngeal

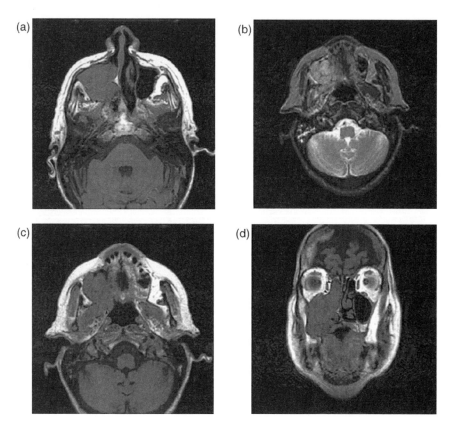

Figure 4   Maxillary sinus carcinoma. (a) Axial T_1-weighted image shows opacification of the maxillary sinus with associated bone destruction. The amount of extension posteriorly is unclear. (b) Axial T_2-weighted image at the same level shows improved muscle/tumor contrast. (c) Axial T_1-weighted image at a higher level shows invasion of the pterygopalatine fossa. (d) Coronal T_1-weighted image better defines the craniocaudal extension of the mass which involves the ethmoid sinus

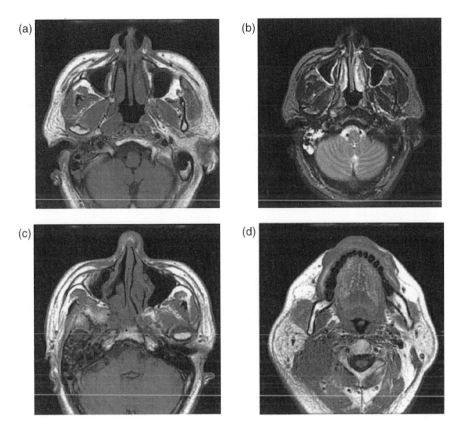

Figure 5   Nasopharynx carcinoma. This man presented with a neck mass. (a) T_1-weighted axial image through the nasopharynx shows a right sided fullness. (b) T_2-weighted image at the same level shows that the mass crosses the normal pharyngobasilar fascia and invades the parapharygeal space. (c) Higher axial T_1-weighted image shows the mass extending cephalad and involving the pterygopalatine fossa. (d) Axial T_1-weighted image shows the large [presenting] posterior triangle lymph node

adenopathy, tumor infiltration beyond the pharyngobasilar fascia, and hypertrophic lymphoid tissue are all better identified on MRI than on CT. The main goal of MRI examination is to evaluate the extension of primary tumor as well as nodal metastases. T_1-weighted images are usually adequate for examining the nasopharynx because of the abundance of loose areolar tissue between various muscle groups and bundles. Tumors can be identified as low signal regions on T_1-weighted images and high signal on T_2-weighted images in these loose areolar planes (Figure 5).

The configuration of the nasopharynx is dominated by a very tough fascial membrane called the pharyngobasilar fascia. This tough fascia represents a continuation of the pharyngeal constrictor muscles and extends from the base of the skull down to the level of the hard palate. Its function is to maintain the airway as an open channel for breathing during normal activities and during chewing. Only malignant tumors and very aggressive inflammatory processes such as mucormycosis will pass from the mucosa of the nasopharynx through the pharyngobasilar fascia to involve the structures within the parapharyngeal space. MRI can routinely show the pharyngobasilar fascia and destruction of this tough fascia which indicates the aggressive nature of mass.

The nasopharynx remains an area that is obscure to casual clinical examination due to gag reflex or poor patient cooperation. Its proximity to the skull base makes cancers in this region particularly devastating. Most malignancies of the nasopharynx are squamous cell carcinoma. The presenting symptoms of nasopharyngeal carcinoma vary widely. The most common complaints of patients presenting with nasopharyngeal tumors are nasal obstruction, local invasion of cranial nerves, serous otitis media, and cervical lymph node metastases. MRI is particularly well suited to complement the clinical examination in this situation. Other nasopharyngeal tumors, such as lymphoma, plasmacytoma, and occasionally rhabdomyosarcomas are also encountered. These tumors tend to be bulkier and infiltrate more widely than squamous carcinoma.

If a patient has symptoms of cranial nerve involvement, the skull base, jugular foramen, and the cavernous sinus should be thoroughly studied with MRI with administration of gadolinium. In particular, the direct coronal and sagittal MR scans are valuable to assess craniocaudal extension of tumor in skull base involvement, which can help to define radiation ports. Abnormalities of the skull base are detected by replacement of the normal low signal cortical bone and normal high signal bone marrow with the invading neoplasm.

The tensor veli palatini muscle is enveloped in a fascial plane of its own which divides the parapharyngeal space into two compartments.[17] The tensor veli palatini muscle fascia passes posterior to the styloid process and divides the space into a lateral compartment which is called the pre-styloid space and a medial compartment which is called the post-styloid

For References see p. 992

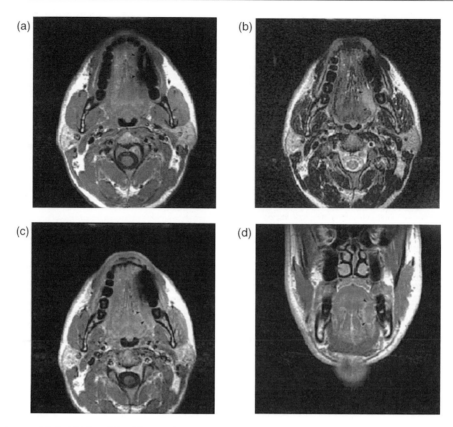

Figure 6   Tongue carcinoma. (a) Axial T_1-weighted image shows an artifact from ferromagnetic dental work. Conventional dental amalgam does not create an artifact. (b) Axial T_2-weighted image shows a small region of increased signal laterally which represents squamous carcinoma. It does not cross the midline nor invade the mandible. (c) Axial T_1-weighted image following contrast shows enhancement of the lesion. The information is similar to that in the T_2-weighted image. (d) Coronal T_1-weighted image following contrast administration again shows the lesion

space. This helps narrow down the list of differential diagnosis, since depending on the anatomical location the possible pathologies are different. For example, the majority of pre-styloid space masses are salivary gland tumors, and most post-styloid masses are of neurovascular origin, such as neuroma and paraganglioma. Metastatic tumor can be seen in both compartments. The capability of multiplanar imaging and the far superior soft tissue resolution make MRI the clear imaging study of choice to evaluate the nasopharynx and the adjacent spaces.

7   TONGUE AND OROPHARYNX

In general, MRI produces superior soft tissue detail in evaluating the tongue and oropharynx compared with CT.[18,19] Lack of artifacts from dental amalgam and beam hardening artifacts from the mandible on MRI also eliminates two major shortcomings of CT in the examination of this area. Moreover, the ability of MRI to obtain direct coronal and sagittal scan planes is a distinct advantage in recognizing intrinsic tongue musculature and assessing tumor volume and spread for treatment planning. MRI is therefore considered the study of choice in this area.

Squamous cell carcinomas account for well over 90% of the malignancies of the tongue. The remaining lesions are lymphomas, rhabdomyosarcomas, adenocarcinoma, adenoid cystic carcinoma originating from the minor salivary gland, and an assortment of benign tumors and inflammatory processes.

Squamous cell carcinomas of the tongue, tonsillar bed and posterior pharyngeal wall are the lesions most likely to require radiological imaging. The main purpose of the MRI study is to evaluate the extent of the primary tumor and nodal spread. Occasionally an infiltrating tumor of these regions may have spread to the nasopharynx or the supraglottic larynx so that additional imaging is needed prior to surgical management.

The tongue is one area in the head and neck where T_2-weighted pulse sequences are extremely valuable (Figure 6). The musculature of the tongue is of low signal on both T_1- and T_2-weighted images. Tumors produce a very high signal on T_2-weighted images, resulting in excellent delineation of tumor margins. Anatomical details of the midline are best obtained by coronal T_1-weighted images. The midline lingual septum appears as high signal fibrofatty tissue separating the two lateral halves of the tongue. If the tumor crosses this midline, the relationship of the tumor margin to the lingual artery and hypoglossal nerve should be accessed.

At least one hypoglossal nerve and one lingual artery must be retained in order to perform a hemiglossectomy. A total glossectomy is a much more involved procedure for selected patients, and ideally requires specialized preoperative planning. The lingual artery is easily visible between the genioglossus muscle and the interdigitation of the styloglossus and hyoglossus muscles. Disruption of the diaphragm of the floor of the mouth (the mylohyoid muscle) in a bilateral fashion clearly indicates a far advanced tumor. Adenoid cystic carcinomas almost invariably extend along perineural lymphatics. In this case, gadolinium enhanced MRI is helpful to demonstrate the enhancement of the involved nerve.

Carcinomas of the base of tongue (posterior to the foramen cecum) are particularly troublesome. Because of their relative inaccessibility, they are usually not discovered until relatively late. Of these tumors, 76% are reported to have metastases at the time of initial examination. Nodal disease is frequently bilateral because of the bilateral lymphatic drainage of the posterior third of the tongue. Tumors of the base of the tongue have a propensity to spread laterally into the glossopharyngeal sulcus and tonsillar bed regions and anteriorly into the vallecula and preepiglottic space.

Since the mucosal surface of the base of tongue varies and asymmetry does not necessary indicate the presence of tumor, the MRI finding which more strongly suggests the presence of an early base of tongue tumor is the disruption of the high signal intrinsic musculature. This is well visualized on axial and sagittal T_1-weighted images. Sagittal images are particularly helpful in showing tumor extension to the supraglottic larynx.

The remaining anterior lesions of the cheek, retromolar trigone, alveolar ridge, and lips are readily visible by inspection or available to the palpating finger, and imaging studies are seldom necessary, except for advanced cases that require the investigation of deep extension.

8 LARYNX AND HYPOPHARYNX

Rarely does any radiological imaging modality play a significant role in reaching a diagnosis of malignancy in the larynx. The larynx is so readily accessible to clinical examination that the combination of biopsy and visual inspection usually strongly indicates the diagnosis of cancer. The very small true cord cancer may not be detected by MRI. While laryngoscopy can show mucosal surfaces and masses involving the lumen, deep extensions are difficult to detect from clinical examination alone, yet, in several areas, these extensions have profound implications on the management of disease.[1,2,20–23] MRI to an even greater extent can define this important deep anatomy.

T_1-weighted images with three scanning planes, axial, coronal, and sagittal, are ideal for the study of the larynx (Figure 7).[2,20] It is valuable to start with sagittal images which show

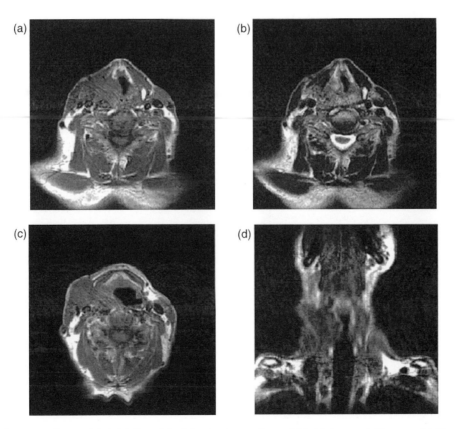

Figure 7 Supraglottic larynx carcinoma. (a) Axial T_1-weighted image shows a large right sided supraglottic mass. (b) The T_2-weighted image at the same level shows similar findings but improved muscle/tumor contrast. (c) More cephalad axial T_1-weighted image shows extralaryngeal spread to the adjacent neck. (d) Coronal T_1-weighted image shows that the mass does not involve the cricoid cartilage of subglottic region

For References see p. 992

the level of true vocal cords. Axial images are obtained parallel to the true cord which is readily visualized in the paramedian sagittal image. Then coronal images are acquired perpendicular to the true vocal cord. This will provide improved delineation of three important anatomical divisions—the glottic, subglottic and supraglottic regions. T_1-weighted sequences maximize contrast between the loose areolar tissue of the parapharyngeal and preepiglottic spaces, and most tumors. In addition, the laryngeal skeleton consists of several cartilages, which contain high signal marrow on T_1-weighted images. The level of the true vocal cord can be identified by the intrinsic vocalis muscle attached to the vocal process of arytenoid cartilage, which rests on top of the cricoid lamina.[24]

The supraglottic larynx consists of the laryngeal ventricle, the false cord, aryepiglottic folds, the epiglottis, and arytenoid cartilages. The supraglottic larynx is embryologically part of the buccopharyngeal anlage, and its lymphatic drainage is shared with the tongue, extending superiorly to the internal jugular chain. The preepiglottic space, anterior to the epiglottis, has a normal high signal which is displaced by low signal when tumors infiltrate this space.

The finding of the vertical spread of supraglottic tumors to the glottic region is even more important. Only advanced lesions spread down the inferior margin of the epiglottis to the anterior commissure or from superior to inferior in the paralaryngeal space. For the treatment of supraglottic tumors involving the tongue base, a partial glossectomy may have to be performed in addition to the primary surgery.[1] Coronal and sagittal MRI scans with T_1-weighted pulse sequences readily demonstrate these spreads. The true vocal cords and subglottic region act more like the trachea with the majority of the lymphatic drainage directed posteriorly and inferiorly.

Planning for any voice conservation laryngeal surgery depends on an accurate pre-operative knowledge of the precise extent of the disease within the larynx. Specifically, all techniques require an intact cricoid cartilage and at least one mobile arytenoid cartilage. MRI can provide this essential information. To plan this type of operation, the direct coronal and direct sagittal imaging capabilities of magnetic resonance far surpass axial CT images in the ability to define critical information regarding the cranial–caudal extent of the tumors. Compared to CT, MRI consistently shows superior soft tissue definition in cooperative patients. The use of direct coronal and sagittal scan planes allows the visualization of cranial–caudal tumor extension.

9 IMAGING GUIDED ASPIRATION CYTOLOGY

Imaging guided aspiration cytology has been applied to the nonpalpable deep sited head and neck lesion and has shown promising results. This was initially performed under CT or ultrasound guidance. Imaging guided aspiration cytology is now a standard procedure in the evaluation of head and neck tumors that cannot be sampled by more blind approaches. Aspiration cytology is now possible with MRI guidance using specially developed MRI compatible needles (E-Z-EM Corporation, Westbury, New York).[25–28] The advantages of using MRI as an imaging guidance are lack of radiation, high soft tissue contrast allowing better delineation and detection of

pathology, and multiplanar capability of showing the needle track in a single plane (see Figure 3).

10 RELATED ARTICLES

Eye, Orbit, Ear, Nose, and Throat Studies Using MRI; Gadolinium Chelate Contrast Agents in MRI: Clinical Applications; Multi Echo Acquisition Techniques Using Inverting Radiofrequency Pulses in MRI; Surface and Other Local Coils for In Vivo Studies; Temperomandibular Joint MRI; Whole Body Machines: NMR Phased Array Coil Systems.

11 REFERENCES

1. R. B. Lufkin, W. N. Hanafee, D. Wortham, and L. Hoover, *Radiology*, 1986, **158**, 747.
2. C. B. McArdle, B. J. Bailey, and E. G. Amparo, *Arch. Otolaryngol. Head Neck Surg.*, 1986, **112**, 616.
3. G. H. Zoarshi, J. R. Maskey, Y. Anzai, W. N. Hanafee, P. S. Melki, R. V. Mulkern, F. A. Jolesz, and R. B. Lughin, *Radiology*, 1993, **188**, 323.
4. J. D. Robinson, S. C. Crawford, M. Teresi, V. L. Schiller, R. B. Lufkin, H. R. Harnsberger, R. B. Dietrich, J. R. Grim, G. R. Duckwiler, and E. M. Spiekler, *Radiology*, 1989, **172**, 165.
5. J. Schwartz and H. Harnsberger, 'Imaging of the Temporal Bone', 2nd edn., Thieme, New York, 1992. pp. 192–246.
6. M. H. Han, B. A. Jabour, J. C. Andrews, R. F. Canelis, F. Chen, Y. Arzai, D. P. Becker, R. B. Lughin, and W. N. Hanafee, *Radiology*, 1991, **179**, 795.
7. R. W. Allen, H. R. Harnsberger, W. L. Davis, B. D. King, J. L. Parkin, and R. I. Affelbaum, *Radiology*, 1993, **189**(P), 141.
8. S. M. Mandelblatt, I. F. Brain, P. C. Davis, S. M. Fry, L. H. Jacobs, and J. Hoffman, Jr., *Radiology*, 1987, **163**, 411.
9. L. M. Teresi, E. Kolin, R. B. Lythin, and W. Hanafee, *Am. J. Neuroradiol.*, 1987, **148**, 995.
10. L. M. Teresi, D. Wortham, E. Abemayor, and W. Hanafee, *Radiology*, 1987, **163**, 405.
11. J. W. Casselman and A. A. Mancuso, *Radiology*, 1987, **165**, 183.
12. D. R. Mirich, C. B. McArelle, and M. V. Kulkarni, *J. Comput. Assist. Tomogr.*, 1987, **11**, 620.
13. D. H. Rice, *Arch. Otolaryngol. Head Neck Surg.*, 1987, **113**, 78.
14. M. P. Som, W. P. Dillon, G. D. Fullerton, R. A. Zimmeroner, B. Ragagopalan, and Z. Maron, *Radiology*, 1989, **172**, 515.
15. M. Shapiro and P. Som, *Radiol. Clin. N. Am.*, 1988, **27**, 447.
16. W. P. Dillon, C. M. Mills, B. Kjos, J. De Groot, and M. Brant-Zawalzki, *Radiology*, 1984, **152**, 731.
17. H. D. Curtin, *Radiology*, 1987, **163**, 195.
18. R. B. Lufkin, G. D. Wortham, R. B. Dietrich, A. L. Hoover, S. G. Larsson, H. Kangarloo, and W. N. Hanafee, *Radiology*, 1986, **161**, 69.
19. J. M. Unger, *Radiology*, 1985, **155**, 151.
20. R. B. Lufkin, S. G. Larsson, and W. N. Hanafee, *Radiology*, 1983, **148**, 173.
21. R. B. Lufkin and W. N. Hanafee, *Am. J. Neuroradiol.*, 1985, **145**, 483.
22. J. A. Castelijns, J. Doomber, B. Verbeeten, Jr., G. J. Vielroye, and J. L. Bloem, *J. Comput. Assist. Tomogr.*, 1985, **9**, 919.
23. H. D. Curtin, *Radiology*, 1989, **173**, 1.
24. C. R. Archer, S. S. Sagel, V. L. Yeager, S. Montin, and W. H. Friedman, *Am. J. Roentgenol.*, 1981, **136**, 571.
25. P. R. Mueller, D. D. Stark, J. F. Simeone, S. Saini, R. J. Butch, R. R. Edelmann, J. Wittenberg, and J. T. Ferrucci, Jr. *Radiology*, 1986, **161**, 605.

26. R. Lufkin and L. Layfield, *J. Comput. Assist. Tomogr.*, 1989, **13**, 1105.

27. R. Lufkin, L. Teresi, and W. Hanafee, *Am. J. Roentgenol.*, 1987, **149**, 380.

28. G. Duckwiler, R. B. Luphin, L. Jeresi, E. Spickler, J. Dion, F. Vinuela, J. Bentson, and W. N. Hanafee, *Radiology*, 1989, **170**, 519.

Head and Neck Studies Using MRA

Paul M. Ruggieri, Jean A. Tkach, and Thomas J. Masaryk

Cleveland Clinic Foundation, Cleveland, OH, USA

1   INTRODUCTION

Magnetic resonance angiography (MRA) is most commonly applied to the evaluation of cerebrovascular disease as neurological examinations continue to be the mainstay of clinical MRI. In addition, the rapid blood flow through the extra-/intracranial cerebral vasculature, the paucity of gross physiological motion, and the existing coil technology make these vessels ideally suited to vascular MRI. MRA is especially appealing since it provides a morphological representation of the vasculature as a relatively rapid, noninvasive alternative to the existing vascular imaging modalities. During the same examination, conventional spin echo images of the brain parenchyma are acquired to assess the consequences of the underlying vascular pathology.

1.1   Carotid MRA: Technical Considerations

Two basic types of MRA sequences are commonly employed to study the cerebral vasculature: phase contrast (PC) and time-of-flight (TOF).[1,2] The PC technique provides optimal suppression of the background stationary tissues via subtraction of alternating flow encoded images, and can be designed to be highly sensitive to blood moving at specific velocities. Appropriately calibrated sequences can also be designed to measure flow velocities. However, this technique demands longer acquisition and postprocessing times than the TOF studies of comparable spatial resolution. The phase sensitivity of this method also makes it more vulnerable to intravascular signal loss due to complex motion (e.g. carotid bifurcation stenosis) and/or local field inhomogeneities (e.g. carotid siphon, densely

calcified atherosclerotic plaque).[3] Consequently, the PC study is not commonly used to evaluate the carotid bifurcations in routine clinical work. A 2D PC study is useful as a quick scout view for subsequent positioning of the 2D or 3D TOF volumes and to confirm vessel patency. The 2D PC technique may also prove to be useful for making flow velocity measurements in the region of a carotid stenosis (similar to duplex ultrasound studies).

The TOF techniques take advantage of the rapid, constant inflow of fresh unsaturated spins to provide the necessary contrast relative to the adjacent stationary tissues. In lieu of subtraction mask acquisitions, angiographic images are rendered using computer postprocessing algorithms which may then create multiple MRA images from a single data set. The TOF methods are capable of minimizing intravoxel phase dispersion (and intravascular signal loss) by incorporating very small voxels and very short echo times with motion refocusing gradients.[4] Alternatively, the TOF sequences are not as sensitive to slow flow as the PC techniques. The TOF images also have less effective background suppression. Because the TOF studies are inherently T_1-weighted, and the commonly used postprocessing method (maximum intensity projection or MIP) cannot necessarily distinguish bright vessels from other tissues with short T_1 relaxation times (e.g. fat, subacute hemorrhage), the vessels in the volume may be obscured by adjacent tissues. The MIP reconstruction technique may introduce additional artifacts in all MRA images. Small vessels may frequently be visible in the original slices but are not represented in the MRA image because of: (1) variation in background intensities; (2) marginal flow-related enhancement; and (3) partial volume averaging artifacts.[5]

Neither the PC nor the TOF technique can provide the contrast and spatial resolution needed to identify very subtle contour abnormalities which would be important in confirming mild vascular dysplasias (e.g. fibromuscular disease) or vasculitis that would only be evident using conventional arteriography. These limitations are predominantly related to signal-to-noise ratio (S/N) issues and the hardware (e.g. gradients, computer memory, processing speed) constraints imposed on the MRA pulse sequences. The two techniques are also unable to provide the same dynamic information as with catheter angiography.

1.2   Carotid MRA: Clinical Significance

MRA has the potential to make its greatest impact on routine clinical work in the evaluation of carotid bifurcation atherosclerotic disease. Justification for imaging the carotid bifurcations is based on the conclusions of the North American Symptomatic Carotid Endarterectomy Trial (NASCET) and the European Carotid Surgery Trial (ECST). These controlled randomized clinical studies support the clinical utility of carotid endarterectomies in symptomatic patients with severe (>70%) carotid bifurcation stenosis, and set well defined standards for the preoperative evaluation of these patients (angiographic demonstration, specific measurements of the stenosis, significant tandem lesions).[6–10]

Ideally, it would be preferable to replace the conventional catheter arteriogram with a noninvasive alternative, thereby avoiding the potential risks of the invasive study. As many of

For References see p. 1001

the symptomatic patients with cerebrovascular disease are evaluated by spin echo MRI to assess the severity of the parenchymal sequelae, MRA of the carotid and intracranial vasculature would be a logical (noninvasive) extension of such studies. However, this makes the assumption that MRA can provide the same information as the invasive study and has similar sensitivity and specificity. The questions that remain to be proven in large scale, prospective clinical trials include the following: (1) can MRA characterize carotid stenoses of 70–99% severity, (2) distinguish complete occlusion from severe stenosis, and (3) exclude tandem stenoses as accurately as catheter angiography? Among these questions, the issue with the least objective data is the ability of MRA to exclude the presence of tandem stenoses, most commonly in the carotid siphon. The presence of distal stenoses have been associated with an increased risk for cerebrovascular and cardiac complications, suggesting a subpopulation with more malignant atherosclerosis.[11,12] Moreover, although 3D TOF and PC MRA continue to improve, there are frequently artifacts mimicking stenoses of the carotid siphons due to complex flow and/or local field inhomogeneities.[13,14] In the evaluation of carotid bifurcation disease, MRA is currently most appropriately assigned the role of a screening study for significant atherosclerotic disease which would be best followed up by conventional angiography, preoperatively.

1.3   Carotid MRA: Technical Considerations

In 2D TOF imaging, the carotid arteries are visualized by acquiring a series of thin 2D axial gradient echo images sequentially (superior to inferior) against the direction of flow. The high signal from the venous structures (particularly the jugular vein) can obscure the carotid bifurcation, but a traveling saturation pulse is placed superior to each excitation slice to eliminate the signal from the caudally directed venous flow.[15] The motion-induced dephasing (intravascular signal loss) is minimized through a combination of constant velocity flow compensation gradients along the slice-select and frequency-encoding directions, the shortest possible echo times (8.0–9.0 ms with conventional gradients), and the smallest possible voxels (1.5–3.0 mm slice thickness and 1.0 mm in-plane resolution).

The major advantage of 2D TOF sequences relate to their sensitivity to slow flow as the spins must only move a short distance (1.5–3.0 mm 2D slice) within each *TR* interval to ensure high intravascular signal in each slice. Consequently, these sequences are readily able to distinguish a severe stenosis from an occlusion. A disadvantage of the 2D technique is the stair-step misregistration artifact or vessel discontinuities in the final MRA arteriogram images that arise due to gross patient motion during the acquisition of the individual slices. More importantly, the 2D sequences demand high gradient amplitudes to define the thin 2D slices, which, in turn, place significant restrictions on the minimum echo time and in-plane voxel dimensions. The high gradient amplitudes accentuate any spin dephasing due to motion which is not corrected by the flow compensating gradients. As a result, severe stenoses are often seen as 'flow voids' or vascular interruptions. In spite of this, 2D TOF MRA has been demonstrated to be a highly sensitive and very acceptable screening study.[16]

3D TOF sequences incorporate the same strategies for vessel visualization as their 2D counterparts, but the rf pulse excites a thick slab of tissue during each *TR* interval and the thin slices within the volume are defined by a second phase-encoding gradient along the slice-select direction. The examination time increases in proportion to the number of phase-encoding steps in the second direction (number of slices), but there is also a proportional increase in the S/N when contrasted to comparable size voxels in 2D sequences. As a result of the reduced gradient demands, the 3D sequences permit shorter echo times (4.5–7.0 ms) and higher spatial resolution (<1.0 mm^3) with comparable refocusing pulses when contrasted to the 2D TOF sequences. These factors combine to minimize the degree of intravoxel phase dispersion and signal loss within the vessel.[17] The phase errors increase quadratically with time in the setting of higher order motion terms so intravascular signal loss is especially likely within and immediately distal to a stenosis, within a focal tortuosity, and in some normal vessel bifurcations. Since the greatest amount of dephasing occurs during the application of the gradients, the phase errors are most effectively suppressed by minimizing the duration, amplitude, and timing of the gradients.[18] Additional flow compensation gradients can be incorporated to correct for the higher order motion terms but only at the expense of significantly stronger and longer gradients before the echo. The signal loss with the longer gradients more than offsets the minimal improvement derived from the supplementary motion compensation. Because these sequences are all gradient echo acquisitions, the intravascular signal is also reduced by T_2^* phenomena. This effect is similarly reduced with small voxels and short echo times.[19]

The rapid unidirectional flow in the carotid arteries is particularly well suited to a 3D TOF acquisition. This technique also has its limitations which primarily relate to the saturation of flowing spins (e.g. slow antegrade flow distal to a severe stenosis). Spatially variable rf pulses (lower flip angle as the spins enter the volume and higher flip angles distally in the volume) have proven particularly helpful in this regard. The vessel/soft tissue contrast is somewhat reduced at the caudal end of the volume, but the contrast increases superiorly and the spins can flow further into the volume before becoming saturated without increasing the *TR*. Higher flip angles at the cranial end also increase the saturation of the stationary tissues for a given *TR*. The advantages of this rf pulse are particularly noticeable with slow flow such as in an elderly patient with poor cardiac output, a very small vessel lumen, or distal to a severe stenosis.

Failure to adequately suppress the background stationary tissues is a limitation of any TOF technique, but particularly the 3D technique in view of the lower signal of the blood (relative to 2D TOF). Excitation with higher flip angles improves the flow-related enhancement but accentuates the problem of spin saturation in 3D acquisitions. Shorter repetition times enhance the background suppression while causing more saturation of the flowing spins. In the neck, fat is the most problematic, as its short T_1 relaxation time and high spin density causes it to appear bright on the final MRA images. Since there tends to be a large amount of perivascular and subcutaneous fat in older patients (the group most likely to be studied for carotid atherosclerotic disease), the fat can largely obscure the underlying vessels. Reconstructing a subvolume that is limited to the region immediately surrounding the vessels excludes much of the stationary tissue from the reconstruction but does not eliminate the problem. Incorporating additional fat suppression pulses into these sequences will reduce the signal of the fat without affecting the vessel signal and can significantly improve the vessel/soft tissue contrast-to-noise level.[20] In practice, these pulses can be unreliable in clinical studies since local field inhomogeneities in the neck may cause inhomogeneous fat suppression or even water (blood) suppression.

In light of the above arguments, a most promising clinical method is the stacked overlapping 3D volume technique, which combines many of the advantages of the 2D and 3D studies.[21] Relatively small volumes minimize the problem of spin saturation while the lower gradient demands of the 3D sequences allow for very short echo times and small voxels. Moreover, the stacked volume approach can be implemented with higher flip angles and/or shorter TR values than are possible with a large single 3D volume so the vessel/soft tissue contrast is enhanced. If the stacked volumes incorporate the specialized rf pulses as discussed above, intermediate sized volumes can be used to improve the postprocessing, increase the anatomical coverage, and reduce the overall examination time.

1.4 Carotid MRA: Clinical Applications—Carotid Bifurcation

A number of studies have evaluated the clinical efficacy of 2D TOF MRA for imaging the carotid bifurcations, the largest of which was a study by Laster et al. in which the images from 101 patients were compared with conventional catheter angiography and, in some cases, Doppler ultrasound.[22,23] As in any MRA technique, the degree of lumenal narrowing tends to be accentuated compared with the conventional arteriogram images. Because of the limitations in echo time and voxel size with the 2D technique, the dephasing caused by higher order motion terms induces more prominent intravascular signal loss and therefore greater apparent narrowing of the lumen than with the 3D technique. In the case of a severe stenosis, there is frequently a flow void or complete absence of signal in the involved segment so the morphology of the stenosis cannot be defined at all. The reader simply concludes that the vessel is severely stenotic because there is complete loss of signal in that segment with reappearance of the vessel distally. Whereas the MRA images agreed with the catheter studies in up to 75% of cases, overestimation of the stenosis was seen in up to 44%.[24,25] As expected, the highest percentage of these errors was evident in those studies with relatively long echo times and large voxel dimensions. The most significant error, however, was seen in those cases of moderate stenoses which were labeled as 'severe' on MRA. This was a problem even with a slice thickness of 1.5 mm and echo times of approximately 9.0 ms.[24] If MRA is used as the sole imaging study preoperatively, a carotid endarterectomy may be recommended inappropriately. It can be argued that Doppler ultrasound could be performed in conjunction with the MRA, but the ultrasound also has potential pitfalls such as operator dependence, machine variability, vessel tortuosity and plaque calcification leading to nondiagnostic studies, overestimation of stenoses, and (although less commonly with color flow systems) misdiagnoses of complete occlusion at the bifurcation.

Although it is difficult to depict the morphology of severe stenoses, the sensitivity of the 2D TOF technique to slow flow generally makes it readily able to distinguish a severe stenosis from an occlusion if the slice thickness is sufficiently small, the TR is sufficiently long, and a long enough segment of the vessel is evaluated to include a segment in which the flow has returned to a laminar profile.[22] Nevertheless, in a few cases, MRA overestimates a critical stenosis as an occlusion which has important clinical implications as the patient is no longer considered a surgical candidate if the vessel is occluded.

The 2D TOF studies have also been compared to Doppler ultrasound examinations of the carotid bifurcations. The only noticeable difference was that the MRA studies tended to be less likely to misdiagnose a severe stenosis as an occlusion, but there was ultimately no statistically significant difference in accuracy between the two modalities.[24,26] Both studies tended to overestimate the degree of carotid stenosis. Even when the results of the two examinations are in agreement, they still do not necessarily correlate with the invasive arteriogram and may not have the accuracy to replace the conventional study.[24–26]

Other groups have primarily relied upon 3D TOF imaging in patient studies because of a lesser degree of flow-related dephasing and, consequently, improved characterization of carotid stenoses. When single-volume 3D TOF MRA was compared to intra-arterial digital subtraction angiography (IADSA) using a caliper technique, ROC analysis showed that technically adequate 3D TOF MRA images can be interpreted consistently, show a very strong correlation with IADSA, and that 3D TOF MRA may be a sensitive screening examination for extracranial carotid stenoses.[27] In spite of the advantages of the 3D sequence parameters, the more severe stenoses may still be accentuated due to higher order motion in the region of the stenosis. As alluded to above, another obvious limitation of the 3D technique is its potential to misrepresent a severe stenosis as an occlusion so 2D and 3D TOF carotid MRA investigations have been combined in patient examinations to avoid this pitfall.[28]

Even when the 2D and 3D TOF studies are combined, it seems unlikely at this time that MRA, duplex ultrasound, or the combination of the two can replace conventional angiography in patients who are being considered for surgery. It would be more appropriate for these modalities to serve as a screening examination, and, if a significant stenosis is detected, an invasive diagnostic study can be performed for further evaluation prior to surgical intervention. Based on the guidelines set by the NASCET study, it is now not only necessary to measure accurately the lumenal narrowing but it is also necessary to exclude the presence of a tandem stenosis elsewhere in the carotid circulation.[6] Moreover, it is important to measure the stenosis in a consistent fashion and to have the capability to distinguish different gradations of severe lumenal compromise since the NASCET study demonstrated a greater benefit of the endarterectomies with increasing severity of the stenoses. While the 3D TOF technique characterizes the morphology of the stenotic segment more accurately than 2D studies, both tend to exaggerate the degree of narrowing and may show complete signal void within the stenosis (particularly in the 2D studies) (Figure 1).

Further improvements should be realized with improved gradient capabilities which will permit shorter echo times,

For References see p. 1001

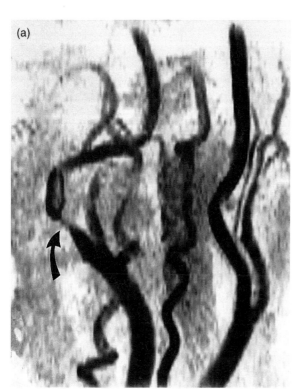

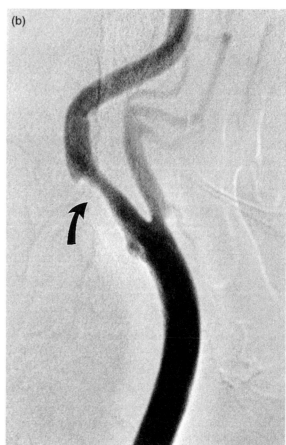

Figure 1 (a) Stacked overlapping volume acquisition of the carotid bifurcations shows no significant saturation despite the tortuosity of the vessels and the proximal stenosis of the internal carotid artery. (b) The degree of stenosis (arrow) is less severe in the conventional arteriogram, likely due to the long gradient duration for the MRA acquisition

higher resolution, and reduced intravoxel dephasing at the site of severe stenoses.[18,29]

1.5 Carotid MRA: Clinical Applications—Carotid Dissection

Dissection of the carotid artery is not an infrequent cause of stroke but tends to occur in younger adults than strokes relating to atherosclerotic disease. If the clinical suspicion is low, a carotid dissection is only one possible explanation for the patient's clinical presentation, and/or there is a relative contraindication to conventional angiography. MRI in conjunction with MRA serves as a noninvasive alternative to evaluate for dissection and to identify simultaneously the extent of parenchymal sequelae. Catheter angiography, however, remains the gold standard, and is the most reliable means by which the underlying etiology can be identified (e.g. fibromuscular dysplasia). The conventional study may also be the only method of fully appreciating the extent of collateral filling of the intracranial circulation if the alternative routes are circuitous with slow flow in small vessels. Alternatively, it is also possible to have a false-negative catheter study if the vessel is completely occluded (nonfilling) or the dissection is subadventitial (normal size and configuration of the lumen) and the MRI/MRA study may be the best way to confirm the diagnosis by demonstrating

the perilumenal blood products.[30] In routine T_1- and T_2-weighted spin echo imaging the dissection is generally detected as a crescentic area of hyperintensity around a narrowed lumen of the cervical internal carotid artery.[31] In particular, an axial T_1-weighted spin echo study using fat saturation and an inferior spatial saturation pulse suppresses the signal of the perivascular fat and intraluminal flow-related enhancement, thereby increasing the conspicuity of the hyperintense (short T_1) periluminal hemorrhage.[32]

MRA is used to complement conventional spin echo imaging in the evaluation of patients with suspected carotid dissections.[33] As in IADSA studies, the internal carotid artery may appear narrowed and there may be poor flow-related enhancement proximally if the outflow through the internal carotid artery is severely impaired. The dissection is often imaged when the periluminal hemorrhage contains methemoglobin, so the vessel lumen can appear artifactually widened on the MRA since the hyperintense thrombus is similar in intensity to the flow-related enhancement in the true lumen. Because of the variability in the state of evolution of the thrombus, the hemorrhagic by-products are either hyperintense or hypointense relative to the lumen. If the distinction between lumen and hemorrhage is not apparent on the projection arteriogram images, the two lumena and the intervening flap are usually evident in the source images for the MRA.

For list of General Abbreviations see end-papers

1.6 Intracranial MRA: Technical Considerations

The intracranial circulation presents problems similar to those seen in the evaluation of the extracranial cerebral vasculature except the studies are further complicated by smaller vessels, field inhomogeneities, and reduced through-plane flow. The 3D sequences seem particularly well suited to this situation as they not only provide the high spatial resolution needed to visualize the smaller vessels but they also provide the reconstruction capabilities (maximum intensity projection) necessary to display the complex intracranial vascular geometry. Intracranially, local field inhomogeneities exist at air–soft tissue and bone–soft tissue interfaces while higher order motion terms occur at focal vessel tortuosities, bifurcations, and lumenal stenoses. Both of those factors are limited by small voxels and short echo times. In comparison to the 2D sequences, the 3D acquisitions are better able to demonstrate the large component of in-plane flow intracranially (with appropriate adjustments in the *TR* and flip angle to avoid saturation).

The reduced flow velocities and large component of in-plane flow are still somewhat limiting in the 3D acquisitions as these factors aggravate the problem of spin saturation, thereby reducing the vessel–soft tissue contrast particularly in the distal volume. Adjustments in the conventional imaging parameters provide only limited benefit. The incorporation of magnetization transfer saturation in all intracranial TOF MRA studies has become routine.[34] Since the off-resonance saturation pulse suppresses the signal of the bound water molecules, and the bound component is largely restricted to the brain parenchyma, the magnetization transfer saturation effectively enhances the vessel–soft tissue contrast. Additional vascular contrast may be obtained by incorporating the spatially varying rf pulses for excitation designed to limit the problem of spin saturation. Because of the slower flow rates in the intracranial arteries, the most promising TOF technique seems to be the stacked volume approach, which combines the advantages of 2D and 3D TOF acquisitions.[21]

Phase contrast sequences are also commonly used to evaluate the intracranial circulation. The vessel geometry and small size of the arteries make the 3D acquisition necessary for most purposes but, for comparable spatial resolution and flow sensitization along all three planes, the acquisition time is longer and the postprocessing more complex for the 3D PC technique. Alternatively, 2D acquisitions may provide similar flow information relatively quickly, but at the expense of limited anatomical coverage. A problem common to both phase contrast techniques is the need for some a priori knowledge of the range of velocities present within a given patient or lesion. Inappropriate choice of the velocity-encoding gradient can cause poor visualization of the small and/or peripheral vessels with slow flow if the chosen velocity-encoding gradient is too weak, or aliasing of flow information and signal loss if the chosen velocity range is too low. On the other hand, the freedom to choose different velocity encodings provides options that are not possible or more cumbersome in a TOF acquisition such as the direction of blood flow (e.g. collateral flow in the circle of Willis).[35,36] Alternatively, directional information is easily attainable in the PC acquisition without performing a separate study. The PC studies also readily provide the capability to measure flow velocities and volume flow rates of individual vessels. Lastly, if 2D PC studies are performed in a cardiac gated cine mode, the flow dynamics in various vascular lesions can be studied at different, specified velocity ranges which is only possible in a very limited fashion with TOF techniques.

1.6.1 Intracranial MRA: Aneurysms

The statistics relative to subarachnoid hemorrhage from a ruptured intracranial aneurysm remain devastating—greater than 50% mortality and major morbidity.[37] Given the low operative morbidity and mortality of surgery on unruptured aneurysms, it is appealing to consider the possibility of using MRA as a screening test. It would be especially applicable in patient groups where there is an increased incidence or a high anxiety level for an aneurysm, such as patients with a strong family history, polycystic kidney disease, aortic coarctation, fibromuscular disease, and collagen vascular disease. Based on the decision analysis of Levy et al., the risk of an aneurysm even in these groups does not warrant the screening of all these patients with conventional arteriography.[38] It is possible, however, to perform an accurate noninvasive pretest that would increase the prevalence in subgroups which could then be studied by conventional arteriography for further evaluation. Based on the results of MRA in the evaluation of intracranial aneurysms, MRA exceeds the criteria of sensitivity and specificity for the noninvasive pretest outlined by Levy et al.[37,39] It is possible to use high resolution contrast enhanced computerized tomography (CT) to study these patients, but the sensitivity and specificity have not been studied, and the requirement of intravenous iodinated contrast and the beam hardening artifact from the skull base are potentially limiting, which make MRA preferable.[40] Currently, TOF MRA is frequently used as a screening test in these groups or in stable patients with a subacute onset of symptoms in which an intracranial aneurysm is only one of a number of potential explanations for the clinical presentation and the suspicion of an aneurysm is not high enough to warrant a conventional arteriogram (Figure 2). However, the patients who present acutely with a subarachnoid hemorrhage are best served by a nonenhanced head CT followed by conventional catheter angiography.

Retrospective studies have been conducted to test the accuracy of the 3D TOF MRA method for the detection of intracranial aneurysms as compared to IADSA examinations.[39,41,42] Aneurysms as small as 3–4 mm have been detected using 3D TOF and 3D PC techniques. In a series of 21 angiographically confirmed aneurysms in 19 patients, the authors demonstrated an increase in sensitivity from 67% when evaluating the 3D TOF MRA projection images alone, to 86% when the MRA arteriogram images were evaluated along with the source images from the 3D data set and the spin echo images. Using the MIP images, the individual slices, and the spin echo study, the sensitivity was calculated to be 95% for the detection of at least one aneurysm in a patient such that the study would lead to the referral for a conventional arteriogram for further evaluation preoperatively.

Recognizing the limitations of these MRA acquisitions becomes particularly important when choosing the method of postprocessing, interpreting the images, and recommending a conventional arteriogram as a follow-up study or in place of MRA. Larger aneurysms or aneurysms compressing the parent vessel may reduce the extent of inflow and washout of fresh

For References see p. 1001

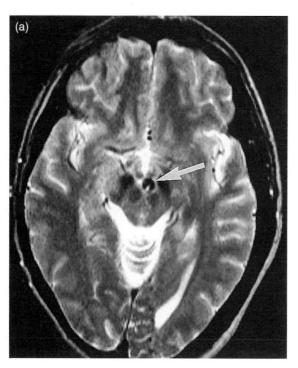

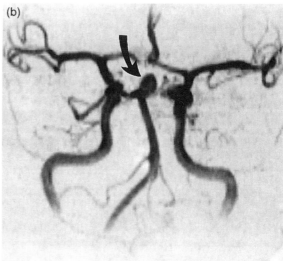

Figure 2 (a) This 55-year-old woman is known to have a basilar tip aneurysm and is presenting for noninvasive follow-up. The aneurysm is seen as a rounded hypointense focus (arrow) in the interpeduncular cistern. (b) The 3D TOF MRA image confirms the aneurysms (arrow), including its size and orientation with respect to the parent vessel

unsaturated spins within the aneurysm lumen, which causes an apparent reduction of the size of the aneurysm on TOF studies.[41] This is less of a problem with a PC acquisition if the preliminary spin echo study reveals a giant intracranial aneurysm, so that a low velocity encoding range can be selected for PC MRA.[42] On the other hand, these lesions are also obvious on the spin echo study owing to the prominent flow artifacts, intimate relationship with the adjacent basilar vessels, and (if present) the contained blood products.

The stagnant flow within a large aneurysm tends to promote thrombus formation which partially fills the aneurysm lumen. Since the TOF sequences are inherently T_1 and spin density weighted, both the thrombus and the patent lumen may be visible on the MIP images, depending on the state of evolution of the thrombus just as in conventional IADSA. This will cause an underestimation of the overall aneurysm size in PC studies because of the strong background suppression. In the case of subarachnoid hemorrhage, the aneurysm and the parent vessel may not be observed altogether.

Although the gross relationships with the parent vessel may be visible, identification of the aneurysm neck may be difficult at times when evaluating the projection images and occasionally with conventional cerebral angiography. In these cases, evaluation of the 3D TOF source images and multiplanar reconstructions of the original slices create 'angiotomograms' to delineate more clearly both the aneurysm and its point of origin. Alternatively, reconstructing vessels from small subvolumes that contain only the vessels of interest provides images with better definition (reduces the artifact inherent to the MIP algorithm) and eliminates the problems relating to vessel overlap.[5] This is especially helpful with aneurysms arising from a tortuous carotid siphon or in determining which branch the aneurysm arises from if it occurs at a bifurcation (e.g. middle cerebral artery bifurcation/trifurcation). More importantly, routine evaluation of the source images and planar reconstructions of the axial slices reduces the likelihood of missing small aneurysms that may otherwise be missed by reviewing only the projection MRA images.

Identifying the relationship of the aneurysm to the parent vessel may also be difficult if the aneurysm arises from an arterial segment that is incompletely visualized due to motion induced dephasing. This is most commonly seen in the carotid siphon, particularly if the supraclinoid segment is immediately apposed to the juxtasellar segment so the turn of the siphon is rather tight. Because of a greater tendency towards motion induced dephasing, this should be more problematic with PC than TOF acquisitions. Shorter gradient times and smaller voxels significantly reduce the phase dispersion and signal loss in these vessels.[19–21]

MRA can also be applied to the long term postoperative follow-up of aneurysm patients treated via an endovascular route (e.g. balloon occlusion) or noninvasive follow-up of a known, untreated aneurysm (Figure 2). Susceptibility artifacts from the nonferromagnetic aneurysm clips preclude the evaluation of aneurysms treated by surgical clipping but would not prohibit the use of MRA to evaluate for intracranial aneurysms elsewhere.

1.6.2   Intracranial MRA: Arterial Occlusive Disease

Patients with symptoms of cerebral ischemia are frequently examined by MRI to assess the extent of parenchymal ischemic disease. The morphology and signal characteristics of the vessel lumena can only suggest stenosis and slow flow or occlusion, whereas MRA can directly identify the site and severity of the underlying stenosis. Not infrequently, the patient has already had a Doppler ultrasound study of the carotid bifurcations, and the ischemic symptoms are not explained by the mild narrowing in the carotid bifurcations. MRA provides an additional noninvasive study which can be performed at the same time as the spin echo study to evaluate the morphology of the intracranial vessels. Alternatively, in a patient with severe stenosis, MRA provides the means by which the physician can evaluate the degree of collateral flow to the circulation that is severely compromised in the extracranial circulation. As compared with transcranial Doppler ultrasound studies, MRA is better able to evaluate the posterior fossa circulation and provides an anatomical display of the intracranial vasculature that is not possible with the transcranial Doppler technique.[43,44] The MRA findings can alert the clinician to the nature and location of the occlusive disease and a conventional angiographic study could then be performed to confirm the full extent of pathology prior to starting therapy if the MR/MRA findings do not explain the clinical situation or if surgical or endovascular therapy is being considered. Alternatively, a negative study might avert the need for an invasive diagnostic examination altogether. The hope for the future is that MRA can effectively evaluate for tandem stenoses in the distal internal carotid arteries in patients being considered for carotid endarterectomies and, in combination with a similar noninvasive study of the carotid bifurcations, can obviate the need for an invasive study preoperatively (i.e. noninvasively meet the NASCET criteria).

Initial studies have been restricted to the evaluation of the larger vessels such as the carotid siphon, the proximal anterior and middle cerebral arteries, the distal vertebral arteries, and the basilar trunk. This is, in part, related to the spatial resolution limitations of these sequences and the small vessel caliber of the intracranial vasculature. Narrowing of the distal peripheral vessels, typical of cerebral vasculitis, cannot be effectively evaluated with MRA, and requires a conventional arteriogram. Flow rates also impose restrictions in the TOF acquisitions as only the larger arteries of the circle of Willis will have flow which is rapid enough that the flow-related enhancement will likely 'opacify' the vessel distal to the stenosis. This is not as problematic with the PC sequences if the physician has a priori knowledge about the expected range of flow velocities in the vessels.

3D TOF MRA is readily able to display normal vessels and to distinguish normal from stenotic vessels. In one study, the sensitivity for identifying stenoses in the distal vertebral and basilar arteries was shown to be 100% whereas the specificity of the study was not nearly as high.[44] When 3D TOF MRA was compared with the IADSA studies, the degree of stenosis was exaggerated in as many as 63% of the MRA images. When correlated with the results of other clinical studies, this error was likely accentuated to some extent by the chosen voxel dimensions, the small but highly variable caliber of the distal vertebral arteries, and the separation of the stenoses into five categories.[44] Reviewing the original 3D source images was found to be helpful for distinguishing occlusion from slow flow in a severely stenotic or hypoplastic vessel. This distinc-

tion was further aided by incorporating selected 2D TOF slices because of their improved sensitivity to slow flow.

Knowledge of the degree of collateral flow between different vascular territories may have a significant impact on the physician's therapeutic decisions. Often times, the method of cross filling is obvious on the standard TOF MRA acquisition. For example, if one internal carotid artery is occluded but there is still good flow to the corresponding anterior circulation, the flow is more likely through the patient's large anterior communicating artery than the diminutive posterior communicating arteries. It is not as easy to draw these conclusions when the potential collateral vessels are more uniform in size or they are all diminutive. The slow flow and small size of leptomeningeal collateral vessels prevents their visualization. Evaluation of the source images from PC studies clearly demonstrates the direction of flow in major vascular trunks, making the pattern of collateral flow obvious.[36] Moreover, PC studies also provide the potential to make flow velocity and flow volume measurements. Alternatively, examining the corresponding phase images from a TOF study can also confirm the direction of blood flow in main vascular trunks. The relative contributions from the other vascular territories can also be identified in a TOF study through the use of appropriately positioned saturation pulses.[35] A narrow saturation slab, which is perpendicular to a 2D TOF slice or narrow 3D TOF volume, can be oriented to eliminate the flow-related enhancement from any one or any combination of the main vascular trunks to identify the source of collateral flow to a given vascular distribution. For example, selectively saturating the contralateral internal carotid artery will demonstrate the relative contribution from the posterior circulation to the remaining internal carotid artery distribution in question.

1.6.3 Intracranial MRA: Vascular Malformations

Intracranial vascular malformations are generally easy to identify on the routine spin echo study of the head. Capillary telangiectasias are usually discovered incidentally at autopsy while cavernous angiomas are most commonly detected as subcortical foci of mottled hyperintensity on the T_1- and T_2-weighted sequences with a peripheral rim of ferritin and hemosiderin deposition on the T_2 images from prior hemorrhage.[45] In both cases, the flow through these lesions is so slow that their abnormal vessels cannot be visualized by any MRA technique. (In fact, TOF MRA images can be misleading since the short T_1 of hemorrhage in a cavernous angioma can mimic the high flow vascular nidus of an arteriovenous malformation.) Venous angiomas are also readily identified on the spin echo study as a caput of veins draining into a large medullary vein extending from a point near the ependymal surface to the cortex. While the sensitivity to slow flow in the 2D TOF or PC studies would make these veins visible with MRA, their characteristic spin echo appearance makes this unnecessary.

High flow arteriovenous malformations are also easily diagnosed on the conventional spin echo parenchymal study. The vascular nidus appears as a cluster of serpiginous flow voids because of the relatively fast flow through these vessels. The factors that are considered important predictors of surgical resectability are also readily appreciated, including the size of the nidus, its relationship to eloquent areas of the brain, and the presence of deep venous drainage.[46] If there is a direct

For References see p. 1001

arteriovenous fistula, there will be no intervening nidus but the enlarged feeding arteries and draining veins will still be easily identified. There may also be associated parenchymal findings related to hemorrhage, ischemic steal, secondary venous thrombosis, or prior surgical, embolization and/or radiation therapy. The arteriovenous malformations can be identified with TOF or PC MRA, but the images generally do little more than re-inforce the 3D spatial relationships of the vascular nidus with the circle of Willis and identify the main feeding arteries and draining veins.

Conversely, one situation in which MRA is essential in the work-up is in the evaluation of suspected dural vascular malformations which make up approximately 15% of the arter-iovenous malformations. Frequently, the dural malformations are not visible at all on the spin echo study because of the small size and lack of contrast between the flow voids and the adjacent hypointense skull base.[47] The only clues to the pre-sence of the malformation may be the clinical history, the presence of enlarged draining cortical veins, or secondary find-ings such as dural venous thrombosis, subdural or parenchymal hematomas, and venous infarcts. On occasion, subtle peripheral flow voids or focal prominence of a dural venous sinus may be present, but these are often identified only in retrospect, once the MRA study makes the diagnosis obvious. MRA is often able to identify the vascular nidus along the dura, enlarged branches of the external carotid and/or vertebral arteries, and possibly enlarged draining veins or dural sinuses if there is a direct arteriovenous fistula present. It is essential to evaluate the source images from the MRA acquisition, as the findings may even be subtle on the projection MRA. Intravenous gado-linium can be used with 3D TOF MRA to evaluate better small feeding arteries and draining veins, but a study of dural fistulas failed to identify venous constriction. A study of dural fistulas with 3D TOF MRA was unable to identify venous con-striction when gadolinium was used to reduce the problems of spin saturation and better identify the slow flow in the small feeding arteries and draining veins.[48]

Ultimately, conventional catheter angiography will be per-formed on any patient with an arteriovenous malformation to characterize it prior to therapy. Only the invasive arterio-gram provides the spatial resolution necessary to identify all of the afferent and efferent vessels. While the MRA acquisition can identify aneurysms along the afferent vessels, these too should be further evaluated by a conventional arteriogram preoperatively. Lastly, only the catheter study provides the temporal resolution to assess the arteriovenous circulation time, to identify individual vessels with the greatest degree of arter-iovenous shunting and, in the case of a direct fistula, to localize the point of communication between the arterial and venous systems.

The MRA acquisitions can be designed to provide a limited degree of physiological information about the vascular malfor-mations. Thin saturation slabs can be applied to a 2D or thin volume 3D TOF acquisition to eliminate the flow-related enhancement from one vascular trunk and monitor the relative intensity and/or size of the malformation in the final image. In this way, it is possible to determine if the malformation is fed by more than one vascular distribution (collateral flow) and to assess the relative contributions from the different vascular territories.[35] The PC acquisitions can also be designed to supply similar information by applying a series of different velocity-encoding gradients in a stepwise fashion to demon-strate different components of the malformation.[35] A low velocity-encoding range (e.g. 10–20 cm s^{-1}) would demon-strate draining veins with low to medium velocities whereas higher velocity-encoding ranges (e.g. 60–100 cm s^{-1}) would preferentially demonstrate the nidus, larger veins, and feeding arteries with more rapid flow.[42] In this fashion, it is possible to provide a velocity flow map of the vascular malformation. Such information may be clinically useful if serial studies indi-cate a change in flow dynamics which may be predictive of hemorrhage, but this remains to be tested.

In spite of the advantages, saturation of flowing spins is per-haps the most significant limitation of the 3D TOF technique. This limits visualization of the small feeding arteries (slower flow) distally in the volume, all but the largest early draining veins, and potentially part of the vascular nidus itself if there is relatively slow flow through the nidus and it is located distally in the volume. The 2D TOF studies are less susceptible to spin saturation and provide better contrast-to-noise levels across the imaging volume. Alternatively, 2D TOF images are more lim-ited by flow-induced dephasing and may not visualize portions of the afferent and efferent vessels depending on the direction along which the slices are acquired relative to the course of these vessels. The problem of spin saturation is reduced with intravenous gadolinium, but this also increases the intensity of the background tissues, which may be quite limiting if the mal-formation is situated along the skull base (e.g. many dural vascular malformations). The best compromise between the TOF techniques is the stacked overlapping volume approach.[21] If the malformation is located within the parenchyma, a para-magnetic contrast medium could be added to improve venous visualization, while the stacked volume approach improves arterial visualization.[49]

1.6.4 Intracranial MRA: Venous Sinus Thrombosis/Occlusion

Dural venous sinus thrombosis and secondary parenchymal findings (e.g. edema, hemorrhagic venous infarct) can be detected on conventional spin echo images although the find-ings are not always obvious and may be misleading. The most sensitive spin echo study to evaluate the venous sinuses is a T_2 weighted sequence in which the slices are oriented perpendicu-lar to the direction of flow in the sinus in question (e.g. coronal for the superior sagittal and straight sinuses). With such a long echo time, blood flowing perpendicular to the slice normally causes a flow void as it is least apt to experience the 90 and 180° pulses and remain in the imaging plane at readout unless the sinus is thrombosed or occluded. An occluded sinus will fail to demonstrate a flow void or may be frankly hyperin-tense on the T_1- and T_2-weighted images. Slow flow may also cause an intermediate to high signal within the sinus and simu-late thrombus. In the case of acute/early subacute thrombus, the hypointense thrombus may simulate a flow void on a T_2-weighted image.[50] In the situations where the spin echo find-ings of thrombosis are ambiguous or an adjacent mass appears to have invaded and occluded the sinus, MRA venograms can be most helpful in confirming or excluding thrombosis/occlu-sion without resorting to an invasive study.

A 2D technique would be the most sensitive TOF study to evaluate the slow flow in the dural sinuses.[51] Coronal slices acquired sequentially, from posterior to anterior, should ensure flow-related enhancement in the major dural sinuses since the flow is largely in the reverse direction except the posterior portion of the superior sagittal sinus. This same strategy can be followed with fewer slices and a shorter examination time if the slices were acquired in an oblique coronal plane (e.g. rotated 75° towards the sagittal plane). In view of the relatively slow normal velocities in these veins, the slice thickness for these 2D slices should be minimized, the *TR* maintained relatively long, and a low-to-intermediate flip angle should be used (e.g. 30°) to avoid spin saturation. Hyperintense thrombus could simulate flow-related enhancement, but this is easily distinguished when correlated with the findings on the spin echo study. The prior administration of intravenous gadolinium would make it difficult to distinguish a normal sinus from contrast enhancement along the periphery of a thrombosed sinus. A more important potential error relates to the limitations of the technique. There may be insufficient flow-related enhancement in a small sinus with slow flow, leading to a false-positive study. Similarly, an oblique coronal acquisition will insure flow-related enhancement in one transverse sinus, but this is not necessarily the case in the opposite transverse sinus. In either case, evaluating the individual slices or reconstructing a small subvolume eliminates the reconstruction artifact and usually makes it possible to distinguish patency from occlusion.

Phase contrast techniques are especially well suited to this application because of strong background suppression and the ability to adjust the sequence so it is sensitive to very slow flow (10–20 cm s^{-1}).[52,53] While the sensitivity of the 2D TOF study is sufficient for most routine clinical work, the sensitivity is somewhat greater with PC, and it may be possible to detect flow in a recanalized sinus when the TOF images would suggest occlusion. The relatively large size of these sinuses, laminar flow profiles, and predominantly unidirectional flow make it easy to study these veins with two quick 2D PC acquisitions. A thick 2D slab can be selected in the midline sagittal plane, with velocity sensitization along the anterior/posterior direction, to assess the patency of the deep and superficial midline veins. A second thick axial slab in the posterior fossa with sensitization in a similar direction will evaluate the patency of the transverse and sigmoid sinuses.

2 RELATED ARTICLES

Abdominal MRA; Peripheral Vasculature MRA; Phase Contrast MRA; Time-of-Flight Method of MRA; Whole Body Magnetic Resonance Angiography.

3 REFERENCES

1. C. L. Dumoulin, *Neuroimaging Clin. N. Am.*, 1992, **2**, 657.
2. G. A. Laub and W. A. Kaiser, *J. Comput. Assist. Tomogr.*, 1988, **12**, 377.
3. C. L. Dumoulin, S. P. Souza, M. F. Walker, and W. Wagle, *Magn. Reson. Med.*, 1989, **9**, 139.
4. G. W. Lenz, E. M. Haacke, T. J. Masaryk, and G. A. Laub, *Radiology*, 1988, **166**, 875.
5. C. M. Anderson, D. Saloner, J. S. Tsuruda, L. G. Shapeero, and R. E. Lee, *Am. J. Radiol.*, 1990, **154**, 623.
6. The North American Symptomatic Carotid Endarterectomy Trial (NASCET) Steering Committee, *Stroke*, 1991, **22**, 711.
7. V. J. Howard, J. Grizzle, H. C. Diener, R. W. Hobson, M. R. Mayberg, and J. F. Toole, *Stroke*, 1992, **23**, 583.
8. European Carotid Surgery Trialists' Collaborative Group, *Lancet*, 1991, **337**, 1235.
9. Editorial, *Lancet*, 1991, **337**, 1255.
10. North American Symptomatic Carotid Endarterectomy Trial Investigators, *Stroke*, 1991, **22**, 816.
11. J. J. Schuler, P. Flanigan, L. T. Lim, T. Keifer, L. R. Williams, and A. J. Behrend, *Surgery*, 1982, **92**, 1058.
12. D. J. Marzewski, A. J. Furlan, P. St. Louis, J. R. Little, M. T. Modic, and G. Williams, *Stroke*, 1982, **13**, 821.
13. T. J. Masaryk, M. T. Modic, J. S. Ross, P. M. Ruggieri, G. A. Laub, G. W. Lenz, E. M. Haacke, W. R. Selman, M. Wiznitzer, and S. I. Harik. *Radiology*, 1989, **171**, 793.
14. J. Huston, D. A. Rufenacht, R. L. Ehman, and D. O. Wiebers, *Radiology*, 1991, **181**, 721.
15. P. J. Keller, B. P. Drayer, E. K. Fram, K. D. Williams, C. L. Dumoulin, and S. P. Souza, *Radiology*, 1989, **173**, 527.
16. J. E. Heiserman, B. P. Drayer, E. K. Fram, P. J. Keller, C. R. Bird, J. A. Hodak, and R. A. Flom, *Radiology*, 1992, **182**, 761.
17. G. W. Lenz, E. M. Haacke, T. J. Masaryk, and G. A. Lamb, *Radiology*, 1988, **166**, 875.
18. A. J. Evans, D. B. Richardson, R. Tien, J. R. MacFall, L. W. Hedlund, E. R. Heinz, O. Boyko, and H. D. Sostman, *Am. J. Neuroradiol.*, 1993, **14**, 721.
19. E. M. Haacke, J. A. Tkach, and T. B. Parrish, *Radiology*, 1989, **170**, 457.
20. J. A. Tkach, P. M. Ruggieri, J. S. Ross, M. T. Modic, J. J. Dillinger, and T. J. Masaryk, *J. Magn. Res. Imaging*, 1993, **3**, 811.
21. D. D. Blatter, D. L. Parker, S. S. Ahn, A. L. Bahr, R. O. Robison, R. B. Schwartz, F. A. Jolesz, and R. S. Boyer, *Radiology*, 1992, **183**, 379.
22. R. E. Laster Jr., J. D. Acker, H. H. Halford III, and T. C. Nauert, *Am. J. Neuroradiol.*, 1993, **14**, 681.
23. B. C. Bowen, R. M. Quencer, P. Margosian, and P. M. Pattany, *Am. J. Roentgenol.*, 1994, **162**, 9.
24. R. L. Mittl, M. Broderick, J. P. Carpenter, H. I. Goldberg, J. Listerud, M. M. Mishkin, H. D. Berkowitz, and S. W. Atlas, *Stroke*, 1994, **25**, 4.
25. T. S. Riles, E. M. Eidelman, A. W. Litt, R. S. Pinto, F. Oldford, and G. W. S. Schwartzenberg, *Stroke*, 1992, **23**, 341.
26. J. Huston III, B. D. Lewis, D. O. Wiebers, F. B. Meyer, S. J. Riederer, and A. L. Weaver, *Radiology*, 1993, **186**, 339.
27. T. J. Masaryk, J. S. Ross, M. T. Modic, G. W. Lenz, and E. M. Haacke, *Radiology*, 1988, **166**, 461.
28. C. M. Anderson, D. Saloner, R. E. Lee, V. J. Griswold, L. G. Shapeero, J. H. Rapp, S. Nagarkar, X. Pan, and G. A. Gooding, *Am. J. Neuroradiol.*, 1992, **13**, 989.
29. J. A. Tkach, P. M. Ruggieri, J. J. Dillinger, J. S. Ross, M. T. Modic, and T. J. Masaryk, *J. Magn. Reson. Imaging*, 1993, **3**, 365.
30. M. Assaf, P. J. Sweeney, G. Kosmorsky, and T. J. Masaryk, *Can. J. Neurol. Sci.*, 1993, **20**, 62.
31. H. I. Goldberg, R. I. Grossman, J. M. Gomori, A. K. Asbury, L. T. Bilaniuk, and R. A. Zimmerman, *Radiology*, 1986, **158**, 157.
32. R. Pacini, J. Simon, L. Ketonen, D. Kido, and K. Kieburtz, *Am. J. Neuroradiol.*, 1991, **12**, 360.
33. C. Levy, J. P. Laissy, V. Raveau, P. Amarenco, V. Servois, M. G. Bousser, and J. M. Tubiana, *Radiology*, 1994, **190**, 97.
34. R. R. Edelman S. S. Ahn, D. Chien, W. Li, A. Goldmann, M. Mantello, J. Kramer, and J. Kleefield, *Radiology*, 1992, **184**, 395.
35. R. R. Edelman, H. P. Mattle, G. V. O'Reilly, K. U. Wentz, C. Liu, and B. Zhao, *Stroke*, 1990, **21**, 56.

For References see p. 1001

36. P. Turski and F. Korosec. *Neuroimaging Clin. N. Am.*, 1992, **2**, 785.
37. T. J. Ingall, J. P. Whisnant, D. O. Wiebers, and W. M. O'Fallon, *Stroke*, 1989, **20**, 718.
38. A. S. Levey, S. G. Pauker, and J. P. Kassirer, *N. Engl. J. Med.*, 1983, **308**, 986.
39. J. S. Ross, T. J. Masaryk, M. T. Modic, P. M. Ruggieri, E. M. Haacke, and W. R. Selman, *Am. J. Neuroradiol.*, 1990, **11**, 449.
40. A. B. Chapman, D. Rubinstein, R. Hughes, J. C. Stears, M. P. Earnest, A. M. Johnson, P. A. Gabow, and W. D. Kaehny. *N. Engl. J. Med.*, 1992, **327**, 916.
41. R. J. Sevick, J. S. Tsuruda, and P. Schmalbrock, *J. Comput. Assist. Tomogr.*, 1990, **14**, 874.
42. J. Huston, D. A. Rufenacht, R. L. Ehman, and D. O. Wiebers, *Radiology*, 1991, **181**, 721.
43. J. E. Heiserman, B. P. Drayer, P. J. Keller, and E. K. Fram, *Radiology*, 1992, **185**, 667.
44. K. U. Wentz, J. Rother, A. Schwartz, H. P. Mattle, R. Suchalla and R. R. Edelman, *Radiology*, 1994, **190**, 105.
45. S. Atlas, *Radiol. Clin. N. Am.*, 1988, **26**, 821.
46. R. F. Spetzler, and N. A. Martin, *J. Neurosurg.*, 1986, **65**, 476.
47. J. K. DeMarco, W. P. Dillon, V. V. Halbach, and J. S. Tsuruda. *Radiology*, 1990, **175**, 193.
48. J. Chen, J. S. Tsuruda, and V. V. Halbach, *Radiology*, 1992, **183**, 265.
49. G. Marchal, H. Bosmans, L. Van Fraeyenhoven, G. Wilms, P. Van Hecke, C. Plets, and A. L. Baert, *Radiology*, 1990, **175**, 443.
50. G. Sze, B. Simmons, G. Krol, R. Walker, R. D. Zimmerman, and M. D. Deck, *Am. J. Neuroradiol.*, 1988, **9**, 679.
51. H. P. Mattle, K. U. Wentz, R. R. Edelman, B. Wallner, J. P. Finn, P. Barnes, D. J. Atkinson, J. Kleefield, and H. M. Hoogewoud, *Radiology*, 1991, **178**, 453.
52. D. J. Rippe, O. B. Boyko, C. E. Spritzer, W. J. Meisler, C. L. Dumoulin, S. P. Souza, and E. R. Heinz, *Am. J. Neuroradiol.*, 1990, **11**, 199.
53. J. S. Tsuruda, A. Shimakawa, N. J. Pelc, and D. Saloner, *Am. J. Neuroradiol.*, 1991, **12**, 481.

Biographical Sketches

Paul M. Ruggieri. *b* 1956. B.S., 1978, M.S., 1979, Bucknell University, Lewisburg, PA, USA. M.D., 1984, UMDNJ-Rutgers Medical School, Piscataway, NJ, USA. Radiology residency at University Hospitals of Cleveland, Neuroradiology fellowship at Cleveland Clinic Foundation including pediatric neuroradiology at Cincinnati Children's Hospital, research fellowship with Siemens Medical Engineering Group in Erlangen, Germany (supervisor Gerhard Laub). 1990–present, Associate Staff, Sections of Neuroradiology and Pediatric Radiology, Cleveland Clinic Foundation, Cleveland, OH, USA. Approx. 25 publications. Current specialties: neuroradiology, pediatric neuroradiology, MRI flow imaging.

Jean A. Tkach. *b* 1961. B.S.E., 1982, M.S., 1985, Ph.D., 1988 (Biomedical Engineering), Case Western Reserve University, Cleveland, Ohio, USA. 1990–present, Head of MR Imaging Research at the Cleveland Clinic Foundation, Cleveland, Ohio, USA. Approx. 30 publications. Current research specialties: MRA, functional MRI, cardiac MRI, flow quantification, magnetization transfer saturation MRI and steady state MRI techniques.

Thomas J. Masaryk. *b* 1955. B.S., 1978, M.D., 1981, Medical College of Ohio, Toledo, Ohio, USA. Radiology residency at the Cleveland Clinic Foundation, Neuroradiology fellowship at the Cleveland Clinic Foundation. 1990–present, Head of Neuroradiology, the Cleveland Clinic Foundation, Cleveland, Ohio, USA. Approx. 92 publications. Current specialties: neuroradiology, interventional neuroradiology, MRI flow imaging.

For list of General Abbreviations see end-papers

Central Nervous System Degenerative Disease Observed by MRI

Frank J. Lexa

Leonard Davis Institute of Health Economics, University of Pennsylvania, Philadelphia, PA, USA

1 INTRODUCTION

The degenerative diseases encompass some of the most interesting and important pathological entities in clinical medicine. In the past decade, MR imaging (MRI) has emerged as the premier radiological method for examining the brain and spinal cord. MRI has many advantages over computerized tomography (CT) for the detection of subtle changes in white and gray matter structures. This has led to the ability finally to detect and characterize some of these diseases early in their course without resorting to invasive techniques. This section will discuss the use of MRI to detect and characterize neural degeneration, beginning with the general case of the processes of Wallerian degeneration, which are common to many types of brain injury, and then move on to discuss the salient clinical and imaging features of the most important and best characterized of the neurodegenerative disorders. In the interest of brevity, only the most important clinical and imaging facets of the diseases will be covered. The reader is referred to two standard texts which are able to provide a more comprehensive discussion and bibliography.[1,2]

2 WALLERIAN DEGENERATION: THE NEURONAL REACTION TO INJURY

Waller is credited for making the first histological description of the series of events which occur in the distal axonal segment following a proximal injury.[3–5] This can occur after a wide variety of insults to the brain or spinal cord and thus represents an important marker of significant central nervous system (CNS) damage. The earliest phase is cessation of axonal transport, and collapse of the axonal segment.[6] The myelin begins to fragment, and the Schmidt–Lanterman incisures open up. This is followed by reaction of the surrounding tissue with cellular proliferation and edema, during which the myelin is chemically degraded and removed. This eventually leads to formation of a gliotic scar.

Until the advent of MRI, it was difficult or impossible to detect degeneration of axonal pathways using antemortem imaging techniques. CT is usually only capable of detecting significant volume loss in end-stage cases. Early reports of the MRI appearance of Wallerian degeneration described T_1 and T_2 prolongation along corticofugal pathways in patients with acquired lesions such as strokes and Schilder's disease.[7–13] In these reports, the lesions were at least several months old when

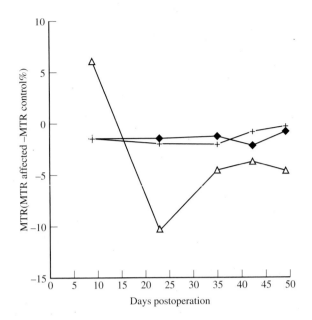

Figure 1 Demonstration of changes in magnetization transfer in a feline model of Wallerian degeneration at 1.5 T. The *x* axis shows the number of days postplacement of a cortical lesion. The *y* axis shows the change in the magnetization transfer ratio (MTR) of the affected side relative to a control in three structures: the degenerating superolateral white matter (△), remote white matter (◆), and the ipsilateral lateral geniculate nucleus (+). The ipsilateral white matter tracts show an early rise and later fall in magnetization transfer as early degeneration progresses. (Reproduced by permission of The American Society of Neuroradiology from Lexa et al.[23])

they had these signal characteristics, although these changes have been reported to be observable earlier.[14] Observations from the first month were reported in which no abnormality could be detected.[15] High signal intensity in the dentatorubral–thalamic tract has been reported after dentate nucleus resection.[16]

Other reports described a more complex pattern with no detectable signal changes in approximately the first month, a transient period of hypointensity on long *TR* images at approximately 1 month, followed by normalization of that hypointensity and later development of the more familiar T_1 and T_2 prolongation effects at approximately the 3–4 month mark.[17–18] Recent work suggests that fast spin echo (FSE) imaging shows this early hypointense phase of Wallerian degeneration more conspicuously than conventional proton density imaging.[19] In clinical medicine, this has important implications for the diagnosis and treatment of brain injury.

Animal models have been used to address the controversies raised by the above studies. Magnetic resonance spectroscopy of peripheral nerves showed significant increases in T_1 and T_2 relaxation times 15 days after sectioning the sciatic nerve.[20] Imaging of the tibial nerve confirmed that the T_2 signal intensity was elevated on long *TR/TE* images at day 15 in a crush injury model.[21] Histological confirmation of Wallerian degeneration was obtained in a feline model of radiation injury. These areas showed high signal intensity on long *TR/TE* images over 200 days after radiation injury.[22]

Newer techniques utilizing magnetization transfer techniques appear very promising for the early detection and separation of some of these changes. Using a cortical ablation model, Lexa et al. were able to demonstrate changes as early as the first week after injury (Figure 1).[23] Magnetization transfer images detected degenerating tracts at a distance from the primary injury site before conventional spin echo images or even routine light microscopy showed significant evidence of injury. Electron microscopy confirmed that the first phase of Wallerian degeneration was underway (Figure 2). Myelin degeneration and cellular degeneration were not detectable by light-microscopic techniques until significantly later. Moreover, the biphasic nature of the early changes in magnetization transfer suggests that it may be possible to separate some of the processes occurring in early axonal injury with magnetization transfer (Figure 3).

3 DEGENERATIVE DEMENTING ILLNESSES: ALZHEIMER'S AND PICK'S DISEASES

Alzheimer's disease is both the commonest cause of dementing illness as well as one of the most common causes of death in the industrialized countries. Both pathological and radiological studies demonstrate extensive atrophy, particularly of the hippocampus.[24,25]

Pick's disease is a much rarer dementing illness with similar cognitive deficits to Alzheimer's disease; however, abnormal behaviors, apathy, abulia, and the Klüver–Bucy syndrome are more common.[26] The disease appears to be transmitted in a dominant, although modified, fashion, and is more common in females.

Neuroradiological studies and gross pathology show findings of marked atrophy affecting the frontal and anterior temporal lobes with relative sparing of the parietal lobes and sparing of the posterior portion of the superior temporal gyrus and the occipital regions.[26] Alzheimer's disease appears to have associations with other degenerative diseases of the CNS–

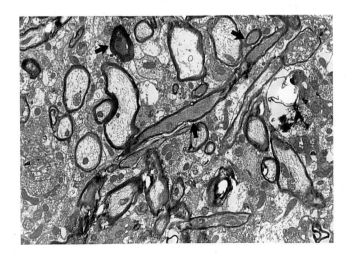

Figure 2 Electron micrograph of white matter immediately superior to the lateral geniculate nucleus 8 days after ablation of the cortical sites of visual processing. Arrows mark examples of early degeneration with axonal collapse and increased cytoplasmic staining. (Reproduced by permission of The American Society of Neuroradiology from Lexa et al.[23])

For References see p. 1006

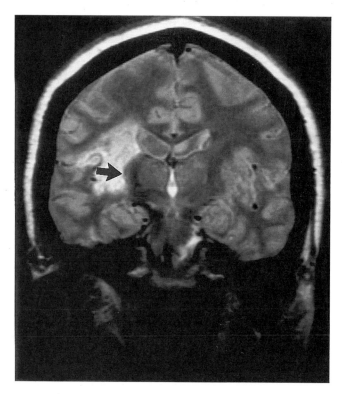

Figure 3 Long *TR* FSE images in a woman 4 weeks after a middle cerebral artery stroke showing evidence of Wallerian degeneration in the ipsilateral corticospinal tract with marked signal loss (arrow) relative to the unaffected side

most notably Parkinson's disease (see below). In some reports, close to half of those patients with Alzheimer's disease have evidence of the degenerative changes associated with Parkinson's disease.[27,28]

4   NORMAL PRESSURE HYDROCEPHALUS

Hydrocephalus can occur secondary to a wide variety of etiologies. This discussion will be limited to the 'degenerative' form of normal pressure hydrocephalus (NPH).

The syndrome of NPH is seen predominately in older patients, and includes a triad of clinical findings—dementia, gait ataxia, and urinary incontinence. The underlying mechanism which leads to NPH is controversial and includes theories related to diminished absorption from prior hemorrhage or inflammatory disease, changes in the physical properties of the ventricular walls, or, rarely, a preexisting congenital form of hydrocephalus. Patients may respond to shunting, and favorable factors include response to a preliminary trial of cerebrospinal fluid removal, presentation with gait disturbance, and short duration of symptoms.[29] NPH and white matter ischemic disease appear to be significantly related, including a direct correlation of the severity of the diseases.[30]

On cross-sectional imaging techniques, NPH may overlap significantly in appearance with other diseases in this age group. In particular, the overlap with atrophy from other causes is problematic. The most useful features are those found in

other forms of hydrocephalus: thinning and elevation of the corpus callosum, and distention of the third ventricle.[31]

MRI may provide an additional clue by allowing detection and quantification of flow in the cerebral aqueduct. The magnitude of flow in this structure is increased in the setting of NPH, and is probably secondary to the diminished compliance of the ventricular system.[32]

5   DEGENERATION OF THE CENTRAL GRAY MATTER STRUCTURES

5.1   Huntington's Chorea

Huntington's chorea is an autosomal dominant degenerative disease which has been localized to chromosome 4.[33] Presentation is in the fourth and fifth decades, with choreathetosis and dementia. Neuroradiological studies confirm characteristic atrophy of the caudate and putamen as well as more generalized atrophy.[34] On MRI, abnormal signal intensity on long *TR/TE* images may be present in the striatum.

In addition to reductions in the caudate and putamen, detailed volumetric studies show volume loss in the thalamus and mesial temporal lobes.[35] In mildly affected individuals, the putaminal measurement appears to be sensitive and specific.[36]

5.2   Wilson's Disease (Hepatolenticular Degeneration)

Wilson's disease is a genetic deficiency of ceruloplasmin, which leads to toxic copper deposition particularly in the liver and the lenticular nuclei of the brain. It is an autosomal recessive disorder which localizes to chromosome 13.[37] Clinical manifestations include liver disease, movement disorders of tremor, ataxia, dysarthria and rigidity, and psychiatric symptoms. The Kayser–Flesicher ring of copper deposition in Descemet's membrane is characteristic, while the definitive diagnosis is made by low serum ceruloplasmin, increased urinary copper, and elevated liver copper levels.

Atrophy, particularly of the gray matter, is a hallmark of Wilson's disease. Focal white matter lesions may occur which are hypointense on short *TR*, but of variable intensity on long *TR*.[38] These lesions appear to have a reversible component during appropriate therapy.[38–40] Clinical neurological symptoms appear to correlate with the presence of radiological abnormalities. The location of lesions can be useful in predicting the type of neurological manifestation.[41]

5.3   Parkinson's Disease

Parkinson's syndrome, or Parkinsonism, is a broad clinical term referring to a cluster of symptoms due to loss of the majority of the dopaminergic efferent projections of the substantia nigra. The predominant clinical symptom is an involuntary tremor with rigidity and akinesia. The syndrome can be caused by a wide variety of etiologies, including stroke, trauma, inflammatory processes (particularly an epidemic which occurred after World War I, von Economo's encephalitis, now rarely with viral illnesses such as Cocksackie B, Japanese B and St. Louis viral encephalitides), drug effects, and associations with other degenerative diseases such as progressive supranuclear palsy, striatonigral degeneration, and Parkinson's disease. The final common pathway is loss of the dopaminergic effects of these nigral projections on the striatal structures.

Parkinson's disease is a true degenerative disease of the CNS with idiopathic loss of nigral–striatal projections. This entity is common, with an age of onset in the 50–60 years range, affecting approximately 1 in 1000 of people over the age of 50 years.[27] The disease is progressive, probably due in part to ongoing loss of nigrostrial function. Early in the course of the disease, patients may respond well to a variety of medications, often with progressive resistance as the degeneration advances.

On pathological examination, there are characteristic cell inclusions called Lewy bodies in the neurons of the substantia nigra, as well as in several related regions of the brain. MRI of Parkinson's disease patients reveals a decrement in the width of the pars compacta of the substantia nigra. This finding is shared with diseases leading to Parkinsonism such as striatonigral degeneration and progressive supranuclear palsy. Atrophy of the brain is also a characteristic that Parkinson's disease shares with other degenerative disorders, although sometimes this can be of a relatively small magnitude.

5.4 Striatonigral Degeneration

This degenerative disease presents with similar symptoms to Parkinson's disease. The primary differences rest with the much greater early involvement of the striatum (the putamen more than the caudate), which may in part explain the poor response to traditional Parkinsonism pharmacotherapy. This putaminal involvement may be detected on MRI with greater than normal hypointensity of the putamen on T_2-weighted sequences. The magnitude of this finding has been reported to correlate with the severity of clinical involvement.[42] A second area of detectable involvement occurs in the substantia nigra, with loss of width of the pars compacta.

5.5 Progressive Supranuclear Palsy

Named for the presence of supranuclear ophthalmoplegia of vertical gaze, this syndrome includes manifestations of pseudobulbar palsy, rigidity, extrapyramidal signs, and dementia, presenting in late middle age and in the elderly population. This degenerative disease localizes predominantly to the midbrain. In particular, the superior colliculi are atrophic, constituting a relatively specific MRI sign of the disease. Neurofibrillary tangles can be detected in the periaqueductal gray matter of the mesencephalon with gliosis–similar to the findings in Alzheimer's disease.[43,44] Changes in this region can be seen on MRI, with high signal intensity on relatively T_2-weighted sequences.[43] As in the other degenerative diseases related to Parkinson's disease, there is diminished size of the pars compacta.

6 DEGENERATION OF THE CEREBELLUM AND BRAIN STEM

There are several degenerative diseases which have a propensity to attack the structures of the posterior fossa (Figure 4).

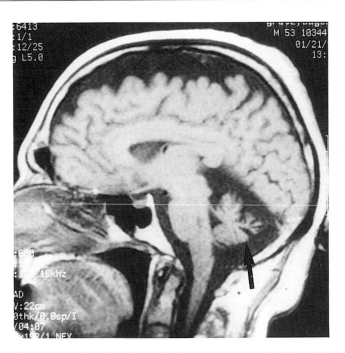

Figure 4 Sagittal T_1-weighted MRI scan demonstrating idiopathic vermian atrophy in a patient with marked cerebellar symptoms

6.1 Down's Syndrome

Down's syndrome is a very important cause of inherited mental retardation. Supranormal amounts of the genetic material on chromosome 21 usually from nondisjunction are the underlying cause. MRI measurements show a global reduction in both gray and white matter. Interestingly, Down's syndrome is a very important risk factor for the ultimate development of Alzheimer's disease, with close to 100% of Down's patients developing a pathology similar to Alzheimer's disease in later life. The gene for the production of the amyloid-β protein which has been linked to some of the pathology in Alzheimer's disease also maps to chromosome 21. Reports of older patients with Down's syndrome (up to 64 years old) show significant widening of the temporal horns on MRI–similar to what has been reported in patients affected by Alzheimer's disease.[45] Finally, NMR spectroscopic analysis has shown a rapid decline in N-acetylaspartate (NAA)/choline in older Down's patients, which correlates well with onset of clinical symptoms and precedes atrophic changes.[46]

6.2 Fragile X and Rett's Syndrome

The fragile X syndrome is a rare genetic cause of mental retardation. The syndrome is characterized by autism, repetitive movements, and hyperactive, self-destructive behaviors. The most profound expression is in males, but female carriers also show mild clinical and neuroradiological features of the disease. MRI reports have shown atrophy of the cerebellar vermis with concomitant enlargement of the fourth ventricle.[47]

In contrast, Rett's syndrome is a disease of girls which manifests also with autism and stereotyped movements with loss of language and motor skills. The brain appears to have a globally

For References see p. 1006

decreased volume, particularly in the frontal lobes. This probably occurs from a combination of congenital hypoplasia and perhaps ongoing degeneration—particularly in the cerebellum. The cerebellar loss is controversial, and other groups have shown greater effects on loss of gray matter volume.

6.3 Autism

Autism has been reported to be associated with findings of focal loss of volume in the superior vermian lobules (decline, folium, and tuber), as well as more generalized volume loss in other parts of the brain.[48] This finding has been contested by others.[49]

6.4 Olivopontocerebellar Degeneration

This rare neurodegenerative disorder can be inherited or sporadic. Ataxia is the presenting sign, and can occur at any age. Because of the nonspecific nature of the presentation, MRI may be particularly useful in providing a specific diagnosis. Pathological and imaging findings include marked loss of volume in the body of the pons, middle cerebellar peduncles, cerebellum, and inferior olive. The pathways which interconnect these structures may show abnormal increased signal intensity on long *TR* sequences.[50] This probably depends on the stage of axonal degeneration.

6.5 Cerebellar Cortical Degeneration

Cerebellar cortical degeneration can present either with an early onset familial form or a late-onset sporadic form. There is a slowly progressive ataxia affecting the limbs, the trunk, and speech. Pathological examination and imaging studies both show marked cerebellar atrophy, particularly in the anterior vermis.

6.6 Acetazolamide-Responsive Familial Paroxysmal Ataxia

This unusual disorder is characterized by recurrent attacks of cerebellar symptoms which are associated with anterior vermian atrophy in the cerebellum.

6.7 Friedreich's Ataxia

Presenting in the second decade of life, this genetic degenerative disorder (localizing to chromosome 9 with multiple inheritance patterns) leads to a progressive ataxia. This leads to gait ataxia and difficulty with speech, as well as kyphoscoliosis, probably on a muscular basis. In addition, to the degeneration of the spinocerebellar tracts, there is loss of the posterior columns of the spinal cord. The spinal cord is atrophic, particularly in the posterior and lateral columns. The majority of subjects also have cerebellar atrophy; however, this is less pronounced than other processes which affect the cerebellum.

7 CONCLUSIONS

MR imaging has revolutionized the detection and diagnosis of many of the neurodegenerative diseases. The patterns of

atrophy and Wallerian degeneration in these entities often allow a specific diagnosis to be made noninvasively. Moreover, in the near future, advances in MRI such as magnetization transfer techniques may allow much earlier detection and characterization of the degenerative processes within the neuron which lead to these tragic diseases.

8 RELATED ARTICLES

Brain Neoplasms Studied by MRI; Hemorrhage in the Brain and Neck Observed by MRI; Intracranial Infections; Magnetic Resonance Imaging of White Matter Disease; Structural and Functional MR in Epilepsy.

9 REFERENCES

1. R. D. Adams and M. Victor, 'Principles of Neurology', McGraw Hill, New York, 1989.
2. S. W. Atlas, 'Magnetic Resonance Imaging of the Brain and Spine', Raven Press, 1991.
3. A. V. Waller, *Philos. Trans. R. Soc. London*, 1850, **140**, 423.
4. R. J. Rossiter, 'Chemical Pathology of the Nervous System', ed. J. Folch-Pi, Pergamon Press, New York, 1961.
5. P. M. Daniel and S. J. Strich, *Acta Neuropathol. (Berlin)*, 1969, **12**, 314.
6. H. Webster and F. De, *Ann. N.Y. Acad. Sci.*, 1964, **122**, 29.
7. L. D. DeWitt, J. P. Kistler, D. C. Miller, E. P. Richardson Jr., and F. S. Buonanno, *Stroke*, 1987, **18**, 342.
8. S. R. Cobb and C. M. Mehringer, *Radiology*, 1987, **162**, 521.
9. M. J. Kuhn, K. A. Johnson, and K. R. Davis, *Radiology*, 1988, **168**, 199.
10. M. Bouchareb, T. Moulin, F. Cattin, J. L. Dretemann, A. Rack, H. Verdot, and J. F. Bonneville, *J. Neuroradiol.*, 1988, **1988**, 238.
11. A. Uchino, K. Onomura, and M. Ohno, *Radiat. Med.*, 1989, **7**, 74.
12. J. E. Pujol, J. L. Mari-Vilata, C. Junque, P. Vendrell, J. Fernandez, and A. Capdevila, *Stroke*, 1990, **21**, 404.
13. T. Orita, T. Tsurutani, A. Izumihara, and T. Matsunaga, *J. Comput. Assist. Tomogr.*, 1991, **15**, 802.
14. Y. Inoue, Y. Matsumura, T. Fukuda, Y. Nemoto, N. Shirahata, T. Suzuki, M. Shakudo, S. Yawata, S. Tanaka, and K. Takemoto, *Am. J. Roentgenol.*, 1990, **11**, 897.
15. A. Danek, M. Bauer, and W. Fries, *Eur. J. Neurosci.*, 1990, **2**, 112.
16. N. P. Bontozoglou, D. W. Chakeres, G. F. Martin, M. A. Brogen, and R. B. McGhee, *Radiology*, 1991, **180**, 223.
17. M. J. Kuhn, D. J. Mikulis, D. M. Ayoub, B. E. Kosofsky, K. R. Davis, and J. M. Taveras, *Radiology*, 1989, **172**, 179.
18. A. Uchino, H. Imada, and M. Ohno, *Neuroradiology*, 1990, **32**, 191.
19. N. Kumar, and F. J. Lexa, in preparation.
20. F. A. Jolesz, J. F. Polak, P. W. Ruenzel, and D. F. Adams, *Radiology*, 1984, **152**, 85.
21. D. S. Titelbaum, J. L. Frazier, R. I. Grossman, P. M. Joseph, L. T. Yu, E. A. Kassab, W. F. Mickey, D. laRossa, and M. J. Brown, *Am. J. Roentgenol.*, 1989, **10**, 741.
22. R. I. Grossman, C. M. Hecht-Leavitt, S. M. Evans, R. E. Lenkinski, G. A. Holland, J. J. van-Winkle, J. T. McGrath, W. J. Curran, A. Shetty, and P. M. Joseph, *Radiology*, 1988, **169**, 305.
23. F. J. Lexa, R. I. Grossman, and A. C. Rosenquist, *Am. J. Roentgenol.*, 1994, **15**, 201.
24. M. L. Schmidt, V. M.-Y. Lee, and J. Q. Trojanowski, *Lab. Invest.*, 1989, **60**, 513.

For list of General Abbreviations see end-papers

25. J. W. Dahlbeck, K. W. McCluney, J. W. Yeakley, M. J. Fenstermacher, C. Bonmati, G. van-Horn, and J. Aldeg, *Am. J. Roentgenol.*, 1991, **12**, 931.

26. J. L. Cummings, and L. W. Duchen, *Neurology*, 1981, **31**, 1415.

27. A. Barbeau, 'Handbook of Clinical Neurology', Elsevier, Amsterdam, 1986, Vol. 49, p. 87.

28. E. B. Larson, B. V. Reifler, S. M. Sumi, C. G. Confield, and N. M. Chinn, *Arch. Intern. Med.*, 1986, **146**, 1917.

29. C. Wikkelso, H. Andersson, C. Blomstrand, M. Matousek, and P. Svenelson, *Neuroradiology*, 1989, **31**, 160.

30. W. G. Bradley Jr., A. R. Whittemore, A. S. Watanabe, S. T. Davis, L. M. Teresi, and M. Komyak, *Am. J. Roentgenol.*, 1991, **12**, 31.

31. T. El Gammal, M. B. Allen, B. S. Brooks, and E. K. Mark, *Am. J. Roentgenol.*, 1987, **8**, 591.

32. W. G. Bradley Jr., K. E. Kortman, and B. Burgoyne, *Radiology*, 1986, **159**, 611.

33. J. F. Gusella, N. S. Wexler, P. M. Conneally, S. L. Naylor, M. A. Anderson, R. E. Tanzi, P. C. Watkins, K. Ottina, M. R. Wallace, and A. Y. Sakaguchi, *Nature*, 1983, **306**, 234.

34. S. T. Grafton, J. C. Mazziotta, J. J. Pahl, P. St. George Hyslop, J. L. Haines, J. Gusella, J. M. Hoffman, L. R. Baxter, and M. E. Phelps, *Arch. Neurol.*, 1992, **49**, 1161.

35. T. L. Jernigan, D. P. Salmon, N. Butters, and J. R. Hesselink, *Biol. Psychol.*, 1991, **29**, 68.

36. G. J. Harris, G. D. Pearlson, C. E. Peyser, E. H. Aylward, J. Roberts, P. E. Barta, G. A. Chose, and S. E. Folstein, *Ann. Neurol.*, 1992, **31**, 69.

37. S. Starosta-Rubinstein, A. B. Young, K. Kluin, G. Hill, A. M. Aisen, T. Gabrielsem, and G. J. Brewer, *Arch. Neurol.*, 1984, **44**, 365.

38. L. Prayer, D. Wimberger, J. Kramer, G. Grimm, K. Oder, and H. Imhof, *Neuroradiology*, 1990, **32**, 211.

39. H. Nazer, J. Brismar, M. Z. Al-Kawi, T. S. Gunasekaran, and K. H. Jorulf, *Neuroradiology*, 1993, **35**, 130.

40. K. A. Thuomas, S. M. Aquilonius, K. Bergstrom, and K. Westermark, *Neuroradiology*, 1993, **35**, 134.

41. W. Oder, L. Prayer, G. Grimm, J. Spatt, P. Ferenci, K. Holleger, B. Schneider, A. Gangl, and L. Deeke, *Neurology*, 1993, **43**, 120.

42. R. Brown, R. J. Polinsky, G. Di Chiro, B. Postakia, L. Wener, and J. J. Simmons, *J. Neurol. Neurosurg. Psychiatry*, 1987, **50**, 913.

43. M. Savoiardo, L. Strada, F. Girotti, L. D'Incerti, M. Sherna, P. Soliveri, and A. Bolzarini, *J. Comput. Assist. Tomogr.*, 1989, **13**, 555.

44. M. L. Schmidt, V. M.-Y. Lee, J. Q. Trojanowski, and H. Hurtig, *Lab. Invest.*, 1988, **59**,

45. M. LeMay and N. Alvarez, *Neuroradiology*, 1990, **32**, 104.

46. T. Murata, Y. Koshino, M. Omori, I. Murati, M. Nishio, T. Horie, Y. Umezawa, K. Isaki, H. Kimura, and S. Itoh, *Biol. Psychol.*, 1993, **34**, 290.

47. E. H. Aylward and A. Reiss, *J. Psychiatr. Res.*, 1991, **25**, 159.

48. E. Courchesne, G. Press, and R. Yeung-Courchesne, *Am. J. Roentgenol.*, 1993, **160**, 387.

49. M. A. Nowell, D. B. Hackney, A. S. Muraki, and M. Coleman, *JMRI*, 1990, **8**, 811.

50. M. Savoiardo, L. Strada, F. Girotti, R. A. Zimmerman, M. Grisoli, D. Testa, and R. Petrillo, *Radiology*, 1990, **174**, 693.

Biographical Sketch

Frank J. Lexa. *b* 1958. A.B. (Biology), Harvard University, USA, 1980. M.S. (Physiology), 1982, M.D. 1985, Stanford University School of Medicine, USA. Postgraduate medical training in internal medicine, Brigham and Women's Hospital, Harvard Medical School, and diagnostic radiology/neuroradiology, Hospital of the University of Pennsylvania, USA. Currently Fellow, Leonard Davis Institute of Health Care Economics and Graduate MBA Candidate, Class of '99, The Wharton School Director, Medical Acquisitions, BTG International. Approx. 30 publications. Current research specialties: early detection and characterization of axonal degeneration using magnetization transfer techniques, MRI contrast agent development for advanced medical applications.

Cranial Nerves Investigated by MRI

Anton N. Hasso
University of California, Irvine, CA, USA

and

Peggy J. Fritzsche
Loma Linda University, Loma Linda, CA, USA

1 INTRODUCTION

The reader of this article will gain an understanding of the normal pathways of the cranial nerves and be able to correlate the function with the course of each nerve. The functional anatomy will also be addressed in conjunction with the clinical manifestations of pathologic entities. Tumors, infections, and inflammations of the cranial nerves will be discussed separately.

2 NORMAL ANATOMY

2.1 Cranial Nerve I (Olfactory)

Smell is consciously perceived in the gray matter covering the medial and lateral striae of the olfactory gyri. The lateral striae travel from the inferior medial aspect of the temporal lobe (pyriform area) and the medial striae from the medial and inferior aspects of the frontal lobe (subcallosal region) to the olfactory trigone located at the anterior perforated substance. The intermediate stria is not significant in humans, but joins the lateral and medial striae at the trigone to extend anteriorly as the olfactory tract. The olfactory tract terminates in the olfactory bulb. The olfactory bulbs are located inferior to the olfactory sulci (Figure 1). The olfactory sulci separate the gyri rectus from the medial orbitofrontal gyri.

Sensory nerve bundles exit the olfactory bulbs via the cribriform plate of the skull to separate into multiple neurosensory

For References see p. 1033

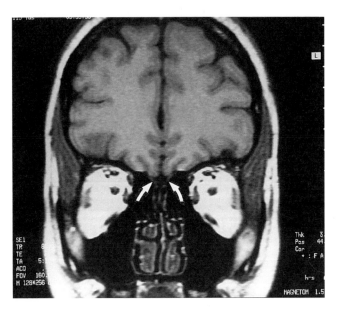

Figure 1 Normal anatomy of the olfactory bulbs. The olfactory bulbs (arrows) are located inferior to the olfactory sulci which separate the gyrus rectus from the medial orbitofrontal gyrus

cells which reside in the nasal epithelium. The function of the olfactory nerve is to provide olfaction. Neuropathy of the first cranial nerve manifests as anosmia, or acute loss of olfaction. Anosmia is usually unilateral, and typically occurs secondary to pathologic processes arising in the nasal cavities. Intracranial neoplasms, infections, arterial disease, or congenital anomalies also can cause anosmia.[1]

2.2 Cranial Nerve II (Optic)

The optic fibers originate in the calcarine portion of the occipital cortex and travel superiorly, inferiorly, and laterally to the occipital horn of the ventricle. The optic fibers then become the optic radiations as they course over the temporal horn of the lateral ventricle and the tail of the caudate nucleus to enter the lateral geniculate body of the thalamus [Figure 2(a)]. Bilateral optic tracts then exit the lateral geniculate bodies and meet in the midline to form the optic chiasm. From the optic chiasm ventrally, the optic fibers then divide into the bilateral optic nerves coursing initially into the cranial openings of the optic canals. The optic nerves exit the orbital openings of the optic canals and extend forward in a sinusoidal manner to enter the globes at the optic nerve heads.

Lesions involving the intraorbital and prechiasmal portions of the optic nerve result in total monocular blindness. The sudden onset of a monocular deficit suggests a possible vascular or demyelinating process. The slow onset of a monocular optic neuropathy may be caused by a lesion intrinsic to the optic nerve sheath or secondary to extrinsic compression by an adjacent mass.

The fibers from the outer (temporal) portion of both visual fields cross at the optic chiasm and thus are carried via the optic tracts to the contralateral calcarine cortex of the occipital lobes. The fibers from the medial (nasal) portion of the visual fields do not cross at the optic chiasm and are projected to the ipsilateral visual cortex. Lesions involving the central portion of the chiasm particularly produce bitemporal

hemianopsias. Lesions along the optic tracts result in loss of vision in the ipsilateral (uncrossed) nasal field and contralateral (crossed) temporal field. Since the retrochiasmal segment of the visual pathway recognizes the opposite visual field, a retrochiasmal lesion leads to a homonomous hemianopsia contralateral to the lesion. Contralateral hemianopsias may be produced by lesions involving any portion of the retrochiasmal optic pathways including the lateral geniculate bodies, the optic radiations, or the occipital cortices.[1–3]

2.3 Cranial Nerve III (Oculomotor)

The motor component of cranial nerve III originates in the motor cortex of the cerebrum. The fibers then travel centrally through the internal capsule to the ipsilateral superior colliculus. The fibers cross between the red nuclei in the dorsal tegmental decussation and continue distally in the paramedian raphe to synapse in the pontine reticular formation. The fibers then ascend in the medial longitudinal fasciculus to the oculomotor nucleus which is situated in the paramedian midbrain tegmentum ventral to the aqueduct of Sylvius at the level of the superior colliculus. At this level, additional fibers from the parasympathetic nucleus, paramedian nucleus, Perlia's nucleus, and Edinger–Westphal nucleus join the motor fibers. These combined fibers bow laterally to extend through the medial aspect of the red nucleus and the cerebral peduncle (Figure 2). Fibers then exit from the brainstem anteriorly at the pontomedullary junction near the midline where the two nerves form parallel structures in a V-configuration as they extend through the interpeduncular fossa (Figure 3).

The oculomotor nerves then travel between the superior cerebellar and posterior cerebral arteries through the prepontine cistern. Each cranial nerve III then penetrates the dura of the cavernous sinus and rises superiorly along the lateral wall of the cavernous sinus adjacent to the fourth cranial nerve and above the sixth cranial nerve and ophthalmic division of the fifth cranial nerve. The parasympathetic and motor fibers are in close proximity, and altogether the nerves penetrate the superior orbital fissure to reach the orbital structures.

The motor fibers provide the somatic motor function to the medial rectus, inferior rectus, superior rectus, and inferior oblique extraocular muscles. Lesions of this portion of the oculomotor nerve lead to a downward abducted eye, due to the unopposed action of the superior oblique and lateral rectus muscles. The oculomotor nerve also supplies the levator palpebrae superioris, which elevates the eyelid. Lesions of this portion of cranial nerve III also result in lid droop or ptosis. The visceral motor fibers from the Edinger–Westphal nucleus provide parasympathetic motor function to the pupillary constrictor and ciliary muscles. Lesions of this portion of the oculomotor nerve result in pupillary dilatation due to unopposed sympathetic input and failure to accommodate. Isolated third-nerve palsies are regarded as either complete (including parasympathetic involvement) or incomplete (due to lack of parasympathetic involvement).[1–3]

2.4 Cranial Nerve IV (Trochlear)

The motor impulses of cranial nerve IV originate in the motor cortex and then extend centrally via the internal cap-

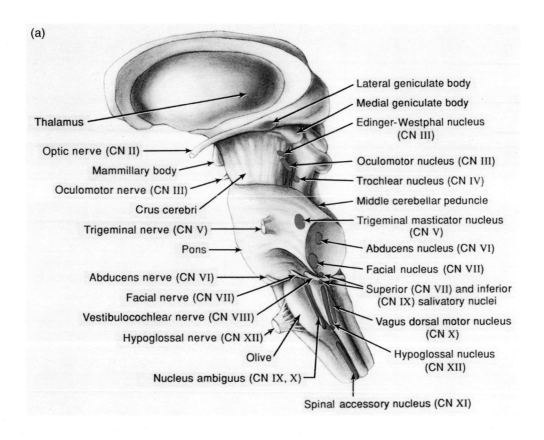

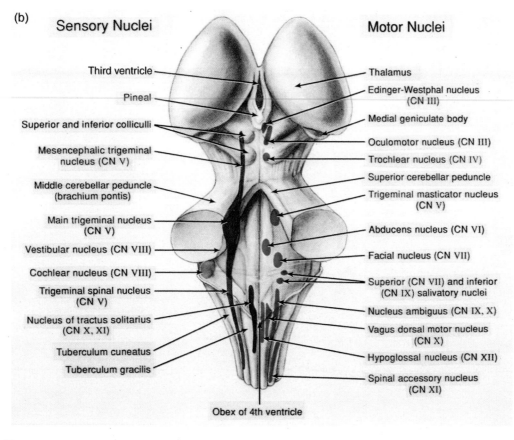

Figure 2   (a, b)

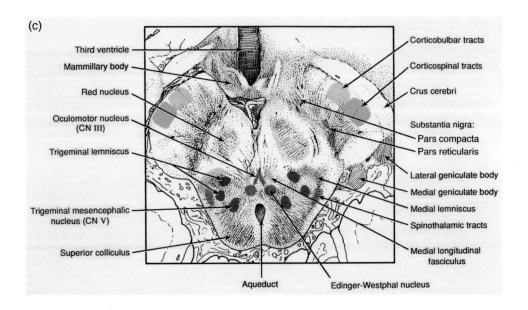

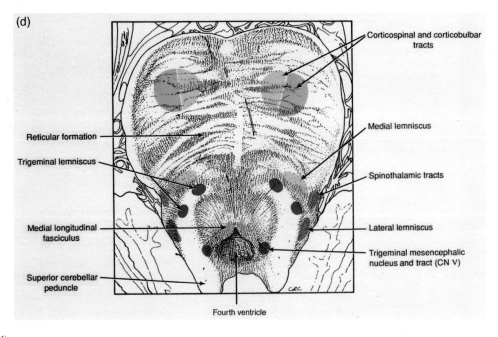

Figure 2 (c, d)

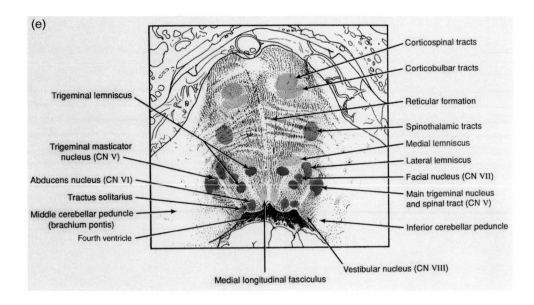

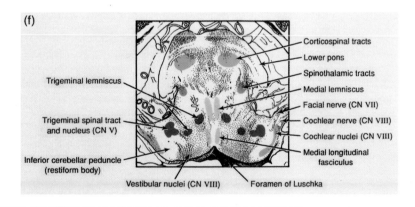

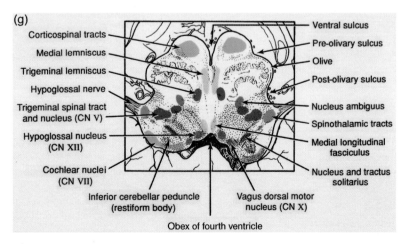

Figure 2 Illustrations of normal anatomy of brainstem and cranial nerves. (a) Lateral view of brainstem with motor nuclei. (b) Posterior view of brainstem with sensory and motor nuclei. (c) Normal axial anatomy of the brainstem at the level of the midbrain. (d) Normal axial anatomy of the brainstem at the level of the pontine isthmus. (e) Normal axial anatomy of the brainstem at the pons. (f) Normal axial anatomy of the brainstem at the upper medulla. (g) Normal axial anatomy of the brainstem at the middle medulla. CN, cranial nerve. Reproduced with permission from W. G. Bradley, Jr, *Radiology*, 1991, **179**, 319

For References see p. 1033

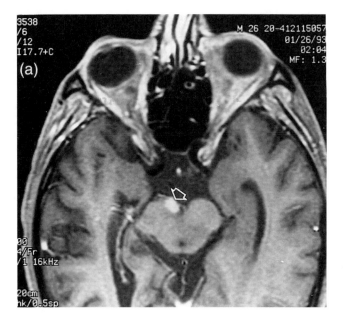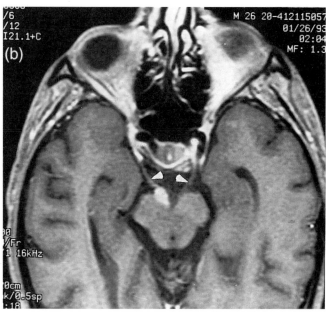

Figure 3 Lymphomatous meningitis in a patient with AIDS. Axial enhanced images with fat saturation pulses. (a) Note the focal nodule in the right side of the interpeduncular cistern (arrow). (b) The characteristic V-shape of the oculomotor nerves in the prepontine cistern is well shown (arrows)

sule to the ipsilateral superior colliculus. The fibers cross between the red nuclei in the dorsal tegmental decussation and travel caudally in the paramedial raphe to synapse in the pontine reticular formation. The fibers then ascend in the medial longitudinal fasciculus to reach the motor nucleus. The fourth cranial nerve nucleus is located in the tegmentum of the mesencephalon at the level of the inferior colliculus immediately caudal to the third nerve nucleus. Fibers from the nucleus loop posteriorly to decussate in the tectum of the lower midbrain beneath the inferior colliculi. This pathway forms a sickle-shaped arch around the aqueduct of Sylvius (Figure 2).

The trochlear nerve exits from the dorsal aspect of the brainstem, emerging from the superior medullary vellum just beneath the inferior colliculus. The nerve extends around the cerebral peduncle between the superior cerebellar and posterior cerebral arteries, just lateral to the third cranial nerve (Figure 4). The nerve then pierces the dura of the cavernous sinus coursing along the superior margin adjacent to the oculomotor nerve and the ophthalmic portion of the trigeminal nerve en route to the superior orbital fissure.

The fourth cranial nerve innervates the superior oblique muscle in the orbit. Lesions of the trochlear nerve or its nucleus result in outward rotation of the affected eye. Because the trochlear nerve has the longest intracranial course (7.5 cm), it has the greatest chance of injury from trauma or surgery in the region of the midbrain.[1,3,4]

2.5 Cranial Nerve V (Sensory Trigeminal)

The sensory impulses from the upper, middle, and lower part of the face extend to the gasserian ganglion via the ophthalmic (V1), maxillary (V2), and mandibular (V3) branches of cranial nerve V. These branches enter the skull via the superior orbital fissure, foramen rotundum and foramen ovale, respectively. The ophthalmic and maxillary divisions extend through the cavernous sinus, while the mandibular branch bypasses the cavernous sinus and directly enters the trigeminal (gasserian) ganglion. This ganglion lies in the inferior portion of Meckel's cavity and contains the cell bodies of numerous afferent sensory fibers. Facial numbness and burning (tic douloureux) represent signs and symptoms caused by sensory trigeminal neuropathies.

Sensory impulses from the face then leave the trigeminal ganglion to enter the midlateral pons through the prepontine cistern. As they enter the brainstem, branches go to the three sensory nuclei: the principal sensory nucleus, the mesencephalic nucleus, and the spinal nucleus. Pain and temperature fibers descend to the C2–C3 cord level where they synapse in the spinal nucleus of cranial nerve V. These fibers then cross the midline ascending in the ventral trigeminal lemmiscus to reach the ventral posteromedial nucleus of the thalamus. The thalamocortical tract then carries the fibers via the internal capsule to the sensory cortex. The sensory impulses that mediate proprioception from the face, particularly of the mandible, synapse in the mesencephalic nucleus of cranial nerve V which extends superiorly into the periaqueductal gray matter of the midbrain. This portion of the sensory nucleus is composed of the sensory neurons that travel with the mandibular nerve. Sensory nerves that mediate touch and pressure from the face synapse in the principal sensory nucleus of cranial nerve V and travel to the ventral posteromedial nucleus of the thalamus via the contralateral dorsal trigeminal lemmiscus. The extension to the sensory cortex is via the thalamocortical tract which passes through the internal capsule (Figure 2).[1–3,5]

2.6 Cranial Nerve V (Motor Trigeminal)

The impulses for the motor components of cranial nerve V originate in the motor cortex of the cerebral hemisphere. The

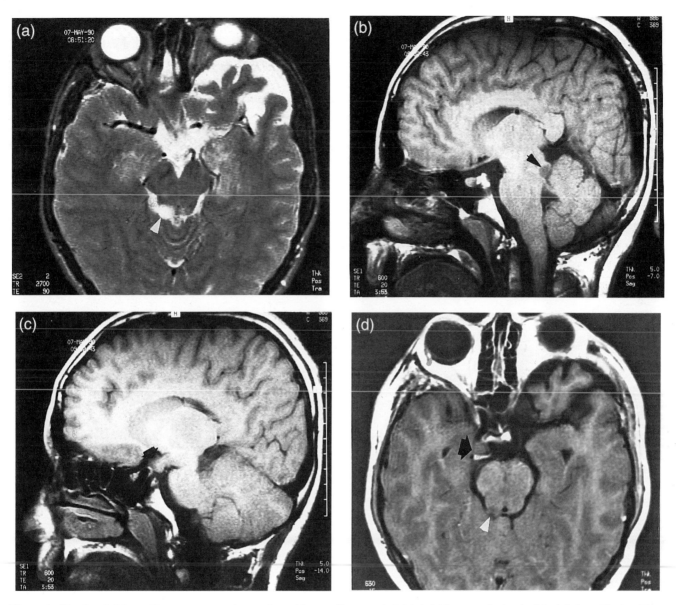

Figure 4 Tectal and right trochlear nerve tumor in a patient with neurofibromatosis I. (a) Axial image at the level of the inferior colliculus. There is a hyperintense mass in the right tectum encompassing the nucleus of cranial nerve IV. Note the incidental dysplasia of the left sphenoid wing which is commonly seen in patients with NF-1. (b and c) Sagittal unenhanced images. There is a hypointense tumor expanding the inferior colliculus [arrow, (b)]. Note the expansion of the right trochlear nerve in the prepontine cistern [broad arrow, (c)]. (d) Axial enhanced image. The tumor foci in the tectal plate (white arrow) and in the right prepontine cistern (black arrow) do not enhance significantly. A hamartoma or low grade glioma is suspected. Note the uninvolved oculomotor nerves in the prepontine cistern

pathway then extends through the genu of the internal capsule and the cerebral peduncle (along the corticobulbar tract) to the motor nucleus which is situated medial to the principal sensory nucleus. The fused motor and sensory components of the fifth nerve emerge from the ventrolateral pons, coursing through the anterior portion of the cerebellopontine angle cistern near the apex of the medial petrous ridge. The nerve fibers then pierce the dura of Meckel's cavity and enter the trigeminal ganglion. At the distal aspect of Meckel's cavity, the motor fibers of the mandibular nerve exit at the inferolateral surface and extend through the foramen ovale (Figure 5). The motor division of the mandibular nerve supplies the muscles of mastication plus other small muscles (tensor villi palatini, mylohyoid, anterior belly of the digastric and tensor tympani).

The exit point of the motor trigeminal nerve from the brainstem is also the entry point of the sensory trigeminal nerve. This area is termed the 'root entry zone' of the sensory trigeminal nerve and is subject to vascular compression by dolichoectatic vessels or aneurysms which may cause tic douloureux or other trigeminal neuropathies. Alternatively, an ipsilateral or contralateral posterior fossa tumor or other mass lesion can torque the brainstem and cause a distention of this root entry zone. If the motor trigeminal portion is affected by a pathologic process, weakness of the muscles of mastica-

For References see p. 1033

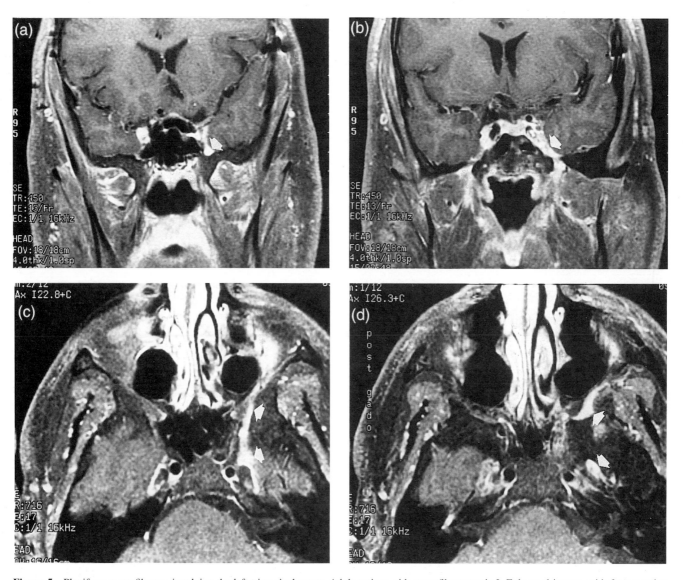

Figure 5 Plexiform neurofibroma involving the left trigeminal nerve. Adult patient with neurofibromatosis I. Enhanced images with fat saturation pulses. (a and b) Coronal images demonstrate the expansion and enhancement in the left maxillary division [arrow, (a)] and mandibular division [arrow, (b)] of the trigeminal nerve. (c and d) Axial images demonstrate the course of the left axillary nerve from the trigeminal ganglion through the foramen rotundum [arrows, (c)]. Note the expansion of the tumor into the left foramen ovale and pterygopalatine fossa [arrows, (d)]

tion (with or without denervation atrophy) can occur (Figure 6).[1–3,5]

2.7 Cranial Nerve VI (Abducens)

The impulses of cranial nerve VI originate in the motor cortex of the cerebral hemisphere. The fibers extend through the internal capsule and the corticobulbar tract to the ipsilateral superior colliculus. At this point, the fibers cross between the red nuclei in the dorsal tegmental decussation. They then extend caudally in the paramedian raphe of the brainstem to synapse in the pontine reticular formation. Both ipsilateral and contralateral fibers ascend in the medial longitudinal fasciculus to reach the brainstem nucleus of the abducens nerve which lies in the pons immediately anterior to the fourth ventricle. These fibers pass through the pontine tegmentum and exits anteriorly from the brainstem at the pontomedullary junc-

tion. The abducens nerve then courses superiorly and slightly laterally along the clivus. At the base of the dorsum sella, the sixth cranial nerve courses anteriorly piercing the dura through Dorello's canal to enter the cavernous sinus. The nerve then extends obliquely through the cavernous sinus just inferior and medial to cranial nerves III and IV, immediately adjacent to the cavernous portion of the internal carotid artery. Cranial nerve VI then exits anteriorly through the cavernous sinus to enter the superior oblique fissure en route to the lateral rectus muscle (Figure 2).

An isolated lateral rectus palsy is the most common lesion of the cranial nerves supplying the extraocular muscles. Injury to the abducens nerve produces paralysis of the ipsilateral lateral rectus muscle, with resultant horizontal diplopia. The unopposed pull of the medial rectus muscle causes the eye to turn inward (adduct), thereby producing internal strabismus.[1–3]

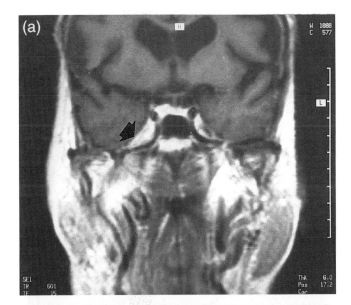

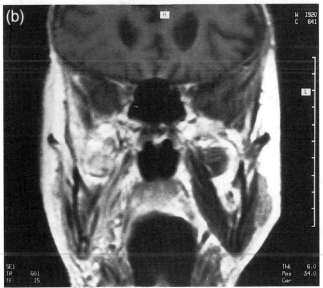

Figure 6 Perineural extension of squamous cell carcinoma. Adult patient with previous known facial cancer. Coronal enhanced images. (a) Note the lateral deviation of the cavernous sinus wall and enhancement along the right trigeminal ganglion (arrow). (b) There is marked denervation atrophy of the right muscles of mastication. The fat which replaces the muscle shows moderate diffuse enhancement

2.8   Cranial Nerve VII (Facial)

The motor impulses of cranial nerve VII originate in the motor cortex of the cerebral hemisphere and extend via the genu of the internal capsule and cerebral peduncle to the brainstem nucleus. The motor nucleus of cranial nerve VII is situated in the caudal one-third of the ventral pontine tegmentum. Fibers from the ipsilateral and contralateral sides meet in this brainstem nucleus. From the nucleus, the fibers extend posteriorly toward the floor of the fourth ventricle and loop around the nucleus of cranial nerve VI (abducens). This portion of the facial nerve contributes to the facial colliculus, a mound of tis-

sue that protrudes internally towards the floor of the fourth ventricle. Extending from the loop, cranial nerve VII emerges from the brainstem at the anterolateral aspect of the pontomedullary junction. The nerve then extends laterally in the cerebellopontine angle cistern reaching the anterior superior quadrant of the internal auditory canal. The facial nerve extends through the temporal bone to the fundus of the internal auditory canal. It then exits to join the geniculate ganglion within the roof of the petrous bone (Figure 7). At this point, there is a synapse with the sensory neurons for taste to the anterior two-thirds of the tongue and the cutaneous fibers for light touch in the region of the external ear.

The superior salivatory nucleus in the pons is the site of origin of the parasympathetic fibers that terminate in and stimulate the lacrimal, sublingual, and submandibular glands. The nucleus solitarius represents the end point of the fibers that convey taste sensation from the anterior two-thirds (via cranial nerve VII) and from the posterior third (via cranial nerve IX) of the tongue. The cell bodies of these taste fibers are found in the geniculate ganglion. The impulses from these nuclei form the intermediate nerve fibers which join the motor root of the facial nerve as it exits the brainstem (Figure 2).

From the region of the geniculate ganglion, the secretomotor fibers project forward as the greater superficial petrosal nerve which joins some parasympathetic fibers from the carotid artery to form the nerve of the pterygoid canal. The pterygoid canal will transmit the nerve to the pterygopalatine fossa from where the parasympathetic components synapse and fibers extend to affect lacrimation and salivation.

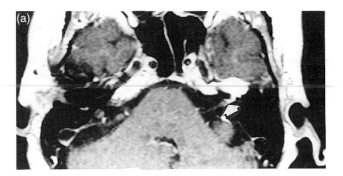

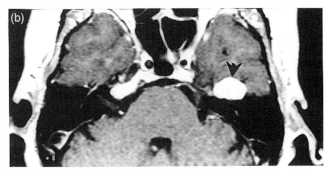

Figure 7 Axial enhanced images of the geniculate ganglion and proximal facial nerve schwannoma. (a) Note the tumor enhancement extending through the fundus of the left internal auditory canal to the geniculate ganglion (arrow). (b) The expansion of the tumor along the surface of the left petrous bone into the middle cranial fossa can be clearly seen (arrow)

For References see p. 1033

From the geniculate ganglion, the facial motor fibers and taste fibers form a 180° turn and are directed posteriorly, at the anterior genu of the facial nerve. Extending posteriorly, these fibers lie at the medial aspect of the tympanic cavity medial to the ossicular chain and below the lateral semicircular canal. At the posterior aspect of the middle ear, the motor and taste fibers extend inferiorly at the posterior genu of the facial nerve to enter the stylomastoid foramen. At this point, the stapedial nerve exits to supply the stapedial muscle which controls the mobility of the ossicles. The chorda tympani fibers for taste also separate from the seventh nerve within the stylomastoid foramen extending via the petrotympanic fissure and eventually joining the lingual branch of the mandibular nerve to participate in the sensory fibers for taste to the anterior two-thirds of the tongue. The seventh nerve then exits the stylomastoid foramen into the parotid gland and extends anteriorly to innervate the muscles of facial expression as well as the posterior belly of the digastric, the stylohyoid, and the platysma muscles.

Lesions of the facial nerve result in a facial palsy, which can be either peripheral or central. Central paralysis refers to a supranuclear injury that becomes manifest as paralysis of the contralateral muscles of facial expression, with sparing of the muscles of the forehead. Peripheral facial nerve function implies injury to the facial nerve from the level of the brainstem nucleus to the end of its motor fibers. All the ipsilateral muscles of facial expression are paralyzed in a peripheral injury. Dysfunction of the facial nerve branches within the greater superficial petrosal nerve is manifested as an impairment of ipsilateral lacrimation. Dysfunction of the stapedius branch of the facial nerve is manifested by hyperacusis. Dysfunction of the chorda tympani nerve causes loss of taste from the anterior two-thirds of the tongue. An impression on the 'root exit zone' of the facial nerve from the brainstem by a vascular anomaly may lead to spasmodic contractures of the ipsilateral muscles of facial expression (facial tic) (Figure 8).[1–3,6]

2.9   Cranial Nerve VIII (Vestibular)

Movement impulses are modulated by the hair cells within the utricle, saccule, and three semicircular canals and transmitted through the superior and inferior vestibular nerves (Figure 9). The combined vestibular nerve courses along with the cochlear nerve and facial nerve within the internal auditory canal. The superior and inferior vestibular divisions enter the brainstem at the pontomedullary junction after crossing the cerebellopontine angle cistern. These fibers extend into the vestibular nuclear complex with postganglionic fibers carrying information regarding equilibrium to many coordination sites including the folliculonodular lobe of the cerebellum, the vestibulospinal tract, both medial longitudinal fasciculi and other pathways which affect the control of the eyes and the muscles of balance (Figure 2).[1–3]

2.10   Cranial Nerve VIII (Cochlear)

Sound waves from the external environment are converted to energy in the inner ear in the form of vibratory impulses at the oval window of the vestibule. The kinetic impulses are propagated as fluid waves in the cochlea. These impulses are then converted to electrical impulses along the spiral organ of Corti. The auditory signals are then transmitted via the spiral ganglion

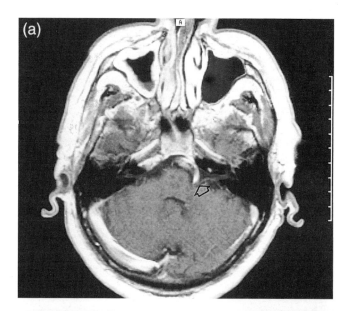

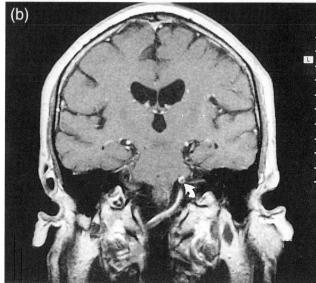

Figure 8 Enhanced images of dolichoectasia of the basilar artery compressing the root exit zone of the left facial nerve. (a) This axial view shows an indentation by the basilar artery on the brainstem at the origin of the left facial nerve (open arrow). (b) This coronal view documents the tortuous course of the left basilar artery, including the brainstem compression at the origin of the facial nerve (arrow)

in the cochlear nerve through the internal auditory canal and cerebellopontine angle to the origin of cranial nerve VIII at the ventrolateral pontomedullary junction. There are two cochlear nuclei (dorsal and ventral) located lateral to the inferior cerebellar peduncles in the upper medulla. The dorsal cochlear nucleus carries high frequencies, while the ventral cochlear nucleus connects low frequencies. Somewhat more than half of the fibers from these nuclei cross to the contralateral superior olivary nucleus to form the trapezoid body. From the trapezoid body, the fibers enter the posteriorly located lateral lemniscus to ascend to the inferior colliculus in the midbrain. From the inferior colliculus, the fibers travel to the medial geniculate

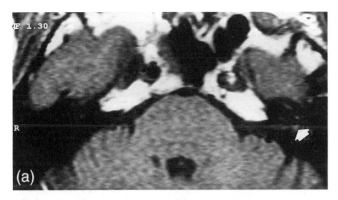

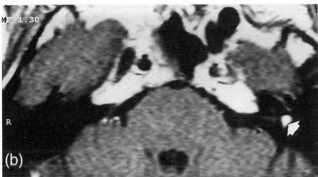

Figure 9 Vestibular schwannoma of the left utricle and saccule. (a) Unenhanced axial image. The small tumor of the membranous labyrinth is difficult to identify within the left petrous temporal bone (arrow). (b) Enhanced axial image. The bulbous tumor in the left vestibule is clearly shown (arrow)

body of the thalamus and then, via the auditory radiations to the superior temporal gyrus (Figure 2).

Clinical manifestations of acoustic pathway injury depend on the level of the injury. Lesions of the cochlear portion of the eighth nerve result in hearing loss and tinnitus (ringing or roaring in the ear). If unilateral sensorineural hearing loss is present, the injury has occurred somewhere between the cochlea and the cochlear nuclei of the brainstem. Unilateral involvement of the auditory pathway above the cochlear nuclei usually causes bilateral hearing loss due to the multiple crossings through the acoustic pathways. The hearing loss is greater in the ear contralateral to the site of the lesion. Cortical acoustic pathway lesions, which rarely cause disruption of auditory function, may result in auditory agnosia.[1–3,7]

2.11 Cranial Nerve IX (Glossopharyngeal)

Impulses from the motor cortex of the cerebral hemisphere extend to the genu of the internal capsule and subsequently follow the corticobulbar tract through the midbrain to the nucleus ambiguus. The fibers extend to the posterior olivary sulcus leaving the medulla at this point to extend anterolaterally to the superior and inferior glossopharyngeal ganglia situated in the jugular foramen. The fibers then travel inferiorly to supply the stylopharyngeus and constrictor muscles of the pharynx. In addition to this motor component to the muscles, there is a solitary nucleus for sensory fibers and a salivatory nucleus for

parasympathetic fibers. These nuclei represent the primary sensory neurons for general visceral and special sensation, including taste, for the posterior third of the tongue (Figure 2).

The sensory portion of the glossopharyngeal nerve joins the motor portion which then extend together into the pars nervosa of the jugular foramen (Figure 10). There is a small tympanic branch (Jacobson's nerve) which supplies sensation to the middle ear and eustachian tube as well as parasympathetic fibers to the parotid gland. A small visceral branch supplies the carotid sinus and carotid body which controls the pressor and chemoreceptor functions. Pharyngeal sensory branches supply the posterior oropharynx and soft palate. The lingual branch

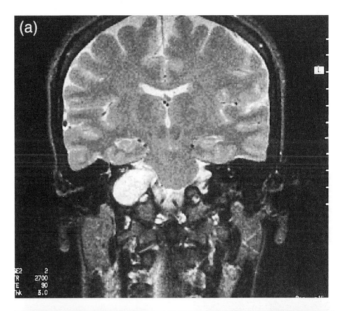

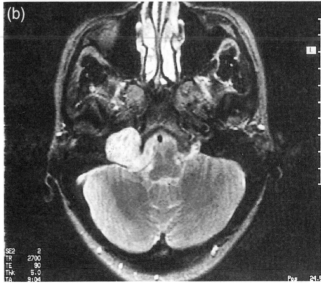

Figure 10 Schwannoma of the right glossopharyngeal nerve in the jugular foramen. (a) Coronal image. The homogeneously hyperintense tumor smoothly expands the right jugular foramen. The apex of the tumor is along the brainstem with its inferior portion extending through the right side of the skull base. (b) Axial image. The homogeneously hyperintense tumor mass in the right side of the skull base is clearly seen on this T_2-weighted image

For References see p. 1033

includes fibers for both sensation and taste for the posterior third of the tongue.

Clinical manifestations of ninth nerve dysfunction include otalgia (referred pain along the tympanic branch to the ear), dysphagia (stylopharyngeus muscle dysfunction), and tachycardia or bradycardia with hypotension (sinus nerve dysfunction). Other symptoms definitely related to ninth nerve injury are loss of the afferent limb of the gag reflex and loss of taste to the posterior third of the tongue.[1–3]

2.12   Cranial Nerve X (Vagus)

Impulses from the motor cortex of the cerebral hemisphere travel through the genu of the internal capsule and cerebral peduncle via the corticobulbar tract to the nucleus ambiguus. The fibers extend to the posterior olivary sulcus slightly more lateral than cranial nerve IX exiting the medulla in the groove between the inferior peduncle and olive. Cranial nerve X continues an anterolateral course with its companion nerves (cranial nerve IX and the bulbar portion of cranial nerve XI) through the perimedullary cistern. The vagus nerve and bulbar portion of the eleventh cranial nerve enter the anterior aspect of the pars vascularis of the jugular foramen. As previously discussed, the ninth cranial nerve enters the pars nervosa of the jugular foramen. After exiting the jugular foramen, the vagus nerve plunges like a plumb line along the posterolateral aspect of the carotid artery to the aortopulmonic window of the mediastinum on the left side and to the clavicle on the right side. The right recurrent laryngeal branch turns cephalad around the right subclavian artery, while the left recurrent laryngeal branch turns cephalad by looping through the aortopulmonic window. Both recurrent laryngeal nerves reach the larynx via the tracheoesophageal groove. Sensory fibers of the vagus nerve begin in the thoracoabdominal viscera and chemoreceptors at the aortic arch. These then synapse in the solitary nucleus (pure sensory fibers) and the dorsal motor nucleus (parasympathetic secretomotor fibers) (Figure 2).

Injury to the vagus nerve below the hyoid bone causes manifestations of recurrent laryngeal nerve malfunction and include hoarseness, aspiration, and cervical dysphagia. Injury to the vagus nerve above the hyoid bone results in the above symptoms in combination with oropharyngeal symptoms of uvular deviation and loss of the efferent limb of the gag reflex secondary to pharyngeal plexus malfunction. Proximal vagal neuropathy usually involves some combination of cranial nerves IX, X, XI, and XII.[1–3,8]

2.13   Cranial Nerve XI (Spinal Accessory)

Impulses from the motor cortex of the cerebral hemisphere extend through the genu of the internal capsule and cerebral peduncle via the corticobulbar tract to the nucleus ambiguus. The motor fibers of the bulbar portion of the spinal accessory nerve are joined by additional fibers which arise from the anterior horn cells of the first five cervical cord segments. The fibers from the spinal nucleus have ascended through the foramen magnum. Together, the combined fibers from the bulbar and spinal portions of the spinal accessory nerve pass anteriorly to exit from the medulla at the posterior olivary sulcus. The eleventh cranial

nerve joins the companion cranial nerves IX and X to pass through the perimesencephalic cistern. The spinal accessory nerve then leaves the skull via the posterior portion of the pars vascularis of the jugular foramen. From the skull base, the eleventh cranial nerve enters the carotid space and then quickly diverges posterolaterally to descend along the medial aspect of the sternocleidomastoid muscle. It then continues its course across the posterior triangle of the neck to terminate in the trapezius muscle.

Clinical manifestations of spinal accessory nerve injury are often found in conjunction with ninth, tenth, and twelfth nerve symptoms. When isolated, the symptoms will consist of shoulder droop and an inability to lift the arm.[1–3]

2.14   Cranial Nerve XII (Hypoglossal)

Impulses from the motor cortex travel via the corona radiata through the genu of the internal capsule and cerebral peduncle via the corticobulbar tract to reach the hypoglossal nucleus, which is found in a paramedian location in the floor of the fourth ventricle. Hypoglossal nerve root fibers as well as fibers from the nucleus ambiguus exit forward from the medulla passing lateral to the medial lemniscus. The fibers exit from the brainstem in the sulcus between the pyramid and olive as multiple small rootlets. The rootlets extend anteriorly to form the hypoglossal nerve extending anterolaterally through the hypoglossal canal. Once exited from the skull base, the fibers descend inferiorly in the carotid space lying medial to cranial nerves IX, X, and XI and passing lateral to the carotid bifurcation. The hypoglossal nerve then continues forward to enter the posterior portion of the sublingual space of the oral cavity. From here it curves upward medially to the submandibular salivary glands to supply the intrinsic and extrinsic muscles of the tongue and the infrahyoid strap muscles.

Clinical manifestations of hypoglossal nerve injury include deviation of the tongue toward the side of the lesion on protrusion. Tongue atrophy, including both the intrinsic and extrinsic muscles will be evident in chronic lesions. Fasciculations of the tongue muscles may also be seen.[1–3]

3   TUMORS OF CRANIAL NERVES

3.1   Neural Tumors

Schwannoma and neurofibroma are the two most common tumors of the cranial and peripheral nerves. Both tumors are derived from Schwann cells, but present quite differently. Schwannomas are usually solitary masses arising from more proximal nerves, while neurofibromas may be multiple, and usually arise in more distal nerves (Figure 11). Histologically, schwannomas show areas of high and low cellularity known as Antoni-A and Antoni-B areas, respectively, while neurofibromas show spindle cells with wavy nuclei not seen in schwannomas. Another differentiation between these two tumors is the presence of a capsule around schwannomas, whereas neurofibromas are usually not encapsulated. In many cases, schwannomas can be removed by surgery without transecting the nerve. Neurofibromas, however, are part of the nerve which has to be excised with the tumor.

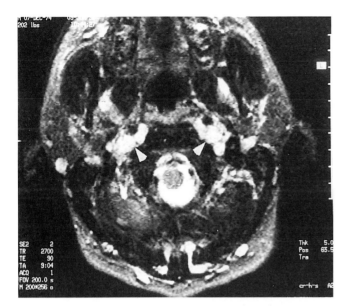

Figure 11 Multiple neurofibromas in a patient with neurofibromatosis I. There are bilateral neurofibromas in the carotid spaces which surround the carotid arteries. The lesions are readily identified by their high signal intensity (arrows). Additional smaller neurofibromas are seen in the posterior scalp region

3.2 Schwannoma (Neurilemoma)

Schwannoma is a benign tumor which arises from the Schwann cells surrounding the cranial nerves, peripheral nerves and sympathetic plexus. The Schwann cells represent structures that both encircle the axons of the nerves and provide the nerves with mechanical support and their myelin sheaths. Histologically, a schwannoma contains a cellular component (formed from the Antoni-A cells) and a myxoid component (formed from the Antoni-B cells). The presence of these components and the absence of nerve fibers in the body of the tumor allow the pathologist to differentiate a schwannoma from a neurofibroma.

Schwannomas of the olfactory system and visual pathway are rare, since cranial nerves I and II are not true cranial nerves, but rather are embryologic invaginations of fiber tracts from the telencephalon and diencephalon. The remaining cranial nerves and the upper spinal nerves are the sites of origin of the majority of schwannomas. These tumors are more common in women aged 30–60 years. The signs and symptoms vary according to the specific site of origin.

Schwannomas of cranial nerves III, IV and VI may be seen along the cisternal, cavernous or intraorbital course of these nerves. Symptoms include diplopia, ophthalmoplegia or proptosis, depending upon the site of greatest mass effect. These tumors may be visualized on MRI as they course through the cranial canals. Bulbous expansion of the canals at the site of a lesion is considered pathognomonic; however, differentiation of schwannomas from various malignancies extending along nerve roots may be difficult.

Schwannomas of the trigeminal ganglion and trigeminal nerve usually cause progressive facial numbness, pain, and paresthesias. Some trigeminal tumors can attain a considerable size in the absence of facial pain or numbness, since they may arise from the nerve sheath and secondarily compress the nerve fibers. Schwannomas involving the trigeminal nerve are divided into three anatomical divisions: preganglionic, ganglionic, and postganglionic. Lesions that originate from the trigeminal ganglion typically straddle the incisura with a component extending into the cavernous sinus (postganglionic) and another portion extending into the prepontine cistern (preganglionic) (Figure 12).[9-12]

Facial nerve schwannomas may present with facial nerve palsy and/or hearing loss, or may be relatively asymptomatic. The development of facial nerve schwannomas from the nerve sheaths allows some of these tumors to enlarge and decompress into the adjacent air-containing cavities of the temporal bone

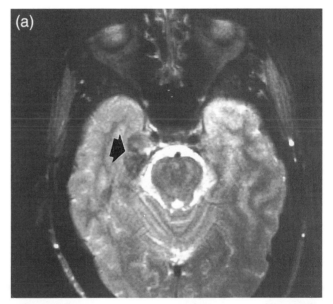

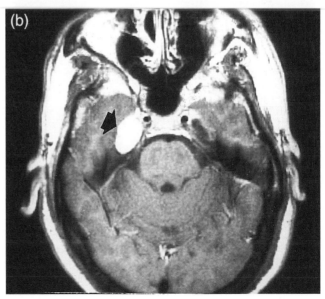

Figure 12 Schwannoma of the right trigeminal ganglion. (a) There is a primarily hypointense lesion in the right Meckel's cave and trigeminal ganglion region (arrow). (b) This enhanced view documents homogeneous enhancement of the tumor mass (arrow)

For References see p. 1033

without producing symptoms. Schwannomas originating from the geniculate ganglion typically become quite large without producing facial nerve dysfunction. These lesions may expand into the superior surface of the temporal bone and present as middle fossa masses (Figures 7 and 13).

Acoustic nerve schwannomas originate from the vestibulo-cochlear nerve along its course from the vestibule and cochlea, internal auditory canal, and cerebellopontine angle cistern. These tumors represent 5–10% of all primary intracranial neoplasms and approximately 85–90% of all cerebellopontine angle tumors. The clinical symptoms of acoustic nerve schwannoma are typically unilateral or asymmetric neurosensory hearing loss. These symptoms are directly related to the size of the tumor. As the tumor enlarges and fills the internal auditory

canal, it compresses the cochlear nerve, causing neurosensory hearing loss and tinnitus, which are the most common initial symptoms (Figure 13). The early hearing loss is of the high-frequency variety, which may be related to the anatomic arrangement of the high frequency fibers around the periphery of the nerve. The hearing loss is typically gradual and progressive, but may have an acute onset. Vestibular dysfunction, such as dizziness and disequilibrium, are relatively uncommon presenting symptoms, even though 85% of acoustic nerve schwannomas arise from the vestibular nerves. Small intracanalicular lesions may be either relatively asymptomatic or may cause progressive symptoms. Some larger neoplasms may develop intramural or extramural hemorrhage with the sudden onset of headache, nausea, and vomiting. True central vertigo

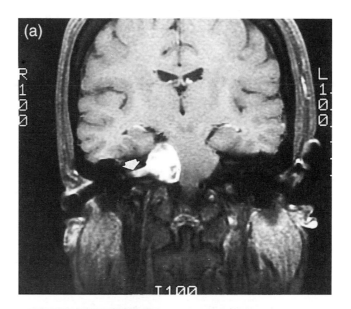

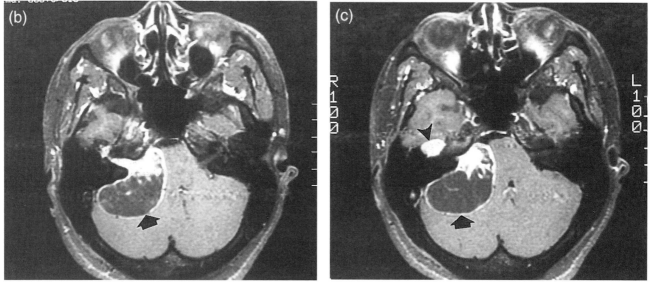

Figure 13 Large schwannomas of the right seventh and eighth cranial nerves extending from the geniculate ganglion to the cerebellopontine angle (CPA). The patient had marked hearing loss, but only minimal facial nerve dysfunction. Enhanced images with fat saturation pulses. (a) This coronal image shows the tumor in the right CPA and internal auditory canal (arrow). (b and c) Axial images. The portion of the tumor in the right CPA shows both solid and cystic components (arrows). The solid component extends through the right internal auditory canal where it merges with a companion tumor in the right geniculate ganglion [arrow, (c)]

may be associated with brainstem involvement. Facial nerve dysfunction may be related to compression of the seventh nerve.

Jugular foramen schwannomas can originate from different sites along cranial nerves IX, X or XI. Early symptoms consist of loss of taste in the posterior third of the tongue (glossopharyngeal nerve), paralysis of the vocal cord and palate (vagus nerve), and paresis of the trapezius and sternocleidomastoid muscles (spinal accessory nerve). Proximal schwannomas may extend into the inferior portion of the posterior fossa, producing hearing loss, vertigo, ataxia, hoarseness, and swallowing difficulties (Figure 10). Schwannomas of the twelfth nerve usually cause hemiatrophy and paresis of the tongue. Some large tumors of the jugular foramen can extend into the hypoglossal canal and large hypoglossal schwannomas may expand into the adjacent jugular foramen. Concomitant involvement of cranial nerves IX, X, XI, and XII is rare, but may be seen with bulky tumors. Additional disturbances of posterior fossa function may include gait abnormalities, swallowing difficulties, dizziness, and vomiting. In the neck, schwannomas are the most common neoplasm of the carotid space and the second most common lesion of the parapharyngeal space. Clinical presentation is a painless, slow growing mass in the anterolateral neck or posterolateral oropharynx. These lesions may become painful with associated neuropathies if they continue to grow and compress the nerves.[11-13]

The MRI appearance of all schwannomas is similar. The key differentiating point is the location of the central portion of the tumor within or near the nerve of origin. Schwannomas are fairly fusiform in appearance and well circumscribed because of their associated capsule. There may be smooth and uniform enlargement of the cranial canal through which the nerve is transmitted. The signal intensities, however, differ dramatically with respect to the size of the tumor. Smaller tumors usually appear homogeneous (since they are predominantly of the Antoni-A cell type), while larger tumors demonstrate heterogeneity and variable signal intensities owing to the presence of hemorrhage, necrosis, and cystic degeneration (Figure 13). Larger tumors typically contain many Antoni-B type cells. These schwannomas can have foci of high and low signal intensities on both the T_1- and T_2-weighted images. The presence of intramural cysts will characteristically produce foci of high signal intensity on the T_2-weighted images (Figure 14).[9,12]

One MRI study suggested that cystic changes in acoustic nerve schwannomas may be related to either intramural cysts (Figures 13 and 14) or extramural arachnoid cysts (Figure 15). The intramural cysts showed high signal intensity on the T_2-weighted images, which was thought to be due to necrotic material, blood, or colloid-rich fluid. The extramural cysts showed high signal intensity related to higher protein and/or colloid contents secreted by the tumor. The extramural cysts were thought to originate from peritumoral adhesions which caused a pseudoduplication by the trapping effect of fluid between the leptomeninges and the schwannoma.[14]

Enhancement following gadolinium contrast administration is also variable, although it is typically rapid due to extravasation of contrast into the extracellular compartment of the often vascular schwannoma. In nearly all cases, there is a sharp demarcation between the margin of the tumor and the adjacent structures. Enhancement of the cranial nerves or their ganglia without an associated soft tissue mass is consistent with cranial

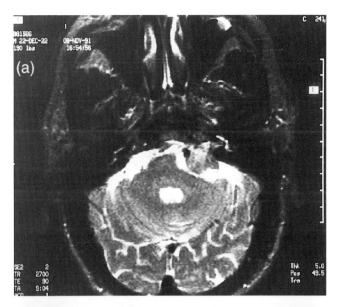

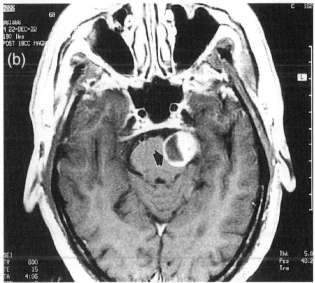

Figure 14 Acoustic nerve schwannoma with solid and intramural cystic components. (a) This axial image shows the heterogeneous mass in the left internal auditory canal and cerebellopontine angle cistern. The cystic component of the tumor merges with the bright signal intensity of the cerebrospinal fluid. (b) This axial enhanced image shows the solid and cystic components. Note that the cystic portion shows peripheral enhancement confirming an intramural cyst (arrow)

nerve neuritis and/or ganglionitis rather than a neoplasm. Sarcomas, hemangiomas, or paragangliomas may have similar MRI enhancement characteristics and need to be considered in the differential diagnosis of schwannomas.[12,15]

3.3 Neurofibroma

A neurofibroma contains neural elements with a fibrous core. This tumor may arise from the cranial, spinal, or peripheral nerves. The patient's symptoms depend upon the nerve of involvement and may be similar to the symptoms associated with a schwannoma. In the extracranial head and neck, neurofi-

For References see p. 1033

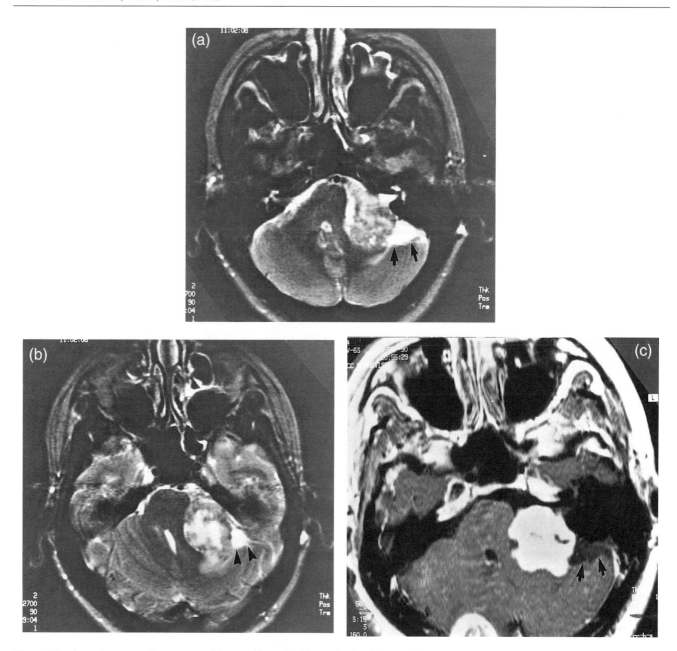

Figure 15 Acoustic nerve schwannoma with a small arachnoid cyst in the left cerebellopontine angle. (a and b) The large tumor is highly heterogeneous and deforms the left side of the brainstem and cerebellum. Note the homogeneously bright signal in the small left arachnoid cyst (arrows). (c) This image shows enhancement of the tumor. The arachnoid cyst, however, is extramural and does not enhance (arrows)

bromas may present as subcutaneous soft tissue nodules in the skin or as a mass lesion in the carotid or parapharyngeal spaces (Figure 11). In such cases, the clinical symptoms result from tumor bulk and may include dysphagia or difficulty turning the neck. Some neurofibromas cause pain or other focal neurologic deficits.

Most neurofibromas are seen in young adult patients. Approximately 10% of patients with neurofibromas have neurofibromatosis type I (NF-1 or von Recklinghausen's disease). Two types of neurofibroma are known to be associated with NF-1. Multiple or single solid lesions may be seen along the cranial, spinal, or peripheral nerves (Figure 11). Another type of lesion presents as an infiltrating soft tissue mass with extension along the cranial nerves through the orbits, skull base, or posterior fossa. This so-called 'plexiform' neurofibroma encases the muscles and soft tissues along the course of the nerves (Figure 5).[16]

The MRI characteristics of neurofibromas are quite varied and simulate schwannomas in many cases. Neurofibromas project an intermediate signal intensity on the T_1- and proton-density-weighted images. They sometimes have a 'salt and pepper' appearance whenever there is intense vascularity within the stroma of the tumor. The T_2-weighted images are variable, depending on whether the tumor is cystic or solid in nature. Variable enhancement is seen following gadolinium administration due to the presence of cystic areas, calcifications, or vascularity. The margins of a neurofibroma appear less distinct than the margins of a schwannoma, because neurofibromas lack a capsule. In many cases, it may not be possible to differentiate a neurofibroma from a schwannoma on MRI characteristics.[12,15]

3.4 Malignant Schwannoma

Many terms have been used to describe a malignant schwannoma, including malignant neurofibroma, neurofibrosarcoma, neurosarcoma, neurogenic sarcoma, malignant neurilemoma, and malignant nerve sheath tumor. Many malignant schwannomas contain interlacing fascicles of spindle cells in a herring-bone pattern, as typically seen in fibrosarcomas. Unlike fibrosarcomas, however, on electron microscopy malignant schwannomas are seen to contain basement membranes. The light microscopic findings include high cellularity, pleomorphism, high incidence of mitotic figures, presence of necrosis, and presence of histologic invasion. There is no capsule around a malignant fibrosarcoma.[16]

Malignant schwannomas are highly malignant lesions that recur following local excision in 50–80% of patients. The histologic grade is the most important factor in determining prognosis and is the primary factor considered in staging. Five-year survival rates in one study of 17 patients over the course of 37 years was 47%. Wide and aggressive excision with or without adjuvant radiation therapy is considered the best treatment alternative. Generally, patients with malignant schwannomas of the head and neck have a poorer prognosis than do those with similar tumors of the extremities, which can be treated with amputation. One factor contributing to the poor prognosis of this tumor is its resistance to radiation therapy.

The incidence of malignant schwannomas is approximately 5–10% of all sarcomas and 2–12% of all nerve sheath tumors. Patients with NF-1 have a 5–30% chance of developing a de novo malignant neurogenic tumor. The typical NF-1 patient with a malignant neural tumor is 20–50 years old. The 5-year survival rate in patients with NF-1 is 15–30%. Recurrences and metastases may occur as late as 5–10 years following treatment.

The tumor initially causes a fusiform, nodular swelling of the involved nerve and ultimately diffusely infiltrates the nerve. It then spreads into the surrounding soft tissues and often contains areas of hemorrhage or necrosis (Figure 16). The tumor may cause neurologic symptoms such as pain, paresthesias, atrophy, and muscle weakness. Occasionally, a symptomless mass along a nerve distribution is seen.[12,15,16]

The MRI appearance is of a tubular mass along the involved nerve with regional extension into the skull base (cranial nerve) or cervical soft tissues. The signal intensity is varied, but is typically hypointense on the T_1-weighted images and hyperintense on the T_2-weighted images. Gadolinium contrast administration helps to outline the margins of the tumor including extension beyond the site of origin into the spinal canal or cranial cavity (Figure 17)[12,15]

3.5 Cranial Nerve Seeding

Both metastatic neoplasms and primary malignant tumors of the central nervous system can disseminate via the subarachnoid cisternal pathways to distant locations. This type of metastatic spread may be manifested as either a diffuse spread along the leptomeningeal surfaces of the brain or loculated deposits of tumor nodules along specific sites within the subarachnoid spaces such as the cranial nerves. The most common dissemination in children is from primitive neuroectodermal tumors and ependymomas. The most common cause of seeding in adults is from a glioblastoma multiforme. The incidence of seeding from a glioblastoma is less than 5% in all age groups. Other tumors that can seed to the leptomeninges or cranial nerves are sarcomas, carcinomas, and adenocarcinomas. Leptomeningeal metastases from various lymphoproliferative tumors are also common in both children and adults. Metastatic lymphomatous foci may be deposited adjacent to the cranial nerves along their courses through the posterior and middle cranial fossae (Figures 3 and 18).[16,17]

Intracranial leptomeningeal metastases tend to be diffuse along the subarachnoid spaces and cranial nerves. Microscopically, the tumor cells infiltrate the leptomeninges as a single layer or as thicker, multilayered aggregates. The earliest infiltration is along cortical vessels and is limited to the perivascular Virchow–Robin spaces. Clinically, patients most often present with a variety of symptoms, which are usually multifocal. Headache, change of mental status, cranial nerve deficits, and gait disturbances are the most common.

These metastatic deposits are visualized best on the T_1-weighted postcontrast images, particularly if fat saturation techniques are utilized. The areas of nodular or linear enhancement follow the contours of the meningeal surfaces or cranial nerves as they traverse through the cerebrospinal fluid

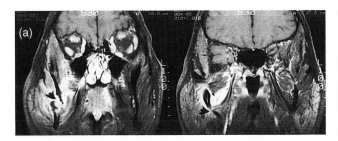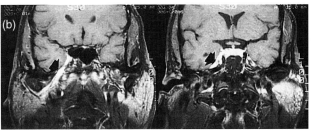

Figure 16 Malignant schwannoma of the right trigeminal and inferior alveolar nerves. Coronal enhanced images with fat saturation pulses. (a) This composite image documents the extensive infiltration of the tumor in the right muscles of mastication. There is a pathologic fracture of the right ramus of the mandible (arrows). (b) This composite view documents the extension of the tumor in a tubular manner through the right foramen ovale into the right trigeminal ganglion (arrows)

For References see p. 1033

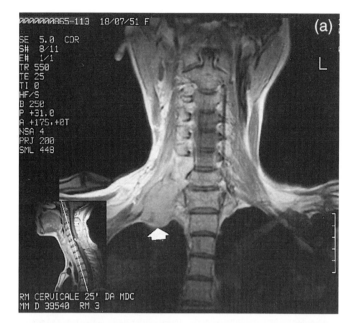

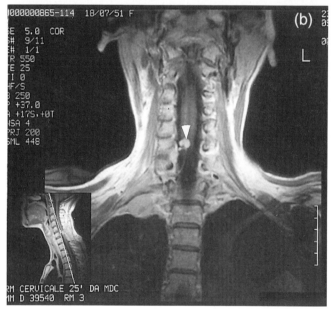

Figure 17 Coronal enhanced images of a malignant schwannoma of the right brachial plexus with intraspinal extension. (a and b) The tumor fills the region of the scalene muscles and brachial plexus. There is downward extrapleural extension [arrow, (a)]. Note a small intraspinal portion along the cervical nerve roots [arrow, (b)]

(Figure 19). The T_2-weighted images are less sensitive because of the surrounding high signal intensity of the cerebrospinal fluid.[12,15]

3.6 Perineural Spread of Malignant Neoplasms

Perineural tumor extension is a form of metastatic disease in which primary tumors spread along neural pathways and gain access to noncontiguous regions. Since the treatment and prognosis are altered when perineural spread is identified, proper

evaluation is critical. The infiltration of the nerves by tumor is best seen on gadolinium enhanced MRI.[18,19]

Perineural tumor spread is visible as smooth thickening of the cranial or spinal nerves and/or concentric expansion of the foramina and fissures through which the nerves traverse. Involvement of the maxillary and mandibular divisions of the trigeminal nerve is common because of the large size of these branches and their strategic passage through the pterygopalatine fossa. In cases of perineural spread to the trigeminal ganglion, there may be replacement of the normal trigeminal cistern hypointensity by an isointense or enhancing soft tissue mass (Figure 20). Other signs of trigeminal ganglion involvement include lateral bulging of the cavernous sinus dural membranes and atrophy of the masticator muscles supplied by the trigem-

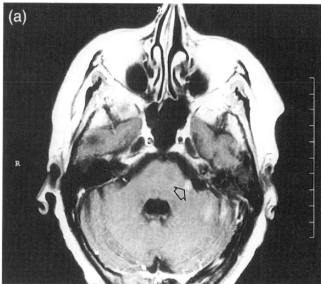

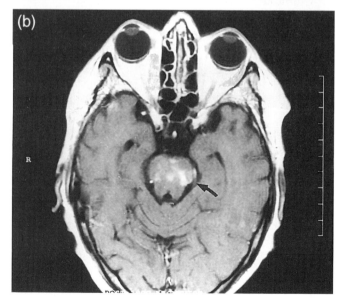

Figure 18 Lymphomatous meningitis of the brainstem and cranial nerves. Axial enhanced images. There is a focal tumor nodule in the left cerebellopontine angle region [arrow, (a)]. There is diffuse infiltration of the midbrain and enhancement along the left trochlear nerve [arrow, (b)]

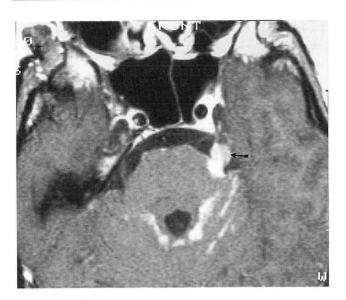

Figure 19 Axial enhanced image of seeding of metastatic adenocarcinoma to the subarachnoid spaces and cranial nerves. There are sheets and nodules of tumor surrounding the midpons and cerebellar folia. A large focal nodule occupies the origin and preganglionic portions of the left trigeminal nerve (arrow)

inal nerve branches (Figure 6). Perineural metastatic disease may occur along the course of the facial nerve, most commonly resulting from neoplasms of the parotid gland, particularly adenoid cystic carcinoma.[15,18,19]

4   INFECTIONS AND INFLAMMATIONS OF CRANIAL NERVES

Infections of the cranial nerves commonly are caused by a latent virus such as herpes simplex I or varicella zoster. These neurotropic viruses tend to affect the cell bodies of the sensory neurons, in which they may reside for many years. Reactivation of the dormant virus results in a swelling of the involved ganglion or of the involved nerve. After contrast medium administration, there may be enhancement in the ganglion or along the course of an infected nerve.[15,20]

The most common viral mononeuritis is Bell's palsy, which is thought to be caused by reactivation of a herpes simplex virus in the geniculate ganglion. Following gadolinium administration, there may be enhancement along the entire course of the facial nerve from the internal auditory canal through the mastoid foramen. More commonly, the portion of the facial nerve in the region adjacent to the geniculate ganglion is enhanced as a result of entrapment of the seventh cranial nerve in the fallopian canal (Figure 21). Herpes simplex trigeminal neuritis and ganglionitis have been described, revealing involvement of the fifth cranial nerve on contrast enhanced MRI (Figure 22).[21,22]

The clinical prevalence of herpetic involvement of the cranial nerves is most likely underestimated. However, in patients infected with the human immunodeficiency virus (HIV), it is fairly common to recognize postgadolinium enhancement along one, two or more cranial nerves. The mechanism is similar to that of herpes simplex virus, since the HIV virus is also neurotropic and directly invades sensory ganglia and their associated nerves.[20,21]

5   OPTIC NERVE AND SHEATH DISORDERS

Primary disorders of the optic nerve and optic nerve sheath typically cause visual disturbances and occasionally proptosis in cases of large neoplasms. The most common primary neoplasms of the optic nerve include gliomas (astrocytomas and hamartomas) or meningiomas. There may be a paraoptic component of optic nerve tumors, which is characteristically seen in patients with neurofibromatosis type I. The paraoptic component results from perineural arachnoidal gliomatosis, consisting of proteinaceous material that seeds and spreads within the optic nerve subarachnoid space and is a portion of the tumor. This process may result in optic nerve elongation and resultant kinking of the nerve just posterior to the posterior portion of the globe. There is no correlation between the size of the lesion and visual loss.

Extension of tumors into the optic chiasm, optic tracts, and lateral geniculate bodies of the thalami is more accurately depicted on MRI than computed tomography (CT). The size and shape of the optic canals are best assessed in the axial projection, while the size and shape of the optic nerves are best appreciated on coronal and oblique sagittal images. Many optic nerve tumors exhibit fusiform homogeneous enhancement (Figure 23). Unenhanced portions of optic nerve tumors may represent the sites of gliomatosis, rather than a true neoplasm. Biopsy is essential to make the correct diagnosis.[23,24]

Meningiomas are the most common optic nerve sheath tumors; they usually arise from the arachnoidal coverings of the optic nerve. Visual loss and optic atrophy are the usual presenting symptoms. Most cases are seen in middle-aged or elderly patients, more often in women. The tumor may be cuff-like surrounding the optic nerves or eccentrically located on only one side of the nerve. CT scans will often demonstrate calcifications and show typical postcontrast enhancement parallel to the length of the optic nerves. MR scans readily depict the irregular thickening along the optic nerves and spread into adjacent meninges on the postcontrast scans (Figure 24). Parallel optic cysts may be identified surrounding the optic nerve immediately distal to the meningioma. The process causes trapping of the cerebrospinal fluid in the subarachnoid space and can add to the mass effect and proptosis.[23,25]

Optic neuritis is seen on MRI as focal or diffuse enlargement of the optic nerve associated with abnormal signal intensities and enhancement (Figure 25). This is caused by immunological inflammation of the optic nerve, which affects the myelin sheaths with relative preservation of the axons. As optic neuritis is the initial manifestation of multiple sclerosis in about 20% of patients and may occur at some point in the disease in approximately 50%, the role of MRI in the evaluation of acute optic neuritis is changing. A multicenter trial was developed to assess the efficacy of corticosteroid treatment for acute optic neuritis. This study showed that the utility of MRI in establishing the diagnosis of optic neuritis was limited; however, with

For References see p. 1033

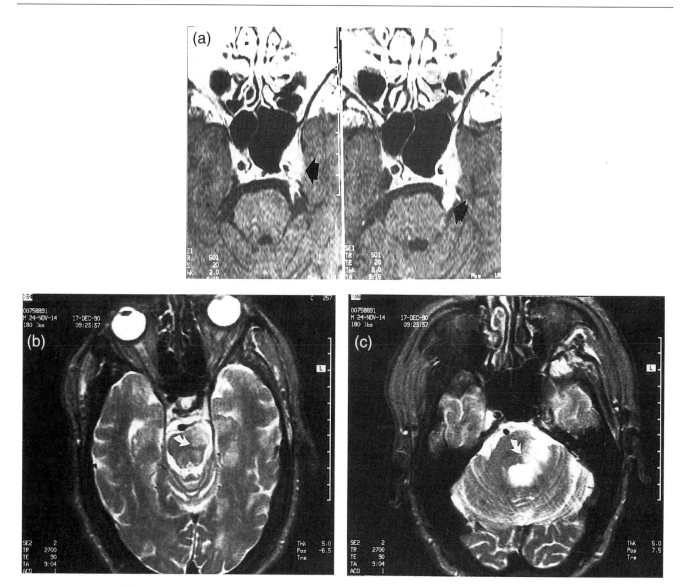

Figure 20   Perineural spread of squamous cell carcinoma. (a) This axial enhanced composite image shows bulging of the left side of the cavernous sinus and tumor extension into the preganglionic portion of the left trigeminal nerve (arrows). (b and c) Axial images. There is well defined edema and tumor infiltration into the left cerebral peduncle, pons, and middle cerebellar peduncle (arrows). The wide extent through the brainstem correlates with the long nuclei of the fifth cranial nerve

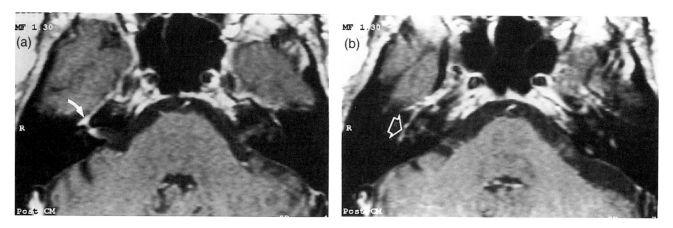

Figure 21   Facial neuritis and geniculate ganglionitis in a patient with Bell's palsy. Axial enhanced images. (a) Note the enhancement in the fundus of the right internal auditory canal with extension into the adjacent geniculate ganglion (arrow). (b) There is enhancement along the tympanic segment of the right facial nerve (arrow)

For list of General Abbreviations see end-papers

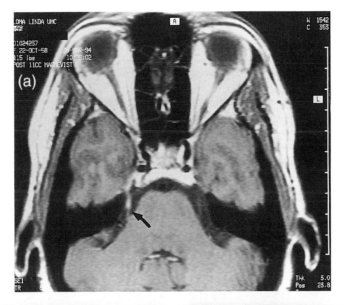

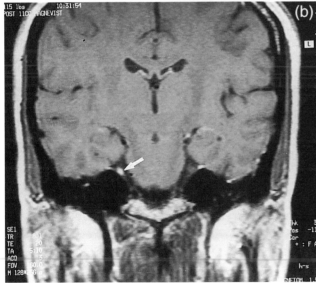

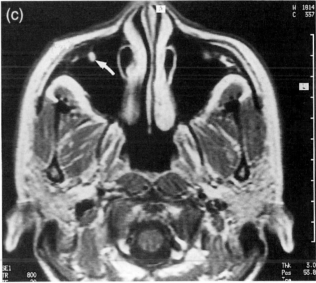

Figure 22   Enhanced images of viral trigeminal neuritis in a patient with trigeminal neuropathy. (a) There is enhancement along the preganglionic portion of the right trigeminal nerve (arrow). (b) This coronal view confirms the enhancement of the cisternal portion of the right trigeminal nerve (arrow). (c) The enhancement extends along the infraorbital division of the right maxillary nerve into the infraorbital groove (arrow)

additional views of the cranial cavity, abnormal MRI scans could differentiate those patients who would later develop multiple sclerosis from those that would not. Thus, MRI is a predictor of multiple sclerosis as it can help to identify a subgroup of patients whose risk of developing the disease appears to be low and it can provide prognostic data about the development of multiple sclerosis after optic neuritis.[23,26]

The papilledema associated with the pseudotumor cerebri may be detected on CT or MRI scans as enlargement of the optic nerve sheaths. If severe, there is a reversal of the optic nerve head with bulging forward into the posterior wall of the globe. This phenomenon is more readily detected on CT than MRI because of the chemical shift artifact inherent to the MR studies. There is some correlation between the severity of the visual loss and the detection of enlarged nerves with reversed nerve heads. Patients with more severe visual loss demonstrate

more frequent and more severe reversal of the optic nerve head.[27]

5.1   Optic Nerve Neuropathies

Isolated visual disturbances may result from a variety of optic nerve neuropathies, caused by radiation, chemotherapy, compressive phenomena, or ischemia.[28–30] Radiation-induced optic neuropathy (RON) is a rare catastrophic complication of radiation therapy regimens used to treat a variety of neoplasms of the skull base, sella, and parasellar regions (Figure 26). There is typically a latency period of 6 to 36 months following treatment. Clinically, the nerve head may appear normal, but gadolinium-enhanced MRI will show patchy, linear, or confluent enhancement along the portions of the optic nerve, chiasm,

For References see p. 1033

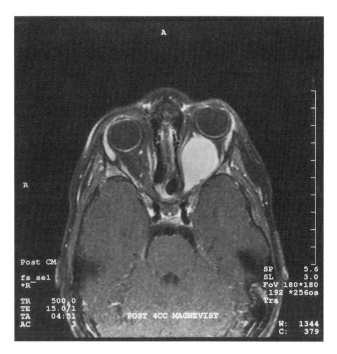

Figure 23 A 4-year-old boy with large left optic nerve glioma. There is a pear-shaped densely enhancing tumor mass in the left orbit causing mild proptosis. Note extension of the tumor into the orbital opening of the left optic nerve canal

or optic tract.[25,28] These findings precisely correlate as the cause of delayed visual loss following radiation therapy. Treatment of RON with corticosteroids may be helpful, although optic atrophy may develop with permanent visual loss.[28]

Compressive optic nerve neuropathies can be seen with a variety of lesions in the orbital apex [including Idiopathic Orbital Pseudotumor (IOP), thyroid ophthalmopathy with muscular hypertrophy, and edema causing compression of the intraocular portion of the optic nerve] or other systemic diseases with common orbital manifestations, such as sarcoidosis or Wegener's granulomatosis.[29–36]

Compression of the optic nerve may occur as a result of cavernous carotid fistulae, arteriovenous malformations, or orbital varices. Such vascular anomalies may produce retrograde flow through the ophthalmic vessels, with subsequent dilatation of the orbital veins and passive congestion of the orbital tissues. This leads to progressive increases in the intraocular pressure and subsequent decreases in visual acuity leading to blindness. Spontaneous thrombosis of the ophthalmic veins occasionally occurs and may aggravate the mass effect within the apex of the orbit and the compressive neuropathy.[23,33]

MRI will demonstrate the dilated ophthalmic veins, facial veins, and other regional venous structures along with enlargement of the cavernous sinus. Large edematous extraocular muscles and other periorbital structures may be identified. These findings are optimally seen with MRI, particularly as the addition of MR angiography allows for flow assessments along with the static morphologic changes. In some cases, conventional angiography may be required to make the definitive diagnosis, although most commonly this is used in conjunction with therapeutic interventional procedures.

6 VERTIGO AND HEARING LOSS

6.1 Dizziness and Vertigo

Dizziness is a common clinical complaint. It accounts for 1% of visits to US office-based physicians. Vertigo is a form of dizziness in which there is an illusion of movement (rotation, tilt, or linear translation). Vertigo occurs through an imbalance of tonic vestibular signals; consequently, it is a hal-

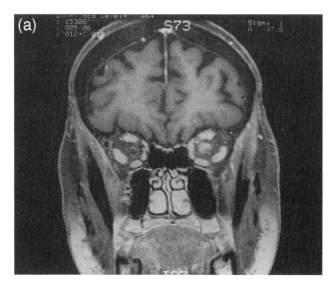
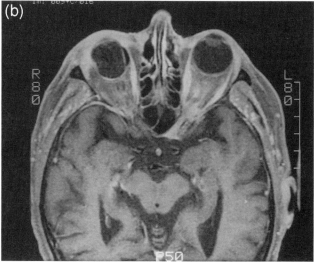

Figure 24 An 80-year-old female with left optic sheath meningioma. (a) Coronal enhanced view shows an enlarged left optic nerve with near complete enhancement of the lesion. (b) Axial enhanced MR scan shows the extent of the tumor in a 'railroad track' manner about the left optic nerve, with extension into the intracanalicular portion of the left optic nerve. Incidental note is made of a retinal banding procedure around the right globe

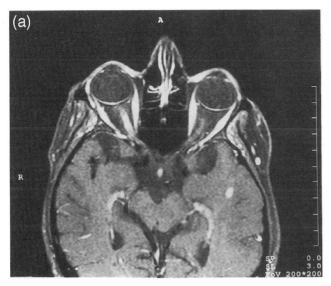 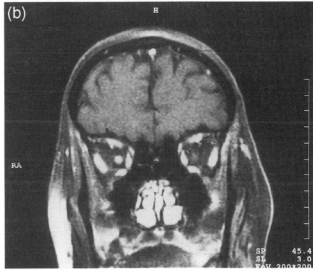

Figure 25 Right optic nerve neuritis in a 40-year-old female. (a) Axial enhanced MR scan showing no enlargement of the right optic nerve. There is enhancement along a major portion of the intraorbital part of the optic nerve. (b) Coronal enhanced scan showing the normal size but enhancing nerve in the right eye

lucination of movement and represents a symptom of a disturbed vestibular system.[37,38]

The complete vestibular system comprises the end organs in the temporal bone, the vestibular components of the VIIIth cranial nerve, and the central connections in the brainstem. The end organs in the temporal bones are the cristae of the three semicircular canals, which respond to movement of the head, and the macula of the utricle, which records the position of the head. The semicircular canals record dynamic actions while the utricle records static function. Vertigo is subdivided into peripheral vertigo (caused by failure of the end organs) or central vertigo (caused by failure of the vestibular nerves or central connections to the brainstem and the cerebellum.[38,39]

6.2 Peripheral Vestibular Disorders

Patients with benign positional vertigo describe episodic vertigo lasting less than a minute, brought on by movements of the head and without other associated symptoms. There are no radiological findings in patients with benign positional vertigo.[37,39]

In Meniere's disease, paroxysmal attacks of whirling vertigo are usually accompanied by nausea; the attacks are transient in nature, lasting a few hours but not days. The severe episodic vertigo is accompanied by tinnitus, fluctuating hearing loss, and a feeling of fullness in the affected ear or ears. Typically, hearing decreases and tinnitus increases during the attack. Hearing may improve between attacks in early stages of the disease. Generally, the hearing loss begins unilaterally and affects primarily the lower frequencies, with mid and high frequencies being affected in later stages of the disease.[37-39]

Meniere's disease is most common in middle age and may become bilateral in up to 50% of the affected patients. The etiology of Meniere's disease is a failure of the mechanism regulating the production and disposal of endolymph, resulting in recurrent attacks of endolymphatic hydrops. Since the endolymphatic duct and sac are the sites of resorption of endolymph, these structures play an important role in the pathogenesis of endolymphatic hydrops. The success of various surgical procedures in relieving the symptoms of Meniere's disease has led to great interest in using CT and/or MRI to evaluate the vestibular aqueduct and the endolymphatic duct and sac.[39-41]

Unfortunately, there is no unanimity on the value of imaging in cases of Meniere's disease. Some investigators have used CT or MRI to evaluate the potential success of shunt surgery, based on showing patency of the vestibular aqueduct.[39,41] Other investigators, however, report that the size, shape, and patency of the vestibular aqueduct are of no value in predicting surgical results in shunt procedures or in predicting occurrence of bilateral disease.[40] MRI with its ability to detect the endolymphatic duct and sac, separate from the bony vestibular aqueduct, may offer more useful information than can CT.[41] The value of CT and MRI may be in their ability to exclude associated infectious or neoplastic disease processes.[38,39]

Vestibular neuronitis is a clinical diagnosis based on an aggregate of specific symptoms. The disease is characterized by an acute onset of severe vertigo, lasting several days, followed by gradual improvement of several weeks. Hearing is typically unaffected. The history includes onset of vertigo following an illness such as an upper respiratory infection. Most patients become completely symptom free following central compensation.[39] Vestibular labyrinthitis is similar in that the disease presents with the acute symptoms of vertigo, but this disease is always associated with hearing loss. Labyrinthitis is usually viral in origin but may result from acute or chronic bacterial middle ear infections. Unlike viral labyrinthitis, labyrinthitis associated with suppurative ear disease may progress to develop partial or complete occlusion of the lumen of the affected labyrinth.[38,39] Early on, the obstructed lumen may

For References see p. 1033

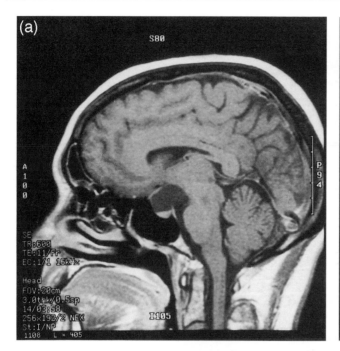

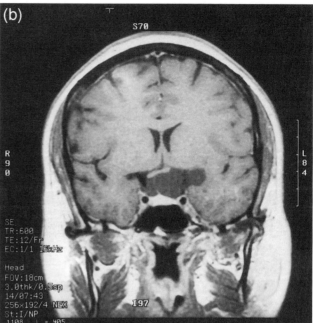

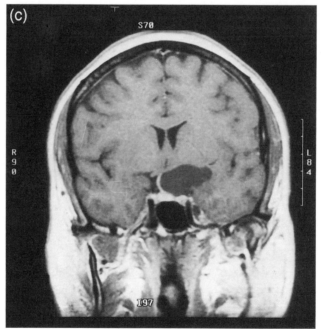

Figure 26 Parasellar and suprasellar Rathke's pouch cyst. (a) Fragile unenhanced scan showing the mass causing elevation of the optic chiasm. (b) Coronal enhanced scan at the level of the chiasm. Note that the non-enhanced cyst causes elevation and deformity of the optic chiasm. (c) Coronal enhanced scan at the level of the pituitary stock shows displacement of the stock to the right by the left parasellar mass

be detected by MRI through loss of the signal intensity of the fluid contents. Later, more complete obliteration of all the labyrinthine structures occurs, leading to labyrinthitis obliterans, which is readily diagnosed on CT or conventional tomography.[42]

With MRI, there may be postgadolinium enhancement of the labyrinthine structures or vestibular nerves during the acute or subacute stages of vestibular neuronitis and/or labyrinthitis.[43,44] Such results must be interpreted with care, since

sudden labyrinthine dysfunction may be caused by spontaneous hemorrhage or injury, which result in abnormal signal intensities within the labyrinthine structures secondary to the presence of blood products.[45]

Diseases of the internal auditory canal and cerebellopontine angle are generally not characterized by severe attacks of vertigo but rather with intermittent dizziness and/or exacerbated periods of dizziness.[37,39] A variety of benign or malignant tumors of the petrous temporal bone, such as paragangliomas,

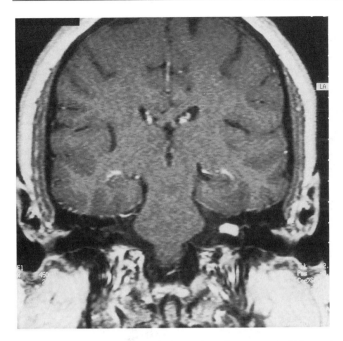

Figure 27 Tumor of the left internal auditory canal. Enhanced coronal image documents a linear enhancing tumor that fills the left auditory canal

carcinomas, or metastatic tumors, may directly involve the labyrinthine structures, causing vertigo. Such processes are readily evaluated with MRI (Figures 27–29).

6.3 Central Vestibular Disorders

Lesions of the brainstem or cerebellum that result in central vertigo can readily be diagnosed by MRI. Vascular insufficiency in the vertebrobasilar circulation is a common cause of vertigo in patients over 50 years of age. Thrombosis of the labyrinthine artery or infarction of the lateral medulla from vertebral or posterior inferior cerebellar artery insufficiency may cause severe vertigo. The subclavian steal syndrome can cause a variety of symptoms, including vertigo.[38,46,47] Such conditions can be carefully evaluated with MR angiography or conventional angiography of the posterior fossa vasculature.

A variety of other central nervous diseases may produce vertigo or dizziness. These include seizure disorders, multiple sclerosis, ataxic diseases, head injuries, or any cause for increased intracranial pressure. Vertigo may result from stroke and transient ischemic attacks may present as episodic dizziness.[39]

Various metabolic disorders may result in dizziness. These include thyroid disorders, hyperlipidemia, diabetes, and hypoglycemia. Autoimmune diseases or diseases that affect the proprioceptive system may be the cause of vertigo. In many cases, the possibility of function neurotic symptoms must be considered in patients in whom no diseases can be found. Finally cervical spondylosis is thought to cause vertigo by disc degeneration and narrowing of the disc space, which affects the nerves in close proximity, or by osteophyte formation, which compresses the blood vessels. In such cases, convention-

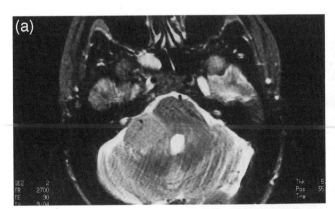

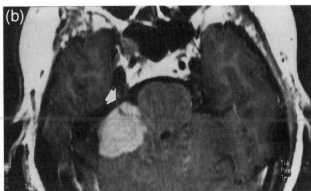

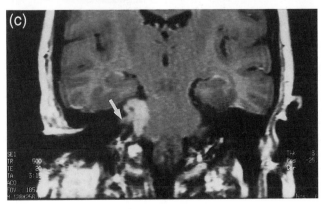

Figure 28 Meningioma of the right cerebellopontine angle. (a) Axial MR scan without contrast. The nearly isointense tumor can be seen to indent the left cerebellopontine angle. (b) Enhanced axial CT scan showing the full extent of the tumor mass. Note that the mass bends up to the level of the left fifth nerve (arrowhead). (c) Enhanced coronal image shows that mass to extend into the internal auditory canal, simulating an acoustic nerve tumor

For References see p. 1033

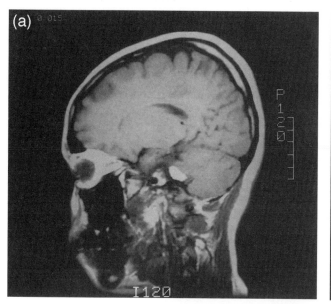
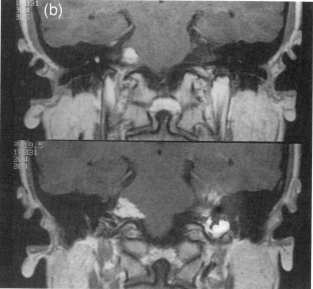
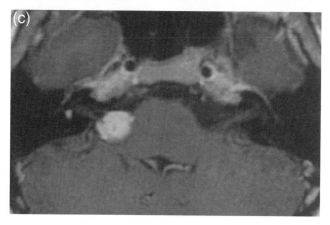

Figure 29 Lipoma of the right cerebellopontine angle and vestibule. (a) Saggital unenhanced image shows the fat intensity mass in the cerebellopontine angle. (b) Coronal unenhanced images document the lesion in the right cerebellopontine angle with a small focus of a high signal in the region of the right vestibule. (c) Axial enhanced CT scan again shows the tumor masses both in the cerebellopontine angle and the vestibule. The patient complained of severe dizziness that was more likely related to the small lesion in the vestibule

al radiographs and/or cross-sectional imaging procedures may be helpful.[38,39,42]

6.4 Sensorineural Hearing Loss

Sensorineural hearing loss may be sudden, fluctuating, or progressive. Sudden sensorineural hearing loss is a manifestation of viral infections, vascular occlusive diseases, or inner ear membrane ruptures.[48–52] As discussed above, there may be vertigo associated with these conditions, which would help to define whether the lesion is peripheral or central. In order to discriminate idiopathic or viral infections from other causes of sensorineural hearing loss, auditory brainstem responses and gadolinium enhanced MRI may be utilized.[48–50] Patients with cochleitis or cochlear nerve neuritis typically have abnormal auditory brainstem responses and may be helped by a tapering course of oral corticosteroids.[49,50] Whether gadolinium enhanced MRI shows enhancement of the cochlear nerve or cochlea is not a helpful indicator for or against using corticos-

teroid therapy. However, one study suggests that sudden deafness associated with MRI changes is more difficult to cure with steroid therapy than that without associated MRI changes.[49]

Fluctuating neurosensory hearing loss is a difficult disease to assess properly. The audiometric examination does, of course, indicate the level of dysfunction but not the likely cause. In patients where MRI indicates large vestibular aqueducts (apertures greater than 4 mm), this may indicate a fluctuating frequency loss more often than low frequency loss. Fluctuating sensorineural hearing loss resulting from an enlarged vestibular aqueduct appears to be more common in children and young adults, which is an important point in differentiating this disease from Meniere's disease in which the majority of patients are middle aged or older. The vestibular aqueduct of patients with Meniere's disease may be small rather than large.[38,39]

There is speculation on the causes of a sudden drop in hearing in patients with large vestibular aqueducts.[53] Two possible causes are reflux of hyperosmolar fluid from the endolymphatic sac to the inner ear and rupture of the membranous labyrinth or a perilymphatic fistula caused by transmission of intracranial pressure to the inner ear through the enlarged vestibular aqueduct. It is well recognized that patients sustaining relative minor head trauma, or patients who are subjected to extreme barotrauma (scuba diving), may have aggravated episodes of hearing loss. Consequently, it may be worthwhile to image the temporal bones in order to detect enlarged vestibular aqueducts and thus advise the patients or their parents of the dangers of contact sports or activities that entail extreme barometric pressure changes. The imaging findings must be correlated with audiometry, since the fluctuating sensorineural hearing loss of large vestibular aqueduct patients does not resemble the low frequency changes characteristic of Meniere's disease which may also be associated with fluctuating hearing loss.[54,55]

The pathophysiologic basis of disease in patients with isolated large vestibular aqueducts may differ from that in patients whose large aqueducts are associated with other inner ear malformations. Patients with complex inner ear malformations may be subject to recurrent episodes of meningitis and/or the 'gusher' syndrome, resulting in a dead ear at the time or surgical intervention such as a stapedectomy.[52,54]

Asymmetric sensorineural hearing loss or gradually declining unilateral sensorineural hearing loss is a common symptom that may be ascribed to a variety of pathologic processes. Initial evaluation attempts to find the site of the lesion (i.e. cochlear or retrocochlear). All retrocochlear lesions are associated with an abnormal auditory brainstem response, which is often obtained prior to an imaging study. Whether auditory brainstem response testing should be eliminated as a cost-saving measure is a subject of considerable debate. It seems unlikely that clinicians will refer patients directly to MRI without at least preliminary audiometric and/or auditory brain response testing.[48,50,56]

A complete MR study of the head should be performed in addition to the studies of the internal auditory canal and temporal bones. The MR examination should include complete evaluation of the central nuclei in the brainstem as well as the auditory pathways extending upwards into the cerebral hemispheres.[7] Whether or not gadolinium contrast enhancement is routinely utilized depends on a variety of factors including coil size, field of view, field strength, and pulse sequences. CT is diagnostic in lesions 1–1.5 cm or greater in diameter but does not readily detect small brainstem lesions such as infarctions or demyelination.[7,56]

7 RELATED ARTICLES

Eye, Orbit, Ear, Nose, and Throat Studies Using MRI; Gadolinium Chelate Contrast Agents in MRI: Clinical Applications; Head and Neck Investigations by MRI.

8 REFERENCES

1. L. A. Hayman, L. A. Rolak, J. M. Gundzik, and R. B. Lufkin, in 'Clinical Brain Imaging', ed. L. Anne Hayman, Mosby, St. Louis, 1992, Chap. 8.
2. W. G. Bradley, Jr, *Radiology*, 1991, **179**, 319.
3. W. M. Kelly, *Neuroimaging Clin. North Am.*, 1993, **3**, 1.
4. L. R. Gentry, R. C. Mehta, R. E. Appen, and J. E. Weinstein, *Am. J. Neuroradiol.*, 1991, **12**, 707.
5. L. G. Hutchins, H. R. Harnsberger, C. W. Hardin, W. P. Dillon, W. R. Smoker, and A. G. Osborn, *Am. J. Neuroradiol.*, 1989, **10**, 1031.
6. S. S. Gebarski, S. A. Telian, and J. K. Niparko, *Radiology*, 1992, **183**, 391.
7. S. S. Gebarski, D. L. Tucci, and S. A. Telian, *Am. J. Neuroradiol.*, 1993, **14**, 1311.
8. C. J. Jacobs, H. R. Harnsberger, R. B. Lufkin, A. G. Osborn, W. R. Smoker, and J. L. Parkin, *Radiology*, 1987, **164**, 97.
9. A. N. Hasso and D. S. Smith, *Semin. Ultrasound, CT, MR*, 1989, **10**, 280.
10. L. G. Hutchins, H. R. Harnsberger, J. M. Jacobs, and R. I. Apfelbaum, *Radiology*, 1990, **175**, 837.
11. N. Martin, O. Sterkers, D. Mompoint, and N. Nahum, *Neuroradiology*, 1992, **34**, 62.
12. A. N. Hasso, in 'MRI Atlas of the Head and Neck', Martin Dunitz, London, 1993, Chaps 1 and 6.
13. M. F. Mafee, C. S. Lachenauer, A. Kumar, P. M. Arnold, R. A. Buckingham, and G. E. Valvassori, *Radiology*, 1990, **174**, 395.
14. E. T. Tali, W. T. C. Yuh, H. D. Nguyen, G. Feng, T. M. Koci, J. R. Jinkins, R. A. Robinson, and, A. N. Hasso, *Am. J. Neuroradiol.*, 1993, **14**, 1241.
15. A. N. Hasso and K. D. Brown *J. Med. Radiol. Imag.*, 1993, **3**, 247.
16. D. S. Russell and L. J. Rubinstein, in 'Pathology of Tumours of the Nervous System', Williams & Wilkins, Baltimore, OH, 1989, Chap. 5.
17. Y. Y. Lee, R. D. Tien, J. M. Bruner, C. A. DePena, and P. Van-Tassel, *Am. J. Roentgenol.*, 1990, **154**, 351.
18. F. J. Laine, I. F. Braun, M. E. Jensen, L. Nadel, and P. M. Som, *Radiology*, 1990, **17**, 65.
19. G. D. Parker and H. R. Harnsberger, *RadioGraphics*, 1991, **11**, 383.
20. R. D. Tien, G. J. Felsberg, and A. K. Osumi, *Am. J. Roentgenol.*, 1993, **161**, 167.
21. A. Vahlne, S. Edstrom, P. Arstila, M. Beran, H. Ejnell, O. Nylen, and E. Lycke, *Arch. Otolaryngol.*, 1981, **107**, 79.
22. R. Tien, W. P. Dillon, and R. K. Jackler, *Am. J. Neuroradiol.*, 1990, **11**, 735.
23. L. T. Bilaniuk, R. A. Zimmerman, and T. H. Newton, in 'Modern Neuroradiology, Vol. 4. Radiology of the Eye and Orbit', ed. T. H. Newton, L. T. Bilaniuk. Clavadell Press, New York, pp. 5.1–5.84.
24. S. R. Kodsi, D. J. Shetlar, R. J. Campbell, J. A. Garrity, and G. B. Bartley, *Am. J. Ophthalmol.*, 1994, **117**, 177.
25. B. J. Goldsmith, S. A. Rosenthal, W. M. Wara, and D. A. Larson, *Radiology*, 1992, **185**, 71.
26. M. C. Brodsky and R. W. Beck, *Radiology*, 1994, **192**, 22.
27. W. A. Gibby, M. S. Cohen, H. I. Goldberg, and R. C. Sergott, *Am. J. Roentgenol.*, 1993, **160**, 143.
28. J. Guy, A. Mancuso, R. Beck, M. Moster, L. A. Sedwick, R. G. Quisling A. L. Rhoton Jr, E. E. Protzko and J. Schiffman, *J. Neursurg.*, 1991, **74**, 426.
29. N. Hosten, B. Sander, and M. Cordes, *Radiology*, 1989, **172**, 759.
30. H. Tonami, H. Tamamura, K. Kimizu, A. Takarada, T. Okimura, I. Yamamoto, and K. Sasaki, *Am. J. Roentgenol.*, 1990, **154**, 385.
31. R. F. Carmody, M. F. Mafee, J. A. Goodwin, K. Small, and C. Haery, *Am. J. Neuroradiol.*, 1994, **15**, 775.
32. N. A. Courcoutsakis, C. A. Langford, M. C. Sneller, T. R. Cupps, K. Gorman, and N. Patronas, *J. Comput. Assist. Tomogr.*, 1997, **21**, 452.
33. A. S. Cytryn, A. M. Putterman, G. L. Schneck, E. Beckman, and G. E. Valvassori, *Ophthal. Plastic Recons. Surg.*, 1997, **13**, 129.

For References see p. 1033

34. P. de Potter, J. A. Shields, and C. L. Shields, *MRI of the Eye and Orbit*, J. B. Lippincott Company, Philadelphia, PA, USA, 1995

35. R. A. Nugent, R. I. Belkin, J. M. Neigel, J, Rootman, W. D. Robertson, J. Spinelli, and D. A. Graeb, *Radiology*, 1990, **177**, 675.

36. T. Ohnishi, S. Noguchi, N. Murakami, J. Tajiri, M. Harao, H. Kawamoto, H. Hoshi, S. Jinnouchi, S. Futami, S. Nagamachi, and K. Watanabe, *Radiology*, 1994, **190**, 857.

37. S. R. McGee, *West. J. Med.*, 1995, **162**, 37.

38. P. D. Phelps and G. A. S. Lloyd in 'Radiology of the Ear', Blackwell Scientific, Boston, MA, 1983, pp. 137–141.

39. J. R. E. Dickins and S. S. Graham, *Ear Hearing*, 1986, **7**, 133.

40. E. M. Kraus and P. J. DuBois, *Arch. Otolaryngol.*, 1979, **105**, 91.

41. F. W. J., Albers, R. V. Van Weissenbruch, and J. W. Casselman, *Acta Otolaryngol.*, 1994, **114**, 595.

42. A. N. Hasso and J. A. Ledington, *Otolaryngol. Clin. North Am.*, 1988, **21**, 219.

43. S. Seltzer and A. S. Mark, *Am. J. Neuroradiol.*, 1991, **12**, 13.

44. A. S. Mark, S. Seltzer, J. Nelson-Drake, J. C. Chapman, D. C. Fitzgerald, and A. J. Gulya, *Ann. Otol. Rhinol. Laryngol.*, 1992, **101**, 459.

45. J. L. Weissman, H. D. Curtin, B. E. Hirsch, and W. L. Hirsch Jr, *Am. J. Neuroradiol.*, 1992, **13**, 1183.

46. S. Kikuchi, K. Kaga, T. Yamasoba, R. Higo, T. O'uchi, and A. Tokumaru, *Acta Otolaryngol.*, 1993, **113**, 257.

47. B. Norving, N. Magnusson, and S. Holtås, *Acta Neurol. Scand.*, 1995, **91**, 43.

48. R. A. Hendrix, R. M. DeDio, and A. P. Sclafani, *Otolaryngol. Head Neck Surg.*, 1990, **103**, 593.

49. K. Kano, T. Tono, Y. Ushisako, T. Morimitsu, Y. Suzuki, and T. Kodama, *Acta Otolaryngol.*, 1994, **514**, 32.

50. N. Y. Busaba and S. D. Rauch, *Otolaryngol. Head Neck Surg.*, 1995, **113**, 271.

51. M.-H. Huang, C.-C. Huang, S. J. Ryu, and N.-S. Chu, *Stroke*, 1993, **24**, 132.

52. J. S. Reilly, *Laryngoscope*, 1989, **99**, 393.

53. G. E. Valvassori and J. D. Clemis, *Laryngoscope*, 1978, **88**, 723.

54. M. F. Mafee, D. Charletta, A. Kumar, and H. Belmont, *Am. J. Neuroradiol.*, 1992, **13**, 805.

55. T. Okumura, H. Takahashi, I. Honjo, A. Takagi, and K. Mitamura, *Laryngoscope*, 1995, **105**, 289.

56. S. H. Selesnick, R. K. Jackler, and L. W. Pitts, *Laryngoscope*, 1993, **103**, 431.

Acknowledgements

The authors sincerely thank W. G. Bradley, Jr, M.D., Ph.D., for reviewing the manuscript and for granting permission to reproduce Figure 2.

Biographical Sketches

Anton N. Hasso. *b* 1940. B.S., 1962, M.D., 1967, Loma Linda University. Faculty, UCLA, 1972–74. Faculty, Loma Linda University, 1974–96, Professor and Chairman, Department of Radiological Sciences, University of California, Irvine, 1996–present. Approx. 130 publications including 15 in NMR; senior author of *MRI Atlas of the Head and Neck* (1993) and *MRI of Brain: Neoplastic Disease* (1991); co-author of *MRI of the Brain, Head and Neck, A Text Atlas* (1985). Research interests include applications of NMR in neuroradiology/head and neck radiology, particularly magnetic resonance angiography and the use of gadolinium contrast agents.

Peggy J. Fritzsche. *b* 1941. B.S., Andrews University, 1962; M.D., Loma Linda University, 1966. Faculty, UCLA, 1973–74. Faculty, Loma Linda University, 1974–91, Medical Director, Riverside MRI, 1991–present. Approx. 85 publications including eight in NMR; co-

author of *MRI of the Body* (1992). Research interests include applications of NMR in genitourinary and abdominal radiology, breast imaging and pelvic imaging.

Pituitary Gland and Parasellar Region Studied by MRI

Richard Farb and Walter Kucharczyk
University of Toronto, Toronto, ON, Canada

1   ANATOMY AND THE MRI TECHNIQUE

The pituitary gland is a small but important organ situated in a small bony depression in the skull base called the sella turcica. Although the average weight of the pituitary gland in the adult is only 0.5 g, it is responsible for the regulation of many of the body's most critical endocrine functions. The gland enlarges slowly with maturation reaching a peak height of no more than 9–10 mm in early adulthood. There are two periods when the gland may transiently enlarge beyond its normal dimensions, those being adolescence and pregnancy. This is principally due to physiological hypertrophy of prolactin-secreting cells.

The pituitary gland is made up of three distinct lobes: anterior, intermediate, and posterior. The intermediate lobe is a vestigial remnant in humans and serves no function; however, it may be the site of an incidental cyst. The anterior lobe, or adenohypophysis, is a true endocrine organ responsible for synthesizing many hormones. Its function is regulated by stimulatory and inhibitory hormones arising from the hypothalamus. The posterior lobe of the pituitary gland, also called the neurohypophysis, is actually a downward extension of the hypothalamus. The hormones of the posterior pituitary gland, vasopressin and oxytocin, are synthesized within the hypothalamus and axonally transported to the posterior pituitary gland where they are stored and from which they are ultimately released.

The anterior and posterior lobes of the pituitary gland are therefore two functionally distinct entities, although they reside in very close association within the sella turcica. The posterior lobe is much smaller than the anterior lobe and takes up only 10–20% of the sella turcica. The anterior lobe of the pituitary gland generally takes up the anterior, central, and lateral aspects of the sella turcica. The posterior lobe resides centrally in the midline just anterior to the dorsum sella. The lateral wings of the adenohypophysis usually extend laterally around the posterior pituitary gland.

The anterior and posterior lobes of the pituitary gland are easily distinguishable on MRI. The anterior lobe of the pituitary gland is normally isointense with cerebral white matter on all pulse sequences, whereas the posterior lobe of the pituitary gland is easily distinguishable by its characteristic hyperintense signal on T_1-weighted images. Following the administration of intravenous contrast material the anterior and posterior lobes, as well as the pituitary stalk, enhance intensely.

The pituitary gland is enclosed within the sella turcica by a dural diaphragm (the diaphragma sella), the center of which has a defect in it allowing for passage of the pituitary stalk. The superior surface of the diaphragma sella is lined with arachnoid, and the suprasellar cistern lies above the diaphragma sella. The suprasellar cistern contains all the vascular structures of the circle of Willis. The optic chiasm is immediately anterior to the pituitary stalk (infundibulum) within the suprasellar cistern.

On either side of the sella turcica lie the cavernous sinuses. These are complex dural venous sinuses passing from the anteromedial aspect of the posterior fossa (Meckel's cave) anteriorly to the superior orbital fissure, transmitting cranial nerves III, IV, V1, and V2 within the lateral walls of the cavernous sinus. Cranial nerve VI is the only cranial nerve that actually passes within the cavernous sinus. The appearance of the cavernous sinuses on MRI is somewhat variable in signal

intensity; however they are usually symmetric in dimensions. Following contrast administration the cavernous sinuses enhance intensely. The normal MRI anatomy is illustrated in Figure 1.

Imaging of the pituitary gland and sella turcica for possible structural abnormalities is a frequent indication for MRI. Exact specifications of the technique are not provided here because the constantly improving capabilities of modern scanners require that technique be continually revised. However high spatial detail is very important in this area and should be achieved through the use of thin slices, a fine matrix size, and a relatively small field-of-view. Requirements for spatial detail must be balanced against the need for adequate signal-to-noise ratio as well as imaging time.

The pulse sequence for best tissue contrast is still a matter of opinion. Several groups have shown that short *TR*, short *TE* spin echo images (i.e. T_1-weighted) generate very good contrast for visualizing pituitary pathology.[1,2,3] Because fast spin echo (FSE) methods are now readily available, and do not incur the time penalty of conventional T_2-weighted spin echo sequences, coronal T_2-weighted FSE imaging is being utilized with increasing frequency as a supplementary sequence for investigation of pituitary adenomas. Several other pulse sequences have been implemented for parasellar imaging and have met with varying degrees of success.[4]

Paramagnetic contrast-enhanced images are widely used and are very useful. Although most adenomas are visible without

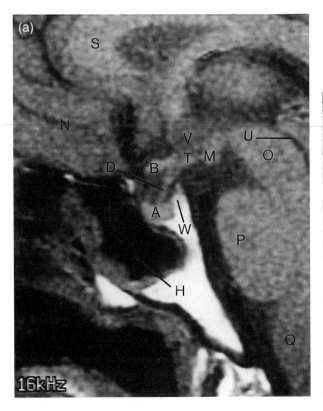
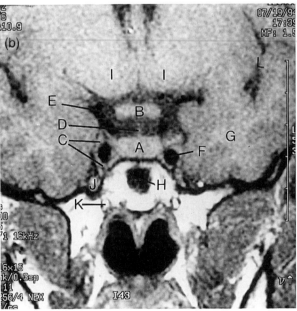

Figure 1 *Normal MR anatomy of the pituitary region.* T_1-weighted sagittal (a), and coronal (b), images. A. Anterior pituitary gland, B. optic chiasm, C. cavernous sinus, D. pituitary stalk, E. right internal carotid artery (supraclinoid portion), F. left internal carotid artery (cavernous portion), G. medial temporal lobe, H. sphenoid sinus, I. hypothalamus, J. trigeminal nerve, K. pterygoid canal, L. sylvian fissure, M. mammillary body, N. frontal lobe, O. midbrain, P. pons, Q. medulla, R. interpeduncular cistern, S. genu of corpus callosum, T. tuber cinereum, U. cerebral aqueduct, V. third ventricle, W. posterior pituitary gland

For References see p. 1042

the injection of intravenous contrast, several studies have shown that small adenomas may become visible only after contrast injection.[5,6] To take better advantage of the differential rates of contrast enhancement between adenomas and normal pituitary gland, dynamic pituitary scanning immediately after bolus contrast injection is also useful.[7]

2   CONGENITAL ABNORMALITIES

2.1   Pituitary Gland Hypoplasia

Congenital abnormalities of the pituitary gland and hypothalamus are usually present in association with anomalies of other midline cranial, orbital, and facial structures. Pituitary gland hypoplasia is usually accompanied by a small sella turcica, and is commonly clinically associated with growth failure and other endocrine abnormalities. A curious form of pituitary hypoplasia has recently been recognized to occur with a slightly higher frequency in patients with a history of breech deliveries.[8-10] These patients have short stature and growth hormone deficiency, as well as the MRI findings of pituitary hypoplasia, hypoplasia of the distal pituitary stalk, and absence of the normal hyperintensity within the posterior pituitary gland.

2.2   Ectopic Posterior Pituitary Tissue

Occasionally, the normal hyperintensity on T_1-weighted images within the posterior pituitary gland and in the posterior aspect of the sella turcica is not visualized. Instead there is a small area of hyperintensity seen along the course of the pituitary stalk. This may be associated with discontinuity of the pituitary stalk. These patients may present with hormonal abnormalities, particularly of the anterior pituitary gland, due to transection of the hypothalamo-pituitary portal anastomosis. It is rare for these patients to present with posterior pituitary gland hormonal deficiencies.

'Ectopic posterior pituitary tissue' is thought to occur by one of two mechanisms, either incomplete embryologic descent of hypothalamic neurons, or possibly due to traumatic transection of the pituitary stalk above the diaphragma sella. In either case, the neurosecretory granules descending down toward the posterior pituitary gland from the hypothalamus accumulate within the proximal stump of the amputated pituitary stalk. This simply becomes the new site or location of posterior pituitary tissue thus accounting for the increased signal intensity on T_1-weighted images as well as the normal hormonal profile of the posterior pituitary gland.

2.3   The Empty Sella Turcica

The terms 'empty sella syndrome' or 'empty sella turcica' refer to the roentgenographic, pneumoencephalographic, or computed tomographic appearance of the sella turcica, which is filled predominantly with cerebrospinal fluid (CSF), with the normal posterior pituitary gland and stalk being crowded posteriorly and inferiorly within the sella turcica. The cause of this displacement is the presence of a large defect in the diaphragma sella allowing the arachnoid membrane to herniate

through the diaphragma sella adjacent to the stalk and thus allowing CSF pulsations to enter the sella turcica and result in the eventual displacement of the pituitary gland and stalk posteriorly. It is actually quite rare for patients with an 'empty sella' to have symptoms referable to the area of the sella turcica. 'The empty sella' is most commonly seen as an incidental finding of little or no clinical significance.

2.4   Cephaloceles

Cephaloceles are herniations of the meninges (meningocele) or of the meninges and the brain (meningoencephalocele) through a congenital cranial defect. Cephaloceles in the sellar region (transphenoidal encephaloceles) are very rare. Most of these are associated with other midline anomalies, particularly agenesis of the corpus callosum.

3   TUMORS AND TUMOR-LIKE CONDITIONS

3.1   Pituitary Adenoma

The most common tumors affecting the pituitary gland are pituitary adenomas. These are benign, slow growing, epithelial adenomas originating in the adenohypophysis. Pituitary adenomas are usually well demarcated lesions that are separated from the normal pituitary gland by a pseudocapsule of compressed tissue. Pituitary adenomas are commonly classified based on size and hormonal activity. Adenomas measuring greater than 1 cm in diameter are referred to as **macro**adenomas; those less than 10 mm are **micro**adenomas. Those adenomas that are hormonally active may also be referred to by the hormone that they secrete. For example, the most common hormonally active adenoma is the 'prolactin-secreting microadenoma', or simply prolactinoma. The various hormonal types of adenomas are indistinguishable from one another by MRI imaging.

On MRI, pituitary microadenomas are usually small, focal hypointensities within the pituitary gland on T_1-weighted images. On T_2-weighted images the corresponding lesion is seen as hyperintense to the surrounding pituitary tissue.[3] Approximately 80–95% of pituitary adenomas present with these characteristic signal intensities (Figure 2). Intratumoral hemorrhage occurs in 20–30% of adenomas. These are usually macroadenomas (Figure 3). The incidence of intratumoral bleeding is higher in patients receiving bromocryptine therapy.[11] We have found that tumors that are hyperintense on T_1-weighted images are always cystic with the cyst containing elements of previous hemorrhage.

Tissue contrast represented by a differential signal intensity between the tumor and normal pituitary gland is the most sensitive and reliable indicator of the presence of a microadenoma (Figure 2). Indirect signs previously utilized for computed tomography imaging of the pituitary gland such as tilt of the pituitary gland, contour of the superior aspect of the pituitary gland, and deviation of the pituitary stalk, all appear to be quite insensitive to the presence of microadenomas and indeed may be misleading (Figure 2).

It is difficult to determine the accuracy of MRI for the diagnosis of pituitary microadenomas. It is estimated that about

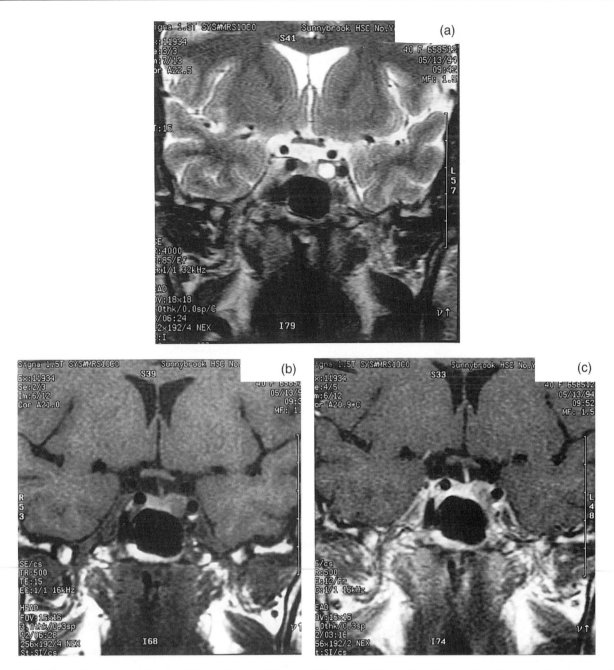

Figure 2 *Pituitary microadenoma.* (a) T_2-weighted, (b) T_1-weighted, and (c) enhanced T_1-weighted images. This 40-year-old female was noted to have marked elevation of cortisol in the blood and urine due to this hormonally active tumor secreting adrenocorticotropic hormone. This tumor demonstrates the typical features of a microadenoma including hyperintensity on T_2-weighted images and hypointensity on T_1-weighted images. Note the differential signal intensity between the intensely enhancing normal, anterior pituitary gland and the less enhancing tumor. Note also how the pituitary stalk deviates toward the tumor, thus stressing the unreliability of contralateral stalk deviation as a lateralizing sign of tumor

90% of microadenomas are detected and accurately localized with MRI.[3] The detection rate of macroadenomas approaches 100%. Their signal intensities are qualitatively similar to their smaller counterparts; however they more commonly demonstrate cystic degeneration or hemorrhage.

An area of continued difficulty in pituitary and parasellar imaging is that of the postoperative examination in search of residual or recurrent pituitary adenoma. In these cases, it may be very difficult to distinguish postoperative scarring, or graft material, from the normal gland or adenomatous tissue. This is

especially true in the first 6 months after surgery. In these cases, progressive growth of a soft tissue mass on sequential postoperative MR scans is the best imaging sign of recurrent tumor. This must also be interpreted in conjunction with endocrinologic markers.

3.2 Craniopharyngioma

Craniopharyngiomas are epithelial-derived neoplasms thought to arise from remnants of Rathke's cleft in the region

For References see p. 1042

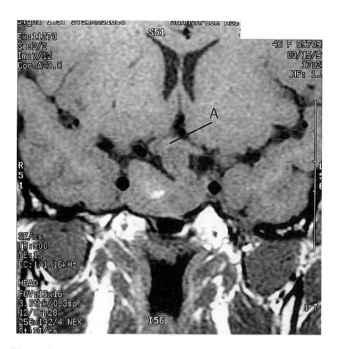

Figure 3 *Pituitary macroadenoma.* Nonenhanced T_1-weighted image demonstrating a large, lobulated tumor occupying the sella and extending superiorly into the suprasellar cistern displacing the optic chiasm (A) superiorly. A focal area of increased signal intensity represents a small hemorrhage within the tumor. The tumor only abuts and does not clearly invade the cavernous sinus regions bilaterally

of the pars tuberalis. They account for 3% of all intracranial tumors and show equal incidence in males and females. Tumors can vary greatly in size from several millimeters to several centimeters in diameter with the epicenter of the tumour usually located in the suprasellar cistern. Craniopharyngiomas typically have both solid and cystic components (Figure 4). Calcification is commonly seen in the solid portion of the tumor. The cystic contents of the tumour can vary in color and viscosity.

On MRI craniopharyngiomas are typically lobulated with heterogeneous signal intensities on T_1- and T_2-weighted images.[12] The cystic component of the tumor is uniform and commonly hyperintense on both T_1- and T_2-weighted images, and contains fluid that has been likened to 'machine oil'.

3.3 Rathke's Cleft Cyst

Rathke's cleft cysts share a common origin with craniopharyngiomas in that they originate from remnants of squamous epithelium of Rathke's cleft. The cysts are common incidental findings at autopsy; however the larger cysts can be symptomatic. The contents of the cysts are typically mucoid. Less commonly they are filled with serous or desquamated cellular debris.[13] The mucoid-containing cysts can be hyperintense on T_1 and T_2-weighted images; the serous-containing cysts have signal intensities that closely match CSF (Figure 5). Rathke's cleft cysts do not enhance following contrast injection, except for perhaps marginal enhancement

around the cyst wall. If nodular enhancement or calcification is seen, craniopharyngioma should be suspected.

3.4 Meningioma

Approximately 10% of meningiomas occur in the parasellar region. These tumors arise from a variety of locations around the sella including the tuberculum sella, clinoid processes, medial sphenoid wing, and cavernous sinus. Meningiomas are most frequently isointense relative to gray matter on unenhanced T_1-weighted images. Approximately 50% remain isointense on T_2-weighted images.[14,15] Meningiomas, as well as other extraaxial tumors, can demonstrate a typical CSF cleft between the tumor and the adjacent brain parenchyma. Also, meningiomas have been noted to demonstrate a peripheral black rim around their margins, thought to represent veins surrounding the tumor.

Meningiomas commonly encase arteries of the parasellar region, particularly the cavernous and supraclinoid portions of the internal carotid artery. Meningiomas will typically narrow the vessels they encase (Figure 6). Adenomas, though they commonly encase vessels, usually do not narrow these vessels. Other distinguishing features of meningiomas include intense

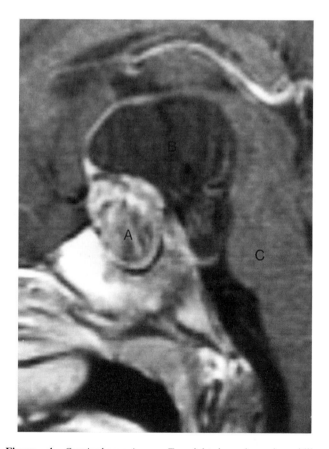

Figure 4 *Craniopharyngioma.* T_1-weighted, enhanced, midline sagittal image of the parasellar region in this 3-year-old male demonstrating the typical findings of a craniopharyngioma. Note the enhancing solid tumor A, and the large cystic portion of the tumor B, extending supero-posteriorly, draping over the dorsum sella into the posterior fossa displacing the pons, C

For list of General Abbreviations see end-papers

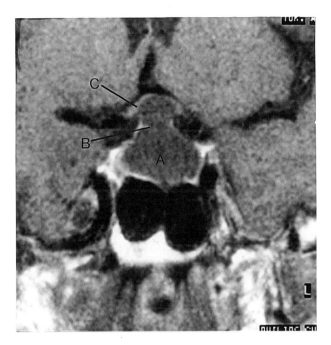

Figure 5 *Rathke's cleft cyst.* T_1-weighted, enhanced coronal image of the sella region demonstrating a somewhat bilobed cystic mass **A**, extending from the sella into the suprasellar cistern through the diaphragma sella **B**, displacing the optic chiasm superiorly **C**. This 28-year-old male presented with bitemporal hemianopsia and normal pituitary function. This typical appearance of the Rathke's cleft cyst has been likened to the appearance of a snowman

enhancement with intravenous contrast, tumor calcification, and hyperostosis of adjacent bone.

3.5 Germinoma and Teratoma

Germinomas account for less than 2% of primary intracranial tumors. Most of these tumors occur in the pineal region. However, approximately 20% of these tumors occur in the suprasellar cistern or pituitary fossa. Suprasellar germinomas occur either as metastatic deposits from a pineal region tumor or arise primarily in the suprasellar cistern. Most patients present between 5 and 25 years of age. Although pineal region germinomas are most commonly seen in males, there is no sex predilection for the primary suprasellar germinoma. Other tumors of germ cell origin occurring in this region, albeit rarely, include yolk sac tumors, choriocarcinoma, embryonal cell carcinoma, and teratoma. Suprasellar germinomas are usually large midline tumors with a propensity to infiltrate and spread by the CSF pathways. In contrast to craniopharyngiomas, germinomas are homogeneous and only rarely have cystic components. Germinomas are mildly hypointense on T_1-weighted images and hyperintense on T_2-weighted images. Marked uniform contrast enhancement is the rule (Figure 7). Teratomas have mixed heterogeneous signal intensities, often demonstrating evidence of fat or calcification.

3.6 Epidermoid and Dermoid

Epidermoids and dermoids are benign, slow-growing 'inclusion tumors' that can occur intracranially or intraspinally

resulting from inclusions of epithelium during the time of neural tube closure. Dermoids are approximately one-fifth as common as epidermoids. Epidermoids account for approximately 1% of all intracranial neoplasms. The cerebellopontine angle is the most frequent site for epidermoids with the parasellar region being the second most common site.

Epidermoids are generally cystic tumors with the walls composed of simple, stratified, squamous epithelium resting on a layer of connective tissue. The interior of the cyst is composed of desquamated keratin products of the cyst wall. These tumors are benign and grow slowly with expansile characteristics. They also typically insinuate within and around structures, or conform to the space within which they are situated. Rarely the cyst can rupture producing a chemical granulomatous meningitis. Epidermoids are typically only slightly hyperintense to CSF on T_1- and T_2-weighted images, making the proton density image the most valuable sequence for identification of the epidermoid tumor. Epidermoids are characteristically hyperintense to brain and CSF on the proton density sequence.[16] Unfortunately epidermoids can also retain signal intensities identical to CSF making them nearly indistinguishable from arachnoid cysts. Enhancement, if it occurs, occurs only along the periphery of the tumor. Dermoids are more heterogeneous. Most dermoid tumors show evidence of fatty component as well as small areas of calcification.

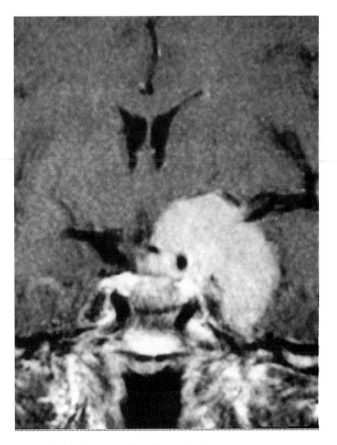

Figure 6 *Meningioma.* This T_1-weighted, enhanced coronal image demonstrates a homogeneously enhancing meningioma arising from the left anterior clinoid. Note the encasement of the left supraclinoid internal carotid artery, which is slightly narrowed when compared with the cavernous portion of the artery

For References see p. 1042

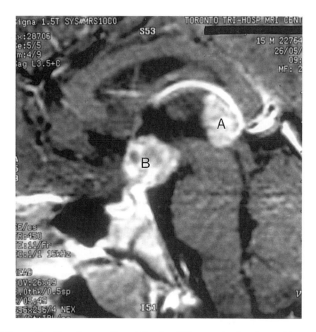

Figure 7 *Germinoma.* T_1-weighted, midline, sagittal, enhanced image demonstrating homogeneously enhancing masses within the pineal region A, and suprasellar region B. The presence of tumors in both the pineal and suprasellar regions is highly characteristic of germinoma

3.7 Metastases

Metastases to the pituitary gland are a common autopsy finding but only about 1–2% of cancer patients have symptomatic metastases to the pituitary gland. These are usually metastases from breast or bronchogenic carcinoma.

3.8 Chordoma

Chordomas are rare neoplasms arising from remnants of the primitive notochord. Most cranial chordomas are found in the midline in relation to the clivus. Chordomas are locally invasive and destructive and commonly result in large, lobulated, destructive lesions extending in an extradural fashion. Portions of the sella and the parasellar regions are commonly involved. Chordomas are characteristically isointense or hypointense on T_1-weighted images and hyperintense on T_2-weighted images[17] with moderate enhancement following intravenous contrast administration.

3.9 Arachnoid Cysts

Approximately 15% of arachnoid cysts occur in the juxtasellar region.[18] Suprasellar arachnoid cysts are thought to be developmental in origin and arise from an imperforate membrane of Liliequist. They have MRI characteristics identical to normal CSF.

3.10 Hamartoma of the Tuber Cinereum

Hamartomas of the tuber cinereum are sessile or pedunculated masses extending inferiorly from the hypothalamus into the suprasellar cistern between the pituitary stalk and mammillary bodies. Histologically they resemble cerebral cortex with little histologic similarity to the normal hypothalamus. Hamartomas of the tuber cinereum can be associated with precocious puberty. On MRI hamartomas are isointense to gray matter on all imaging sequences and should not enhance with contrast material (Figure 8).[19]

3.11 Eosinophilic Granulomatosis (Histiocytosis X, Langerhan's Cell Granulomatosis)

Histiocytosis can involve the pituitary gland, pituitary stalk, or hypothalamus. Twenty-five per cent of cases develop the classical clinical triad of diabetes insipidus, exophthalmos, and lytic bone lesions (the Hand–Schuller–Christian syndrome). In these cases granulomas can be found in the hypothalamus or the pituitary stalk. With MRI there is thickening of the pituitary stalk, hyperintensity on T_2-weighted images, and intense enhancement following contrast administration.[20,21]

4 INFECTIONS AND INFLAMMATORY CONDITIONS

4.1 Infections

Infections of the pituitary gland are a rare occurrence. Previous viral infection has been proposed as an etiology for diabetes insipidus. Leptomeningeal tuberculosis at the base of the brain can involve the pituitary gland. Bacterial infection of the pituitary gland is usually unnoticed until it manifests as a

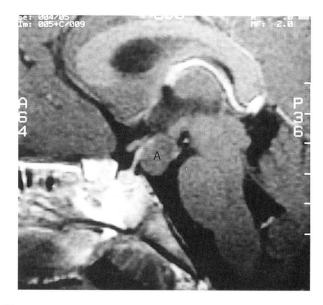

Figure 8 *Hamartoma of the tuber cinereum.* T_1-weighted, enhanced, midline sagittal image demonstrating a homogeneous mass located posterior to the pituitary stalk and anterior to the mammillary bodies, extending from the region of the tuber cinereum into the posterior aspect of the suprasellar cistern. This mass did not enhance and remained isointense to gray matter on all sequences, demonstrating the typical findings of hamartoma of the tuber cinereum

For list of General Abbreviations see end-papers

pituitary abscess, an entity which is also exceedingly rare. Pituitary abscesses are most frequently seen in association with preexisting conditions such as craniopharyngioma or pituitary adenoma. Transphenoidal surgery is rarely complicated by infection.

Infection of the parasellar regions, in particular the cavernous sinus is also very rare. Cavernous sinus thrombophlebitis is thought to be secondary to venous spread of a septic embolus from a periorbital or perinasal source.

4.2 Lymphocytic Hypophysitis

Lymphocytic hypophysitis is characterized by diffuse infiltration of the adenohypophysis by lymphocytes. It is most commonly seen in late pregnancy or the postpartum period, and has been associated with other autoimmune disorders such as Hashimoto's thyroiditis. Clinically the woman complains of headache, visual loss, and failure to resume normal menstrual cycle. MRI demonstrates diffuse enlargement of the anterior pituitary gland without focal abnormality.

4.3 Sarcoidosis

Sarcoidosis most commonly occurs in an intrathoracic location. Any part of the central nervous system however can also be involved and 5% of patients with sarcoidosis will have neurologic complaints. The most typical findings for neurosarcoidosis include a basal meningitis with an abnormal, thick, enhancement pattern involving particularly the leptomeninges at the base of the brain characteristically in the midline. Abnormal parenchymal lesions are also noted within the brain. Neurosarcoidosis can be indistinguishable from other causes of leptomeningeal disease, in particular tuberculosis and leptomeningeal carcinomatosis.

4.4 Tolosa–Hunt Syndrome

This syndrome consists of painful ophthalmoplegia associated with a lesion within the cavernous sinus which is responsive to steroid therapy. Pathologically the lesion is similar to orbital pseudotumor.

5 VASCULAR LESIONS AND INFARCTION

5.1 Aneurysms

The large arteries in the suprasellar region are a common site of aneurysms. Aneurysms are for obvious reasons extremely important lesions to identify correctly. Fortunately their MRI appearance is distinctive and easily appreciated.[22,23] Aneurysms centrally contain signal void created by rapidly flowing blood. This flowing blood may also cause a substantial ghosting artifact in the phase-encoding direction, which is a useful diagnostic sign. Thrombus within the aneurysm is laid down against the wall in a sequential fashion over a long period of time and results in a typical laminated appearance. Hemosiderin staining of the adjacent brain can also be seen. If confusion persists regarding the diagnosis, conventional angio-graphy remains the gold standard for diagnosis of intracranial aneurysms.

5.2 Carotid Cavernous Fistulas

Carotid cavernous fistulas (direct or indirect type) can be spontaneous, posttraumatic, or atherosclerotic. Dural arteriovenous malformations of the cavernous sinus are another form of abnormal venous communication in this region. On MRI, dilatation of the venous structures, in particular the ophthalmic vein and cavernous sinus, is usually clearly visible. The carotid artery can also be somewhat dilated.

5.3 Postpartum Pituitary Necrosis

During pregnancy the pituitary gland increases in size thus making it vulnerable to circulatory disturbances. Postpartum ischemic necrosis of the pituitary gland is a well-known entity, which can follow complicated deliveries associated with hemorrhage and shock.

6 MISCELLANEOUS CONDITIONS

6.1 Diabetes Insipidus

Diabetes insipidus results from the lack of appropriate release of antidiuretic hormone (vasopressin) from the posterior pituitary gland. This results in the inability to concentrate urine and causes polydipsia and polyuria. The inability to release vasopressin in response to normal stimuli can have several possible etiologies: (1) primary failure to synthesize the hormone in the periventricular nuclei of the hypothalamus, (2) destruction of these nuclei or proximal transport pathways in the hypothalamus and proximal pituitary stalk, (3) dysfunction of hypothalamic osmoreceptors such that despite adequate vasopressin reserves, the hormone is not released appropriately. The causes of diabetes insipidus include neoplasms, infiltrative disorders, surgery, and head trauma. Approximately 30% of cases are idiopathic. MRI is valuable in demonstrating not only the causative lesions, but also the status of the posterior pituitary gland, the storage site for the neurosecretory vesicles containing vasopressin. It is a unique feature of this gland that it is hyperintense on T_1-weighted images and regardless of the etiology this signal disappears in cases of diabetes insipidus.

6.2 Hemochromatosis

Hemochromatosis is a metabolic disorder in which excess iron is deposited throughout the body particularly in the solid abdominal viscera. For unknown reasons the pituitary gland is frequently involved. There is relative preferential deposition of iron within the gonadotrophic cells of the adenohypophysis. Dysfunction commonly manifests as loss of libido and hypogonadism. The characteristic finding on MRI is marked hypointensity emanating from the anterior pituitary gland on T_2-weighted images. This is due to the high levels of iron

For References see p. 1042

deposited within the region resulting in susceptibility effects and loss of signal.[24]

7   RELATED ARTICLES

Cranial Nerves Investigated by MRI; Eye, Orbit, Ear, Nose, and Throat Studies Using MRI; Head and Neck Investigations by MRI; Intracranial Infections; Temperomandibular Joint MRI.

8   REFERENCES

1. D. W. Chakeres, A. Curtin, and G. Ford, *Radiol. Clin. North Am.*, 1989, **27**, 265.
2. W. W. Peck, W. P. Dillon, D. Norman, T. H. Newton, and C. B. Wilson, *Am. J. Neuroradiol.*, 1988, **9**, 1085.
3. W. Kucharczyk, D. O. Davis, W. M. Kelly, G. Sze, D. Norman, and T. H. Newton, *Radiology*, 1986, **161**, 761.
4. T. Stadnik, A. Stevenaert, A. Beckers, R. Luypaert, T. Buisseret, and M. Osteaux, *Radiology*, 1990, **176**, 419.
5. J. L. Doppman, J. A. Frank, A. J. Dwyer, E. H. Oldfield, D. L. Millor, L. R. Nieman, G. P. Chrousos, G. B. Cutler, Jr., and D. L. Loriaux. *J. Comput. Assist. Tomogr.*, 1988, **12**, 728.
6. D. R. Newton, W. P. Dillon, D. Norman, T. H. Newton, and C. B. Wilson. *Am. J. Neuroradiol.*, 1989, **10**(5), 949.
7. W. Kucharczyk, J. E. Bishop, D. B. Plewes, M. A. Keller, and S. George. *Am. J. Roentgenol.*, 1994, **163**, 671.
8. W. M. Kelly, W. Kucharczyk, J. Kucharczyk, B. Kjos, W. W. Peck, D. Norman, and T. H. Newton. *Am. J. Neuroradiol.* 1988, **9**, 453.
9. I. Fujisawa, K. Kikuchi, K. Nishimura, K. Togashi, K. Itoh, S. Noma, S. Minami, T. Sagoh, T. Hiraoka, and T. Nomaoi, *Radiology*, 1987, **165**, 487.
10. M. Maghnie, F. Triulzi, D. Larizza, G. Scotti, G. Beluffi, A. Cecchini, and F. Severi. *Pediatr. Radiol.*, 1990, **20**, 229.
11. D. M. Yousem, J. A. Arrington, S. J. Zinreich, A. J. Kumar, and R. N. Bryan, *Radiology*, 1989, **170**, 239.
12. E. Pusey, K. E. Kortman, B. D. Flannigan, J. Tsuruda, and W. G. Bradley, *Am. J. Neuroradiol.*, 1987, **8**, 439.
13. W. Kucharczyk, W. W. Peck, W. M. Kelly, D. Norman, and T. H. Newton, *Radiology*, 1987, **165**, 491.
14. M. Castillo, P. C. Davis, W. K. Ross, and J. C. Hoffman, Jr., *J. Comput. Assist. Tomogr.*, 1989, **13**, 679.
15. J. W. Yeakley, M. V. Kulkarni, C. B. McArdle, F. L. Hoar, and R. A. Tang, *Am. J. Neuroradiol.*, 1988, **9**, 279.
16. D. Tampieri, D. Melanson, and R. Ethier, *Am. J. Neuroradiol.*, 1989, **10**, 351.
17. G. Sze, L. S. Uichanco, M. N. Brant-Zawadzki, R. L. Davis, P. M. Gutin, C. B. Wilson, D. Norman, and T. H. Newton, *Radiology*, 1988, **166**, 187.
18. S. N. Wiener, A. E. Pearlstein, and A. Eiber, *J. Comput. Assist. Tomogr.*, 1987, **11**, 236.
19. E. M. Burton, W. S. Ball, Jr., K. Crane, and L. M. Dolan, *Am. J. Neuroradiol.*, 1989, **10**, 497.
20. J. B. Moore, R. Kulkarn, D. C. Crutcher, and S. Bhimani, *Am. J. Pediatr. Hematol. Oncol.*, 1989, **11**, 174.
21. R. D. Tien, T. H. Newton, M. W. McDermott, W. P. Dillon, and J. Kucharczyk, *Am. J. Neuroradiol.*, 1990, **11**, 703.
22. S. W. Atlas, A. S. Mark, E. K. Fram, and R. Grossman, *Radiology*, 1988, **169**, 455.
23. A. Biondi, G. Scialfa, and G. Scotti, *Neuroradiology*, 1988, **30**, 214.
24. I. Fujisawa, M. Morikawa, Y. Nakano, and J. Konishi, *Radiology*, 1988, **168**, 213.

Biographical Sketches

Richard I. Farb. *b* 1959. B.Sc., 1981, University of Toronto. MD, 1985, University of Toronto, Canada. Radiology residency, Wayne State University Medical Center, Michigan 1991. Certification by examination by the Royal College of Physicians (Canada) and the American Board of Radiology 1991. Two year Neuroradiology Fellowship at University of Toronto completed 1993. Currently staff neuroradiologist at Sunnybrook Health Science Centre and lecturer at University of Toronto.

Walter Kucharczyk. *b* 1955. MD, 1979, University of Toronto, Canada. Radiology residency, University of Toronto, completing in 1983. Certified by examination as a Fellow of the Royal College of Physicians (Canada) in 1983, followed by two years of Neuroradiology MRI post residency training at UCSF and University of Toronto. Successively, visiting assistant professor (UCSF), assistant professor, full professor (University of Toronto). Currently Professor and Chair of the Department of Medical Imaging at the University of Toronto. Main interests: MRI of pituitary gland and parasellar area, tissue characterization using MR techniques, interventional methods with MRI.

Pediatric Brain MRI: Applications in Neonates and Infants

Jacqueline M. Pennock

Royal Postgraduate Medical School, Hammersmith Hospital, London, UK

1   INTRODUCTION

The ability of magnetic resonance imaging (MRI) to provide physiological, anatomical and functional information without posing any biological hazard makes it particularly suitable for studying the central nervous system of children where repeated examinations may be critical for diagnosis.

Clinical MRI systems are almost always designed for adults and sequences are optimized to maximize contrast in adults. The first task in examining infants is often to modify the equipment to optimize it for small infants and to develop imaging patterns which provide equivalent contrast to that seen in adults. This article covers particular techniques used to examine preterm babies and infants, the normal appearances of the developing brain, and clinical conditions encountered in infants and children.

2   PATIENT PREPARATION

For a successful examination, both the child and the parent must be considered. We actively encourage parents to accompany their infants to the imaging unit so that they do not feel excluded and can see that their child is asleep and comfortable. If it is their wish, we invite the parents to watch us putting their baby into the machine, bearing in mind that for many parents seeing their sleeping child enclosed in a head coil and being slid into the machine may be a stressful experience. With older children, parents are asked to stay with them during the examination, although talking is limited to the time between scans.

2.1   Sedation

Diagnostic images can only be obtained in a quiet, immobile child, so if possible appointments are made to coincide with the natural sleeping pattern of the child. Children come to a children's day ward where they are seen by a pediatrician who goes through the metal check with the parent and ensures that there are no contraindications to sedation. This is given to children under 2 years of age 20–30 min before the scan time. The child stays on the ward with the supervision of a nurse until she or he is asleep and is then brought to the scanning unit. Once the scan is over, the child returns to the ward until she or he is fully recovered.

Preterm infants and neonates are fed prior to the examination and scanned during natural sleep. However, in the irritable stable neonate, oral chloral hydrate via a nasal gastric tube or per rectum (50–80 mg kg^{-1}) is used. For infants aged 6 weeks and over, oral chloral hydrate (70–100 mg kg^{-1}) is given. This dose is successful in most children up to the age of 3 years or about 15 kg. Once the child is asleep, further immobilization is achieved by partly surrounding the head with a plastic globe filled with tiny polystyrene balls which is then evacuated. This minimizes motion, helps to keep the baby warm and provides insulation from sound. Swaddling the infant also prevents movement and provides extra security and warmth. Neonates usually sleep better on their sides, and this position also reduces the risk of inhalation of regurgitated milk or vomit.

MRI of the critically ill intubated infant is feasible as long as the imaging unit has the appropriate facilities to offer the same care for the child as does the neonatal intensive care unit (NICU). These include suitably trained staff, gases and suction, and pediatric-sized resuscitation equipment.

In our hospital the baby is brought from the NICU in a standard Vickers transport incubator. The incubator is parked and locked in a corner of the room well beyond the 5 G (5 × 10^{-4} T) line and, as an extra precaution, attached to a wall by a plastic chain. Infants are ventilated from the Vicker neovent unit via long extension tubes from the incubator to the baby, with appropriate adjustment of the inspiratory and expiratory pressure to take account of the large dead space.

2.2   Monitoring

Monitoring of the sedated and naturally sleeping infant is mandatory. A variety of magnetic resonance (MR) compatible physiological monitoring devices is available for monitoring blood pressure, heart rate, respiration, and blood oxygen saturation. We use electrocardiogram (ECG) monitoring with fiber optic leads and infant-sized MR-compatible electrodes, as well as pulse oximetry with neonatal probes attached to the infant's foot to display blood oxygen saturation and pulse rate. Further details and information on the sedation and anesthesia of the critically ill infant is available in the radiological literature.[1–3]

2.3   General Safety

Baby clothing with metal poppers should be removed. Although the fastenings are not ferromagnetic they may conduct eddy currents which can lead to artifacted images. An infant is almost always accompanied by medical staff and parents unaccustomed to an MR unit, and constant care has to be taken to check for loose metal objects on persons in a magnetic field. Pacemakers and aneurysm clips are usually associated with an older population; however, young mothers may also have aneurysm clips. A strict routine and protocol for access is necessary.

Scheduling of sedated infants and sleeping children is difficult and requires considerable flexibility. It is much easier to have exclusively pediatric sessions and not to try to mix sedated and sleeping children with ambulant adults. Prior knowledge of the clinical condition with a presumptive diagnosis is essential so that suitable protocols can be set up before each scan. Protocols which we have found useful in neonatal imaging are given in Table 1.

3   TECHNICAL CONSIDERATIONS

The signal-to-noise ratio is an important factor in producing high quality diagnostic images at all field strengths. Image quality can be improved by using the smallest diameter coil available. Adult knee coils with an internal diameter of approximately 19 cm have proved useful for imaging premature infants (head circumference approximately 30 cm) as well as infants up to 45 weeks gestational age. For children up to 18 months of age we use a receiver coil with an internal diameter of 24 cm. This coil is made in two halves of clear Perspex, making positioning and observation of the baby easier.

4   PULSE SEQUENCES

The neonatal brain contains a higher proportion of water (92–95%) than the adult brain (82–85%) and this is associated with a marked increase in T_1 and T_2.[4] The amount of water in the brain falls with increasing age, and by 2 years of age it has almost reached adult levels.[5]

A brief description of the standard pulse sequences and useful variations for imaging neonates and children are given below. The major pulse sequences and their specific variations for neonatal practice are shown in Table 2.

For References see p. 1055

Table 1 Suggested Protocol for Imaging Neonates at 1.0 T

	First scan (all ages, short *TR/TE* SE)	Inversion recovery	T_2-weighted spin echo
Slice orientation	Transverse	Transverse	Transverse
Phase-encoding axis	Left–right	Left–right	Left–right
Number of slices	24	18	24
Echo time (*TE*)	20	30	200–120[a]
Inversion time (*TI*)	–	1500–700[a]	–
Repeat time (*TR*)	860	6000–3500	3500
Field-of-view	22–25	22–25[a]	22–25
Phase resolution	192	128	128
Frequency encode resolution	256	256	256
Number of signals (average)	2	1 or 2	1 or 2
Slice thickness (mm)	4	5–6	5
Slice gap	0	0	0
Receive coil	Small diameter	Small diameter	Small diameter
Flip angle (degrees)	90	90	90
Estimated time of scan	4 min	Variable	Variable

[a]Choice will depend on age of infant (see Table 2).

4.1 The Inversion–Recovery (IR) Sequence

The short inversion time (*TI*) inversion–recovery (STIR) sequence is of value in demonstrating myelination and pathological periventricular changes where the cerebrospinal fluid (CSF) signal can be kept less than that of brain.[6] This sequence has many features in common with the T_2-weighted spin echo sequence, but gives greater gray–white matter contrast (Figure 1).

The medium *TI* version of the IR sequence provides excellent gray–white matter contrast and displays white matter with high signal intensity and gray matter with a lower signal intensity. This sequence is particularly useful for scanning the developing brain, especially for assessment of myelination, but the *TI* requires adjustment to accommodate the changing water content of the brain with age (see Figure 4 and Section 5).

The long *TI*, long *TE* version of this sequence is designed to display the CSF with low signal intensity and heavy T_2-weighting. The acronym for this sequence is FLAIR (fluid atte-

nuated inversion–recovery sequence) and it provides better lesion conspicuity than conventional T_2 spin echo sequences (Figure 2).[7] It is not always helpful in the immature brain, but in older children we have found it of great help for lesions with a periventricular distribution.

4.2 The Spin Echo Sequence

The short *TR TE* T_1-weighted spin echo sequence has less gray–white matter contrast than the medium *TI* inversion recovery sequence. However, the former sequence is quicker and is recommended for use as the first sequence in all studies on sedated infants under 2 years of age. It is extremely useful for the assessment of brain swelling, hemorrhage, cystic change, and contrast enhancement with gadopentetate dimeglumine.

With the T_2-weighted spin echo sequence it is necessary to increase the *TE* to 120–200 ms in order to identify long T_2 lesions against the background of the long T_2 of the immature

Table 2 Inversion Recovery and Spin Echo (SE) sequences at 1.0 T

Age	Type	*TR* (ms)	*TE* (ms)	*TI* (ms)
Inversion–recovery sequences				
29–34 weeks	IR medium *TI*	7000	30	1500
35–39 weeks		6500	30	1200
40 weeks to 3 months		3800	30	950
>3 months to 2 years		3500	30	800
>2 years		3200	30	700
<3 months	FLAIR long *TI*	6000	240	2100
>3 months		6000	160	2100
All ages	STIR short *TI*	3800	30	130–150
Spin echo sequences				
All ages	T_1-weighted	860	20	
⩽2 years	T_2-weighted	3500	120	
>2 years	T_2-weighted	2500	20/80	

For list of General Abbreviations see end-papers

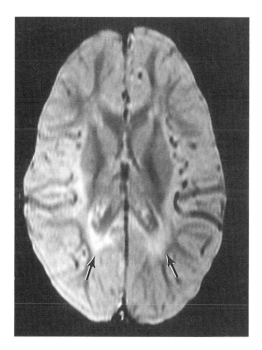

Figure 1   Female infant aged 9 months. Short *TI* inversion recovery (STIR) sequence (IR 3000/30/125). Pathological white matter adjacent to the posterior horns of the lateral ventricles is seen (arrows). The CSF is displayed as moderate signal intensity and the myelin as low signal intensity

brain.[8] However, the extended *TE* results in high signal intensity of CSF, which makes it difficult to detect lesions in a periventricular distribution.

4.3   Diffusion Weighted Imaging

Image contrast in diffusion weighted sequences depends on the molecular motion of water. These sequences are particularly useful for demonstrating early myelination before it can be seen with conventional sequences.[9] It also shows regions of focal infarction and diffuse ischemic anoxic brain injury which are not visible on spin echo and inversion–recovery images in the early phase (see Section 9).

4.4   Magnetization Transfer Techniques

These sequences have proved useful for improving lesion contrast both for short and long T_1 lesions[10] and in newborn infants the conspicuity of some cystic lesions is improved. It is also a useful technique for demonstrating myelination.

4.5   MR Angiography

MR angiography (MRA) is a noninvasive method of examining cerebral blood flow, although its use in pediatric practice has so far been limited. Cerebral blood flow is slow in infants and the blood vessels are smaller than in adults. A recent innovation in this technique has been the implementation of magnetization transfer MRA which involves the application of a specialized frequency-selective pulse followed by the desired MRA sequence. This results in a darkening of the background static tissue with an increase in contrast with the flowing blood. Our initial studies using two- and three-dimensional time-of-flight angiography have been of value in showing the internal

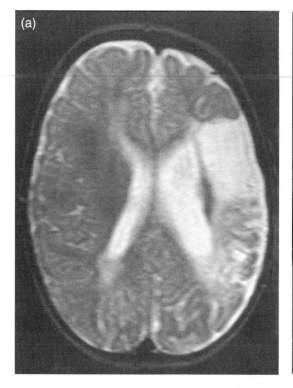

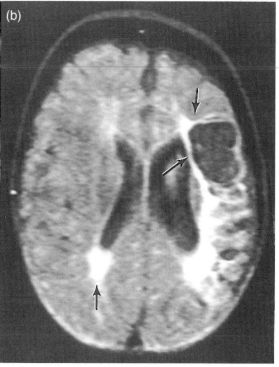

Figure 2   Male infant aged 1 year. (a) T_2-weighted spin echo (SE 2700/120) and (b) FLAIR (IR 7203/240/2100) sequences. A porencephalic cyst is seen in both images in the left hemisphere. Gliosis is shown as high signal intensity in both sequences; however, it appears more extensive and is seen with greater conspicuity in (b) (arrows). Note the low signal intensity of CSF within the porencephalic cyst with the FLAIR sequence

For References see p. 1055

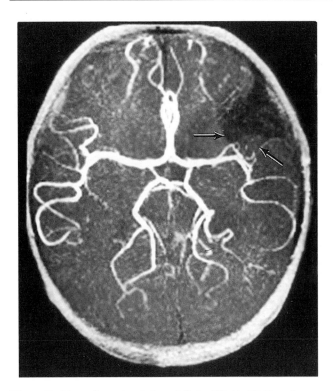

Figure 3 Three-dimensional time-of-flight (*TR* 44, *TE* 8) angiogram: an 18-month-old male infant with known infarction of the left middle cerebral artery. The anterior, posterior, and middle cerebral arteries are seen, but there is a paucity of vessels on the left in the anterior branches from the middle cerebral artery (arrows)

carotid and basilar arteries and the proximal regions of the anterior, middle, and posterior cerebral arteries in the term infant (Figure 3). Large veins are also well shown.

4.6 The Gradient Echo (GE) and Partial Saturation (PS) Sequences

The gradient echo and partial saturation sequences have a specific role in pediatric practice, especially in the detection of neonatal hemorrhage. Low signal regions in and around hematomas are seen with higher sensitivity with partial saturation sequences than with spin echo sequences, indicating a significant contribution from susceptibility effects.[11] Phase maps can be derived from two partial saturation sequences with different values of *TE*, and are particularly useful for looking at the susceptibility effects of intracerebral hemorrhage.[12]

The use of these sequences in fast imaging techniques increases the speed of MR examinations, which is of particular interest when scanning infants; however, high signal from CSF remains a problem.[13] A GE sequence with a *TE* of 29 ms and a flip angle that corresponds to the choice of *TR* is recommended at 1.5 T to optimize the detection of calcification in the brain.[14]

4.7 Volume Imaging

This may be of particular value in infants and children because reconstruction and reslicing of images can provide a

precise correction for differences in registration of follow-up examinations so that subtle differences in growth and development can be recognized.

5 NORMAL DEVELOPMENT AND MRI APPEARANCES

During the first 2 years of life the pediatric brain changes rapidly as physiological myelination takes place. This continues at a slower rate into the second decade. The process of myelination begins in utero and is first seen in the cerebellum and brainstem and then spreads from the posterior limb of the internal capsule to the postcentral gyrus. After birth, white matter develops in a predictable manner with sensory tracts myelinating before motor tracts from dorsal to ventral and from central to the peripheral areas of the brain.[15,16] Several descriptions of this process with MRI using a variety of sequences are available in the literature.[2,3,17–20] The pattern of development from 29 weeks gestational age to 1 year of age is shown by IR sequences in Figure 4.

5.1 Delays or Deficits in Myelination

Delays or deficits in myelination are difficult to recognize before 3–6 months of age, since relatively little myelin is present. Conversely, after 2 years of age there is time for cases of delayed myelination to 'catch up'. As a result, delays or deficits are most obvious from 6 to 24 months of age. With only limited information about the normal range we have preferred to use age-matched controls and diagnose delays only in the absence of or marked reduction in myelination of named tracts or commissures relative to controls, with both examinations performed using the same technique.

Delays or deficits in myelination have been recognized following probable intrauterine rubella infection, in posthemorrhagic hydrocephalus, after hypoxic ischemic encephalopathy, infarction, periventricular cystic leukomalacia, and metabolic disease.

6 INTRACRANIAL HEMORRHAGE

Intraventricular/periventricular hemorrhage (IVH/PVH) is the most important and most common type of hemorrhage in infants of less than 32 weeks gestation, with 90% of the hemorrhage occurring in the germinal matrix adjacent to the heads of the caudate nuclei.[21,22]

The basal ganglia are the most vulnerable site for hemorrhagic lesions in term infants suffering from severe birth asphyxia.[22] Ultrasound is used to detect and monitor the progression of IVH/PVH in the early stages in infants too sick to transport to the MR unit. However, MR is useful in the stable neonate[23] and for long-term follow-up. The MR appearance of hemorrhage parallels that seen in adults; however, in neonates hemorrhage occurs against a background of long normal values of brain T_1 and T_2, so that the T_2 of hematoma may appear distinctly shorter than that of the surrounding brain (Figure 5).

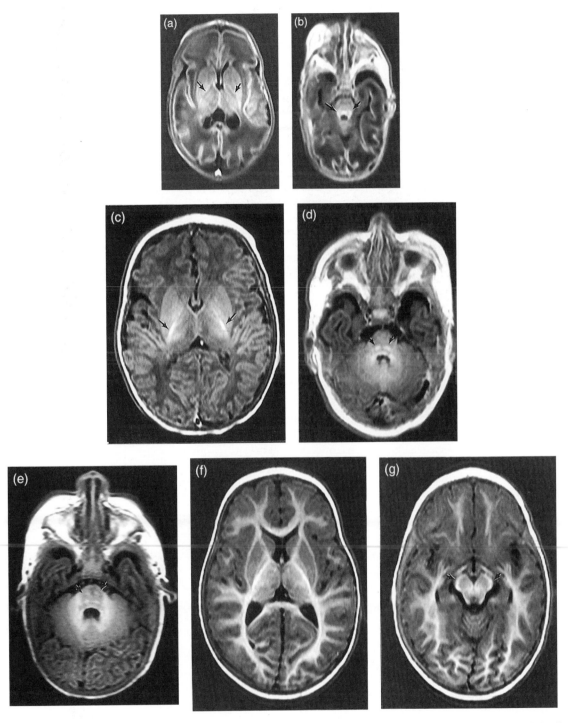

Figure 4 Infant born at 29 weeks gestational age and examined at 31 weeks gestational age (IR 7096/30/1500). At the level of the diencephalon (a) no myelin is seen; however, the posterior limb of the internal capsule is seen as low signal intensity (arrows), the remaining white matter is featureless. The cortex is of high signal intensity and the gyral pattern is underdeveloped. At the level of the pons (b) the corticospinal tracts are of low signal intensity (SI); however, myelin is seen in the medial longitudinal fasciculus and the inferior cerebellar peduncle (arrows). Term infant (IR 3500/30/950) (c–d). At the level of the diencephalon (c) the cortex remains of higher SI than unmyelinated white matter, but more structure is seen than in (a). Myelination is present in the ventrolateral nuclei in the thalami and in the posterior limb of the internal capsule (arrows). The lentiform nuclei have relatively high signal compared to the caudate nuclei and the thalami. At the level of the pons (d) myelin is seen in the medial lemnisci (arrows). At the age of 2 months at the level of the pons (e) myelin is present in the corticospinal tracts (arrows) anterior to the medial leminisci and further myelination is seen in the cerebellum. (f, g) Infant aged 12 months (IR 3200/30/800). At the level of the diencephalon (f) myelination is now seen throughout the internal and external capsule, along the occipitothalamic pathways and the corpus callosum. At the level of the mesencephalon (g) myelin is seen in the crus cerebri (arrows)

For References see p. 1055

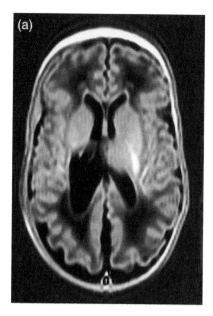
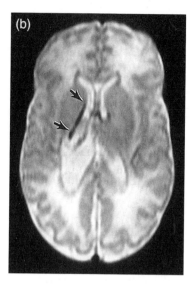

Figure 5 Female infant with IVH/PVH born at 26 weeks gestational age and examined at 40 weeks gestational age. (a) Inversion–recovery (IR 6000/30/1200) and (b) T_2-weighted spin echo (SE 2700/120) sequences. Hemorrhage is seen as high signal intensity on the IR image (a), but is seen more clearly on (b) as low signal intensity (arrows). Myelin in the posterior limb of the internal capsule is better seen in (a) than in (b)

The signal characteristics of subdural and extradural hematomas are similar to those seen in adults. Subarachnoid blood, especially along the Rolandic fissure, is a common finding in the newborn term infant (Figure 6).

The ability to demonstrate methemaglobin and hemosiderin by a shortening of T_2^* and susceptibility effects provides a high degree of sensitivity, and may function as a marker of early hemorrhage years after the event.[24] The late sequelae of intracranial hemorrhage may have important consequences such as hydrocephalus.

7 PERIVENTRICULAR LEUKOMALACIA

The second most frequent condition to afflict premature infants is periventricular leukomalacia (PVL),[21,22] which occurs in a characteristic distribution in white matter around the lateral ventricles. Ultrasound scanning is generally used to make the diagnosis of leukomalacia; however, periventricular cysts and subcortical cysts are clearly seen with MRI in the early neonatal period and the changes can be quite extreme. Sometimes the cysts coalesce and become continuous with the adjacent ventricles producing hydrocephalus. The presence of severe or moderately severe cysts in infancy is frequently associated with a delay or deficit in myelination (Figure 7).

The MRI features are closely correlated with clinical outcome.[25] For example, infants with mild periventricular change on MRI have a mild spastic diplegia and normal intellectual development, whereas infants with subcortical cysts, diminution of white matter and microcephaly are severely mentally retarded with quadriplegia and seizures. In older children PVL is seen, with a T_2-weighted spin echo sequence, as an increased signal intensity usually in white matter in the centrum semiovale and adjacent to the anterior and/or the

posterior horns of the lateral ventricles. We have found greater conspicuity of the lesions with the FLAIR and the short TI inversion–recovery sequences. To date, MRI has been most useful for studying the late stages of cystic leukomalacia.[26] MRI also permits retrospective diagnosis of the condition in infants who were not scanned either during the neonatal period

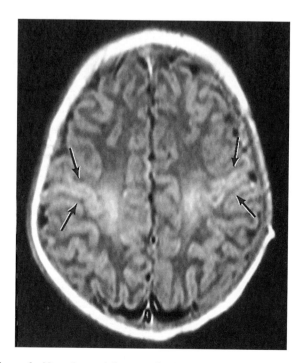

Figure 6 Normal term infant examined at 24 hours of age (IR 3800/30/950). Note the high signal intensity of the subarachnoid blood along the Rolandic fissure (arrows)

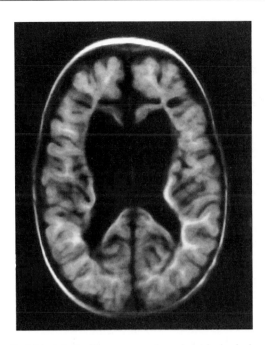

Figure 7 Male infant with severe cystic periventricular leukomalacia born at 33 weeks and examined aged 11 months (IR 1800/44/600). There is a severe delay in myelination compared with that seen in Figure 4(f)

or were scanned at a stage of evolution when the cysts were not apparent.

8 INFARCTION

Cerebral infarction is most commonly seen in term infants and appears as a region of increased T_1 and T_2 which may be difficult to distinguish from areas of unmyelinated white matter normally present in the brain. There may be a loss of gray–white matter contrast and loss of the normal gyral pattern in a focal area, with or without a slight increase in T_1. The use of diffusion weighted imaging, where contrast is mainly determined by differences in the molecular motion of water (and not T_1 and T_2), may show the extent of the lesion within hours of the onset of injury (Figure 8).[27]

However, 1–4 days after the onset of symptoms, changes in T_1 and T_2 occur and the region of infarction may become visible with standard imaging techniques. With increasing age there is some apparent shrinking of the lesion and a porencephalic cyst may develop. Some porencephalic cysts decrease in size, some remain the same size, and others appear smaller due to increasing head size (Figure 9). Associated hydrocephalus with porencephalic cysts communicating with the ventricles may produce an apparent increase in the size of cysts. Waller-

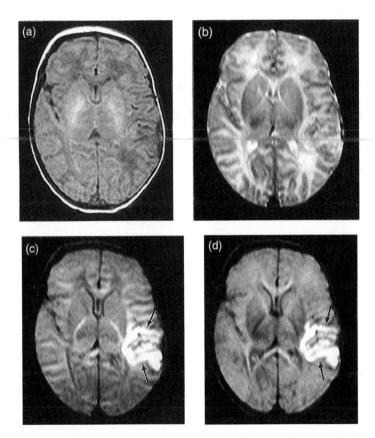

Figure 8 Cerebral infarction in a male infant born at 40 weeks gestational age. (a) Transverse T_1-weighted spin echo (SE 720/20), (b) T_2-weighted spin echo (SE 3000/120), (c) diffusion weighted (SE pulse interval/200 ms, A-P sensitization, $b = 600$ s mm^{-2}), and (d) diffusion weighted (SE pulse interval/200 ms through plane sensitization, $b = 600$ s mm^{-2}) images. The infarction is difficult to recognize on (a) or (b) but is readily apparent as a high signal region on (c) and (d) (arrows)

For References see p. 1055

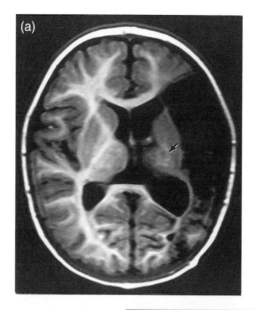

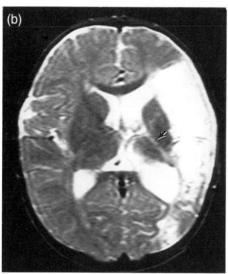

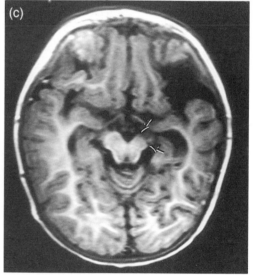

Figure 9 Infant aged 9 months with a left middle cerebral infarct. (a, c) Inversion–recovery (IR 3400/30/800) and (b) T_2-weighted spin echo (SE 2700/120) images. A large porencephalic cyst is seen in the left hemisphere. Increased signal intensity is seen along the length of the posterior limb of the internal capsule in (b) and low signal intensity is seen in (a) (arrows). The mesencephalon on the left is smaller than on the right (c) (arrows). Less myelin is seen on the left hemisphere than on the right

ian degeneration within the corticospinal tracts can be seen as early as 7 days after the insult in newborn infants, with atrophy of the brainstem occurring by 3 months of age. These findings occur much earlier in children than in adults.[28,29]

9 HYPOXIC ISCHEMIC ENCEPHALOPATHY

Hypoxic ischemic encephalopathy (HIE) is most frequently seen in term infants[22] and the pattern of damage and its evolution is very variable even in children who appear to be seriously affected immediately after birth. Specific early MR features include brain swelling and increased signal intensity on the inversion–recovery sequence in the cortex which is usually the most marked around the Rolandic fissure with decreased signal intensity on the T_2-weighted spin echo sequence. Loss of the normal signal intensity in the posterior limb of the internal capsules as well as focal hemorrhagic lesions in the basal ganglia, which are most commonly located in the lentiform nuclei, are also seen (Figure 10).

Brain swelling is not seen after the first 4 days and the evolution of the early MR findings may include breakdown of white matter into subcortical cysts within the first 2 weeks of life, with diminution of white matter and a severe delay in myelination by 3 months of age. The basal ganglia lesions become less obvious with time and may regress completely.

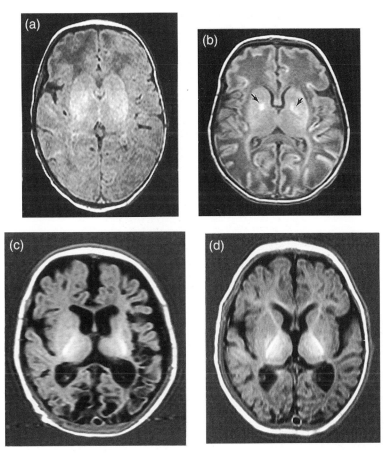

Figure 10 Male infant born at 40 weeks with severe birth asphyxia: T_1-weighted spin echo (SE 860/20). Images at age 2 days (a) and 10 days (b); (c) inversion–recovery (IR 3797/30/95) image at age 5 weeks; (d) IR 3405/30/800 at age 6 months. At age 2 days brain swelling is present and there is a loss of sulcal patterns and gray–white matter contrast. The ventricles and the interhemispheric fissure are small (a). At 10 days, the brain swelling is no longer present. High signal intensity is seen throughout the cortex and the white matter is featureless. Hemorrhagic lesions are seen with the globus pallidus (arrows). There has been loss of the normal increased signal intensity in the posterior limb of the internal capsule (b). At 5 weeks (c) there has been a diminution in white matter and cystic change is noted in the posterior lobes. Myelin is now present in the posterior limb of the internal capsule. At 6 months (d), there has been further loss of white matter relative to normal. Abnormal signal intensity is seen in the thalami and globus pallidus. Further myelination has taken place

Development of myelination in the posterior limbs of the internal capsules may occur by 3 months of age. Alternatively, the lesions in the basal ganglia and thalami become cystic and this may occur as early as 17 days of age with atrophy of the basal ganglia by 6 weeks of age. The degree of cortical highlighting is also variable and diminishes with time; however, in our experience it may still be present up to 6 months of age.

Early diffusion weighted imaging is important in these children (Figure 11) and correctly predicts the sites of injury which become more obvious on standard imaging at a later stage. These striking early MR findings closely correlate with the location of the selective neuronal necrosis[21,22] seen at post mortem in asphyxiated infants.

The prognosis for infants with global hypoxia can be devastating. However, new cerebroprotective drugs are at present undergoing investigation in stroke models, but these therapies are only effective in the first few hours after birth before the onset of secondary energy failure.[30] It is hoped that these MR findings may be used to monitor such treatment in the future.

10 HYDROCEPHALUS

Hydrocephalus can arise in a number of circumstances in children.[31] The ventricular size can be readily assessed, and MRI has obvious advantages in the long-term follow-up of children with shunts. It is also possible to recognize periventricular edema both with spin echo and inversion–recovery sequences; this condition may regress following satisfactory ventricular shunting. The periventricular changes are displayed as increased signal intensity in the periventricular regions in some cases of hydrocephalus, probably indicating transependymal spread of fluid. We have found the STIR sequence (CSF displayed as moderate signal intensity) and the FLAIR sequence (CSF displayed as low signal intensity) are better

For References see p. 1055

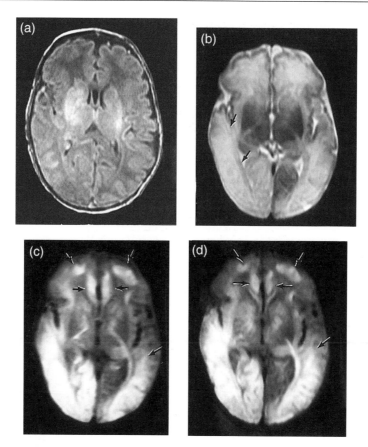

Figure 11 Hypoxic ischemic encephalopathy on day 1: (a) transverse T_1-weighted spin echo (SE 720/20), (b) T_2-weighted spin echo (SE 3000/120) (c) diffusion weighted (SE pulse interval/200 ms, left–right sensitization, $b = 600$ s mm^{-2}), and (d) diffusion weighted (SE pulse interval/200 ms, through-plane sensitization, $b = 600$ s mm^{-2}) images. High intensity signal is only seen in the medial right occipital lobe in (b) (arrows), suggesting the diagnosis of infarction; much more extensive abnormalities are seen bilaterally in the frontal, temporal, and occipital lobes using diffusion weighted imaging (c and d) (arrows showing some of the high intensity signal areas)

than the conventional spin echo sequence for looking at periventricular change in children with hydrocephalus (Figure 12). However, similar changes may be seen in other diseases such as periventricular leukomalacia, and the changes are not specific. It is not always possible to distinguish between acute and chronic hydrocephalus.

Hydrocephalus caused by aqueduct stenosis and other obstructive lesions is generally well displayed with MRI and the ability to scan in more than one plane is very useful in these conditions.

11 CONGENITAL MALFORMATIONS

Brain development follows a well defined sequence. A disturbance at any particular time may affect one or more stages and result in a developmental anomaly.[2,3,32–34] Dorsal induction occurs during week 3–4 of gestation when the nasal plate folds to form the nasal tube. Failure to close caudally results in myelomeningocoele, and failure to close at the cephalad end may result in anencephaly encephalocele, etc. In the next stage the mesencephalon divides to form the telencephalon. Failure at this stage produces prosencephaly. Normal proliferation then follows in which the germinal matrix forms the neurons that

form the cortex. Neurons may fail to form the normal cortical layers, or stop along their path resulting in multiple cortical abnormalities including heterotopia.

Neurofibromatosis, tuberous sclerosis, and Sturge–Weber disease are the common neurocutaneous diseases occurring in children.[35] However, the MRI features may be difficult to define in the neonate, before some degree of myelination has taken place. MRI is of considerable value in demonstrating tumors associated with the phakomatoses as well as MRI regions of gliosis, hematomas, and cerebral atrophy.[36] Calcification in tuberous sclerosis is poorly shown, but may be seen.[37] Computed tomography (CT) or plain skull X-rays may be more useful in this situation.

Obvious anatomical deformations are readily shown, e.g. anencephalopathy, holoprosencephaly, and Dandy–Walker syndrome. The sagittal plane lends itself to demonstration of many of these conditions including, for example, agenesis of the corpus callosum.

12 WHITE MATTER DISEASE

The most common white matter disease to affect infants is periventricular leukomalacia (see Section 7). Recognition of

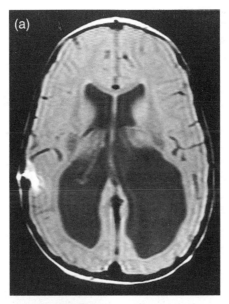

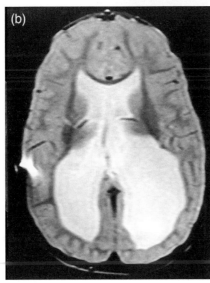
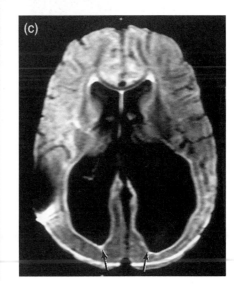

Figure 12 Male infant aged 2 years with posthemorrhagic hydrocephalus: (a) mildly T_2-weighted (SE 2500/20) and (b) moderately T_2-weighted (SE 2500/80) spin echo sequences; (c) fluid attenuated inversion–recovery sequence (FLAIR 6500/160/2100). A shunt artifact is seen. There is an increase in signal intensity around the lateral ventricles in (a), which is less well seen in (b) and best seen in (c) (arrows)

other white matter disease is difficult in early life because of the lack of myelin present at birth and in the first 6 months of life. Once myelin has been laid down it can disappear in two principal ways, namely demyelination and dysmyelination.[38] In demyelination the breakdown of myelin is caused by extrinsic factors (e.g. infection, trauma, chemotherapy), and in dysmyelination there is a genetic disorder of myelin formation which is a feature of metabolic disorders such as the leukodystrophies. Diffuse abnormalities are seen within white matter in leukodystrophy. The changes are usually extensive and not confined to the periventricular region. In other forms of white matter disease, such as Alexander's disease, changes may be confined to the frontal lobes. A variety of other abnormalities have been described in different forms of leukodystrophy.[2,38] We have also seen periventricular abnormalities associated with intrathe-

cal methotrexate therapy in leukemia. Along with delays in myelination, both demyelination and dysmyelination are readily recognized on MRI in children.[2,38,39]

13 INFECTION

Excellent reviews on inflammatory diseases of the brain in childhood are available in the literature.[2,40] Cerebral abscess displays an increase in T_1 and T_2. Edema is well displayed but the exact margins of the abscess may be difficult to define.[40] However, ring enhancement after the intravenous injection of gadolinium diethylenetriaminepentaacetic acid (Gd-DTPA) may help in the differential diagnosis of cerebral abscess. Calcifi-

For References see p. 1055

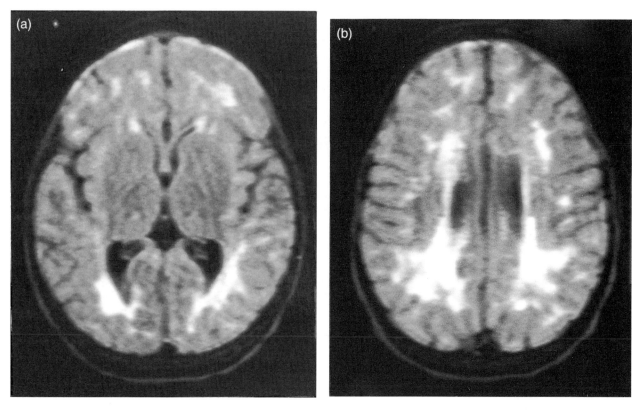

Figure 13   Female infant aged 2 years with cytomegalovirus; fluid attenuated inversion–recovery (FLAIR 6500/160/2100) sequences. Areas of long T_2 are seen throughout the white matter

cation associated with abscess is poorly demonstrated in comparison with CT.

In two cases of brainstem encephalitis, changes have been seen with very little associated mass effect. This has been the main distinction between tumors at the initial examination and on regression on follow-up examination. It provides strong support for the diagnosis, although a certain amount of caution is necessary as patients are frequently treated with steroids which may result in some regression of edema associated with a tumor. In a case of cytomegalovirus, marked white matter change has been observed in a child of 2 years without any overt clinical signs (Figure 13). In neonatal meningitis, contrast enhancement may be seen in the meninges (Figure 14).

14   TUMORS

In general, the features of tumors in children parallel those in adults; however, there is a higher incidence of tumors in the posterior fossa, and embryological tumors are more common.[2] The high incidence of midline tumors lends itself to sagittal imaging, and the clarity with which the posterior fossa is seen is also an advantage.

Most tumors display an increased T_1 and T_2 providing high contrast with long TE long TR spin echo sequences, although distinction between tumor and edema may be difficult. Differentiation between brainstem and cerebellar sites is reasonably easy. Craniopharyngiomas and various other lipid-containing tumors may show characteristic features. Hamartomas may not display a significant change in T_1 and T_2 and may then need to

be recognized by their indirect signs. Hypothalamic tumors which may be poorly seen with CT are well recognized with MR.

15   OTHER DISEASE

Certain other conditions are worth reviewing, although they are quite rare. Delays or deficits in myelination have been recognized in Hurler's disease, and these may be reversed following successful bone marrow transplantation.[41]

Hallervorden–Spatz disease is of particular theoretical interest as a condition in which there is abnormal iron deposition in the brain, and in one case abnormalities have been seen in the basal ganglia.

In Wilson's disease abnormalities are seen in the lentiform nucleus and within the thalamus;[42] however, the findings are not specific as similar changes are seen in children with Leigh's disease.[2]

16   FOLLOW-UP EXAMINATIONS

This is an important aspect of pediatric practice. The normal appearances including the value of T_1 and T_2, the presence of periventricular long T_1 areas, the degree of myelination, as well as the size and the shape of the brain, all change. Pathological changes must be assessed against this changing background.

The lack of known hazard is a strong incentive for pediatric MRI. Follow-up examinations in conditions in which long-term

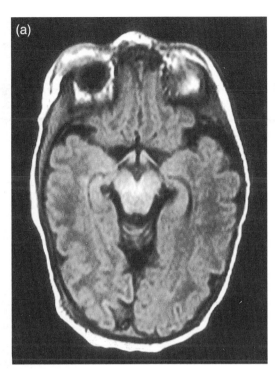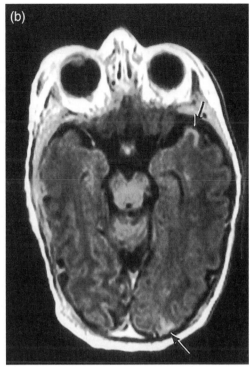

Figure 14 Male infant aged 6 weeks with meningitis. T_1-weighted spin echo (SE 720/20) sequence before (a) and after intravenous injection of Gd-DTPA (b). Meningeal enhancement is seen in (b) (arrows)

survival is expected without accumulating significant X-ray dosage is important. Nevertheless, there are problems in achieving MRI scans at the same level and angulation as in the initial studies. There is also a theoretical problem in using age-adjusted sequences since the machine parameters are different. Genuine advances in technique can also make comparison difficult.

17 CONCLUSIONS

Developments in pediatric MRI lag behind those of adults, but it is possible to extrapolate findings in adults to children. Clinical correlation has been progressing, but the correlation is not precise and some children may have very large lesions with relatively small clinical deficits. Large unsuspected lesions have been found where the clinical signs are quite subtle. The capacity for repeated examination without cumulative radiation dosage problems has been of value in studying the natural history of a variety of neonatal insults.

The versatility of MRI with its basic image parameters ρ, T_1, T_2, chemical shift, flow, susceptibility, and diffusion effects provides a wide variety of options for the various problems encountered in clinical practice. Only a small number of these options has yet been employed in pediatric practice, and a growing role for MRI in this area is certain. The application of MR may be greatly expanded by the installation of suitable systems in neonatal intensive care units. To date only relatively large and stable infants have been examined, but the impact of

MRI on the management of the critically ill and very premature infants may be much expanded in the future.

18 RELATED ARTICLES

Brain MRS of Infants and Children; Diffusion: Clinical Utility of MRI Studies; Intracranial Infections; MRI in Clinical Medicine.

19 REFERENCES

1. R. S. Boyer, *Am. J. Neuroradiol.*, 1992, **13**, 777.
2. A. J. Barkovich, in 'Contemporary Neuroimaging', ed. D. Norman, Raven, New York, 1990, Vol. 1.
3. M. D. Cohen and M. K. Edwards (eds), 'Magnetic Resonance Imaging of Children', B. C. Decker, Philadelphia, 1990.
4. M. A. Johnson, J. M. Pennock, G. M. Bydder, R. E. Steiner, D. J. Thomas, R. Hayward, D. J. Bryant, J. A. Payne, M. I. Levene, A. Whitelaw, L. M. S. Dubowitz, and V. Dubowitz, *Am. J. Roentgenol.*, 1983, **141**, 1005; *Am. J. Neuroradiol.*, 1983, **4**, 1013.
5. J. Dobbing and J. Sands, *Arch. Dis. Child.*, 1973, **48**, 757.
6. L. S. de Vries, L. M. S. Dubowitz, V. Dubowitz, and J. M. Pennock, 'Colour Atlas of Brain Disorders in the Newborn', Wolfe Medical, Chicago, 1990.
7. B. De Coene, J. V. Hajnal, P. Gatehouse, D. B. Longmore, S. J. White, A. Oatridge, J. M. Pennock, I. R. Young, and G. M. Bydder, *Am. J. Neuroradiol.*, 1992, **13**, 1555.
8. M. A. Nowell, D. B. Hackney, R. A. Zimmerman, L. T. Bilaniuk, R. I. Grossman, and H. I. Goldberg, *Radiology*, 1987, **162**, 272.

For References see p. 1055

9. M. A. Rutherford, F. M. Cowan, A. Y. Manzur, L. M. S. Dubowitz, J. M. Pennock, J. V. Hajnal, I. R. Young, and G. M. Bydder, *J. Comput. Assist. Tomogr.*, 1991, **15**, 188.

10. J. V. Hajnal, C. J. Baudouin, A. Oatridge, I. R. Young, and G. M. Bydder, *J. Comput. Assist. Tomogr.*, 1992, **16**, 7.

11. R. R. Edelman, K. E. Johnson, R. Buxton, G. Shoukimos, B. R. Rosen, K. R. Davis, and T. J. Brady, *Am. J. Neuroradiol.*, 1986, **7**, 751.

12. I. R. Young, S. Khenia, D. G. T. Thomas, C. H. Davis, D. G. Gadian, I. J. Cox, B. D. Ross, and G. M. Bydder, *J. Comput. Assist. Tomogr.*, 1987, **11**, 2.

13. F. W. Wehrli, 'Fast-Scan Magnetic Resonance: Principles and Applications', Raven, New York, 1991.

14. R. M. Henkelman and W. Kucharczyk, *Am. J. Neuroradiol.*, 1994, **15**, 465.

15. A. Feess-Higgins and J.-C. Larroche, 'Le Developpement du Cerveau Foetal Humain: Atlas Anatomique', Masson, Paris, 1987.

16. P. I. Yakovlev and A. R. Lecours, in 'Regional Development of the Brain in Early Life', ed. A. Minkowski, Blackwell Scientific, Oxford, 1967, p. 3.

17. C. B. McArdle, C. J. Richardson, D. A. Nicholas, M. Mirfakhraee, C. F. Hayden, and E. G. Amporo, *Radiology*, 1987, **162**, 223.

18. A. J. Barkovitch and C. L. Truwit, 'Practical MRI Atlas of Neonatal Brain Development', Raven, New York, 1990.

19. E. C. Prenger, W. W. Beckett., S. S. Koleias, and W. S. Ball, *JMRI*, 1994, 179.

20. S. M. Wolpert and T. D. Barnes, 'MRI in Pediatric Neuroradiology', Mosby, St Louis, 1992.

21. K. E. Pape and J. S. Wigglesworth, 'Hemorrhage, Ischemia and the Perinatal Brain', Lippincott, Philadelphia, 1979.

22. J. J. Volpe, 'Neurology of the Newborn', Saunders, Philadelphia, 1987.

23. C. B. McArdle, C. J. Richardson, C. K. Hayden, D. A. Nicholas, M. J. Crofford, and E. G. Amparo, *Radiology*, 1987, **163**, 387.

24. J. M. Gomori, R. I. Grossman, H. I. Goldberg, D. B. Hackney, R. A. Zimmerman, and L. T. Bilaniuk, *Neuroradiology*, 1987, **29**, 339.

25. L. S. de Vries, L. M. S. Dubowitz, J. M. Pennock, and G. M. Bydder, *Clin. Radiol.*, 1989, **40**, 158.

26. L. L. Baker, D. K. Stevenson, and D. R. Enzmann, *Radiology*, 1988, **168**, 809.

27. F. M. Cowan, J. M. Pennock, D. D. Hanrahan, K. Manji, J. D. Hanrahan, and A. D. Edwards, *Neuropediatrics*, 1994, **25**, 172.

28. J. M. Pennock, M. A. Rutherford, F. M. Cowan, and G. M. Bydder, *Clin. Radiol.*, 1993, **47**, 311.

29. M. J. Kuhn, D. J. Mikulis, D. M. Ayoub, B. E. Kosofsky, K. R. Davis, and J. M. Taveras, *Radiology*, 1989, **172**, 179.

30. D. Azzopardi, J. S. Wyatt, E. B. Cady, D. T. Delby, T. Boudin, A. L. Stewart, P. L. Hope, P. A. Hamilton, and E. O. Reynolds, *Pediatr. Res.*, 1989, **25**, 445.

31. C. R. Kitz, *Semin. US, CT MR*, 1988, **9**, 216.

32. S. E. Byrd, R. E. Osborn, M. A. Radkorvoski, C. B. McArdle, E. C. Prengen, and T. P. Naidich, *Semin. US, CT MR*, 1988, **9**, 201.

33. S. R. Pollei, R. S. Boyer, S. Crawford, H. R. Harnsberger, and A. J. Barkovich, *Semin. US, CT MR*, 1988, **9**, 231.

34. M. S. Van der Knaap, and J. Valk, *Am. J. Neuroradiol.*, 1988, **9**, 315.

35. S. C. Crawford, R. S. Boyer, H. R. Harnsberger, S. R. Pollei, W. R. T. Smoker, and A. G. Osborn, *Semin. US, CT MR*, 1988, **9**, 247.

36. J. G. Smirniotopoulos and F. M. Murphy, *Am. J. Neuroradiol.*, 1992, **13**, 725.

37. B. H. Braffman, L. T. Bilaniuk, and R. A. Zimmerman, *Radiol. Clin. North Am.*, 1988, **26**, 773.

38. J. Valk, 'MRI of the Brain, Head, Neck and Spine: A Teaching Atlas of Clinical Application', Martinus Nijhoff, Dordrecht, 1987.

39. M. A. Nowell, R. I. Grossman, D. B. Hackney, R. A. Zimmerman, H. I. Goldberg, and L. T. Bilaniuk, *Am. J. Roentgenol.*, 1988, **151**, 359.

40. C. R. Fitz, *Am. J. Neuroradiol.*, 1992, **13**, 551.

41. M. A. Johnson, S. Desai, K. Hugh-Jones, and F. Starer, *Am. J. Neuroradiol.*, 1984, **5**, 816.

42. G. A. Lawler, J. M. Pennock, R. E. Steiner, W. J. Jenkins, S. Sherlock, and I. R. Young, *J. Comput. Assist. Tomogr.*, 1983, **7**, 1.

Biographical Sketch

Jacqueline M. Pennock. *b* 1940. M.Phil., 1976. Department of Diagnostic Radiology, Hammersmith Hospital, London 1963–present. Currently, senior scientific officer. Approx. 80 publications. Metabolic and endocrine disease (with Professor Frank Doyle). Current research interests: pediatric magnetic resonance, specifically related to normal development and critically ill preterm and term infants (with Frances Cowan, Mary Rutherford, Lilly Dubowitz and Graeme Bydder).

Temporomandibular Joint MRI

Steven E. Harms

Baylor University Medical Center, Dallas, TX, USA

1 INTRODUCTION

The value of medical imaging is determined by its ability to define treatment decisions. If those decisions are separated by treatments that greatly differ in cost or morbidity, the value of the diagnostic test is high. On the other hand, if the treatment is safe and inexpensive, then it may be more cost-effective to treat without diagnosis. As the treatment of temporomandibular joint (TMJ) disease evolves, the role of MR imaging as a determinant for clinical management decisions becomes better defined. In this chapter, the MR imaging of the TMJ is reviewed with its relationship to the current concepts for clinical management of TMJ disorders.

2 ANATOMY AND PHYSIOLOGY

The TMJ is a compound joint with the three components being the temporal bone, the mandibular condyle, and the

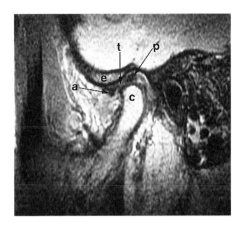

Figure 1 Normal temporomandibular joint. The bone landmarks are clearly visible on the oblique sagittal SE 15/2000 image of the temporomandibular joint. The mandibular condyle, c, and eminence, e, are labeled. Normal disk architecture is demonstrated with good demonstration of the anterior band, a, posterior band, p, and the thin zone, t

articular disk (Figure 1). There are two joint spaces: the superior joint space lying between the disk and the temporal bone and the inferior joint space lying between the disk and the mandibular condyle.[1,2]

The articular disk is a lens-shaped structure divided into three parts: the posterior band, the anterior band, and the thin zone. In the young asymptomatic population without previous orthodontic manipulation, the posterior band normally lies between 11 o'clock and 1 o'clock relative to the superior aspect of the mandibular condyle. As opposed to what was originally thought, the disk can be quite variable in location without producing symptoms.[3]

The articular disk of the TMJ is held in place by the discal ligaments which extend from the medial and the lateral poles of the condyle to the medial and lateral margins of the disk. As a result of the strong attachment to the condyle by the diskal ligaments, the disk and condyle move together as a unit called the disk–condyle complex. The posterior band of the disk is connected to the posterior aspect of the temporal fossa by elastic tissues called the retrodiskal attachments. Since the retrodiskal attachment is composed of two major elastic structures, this area is sometimes called the bilaminar zone. The anterior band of the disk is attached to the superior belly of the lateral pterygoid muscle.[1,2]

There are two movements that occur with opening: translation and rotation. The first movement is rotation which is the swinging of the condyle relative to the disk. The next movement is the translation of disk–condyle complex anteriorly from the articular fossa to a position beneath the articular eminence.[1,2]

The disk position is primarily determined by weight bearing, with the thin zone being the point of maximum load. During mastication, weight bearing is transferred to the food between the teeth and the TMJ is unloaded. Without weight bearing to hold the thin zone in position, there is a tendency for the elastic retrodiskal tissues to pull the disk posteriorly. This force is opposed by a contraction of the anterior belly of the lateral pterygoid muscle. Disk displacement is most com-monly anterior due to laxity or disruption of the elastic retrodiskal tissues that hold the disk posteriorly.[1,2]

3 TREATMENT

Currently, most treatment teams accept the goal of TMJ treatment as relief of pain and/or restoration of function. This philosophy is a departure from the early years of TMJ treatment when restoration of normal anatomical relationships was thought to be the treatment goal. This distinction is important since relief of symptoms may be achieved in the face of abnormal anatomical relationships.

Conservative TMJ therapy may consist of approaches such as: biofeedback, nonsteroidal antiinflammatory agents, muscle relaxers, bedtime sedatives, splints, and/or any combination of these approaches. Conservative therapy generally refers to the fact these treatments are generally noninvasive and reversible. In terms of cost, however, some conservative therapy may be more expensive than surgery. It is generally accepted that a trial of some form of conservative therapy should be attempted before invasive therapy is considered. Since the decision for conservative therapy approach is based upon the symptoms and clinical settings, MRI is not typically indicated in this management decision.[4,5]

If conservative therapy fails, then surgery may be required to relieve pain and/or restore function. There are a number of possible surgical treatments that are designed to address a particular physical defect. Joint adhesions may be treated by therapies as simple as joint lavage or as complex as arthroscopy.[6] Although attempts to correct disk displacement are made by TMJ arthroscopists, it is generally recognized that the major treatment effect of arthroscopy is the lysis of joint adhesions.[7] The lysis of adhesions is often highly effective in relieving joint symptomology without any correction in disk displacement. This discovery is a significant change from the original concept of TMJ disorders which attributed TMJ pain and dysfunction to disk displacement alone. The success of adhesion lysis therapy has made the imaging diagnosis of adhesions an important part of the MR evaluation.

Internal derangements of the articular disk can be approached with a variety of conventional surgical procedures. If the disk architecture is relatively well preserved, then disk plication may be used. The plication procedure pulls the usually anteriorly displaced disk back into position over the condyle and repairs the posterior attachment. Disk plication was once the most common surgical procedure but has fallen out of favor with many surgeons due to the common complication of adhesions in the postoperative period. In addition, abnormal disk architecture is often seen in association with displacement. Repositioning of a torn or otherwise deformed disk by plication is not possible. Diskectomy is often accompanied by replacement with a dermal graft, elastic cartilage graft, or alloplastic material.[8] The common complication of giant cell foreign body reaction to Proplast TMJ implants has greatly reduced the use of alloplastic disk replacements.[9,10] The poor outcome experienced with alloplastic implants has increased the popularity of autologous tissues as disk replacements. Postoperative adhesions remain a problem and despite attempts the ideal

For References see p. 1060

replacement material for severely degenerated or deformed disks has not yet been found.

For end-stage TMJ disorders where degeneration changes in the condyle and temporal bone predominate, the surgical treatment is often total joint replacement. A variety of appliances have been designed for total TMJ replacement. All of the devices comprise some form of artificial fossa and condyle. Because metallic components are used, MRI evaluations of the TMJ in postoperative total joint patients is not possible. As with total joint replacement in other joints, the long-term durability of the appliance is a problem. As a result, surgeons tend to restrict this approach to older individuals or to patients with severe degenerative change.[11]

Avascular necrosis of the mandibular condyle has been described by some workers as a common cause of TMJ disorders.[12,13] These workers favor corrections of the problem by condylar decompression or replacement. This diagnosis and treatment is highly controversial. Many in the TMJ treatment field refute this theory.

4 TECHNIQUE

The TMJ examination should consist of the following components:
1. static sagittal high-resolution images in the closed mouth position;
2. static coronal views in the closed mouth position;
3. kinematic or real time images during opening;
4. at least one sagittal T_2-weighted series.

A dedicated TMJ surface coil is now considered an essential feature for adequate TMJ imaging. The coil may be as simple as a single 3-inch-diameter loop.[14] Both sides must be evaluated simultaneously in order to demonstrate asymmetry. About 80% of TMJ disorders are bilateral. These coils can be combined either with a special combiner box or with signals separately acquired with phased array coils and receivers.[15,16] The phased array approach will result in a theoretical square root of 2 S/N improvement but the technology is considerably more costly.

The anatomical diagnosis of most TMJ disorders is made from the high-resolution sagittal images.[14,17–23] A short TE proton density or T_1-weighted pulse sequence, similar to the technique used for knee imaging, is required for optimal S/N and resolution. Because volume averaging effects are an important cause of image degradation, the slice thickness should be around 3 mm. An oblique image plane is performed using the angle of the mandibular body and condyle as determined on axial scout views as a reference. The short echo anatomical images are most important in TMJ diagnosis. As a result, traditional spin echo (SE) pulse sequences are preferred over fast SE or RARE sequences due to the blurring of short, effective TE echo images.

T_2 weighting can be achieved in both a dual echo SE sequence and a separate, long, effective TE fast SE (FSE) sequence. The FSE will result in improved image resolution but the fat signal will be more intense. The fat signal can be addressed with the addition of a fat saturation prepulse which lengthens the TR. Since the objective of T_2 weighting is the identification of fluid and not anatomical definition, there is

little advantage in the use of FSE over a dual echo traditional SE.[14,17–23]

Coronal T_1-weighted imaging is used to localize the disk in the medial-lateral direction. A faster sequence with diminished resolution compared with the sagittal images will usually be satisfactory.[24,25]

Kinematic or real-time motion images are needed to demonstrate TMJ physiology (Figure 2). A fast T_1-weighted SE is usually satisfactory. Gradient echo sequences may be employed if the timing parameters are adjusted to demonstrate good diskal anatomy. Often gradient echo images for cine TMJ imaging are inadequate to demonstrate this anatomy and are not the favored method for this reason. Because multiple slices are needed, there is no time saving associated with the use of gradient echo sequences. Scan time on the order of one minute per view are satisfactory. At least two slices per side should be performed in case the disk moves medially or laterally out of the view of one slice. Since the condyle is the reference for disk reduction, the slices are oriented sagittally, not oblique sagittally, as in the anatomical images. The condyle is a rigid structure that must travel in a straight line in the sagittal plane. The disk, however, will tend to move anteromedially. If oblique images are performed, the late opening images will display a disk and no condyle for reference. The determination of

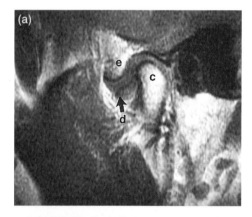

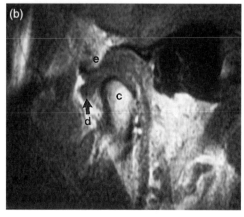

Figure 2 Kinematic imaging. Two images from a kinematic display are shown: the closed mouth (a) and the maximum open mouth (b) views. The closed mouth view demonstrates an anteriorly displaced disk, d, relative to the mandibular condyle, c. With opening, there is poor translation of the condyle which only translates to mid-eminence, e. The disk remains anteriorly displaced throughout opening

disk reduction will require the identification of the disk relative to the condyle. For kinematic imaging, the TMJ will need to be incrementally opened at predetermined intervals. We prefer to open in 3 mm increments until the patient can no longer proceed either due to pain or inability to open. One of several commercially available patient-controlled positioning imaging devices may be employed for establishing consistent incrementation. A reliable device is needed in order to quantitate the interincisor opening at the time of disk reduction.[26–29]

Most commercially available instruments have this kinematic display capability. Unfortunately, some manufacturers offer this capability only at additional cost. The lack of kinematic display will greatly limit the ability to make clinically significant TMJ diagnoses.

A more sophisticated and more physiologic approach is the gating of the MR imaging acquisition. Gating similar to cardiac gating is used during repetitive opening of the TMJ. It remains to be seen if this approach provides a clinically significant improvement in examination quality.

Echo planar methods may be used in the future for real-time TMJ imaging. Some believe that this approach would be more physiologic. Imaging during mastication would be possible with echo planar imaging.

5 ELEMENTS OF A TMJ INTERPRETATION

5.1 Bone

The landmarks for TMJ anatomy are the mandibular condyle and temporal bone. These structures should themselves be evaluated for anatomic defects. Osteophytes and beaking of the mandibular condyle are common. A steep articular eminence is associated with an increased incidence of TMJ disorders. Displacement of the condyle in the fossa may be a clue as to more subtle soft tissue defects. Hypointensity of the condylar marrow may indicate avascular necrosis.[12,13]

End-stage TMJ disorders result in severe degenerative changes. Erosions and large osteophytes are common. The eminence and condyle may be flattened. The joint space may be almost totally obliterated. If these findings are detected, a total joint replacement may be indicated.

In patients with a history of alloplastic implants, destruction of bone should be recognized as an effect associated with giant-cell foreign-body reaction. If large erosions are seen in these patients, look for an associated soft tissue mass that is hypointense on T_2-weighted images which is the hallmark of giant-cell reaction. Fragments of implant material that are hypointense on all sequences may be identified. Giant-cell foreign-body reactions may be very severe and on occasion can penetrate through the temporal bone into the middle cranial fossa.[8–10]

5.2 Disk Position and Morphology

The common method for recording disk position in the closed mouth sagittal views is the relationship to the condyle. Imagine the round mandibular condyle as a clock face and then locate the posterior band of the disk. The posterior band will usually be the thickest part of the posterior aspect of the disk. Sometimes the posterior band will be flattened or the disk may be deformed with an unidentifiable posterior band. In these cases record position as the posterior aspect of the disk rather than the posterior band. The normal posterior band position is 11 o'clock to 1 o'clock; anything else is displaced. If the disk is displaced greater than the length of the disk from the fossa, then this severe displacement should be noted. Anterior displacements are the most common. Posterior displacements are rare.[3–5]

The position of the disk in the medial-lateral direction is demonstrated on the coronal views. If the disk is difficult to identify on the sagittal view, look on the coronal view where a pure medial or lateral displacement may be seen. Antero-medial displacements are common. Antero-lateral displacements are less commonly seen.[24,25]

Disk morphology should be described. Are the anterior band, posterior band, and thin zone well demonstrated? If not, the disk morphology is abnormal. The disk may be thickened or distorted. A thin posterior band is a common finding in early TMJ disease.

End-stage TMJ disease results in fragmentation and degeneration. The disk may not be identifiable. Fragments of disk-like intensity may be seen. These areas commonly represent a combination of disk fragments, fibrosis, and metaplasia of the retrodiscal tissues.

Tears or perforations are findings described in TMJ arthrography as a communication of contrast from one joint space to another. Most of these perforations actually involve the retrodiscal tissues, not the disk itself. Perforations of the disk itself are rare and are almost always associated with other, more significant, abnormalities. Temporomandibular joint perforations are difficult to diagnose with MRI. With more experience, MR imaging has demonstrated more detail of the defects that better define appropriate treatment requirements.[17]

5.3 Hyperintense Signal on T_2-Weighted Images

Fluid in either the superior or interior joint spaces is best seen on the T_2-weighted images. As with other joints, fluid in the TMJ is a nonspecific sign of inflammation. Fluid may also be seen around the pterygoid muscle as a sign of possible myofascial disease.

Edema in the retrodiskal area indicates possible retrodiskitis. Retrodiskitis is acutely painful and often difficult to separate from an internal derangement on physical examination. The MR diagnosis of retrodiskitis is important since this disorder may be effectively treated with antiinflammatory agents. Hyperintensity in retrodiskitis begins at the posterior margin of the disk and extends posteriorly in the fossa. Often the condyle is displaced antero-inferior by the soft tissue mass. This appearance should be distinguished from normal retrodiskal veins that are hyperintense on T_2-weighted images. These veins can be distinguished by their location, posterior and inferior compared with retrodiskitis. In contrast to retrodiskitis which is confluent, retrodiskal veins are seen as multiple, small spots typical of vessels in cross-section.[17]

5.4 Kinematic Evaluation

Kinematic imaging is needed to describe accurately abnormalities in joint function. Disk rotation and disk–condyle

For References see p. 1060

complex translation should be observed. The right and left sides should be compared side-by-side using a kinematic display.[26–29]

What is the relationship of the disk to the condyle during the opening cycle? If the disk is displaced in the closed mouth view, does it reduce upon opening? When does it reduce? Early or late? By knowing the interincisor distance associated with each step, the extent of opening at the point of disk reduction can be determined.

Kinematic imaging is used for the diagnosis of joint adhesions. Adhesions between the disk and the condyle are demonstrated by fixation of the disk to the condyle during the entire opening cycle. Adhesions to the eminence are shown as lack of separation between the disk and eminence as well as fixation to the eminence during opening. Asymmetric translation is indirect evidence of adhesions.

Very poor translation on physical examination is often called a closed locked joint. The classic description of closed locked physiology is an anteriorly displaced disk without reduction. Theoretically the displaced disk impairs condylar translation and locks the joint in the closed mouth position. MR studies, however, reveal that an anteriorly displaced disk, without reduction on opening, uncommonly produces locking. Other causes for a closed locked joint on physical examination should be considered. Severe adhesions can more commonly cause the clinical picture of a locked joint. When these adhesions totally encompass the condyle and prevent any translation, the condition is called fibrous ankylosis. Previous surgery with dislocated or adhesed implant is another common cause of poor translation in the postoperative patient. As mentioned previously, adhesions can be managed with arthroscopy, whereas a closed locked joint, due to disk displacement, usually requires surgery (an open surgical procedure).[26–29]

6 SUMMARY

There is a multifaceted spectrum of causes for TMJ-associated pain and dysfunction. As the understanding of TMJ disorders increases, the methods for treatment become more focused and more effective. Accurate diagnosis of these entities is essential in designing treatment protocol that is appropriate for a particular patient. Modern MRI methods can make the most of the diagnostic determinations needed for these treatment decisions.

7 RELATED ARTICLES

Cardiac Gating Practice; Eye, Orbit, Ear, Nose, and Throat Studies Using MRI; Surface and Other Local Coils for In Vivo Studies; Whole Body Machines: NMR Phased Array Coil Systems.

8 REFERENCES

1. J. P. Okeson, 'Fundamentals of Occlusion and Temporomandibular Disorders', Mosby Year Book, St. Louis, MO, 1985.
2. W. E. Bell, 'Temporomandibular Disorders: Classification, Diagnosis, Management', 2nd edn., Mosby Year Book, Chicago, 1986.
3. L. T. Kircos, D. A. Ortendahl, A. S. Mark, and M. Arakawa, J. Oral Maxillofac. Surg., 1987, 45, 852.
4. F. M. Bush, J. H. Butler, and D. M. Abbott, in 'Advances in Occlusion: Diagnosis and Treatment', eds. H. C. Lundeen and C. H. Gibbs, Publishing Sciences Group, Littleton, MA, 1982.
5. M. F. Dolwick, in 'Internal Derangements of the Temporomandibular Joints,' eds. C. A. Helms, R. W. Katzberg, and M. F. Dolwick, Radiology Research and Education Foundation, San Francisco, 1983.
6. J. J. Moses, D. Sartoris, R. Glass, T. Tanaka, and I. Toker, J. Oral Maxillofac. Surg., 1989, 47, 674.
7. M. T. Montgomery, J. E. Van Sickels, S. E. Harms, and W. J. Thrash, J. Oral Maxillofac. Surg., 1989, 47, 1263.
8. D. P. Timmis, S. B. Aragon, J. E. Van Sickels, and T. B. Aufdemorte, J. Oral Maxillofac. Surg., 1986, 44, 541.
9. P. A. Kaplan, J. D. Ruskin, H. K. Tu, and M. A. Knibbe, Am. J. Roentgenol., 1988, 151, 337.
10. K. P. Schellhas, C. H. Wilkes, M. el Deeb, L. B. Lagrotteria, and M. R. Omlie, Am. J. Roentgenol., 1988, 151, 731.
11. R. W. Bessette, R. Katzberg, J. R. Natrella, and M. J. Rose, Plast. Reconstr. Surg., 1985, 75, 192.
12. K. P. Schellhas, C. H. Wilkes, H. M. Fritts, M. R. Omlie, and L. B. Lagrotteria, Am. J. Roentgenol., 1989, 152, 551.
13. C. H. Wilkes, Arch. Otolaryngol. Head Neck Surg., 1989, 115, 469.
14. S. E. Harms and R. W. Wilk, Radiographics, 1987, 7, 521.
15. C. J. Hardy, R. W. Katzberg, R. L. Frey, J. Szumouski, S. Totterman, and O. M. Mueller, Radiology, 1988, 167, 835.
16. J. S. Hyde, A. Jesmanowicz, W. Froncisz, J. B. Kneeland, T. M. Grist, and N. F. Campagna, J. Magn. Reson., 1986, 70, 512.
17. S. E. Harms, R. M. Wilk, L. M. Wolford, D. G. Chiles, and S. B. Milan, Radiology, 1985, 157, 133.
18. C. A. Helms, L. B. Kaban, C. McNeill, and, T. Dodson, Radiology, 1989, 172, 817.
19. C. A. Helms, T. Gillespy III, R. E. Sims, and M. L. Richardson, Radiol. Clin. North Am., 1986, 24, 189.
20. T. L. Miller, R. W. Katzberg, R. H. Tallents, R. W. Bessette, and K. Hayakawa, Radiology, 1985, 154, 121.
21. R. W. Katzberg, R. W. Bessette, R. H. Tallents, D. B. Plewes, and J. V. Manzione, Radiology, 1986, 158, 183.
22. R. W. Katzberg, Radiology, 1989, 170, 297.
23. M. T. Cirbus, M. S. Smilack, J. Beltran, and D. C. Simon, J. Prosthet. Dent., 1987, 57, 488.
24. R. W. Katzberg, P.-L. Westesson, R. H. Tallents, R. Anderson, K. Kurita, J. V. Manzione Jr., and S. Totterman, Radiology, 1988, 169, 741.
25. M. B. Khoury and E. Dolan, American Journal of Neuroroentgenography, 1986, 7, 869.
26. K. R. Burnett, C. L. Davis, and J. Read, Am. J. Roentgenol., 1987, 149, 959.
27. W. F. Conway, C. W. Hayes, and R. L. Campbell, J. Oral Maxillofac. Surg., 1988, 46, 930.
28. W. F. Conway, C. W. Hayes, R. L. Campbell, and D. M. Laskin, Radiology, 1989, 172, 821.
29. J. M. Fulmer and S. E. Harms, Top. Magn. Reson. Imag., 1989, 1, 75.

Biographical Sketch

Steven E. Harms. M.D., 1978, University of Arkansas for Medical Sciences. Postdoctoral research associate, physical chemistry, Stony Brook, 1977–78; assistant radiologist, Depts. of Diagnostic Radiology and Biophysics; M.D., Anderson Cancer Center, Houston, 1982–83 TX. Director of Magnetic Resonance, Dept. of Radiology, Baylor University Medical Center, Dallas, 1983–present. Approx. 80 papers. Current research interests: three-dimensional imaging applications, development of new fast scanning sequence, and musculoskeletal, temperomandibular joint, breast, body, and ophthalmological applications of MRI.

Degenerative Disk Disease Studied by MRI

Michael T. Modic
Cleveland Clinic Foundation, OH, USA

1   INTRODUCTION

Degeneration of the intervertebral disk complex is a process that begins early in life and is a consequence of a variety of environmental factors as well as normal aging. The pathophysiology of this disorder is complicated and poorly understood. It has been stated[1] that degeneration as commonly applied to the intervertebral disk covers such a wide variety of clinical, radiologic, and pathologic manifestations as to be really 'only a symbol of our ignorance'.

It is estimated that low back pain affects 5% of the adult population each year and that the lifetime incidence of LBP in adults is 60–80%.[2] Fortunately, the natural history of back pain is such that the vast majority of back pain sufferers improve with little or no medical intervention; only 14% of patients have episodes of pain lasting more than 2 weeks.[3] Despite the self-limited nature of most LBP, the total societal costs of LBP have been estimated at US$16–100 billion annually,[2] and, as back pain is the second leading cause of physician visits among American adults, it is estimated to account for more than US$10 billion in direct medical care costs each year.[4] Much of the direct cost of treating patients with LBP is related to diagnostic tests;[5] it is estimated that MRI alone contributes US$2 billion dollars in direct medical care costs (Modic MT, 1993, projected estimate).

The sequelae of disk degeneration remain among the leading causes of functional incapacity in both sexes and are an all too common source of chronic disability in the working years. However, by the age of 50 years, 85–95% of adults show evidence of degenerative disk disease at autopsy;[6] thus the jump from identifying an anatomic derangement to proposing a symptom complex must be made with caution, since to date only a moderate correlation has been found between imaging evidence of disk degeneration and symptomatology.[7] There is no unifying theory regarding the cause of disk degeneration. It is likely that degeneration and aging are multifactorial processes that encompass a wide spectrum of changes and sequelae. Nevertheless, in patients in whom surgical therapy is planned, or where the diagnosis remains elusive and patient treatment and outcome is dependent upon a more definitive diagnosis, imaging studies are an important component in the evaluation of patients with symptoms of disk disease.

2   MORPHOLOGY

The contrast sensitivity and multiplanar imaging capability of proton magnetic resonance (MR) place this modality in a position to provide a unique noninvasive means of imaging intervertebral disks. The implementation of surface coil technology,[8] cardiac gating,[9] gradient refocusing,[10] paramagnetic contrast agents,[11] saturation pulses,[12] gradient echo volume imaging, turbo (fast) T_2-weighted spin echo sequences, and magnetization transfer techniques continue to refine the capability of MRI for degenerative disk disease. When a combination of imaging planes and pulse sequence parameters is used, the anatomy of the intervertebral disk (Figure 1), spinal nerves, dural sac, and adjacent structures can be clearly depicted. From a morphological aspect, MRI may be the most accurate means of evaluating the intervertebral disk. Accordingly, most research to date has been directed clinically towards optimizing anatomic image display in a fashion similar to that of computerized tomography (CT) for assessment of disk contour.[13–19] Unlike CT and conventional radiography, however, which are dependent on information related to electron density, proton MR signals are influenced by the T_1 and T_2 relaxation times, and by proton density, providing greater tissue contrast. Thus the role of this technique may go beyond gross anatomic appraisal to actual tissue characterization of pathology and biochemical change.[20]

The relationship among the vertebral body, endplate, and disk has been studied[21–24] using both degenerated and chymopapain-treated disks as models. Signal intensity changes in vertebral body marrow adjacent to the endplates of degenerated disks are a common observation in MRI. Work using T_2-weighted spin echo sequences[20] further suggests that MRI is capable of detecting changes in the nucleus pulposus and anulus fibrosus relative to degeneration and aging based on a loss of signal presumed to be secondary to known changes in hydration that occur within the intervertebral disk (Figure 2). However, the correlation is not straightforward, since differences in signal intensity appear to be somewhat exaggerated for the degree of water loss noted with degeneration (about 15%).[25] At present the role that specific biochemical changes (proteoglycan ratios, aggregating of complexes) play in the altered signal intensity is not well understood. In fact, the important factor may not be the total quantity of water but the state that the water is in. Sodium images suggest that the T_2 signal intensity in the disk tracks the concentration and regions of high glycosamino glycans (GAG) percentages. Thus it seems likely that the health and status of the proteoglycans determine the signal intensity.

The vacuum phenomenon within a degenerative disk is represented on spin echo images as areas of signal void.[26] Gradient echo sequences demonstrate this better than do conventional spin echo sequences, plain radiographs and CT. This is due to the magnetic susceptibility effects caused by the intradiskal gas collection. Whereas the presence of gas within the

For References see p. 1069

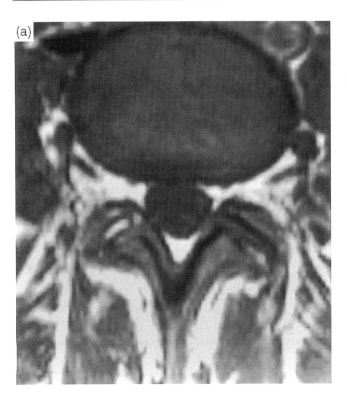
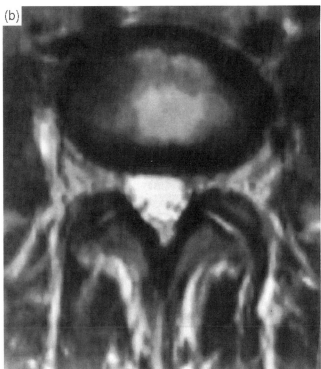

Figure 1   Axial L3–L4 disk: (a) and (b) are normal. Axial T_1-weighted spin echo (500/17), and T_2-weighted spin echo (2000/90) images through the axial midplane of the intervertebral disk. On the T_1-weighted axial image (a) the disk has a homogeneous soft tissue signal intensity. On the T_2-weighted image (b) three distinct components of the disk can be identified: first, a decreased signal intensity is noted in the outer annulus and ligamentous region; second, the inner annulus has a slightly increased signal intensity, and third, the region of the nucleus pulposus has the highest signal intensity consistent with the highest concentration of proteoglycans

disk is usually suggestive of degenerative disease rather than an infected process, spinal infection may (rarely) be accompanied by intradiskal or intraosseous gas.[27] A gas density cleft within a transverse separation of the vertebral body appearing in extension and disappearing in flexion is characteristic of the vacuum phenomenon within a region of ischemic vertebral collapse. Rarely, the phenomenon has been identified with vertebral body neoplasms such as multiple myeloma.[28–32]

Hyperintense intervertebral disks are not an infrequent finding on T_1-weighted MR images of the spine, and it has been suggested[33] that the relatively 'bright' intervertebral disk may reflect diffuse abnormality and loss of normal signal in the marrow of the adjacent vertebral bodies. Calcification has usually been described on MR images as a region of decreased or absent signal. The loss of signal is attributed to a low mobile proton density as well as, in the case of gradient echo imaging, to its sensitivity to the heterogeneous magnetic susceptibility found in calcified tissue.

There is, however, variability of signal intensity of calcium on various sequences; the type and concentration of calcification are probably important factors. Multiple examples of a hyperintense signal on T_1-weighted spin echo images in areas that contain calcification on CT have been reported in the literature. These hyperintensities have been attributed to the paramagnetic effects of methemoglobin,[34–36] melanin,[37] and trace elements,[38,39] as well as to the T_1-shortening effects of lipids and cholesterol,[40] proteins,[41,42] and laminar necrosis as-

sociated with infarction and calcification.[43–47] Focal or diffuse areas of hyperintensity on T_1-weighted spin echo sequences may also be encountered in the intervertebral disk. Hyperintensities that are affected by fat-suppression techniques have also been noted within intervertebral disks.[48] These are presumably related to areas of ossification with formation of a lipid marrow in the ossified disk space; they also appear calcified on conventional studies.

3   SEQUELAE OF DISK DEGENERATION

Disk herniation, especially in the lumbar and thoracic regions, is probably better depicted by MRI than by other modalities.[14–19,49,50] It is, again, important to bear in mind that not all morphological changes are a cause of symptoms. In a prospective study of individuals who never had low back pain, sciatica or neurogenic claudication,[51] one-third of the subjects were found to have substantial abnormalities. In patients who were less than 60 years of age, 20% had a herniated nucleus pulposus. In the group that was 60 years or older, findings were abnormal on 57% of the scans. Thirty-six per cent of the subjects had a herniated nucleus pulposus, and 21% had spinal stenosis. More recent studies[52,53] have confirmed the observation that disk bulges and protrusions are common findings on MRI in the asymptomatic population. These latter studies,

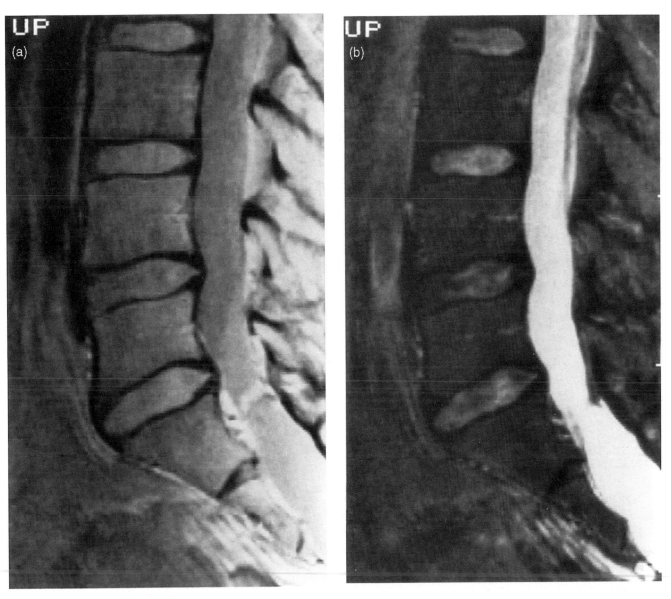

Figure 2 Spin echo and gradient sagittal images. (a) Intermediate spin echo (2000/20). (b) T_2 weighted spin echo (2000/90). (c) FLASH (50/13/10 degrees). (d) FLASH (50/13/60 degrees). (b) There is decreased signal intensity in the L4–L5 and L5–S1 disk, consistent with degeneration. The L2–L3 and L3–L4 disks have a normal signal intensity. The signal intensity changes consistent with degeneration are not as well appreciated from sequences (a), (c) and (d). (Reproduced with permission from *Magnetic Resonance Imaging of the Spine*, 2nd edn, Chap. 2, p. (52, Fig. 2-14)

however, suggest that frank disk extrusion is rare; a disk bulge was observed to be age related, but disk protrusion was not.

Multidimensional imaging allows the direct acquisition of orthogonal views covering long segments of the spine, without requiring secondary reconstructions. Alternatively, three-dimensional datasets can be acquired with or without contrast, which allows multiplanar reformation with variable partition thicknesses for improved overall spatial resolution and reduced examination times. The outer anulus–posterior longitudinal ligament (PLL) complex can usually be seen as an area of decreased signal relative to the inner anulus–nucleus pulposus, which helps in characterizing the type of herniation (protrusion, extrusion and/or sequestration).[54] This ability to characterize and differentiate the various subgroups

of disk herniation has been proposed to have certain diagnostic and therapeutic ramifications, particularly in the lumbar region.

Abnormal disks can be classified as an anular bulge or herniated (protruded, extruded, or free fragment). Concurrently, herniated disk disease should be described by contour, size, location, and presence or absence of enhancement.

4 BULGE

An anular bulge is a result of disk degeneration with a grossly intact, albeit lax, anulus, usually recognized as a gener-

For References see p. 1069

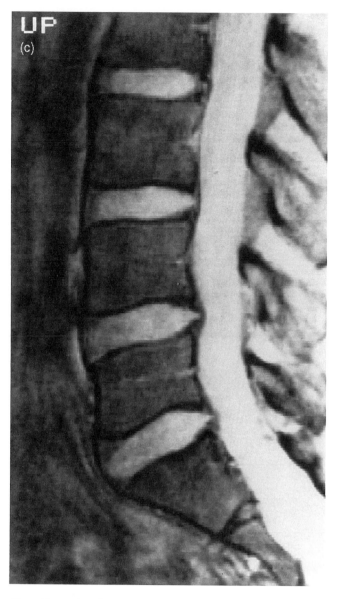

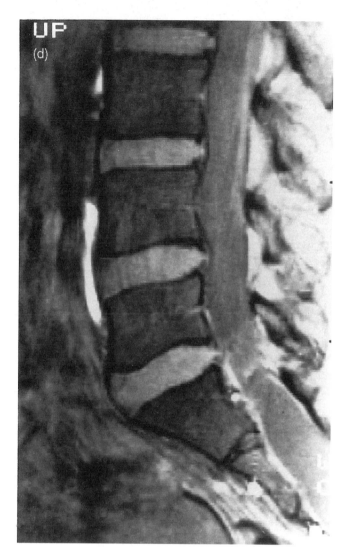

Figure 2 continued

alized extension of the disk margin beyond the margins of the adjacent vertebral endplates, regardless of the signal of the interspace. An index of disk bulging using sagittal anatomic sections from 149 lumbar disks[55] has been studied. The largest disk bulgings were always associated with radial tears of the anulus, which contradicts the previously held concept that the anulus fibrosus remains intact with a bulging disk.

Some investigators would recognize the anular tear as an intermediate category between bulge and herniation. The significance of the anular tear, however, remains extremely unclear; it is perhaps best considered as a sequela of disk degeneration rather than as a discrete category that may imply some type of clinical significance. On nonenhanced T_2-weighted MR images anular tears are best appreciated as an area of high signal extending into the area of decreased signal of the anulus–ligamentous complex (Figure 3).

5 LUMBAR DISK HERNIATIONS

Obviously some degree of anular disruption is an intrinsic component of any disk herniation.

A protrusion represents herniation of nuclear material through a defect in the anulus, producing a focal or broad-based extension of the disk margin. The extension is less than that which occurs with an extrusion. At least some fibers of the overlying anulus and PLL remain intact, and the disk is described as 'contained'. The signal intensity of the parent nucleus is usually decreased.

The next two categories, extrusion and sequestration, represent herniations that are no longer contained by the overlying anulus and ligament (Figure 4). In an extrusion, nuclear material becomes an anterior extradural mass that remains attached to the nucleus of origin, often via a high-signal pedi-

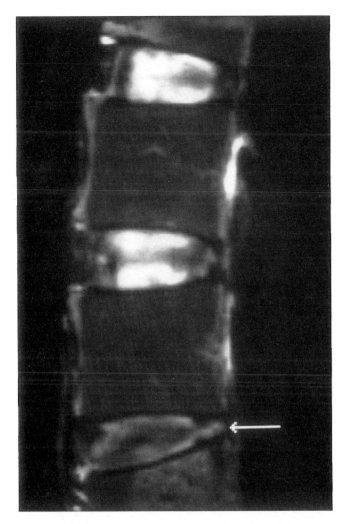

Figure 3 Sagittal T_2-weighted spin echo image (2000/90) through a cadaveric spine. Note the loss of signal intensity at the L4–L5 disk and the region of high signal extending into the anulus–ligament complex (arrow). This represents an annular tear. (Reproduced with permission from *Magnetic Resonance Imaging of the Spine*, 2nd edn, Chap. 3, p. 96, Fig. 3-16)

cle on the T_2-weighted image. The signal intensity of the extruded portion may be increased or decreased. The disk usually appears contained by the PLL and remaining contiguous portions of the anulus, which show up as curvilinear areas of decreased signal.

The term 'free fragment' or 'sequestrated disk' refers to disk material external to the anulus fibrosus and no longer contiguous with the parent nucleus. A frequent finding with both extruded and free fragments is the presence of a high signal intensity extradural defect often surrounded by a curvilinear area of decreased signal that is distinct from the interspace of origin.

Whereas all of the foregoing has become a part of our clinical lore, it is not at all clear whether these findings are clinically relevant. It has been proposed, for example, that differentiating between various degrees of herniation is critically important; yet the reality of the situation is that any disk herniation most likely represents a spectrum or continuum rather

than a discrete entity with specific clinical relationships. One may view the continuum of herniated disk disease as starting with anular disruption, proceeding to a small focal herniation that does not break through the anulus–ligamentous complex, and winding up as a frank herniation (extrusion) that does indeed dissect through the anulus–posterior ligamentous complex.

Traversing nerve roots within the thecal sac above a particular level of impingement from a herniated disk or stenosis have been noted to be enlarged compared to the contralateral side. This may be a reflection of nerve root edema. Enhancement of traversing nerve roots within the thecal sac has also been reported[56] on MRI, the majority of cases being associated with focally protruding disk pathology. A significant number of these patients, however, had isolated enhancement of multiple nerve roots without significant associated anatomic pathology. The mechanism of such enhancement may be related to the blood–nerve barrier of this spinal nerve root, which is altered by compression.[57] It has also been suggested that the enhancement may be related to the vascular anatomy as a variation of normal, rather than as a pathologic reaction.[58]

In a prospective evaluation of surface coil MRI, CT, and myelography for lumbar herniated disk disease and canal stenosis[16] there was 82.6% agreement between MRI and surgical findings regarding type and location of disease, 83% agreement between CT and surgical findings, and 71.8% agreement between myelography and surgical findings. In another study,[59] the accuracy of CT, myelography, CT-myelography and MRI for lumbar herniated nucleus pulposus was compared prospectively in 59 patients, all of whom underwent surgical exploration. MRI was the most accurate (76.5%), CT-myelography next (76%), and then CT (73%) and myelography (71%). The false-positive rate was lowest for MRI, at 13%, followed by myelography and CT.[59] Similar data from a carefully controlled prospective study in Wisconsin[60] suggest the same equivalence for plain CT, CT-myelography, and MRI.

The natural history of lumbar disk herniation has engendered recent interest and MRI has been an excellent tool for these investigations. Multiple studies have demonstrated that the size of the disk herniation can change with time, and studies of patients treated conservatively have demonstrated that the majority will show a reduction in disk herniation size of 30–100%. The larger herniations show the most significant decrease in size. A resulting impression has also been that patients who do well conservatively will show this reduction in herniation size, yet the correlation of disk changes with time and symptoms resolution has not yet been carefully worked out.

As part of a long term natural history study, patients with acute radiculopathy were evaluated clinically and with contrast enhanced MRI to determine if there was a correlation between presenting symptoms and the type, size, location and enhancement of the disk herniations at presentation. In this group, 72% had a herniated disk at one or more levels at presentation. There was excellent agreement between MRI and clinical findings for level and size, but no correlation of pain and disability with disk size, type or enhancement. At the 6-week follow-up, 36% of large herniated disks demonstrated a significant reduction in size (Figure 5), but there was no difference in

For References see p. 1069

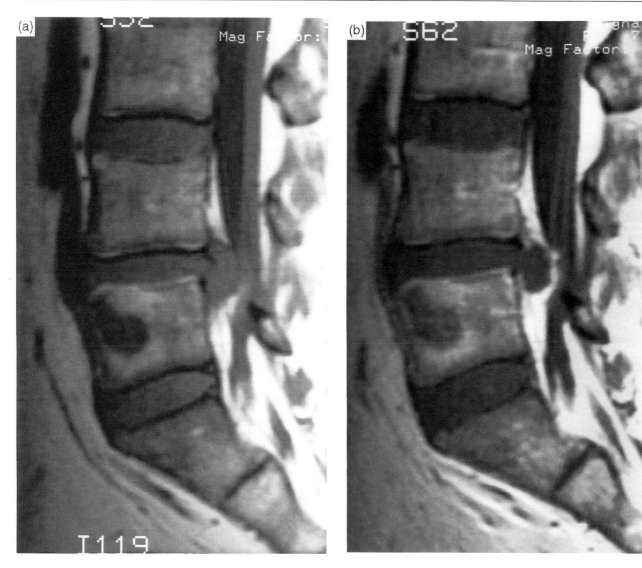

Figure 4 L4–L5 disk herniation. (a) Pre- and (b) post-contrast sagittal T_1-weighted spin echo images demonstrate an extradural mass at the L4–L5 disk level. Axial T_1-weighted spin echo (500/17 images) before (c) and after (d) contrast show the peripheral enhancement of disk material, which is also appreciated on the sagittal images. The enhancing areas of granulation tissue contribute to the overall size of the mass. (Reproduced with permission from *Magnetic Resonance Imaging of the Spine*, 2nd edn, Chap. 3, p. 102, Fig. 3-26)

clinical outcome based on disk size, change, or type. At 6 months, 60% of herniated disks had reduced significantly in size, but again there was no correlation of pain and disability with disk size, change, or type.

Enhancement of herniated disk was almost a constant feature. The degree of enhancement was variable and probably related to granulation tissue which develops around and through herniated disk. The granulation tissue response itself may comprise a fair percentage of the herniated disk mass. One possible explanation for the change of a disk with time then is that the granulation tissue undergoes cicatrization and retraction, a normal phenomenon for reparative tissue.

Although less common than lumbar or cervical herniations, thoracic herniations have been noted more frequently with the advancement of imaging techniques.[19] A higher prevalence of asymptomatic morphologic change is also present in the thoracic spine.

Symptomatic cervical disk herniations are most common in young patients (third and fourth decades) and frequently occur without recognized trauma. In a prospective study of patients with no symptoms of cervical disease,[61] 10% of subjects less than 40 years of age had a herniated nucleus pulposus and 4% had foraminal stenosis. Of the subjects who were older than 40 years of age, 5% had a herniated nucleus pulposus, 3% a bulging disk, and 20% foraminal stenosis. Narrowing of the disk space, degeneration of a disk, spurs, or compression of the cord were noted at one or more levels in 25% of the subjects less than 40 years of age and in almost 60% of those who were older than 40.

Cervical disk herniation, especially when central or large, is well appreciated on routine sagittal and axial MR images.

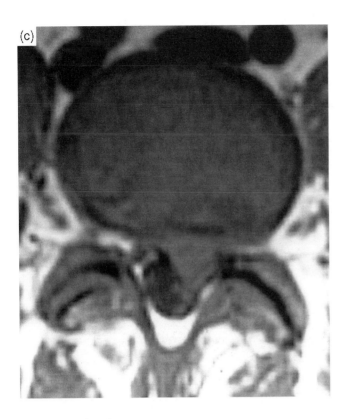

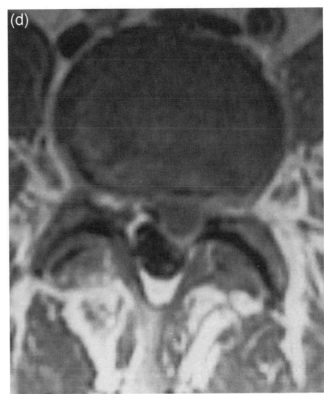

Figure 4   continued

Again, thin slices (3 mm or less) are critical for accurate diagnosis. This may necessitate the use of three-dimensional volume sequences with partitions of 2 mm or less and/or reconstructions in other planes. The T_2 signal intensity of intervertebral disks in the cervical region is not as helpful as that in the lumbar region for identifying the presence or absence of degeneration. Cervical disk herniation is usually identified on sagittal images as an anterior or anterior–lateral extradural defect that may indent or compress the cervical cord. In a prospective study to compare the accuracy of surface coil MRI with metrizamide myelography (MM) and CT with metrizamide (CTM),[17] there was surgical agreement in 74% of patients with surface coil MRI, 85% with CTM, and 67% with MM. When surface coil MRI and CTM were used jointly, 90% agreement with surgical findings was seen, and when CTM and MM were used jointly there was 92% agreement. In general, surface coil MRI was as sensitive as CTM for identifying disease level, but not as specific for type of disease.

In another study comparing MRI and CT, plain film myelography, and CT-myelography in 35 patients operated on for cervical radiculopathy and myelopathy,[62] MRI correctly predicted 88% of the surgically proved lesions. The corresponding rates were 81% for CT myelography, 58% for plain myelography, and 50% for CT.

Although we would maintain that most evaluations of the cervical spine for extradural disease can be done in an adequate fashion using spin echo T_1-weighted sagittal and axial images, the introduction of fast gradient echo images with small flip angles (less than 15°) has improved the accuracy of the examination by increasing the conspicuousness of extradural defects. Recent work[63–65] has confirmed the value of this technique. This capability, coupled with the potential of utilizing very short TRs (50 ms or less), has provided the stimulus for using volume imaging in the evaluation of extradural disease, in the hope of shortening the examination time and decreasing slice thickness.

The disadvantages of gradient echo imaging relate to problems with field inhomogeneity and contrast detectability of pathologic processes within the cord itself. Furthermore, depending upon the echo time, the ability accurately to characterize morphologic defects, foraminal narrowing, and bony ridges is inferior to that for viewing bone by CT.

An adjunct to conventional MRI in evaluating degenerative disk disease is the utilization of gadolinium diethylenetriaminepentaacetic acid (Gd-DTPA). Studies with virgin disks and postoperative disk herniation[11] indicate that it is the most accurate means of separating epidural fibrosis and disk tissue. There is consistent enhancement of peridiskal fibrosis early (less than 15 min after injection), and variable enhancement occurs in herniated or degenerated parent disks late (30 min after injection). Enhancement of intervertebral disks does not appear to occur in the normal state and is probably a sequela of the degenerative process.[11,66] Thus it seems likely that Gd-DTPA may play a role in identifying the sequelae of degenerative disk disease by enhancing the reactive granulation tissue that forms secondary to disruption of the disk and associated structures.

For References see p. 1069

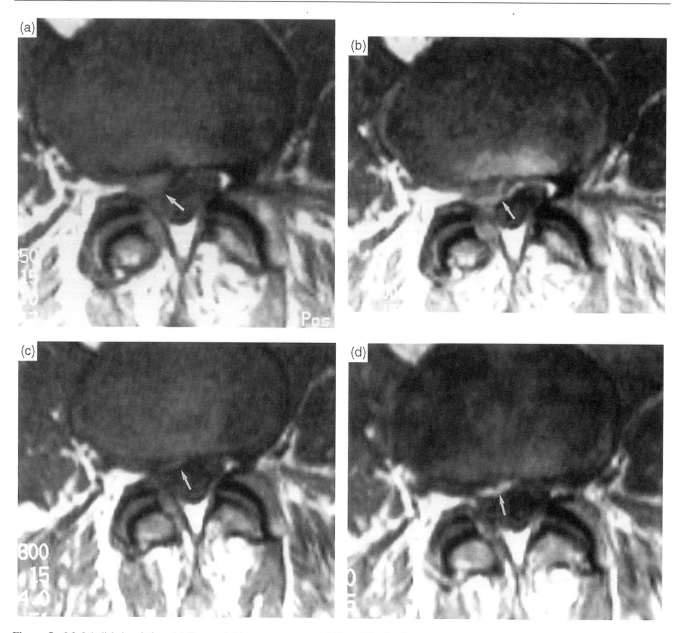

Figure 5   L3–L4 disk herniation. (a) Pre- and (b) post-contrast axial T_1-weighted spin echo images through the L3–L4 disk in a patient who presented with the onset of acute right-sided radiculopathy. A moderate sized extradural mass is identified on the right which shows peripheral enhancement following the administration of contrast. (c) (d) Axial T_1-weighted spin echo (500/17) images at the same level as (a) and (b) following 6 weeks of conservative management. There is a marked reduction in the size of the extradural mass, although there is still some enhancement in this region

In addition to changes within the disk, including hernia-tion, secondary changes are noted both in animal models and in humans following disk degeneration. The stability of a motion segment depends on the integrity of all its com-ponents. Diseases occurring in this area are circumferential, with that in one joint affecting another, and degeneration of one disk leads to a loss of disk height and forces the facet joints into malalignment, so-called 'rostrocaudal subluxation'. This leads to increased biomechanical forces at the facet joint with increasing joint relaxation and instability, secondary facet and arthritic changes, and potential fractures. Similarly, abnor-mal movements allowed by disk degeneration and facet changes add stress to the posterior ligaments and can result in hypertrophy. A vicious degenerative cycle is established which includes degenerative disk disease, facet arthrosis, liga-mentous and capsular hypertrophy, spinal instability, and lumbar stenosis.[67]

6   SPINAL STENOSIS

Spinal stenosis results from an overall diminution of the spinal canal, lateral recesses, or neural foramina. It occurs more commonly in the lumbar and cervical regions. The symptoms are usually a reflection of the compressive pathology.

In the lumbar spine, central stenosis tends to occur at multiple levels but is most frequent and usually most severe at the L4–L5 level, where it may occur alone.

The transverse and sagittal dimensions of the central neural canal are best depicted by integrating orthogonal (e.g. axial and sagittal) planes. Gradient echo images with low flip angles or more T_2-weighted spin echo sequences provide the best views of thecal sac dimensions by furnishing a gray-scale inversion of the cerebrospinal fluid and extradural elements.[63,64]

Peripheral stenosis is best appreciated on T_1-weighted images, which maintain the separation between neural structures and epidural fat quite well. The signal intensity between neural elements and fat is less well seen on more T_2-weighted and gradient echo images, although bony overgrowth can be identified.

In the cervical spine degenerative changes affect all major spinal articulations, including the intervertebral disks, apophyseal joints, ligamentous connections between vertebral bodies, and the vertebral bodies themselves.

The suggestion has been made[68] that myelopathy or stenosis is associated with an average canal diameter of less than 12 mm. Care must be taken, however, when using gradient echo images since the degree of spinal narrowing can be overestimated because of susceptibility effects. Another method suggested for determining cervical spinal stenosis[69] is the vertebral body/canal ratio, which has been suggested to correct for body size. A spinal canal/vertebral body ratio of less than 0.82 would indicate significant cervical spinal stenosis. MRI and myelographic CT studies appear to be equivalent in measuring the degree of cord compression. There may be a correlation between compression and the amount of neurologic dysfunction.[70]

Although there are no good studies to indicate that MRI has led to improved therapeutic choices, there is at present sufficient evidence to suggest that MRI is adequate for most therapeutic planning to allow physicians to stop further testing, obviating the need for a myelogram or CT scan.[71,72]

7 FACET JOINTS

Although facet arthrosis constitutes an important cause of acquired lumbar stenosis, degenerative change can occur independently of it and may be a cause of low back pain and radiculopathy.[67,73]

On MRI the facet joints are best evaluated by visually integrating sagittal and axial images. Again, osteophytes are easier to identify on gradient echo or spin echo images with long relaxation times. Those that contain marrow are also better demarcated than are those that are sclerotic, which can be confused with the adjacent capsular ligamentous structures. Altered signal intensity in bone marrow consistent with fatty replacement has also been reported to be commonly associated.[74] The articular cartilage can be directly visualized on both T_1 and T_2-weighted spin echo images, but thinning is difficult to measure accurately because of variable axial obliquity and chemical shift artifact from the adjacent facet. Gradient echo examinations have demonstrated that they provide better conspicuousness of the articular cartilage itself as distinct from adjacent osseous structures.

In summary, degenerative disk changes appear to be a normal consequence of the aging process. MRI is an excellent modality for depicting them. However, the clinical relevance of these morphologic changes remains to be established and the value of MRI as a prognostic indicator for patient outcome needs to be studied more fully. At the present time, one can clearly recommend MRI as the single best presurgical decision-making tool, but its role in the evaluation of patients who are going to be treated in a conservative fashion appears to be much more limited.

8 RELATED ARTICLES

Contrast Agents in Magnetic Resonance: Operating Mechanisms; Contrast Agents in Whole Body Magnetic Resonance: An Overview; Gadolinium Chelate Contrast Agents in MRI: Clinical Applications; Head and Neck Investigations by MRI; Lung and Mediastinum MRI.

9 REFERENCES

1. K. P. H. Pritzker, *Orthop. Clin. North Am.* 1977, **8**, 66.
2. J. W. Frymoyer and W. L. Cats-Baril, *Orthop. Clin. North Am.*, 1991, **22**, 263.
3. R. A. Deyo and Y. J. Tsui-Wu, *Spine*, 1987, **12**, 264.
4. B. K. Cypress, *Am. J. Public Health*, 1983, **73**, 389.
5. R. A. Deyo, *J. Gen. Intern. Med.* 1986, **1**, 328.
6. R. J. Quinet and N. M. Hadler, *Semin. Arthritis Rheum.* 1979, **8**, 261.
7. A. A. White and S. L. Gordon, *Spine*, 1982, **7**, 141.
8. L. Axel, *J. Comput. Assist. Tomogr.*, 1984, **8**, 381.
9. D. R. Enzmann, J. B. Rubin, and A. Wright, *Radiology*, 1987, **162**, 763.
10. E. M. Haacke and G. Lenz, *Am. J. Roentgenol.* 1987, **148**, 1251.
11. M. G. Hueftle, M. T. Modic, J. S. Ross, T. J. Masaryk, J. R. Carter, R. G. Wilber, H. H. Bohlman, P. M. Steinberg, and R. B. Delamartor, *Radiology* 1988, **167**, 817.
12. R. R. Edelman, D. J. Atkinson, M. S. Silver, F. L. Hoaiza, and W. S. Warren, *Radiology* 1988, **166**, 231.
13. N. I. Chafetz, H. K. Genant, K. L. Moon, C. A. Helms, and J. M. Morris, *Am. J. Roentgenol.*, 1983, **141**, 1153.
14. R. R. Edelman, G. M. Shoukimas, D. D. Stark, R. R. Davis, P. F. New, S. Saini, D. I. Rosenthal, G. L. Wismer, and T. J. Brady, *Am. J. Roentgenol.*, 1985, **144**, 1123.
15. K. R. Maravilla, H. P. Lesh, J. C. Weinreb, D. K. Selby, and V. Mooney, *Am. J. Neuroradiol.*, 1985, **6**, 237.
16. M. T. Modic, T. J. Masaryk, F. Boumphrey, M. Goormastic, and G. Bell, *Am. J. Roentgenol.*, 1986, **147**, 757.
17. M. T. Modic, T. J. Masaryk, G. P. Mulopolos, C. V. Bundschuh, J. J. Han, and H. Bohlman, *Radiology*, 1986, **161**, 753.
18. M. T. Modic, T. J. Masaryk, J. S. Ross, G. P. Mulopolos, C. V. Bundschuh, and H. Bohlman, *Radiology*, 1987, **163**, 227.
19. J. S. Ross, N. Perez-Reyes, T. J. Masaryk, M. Bohlman, and M. T. Modic, *Radiology*, 1987, **165** 511.
20. M. T. Modic, W. Pavlicek, M. A. Weinstein, F. Boumphrey, F. Ngo, R. Hardy, and P. M. Duchesneau, *Radiology*, 1984, **152**, 103.
21. A. deRoos, H. Kressel, C. Spritzer, and M. Dalinka, *Am. J. Roentgenol.*, 1987, **149**, 531.

22. M. T. Modic, P. M. Steinberg, J. S. Ross, T. J. Masaryk, and J. R. Carter, *Radiology*, 1988, **166**, 193.

23. J. Aoki, I. Yamamoto, N. Kitamura, T. Sone, H. Iroh, K. Torizuka, and K. Takasu, *Radiology*, 1987, **164**, 411.

24. T. J. Masaryk, F. Boumphrey, M. T. Modic, C. Tamborrello, J. S. Ross, and M. D. Brown, *J. Comput. Assist. Tomogr.* 1986, **10**, 917.

25. V. M. Haughton, *Radiology*, 1988, **166**, 297.

26. N. Grenier, R. I. Grossman, M. L. Schiebler, B. A. Yeager, H. I. Goldberg and H. Y. Kressel, *Radiology*, 1987, **164**, 861.

27. D. K. Bielecki, D. Sartoris, D. Resnick, K. Van Lom, J. Fierer, and P. Haghighi, *Am. J. Roentgenol.*, 1986, **147**, 83.

28. W. Kumpan, E. Salomonowitz, G. Seidl, and G. R. Wittich, *Skeletal Radiol.* 1986, **15**, 444.

29. D. Resnick, G. Niwayama, J. Guerra, V. Vint, and J. Usselman, *Radiology*, 1981, **139**, 341.

30. L. G. Naul, G. J. Peet, and W. B. Maupin, *Radiology*, 1989, **172**, 219.

31. B. E. Maldague, H. M. Noel, and J. J. Malghem, *Radiology*, 1978, **129**, 23.

32. F. Gagnerie, B. Taillan, L. Euller-Ziegler, and G. Ziegler, *Clin. Rheumatol.*, 1987, **6**, 597.

33. M. Castillo, J. A. Malko, and J. C. Hoffman, *Am. J. Neuroradiol.*, 1990, **11**, 23.

34. J. M. Gomori, R. I. Grossman, H. I. Goldberg, R. A. Zimmerman, and L. T. Bilaniuk, *Radiology*, 1985, **157**, 87.

35. J. M. Gomori, R. I. Grossman, D. B. Hackney, H. I. Goldberg, R. A. Zimmerman, and L. T., Bilaniuk *Am. J. Neuroradiol.*, 1987, **8**, 1019; *Am. J. Roentgenol.*, 1988.

36. H. Nabatame, N. Fujimoto, K. Nakamura, Y. Imura, Y. Dodo, H. Fukuyama, and J. Kimura, *J. Comput. Assist. Tomogr.*, 1990, **14**, 521.

37. S. A. Mirowitz, K. Sartor, and M. Gado, *Am. J. Neuroradiol.*, 1989, **10**, 1159; *Am. J. Roentgenol.*, 1990, **154**, 369.

38. S. A. Mirowitz, T. J. Westrich, and J. D. Hirsch, *Radiology*, 1991, **181**, 117.

39. S. A. Mirowitz and T. J. Westrich, *Radiology*, 1992, **185**, 535.

40. P. P. Maeder, S. L. Holtas, L. N. Basibuyuk, L. A. Salford, V. A. Tapper, and A. Bruh, *Am. J. Neuroradiol.*, 1990, **11**, 575.

41. P. M. Som, W. P. Dillon G. D. Fullerton, R. A. Zimmerman, B. Rajagopalan, and Z. Marom, *Radiology*, 1989, **172**, 515.

42. K. Abe, H. Hasegawa, Y. Kobayashi, H. Fujimura, S. Yorifuji, and S. Biton, *Neuroradiology*, 1990, **32**, 166.

43. O. B. Boyko, P. C. Burger, J. D. Shelburne, and P. Ingram, *Am. J. Neuroradiol.*, 1992, **13**, 1439.

44. R. M. Henkelman, J. F. Watts, and W. Kucharczyk, *Radiology*, 1991, **179**, 199.

45. L. A. Dell, M. S. Brown, W. W. Orrison, C. G. Eckel, and N. A. Matwiyoff. *Am. J. Neuroradiol.*, 1988, **9**, 1145.

46. R. D. Tien, J. R. Hesselink, and A. Duberg, *Am. J. Neuroradiol.*, 1990, **11**, 1251.

47. Y. Araki, T. Furukawa, K. Tsuda, T. Yamamoto and I. Tsukaguchi, *Neuroradiology*, 1990, **32**, 325.

48. B. A. Bangert, M. T. Modic, J. S. Ross, N. A. Obuchowski, J. Perl, P. M. Ruggieri, and T. J. Masaryk, *Radiology*, 1995.

49. R. P. Jackson, J. E. Cain, R. R. Jacobs, B. R. Cooper, and G. E. McManus, *Spine*, 1989, **14**, 1362.

50. J. S. Ross, M. T. Modic, T. J. Masaryk, J. Carter, R. E. Marcus, and H. Bohlman, *Am. J. Neuroradiol.*, 1989, **10**, 1243.

51. S. D. Boden, D. O. Davis, T. S. Dina, N. J. Patronas, and S. W. Wiesel, *J. Bone Joint Surg (Am.)*, 1990, **72**, 403.

52. M. C. Jensen et al., *N. Engl. J. Med. 2*, 1994, **331**, 69.

53. D. Fardon, S. Pinkerton, R. Balderston, S. Garfin, R. Nasca, and R. Salib, *Spine*, 1993, **18**, 274.

54. T. J. Masaryk, J. S. Ross, M. T. Modic, F. Boumphrey, H. Bohlman, and G. Wilberg, *Am. J. Neuroradiol.*, 1988, **150**, 1155.

55. S. W. Yu, V. M. Haughton, L. A. Sether, and M. Wagner, *Radiology*, 1988, **169**, 761.

56. J. R. Jinkins, *Am. J. Neuroradiol.*, 1993, **14**, 193.

57. S, Kobayashi, H. Yoshizawa, and Y. Hachiya in '18th Annual Meeting of the International Society for the Study of the Lumbar Spine, 1991'.

58. R. Quencer *Am. J. Neuroradiol.*, 1994,

59. R. P. Jackson, J. E. Cain, R. R. Jacobs, B. R. Cooper, and C. E. McManus, *Spine* 1989, **14**, 1362.

60. J. R. Thornbury, D. G. Fryback, P. A. Turski, M. J. Jarid, J. V. McDonald, B. R. Bernlich, L. R. Gentry, J. F. Saekott, E. J. Dosbach, and P. A. Martin, *Radiology*, 1993, **186**, 731.

61. S. D. Boden, P. R. McCowin, D. O. Davis, T. S. Dina, A. S. Mark, and S. Wiesel, *J. Bone Joint Surg. (Am.)*, 1990, **72**, 1178.

62. B. M. Brown, R. H. Schwartz, E. Frank, and N. K. Blank, *Am. J. Neuroradiol.*, 1988, **9**, 859.

63. M. C. Hedberg, B. P. Drayer, R. A. Flom, J. A. Modak and C. R. Bird, *Am. J. Roentgenol.*, 1988, **150**, 683.

64. D. R. Enzmann, and J. B. Rubin, *Radiology*, 1988, **166**, 467.

65. D. M. Yousem, S. W. Atlas, H. I. Goldberg, and R. I. Grossman, *Am. J. Neuroradiol.*, 1991, **12**, 229.

66. E. E. Awwad, D. S. Martin, K. R. Smith, and R. D. Bucholz, *J. Comput. Assist. Tomogr.*, 1990, **14**, 415.

67. D. Schellinger, L. Wener, B. D. Ragsdale, and N. J. Patronas, *RadioGraphics*, 1987, **7**, 944.

68. T. B. Freeman and C. R. Martinez, *Perspect. Neurol. Surg.*, 1992, **3**, 34.

69. H. Pavlov, J. S. Torg, B. Robie, and C. Jahre, *Radiology*, 1987, **164**, 771.

70. T. Fukushima, T. Ikata, Y. Taoka, and S. Takata, *Spine*, 1991, **16**, (Suppl. 10), 5534.

71. P. F. Statham, D. M. Hadley, P. MacPherson, R. A. Johnston, I. Bone, and G. M. Teasdale, *J. Neurol. Neurosurg. Psychiatry.*, 1991, **54**, 484.

72. B. M. Brown, R. H. Schwartz, E. Frank, and N. K. Blank, *Am. J. Roentgenol.*, 1988, **151**, 205.

73. R. I. Harris and I. McNab, *J. Bone Joint Surg. (Br.)*, 1954, **36**, 304.

74. N. Grenier, H. Y. Kressel, M. L. Schiebler, R. I. Grossman, and M. K. Dalinka, *Radiology*, 1987, **165**, 517.

Biographical Sketch

Michael T. Modic received his M.D. degree from Case Western Reserve School of Medicine in 1975. He completed his residency in radiology and fellowship in neuroradiology at the Cleveland Clinic Foundation. He spent a year as assistant professor of Radiology and as a staff neuroradiologist at University Hospitals in Cleveland from 1979–1980 and in 1980 returned to the Cleveland clinic as a staff neuroradiologist. In 1982 he was appointed head of the section of Magnetic Resonance. In 1985 he returned to Case Western Reserve School of Medicine/University Hospitals as Director of Magnetic Resonance and Neuroradiology, positions he held through 1989. During that time he also held the rank of Professor of Radiology, Neurology, and General Medical Sciences. In 1988 he was given a joint appointment as Professor of Neurosurgery as well. In 1989, Dr. Modic returned to the Cleveland Clinic Foundation as Chairman of Radiology and in 1993 was appointed as Professor of Radiology, Ohio State University. Dr. Modic has served on the Editorial Boards of the journals Radiology, American Journal of Neuroradiology, Neurology, Magnetic Resonance in Medicine, and Magnetic Resonance Imaging. He has served as a member of the Board of Trustees of the Society of Magnetic Resonance in Medicine and Board of Directors of the Society of Magnetic Resonance Imaging. He was President of the Society of Magnetic Resonance in Medicine for the 1992–1993 year and was the recipient of the Society Gold Medal in Clinical Science for his research activities related to the spine in 1991. He is co-author of the text 'Magnetic Resonance Imaging of the Spine' which is in its second edition and is the author or co-author of over 125 peer reviewed articles related to neuroradiology.

Brain MRS of Human Subjects

James W. Prichard

Yale University, New Haven, CT, USA

1 INTRODUCTION

A biomedical revolution based on NMR technology is under way. Many of the NMR methods described in this Encyclopedia are already in use for medical research and diagnosis, and many others will find such application soon. As the full impact of mature NMR methods becomes evident in the latter 1990s, it will be measured on the same scale with such things as microscopy and genetics. Two well-established characteristics of biomedical NMR technology ensure this outcome: it is remarkably noninvasive (see **Health and Safety Aspects of Human MR Studies**) and it is remarkably versatile, as attested in many other chapters. In each respect separately, it exceeds most other technologies available for the study of living tissue. By their combination, it stands alone.

This chapter deals with human brain research done by the branch of biomedical NMR technology designated magnetic resonance spectroscopy (MRS) to distinguish it from magnetic resonance imaging (MRI), which makes pictures of anatomical structure from the water proton signal. MRS obtains chemically specific information from biological tissue by the measurement of much smaller signals from ^1H, ^{31}P, ^{13}C, ^{17}O, and other magnetic nuclei in a variety of compounds. It is extending the range of biomedically useful NMR measurements to practical applications that would have courted dismissal as visionary a decade ago. In 1994, signals from more than two dozen compounds can be detected routinely in the human brain, and many more are in prospect. All report on biochemical phenomena previously accessible only by surgical removal of tissue or at autopsy. As various forms of MRS supported by MRI reach technical maturity, their collective contributions to an understanding of normal and deranged biochemistry in the brain will amount to nothing less than a new neurobiological discipline comparable in scope to today's neurochemistry, neuroanatomy, and neuropathology combined. All stages of normal human brain development and disease processes—including intrauterine stages—will be open to inspection along the many axes MRS can measure.

The MRS studies mentioned in this chapter are intended to help scientists and physicians interested in the brain to evaluate the above assertions, which may appear extravagant to readers who are exploring modern NMR for the first time. Many more studies appropriate for that purpose have been published than can be cited. Selection among them was intended to favor efforts that illuminate each other or set new problems by failing to do so, but it cannot have escaped all of the author's biases.

2 ^1H AND ^{31}P SPECTRA

Most human brain MRS research to date has used these two nuclei. Figures 1 and 2 illustrate the normal features and some disease-related abnormalities of brain ^1H and ^{31}P spectra with simulations that are free of the wide technical variation among spectra in original publications. Differences in acquisition parameters and data processing can cause spectra that are biologically equivalent to look quite different from each other. For readers new to either NMR or neuroscience, the variation can easily distract attention from features of biological importance.

The simulated spectra in Figures 1 and 2 were generated by Mathematica© and annotated in Adobe Illustrator©. Within each figure, random noise is of the same intensity in all spectra. Resonance variables were adjusted to mimic the appearance typical of many published spectra, in which less intense resonances are not evident because of such factors as long spin echo times, J coupling, and processing methods. The resonances shown are usually identifiable in all spectra from normal adult brain. The disease-related variations are large ones that have been reported by more than one group. They are presented together to emphasize their potential for metabolic fingerprinting of different disease processes.

3 NORMAL BRAIN FUNCTION: LACTATE IN HUMAN SENSORY SYSTEMS

A signal from lactate is detectable in ^1H spectra from the normal human brain.[1] Stimulation of the human visual[2-4] and auditory[5] systems can cause increases in lactate observable by ^1H MRS in the primary cortical receiving areas. These findings are consistent with positron emission tomography (PET) data showing that similar stimulation can increase glucose uptake more than oxygen extraction in the regions affected.[6] The PET workers interpreted their observations as evidence for preferential activation of nonoxidative glycolysis. The MRS data strengthen that view.

The notable aspect of this research area is the evidence from at least five different PET and NMR groups showing that glycolysis can increase more than respiration in response to some kinds of activation within the normal range of brain function. The phenomenon is well known in skeletal muscle. Selective activation of muscle cells with differing capacities for glycolysis and respiration is a normal feature of muscle function, and heterogeneous distribution of glycolytic and respiratory enzymes has long been employed for histochemical characterization of individual muscle cells. Evidence for similarly heterogeneous enzyme distribution already exists at the level of cell groups in

For References see p. 1077

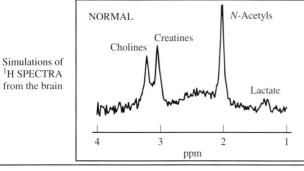

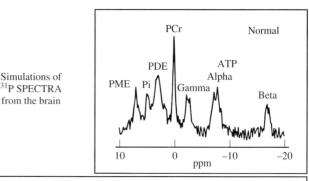

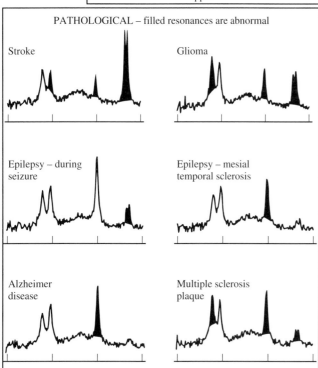

Simulations of
¹H SPECTRA
from the brain

Simulations of
³¹P SPECTRA
from the brain

PATHOLOGICAL – filled resonances are abnormal

PATHOLOGICAL – filled resonances are abnormal

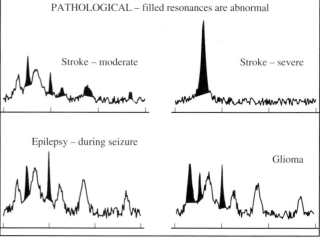

Figure 2 Simulated ³¹P spectra from normal and diseased brain, created and presented by the same procedures used for Figure 1. The resonances represented are from phosphomonoesters (PME), inorganic phosphate (Pi), phosphodiesters (PDE), phosphocreatine (PCr) and the α, β, and γ phosphates of adenosine triphosphate (ATP). In the 'stroke - severe' example, all of the signal present is Pi, indicating total loss of high-energy phosphate compounds

Figure 1 Simulated ¹H spectra from brain, created in Mathematica© and annotated in Adobe Illustrator©. Alterations typically caused by six pathological conditions are contrasted with each other and to normal. Spectral variables were adjusted to mimic the appearance common in published spectra. Random noise and each unfilled resonance are of the same intensity in all spectra. The 'normal' spectrum represents resonances from choline-containing compounds (cholines), creatine and phosphocreatine (creatines), N-acetyl-containing compounds, mostly N-acetylaspartate (N-acetyls), and lactate methyl protons (lactate). Glutamate and glutamine signals are represented as unresolved low-intensity resonances between 2 and 3 ppm. The 'pathological' spectra were created by varying the intensities of signals reported to be affected by the disease states indicated by labels on each spectrum. The abnormalities shown are the ones most often associated with each condition, but they are not all present in every case. The elevated lactate signal in four spectra displays the characteristic 7 Hz splitting caused by J coupling to the C-2 proton

the animal brain.[7] Like muscle cells, neurons and possibly glia may well have variable capacity for glycolysis and respiration related to their physiological functions.

These PET and MRS data prompt the question of what adaptive utility lactate elevation resulting from brain activity

might have. Two novel possibilities that can be tested by experiment are the following:

(1) *Lactate may be a local energy storage medium that is increased in anticipation of sudden high energy demand.* Ammunition stacked beside cannons defending a fort is more useful than ammunition in a remote storehouse if the fort is suddenly attacked. Suspicion of imminent attack would certainly cause the fort's defenders to move extra ammunition from the storehouse to the cannons, to shorten the response time and lengthen the duration of maximum firing. In the brain, neurons can suddenly be called upon to fire hundreds of action potentials a second and surrounding cells to recycle transmitter at proportional rates. The immediate metabolic burden of this intense activity falls principally on synaptic terminals, which are quite small and in consequence have low ratios of cytoplasmic volume to the area of excitable surface membrane that they must support. In such a structure, the Krebs cycle can respond to sudden maximum energy demand more quickly if it is not limited by the time necessary for activation of glycolysis, which is long on the scale of intense neuronal discharge and transmitter release. Local lactate elevation by previous activity would prime synaptic terminals to deliver their maximum response to sudden demand while glycolytic rate is increasing.

Existing techniques for the measurement of glucose and oxygen uptake and lactate concentration would not detect most instances of this process, because the ones that can be used in vivo, including MRS, report on overall changes in volumes of tissue much larger than the scattered locations at which such priming would ordinarily occur, and the others are bedeviled by the possibility of agonal artifact. Only when it occurs in many cells at the same time—as during sensory barrages and seizure discharge—could such a process be detected by present methods.

(2) *Lactate may have a signaling function in the brain.* Lactate is well suited to be a neuromodulator—a substance which alters the excitability of local neural ensembles. It is produced at increased rates whenever glycolysis accelerates in response to neural activity. It accumulates during pathological activation caused by ischemia or lack of oxygen, and, as described above, during some kinds of activation in the physiological range. If—upon escape from the active cell that produced it—it reduced the excitability of adjacent cells, it could both inhibit seizure discharge and 'sharpen' transmission along multifiber pathways, a process known to occur by other mechanisms. A finding consistent with this idea is available: in *Torpedo* electroplax, lactate in concentrations that can be produced by brain activation inhibited release of acetylcholine.[8]

4 STROKE

Stroke is the commonest major brain disease and a principal medical burden on society through the lost productivity it causes and its consumption of resources by extended disability. It is a natural place to test the benefits of new biomedical technology.

MRS observations on human stroke are far enough advanced to allow initial assessment of the role MRS will come to have in patient management. Lactate and *N*-acetyl (NA) resonances in the ^{1}H spectrum and intracellular pH (pHi) derived from the ^{31}P spectrum appear to have the best prospects for clinical utility.

4.1 ^{1}H MRS in Stroke

Figure 1 illustrates the kind of change commonly observed in ^{1}H spectra of human brain infarction: loss of the NA signal and elevation of the signal from lactate. Loss of NA signal should ensue from death of neurons, which are the only cells in mature brain thought to contain *N*-acetylaspartate (NAA), the principal source of the NA resonance. The NA signal was depressed both in the first few days after stroke onset and several weeks later in six patients studied at least twice.[9] In 10 patients studied within 60 h of stroke onset, it was reduced relative to the homologous region of the contralateral hemisphere in all but two; in seven of the same patients, repeat examination 1–2 weeks later showed additional decline at an overall rate calculated at $-29\% \pm 9\%$ per week.[10] Similar findings have been reported by other groups.[11–13]

Decline of the NA signal in the first days after a stroke surely reflects, in part, clearing of debris from neurons killed at the outset, but it may also be evidence of continuing neuronal loss after the acute period, possibly in the ischemic penumbra.[14] Therapy which prevents delayed neuronal loss would presumably also reduce eventual fixed deficit. Monitoring of the NA signal may prove useful as a surrogate endpoint for evaluation of therapies intended to reduce continuing neuronal loss in the subacute period.

Lactate elevation by acute infarction is readily detectable by ^{1}H MRS (Figure 1). However, the first report of increased lactate associated with stroke[14] was on patients studied months after the stroke occurred. Many observers including the present writer doubted that stroke-associated lactate elevation would persist so long, but subsequent observations by several groups have shown that it does. Serial study of individual patients demonstrated continuous lactate elevation for weeks after stroke in most and for several months in some.[9] Again, similar observations have been made by other groups.[11–13]

Multiple mechanisms of lactate elevation must be at work over so long a period. A newly infarcted region of brain accumulates lactate to a concentration of 15–30 mM within minutes of losing its blood supply, as glucose and glycogen in unperfused tissue are metabolized in the absence of oxygen, which is exhausted in the first few seconds. If the region is never reperfused, the lactate in it can dissipate only by diffusion, but the process would not take weeks to months in the presence of active tissue repair processes. Other possible sources of elevated lactate associated with infarction include adjacent regions of surviving but impaired tissue referred to as an 'ischemic penumbra',[15] infiltrating cells involved in the tissue's reaction to injury, and altered metabolism of surviving brain. MRS has provided evidence for the second of these:[16] the brain of a patient studied two weeks after a stroke and autopsied a week later had large numbers of macrophages in the regions where ^{1}H MRS had shown elevated lactate. The role of altered metabolism is conjectural at present; the long-term metabolic response of brain tissue that survives ischemic insult is not well understood.

Lactate that accumulates in the killed core of an infarct is metabolically inert, while that in an ischemic penumbra turns over, albeit probably at a rate different from that of normal brain tissue. The anatomical boundary between pools of lactate in these two states need not be sharp; the pools may even be anatomically interdigitated. Their relative sizes reflect the proportion of killed to surviving tissue early in the history of the lesion, and may therefore indicate how a particular lesion is likely to evolve. An advanced form of combined ^{1}H/^{13}C MRS that has been shown to be feasible in humans can make a pertinent measurement. Inert and metabolically active lactate associated with a human stroke can be distinguished from each other by determining how much of the total stroke-associated lactate pool can be labeled with ^{13}C from blood glucose. Explanation of this remarkable possibility requires a brief description of the NMR properties of [13]C.

4.2 ^{13}C MRS

Carbon in Nature is nearly 99% [12]C, which is not magnetic and hence gives no NMR signal. The stable magnetic isotope ^{13}C is 1.1% naturally abundant. Its potential for biological studies was appreciated early in the second decade of empirical

For References see p. 1077

NMR work. In 1958, the following passage was written by P. C. Lauterbur:[17]

'Most practical applications of ^{13}C spectra must await some improvement in the signal-to-noise ratio. *One application that might be made immediately is in the use of ^{13}C as an isotopic tracer in reactions.* Various chemical forms of carbon could be identified by their chemical shifts and fine structures, even in complex mixtures, after the introduction of a compound enriched in ^{13}C. *Even some biological systems might be studied by this technique.*' (Emphases added.)

These words are the conceptual origin of biomedical ^{13}C MRS. In 1957, their author had published the first report of ^{13}C chemical shifts,[18] which was followed shortly by another.[19] Years became decades while NMR technology evolved to a stage at which Lauterbur's idea could be explored by ^{13}C labeling studies, first on enzyme systems and functioning cells,[20] later on intact animal liver,[21] animal brain,[22] and human brain.[23,24] The potential adumbrated by these studies is so great that even the most advanced of them belong to what later generations of biomedical scientists will regard as the early history of ^{13}C MRS. Lauterbur was remarkably prescient.

In living subjects including humans, natural abundance ^{13}C MRS has much still unexploited potential for characterization of normal and pathological variation among tissues, but the most inviting opportunity that ^{13}C MRS presents to biomedical workers is ^{13}C enrichment of molecules observable in vivo. As demonstrated more than two decades ago in *Candida utilis*,[25] feeding a ^{13}C-enriched nutrient to a living organism provides the simultaneous advantages of increased signal-to-noise ratio for observation of molecules that receive the ^{13}C through metabolic processes and a nondestructive means of measuring the rates of those processes. The adult brain normally derives nearly all of its energy from glucose, most of which it converts to two molecules of lactate; if the glucose was enriched in ^{13}C at the C-1 position, one of the lactates receives ^{13}C glucose in its methyl position. Further metabolism in the Krebs cycle creates several other ^{13}C-labeled molecules observable in vivo, the most prominent of which is 4-^{13}C-glutamate. After animal experiments demonstrated that 1-^{13}C-glucose, 3-^{13}C-lactate, and ^{13}C-labeled amino acids could be detected in living brain by ^{13}C MRS,[22] similar observations were made in the human.[24]

4.3 Detection of ^{13}C by ^{1}H MRS

Direct ^{13}C MRS has the disadvantage for in vivo work of long acquisition times due to the low sensitivity of ^{13}C compared to ^{1}H. In samples containing enough ^{1}H–^{13}C bonds, the presence of the ^{13}C can be detected in ^{1}H spectra by the characteristic way in which it splits the signal from protons bonded to it. Exploitation of this phenomenon confers proton sensitivity on NMR measurement of ^{13}C in some positions of intact, functioning molecules. The principle was used in studies of organic molecular structure as early as the 1960s.[20] Later, technological advances and diligent effort allowed its successful adaptation to observation of ^{13}C-labeled compounds in the brains of living animals[26] and humans.[23] By the use of appropriate models, cerebral metabolic rates can be calculated from the time course of ^{13}C accumulation in observable metabolic pools.[27] These and related NMR methods are still in the early

stages of development; in their fully mature forms, they are likely to be the preeminent means of measuring metabolic rates in the living human brain.

4.4 Labeling of Stroke-Elevated Lactate with ^{13}C

In stroke, ^{13}C labeling can make the important distinction between lactate which is trapped in a nonmetabolizing compartment and lactate that is persistently elevated in the presence of competent metabolic machinery. This strategy was used to show that shock-elevated lactate in rabbit brain[28] is all metabolically active,[29] and it is equally applicable to assessment of stroke-elevated lactate in the human, as has been demonstrated.[30]

However, the prospect for the routine use of this very sophisticated technique in emergency stroke evaluation depends on other things. Carbon-13 labeling of the kind described can certainly provide information about how much of a fresh stroke is an ischemic penumbra,[14] but rapid evolution of MRI may provide equivalent information from simpler measurements. The best candidates are diffusion weighted imaging (DWI)[31] (see *Diffusion: Clinical Utility of MRI Studies*) and magnetization transfer contrast imaging (see *Magnetization Transfer Contrast: Clinical Applications*). The point here is that if ^{13}C labeling of stroke-elevated lactate proves to be a sufficiently novel predictor of a fresh infarct's later course to justify its relative complexity, its minor risk, and a few hundred dollars for each patient, it can be widely implemented.

4.5 ^{31}P MRS in Stroke

Figure 2 illustrates two degrees of loss of energy stores and elevation of inorganic phosphate that can be caused by stroke. Intracellular acidosis is measurable from the resonant frequency of the inorganic phosphate signal, but the frequency difference is too small to be evident in the figure.

The first ^{31}P MRS study of human stroke found normal metabolite ratios with reduced total phosphorus signal in chronic stroke, consistent with the replacement of infarcted tissue by cerebrospinal and interstitial fluid.[32] After developing the capability to study acutely ill patients in a magnet—a decidedly nontrivial acheivement—a group at Henry Ford Hospital in Detroit used it to monitor ^{31}P changes from the acute to the subacute period.[33] Phosphocreatine and ATP were reduced in the acute period, and inorganic phosphate was elevated, as shown in Figure 2. Intracellular pH was acidotic. All of these changes are consistent with the traditional understanding of stroke pathophysiology, but they did not correlate well with clinical measures. Alkalosis replaced acidosis within a few days. The authors suggested that effective therapy might have to be instituted during the period of acidosis.

An estimate of tissue Mg^{2+} concentration can be made from information in ^{31}P spectra.[34] In stroke, Mg^{2+} was elevated in the acidotic period; it might be a pathophysiological factor or a marker of cellular injury.[35]

^{1}H and ^{31}P spectra can both be obtained in the same session. Combined ^{1}H/^{31}P observations are a powerful way of analyzing relationships among a wide range of brain metabolites, as was first demonstrated in animal work on hypoglycemia.[36] Despite rather considerable technical obstacles to doing this in human stroke patients, two groups have accomplished it,[37,38] and the results are illuminating. Both groups found that stroke-associated lactate and pH were usually not inversely correlated. In combined ^{1}H/^{31}P observations ranging from the acute to the chronic period, the more common association was of elevated lactate with alkalosis, rather than acidosis or normal pH. Because ^{31}P and ^{1}H spectra come from tissue volumes that are not the same size and are both large compared with the dimensions of any of the several metabolic compartments they contain, MRS-observable metabolites should not be expected to behave as though they were in a well-stirred test tube. Dissociation of lactate from pH has been observed in experimental status epilepticus by NMR[39] and, by biochemical techniques, in tumors and one day after global ischemia.[40]

5   EPILEPSY

5.1 ^{31}P MRS in Epilepsy

Several animal studies on seizure phenomena in the early years of in vivo MRS demonstrated that the PCr/Pi (Figure 2) decline well established by traditional biochemical research could be observed in the the living brain.[41] Due to the limited bore size of available spectrometers, the first MRS observations on the epileptic human brain were ^{31}P spectra from infants.[42] Findings during seizures were as expected: the PCr/Pi ratio was decreased about 50%, as shown in Figure 2, and it returned to normal after seizure discharge ceased. Infants who had the lowest ratios during seizures developed long-term neurological sequelae.

The development of spectrometers with bores large enough for adults was quickly followed by studies of chronic temporal lobe epilepsy, which is a major problem in modern epilepsy management. Two groups have published data on enough patients to allow comparison of results, which are somewhat different. One found alkaline pHi associated with the seizure focus in eight patients, seven of which also had increased Pi and decreased PME.[43,44] The other group reported low PCr/Pi ratios without other changes.[45] Further work will be necessary to determine whether the discrepancy reflects technical factors of differences between patient populations.

5.2 ^{1}H MRS in Epilepsy

The principal component of the NA signal in ^{1}H spectra of normal brain is from NAA, which occurs mainly if not exclusively in neurons. Reduction of the NA signal implying loss of neurons is a common feature of chronically epileptogenic brain tissue, having been documented in two papers,[46,47] and in six preliminary reports by these groups and others at the Annual Meeting of The Society of Magnetic Resonance in Medicine in 1993. The phenomenon is illustrated in Figure 1.

These data have the important implication that chemically specific ^{1}H MRS abnormalities may be detectable in vivo before structural changes in patients with chronic temporal lobe epilepsy, which is commonly associated with a neuropathological state known as mesial temporal sclerosis and can often be relieved by surgery. MRI techniques that appear to be especially sensitive to this pathology have been reported,[48,49] but even if they become routine, noninvasive preoperative detection of chemical abnormality is likely to improve accurate selection of the tissue to be resected for relief of complex partial epilepsy. MRS will be especially important for that purpose if the NA signal proves to be the first NMR quantity to change as mesial temporal sclerosis develops. Noninvasive electroencephalographic and NMR techniques together may soon eliminate the need for implantation of intracranial electrodes to determine which patients with this kind of intractable epilepsy can benefit from surgery.

Observation by ^{1}H MRS of two patients with a form of chronic localized epileptogenic encephalitis known as Rasmussen's syndrome produced the first report of elevated lactate associated with seizure discharge in human brain.[46] Lactate elevation caused by seizure discharge is illustrated in Figure 1.

Monitoring of signals from γ-aminobutyric acid and glutamine in the human brain by newly developed ^{1}H MRS methods[50] has allowed direct observation of the effects of vigabatrin, a new antiepileptic drug.[51] Such observation of drug effects directly in the living target organ opens a new era in neuropharmacology.

6   DEMENTIA

Primary dementia—decline of mental function not secondary to tumors, trauma, drugs, or other obvious causes—is about one-half dementia of the Alzheimer type (hereinafter 'Alzheimer disease'). Vascular, infectious, and other dementias are much less common. Alzheimer disease is a condition of unknown etiology that is defined by its characteristic neuropathology, although the diagnosis can usually be made correctly in life from its distinctive constellation of clinical and laboratory abnormalities. The protracted disability that it causes places a large burden on society, motivating intense research effort which now includes MRS.

6.1 ^{31}P Studies of Alzheimer Disease

Published ^{31}P data on Alzheimer patients do not agree with each other. Conflicting claims about what changes, if any, are present have persisted for several years. A series of studies recently summarized[52] reported that PME and Pi were above normal in the brains of Alzheimer patients, while another group found no clear ^{31}P changes associated with the disease.[53,54] No obvious difference in patient selection or characterization accounts for the difference. Technical factors might; the two studies used somewhat different ^{31}P acquisition methods.

6.2 ^{1}H Studies of Alzheimer Disease

In contrast to the ^{31}P findings, ^{1}H MRS done by several groups has produced general agreement that a reduced NA signal is characteristic of Alzheimer disease (see Figure 1). Preliminary reports of nine studies of living Alzheimer patients all describe decreased NA signals (*Proc. 12th Ann Mtg. (Int) Soc. Magn. Reson. Med.*, New York, 1993). These very consistent data urge the conclusion that NA is characteristically reduced in Alzheimer disease, apparently in proportion to the severity of neuron loss.

For References see p. 1077

7 BRAIN TUMORS

The cellular heterogeneity of brain tumors is a substantial impediment to progress in understanding neoplasia in the nervous system by in vivo techniques, as PET workers have known for years. The heterogeneity is far below the anatomical resolution of current MRS techniques, and technical improvements that appear feasible offer no hope that the gap can be closed. Its effects were evident in an early ^{31}P MRS study of brain tumors, which commented on the 'striking diversity' of the metabolic patterns observed.[55] Later studies associated decreased PCr, alkaline pH, increased PME, and altered PDE signals with heightened aggressivity of gliomas.[56,57] One constellation of ^{31}P changes that can occur in gliomas is illustrated in Figure 2.

More detailed observation of metabolic properties of brain tumors is possible by ^{1}H MRS due to its finer anatomical resolution. In the form of chemical shift imaging (CSI), it allows variations within single lesions to be detected. A study that combined ^{1}H CSI with PET found that regions of high lactate tended to coincide with regions of high glucose uptake.[58] Later work with improved techniques by the same group revealed a more complicated situation: high lactate was also associated with loculations of extracellular fluid.[59] Another group able to study patients with both ^{1}H CSI and PET reported similar variability and metabolic findings.[60,61] Variable reductions in NA and increased choline signals were observed by both groups. These changes and the elevated lactate seen in some brain tumors are illustrated in Figure 1.

Metabolic maps made from ^{31}P have lower anatomical resolution than ^{1}H maps, due to the lower sensitivity of ^{31}P, but they are capable of showing metabolite distributions across major brain structures and defects caused by large lesions (also see the related *Chemical Shift Imaging*).[62,63] The information about energy state, pH, and phospholipids available in ^{31}P spectra is so valuable for the understanding of many disease states that continued vigorous efforts to refine ^{31}P CSI are certain.

8 MULTIPLE SCLEROSIS

Demonstration of the extensive, clinically silent pathology of cerebral white matter in multiple sclerosis was among the first major new findings of MRI in the nervous system. MRS of the disease was not practical until several years later, for the usual reason that MRS signals are of much lower intensity. By 1993, NMR technology had advanced to the point that nine preliminary reports on ^{1}H MRS studies of multiple sclerosis appeared in that year's *Proc. 12th Ann Mtg. (Int) Soc. Magn. Reson. Med.* Reduced NA and increased choline signals associated with plaques were the most common findings. Elevated lactate was also observed; it may reflect the presence of highly glycolytic white cells, as in subacute cerebral infarction.[16] All three changes are illustrated in Figure 1.

These metabolic abnormalities were followed over a period of months in a single multiple sclerosis patient with an unusually large cerebral plaque.[64] Metabolic aspects of plaque evolution have not previously been accessible for study in human patients. The combination of MRI and chemically specific MRS will produce new understanding of the pathophy-

siology of multiple sclerosis, as they are used together to obtain new information on the natural history and therapeutic responsiveness of the disease in individual patients. All of the data reported by Arnold and colleagues[64] are new information bearing on the underlying cellular and molecular biology of episodic demyelination. The opening of so wide a window on a pathophysiological process is certain to improve understanding of it.

9 THE FUTURE OF HUMAN BRAIN MRS

The chemical specificity of MRS guarantees a major role for it in neurobiology. That property, together with the noninvasiveness that it shares with all NMR methods, offers scope for the investigation of the normal human brain that has no close precedent in the history of any earlier technology. Detailed biochemical characterization of the living human brain at all its stages of development and decline is coming within the reach of noninvasive, chemically specific MRS. Investigation of how the human brain works when it is normal and when it is diseased will move more rapidly, and in new directions, with abundant benefit to both science and medicine.

9.1 Science

Neurochemistry is a difficult discipline, because nervous tissue is well protected and highly intolerant of the kind of disruption required for study by standard chemical techniques. MRS can reduce that barrier considerably by providing abundant neurochemical data from the living organ, free of agonal artifact and remeasurable as often as necessary in the same individual. The MRS studies of normal function and metabolic rates mentioned above are early examples of work that will grow into a new dimension of neurochemistry touching nearly every aspect of human brain biology. While MRS can measure only a small fraction of the compounds present in living brain, information about that fraction is unique because previously it was not available at all. Neurobiologists can now look through the window of several dozen MRS-measurable compounds at the biochemical milieu of which they are part. As the number of compounds observable in vivo grows, MRS will become an increasingly powerful complement to cellular and molecular methods in neurobiological research.

9.2 Medicine

In the late 1990s, no routine clinical application of MRS is yet standard practice of the kind that every hospital must provide, like X-ray equipment and electrocardiographs, but MRI has not reached that point either. Both will. Diagnostic MRI is so much more versatile and efficient than earlier technologies that its emergence as the premier medical imaging method of the latter 1990s is certain. Implementation of MRS (small signal capability) on standard clinical MRI machines is no longer a large step in either technique or money. The widespread availability of MRI machines needed for efficient medical diagnosis will facilitate the introduction of MRS into routine clinical practice as rapidly as MRS research demonstrates useful applications.

The most important prospect that MRS offers clinical medicine is chemically specific characterization of disease processes at all of their stages. Noninvasive longitudinal MRS data that provide new understanding of how pathophysiological processes evolve are unique. They will affect medicine no less than the data from microscopic and chemical study of removed brain tissue that are much of the basis for modern conceptions of disease. MRS-defined biochemical profiles that distinguish ischemic, neoplastic, inflammatory, degenerative, and other pathophysiological categories from each other in vivo will emerge, as will profiles that identify specific diseases. Many MRS measurements will be useful in monitoring the effects of therapy. As this body of knowledge grows, the use of specific MRS measurements in the management of individual patients will become routine.

10 RECENT PROGRESS

This article was first written in 1994. In the four intervening years, MRS of the human brain has advanced rapidly, as have nearly all biomedical applications of NMR technology. Most of the advances in MRS have been along paths predictable from the work described in the article, which continues to provide useful orientation to the origins of a field that is both revolutionary and still young.

However, in several areas recent progress is either novel or extensive to a degree that the article does not adequately indicate. The following recent citations, which are mostly reviews, will help the interested reader find relevant literature.

10.1 Spectroscopic Imaging

This procedure, also known as chemical shift imaging, allows mapping of metabolites in two dimensions. The major technical problems that it presents have been the object of intense development efforts in recent years,[65,66] and it has been used by several groups in clinical research studies on brain tumors,[67] multiple sclerosis,[68] and various aspects of brain metabolism.[69,70]

10.2 High-field Magnets

Spectrometers suitable for human studies at fields as high as 4.1 T have been used in research for several years, and instruments with fields up to 8 T are under development. Notable examples of the neurobiologic opportunities opened by work at 4 T include improved observation of glucose[71] and amino acid[69] resonances in the human brain.

10.3 Measurement of Brain pH in ¹H Spectra

Human ^{1}H MRS at 2.1 T has shown that a titrating signal from homocarnosine can provide a measure of cytosolic pH in neurons with high concentrations of that compound, probably a subset specialized for synaptic release of γ-aminobutyric acid.[72] A general review of brain pH measurements by MRS has appeared.[73]

11 RELATED ARTICLES

Anisotropically Restricted Diffusion in MRI; Brain Infection and Degenerative Disease Studied by Proton MRS; Brain MRS of Infants and Children; Brain Neoplasms Studied by MRI; Chemical Shift Imaging; Diffusion: Clinical Utility of MRI Studies; Echo-Planar Imaging; Health and Safety Aspects of Human MR Studies; Localization and Registration Issues Important for Serial MRS Studies of Focal Brain Lesions; Single Voxel Localized Proton NMR Spectroscopy of Human Brain In Vivo; Sodium-23 Magnetic Resonance of Human Subjects; Structural and Functional MR in Epilepsy; Systemically Induced Encephalopathies: Newer Clinical Applications of MRS; Whole Body Studies: Impact of MRS.

12 REFERENCES

1. C. C. Hanstock, D. L. Rothman, J. W. Prichard, T. and R. G. Shulman, *Proc. Natl. Acad. Sci. USA*, 1988, **85**, 1821.
2. J. Prichard, D. Rothman, E. Novotny, O. Petroff, T. Kuwabara, M. Avison, A. Howseman, C. Hanstock, and R. Shulman, *Proc. Natl. Acad. Sci. USA*, 1991, **88**, 5829.
3. D. Sappey-Marinier, G. Calabrese, G. Fein, J. Hugg, C. Biggins, and M. Weiner, *J. Cereb. Blood Flow Metab.*, 1992, **12**, 584.
4. B. G. Jenkins, J. W. Belliveau, and B. R. Rosen, *Proc. 11th Ann Mtg. (Int) Soc. Magn. Reson. Med.*, Berlin, 1992, 2145.
5. M. Singh, *IEEE Trans. Nucl. Sci.*, 1992, **39**, 1161.
6. P. T. Fox, M. E. Raichle, M. A. Mintun, and C. Dence, *Science*, 1988, **241**, 462.
7. I. W. Borowsky and R. C. Collins, *J. Comp. Neurol.*, 1989, **288**, 401.
8. Y. M. Gaudry-Talarmain, *Eur. J. Pharmacol.*, 1986, **129**, 235.
9. G. D. Graham, A. M. Blamire, A. M. Howseman, D. L. Rothman, P. B. Fayad, L. M. Brass, O. A. Petroff, R. G. Shulman, and J. W. Prichard, *Stroke*, 1992, **23**, 333.
10. G. D. Graham, A. M. Blamire, D. L. Rothman, L. M. Brass, P. B. Fayad, O. A. C. Petroff, and J. W. Prichard, *Stroke*, 1993, **24**, 1891.
11. J. H. Duijn, G. B. Matson, A. A. Maudsley, J. W. Hugg, and M. W. Weiner, *Radiology*, 1992, **183**, 711.
12. C. C. Ford, R. H. Griffey, N. A. Matwiyoff, and G. A. Rosenberg, *Neurology*, 1992, **42**, 1408.
13. P. Gideon, B. Sperling, P. Arlien-Soborg, T. S. Olsen, and O. Henriksen, *Stroke*, 1994, **25**, 967.
14. J. W. Berkelbach van der Sprenkel, P. R. Luyten, P. C. van Rijen, C. A. Tulleken, and J. A. den Hollander, *Stroke*, 1988, **19**, 1556.
15. J. W. Prichard, in 'Molecular and Cellular Approaches to the Treatment of Brain Disease', ed. S. G. Waxman, Raven Press, New York, 1993, pp. 153–174.
16. O. A. Petroff, G. D. Graham, A. M. Blamire, R. M. al, D. L. Rothman, P. B. Fayad, L. M. Brass, R. G. Shulman, and J. W. Prichard, *Neurology*, 1992, **42**, 1349.
17. P. C. Lauterbur, *Ann. N.Y. Acad. Sci.*, 1958, **70**, 841.
18. P. C. Lauterbur, *J. Chem. Phys.*, 1957, **26**, 217.
19. C. H. Holm, *J. Chem. Phys.*, 1957, **26**, 707.
20. N. A. Matwiyoff, and D. G. Ott, *Science*, 1973, **181**, 1125.
21. J. R. Alger, L. O. Sillerud, K. L. Behar, R. J. Gillies, R. G. Shulman, R. E. Gordon, D. Shae, and P. E. Hanley, *Science*, 1981, **214**, 660.
22. K. L. Behar, O. A. C. Petroff, J. W. Prichard, J. R. Alger, and R. G. Shulman, *Magn. Reson. Med.*, 1986, **3**, 911.
23. D. L. Rothman, E. J. Novotny, G. I. Shulman, A. M. Howseman, O. A. C. Petroff, G. Mason, T. Nixon, C. C. Hanstock, J. W.

For References see p. 1077

Prichard, and R. G. Shulman, *Proc. Natl. Acad. Sci. USA*, 1992, **89**, 9603.

24. R. Gruetter, E. J. Novotny, S. D. Boulware, D. L. Rothman, G. F. Mason, G. I. Shulman, R. G. Shulman, and W. V. Tamborlane, *Proc. Natl. Acad. Sci. USA*, 1992, **89**, 1109.

25. R. T. Eakin, L. O. Morgan, C. T. Gregg, and N. A. Matwiyoff, *FEBS Lett.*, 1972, **28**, 259.

26. D. L. Rothman, K. L. Behar, H. P. Hetherington, J. A. den Hollander, M. R. Bendall, O. A. C. Petroff, and R. G. Shulman, *Proc. Natl. Acad. Sci. USA*, 1985, **82**, 1633.

27. G. F. Mason, D. L. Rothman, K. L. Behar, and R. G. Shulman, *J. Cereb. Blood Flow Metab.*, 1992, **12**, 434.

28. J. W. Prichard, O. A. Petroff, T. Ogino, and R. G. Shulman, *Ann. N.Y. Acad. Sci.*, 1987, **508**, 54.

29. O. A. C. Petroff, E. J. Novotny, M. Avison, D. L. Rothman, J. R. Alger, T. Ogino, G. I. Shulman, and J. W. Prichard, *J. Cereb. Blood Flow Metab.*, 1992, **12**, 1022.

30. D. L. Rothman, A. M. Howseman, G. D. Graham, O. A. C. Petroff, G. Lantos, P. B. Fayad, L. M. Brass, G. I. Shulman, R. G. Shulman, and J. W. Prichard, *Magn. Reson. Med.*, 1991, **21**, 302.

31. S. Warach, D. Chien, W. Li, M. B. Ronthal, and R. R. Edelman, *Neurology*, 1992, **42**, 1717.

32. P. A. Bottomley, B. P. Drayer, and L. S. Smith, *Radiology*, 1986, **160**, 763.

33. S. R. Levine, J. A. Helpern, K. M. Welch, A. M. Vande Linde, K. L. Sawaya, E. E. Brown, N. M. Ramadan, R. K. Deveshwar, and R. J. Ordidge, *Radiology*, 1992, **185**, 537.

34. H. R. Halvorson, A. M. Vande Linde, J. A. Helpern, and K. M. Welch, *NMR Biomed.*, 1992, **5**, 53.

35. J. A. Helpern, A. Vande Linde, K. M. Welch, S. R. Levine, L. R. Schultz, R. J. Ordidge, H. R. Halvorson, and J. W. Hugg, *Neurology*, 1993, **43**, 1577.

36. K. L. Behar, J. A. den Hollander, O. A. C. Petroff, H. Hetherington, J. W. Prichard, and R. G. Shulman, *J. Neurochem.*, 1985, **44**, 1045.

37. J. W. Hugg, J. H. Duijn, G. B. Matson, A. A. Maudsley, J. S. Tsuruda, D. F. Gelinas, and M. W. Weiner, *J. Cereb. Blood Flow Metab.*, 1992, **12**, 734.

38. P. Gideon, B. Sperling, O. Henriksen, N. A. Lassen, T. Skyhøj, T. S. Olsen, P. Sidenius, and P. Arlien-Soborg, *Proc. 12th Ann Mtg. Soc. Magn. Reson. Med.*, New York, 1993, 1484.

39. O. A. C. Petroff, J. W. Prichard, T. Ogino, M. J. Avison, J. R. Alger, and R. G. Shulman, *Ann. Neurol.*, 1986, **20**, 185.

40. W. Paschen, B. Djuricic, G. Mies, R. Schmidt-Kastner, and F. Linn, *J. Neurochem.*, 1987, **48**, 154.

41. J. W. Prichard, *Epilepsia*, 1994, **35**(Suppl 6), S14.

42. D. P. Younkin, P. M. Delivoria, J. Maris, E. Donlon, R. Clancy, and B. Chance, *Ann. Neurol.*, 1986, **20**, 513.

43. J. W. Hugg, G. B. Matson, D. B. Tweig, A. A. Maudsley, D. Sappey-Marinier, and M. W. Weiner, *Magn. Reson. Imag.*, 1992, **10**, 227.

44. K. D. Laxer, B. Hubesch, M. D. Sappey-Marinier, and M. W. Weiner, *Epilepsia*, 1992, **33**, 618.

45. R. Kuzniecky, G. A. Elgavish, H. P. Hetherington, W. T. Evanochko, and G. M. Pohost, *Neurology*, 1992, **42**, 1586.

46. P. M. Matthews, F. Andermann, and D. L. Arnold, *Neurology*, 1990, **40**, 985.

47. G. Layer, F. Traber, U. Muller-Lisse, J. Bunke, C. E. Elger, and M. Reiser, *Radiologe*, 1993, **33**, 178.

48. G. D. Jackson, S. F. Berkovic, J. S. Duncan, and A. Connelly, *Am. J. Neuroradiol.*, 1993, **14**, 753.

49. G. D. Jackson, A. Connelly, J. S. Duncan, R. A. Grunewald, and D. G. Gadian, *Neurology*, 1993, **43**, 1793.

50. D. L. Rothman, O. A. Petroff, K. L. Behar, and R. H. Mattson, *Proc. Natl. Acad. Sci. USA*, 1993, **90**, 5662.

51. O. Petroff, D. Rothman, K. Behar, and R. Mattson, *Proc. 12th Ann Mtg. Soc. Magn. Reson. Med.*, New York, 1993, 434.

52. J. N. Kanfer, J. W. Pettegrew, J. Moossy, and D. G. McCartney, *Neurochem. Res.*, 1993, **18**, 331.

53. P. A. Bottomley, J. P. Cousins, D. L. Pendrey, W. A. Wagle, C. J. Hardy, F. A. Eames, R. J. McCaffrey, and D. A. Thompson, *Radiology*, 1992, **183**, 695.

54. D. G. Murphy, P. A. Bottomley, J. A. Salerno, C. DeCarli, M. J. Mentis, C. L. Grady, D. Teichberg, K. R. Giacometti, J. M. Rosenberg, C. J. Hardy, M. B. Schapiro, S. I. Rapoport, J. R. Alger, and B. Horwitz, *Arch. Gen. Psychiatr.*, 1993, **50**, 341.

55. R. D. Oberhaensli, J. D. Hilton, P. J. Bore, L. J. Hands, R. P. Rampling, and G. K. Radda, *Lancet*, 1986, **2**, 8.

56. J. Jeske, K. Herholz, W. Heindel, and W. D. Heiss, *Onkologie*, 1989, **1**, 42.

57. W. D. Heiss, W. Heindel, K. Herholz, J. Rudolf, J. Bunke, J. Jeske, and G. Friedmann, *J. Nucl. Med.*, 1990, **31**, 302.

58. P. R. Luyten, A. J. Marien, W. van G. P. Heindel, K. den H. J. Herholz, G. Friedmann, and W. D. Heiss, *Radiology*, 1990, **176**, 791.

59. K. Herholz, W. Heindel, P. R. Luyten, J. A. denHollander, U. Pietrzyk, J. Voges, H. Kugel, G. Friedmann, and W. D. Heiss, *Ann. Neurol.*, 1992, **31**, 319.

60. J. R. Alger, J. A. Frank, A. Bizzi, M. J. Fulham, B. X. DeSouza, M. O. Duhaney, S. W. Inscoe, J. L. Black, P. C. Van Zijl, C. T. Moonen, and G. Di Chiro, *Radiology*, 1990, **177**, 633.

61. M. J. Fulham, A. Bizzi, M. J. Dietz, H. H. Shih, R. Raman, G. S. Sobering, J. A. Frank, A. J. Dwyer J. R. Alger, and G. Di Chiro, *Radiology*, 1992, **185**, 675.

62. T. R. Brown, *NMR Biomed.*, 1992, 5, 238.

63. B. J. Murphy, R. Stoyanova, R. Srinivasan, T. Willard, D. Vigneron, S. Nelson, J. S. Taylor, and T. R. Brown, *NMR Biomed.*, 1993, **6**, 173.

64. D. L. Arnold, M. D. Matthews, G. S. O. C. J. Francis, and J. P. Antel, *Ann. Neurol.*, 1992, **31**, 235.

65. S. J. Nelson, D. B. Vigneron, J. Star-Lack, and J. Kurhanewicz, *NMR Biomed.*, 1997, **10**, 411.

66. R. V. Mulkern, H. Chao, J. L. Bowers, and D. Holtzman, *Ann. N. Y. Acad. Sci.*, 1997, **820**, 97.

67. M. C. Preul, Z. Caramanos, R. Leblanc, J. G. Villemure, and D. L. Arnold, *NMR Biomed.*, 1998, **11**, 192.

68. D. L. Arnold, J. S. Wolinsky, P. M. Matthews, and A. Falini, *J. Neurol. Neurosurg. Psychiat.*, 1998, **64**(Suppl. 1), S94.

69. H. P. Hetherington, J. W. Pan, W. J. Chu, G. F. Mason, and B. R. Newcomer, *NMR Biomed.*, 1997, **10**, 360.

70. P. C. van Zijl and P. B. Barker, *Ann. N. Y. Acad. Sci.*, 1997, **820**, 75.

71. R. Gruetter, M. Garwood, K. Ugurbil, and E. R. Seaquist, *Magn. Reson. Med.*, 1996, **36**, 1.

72. D. L. Rothman, K. L. Behar, J. W. Prichard, and O. A. Petroff, *Magn. Reson. Med.*, 1997, **38**, 924.

73. J. W. Prichard, D. L. Rothman, and O. A. C. Petroff, in 'pH and brain function', eds K. Kaila and B. R. Ransom, John Wiley, New York, 1998, p. 149.

Acknowledgements

The author's own work was supported by USPHS Grants NS 27883, DK 34576, and NS 21708.

Biographical Sketch

J. W. Prichard. *b* 1934. A.B., 1955, Philosophy, Washington University, St. Louis, M.D., 1959 Harvard Medical School, Boston. Clinical and postdoctoral training at Bellevue Hospital (New York), The National Hospital for Nervous Diseases (Queen Square, London), Yale (New Haven), and The National Institutes of Health (Bethesda).

For list of General Abbreviations see end-papers

Research career in neurology electrophysiology prior to 1981. Entered NMR research then through collaboration with Yale spectroscopists in development and validation of NMR methods for study of the brain in vivo. Approx. 110 publications, half in NMR. Principal research interests: application of NMR methods to clinical neurology and their relationship to cerebral electrophysiology.

Brain Infection and Degenerative Disease Studied by Proton MRS

Robert E. Lenkinski, Dolores López-Villegas, and Supoch Tunlayadechanont

University of Pennsylvania, Philadelphia, PA, USA

1 INFECTIONS

1.1 Creutzfeldt–Jakob Disease (CJD)

Bruhn et al.[1] showed in an initial case report of CJD that there was significant loss of *N*-acetylaspartate (NAA) in both white matter (40%) and gray matter (30%). The level of inositol, presumably myo-inositol (MI), was higher than normal (30%) in white matter. The conventional magnetic resonance (MRI) images of this patient showed only mild cortical atrophy and hyperintensities in the lentiform nuclei. These authors suggested that proton MRS might be useful in the early detection of CJD.

Graham et al.[2] have studied two patients with biopsy proven CJD using in vivo MRS. High-resolution NMR spectroscopy was also carried out on an extract of a biopsy specimen of a third patient with CJD.[2] These NMR results were compared with neuronal cell counts. Marked decreases in the levels of NAA (detected in vivo) were observed at later stages of disease in two patients. However in the early stages of CJD smaller decreases were observed (15% and 27%) in the level of NAA. The NMR spectrum obtained from the biopsy sample taken from a third patient showed little or no reduction in the levels of metabolites. In contrast with the previous case report of Bruhn et al.[1] Graham et al. suggested that there should be little or no change in the levels of metabolites in the early stages of CJD. This suggestion was supported by referring to the known pathophysiology of CJD which indicates that neurons are lost relatively late in the disease process.

1.2 Herpes Simplex Encephalitis (HSE)

Menon et al.[3] showed in an early case report that the ratio of NAA to choline (NAA/Cho) was reduced in an 11 year old boy. This reduction observed at eight weeks after the onset of

symptoms was found to be unaltered on follow-up examination (16 weeks), indicating that there was no progressive neuronal loss. This observation was consistent with the clinical evaluation of the patient.

As part of an MRI study of HSE involving eight patients, Demaerel et al.[4] obtained proton spectra from two patients with HSE. The proton spectra from one of the HSE subjects are shown in Figure 1.

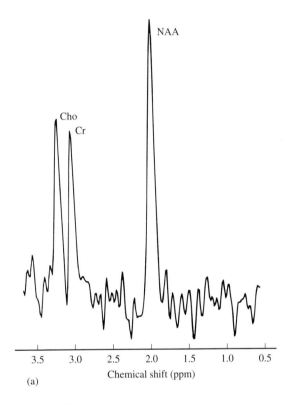

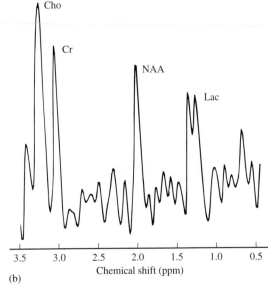

Figure 1 Solvent-suppressed proton spectra obtained using the STEAM sequence at 1.5 T (*TR* = 3000 ms, *TE* = 270 ms) from a 27 cm^3 voxel located at (a) the left and (b) the right temporal lobe. Each spectrum is the sum of 256 acquisitions. Reproduced with permission from Demaerel et al.[4]

For References see p. 1085

Note the reduction in NAA observed in the affected region as compared with the contralateral temporal lobe. There is also a significant amount of lactate observed. The authors interpret the observation of lactate to indicate the presence of an infarct.

1.3 Intracranial Tuberculomas

Gupta et al.[5] have reported the results of localized proton MRS carried out on two patients with intracranial tuberculomas. A large resonance was observed between 0.7–1.6 ppm (see Figure 2) which was assigned to the $(CH_2)_n$ group of saturated fatty acids. The presence of this peak was attributed to the large lipid fraction present in the tubercule bacillus.

1.4 Human Immunodeficiency Virus (HIV) Infection

Menon et al.[6] reported the first spectroscopic study of the brains in patients infected by Human Immunodeficiency Virus (HIV). The authors examined two patients with AIDS [Centers for Disease Control classification (CDC) Group IV] and evidence of focal CNS pathology. A multiple-echo spectroscopy acquisition (MESA) sequence with a *TR* of 2000 ms and *TE* of 270 ms was employed. The spectra were collected from a 64 cm³ voxel in the right parietal region. Although these patients had abnormal MRI, the spectra were taken in areas of normal appearing white matter. In both cases the spectra showed a marked reduction in the ratio of NAA/Cho and NAA/creatine (NAA/Cr) when compared with control subjects. The authors proposed that the spectral changes in these regions which appear normal on MRI may be attributed to other causes but is most likely to be due to primary HIV infection. The authors first demonstrated the ability of proton MRS to detect brain abnormalities in HIV-infected patients at a stage when it is undetectable by imaging and proposed a possible role for the early selection of patients for treatment with antiviral drugs.

The same group examined 11 patients with HIV infection and varying stages of AIDS dementia complex (ADC) (actually known as HIV-1-associated cognitive/motor complex).[7] In this study, patients with pathology seen on MR imaging that could not be directly attributed to HIV infection of the brain were excluded from the study. Spectra were obtained from 27 to 64 cm³ voxels in the parieto–temporal regions using the same sequence as in the previous study. Spectra from patients with moderate (Stage 2) to severe ADC (Stage 3), when compared with spectra from normal volunteers, exhibited significant reductions in NAA/Cr ratio and a tendency to increased Cho/Cr ratio, although this last trend did not reach statistical significance. Spectra from patients with no ADC (Stage 0) or early ADC (Stage 1) were not significantly different from normal volunteers (see Figures 3 and 4).

Many of the patients in this study exhibited abnormalities on MRI, but apart from the presence of atrophy, which was seen in all of the patients with moderate to severe ADC, MRI did not seem to discriminate between patients with and without ADC. The authors concluded that although the NAA/Cr ratio may not be an early or sensitive marker of ADC, it may be relatively specific since all of the patients with significantly low values of this ratio had a clinical diagnosis of ADC.

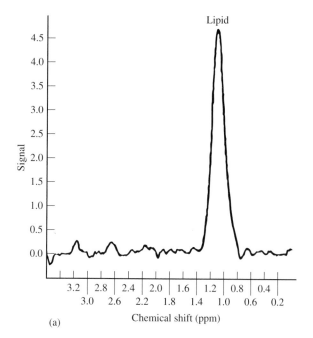

(a)

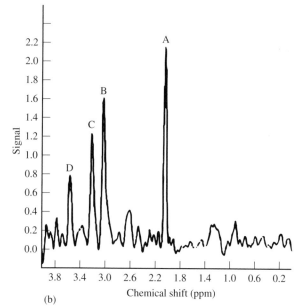

(b)

Figure 2 Solvent-suppressed proton spectra obtained using the STEAM sequence at 2.0 T (*TR* = 5000 ms, *TE* = 20 ms) from an 8 cm³ voxel located (a) in a tuberculoma and (b) in a contralateral normal brain. Each spectrum is the sum of 128 acquisitions. Reproduced with permission from Gupta et al.[5]

Meyerhoff et al.[8] examined 14 HIV seropositive patients, 10 with varying degrees of cognitive impairment and four who were cognitively asymptomatic. Spectra were obtained from nine 2.5 cm³ volumes in the centrum semiovale and the mesial cortex in each patient. Significantly reduced NAA/Cho and NAA/Cr ratios were observed in cognitively impaired subjects versus normal controls, without significant regional differences between the voxels studied. No significant differences were found between groups with cognitive impairment and asymptomatic groups or between asymtomatic and control groups.

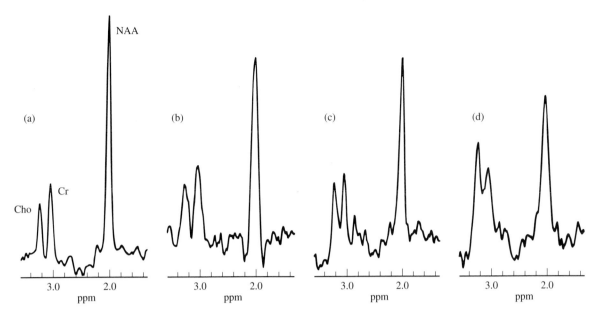

Figure 3 Solvent-suppressed proton spectra obtained from a 27 cm³ voxel located in the parietal lobe from a healthy volunteer and three patients with HIV infection. The spectra were obtained with a double-echo sequence, TR = 2000 ms, TE = 270 ms. Each spectrum is the sum of 128 acquisitions. The spectra are displayed with creatine scaled similarly. The spectra are from (a) a healthy volunteer, (b) a patient with ADC Stage 0, (c) a patient with ADC Stage 1, and (d) a patient with ADC Stage 3. Reproduced with permission from Menon et al.[7]

This study reported diffuse reductions in NAA in individuals with cognitive impairment due to HIV. Contrary to previous studies, most of the patients in this study had normal appearing MRI (80%) suggesting that proton MRS is more sensitive than imaging in assessing the effects of HIV infection on the brain.

Jarvik et al.[9] examined 11 HIV seropositive patients without clinical, radiological, or laboratory evidence of CNS infections other than HIV. Proton spectra were acquired with the stimulated echo acquisition mode (STEAM), TR of 2000 ms, TE of 19 ms, and TM of 10.6 ms. Voxels of 3.4–8 cm³ were chosen to cover areas of abnormal white matter signal intensity if present, or centrum semiovale if the white matter appeared normal at imaging. Analysis of the images showed that there was a significant difference between the patients and control subjects with respect to atrophy, although no significant difference was found between the appearance of the white matter in patients and control subjects. Analysis of the spectra showed that the NAA/Cr ratio was significantly lower and Cho/Cr and marker peak/Cr ratios significantly higher in patients as compared with control subject. The authors calculated an aggregate score that combined these three ratios. This aggregate score proved to be a good discriminator between the patient and control populations (P = 0.001) (see Figure 5).

The aggregate scores were abnormal (>2 SDs from the mean of the control subjects) in 13 out of 15 patient spectra (87%), while 8 out of 11 patients (73%) had abnormal MR images. Moreover, only one out of 10 control spectra (10%) was abnormal while four out of 11 controls (36%) had abnormal imaging. These results suggested that MRS may be more sensitive and specific than MR imaging in detecting CNS involvement in HIV-infected patients.

Chong et al.[10] reported the largest study in which proton MRS was performed in 103 HIV seropositive patients and 23 control subjects. Spectra were collected from an 8 cm³ voxel placed in a normal parieto–occipital region of the brain using a 90°, 180°, 180° spin echo sequence (PRESS), TR of 1600 ms and TE of 135 ms. In the first part of the study, the spectra of HIV seropositive patients were compared and correlated with clinical, inmunologic, and radiologic measures of HIV infection. A significant reduction in the NAA/Cr ratio was seen in patients with late-stage disease (CDC Group IV). The NAA/Cr and NAA/Cho ratios were also reduced in patients with CD4 counts <200 mm⁻³ and in patients with neurologic signs. Significant increases in Cho/Cr ratios were seen in patients with low CD4 counts and abnormal MR images. Reduced NAA

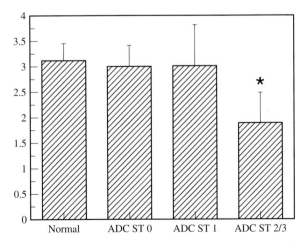

Figure 4 A comparison of the NAA/Cr ratios in normals and patients with different stages of ADC. The bars represent 1 SD. The NAA/Cr ratio is significantly reduced in the ADC Stage 2 and 3 as compared with any of the other groups (P < 0.05). Reproduced with permission from Menon et al.[7]

For References see p. 1085

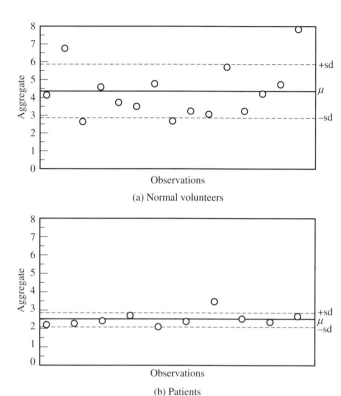

Figure 5 Aggregate spectral scores (Cr/NAA + Cho/Cr + marker/Cr) obtained from normal volunteers and patients with HIV infection. The mean and 1 SD lines are indicated. Note the almost complete separation between the two groups. See Jarvik et al. for experimental details[9]

ratios correlated with diffuse but not focal MR imaging abnormalities. In the second part of the study, the authors evaluated the utility of combining the results of MR imaging and spectroscopy, finding that the combination of both modalities provides closer relationships to clinical and immunologic measures of disease than either modality alone. Moreover, abnormal spectra correlated more with abnormal neurologic finding than did abnormal MR images, suggesting that spectroscopy may be sensitive to changes in cerebral chemistry that are clinically relevant and are not apparent on MR images.

The same authors[11] studied 43 HIV seropositive patients, including 26 who had clinical or radiologic evidence of late-stage disease (CDC Group IV) and 17 in the early stages of infection (CDC Groups II and III). Using the same sequence as described above, spectra were obtained from a 8 cm^3 voxel placed in a normal parieto-occipital region of the brain. When patients grouped by different criteria were compared, a significant reduction in the NAA/(NAA + Cho + Cr) ratio was found in patients in the late stage of the disease as compared with those in early stage disease. This ratio was lower in patients with HIV-1-associated cognitive/motor complex as compared with patients who were neurologically healthy. Also this ratio was decreased in patients with abnormal MR images (diffuse white matter abnormalities) as compared with those with normal appearing white matter. When patients were compared with healthy control subjects, the NAA/(NAA + Cho + Cr) ratios were significantly higher in control subjects than in CDC Group II and III patients and also significantly higher than in all seropositive patients with

normal MR imaging. The authors noted, however, that the control group was younger and so these last results may be influenced by the expected decrease in NAA levels which parallel neuronal loss with age. The authors reported the first follow-up spectroscopic study performed in 15 patients between three and eight months after their initial studies. A significant reduction in the NAA/(NAA + Cho + Cr) ratio was observed at follow-up study when compared with the initial examinations.

The results of all of the MRS studies of HIV infection are summarized in Table 1.

A common finding in all of the studies on HIV infection of the brain is the reduced level of NAA. Although NAA, which resonates at 2.0 ppm, is the most prominent resonance in the brain proton spectrum, its precise biochemical function is not clear. It has been suggested that NAA is involved in the regulation of lipid synthesis, the regulation of protein synthesis, or as a storage buffer for aspartate. Since NAA is largely confined to neurons it has been proposed as a neuronal marker. Hence, decreases in the level of NAA may be interpreted as a sign of neuronal loss or injury. This finding is consistent with the pathologic evidence of neuronal loss reported in HIV infected patients. Another common finding, although not present in all of the studies, is the increase in Cho/Cr ratio. Choline, which resonates at 3.2 ppm, is an important precursor of cell membrane synthesis and is often found to be elevated in tissues that are rapidly regenerating or undergoing membrane disruption. The significance of the elevation in Cho/Cr ratio remains uncertain. A possible interpretation for this increase in HIV infected patients is that the increased Cho may not directly arise from changes in neurons but may result from metabolic alterations in glial or inflammatory cells.[7] An alternative interpretation is that the increase in Cho/Cr ratio combined with the increase in marker/Cr ratio, also observed in these patients, may reflect myelin damage.[9] Choline phosphoglycerides contribute 11.2% of myelin lipids, with phosphatidylcholine being the most abundant. Therefore, as myelin damage occurs, free choline may be released increasing the choline resonance detected by MRS. The 'marker peak' region between 2.1–2.6 ppm may be a combination of nonspecific amino acids and possibly myelin catabolites which may increase with myelin breakdown. Both explanations for the increase in Cho/Cr ratio are consistent with the neuropathologic finding of inflammatory infiltrates and white matter abnormalities in the brain of HIV infected patients. An alternative explanation proposed for the increase in this ratio is the decrease in Cr resonance.[10] This resonance at 3.0 ppm reflects the concentration of the total creatine pool (PCr and Cr). The level of this peak remains stable under many conditions in the brain. The decrease in creatine pool may indicate impairment of cellular metabolism in the brain of these patients and is supported by the decrease in PCr peak observed in phosphorus MRS studies.

Abnormal proton spectra may be found in normal-appearing white matter on imaging, suggesting that MRS may be more sensitive than MRI in detecting CNS involvement in HIV infected patients. Moreover, MRS seems to be a better discriminator between the patient and control populations than MRI, suggesting that MRS may be also more specific. When individuals displaying different manifestations of illness were compared, the abnormalities in the spectra correlated with the presence and the severity of the cognitive impairment and with clinical and immunologic measures of late-stage disease. These

Table 1 A Summary of the Proton MRS Findings in HIV infection

Study	Patients (p) Controls (c)	Stage[a]	n	Clinical evidence of CNS involvement	n	MRI findings[c]	n	MRS sequence and parameters	MRS findings	Groups statistically distinguished by MRS
Menon et al. (1990)	2p/6c	late-stage	2	cogn. imp. + focal signs	1	focal lesions not attributed to HIV	1	MESA *TR* 2000 ms *TE* 270 ms V = 64 cm^3 normal parietal region	↓NAA/Cho ↓NAA/Cr	patients/controls
Menon et al. (1992)	11p/8c	late-stage	10	cogn. imp. (ADC)	7	abnormal	9	MESA *TR* 2000 ms *TE* 270 ms	↓NAA/Cr	ADC Stage 2/3/ controls
		early-stage	1	no cogn. imp.	4	normal	2	V = 27–64 cm^3 parieto-temporal region	↑Cho/Cr (NS)	
Meyerhoff et al. (1993)	14p/7c	late-stage	4	cogn. imp.	10	abnormal		CSI V = 2.5 cm^3 centrum semiovale and mesial cortex	↓NAA/Cr	cogn. impair/ controls
		early-stage	10	no cogn. imp.	4	normal			↓NAA/Cho	
Jarvik et al. (1993)	11p/8c	late-stage	4	cogn. imp.	9	abnormal	8	STEAM *TR* 2000 ms *TE* 19 ms 10.6 ms V = 3.4–8 cm^3 abnormal white matter or centrum semiovale	*TM*↓NAA/Cr ↑Cho/Cr ↑ Marker peak/Cr	patients/ controls
		early-stage	7	no cogn. imp.	2	normal	3			
Chong et al. (1993)[d]	103p/23c	late-stage	70	neurologic signs	19	abnormal	34[e]	PRESS *TR* 1600 ms *TE* 135 ms V = 8 cm^3 normal parieto-occipital region	↓NAA/Cr	late-stage/early-stage; <CD4 count/>CD4 count; neurologic signs/no neurologic signs; abnormal MRI/ normal MRI
		early-stage	22	no neurologic signs	31	normal	36		↓NAA/Cho ↑Cho/Cr	
Chong et al. (1994)[d]	43p/8c	late-stage	26	cogn. imp. (HIV-1-associated cognitive/motor complex)	6	abnormal	9	PRESS *TR* 1600 ms *TE* 135 ms V = 8 cm^3 normal parieto-occipital region	↓NAA/NAA + Cho + Cr	late-stage/early stage; cogn imp/no cogn imp; abnormal MRI/ normal MRI; early-stage/controls; normal MRI; patients/controls
		early-stage	17	no cogn. imp.	14	normal	34			

[a]Stage: late stage refers to CDC Group IV, early stage to CDC Groups II and III. [b]ADC refers to the aids dementia complex, actually known as HIV-1-associated cognitive/motor complex. Cogn. imp. is an abbreviation for cognitive impairment. [c]Abnormal MRI: abnormalities were either GM atrophy or WM signal changes. [d]Not all of the subjects had complete neurological examinations. [e]The remaining patients (33) were excluded because they exhibited focal lesions on MRI.

For References see p. 1085

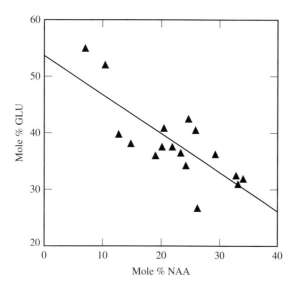

Figure 6 The variation in the mol% of glutamate with the mol% of NAA in the NMR spectra of extracts of brain samples from patients with AD. See Klunk et al. for experimental details[12]

findings suggest that MRS may serve as an indicator of the degree of CNS involvement. Therefore, MRS may serve to monitor the course of disease progression and may have a prognostic role regarding CNS involvement. This modality may provide a sensitive method for the early detection of HIV migration to the CNS. Finally, MRS can be employed in longitudinal studies to monitor the response to therapy and thus may lead to individual optimized treatment effectiveness.

2 DEGENERATIVE DISEASES

2.1 Alzheimer's Disease (AD)

2.1.1 In Vitro Studies of AD

Klunk et al.[12] determined the mole ratios of several amino acid metabolites in perchloric extracts of 12 AD and five control brain samples from either the junction of the superior and middle frontal cortex or the superior temporal cortex. NAA and γ-aminobutyric acid (GABA) were found to be lower in AD. The NAA levels also had a negative correlation to the numbers of senile placques (SP) and neurofibrillary tangles (NFT). As shown in Figure 6, glutamate levels were greater in AD and had an inverse correlation with NAA levels. The authors suggested that these findings applied to in vivo studies and reflected neuronal loss, while the remaining neurons were exposed to the potentially excitatotoxic effect of glutamate.

2.1.2 In Vivo Studies of AD

Miller et al.[13] used the STEAM sequence, at field strength of 1.5 T and *TE* 30 ms, to acquire proton spectra from 10–15 ml voxels in white matter located in the parietal area (WM) and gray matter in the occipital cortex (GM). Summed spectra are shown in Figure 7. They reported an increased myoinositol/creatine (MI/Cr) ratio, which suggested abnormalities in the inositol polyphosphate messenger pathway, and reduced NAA/

Cr in AD compared to controls. The difference in NAA/Cr level was small but statistically significant in both WM and GM voxels. The increase in MI/Cr was more prominent, especially in GM. However, there was no correlation between the severity of the disease and spectroscopic findings.

Shiino et al.[14] studied a patient group (*n* = 9) with primary degenerative dementia (PDD) which included seven patients with probable AD, three normal pressure hydrocephalus (NPH), and healthy controls. The DRY STEAM technique was applied with a *TR* of 2500 ms, *TE* of 19 ms, and *TM* of 5.7 ms. The NAA/Cr was significantly reduced in patients with PDD, with no significant brain atrophy or reduction in regional blood flow detected by SPECT. There was no reduction of NAA/Cr in PH. These authors concluded that, in the appropriate clinical setting, proton MRS was a useful measure for the early detection and study of PDD (see Figure 8).

2.2 Parkinson's Disease (PD)

Shiino et al.[14] examined two patients with PD using proton MRS. Spectra were acquired from a 27 cm³ voxel located in the insular area of the brain. A decrease in the NAA/Cr ratio was observed in the two patients with PD when compared with control subjects. Even though the mean age in the control group was lower, the authors observed that there were no age-related changes in the mean area ratio of NAA/Cr in this group. Both PD patients examined in this study exhibited marked atrophy on MRI, and since the spectra were obtained from voxels placed in the insular area these results should be interpreted with caution.

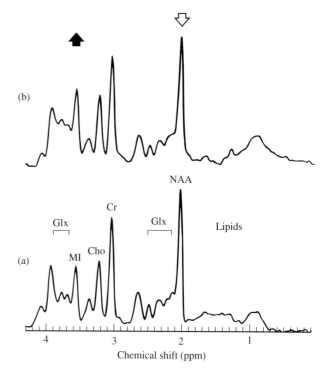

Figure 7 Representative summed solvent-suppressed proton spectra obtained using the STEAM sequence (*TR* = 1500 ms, *TE* = 30 ms) from (a) normal volunteers (*n* = 8) and (b) patients with AD (*n* = 8). Reproduced with permission from Miller et al.[13]

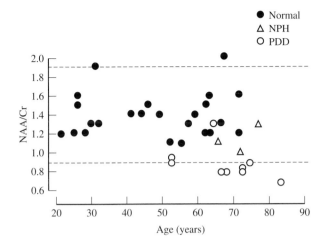

Figure 8 A correlation plot of NAA/Cr versus age. Normals are designated by filled circles. Patients with primary degenerative dementia (PDD) are designated with open circles and patients with normal pressure hydrocephalus are designated by triangles. See Shiino et al.[14] for experimental details

2.3 Huntington's Disease (HD)

Jenkins et al.[15] reported the results of a proton MRS study in a series of HD patients. Sixteen patients with clinical signs of definite HD and confirmed family history and two individuals without neurologic affectation but with linked DNA markers, indicating a high probability to have inherited the HD gene, were examined. Proton spectroscopy was performed using a STEAM sequence with *TR* of 2000 ms, *TE* of 272 ms, and *TM* of 10 ms. Spectra were collected from a 15.6 cm³ voxel placed over the visual cortex and from a 5.4–20 cm³ voxel placed in basal ganglia. A 5 inch surface coil was used to study the occipital cortex and the quadrature head coil was used for studies of the basal ganglia. Elevated lactate levels were observed in the occipital cortex in all of the symptomatic patients when compared with normal controls. The lactate level correlated with the duration of illness. Lactate levels were normal in the two asymptomatic subjects with high probability of having the HD gene. In the basal ganglia, the levels of NAA were decreased and the levels of Cho elevated relative to creatine. Several patients also showed elevated lactate levels in the basal ganglia.

The authors suggested that the increase in lactate may be explained by a defect in oxidative phosphorylation in HD. This possibility is supported by biochemical studies that have reported reduced mitochondrial enzyme activity in HD patients, and ultrastructural studies that have shown abnormalities in the mitochondria of these patients. The reduced NAA and increased Cho were interpreted in these patients as resulting from neuronal loss and gliosis. The authors suggested that lactate likely precedes neuronal death and that this may be the explanation for the more variable lactate elevation in the basal ganglia where evidence of more neuronal loss and gliosis was found. On this basis, they proposed that the elevated lactate may provide a simple marker to monitor the progression of the disease and possible therapies for HD patients.

High-resolution proton MRS was employed by Nicoli et al.[16] to study CSF and serum metabolic samples obtained from patients with HD. Serum and CSF samples were collected from 11 patients suffering HD and 12 reference patients suffering miscellaneous neurological diseases. A significant increase in the pyruvate concentration in CSF was found in patients with HD. This finding may be explained by the decrease of pyruvate-dehydrogenase and Kreb's cycle enzyme activity observed in HD patients. This observation is consistent, as was the finding of elevated lactate reported by Jenkins et al.,[15] with the hypothesis of the existence of a defect in oxidative phosphorylation present in this disease.

3 RELATED ARTICLES

Brain MRS of Human Subjects; Brain Neoplasms in Humans Studied by Phosphorus-31 NMR Spectroscopy; Localization and Registration Issues Important for Serial MRS Studies of Focal Brain Lesions; Systemically Induced Encephalopathies: Newer Clinical Applications of MRS.

4 REFERENCES

1. H. Bruhn, T. Weber, V. Thorwirth, and J. Frahm, *Lancet*, 1991, **337**, 1610.
2. G. D. Graham, O. A. Petroff, A. M. Blamire, G. Rajkowska, P. Goldman-Rakic, and J. W. Prichard, *Neurology*, 1993, **43**, 2065.
3. D. K. Menon, J. Sargentoni, C. J. Peden, J. D. Bell, I. J. Cox, G. A. Coutts, C. Baudouin, and C. G. Newman, *J. Comput. Assist. Tomogr.*, 1990, **14**, 449.
4. Ph. Demaerel, G. Wilms, W. Robberecht, K. Johannik, P. Van Hecke, H. Carlton, and A. L. Baert, *Neuroradiology*, 1992, **34**, 490.
5. R. K. Gupta, R. Pandey, E. M. Kahn, P. Mittal, R. B. Gujral, and D. K. Chhabra, *Magn. Reson. Imag.* 1993, **11**, 443.
6. D. K. Menon, C. J. Baudouin, D. Tomlinson, and C. Hoyle, *J. Comput. Assist. Tomogr.*, 1990, **14**, 882.
7. D. K. Menon, J. G. Ainsworth, I. J. Cox, R. C. Coker, J. Sargentoni, G. A. Coutts, C. J. Baudouin, A. E. Kocsis, and J. R. Harris, *J. Comput. Assist. Tomogr.*, 1992, **16**, 538.
8. D. J. Meyerhoff, S. Mackay, L. Bachman, N. Poole, W. P. Dillon, M. W. Weiner, and G. Fein, *Neurology*, 1993, **43**, 509.
9. J. G. Jarvik, R. E. Lenkinski, R. I. Grossman, J. M. Gomori, M. D. Schnall, and I. Frank, *Radiology*, 1993, **186**, 739.
10. W. K. Chong, B. Sweeney, I. D. Wilkinson, M. Paley, M. A. Hall-Craggs, B. E. Kendall, J. K. Shepard, M. Beecham, R. F. Miller, I. V. D. Weller, S. P. Newman, and M. J. G. Harrison, *Radiology*, 1993, **188**, 119.
11. W. K. Chong, M. Paley, I. D. Wilkinson, M. A. Hall-Craggs, B. Sweeney, M. J. G. Harrison, R. F. Miller, and B. E. Kendall, *AJNR*, 1994, **15**, 21.
12. W. E. Klunk, K. Panchalingam, J. Moossy, R. J. McClure, J. W. Pettigrew, *Neurology*, 1992, **42**, 1578.
13. B. L. Miller, R. A. Moats, T. Shonk, T. Ernst, S. Woolley, and B. D. Ross, *Radiology*, 1993, **187**, 433.
14. A. Shiino, M. Matsuda, S. Morikawa, T. Inubushi, I. Akiguchi, and J. Handa, *Surg. Neurol.*, 1993, **39**, 143.
15. B. G. Jenkins, W. J. Koroshetz, M. F. Beal, and B. R. Rosen, *Neurology*, 1993, **43**, 2689.
16. F. Nicoli, J. Vion-Dury, J. M. Maloteaux, C. Delwaide, S. Confort-Gouny, M. Sciaky, and P. Cozzone, *Neurosci. Lett.*, 1993, **154**, 47.

Acknowledgements

Supported in part by NIH grant NS 31464 (Robert E. Lenkinski). Dr. López-Villegas would like to acknowledge support by the Ministry of Health of Spain (FIS: BAE 94/5033). Dr. Tunlayadechanont would like to acknowledge support from Ramathibodi Hospital in Thailand.

Biographical Sketches

Robert E. Lenkinski. *b* 1947. B.Sc., 1969, University of Toronto, Ph.D., 1973, University of Houston. Postdoctoral Fellow, Weizmann Institute of Science, Isotope Department, Rehovot, Israel, 1973–75. Faculty at University of Houston, 1975–76; Comprehensive Cancer Center, University of Alabama in Birmingham, 1976–80; Department of Chemistry, University of Guelph, Ontario, Canada, 1980–86; University of Pennsylvania, 1986–present. Approx. 92 publications. Research interests include clinical MRS particularly in the diagnosis and staging of disease.

Dolores López-Villegas. *b* 1964. M.D., 1988, University of Barcelona, Medical School. Intern in Medicine, Hopital Clinic, University of Barcelona, 1988. Resident in Medicine, Hopital de la Santa Creu i Sant Pau, Autonomous University of Barcelona, 1989. Resident in Neurology in the same hospital, 1990–92. Research Fellow in MRS (supported by the Ministry of Health of Spain), University of Pennsylvania, Department of Radiology, 1993–present. Approx. 10 publications. Research interests include application of MRS to investigation of neurological diseases.

Supoch Tunlayadechanont. *b* 1962. M.D., 1986, Ramathibodi Hospital, Mahidol University. Residency training in Neurology, Ramathibodi Hospital, 1989–92; Graduate Diplomate in Clinical Science, 1990, Mahidol University; Diplomate the Thai Board of Neurology, 1992, Thai Medical Council. Training in MRI and MRS, Institute of Magnetic Resonance, University of Innsbruck, Austria, 1993. Research Fellow in MRS, University of Pennsylvania, 1993–present. Approx. 10 publications. Research interests include clinical MRS particularly in neurological disease.

Brain MRS of Infants and Children

Ernest B. Cady

University College London Hospitals, UK

and

E. Osmund R. Reynolds

University College London Medical School, UK

1 INTRODUCTION

Magnetic resonance spectroscopy (MRS) was first introduced into pediatrics in the early 1980s as a method for the

For list of General Abbreviations see end-papers

noninvasive investigation of perinatal brain injury.[1] This injury is often due to cerebral periventricular hemorrhage (PVH) in preterm infants (born before 37 weeks of gestation) or to hypoxia-ischemia in both preterm and term infants. It is responsible for permanent neurodevelopmental impairments in about 10% of disabled children of school age. In the 1970s, X-ray computerized tomography (CT) and, more especially, ultrasound imaging using portable equipment at the cotside, were found to be useful for investigating the incidence, pathogenesis, and prognostic significance of PVH, but long-term follow-up studies indicated that hypoxia-ischemia, often causing periventricular leukomalacia, was a more important cause than PVH of permanent disabilities in preterm infants.[2] In term infants, hypoxia-ischemia was the mechanism by which 'birth asphyxia' (critically impaired gas exchange during labor) damaged the brain. Overall, hypoxia-ischemia turned out to be the most prevalent cause of serious perinatal brain injury, so there was an obvious need for noninvasive investigation of cerebral oxidative metabolism, and hence of the pathogenesis and evolution of hypoxic-ischemic brain injury. MRS was thought to provide a suitable, noninvasive, method for obtaining biochemical information from the otherwise inaccessible neonatal brain.[3-5]

In the early 1980s, ^{31}P MRS had already been used to investigate oxidative metabolism in human skeletal muscle in vivo and the first in vivo spectra had been obtained from mammalian brain.[6] However, studies of newborn infants had to await the development of superconducting magnets with a sufficiently wide bore (approximately 20 cm) and a high enough field strength for spectroscopic work. These magnets (intended primarily for studies of human limb muscles), became available in the early 1980s, and investigations of the brain in newborn infants started shortly thereafter.[1,7] The first in vivo spectra from the human brain were obtained from a preterm baby at University College London in 1982 using the ^{31}P nucleus.[1] It rapidly became clear that metabolites involved in energy metabolism [e.g. adenosine triphosphate (ATP), phosphocreatine (PCr), and inorganic phosphate (Pi)] as well as those related to more fixed cellular components, namely phosphomonoesters (PMEs) and phosphodiesters (PDEs), were easily detected. Intracellular pH (pH$_i$) could also be measured. Obvious spectral abnormalities were soon found in the brains of infants with hypoxic ischemic injury.[1,7] ^{31}P spectra have also been obtained from the brains of older children and developmental changes have been elucidated. In addition, ^{1}H spectra have demonstrated products of anerobic glycolysis such as lactate (Lac), various amino acids (e.g. glutamate and glutamine), and other brain metabolites, including those containing choline (Cho; related to membrane metabolism) and *N*-acetylaspartate (NAA; a putative neuronal marker). Localization of spectra to specific regions of the brain was initially crude, relying solely on the sensitive volume of a surface coil. Recently, techniques have been developed that enable spectra to be acquired from well defined volumes of tissue, so deeper and focal lesions can be investigated.

The purposes of this article are to summarize the methods used for MRS of the brain in infants and children, to describe what has been learnt about normal brain development, and then to consider the role of MRS, particularly in the investigation of hypoxic ischemic brain injury.

2   PATIENT MANAGEMENT

MRS studies of the brains of infants and children are safe provided that current recommendations are observed,[8] though the requirements for physiological support and monitoring are often much greater than for adults.

Newborn infants, who may be in an unstable condition, have to be conveyed and studied in a specially designed transport incubator incorporating a nonmagnetic, cylindrical pod that encloses the baby and is suitable for insertion into the magnet. Conventional facilities for temperature control, monitoring, and mechanical ventilation are provided.[9,10] Monitoring of heart rate [by electrocardiogram (ECG) or ultrasound Doppler flow], blood pressure, arterial oxygen saturation (by pulse oximetry), end-tidal CO_2, transcutaneous pO_2 and pCO_2, and body temperature (by skin and/or rectal sensors) should be available. Intravenous infusions may have to be provided. Newborn infants can often be studied satisfactorily whilst naturally asleep following a feed. However, sedation with, for example, chloral hydrate is often required for studies of older children. Especially for the newborn, the head must be gently immobilized during studies using, for example, a polythene bag containing expanded polystyrene beads and which can be evacuated. Clinical apparatus must be carefully designed to take into account the large range of size of infants and children. It is important that the data acquisition protocol is such that useful information can be obtained even if small movements take place.

Interactions between magnetic fields (both static and transient) and the pulsed rf, and the monitoring equipment needed for patient care, require consideration. Some sensors contain ferromagnetic components that can impair the homogeneity of the magnetic field. Many manufacturers of monitoring equipment can supply special magnetic resonance compatible devices. Alternatively, it may be possible to position ferromagnetic components so that their effects are unimportant. Some sensors may pick up rf interference, thereby decreasing the signal-to-noise ratio and transient rf and pulsed magnetic fields may overload ECG and other devices. The accidental formation of conductive loops by sensor cables must be avoided. This is in order to eliminate the risk of burns that could be caused if currents are induced by transient rf and pulsed gradient fields.[11] Microprocessors, included in many newer monitors, emit rf interference and thus special rf Faraday shields and filters are required. Also, the spectrometer must be safe in the presence of the high inspired oxygen concentrations needed by some patients. Probes (rf) must be tested to ensure that transmitter pulses do not cause arcing; they must be in sealed containers which, as a further precaution, can be flushed with a relatively inert gas such as N_2.

3   DATA ACQUISITION

Using surface coils with diameters in the range 5–7 cm and without further localization techniques, good quality ^{31}P spectra can be obtained from 10–50 mL of the cerebral cortex of the newborn infant. These spectra have very little contamination from other tissues since skin and fat contain little ^{31}P, cranial muscle is sparse, and the cranial bones are poorly mineralized and have relatively immobile ^{31}P nuclei. For basic studies of the normal brain and abnormalities associated with nonfocal brain injury, valuable information has been obtained without better spectroscopic localization. However, the sensitive volume provided solely by a surface coil is not well defined. For investigations of focal metabolism, particularly deep in the brain, and in older infants and children who have better developed cranial musculature, and also for ^{1}H studies, in which fat contamination can be a problem, spectroscopic localization techniques are essential. Many volume localization methods exist which have been used successfully for studies of infants and children. These techniques all employ pulsed magnetic field gradients to obtain low-contamination spectra from a well defined voxel positioned with reference to MRI scout images. In these multipulse techniques, rf power is higher than for 'single pulse' surface coil methods and consequently greater attention to safety is mandatory.

Image-selected in vivo spectroscopy (ISIS; see *Single Voxel Whole Body Phosphorus MRS*)[12] has been used for ^{31}P studies of the developing brain[13] and, in order to combine MRI with localized ^{31}P MRS, double-tunable probes dedicated to the examination of children's brains have been developed.[14] For neonatal applications, standard probeheads are usually too large and give poor signal-to-noise ratio with the tiny voxels (about 1 mL) often necessary when studying the smaller anatomy. Localized ^{1}H spectra from infants and older children have been acquired using both the point resolved spectroscopy (PRESS) and stimulated echo amplitude mode (STEAM) techniques (see *Single Voxel Localized Proton NMR Spectroscopy of Human Brain In Vivo*).[15] MR spectroscopic imaging[16] (see *Chemical Shift Imaging*) has not been as widely applied as the single voxel methods. Absolute concentrations (i.e. in millimoles per kilogram wet weight, or millimoles per liter) of ^{31}P metabolites in developing brain have been estimated both by PRESS with brain water as an internal reference,[17] and by ISIS with an external reference.[18] The absolute concentrations of ^{1}H metabolites are more difficult to estimate, due largely to the dependences of peak areas on T_2 relaxation and, in the case of multiplets, on phase modulation. However, some information about developmental changes has been acquired using an external reference[19] or brain water as an internal reference.[17,20]

4   NORMAL BRAIN DEVELOPMENT IN INFANTS AND CHILDREN

Information about the relative and absolute concentrations of metabolites detectable by ^{31}P and ^{1}H MRS in the developing human brain has been acquired at several centers worldwide with good agreement of results.

4.1   The Neonatal Brain

The major ^{31}P energy related resonances in spectra from skeletal muscle had been securely assigned when the first studies of the neonatal brain were carried out. Peaks attributable to γ-, α-, and β-nucleotide triphosphate (NTP), PCr, and Pi, all of which are involved in energy metabolism, were also detected in the brains of newborn infants (see Figure 1). Magnesium complexed ATP is the main contributor to cerebral

NTP, with guanosine, uridine, and cytidine triphosphates making up about a quarter of the remaining NTP pool;[21] the γ- and α-NTP resonances have small contributions from nucleotide diphosphates; and an often unresolved nicotinamide adenine dinucleotide (NAD + NADH) quartet resonates on the right-hand shoulder of α-NTP.[22] Within the physiological range, the PCr resonant frequency is independent of both pH_i (it only shifts significantly at very low pH) and metal ion concentration; it is often used as a chemical shift reference (assigned as 0 ppm).

Only a single Pi resonance (peak 2) is visible in the ^{31}P spectra shown in Figure 1. Its chemical shift relative to PCr (δ_{Pi}) depends strongly on the concentration ratio of its acidic and basic species ($pK_a \approx 6.8$) and hence pH_i can be estimated from the Henderson–Hasselbalch equation:[23]

$$pH_i = 6.77 + \log_{10}[(\delta_{Pi} - 3.29)/(5.68 - \delta_{Pi})] \quad (1)$$

Estimation of pH_i is accurate to about 0.1 pH units. At high field strengths ($\geqslant 4.7$ T), the Pi peak can often be resolved into

extra- and intracellular[24] and perhaps also mitochondrial components,[25] implying the presence of compartments of different pH.

When the first ^{31}P spectra were acquired from the neonatal brain, problems arose with the interpretation and assignment of prominent resonances other than those particularly involved in energy metabolism. The PME and the moderately broad, unresolved PDE peaks were both surprisingly large, and surface coil spectra were superimposed on a broad hump [see Figure 1(c)]. Chromatography and ^{31}P and ^{1}H MRS of extracts of neonatal mammalian brain have identified phosphoethanolamine (PEt) as the main contributor to the PME peak[26] which also includes phosphocholine (PC) and several smaller resonances, due for example to phosphoserine and phosphoinositol.[22] These PMEs are involved in phospholipid metabolism, including the synthesis of myelin.

The moderately broad resonance at about 2.9 ppm [see Figures 1(a) and 1(b), peak 3] is largely attributable to cross linked PDEs in membrane bilayer phospholipids and also

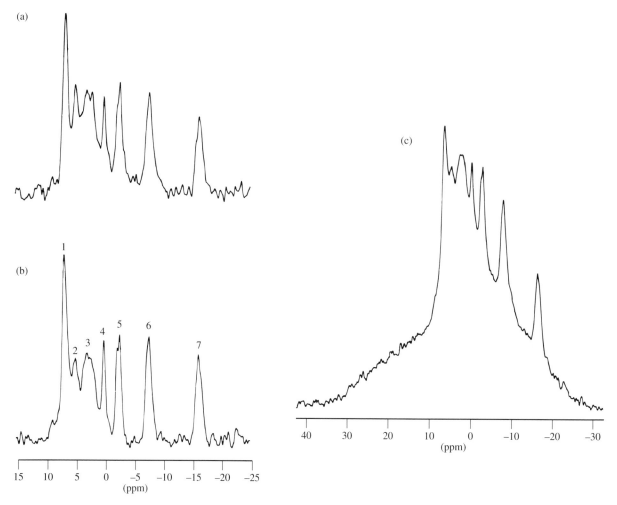

Figure 1 ^{31}P spectra acquired from the temporo-parietal cortex of (a) a preterm infant of gestational plus postnatal age (GPA) 31 weeks; (b) an infant born and studied at full term. Pi is relatively smaller and PCr larger in spectrum (b) as compared with spectrum (a). Spectrum (c) is from the same infant as (b) but without removal by postacquisition processing of the broad underlying hump which was due to relatively immobile ^{31}P in bone and membrane phospholipids. Peak identifications: (1) PME; (2) Pi; (3) PDE; (4) PCr; (5) γ-NTP; (6) α-NTP; (7) β-NTP. (Reproduced by permission of Plenum Press from E. B. Cady, 'Clinical Magnetic Resonance Spectroscopy', New York, 1990, p. 86)

For list of General Abbreviations see end-papers

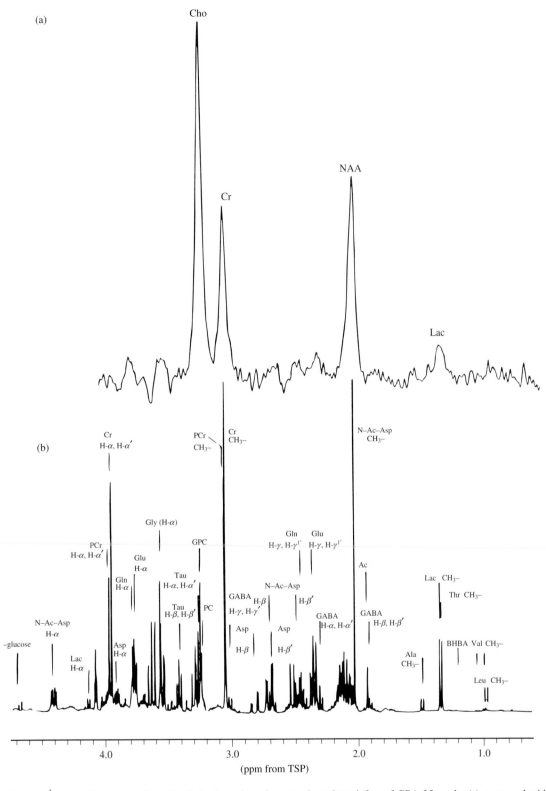

Figure 2 An in vivo ¹H PRESS spectrum from the thalamic region of a normal newborn infant of GPA 35 weeks (a) compared with an in vitro spectrum of an extract of adult rat brain (b). Spectrum (a) was collected at 100 MHz and the acquisition conditions were: *TE* 270 ms; repetition time (*TR*) 2 s, 128 averaged echoes; and an 8 mL voxel. Resonance identifications: Cho, choline containing compounds; Cr, creatine + phosphocreatine; NAA, *N*-acetylaspartate; Lac, lactate. Spectrum (b) was obtained at 360 MHz with *TR* 3 s and a 90° flip angle (water-suppressed single pulse sequence). [Spectrum (a) was obtained in collaboration with A. Lorek, J. Penrice, J. S. Wyatt, R. Aldridge, and M. Wylezinska. Spectrum (b) is reproduced by permission of Elsevier Science Publishers from S. Cerdan, R. Parrilla, J. Santoro, and M. Rico, *FEBS Lett.*, 1985, **187**, 167]

For References see p. 1099

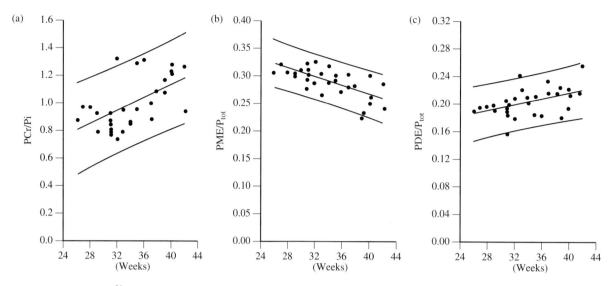

Figure 3 Relations between ^{31}P metabolite concentration ratios in the temporo-parietal cortex and GPA in 30 infants of appropriate weight for gestational age. Regression lines and 95% confidence limits are shown. (Reproduced by permission of International Pediatric Research Foundation Inc. from D. Azzopardi, J. S. Wyatt, P. A. Hamilton, E. B. Cady, D. T. Delpy, P. L. Hope, and E. O. R. Reynolds, *Pediatr. Res.*, 1989, **25**, 440)

to a small pool of mobile phospholipid breakdown products [glycerolphosphorylethanolamine (GPE) and glycerolphosphorylcholine (GPC)]. The membrane bilayer phospholipids exhibit chemical shift anisotropy leading to a reduced signal-to-noise ratio at higher field strengths (e.g. 4.7–7 T). The very broad underlying signal, shown in Figure 1(c), is due to more rigidly bound ^{31}P nuclei mainly in membrane phospholipids and myelin.

Well resolved ^{1}H spectra can be acquired from the brains of newborn infants if long echo times (*TE*) are used (e.g. 270 ms). In Figure 2 an in vivo spectrum from a normal newborn infant is compared with an in vitro high-resolution spectrum of adult rat brain extract.[27] The major in vivo peaks originate from Cho, Cr, and NAA; a smaller but clearly detectable Lac methyl resonance is also present. In the extract spectrum, Cho is seen to include GPC, PC, and taurine, and NAA is adjacent to glutamate, glutamine, and γ-aminobutyric acid (GABA). At shorter *TE* times (e.g. about 25 ms), T_2 relaxation and phase modulation are less important and numerous peaks are easily detected; however, ^{1}H spectra become more difficult to analyze. This problem is further complicated by the presence of many broad overlapping resonances from fatty acids (from extravoxel contamination and perhaps also free fatty acids) and macromolecules (e.g. proteins). Unlike ^{31}P metabolites, whose biochemical pathways are largely known, many ^{1}H resonances other than Lac still pose problems of interpretation.[28] For

Table 1 Relationships between Metabolite Concentration Ratios and pH_i, in the temporo-parietal cortex and Gestational plus Postnatal age in 30 Infants of Appropriate Weight for Gestational Age[a]

	x	c	r	p	28 weeks	42 weeks
PCr/Pi	0.024	0.190	0.58	<0.001	0.85 ± 0.33	1.18 ± 0.33
NTP/P_{tot}	0.001	0.075	0.19	NS	0.09 ± 0.03	0.10 ± 0.03
PCr/P_{tot}	0.001	0.051	0.56	<0.01	0.09 ± 0.02	0.11 ± 0.02
Pi/P_{tot}	−0.001	0.128	−0.40	<0.05	0.10 ± 0.02	0.09 ± 0.02
PME/P_{tot}	−0.004	0.429	−0.68	<0.001	0.32 ± 0.04	0.26 ± 0.04
PDE/P_{tot}	0.002	0.131	0.47	<0.01	0.19 ± 0.04	0.22 ± 0.04
PCr/NTP	0.009	0.717	0.20	NS	0.79 ± 0.42	1.09 ± 0.42
Pi/NTP	−0.013	1.482	−0.24	NS	1.13 ± 0.50	0.95 ± 0.50
PME/NTP	−0.050	4.887	−0.37	<0.05	3.49 ± 1.23	2.79 ± 1.25
PDE/NTP	0.016	1.658	0.15	NS	2.11 ± 1.03	2.34 ± 1.05
Hump/NTP[b]	1.011	−7.790	0.49	<0.05	20.5 ± 20.5	34.7 ± 21.6
pH_i[c]	−0.003	7.233	0.12	NS	7.14 ± 0.28	7.09 ± 0.28

[a]x, Slope; c, ordinate intercept from the regression; p, significance of the correlation coefficient r. Data for 28 and 42 weeks are mean values ±95% confidence intervals from the regressions. [b]n = 24. [c]n = 23; values for the remaining infants were not calculated, because spectra were not well resolved. (Reproduced by permission of International Pediatric Research Foundation Inc. from D. Azzopardi, J. S. Wyatt, P. A. Hamilton, E. B. Cady, D. T. Delpy, P. L. Hope, and E. O. R. Reynolds, *Pediatr. Res.*, 1989, **25**, 440)

For list of General Abbreviations see end-papers

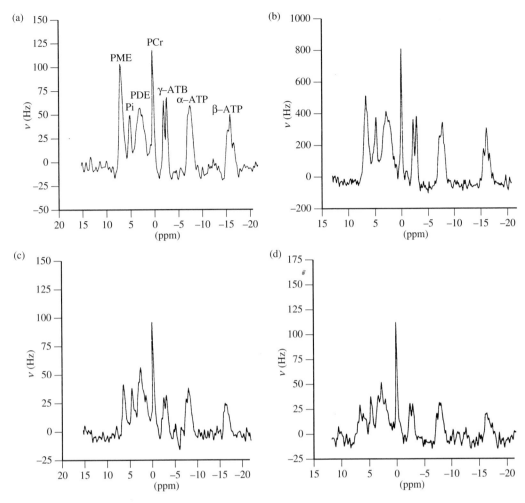

Figure 4 ^{31}P ISIS spectra from the paraventricular region of the human brain at postnatal ages of 1 month (a), 4 months (b), 30 months (c), and 15 years (d). The spectra were obtained at 1.5 T using a sensitive volume varying from 54 mL in infants to 72 mL in older children. *TR* was 3.75 s and 256 FIDs were averaged. Significant developmental changes occurred within the first few years of life. (Reproduced by permission of The Radiological Society of North America from M. S. Van der Knaap, J. van der Grond, P. C. van Rijen, J. A. J. Faber, J. Valk, and K. Willemse, *Radiology*, 1990, **176**, 509)

example, although NAA is widely regarded as an intraneuronal marker, its biochemical role remains unclear and NAA has also been found in oligodendrocyte progenitor cells.[29]

4.2 Changes with Age

Age-dependent changes in the concentration ratios and absolute concentrations of ^{31}P and ^{1}H metabolites have been defined in several studies. A major reason for doing these investigations has been to establish normal age-related values so that metabolic abnormalities may be securely identified. Attention has focused on the rapid changes taking place in the last 3 months of gestation as determined in preterm and term infants studied during the first days of life. Investigations of development during later childhood have also been undertaken. Figure 3 and Table 1 give ^{31}P surface coil data from the temporo-parietal cortex of 30 infants of appropriate weight for gestational age (AGA) studied at a median postnatal age of 4 days.[30] It can be seen that, with increasing gestational plus postnatal age (GPA), [PCr]/[Pi] and [PDE]/[total mobile phosphorus (P_{tot})] increased, whereas [PME]/[P_{tot}] fell. These alterations, together with an increase in [PCr]/[NTP], continue during the first months of life, but then the changes slow down as adulthood is approached (see Figures 4 and 5).[13] Adult values are attained by about 4 years. Values from normal, small-for-gestational-age (SGA; i.e. with birthweights below the 3rd centile for gestation) infants were similar to those from AGA infants.[30] [PCr]/[Pi] is directly related to the phosphorylation potential and to the free energy change of hydrolysis of ATP. The increase in [PCr]/[Pi] with age may be attributable to a greater need for an energy reserve in the mature brain and perhaps also to less efficient oxidative phosphorylation in the immature one. In the newborn human infant, [PCr]/[Pi] is similar to that in other altricious species, such as the newborn rat,[31] but smaller than in precocious animals such as the lamb.[32]

The large amplitude of the PME resonance in the neonatal period is due to high [PEt].[33,34] This metabolite may have an important role as a phospholipid precursor related to increased synthesis of membrane phospholipids and myelin at this time.

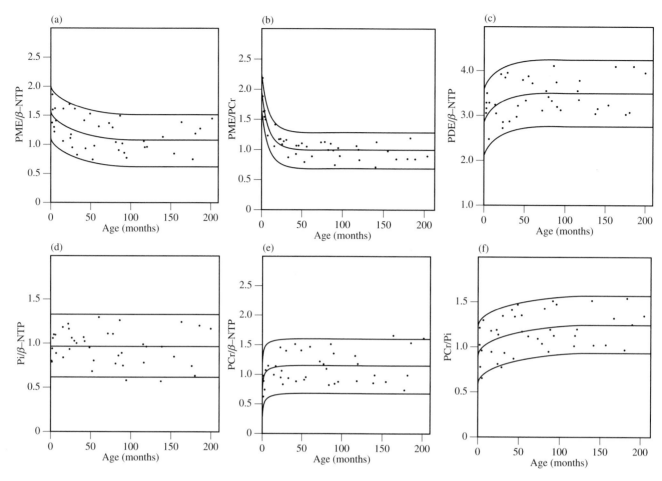

Figure 5 Relationships between age and PME/β-NTP (a), PME/PCr (b), PDE/β-NTP (c), Pi/β-NTP (d), PCr/β-NTP (e), and PCr/Pi (f) in the paraventricular region. Regression lines and 2.5 and 97.5 percentile lines are shown. (Reproduced by permission of The Radiological Society of North America from M. S. Van der Knaap, J. van der Grond, P. C. van Rijen, J. A. J. Faber, J. Valk, and K. Willemse, *Radiology*, 1990, **176**, 509)

Of the ^{31}P resonances only PME shows a change in relaxation with development; T_1 decreases from a neonatal value of about 1.8 ms to about 1.2 ms in adulthood.[18] This fall is probably due to alterations in the constituents of the peak, notably the reduction in PEt. No change has been detected in pH$_i$ during maturation (pH$_i \simeq 7.10$).[13,30]

Measurements of the absolute concentrations of ^{31}P metabolites in the newborn human brain obtained by PRESS[17] and ISIS[18] techniques have shown disagreements. For example, with the ISIS technique, the apparent [NTP] was much less than estimated with the PRESS method. These discrepancies could be due to a number of factors, including the acquisition of data from different parts of the brain. Furthermore, ISIS and PRESS estimates of [PDE] may be affected by relaxation during the delays and the pulses of the acquisition sequence; the cross-linked membrane bilayer fraction of PDE has a T_2 of about 3 ms.[35]

As with ^{31}P metabolites, the relative peak areas[13,19,36] and absolute concentrations[17,19] of ^{1}H metabolites change with increasing age (see Figures 6–8). NAA/Cr increases and Cho/Cr decreases up to about 3 years when the rate of change declines. [NAA] and [Cr] both appear to increase with age, whereas both [Cho] and [myoinositol] decrease.[17,19,37] The

relatively low NAA signal seen in younger subjects and the increase with age are supported by biochemical studies, indicating that [NAA] increases greatly during mammalian brain development.[38] Since NAA is mainly located in neurons, this increase may reflect neuronal maturation. The Cho peak contains signals from PCh, GPC and other metabolites; changes in its amplitude probably relate to alterations in the PME and PDE peaks in ^{31}P spectra. Recent evidence suggests that Lac is detectable in the brains of normal newborn infants [see Figure 2(a)] and is higher in preterm and SGA infants. The implied greater reliance on anerobic glycolysis in preterm and SGA infants may relate to their reduced values for [PCr]/[Pi].[17,36]

5 CEREBRAL PATHOLOGY

^{31}P MRS has proved particularly valuable for the investigation of hypoxic-ischemic brain injury, especially in newborn infants. Most of the data have been acquired from infants who have suffered birth asphyxia, but similar results have been obtained in those with periventricular leukomalacia and other forms of hypoxic ischemic injury. More recently, ^{1}H MRS has been used for monitoring abnormal glycolytic pathway activity

For list of General Abbreviations see end-papers

(a)

(b)

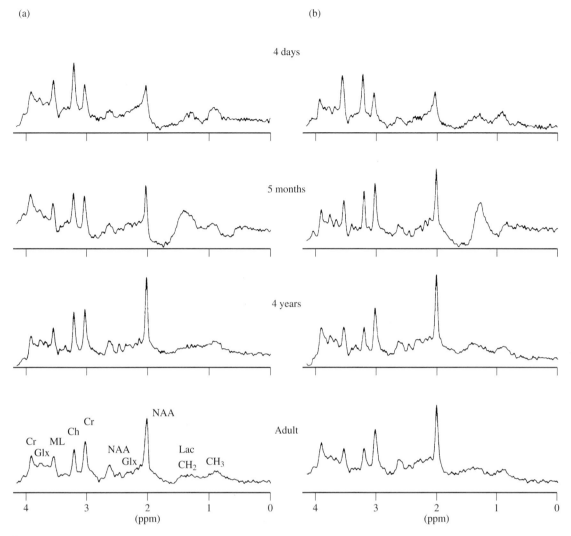

Figure 6 ^{1}H STEAM spectra obtained from (a) the parietooccipital cortex (predominantly white matter) and (b) the midline occipital cortex (mostly gray matter) of human brain at the ages of 4 days, 5 months, 4 years and from an adult. Amplitude scales have been adjusted so that absolute differences depending on both development and location can be visualized directly. Acquisition conditions: *TR* 2.87 s; mixing time (*TM*) 13 ms; *TE* 272 ms. Peak assignments: Ch, choline containing compounds; Cr, creatine + phosphocreatine; NAA, *N*-acetylaspartate; Glx, glutamate and glutamine; MI, myoinositol; Lac, lactate; CH$_2$/CH$_3$, lipid. (Reproduced by permission of Williams and Wilkins from R. Kreis, T. Ernst, and B. D. Ross, *Magn. Reson. Med.*, 1993, **30**, 424)

(via increased Lac) and neuronal degeneration (via decreased NAA).

5.1   Perinatal-Hypoxia Ischemia

Studies of babies with suspected cerebral hypoxic-ischemic injury were initiated on the premise that in conditions where the supply of oxygen to the brain was curtailed, or the mitochondrial mechanisms for its consumption were damaged, [ATP] would tend to fall. It was expected that the decrease in [ATP] would initially be very small because, unless [PCr] is grossly depleted, [ATP] is buffered by the creatine kinase reaction[39] which maintains it close to normal, but at the expense of

a fall in [PCr] and a reciprocal rise in [Pi]. It was hypothesized that [PCr]/[Pi] in the brains of infants with hypoxic ischemic injury would be reduced and that, in extreme circumstances, [ATP] might also be low. Surprisingly, studies of severely birth-asphyxiated babies have shown that cerebral ^{31}P spectra were often normal on the first day of life.[40] Subsequently [PCr]/[Pi] gradually fell, with a nadir at 2–4 days of age (see Figure 9 and Table 2), in spite of normal arterial oxygen saturation, blood pressure, and blood glucose. In the most affected infants, [NTP] then fell and death ensued. In contrast to the profound intracellular acidosis seen in experimental hypoxia-ischemia,[32] pH$_i$ tended to be slightly alkaline. (It is of interest that PCr and NTP were almost undetectable in two infants with inborn metabolic errors, propionic acidemia and arginosuccinic aciduria.[41] However, cerebral pH$_i$ was profoundly acidic in

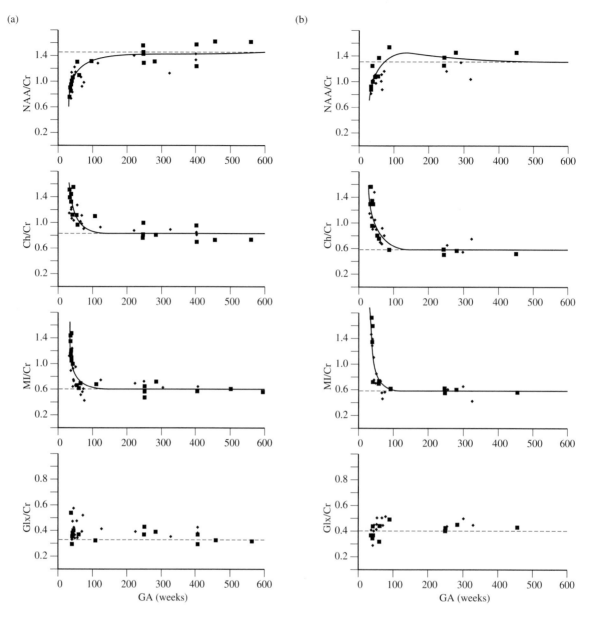

Figure 7 Relations between gestational age (GA) and cerebral peak-amplitude ratios at the same anatomical locations as in Figure 6. Acquisition conditions and resonance identifications are as in Figure 6. Cho/Cr and MI have been fitted to a monoexponential function, whereas the NAA/Cr fit is biexponential. (Reproduced by permission of Williams and Wilkins from R. Kreis, T. Ernst, and B. D. Ross, *Magn. Reson. Med.*, 1993, **30**, 424)

these infants.) In survivors, the spectra usually returned to normal within about 2 weeks, but the overall ^{31}P signal amplitude was often reduced, indicating loss of brain cells.

Since the lowest observed [PCr]/[Pi] values were strongly related to long-term prognosis (see below) and because of the long timescale of the postnatal biochemical changes, in order not to miss useful information it is important that spectra are obtained on several occasions during the first 2–4 days of life during which metabolic abnormalities are at their greatest. An explanation for the observed sequence of spectral changes is that the brain suffered an acute episode of hypoxia-ischemia before birth causing 'primary' energy failure, which was reversed by resuscitation at delivery, so that the energy state of

the brain returned to normal. The subsequent changes in ^{31}P spectra have been attributed to a 'secondary' impairment or failure of energy generation set off by the primary phase or events associated with it.[41] Studies using near-infrared spectroscopy (NIRS) in infants developing secondary energy failure have shown that both cerebral blood flow and blood volume are increased; continuing inadequate oxygen or substrate supply is therefore not likely to be causal.

Many mechanisms have been suggested to account for the progression from primary to secondary energy failure, which is associated with delayed neuronal death. For example, one possibility is that excitatory neurotransmitters, particularly glutamate, which are released at the synapses in response to acute

For list of General Abbreviations see end-papers

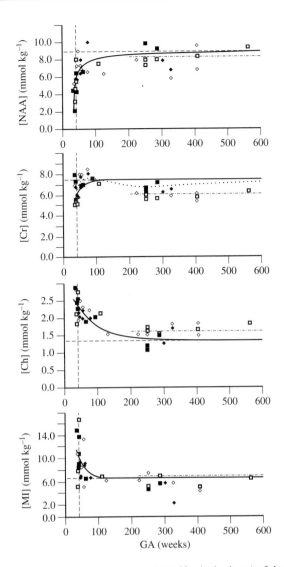

Figure 8   Absolute concentrations (mmol/kg brain tissue) of the main cerebral metabolites detectable by in vivo ^{1}H MRS. Unfilled symbols are data from parietal white matter [location (a) in Figure 6]; filled symbols are data from occipital gray matter [location (b) in Figure 6]. ($\Box$, $\blacksquare$) Data points from both normal infants and subjects with no primary cerebral pathology; ($\Diamond$, $\blacklozenge$) data points from infants with essentially normal neurological status for whom the eventual diagnosis was considered immaterial to the cerebral metabolite profile. The solid lines are best fits to the data; biexponential for [NAA] and monoexponential for [Cr], [Cho], and [MI]. (The dotted curve in the [Cr] plot is triexponential and appears to fit the data better.) The vertical dashed lines indicate term at 40.5 weeks gestational age. The horizontal dashed lines show the mean adult values measured at location (a) in Figure 6. (Reproduced by permission of Williams and Wilkins from R. Kreis, T. Ernst, and B. D. Ross, *Magn. Reson. Med.*, 1993, **30**, 424)

hypoxia ischemia may, by stimulating *N*-methyl-D-aspartate (NMDA) and other receptors, cause massive calcium entry to cells and damage to the mitochondrial electron transport chain. Other possible mechanisms include those involving prostanoids, nitric oxide, free radicals, immune mechanisms, phagocytes, growth factors, and impaired protein synthesis.

Because of the relation between secondary energy failure and long-term prognosis, much effort is being made to explore the mechanisms involved and how to interrupt them, with a view to the cerebroprotective treatment of asphyxiated infants. The same mechanisms are likely to be involved in stroke in adults.

Some evidence is emerging from animal studies of the feasibility of cerebroprotection. The development of secondary energy failure has been modeled in the newborn piglet, giving ^{31}P MRS results closely resembling those seen in birth-asphyxiated human infants.[42] This model is suitable for testing clinically feasible cerebroprotective strategies, with the amelioration of secondary energy failure, as measured by ^{31}P and ^{1}H MRS, providing an initial end-point. Recently, mild hypothermia, applied after a severe cerebral hypoxic–ischemic insult, was shown to be markedly cerebroprotective in this model,[43] whereas intravenous magnesium sulfate was ineffective.[44]

^{1}H MRS studies of infants with hypoxic-ischemia brain injury[36,45,46] show that increased Lac (due to failure of oxidative phosphorylation and consequent anaerobic glycolysis), and occasionally also reduced NAA (due to neuronal death), may be detected (see Figure 10). The Lac/NAA peak area ratio is often high and this index may prove a sensitive marker of cerebral injury. Reduced [NAA] has also been seen in newborn infants with a variety of nonfocal pathologies,[19] and in children with generalized demyelination.[47] Abnormal ^{1}H spectra have been found in older children with a variety of inborn errors of metabolism, including increased NAA in Canavan's disease (see Figure 11), and increased Lac in Leigh, Schilder, and Cockayne disease, and also in neuroaxonal dystrophy.[47]

5.2   Prognosis

The prognostic significance of secondary energy failure has been investigated by Azzopardi et al.[40] They studied 61 infants recruited because of evidence or suspicion of hypoxic ischemic brain injury, many of whom had sustained birth asphyxia. Clear evidence was found that the chances of survival and the severity of neurodevelopmental impairments at 1 year of age were directly related to the maximum detected extent of energy failure during the first days of life, quantified as the minimum observed value for [PCr]/[Pi] (see Table 2 and Figure 12). Furthermore, if [NTP]/[P$_{tot}$] fell, death was almost inevitable.

As a continuation of the same study, Roth et al.[48] recruited a further group of birth-asphyxiated infants, so that data from a total of 52 such infants were available for analysis. These infants had been in very poor condition at birth; their mean arterial base excess in cord blood shortly after delivery was -20 mmol L^{-1}, 38 had fits, and 26 required mechanical ventilation. The relation between minimum recorded [PCr]/[Pi] in the first days of life and neurodevelopmental outcome is illustrated in Figure 13. These data extend those of Azzopardi et al. and confirm the bad prognosis of secondary energy failure. The sensitivity, specificity and positive predictive value for death or multiple disabling impairments of values of [PCr]/[Pi] more than 2 SD below normal were 72%, 92%, and 91%, respectively. Roth et al. also found a relationship between minimum

For References see p. 1099

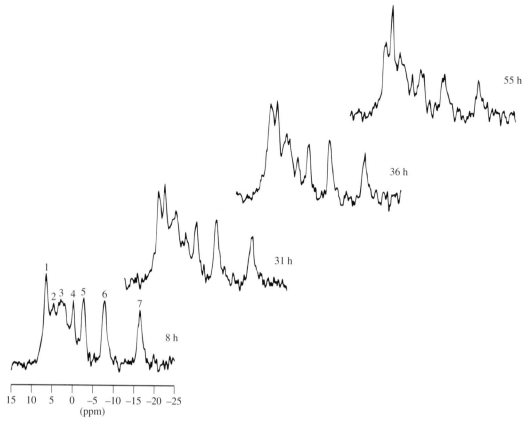

Figure 9 ^{31}P spectra from the temporo-parietal cortex of a birth-asphyxiated infant born at 37 weeks gestation. The postnatal ages at the times of study are given. Peak assignments are as in Figure 1. At age 8 h, [PCr]/[Pi] was 0.99, [NTP]/[P$_{tot}$] was 0.09 and pH$_i$ was 7.06; pH$_i$ increased to a maximum of 7.28 at 36 h. The minimum [PCr]/[Pi] was 0.32 at 55 h, when [NTP]/[P$_{tot}$] was 0.04 and pH$_i$ was 6.99. The infant died aged 60 h. (Reproduced by permission of International Pediatric Research Foundation Inc. from D. Azzopardi, J. S. Wyatt, E. B. Cady, D. T. Delpy, J. Baudin, A. L. Stewart, P. L. Hope, P. A. Hamilton, and E. O. R. Reynolds, *Pediatr. Res.*, 1989, **25**, 445)

Table 2 Age-dependent Standard Deviation Scores (SDS) for Phosphorus Metabolite Concentration Ratios and pH$_i$ in the temporo-parietal cortex of Newborn Infants Suspected of Hypoxic–ischemic Brain Injury[a]

	All infants (n = 61)	Birth asphyxia (n = 40)	Postnatal asphyxia (n = 5)	Increased cerebral echodensities (n = 16)
PCr/Pi	-2.14 ± 2.10^b	-2.14 ± 2.19^b	-0.03 ± 0.76	-2.04 ± 1.78^b
NTP/P$_{tot}$	-0.98 ± 1.86^c	-1.17 ± 1.82^c	-0.01 ± 1.03	-0.83 ± 1.99
PCr/P$_{tot}$	-1.72 ± 2.58^b	-1.79 ± 2.59^b	0.33 ± 1.04	-2.19 ± 2.52^b
Pi/P$_{tot}$	5.40 ± 8.77^b	5.97 ± 8.91^b	0.69 ± 0.99	5.51 ± 9.06^c
PME/P$_{tot}$	0.42 ± 1.69	0.63 ± 1.58	-0.79 ± 1.26	0.30 ± 1.83
PDE/P$_{tot}$	0.44 ± 1.43	-0.60 ± 1.44^d	0.39 ± 0.50	-0.32 ± 1.47
PCr/NTP	0.52 ± 2.98	0.65 ± 3.14	0.15 ± 0.66	0.31 ± 2.91
Pi/NTP	7.21 ± 16.41^b	8.96 ± 17.94^b	0.35 ± 1.06	4.94 ± 13.27^c
PME/NTP	1.94 ± 4.40^d	2.49 ± 4.71^c	-0.26 ± 1.11	1.25 ± 3.72
PDE/NTP	1.02 ± 3.84	1.43 ± 4.48	0.17 ± 0.77	0.23 ± 1.81
pH$_i$	0.05 ± 1.47	0.23 ± 0.97	-0.28 ± 0.22	0.00 ± 2.21
	(n = 55)	(n = 36)	(n = 5)	(n = 14)

[a]Values are those obtained when PCr/Pi was at its lowest. Mean values for SDS $\pm$ SD vs the normal control infants (see Table 1) are given. [b]$p < 0.001$. [c]$p < 0.01$. [d]$p < 0.05$. (Reproduced by permission of International Pediatric Research Foundation Inc. from D. Azzopardi, J. S. Wyatt, E. B. Cady, D. T. Delpy, J. Baudin, A. L. Stewart, P. L. Hope, P. A. Hamilton, and E. O. R. Reynolds, *Pediatr. Res.*, 1989, **25**, 445)

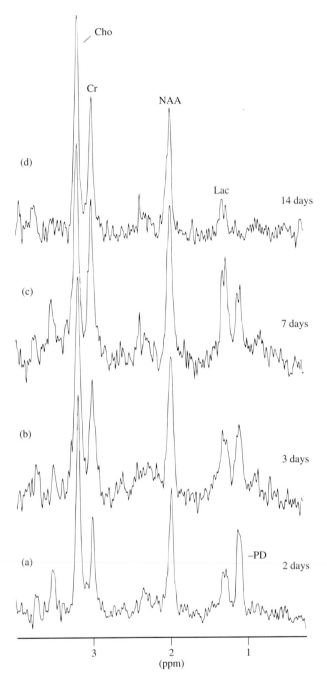

Figure 10 ^{1}H PRESS spectra acquired at 100 MHz from the thalamus of a severely birth-asphyxiated term infant. The ages at which the spectra were acquired are shown. Acquisition conditions: *TE* 270 ms; *TR* 2 s; 128 averaged echoes; 8 mL voxel. The lactate peak at about 1.3 ppm is greatly increased when compared with normal infants [see Figure 2(a)], although this may be partly due to raised free fatty acids. The peak at about 1.1 ppm is due to propan-1,2-diol (PD; the injection medium for phenobarbitone and phenytoin preparations). (Reproduced by permission of Williams and Wilkins from E. B. Cady, A. Lorek, J. Penrice, E. O. R. Reynolds, R. A. Iles, S. P. Burns, G. A. Coutts, and F. M. Cowan, *Magn. Reson. Med.*, 1994, **32**, 764)

observed [PCr]/[Pi] and head growth. Although the infants had normal head circumferences at birth, head growth was slowest, leading to microcephaly, in those whose values for [PCr]/[Pi] fell the most. Further follow-up indicated that the severities of

adverse outcomes at age 4 years were also closely related to the extent of cerebral energy impairment in the first week of life.[49]

The results of these studies may be summed up as showing that secondary energy failure leads both to neurodevelopmental disabilities and to microcephaly. The worse the energy failure, the worse the outcome is likely to be.

^{31}P spectra from the brains of infants with increased echo densities on ultrasound scans are also often abnormal, showing reduced values for [PCr]/[Pi] and sometimes [NTP]/[P$_{tot}$] (see Table 2).[40,50] Such echo densities are usually due to hypoxic ischemic injury or to hemorrhage. ^{31}P spectroscopy can be useful for segregating infants whose echo densities carry a relatively good, from those implying a bad, prognosis.

Data are becoming available which indicate that abnormal ^{1}H spectra also imply a bad prognosis.[36,45] Observation of raised Lac and of reduced NAA appear predictive of an unfavorable outcome. Further investigations, including long-term follow-up, are required before definitive statements can be made.

6 CONCLUSIONS

The first in vivo MRS studies of the human brain were in newborn infants. Since these studies were carried out, investigations of older children and adults have been undertaken and developmental changes have been defined for ^{31}P and ^{1}H metabolites. For example, [PCr]/[Pi] rises with age, indicating increased phosphorylation potential, and [PME]/[PDE] falls. ^{31}P MRS has proved particularly useful for investigating the cerebral metabolic consequences of severe birth-asphyxia. Spectra are often normal shortly after birth, but then [PCr]/[Pi] and, in severe cases, [NTP]/[P$_{tot}$] gradually falls, reaching minimum values at 2–4 days of age. The extent of the falls in [PCr]/[Pi] and [NTP]/[P$_{tot}$] are strongly related to the likelihood of long-term neurodevelopmental impairments or death. These changes in ^{31}P metabolite concentrations have been termed 'secondary' cerebral energy failure, on the hypothesis that they were initiated by an episode of 'primary' energy failure taking place shortly before or during delivery. The prognostic information acquired by ^{31}P MRS is valuable for guiding the clinical care of infants suspected of hypoxic ischemic brain injury. In the future, the prevention of secondary energy failure, as measured by MRS, may be used to test the efficacy of cerebroprotective strategies. ^{1}H spectroscopy has demonstrated increased Lac/NAA peak area ratios in the brains of asphyxiated infants and this may also have prognostic significance.

Recent technological developments will enhance the value of MRS to clinical pediatrics. Five major areas have potential for immediate improvement: the development of special neonatal head probes to increase the signal-to-noise ratio; optimized acquisition methods to enhance the quality and variety of spectroscopic information; spectroscopic quantification; and spectral editing to detect resonances that would otherwise be difficult to resolve. Spectroscopic imaging for metabolite mapping and direct comparison of MRI and spatial MRS data should help to improve the quality of the information available

For References see p. 1099

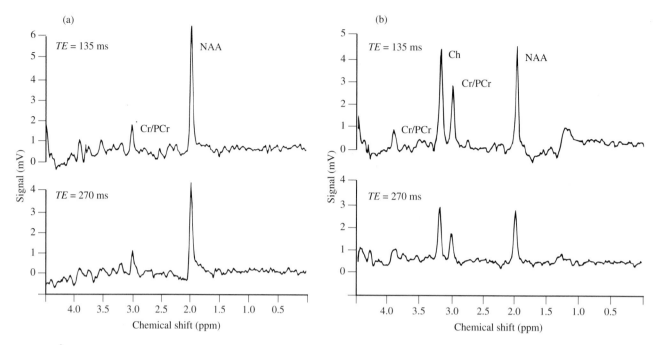

Figure 11 ¹H PRESS spectra from occipital voxels in the brain of a 7-month-old child with Canavan disease (a) and in that of an age-matched, normal control (b). Spectrum (a) shows increased *N*-acetylaspartate (NAA) signal, and the resonance from choline-containing compounds (Cho) is almost undetectable. Canavan disease has been linked with high body fluid levels of NAA caused by aspartoacylase deficiency. The disease also affects myelin sheaths due to a defect of myelin formation and/or glial metabolism, and this may be the reason for the low Cho signal. (Reproduced by permission of The Radiological Society of North America from W. Grodd, I. Krageloh-Mann, U. Klose, and R. Sauter, *Radiology*, 1991, **181**, 173)

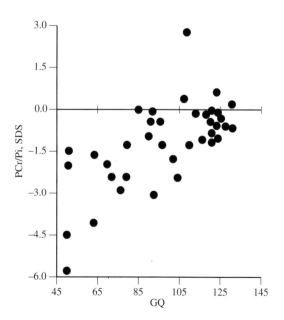

Figure 12 The relationship between minimum observed PCr/Pi in the temporo-parietal cortex expressed as standard deviation score (SDS) vs normal infants and the Griffiths general quotient (GQ) assessed at age 1 year in 38 infants surviving after hypoxic ischemic brain injury. Scores below 50 are recorded as 50 ($r = 0.67$, $p < 0.001$). (Reproduced by permission of International Pediatric Research Foundation Inc. from D. Azzopardi, J. S. Wyatt, E. B. Cady, D. T. Delpy, J. Baudin, A. L. Stewart, P. L. Hope, P. A. Hamilton, and E. O. R. Reynolds, *Pediatr. Res.*, 1989, **25**, 445)

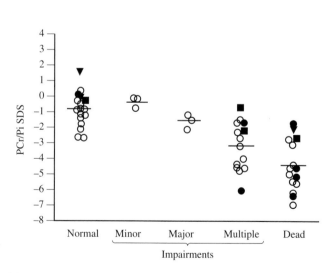

Figure 13 Minimum observed age-related PCr/Pi standard deviation score (SDS) in the temporo-parietal cortex of birth-asphyxiated newborn infants and neurodevelopmental outcome at age 1 year: (○) term AGA infants; (●) term SGA infants; (▼) preterm AGA infants; (■) preterm SEA infants. (Reproduced by permission of Mac Keith Press from S. C. Roth, D. Azzopardi, A. D. Edwards, J. Baudin, E. B. Cady, J. Townsend, D. T. Delpy, A. L. Stewart, J. S. Wyatt, and E. O. R. Reynolds, *Dev. Med. Child Neurol.*, 1992, **34**, 285)

and further develop MRS as an acceptable and valuable clinical tool.

7 RELATED ARTICLES

Brain Infection and Degenerative Disease Studied by Proton MRS; Brain MRS of Human Subjects; Localization and Registration Issues Important for Serial MRS Studies of Focal Brain Lesions; MRI in Clinical Medicine; Patient Life Support and Monitoring Facilities for Whole Body MRI; Single Voxel Localized Proton NMR Spectroscopy of Human Brain In Vivo; Single Voxel Whole Body Phosphorus MRS; Tissue and Cell Extracts MRS.

8 REFERENCES

1. E. B. Cady, A. M. de L. Costello, M. J. Dawson, D. T. Delpy, P. L. Hope, E. O. R. Reynolds, P. S. Tofts, and D. R. Wilkie, *Lancet*, 1983 **i**, 1059.
2. A. L. Stewart, R. J. Thorburn, P. L. Hope, M. Goldsmith, A. P. Lipscomb, and E. O. R. Reynolds, *Arch. Dis. Child.*, 1983, **58**, 598.
3. E. B. Cady, J. Hennig, and E. Martin, in 'Imaging Techniques of the CNS of Neonates', ed. J. Haddad, D. Christmann, and J. Messer, Springer, Berlin, 1991, Chap. 6.
4. E. B. Cady, in 'Magnetic Resonance Spectroscopy in Biology and Medicine', ed. J. D. de Certaines, W. M. M. J. Bovee, and F. Podo, Pergamon, Oxford, 1992, Chap. 23.
5. E. B. Cady, C. Boesch, and E. Martin, in 'Perinatal Asphyxia', ed. J. Haddad and E. Saliba, Springer, Berlin, 1993, Chap. 13.
6. A review of this early in vivo work is given in: E. B. Cady, 'Clinical Magnetic Resonance Spectroscopy', Plenum, New York, 1990.
7. D. P. Younkin, M. Delivoria-Papadopoulos, J. C. Leonard, V. H. Subramanian, S. Eleff, J. S. Leigh, and B. Chance, *Ann. Neurol.*, 1984, **16**, 581.
8. Department of Health (United Kingdom), 'Guidelines for Magnetic Resonance Diagnostic Equipment in Clinical Use', HMSO, Norwich, 1993.
9. A. Chu, D. T. Delpy, and S. Thalayasingam, in 'Fetal and Neonatal Physiological Measurements', ed. P. Rolfe, Butterworths, London, 1986, Chap. 59.
10. C. Boesch and E. Martin, *Radiology*, 1988, **168**, 481.
11. F. G. Shellock and G. L. Slimp, *Am. J. Roentgenol.*, 1989, **153**, 1105.
12. R. J. Ordidge, A. Connelly, and J. A. B. Lohman, *J. Magn. Reson.*, 1986, **66**, 283.
13. M. S. Van der Knaap, J. van der Grond, P. C. van Rijen, J. A. J. Faber, J. Valk, and K. Willemse, *Radiology*, 1990, **176**, 509.
14. R. Gruetter, C. Boesch, M. Muri, E. Martin, and K. Wuthrich, *Magn. Reson. Med.*, 1990, **15**, 128.
15. C. T. W. Moonen, M. von Kienlin, P. C. M. van Zijl, J. Cohen, J. Gillen, P. Daly, and G. Wolf, *NMR Biomed.*, 1989, **2**, 201.
16. J. H. Duyn and C. T. W. Moonen, *Magn. Reson. Med.*, 1993, **30**, 409.
17. E. B. Cady, J. Penrice, P. N. Amess, A. Lorek, M. Wylezinska, R. F. Aldridge, F. Franconi, J. S. Wyatt, and E. O. R. Reynolds, *Magn. Reson. Med.*, 1996, **36**, 878.
18. R. Buchli, E. Martin, P. Boesiger, and H. Rumpel, *Pediatr. Res.*, 1994, **35**, 431.
19. R. Kreis, T. Ernst, and B. D. Ross, *Magn. Reson. Med.*, 1993, **30**, 424.
20. P. Christiansen, O. Henriksen, M. Stubgaard, P. Gideon, and H. B. W. Larsson, *Magn. Reson. Imaging*, 1993, **11**, 107.
21. A. G. Chapman, E. Westerberg, and B. K. Siesjö, *J. Neurochem.*, 1981, **36**, 179.
22. T. Glonek, S. J. Kopp, E. Kot, J. W. Pettegrew, W. H. Harrison, and M. M. Cohen, *J. Neurochem.*, 1982, **39**, 1210.
23. O. A. C. Petroff, J. W. Prichard, K. L. Behar, J. R. Alger, J. A. den Hollander, and R. G. Shulman, *Neurology*, 1985, **35**, 781.
24. R. C. Robbins, R. S. Balaban, and J. A. Swain, *J. Thorac. Cardiovasc. Surg.*, 1990, **99**, 878.
25. P. B. Garlick, S. Soboll, and G. R. Bullock, *NMR Biomed.*, 1992, **5**, 29.
26. J. W. Pettegrew, S. J. Kopp, J. Dadok, N. J. Minshew, J. M. Feliksik, T. Glonek, and M. M. Cohen, *J. Magn. Reson.*, 1986, **67**, 443.
27. S. Cerdan, R. Parrilla, J. Santoro, and M. Rico, *FEBS Lett.*, 1985, **187**, 167.
28. Reviews of biochemical aspects of the resonances seen in ^{1}H brain spectra are given, in: B. L. Miller and B. D. Ross, *NMR. Biomed.*, 1991, **4**, 47, 59.
29. J. Urenjak, S. R. Williams, D. G. Gadian, and M. Noble, *J. Neurosci.*, 1993, **13**, 981.
30. D. Azzopardi, J. S. Wyatt, P. A. Hamilton, E. B. Cady, D. T. Delpy, P. L. Hope, and E. O. R. Reynolds, *Pediatr. Res.*, 1989, **25**, 440.
31. P. Tofts and S. Wray, *J. Physiol. (London)*, 1985, **359**, 417.
32. P. L. Hope, E. B. Cady, A. Chu, D. T. Delpy, R. M. Gardiner, and E. O. R. Reynolds, *J. Neurochem.*, 1987, **49**, 75.
33. H. C. Agrawal and W. A. Himwich, in 'Developmental Neurobiology', ed. W. A. Himwich, C. C. Thomas, Springfield, IL, 1970, Chap. 9.
34. N. Okumura, S. Otsuki, and A. Kameyama, *J. Biochem.*, 1960, **47**, 315.
35. W.-I. Jung, S. Widmaier, M. Bunse, U. Seeger, K. Straubinger, F. Schick, K. Kuper, G. Dietze, and O. Lutz, *Magn. Reson. Med.*, 1993, **30**, 741.
36. J. Penrice, E. B. Cady, A. Lorek, M. Wylezinska, P. N. Amess, R. F. Aldridge, A. L. Stewart, J. S. Wyatt, and E. O. R. Reynolds, *Pediatr. Res.*, 1996, **40**, 6.
37. P. S. Huppi, S. Posse, F. Lazeyras, R. Burri, E. Bossi, and N. Herschkowitz, *Pediatr. Res.*, 1991, **30**, 574.
38. H. H. Tallan, *J. Biol. Chem.*, 1957, **223**, 41.
39. B. K. Siesjö, 'Brain Energy Metabolism', Wiley, New York, 1978.
40. D. Azzopardi, J. S. Wyatt, E. B. Cady, D. T. Delpy, J. Baudin, A. L. Stewart, P. L. Hope, P. A. Hamilton, and E. O. R. Reynolds, *Pediatr. Res.*, 1989, **25**, 445.
41. P. L. Hope, A. M. de L. Costello, E. B. Cady, D. T. Delpy, P. S. Tofts, A. Chu, P. A. Hamilton, E. O. R. Reynolds, and D. R. Wilkie, *Lancet*, 1984, **ii**, 366.
42. A. Lorek, Y. Takei, E. B. Cady, J. S. Wyatt, J. Penrice, A. D. Edwards, D. Peebles, M. Wylezinska, H. Owen-Reece, V. Kirkbride, C. Cooper, R. F. Aldridge, S. C. Roth, G. Brown, D. T. Delpy, and E. O. R. Reynolds, *Pediatr. Res.*, 1994, **36**, 699.
43. M. Thoresen, J. Penrice, A. Lorek, E. B. Cady, M. Wylezinska, V. Kirkbride, C. E. Cooper, G. C. Brown, A. D. Edwards, J. S. Wyatt, and E. O. R. Reynolds, *Pediatr. Res*, 1995, **37**, 667.
44. J. Penrice, P. N. Amess, S. Punwani, M. Wylezinska, L. Tyszczuk, P. D'Souza, A. D. Edwards, E. B. Cady, J. S. Wyatt, and E. O. R. Reynolds, *Pediatr. Res.*, 1997, **41**, 1.
45. E. B. Cady, *Neurochem. Res.*, 1966, **21**, 1043.
46. E. B. Cady, P. Amess, J. Penrice, M. Wylezinska, V. Sams, and J. S. Wyatt, *Magn. Reson. Imaging*, 1997, **15**, 605.
47. W. Grodd, I. Krageloh-Mann, U. Klose, and R. Sauter, *Radiology*, 1991, **181**, 173.

48. S. C. Roth, A. D. Edwards, E. B. Cady, D. T. Delpy, J. S. Wyatt, D. Azzopardi, J. Baudin, J. Townsend, A. L. Stewart, and E. O. R. Reynolds, *Dev. Med. Child Neurol.*, 1992, **34**, 285.

49. S. Roth, J. Baudin, E. B. Cady, K. Johal, J. Townsend, J. S. Wyatt, E. O. R. Reynolds, and A. L. Stewart, *Dev. Med. Child Neurol.*, 1997, **39**, 718.

50. P. A. Hamilton, P. L. Hope, E. B. Cady, D. T. Delpy, J. S. Wyatt, and E. O. R. Reynolds, *Lancet*, 1986, **i**, 1242.

Biographical Sketches

Ernest B. Cady. *b* 1952. B.Sc. (astronomy), University College London, 1973; Dip. Adv. Sc. Studies (radio astronomy), Manchester University, 1974. Introduced to NMR by Prof. D. R. Wilkie FRS. Medical Physicist, University College London Hospitals, 1978–present. Approx. 100 publications, including the book *Clinical Magnetic Resonance Spectroscopy*. Research interests: probe design, data analysis, absolute quantification, and neonatal brain metabolism.

E. Osmund R. Reynolds. *b* 1933. B.Sc. (physiology), 1955, M.B., B.S., 1958, M.D., 1965, London University; M.R.C.P., 1966, F.R.C.P., 1975, F.R.C.O.G. (*ad eundem*), 1983; F.R.S., 1993; C.B.E., 1995; Hon. F.R.C.P.C.H., 1997; Professor of Neonatal Pediatrics, University College London Medical School, 1976–96 (emeritus). Approx. 300 publications. Research interests: noninvasive investigation of perinatal brain injury, in particular employing ultrasound imaging, NMR spectroscopy, and near-infrared spectroscopy.

Brain Neoplasms in Humans Studied by Phosphorus-31 NMR Spectroscopy

Wolfhard Semmler

Institut für Diagnostikforschung (IDF) an der Freien Universität, D-14050 Berlin, Germany

and

Peter Bachert

Forschungsschwerpunkt Radiologische Diagnostik und Therapie, Deutsches Krebsforschungszentrum (DKFZ), D-69120 Heidelberg, Germany

1 INTRODUCTION

1.1 Value of ^{31}P NMR Spectroscopy for Studies of the Human Brain

In vivo ^{31}P NMR spectroscopy in humans was first applied in studies of muscle diseases and soft tissue tumors. Subsequently, metabolic, ischemic, and neoplastic disorders of the human brain have been studied by means of this technique. In vivo ^{31}P NMR spectroscopy offers a window to study high-energy phosphate and membrane phospholipid turnover in human neurometabolism. It also permits the noninvasive measurement of the intracellular pH in the tissue.

1.2 Problems of Tumor Diagnosis

The major diagnostic problems in oncology are: (i) early tumor detection, (ii) tumor grading, and (iii) tumor staging. After diagnosis of a malignant tumorous disease and selection of an adequate therapy, monitoring is important for the optimization of the treatment and for the prediction of the response to treatment.

The value of ^{31}P NMR spectroscopy for solving these clinical problems is not yet established, although animal experiments could clearly demonstrate the high potential of this noninvasive technique for monitoring tumor biochemistry and the response of tumors to radiation therapy and chemotherapy.[1–6]

1.3 Heterogeneity of Neoplastic Tissue

Tumor tissue is heterogeneous on both macroscopic and microscopic scales. Vital and hypoxic tissue as well as necrotic cell masses and hemorrhage may coexist in tumors, reflecting regional differences in oxygen supply, vascularization, and nutrition of the cells.[1,5] Because of the poor spatial resolution, ^{31}P NMR spectroscopy can provide only an overall picture of the metabolic state of the tumor, being completely insensitive to tissue heterogeneity below the centimeter range.

This is a particular problem when high-grade gliomas are examined in vivo by ^{31}P NMR spectroscopy. Histopathological studies show heterogeneities in these tumors, but often they are also heterogeneous macroscopically as can be detected by MR imaging. ^{31}P NMR spectra will therefore comprise different signal contributions from vital, hypoxic, and necrotic tissue within the selected voxel rather than give a metabolic fingerprint of this tumor entity. The problem is less serious in the case of the relatively homogeneous meningioma and pituitary adenoma. In any case, however, MR imaging is mandatory before each NMR spectroscopic examination in order to detect gross inhomogeneities within the tumor mass.

1.4 NMR Spectroscopic Techniques

Adequate performance of the method requires the acquisition of localized NMR spectra with high signal-to-noise ratio (SNR) and high frequency resolution. A variety of localization techniques has been developed for ^{31}P NMR spectroscopy in humans. Localization means that NMR signals are obtained from a volume of interest (VOI) in the examined organ, while signal contributions of regions outside the VOI are suppressed. Single-voxel localization methods, e.g. ISIS,[7] and STEAM,[8] are available that employ selective rf pulses in the presence of orthogonal magnetic field gradients. A frequently used localization technique for in vivo ^{31}P NMR spectroscopy studies of the human brain is chemical shift imaging (CSI), also called spectroscopic or metabolic imaging, which employs phase-encoding gradients in two or three spatial directions to generate a 2D or 3D array of localized spectra in one measurement process.[9–14] These techniques permit the selective acquisition of ^{31}P NMR signals from regions with volumes >30 cm^3 within the human brain.

Frequency resolution and SNR of ^{31}P NMR spectra can be improved significantly using ^{31}P-{^{1}H} double resonance. By this means multiplet splittings resulting from scalar ^{31}P–^{1}H spin–spin couplings are removed (proton-decoupling[15]) and ^{31}P NMR signal intensities are enhanced owing to dipolar relaxation of phosphorus nuclei interacting with close protons (^{31}P–{^{1}H} NOE[16,17]).

Luyten and co-workers obtained localized (ISIS with 200 cm^3 voxel) proton-decoupled ^{31}P–{^{1}H} NMR spectra with resolved phosphomonoester (PME) and phosphodiester (PDE) resonance bands from the human brain.[18] The 2D ^{31}P CSI combined with ^{1}H irradiation to induce NOE signal enhancements yields in vivo brain spectra of good quality from 8×8 voxels with 3 cm × 3 cm × 4 cm volume each in a measurement time of 19 min.[19]

In general, comparison of in vivo NMR spectra without quantitative analysis is of limited value. Quantification must include the fit of the signals in the time or frequency domain to obtain chemical shifts, line widths, and a measure of signal intensity. Determination of absolute metabolite concentrations in tissue from in vivo ^{31}P NMR data would be of great value. However, this is a complicated problem and, therefore, ratios of integrated peak areas are often used for data analysis.

2   ASSESSMENT OF NEOPLASTIC BRAIN DISEASE WITH THE USE OF ^{31}P NMR SPECTROSCOPY

2.1   Detection and Differentiation of Brain Tumors

Tumor diagnosis was one of the primary issues to which ^{31}P NMR spectroscopy has been directed. The central diagnostic problem is the detection, localization, and differentiation of a tumor in its *early* stage when the neoplastic tissue mass is still very small. Because of its poor spatial resolution, ^{31}P NMR can hardly contribute to the solution of this problem. To our knowledge there is no case reported which clearly demonstrates the value of ^{31}P NMR spectroscopy for early cancer detection and where this modality was indispensable for clinical decisions.

On the other hand, the potential of ^{31}P NMR spectroscopy for tumor grading[19] is believed to be high, because changes of signal intensities and linewidths of the various resonances, and of the inorganic phosphate (Pi) chemical shift (a measure of intracellular pH) are observed in tumor spectra in comparison to spectra from normal brain tissue. These effects reflect metabolic differences of tumor and surrounding unaffected tissue as well as physiological conditions related to tumor oxygenation, perfusion, microcirculation, and angiogenic activity.[1,5,20,21]

Spectral differences are also expected when human brain tumors are examined at different growth stages. Animal exper-

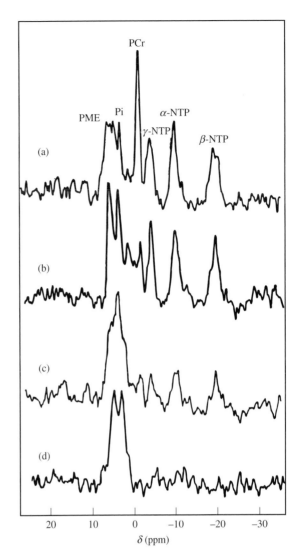

Figure 1 Metabolic changes during tumor growth reflected in in vivo ^{31}P NMR spectra, showing progressive decline of high-energy phosphate signals (NTP, nucleoside 5′-triphosphate, mainly ATP, adenosine 5′-triphosphate; PCr, phosphocreatine) and an increase of inorganic phosphate (Pi) signals. The concentration of phosphomonoesters (PME), which are attributed to precursors of major membrane phospholipids, is elevated. (a) Base line spectrum and subsequent spectra recorded at (b) day 3, (c) day 7, and (d) day 11. The pH value of the tumor tissue dropped from 7.1 for (a) to 6.6 for (c). Experimental tumor: subcutaneously implanted MOPC 104E myeloma (modified from Evanochko et al.[2])

iments show strong changes of ^{31}P NMR spectra during growth (Figure 1). A progressive decrease of high-energy phosphate levels is attributed to an increase of the fraction of hypoxic cells,[22] while an increase of PME and P$_i$ intensities may result from the emergence of necrotic cells.[1] The frequently observed alkalinic shift of pH in tumor tissue[20] is explained by necrosis.[5]

For tumor characterization, the presence of an NMR-visible compound unique to a specific type of tumor cells or tumor

For References see p. 1106

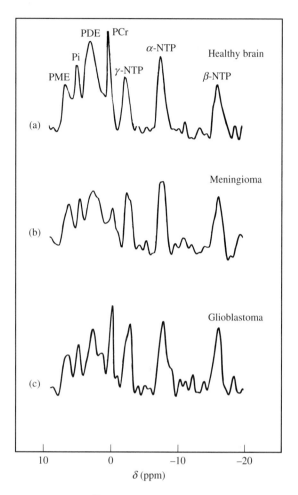

Figure 2 Localized ^{31}P NMR spectra of (a) healthy human brain tissue, (b) meningioma, and (c) glioblastoma obtained with the ISIS localization technique (modified from Heindel et al.[24])

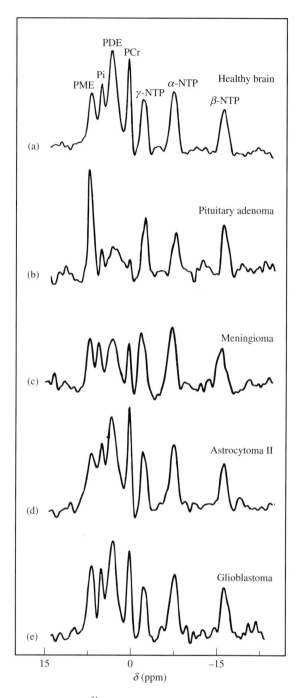

Figure 3 Localized ^{31}P NMR spectra in vivo of (a) normal human brain tissue and different tumors of the brain: (b) pituitary adenoma, (c) meningioma, (d) grade II astrocytoma, and (e) glioblastoma. Spectra were obtained by means of the ISIS localization technique with voxel sizes of 41–220 cm^3 (modified from Arnold et al.[26])

species would be highly important. In fact, no tumor-specific resonance has been found in in vivo ^{31}P NMR spectra of all tumors studied so far. Neuroepithelial and mesodermal tumors and cerebral metastases, for example, show the same resonances as spectra obtained from healthy brain tissue.[23–25]

The ^{31}P NMR spectrum of a glioblastoma, a neuroepithelial tumor, shows a reduction of PME, Pi, PDE, and PCr relative to nucleoside 5′-triphosphate (NTP) signals [Figure 2(c)]. Heiss et al. found a similar spectral pattern in their ^{31}P NMR study of glioblastomas except for a cystic glioblastoma which produced only poor spectral SNR.[23] In contrast, Arnold et al. observed enhanced PME and barely reduced PDE levels in glioblastomas[23,26] (Figure 3(e)). Resonance intensities from high- and low-grade astrocytomas, which were also examined in this study, did not differ significantly. In pituitary adenomas, reduced PDE and exceedingly high PME signals were found (Figure 3(b)). Segebarth et al. observed elevated PME and reduced PCr intensities in the ^{31}P NMR spectrum of a prolactinoma[27] (Figure 4(a)) in comparison to the spectrum from the unaffected brain region of the same patient (Figure 4(b)). Hwang et al. found reduced PDE/NTP and PCr/NTP

signal intensity ratios and elevated pH values in high-grade gliomas and in a meningioma.[28] Similar findings were reported from other studies.[29,30] The localized in vivo ^{31}P NMR spectra in Figure 2 show increased PME and decreased PDE and PCr signal intensities for the meningioma (Figure

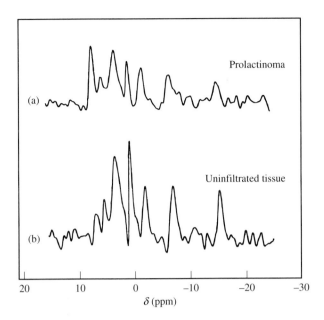

Figure 4 Localized ^{31}P NMR spectra in vivo of (a) a prolactinoma and (b) the unaffected brain tissue in the same patient (modified from Segebarth et al.[27])

2.2 Therapy Monitoring of Brain Tumors by Means of ^{31}P NMR

Studies of experimental tumors in animals by means of in vivo ^{31}P NMR spectroscopy showed significant changes of levels of phosphorus-containing metabolites during the course of radiation therapy and chemotherapy.[2,3] A common feature is the decrease of high-energy phosphate signals (PCr, NTP) and of pH and the increase of Pi, PME, and PDE resonances; however, the opposite effects have also been observed.[2,3,6] A strong increase in the Pi resonance is often accompanied by a collapse of high-energy phosphate signals, as demonstrated in animal studies (Figure 5). This effect has also been observed in a clinical study monitoring the hyperthermic regional perfusion therapy of the recurrence of a malignant melanoma.[34]

The majority of studies on therapy monitoring by means of in vivo ^{31}P NMR spectroscopy have been performed in

2(b)),[24] a mesodermal tumor that originates from meningial tissue, compared with the spectrum from healthy brain tissue (Figure 2(a)).

Negendank reviewed in vivo NMR data of human tumors.[31] His evaluation shows that about 80% of all examined high-grade glioma (Kernohan III–IV) showed reduced PDE and about 50% reduced PCr signal intensities. In 96% of all examined meningioma, low PDE and low PCr signals were observed. Intracellular pH values were higher in high-grade glioma and meningioma than in normal brain tissue.

Benign pituitary adenomas showed reduced PDE levels in 75% and high PME and low PCr levels in all patients examined.[31] In all cases the observed pH was comparable to the value measured in normal brain tissue. Low PDE/NTP signal intensity ratios and elevated pH values were found in ependymoma.

In vivo ^{31}P NMR spectra from neuroepithelial tumors (glioma) and mesodermal tumors (meningioma) are similar, so this technique does not allow the differentiation between these two entities. On the other hand, the pH, which was found in the normal range for low-grade and elevated in high-grade glial tumors, may help to grade gliomas. However, discrepant pH values have also been reported.[23,25] High pH values in post-therapy stages of soft tissue sarcomas were found to be related to necrosis.[5,32]

In a recent study, Rutter et al. employed 1D ^{31}P CSI in vivo and detected statistically significant differences in NMR parameters of brain hemispheres of patients with untreated brain tumors (astrocytomas, glioblastomas, meningiomas).[33] PME/NTP and PDE/NTP signal intensity ratios were higher in glioblastomas and astrocytomas than in healthy brain tissue. The P$_i$/NTP and PCr/NTP ratios of astrocytomas were higher compared with glioblastomas and normal brain.

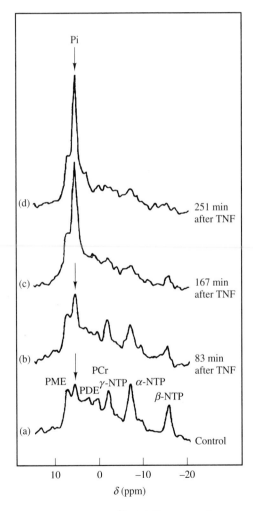

Figure 5 Changes of in vivo ^{31}P NMR spectra of an experimental tumor (murine methylcholanthrene-induced [Meth-A] sarcoma) upon application of recombinant human tumor necrosis factor α (rHuTNF-α). The series of spectra shows a progressive decline of high-energy phosphate signals (NTP, PCr) and a strong increase in the inorganic phosphate (Pi) level (modified from Shine et al.[6])

For References see p. 1106

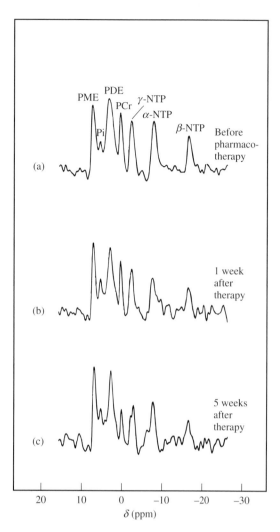

Figure 6 Localized ^{31}P NMR spectra of prolactinoma obtained in a patient before and one and five weeks after the beginning of bromocriptine pharmacotherapy. The spectra were rescaled to take into account differences of measured voxel sizes (modified from Segebarth et al.[27])

More pronounced ^{31}P spectral effects were found after two weeks of radiation therapy in a patient with an intracranial lymphoma. The enhanced PME and the reduced PCr resonances before treatment changed to the intensities of normal tissue after a radiation dose of 24.2 Gy (Figure 7). The pH of the tumor tissue did not change. The patient improved clinically during therapy. Similar results were obtained in the examination of a patient with a grade II astrocytoma treated with radiation therapy of 60 Gy total dose after subtotal surgical removal of the tumor.

In their ^{31}P NMR follow-up study of tumor radiation therapy, Heindel et al. examined patients with meningiomas, glioblastomas, astrocytomas, and other cerebral tumors.[24] NMR signal intensity changes similar to those in Figure 7, in particular a PME signal reduction, were observed in the spectra of grade II astrocytoma upon radiation therapy (Figure 8).

Arnold et al. examined four patients with malignant gliomas after intraarterial (Figure 9) and three patients after intravenous 1,3-bis(2-chloroethyl)-1-nitrosourea (BCNU) therapy.[44,45] They found statistically significant differences of pH values upon evaluation of ^{31}P NMR spectra acquired before and after chemotherapy: i.v. application of BCNU resulted in a transient acidosis and i.a. administration of the same drug in an alkalosis in the tumor tissue. These pH changes were detected before any effects were seen by imaging modalities. The authors discuss the alkalosis after i.a. administration as being a result of cell membrane damage due to the high local drug concentration in this route of application. In soft tissue sarcomas, alkalosis was found to be associated with necrosis[32] which may also be true for brain tumors.

soft tissue tumors[32,34–41] (for reviews, see Steen[21] and Negendank[31]). Only a few studies focused on monitoring treatment of brain tumors, among them monitoring of radiation therapy and/or chemotherapy,[23,24,26,27,29,42] immunotherapy,[43] and embolization.[18]

One of the first reports in this field was published in 1987 by Segebarth et al.[27] Figure 6 shows the ^{31}P NMR spectra obtained in a patient with invasive prolactinoma before and one and five weeks after pharmacotherapy with bromocriptine [the same patient as in Figure 4(a)]. Upon therapy the patient improved clinically while no morphological changes of the tumor were found by MR imaging. After five weeks, a small decrease of the NTP level was detected, but no pH change. Elevated pH values were observed in the tumor tissue before and one and five weeks after therapy when compared with the pH of the unaffected brain tissue.

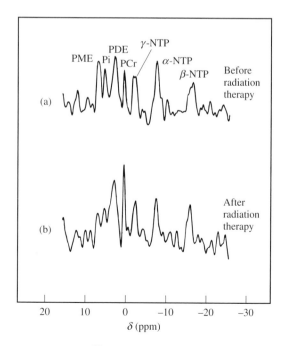

Figure 7 Localized ^{31}P NMR spectra in vivo of intracranial lymphoma obtained (a) before and (b) after radiation therapy with a dose of 24.2 Gy (modified from Segebarth et al.[27])

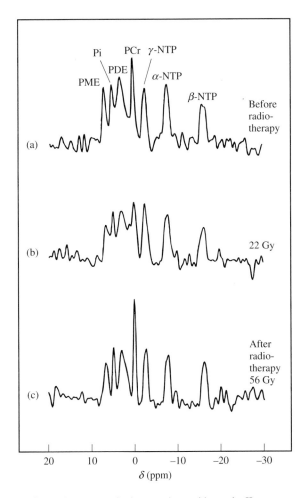

Figure 8 Follow-up study in a patient with grade II astrocytoma. Localized ^{31}P NMR spectra obtained (a) before, (b) after 22 Gy, and (c) after completion of the radiation therapy (56 Gy) show intensity changes of phosphomonoester (PME) and phosphocreatine (PCr) resonances (modified from Heindel et al.[24])

potential of ^{31}P NMR spectroscopy for monitoring tumor therapy response in patients.

3   CONCLUSIONS

At present, clinical ^{31}P NMR spectroscopy of human brain tumors can be rated as follows:
1. The method allows an assessment noninvasively of bioenergetic status, levels of intermediates of phospholipid metabolism, intracellular pH, and the extent of necrosis within macroscopic regions of proliferating tissue in the brain.
2. Limitations in spatial localization, sensitivity, and spectral resolution prevent differential diagnosis of human brain tumors by means of ^{31}P NMR spectroscopy.
3. Tumor therapy monitoring is feasible with the use of ^{31}P NMR due to the observation of cellular energy status and membrane turnover rate, and the quantitative comparison with spectral data obtained before the beginning of therapy and from unaffected tissue.

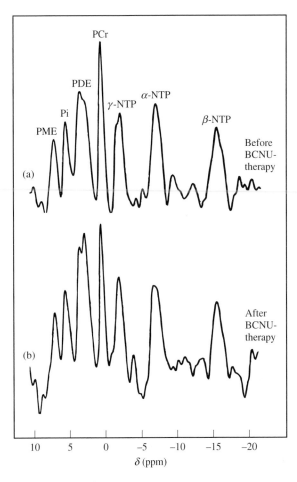

Figure 9 Localized ^{31}P NMR spectra of a glioblastoma in a patient (a) before and (b) after superselective intraarterial infusion of 1,3-bis(2-chloroethyl)-1-nitrosourea (BCNU) over 3 h (modified from Arnold et al.[45])

Superselective catheter embolization of meningiomas with poly(vinyl alcohol) particles is performed to minimize bleeding during subsequent neurosurgery. This treatment form is suitable to validate ^{31}P NMR spectroscopy for clinical application. Since the spectra reflect the strong ischemia caused by the embolization, in vivo NMR spectroscopy should help to assess response to therapy. Employing two-dimensional CSI, increased P_i/β-NTP signal intensity ratio was detected in meningioma the day following embolization.[18] A complete depletion of the NTP pool was not observed (Figure 10).

In studies of therapy monitoring of human brain tumors by means of ^{31}P NMR spectroscopy, different tumor entities have been examined and also different treatment protocols applied. The comparability of results of independent studies can be compromised when different measurement techniques are used. Besides further progress in experimental techniques and quantification, a generally applicable examination protocol must be established. Notwithstanding these current limitations, the clinical studies presented on cerebral neoplasms demonstrate the

For References see p. 1106

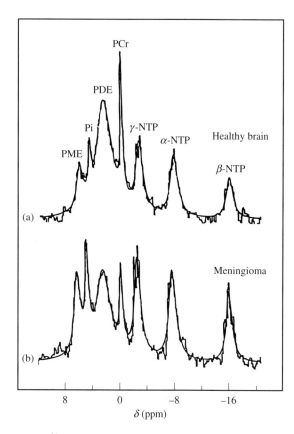

Figure 10 ^{31}P spectroscopic imaging of the brain of a patient with meningioma after presurgical superselective catheter embolization. Localized spectra (voxel size 3 cm × 3 cm × 3 cm) from (a) unaffected hemisphere and (b) meningioma. For quantitative evaluation, Lorentzian line-fits are superimposed on the measured spectra (modified from Knopp et al.[18])

4 ABBREVIATIONS

CSI chemical shift imaging
ISIS image-selected in vivo spectroscopy
NOE nuclear Overhauser effect
NTP nucleoside 5′-triphosphate
PCr phosphocreatine
PDE phosphodiester
P_i inorganic phosphate
PME phosphomonoester
SNR signal-to-noise ratio
STEAM stimulated echo acquisition mode
VOI volume of interest

5 RELATED ARTICLES

Chemical Shift Imaging; Localization and Registration Issues Important for Serial MRS Studies of Focal Brain Lesions; Proton Decoupling During In Vivo Whole Body Phosphorus MRS; Single Voxel Localized Proton NMR Spectroscopy of Human Brain In Vivo; Single Voxel Whole Body Phosphorus MRS; Whole Body Studies: Impact of MRS.

6 REFERENCES

1. T. C. Ng, W. T. Evanochko, R. N. Hiramoto, V. K. Ganta, M. B. Lilly, A. J. Lawson, T. H. Corbett, J. R. Durant, and J. D. Glickson, *J. Magn. Reson.*, 1982, **49**, 271.
2. W. T. Evanochko, T. C. Ng, and J. D. Glickson, *Magn. Reson. Med.*, 1984, **1**, 508.
3. S. Naruse, K. Hirakawa, Y. Horikawa, C. Tanaka, T. Higuchi, S. Ueda, H. Nishikawa, and H. Watari, *Cancer Res.*, 1985, **45**, 2429.
4. P. F. Daly, R. C. Lyon, P. J. Faustino, and J. S. Cohen, *J. Biol. Chem.*, 1987, **262**, 14875.
5. P. Vaupel, F. Kallinowski, and P. Okunieff, *Cancer Res.*, 1989, **49**, 6449.
6. N. Shine, M. A. Palladino, J. S. Patton, A. Deisseroth, G. S. Karczmar, G. B. Matson, and M. W. Weiner, *Cancer Res.*, 1989, **49**, 2123.
7. R. J. Ordidge, A. Connelly, and J. A. B. Lohman, *J. Magn. Reson.*, 1986, **66**, 283.
8. K. D. Merboldt, D. Chien, W. Hänicke, M. L. Gyngell, H. Bruhn, and J. Frahm, *J. Magn. Reson.*, 1990, **89**, 343.
9. P. C. Lauterbur, D. M. Kramer, W. V. House, and C.-N. Chen, *J. Am. Chem. Soc.*, 1975, **97**, 6866.
10. T. R. Brown, B. M. Kincaid, and K. Ugurbil, *Proc. Natl. Acad. Sci. USA*, 1982, **79**, 3523.
11. A. A. Maudsley, S. K. Hilal, W. H. Perman, and H. E. Simon, *J. Magn. Reson.*, 1983, **51**, 147.
12. A. A. Maudsley, S. K. Hilal, H. E. Simon, and S. Wittekoek, *Radiology*, 1984, **153**, 745.
13. D. B. Vigneron, S. J. Nelson, J. Murphy-Boesch, D. A. Kelly, H. B. Kessler, T. R. Brown, and J. S. Taylor, *Radiology*, 1990, **177**, 643.
14. J. W. Hugg, G. B. Matson, D. B. Twieg, A. A. Maudsley, D. Sappey-Marinier, and M. W. Weiner, *Magn. Reson. Imaging*, 1992, **10**, 227.
15. P. R. Luyten, G. Bruntink, F. M. Sloff, J. W. A. H. Vermeulen, J. I. van der Heijden, J. A. den Hollander, and A. Heerschap, *NMR Biomed.*, 1989, **1**, 177.
16. P. Bachert-Baumann, F. Ermark, H.-J. Zabel, R. Sauter, W. Semmler, and W. J. Lorenz, *Magn. Reson. Med.*, 1990, **15**, 165.
17. P. Bachert and M. E. Bellemann, *J. Magn. Reson.*, 1992, **100**, 146.
18. M. V. Knopp, P. Bachert, G. Ende, M. Blankenhorn, H. Kolem, W. Semmler, T. Hess, M. Forsting, K. Sartor, W. J. Lorenz, and G. van Kaick, *Proc. 11th Ann Mtg. Soc. Magn. Reson. Med.*, Berlin, 1992, p. 1954.
19. D. L. Arnold, E. A. Shoubridge, J. G. Villemure, and W. Feindel, *NMR Biomed.*, 1990, **3**, 184.
20. R. D. Oberhänsli, D. Hilton-Jones, P. J. Bore, L. J. Hands, R. P. Rampling, and G. K. Radda, *Lancet*, 1986, **ii**, 8.
21. R. G. Steen, *Cancer Res.*, 1989, **49**, 4075.
22. W. T. Evanochko, T. C. Ng, M. B. Lilly, A. J. Lawson, T. H. Corbett, J. R. Durant, and J. D. Glickson, *Proc. Natl. Acad. Sci. USA*, 1983, **80**, 334.
23. D. L. Arnold, E. A. Shoubridge, J. G. Villemure, and W. Feindel, *Proc. 7th Ann Mtg. Soc. Magn. Reson. Med.*, San Francisco, 1988, p. 333.
24. W. Heindel, J. Bunke, S. Glathe, W. Steinbrich, and L. Mollevanger, *J. Comput. Assist. Tomogr.*, 1989, **12**, 907.
25. W. D. Heiss, W. Heindel, K. Herholz, J. Rudolf, J. Bunke, J. Jeske, and G. Friedmann, *J. Nucl. Med.*, 1990, **31**, 302.
26. D. L. Arnold, J. Emrich, E. A. Shoubridge, J.-G. Villemure, and W. Feindel, *J. Neurosurg.*, 1991, **74**, 447.

27. C. M. Segebarth, D. F. Baleriaux, D. L. Arnold, P. R. Luyten, and J. A. den Hollander, *Radiology*, 1987, **165**, 215.

28. Y. C. Hwang, J. Mantil, M. D. Boska, D.-R. Hwang, W. Banks, M. Jacobs, and C. Peterson, *Proc. 11th Ann Mtg. Soc. Magn. Reson. Med.*, Berlin, 1992, p. 3610.

29. B. Hubesch, D. Sappey-Marinier, K. Roth, D. J. Meyerhoff, G. B. Matson, and M. W. Weiner, *Radiology*, 1990, **174**, 401.

30. T. A. D. Cadoux-Hudson, M. J. Blackledge, B. Rajagopalan, D. J. Taylor, and G. K. Radda, *Br. J. Cancer*, 1989, **60**, 430.

31. W. Negendank, *NMR Biomed.*, 1992, **5**, 303.

32. D. Sostman, M. Dewhirst, C. Charles, K. Leopold, D. Moore, R. Burn, A. Tucker, J. Harrelson, and J. Oleson, *Proc. 9th Ann Mtg. Soc. Magn. Reson. Med.*, New York, 1990, p. 319.

33. A. Rutter, H. Hugenholtz, J. K. Saunders, and I. C. Smith, *Invest. Radiol.*, 1995, **30**, 359.

34. W. Semmler, G. Gademann, P. Schlag, P. Bachert-Baumann, H.-J. Zabel, W. J. Lorenz, and G. van Kaick, *Magn. Reson, Imaging*, 1988, **6**, 335.

35. J. M. Maris, A. E. Evans, A. D. McLaughlin, G. J. D'Angio, L. Bolinger, H. Manos, and B. Chance, *N. Engl. J. Med.*, 1985, **312**, 1500.

36. T. C. Ng, S. Vijayakumar, A. W. Majors, F. J. Thomas, T. F. Meaney, and N. J. Baldwin, *Int. J. Radiat. Oncol. Biol. Phys.*, 1987, **13**, 1545.

37. B. D. Ross, J. T. Helper, I. J. Cox, I. R. Young, R. Kempf, A. Makepeace, and J. Pennock, *Arch. Surg.*, 1987, **122**, 1464.

38. W. Semmler, G. Gademann, P. Bachert-Baumann, H. J. Zabel, W. J. Lorenz, and G. van Kaick, *Radiology*, 1988, **166**, 533.

39. O. M. Redmond, J. Stack, M. Scully, P. Dervan, D. Carney, and J. T. Ennis, *Proc. 7th Ann Mtg. Soc. Magn. Reson. Med.*, San Francisco, 1988, p. 432.

40. G. S. Karczmar, D. J. Meyerhoff, M. D. Boska, B. Hubesch, J. Poole, G. B. Matson, F. Valone, and M. W. Weiner, *Radiology*, 1991, **179**, 149.

41. O. M. Redmond, J. P. Stack, N. G. O'Connor, D. N. Carney, P. A. Dervan, B. J. Hurson, and J. T. Ennis, *Magn. Reson. Med.*, 1992, **25**, 30.

42. S. Naruse, Y. Horikawa, C. Tanaka, T. Higuchi, H. Sekimoto, S. Ueda, and K. Hirakawa, *Radiology*, 1986, **160**, 827.

43. B. D. Ross, J. Tropp, K. A. Derby, S. Sugiura, C. Hawryszko, D. B. Jacques, and M. Ingram, *J. Comput. Assist. Tomogr.*, 1989, **13**, 189.

44. D. L. Arnold, E. A. Shoubridge, W. Feindel, and J.-G. Villemure, *Can. J. Neurol. Sci.*, 1987, **14**, 570.

45. D. L. Arnold, E. A. Shoubridge, J. Emrich, W. Feindel, and J.-G. Villemure, *Invest. Radiol.*, 1989, **24**, 958.

Biographical Sketches

Wolfhard Semmler. *b* 1944. M.Sc. (Dipl.-Phys.), 1972, Ph.D. (Dr. rer. nat.), 1976, Free University of Berlin, M.D. (Dr. med.), 1990, University of Heidelberg. University of Aarhus (Denmark), 1976; Rutgers University and Bell Laboratories, New Jersey, USA, 1977–79; Hahn–Meitner-Institute, Berlin, 1979–83; Department of Radiology, Free University of Berlin, 1983–85; German Cancer Research Center (DKFZ), Heidelberg, 1985–91; Director, Institute of Diagnostic Research, Free University of Berlin, 1992–present. Approx. 120 publications. Research specialties: MR imaging and spectroscopy, development of contrast media for MRI, basic principles of contrast media for imaging modalities.

Peter Bachert. *b* 1954. M.Sc., 1979, Ph.D. (Dr. rer. nat), 1983, University of Heidelberg. Introduced to NMR by K. H. Hausser and U. Haeberlen. Application scientist, Siemens, 1984–86; German Cancer Research Center (DKFZ), Heidelberg, 1986–present. Approx. 40 publications. Research specialties: applications of NMR to problems in biophysics and biomedicine.

Systemically Induced Encephalopathies: Newer Clinical Applications of MRS

Brian D. Ross, Stefan Bluml, Kay J. Seymour, Jeannie Tan, Jong-Hee Hwang and Alexander Lin

Huntington Medical Research Institutes, Pasadena, CA, USA

1   SYSTEMICALLY INDUCED ENCEPHALOPATHIES

1.1   Background: Biochemistry of Coma

Metabolic disturbances of liver, kidney, endocrine or other systems have remote effects; those on the brain result in a variety of well-defined encephalopathies. When severe, these disorders present as coma. Posner and Plum published a comprehensive account of human coma.[1,2] Many were the result of presumed metabolic events with normal brain anatomy, setting the stage for noninvasive elucidation by means of biochemically based techniques. Among these are MRA, diffusion imaging, positron emission tomography (PET), single-photon emission tomography (SPECT), and multinuclear magnetic resonance spectroscopy (MRS). MRS has increasingly been used to identify specific biochemical changes in the brain, from which information on diagnosis and pathogenesis of these poorly understood disorders is beginning to emerge.

As a model for this group of disorders, we discuss hepatic encephalopathy, with additional remarks about diabetic, hyperosmolar, and hypoxia-induced encephalopathies. Renewed interest in the use of MRS for differential diagnosis of focal brain pathologies is reflected in studies of stroke and adrenoleukodystrophy.

1.2   Neurochemical Pathology of Hepatic Encephalopathy

Hepatic encephalopathy (HE) is an excellent example of metabolic encephalopathy,[3] with identifiable neurotoxins originating in a systemic disease. In animal studies, several distinct neurotoxins have been identified. The earliest of these was ammonia;[4,5] ammonia is normally fully removed from portal blood by hepatic urea synthesis.[6,7] The combination of the loss of biosynthetic liver function and the diversion of nondetoxified blood to the brain by so-called portal systemic shunts (PSSs) accounts for the frequently demonstrated excess of cerebral and cerebrospinal fluid glutamine[8] [Equation (1)].

$$NH_4^+ + glutamate \underset{glutaminase}{\overset{GS}{\rightleftharpoons}} glutamine \qquad (1)$$

The enzyme glutamine synthetase (GS) responsible for this reaction is located exclusively in astrocytes.[9] Resynthesis of

For References see p. 1123

glutamate and of γ-aminobutyrate (GABA) from glutamine (both vital neurotransmitter amino acids) occurs principally in neurons.[10]

Lest it be thought that ammonia 'toxicity' accounts for all of the clinical syndromes covered by the term HE, the interested reader is referred to Butterworth[11] for an extensive review of several other well-documented alternatives. Metabolic theories abound: failure of oxidative energy metabolism (a corollary of glutamate and 2-oxoglutarate depletion from the Krebs cycle), tryptophan and serotonin (5-hydroxytryptamine) accumulation, branched-chain amino acid deficits, endogenous benzodiazepine agonists which modify access of GABA to its inhibitory receptors, and neurotoxic fatty acids, octanoate in particular, have been proposed.

An attractive unifying theory proposed by Zieve[12] is that multiple neurotoxins derived from liver, blood, or from the diet, gain access via PSSs to a previously 'sensitized' brain. No mechanism of sensitization is known, but with the advent of MRS, a candidate has been proposed in the form of cerebral *myo*-inositol (mI) depletion.[13]

1.3 Animal Studies in Hepatic Encephalopathy Using Multinuclear MRS

Three groups[14–16] independently perfected methods for the noninvasive determination of cerebral glutamine (including a variable contribution from glutamate in each assay) with ^{1}H MRS and confirmed the elevation of this metabolite in a variety of animal models of acute liver 'failure' with HE. One of these groups[14] also demonstrated a hitherto unrecognized abnormality, the significant reduction of choline (Cho; cerebral choline-containing compounds) in the ^{1}H spectra of rats with acute HE.

More recently, ^{15}N NMR[17–19] and [1]H–[15]N heteronuclear multiple quantum coherence (HMQC)[20,21] identified cerebral glutamine unequivocally in vivo in HE produced by ammonia infusion in the normal and portocaval shunted rat, respectively.

The glutamine protons coupled to amide nitrogen (termed H_Z and H_E) provide additional information regarding the compartmentalized pH adaptation during severe ammonia-induced coma in rats.[22] Alkalinization of the astrocyte (glial) cell cytoplasm can be inferred from alterations in [1]H–[15]N HMQC spectral line-widths in vivo, whereas whole brain pH, determined from chemical shift of inorganic phosphate (^{31}P MRS, *pH Measurement In Vivo in Whole Body Systems*) is apparently unaltered.[23]

In extracts of portacaval shunt rat brain, high-performance liquid chromatography (HPLC) confirms the accumulation of glutamine and the depletion of glycerophosphorylcholine (GPC). The latter has also been demonstrated in human subjects using a technique of quantitative proton-decoupled ^{31}P MRS.[24] It partly explains the reduced Cho in the prior ^{1}H NMR study as well as demonstrating the expected depletion by 50% or more of the cerebral mI and *scyllo*-inositol (sI) content.[25] This result in rats confirmed the observations that first emerged from human studies with short-echo time ^{1}H MRS[13,26,27] and establishes the validity of the portacaval shunt rat for future experimental studies.

1.4 Pathogenesis, Diagnosis, and Therapeutic Management of Hepatic Encephalopathy in Humans: The Emerging Role of Proton MRS

Studies using stimulated echo acquisition mode (STEAM) localized, short echo time ^{1}H MRS defined the changes in 10 patients with clinically confirmed chronic HE. The average increase in cerebral glutamine was estimated as +50%, Cho decreased 14% and mI decreased by 45%[21] (Figure 1). Very similar findings have now been reported at 2 T by Bruhn (Figure 2)[26,29] and by McConnell[30] using PRESS at short or long echo times. An early study which used long echo time PRESS-localized ^{1}H MRS failed to identify the depletion of mI but was the first clearly to show 'Cho' depletion in human brain.[31]

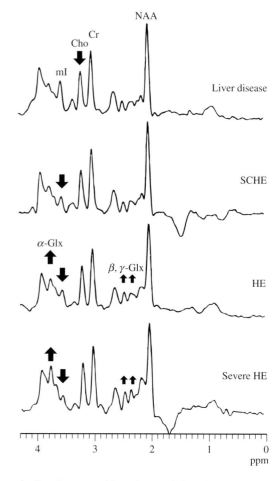

Figure 1 Development of hepatic encephalopathy in human subjects (1.5 T spectra). A series of spectra of parietal cortex (white matter) acquired under closely similar conditions (GE Signa 1.5 T, STEAM localization *TR* 1.5 s, *TE* 30 ms, NEX 128) from different patients is presented. A normal spectrum for comparison is that in Figure 7(b). In liver disease (top) there is a relatively normal spectrum with a slight decrease in choline (Cho). In subclinical hepatic encephalopathy (SCHE) there is a definite decrease in *myo*-inositol (mI) with a minor increase in the glutamine (Gln) regions (glutamine plus glutamate, Glx). There is a very significant increase in the Glx regions in HE (grade 1) and mI is further depleted. The spectrum of grade 3 HE shows more severe changes in the biochemical markers of this disease, most notably glutamine. Cr, creatine

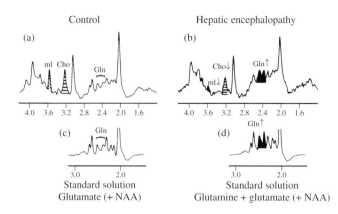

Figure 2 Identification of glutamine in the proton spectra at 2.0 T. A 51-year-old patient with HE resulting from surgical portacaval shunt (b) is compared with a normal control (a). Spectra were obtained from an occipital gray matter location and show the expected changes of HE: increased glutamine, decreased Cho/Cr and mI/Cr. With the improved resolution at 2 T, separate analysis of glutamine and glutamate is possible. Inset are spectra from model solutions (c) 5 mM glutamate + 5 mM N-acetylaspartate (NAA); (d) 5 mM glutamine + 5 mM glutamate + 5 mM NAA, indicating that in this patient the increase in glutamine occurs without obvious depletion of cerebral glutamate. Spectra were acquired on a Siemens 2.0 T clinical spectrometer, with STEAM localization; *TR* 3 s, *TE* 20 ms. Abbreviations as in Figure 1. (Modified from Bruhn et al.,[26] with permission from J. Frahm, T. Michaelis and colleagues)

This information considerably alters our understanding of the pathogenesis of HE, emphasizing an underlying defect in cerebral osmoregulation in addition to the clear relationship to ammonia toxicity and glutamine accumulation by the cerebral astrocytes. Proton-decoupled ^{31}P MRS revealed additional osmotic and metabolic disturbances in patients with HE.[24] Quantitative analyses in 16 patients with liver disease, ten with and six without chronic hepatic encephalopathy, in four patients with hyponatremia, and in 20 age-matched normal subjects were reported (Figure 3). Patients with HE were distinguished from controls by significant reduction in cerebral nucleoside triphosphate (NTP = ATP) (2.45 ± 0.20 versus 2.91 ± 0.21 mmol kg^{-1} brain; $P < 0.0003$), inorganic phosphate ($P < 0.03$) and phosphocreatine ($P < 0.04$).

In addition to increased cerebral levels of glutamate plus glutamine (Glx) and decreased concentrations of mI, patients with HE showed reduction of total visible Cho (in ^{1}H MRS), GPC (0.67 ± 0.13 versus 0.92 ± 0.20 mmol kg^{-1} brain in controls; $P<0.005$), and glycerophosphorylethanolamine (GPE) (0.40 ± 0.12 versus 0.68 ± 0.12 mmol kg^{-1} brain in controls; $P < 0.0003$) (in proton-decoupled ^{31}P MRS). Of the reduction of 'total Cho', 61% was accounted for by GPC, a cerebral osmolyte. Similar metabolic abnormalities were seen in hyponatremic patients.

The results are consistent with disturbances of cerebral osmoregulation and energy metabolism in patients with chronic HE.[24]

1.5 Energetics of the Human Brain in Hepatic Encephalopathy

The predicted cerebral energy deficit (reduction in [ATP]) has previously only been convincingly shown in mice and tree shrews.[32] Phosphorus-31 MRS should be the easiest tool with which to establish any energy deficit in HE (as predicted) but Ross,[33] Tropp,[34] Chamuleau,[35] Barbiroli,[36] and Morgan[37] have obtained conflicting and hence unconvincing data with ^{31}P MRS on such effects in humans. Using quantitative ^{1}H MRS, however, Geissler and colleagues showed a small but significant increase in cerebral creatine concentration [Cr] following liver transplantation in humans.[38] Since [Cr] is the sum of phosphocreatine (PCr) and creatine, this may indicate that patients with HE have an 'energy deficit'. A 10% reduction of PCr and a 16% reduction of NTP (of which perhaps up to 80% of the ^{31}P signal arises from ATP) is reported by Bluml et al. in patients with HE[24] and is hypothesized to be a contributing factor in the complex pathogenesis of the disorder. Because ATP depletion occurred without a marked increase in glutamine accumulation, the depletion of TCA cycle intermediates, hypothesized to limit energy metabolism,[5] is deemed an unlikely explanation. Instead, the hypothesis proposed is that ATP depletion may occur in the astrocytes as part of a failure of astrocyte volume regulation, which is reflected in reduced concentrations of several cerebral osmolytes. In the astrocyte, a possible rate-limiting role of ATP as a co-factor for GS would be most relevant. The difficulty which has been confronted in demonstrating such an effect by ^{31}P MRS, where sensitivity is 10% that of ^{1}H MRS, becomes obvious. Since reduced cerebral oxygen consumption and blood flow are recognized abnormalities in HE, the ^{13}C NMR techniques pioneered by Shulman may be best suited to demonstrate the anticipated reduction in the overall rate of the Krebs cycle.[39]

From these clinical studies, it appears that ^{1}H MRS, particularly if applied with effective water suppression to reveal mI, is currently most efficient for the elucidation of pathogenesis of HE, and also, as will be shown, for its early clinical diagnosis. Direct measurement of mI by ^{13}C NMR may become a valuable adjunct.[40-42]

2 SUBCLINICAL HEPATIC ENCEPHALOPATHY

As indicated, HE is a group of diseases presumed to have identical etiology but presenting a great variety of distinct clinical pictures. Underlying all of them is believed to be the entity of 'subclinical HE' (SCHE), although it is by no means clear that the HE and coma of fulminant (very acute) liver failure, goes through any truly 'subclinical' phase.

Proton MRS accurately reflects the entity of SCHE,[43,44] and in preliminary studies also appears to mirror the progressive and increasingly severe syndromes of overt HE defined by Parsons-Smith as grades 1–4 (Figure 1). Elevation of glutamine is rather more obvious at 2 T (Figure 3) but presents little difficulty at 1.5 T.

Figure 1 should not be taken as proof, however, that in the individual patient such orderly progression of neurochemical dysfunction occurs. Longitudinal studies have not been performed in sufficient numbers to be certain of this. Nevertheless, it is tempting to suggest, as Figure 1 appears to indicate, that in the brain exposed to liver 'toxins', Cho depletion (principally GPC) precedes the loss of mI and sI, with later accumulation of glutamine. If this sequence is correct, then perhaps 'sensitization' of the brain is the result of mI or

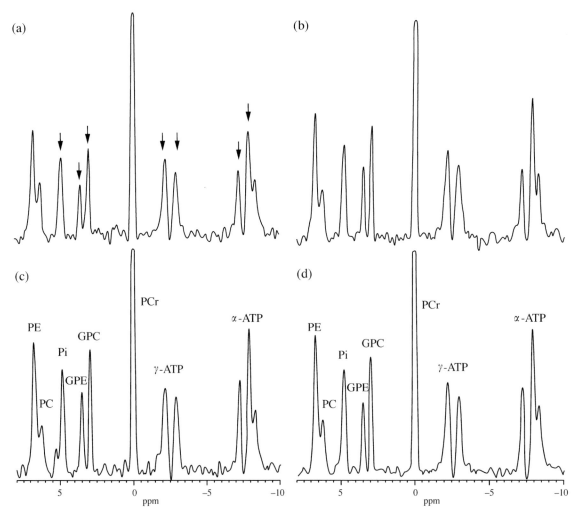

Figure 3 Osmotic and metabolic disturbances in hepatic encephalopathy (HE). Averaged proton-decoupled ^{31}P MRS from patients with (a) HE, (b) liver disease without HE, (c) elderly controls and (d) young controls. The averaged spectra calculated from all patients with HE can readily be distinguished from averaged spectra obtained from liver-diseased patients without HE and controls by reduced concentrations of phosphoethanolamine (PE), inorganic phosphate (Pi), glycerophosphorylethanolamine (GPE), glycerophosphorylcholine (GPC), and ATP. Furthermore, phosphocreatine (PCr) content (not obvious here owing to the expanded scale) was significantly decreased in HE. Phosphorylcholine (PC) remained normal

Cho depletion, or both. In keeping with the theory of many years standing, increasing cerebral glutamine underlies the neurological syndromes of grades 1–4, severe, overt chronic HE, as well as, perhaps, that of acute, fulminant HE and coma. Figure 4 shows the similar but more severe neurochemical changes of Reye's syndrome, giving an effective indication of what may be seen in acute HE.[45] Unfortunately, published spectra from patients with fulminant HE are limited and difficult to interpret.[46]

2.1 Depletion of *myo*-Inositol and the Induction of Hepatic Encephalopathy

A human 'experiment' which goes some way towards verifying this sequence is the new interventional procedure known as TIPS (transjugular intrahepatic portal systemic shunt), which is used as a life-saving procedure in cirrhosis-induced hematemesis. Not surprisingly, TIPS induces clinical HE in up to 90% of survivors.[47–49] Proton MRS performed both before and after TIPS in 10 patients showed a universal increase in mean intracerebral glutamine following the procedure. More importantly, in a small number of individuals in whom mI was normal prior to TIPS (these patients often show SCHE), a progressive reduction in mI/Cr and development of subclinical and clinical HE follows the introduction of the shunt (Figure 5).

2.2 Restoration of Cerebral *myo*-Inositol and Choline Accompanies Reversal of Hepatic Encephalopathy

Yet another common human 'experiment', that of orthotopic liver transplantation, allows the reverse process to be unequivocally demonstrated, thereby establishing a firm, albeit

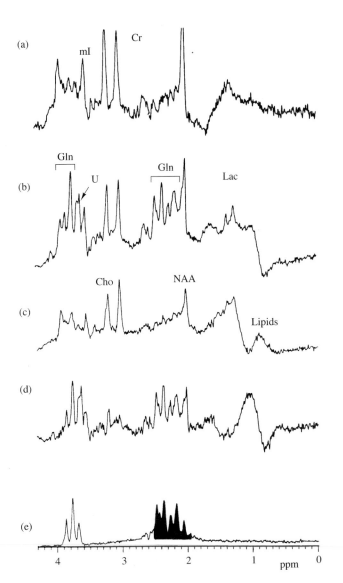

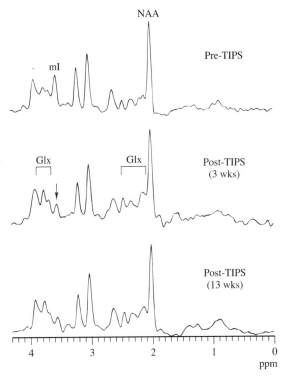

Figure 5 Effect of a transjugular intrahepatic portal systemic shunt (TIPS) on the proton MR spectrum. Spectra are from a 30-year-old female, 5 days after a hematemesis caused by esophageal varices and chronic alcoholic liver disease. Spectra acquired from a parietal white matter location (GE Signa 1.5 T, volume 12.5 cm^3, STEAM *TR* 1.5 s *TE* 30 ms, NEX 128) before a TIPS procedure shows no abnormalities apart from a small but significant reduction in Cho/Cr, attributable to liver disease (top). Three weeks after TIPS, changes are seen in the Glx and mI regions and a further reduction on Cho/Cr (middle). The bottom spectrum shows the progression of changes seen (13 weeks post-TIPS); Glx is markedly increased and mI is significantly reduced. Abbreviations as in Figure 1. (With permission from Shonk et al.[48])

Figure 4 Proton MRS in acute hepatic encephalopathy caused by Reye's syndrome. In vivo proton MRS spectra acquired from a parietal white matter region in infant brain from: (a) 10-month-old normal subject; (b) patient, day 2 after admission; (c) patient, day 8 after admission; (d) difference (b) − (c) between days 2 and 8 in patient; (e) solution with 15 mM glutamine. All spectra are scaled. Spectral assignments: Lac, lactate (1.3 ppm); NAA, *N*-acetylaspartate (2.02 ppm); Gln, glutamine (2.10–2.50 ppm, and 3.65–3.90 ppm); Cr + PCr, creatine + phosphocreatine (3.03 ppm); Cho, choline-containing compounds (3.23 ppm); mI, *myo*-inositol (3.56 ppm). U, Unassigned (3.62 ppm). Notable abnormalities concern a huge accumulation of cerebral Gln, reduced Cho, and, later, reduced mI, all of which reflect liver failure. In addition there is a decrease in NAA and [Cr] and appearance of the unassigned peak. Reye's syndrome, a toxic viral disease associated with aspirin intake, is known to produce severe neuronal damage, and may not, therefore, completely correspond to the picture of acute liver failure. An occipital gray matter region gave almost identical results. (Reproduced with permission from Ernst et al.)[45]

circumstantial link between mI depletion and the syndromes of SCHE and overt HE (Figure 6). There was also a recovery in Glx and Cho also recovered. Indeed, an overshoot of cerebral Cho is consistently observed, perhaps linking the earlier Cho depletion with deficient hepatic biosynthesis of some relevant precursor of GPC.

2.3 Ornithine Transcarbamylase Deficiency

A description of MRS in HE would be incomplete without consideration of a rare but informative inborn error of hepatic urea synthesis, ornithine transcarbamylase (OTC) deficiency. The single known biochemical consequence is hyperammonemia. Gadian and colleagues[50] demonstrated the inevitable elevation of cerebral glutamine in two such patients, while Ross[51] showed the extraordinary parallel with HE, in depletion of cerebral mI (Figure 7).

For References see p. 1123

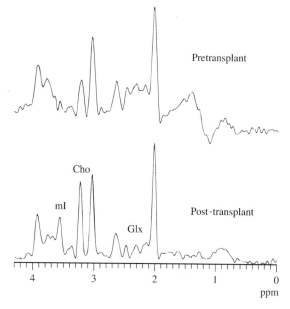

Figure 6 Restoration of biochemical abnormalities post-liver transplant. The patient is a 30-year-old male with acute-on-chronic hepatic encephalopathy secondary to hepatitis, and subsequently successfully treated by liver transplantation. Spectra were acquired 6 months apart from the same parietal white matter location, (15.0 cm³ STEAM *TR* 1.5 s, *TR* 30 ms, NEX 128) and scaled to the same Cr intensity for comparison. The obvious abnormalities before liver transplantation, increased α-, β-, and γ-glutamine, reduced Cho/Cr and mI/Cr (upper spectrum) were completely reversed 3 months after transplantation, and Cho/Cr exceeded normal (lower spectrum). Abbreviations as in Figure 1

3   CONTRIBUTION OF NMR TO ASSESSMENT OF CLINICAL HEPATIC ENCEPHALOPATHY

3.1   Pathogenesis

Both experimentally and clinically, NMR (particularly ¹H MRS) supports the classical concept of HE as a disorder of cerebral ammonia metabolism, be it by ammonia toxicity or glutamine synthesis (as a newer variant of the theory would have it).[52] The concept of an underlying brain 'sensitization' receives substantial new impetus deserving of further research in animal models. Either Cho (GPC) depletion, caused perhaps by failure of hepatic synthesis of a necessary precursor, or cerebral mI depletion could fulfill the role of 'sensitizer'. In neither case is there a precedent; as a result basic research is urgently required. It is likely that NMR will play a crucial role in such investigations, and the portacaval shunt rat is a convenient model. Carbon-13 NMR is the only method of unequivocally determining mI as distinct from the lower concentrations of inositol-1-phosphate (Inos-1-P) and glycine with which mI coresonates in the ¹H MR spectrum.

3.2   Osmoregulation in Human Hepatic Encephalopathy

A new concept in HE can justifiably be attributed to these endeavors with in vivo MRS. This is related to cell-volume

regulation in the brain and is conveniently termed osmolyte disorder.

GPC was first identified as an osmolyte in the kidney, while mI, GPC and taurine (in rats) were proposed for that role in brain,[53] Haussinger et al.[54] first used the term hypo-osmolarity to link mI with HE. The importance of the concept may lie in the hitherto obscure connection between hepatic coma (in fulminant liver failure), a disease treated as an emergency because of the massive edema that occurs (dysregulation of cerebral osmolytes perhaps?), and the more mundane and relatively slow-evolving biochemical disturbance outlined above as chronic HE and SCHE.

3.3   Proton MRS for Diagnosis

Early in these investigations, it became apparent that mI depletion and Glx accumulation occurred when clinical HE was absent. Other systemic or metabolic diseases (apart from OTC deficiency already discussed) did not result in mI depletion; consequently ¹H MRS offers an unique opportunity for early,

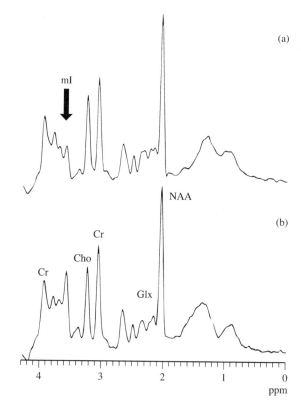

Figure 7 Treated ornithine transcarbamylase (OTC) deficiency versus healthy age-matched subject. Both spectra were acquired from a similar location in parietal white matter (STEAM *TR* 1.5 s, *TE* 30 ms, NEX 128) and processed as described in the literature.[47] Single-enzyme defects in the urea cycle result in severe hyperammonemia and 'hepatic encephalopathy'. Because the patient was receiving treatment with sodium benzoate, no excess of cerebral glutamine is present. However, as anticipated by studies in the commoner condition of hepatic encephalopathy caused by portosystemic shunting, a decrease in mI was noted in the proton spectrum of a 14-year-old with OTC deficiency (a) when compared with a normal age-matched control (b). NAA, *N*-acetylaspartate; other abbreviations as in Figure 1

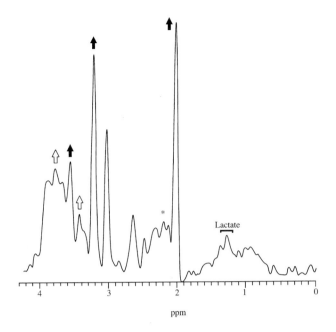

Figure 8 Proton MRS of occipital gray matter in recovering nonketotic hyperglycemia. Excess Cho/Cr, NAA/Cr, mI/Cr (▲) as well as abnormal intensities for glucose (⇧), lactate and (possibly) acetone (*) are indicated. Abbreviations as in Figure 1

specific diagnosis of this still perplexing condition. Paradoxically, there is at present little enthusiasm for this, most probably because prevention (with lactulose or neomycin) and treatment (by liver transplantation) of overt or severe HE is relatively straightforward (albeit rather costly in the case of orthotopic liver transplantation (OLT) at $150 000–200 000 per patient in USA). If a ready medical means of restoring cerebral mI were to be discovered, the value of ^{1}H MRS in diagnosis might increase. This point is covered in a recent report in which treatment with lactulose for 7 days in overt HE significantly increased mI/Cr.[55,56]

3.4 Unanswered Questions in Neurology of Hepatic Encephalopathy

Wilson's disease, caused by excess copper deposition, is believed to result in an encephalopathy analogous to HE. However, the ^{1}H MRS findings are not surprisingly rather different, lacking either mI depletion or glutamine accumulation (unpublished study from this laboratory).

Myelopathy is an unusual form of chronic HE. It presents with paraplegia. Athough neurological considerations would suggest cord involvement, the ^{1}H MRS findings in the parietal cortex are typical of other patients with the more classical clinical presentation (see Fig. 4 in Ross et al.).[51]

Often noted in MRI, the basal ganglia may be 'bright' in inversion recovery (IR) images of patients with known HE. There is no consistent relationship with ^{1}H MRS findings, and increasingly this MRI finding is recognized as nonspecific. Nevertheless, the extrapyramidal signs, the changes on postmortem, and these albeit inconsistent MRI findings continue to suggest that there may be as yet unrecognized underlying neurochemical changes in the basal ganglia in HE.

4 OTHER SYSTEMIC ENCEPHALOPATHIES

4.1 Diabetes Mellitus and Ketogenesis

Like HE, diabetic encephalopathy is common and obviously 'metabolic' in origin. Three principal syndromes are recognized: diabetic ketoacidosis, lactacidosis, and nonketotic hyperosmolar coma.

Elevated cerebral glucose,[57] significant excess of mI, a reversible accumulation of 'Cho', and the presence of ketone bodies have been detected in various patients with varying severity of diabetic encephalopathy.[58] The hyperosmolar state, which is discussed in more detail below, is well known in diabetes mellitus as so-called nonketotic hyperglycemia (NKH). NKH is responsible for coma and neurological symptoms more frequently even than diabetic ketoacidosis. MRS findings in a young adult with NKH-induced coma[59] are consistent with many years of experimental work in demonstrating altered cer-

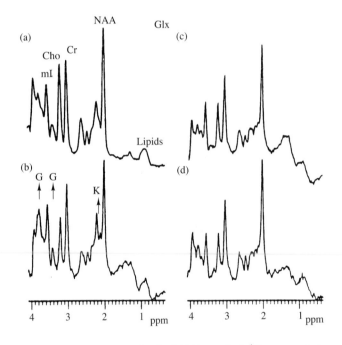

Figure 9 Diabetic ketoacidosis (DKA): cerebral ^{1}H spectra during two DKA episodes and recovery. (a) Episode 1, acquired 3 days after admission to hospital, at a time when the patient had supposedly totally recovered and was ready to be discharged. A peak characteristic for the presence of ketone (K) bodies was noted at 2.22 ppm, obtained from an 18 cm³ volume in the left parietal lobe. The patient relapsed into DKA 2 days later. (b) Episode 2, acquired 5 months later from an occipital gray matter location (10.3 cm³) during a second episode of DKA. In addition to the ketone peak, peaks for glucose (G) were seen. (c) Recovery, 6 days after episode 2 (from the same occipital location), when no more ketone bodies were detected in the urine. (d) Occipital cortex of an age- and sex-matched healthy subject. Acquisition conditions: GE Signa 1.5 T, 4X software; STEAM *TR* 1.5 s, *TE* 30 ms, NEX 128. More detailed analysis of peak K indicates the resonance frequency to be that of acetone, rather than acetoacetate, the ketone body more commonly identified in blood and urine of diabetics in coma. Abbreviations as in Figure 1. (Reproduced by permission of the Radiological Society of North America from R. Kreis and B. D. Ross, *Radiology*, 1992, **184**, 123)

For References see p. 1123

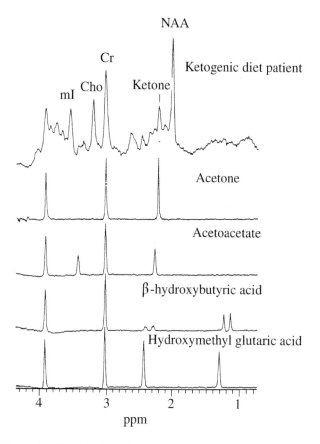

NAA

Cr

mI   Cho   Ketone   Ketogenic diet patient

Acetone

Acetoacetate

β-hydroxybutyric acid

Hydroxymethyl glutaric acid

4   3   2   1

ppm

Figure 10 Identification of the ketone body observed in patients on the ketogenic diet. Top trace is the summed ^{1}H MRS of three examinations of a patient on ketogenic diet. Lower four traces show ^{1}H MRS of model solutions containing 5 mmol l^{-1} of acetone, β-hydroxybutyric acid and hydroxymethyl glutaric acid, and 10 mmol l^{-1} acetoacetate. Each model solution also contained 10 mmol l^{-1} creatine as a chemical shift reference at 3.02 and 3.94. Abbreviations as in Figure 1. (Reproduced with permission from Seymour et al.)[60]

ebral osmolytes, Cho, mI and even *N*-acetylaspartate (NAA) (Figure 8).

The most surprising finding of proton MRS, which requires further study, is the identification of acetone (rather than the more widely anticipated β-hydroxybutyrate and acetoacetate) as the ketone of human diabetic ketoacidotic encephalopathy (Figure 9).[58]

A therapeutic idea dating from the 1920s has reemerged in the use of a so-called 'ketogenic diet' to provide excellent control of seizures in drug-resistant patients with epilepsy. Perhaps not surprisingly, the cerebral ketone body that accumulates in these patients is also acetone, rather than acetoacetate or β-hydroxybutyrate (Figure 10).[60] In epilepsy, we suggest that this finding is more than of academic importance and may ultimately lead to more focused dietary or pharmacological treatments. At the very minimum, ^{1}H MRS should be considered for clinical monitoring in iatrogenic ketosis.

Long-term neurological and cerebral 'complications' of diabetes contribute to the much increased mortality. The biochemical basis of these conditions may lie in those changes in Cho, NAA, mI, ketones, and glucose, now recognized by ^{13}C and ^{1}H MRS to be present in the acutely and chronically diabetic brain.

4.2 The Hyperosmolar State: Identification of Idiogenic Organic Osmolytes by Proton MRS

Lien et al. first used ^{1}H MRS to investigate a 'new' family of cerebral metabolites collectively known as organic osmolytes because of their believed role in the maintenance of cerebral osmotic equilibrium.[61] Such molecules were also first thoroughly researched in the papilla of the kidney with the help of in vitro NMR.[62]

First in sporadic cases of diabetic hyperosmolar coma[58] and in a patient after closed head injury,[63,64] and then most convincingly in a single infant with holoprosencephaly and deficient thirst mechanisms, these concepts were confirmed as contributing to human encephalopathy by the use of ^{1}H MRS (Figure 11). mI was three times normal, and other resonances, also markedly affected, returned toward normal with treatment. Difference spectroscopy is particularly helpful in identifying

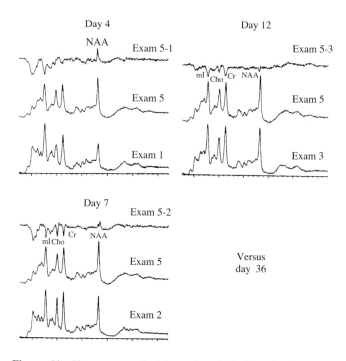

Day 4
NAA
Exam 5-1

Exam 5

Exam 1

Day 7
Exam 5-2

mI Cho   Cr   NAA

Exam 5

Exam 2

Day 12
Exam 5-3

mI   Cho   Cr   NAA

Exam 5

Exam 3

Versus
day 36

Figure 11 Time course of changes in principal cerebral organic osmolytes during correction of severe dehydration. A series of spectra were obtained from the same (occipital gray matter) brain location, in a 14-month-old child, recovering from severe dehydration and hypernatremia (plasma sodium 195 mmol l^{-1}; normal range 135–142 mmol l^{-1}) using identical acquisition conditions. Spectra were processed and scaled identically to permit subtraction of sequential spectra (difference spectroscopy). The day of examination refers to interval since admission to hospital. Examinations are numbered sequentially, from the first (Exam. 1) to last (Exam. 5) on day 36. Compared with a relevant normal [spectrum (b), Figure 7], the principal abnormality appears to be a reversal of the intensities of *N*-acetylaspartate (NAA) (reduced) and mI (increased); as a result mI dominates the spectrum. These changes slowly reverse to be nearly normal on day 36. The resultant difference spectra more clearly identify the progressively falling concentrations of several metabolites, with a possible elevation in the resonance peak assigned to the neuronal marker NAA. Quantitative MRS defines the principal abnormality as a threefold increase in the concentration of the cerebral osmolyte mI. Abbreviations as in Figure 1

these changes.[65,66] The converse, or hypoosmolar state of hyponatremia has been identified by ^{1}H MRS in significant numbers of patients.[67]

4.3   The Hypoosmolar State: Hyponatremia and Pituitary Failure

Hyponatremia is so common that its impact on the human brain spectrum needs to be included in all discussion of clinical MRS interpretations. The characteristic features are reduction of mI/Cr and Cho/Cr in the ^{1}H MRS examination[67] and reduced GPE and GPC in the proton-decoupled ^{31}P spectrum.[24] Similarities to the findings in HE have already been discussed. Unexpected reduction of mI/Cr or Cho/Cr, in, for example, screening MRS examinations for dementia might be an indicator of hyponatremia.[59]

Central pontine myelinolysis (CPM) is a rare clinical disorder believed to follow inappropriate corrections of severe hyponatremia in patients. This link led Lien and colleagues to investigate the value of in vitro MRS of the newly discovered cerebral osmolytes in rats.[68]

The correction of cerebral osmolytes disturbed by hyponatremia is even more striking than the initial discovery of their depletion because of its relevance to clinical management. Slow restoration of plasma sodium is advocated and is accompanied by even slower restoration of cerebral osmolytes. In one human study, a mean interval of 10.5 weeks elapsed before cerebral osmolyte correction in three of four patients.[67] Now we have identified the MRS pattern of CPM in a patient in whom hyponatremia was not apparent, and in whom cerebral osmolytes were normal (Figure 12). An illuminating study of hyponatremia secondary to pituitary failure suggests that some cerebral osmolytes may also be markers of hormone activity. A

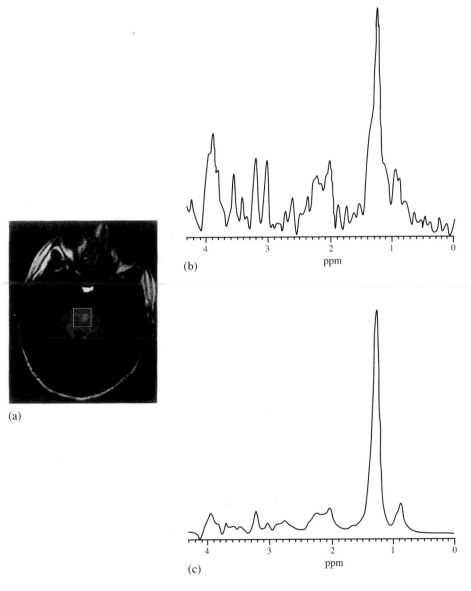

(a)

(b)

ppm

(c)

ppm

Figure 12 Proton MRS of the pons in central pontine myelinolysis (CPM). (a) A 29-year-old female with characteristic MRI changes that developed in the absence of hyponatremia. The spectra [9 weeks post-insult (b); 1 month later (c)] shows marked elevation in lipid, which appropriately reflects the histology of this lesion in autopsy-proven cases of CPM. (Illustration courtesy of David Dubowitz M.D.)

For References see p. 1123

clearer understanding of the neuroendocrine axis may emerge from systematic use of MRS (either ${}^1H$ or ${}^1H–{}^{31}P$) in these very common disorders (Figure 13).

5 HYPOXIC ENCEPHALOPATHY AND 'BRAIN ATTACK'

The prime example of a systemically induced encephalopathy is that resulting from insufficient oxygenation of blood, with resultant failure of cerebral oxygen delivery. Hypoxic encephalopathy is best understood in the context of energy failure and altered redox state through all cerebral metabolic pathways. Loss of ATP and PCr, accumulation of ADP, AMP, inorganic phosphate, and Cr are obvious consequences; H^+ accumulates, both from failure to remove CO_2 and from formation of lactate. Reduced partners of the equilibrium enzymes lactate and glutamate dehydrogenase accumulate. In practice, glutamate is probably equally rapidly converted to glutamine. Innumerable animal studies, using principally ${}^{31}P$ NMR, but also ${}^1H$ NMR, confirm these principles and document other previously unsuspected changes, notably the loss of the neuronal marker NAA.

The very common occurrence of hypoxic encephalopathy in humans has given ample opportunity for verification of these events in the human brain. 'Recovery' from nonlethal hypoxic encephalopathy gives yet another view of the metabolic process. From ${}^{31}P$ MRS in neonates,[69] the expected changes emerged.

More recently, ${}^1H$ MRS has been used to quantify and plot the time course of changes which occur after oxygen deprivation, applying the information to defining the degree of irreversible hypoxic neuronal damage—and hence prognosis. 'Near-drowning' is the term applied when virtually total oxygen deprivation occurs owing to submersion in water. It must be presumed that all ATP and PCr is lost, and all glucose converts to lactate in this period of anoxia. Yet at the first clinical examinations, 24 hours after rescue and artificial life support, MRS demonstrates more often than not the absence of lactate, normal Cr (plus PCr), and NAA at nearly full concentration. Only subsequently do NAA and total Cr fall and lactate appears. The severity of the changes gives a fairly good guide to outcome.[70]

The classic events of anoxia described by Lowry[71] are probably reversible, but secondary damage results in progressive cell death, the consequences of which are loss of NAA (in the case of dying neurons), loss of PCr and oxidative function and reaccumulation of lactate. It is unclear why the large accumulations of glutamine occur (Figure 14). A systematic analysis of the MRS changes and their impact on outcome after global hypoxia[72] should go some way to clarifying the MRS literature in focal ischemia of the human brain in stroke and provide much needed noninvasive guidance in treatment of hypoxic encephalopathy. This includes tissue plasminogen activator (TPA), neuroprotectives and other innovations in human 'brain attack'.

5.1 Chemical Shift Imaging in Acute Stroke

Most clinical questions can be quickly answered using single-voxel technique discussed in the previous section, par-

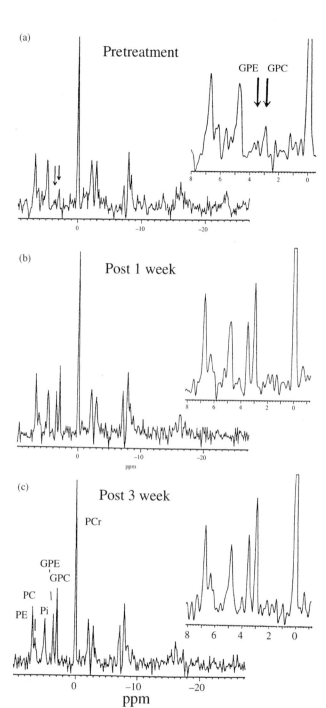

Figure 13 Endocrine-responsive hyponatremia. Proton-decoupled ${}^{31}P$ MRS shows almost complete depletion of GPE and GPC [(a) pretreatment] that recovered rapidly after initiation of pituitary replacement therapy [after 1(b) and 3(c) weeks of treatment]. Inset are expanded portions of the spectra. Abbreviations as in Figure 3. (Adapted from J. Tan et al., unpublished work in this laboratory)

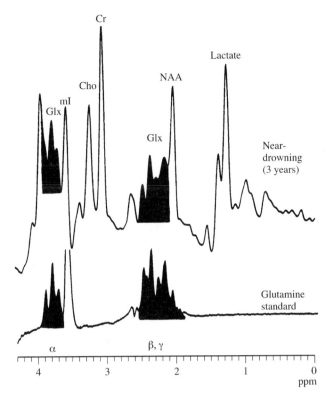

Figure 14 Near-drowning with fatal outcome. Brain spectrum from occipital cortex of a severe near-drowning victim (3 years old), examined 48 hours post-immersion (top), and the glutamine standard (bottom). The patient died on the 5th day post-injury. The ^{1}H spectrum was acquired from a volume of 12 cm^3 (STEAM *TR* 3.0 s, *TE* 30 ms, data processing including correction for residual water, line fitting, and quantification as described by Kreis et al.).[63] The spectrum of the glutamine standard (10 mmol l^{-1}) was acquired in the same way and scaled appropriately. Notable abnormalities in the patient spectrum are the very much reduced *N*-acetylaspartate (NAA) concentration and NAA/Cr ratio, excess of lactate (doublet at 1.3 ppm), and of the strongly coupled resonances of α, β, and γ glutamine protons. A 25% reduction in [Cr] was apparent on quantitative examination. Abbreviations as in Figure 1. (Thanks to R. Kreis, T. Ernst, and E. Arcinue)

ticularly since the arrival of complete automation of voxel selection, localization, shimming, water suppression, data acquisition, phasing and semiquantitative print-out of the salient metabolite ratios.[73] Nevertheless, chemical shift imaging (CSI), a much more efficient use of MR signal,[74] is re-emerging for clinical questions in which multivoxel information is essential. In our opinion, only stroke (in adults) and adrenoleukodystrophy (ALD) in children completely fulfill this criterion. In tumor MRS, for example, single voxel[75] matches CSI with 96–97% sensitivity for diagnosis,[76] making it generally unnecessary to use the more time-consuming CSI.

The MRS changes in chronic stroke are well documented. More recent interest in reversal therapies for acute stroke has emphasized the need for very early differentiation of penumbra, infarcts and transient ischemic attacks. When cerebral artery flow falls below 100 ml min^{-1} (normal ~130 ml min^{-1}), lac-

tate was observed in the symptomatic cerebral cortex with increasing frequency.[77] This single-voxel study was excellent in determining differences between the affected and unaffected hemisphere and in distinguishing borderzone infarcts, territory infarcts and no infarct.[77] However, obvious regional heterogeneity could not be determined. The power of a multivoxel (single-slice) ^{1}H MRS acquisition in distinguishing three degrees of neurochemistry after stroke is illustrated with spectra obtained for a patient who had suffered two strokes 7 days apart (Figure 15). These are seen in T_2 MRI and in diffusion-weighted images (DWI) as bright areas. While it was readily apparent which of the two infarcts occurred first, DWI could not immediately separate the acute from the subacute lesion. The synchronous acquisition of spectra from each shows subtle differences between them, and from normal as determined in the unaffected hemisphere. Intuitively, it is attractive to consider lactate (present as an inverted doublet at 1.3 ppm only in the more recent 'infarct') to be the much needed marker of viable tissue. This example is merely a pilot to the much larger studies now in progress in which CSI will be performed rapidly and as part of a broader MRI protocol of DWI, MRS and perfusion imaging.

5.2 Adrenoleukodystrophy

The most common of a very rare group of inherited brain disorders of white matter, the peroxisomal disease adrenoleukodystrophy, has now been studied exhaustively with single-voxel MRS. MRS is consistently more sensitive than MRI in detection of neurologic abnormality,[78] which leads to the next question. If MRI is normal in a leukodystrophy, where is the single voxel to be sited to achieve reliable preclinical diagnosis in this disorder? Currently we use the posterior horn of the lateral ventricle as a landmark to future disease. It would be safer to apply a robust multivoxel technique.[79]

6 OTHER FOCAL BRAIN DISORDERS

There is now a huge published database on which to found the clinical practice of MRS (neurospectroscopy). In papers where original data are provided (i.e. actual spectra with details of the processing applied after acquisition), this can be reliably assumed to be transferable from one instrument to another, and even across field-strength (1.0–4.02).[59] In the USA, a CPT code (76390) signals release of MRS into the day-to-day clinical environment by virtue of its efficacy in a number of common diagnoses. Limited use has been made of the discoveries reviewed in Sections 1–4, perhaps because MRI is often entirely normal! By and large, radiologists have used MRS to add detail in the differential diagnosis of lesions identified first by MRI: tumor recurrence versus necrosis, primary versus secondary tumor, stroke versus venous infarction, tumor versus stroke, abcess versus tumor in the immunocompromised patient.[80]

Single-voxel ^{1}H MRS readily distinguishes between lymphoma, toxoplasmosis, progressive multifocal leuko-

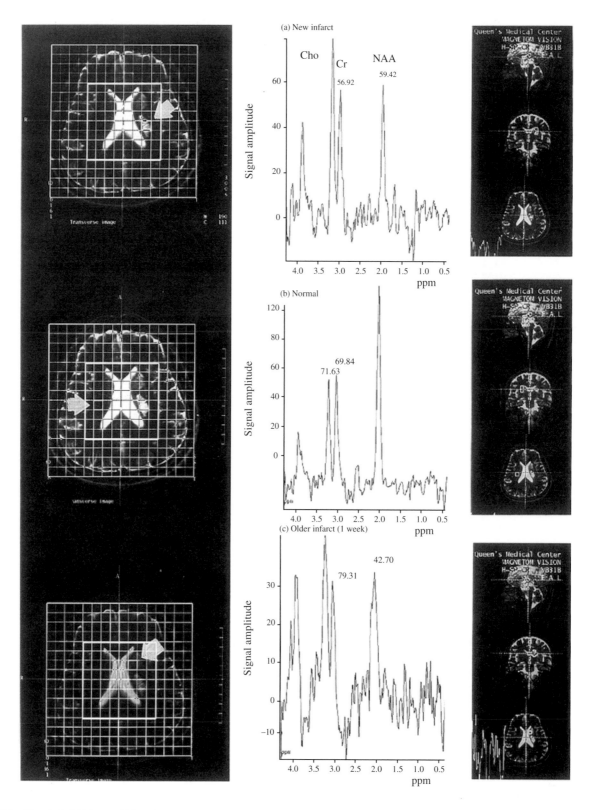

Figure 15 Chemical shift image (CSI) of acute stroke. The patient suffered a series of focal infarcts in the left internal capsule, seen in diffusion-weighted images (DWI) in left of figure. Three representative spectra (CSI, STEAM *TE* 135 ms) illustrates a new infarct (a), normal contralateral brain region (b) and a slightly older infarct, possibly 1 week earlier (c). Note that reduced *N*-acetylaspartate (NAA) and increased choline (Cho) are present in the early and later stages of evolving stroke. However, lactate (inverted doublet at 1.33 ppm) was present only in the more recent stroke. In this case, DWI and CSI give similar information. (Reproduced with kind permission of Stephen Holmes M.D., Queens Medical Center, Honolulu, HI)

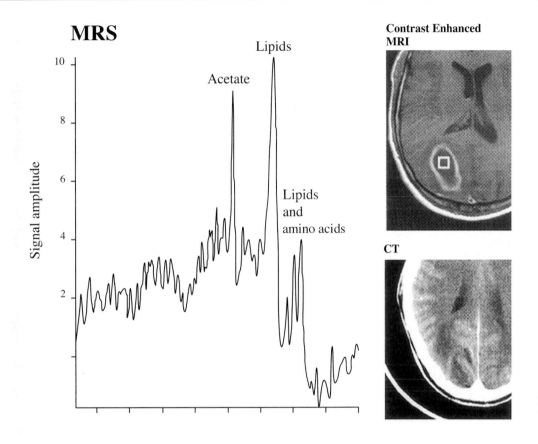

Figure 16   Diagnostic ^{1}H MRS of brain abscess. Computed tomographic scan (CT) was misleadingly diagnosed as malignant tumor in a young man. MRI and ^{1}H MRS confirms the presence of an abscess, based upon high concentrations of bacterial breakdown products within a ring-enhancing cyst. (Reproduced with kind permission of Else R. Danielsen, University of Copenhagen, Copenhagen, Denmark)

encephalopathy and coccidiomycosis in patients with AIDS.[81] Bacterial abcesses can no longer be confused with tumor, stroke, or multiple sclerosis plaque, since ^{1}H MRS identifies the abcess from its unique bacterial metabolic products (E. R. Danielsen, personal communication; Figure 16).

6.1   Central Pontine Myelenolysis

A rare focal lesion known pathologically to consist of lipid-filled macrophages similarly yields an unmistakable ^{1}H MR spectrum of triglyceride (see Figure 12). Lipid pathologies in the brain can be reliably studied only with short echo times. One example of a useful clinical MRS examination, prognosis after non accidental trauma (shaken-baby syndrome) hinges on identification of increasing concentrations of free lipid or 'macromolecules'[82] in ^{1}H MRS.[83,84]

6.2   Neurotransplantation in Humans

Among the neurodegenerative diseases regularly examined by ^{1}H MRS, only Alzheimer disease appears to have a diag-

nostic pattern;[85–87] Parkinson's disease and Huntington's disease do not. In particular, earlier demonstrations of lactate in brain spectra of patients with Huntington's disease, or of excess glutamate in basal ganglia of patients with Parkinson's disease, have not been confirmed.[88]

A recent study of patients who had received fetal cell transplants into the diseased putamen and caudate demonstrates a new clinical role of ^{1}H MRS in focal 'lesions'. The maturation, viability, partial rejection and subsequent recovery of the grafts can all be tracked in repeated ^{1}H MRS examinations (Figure 17).

7   ADVANCES IN MULTINUCLEAR CLINICAL NEUROSPECTROSCOPY

Human brain research with ^{31}P MRS and ^{13}C MRS are dealt with elsewhere in this series (*MRI and MRS of Neuropsychiatry*; *Localization and Registration Issues Important for Serial MRS Studies of Focal Brain Lesions*; *Central Nervous System Degenerative Disease Observed by MRI*; *Brain*

For References see p. 1123

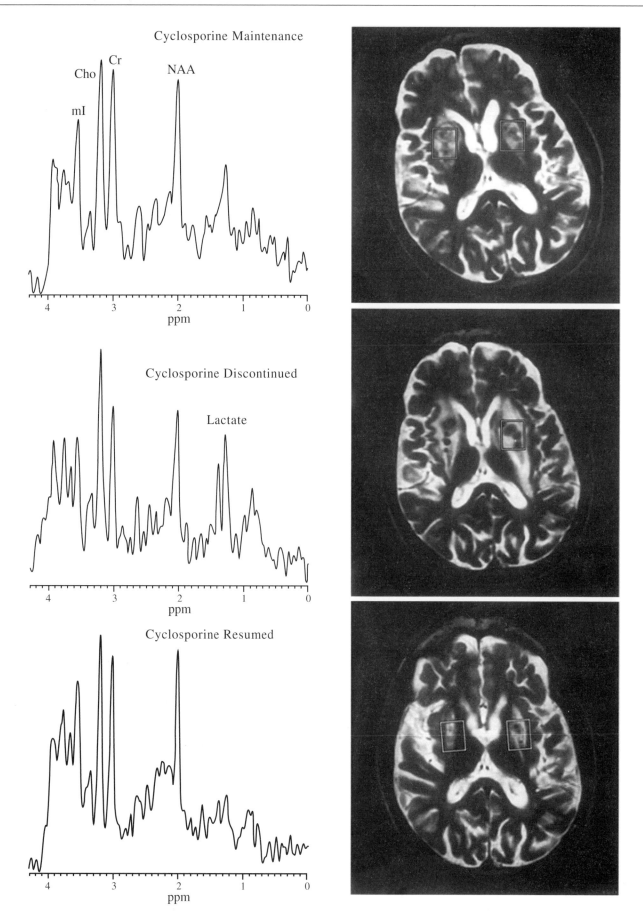

Cyclosporine Maintenance

Cho
Cr
mI
NAA

ppm

Cyclosporine Discontinued

Lactate

ppm

Cyclosporine Resumed

ppm

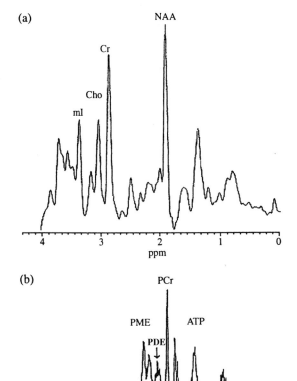

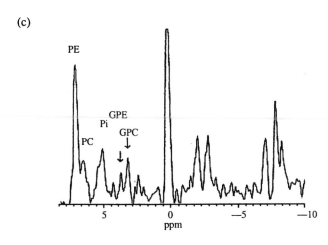

Neoplasms in Humans Studied by Phosphorus-31 NMR Spectroscopy; *Brain Infection and Degenerative Disease Studied by Proton MRS* and *Brain MRS of Human Subjects*). In the purely clinical setting, ^{31}P MRS remains disappointing. With the exception of the discovery of a new syndrome characterized by absence of PCr from the brain spectrum[89] and the characterization of brain death as spectra lacking both PCr and ATP, few diagnostic ^{31}P MR brain spectra have been identified.

However, proton-decoupling of ^{31}P spectra has brought significant clinical advantages and several diagnostic uses to this modality. In particular, proton-decoupled ^{31}P MRS now permits accurate evaluation of the state of myelination in the normally developing and dysmyelinating infant brain (Figure 18).[90]

The identification of other membrane or myelin disorders in adults is also gaining some specificity through the application of proton-decoupled ^{31}P MRS, since GPC, GPE, phosphorylcholine (PC) and phosphoethanolamine apparently vary almost independently in different pathological settings.[91] New insights into osmolyte disturbances of hepatic encephalopathy and hyponatremia were discussed above.

From being a research tool in the 1980s, natural abundance ^{13}C MRS has achieved clinical status through identification of mI, glutamate and several other useful neurometabolites in individual patients (Figure 19).[41]

FDA-IND approval of [1-^{13}C]-glucose (FDA 56,510; Figure 20)[91] infusion allows exploration in a clinical setting of the very promising techniques developed in the 1980s at Yale.[92] This is a dynamic technique, with many parallels to functional MRI. Carbon-13 MRS has particular relevance to the diagnosis and elucidation of the encephalopathies discussed at the beginning of this article.[93–95]

Figure 18 Identification of abnormal or delayed myelination with proton-decoupled ^{31}P MRS. MRS of a child with an unidentified dysmyelination syndrome who appears to lack significant amounts of white matter. (a) ^{1}H MRS of parietal 'white matter' shows significant reduction of Cho. An unidentified peak at 1.4 ppm is possibly lipid. (b) ^{31}P MRS with low–normal phosphodiester (PDE) for age. (c) Proton-decoupled ^{31}P MRS is markedly abnormal for a 2-year-old. GPE and GPC are reduced to the neonatal MRS level. PE and PC, however, are normal for age. Abbreviations as in Figures 1 and 3

8   CONCLUSIONS

Diffuse metabolic changes in brain biochemistry are the result of complex interactions of disordered biochemistry in many other organs and tissues. Hepatic encephalopathies, diabetic coma, hypo- and hyperosmolar states, endocrine and hypoxic encephalopathy are examples of such conditions in which accurate application of quantitative NMR spectroscopy sheds new light. A thorough knowledge of the clinical MRS of coma will almost certainly be of crucial importance to health care in these difficult cases. Clinical uses of ^{1}H MRS has focused on lesions already identified by MRI, but its uses are likely to expand. Routine clinical MRI scanners can now provide excellent informative proton-decoupled ^{31}P spectroscopy and ^{13}C brain spectra; consequently these tools too are now entering clinical use.

Figure 17   MRI and ^{1}H MRS in human neurotransplantation. MRI and MRS of an individual patient with Huntington's disease showing CNS immune response to resuming immunosuppression after halting for 6 months. Significant decrease of *N*-acetylaspartate (NAA/Cr) ratio and presence of lactate strongly suggests deterioration of graft growth without immunosuppression (middle). Abbreviations as in Figure 1

For References see p. 1123

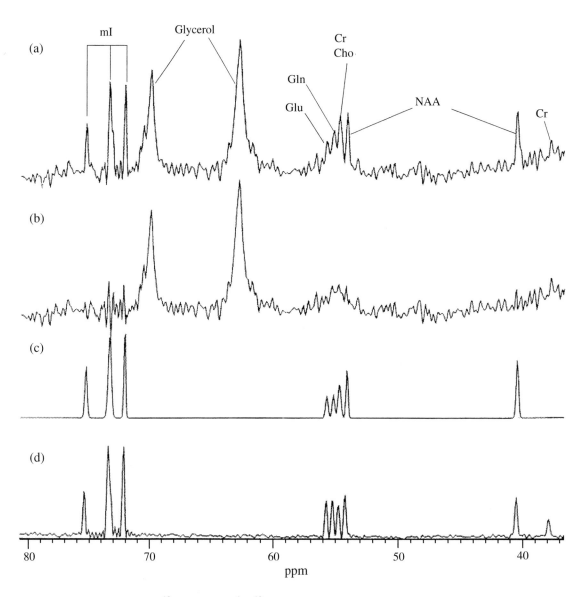

Figure 19 Clinical neurospectroscopy with ^{13}C MRS. (a) A ^{1}H–^{13}C spectrum acquired in Canavan disease. Metabolite peak areas were fitted and subtracted from the spectrum (b). From the fitted curves (c) the peak ratios of Glu, Gln, and N-acetylaspartate (NAA) relative to mI were calculated and compared with the peak ratios obtained from a model solution containing equal concentrations of NAA, Cr, mI, glutamate, and glutamine (d). Absolute concentrations in mmol per kilogram brain tissue were calculated using [mI] quantified with ^{1}H MRS in the occipital region of the brain as an internal reference. Abbreviations as in Figure 1

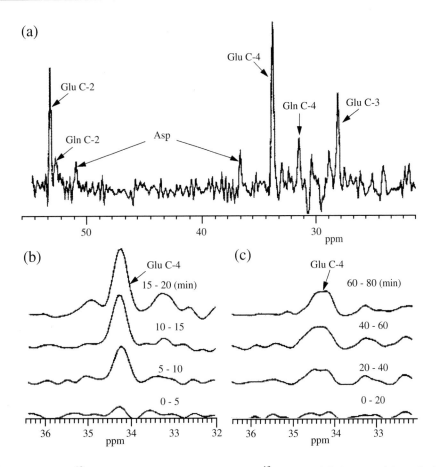

Figure 20   Dynamic ^{13}C MRS after [1-^{13}C]-glucose infusion. Using intravenous ^{13}C glucose infusion to enrich cerebral metabolite pools several fold higher than the 1% natural abundance (illustrated in Figure 19) provides a measure of glutamate (Glu) synthesis and metabolism. (a) Difference spectrum from a fed adult acquired 50–70 min after infusion start. Incorporation of the ^{13}C label into Glu C-4 in fasted mature (b) and immature brain (c) is clearly different and has diagnostic value in the premature infant (S. Bluml, J.-H. Wang, and B. D. Ross[91])

9   RELATED ARTICLES

Animal Methods in MRS; Brain MRS of Human Subjects; In Vivo Hepatic MRS of Humans; Single Voxel Localized Proton NMR Spectroscopy of Human Brain In Vivo; Water Suppression in Proton MRS of Humans and Animals.

10   REFERENCES

1. F. Plum and J. B. Posner, 'The Diagnosis of Stupor and Coma', Davis, Philadelphia, 1986.
2. G. B. Young, A. H. Ropper, and C. F. Bolton, 'Coma and Impaired Consciousness', McGraw-Hill, New York, 1998.
3. S. Sherlock, W. H. J. Summerskill, L. P. White, and E. A. Phear, *Lancet*, 1954, **2**, 453.
4. S. Bessman and A. Bessman, *J. Clin. Invest.*, 1955, **34**, 622.
5. S Bessman and N. Pal, in 'The Urea Cycle', eds. S. Grisolia, R. Baguena, and F. Mayor, Wiley, Chichester, 1976, p. 83.
6. H. A. Krebs and K. Henseleit, *Hoppe-Seyler's Z. Physiol. Chem.*, 1932, **210**, 33.
7. A. Geissler, K. Kanamori, and B. D. Ross, *Biochem. J.*, 1992, **287**, 813.
8. B. T. Hourani, E. M. Hamlin, and T. B. Reynolds, *Arch. Intern. Med.*, 1971, **127**, 1033.
9. M. D. Norenberg and A. Martinez-Hernandez, *Brain Res.*, 1979, **161**, 303.
10. J-C Reubi, C. van den Berg, and M. Cuenod, *Neurosci. Lett.*, 1978, **10**, 171.
11. R. F. Butterworth and G. P. Layrargues, 'Hepatic Encephalopathy: Pathophysiology and Treatment', Humana Press, Clifton, 1989.
12. L. Zieve, in 'Diseases of the Liver', eds. L. Schiff and E. R. Schiff, Lippincott, Philadelphia, 1987, p. 925.
13. R. Kreis, N. A Farrow, and B. D. Ross, *Lancet*, 1990, **336**, 635.
14. N. E. P. Deutz, A. A. de Graaf, J. B. de Haan, W. M. M. J. Bovee and R. A. F. M. Chamuleau, *J. Hepatol.*, 1987, **4**(1), S13.
15. T. E. Bates, S. R. Williams, R. A. Kauppinen, and D. G. Gadian, *J. Neurochem.*, 1989, **53**, 102.
16. S. M. Fitzpatrick, H. P. Hetherington, K. L. Behar, and R. G. Shulman, *J. Cereb. Blood Flow Metab.*, 1990, **10**, 170.
17. K. Kanamori and B. D. Ross, *Biochem. J.*, 1993, **293**, 461.
18. K. Kanamori, F. Parivar, and B. D. Ross, *NMR Biomed.*, 1993, **6**, 21.

19. J. Shen, N. R. Sibson, G. Cline, K. L. Behar, D. L. Rothman, and R. G. Shulman, *Devel. Neurosci.*, 1998, **20**, 434.

20. K. Kanamori, B. D. Ross, and E. L. Kuo, *Biochem. J.*, 1995, **311**, 681.

21. K. Kanamori, B. D. Ross, and J. Tropp, *J. Magn. Reson. B*, 1995, **107**, 107.

22. K. Kanamori and B. D. Ross, *J. Neurochem.*, 1997, **68**, 1209.

23. S. H. Fitzpatrick, H. P. Hetherington, K. L. Behar, and R. G. Shulman, *J. Neurochem.*, 1989, **52**, 741.

24. S. Bluml, E. Zuckerman, J. Tan, and B. D. Ross, *J. Neurochem.*, 1998, **71**, 1564.

25. R. A. Moats, Y-H. H. Lien, D. Filippi, and B. D. Ross, *Biochem. J.*, 1993, **295**, 15.

26. H. Bruhn, J. Frahm, T. Michaelis, K.-D Merboldt, W. Hänicke, M. L. Gyngell, P. Brunner, J. Frohlich, D. Haussinger, P. Schauder, and B.D. Ross, *Hepatology*, 1991, **14**, 121.

27. R. Kreis, N. A. Farrow, and B. D. Ross, *NMR Biomed.*, 1991, **4**, 109.

28. R. Kreis, B. D. Ross, N. A. Farrow, and Z. Ackerman, *Radiology*, 1992, **182**, 19.

29. H. Bruhn, K.-D. Merboldt, T. Michaelis, M. L. Gyngell, W. Hanicke, J. Frahm, P. Schauder, K. Held, G. Brunner, J. Frolich, D. Haussinger, and B. D. Ross, *Proc. Xth Annu. Mtg Soc. Magn. Reson. Med.*, San Francisco, 1991, p. 400.

30. J. R. McConnell, C. S. Ong, W. K. Chu, M. F. Sorrell, B. W. Shaw, and R. K. Zetterman, *Proc. XIth Annu. Mtg Soc. Magn. Reson. Med.*, Berlin, 1992, p. 1957.

31. R. A. F. M. Chamuleau, D. K. Bosman, W. M. M. J. Bovee, P. R. Luyten, and J.A. den Hollander, *NMR Biomed.*, 1991, **4**, 103.

32. S. Schenker, K. J. Breen, and A. M. Hoyumpa, *Gastroenterology*, 1974, **66** 121.

33. B. D. Ross, M. R. Morgan, I. J. Cox, K. E. Hawley, and I. R. Young, *J. Cereb. Blood Flow Metab.*, 1987, **7**, 5396.

34. B. D. Ross, J. P. Roberts, J. Tropp, K. Derby, N. Bass, and C. Hawryszko, *Magn. Reson. Imag.*, 1989, **7**, 82.

35. P. R. Luyten, J. A. den Hollander, W. M. M. J. Bovée, B. D. Ross, D. K. Bosman, and R. A. F. M. Chamuleau, *Proc. VIIIth Annu. Mtg Soc. Magn. Reson. Med.*, Amsterdam, 1989, p. 375.

36. L. Barbara, B. Barbiroli, S. Gaiani, L. Bolondi, S. Sofia, G. Zironi, R. Lodi, S. Iotti, P. Zaniol, C. Sama, and S. Brillanti. *Eur. J. Hepatol.*, 1993, **2**, 60.

37. S. Taylor-Robinson, R. J. Mallalieu, J. Sargentoni, J. D. Bell, D. J. Bryant, G. A. Coutts, and M. Y. Morgan, *Proc. XIIth Annu. Mtg Soc. Magn. Reson. Med.*, New York, 1993, p. 89.

38. A. Geissler, N. Farrow, F. Villamil, L. Makowka, T. Ernst, R. Kreis, and B. Ross, *Proc. XIth Annu. Mtg Soc. Magn. Reson. Med.*, Berlin, 1992, p. 647.

39. R. Gruetter, E. J. Novotny, S. D. Boulware, G. F. Mason, D. L. Rothman, G. I. Shulman, J. W. Prichard, and R. G. Shulman, *J. Neurochem.*, 1994, **63**, 1377.

40. R. Gruetter, D. L. Rothman, E. J. Novotny, and R. G. Shulman, *Magn. Reson. Med.*, 1992, **25**, 204.

41. S. Bluml, *J. Magn. Reson.*, 1999, **136**, 219.

42. S. Bluml and B. D. Ross, in 'Impact of Molecular Biology and New Technical Developments in Diagnostic Imaging', eds. W. Semmler and M. Schwaiger, Springer-Verlag, Heidelberg, 1998, p. 43.

43. B. D. Ross, S. Jacobson, F. G. Villamil, R. A. Moats, T. Shonk, and J. Draguesku, *Hepatology*, 1993, **18**, 105A.

44. B. D. Ross, S. Jacobson, F. Villamil, J. Korula, R. Kreis, T. Ernst, T. Shonk, and R.A. Moats, *Radiology*, 1994, **193**, 457.

45. T. Ernst, B. D. Ross, and R. Flores, *Lancet*, 1992, **340**, 486.

46. R. K Gupta, V. A Saraswat, H. Poptani, R. K. Dhiman, A. Kohli, R. B Gujral, and S. R. Naik, *Am. J. Gastroenterol.*, 1993, **88**, 670.

47. B. D. Ross, T. Shonk, R. A. Moats, S. Jacobson, J. Draguesku, T. Ernst, J. H. Lee, and R. Kreis, *Proc. XIIth Annu. Mtg Soc. Magn. Reson. Med.*, New York, 1993, p. 131.

48. T. Shonk, R. Moats, J. H. Lee, J. Korula, T. Ernst, R. Kreis, J. Draguesku, and B. D. Ross, *Gastroenterology*, 1993, **104**, A-449, 1793

49. J. Korula, D. Kravetz, M. Katz, T. Shonk, S. Hanks, and B. D. Ross, *Hepatology*, 1993, **18**, 282A, 903.

50. D. G. Gadian, A. Connelly, J. H. Cross, S. Burns, R. A Iles, and J. V. Leonard, *Proc. Xth Annu. Mtg Soc. Magn. Reson. Med.*, San Francisco, 1991, p. 193.

51. B. D. Ross, R. Kreis, and T. Ernst, *Eur. J. Radiol.*, 1992, **14**, 128.

52. R. A. Hawkins, J. Jessy, A. M. Mans, and M. R. De Joseph, *J. Neurochem.*, 1993, **60**, 1000.

53. B. D. Ross and S. Bluml, *NMR Biomed.*, 1996, **9**, 279.

54. D. Haussinger, J. Laubenberger, S. V. Dahl, T. Ernst, S. Bayer, M. Langer, W. Gerok, and J. Hennig, *Gastroenterology*, 1994, **107**, 1475.

55. L. J. Haseler, W. L. Sibbit Jr, H. N. Mojtahedzadeh, S. Reddy, V. P. Agarwal, and D. M. McCarthy, *Am. J. Neuroradiol*, 1998, **19**, 1681.

56. R. N. Bryan and P. Barker, *Am. J. Neuroradiology*, 1998, **19**, 1593.

57. H. Bruhn, T. Michaelis, K-D. Merboldt, W. Hanicke, M. L. Gyngell, and J. Frahm, *Lancet*, 1991, **337**, 745.

58. R. Kreis, and B. D. Ross, *Radiology*, 1992, **184**, 123.

59. E. R. Danielsen and B. D. Ross, 'Magnetic Resonance Spectroscopy Diagnosis of Neurological Diseases', Marcel Dekker, New York, 1999.

60. K. J. Seymour, S. Bluml, J. Sutherling, W. Sutherling, and B. D. Ross, *MAGMA*, 1999, **8**, 33.

61. Y-H. H. Lien, J. I. Shapiro, and L. Chan, *J. Clin. Invest.*, 1990, **85**, 1427.

62. G. G. Wong D.Phil. Thesis, University of Oxford, 1981.

63. R. Kreis, T. Ernst, and B. D. Ross, *J. Magn. Reson.*, 1993, **102**, 9.

64. B. D. Ross, T. Ernst, R. Kreis, L. J. Haseler, S. Bayer, E. Danielsen, S. Bluml, T. K. Shonk, J. C. Mandigo, W. Caton, C. Clark, S. Jensen, N. Lehman, E. Arcinue, R. Pudenz, and C. H. Shelden, *JMRI*, 1998, **8**, 829.

65. J. H. Lee and B. D. Ross, *Proc. XIIth Annu. Mtg Soc. Magn. Reson. Med.*, New York, 1993, p. 1553.

66. J. H. Lee, E. Arcinue, and B. D. Ross, *N. Engl. J. Med.*, 1994, **331**, 439.

67. J. S. Videen, T. Michaelis, P. Pinto, and B. D. Ross, *J. Clin. Invest.*, 1995, **95**, 788.

68. Y.-H. H. Lien, J. I. Shapiro, and L. Chan, *J. Clin. Invest.*, 1991, **88**, 303.

69. P. L. Hope, E. B. Cady, P. S. Tofts, P. A. Hamilton, A. M. de Costello, D. T. Delpy, A. Chu, E. O. R. Reynolds, and D. R. Wilkie, *Lancet*, 1984, 366.

70. R. Kreis, T. Ernst, E. Arcinue, R. Flores, and B. D. Ross, *Proc. XIth Annu. Mtg Soc. Magn. Reson. Med.*, Berlin, 1992, p. 237.

71. O. H. Lowry in 'Neurology of the Newborn,' ed. J. J. Volpe, Saunders, Philadelphia, 1987, p. 33.

72. R. Kreis, E. Arcinue, T. Ernst, T. Shonk, R. Flores and B. D. Ross, *J. Clin. Invest.*, 1996, **97**, 1.

73. G. A. Webb (ed.), 'Annual Reports on NMR Spectroscopy', Vol. 25, Academic Press, London, 1993, p. 480.

74. D. R. Bailes, D. J. Bryant, H. A. Case, A. G. Collins, I. J. Cox, A. S. Hall, R. R. Harman, S. Khenia, P. McArthur, B. D. Ross, and I. R. Young, *J. Magn. Reson.*, 1988, **77**, 460.

75. A. Lin, S. Bluml, W. Caton, C. Duma, A. Mamelak, R. Rand, S. Wiseman, and B. D. Ross, *Proc. VIIth Annu. Mtg (Int.) Soc. Magn. Reson. Med.*, Philadelphia, 1999, p. 1394.

76. M. C. Preul, Z. Caramanos, D. L. Collins, J.-G. Villemure, R. Leblanc, A. Olivier, R. Pokrupa, and D. L. Arnold, *Nature Med.*, 1996, **2**, 323.

77. J. van der Grond, K. J. van Everdingen, B. C. Eikelboom, J. Kenez, and W. P. T. M. Mali, *JMRI*, 1999, **9**, 1.

78. P. Dechent, P. J. W. Pouwels, F. Hanefield, and J. Frahm, *Proc. VIth Annu. Mtg Soc. Magn. Reson. Med.*, Sydney, 1998, **3**, p. 1754.

79. B. Kruse, P. B. Barker, P. C. M. van Zigl, J. H. Duyn, C. T. W. Moonen, and H. W. Moser, *Ann. Neurol.*, 1994, **36**, 595.

80. M. Castillo (ed.), 'Proton MR Spectroscopy of the Brain', Vol. 8, Saunders, Philadelphia, PA, 1998, p. 926.

81. L. Chang, B. L. Miller, D. McBride, M. Cornford, G. Oropilla, S. Buchthal, F. Chiang, H. Aronow, and T. Ernst, *Radiology*, 1995, **197**, 525.

82. K. L. Behar and T. Ogino, *Magn. Reson. Med.*, 1993, **30**, 38.

83. L. J. Haseler, E. Arcinue, E. R. Danielsen, S. Bluml, and B. D. Ross, *Pediatrics*, 1997, **99**, 4.

84. T. Chen, N. Farrow, S. Bluml, C. Huang, and B. D. Ross, *Proc. VIth Annu. Mtg (Int.) Soc. Magn. Reson. Med.*, Sydney, 1998, **1**, 537.

85. B. L. Miller, R. Moats, T. Shonk, T. Ernst, S. Woolley, and B. D. Ross, *Radiology*, 1993, **187**, 433.

86. R. A. Moats and T. Shonk, *Am. J. Neuroradiol.*, 1995, **16**, 1779.

87. T. K. Shonk, R. A. Moats, P. Gifford, T. Michaelis, J. C. Mandigo, J. Izumi, and B. D. Ross, *Radiology*, 1995, **195**, 65.

88. T. Hoang, D. Dubowitz, S. Bluml, O. V. Kopyov, D. Jacques, and B. D. Ross, *Proc. 27th Meeting of the Society for Neuroscience*, New Orleans, FL, 1997, **23**(2), 1682.

89. S. Stockler, F. Hanefeld, and J. Frahm, *Lancet*, 1996, **348**, 789.

90. K. Seymour, S. Bluml, G. McComb, and B. D. Ross, *Proc. VIth Annu. Mtg (Int.) Soc. Magn. Reson. Med.*, Sydney, 1998, **3**, p. 1807.

91. S. Bluml, J. H. Hwang, and B. D. Ross, *Proc. VIIth Annu. Mtg (Int.) Soc. Magn. Reson. Med.*, Philadelphia, 1999, p. 335.

92. G. F. Mason, R. Gruetter, D. L. Rothman, K. L. Behar, R. G. Shulman, and E. J. Novotny, *J. Cereb. Blood Flow Metab.*, 1995, **15**, 12.

93. S. Bluml, *J. Magn. Reson.*, in press.

94. S. Bluml, J. H. Wang, A. Moreno, L. Lim, J. Tam, and B. D. Ross, *Radiology*, 1999, **213**, 330P.

95. S. Bluml, A. Moreno, and B. D. Ross, *Lancet* submitted.

Acknowledgements

Work reported was largely funded by the Rudi Schulte Research Institute, Santa Barbara, the Jameson Foundation and by funds from the HMRI MRS Program. BDR is grateful to the following colleagues: K. Kanamori, E. Rubaek-Danielsen, J.D. Roberts, N. Kumar, M. Linsey and C. Sharp.

Biographical Sketch

Brian D. Ross. *b* 1938; B.Sc., 1958, University College, London. D.Phil., 1966, University of Oxford. M.B., 1961, University College Hospital, London. F.R.C.S., 1973, Royal College of Surgeons, London. M.R.C.Path., 1976, Royal College of Pathologists, London. 1989, F.R.C.Path. University of Oxford lecturer, Metabolic Medicine. Director, Renal Metabolism Unit and Consultant Chemical Pathologist, Radcliffe Infirmary, Oxford, 1976–84. Director of Clinical Spectroscopy Programs at Radcliffe Infirmary, Oxford, 1981–84, Hammersmith Hospital, London, 1986–88, and Huntington Medical Research Institutes, Pasadena, CA, 1986–present. Visiting Associate, California Institute of Technology, 1986–present.

Structural and Functional MR in Epilepsy

Graeme D. Jackson

Brain Imaging Research Institute, Austin and Repatriation Medical Centre, Heidelberg, West Australia and Howard Florey Institute, University of Melbourne, Australia

and

Alan Connelly

Institute of Child Health and Great Ormond Street Hospital for Children, NHS Trust, London, UK

1 SUMMARY

Recently, developments in magnetic resonance imaging (MRI), magnetic resonance spectroscopy (MRS), and functional magnetic resonance imaging (fMRI) have opened up new opportunities for the noninvasive investigation of the brain. In epilepsy, these noninvasive techniques play a major role in the clinical investigation of patients with epilepsy.

MRI can noninvasively detect virtually all foreign tissue lesions (tumors) such as hamartomas, gliomas, oligodentrocytomas, dysembryoplastic neuroepithelial tumors and other developmental lesions. It has been able to define these lesions with a great deal of anatomical accuracy. This in itself is a tremendous advance which can now easily be taken for granted. Perhaps even more impressive has been the ability of optimized structural imaging techniques to detect smaller abnormalities of gray matter, in particular lesions like subtle cortical dysplasias, minor abnormalities of gray matter, and especially hippocampal sclerosis. In the important lesion of hippocampal sclerosis, quantitative measures of both the abnormal morphology (volume) and abnormal signal (T_2 relaxation measurement) have allowed this diagnosis to be made objectively. The clinical impact of this noninvasive information cannot be overstated. A major challenge now is to be able to reliably detect subtle areas of dysgenesis in the cortical gray matter.

MR spectroscopy provides information about brain metabolites that appears to provide objective information about regional damage in the temporal lobe which is not readily apparent with conventional imaging. Both these methods have enabled the detection of bilateral temporal lobe abnormalities, and the consequences of this bilateral damage on cognitive and seizure outcome is being explored. These techniques show fixed brain pathology. With the development of functional MRI it is possible to see signal change associated with activated neurons. As well as detecting normal activation (such as hand movement), it has been possible to image seizures. This gives information about both the spatial and the temporal location of signal changes during seizures. It appears that interictal activations may also be detectable.

For References see p. 1133

This abundance of new MR techniques that allows so many aspects of brain structure, function and biochemistry to be investigated in the clinical setting has revolutionized the ability to detect brain abnormalities which underlie the epilepsy condition. In surgical programs for the treatment of epilepsy, MR has become at least as important as the EEG. The information from MR will have a major impact on the classification and understanding of epilepsy syndromes.

2   EPILEPSY, THE CLINICAL PROBLEM

Epilepsy is a common problem occurring in up to 1% of the population.[1,2] Intractable epilepsy which is a debilitating disorder may occur in up to 0.25% of the population. It is an extremely important neurological condition because there are severe social and medical consequences of the disorder[3,4] (up to 1% of patients die per year and many more are severely affected). Individuals otherwise often have the capacity to live normal lives, and complete cure is possible in a large number of these if a seizure focus exists, can be identified, and can be surgically removed.

Epilepsy is a disorder predominantly of the gray matter. This includes a wide variety of pathologies such as tumors (some of which may be small), subtle lesions such as cortical dysplasia, and often minor abnormalities of development of subtle degrees of brain injury such as hippocampal sclerosis or cortical gliosis. The predominant abnormality in these patients varies from the macroscopic to the cellular level, and may be characterized by predominantly biochemical or metabolic abnormalities.

3   CLASSIFICATION OF EPILEPSY CONDITIONS

In general terms epilepsy can be divided into two main groups: generalized and partial epilepsy.[5] In generalized epilepsy the seizure arises almost simultaneously in all parts of the brain. In partial epilepsy, by comparison, the seizure begins in one distinct part of the brain, and may then spread to other parts of the brain. There is an implication that generalized epilepsy involves an abnormality in multiple, or all, parts of the brain, while partial epilepsy implies that the abnormality is confined to a limited portion of the brain. The international classification which is most widely accepted[6] further divides the epilepsy conditions into those that are 'symptomatic' (that is have a defined pathological cause such as a tumor) and those that are 'idiopathic' (have no defined lesional cause). The recent development of MR tools for the investigation of brain structure, biochemistry, and function have had a major impact on the thinking in clinical epilepsy. The ability to define lesions (such as hippocampal sclerosis and cortical dysplasia), which were previously 'cryptogenic' and only detectable when pathological studies had been performed, means that the classification and principles upon which clinical epilepsy is based are undergoing major revision at the present time as a direct consequence of the information provided by MR studies.

4   THE CHALLENGE FOR MR STUDIES: THE INFORMATION THAT IS SOUGHT IN EPILEPSY PATIENTS

The problem of epilepsy necessitates understanding of brain structure, function, and biochemistry in normal and pathological states. The central problems in the understanding and management of intractable epilepsy are as follows.

1. To determine whether the epilepsy syndrome is generalized (no defined site of seizure onset, i.e. onset almost simultaneous in all or many parts of the brain) or partial (focal or localized seizure onset, with or without subsequent generalized spread).
2. To define whether a structural abnormality of the brain exists which may give rise to the epilepsy disorder (if so, known as symptomatic epilepsy; if not, then defined as idiopathic epilepsy).
3. If partial, to define the location, and extent, of the region (or regions) responsible for the generation of the seizures, and how functional events relate to underlying structural abnormality.
4. To understand which lesions are epileptogenic, and what abnormalities of structure and function define such areas.
5. To determine the effects of seizures on the brain. Do seizures cause damage (e.g. cellular damage, neuronal loss, hippocampal sclerosis) or is it the disease or condition that gives rise to the epilepsy disorder which causes all of the damage to the brain?
6. To identify important functional areas of cortex (movement, speech, memory) which must be preserved if a neurosurgical procedure is to be performed for treatment of intractable seizures.

Therefore, noninvasive investigations contribute to the solution of these problems by identifying gray matter lesions such as hippocampal sclerosis, replacing the use of invasive methods used to localize the site of seizure origin, defining the nature and extent of the structural, functional and metabolic abnormalities of the seizure focus, and determining preoperatively factors which influence the likely seizure and functional outcome from surgical treatment.

5   STRUCTURAL ABNORMALITIES SHOWN BY MR IMAGING

5.1   Tumors

Approximately 20% of all patients with intractable epilepsy will have a relatively large lesion (tumor) as the basis of their epilepsy. Before MRI only about 50% of these lesions were detected preoperatively (with CT) when located in the temporal lobes.[7] As these patients generally have an excellent outcome after surgery it is essential to identify them. Many series have now established that MR detects virtually all tumors, including dysembryoblastic lesions, hamartomas, and gliomas.

5.2   Hippocampal Sclerosis

Hippocampal sclerosis is one of the most common lesions found in the brains of patients with intractable epilepsy.[8-10] It is important for several reasons. It is a highly epileptogenic lesion. The side of the most affected hippocampus is almost always the side from which the majority of temporal lobe seizures originate. The detection of hippocampal sclerosis by MRI may obviate the need for invasive EEG monitoring with its attendant morbidity in patients being considered for surgery. Until new MR techniques arose, hippocampal sclerosis was considered as nonlesional epilepsy: the reliable detection of hippocampal sclerosis changes the clinical perspective of these patients.

The first reports of the MR detection of hippocampal sclerosis were rather confusing with encouraging results reported in some small studies, confusion with artifacts in others, and inadequate pathological verification in many. There were even studies that failed to detect any abnormality. In 1987, Kuzniecky and colleagues, from the Montreal Neurologic Institute, published the first major series of papers in which hippocampal sclerosis was detected in a systematic way, and shown to be related to pathological findings.[11] This study relied almost entirely on T_2-weighted signal changes. Using optimally oriented imaging planes, and heavily T_1-weighted inversion recovery sequences in addition to T_2-weighted sequences, hippocampal sclerosis was shown to be reliably detected by visual inspection of images acquired at 0.3 tesla using the criteria of hippocampal atrophy and T_2-weighted signal change in hippocampal gray matter.[7,12] Since then optimized imaging at 1.5 tesla has shown that there are four main features of hippocampal sclerosis visible in MRI[13,14] (Table 1).

The criteria for assessing these optimized images include both morphological and signal intensity changes. The morphological features are atrophy and disruption of internal hippocampal structure. Abnormal signal in the hippocampus can be seen on both inversion–recovery (T_1) and T_2-weighted imaging. The imaging of the fine anatomical structure of the hippocampus, to a level of detail previously possible only with microscopic examination (Figure 1 shows details of this) may become an even more important method of detecting hippocampal sclerosis with improvements in imaging resolution.

5.3 Visual Analysis of the Hippocampus

The visual diagnosis of hippocampal sclerosis can be highly subjective, and even experienced neuroradiologists may have difficulty detecting this subtle lesion. It must be emphasized that not all features are seen in all cases. Unequivocal signal change (as long as the hippocampus is not enlarged) or atrophy may be accepted as diagnostic of hippocampal sclerosis, with the certainty increased if both are present. The addition of internal structure loss can be very helpful if the abnormality is subtle. Several series which compare blinded visual reports to pathological material have shown beyond doubt that hippocam-

pal sclerosis can be reliably diagnosed by visual analysis. In experienced hands, and with optimized images, high sensitivity and specificity can be achieved. In order to do this optimized images must be performed[13] and all features of hippocampal sclerosis must be appreciated and searched for. It is our experience that no single feature (such as atrophy) on its own is sensitive enough for reliable routine visual diagnosis, although some expert centers achieve an accurate diagnosis in about 80% of cases.[15,16]

5.3.1 Atrophy

The assessment of the cross-sectional size of the hippocampus must be made in images obtained in the modified (tilted) coronal axis that transects the hippocampus at right angles. A smaller hippocampus as detected in this plane either qualitatively, or by quantitative methods, reliably predicts the side of the epileptogenic focus in the case of temporal lobe epilepsy but absolute measures of hippocampal size must be interpreted with caution.[17] Quantitation of atrophy (hippocampal volume measurement) is slightly more sensitive than visual assessment of hippocampal size. But with the addition of other visual features of hippocampal sclerosis this difference is less marked, or even reversed.

5.3.2 Loss of the Normal Internal Morphological Structure of the Hippocampus

Normal internal morphological structure of the hippocampus is produced by the alveus, the molecular cell layer of the dentate gyrus, and the pyramidal cell layer of the cornu ammonis, and can be seen on optimized coronal MR images [Figure 2(a) and (b)]. In hippocampal sclerosis, the loss of this normal internal structure is a consequence of neuronal cell loss and replacement of normal anatomical layers with gliotic tissue [Figure 2(c) and (d)]. This feature of hippocampal sclerosis is potentially very important as, with increasing spatial resolution, thinning of the CA1 region of the cornu ammonis may prove to be the most sensitive and specific means of diagnosing hippocampal sclerosis. Attempts have been made to define this with specially designed coils; however, increased resolution which will routinely show this microanatomy is becoming possible using standard equipment.

5.3.3 Signal Hyperintensity on T_2-Weighted Images

Increased T_2-weighted signal when localized imprecisely to the 'mesial temporal region' may be due to foreign tissue such as a glioma or hamartoma, to gliotic tissue in the hippocampus, to increased cerebrospinal fluid in the atrophied region, to flow artifacts and occasionally from a developmental cyst in the hippocampal head stemming from failure of closure of the lateral aspect of the hippocampal fissure. Careful determination of the exact location of this signal change

Table 1 Features of hippocampal sclerosis

MRI feature of hippocampal sclerosis	Suggested histopathological correlate of the MR imaging abnormality
• Unilateral atrophy (right cf. left)	• hippocampal atrophy
• Loss of internal morphological structure on IR images	• loss of neurones in CA1, CA2, and CA4 and replacement gliosis [Figure 6(d)]
• Increased signal on T_2-weighted images	• gliosis
• Decreased signal on T_1-weighted images (IR)	• gliosis

For References see p. 1133

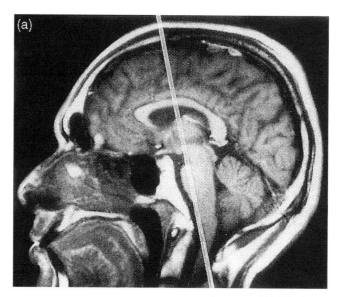

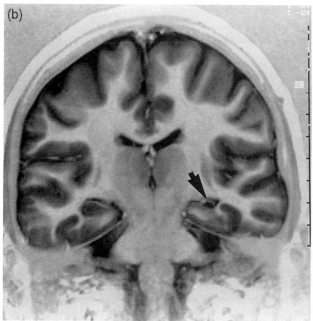

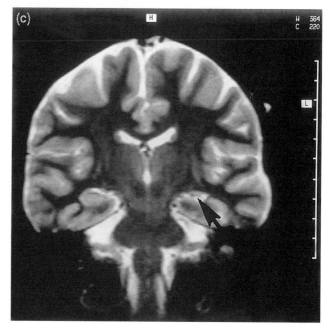

Figure 1   The features of hippocampal sclerosis on optimized imaging are shown here in the imaging plane which transects the hippocampus at right angles (a). These images show atrophy, and reduced T_1 signal (b) and increased T_2 signal (c)

by detailed examination of the anatomical features enables the correct diagnosis to be made. It is important to have sufficient knowledge of both the hippocampal anatomy and the easily recognizable artifacts that can occur in this region so that artifacts are not confused with significant signal abnormalities in the hippocampal gray matter.

5.3.4   Signal Hypointensity on T_1-Weighted IR Images

At 1.5 tesla, using a *TR* of 3500 and a *TI* of 300 ms, a sclerotic hippocampus appears small and dark, and the internal features are obscured (Figure 1). The presence of three features in a single coronal image makes the visual diagnosis of hippocampal sclerosis much easier, and makes it possible to detect mild degrees of abnormality.

An abnormal signal on T_1- or T_2-weighted images arising from an atrophic hippocampus almost always represents hippocampal sclerosis. An abnormal signal arising from an apparently enlarged hippocampus may represent a hamartoma or glioma. If one relied only on a single feature such as atrophy, then these cases of the larger hippocampus being abnormal would be incorrectly lateralized.

5.4   Visual Analysis: Findings Other than Hippocampal Sclerosis

Detailed analysis of MRI images reveals a high percentage of abnormalities which can be detected in the brains of patients with intractable partial epilepsy. As MR techniques improve, it is becoming clear that most patients with intract-

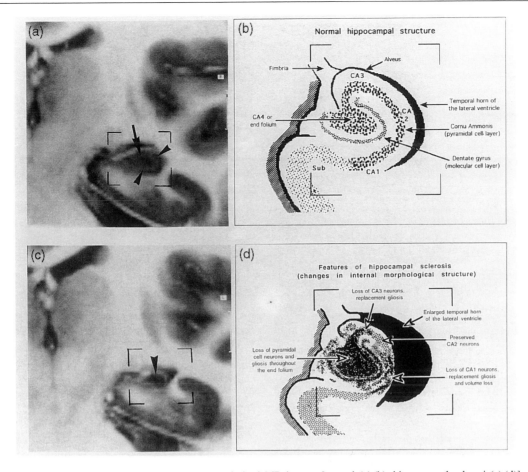

Figure 2 Internal structure of the hippocampus as seen on optimized MR images [normal (a),(b), hippocampal sclerosis(c),(d)]

able epilepsy have detectable imaging abnormalities. Table 2 shows the range of abnormalities which were detected in a combined series of patients from several centers. While often, in the past, no cause could be identified for many cases of intractable partial epilepsy, it is now becoming clear that most adults and children with partial epilepsy will have defined brain abnormalities visualized on appropriate optimized imaging.[7,13,16,18–26]

6 QUANTITATIVE DIAGNOSIS OF HIPPOCAMPAL SCLEROSIS

There has been a widespread problem in achieving the best results using visual analysis (and routine reporting of MR studies) in the clinical environment. It is clear that the results of visual inspection can be replicated in different centers, and that the findings are specific and sensitive when optimized images

Table 2 340 Patients, pathological verification of MRI diagnosis in 149

MRI diagnosis	%	Number/location
Hippocampal sclerosis	57	194
Foreign tissue lesion (glioma, astrocytoma or dysembryoplastic tumor)	13.5	46 (36 temporal, 10 extratemporal)
Cortical dysplasia	10.5	35 (12 temporal, 23 extratemporal)
Vascular malformations (12 cavernous hemangiomas, two high flow lesions)	4	14 (6 temporal, 8 extratemporal)
Cystic lesions	1.5	5 (4 temporal, 1 extratemporal)
Miscellaneous	5	17 (5 trauma, 1 tuberous sclerosis, 2 epidermoid, 4 extensive white matter lesions, 1 cerebellar atrophy, 4 uncertain)
No lesion demonstrated	8.5	29
Total	100	340

For References see p. 1133

are interpreted in expert hands. It is equally clear that this expertise is hard to come by. This is probably because the lesion of hippocampal sclerosis differs from a normal hippocampus by a degree that might have previously been attributed to artifact or normal variation. Therefore considerable experience is required to gain expertise in this diagnosis. For this reason, for research purposes, and for the detection of abnormalities beyond the sensitivity of the eye, quantitative diagnosis of hippocampal abnormalities has been essential. The features of hippocampal sclerosis are the same as when assessed visually. Much initial attention has been given to the quantitation of hippocampal atrophy but quantitation of signal characteristics is also possible.

6.1 Volumetric Analysis

In 1990 Jack and co-workers[17,27,28] published their method of quantifying hippocampal atrophy which for the first time enabled an objective measurement of hippocampal pathology. The use of volumetric measurement to assess hippocampal size had been successfully adopted (and adapted) by many centers.[7,13,16,22–26,29–33] Despite different protocols, this has proven to be a reliable means of determining the lateralization of hippocampal pathology in up to 90% of cases with hippocampal sclerosis. It is our impression that visual analysis in the most expert centers detects virtually all cases that are detected by volumetric analysis, although volume techniques are more sensitive than visual analysis for the detection of the single feature of hippocampal atrophy. There are some cases, because features in addition to atrophy are considered, which can be detected by visual analysis and not by volumetric analysis.[34] Because of the range of normal variation, and measurement error, the most reliable use of the volume measurement technique (and indeed visual analysis) has been in lateralization. It has not been easy to determine bilateral hippocampal abnormalities or to determine abnormality without comparison between sides. The strength of the volumetric technique is that the anterior–posterior distribution of the atrophy can be determined[35] and the quantitative analysis removes the sometimes subjective nature of the analysis. The greatest weakness is that the only well verified and reliable technique (because of normal variation) is the use of side to side ratios. This means that bilateral disease is not usually detectable.

6.2 T_2 Relaxometry

As well as quantifying the hippocampal atrophy, one can quantify the T_2 signal in the hippocampus by measuring the T_2 relaxation time in the hippocampal gray matter (Figure 3).[36] The T_2 relaxation time can be measured quantitatively by measuring the decay in signal intensity at different echo times in a series of T_2-weighted images acquired in the same slice. Each pixel of the resulting T_2 map is derived from the intensity in each of 16 images in that same slice.

This objective measurement has a small range of values in normal subjects. The T_2 relaxation time appears to be very precise in normal tissue. This enables the detection of pathology without requiring comparison between two hippocampi. Therefore, as well as being sensitive, it permits the detection of pathology in the contralateral hippocampus.

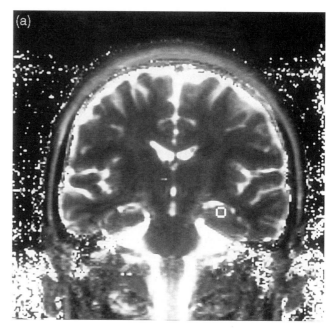

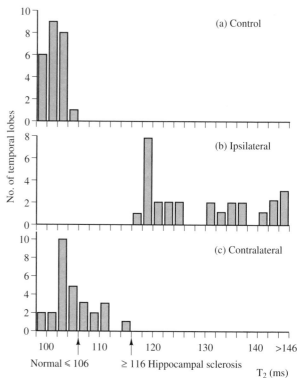

Figure 3 The T_2 map (a) is constructed from the T_2 relaxation times measured for each pixel. The relaxation time is presented as intensity, and can be measured directly for any region of interest (shown for the hippocampus). The histogram (b) shows the distribution of measured T_2 relaxation times in those patients with hippocampal sclerosis (labeled ipsilateral and contralateral to the side of seizure onset) and controls

In our experience, the T_2 relaxation time within the hippocampus is a robust and reliable objective measurement of hippocampal pathology, providing a means of assessing the hippocampus which is as good as our most skilled visual interpretation of hippocampal abnormality in optimized scans. In contrast to both visual interpretation and volumetric analysis of hippocampal atrophy, the definition of a normal hippocampal T_2 relaxation time is very precise. Therefore, T_2 quantification has the ability to detect very mild, bilateral and progressive hippocampal abnormalities. Moreover, T_2 values can be interpreted in terms of hippocampal pathology even when the other hippocampus is incomplete or distorted, such as when a lesion is present or following temporal lobe surgery. In these difficult cases, pathology of the residual ipsilateral hippocampus and the contralateral hippocampus may still be diagnosed. It has recently been shown that there is a very close correlation between hippocampal atrophy, hippocampal T_2 abnormality and pathological findings. Therefore, findings from hippocampal volume studies (such as outcome, and pathology correlations) should apply to T_2 abnormalities, while the latter has the advantage of detecting bilateral disease.

7 MR SPECTROSCOPY

As discussed elsewhere in this volume, several lines of evidence suggest that almost all the *N*-acetylaspartate (NAA) within the brain is neuronal,[37–40] and so a reduction in the NAA signal is commonly interpreted in terms of neuronal loss or damage. Such a case is shown for the temporal lobe in Figure 4. Here the NAA signal is reduced and the signals from choline containing compounds (Cho) and creatine + phosphocreatine (Cr) are increased. While interest focuses on the NAA signal as a marker of neuronal damage or loss, it is often impractical to

quantitate this as an absolute quantity. Usually a marker of abnormality is being sought, so the ratio of NAA to Cho or Cr is often used as such a marker. In the temporal lobe it may often be difficult to resolve the Cho and Cr signals unequivocally, therefore we recommend that the ratio NAA/(Cho + Cr) is an appropriate marker of pathology for clinical use.

MR spectroscopy studies have shown overall abnormalities in the NAA, Cr and Cho signals in the temporal lobes of patients with well localized intractable temporal lobe epilepsy.[41,42] In comparison with controls, the temporal lobes ipsilateral to the seizure focus show a 22% reduction in NAA signal intensity, a 15% increase in the Cr signal, and a 25% increase in the Cho signal. The NAA change is interpreted in terms of neuronal loss (or damage). The interpretation of the increase in Cho and Cr signals is not yet clearly defined. However, studies of neural cells show that the concentrations of choline-containing compounds and of creatine + phosphocreatine are much higher in astrocyte and oligodendrocyte preparations than in cerebellar granule neurons.[37] Thus, increases in Cr and Cho may reflect reactive astrocytosis suggesting that both neuronal loss and astrocytosis may be identified by the noninvasive measurement of metabolites by MRS.[37,43–45]

Using the ratio NAA/(Cho + Cr), MRS has been used to lateralize up to 70 patients with intractable temporal lobe epilepsy. In about 40 of these cases the abnormality was bilateral, and like T_2 relaxation time measurements could be used to detect bilateral abnormalities. MRS is sensitive to bilateral and diffuse pathology and it is an objective measure of metabolite abnormalities which cannot be visualized directly with MR imaging. We view these MR spectroscopic abnormalities as a marker of regional abnormalities of the temporal lobe in these patients. We do not believe that it is a marker simply of hippocampal sclerosis, but it provides additional information about pathology of the temporal lobes which is not available by other MR methods.

Chemical shift imaging (CSI) or magnetic resonance spectroscopic imaging has the advantages of being able to determine the regional distribution of metabolites and to identify areas of maximal abnormality. In a study of ten patients with temporal lobe epilepsy and five controls the left–right asymmetry of NAA/Cr ratios was found to be significantly different from controls in all cases.[46] The use of such an asymmetry index alone precludes the detection of bilateral abnormalities. However, comparison of NAA/Cr ratios in patients and controls indicated that two patients had bilateral reduction in the NAA/Cr ratio in the posterior temporal region, and the greatest reduction was ipsilateral to the maximum EEG disturbance. A further CSI study, using a 2T Philips system with a 4 mL effective voxel size, was performed on eight patients with unilateral complex partial seizures and eight controls.[47] Significant asymmetry in the intensity of the NAA signal was found in all patients. In each case the lower NAA corresponded to the side of seizure focus as determined by EEG. No significant changes in Cho or Cr were observed.

It is apparent that CSI has distinct advantages over single voxel techniques in terms of coverage of the brain, and it is becoming the method of choice in a number of centers for the study of epilepsy. However, it is technically more demanding than single voxel MRS, particularly with respect to magnetic field homogeneity (shimming), water suppression, and infiltration of subcutaneous fat signal into voxels other than just those

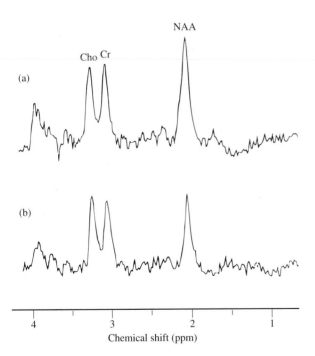

Figure 4 Spectra from (a) the temporal lobe contralateral to the seizure focus in a patient with temporal lobe epilepsy and (b) from the ipsilateral temporal lobe. Note the decrease in NAA in the ipsilateral temporal lobe

adjacent to the scalp. Cendes et al.[46] noted that anterior temporal lobe structures were more accessible to single voxel methods, and reported only posterior and midtemporal results from their CSI study and Xue et al.[48] have reported problems with suboptimal shimming when performing CSI in a large region including both temporal lobes. They have therefore adopted the strategy of acquiring CSI volumes from each temporal lobe separately.

8 FUNCTIONAL MRI (fMRI)

The clinical potential of fMRI is enormous and will be dealt with in many other sections of this Encyclopedia. For clinical epilepsy the following are major applications.

8.1 Mapping of Eloquent Areas of Cortex

Epilepsy surgery entails resection of abnormal areas of cortex in order to relieve the epilepsy condition. In many cases it is important to determine the location of important functions which must not be affected by this surgery such as movement, speech, and memory. The ability of fMRI to demonstrate this eloquent cortex helps in the preoperative assessment of these patients. At present this is largely limited to the motor strip[49] (Figure 5), but demonstration of speech areas will also be important.

At the moment, many groups have produced images of signal changes during 'speech activation'. While some of these are compelling, the problem of speech activation and localization is complex, and a great deal of validation and careful interpretation of these signal changes will be required before they assume their potential importance in clinical practice.

8.2 Functional Imaging of Seizures

Functional MRI can detect cortical activation which occurs during partial motor seizures.[50] Activation can be seen in the region which is activated during seizures even when no clinical seizure occurs. Also, quite remarkably, activation could first be seen up to 1 min before the onset of clinical or EEG changes during similar seizures [Figure 6(a) and (b)]. The implication of these observations is that the vascular and oxygenation changes may precede, or at least be detectable earlier than, the EEG or clinical events which are associated with seizures. It also provides compeling evidence that subclinical activation can be identified using fMRI, and this may enable precise localization of the seizure focus in some cases. These observations allow structural and dynamic functional information to be obtained in a single, integrated, totally noninvasive, MR examination, and point the way to the future role of MR as a means of imaging neurophysiology.

9 CONCLUSION

The impact of MR and its application to clinical epilepsy is akin to the impact of the EEG in the 1940s. It is enabling abnormalities of the brain to be demonstrated by noninvasive techniques in many of these patients with epilepsy. The impact has been great in the field of epilepsy surgery where, already, patients who had previously required invasive depth electrodes, are now able to go to resective surgery directly based on noninvasive studies which include noninvasive EEG, routine clinical evaluation, and these new MR techniques. In the broader field of clinical epilepsy problems, MR findings are beginning to have a large impact on the view of epilepsy, and of the syndromes that can be defined in individual patients. This will ultimately affect the classification of epilepsy, with

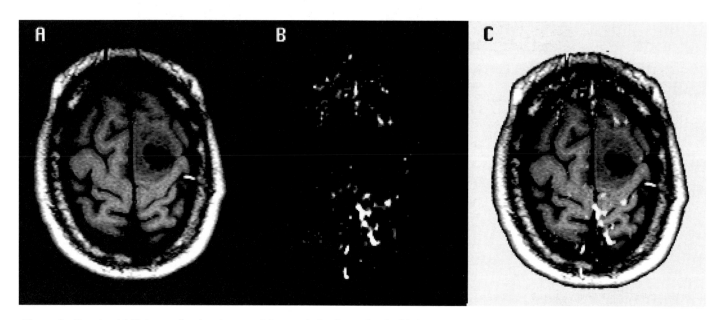

Figure 5 Functional MR image showing the area of increased signal associated with leg movement near a tumor thought to involve the motor strip: (a) baseline image; (b) activation image; (c) superimposition image

For list of General Abbreviations see end-papers

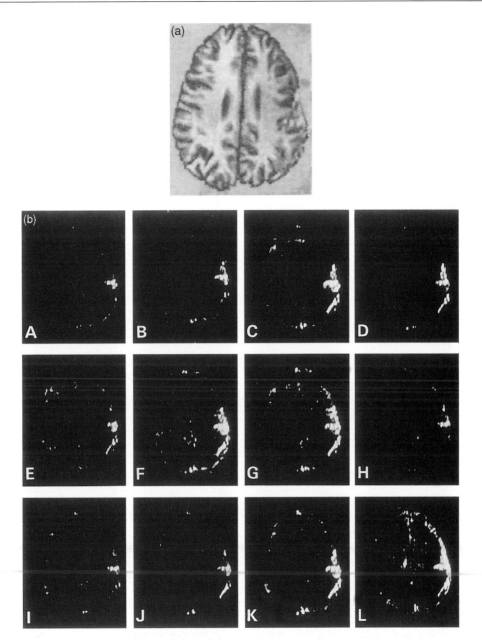

Figure 6 (a) Signal change associated with the onset of a focal motor seizure in the same region in a 4 year old boy with intractable epilepsy. This activation was seen 20 s before the onset of the seizure. (b) The upper two rows (A–D and E–H) show activation seen using fMRI during two clinical seizures. After activity is first seen, images 20, 60 and 100 s later are shown (all rows). In the first row a clinical motor seizure began at the time of the third image. The lower row is a period of activation not associated with a seizure. The surface EEG suggested seizure onset in this region, and ictal single photon emission computed tomography (SPECT) localized it to this same region. The base anatomy image is the same for all these figures

new syndromes such as temporal lobe epilepsy with hippocampal sclerosis being defined. The consequence of these new MR techniques will be to benefit those who are most affected by this disease: patients with epilepsy.

10 RELATED ARTICLES

Brain MRS of Human Subjects; Brain Neoplasms in Humans Studied by Phosphorus-31 NMR Spectroscopy; Localization and Registration Issues Important for Serial MRS Studies of Focal Brain Lesions.

11 REFERENCES

1. W. A. Hauser, J. F. Annegers, and L. T. Kurland, *Epilepsia*, 1991, **32**, 429.
2. W. A. Hauser, 'Epilepsy Surgery'; ed. H. A. Luders, Raven Press, New York, 1992, pp. 133–141.

For References see p. 1133

3. P. Klenerman, J. W. Sander, and S. D. Shorvon, *J. Neurol. Neurosurg. Psychiatry* 1993, **56**, 149.

4. D. C. Taylor 'Surgical treatment of the epilepsies', ed. J. Engel, Jr. 2nd edn., Raven Press, New York, 1993.

5. J. J. Engel 'Seizures and epilepsy' Contemporary Neurology series, Davis, Philadelphia, 1989.

6. Commission on classification and terminology of the international league against epilepsy, *Epilepsia*, 1989, **30**, 389.

7. G. D. Jackson, S. F. Berkovic, B. M. Tress, R. M. Kalnins, G. Fabinyi, and P. F. Bladin, *Neurology*, 1990, **40**(12), 1869.

8. J. H. Margerison and J. A. N. Corsellis, *Brain*, 1966, **89**, 499.

9. M. A. Falconer, E. A. Serafetinides, and J. A. N. Corsellis, *Arch. Neurol.*, 1964, **10**, 233.

10. M. A. Falconer, *Lancet*, 1974, **2**, 767.

11. R. Kuzniecky, V. De La Sayette, R. Ethier, D. Melanson, F. Andermann, S. Berkovic, Y. Robitaille, A. Olivier, T. Peters, and W. Feinder, *Ann. Neurol.*, 1987, **22**(3), 341.

12. S. F. Berkovic, F. Andermann, A. Olivier, R. Ethier, D. Melanson, Y. Robitaille, R. Kuzniecky, T. Peters, and W. Feindel, *Ann. Neurol.*, 1991, **29**, 175.

13. G. D. Jackson, S. F. Berkovic, J. S. Duncan, and A. Connelly, *Am. J. Neurorad.*, 1993, **14**, 753.

14. R. Kuzniecky, E. Faught, and R. Morawetz, *Epilepsia*, 1993, **34**(6), 141.

15. C. R. J. Jack, F. W. Sharbrough, G. D. Cascino, K. A. Hirschorn, P. C. O'Brien, and W. R. Marsh, *Ann. Neurol.*, 1992, **31**(2), 138.

16. G. D. Cascino, C. R. Jack, Jr., K. A. Hirschorn, and F. W. Sharbrough, *Epilepsy Res.*, 1992, (suppl. 5), 95.

17. C. J. Jack, M. D. Bentley, C. K. Twomey, and A. R. Zinsmeister, *Radiology*, 1990, **176**(1), 205.

18. J. H. Cross, G. D. Jackson, B. G. R. Neville, A. Connelly, F. J. Kirkham, S. G. Boyd, M. C. Pitt, and D. G. Gadian, *Arch. Dis. Child.*, 1993, **69**, 104.

19. G. D. Cascino, C. R. Jack, Jr., J. E. Parisi, W. R. Marsh, P. J. Kelly, F. W. Sharbrough, K. A. Hirschorn, and M. R. Trenerty, *Epilepsy Res.*, 1992, **11**(1), 51.

20. R. Duncan, J. Patterson, D. M. Hadley, P. Macpherson, M. J. Brodie, I. Bone, A. P. McGeorge, and D. J. Wyper, *J. Neurol. Neurosurg. Psychiatry*, 1990, **53**(1), 11.

21. P. Gulati, A. Jena, R. P. Tripathi, and A. K. Gupta, *Indian Pediatr.*, 1991, **28**(7), 761.

22. B. Jabbari, A. D. Huott, G. DiChiro, A. N. Martins, and S. B. Coker, *Surg. Neurol.*, 1978, **10**, 319.

23. R. Kuzniecky, A. Murro, D. King, R. Morawetz, J. Smith, K. Powers, F. Yaghmai, E. Faught, B. Gallagher, and O. C. Snead, *Neurology*, 1993, **43**, 681.

24. C. J. Kilpatrick, B. M. Tress, C. O'Donnell, S. C. Rossiter, and J. L. Hopper, *Epilepsia* 1991, **32**(3), 358.

25. K. Miura, M. Kito, F. Hayakawa, M. Maehara, T. Negoro, and K. Watanabe, *J. Jpn. Epilepsy Soc.*, 1990, **8**(2), 159.

26. S. S. Spencer, G. McCarthy, and D. D. Spencer, *Neurology*, 1993.

27. C. J. Jack, C. K. Twomey, A. R. Zinsmeister, F. W. Sharbrough, R. C. Petersen, and G. D. Cascino, *Radiology*, 1989, **172**(2), 549.

28. C. J. Jack, F. W. Sharbrough, C. K. Twomey, G. D. Cascino, K. A. Hirschorn, W. R. Marsh, A. R. Zinsmeister, and B. Scheithauer, *Radiology* 1990, **175**(2), 423.

29. M. Ashtari, W. B. Barr, N. Schaul, and B. Bogerts, *Am. J. Neurorad.*, 1991, **12**, 941.

30. M. Baulac, O. Granat, X. Gao, and D. Laplane, *Epilepsia*, 1991, **3**(2), 2.

31. G. Castorina and D. L. McRae, *Acta Radiol.*, 1963, **1**, 541.

32. T. Lencz, G. McCarthy, R. A. Bronen, T. M. Scott, J. A. Inserni, K. J. Sars, R. A. Novelly, J. H. Kim, and D. D. Spencer, *Ann. Neurol.*, 1992, **31**(6), 629.

33. K. Matsuda, K. Yagi, T. Mihara, T. Tottori, Y. Watanabe, and M. Seino, *Jpn. J. Psychiatry Neurol.*, 1989, **43**(3), 393.

34. G. D. Jackson, R. I. Kuzniecky, and G. D. Cascino, *Neurology*, 1994, **44**, 42.

35. M. J. Cook, D. R. Fish, S. D. Shorvon, K. Straughan, and J. M. Stevens, *Brain*, 1992, **115**, 1001.

36. G. D. Jackson, A. Connelly, J. S. Duncan, R. A. Grünewald, and D. G. Gadian, *Neurology*, 1993, **43**, 1793.

37. J. Urenjak, S. R. Williams, D. G. Gadian, and M. Noble, *J. Neurochem.*, 1992, **59**, 55.

38. J. Urenjak, S. R. Williams, D. G. Gadian, and M. Noble, *Neurosci.*, 1993, **13**, 981.

39. J. W. Hugg, K. D. Laxer, G. B. Matson, A. A. Maudsley, C. A. Husted, and M. W. Weiner, *Neurology*, 1992, **42**, 2011.

40. J. K. Joller, R. Zaczek, and J. T. Coyle, *J. Neurochem.*, 1984, **43**, 1136.

41. J. W. Hugg, K. D. Laxer, G. B. Matson, A. A. Maudsley, and M. W. Weiner, *Proc. XIth Ann Mtg. Soc. Magn. Reson. Med.*, Berlin, 1992, 1913.

42. J. Peeling, G. Sutherland, *Neurology*, 1993, **43**, 589.

43. D. G. Gadian, A. Connelly, J. S. Duncan, J. M. Cross, F. J. Kirkham, C. L. Johnson, F. Vargha-Khadom, B. G. Nevik, and G. D. Jackson, *Acta Neurol. Scand.*, 1994, suppl. 152, 116.

44. D. G. Gadian, A. Connelly, J. H. Cross, G. D. Jackson, F. J. Kirkham, J. V. Leonard, B. G. R. Neville and F. Vargha-Khadom, 'New Trends in Pediatric Neurology', eds. N. Fejerman and N.A. Chamoles, Amsterdam 1993, pp. 23-32.

45. A. Connelly, D. G. Gadian, D. G. Jackson, J. H. Cross, M. D. King, J. S. Duncan, and F. J. Kirkham, 'Proton Spectroscopy in the Investigation of Intractable Temporal Lobe Epilepsy', *Proc. XIth Ann Mtg. Soc. Magn. Reson. Med.*, Berlin, 1992, p. 234.

46. F. Cendes, F. Andermann, P. C. Preul, and D. L. Arnold, *Ann. Neurol.*, 1994, **35**, 211.

47. J. W. Hugg, K. D. Laxer, G. B. Matson, G. B. Maudsley, and M. W. Weiner, *Ann. Neurol.*, 1993, **34**, 788.

48. M. Xue, T. C. Ng, M. Modic, Y. Comair, and H. Kolem, *Proc. XIIth Ann Mtg. Soc. Magn. Reson. Med.*, New York, 1993, 435.

49. C. R. Jack, R. M. Thompson, R. C. Botts, R. R. Butts, F. W. Sharbrough, P. J. Kelly, D. P. Hanson, S. J. Rieslerer, R. L. Ehman, N. J. Hangrandrean, and G. D. Cascino, *Radiology*, 1994, **190**(1), 85.

50. G. D. Jackson, A. Connelly, J. H. Cross, I. Gordon, and D. G. Gadian, *Neurology*, 1994, **44**, 850.

Acknowledgements

Dr Jackson was supported, in part, by grants from the Wellcome Trust and Action Research. We would like to thank David Gadian, Cheryl Johnson, Brian Neville, John Duncan, Richard Grünewald, Wim Van Paeschen, and Simon Robinson.

Biographical Sketches

Graeme D. Jackson. *b* 1956. B.Sc. (Hons) Psychology, 1977; MB.BS. Monash University, Melbourne, Australia, 1982.; FRACP (Neurology), 1992; M.D. thesis 'Magnetic Resonance in Intractable Epilepsy', Monash University, 1995. Introduced to epileptology by Drs. Peter Bladin and Sam Berkovic, Austin Hospital, Melbourne, 1988–90. Then a research registrar, National Hospital for Neurology and Neurosurgery, 1990–92. Subsequently lecturer then senior lecturer and Honorary Consultant in Paediatric Neurology at the Institute of Child Health, and Great Ormond Street Hospital and NHS Trust, London, 1992–1996. Director of the Brain Imaging Research Institute, Austin and Repatriation Medical Centre, Heidelberg, West Australia 1996–current. Howard Florey Institute, University of Melbourne 1998–current. Current research specialities: MR applications in epilepsy, neurotoxicity.

Alan Connelly. *b* 1955. B.Sc. (Hons) Chemistry, University of Glasgow, Scotland, 1977; Ph.D., University of East Anglia, Norwich, UK.

Ph.D. in high resolution NMR under the supervision of Prof. Robin Harris. Began work on in vivo spectroscopy and imaging as NMR development scientist at Oxford Research Systems, UK, 1983–88. Subsequently lecturer (1988–93) then senior lecturer (1993–present) at the Institute of Child Health, London, UK. Current research specialties: imaging and spectroscopy applications in stroke and epilepsy.

MRI and MRS of Neuropsychiatry

Basant K. Puri

MRC Clinical Sciences Centre, Imperial College School of Medicine, London, UK

1 INTRODUCTION

MRI and MRS studies are becoming increasingly important in neuropsychiatry. In this chapter, the contributions will be considered of such studies to our understanding of schizophrenia, mood disorders, anxiety disorders, obsessive-compulsive disorder, eating disorders, attention deficit hyperactivity disorder (ADHD), psychoactive substance use, Alzheimer's disease, Lewy body disease and Binswanger's disease, Huntington's disease, autism, electroconvulsive therapy, dyslexia, brain changes following incomplete spinal injury in humans, and drug monitoring.

2 SCHIZOPHRENIA

Many studies using MRI have been carried out on patients with schizophrenia since 1983. In a well-researched critical review of these by Chua and McKenna,[1] it was found that the only well-established structural abnormality in schizophrenia is lateral ventricular enlargement; this is modest and overlaps with ventricular size in the normal population. The authors of the review came to the following conclusions: 'there is no consistent evidence from MRI studies for a global reduction in brain size in schizophrenia, and only a minority of studies have pointed to a focal reduction in the size of the frontal lobes. However, the numbers of positive and negative replications are approximately equal for the finding of reduced temporal lobe size, and when the hippocampus and amygdala (and perhaps also the parahippocampal gyrus) are specifically considered this turns into a slight majority in favour of reduced size. A reasonable conclusion might therefore be that, while not yet established beyond reasonable doubt, it is likely that any brain substance abnormality in schizophrenia will be found to be localised to the temporal lobe, where it will be predominantly subcortical and perhaps also predominantly left-sided.'

Recently developed techniques of subvoxel registration of high-resolution three-dimensional (3D) serial MR scans[2,3] and quantification of changes thereby discovered[4] have just started to be applied to various aspects of this disorder. For example, when first-episode schizophrenic patients were classified according to Gruzelier's syndromal model,[5,6] it was found that, compared with normal controls, over an 8-month period patients who were 'withdrawn' showed progressive ventricular enlargement, with an increase in ventricle-to-brain volume ratio. In contrast a group of 'active' patients showed a reduction in ventricle-to-brain volume ratio, with a change that was opposite in sign and smaller in magnitude.[7] These findings suggest that opposite patterns of functional hemispheric activation early in the course of schizophrenia may be associated with strikingly different structural cerebral changes. These techniques have also found application in testing specific predictions of Horrobin's neuronal membrane phospholipid model of schizophrenia.[8,9] In the first example of this, it has been found that in a patient with long-standing disease not being treated with conventional medication, sustained remission of positive and negative symptoms of schizophrenia associated with treatment with the omega-3 fatty acid eicosapentaenoic acid (EPA; Kirunal) was accompanied by a reversal of cerebral atrophy (Figure 1).[10]

Using ^{31}P MRS to study the prefrontal cortex in schizophrenia, a number of groups, including Pettegrew and colleagues[11] and Stanley and colleagues,[12] have reported changes in membrane phospholipid metabolism, irrespective of antipsychotic medication status, with reduced levels of phosphomonoesters (precursors of phospholipid biosynthesis) and increased levels of both phosphodiesters (phospholipid breakdown products) and intracellular magnesium ions. It has been suggested that these findings may be fundamentally related to the pathophysiology of schizophrenia,[11–13] with the reduced levels of phosphomonoesters being caused by reduced biosynthesis or altered degradation and the elevated levels of phosphodiesters being associated with increased activity of phospholipase A_2 or A_1, or perhaps decreased phosphodiesterase activity. An alternative explanation involves a putative disturbance of metabolic compartmentation of phosphatidylcholine biosynthesis.[14]

Many studies using proton MRS have demonstrated a reduction in the neuronal marker N-acetylaspartate, particularly in the left temporal lobe. In a recent study combining this technique with MRI, the volume of cortical gray matter was found to be reduced in patients with schizophrenia, while the N-acetylaspartate signal intensity from a comparable region was normal; by comparison, the volume of cortical white matter was normal while the N-acetylaspartate signal intensity from a comparable region was reduced.[15] The lack of reduction in gray matter N-acetylaspartate signal intensity suggests that the cortical gray matter deficit involved both neuronal and glial compartments, rather than a neurodegenerative process in which there is a decrease in the neuronal relative to the glial compartment. The reduced white matter N-acetylaspartate signal intensity without a white matter volume deficit may reflect abnormal axonal connections.[15]

For References see p. 1142

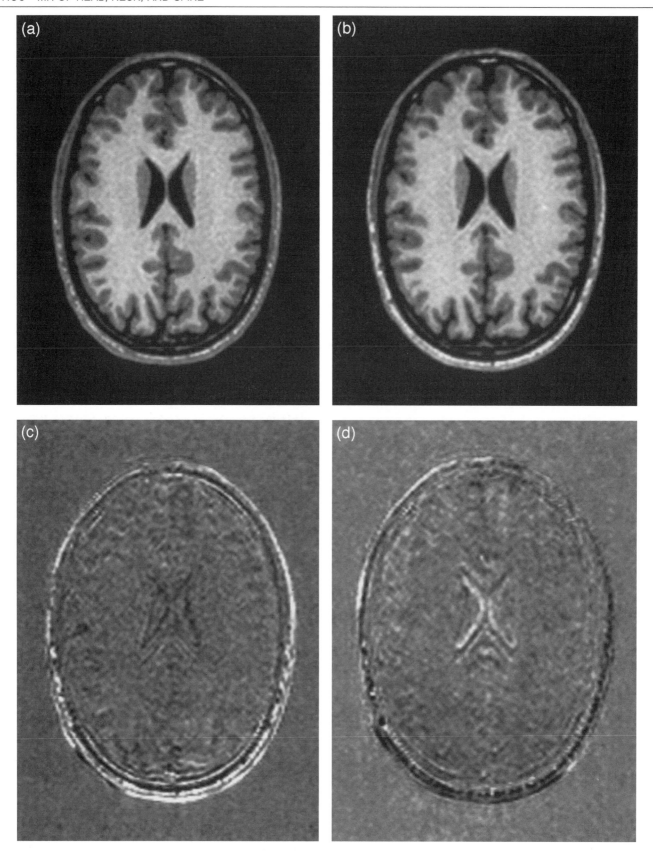

Figure 1   MRI in a patient with long-term schizophrenia. (a) Transverse image of the brain 12 months prior to commencing treatment with eicosapentaenoic acid (EPA). (b) Transverse image of the brain at baseline (0 months) with respect to EPA treatment. (c) Registered difference image of the baseline scan minus the scan at −12 months ((b) minus (a)). The dark lines around the ventricles are caused by a decrease in brain size. (d) Registered difference image of the scan at 6 months minus the scan at baseline (0 months). The white lines around the ventricles are caused by an increase in brain size. Changes are also seen in the cerebral cortex, with narrowing evident in some sulci and increased volume evident in some gyri

For list of General Abbreviations see end-papers

3 MOOD DISORDERS

To date there have been relatively few MR studies of mood disorders and the findings are not consistent. For example, ventricular enlargement is an inconsistent finding in depression (using MRI or computed tomography (CT)); when it has been found it has sometimes been shown to be positively correlated with the length of illness. Neither first-episode bipolar disorder nor first-episode major depression appear to be associated with ventricular enlargement, however.[16] Another inconsistent finding is the possibility of an increased frequency of signal hyperintensities on T_2-weighted scans in elderly depressed patients, which may be associated with poor cognitive performance.[17,18] Although such hyperintensities may be a marker of underlying pathology, they are by no means specific to depression and indeed may also occur in older normal controls. It has been reported that the presence of such hyperintensities in both the basal ganglia and the pontine reticular formation in patients aged 65 years and over is associated with a poor response to antidepressant monotherapy.[19] It has also been suggested that treatment-resistant chronic unipolar depression is associated with reduced gray matter density in the left temporal cortex, including the hippocampus.[20] Studies using ^{31}P MRS and ^1H MRS have indicated possible abnormalities in membrane phospholipid metabolism, high-energy phosphate metabolism, and intracellular pH in mood disorders.[21]

4 ANXIETY DISORDERS

There have been very few MR studies of anxiety disorders, perhaps because anxiety and claustrophobic symptoms constitute a recognized cause of incomplete or cancelled MR examinations.[22] In one ^{31}P MRS study of the frontal lobes in panic disorder, no significant differences were found between patients and controls in ^{31}P metabolite levels, although a significant asymmetry (left greater than right) of phosphocreatine concentration was found in the patients; raised intracellular pH in 2 out of 18 of the patients may have resulted from respiratory alkalosis secondary to hyperventilation in the anxiety state.[23]

5 OBSESSIVE-COMPULSIVE DISORDER

Structural neuroimaging studies indicate that at least a subgroup of patients with obsessive-compulsive disorder may have abnormal basal ganglia development.[24] Although not all such studies demonstrate reduced volumes of these structures, it is noteworthy that a reduced level of the neuronal marker N-acetyl-aspartate has been found in either the left[25] or right[26] corpus striatum in obsessive-compulsive disorder using proton MRS, even when volumetric MRI studies of the same patients do not show reduced volumes.[25] Hence the inconsistent volumetric findings may reflect the relatively poorer sensitivity of MRI morphometry for detecting neuronal loss compared with proton MRS measurement of N-acetylaspartate.

6 EATING DISORDERS

The CT finding of cerebral atrophy in patients with eating disorders has been replicated using MRI.[27,28] Female patients with anorexia nervosa and bulimia nervosa have been reported to have smaller pituitary glands than matched controls.[29,30] In the absence of any other pituitary pathology, this atrophy is likely to be secondary to nutritional or endocrine alterations. Other reported structural abnormalities include enlarged lateral ventricles with dilated cortical and cerebellar sulci,[31] and subcortical signal hyperintensities on T_2-weighted scans.[32]

In a small cerebral ^{31}P MRS study of anorexia nervosa before treatment, increased levels of phosphodiesters were found compared with controls, while decreased phosphomonoesters were found that were associated with malnutrition reflected by endocrinological abnormalities.[33] These data suggest that severe malnutrition in patients with anorexia nervosa may result in an abnormality in membrane phospholipid metabolism, which might be related etiologically to the cerebral atrophy of anorexia nervosa. In another study of patients with anorexia nervosa recording proton MR spectra from parieto-occipital white matter immediately following an interval of excessive loss of body mass, higher signal intensity ratios of choline-containing compounds relative to total creatine and lower ratios of N-acetylaspartate relative to choline-containing compounds were found compared with controls,[34] suggesting that starvation may be associated with an abnormal neuronal membrane turnover in the white matter of the brain.

7 ATTENTION DEFICIT HYPERACTIVITY DISORDER

Recent MRI studies have shown that some regions of the frontal lobes (anterior superior and inferior) and basal ganglia (caudate nucleus and globus pallidus) are about 10% smaller in ADHD groups than in control groups of children,[35] with the right caudate nucleus being larger,[36] or left caudate being smaller,[37] in children with ADHD. These findings are consistent with theories implicating frontal-striatal circuit abnormalities in this disorder. Also in harmony with this theory is the fact that the corpus callosum has been found relatively consistently to be smaller in children with ADHD, particularly in the region of the genu and splenium.[38] Recently, the cerebellum has been systematically studied in this disorder; the vermal volume was found to be significantly smaller in a large sample of boys with ADHD than in matched controls.[39] This reduction involved mainly the posterior inferior lobe (lobules VIII to X) but not the posterior superior lobe (lobules VI to VII) and suggests that perhaps cerebello-thalamo-prefrontal circuit dysfunction may subserve the motor control, inhibition, and executive function deficits seen in this disorder.

Advances in genetic studies of ADHD have occurred while these advances in structural neuroimaging have been taking place. An important example of how both of these investigative techniques can complement each other relates to polymorphisms of the D_4 dopamine receptor (DRD4). One allele with seven tandem repeats in exon 3 (DRD4*7R) has been associated with ADHD, and when this putative association was investigated by Castellanos and colleagues, it was found that

For References see p. 1142

cerebral MRI measures, previously found to discriminate ADHD patients from controls, did not differ significantly between subjects having and those lacking a *DRD4*7R* allele.[40] Hence the MRI results did not support the reported association between *DRD4*7R* and the behavioral or brain morphometric phenotype associated with ADHD.

8   PSYCHOACTIVE SUBSTANCE ABUSE

Chronic alcoholism is associated with MRI-detectable atrophic changes in many regions of the brain, including the cerebral cortex, cerebellum,[41] and corpus callosum.[42] Hippocampal volume reduction is proportional to the reduction in volume of the brain as a whole.[43] It has been found that over a 5-year period brain volume shrinkage is exaggerated in the prefrontal cortex in normal aging but with additional loss occurring in the anterior superior temporal cortex in alcoholism.[44] This association of cortical gray matter volume reduction with alcohol consumption over time suggests that continued alcohol abuse results in progressive cerebral tissue volume shrinkage.

MRI, but not CT, has been shown to be useful in confirming the diagnosis of acute Wernicke's encephalopathy. In one recent study, increased T_2 signal of the paraventricular regions of the thalamus and the mesencephalic periaqueductal regions was observed in patients with Wernicke's encephalopathy compared with both controls and asymptomatic chronic alcohol abusers, with the sensitivity of MRI in revealing evidence of this disease being 53% and the specificity 93%.[45] It should be borne in mind, however, that the absence of abnormalities on MRI does not exclude this diagnosis.

With MRI, widespread cerebral atrophy is seen in alcoholic Korsakoff patients;[46] this is largely subcortical and does not develop independently of the diencephalic pathology. It should be noted that while chronic alcohol abuse is associated with mammillary body and cerebellar tissue volume loss, these markers do not distinguish accurately between amnesic and nonamnesic patients; mammillary body atrophy that is detectable on MRI is not necessary for the development of amnesia in alcoholic patients.[47]

It has recently been shown that cerebral MRI may be of use clinically in the differential diagnosis of chronic alcohol abuse and schizophrenia.[48] In this study, patients with both disorders showed widespread cortical gray matter volume deficits compared with controls, but only those with chronic alcoholism showed white matter volume deficits. The patients with schizophrenia had significantly greater volume deficits in prefrontal and anterior superior temporal gray matter than in more posterior cortical regions. By contrast, the deficits in the patients with alcoholism were relatively homogeneous across the cortex. For white matter, the deficits in the patients with alcoholism were greatest in the prefrontal and temporoparietal regions. Although both patient groups had abnormally larger cortical sulci and lateral and third ventricles than the controls, the patients with alcoholism had significantly larger sulcal volumes in the frontal, anterior, and posterior parieto-occipital regions than those with schizophrenia.

Reduced levels of *N*-acetylaspartate and choline have been found using cerebral proton MRS in chronic alcoholism.[49] The reduction of *N*-acetylaspartate is consistent with neuronal loss while the reduction in choline may be related to neuronal membrane lipid changes.

In a recent large MRI and proton MRS study comparing asymptomatic abstinent cocaine users with matched controls, it was found that while the ventricle-to-brain ratio and level of white matter lesions did not differ significantly between the two groups, elevated creatine and *myo*-inositol in the white matter were associated with cocaine use.[50] The *N*-acetylaspartate level was normal in the cocaine users, suggesting that there was no neuronal loss or damage in the brain regions examined. It was, therefore, concluded that the neurochemical abnormalities observed might result from alterations in nonneuronal brain tissue.

MRI changes associated with chronic toluene abuse include cerebral atrophy, cerebral and cerebellar white matter T_2 hyperintensity, T_2 hyperintensity involving the middle cerebellar peduncle and the posterior limb of the internal capsule, and T_2 hypointensity involving the basal ganglia and thalamus.[51,52] Chronic solvent abusers who have white matter MRI changes have been found have a lower performance intelligence quotient, as measured by the Weschler Adult Intelligence Scale – Revised, with a particularly low score on the digit symbol subtest.[53]

In polysubstance abusers, there is MRI evidence of reduced volume of the prefrontal cortex (both left and right) consistent with either atrophy or hypoplasia.[54] Ventriculomegaly has not been found to be a feature.[55] Abnormal cerebral metabolism has been found using ^{31}P MRS in male polysubstance abusers during early withdrawal: increased phosphomonoesters and decreased β-nucleotide trisphosphates were found in the abusers compared with controls, indicating that cerebral high-energy phosphate and phospholipid metabolite changes result from long-term drug abuse and/or withdrawal.[56]

9   ALZHEIMER'S DISEASE, LEWY BODY DISEASE, AND BINSWANGER'S DISEASE

The first part of this section considers recent studies focusing on the use of MRI in differentiating Alzheimer's disease from both normal aging and other causes of dementia.

While the finding of cortical or subcortical atrophy on MRI or CT is not pathognomonic of Alzheimer's disease, hippocampal atrophy provides a useful early marker of the disorder, although further longitudinal and neuropathological study is required.[57] CT- and MRI-based measurements of hippocampal atrophy may provide useful diagnostic information for differentiating patients with probable Alzheimer's disease from normal elderly individuals.

A recent pilot study has indicated that MRI may have a role in assisting with the clinical differentiation between dementia with Lewy bodies and Alzheimer's disease.[58] Subjects with known or probable Alzheimer's disease were found to have significantly smaller left temporal lobes and parahippocampal gyri than those with known or probable Lewy body disease. Medial temporal atrophy was present in 9 out of 11 patients with Alzheimer's disease and absent in six out of nine patients with Lewy body disease. While two patients with neuropathologically confirmed Lewy body disease had severe medial temporal atrophy, in both concurrent Alzheimer's disease-type pathology was present in the temporal lobe. Therefore, this pilot study supports the hypothesis that a greater burden of pathology centers on the temporal lobes in Alzheimer's disease compared with Lewy body disease, unless Lewy body disease occurs with concurrent Alzheimer pathology.

Another recent study has suggested that diffusion-weighted MRI may be useful in the differential diagnosis of subcortical arteriosclerotic encephalopathy (vascular dementia of the Binswanger type) and Alzheimer's disease with white matter lesions.[59] Apparent diffusion coefficients in the anterior and posterior white matter and the genu and splenium of the corpus callosum were significantly higher in patients with both these disorders compared with age-matched controls, with apparent diffusion coefficient values in the groups with Binswanger's disease and those with Alzheimer's disease being almost the same. Apparent diffusion coefficient ratios, defined as diffusion-restricted perpendicular to the direction of nerve fibers, were also significantly higher in both groups of patients than in the controls. However, there were regional differences in these ratios in the two disorders, with ratios in Binswanger's disease being higher in the anterior portions of the white matter while ratios in Alzheimer's disease were higher in the posterior portions.

In vitro and in vivo ^{31}P MRS studies of the brain in Alzheimer's disease show alterations in membrane phospholipid metabolism and high-energy phosphate metabolism: compared with control subjects, mildly demented patients with Alzheimer's disease have increased levels of phosphomonoesters, decreased levels of phosphocreatine and probably adenosine diphosphate, and an increased oxidative metabolic rate; as the dementia worsens, levels of phosphomonoesters decrease and levels of phosphocreatine and adenosine diphosphate increase.[60] The changes in oxidative metabolic rate suggest that the brain in Alzheimer's disease is under energetic stress while the phosphomonoester findings implicate basic defects in membrane metabolism in the brain.[60] Thus, in addition to aiding with diagnosis, ^{31}P MRS may provide a noninvasive tool to follow both the progression of this disorder and any response to putative therapeutic interventions.

Proton MRS studies of occipital gray matter show that reduced levels of N-acetylaspartate (presumably reflecting neuronal loss) and increased levels of myo-inositol characterize Alzheimer's disease.[61,62] Studies using proton MRS to measure cerebral amino acids have tended to demonstrate increased glutamate levels and sometimes reduced γ-aminobutyric acid (GABA); following neuronal loss, the remaining neurons might be exposed to excess glutamate and relatively low levels of GABA, an imbalance that might be neurotoxic.[63,64]

10 HUNTINGTON'S DISEASE

Initially, structural neuroimaging studies showed atrophy of the caudate and loss of definition between the caudate and the adjacent ventricle as Huntington's disease progresses; however more recent studies have also shown cortical atrophy, particularly in the frontal lobes.[65]

Using proton MRS, Jenkins and colleagues found that lactate concentrations were increased in the occipital cortex of patients with symptomatic Huntington's disease compared with normal controls, with the lactate level correlating with duration of illness.[66] Several patients in the same study also showed highly elevated lactate levels in the basal ganglia, while basal ganglia levels of N-acetylaspartate were lowered and choline dramatically elevated, relative to creatine, reflecting neuronal loss and gliosis in this brain region. The authors of this study suggested that these findings are consistent with a possible defect in energy metabolism in Huntington's disease, which could contribute to the pathogenesis of the disease, and that the presence of elevated lactate might provide a simple marker that could be followed over time noninvasively and repeatedly to aid in devising and monitoring possible therapies.

A more recent proton MRS study by Taylor-Robinson and colleagues found an elevated ratio of glutamine and glutamate relative to creatine in the striatum compared with healthy controls, suggesting disordered striatal glutamate metabolism and possibly supporting the theory of glutamate excitotoxicity in Huntington's disease.[67]

Huntington's disease is now known to result from expanded CAG repeats in a gene on chromosome 4, a possible consequence of which might be progressive impairment of energy metabolism. Jenkins and colleagues have recently extended their previous studies to examine correlations between proton MRS changes and CAG repeat number.[68] The spectra in three presymptomatic gene-positive patients were found to be identical to normal control subjects in cortical regions, but three in eight showed elevated lactate in the striatum. Similar to recently reported increases in task-related activation of the striatum in the dominant hemisphere, they found that striatal lactate levels in patients with Huntington's disease were markedly asymmetric (left greater than right). Markers of neuronal degeneration, decreased N-acetylaspartate to creatine and increased choline to creatine ratios, were symmetric. Both decreased N-acetylaspartate and increased lactate in the striatum significantly correlated with duration of symptoms. When divided by the patient's age, an individual's striatal N-acetylaspartate loss and lactate increase were found to correlate with the subject's CAG repeat number, with correlation coefficients of 0.8 and 0.7, respectively. Similar correlations were noted between postmortem cell loss and age versus CAG repeat length. Together, these data provide further evidence for an interaction between neuronal activation and a defect in energy metabolism in Huntington's disease that may extend to presymptomatic subjects.[68]

11 AUTISM

MRI studies of individuals with autism have variously and inconsistently shown evidence of hypoplasia of the cerebellum and brainstem with increased size of the fourth ventricle, increased brain volume (though with relative hypofrontality), and smaller size of the body and posterior subregions of the corpus callosum; in addition, previous pneumoencephalographic and CT studies have described lateral ventricular enlargement while MRI studies in general have failed to show abnormality in limbic structures.[69] The degree of cerebellar hypoplasia is significantly correlated with the degree of slowed attentional orienting to visual cues in both children and adults with autism.[70] It should be noted that even in the absence of

For References see p. 1142

abnormal MRI findings, autism may be associated with focal areas of decreased perfusion.[71]

The finding that autism is not necessarily associated with MRI abnormalities is consistent with the results of a recent cerebral proton MRS study comparing 28 patients with autism with both 28 age-matched patients with unclassified mental retardation and 25 age-matched healthy children. The ratio of N-acetylaspartate to choline was lower in the nonautistic patients with mental retardation than in the patients with autism and the controls, and, interestingly, there were no significant differences in this ratio between patients with autism and controls.[72]

A ^{31}P MRS study of the dorsal prefrontal cortex of 11 high-functioning autistic adolescent and young adult men and 11 matched normal controls found that the autistic group had decreased levels of phosphocreatine, α-ATP, α-ADP, dinucleotides, and diphosphosugars compared with the controls.[73] When the metabolite levels were compared within each subject group with psychological and language test scores, a common pattern of correlations was observed across measures in the autistic group, but not in the control group. As test performance declined in the autistic subjects, levels of the most labile high-energy phosphate compound and of membrane-building blocks decreased, and levels of membrane breakdown products increased. No significant correlations were present with age in either group or with IQ in the control group, suggesting that these findings were not the consequence of age or IQ effects. This study provides some evidence of alterations in brain energy and phospholipid metabolism in autism that correlate with psychological and language deficits.

12 ELECTROCONVULSIVE THERAPY

For many years clinicians have been concerned that electroconvulsive therapy may result in acute cerebral structural changes. Indeed, some retrospective imaging studies using MRI and CT have reported an association between a history of electroconvulsive therapy and cerebral change, particularly affecting the lateral ventricles and/or cerebral cortex. However, recently, a prospective MRI study of four electroconvulsive therapy-naïve depressed patients in which they underwent scanning 1 week prior to their first treatment with electroconvulsive therapy and then again following this treatment showed that, using accurate subvoxel registration and subtraction of serial MR images,[2,3] there was no significant difference in cerebral structure following electroconvulsive therapy, either within 24 h or after 6 weeks (Figure 2).[74]

A proton and ^{31}P MRS study of three patients found no evidence of changes in lactate or in cerebral energy metabolism following electroconvulsive therapy.[75] However, Woods and Chiu have found, using proton MRS, that electroconvulsive therapy reliably induces an elevation in the lipid signal that resonates at approximately 1.2 ppm and observed a similar increase in brain lipids in a patient with temporal lobe epilepsy temporarily off medication, the signal disappearing following restarting medication.[76] This is of interest given that elevations of brain concentrations of arachidonic acid and other free fatty acids have been demonstrated to occur after seizures induced in animals. Large shifts of potassium ions from the intra- to the extracellular space occur during seizure activity, and free fatty

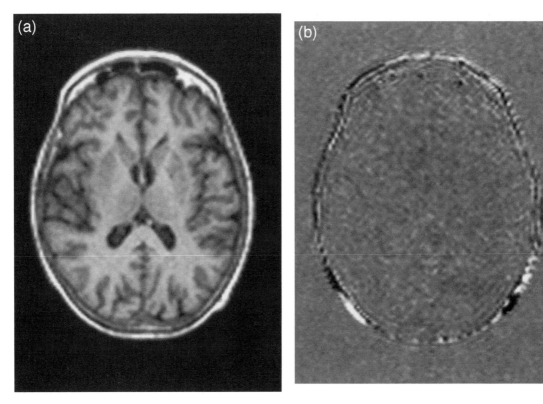

Figure 2   MRI in patients receiving electroconvulsive therapy. (a) Transverse T_1-weighted MR baseline scan showing the anatomy. (b) Difference image obtained by subtracting the baseline scan from the registered follow-up scan showing no evidence of acute structural changes in the brain following electroconvulsive therapy

For list of General Abbreviations see end-papers

acids have a direct effect on membrane potassium ion conductance, suggesting that free fatty acids may play a primary role in seizure evolution in brain tissue.[76]

13   DYSLEXIA

MRI studies have inconsistently shown reversed or diminished asymmetry, compared with normal, in the brain in children with dyslexia, including loss of the usual left greater than right asymmetry of the lateral ventricles and right greater than left asymmetry of the temporal lobes; loss of the normal left greater than right asymmetry of the planum temporale in adolescents, which correlates with the degree of phonological decoding deficits; reversal of the normal left greater than right asymmetry of the angular gyrus in familial dyslexia; and loss of normal right greater than left asymmetry of the frontal cortices and bilaterally smaller size of the frontal cortices. Inconsistent corpus callosum changes have also been reported.[77]

In the first ^{31}P MRS study of dyslexia, Richardson and colleagues found elevated phosphomonoesters in the brain in dyslexia compared with that in controls.[78] This finding is consistent with the hypothesis that neuronal membrane phospholipid metabolism is abnormal in dyslexia, with reduced incorporation of phospholipids into neuronal membranes occurring.[79]

The first proton MRS study of dyslexia showed lateral differences in the ratios of choline to *N*-acetylaspartate and of creatine to *N*-acetylaspartate in the temporo-parietal region and the cerebellum in dyslexic subjects but not in controls.[80]

14   SPINAL INJURY

The first proton MRS study of the human motor cortex following incomplete spinal cord injury showed elevation of *N*-acetylaspartate in this part of the brain compared with normal controls.[81] The authors suggested that this might reflect neuronal adaptation to injury, the finding being consistent with the hypothesis that dendritic sprouting occurs in the motor cortex following recovery from incomplete spinal injury in humans. Clinically, this finding also suggests that MRS might provide a noninvasive method for monitoring such patients.

15   DRUG MONITORING

It is possible to use ^{7}Li MRS directly to measure the cerebral concentration of lithium while ^{19}F MRS can be used to measure the cerebral concentrations of psychotropic drugs containing fluorine, for example the selective serotonin re-uptake inhibitor fluoxetine and the antipsychotics trifluoperazine and fluphenazine.[82]

16   FUTURE DIRECTIONS

As mentioned above, the recently developed MRI techniques of subvoxel registration of high-resolution 3D serial MR scans[2,3] and quantification of the changes thereby discov-

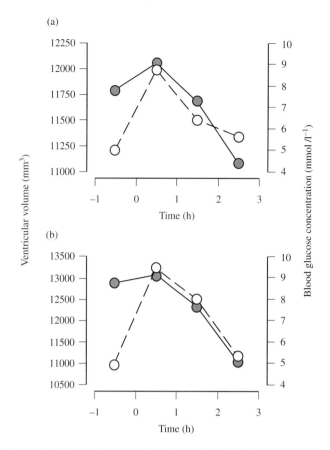

Figure 3   Volume changes in lateral ventricles following oral glucose loading. Blood glucose concentrations (○) and lateral ventricular volumes (●) for two human subjects (a and b) who each ingested 53.7 g glucose at time zero. (After Puri et al.[85])

ered[4] are only just starting to be applied in neuroimaging studies. The sensitivity and accuracy of these techniques hold great promise for neuroimaging applications and the discovery of important new facts concerning the central nervous system.[83,84] For example, they have been used recently to show that volumetric change takes place in the lateral ventricles in the human brain following oral glucose loading (Figure 3).[85]

Cerebral MRS is currently used primarily as a research tool in neuropsychiatry; in due course it is likely to become more widely used diagnostically and prognostically.

It seems probable that MRI and MRS will interface more often with other disciplines (for instance molecular genetics, as in ADHD) and other investigative tools (such as transcranial magnetic stimulation).

In summary, MRI and MRS are proving to be extremely useful in neuroscientific and neuropsychiatric research. These powerful noninvasive tools are likely to continue to grow in importance in these fields and to gain ever more important clinical applications.

17   RELATED ARTICLES

Brain Infection and Degenerative Disease Studied by Proton MRS; Brain MRS of Human Subjects; Brain Neoplasms in

For References see p. 1142

Humans Studied by Phosphorus-31 NMR Spectroscopy; Brain Neoplasms Studied by MRI; Brain Parenchyma Motion Observed by MRI; Hemodynamic Changes owing to Sensory Activation of the Brain Monitored by Echo-Planar Imaging; Central Nervous System Degenerative Disease Observed by MRI; Chemical Shift Imaging; CSF Velocity Imaging; Intracranial Infections; Localization and Registration Issues Important for Serial MRS Studies of Focal Brain Lesions; Magnetic Resonance Imaging of White Matter Disease; Structural and Functional MR in Epilepsy.

18   REFERENCES

1. S. E. Chua and P. J. McKenna, *Br. J. Psychiatry*, 1995, **166**, 563.
2. J. V. Hajnal, N. Saeed, E. J. Soar, A. Oatridge, I. R. Young, and G. M. Bydder, *J. Comput. Assist. Tomogr.*, 1995, **19**, 289.
3. J. V. Hajnal, N. Saeed, A. Oatridge, E. J. Williams, I. R. Young, and G. M. Bydder, *J. Comput. Assist. Tomogr.*, 1995, **19**, 677.
4. N. Saeed, B. K. Puri, A. Oatridge, J. V. Hajnal, and I. R. Young, *Magn. Reson. Med.*, 1998, **16**, 1237.
5. J. H. Gruzelier and R. Manchanda, *Br. J. Psychiatry*, 1982, **141**, 488.
6. J. H. Gruzelier 'Handbook of Schizophrenia', Vol. 5, ed. S. R. Steinhauer, J. H. Gruzelier, and J. Zubin, Elsevier, London, 1991, p. 599.
7. B. K. Puri, N. Saeed, A. J. Richardson, A. Oatridge, S. B. Hutton, J. V. Hajnal, and G. M. Bydder, *Schizophr. Res.*, 1999.
8. D. F. Horrobin, *Lancet*, 1977, **i**, 936.
9. D. F. Horrobin, A. I. M. Glen, and K. Vaddadi, *Schizophr. Res.*, 1994, **13**, 195.
10. B. K. Puri, A. J. Richardson, T. Easton, N. Saeed, A. Oatridge, J. V. Hajnal, D. F. Horrobin, and G. M. Bydder, *Schizophr. Res.*, 1999, **36**, 312.
11. J. W. Pettegrew, M. S. Keshavan, K. Panchalingam, S. Strychor, D. B. Kaplan, M. G. Tretta, and M. Allen, *Arch. Gen. Psychiatry*, 1991, **48**, 563.
12. J. A. Stanley, P. C. Williamson, D. J. Drost, T. J. Carr, R. J. Rylett, A. Malla, and R. T. Thompson, *Arch. Gen. Psychiatry*, 1995, **52**, 399.
13. D. F. Horrobin, *Prostaglandins Leukot. Essent. Fatty Acids*, 1996, **55**, 3.
14. T. P. George and M. W. Spence, *Arch. Gen. Psychiatry*, 1996, **53**, 1065.
15. K. O. Lim, E. Adalsteinsson, D. Spielman, E. V. Sullivan, M. J. Rosenbloom, and A. Pfefferbaum, *Arch. Gen. Psychiatry*, 1998, **55**, 346.
16. W. G. Iacono, G. N. Smith, M. Moreau, M. Beiser, J. A. Fleming, T. Y. Lin, and B. Flak, *Am. J. Psychiatry*, 1988, **145**, 820.
17. C. E. Coffey, G. S. Figiel, W. T. Djang, W. B. Saunders, and R. D. Weiner, *J. Neuropsychiatry Clin. Neurosci.*, 1989, **1**, 135.
18. E. Kramer-Ginsberg, B. S. Greenwald, K. R. Krishnan, B. Christiansen, J. Hu, M. Ashtari, M. Patel, and S. Pollack, *Am. J. Psychiatry*, 1999, **156**, 438.
19. S. Simpson, R. C. Baldwin, A. Jackson, and A. S. Burns, *Psychol. Med.*, 1998, **28**, 1015.
20. P. J. Shah, K. P. Ebmeier, M. F. Glabus, and G. M. Goodwin, *Br. J. Psychiatry*, 1998, **172**, 527.
21. T. Kato, T. Inubushi and N. Kato, *J. Neuropsychiatry Clin. Neurosci.*, 1998, **10**, 133.
22. S. A. Sarji, B. J. Abdullah, G. Kumar, A. H. Tan, and P. Narayanan, *Australas. Radiol.*, 1998, **42**, 293.
23. T. Shioiri, T. Kato, J. Murashita, H. Hamakawa, T. Inubushi, and S. Takahashi, *Biol. Psychiatry*, 1996, **40**, 785.
24. S. Saxena, A. L. Brody, J. M. Schwartz, and L. R. Baxter, *Br. J. Psychiatry*, 1998, **35 (Suppl.)**, 26.
25. R. Bartha, M. B. Stein, P. C. Williamson, D. J. Drost, R. W. Neufeld, T. J. Carr, G. Canaran, M. Densmore, G. Anderson, and A. R. Siddiqui, *Am. J. Psychiatry*, 1998, **155**, 1584.
26. D. Ebert, O. Speck, A. Konig, M. Berger, J. Hennig, and F. Hohagen, *Psychiatry Res.*, 1997, **74**, 173.
27. G. W. Hoffman Jr, E. H. Ellinwood Jr, W. J. Rockwell, R. J. Herfkens, J. K. Nishita, and L. F. Guthrie, *Biol. Psychiatry*, 1989, **25**, 894.
28. G. W. Hoffman Jr, E. H. Ellinwood Jr, W. J. Rockwell, R. J. Herfkens, J. K. Nishita, and L. F. Guthrie, *Biol. Psychiatry*, 1989, **26**, 321.
29. P. M. Doraiswamy, K. R. Krishnan, G. S. Figiel, M. M. Husain, O. B. Boyko, W. J. Rockwell, and E. H. Ellinwood Jr, *Biol. Psychiatry*, 1990, **28**, 110.
30. P. M. Doraiswamy, K. R. Krishnan, O. B. Boyko, et al., *Prog. Neuropsychopharmacol. Biol. Psychiatry*, 1991, **15**, 351.
31. K. Kingston, G. Szmukler, D. Andrewes, B. Tress, and P. Desmond, *Psychol. Med.*, 1996, **26**, 15.
32. K. G. Sieg, M. S. Hidler, M. A. Graham, R. L. Steele, and L. R. Kugler, *Int. J. Eat. Disord.*, 1997, **21**, 391.
33. T. Kato, T. Shioiri, J. Murashita, and T. Inubushi, *Prog. Neuropsychopharmacol. Biol. Psychiatry*, 1997, **21**, 719.
34. H. P. Schlemmer, R. Mockel, A. Marcus, F. Hentschel, C. Gopel, G. Becker, J. Kopke, F. Guckel, M. H. Schmidt, and M. Georgi, *Psychiatry Res.*, 1998, **82**, 171.
35. J. Swanson, F. X. Castellanos, M. Murias, G. LaHoste, and J. Kennedy, *Curr. Opin. Neurobiol.*, 1998, **8**, 263.
36. M. Mataro, C. Garcia-Sanchez, C. Junque, A. Estevez-Gonzalez, and J. Pujol, *Arch. Neurol.*, 1997, **54**, 963.
37. P. A. Filipek, M. Semrud-Clikeman, R. J. Steingard, P. F. Renshaw, D. N. Kennedy, and J. Biederman, *Neurology*, 1997, **48**, 589.
38. G. W. Hynd, M. Semrud-Clikeman, A. R. Lorys, E. S. Novey, D. Eliopulos, and H. Lyytinen, *J. Learn. Disabil.*, 1991, **24**, 141.
39. P. C. Berquin, J. N. Giedd, L. K. Jacobsen, S. D. Hamburger, A. L. Krain, J. L. Rapoport, and F. X. Castellanos, *Neurology*, 1998, **50**, 1087.
40. F. X. Castellanos, E. Lau, N. Tayebi, P. Lee, R. E. Long, J. N. Giedd, W. Sharp, W. L. Marsh, J. M. Walter, S. D. Hamburger, E. I. Ginns, J. L. Rapoport, and E. Sidransky, *Mol. Psychiatry*, 1998, **3**, 431.
41. K. Hayakawa, H. Kumagai, Y. Suzuki, N. Furusawa, T. Haga, T. Hoshi, Y. Fujiwara, and K. Yamaguchi, *Acta Radiol.*, 1992, **33**, 201.
42. R. Estruch, J. M. Nicolas, M. Salamero, C. Aragon, E. Sacanella, J. Fernandez-Sola, and a Urbano-Marquez, *J. Neurol. Sci.*, 1997, **146**, 145.
43. I. Agartz, R. Momenan, R. R. Rawlings, M. J. Kerich, and D. W. Hommer, *Arch. Gen. Psychiatry*, 1999, **56**, 356.
44. A Pfefferbaum, E. V. Sullivan, M. J. Rosenbloom, D. H. Mathalon, and K. O. Lim, *Arch. Gen. Psychiatry*, 1998, **55**, 905.
45. E. Antunez, R. Estruch, C. Cardenal, J. M. Nicolas, J. Fernandez-Sola, and A. Urbano-Marquez, *Am. J. Roentgenol.*, 1998, **171**, 1131.
46. R. Emsley, R. Smith, M. Roberts, S. Kapnias, H. Pieters, and S. Maritz, *Alcohol Alcohol.*, 1996, **31**, 479.
47. P. K. Shear, E. V. Sullivan, B. Lane, and A. Pfefferbaum, *Alcohol. Clin. Exp. Res.*, 1996, **20**, 1489.
48. E. V. Sullivan, D. H. Mathalon, K. O. Lim, L. Marsh, and A. Pfefferbaum, *Biol. Psychiatry*, 1998, **43**, 118.
49. N. R. Jagannathan, N. G. Desai, and P. Raghunathan, *Magn. Reson. Imag.*, 1996, **14**, 553.
50. L. Chang, C. M. Mehringer, T. Ernst, R. Melchor, H. Myers, D. Forney, and P. Satz, *Biol. Psychiatry*, 1997, **42**, 1105.

51. S. Kojima, K. Hirayama, H. Furumoto, T. Fukutake, and T. Hattori, *Rinsho Shinkeigaku*, 1993, **33**, 477.

52. K. S. Caldemeyer, S. W. Armstrong, K. K. George, C. C. Moran, and R. M. Pascuzzi, *J. Neuroimaging*, 1996, **6**, 167.

53. N. Yamanouchi, S. Okada, K. Kodama, T. Sakamoto, H. Sekine, S. Hirai, A. Murakami, N. Komatsu, and T. Sato, *Acta Neurol. Scand.*, 1997, **96**, 34.

54. X. Liu, J. A. Matochik, J. L. Cadet, and E. D. London, *Neuropsychopharmacology*, 1998, **18**, 243.

55. X. Liu, R. L. Phillips, S. M. Resnick, V. L. Villemagne, D. F. Wong, J. M. Stapleton, and E. D. London, *Acta Neurol. Scand.*, 1995, **92**, 83.

56. J. D. Christensen, M. J. Kaufman, J. M. Levin, J. H. Mendelson, B. L. Holman, B. M. Cohen, and P. F. Renshaw, *Magn. Reson. Med.*, 1996, **35**, 658.

57. P. Scheltens, *J. Neurol.*, 1999, **246**, 16.

58. G. T. Harvey, J. Hughes, I. G. McKeith, R. Briel, C. Ballard, A. Gholkar, P. Scheltens, R. H. Perry, P. Ince, and J. T. O'Brien, *Psychol. Med.*, 1999, **29**, 181.

59. H. Hanyu, Y. Imon, H. Sakurai, T. Iwamoto, M. Takasaki, H. Shindo, D. Kakizaki, and K. Abe, *Eur. J. Neurol.*, 1999, **6**, 195.

60. J. W. Pettegrew, W. E. Klunk, K. Panchalingam, R. J. McClure, and J. A. Stanley, *Ann. N. Y. Acad. Sci.*, 1997, **826**, 282.

61. R. A. Moats, T. Ernst, T. K. Shonk, and B. D. Ross, *Magn. Reson. Med.*, 1994, **32**, 110.

62. T. K. Shonk, R. A. Moats, P. Gifford, T. Michaelis, J. C. Mandigo, J. Izumi, and B. D. Ross, *Radiology*, 1995, **195**, 65.

63. W. E. Klunk, K. Panchalingam, J. Moossy, R. J. McClure, and J. W. Pettegrew, *Neurology*, 1992, **42**, 1578.

64. R. J. McClure, J. N. Kanfer, K. Panchalingam, W. E. Klunk, and J. W. Pettegrew, *Neuroimaging Clin. N. Am.*, 1995, **5**, 69.

65. J. M. Hoffman 'Brain Imaging in Clinical Psychiatry', ed. K. R. R. Krishnan and P. M. Doraiswamy, Marcel Dekker, New York, 1997, p. 544.

66. B. G. Jenkins, W. J. Koroshetz, M. F. Beal, and B. R. Rosen, *Neurology*, 1993, **43**, 2689.

67. S. D. Taylor-Robinson, R. A. Weeks, D. J. Bryant, J. Sargentoni, C. D. Marcus, A. E. Harding, and D. J. Brooks, *Mov. Disord.*, 1996, **11**, 167.

68. B. G. Jenkins, H. D. Rosas, Y. C. Chen, T. Makabe, R. Myers, M. MacDonald, B. R. Rosen, M. F. Beal, and W. J. Koroshetz, *Neurology*, 1998, **50**, 1357.

69. S. Deb and B. Thompson, *Br. J. Psychiatry*, 1998, **173**, 299.

70. N. S. Harris, E. Courchesne, J. Townsend, R. A. Carper, and C. Lord, *Brain Res. Cogn. Brain Res.*, 1999, **8**, 61.

71. Y. H. Ryu, J. D. Lee, P. H. Yoon, D. I. Kim, H. B. Lee, and Y. J. Shin, *Eur. J. Nucl. Med.*, 1999, **26**, 253.

72. T. Hashimoto, M. Tayama, M. Miyazaki, Y. Yoneda, T. Yoshimoto, M. Harada, H. Miyoshi, M. Tanouchi, and Y. Kuroda, *J. Child Neurol.*, 1997, **12**, 91.

73. N. J. Minshew, G. Goldstein, S. M. Dombrowski, K. Panchalingam, and J. W. Pettegrew, *Biol. Psychiatry*, 1993, **33**, 762.

74. B. K. Puri, A. Oatridge, N. Saeed, J. E. Ging, H. M. McKee, S. K. Lekh, and J. V. Hajnal, *Br. J. Psychiatry*, 1998, **173**, 267.

75. S. R. Felber, R. Pycha, M. Hummer, F. T. Aichner, and W. W. Fleischhacker, *Biol. Psychiatry*, 1993, **33**, 651.

76. B. T. Woods and T. M. Chiu, *Adv. Exp. Med. Biol.*, 1992, **318**, 267.

77. J. N. Giedd and F. X. Castellanos, 'Brain Imaging in Clinical Psychiatry', ed K. R. R. Krishnan and P. M. Doraiswamy, Marcel Dekker, New York, 1997, p. 126.

78. A. J. Richardson, I. J. Cox, J. Sargentoni, and B. K. Puri, *NMR Biomed.*, 1997, **10**, 309.

79. D. F. Horrobin, A. I. M. Glen, and C. J. Hudson, *Med. Hypotheses*, 1995, **45**, 605.

80. C. Rae, M. A. Lee, R. M. Dixon, A. M. Balmire, C. H. Thompson, P. Styles, J. Talcott, A. J. Richardson, and J. F. Stein, *Lancet*, 1998, **351**, 1849.

81. B. K. Puri, H. C. Smith, I. J. Cox, J. Sargentoni, G. Savic, D. W. Maskill, H. L. Frankel, P. H. Ellaway, and N. J. Davey, *J. Neurol. Neurosurg. Psychiatry*, 1998, **65**, 748.

82. H. C. Charles, T. B. Snyderman, and E. Ahearn, 'Brain Imaging in Clinical Psychiatry', ed. K. R. R. Krishnan and P. M. Doraiswamy, Marcel Dekker, New York, 1997, p. 20.

83. G. M. Bydder and J. V. Hajnal, 'Advanced MR Imaging Techniques', ed. W. G. Bradley and G. M. Bydder, Martin Dunitz, London, 1997, p. 239.

84. G. M. Bydder and J. V. Hajnal, 'Advanced MR Imaging Techniques', ed. W. G. Bradley and G. M. Bydder, Martin Dunitz, London, 1997, p. 259.

85. B. K. Puri, H. J. Lewis, N. Saeed, and N. J. Davey, *Exp. Physiol.*, 1999, **84**, 223.

Biographical Sketch

Basant K. Puri. *b* 1961. B.A. 1982, B.Chir. 1984, M.B. 1985, M.A. 1986, University of Cambridge; M.R.C. Psych. 1989; Dip. Math. 2000. Residency in psychiatry, Addenbrooke's Hospital, Cambridge 1986–88. Research fellow in molecular genetics, MRC Molecular Neurobiology Unit, Cambridge and Peterhouse, Cambridge University 1988–89. Residency in psychiatry, Charing Cross & Westminster Medical School, London 1990–96. Senior Lecturer and Consultant Psychiatrist, MRI Unit, MRC Clinical Sciences Centre, Imperial College School of Medicine, Hammersmith Hospital, London University, and Honorary Consultant, Department of Radiology, Hammersmith Hospital, London 1997–present. Approx. 13 books (psychiatry, neuroscience, statistics) and 60 papers. Research interests: central nervous system MRI and MRS.

Neurosurgical Procedures Monitored by Intraoperative MRI

Terence Z. Wong
Duke University, Durham, NC, USA

and

Richard B. Schwartz, Arya Nabavi, Richard S. Pergolizzi Jr, Peter M. Black, Eben Alexander III, Claudia H. Martin, and Ferenc A. Jolesz
Brigham and Women's Hospital, Boston, MA, USA

1 INTRODUCTION

Image guidance has historically been an integral part of neurosurgery. The introduction of the operating microscope

For References see p. 1153

was a major development that enabled the surgeon to decrease the size of the craniotomy and provided superior surface visualization through magnification and perfect illumination. Nonetheless, its advantages for surgical navigation were confined to the visible surface, and the distance to underlying structures had to be estimated. Subsequently, interactive stereotactic image-guided systems were introduced to yield information on the subsurface structures. This allowed the position of the surgeon's tools or visual field to be interactively correlated anatomically with preoperative computed tomographic (CT) or MR images.[1–8] However, image-guided systems that are based on preoperative imaging cannot account for tissue shifts that occur during the course of surgery.[9] This has ultimately led to the development of intraoperative imaging techniques. One modality that can be used is ultrasound. Although the image quality of intraoperative ultrasound has improved in recent years, it is still inadequate as the sole means for guiding neurosurgical procedures.

Intraoperative MRI would appear to be an ideal imaging modality for guiding neurosurgical procedures. The inherent multiplanar capability of MRI facilitates interactive selection of imaging planes in near-realtime. Conventional T_1- and T_2-weighted images provide exquisite sensitivity for identifying abnormalities within the brain. In addition, specialized imaging sequences are available. These include MR angiographic techniques for mapping blood vessels, dynamic gadolinium-enhanced imaging, functional imaging to map cortical activity, and optimized gradient-echo sequences for early identification of blood products.[10,11]

The design of new open-configuration MRI systems combined with the development of MR-compatible surgical instrumentation, anesthesia equipment, and monitoring devices has made intraoperative MRI-guided neurosurgery a reality. At Brigham and Women's Hospital, over 400 neurosurgical procedures have been performed using intraoperative MRI guidance. Design and operational features of the intraoperative MRI system will be described, and our clinical experience with neurosurgical procedures summarized.

2 INTERVENTIONAL MRI DESIGN

Several magnet configurations are currently used for MRI-guided procedures. These include modification of the conventional long-bore configuration, a short-bore design, a horizontal gap open configuration, and a vertical gap open configuration. While the conventional closed long-bore configuration will provide the highest image quality, it offers the least patient access. However, noninvasive energy sources such as high-energy focused ultrasound can be integrated into long-bore MRI systems; the imaging capabilities of the 1.5 T magnet allow prelocalization with subtherapeutic temperatures prior to high-energy ablation.[12,13] The closed short-bore configuration with flanged bore openings provides improved patient access and maintains the high image quality of a high-field system; however it is not easily amenable for perioperative applications. The open-configuration horizontal gap MRI design allows access to the patient from the sides, permitting most radiologic percutaneous procedures to be performed. The low field minimizes the need for shielding of equipment but also potentially limits imaging capability. Currently, the overlying magnet

structure precludes open surgical procedures while the patient is within the imaging field.

Both conventional closed-bore and horizontal open configuration MRI designs can be used intraoperatively by performing surgery outside of the magnet and physically moving the patient into the imaging field for scanning. This must be accomplished while maintaining sterility and is further complicated by the presence of equipment for anesthesia and monitoring the patient. In 1998, Hall et al. reported on perioperative MRI using a conventional 1.5 T system for guiding pediatric neurosurgery.[14] Although the patient had to be repositioned in the magnet for each set of images, these authors point out that the use of the high-field imager allowed state-of-the-art pulse sequences to be used and enabled images of the highest quality to be obtained. In 1997, Tronnier et al. reported on the use of a 0.2 T horizontal-gap open-configuration magnet for intraoperative neurosurgical guidance on 27 patients.[15] The advantages of direct MRI visualization during biopsies, cyst aspiration, and catheter placement procedures were realized. For open craniotomy procedures, however, the patient had to be transferred into the magnet for each imaging set. Newer horizontal-gap MRI units are being designed with improved access, which may permit intraoperative imaging without moving the patient.

In 1993, a vertical gap mid-field strength MRI system was developed as a collaborative effort by General Electric (Signa SP, Milwaukee, WI, USA) and Brigham and Women's Hospital (Figure 1). By allowing vertical access to the patient, this is currently the only interventional MRI design that permits imaging during surgical procedures without moving the patient. The MRI system has been previously described in detail,[11,16–22] and consists of two vertically oriented superconducting coils (55 cm inner bore diameter) in a modified Helmholtz configuration separated by a 56 cm gap. This spacing permits access to the patient from both sides and allows the surgeon and assistant to work simultaneously in the operating field. Other design features included the use of niobium tin as the superconducting material, which has a higher transition temperature and allows the static field to be maintained without a liquid helium bath, thus permitting a wider gap for patient access. The design results in an imaging volume having a 0.5 T static field defined by a 30 cm sphere in the center of the open magnet bore. Flexible transmit/receive coils have been designed to be placed on the patient in close proximity to the imaging volume, while simultaneously allowing surgical and interventional access. The MRI system is maintained in an operating room environment, with adjacent clean and scrub areas. Disposable sterile drapes have been designed to fit within the surgical space of the magnet (Baxter, Deerfield, IL, USA). The design considerations for the intraoperative MRI interventional suite have been previously reported in detail.[20] Considerations for safety and imaging compatibility in the MRI environment have been reviewed,[20,23] and appropriate materials are available that have low magnetic susceptibility and minimal effect on local MRI.[24] Anesthesia equipment, surgical instruments, patient monitors, and electrocautery devices have already been modified by the manufacturers to be MRI compatible. Specialized devices such as the neurosurgical head holder (Mayfield; Ohio Medical Instrument Co., Cincinnatti, OH, USA), flexible Bookwalter arm (Codman Inc., Burlington, MA, USA), pneumatic neurosurgical drill (Midas Rex, Ft Worth, TX, USA), ultrasonic

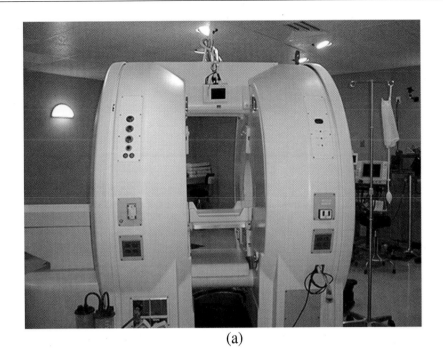

(a)

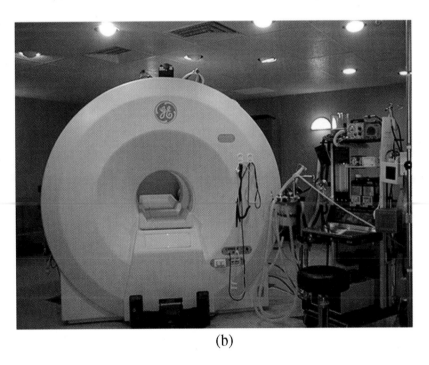

(b)

Figure 1   Interventional/intraoperative MRI system (GE Signa SP) features a vertical gap for surgical access, and is sited within an operating room. (a) Side view: liquid crystal display monitors mounted at eye level on either side of the magnet allow interactive visualization of the MR images. Connections for headlights, suction, navigation, and other instrumentation can be seen on the side of the magnet structure. (b) End-on view: MRI-compatible anesthesia and patient monitoring equipment are shown to the right

aspirator for neurosurgical resection (Selector; Elekta, Atlanta, GA, USA), and MR-compatible microscope (Studor Medical Engineering, Rhenfaal, Switzerland) have also been developed and are used routinely at our institution.[16]

The environment of the MRI system permits it to be used for both *interventional* and *intraoperative* applications.[20]

Although open surgery is by definition not a minimally invasive procedure, image guidance can, with accurate targeting and presurgical planning, minimize the damage to healthy tissue and maximize removal of the abnormality; this reduces the invasiveness of the surgical intervention. We summarize below our experience with MRI-guided interventional and intraopera-

For References see p. 1153

tive intracranial procedures using the vertical gap 0.5 T MRI system. To date, approximately 300 tumor resection procedures and 100 intracranial biopsies have been performed by our group with intraoperative MRI guidance. Other procedures have included evaluation of cystic lesions by injection of dilute gadopentetate dimeglumine,[25] laser ablation of intracranial tumors,[26] and endoscopic sinus surgery.[27,28]

2.1 Interactive Image Guidance

During the MRI-guided neurosurgical procedures, the patient's head is immobilized with a head frame and maintained in the same position throughout the procedure. This provides several important advantages for image guidance. First, patient movement is minimized during imaging. Second, identical slice positions can be obtained and compared during the procedure. Finally, other three-dimensional imaging data sets, such as preoperative MRI, SPECT (single photon emission computed tomography), or PET (positron emission tomography) scans can be co-registered with the MR images and utilized for planning approach and therapy.[29–35] It should be emphasized, however, that although the skull is fixed in position, we have observed that spatial shifts and deformations occur within the brain during the course of surgery, and anatomic co-registration to preoperative imaging studies alone can be subject to significant errors. This problem of tissue shifts may be minimized if access for biopsy or therapeutic probes is limited to small burr holes. In addition, computer models are under development to account for this problem.[36,37]

2.2 Interactive Imaging Tools

Optical, acoustic, and rf techniques have been used to establish the location of biopsy needles or interventional probes in three-dimensional space, and this technology is commonly applied to frameless stereotactic surgery where target coordinates are defined based on preoperative imaging.[1–8]. The intraoperative MRI unit features built-in optical (Flashpoint; Image Guided Technologies Inc., Boulder, CO, USA) and rf (CRD; GE Medical Systems, Inc., Schenectady, NY, USA) tracking capabilities.

Tracking by rf can be implemented using miniature coils mounted on the surgical device.[38,39] Unlike optical or acoustic tracking, intervening tissues are not a problem with rf localization, and this technique is ideal for endoscopic or intravascular procedures.

Currently, the optical tracking system is used for most procedures at Brigham and Women's Hospital. Three charge-coupled device (CCD) cameras are mounted over the imaging field in the MRI unit. These detect infrared light-emitting diodes (LEDs) that are, in turn, attached to the interventional probe. Since the spatial relationship between the LEDs and interventional probe is fixed, the position of the LEDs can be used to define an imaging plane either containing or perpendicular to the probe. Knowledge of the length of the probe allows localization of the probe tip or its projected location. Coordinates of the probe are determined through an interactive workstation (Sun Microsystems, Mountain View, CA, USA), which, in turn, is used to prescribe imaging planes to the SP scanner. Images are subsequently acquired and displayed to the surgeon or interventionalist on liquid crystal display (LCD)

monitors which are mounted at eye level on either side of the work area for the MRI (Figure 1). Images are updated every few seconds, reflecting adjustments in probe position and providing visual feedback in near-realtime. For *interactive imaging*, the interventionalist or surgeon uses the interactive probe to navigate through the region of interest, in a manner similar to that utilizing ultrasonography. This technique is particularly useful for establishing the relationship between surface landmarks and underlying structures, and for planning the approach for biopsy or surgical exposure. For needle biopsies or placement of therapeutic probes, the *targeting* mode is particularly useful. Annotation is added to the acquired images in the form of a 'virtual needle', which enables the operator to view the proposed needle path and tip. The trajectory and target can then be verified by imaging in three orthogonal planes, and the operator is provided with near-realtime visual feedback. When the desired trajectory and target point has been established, the interventional probe can be advanced in a stepwise fashion and observed on the MR images as they are acquired in near-realtime; this is termed the *tracking* mode and confirms satisfactory positioning of the biopsy needle or therapeutic probe.

Two optical tracking handpieces are available for attachment to interventional probes (Figure 2): one has three localization LEDs in a triangular configuration that form a plane perpendicular to the probe. This device can be fixed to the surgical table with the Bookwalter clamp and is the handpiece of choice for intracranial biopsies. One of the three LEDs is offset relative to the other two, and this LED along with the biopsy needle defines a reference imaging plane that contains the needle trajectory. Other planes containing the needle can be selected by physically rotating the handpiece and/or by specifying a relative rotation through the interactive navigation software. This is done at the workstation under the direction of the radiologist and MR technologist. In addition, the software allows the plane perpendicular to the needle to be imaged at any desired depth relative to the needle tip. The needle length is specified, and software annotation provides a projected tract, or 'virtual needle' that can be interactively directed to the target lesion by positioning the handpiece.

The second handpiece for optical navigation (Figure 2, left) features two LEDs mounted on a handle that defines the trajectory of the interventional probe. This device is used most frequently for non-neurosurgical applications, such as interventions in the abdomen, pelvis, or soft tissues.[19] With the two-LED handpiece, the reference imaging plane is defined to be the plane containing the interventional probe and perpendicular to the ground. Other planes can be selected through interactive software prescription.[19]

Although used routinely for biopsy procedures, the optical navigation system can be similarly utilized to position other catheters or needles (e.g. for cyst evaluation or drainage[25]) or to place thermal ablation probes.[26]

3 MRI-GUIDED BIOPSY PROCEDURES

Interactive MRI guidance offers several advantages over conventional stereotactic biopsy techniques. Conventional stereotactic procedures require the placement of a head frame

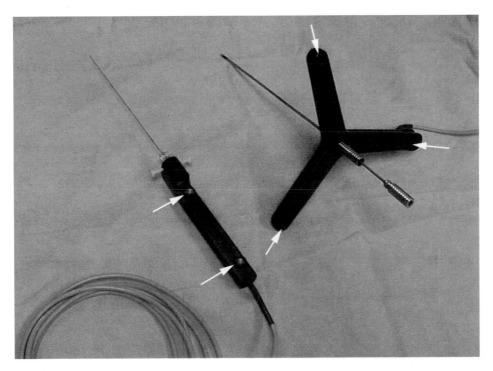

Figure 2 Navigational tools feature either two or three LEDs (white arrows) for optical tracking by three cameras mounted overhead in the MRI unit. The two-LED device (left) is used to establish the trajectory of a biopsy needle attached to the end of the handle and is generally used for interventions involving the abdomen, pelvis, or extremities. In the three-LED handpiece (right), the probe is positioned perpendicular to the handle and is typically immobilized using a Bookwalter clamp during neurosurgical procedures. Near-realtime annotation is used to establish the needle trajectory (Figure 3), after which the biopsy needle is advanced under MRI observation

on the patient followed by a set of preprocedure CT or MR images to define target coordinates. A significant advantage of interactive guidance is that it eliminates the need for this two-step process. The MRI sequence best defining the target tissue can be selected for image guidance; gadolinium contrast can be injected prior to biopsy; this is usually done after the burr hole or craniotomy has been made and the dura exposed. Perhaps the most important advantage of interactive MRI guidance is that the actual path of the biopsy needle can be followed as it is advanced toward the lesion, and small compensatory changes can be made until the sampling port is positioned within the target. Imaging can then confirm in all three dimensions that the biopsy site lies within the tissue having the abnormal MR signal. The ability to make realtime adjustments of the biopsy needle, along with confirmation of biopsy location, may significantly reduce sampling error, as well as decrease the number of needle passes. This can lead to higher diagnostic yield with reduced morbidity.

Of the nearly 100 MRI-guided intracranial biopsies performed to date at Brigham and Women's Hospital, approximately two thirds were performed under general anesthesia with the remaining one third performed using monitored conscious sedation. Average procedure time was less than 2 h if general anesthesia was used, and approximately 1.5 h when conscious sedation was utilized. For navigational MRI, sequences were selected to define the target best. For enhancing lesions, 10–20 ml intravenous gadopentetate dimeglumine was injected, and T_1-weighted fast spin-echo (FSE) images used (updated every 14 s). For nonenhancing lesions, T_2-

weighted FSE imaging was usually used (updated every 15–30 s). More recently, a SSFSE (single-shot fast spin-echo) protocol has been developed that allows T_2-weighted images to be acquired every 4 s. An example of a biopsy performed using MRI-based navigation is illustrated in Figure 3. In this case, contrast-enhanced T_1-weighted imaging was used for guidance. After the patient has been positioned in the headholder on the MRI operating table, the handpiece may be used in the interactive imaging mode to plan the incision location and biopsy trajectory. Prior to inserting the biopsy needle, the targeting mode is utilized, with annotation used to define a virtual needle. The target position can be confirmed in three orthogonal planes (Figure 3). With the handpiece immobilized with the Bookwalter arm, the titanium biopsy needle is advanced in a stepwise manner as MR images are acquired. The needle is observed as a signal void overlying the virtual needle annotation (Figure 3c,d). It should be noted that the tip of the virtual needle is set to correspond to the location of the sideport of the biopsy needle; therefore, the actual needle tip extends several millimeters beyond the virtual needle annotation. This confirms that the biopsy needle is traveling along the expected trajectory and to the desired target depth. If necessary, the annotation can be removed to allow the actual needle location to be seen more clearly and reconfirmed in all three planes. The ability to confirm that the biopsy specimen was obtained within the MR-defined abnormality minimizes sampling error and potentially reduces the number of needle passes and associated complications.

For References see p. 1153

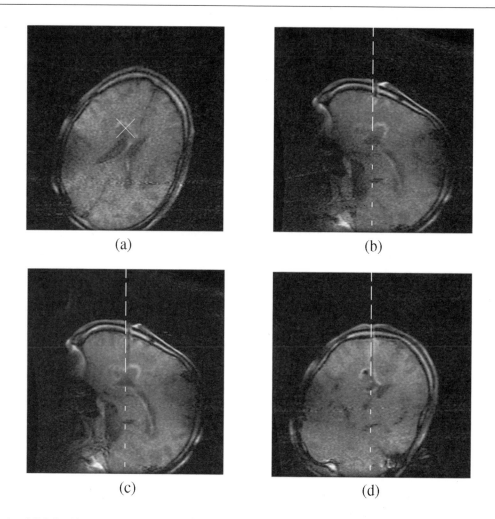

Figure 3 Interactive MRI for biopsy of a right frontal lesion using the 3-LED handpiece and T_1-weighted images following intravenous gadolinium enhancement. (a) Para-axial image perpendicular to needle in the plane of the biopsy port allows targeting of enhancing margin. (b) Para-sagittal image in the plane of the needle shows the 'virtual needle' depicted as dashed lines directed at the target. Note the small signal void at the cortical surface from the biopsy needle tip prior to insertion. (c) As the titanium needle is advanced, near-realtime imaging allows observation of the needle tract along the virtual needle annotation. This allows fine adjustments to be made accordingly and enables the position of the actual biopsy needle to be confirmed prior to sampling. (d) Actual needle tract is confirmed in the perpendicular para-coronal plane. The tip of the virtual needle corresponds to the level of the sampling port of the biopsy needle; as a result the actual needle tip extends a few millimeters beyond the dashed annotation. Note focal signal void within the enhancing rim from biopsy sample

4 MRI GUIDANCE FOR OPEN CRANIOTOMIES

Since 1996, approximately 270 open craniotomies have been performed at Brigham and Women's Hospital for MRI-guided tumor resection. Patient selection, demographics, and preliminary clinical results have been previously reported.[11,16,18,22] In preparation for surgery, the patient's head is positioned in the MRI with the head holder to allow the best surgical access. As a result, the patient's anatomic frame of reference no longer corresponds to conventional MRI planes. To overcome this problem, a new frame of reference is established to allow true axial, sagittal, and coronal images to be obtained. This can be easily accomplished using the navigational tools previously described; interactive imaging is used to define an appropriate reference imaging plane from which conventional axial, sagittal, or coronal images can be prescribed.

Since the patient is immobilized, the reference imaging procedure need only be performed once and is typically done immediately after the patient has been positioned within the intraoperative MRI. The use of conventional imaging planes facilitates accurate localization and identification of critical anatomic structures as the surgery progresses. During the procedure, the radiologist prescribes the appropriate imaging planes and sequences that will best address specific questions of the surgeon as the surgery progresses. For example, for initial identification of tumor boundaries, either conventional spin-echo gadolinium-enhanced T_1-weighted images or T_2-weighted images may be most useful. Other specialized imaging sequences have been developed, including SSFSE for rapid T_2-weighted imaging, dynamic contrast-enhanced imaging, susceptibility sensitive sequences to evaluate early hemorrhage,[10,11] and rapid volumetric image acquisition techniques.

In patients who have been previously treated with high-dose radiotherapy (stereotactic radiosurgery), it is not possible to distinguish active tumor recurrence from postradiation necrosis on conventional delayed gadolinium-enhanced images. Dual-isotope ^{201}Tl/^{99m}Tc-HMPAO (hexamethylpropyleneamine oxime) SPECT scanning has been found to be useful for making this important distinction,[33,34] and these images can be co-registered with intraoperative MR images. Alternatively, a dynamic contrast-enhanced MR study can be performed. These techniques have been useful for estimating tumor grade as well as for distinguishing recurrent tumor from post-treatment change.[40–42] Our intraoperative technique is to obtain two-dimensional fast spoiled gradient-echo images through the tumor volume, with each slice sampled every 11–15 s for 2–3 min following rapid bolus contrast injection. Tumor recurrence, with associated neovascularily, will tend to enhance early, while post-treatment necrosis, related primarily to blood–brain barrier breakdown, will tend to show delayed enhancement. Distinguishing active recurrence from radionecrosis is valuable both for selecting appropriate biopsy sites and for planning the surgical resection. This has been very useful in the intraoperative setting. In an evaluation of a series of 24 patients with intraoperative dynamic MRI,[43] 14 of 15 lesions demonstrating early enhancement had recurrent tumor by pathology; out of the nine lesions that did not show early enhancement, eight had only reactive post-treatment changes. In this study, early enhancement was determined visually when the appearance of contrast within the lesion occurred at the same time as in vascular structures or the choroid plexus.

For all intraoperative MR, an initial baseline set of two-dimensional slices is obtained through the volume of interest. Imaging is repeated at intervals during the surgery to verify the present location in relation to the remaining tumor to be excised. Additionally, the location of the surgical resection cavity relative to adjacent eloquent cortical pathways is always of major importance. Once the most appropriate imaging sequence and imaging plane have been preloaded into the MR console, identical image slices can be acquired simply by rescanning. This allows direct serial comparison in the same imaging plane. Acquisition time for each imaging data set varies depending on the imaging sequence but takes approximately 1–2 min. Prior to imaging, electrical equipment and metallic instruments are removed from the surgical field. Although these devices are MRI-compatible they can produce significant imaging artifacts if they are left in place. Imaging is possible, however, with the operating microscope in position. In our experience, intraoperative MRI guidance enables the surgeon to verify his or her planned approach and surgical procedure and also allows him or her to change the strategy during the procedure according to the newly acquired information.

Figure 4 illustrates an MRI-guided tumor resection. In this case, T_2-weighted images were used for surgical guidance. Serial images from a single axial slice are shown in this example. Early in the procedure, the resection site was defined by a locator. The tissue anterior to this (white arrow, Figure 4(a)) was felt to represent non-neoplastic changes after previous surgery, and this was confirmed by intraoperative histopathology. The remaining tumor was represented by a high T_2 signal posterior to the pointer and just anterior to the motor strip. Stepwise resection of the tumor is illustrated in Figure 4b,c, with complete resection shown in Figure 4d. Prior

to closing, the cavity was irrigated with saline (Figure 4e), and multi-echo gradient-echo images were obtained to evaluate for significant hemorrhage. The MRI characteristics of hyperacute intracranial hemorrhage have been recently described.[25] In our experience, T_2-weighted images (FSE, TR/TE=4000/100 ms; Figure 4e) in combination with gradient-echo sequences (30° flip angle, TR=600 ms) is sensitive for identifying hyperacute hemorrhage and for distinguishing this from postsurgical fluid. Figure 4f,g shows postsurgical gradient-echo images with TE of 9 and 60 ms, respectively. A typical postoperative appearance is illustrated here, with irrigation saline (bright T_2) within the resection cavity, and a thin rim of low signal bordering the resection cavity on the short TE image (Figure 4f) that 'blooms' on the second echo (TE=60 ms; Figure 4g). This rim is felt to represent deoxyhemoglobin content within acute blood products at the resection border. No hematoma, mass effect, or extra-axial collection was noted in this patient. Hematomas tend to have a nonfluid central area (lower T_2 signal), with a rim of susceptibility that also blooms on gradient-recalled echo imaging.

Intraoperative MRI-guided tumor resection provides several advantages over conventional techniques. Image guidance is interactive. Specific sequences and imaging planes can be applied during the procedure to optimize selection of involved tissue. For example, dynamic or steady-state contrast-enhanced images can be used to delineate active tumor, and T_2-weighted images can be used to identify nonenhancing tumor. Neoplastic tissue having abnormal MR signal characteristics but normal appearance visually can be identified and resected. Tissue shifts during surgery have been observed and are easily compensated for with intraoperative imaging. Critical anatomical and functional structures can be identified during resection and, at crucial points in the procedure, images can be obtained more frequently in multiple planes if necessary to refine the resection margin. Finally, imaging capability allows early detection of any unexpected complication, such as hemorrhage. Out of some 400 neurosurgical procedures, postsurgical hematomas were observed in four patients. In all cases, imaging permitted immediate identification, and facilitated removal in three patients. In the fourth patient, serial MR images were used to confirm stability of a small hematoma, and removal was unnecessary. In one patient, a hematoma was noted in the hemisphere contralateral to the surgery and was felt to represent a spontaneous hematoma, possibly from blood pressure instability. This would not have been identified without immediate postoperative imaging; a second craniotomy was performed using MRI guidance to evacuate it.

5   FUTURE PERSPECTIVES

Specialized MRI sequences are available on higher-field MRI units that are not yet available on the intraoperative MRI system, and sophisticated interactive methods have been developed for preoperative surgical planning.[44] In addition, other three-dimensional imaging modalities can provide important information not available by MR techniques. Therefore, we have developed co-registration software to integrate preoperatively acquired multimodality information with the intraoperative ima-

For References see p. 1153

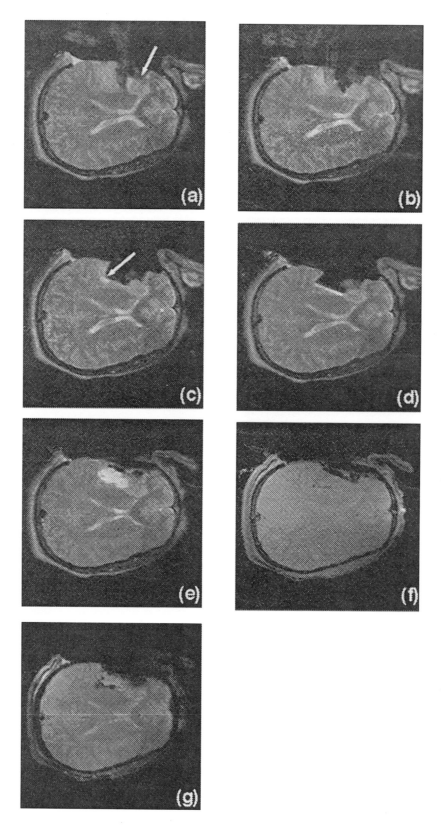

Figure 4   Intraoperative imaging during neurosurgical resection: serial T_2-weighted axial images. (a) Early in the surgery, the neoplastic tissue is posterior to the surgeon's finger. Anterior tissue (arrow) represents postoperative change from previous surgery and was present on prior MR study (b,c) Serial resection, (d) complete resection of the T_2-abnormal tissue. Some tissue shift is evident. (e) Following resection, the cavity is filled with saline. (f,g) Postsurgical gradient-echo images allow evaluation of hyperacute hemorrhage. These show typical postsurgical changes. (Reprinted with permission from Wong et al.[22])

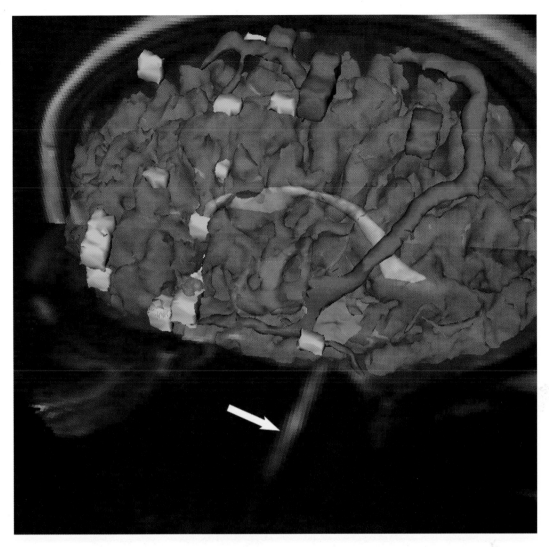

Figure 5   Intraoperative parasagittal image (note intraoperative pointer, white arrow) co-registered with the patient's preoperative three-dimensional brain model. Operative skin flap in the lower portion of the image on the left. The preoperative model includes conventional MRI with segmentation of tumor (green) and ventricles (light blue). In addition, MR angiography was performed to display large vessels (red). Functional MRI was used to identify cortical activation during complex speech tasks (turquoise and yellow) and motor tasks (magenta)

ging. Examples of specialized studies that can be performed prior to surgery include MR angiography, PET, SPECT, and functional MRI. MR angiography can define the relationship of the tumor to major vascular structures during the surgical procedure. PET imaging with [^{18}F]-fluorodeoxyglucose reflects metabolic activity of gliomas, and areas of higher uptake correlate with higher tumor grade and poorer prognosis.[29–31] Dual-isotope ^{201}Tl/^{99m}Tc-HMPAO SPECT is useful for distinguishing tumor recurrence from radiation necrosis,[33] and findings also correlate with histopathological findings and survival.[34] Image registration of MRI with PET or SPECT images can, therefore, be useful for directing biopsy to regions most representative of tumor behavior; this may reduce sampling error. For image-guided tumor resection, the addition of PET or SPECT data can help the neurosurgeon to select most aggressive portions of tumor for resection. Functional MRI can be used to map areas of cortical activation during motor tasks,

enabling critical sensorimotor areas to be identified and allowing the neurosurgeon to avoid eloquent areas during tumor resection.[45]

The preoperatively acquired information is combined with the intraoperative imaging data set using registration algorithms, thus maximizing the information available during surgery. The grayscale images and three-dimensional models of specific structures are displayed to the surgeon and correspond spatially to the actual patient as positioned in the intraoperative MRI (Figure 5). The combination of the preoperative information and the intraoperative renewal of the grayscale information allows updated neuronavigation. With progress of the surgical intervention and increasing brain deformation, the validity of the preoperative information decreases. We are currently evaluating the definition of a deformation field from the intraoperatively acquired grayscale images to align the three-dimensional models and their preoperative information to the

For References see p. 1153

updated situation. Other models to account for brain deformation are being developed.[36,37] The combination of pre- and intraoperative information, their alignment and the tracking of instruments in this defined space comprise the capabilities of this system for realtime interactive image-guided surgery (Figure 6).

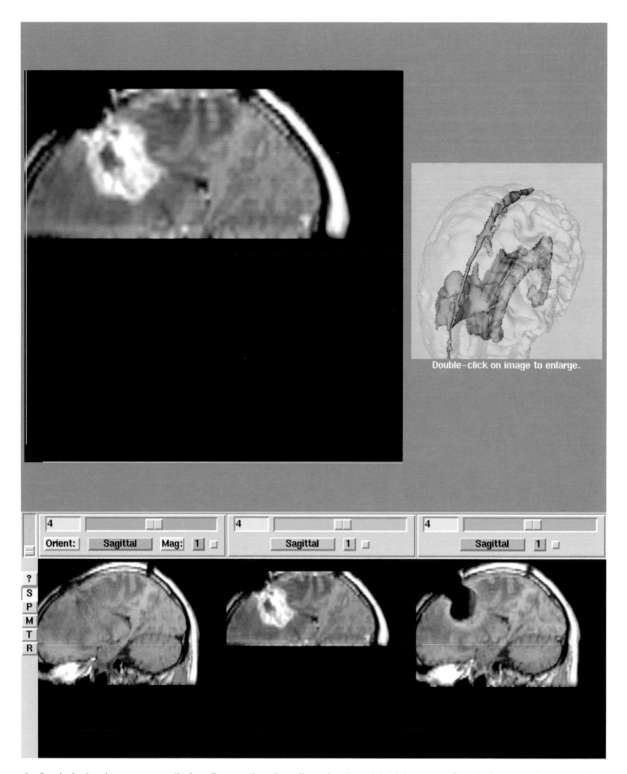

Figure 6 Surgical planning computer display. Preoperative three-dimensional model (right, upper frame) demonstrating tumor (green) and anatomic relationship to superior sagittal sinus (red), corpus callosum (brown), and ventricle (blue). The bottom row displays three intraoperative sagittal images in the same imaging plane. On the left is the unenhanced T_1-weighted image prior to craniotomy. The middle image demonstrates enhancing tumor prior to resection and is reproduced in the large upper left frame. The image on the bottom right demonstrates the surgical resection cavity following tumor removal

For list of General Abbreviations see end-papers

6 CONCLUSIONS

Intraoperative MRI-guided neurosurgical procedures are a collaborative effort, involving neurosurgeons, neuroradiologists, anesthesiologists, MR technologists, nurses, and engineers. MRI provides excellent contrast between abnormal and healthy brain tissue, and intraoperative imaging permits these areas to be defined relative to the surgeon's field of view. This improves the accuracy of intracranial biopsies and surgical resection of tumors. Early identification of unexpected complications is another benefit. To date, MRI guidance has been utilized in selected complex cases or repeat surgery. Based on our current experience, there is little question that MRI guidance for neurosurgical procedures augments the technical capability of the neurosurgeon to remove neoplastic tissue selectively. However, evaluation of cost effectiveness and clinical outcomes relative to conventional neurosurgical procedures will require a prospective, randomized clinical trial.

In the future, technological developments will continue to reduce the degree of invasiveness required for tumor resection. Definition of the tissue targeted for resection and its relationship to critical functional structures will continue to be improved through multimodality presurgical planning. Improvements to MRI will include faster acquisition times and refinement of imaging sequences for neurosurgical guidance. Currently available 'near-realtime' imaging will become closer to realtime. Image processing will continue to become faster, and techniques to compensate for intraoperative tissue shifts and removal are under development. Surgical image guidance will ultimately be a synthesis of anatomical/functional preoperative images with updated co-registered intraoperative imaging.

Image guidance will allow neurosurgeons to treat benign and malignant lesions using minimally invasive tools. Thermal ablation techniques using laser, rf, ultrasound, and microwave energy sources are currently being investigated, as is cryotherapy. Central to the success of these minimally invasive therapies is image guidance to position the ablation probes and monitor the thermal therapy. The navigational tools available for MRI-guided biopsies described above can be used to position thermal ablation probes accurately. In addition, MRI is uniquely suited to reveal subtherapeutic as well as irreversible thermal effects, and development of improved, faster imaging sequences to monitor these tissue effects will parallel the refinement of energy sources to deliver controlled, directed thermal ablation.

7 RELATED ARTICLES

ESR Probes as Field Detectors in MRI; MR-Guided Biopsy, Aspiration, and Cyst Drainage; MR-Guided Therapy in the Brain; Thermal Therapies in the Body Monitored by MRI.

8 REFERENCES

1. D. W. Roberts, J. W. Strohbehn, J. F. Hatch, W. Murray, and H. A. Kettenberger, *J. Neurosurg.*, 1986, **65**, 545.
2. R. J. Maciunas, M. S. Berger, B. Copeland, M. R. Mayberg, R. Selker, and G. S. Allen, *Neurosurg. Clin. North Am.*, 1996, **7**, 245.
3. D. Kondziolka and L. D. Lunsford, Intraoperative navigation during resection of brain metastases, *Neurosurg. Clin. North Am.*, 1996, **7**, 267.
4. P. K. Doshi, L. Lemmieux, D. R. Fish, S. D. Shorvon, W. H. Harkness, and D. G. Thomas, *Acta Neurochir. Suppl. (Wien)*, 1995, **64**, 49.
5. C. R. Maurer Jr, J. M. Fitzpatrick, M. Y. Wang, R. L. Galloway Jr, R. J. Maciunas, and G. S. Allen, *IEEE Trans. Med. Imag.*, 1997, **16**, 447.
6. D. G. Thomas, P. Doshi, A. Colchester, D. J. Hawkes, D. L. Hill, J. Zhao, N. Maitland, A. J. Strong, and R. I. Evans, *Sterotact. Funct. Neurosurg.*, 1996, **66**, 81.
7. E. Watanabe, Y. Mayanagi, Y. Kosugi, S. Manaka, and K. Takakura, *Neurosurgery*, 1991, **28**, 792.
8. C. Schaller, B. Meyer, D. van Roost, and J. Schramm, *Comput. Aided Surg.*, 1997, **2**, 162.
9. C. R. Maurer Jr, D. L. Hill, A. J. Martin, H. Liu, M. McCue, D. Rueckert, D. Lloret, W. A. Hall, R. E. Maxwell, D. J. Hawkes, and C. L. Truwit, *IEEE Trans. Med. Imag.*, 1998, **17**, 817.
10. S. W. Atlas and K. R. Thulborn, *Am. J. Neuroradiol.*, 1998, **19**, 1471.
11. R. B. Schwartz, L. Hsu, T. Z. Wong, D. F. Kacher, A. A. Zamani, P. M. Black, E. Alexander, P. E. Stieg, T. M. Moriarty, C. A. Martin, R. Kikinis, and F. A. Jolesz, *Radiology*, 1999, **211**, 477.
12. K. Hynynen, N. I. Vykhodtseva, A. H. Chung, V. Sorrentino, V. Colucci, and F. A. Jolesz, *Radiology*, 1997, **204**, 247.
13. K. Hynynen, W. R. Freund, H. Cline, A. H. Chung, R. D. Watkins, J. P. Vetro, and F. A. Jolesz, *Radiographics*, 1996, **16**, 185.
14. W. A. Hall, A. J. Martin, H. Liu, C. H. Pozza, S. O. Casey, E. Michel, E. S. Nussbaum, R. E. Maxwell, and C. L. Truwit, *Pediatr. Neurosurg.*, 1998, **29**, 253.
15. V. M. Tronnier, C. R. Wirtz, M. Knauth, G. Lenz, O. Pastyr, M. M. Bonsanto, F. K. Albert, R. Kuth, A. Staubert, W. Schlegel, K. Sartor, and S. Kunze, *Neurosurgery*, 1997, **40**, 891.
16. P. M. Black, T. Moriarty, E. A. Alexander III, P. Stieg, E. J. Woodard, P. L. Gleason, C. H. Martin, R. Kikinis, R. B. Schwartz, and F. A. Jolesz, *Neurosurgery*, 1997, **41**, 831.
17. J. F. Schenck, F. A. Jolesz, P. B. Roemer et al., *Radiology*, 1995, **195**, 805.
18. T. M. Moriarty, R. Kikinis, F. A. Jolesz, P. M. Black, and E. Alexander III, *Neurosurg. Clin. North Am.*, 1996, **7**, 323.
19. S. G. Silverman, B. D. Collick, M. R. Figueira, R. Khorasani, D. F. Adams, R. W. Newman, G. P. Topulos, and F. A. Jolesz, *Radiology*, 1995, **197**, 175.
20. S. G. Silverman, F. A. Jolesz, R. W. Newman, P. R. Morrison, A. R. Kanan, R. Kikinis, R. B. Schwartz, L. Hsu, S. J. Koran, and G. P. Topulos, *Am. J. Roentgenol.*, 1997, **168**, 1465.
21. F. A. Jolesz, *Radiology*, 1997, **204**, 601.
22. T. Z. Wong, R. B. Schwartz, PMcL Black, E. Alexander III, and F. A. Jolesz, *Semin. Intervent. Radiol.*, 1999, **16**, 23.
23. J. F. Schenck, *Med. Phys.*, 1996, **23**, 815.
24. F. A. Jolesz, P. R. Morrison, S. J. Koran, R. J. Kelley, S. G. Hushek, R. W. Newman, M. P. Fried, A. Melzer, R. M. Seibel, and H. Jalahej, *JMRI*, 1998, **8**, 8.
25. R. B. Schwartz, L. Hsu, P. M. Black, *JMRI*, 1998, **8**, 807.
26. J. Kettenbach, S. G. Silverman, N. Hata, K. Kuroda, P. Saiviroonporn, G. P. Zientara, P. R. Morrison, S. G. Hushek, P. M. Black, R. Kikinis, and F. A. Jolesz, *JMRI*, 1998, **8**, 933.
27. M. P. Fried, G. Topulos, L. Hsu, H. Jalahej, H. Gopal, A. Lauretano, P. R. Morrison, and F. A. Jolesz, *Otolaryngol. Head Neck Surg.*, 1998, **119**, 374.
28. L. Hsu, M. P. Fried, and F. A. Jolesz, *Am. J. Neuroradiol.*, 1998, **19**, 1235.

For References see p. 1153

29. M. W. Hanson, M. J. Glantz, J. M. Hoffman, A. H. Friedman, P. C. Burger, S. C. Schold, and R. E. Coleman, *J. Comput. Assist. Tomogr.*, 1991, **15**, 796.

30. J. B. Alavi, A. Alavi, J. Chawluk, M. Kushner, J. Powe, W. Hickey, and M. Reivich, *Cancer*, 1988, **62**, 1074.

31. M. Levivier, S. Goldman, B. Pirotte, J. Brucher, D. Baleriaux, A. Luxen, J. Hildebrand, and J. Brotchi, *J. Neurosurg.*, 1995, **82**, 445.

32. T. G. Turkington, R. J. Jaszczak, C. A. Pelizzari, C. C. Harris, J. R. MacFall, J. M. Hoffman, and R. E. Coleman, *J. Nucl. Med.*, 1993, **34**, 1587.

33. R. B. Schwartz, P. A. Carvalho, E. A. Alexander III, J. S. Loeffler, R. Folkerth, and B. L. Holman, *Am. J. Neuroradiol.*, 1991, **12**, 1187.

34. R. B. Schwartz, B. L. Holman, J. F. Polak, B. M. Garada, M. S. Schwartz, R. Folkerth, P. A. Carvalho, J. S. Loeffler, D. C. Shrieve, P. M. Black, and E. Alexander III, *J. Neurosurg.*, 1998, **89**, 60.

35. B. L. Holman, R. E. Zimmerman, K. A. Johnson, P. A. Carvalho, R. B. Schwartz, J. S. Loeffler, E. Alexander, C. A. Pelizzari, and G. T. Chen, *J. Nucl. Med.*, 1991, **32**, 1478.

36. P. J. Edwards, D. L. Hill, J. A. Little, and D. J. Hawkes, *Med. Image Anal.*, 1998, **2**, 355.

37. K. D. Paulsen, M. I. Miga, F. E. Kennedy, P. J. Hoopes, A. Hartov, and D. W. Roberts, *IEEE Trans. Biomed. Eng.*, 1999, **46**, 213.

38. D. A. Leung, J. F. Debatin, S. Wildermuth, N. Heske, C. L. Dumoulin, R. D. Darrow, M. Hauser, C. P. Davis, and G. K. Schulthess, *Radiology*, 1995, **197**, 485.

39. S. Wildermuth, J. F. Debatin, D. A. Leung, C. L. Dumoulin, R. D. Darrow, G. Uhlschmid, E. Hoffman, J. Thyregod, and G. K. von Schulthess, *Radiology*, 1997, **202**, 578.

40. H. J. Aronen, I. E. Gazit, and D. N. Louis et al., *Radiology*, 1994, **191**, 41.

41. H. F. Aronen, J. Glass, F. S. Pardo et al., *Acta Radiol.*, 1995, **36**, 520.

42. J. D. Hazle, E. F. Jackson, D. F. Schomer, and N. E. Leeds, *JMRI*, 1997, **7**, 1084.

43. R. B. Schwartz, L. Hsu, D. F. Kacher et al., *JMRI*, 1998, **8**, 1085.

44. R. Kikinis, P. L. Gleason, T. M. Moriarty, M. R. Moore, E. Alexander III, P. E. Stieg, M. Matsumae, W. E. Lorensen, H. E. Cline, P. M. Black, and F. A. Jolesz, *Neurosurgery*, 1996, **38**, 640.

45. M. F. Nitschke, U. H. Melchert, C. Hahn, V. Otto, H. Arnold, H. D. Herrmann, G. Nowak, M. Westphal, and K. Wessel, *Acta Neurochir. (Wien)*, 1998, **140**, 1223.

Biographical Sketches

Terence Z. Wong. *b* 1955. B.S.E. biomedical engineering, 1977, Duke University, M.D. and Ph.D., 1990, Dartmouth College and Dartmouth Medical School. Residency in Diagnostic Radiology, Deaconess Hospital, Harvard Medical School, 1992–96, Fellowship in Nuclear Medicine, Deaconess Hospital 1995–96, Fellowship in Cross-sectional Imaging, Beth Israel Deaconess Medical Center 1996–97, Fellowship in Interventional MRI, Brigham and Women's Hospital 1997–98. Assistant Professor of Radiology, Assistant Professor of Radiation Oncology, and Assistant Professor of Biomedical Engineering, Duke University Medical Center 1998–present. Current research interests: image-guided therapy, multimodality metabolic and functional imaging, selective oncologic imaging and therapy, thermal therapy.

Richard B. Schwartz. *b* 1956. B.S., 1978, M.D., 1982, Ph.D., 1985, Neuroanatomy, University of Pennsylvania, USA. Neuropathology internship, Tufts-New England Medical Center, Boston MA, 1984–85. Radiology residency, Brigham and Women's Hospital, Boston MA, 1985–89. Neuroradiology fellowship, Brigham and Women's Hospital, Boston MA, 1989. Certificate of Additional Qualification, 1996.

Instructor, assistant professor, and associate professor of radiology, Harvard University, 1989–present. Assistant Director of the Neuroradiology division, Brigham and Women's Hospital, Boston MA, 1992–present. Director of Functional and Navigational Neuroimaging, 1999. Approx. 80 publications in CT, MR and SPECT scanning of the brain. Current research interests: intracranial intervention under MR guidance, acute hypertensive disorders of the brain, neurovascular applications of spiral CT, and functional imaging of brain tumors using MR and SPECT.

Arya Nabavi. *b* 1966. M.D. University of Kiel/Germany. Neurosurgeon, Research Fellow at the Department of Neurosurgery, Brigham and Women's Hospital, Harvard Medical School. Current research interests: application of MR in neurosurgery, image-guided surgery, intraoperative brain deformations.

Richard S. Pergolizzi Jr. *b* 1964. B.Sc. Chemistry, 1987, The State University of New York at Stony Brook, M.D. 1993, Northeastern Ohio Universities College of Medicine. Residency in Diagnostic Radiology at University of Cincinnati, University Hospital 1993–97, Fellowship in Neuroradiology, University Hospital Health Science Center, Syracuse NY 1997–98, Fellow in Neurointerventional Radiology, Brigham and Women's Hospital and Massachusetts General Hospital, Harvard University, 1998–present. Current research interests: clinical applications for MR image-guided therapy.

Peter M. Black. *b* 1944, A.B., 1966, Harvard College, M.D. 1970 McGill University, Montreal, Canada, Ph.D., 1978, Georgetown University, USA. Research fellow in Neuro-oncology, Massachusetts General Hospital, 1976, Asst. Professor of Surgery, Harvard Medical School, 1980–84, Associate Professor of Surgery, Harvard Medical School, 1984–86. Franc D. Ingraham Professor of Neurosurgery, Harvard Medical School, 1987–present. Neurosurgeon-in-Chief, Brigham and Women's and Children's Hospitals, 1987–present. Chief of Neurosurgical Oncology, Dana Farber Cancer Institute, 1987–present. Chairman of the Board, Brain Tumor Center, BWH, CH, DFCI, 1988–present. Chairman, Department of Surgery, CH, 1994–98. More than 40 visiting professorships and lectureships. Approx. 155 original articles, 7 books, 150 reviews and book chapters, and several clinical communications. Current major research interests: advanced imaging techniques for surgery including the intraoperative MRI and brain mapping, and in the laboratory, molecular biology of brain tumors, specifically the relation between glial development and glioma formation, and hormonal influences on meningioma progression.

Eben Alexander III. *b* 1953. A.B., 1975, Chemistry, University of North Carolina, Chapel Hill. M.D., 1980, Duke University School of Medicine, Durham, North Carolina, Intern in General Surgery, Duke University Medical Center, 1980–81. Resident, Neurological Surgery, Duke University Medical Center, 1981–83. Research Fellow in Neurosurgery, Massachusetts General Hospital, Harvard Medical School, 1983–84. Acting Resident in Neurology, Massachusetts General Hospital, Boston, Massachusetts, 1985. Resident in Neurological Surgery, Duke University Medical Center, 1985–87. Senior Registrar and Cerebrovascular Fellow, Neurosurgical Service, Newcastle General Hospital, Newcastle-Upon-Tyne, England, U.K., 1987. Instructor in Surgery, Brigham and Women's Hospital, Harvard Medical School, Boston, Massachusetts 1988–90. Assistant Professor of Surgery (Neurosurgery), 1990–94, and Radiation Oncology, Harvard Medical School, 1990–99. Associate Professor of Surgery (Neurosurgery), Harvard Medical School, 1994–99. Director of the Stereotastic and Functional Neurosurgery Program, Brigham and Women's Hospital. Approx. 150 publications in neurosurgery, especially radiosurgery and image-guided surgery. Current research speciality: development of advanced synergistic radiation and surgical techniques through image-guidance.

Claudia H. Martin. *b* 1959. B.Mus., 1984, M.D., 1988, McGill University, Canada: Surgical Internship. Yale University, New Haven, 1989–90, NIH postdoctoral fellow in neurosurgery, Bethesda, Maryland, 1990–91, Neurosurgery resident, Yale University, New Haven, 1991–

95, Neurosurgery clinical fellow, Brigham and Women's Hospital, Harvard University, Boston, 1996–99. Chief resident Brigham and Women's Hospital, Boston, 1999–2000. Approximately 10 publications in image guided neurosurgery.

Ferenc A. Jolesz. b 1946. M.D., 1971, Semmelweis Medical School, Hungary. Resident in Neurosurgery, Medical School Pecs, Hungary, 1971–72. Research Fellow, Biomedical Engineering Computer Sciences, K. Kando College, Hungary, 1972–73. Resident in Neurosurgery, Institute of Neurosurgery, Hungary, 1975–79, Research Fellow, Neurology, Massachusetts General Hospital, Boston, MA, 1979–80, Milton Research Fellow, Physiology, Harvard Medical School, Boston, MA, 1980, Research Fellow, Physiology, Harvard Medical School, Boston, MA, 1980–81, Research Associate, Physiology, Harvard Medical School, Boston, MA, 1981–82, Resident, Radiology, Brigham and Women's Hospital, Boston, MA, 1982–85, Assistant Professor, Radiology, Harvard Medical School, Boston, MA, Director, Neuro MR Imaging Section, Brigham and Women's Hospital, Boston MA, 1987–88, Director, Division of Magnetic Resonance Imaging, Brigham and Women's Hospital, Boston, MA, 1988, Associate Professor of Radiology, Harvard Medical School, Boston, MA, 1989–96, Director, Image-guided Therapy Program, Brigham and Women's Hospital, Boston, MA, 1993–present. Professor of Radiology, Harvard Medical School, Boston, MA, 1996. B. Leonard Holman Professor of Radiology, Harvard Medical School, Boston, MA, 1998. Member of the Institute of Medicine, National Academy of Sciences. Approx. 200 publications. Current research speciality: interventional and intraoperative MRI, quantitative neuroimaging, focused ultrasound surgery.

MR-Guided Therapy in the Brain

Volker Tronnier
University of Heidelberg, Germany

Antonio A. F. De Salles
University of California at Los Angeles, CA, USA

Yoshimi Anzai
University of Michigan, USA

Keith L. Black
University of California at Los Angeles, CA, USA

and

Robert B. Lufkin
University of California at Los Angeles, CA, USA

1   INTRODUCTION

Through the technological and clinical advances of the 1990s, MRI has become one of the most powerful noninvasive diagnostic tools in the evaluation of the brain. The superior soft-tissue contrast, the oblique and multiplanar imaging capability, and the absence of ionizing radiation have made MRI the imaging modality of choice for most diagnostic applications of the brain.

These advantages also make MRI an ideal guide for invasive and interventional procedures in the brain. The high sensitivity of MRI to flow enables the delineation of blood vessels without the use of contrast agents. The absence of beam-hardening artifacts from bone and metal devices allows complex instrumentation approaches to anatomic regions that may be difficult or impossible to reach under conventional computed tomography (CT) guidance. Nearly a decade of experience with MR-guided stereotaxis has proven MRI to be an effective and reliable means to visualize and localize neurosurgical targets and a valuable aid in optimizing surgical approaches.[1-4]

Recently, MRI has been applied to treatment monitoring procedures in the brain, such as interstitial therapy and even open surgery.[5] New developments in MR imaging techniques and MR-compatible instrumentation and the availability of open magnet designs are revolutionizing the role of MRI in guiding and monitoring interventional procedures.

2   MR-GUIDED STEREOTAXIS

Since the mid 1980s, MRI stereotaxis has been successfully employed in neurosurgery to guide biopsies, in tumor resection, deep electroencephalographic (EEG) electrode placement, functional target development and resection, and other interventional procedures.[6-14] The noninvasive visualization and localization provided by MRI greatly reduce the surgical invasion necessary to access the target and significantly lower patient morbidity. Several stereotactic systems with localizer frames designed for MRI localization are commercially available: the BRW and CRW (Radionics, Burlington, MA), Leksell (Elekta Instruments Inc., Atlanta, GA), Laitinen (Sandström Trade & Technology Inc., Welland, Ontario, Canada), and others.

Since MRI is more prone to geometric distortion than CT, the accuracy of MRI stereotaxis must be evaluated. The maximum localization error reported in the literature ranges from 1 to 5 mm.[15-18] The variation in the maximum error may be accounted for partly by the use of different stereotactic systems and different reference imaging modalities (CT, ventriculography, etc.) with which MRI is compared. The spatial accuracy of MRI depends on the linearity and calibration of magnetic field gradients and on magnetic susceptibility artifacts, which manifest as spatial distortions. Meticulous quality control should decrease errors in magnetic field linearity and calibration. MRI sequences should use high bandwidth signal acquisition to reduce spatial distortion from susceptibility effects. Several researchers have proposed practical correction algorithms to improve the spatial accuracy of MRI further.[19,20]

Although accurate and reliable, conventional stereotactic systems require the fixation of a frame to the patient's skull. Recent developments in frameless stereotactic systems will improve patient comfort, localization efficiency, and surgical access in open procedures.[21-23] Projects are underway to

For References see p. 1163

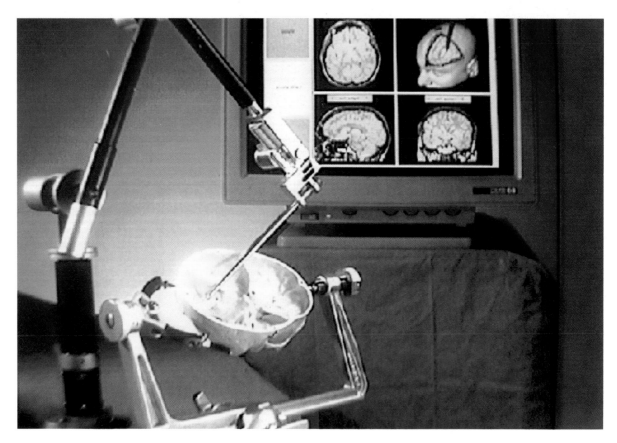

Figure 1 Mechanical articulated arm systems (Operating Arm System Radionics Software Applications, Inc., Burlington, MA) provides intraoperative neuronavigation using a previously acquired three-dimensional CT or MR data set

develop MR-compatible navigational systems that use an arm-based three-dimensional digitizer or light-emitting diode (LED)-based optical digitizers registered to the coordinate system of the interventional MR scanner for interactive scan plane control and near-realtime MRI monitoring of the interventional procedure (Figure 1).

3 MR-GUIDED BIOPSY

One of the earliest applications of intracranial interventional MR was to obtain biopsy specimens for tissue diagnosis. As with biopsies elsewhere in the body, the approach utilized for accessing an intracranial lesion is chosen to limit the risk of injury to surrounding vital structures. Either fine needle aspirates or core biopsies can be obtained by utilizing various gauge MR-compatible needles. Also, either frame-based or frameless stereotaxis may be used to guide the approach, depending on the location and size of the target lesion. The primary goal of an intervention within the MR scanner, however, will be a frameless procedure with on-line imaging control of the needle trajectory. The point of entry is determined with an optical tracking device, a gadolinium-filled wand, or simply with the surgeon's finger. Fast image sequences (e.g., two-dimensional fast imaging, with steady-state precession, FISP) allow a near realtime image while advancing the biopsy needle to the desired target. An obvious advantage compared with intraoperative CT is the possibility of taking specimens from

different tumor areas best displayed by MRI under direct visual control (Figure 2).

4 MR-GUIDED ELECTRODE PLACEMENT

Another application of interventional MR is to assist in the placement and in post-placement evaluation of intracerebral depth electrodes.[24–27] Intracerebral depth electrodes are used to localize the seizure focus in those patients who are refractory to medical therapy. Prior to the development of this technique, such electrodes were placed by inferring their position relative to previously acquired imaging studies. Postplacement examination of electrode placement could not be accurately performed because of the extensive beam-hardening artifact produced in CT by the stainless steel composition of the electrodes. Previously, MR imaging was not feasible because of the ferromagnetic properties of stainless steel and concern for heat and torque production, which could result in local brain injury. With the manufacture of electrodes with different metal compositions, such as platinum alloys, MR imaging guidance can be employed during electrode placement to confirm position without risk of brain injury. Sufficient anatomic detail is obtained to evaluate electrode placement adequately and to avoid possible procedural complications (Figure 3). This should result in improved anatomic–physiologic correlation and surgical outcome for these patients. Intraoperative documentation of long-term implanted depth electrode placed for treatment of chronic

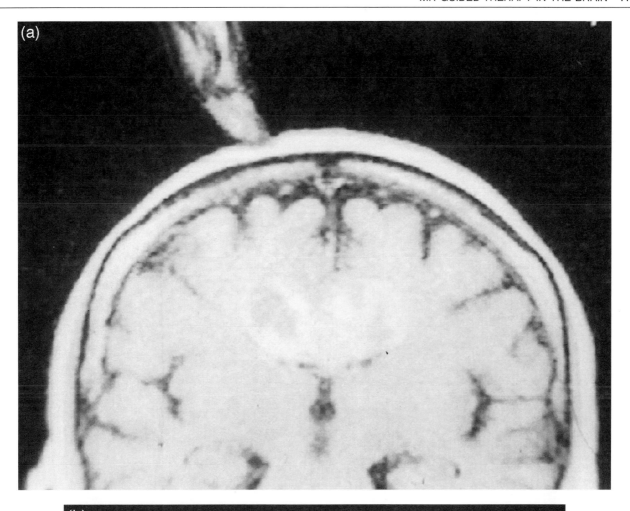

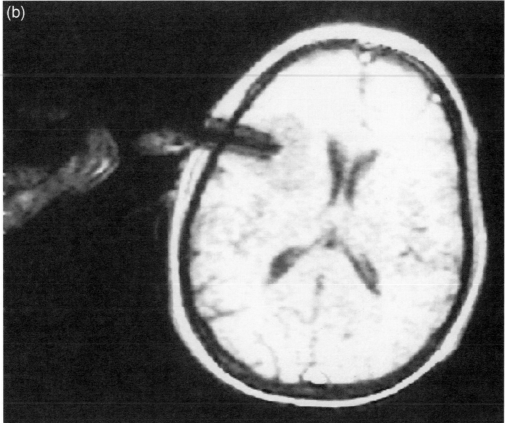

Figure 2 MR-guided biopsy. (a) Direct imaging allows rapid localization of a glioma. (b) MR-guided approach for a deeper lesion

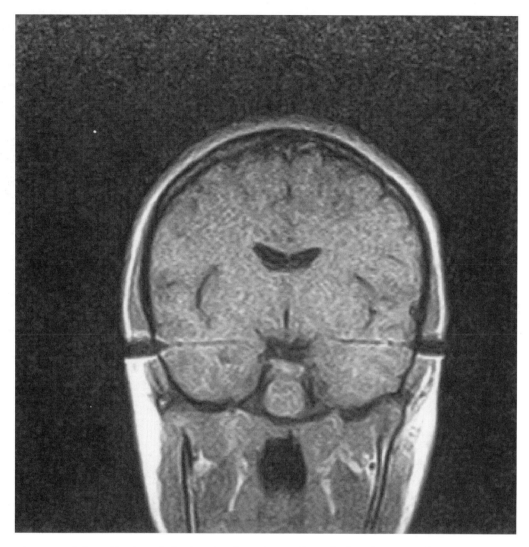

Figure 3 Deep electroencephalographic electrode implantation for patients with medically refractory partial complex seizures who are being considered for possible temporal lobectomy. Coronal MRI was obtained with MR-compatible electrodes in satisfactory position in the hippocampal region

pain or movement disorders is another indication in which the anatomical position can be correlated with intraoperative neurophysiological target determination.

5   MR-GUIDED CLOSED MINIMALLY INVASIVE THERAPEUTIC PROCEDURES

In the effort to reduce patient morbidity, hospitalization length, and recovery time associated with conventional craniotomy, several researchers are pioneering new, minimally invasive therapeutic techniques that may achieve results comparable to surgical resection or serve as effective palliative measures. In these minimally invasive procedures, it is particularly appealing to employ MRI beyond mere stereotactic localization and begin to use MRI to monitor treatment administration directly. The sensitivity of MRI to temperature change and other acute tissue effects (e.g. edema) enables the physician to adjust treatment parameters interactively to maximize

the destruction of pathologic tissue and minimize undesirable injury to normal tissue.[28–30]

5.1   Cryoblation

Freezing has been explored as a minimally invasive technique for tissue ablation in the brain. Postoperative MRI of patients who underwent cryothalamotomy demonstrated excellent depiction of the lesions created (Figure 4).[31] Intraoperative MRI guidance and monitoring of the cryoablation procedure awaits the development of practical MR-compatible freezing devices.

5.2   Ablation by Interstitial Laser Photocoagulation

Traditionally, the use of lasers in neurosurgery has been limited to applications in which direct visualization was possible. Now, MRI guidance allows stereotactic placement of an interstitial laser probe into a deep tumor through a small burr

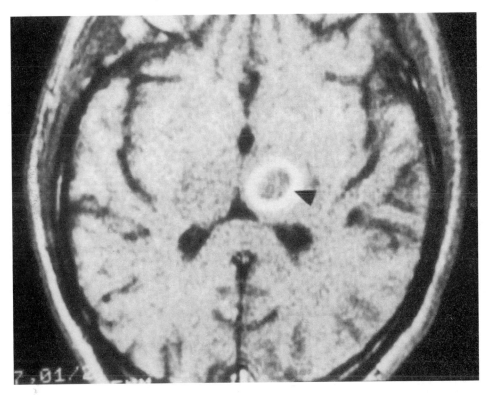

Figure 4 A T_1-weighted image obtained several days after stereotactic cryothalamotomy shows the lesion well circumscribed by a hyperintense ring (arrowhead)

hole in the skull. Laser energy travels through the optic fiber directly into the tumor and produces heat-induced protein denaturation and coagulation necrosis. However, since the penetration and absorption of laser energy are complex functions of tissue characteristics, it is difficult to predict a priori the extent of tissue damage as a function of dose. Therefore, a monitoring technique is required to follow the destructive process in tissue as laser energy is applied. Although CT can show laser-induced tissue changes, these are relatively late findings. Early tissue reaction to laser is missed by CT because of its low soft-tissue contrast.[32] Fortunately, many researchers have demonstrated the ability of MRI to detect acute tissue changes secondary to the application of laser energy.[33–37]

Animal studies have characterized the acute and chronic tissue effects of the Nd:YAG (neodymium:yttrium–aluminum–garnet) laser and correlated the histopathologic findings with their MRI appearance.[33,35,38] It has been shown that T_2-weighted images are particularly sensitive to tissue changes and can clearly demonstrate concentric rings of central cavitation, coagulation necrosis, and edema around the point of laser application. Higuchi et al. detected similar patterns in T_2-weighted images 5 min after laser irradiation in the live rabbit brain.[33]

Since the primary tissue effect of Nd:YAG laser irradiation is thermal damage,[39,40] the thermal sensitivity of MRI can also be exploited to monitor the progress of the laser therapy. The dependence of T_1-time, diffusion coefficient, water proton chemical shift, and other MR parameters on temperature is well known and has been utilized for noninvasive thermal map-

ping during the application of laser and other thermal ablation techniques.[28–30,41–43] Knowledge of the temperature distribution in the treatment area will allow the surgeon to confirm the destruction of the target unambiguously while avoiding irreversible thermal damage to vital structures.

Interstitial laser therapy is particularly well suited for intraoperative MRI guidance because the laser-delivering optic fiber is intrinsically MR compatible and because the application of laser energy does not interfere with MR signal acquisition. Initial clinical experience with MR-guided interstitial laser therapy has yielded promising results in the treatment of brain tumors.[24] This procedure can be executed under local anesthesia, which allows the neurologic status of the patient to be monitored continuously. MR guidance enables the physician to minimize damage to normal brain tissue, while the absence of systemic toxicity permits repetition of the procedure as necessary.

The Dusseldorf group performed a clinical pilot study of MR-guided Nd:YAG laser ablation on 31 patients with brain tumors. Three-dimensional turbo fast low angle shot (FLASH) sequences were used to determine the position of light guide. The time of irradiation varied from 10 to 20 min. Phase-sensitive two-dimensional FLASH and echo-shifted turbo FLASH were used to generate a color-coded heat map. The result is summarized in Table 1. The study shows that MRI is well suited to monitor laser-induced thermotherapy. A typical laser lesion has a central and peripheral zone that make up the total lesion, and it is circumscribed by an enhancing rim demarcat-

For References see p. 1163

Table 1 Results of a Clinical Pilot Study of MR-guided Laser Ablation of Brain Tumors: the Dusseldorf Study

	No.
Study population	31
Neurologic deterioration during ablation	2
Transient deficit after ablation (vasogenic edema)	4
Persistent deficit after ablation	1
Patients with astrocytoma WHO II	24
Neurologic status unchanged after ablation	6
Neurologic status improved after ablation	18

ing the outer border of the irreversible damaged lesion (Figure 5).

5.3 Radiofrequency-induced Thermal Ablation

Thermal ablation induced by rf is a technique of tissue destruction by resistive heating using rf (~500 kHz) electrical current. A probe with a conductive tip is inserted into the target, and rf current is then delivered from the tip to cause focal heating; this results in protein denaturation and coagulation necrosis. Over three decades of extensive neurosurgical experi-

ence with rf has established the efficacy and safety of this ablation technique.[26,27,44–52] Ablation with rf has been used primarily for functional neurosurgical procedures including thalamotomy and pallidotomy for parkinsonism and other movement disorders, leucotomy for intractable pain, and rhizotomy for trigeminal neuralgia. Postoperative MRI has demonstrated excellent depiction of lesion details in patients who underwent rf thalamotomy.[44,53,54]

One group has recently completed an initial study of MR-guided rf ablation of primary and metastatic brain tumors.[5,55] The procedure used MR-guided stereotaxis to place an rf probe (Radionics, Burlington, MA) into the tumor through a 2 mm twist-drill hole in the skull (Figure 6). The rf power was then applied to achieve an intratumoral temperature of 80 °C for 1 min. Depending on the shape and size of the tumor, the probe position was adjusted and additional doses were administered. The entire procedure was performed under light sedation and local anesthesia, which allowed the surgeon to monitor the neurologic status of the patient throughout.

Pretreatment T_1- and T_2-weighted MR images provided clear depiction of tumor morphology and anatomic relationships with surrounding vital structures and proved critical in planning the best trajectory for the rf probe. MR angiography with two-dimensional time-of-flight sequences also proved useful to visualize vascular structures surrounding the planned trajectory

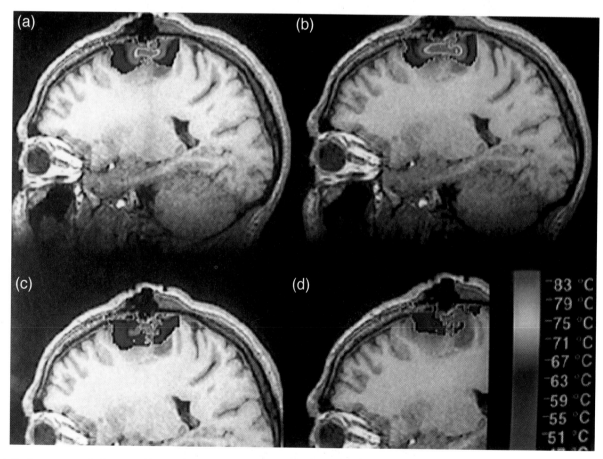

Figure 5 Post-processed phase-sensitive two-dimensional fast low angle shot images acquired and displayed during and after laser-induced thermotherapy (LITT). The calculated temperatures are color-coded (bottom right), gradually increasing the heat-affected zone with increasing maximum temperatures. Images acquired during treatment; (a) 1; (b) 4; (c); 8; and (d) 10 min after starting the laser therapy. (Courtesy of T. Kahn. In R. Lufkin: 'Interventional MRI', St Louis, 1999, Mosby)

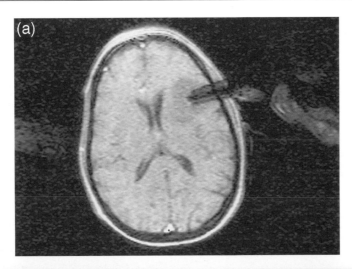

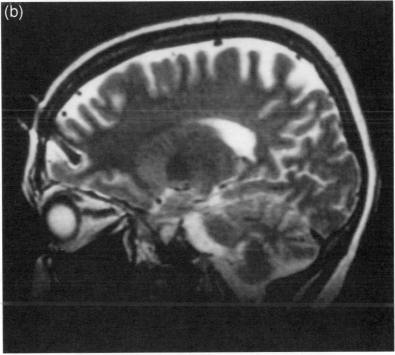

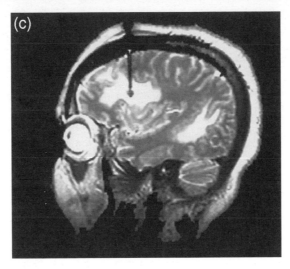

Figure 6 MR-guided rf ablation. (a) The therapy probe is placed into the patient's brain through a 2 mm twist-drill hole using MR-guided stereotaxis. The entire procedure is done under light sedation and local anesthesia in the MR suite. (b,c) Images from two other patients treated in the same way

For References see p. 1163

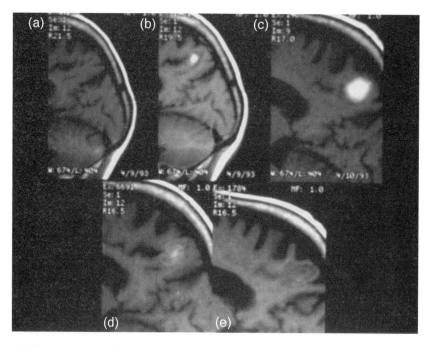

Figure 7 Contrast-enhanced MR images of a patient who underwent MR-guided rf ablation of a metastatic adenocarcinoma. (a) Image before treatment and (b) immediately after rf ablation. Further images were obtained at (c) 1 week, (d) 5 months, and (e) 1 year after treatment

of the rf probe. The rf lesion was detected in post-treatment MRI as a region of low signal on T_1-weighted images and as a region of high signal on T_2-weighted images within 20 to 30 min of the rf ablation. Gadolinium-enhanced T_1-weighted images demonstrate a rim of enhancement that outlines the zone of tissue necrosis. A total of 15 tumors were treated in 11 patients. During the 2-year follow-up period, seven tumors were controlled by treatment while eight tumors recurred. Five tumors had a durable response to treatment (11.9–23.8 months of follow up; median 15.1 month) (Figure 7). Neurologic complications were few and transient except for one patient, who still had mild expressive speech difficulty at 10 weeks after treatment.

MR-guided rf thermal ablation promises to be a safe, minimally invasive treatment of brain tumors. This technique may become an effective palliative measure and may serve as an adjuvant to radiation therapy. However, unlike radiation therapy, which is contraindicated by accumulative toxicity, rf ablation can be repeated as many times as necessary to treat recurrences. Greater efficacy is anticipated as improved MR-compatible rf generator and therapy probes are developed to allow real-time artifact-free MR imaging of the ablation procedure. As with interstitial laser therapy, the heating effects of the rf ablation can be monitored by temperature-mapping MRI sequences during the application of rf power to confirm the destruction of target tissue before physiologic changes become apparent.

5.4 High-intensity Focused Ultrasound Ablation

High-intensity focused ultrasound (HIFU) is a new technique of thermal ablation by remote heating. An extracorporeal ultrasound transducer focuses vibrational energy to a point inside the patient's body so that only the tissue within a small focal zone is heated to lethal temperatures (60 °C or above for thermal coagulation and protein denaturation) while the surrounding tissue experiences only a slight temperature increase. HIFU ablation under ultrasound imaging guidance has been used in clinical trials to treat glaucoma, benign prostatic hyperplasia, tumors of the brain, the breast, the liver, the kidney, and the bladder.[56–60] The greatest advantage of HIFU over other thermal ablation techniques is that no surgical invasion is required to access the target provided there is a sonic path clear of bone and air. However, noninvasive methods to measure the temperature of the focal heating zone must be developed to provide adequate control.

By combining the high soft-tissue contrast and the temperature sensitivity of MRI with HIFU ablation, a powerful minimally invasive therapy technique can be developed. Several researchers have demonstrated the feasibility of using MRI to guide and monitor HIFU ablation procedures.[41,43,61,62] Fast T_1-weighted gradient echo sequences can be used to track in realtime the high-temperature focal zone, which appears as a hypointense region, to guide the positioning of the ultrasound transducer. Tissue changes secondary to HIFU heating can be detected easily on T_2-weighted images, as with interstitial laser therapy. Although practical application of HIFU to the intracranial region necessitates the removal of a skull bone flap to expose the dura, surgical invasion of the brain parenchyma can be avoided and the potential for infection and other complications greatly reduced. Clinical application of HIFU in neurosurgery awaits the further development of MRI-compatible HIFU therapy systems and MRI temperature-mapping techniques.

6 CONCLUSION

MRI has traditionally been criticized as an expensive diagnostic technology. However, when compared with more costly open surgical procedures and the associated higher patient morbidity, hospitalization costs, and lost time from work, MRI-guided minimally invasive interventions can potentially lower the final cost of medical care in selected patients. More work is clearly warranted in this exciting field to determine its ultimate areas of application in neurosurgery.

7 RELATED ARTICLES

MR-Guided Biopsy, Aspiration, and Cyst Drainage; Temperature Measurement Using In Vivo NMR; Thermal Therapies in the Body Monitored by MRI.

8 REFERENCES

1. P. Mueller, D. Stark, J. Simeone, et al. *Radiology*, 1986, **161**, 605.
2. R. Lufkin, L. Teresi, and W. Hanafee, *Am. J. Roentgenol.*, 1987, **149**, 380.
3. R. Lufkin, and L. Layfield, *J. Comput. Assist. Tomogr.*, 1998, **13**, 1105.
4. S. To, R. Lufkin, and L. Chiu, *Comput. Med. Imaging Graph.*, 1989, **13**, 469.
5. Y. Anzai, K. Black, A. DeSalles, et al. *Am. J. Neuroradiol.*, 1999, in press.
6. G. Kratimenos, and D. Thomas, *Br. J. Neurosurg.*, 1993, **7**, 155.
7. D. Wen, W. Hall, D. Miller, et al. *Neurosurgery*, 1993, **32**, 407.
8. P. Kelly, *Adv. Tech. Stand. Neurosurg.*, 1990, **17**, 77.
9. R. Lufkin, S. Jordan, P. Lylyck, et al. *Am. J. Neuroradiol.*, 1988, **9**, 953.
10. L. Lunsford, *Neurosurgery*, 1988, **23**, 363.
11. P. Kelly, F. Sharbrough, K. Ball, et al. *Mayo. Clin. Proc.*, 1987, **62**, 103.
12. S. Uematsu, A. Rosenbaum, M. Delong, et al. *Acta Neurochir. Suppl.*, 1987, **39**, 21.
13. L. Lunsford, A. Martinez, and R. Latchaw, *J. Neurosurg.*, 1986, **64**, 872.
14. L. Leksell, D. Leksell, and J. Schwebel, *J. Neurol. Neurosurg. Psychiatry*, 1985, **48**, 14.
15. K. Burchiel, *Clin. Neurosurg.*, 1992, **39**, 314.
16. D. Kondziolka, P. Dempsey, L. Lunsford, et al. *Neurosurgery*, 1992, **30**, 402.
17. C. Derosier, G. Delegue, T. Munier, et al. *J. Radiol.*, 1991, **72**, 349.
18. J. Villemure, E. Marchand, T. Peters, et al. *Appl. Neurophysiol.*, 1987, **50**, 57.
19. C. Bakker, M. Moerland, R. Bhagwandien, et al. *Magn. Reson. Imag.*, 1992, **10**, 597.
20. J. Fitzpatrick, H. Chang, M. Willcott, et al. *Proc. IXth Annu. Mtg. Soc. Magn. Reson. Med.*, New York, 1990, p. 426.
21. T. Takizawa, *Stereotact. Funct. Neurosurg.*, 1993, **60**, 175.
22. R. Maciunas, R. J. Galloway, J. Fitzpatrick, et al. *Stereotact. Funct. Neurosurg.*, 1992, **58**, 108.
23. A. Kato, T. Yoshimine, T. Hayakawa, et al. *J. Neurosurg.*, 1991, **74**, 845.
24. P. Ascher, E. Justich, and O. Schrottner, *Acta Neurochir.*, 1991, **52**, 78.
25. T. H. Khan, et al. *JMRI*, 1998, **8**, 160.
26. S. Blond, D. Caparros-Lefebvre, and F. Parker, et al. *Neurosurg.*, 1992, **77**, 62.
27. L. Laitinen, M. Harriz, and T. Bergenheim, *J. Neurosurg.*, 1992, **76**, 53.
28. D. LeBihan, J. Delannoy, and R. L. Levin, *Radiology*, 1989, **161**, 401.
29. R. Dickinson, A. Hall, A. Hind, et al. *J. Comput. Assist. Tomogr.*, 1986, **10**, 468.
30. D. Parker, *IEEE Trans. Biomed. Eng.*, 1984, **31**, 161.
31. E. Vining, G. Duckwiler, R. Udkoff, et al. *J. Neuroimag.*, 1991, **1**, 146.
32. R. A. Gatenby, W. H. Hartz, P. F. Engstrom, et al. *Radiology*, 1987, **163**, 172.
33. N. Higuchi, A. Bleier, F. Jolesz, et al. *Invest. Radiol.*, 1992, **27**, 814.
34. J. Robinson, R. Lufkin, D. Castro, et al. *Eur. Radiology*, 1992, **2**, 24.
35. Y. Anzai, R. Lufkin, D. Castro, et al. *JMRI*, 1991, **1**, 553.
36. D. J. Castro, R. E. Saxton, L. J. Layfield, et al. *Laryngoscope*, 1990, **100**, 541.
37. F. A. Jolesz, A. R. Bleier, P. Jokab, et al. *Radiology*, 1988, **168**, 249.
38. Y. Anzai, R. Lufkin, S. Hirschowitz, et al. *JMRI*, 1992, **1**, 671.
39. F. Jolesz, G. Moore, R. Mulkern, et al. *Invest. Radiol.*, 1989, **24**, 1024.
40. K. Matthewson, P. Coleridge-Smith, J. P. O'Sullivan, et al. *Gastroenterology*, 1987, **93**, 550.
41. R. Matsumoto, R. Mulkern, S. Hushek, et al. *JMRI*, 1994, **4**, 65.
42. R. Stollberger, M. Fan, E. Ebner, et al., *Proc. XIIth Annu. Mtg. Soc. Magn. Reson. Med.*, New York, 1993, p. 156.
43. H. Cline, J. Schenck, K. Hynynen, et al. *J. Comput. Assist. Tomogr.*, 1992, **16**, 956.
44. F. Tomlinson, C. Jack, and P. Kelly, *J. Neurosurg.*, 1991, **74**, 579.
45. S. Hassenbusch and P. Pillary, *Neurosurgery*, 1990, **27**, 220.
46. K. Matsumoto, F. Shichijo, and T. Fukami, *J. Neurosurg.*, 1984, **60**, 1033.
47. P. Kelly and F. Gillingham, *J. Neurosurg.*, 1980, **53**, 322.
48. J. Siegfried, *Surg. Neurol.*, 1977, **8**, 126.
49. L. Organ, *Appl. Neurophysiol.*, 1976, **39**, 69.
50. H. Rosomoff, C. Brown, and P. Sheptak, *J. Neurosurg.*, 1965, **23**, 639.
51. S. Aronow, *J. Neurosurg.*, 1960, **17**, 431.
52. W. Sweet, V. Mark, and H. Hamlin, *J. Neurosurg.*, 1960, **17**, 213.
53. Y. Anzai, A. Desalles, K. Black, et al. *Radiographics*, 1993, **13**, 897.
54. S. Matsumoto, F. Shima, K. Hasuo, et al. *Nippon Igaku Hoshasen Gakkai Zasshi*, 1992, **52**, 1559.
55. K. Black, A. DeSalles, Y. Anzai, et al. *J. Clin. Oncol.*, 1999, in press.
56. R. Foster, R. Bihrle, N. Sanghvi, et al. *Eur. Urol.*, 1993, **23**(Suppl. 1), 29.
57. G. Vallancien, M. Harouni, B. Veillon, et al. *J. Endourol.*, 1992, **6**, 173.
58. R. Silverman, B. Vogelsang, M. Rondeau, et al. *Am. J. Ophthalmol.*, 1991, **111**, 327.
59. F. Fry, and L. Johnson, *Ultrasound Med. Biol.*, 1978, **4**, 337.
60. A. Burov, *Dokl. Akad. Nauk. SSSR*, 1956, **106**, 239.
61. H. Cline, K. Hynynen, C. Hardy, et al. *Magn. Reson. Med.*, 1994, **31**, 628.
62. K. Hynynen, A. Darkazanli, E. Unger, et al. *Med. Phys.*, 1993, **20**, 107.

For References see p. 1163

Part Thirteen
MR of the Pelvis, Abdomen, and Thorax

Heart: Clinical Applications of MRI

Scott D. Flamm and Charles B. Higgins
University of California at San Francisco, San Francisco, CA, USA

1   INTRODUCTION

The first electrocardiographically gated magnetic resonance images of the human heart were shown in 1984. In the decade that followed several important techniques have been developed which facilitate the use of MRI for the evaluation of cardiovascular morphology, contractile function, myocardial tissue characterization, and blood flow. With these techniques, MRI has been found to be effective for the investigation of a variety of acquired and congenital cardiovascular diseases. This role of MRI in relation to preexisting cardiac imaging techniques continues to evolve.

The clearly defined indications for the effective use of MRI in acquired heart disease include the evaluation of pericardial disease, right ventricular dysplasia, thoracic aortic disease, intra- and extracardiac masses, cardiomyopathies, valvular heart disease, ischemic myocardial disease, and the morphology of the coronary arteries.

The indications for the application of MRI in congenital heart disease include the assessment of pulmonary artery morphology, coarctation of the aorta, vascular rings, complex congenital heart disease, and postoperative evaluation.

extracardiac structures (e.g. the great vessels). Anatomical information also can be obtained using electrocardiogram (ECG)-gated cine gradient echo (cine GRE) and newer, fast gradient echo techniques that sample segmented k-space during a single breath-hold (breath-hold cine GRE). These latter two techniques are also useful for acquiring multiple, ECG referenced images of the heart, in either long or short axis projections similar to echocardiography, which may be electronically 'looped' and viewed in a cinematic format to evaluate global ventricular function and segmental wall motion. Velocity-encoded cine MRI techniques can assess quantitatively blood flow velocity and flow volume within the heart to evaluate intracardiac shunts, and valvular pathology and function. This same technique may also be used in the great vessels to determine pulmonary and aortic blood flow, including differential flow to the right and left pulmonary arteries, as well as to estimate gradients across areas of stenosis such as in coarctation of the aorta.

For multislice spin echo MRI, ECG gating is required to 'freeze' the motion of the heart, and respiratory compensation further improves image quality (see also **Cardiac Gating Practice** and **Motion Artifacts: Mechanism and Control**). In adults, typical sequences include ECG gating T_1-weighted images of 7–10 mm slice thickness obtained in the axial plane, with supplemental images acquired in the sagittal, and oblique sagittal, axial, and coronal planes, as dictated by the particular clinical concern.

In adults and children, body coils are used most commonly. Flexible phased array surface coils are gaining in use for both general and special purposes, as they provide greater resolution and image contrast as a result of improved signal-to-noise ratios. For some infants and small children a head coil may be practical, and will result in the greatest resolution and signal-to-noise ratio for the relatively small areas of interest.[1]

2   IMAGING TECHNIQUES

A multitude of sequences are available for cardiac imaging (Table 1). These include spin echo and multiplanar multiphasic techniques for morphological information of both cardiac and

3   LEVEL OF CURRENT CLINICAL ACTIVITY

Cardiovascular MRI is used primarily for detection and delineation of anatomical abnormalities according to a recent survey of members of the Society for Magnetic Resonance

Table 1   Techniques for MRI of the Heart

Technique	Application
ECG-gated spin echo	Anatomy and morphology
Cine GRE	Anatomy; atrial and ventricular dimensions and volumes; global and regional ventricular function
Fast (turbo) GRE	Anatomy; myocardial perfusion; rapid imaging of great vessels
Breath-hold, segmented k-space GRE	Anatomy including coronary arteries; myocardial perfusion; global and regional ventricular function
Velocity encoded cine MRI	Valvular function (quantification of regurgitation, and estimation of peak valvular gradient), measurement of stroke volume and cardiac output of each ventricle, quantification of shunts, measurements of differential flow in right and left pulmonary arteries, quantification of gradients at stenoses of the pulmonary arteries or coarctation of the aorta
Echo planar	Global and regional function of left and right ventricles, quantification of regional myocardial blood volume and perfusion

For References see p. 1176

Imaging. Of the responding centers, 46% reported exclusively clinical work, 3% exclusively research, and 51% both clinical and research. The majority of MRI studies are performed in the evaluation of thoracic aortic disease, congenital heart disease, and cardiac masses. While most studies are performed for morphological information, half of the centers also accomplish functional studies, usually for ventricular and/or valvular function, in conjunction with anatomical studies.[2]

4 ACQUIRED HEART DISEASE

4.1 Pericardial Disease

Differentiating constrictive pericarditis from restrictive pericarditis is the most frequent indication for applying MRI in pericardial disease. This is an important distinction because both diseases cause similar hemodynamic abnormalities, and are usually indistinguishable on cardiac catheterization. Only constrictive pericarditis can be alleviated by surgery. Constrictive pericarditis is seen most commonly after cardiothoracic surgery, and radiation therapy. On spin echo sequences it is manifested by local or generalized thickening of the pericardium to 4 mm or greater. This thickening is more readily evident over the right ventricle and right atrium. Associated findings include enlargement of the right atrium, inferior vena cava and hepatic veins, pericardial effusion, and tubular narrowing of the right ventricle. The reported diagnostic accuracy of MRI in establishing the diagnosis is 93%.[3]

MRI also can be used to examine the pericardium and its contents for evidence of effusion or hemorrhage, and mass lesions. Echocardiography is a fast, effective technique to determine the presence and general quantity of pericardial effusions, but can be less accurate in discerning hemorrhagic from nonhemorrhagic effusion, or determining the extent and location of loculated effusions, particularly in the posterior and basal portions of the heart. Spin echo MRI can easily determine the anatomical location and size of effusions, as well as the extent of loculations in all portions of the heart. It also has the advantage of more easily identifying the presence of pericardial masses, which may represent metastatic tumors responsible for hemorrhagic and/or recurrent effusion. The hemorrhagic nature of effusions can be identified with T_1-weighted spin echo sequences as high signal intensity, in contrast to the very low signal intensity of nonhemorrhagic pericardial effusions.

4.2 Right Ventricular Dysplasia

Right ventricular dysplasia is a relatively uncommon disorder generally occurring in young people, and manifesting as potentially life threatening arrhythmias originating from the right ventricle. The diagnosis is predicated on the identification of fatty infiltration or fibrosis replacing the right ventricular myocardium. The diagnosis of right ventricular dysplasia is difficult to establish even with electrophysiological studies and myocardial biopsy.

The primary diagnostic feature on MRI is transmural fat within the free wall of the right ventricle, but the spectrum of

findings includes 'islands' of fat within, or abnormal thinning of, the right ventricular free wall, regional or global contraction abnormalities of the right ventricle, and aneurysmal outpouching of the right ventricular outflow tract during systole.[4] Axial spin echo images focused on the right ventricular free wall are used to demonstrate myocardial fat, and cine GRE sequences obtained in the axial or horizontal long axis are useful to evaluate regional right ventricular contraction abnormalities. The utility of MRI in establishing the diagnosis noninvasively was recently documented in a group of patients with biopsy proven right ventricular dysplasia. Those patients with inducible right ventricular arrhythmias were significantly more likely to have identifiable fat within the myocardium, contraction abnormalities of the right ventricle, and/or a lower ejection fraction.[5]

The course of patients with right ventricular dysplasia typically has been followed with periodic right heart angiography and myocardial biopsy, which entails a significant degree of risk. MRI appears to be an effective technique for identifying at least a subset of patients with right ventricular dysplasia, and may become the procedure of choice for follow-up.

4.3 Thoracic Aortic Disease

Aneurysm and dissection are the most common indications for MRI of the thoracic aorta. Spin echo sequences are used to demonstrate the morphology of thoracic aortic aneurysms, which may occur as a result of atherosclerosis, trauma, or Marfan's syndrome, or may be mycotic or syphilitic in origin. In aortic dissection the intimal flap can often be identified on spin echo images, but is more readily seen with GRE and cine GRE sequences (Figure 1). These techniques also are effective in correctly identifying the true and false lumens, and differentiating slow flow versus thrombus within the false lumen. Additionally, MRI can identify the presence of hematoma within the wall of the aorta.[6] This cannot be defined with X-ray angiography.

MRI is a highly sensitive and specific technique for the detection of aortic dissection that can identify both the intra- and extravascular space, and the walls of the aorta. Conventional angiography has been considered the gold standard, but requires an invasive procedure, may incompletely identify the false lumen and its extent, and is unable to evaluate fully the walls of the aorta, and the extravascular space. MRI has proven to be superior to X-ray angiography and transthoracic echocardiography.

MRI has been compared to transesophageal echocardiography, which is also a highly sensitive technique, and has the advantage of being portable, permitting its use in critical and unstable patients unable to leave the intensive care setting. Multiple studies have demonstrated a similar high sensitivity (98–100%) for both methods, but MRI has a significantly higher specificity (98–100%) than transesophageal echocardiography (68–77%) in high risk populations.[7,8]

Evaluation of the thoracic aorta usually requires approximately 20–30 min using standard cine GRE sequences. Newer sequences, particularly segmented k-space breath-hold cine GRE sequences, can acquire each slice in 12–16 s, providing images of the entire thoracic aorta in approximately 5 min. The study can be extended to the abdominal aorta to evaluate for potential involvement of the mesenteric vessels with the addition of a few minutes of imaging time.

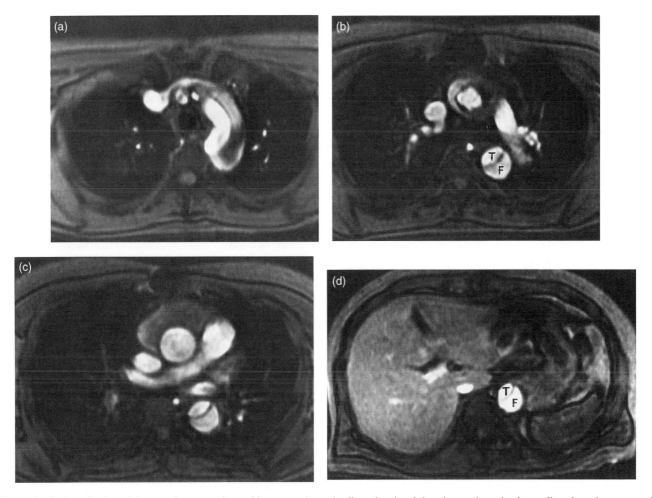

Figure 1 Series of selected images from a patient with a type A aortic dissection involving the aortic arch, descending thoracic aorta, and abdominal aorta. These are images acquired using a breath-hold segmented *k*-space cine MRI sequence acquired at the level of (a) the aortic arch, (b) the left pulmonary artery, (c) the main and right pulmonary artery, and (d) the upper abdomen. Note the compression of the true lumen by the false lumen. The bright signal within the false lumen confirms the presence of flow. T, true lumen; F, false lumen

4.4 Intracardiac Masses

Tumor, thrombus, and valvular vegetations (as occur in bacterial endocarditis) are the primary considerations when there is concern for an intracardiac mass. There are no studies demonstrating the benefit of MRI for valvular vegetations; echocardiography remains the most effective noninvasive technique for this determination.

MRI is effective for evaluating intracardiac masses, and in most instances can differentiate between tumor and thrombus.[9] Spin echo sequences are usually adequate to identify the size and location of intracardiac masses. Both tumors and thrombus tend to be isointense to myocardium on spin echo images; however, some thrombi may have a low signal intensity and be indistinguishable from the ventricular cavity without the additional application of cine GRE sequences.[10] Cine GRE sequences are particularly useful in that they permit identification of the majority of masses, and typically are successful in differentiating thrombus from tumor.[11] The most common tumors include myxomas (usually left atrial), rhabdomyomas (usually right ventricular), and metastases from primary carcinoma of the breast or lung, or melanoma. On cine GRE images tumors tend to be intermediate in signal intensity, with

myocardium being low–intermediate in intensity, while thrombus is frequently very low in signal intensity. In the case of relatively fresh thrombus, high signal intensity may be present on cine GRE images. These signal characteristics hold true in the majority of instances; however, some tumors, particularly myxomas which have bled internally, may contain increased amounts of hemosiderin, making them very low signal intensity on GRE images.

4.5 Cardiomyopathy

There are four recognized types of cardiomyopathy—hypertrophic, congestive (or dilated), restrictive, and obliterative—each of which has a distinct constellation of morphological and functional features. The etiology of the myocardial abnormality is varied. Most forms are idiopathic in nature, but chronic injury from ischemia or toxins, or infiltrative disorders are responsible for a minority of cases. The hypertrophic and congestive forms have been studied extensively with MRI; less is known about the obliterative form as it is very rare, and encountered primarily in Africa.

Hypertrophic cardiomyopathy usually manifests as global hypertrophy of the left ventricular myocardium, though many

For References see p. 1176

variants have been documented with MRI, including focal or diffuse septal hypertrophy, and midventricular and apical forms.[12] Cine GRE, acquired in either the long or short axis, has been used to quantify left ventricular mass, volumes, and ejection fraction in this disease. MRI has shown high accuracy and interstudy reproducibility for quantifying left ventricular volumes and mass, providing a precise technique for serial assessment during treatment.[13,14] Patients with hypertrophic cardiomyopathy also demonstrate abnormal diastolic and systolic function. The overall heart size remains normal, but the end-diastolic volume and filling rates are reduced in both the left and right ventricles.[15] Systolic function is hyperdynamic, and near obliteration of the ventricular cavity, particularly at the apex, is a characteristic feature (Figure 2).

In congestive cardiomyopathy the overall cardiac dimensions are increased, and there is left ventricular dilation. The idiopathic forms have normal myocardial thickness, and as a result the myocardial mass is greater than normal. In patients with chronic injury or ischemic etiology the myocardium may be globally or focally thinned. As in the hypertrophic form, cine MRI in congestive cardiomyopathy can quantify the ventricular volumes and mass, and assess ventricular function and wall stress.[16] Recent studies have demonstrated the effectiveness of using serial MRI examinations to follow patients with congestive cardiomyopathy undergoing treatment with angiotensin-converting enzyme inhibitor therapy.[17,18]

Restrictive cardiomyopathy may have either normal or thickened myocardium, and normal, increased, or decreased left ventricular end-diastolic volume. The salient characteristic functional feature is markedly reduced diastolic compliance, which is similar to that seen in constrictive pericarditis. Constrictive pericarditis, however, may be differentiated by the

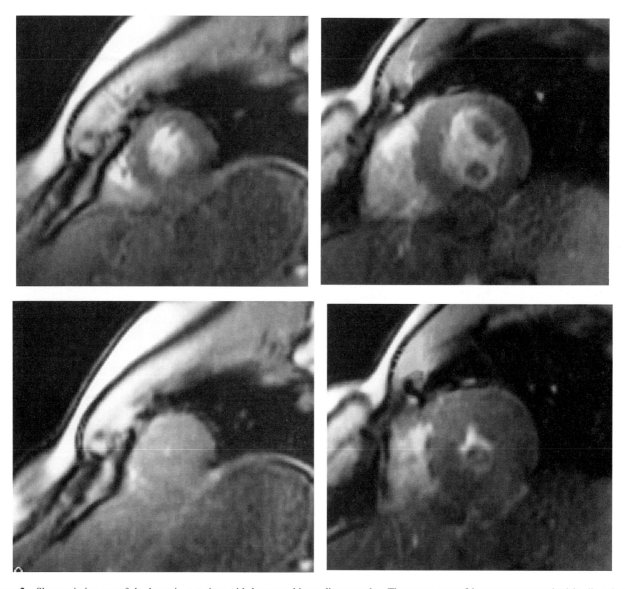

Figure 2   Short axis images of the heart in a patient with hypertrophic cardiomyopathy. The upper row of images were acquired in diastole (the images on the left are from near the apex, and the images on the right from the mid-venticular region), and directly below each is the respective systolic image, demonstrating marked, global thickening of the left ventricular myocardium, and obliteration of the apical cavity during systole. The sequence used is a breath-hold segmented *k*-space cine MRI

abnormally thickened pericardium (>4 mm), usually noted over the right side of the heart. Patients who have restrictive cardiomyopathy with thickened myocardium (such as in amyloidosis) may be differentiated from those with hypertrophic cardiomyopathy based on the reduced systolic function in the former, compared with the hyperdynamic function in the latter.[19]

4.6 Valvular Heart Disease

MRI using cine GRE sequences can identify the severity, extent, and direction of the turbulent jet of blood across stenoses of the aortic, pulmonic, mitral, and tricuspid valves. Turbulence, on cine GRE images is shown as a signal void against the background of bright signal intensity blood in the great vessels or ventricle. Using these images as a reference, the peak velocity of the turbulent jet may be determined with velocity-encoded cine MRI images obtained perpendicular to the long axis of the turbulent jet near the valve's orifice. Velocity-encoded cine MRI can determine the peak velocity of stenotic jets with accuracy up to 6 m s^{-1}.[20] By using the modified Bernoulli equation [gradient (mmHg) = 4 × (peak velocity (m s^{-1}))^2], the peak gradient across the stenosis can be calculated; this is one of the important clinical indicators used in determining the effectiveness of medical therapy, and, when necessary, timing of valve replacement surgery. Recent work has demonstrated the usefulness of this technique in both the aortic[21] and mitral[22] valves.

Cine GRE sequences also have been used to measure the extent and direction of the regurgitant jet in aortic and mitral insufficiency. Quantification of the volume of regurgitant flow may be acquired using velocity-encoded cine MRI sequences where the cardiac cycle (each R–R interval) is divided into 16 phases, and flow is measured at each phase of the cycle. In the case of aortic regurgitation, an imaging slice is acquired perpendicular to the long axis of the aorta proximal to the origin of the innominate artery. The data at each of the 16 phases can be plotted versus time, and the forward, reverse, and total flow calculated.[23] This method has the advantage of determining absolute flow, and has high interstudy reproducibility.[24] Regurgitant volume across the mitral valve has been calculated by measuring the amount of left ventricular inflow during diastole (velocity-encoded cine MRI slice across the mitral valve) and subtracting the left ventricular outflow during systole (imaging slice in the ascending aorta).[25]

4.7 Ischemic Heart Disease

MRI has the potential to become an important technique to differentiate normal from ischemic or infarcted myocardium by combining information about morphology, function, and perfusion in a single study. Of these three the largest hurdle has been determining global and regional myocardial perfusion in a rapid, reproducible manner under conditions reflecting both the basal state, and stress.

The earliest attempts at differentiating ischemic from normal myocardium focused on acute infarctions because of the edema that occurs secondary to vascular leak at the site of damage. Increased water in tissue results in increased signal on T_2-weighted scans in areas of acute infarction due to T_2 prolongation.[26] Subsequently, MRI contrast agents have been used to improve the distinction between acute and subacute infarc-

tions, and normal myocardium. Contrast enhancement (greater than the adjacent normal myocardium) has been noted uniformly in many reperfused areas of myocardium, or only peripherally in areas of infarction in which the central zones had necrosis, hemorrhage, or marked edema that limited or ceased blood flow centrally.[27,28]

Detecting areas of stress induced ischemic myocardium has been approached using contrast-prepared fast imaging techniques. Recent studies have demonstrated the feasibility of monitoring the passage of a bolus of MRI contrast agent through the myocardium using rapid, sequential images.[29–32] Myocardium supplied by coronary arteries with angiographically proven significant stenoses demonstrates lower peak signal intensity after contrast administration, and a lower rate of signal increase than normal myocardium.[29] Ischemic myocardium, induced with the administration of dipyridamole, a pharmacological stress agent, also demonstrates lower peak signal intensity as compared with normal myocardium.[30] At present, imaging is limited to a few imaging planes for each study, but with the development of echo planar imaging, dynamic acquisition of the entire heart during the first pass of MRI contrast media will be feasible (see Figure 3; also ***Echo-Planar Imaging***).

Ischemic myocardium also may be identified by functional wall motion abnormalities observed on cine GRE sequences acquired in the basal state or during pharmacological stress. Areas of threatened myocardium demonstrate decreased contractility during dipyridamole infusion.[33] One study found the sensitivity/specificity (percent values) for localization of stenotic vessels at 78/100 for the left anterior descending, 73/100 for the left circumflex, and 88/87 for the right coronary artery.[34] This technique has also been compared to radionuclide scanning using ^{99m}Tc–methoxyisobutyl isonitrile SPECT (MIBI SPECT), with no significant differences detected in sensitivity for lesion localization.[35] Dobutamine is another pharmacological stressor that has been used to assess wall motion abnormalities with cine GRE sequences. Dobutamine MRI studies have demonstrated close agreement with findings on dobutamine thallium-201 SPECT,[36] and a sensitivity and specificity of 81 and 100%, respectively, using coronary arteriography as the gold standard.[37]

4.8 Coronary Arteries

MRA techniques have recently been used for imaging of the coronary arteries.[38–40] The coronary arteries present a formidable challenge for MRI, because they are small structures whose motion within the thorax is complex, being influenced by the spiral contraction of the heart, expansion of the chest, and descent of the diaphragm.

The proximal and midportions of the coronary arteries have been demonstrated by MRA with a fat-suppressed, cine GRE technique using segmented k-space and breath-holding. In a group of 19 normal volunteers, and six patients with recent angiograms, the left anterior descending and right coronary arteries were identified in 100%, the left main in 96%, and the left circumflex in 76%.[41] A subsequent study using a similar MRA technique in 39 patients with angiographically identified coronary stenoses demonstrated a sensitivity and specificity of 90 and 92% for detection of significant (≥50% narrowing) lesions.[42]

For References see p. 1176

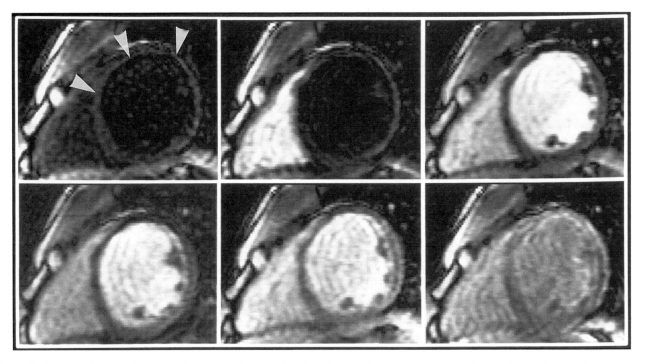

Figure 3   Series of short axis images of the heart acquired repetitively at the same level over a 90 s period during bolus administration of contrast media. The transit of contrast material increases signal intensity first in the right ventricular blood pool, then left ventricular blood pool, and finally the myocardium of the right and left ventricles. The patient has had an infarction in the anterior and anteroseptal distribution (arrowheads); note the delayed rise in signal intensity compared to the adjacent normal myocardium

The use of complex, oblique planes has permitted visualization of more than 8 cm lengths of the right coronary artery and left anterior descending coronary artery (Figure 4).[38] Other recent advances include non-breath-hold three-dimensional

imaging using fat saturation and magnetization transfer contrast techniques. Initial work has demonstrated the first 3–10 cm of the left and right coronary arteries from a single 7–10 min acquisition.[43] Further improvements in scanning time, image

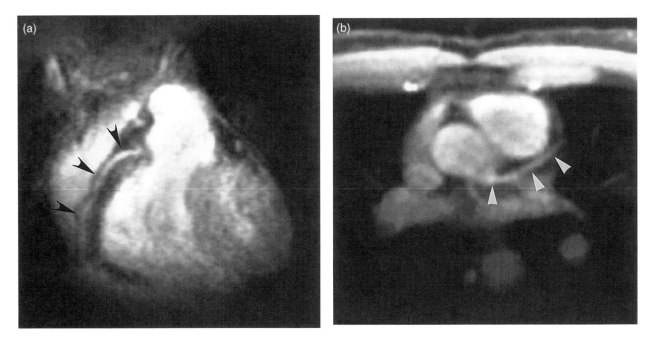

Figure 4   MRA of the coronary arteries using a breath-hold segmented *k*-space cine sequence. (a) An oblique coronal image demonstrating the proximal and midportions of the right coronary artery (arrowheads). (b) An axial image showing the left main and proximal left anterior descending coronary arteries (arrowheads)

resolution, and surface coil technology may result in a complete magnetic resonance angiogram of the coronary arteries.

5   CONGENITAL HEART DISEASE

5.1   Coarctation of the Aorta

MRI can provide excellent anatomic visualization of the aorta. Images are acquired first in the axial plane using a spin echo sequence, and then using these images as a reference a set of oblique sagittal images are acquired parallel to the plane of, and including, the ascending and descending aorta. These two sets of images can adequately describe the anatomical location and severity of the stenosis, including involvement of arch vessels, and degree of poststenotic dilation. A cine GRE sequence in the same oblique sagittal plane can demonstrate the jet of turbulent flow along the descending thoracic aorta just distal to the site of coarctation.[44] The gradient across the stenosis can be estimated from the peak velocity (using the modified Bernoulli equation described previously); measurements performed with velocity-encoded cine MRI reveal a close correlation ($r = 0.95$) with values obtained by continuous wave Doppler ultrasound.[45] The length of the turbulent jet has been correlated with the severity of stenosis, a factor important in gauging the appropriate timing of initial corrective surgical procedure, or subsequent repair in recoarctation.

At present, determining the optimal time for corrective surgery has been based on a combination of clinical findings and the echocardiographically determined gradient across the coarctation, both of which are only loosely correlated with the severity of stenosis. Recent work has revealed a unique ability of MRI in the evaluation of coarctation of the aorta. Velocity-encoded cine MRI was used to measure blood flow in the descending thoracic aorta just distal to the coarctation and at the level of the diaphragm, with the difference in flow representing the contribution from collateral vessels (Figure 5). While normal individuals have an average 8% decrease in blood flow along the descending thoracic aorta, patients with significant coarctations have an average 80% increase in flow.[46] This technique represents the only currently available noninvasive method to measure the extent of collateral development, and their contribution in supplying blood to the abdominal aorta.

5.2   Pulmonary Arteries

The evaluation of pulmonary arterial anatomy, orientation, size, and flow is critical to the management of children with lesions affecting the pulmonary arteries such as tetralogy of Fallot, and pulmonary atresia and stenosis.[47]

The pulmonary arteries are best visualized using 3–5 mm ECG-gated spin echo images in the axial plane, followed by oblique sagittal images through the pulmonary infundibulum, and/or right and left pulmonary arteries. MRI can define accurately the location and severity of narrowing in the right ventricular outflow tract and main pulmonary artery, as well as associated poststenotic dilation.[48,49] In a series of 12 patients who had undergone the arterial switch operation MRI correctly identified

92% of the pulmonary artery stenoses, while echocardiography identified only 58%. Moreover, MRI was superior in identifying peripheral pulmonary stenoses, as echocardiography visualized only seven of the 17 peripheral stenoses seen by MRI.[50]

Pulmonary atresia is identified by a fibrous and/or muscular band at the expected junction of the main pulmonary artery and the right ventricular outflow tract, on both axial and sagittal images. Determining the presence or absence of the main pulmonary artery and continuity of the right and left pulmonary arteries is crucial information for surgical planning. At least two studies, using angiography as the gold standard, have confirmed the ability of MRI to determine correctly the status of main, left and right pulmonary arteries, without the need for an invasive procedure.[51,52]

Quantitative, physiological information regarding flow within the pulmonary arteries can be obtained using velocity-encoded cine MRI images. This technique can accurately quantify both total, and right and left lung differential blood flow,[53] and correlates well with radionuclide lung scanning for the evaluation of relative blood flow to the left and right lungs.[54] This information is useful for determining the severity of stenoses and their effect on total and differential pulmonary blood flow (Figure 6). In addition, it may prove particularly valuable in serially assessing flow at sites of stenosis after balloon valvuloplasty, or surgical repair.

5.3   Vascular Rings

MRI is ideal to define the aortic arch and great vessel relationships because of its multiplanar capabilities and ability to image the adjacent air-filled trachea and esophagus, both of which may be compressed by the vascular structures. Spin echo sequences are typically employed in the axial and sagittal planes to demonstrate the commonly encountered abnormalities which include double aortic arch, right aortic arch with mirror image branching, right aortic arch with aberrant left subclavian artery, left aortic arch with aberrant left subclavian artery, and pulmonary sling. The sagittal images often best demonstrate the compression on both the trachea and esophagus. Coronal images may provide further information regarding double aortic arches.

Imaging may also be performed using magnetic resonance angiography, with either time-of-flight or phase contrast techniques.[55,56] Both techniques produce signal primarily from moving blood resulting in images that reflect the intravascular space, similar to conventional contrast angiograms, but without the need for contrast or an invasive procedure. These sequences have a further advantage in that the images may be reconstructed into a three-dimensional format, and viewed from any desired angle. Further postprocessing permits close examination of a particular area of interest for detailed analysis.

5.4   Complex Congenital Heart Disease

Complex congenital heart disease requires definition of the complicated anatomical relationships of the heart. MRI permits segmental analysis of complex congenital heart disease by providing tomograms encompassing the great vessels, heart, and infradiaphragmatic structures. Axial spin echo sequences are

For References see p. 1176

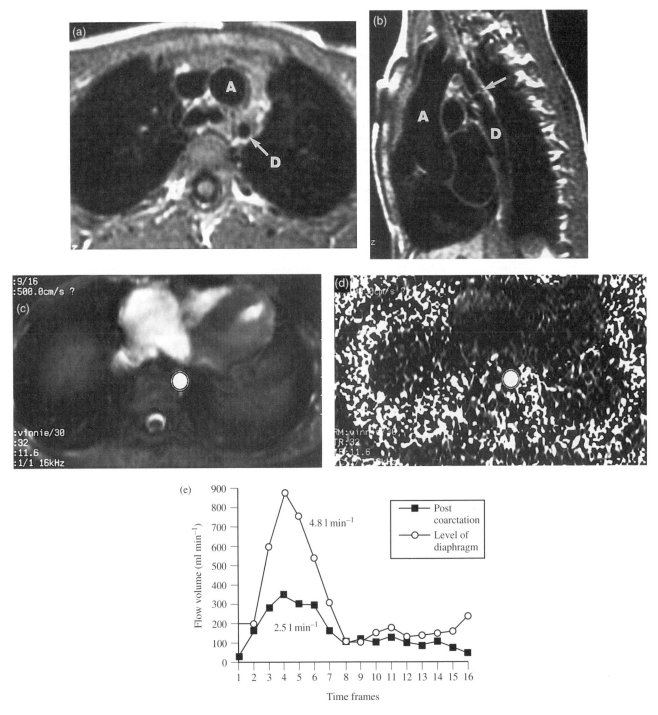

Figure 5 Patient with coarctation of the aorta. (a, b) ECG-gated spin echo images acquired in the axial and oblique sagittal projections, respectively. (c, d) Magnitude and phase images, respectively, from a VEC sequence acquired at the level of the diaphragm. Regions-of-interest have been placed over the aorta. (e) A flow curve can be derived, and blood flow at that level calculated (as described in the text). Similar information is obtained from a slice positioned just distal to the site of coarctation. The difference in blood flow between the sites in the descending thoracic aorta is a direct measure of the contribution of collateral blood flow to the abdominal aorta. A, ascending aorta; D, descending aorta; arrow, coarctation

usually sufficient to define the anatomy, with sagittal, coronal, and oblique imaging planes used to further complement the anatomical information.

A segmental approach is used to define the anatomy beginning with determination of the abdominal situs, and the relationship of the inferior vena cava and aorta. After this, the pattern of blood flow through the heart and great vessels is determined starting with the systemic and pulmonary venous connections, and then identifying the atrial situs, and presence and position of the atrioventricular valves. Next, the ventricular anatomy, and ventriculoarterial connections and great artery relationship are determined.

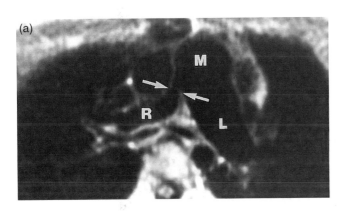
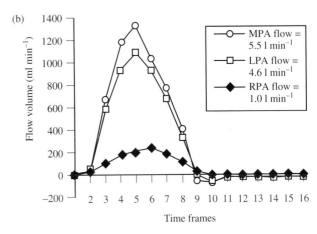

Figure 6 (a) Axial spin echo image of a patient with stenosis (between arrows) of the proximal right pulmonary artery. VEC MRI can be performed perpendicular to the main, and proximal right and left pulmonary arteries, and the total blood flow to the lungs, and differential blood flow to each lung calculated. (b) The different flows calculated for each of the respective arteries are demonstrated here. R, right pulmonary artery; L, left pulmonary artery; M, main pulmonary artery

The absolute and relative sizes of the ventricular system including quantification of ventricular mass, and the quantity and spatial extent of septal tissue are all well defined by MRI.[57–59] Many complex lesions have been evaluated correctly with MRI including double-outlet right ventricle,[60,61] double-outlet left ventricle,[62] double-inlet ventricle,[63] single ventricle and atrioventricular canal,[64] partial and total anomalous pulmonary venous connection,[65,66] and Ebstein's anomaly.[67] In a study of 29 patients with complex lesions, MRI was shown to be more informative for defining complex congenital heart disease than conventional angiography.[64] Receiver operating characteristic curves have been derived for MRI at a specificity level of 90%, demonstrating 100% sensitivity for evaluating great vessel relationships, ventricular septal defects, visceroatrial situs, and ventricular loop, 95% for right ventricular outflow tract obstructions, 94% for thoracic aortic anomalies, and 91% for atrial septal defects.[68]

5.5 Postoperative Evaluation

The metallic clips, wires, stents, shunts, and prosthetic valves used in interventional and surgical correction of congenital heart disease are sometimes problematic for MRI. On magnetic resonance images these devices may result in paramagnetic artifacts that degrade image quality, but fortunately, in most instances, they do not affect significantly the diagnostic utility of the study. The characteristics of noninvasiveness, lack of ionizing radiation, and good interstudy reproducibility are all distinct advantages for an imaging study intended to be used serially in children.

Congenital obstruction at the level of the main pulmonary artery, right ventricular outflow tract, or tricuspid valve requires a corrective procedure that involves construction of an atriopulmonary connection (Fontan procedure) or ventriculopulmonary conduit (Rastelli procedure) using native tissue and/or homograft material. These procedures can be complicated by insidious stenosis of the conduit, or the anastomotic site(s), and will benefit from periodic imaging examinations in addition to clinical follow up. The Fontan procedure has been evaluated

with the spin echo technique alone.[69] Addition of velocity-encoded cine MRI can provide further information regarding quantification of blood flow through the connection into the pulmonary arteries. This technique has demonstrated the normal biphasic pattern of blood flow, similar to venous flow, seen in atriopulmonary Fontans, or the normal monophasic pattern, seen in atrioventricular Fontans, thus confirming the success of the operation.[70] Rastelli conduits may also be complicated by stenosis, which is often difficult to evaluate with echocardiography because of the anterior location of the conduit in the chest. The velocity-encoded cine MRI technique has been applied in these patients to measure the gradient and to localize accurately the site of obstruction, providing a measure of the severity of the lesion which was superior to echocardiography.[71] The Norwood[72] and Jatene[50,73] procedures have been evaluated similarly, and MRI has been shown to be equivalent to angiography, and superior to echocardiography, respectively.

Extracardiac conduits such as the Glenn (superior vena cava to the right pulmonary artery) and Blalock–Taussig (subclavian artery to the ipsilateral pulmonary artery) shunts also can be evaluated easily with spin echo techniques, often in a single imaging plane. Serial imaging studies are valuable to evaluate for size, course, and patency of these shunts, as well as growth of the pulmonary arteries.[63,74]

6 THE FUTURE

The technology relevant to MRI of the heart continues to evolve. Consequently, the application of MRI in cardiovascular disease can be expected to increase further in the next several years.

7 RELATED ARTICLES

Cardiac Gating Practice; Cardiovascular NMR to Study Function; Echo-Planar Imaging; Lung and Mediastinum MRI;

For References see p. 1176

Motion Artifacts: Mechanism and Control; NMR Spectroscopy of the Human Heart.

8 REFERENCES

1. M. Erlichman, *Health Technol. Assess. Report*, 1990, **4**, 1.
2. R. D. White, R. L. Ehman, and J. C. Weinreb, *JMRI*, 1992, **2**, 365.
3. T. Masui, G. R. Caputo, J. C. Bowersox, and C. B. Higgins, *J. Comput. Assist. Tomogr.*, 1995.
4. L. M. Blake, M. M. Scheinman, and C. B. Higgins, *Am. J. Roentgenol.*, 1994, **162**, 809.
5. W. Auffermann, T. Wichter, G. Breithardt, K. Joachimsen, and P. E. Peters, *Am. J. Roentgenol.*, 1993, **161**, 549.
6. E. K. Yucel, F. L. Steinberg, T. K. Egglin, S. C. Geller, A. C. Waltman, and C. A. Athanasoulis, *Radiology*, 1990, **177**, 779.
7. C. A. Nienaber, Y. von Kodolitsch, V. Nicolas, V. Siglow, A. Piepho, C. Brockhoff, D. H. Koschyk, and R. P. Spielmann, *N. Engl. J. Med.*, 1993, **328**, 1.
8. C. A. Nienaber, R. P. Spielmann, Y. von Kodolitsch, V. Siglow, A. Piepho, T. Jaup, V. Nicolas, P. Weber, H. J. Triebel, and W. Bleifeld, *Circulation*, 1992, **85**, 434.
9. N. Fujita, G. R. Caputo, and C. B. Higgins, *Am. J. Card. Imaging*, 1994, **23**, 959.
10. M. Jungehulsing, U. Sechtem, P. Theissen, H. H. Hilger, and H. Schicha, *Radiology*, 1992, **182**, 225.
11. K. C. Seelos, G. R. Caputo, C. L. Carrol, H. Hricak, and C. B. Higgins, *J. Comput. Assist. Tomogr.*, 1992, **16**, 169.
12. J. H. Park, Y. M. Kim, J. W. Chung, Y. B. Park, J. K. Han, and M. C. Han, *Radiology*, 1992, **185**, 441.
13. J. D. Allison, F. W. Flickinger, J. C. Wright, D. G. Falls, L. M. Prisant, T. W. VonDohlen, and M. J. Frank, *Magn. Reson. Imag.*, 1993, **11**, 329.
14. R. C. Semelka, E. Tomei, S. Wagner, J. Mayo, G. Caputo, M. O'Sullivan, W. W. Parmley, K. Chatterjee, C. Wolfe, and C. B. Higgins, *Am. Heart J.*, 1990, **119**, 1367.
15. J. Suzuki, J. M. Chang, G. R. Caputo, and C. B. Higgins, *J. Am. Coll. Cardiol.*, 1991, **18**, 120.
16. J. Suzuki, G. R. Caputo, T. Masui, J. M. Chang, M. O'Sullivan, and C. B. Higgins, *Am. Heart J.*, 1991, **122**, 1035.
17. N. E. Doherty, K. C. Seelos, J. Suzuki, G. R. Caputo, M. O'Sullivan, S. M. Sobol, P. Cavero, K. Chatterjee, W. W. Parmley, and C. B. Higgins, *J. Am. Coll. Cardiol.*, 1992, **19**, 1294.
18. N. Fujita, J. Hartiala, M. O'Sullivan, D. Steiman, K. Chatterjee, W. W. Parmley, and C. B. Higgins, *Am. Heart J.*, 1993, **125**, 171.
19. T. Masui, S. Finck, and C. B. Higgins, *Radiology*, 1992, **182**, 369.
20. P. J. Kilner, D. N. Firmin, R. S. Rees, J. Martinez, D. J. Pennell, R. H. Mohiaddin, S. R. Underwood, and D. B. Longmore, *Radiology*, 1991, **178**, 229.
21. A. C. Eichenberger, R. Jenni, and G. K. von Schulthess, *Am. J. Roentgenol.*, 1993, **160**, 971.
22. P. A. Heidenreich, J. C. Steffens, N. Fujita, M. O'Sullivan, G. R. Caputo, E. Foster, and C. B. Higgins, *Am. J. Cardiol.*, 1994, **75**, 365.
23. N. Honda, K. Machida, M. Hashimoto, T. Mamiya, T. Takahashi, T. Kamano, A. Kashimada, Y. Inoue, S. Tanaka, and N. Yoshimoto, *Radiology*, 1993, **186**, 189.
24. M. C. Dulce, G. H. Mostbeck, M. O'Sullivan, M. Cheitlin, G. R. Caputo, and C. B. Higgins, *Radiology*, 1992, **185**, 235.
25. N. Fujita, A. F. Chazouilleres, J. J. Hartiala, M. O'Sullivan, P. Heidenreich, J. D. Kaplan, H. Sakuma, E. Foster, G. R. Caputo, and C. B. Higgins, *J. Am. Coll. Cardiol.*, 1994, **23**, 951.
26. M. T. McNamara, C. B. Higgins, N. Schechtmann, E. Botvinick, M. J. Lipton, K. Chatterjee, and E. G. Amparo, *Circulation*, 1985, **71**, 717.
27. A. de Roos, A. C. van Rossum, E. van der Wall, S. Postema, J. Doornbos, N. Matheijssen, P. R. van Dijkman, F. C. Visser, and A. E. van Voorthuisen, *Radiology*, 1989, **172**, 717.
28. M. C. Dulce, A. J. Duerinckx, J. Hartiala, G. R. Caputo, M. O'Sullivan, M. D. Cheitlin, and C. B. Higgins, *Am. J. Roentgenol.*, 1993, **160**, 963.
29. W. J. Manning, D. J. Atkinson, W. Grossman, S. Paulin, and R. R. Edelman, *J. Am. Coll. Cardiol.*, 1991, **18**, 959.
30. S. Schaefer, R. van Tyen, and D. Saloner, *Radiology*, 1992, **185**, 795.
31. M. A. Klein, B. D. Collier, R. S. Hellman, and V. S. Bamrah, *Am. J. Roentgenol.*, 1993, **161**, 257.
32. F. P. van Rugge, J. J. Boreel, E. E. van der Wall, P. R. van Dijkman, A. van der Laarse, J. Doornbos, A. de Roos, J. A. den Boer, A. V. Bruschke, and A. E. van Voorthuisen, *J. Comput. Assist. Tomogr.*, 1991, **15**, 959.
33. D. J. Pennell, S. R. Underwood, P. J. Ell, R. H. Swanton, J. M. Walker, and D. B. Longmore, *Br. Heart J.*, 1990, **64**, 362.
34. F. M. Baer, K. Smolarz, M. Jungehulsing, P. Theissen, U. Sechtem, H. Schicha, and H. H. Hilger, *Am. J. Cardiol.*, 1992, **69**, 51.
35. F. M. Baer, K. Smolarz, P. Theissen, E. Voth, H. Schicha, and U. Sechtem, *Int. J. Card. Imaging*, 1993, **9**, 133.
36. D. J. Pennell, S. R. Underwood, C. C. Manzara, R. H. Swanton, J. M. Walker, P. J. Ell, and D. B. Longmore, *Am. J. Cardiol.*, 1992, **70**, 34.
37. F. P. van Rugge, E. E. van der Wall, A. de Roos, and A. V. Bruschke, *J. Am. Coll. Cardiol.*, 1993, **22**, 431.
38. H. Sakuma, G. R. Caputo, J. C. Steffens, A. Shimakawa, T. K. F. Foo, and C. B. Higgins, *Radiology*, 1993, **189**(P), 278 (abstract).
39. R. R. Edelman, W. J. Manning, D. Burstein, and S. Paulin, *Radiology*, 1991, **181**, 641.
40. R. R. Edelman, W. J. Manning, J. Pearlman, and W. Li, *Radiology*, 1993, **187**, 719.
41. W. J. Manning, W. Li, N. G. Boyle, and R. R. Edelman, *Circulation*, 1993, **87**, 94.
42. W. J. Manning, W. Li, and R. R. Edelman, *N. Engl. J. Med.*, 1993, **328**, 828.
43. D. Li, C. B. Paschal, E. M. Haacke, and L. P. Adler, *Radiology*, 1993, **187**, 401.
44. S. Rees, J. Somerville, C. Ward, J. Martinez, R. H. Mohiaddin, R. Underwood, and D. B. Longmore, *Radiology*, 1989, **173**, 499.
45. R. H. Mohiaddin, P. J. Kilner, S. Rees, and D. B. Longmore, *J. Am. Coll. Cardiol.*, 1993, **22**, 1515.
46. J. C. Steffens, M. W. Bourne, H. Sakuma, M. O'Sullivan, G. R. Caputo, and C. B. Higgins, *Radiology*, 1993, **189**(P), 302 (abstract).
47. A. S. Gomes, *Radiol. Clin. North Am.*, 1989, **27**, 1171.
48. P. R. Julsrud, R. L. Ehman, D. J. Hagler, and D. M. Ilstrup, *Radiology*, 1989, **173**, 503.
49. A. S. Gomes, J. F. Lois, and R. G. Williams, *Radiology*, 1990, **174**, 51.
50. F. Blankenberg, J. Rhee, C. Hardy, G. Helton, S. S. Higgins, and C. B. Higgins, *J. Comput. Assist. Tomogr.*, 1994, **18**, 749.
51. D. A. Lynch and C. B. Higgins, *J. Comput. Assist. Tomogr.*, 1990, **14**, 187.
52. J. M. Parsons, E. J. Baker, A. Hayes, E. J. Ladusans, S. A. Qureshi, R. H. Anderson, M. N. Maisey, and M. Tynan, *Int. J. Cardiol.*, 1990, **28**, 73.
53. G. R. Caputo, C. Kondo, T. Masui, S. J. Geraci, E. Foster, M. M. O'Sullivan, and C. B. Higgins, *Radiology*, 1991, **180**, 693.
54. J. M. Silverman, P. J. Julien, R. J. Herfkens, and N. J. Pelc, *Radiology*, 1993, **189**, 699.
55. K. S. Azarow, R. H. Pearl, M. A. Hoffman, F. Zurcher, F. H. Edwards, and A. J. Cohen, *Ann. Thorac. Surg.*, 1992, **53**, 882.
56. R. B. Jaffe, *Semin. Ultrasound. CT MR*, 1990, **11**, 206.

For list of General Abbreviations see end-papers

57. E. J. Baker, V. Ayton, M. A. Smith, J. M. Parsons, E. J. Ladusans, R. H. Anderson, M. N. Maisey, M. Tynan, N. L. Fagg, and P. B. Deverall, *Br. Heart J.*, 1989, **62**, 305.

58. G. R. Caputo, J. Suzuki, C. Kondo, H. Cho, R. A. Quaife, C. B. Higgins, and D. L. Parker, *Radiology*, 1990, **177**, 773.

59. N. E. Doherty, N. Fujita, G. R. Caputo, and C. B. Higgins, *Am. J. Cardiol.*, 1992, **69**, 1223.

60. J. R. Mayo, D. Roberson, B. Sommerhoff, and C. B. Higgins, *J. Comput. Assist. Tomogr.*, 1990, **14**, 336.

61. J. M. Parsons, E. J. Baker, R. H. Anderson, E. J. Ladusans, A. Hayes, N. Fagg, A. Cook, S. A. Qureshi, P. B. Deverall, and M. N. Maisey, *J. Am. Coll. Cardiol.*, 1991, **18**, 168.

62. S. A. Rebergen, G. L. Guit, and A. de Roos, *Br. Heart J.*, 1991, **66**, 381.

63. I. C. Huggon, E. J. Baker, M. N. Maisey, A. P. Kakadekar, P. Graves, S. A. Qureshi, and M. Tynan, *Br. Heart J.*, 1992, **68**, 313.

64. B. A. Kersting-Sommerhoff, L. Diethelm, P. Stanger, R. Dery, S. M. Higashino, S. S. Higgins, and C. B. Higgins, *Am. Heart J.*, 1990, **120**, 133.

65. T. M. Vesely, P. R. Julsrud, J. J. Brown, and D. J. Hagler, *J. Comput. Assist. Tomogr.*, 1991, **15**, 752.

66. B. Kastler, A. Livolsi, P. Germain, A. Gangi, A. Klinkert, J. L. Dietemann, D. Willard, and A. Wackenheim, *Pediatr. Radiol.*, 1992, **22**, 262.

67. B. Kastler, A. Livolsi, H. Zhu, E. Roy, G. Zollner, and J. L. Dietemann, *J. Comput. Assist. Tomogr.*, 1990, **14**, 825.

68. B. A. Kersting-Sommerhoff, L. Diethelm, D. F. Teitel, C. P. Sommerhoff, S. S. Higgins, S. S. Higashino, and C. B. Higgins, *Am. Heart J.*, 1989, **118**, 155.

69. B. A. Kersting-Sommerhoff, K. C. Seelos, C. Hardy, C. Kondo, S. S. Higgins, and C. B. Higgins, *Am. J. Roentgenol.*, 1990, **155**, 259.

70. S. A. Rebergen, J. Ottenkamp, J. Doornbos, E. E. van der Wall, J. G. Chin, and A. de Roos, *J. Am. Coll. Cardiol.*, 1993, **21**, 123.

71. J. E. Martinez, R. H. Mohiaddin, P. J. Kilner, K. Khaw, S. Rees, J. Somerville, and D. B. Longmore, *J. Am. Coll. Cardiol.*, 1992, **20**, 338.

72. C. Kondo, C. Hardy, S. S. Higgins, J. N. Young, and C. B. Higgins, *J. Am. Coll. Cardiol.*, 1991, **18**, 817.

73. C. E. Hardy, G. J. Helton, C. Kondo, S. S. Higgins, N. J. Young, and C. B. Higgins, *Am. Heart J.*, 1994, **128**, 326.

74. B. Kastler, A. Livolsi, P. Germain, G. Zollner, and J. L. Dietemann, *Int. J. Card. Imaging*, 1991, **7**, 1.

Biographical Sketches

Scott D. Flamm. *b* 1960. B.A., 1982, University of California, Berkeley. M.D., 1988, George Washington University, USA. Residency training in diagnostic radiology, University of California, Los Angeles. Fellowship training in cardiovascular imaging and MRI (supervisor Charles B. Higgins), University of California, San Francisco. Currently clinical instructor in cardiovascular imaging/MRI, University of California, San Francisco, Approx. 10 publications. Research specialties: MRI of acquired and congenital cardiac disease, MRA, and quantitative vascular flow measurements.

Charles B. Higgins. *b* 1944. B.S., 1963, Villanova University. M.D., 1967, Jefferson Medical College, USA. Residency training in surgery, University of California, Los Angeles, and in diagnostic radiology, University of California, San Diego. Fellowship training in cardiovascular radiology, Stanford University. Research training at the Cardiovascular Research Laboratory (supervisor Eugene Braunwald), University of California, San Diego. Currently Professor of Radiology, and Chief, Magnetic Resonance Imaging and Cardiac Imaging Sections, University of California, San Francisco. Approx. 450 publica-

tions. Research specialties: MRI of acquired and congenital cardiac disease, and strategies for detection of myocardial ischemia using MR contrast media.

Cardiovascular NMR to Study Function

Donald B. Longmore
Royal Brompton Hospital, London, UK

1 INTRODUCTION

Approximately half of those reading this volume will die of cardiovascular disease, a further quarter will die of cancer, and 12% of lung disease. The diagnosis of these diseases are all now within the capabilities of NMR. However, vast intellectual and financial resources have been, and are being, applied to magnetic resonance technology, ignoring these major causes of death. Only a handful of the world's 9800 NMR machines are available to cardiologists, oncologists, and lung specialists. Presently NMR is used to study less than 3.5% of all disease. Open access rapid cardiovascular and lung imagers with facilities for intervention including laser and ultrasound surgery are technically feasible and needed urgently. Furthermore, the diagnostic power and, from the patient's point of view, the simplicity of a functional cardiovascular NMR examination, means that, for the first time, the ultimate medical objective of preventing the commonest diseases could be realized. The potential of functional evaluation of the cardiovascular system is for population screening to enable an understanding of the natural history of atheromatous disease, the commonest cause of death in the western world. Detection of presymptomatic cardiovascular disease enables both secondary prevention monitoring the efficacy of therapeutic and preventive measures using repeated functional NMR studies. Demographic evidence reveals the changing incidence of the disease in the same population over time and the varying incidence from place to place in the same ethnic groups (see section 8). This, coupled with experimental evidence, shows atheroma to have causative factors—and therefore to be preventable.[1-4] While the heart and circulation are being studied, it is only a small step to include screening for breast cancer[5] (*Breast MRI*) and, with the new ultra-short (50 μs to 500 μs) *TE* sequences, to study the lung (*Lung and Mediastinum MRI*).

Contemporary cardiovascular diagnosis relies on an expensive and time-consuming combination of clinical examination, chest radiography, electrocardiography (ECG), stress ECG, echo cardiography, nuclear medicine studies with and without pharmacological stress and invasive cardiac catherization,

For References see p. 1184

which carries a small but important mortality and morbidity. Unsuccessful attempts have been made to evaluate the cardiovascular system avoiding cardiac catherization using external recording devices such as apex cardiography, ballisto cardiography, ECG, impedance cardiography, etc. Now the balance has been profoundly altered following advances in nuclear medicine techniques, X-ray computed tomography, and echo cardiography. NMR is the most powerful diagnostic instrument yet conceived and this, combined with echo cardiography, means that virtually every cardiac diagnosis that previously required invasion can now be made noninvasively and cost-effectively, without mortality or morbidity. Until recently the exception was coronary artery disease. The NMR alternative to the coronary angiogram is dealt with elsewhere.

2 TECHNICAL REQUIREMENTS FOR CARDIOVASCULAR NMR

In order to study cardiovascular function, magnetic resonance techniques additional to those for imaging motionless parts of the body are required. Essential enhancements are cardiac gating, velocity mapping (*Cardiac Gating Practice*), rapid imaging, and velocity encoded rapid imaging (*Whole Body MRI: Strategies for Improving Imaging Efficiency*) in all planes.[6-8]

2.1 Imaging Strategy

A comprehensive heart NMR study is achieved in minimal time using a simple strategy. A transverse (transaxial) image of the chest is used to measure the angle of the heart to the left. A scan in the plane through the long axis of the left ventricle produces a vertical long axis (VLA) image through the left atrium and the left ventricle. From this, the downward angle of the heart is measured and a double oblique image from below produced to provide a horizontal long axis (HLA) view [Figure 1(a)] showing the five functional chambers of the heart and the coronary arteries in cross-section. Image planes can be selected perpendicular to the HLA to produce the short axis (SA) image [Figure 1(b)].

2.2 Rapid Imaging applied to the Cardiovascular System

Two rapid imaging techniques are currently available[9,10] (dealt with in detail in *Blood Flow: Quantitative Measurement by MRI*). Orders of magnitude of reduction of acquisition times for NMR imaging were first achieved by Mansfield and Pykett[8] with the one-shot echo planar 'Snapshot' (EPI) technique. The advantage of EPI for the cardiovascular system is that data acquisition can follow preparation pulses required to study cardiovascular function for diffusion, chemical shift imaging, and velocity encoding. The alternative to EPI with velocity mapping is the fast low angle shot (FLASH) technique with velocity mapping. EPI has a better signal-to-noise ratio (SNR) and can freeze movement, which is vital in the neonate. Less than 50 ms acquisition time is needed to obtain a velocity encoded image. The FLASH technique uses a total acquisition time of over 300 ms. There is no stage in the cardiac cycle of

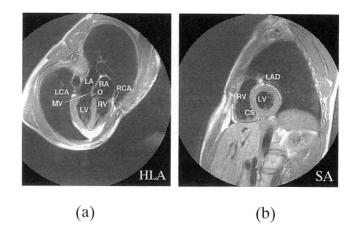

(a)   (b)

Figure 1 (a) The spin echo HLA image shows the right atrium (RA), the right ventricle (RV), the left atrium (LA), the left ventricle (LV) and the left ventricular outflow tract (O). The tricuspid valve which is not indicated is at the same level as the mitral valve marked (MV). The right coronary artery (RCA) and the left coronary artery (LCA) are also seen in cross-section. Fat at the apex of the heart is prominent. (b) The spin echo SA image shows the 'doughnut'-shaped left ventricular muscle (LV) with the right ventricle (RV) anterior to it. The left anterior descending coronary artery (LAD) is seen in the upper interventricular groove and the coronary sinus (CS) in the lower interventricular groove

an infant when the heart is relatively still for this long. Furthermore, at all stages throughout life the filling and contraction of the heart varies from beat to beat. This variability is never more marked than in the neonate with congenital disease. EPI is more suitable for the measurement of aperiodic movements (irregularly irregular) than FLASH, which relies on periodicity and the heart being in the same position and at the same stage of contraction for 16 or more heart beats. However, despite these disadvantages, because the gross appearance of a FLASH image is mainly determined by the data from a relatively few central phase encoding steps acquired in the first 50 ms or so of a total acquisition of over 300 ms, acceptable FLASH velocity encoded imaging is possible; FLASH is also less demanding on the gradients, the rf, data collection, etc., and it can be implemented on standard machines with relatively simple modification.

Either technique can be used to find a jet, measure its velocity and, from the velocity measurement, calculate the gradient and therefore the severity of the lesion using a modified Bernoulli equation.[11] Velocity mapping combined with good anatomical imaging usually reveals more useful functional information about congenital and acquired disease than can be obtained from a cardiac catheter angiogram.

2.3 Volume Measurements

The volumes of all the cardiac chambers, notably the ventricles, can be measured accurately from two multislice sets of transverse, or HLA, or SA images, one set taken at end systole (when the volumes are least) and the other at end diastole (when the heart is fullest). Usually 10–12 slices are sufficient to cover the whole of the heart. The areas of the cardiac cavities can be measured manually or by automatic edge detection or area growth techniques. Ventricular volume measurements are obtained by summing the areas of the images covering the ventricles in systole or diastole. The thickness of the slices is known and the volume of each slice calculated. Summation of the various slice volumes gives the total cavity volume.[12]

A 'shorthand' method of measuring stroke volume uses area length calculations of VLA and HLA systolic and diastolic images assuming the ventricles to approximate to a cone. The end diastolic volume is calculated from the formula:

$$\frac{\text{area of diastolic VLA} \times \text{area of diastolic HLA}}{\text{minimum diastolic length apex to valve plane}} \times 0.85 \quad (1)$$

and the end systolic volume from the formula:

$$\frac{\text{area of systolic VLA} \times \text{area of systolic HLA}}{\text{minimum length apex to valve plane}} \times 0.85 \quad (2)$$

This method is only valid if the ventricles are contracting evenly.

2.3.1 Clinical Significance of Volume Measurements

Subtraction of the systolic volume from the diastolic volume produces an accurate measure of stroke volume, which when multiplied by the heart rate gives the cardiac output. Continued life depends on the way the heart is functioning rather than on the output at a moment in time. An efficient heart retains a small residual volume of blood which when compared with the stroke volume gives the ejection fraction. A low ejection fraction carries a poor prognosis.

2.3.2 The Four Way Comparison

Volume measurements have been validated by comparing the stroke volumes of the right and left ventricles which in normal subjects is unity. The stroke volumes have also been compared with the aortic and pulmonary artery flows. Three measurements (right ventricular output, pulmonary artery flow, and left ventricular output) correspond to within 2%. Aortic flow is 5–10% below the other measurements due to the 'take off' of coronary flow.

Functional assessment of the cardiovascular system can be applied to all types of disease. In this article the classical 'surgical sieve' is used, dividing disease into congenital and acquired, and further dividing acquired disease into traumatic, inflammatory, neoplastic, etc. There is a 'gray area' where it is difficult to distinguish between diseases we are born with and those we acquire, possibly because we are genetically vulnerable.

3   FUNCTIONAL ASSESSMENT OF CONGENITAL HEART DISEASE

Approximately three in every 1000 live births manifest one of the commoner congenital heart diseases, pulmonary artery stenosis, atrial septal defect, patent ductus arteriosis, and tetralogy of Fallot, though there is a significant incidence of many other complex forms of congenital disease. NMR functional imaging can detect and quantify the threat to life and form a vital part of the management in all these conditions. Three discrete groups of patients require functional assessment.

These are: neonates and infants; children, when the less catastrophic abnormalities begin to manifest themselves; and 'grown up' congenital heart disease usually to follow up the results of surgical intervention.

3.1   Neonates and Infants

The small size of an infant and its inability to lie still, combined with the complexity of many congenital abnormalities, necessitates a rapid imaging technique and the use of special surface coils or a head coil.

3.2   Functional Imaging in Children

Functional techniques have given an insight into the compensatory mechanisms which can allow an infant with gross cardiac abnormalities to survive to childhood when operative intervention is easier and safer. The ideal diagnostic techniques for children are a combination of rapid imaging used in neonates and infants and the standard adult functional imaging methods.

3.3   'Grown Up' Congenital Heart Disease (GUCH)

NMR is the only available technique for the repeated long-term follow-up of adolescents and young adults who have undergone successful surgery. It is vital in postoperative cases to include velocity mapping searching for deterioration, because calcified obstructions are not visualized on anatomical images and can only be demonstrated by blood velocity mapping.[13]

NMR is uniquely able to detect the afferent and efferent vessels to congenital arteriovenous (AV) malformations, helping the surgeon to plan an operation. In cases with multiple AV malformations (Klippel, Tralawney syndrome) the measurement of the total blood flow through all the afferent or efferent vessels to the AV malformations can predict the impending onset of high-output heart failure.[14]

3.4   The Late Effects of Congenital Heart Disease

3.4.1   Valve Disease

Valve disease may be acquired or present as a late manifestation of congenital heart disease. A small proportion of the population are born with a bicuspid aortic valve which will usually function normally until the patient reaches the fourth decade of life, by which time calcification causing stenosis and regurgitation results from the abnormal vibrations and stresses on the two cusps. NMR can be used to measure the jet velocity and to calculate the gradient and to measure the regurgitant fraction. This regurgitation can be quantified directly using velocity mapping or by comparing the ratio of the stroke volumes of the right and left ventricles.

3.4.2   Septal Defects

Pulmonary valve stenosis and atrial septal defect are the commonest congenital deformities of the heart, but they may not prove to be significant until adolescence or early adult life when the gradients across the pulmonary valve become significant and can be measured from the jet velocity. The shunt

For References see p. 1184

across the septal defect can be measured by the volume method or by velocity mapping. It is important to determine the amount of blood shunting from the left (higher pressure side of the heart) to the right (lower pressure side). It is not uncommon for the shunt ratio to be as high as 4:1, with four times as much blood flowing through the lungs as the body. This shunt gradually and irreversibly damages the lungs, increasing their resistance to flow and eventually reversing the shunt and making closure of the defect irrelevant and dangerous. If only one defect is present in the absence of valve regurgitation, the shunt can be measured by comparison of the outputs of the two sides of the heart. Velocity mapping measuring the shunt directly is more reliable.

4   FUNCTIONAL ASSESSMENT OF ACQUIRED HEART DISEASE

4.1   Traumatic Cardiovascular Disease

The commonest traumatic cardiovascular lesion due to road traffic and other accidents is a partial tear of the aorta causing a dissecting aneurysm. These can be detected with a computed tomography (CT) scan or an anatomical MR image which reveal the presence of the dissection. A dissection usually travels down the length of the aorta and sometimes it extends back towards the origin of the coronary arteries. Before any repair can be attempted it is essential to use velocity mapping to determine where the blood is flowing, and which branches to vital organs arise from the true lumen and which from the false lumen.[15]

Tears and penetrating injuries such as stab wounds to the heart and the associated blood in the pericardium can be seen on spin echo and better on gradient echo images. Blood flowing in and out of false aneurysms indicates their size, and more significantly, differentiates them from pericardial cysts.

4.2   Inflammatory and Toxic Damage to the Heart

4.2.1   Valve Disease

The mitral and aortic valves may be damaged by bacterial infection on an already damaged or abnormal valve, or by rheumatic fever which for the past decades has been uncommon in the western world. Hypersensitivity to *Streptococcus* spp. still commonly causes valve disease in developing countries and there is now a resurgence, notably in the USA. Thickened valve cusps can be seen on 'black blood' images or as filling defects on white blood images. However, life depends on the normal function of valves, and appearance may not correlate with the severity of their malfunction. Turbulent and regurgitant flow associated with stenosis and regurgitation due to rheumatic fever, etc. can be visualized as areas of signal loss on gradient echo images. If velocity mapping is not available, a remarkably accurate assessment of the volume of a regurgitant jet can be made from the proportion of the chamber occupied by the jet and the duration of the back flow.[16] Stenosis and regurgitation are best measured directly using velocity mapping with short echo times. A *TE* of 3 ms to the second echo can capture velocities of up to 6 m s^{-1}: the highest vel-

ocity attained through a stenotic valve. Multivalve disease invalidates comparison of stroke volumes from the two sides of the heart. The same multislice technique can be used to measure ventricular mass, which is increased in congenital cardiomyopathies.

4.2.2   Heart Muscle Disease

Heart muscle can be affected directly by viral illnesses such as influenza, the ME virus and glandular fever. Conditions such as post partum cardiomyopathy and inflammatory/allergic infiltrative processes such as rheumatic fever also impair muscle function. Alcohol, drugs, and industrial toxic substances can also cause generalized dysfunction of the myocardium. In florid cases, cine NMR imaging is sufficient to demonstrate the dilated poorly contracting heart with a low ejection fraction. In less obvious cases, measurement of global ventricular function, as described below, is required.

4.3   Neoplastic Disease

Primary malignant tumors of the heart muscle are uncommon. They can be seen invading the muscle on multislice NMR images. Their effects on cardiac function can be determined by measuring global and regional function (see sections 4.4.2 and 4.4.3). Benign tumors and developmental cysts are commoner and do not usually have a significant effect on heart function, but repeated studies are necessary to check them for malignant change. Tumors in the atria can float through the atrioventricular valves during artrial systole and be blown back into the atria during ventricular systole, causing cardiac dysfunction by obstructing the valves and by displacing blood in a pumping chamber. Anatomical images do not distinguish well between intracardiac blood clot and intracavity tumor on spin echo images and gradient echo, and velocity mapping techniques are needed to reveal the presence of clot. Tumors in the right atrium may appear to have an obvious connection to the atrial wall but are frequently extensions of tumors in other parts of the vascular system which have grown along the great veins into the atrium. An 'NMR search' can locate the origin of the tumor. The hemodynamic significance of intracavity space-occupying lesions can be determined by velocity mapping. More frequently the heart is invaded by a local tumor such as a lung cancer with hilar glandular involvement. In these cases cardiac function can be impaired in three ways: by stiffening of the wall, by reduction of chamber size due to the mass of tumor tissue, or directly by tumor replacing contractile elements of the heart.

4.4   Degenerative Heart Disease

Most cardiovascular deaths, and hence about 40% of all deaths, are due to various forms of degenerative heart disease of which the commonest are hypertension and atherosclerotic disease.

4.4.1   Hypertension and Ventricular Mass

The reduction of ventricular mass following effective treatment of hypertension can be monitored using the multislice method (section 2.3). The muscle hypertrophy caused by high blood pressure can be quantified using the same multislice technique as for ventricular volumes.[17]

Studies of the caval or pulmonary venous flow which normally exhibit two similar flow peaks, one in systole and one in diastole, can be used to distinguish between ventricular function which is abnormal due to valvar regurgitation which causes an enhanced second peak 'constriction' of the heart by pericardial effusion or scar tissue around it. 'Restriction' due to stiffness of the heart itself, usually as a result of an inadequate blood supply due to coronary artery disease, correlates with an enlarged first peak.

Ventricular function can be studied globally or regionally.

4.4.2 Global Ventricular Function

Ventricular function measurements are clinically most relevant, but the heart has four chambers and the techniques described here are equally relevant to the atria. Global ventricular function measured in the resting state is frequently within normal limits, but the patient's disability indicates that there is inadequate cardiac output during exercise associated with normal activities.

Pharmacological stress (section 4.4.4) can reveal occult abnormalties of global function in two ways: by changes in velocity and acceleration of blood in the aorta, or by changes of stroke volume and ejection fraction.[18]

In the normal heart, increasing doses of a pharmacological stress (section 4.4.4) agent such as dobutamine cause an increase in aortic blood flow, velocity, and acceleration. In the abnormal heart, increasing the dose produces increased performance up to a much lower limit than in the normal heart, which has large reserves of power, and there is a 'fall off' in performance at dose levels which in the normal heart would still increase velocity. The dose level at which the 'fall off' occurs is an indication of the extent of the myocardial damage.

When the resting global function is at the lower end of normal limits, and pharmacological stress (section 4.4.4) indicates myocardial dysfunction with exercise, this may be because regions of the ventricle are contracting suboptimally, not contracting, or even moving paradoxically, bulging during systole thereby further disadvantaging the remaining normal heart muscle. In these cases it is necessary to study regional ventricular function.

4.4.3 Regional Ventricular Function

Multislice VLA, HLA, or, most usefully, SA views of the heart can be used to assess regional thickening and movement of the heart wall during contraction. If these appear normal but the clinical history suggests otherwise, then pharmacological stress is required.

The inner surface of the heart muscle is under maximum stress during cardiac contraction and is remote from the blood supply which enters the heart from its outer surface. The earliest sign that the heart is becoming injured is the failure of the subendocardial muscle fibers to 'actively relax' to enable filling at the beginning of diastole. Later in the disease process, contraction of these muscle fibers is also impaired. Velocity mapping of the inner surface of the heart in the direction of the long axis of the heart provides a sensitive measure of the elas-

ticity and capability of the heart to fill actively. Later in the disease process, the same velocity mapping technique can be used as an accurate measure of the local failure of contraction.[19]

4.4.4 Pharmacological Stress

To mimic normal exercise, pharmacological stress using inotropic agents to increase cardiac work are used. Intravenous dobutamine, which has a half-life of less than 100 s, is an example of a safer, effective agent than the long-acting coronary vasodilator dipyramidole (Persantin), which is more commonly used. Any angina caused by the dobutamine infusion passes off within 90 s of stopping the infusion.[20]

Pharmacological stress agents highlight areas which fail to contract normally, and indicate which coronary artery territory is receiving an inadequate blood supply. During simulated exercise there is good correlation between areas which do not thicken and move during systole and bolus tracking of a 'pseudo blood pool' magnetic resonance contrast agent. Both correlate well with nuclear medicine techniques.[21]

Measurements such as global and regional ventricular function demonstrate the effect, rather than the cause, of reduced ventricular function, which is most commonly a result of a diminished blood supply caused by atheromatous disease partially blocking the coronary arteries.

4.5 Atheromatous Disease

Atheromatous disease is progressive, and from demographic studies and animal experimentation is known to be reversible in its early presymptomatic stages (see section 8). The reader, hopefully a normal subject, requires answers to three fundamental questions about atheromatous disease. (a) Have I *got* it? (b) Are the *atheromatous* plaques of a type which can ulcerate and cause sudden death? (c) Can MR monitor the *efficacy* of preventive and therapeutic measures? An NMR machine capable of chemical shift imaging and velocity mapping is capable of answering all these questions.

4.5.1 Arterial Compliance

(a) *Have I got arterial disease?*: Atheromatous disease, mainly in the coronary arteries in the male and the carotid arteries in the female, appears not to coexist with a normal compliant arterial system. For this reason, measurements of aortic compliance, or its reciprocal the velocity of the onset of the arterial pulse wave along the aorta, can be used as a screening test to determine whether the subject is normal or an arteriopath.[22] Aortic compliance can be determined by NMR directly by measuring the volume of the arch of the aorta at peak systole when it is most dilated, and at the end of diastole when its elastic recoil reduces it to its smallest volume. It can be derived indirectly with similar accuracy by measuring the onset of the arterial pulse wave in the ascending aorta and at the end of the arch of the aorta during peak systolic flow. In a normal subject under the age of 55, the pulse wave takes an average of 32 ms to traverse the arch, whereas if the transit time of the leading edge of the pulse wave takes less than 12 ms diffuse arterial disease is present. This simple screening test divides subjects under the age of 55 into three groups: athletes

For References see p. 1184

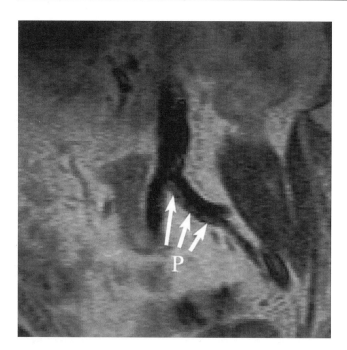

Figure 2 Atheromatous plaques (P) in the aorta of a 59-year-old male seen on a black blood spin echo image

who have remained fit; the average normal; and arteriopaths. Over this age the normal age-related aortic dilatation and reduction of arterial compliance is such that all subjects appear to be arteriopaths.

4.5.2 Chemical Shift Imaging

(b) *Are the atheromatous plaques of a type which can ulcerate and cause sudden death?*: If the patient is shown to be an arteriopath by the screening tests outlined above, arterial plaques in the larger vessels can be found on spin echo images of the vessel walls (Figure 2). The presence of the plaque can be confirmed and artefacts excluded by detecting acceleration of blood upstream of the obstruction, increased velocity across it, and deceleration downstream of it.

Water and fat images of the plaque can be acquired by using Dixon's selective reading of fat and water or Hinks' selective excitation of fat and water. A subtraction image indicates the percentage of lipid present.[23] Figures 3(a) and (b) show an atheromatous plaque with low lipid content. The range of lipid content in a plaque varies from zero up to a maximum of 28%. Figures 4(a) and (b) show a plaque with a high lipid content. A fibrous plaque [Figures 5(a) and (b)] with no or very low lipid content, though it impairs the circulation, is unlikely to cause sudden death. A plaque with a high lipid content puts the subject at risk because the plaque may rupture and ulcerate, precipitating acute thrombosis on its surface and, frequently, sudden death.

At the present stage of development, chemical shift imaging is only applicable to larger vessels, and the resolution is inadequate for NMR studies of coronary arteries based on single breathhold imaging acquiring 16 turbo-FLASH images. Furthermore, fat suppression techniques are needed in order to produce accurate images of the coronary arteries and velocity

measurements in them, because phase changes due to the presence of fat surrounding the vessels invalidate the Dixon technique. Coronary velocity mapping is however adequate for the determination of the hemodynamic significance of coronary plaques by measuring the increased velocity across the plaque taking 24 heart beats (a breathhold duration within the capabilities of most subjects). The way ahead for coronary imaging depends on the development of velocity encoded echo planar and spiral imaging, both techniques with potentially sufficient

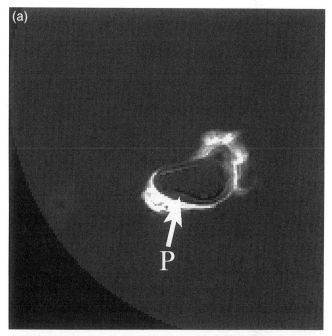

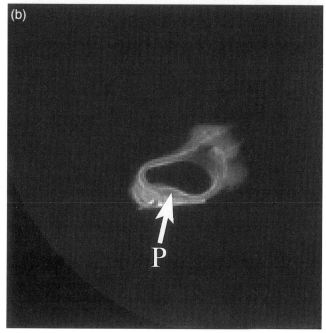

Figure 3 (a) A spin echo image of an atheromatous plaque (P) in a cadaveric specimen of aorta. (b) The image produced by subtraction of the fat and water images. There is no signal in the plaque which contains no lipid, confirmed by histology

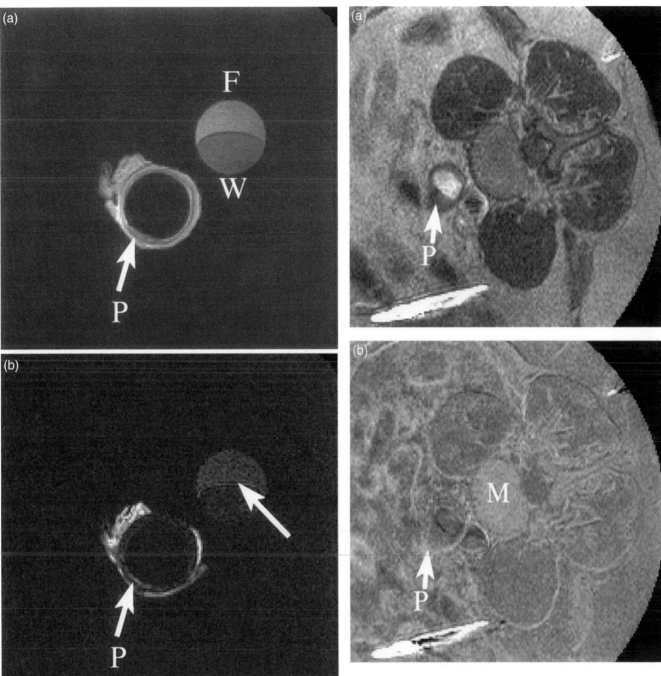

Figure 4 (a) Spin echo cross-sectional images of a cadaveric specimen of human aorta with an atheromatous plaque (P) and a test tube with fat floating on water (W + F). (b) The images produced by subtraction of fat and water images. The plaque (P) and the periaortic fat have high signals corresponding with the high lipid content in the histology specimen. There is a line of high signal on the interface between the fat and water in the test tube (arrowed)

Figure 5 (a) A spin echo image of the abdominal aorta in a 61-year-old male showing a typical asymmetric plaque (P). (b) The image produced by subtraction of fat and water images. The plaque (P) has a lower signal than the marrow (M) in the vertebral body which contains fat in a similar proportion to a lipid-rich plaque; therefore this is a fibrous plaque which will not rupture and cause sudden death

resolution to enable analysis of coronary plaques. At present the inference has to be made that the lipid content of plaques in the coronary arteries is similar to that in plaques elsewhere in the circulation, as is usually the case.

4.5.3   Monitoring Atheroma using Functional NMR

(c) *Can NMR monitor the efficacy of preventive and therapeutic measures?*: Experiments with artificially induced atheroma in animals show that plaques regress with starvation (see section 8). Simple preventive measures including strict

For References see p. 1184

diet and exercise have also been shown to cause regression of atheromatous plaques. In subjects with very high cholesterol and other lipid levels, lipid-lowering agents may be necessary to augment the simple preventive regime. The advantages of NMR are that functional studies can be repeated regularly to monitor the progress of preventive measures.

5 THE FUTURE

More versatile magnetic resonance machines capable of echo planar imaging and spiral velocity mapping will become available. These machines will use a new generation of open-access magnets in which ultrasound and laser intervention will be done using direct NMR vision. Ultrasound surgery with NMR vision has the advantage that the ultrasound energy can be applied at a low level using the changes to the NMR signal with increased temperature to confirm that the ultrasound transducers are focused on the area of interest. The ultrasound energy can then be increased to ablate the lesion. The massive intellectual pool of knowledge of NMR will be switched from less common diseases to cardiology, oncology, and to the lung. Proper training in cardiovascular NMR is required, not so much for imagers, as for cardiologists and experts in vascular disease.

6 CONCLUSIONS

Functional NMR evaluation of the cardiovascular system using volume measurements, velocity encoded rapid imaging, and chemical shift imaging is capable of making the diagnosis in congenital and acquired cardiovascular disease which, between them, cause the largest number of deaths of any disease in the western world and massive morbidity and suffering. Furthermore, for the first time in the history of medicine, there is the opportunity to apply preventive measures to eradicate the epidemic of preventable arterial disease.

7 RELATED ARTICLES

Blood Flow: Quantitative Measurement by MRI; Breast MRI; Echo-Planar Imaging; Heart: Clinical Applications of MRI; Lung and Mediastinum MRI; Whole Body Magnetic Resonance: Fast Low-Angle Acquisition Methods.

8 BACKGROUND NOTE

Atheromatous disease was unknown in Central Europe after World War I. Autopsy studies at the London Hospital from 1908–40 showed a steady increase in coronary artery disease. Casualties in the starving population of the siege of Leningrad in World War II showed no arterial disease; likewise there was no atheromatous disease in the survivors of concentration camps. Young soldiers killed in the Korean and Vietnam Wars showed massive diffuse disease. Experimental atheroma in laboratory animals regresses with starvation. Atheromatous disease does not coexist with wasting diseases. It also varies not only from time to time but from place to place. There are wide vari-

ations in its incidence in apparently genetically similar populations in different countries. There is no relationship between the 'spend' on health care and the incidence of disease. The USA spends the highest proportion of its GDP on health care and the UK nearly the lowest. Both countries are amongst those with the highest incidence. France, a high spender, and Japan, the lowest, share the lowest incidence of occlusive vascular disease.

9 REFERENCES

1. D. B. Longmore, *Diagn. Imag.*, 1988, 399.
2. Physiology and Treatment of Starvation: Experiences in War-starved Europe, Report of Societies 818, *Br. Med. J.*, June 9, 1945.
3. J. Brozek, S. Wells, and A. Keys, 'Medical Aspects of Semistarvation in Leningrad (Seige 1941–42)'.
4. P. Mollison, *Br. Med. J.*, January 1946.
5. B. A. Porter and J. P. Smith, *Magn. Reson.*, 1993.
6. G. L. Nayler, D. N. Firmin, and D. B. Longmore, *J. Comput. Assist. Tomogr.*, 1986, **10**, 715.
7. D. J. Bryant, J. A. Payne, D. N. Firmin, and D. B. Longmore, *J. Comput. Assist. Tomogr.*, 1984, **8**, 588.
8. P. Mansfield and I. L. Pykett, *J. Magn. Reson.*, 1978, **29**, 355.
9. J. Frahm, A. Haase, and D. Matthaei, *Magn. Reson. Med.*, 1986, **3**, 321.
10. D. N. Firmin, P. D. Gatehouse, and D. B. Longmore, *Proc. XIth Annu. Mtg. Soc. Magn. Reson. Med.*, Berlin, 1992, 2915.
11. P. J. Kilner, C. C. Manzara, R. H. Mohiaddin, D. J. Pennell, M. G. St. J. Sutton, D. N. Firmin, S. R. Underwood, and D. B. Longmore, *Circulation*, 1993, **87**, 1239.
12. D. B. Longmore, R. H. Klipsten, S. R. Underwood, D. N. Firmin, G. N. Hounsfield, M. Watanabe, C. Bland, K. Fox, P. A. Poole-Wilson. R. S. O. Rees, D. N. Denison, A. M. McNeilly, and E. D. Burman, *Lancet*, 1985, **i**, 1360.
13. P. J. Kilner, D. N. Firmin, J. Martinez, J. Somerville, S. R. Underwood, R. S. O. Rees, and D. B. Longmore, *Eur. Heart J.*, 1991, **12**(Suppl.), 284.
14. M. Huber, D. B. Longmore, D. N. Firmin, J. Assheuer, H. Bewermeyer, and W. D. Heiss, *Digit. Bilddagnos.*, 1989, **9**, 1.
15. H. G. Bogren, S. R. Underwood, D. N. Firmin, R. H. Mohiaddin, R. H. Klipstein, R. S. O. Rees, and D. B. Longmore, *Br. J. Radiol.*, 1988, **61**, 456.
16. R. H. Mohiaddin and D. J. Pennell, in 'Percutaneous Balloon Valvuloplasty', ed. T. O. Cheng, Igaku-Shoin Medical Publishers, New York, 1992, pp. 185–213.
17. S. M. Forbat, S. P. Karwatowski, P. D. Gatehouse, D. N. Firmin, S. R. Underwood, and D. B. Longmore, *Br. J. Radiol*, 1993, **66**, 957.
18. D. J. Pennell, S. R. Underwood, and D. B. Longmore, *J. Comput. Assist. Tomogr.*, 1990, **14**, 167.
19. S. P. Karwatowski, R. H. Mohiaddin, G. Z. Yang, D. N. Firmin, M. G. St. J. Sutton, S. R. Underwood, and D. B. Longmore, *JMRI*, 1994, **4**, 151.
20. D. J. Pennell, S. R. Underwood, C. C. Manzara, R. H. Swanton, J. M. Walker, P. J. Ell, and D. B. Longmore, *Am. J. Cardiol.*, 1992, **70**, 34.
21. D. J. Pennell, S. R. Underwood, P. J. Ell, R. H. Swanton, J. M. Walker, and D. B. Longmore, *Br. Heart J.*, 1990, **64**, 362.
22. R. H. Mohiaddin, S. R. Underwood, H. G. Bogren, D. N. Firmin, R. H. Klipstein, R. S. O. Rees, and D. B. Longmore, *Br. Heart J.*, 1989, **62**, 90.
23. R. H. Mohiaddin, D. N. Firmin, S. R. Underwood, A. K. Abdulla, R. H. Klipstein, R. S. O. Rees, and D. B. Longmore, *Br. Heart J.*, 1989, **62**, 81.

NMR Spectroscopy of the Human Heart

Paul A. Bottomley

Johns Hopkins University, Baltimore, MD, USA

1 INTRODUCTION

The heart is the largest consumer of energy per gram of tissue, and defects involving energy metabolism, including energy supply and demand, are central to much disease of the heart. It is inevitable, therefore, that human cardiac spectroscopy has focused on the energy metabolites that are NMR detectable. Phosphorus (^{31}P) NMR spectroscopy (MRS) can see adenosine triphosphate (ATP), the fundamental energy currency of the body, and phosphocreatine (PCr), a reservoir of cellular energy, in the heart, as well as the metabolic by-product inorganic phosphate (P_i). The signal-to-noise ratio is about 1×10^{-5} or 2×10^{-5} of that of the proton (^{1}H) NMR signal from muscle water (from the same volume), at the same magnetic field strength.[1,2] PCr and ATP cannot therefore be imaged with the same spatial resolution, signal-to-noise ratio, and scan time as tissue water protons in the same magnetic field. However, with compromises in spatial resolution and scan time, PCr and ATP can be seen in anterior wall volume elements (voxels) as small as 8 mL in the anterior myocardium.[3]

PCr and ATP are linked via the creatine kinase (CK) reaction:

$$PCr + ADP \rightleftharpoons ATP + creatine \qquad (1)$$

whereupon inorganic phosphate (P_i) may be formed as

$$ATP \rightarrow ADP + P_i + energy \qquad (2)$$

The total creatine pool (CR) comprising unphosphorylated (Cr) plus PCr is also detectable in the heart via its *N*-methyl resonance at 3.0 parts per million (p.p.m.) using ^{1}H MRS, with a current resolution of 3–9 ml in scan times of <10 min.[4] Consequently, a combination of ^{1}H and ^{31}P MRS could potentially provide a near-complete picture of CK metabolism in the heart. Under anaerobic conditions, such as may occur in ischemic heart disease in regions of the heart supplied by blocked arteries,[5] a transient mismatch between oxygen supply and demand can cause excess PCr consumption to maintain adequate ATP, along with possible elevations in P_i. This is manifested as reductions in PCr/ATP, PCr/P_i, and/or ATP/P_i in the corresponding ^{31}P NMS during the ischemic episode. The rapid depletion of ATP and PCr levels assayed after coronary occlusion or low-flow ischemia has been demonstrated in animal models.[6] Biochemical assays of biopsies taken at surgery have also demonstrated chronic reductions in the levels of PCr, Cr, and creatine kinase activity in patients with dilated cardiomyopathy (DCM), hypertrophy, and coronary artery disease (CAD).[7,8] These findings link changes in energy metabolism to contractile dysfunction, but in the absence of evidence of active ischemia.[9] Thus, the levels and ratios of PCr, ATP, CR, Cr, and P_i measured by ^{1}H and ^{31}P MRS may be useful indices of supply-side myocardial energy metabolism.

To date, ^{31}P NMR has been applied extensively in studies of patients with cardiomyopathies, including DCM, hypertrophic cardiomyopathy (HCM), and pressure overload left ventricular hypertrophy (LVH); patients with transplanted hearts, in an attempt to assess histological rejection; and patients with CAD and myocardial infarction, including exercise stress testing of patients with reversible ischemia. The literature on ^{31}P NMR of the heart has recently been reviewed in detail.[10] Abnormalities of varying extents in the phosphate metabolites have been reported in all of these disease states. In particular, the variability of findings in cardiomyopathy, and the results from transplanted hearts, suggest the existence of confounding variables that are as yet incompletely understood. However, quantitative studies are approaching a consensus on the true PCr/ATP ratio for human heart, and there are preliminary values reported for the PCr and ATP concentrations, the intracellular pH, and the forward CK reaction rate, and flux. Quantification of P_i has proved difficult because of the presence of neighbouring blood 2,3-diphosphoglycerate (DPG) resonances. Partial saturation effects and contamination from blood and chest muscles are probably the main sources of discrepancy amongst reported myocardial ^{31}P NMR PCr/ATP data.

^{1}H NMR spectra from the human heart are dominated by water and fat resonances, but metabolites such as CR can be measured either relative to water, or as concentrations.[4] CR appears to be depleted in myocardial infarction,[4] and alterations are also anticipated in cardiomyopathy and heart failure.[9]

NMR spectroscopy studies of the human heart with nuclei other than ^{31}P or ^{1}H are scarce. Carbon (^{13}C) NMR potentially provides access to glycolytic and citric acid cycle metabolites such as lactate and glutamate but has the disadvantage of low NMR sensitivity and a 1.1% natural isotopic abundance. This effectively prevents their detection at natural abundance with current technology without resorting to ^{13}C-enriched substrates.[2] Fatty acid resonances are the main contributors to the natural abundance ^{13}C human heart spectrum, and pericardial fat is probably a dominant contaminant, but glycogen may be discernible with ^{1}H decoupling and NOE.[11] While strictly not spectroscopy since there is usually only a single resonance, sodium (^{23}Na) NMR has been performed in human heart[12,13] and has potential for detecting sodium increases in reperfused infarction where sodium–potassium pump function is lost.[14]

2 METHODS

2.1 Localization

Localized ^{31}P NMR spectroscopy of the human heart was first reported in 1985.[1] That study, and all those since, employed surface detection coils placed on the chest closest to the anterior myocardium. Myocardium was distinguished from chest wall using MRI slice selection (the depth resolved surface coil spectroscopy, or DRESS, technique), and ^{1}H MRI was used for tissue identification. While an unlocalized surface coil

For References see p. 1193

study of the chest of an infant with congenital cardiomegaly was also reported in 1985,[15] the first studies of heart patients with localized spectroscopy protocols were not published until 1987.[16,17] These were limited to two to four patients with hypertrophic cardiomyopathy or recent myocardial infarction, and localization was afforded by rotating frame zeugmatography (RFZ), which uses B_1 gradients, or by the DRESS technique.

Full three-dimensional MRI gradient localization of ^{31}P NMR spectra to single voxels in the heart via the image selected in vivo spectroscopy (ISIS) technique appeared in 1988.[18] Multiple voxel, chemical shift imaging (CSI) studies of the heart using MRI phase-encoding gradients in one or more dimensions were published around 1990.[5,18,19] These spectro-

scopy localization techniques are reviewed in greater detail elsewhere (see *Chemical Shift Imaging*, *Selective Excitation in MRI and MR Spectroscopy*, and *Single Voxel Whole Body Phosphorus MRS*), and have various advantages and disadvantages. For example, a problem with three-dimensional CSI is that the large number of phase-encoding steps required to encode the entire sample volume sets a minimum scan time, which may not be compatible with clinical study protocols of limited duration such as in stress testing for myocardial ischemia. One-dimensional CSI and RFZ can be completed faster, but at the expense of the fuzzy localization afforded by the surface coil in the two dimensions that are not gradient encoded. An example of a one-dimensional CSI data set and the corresponding annotated cardiac image is shown in Figure 1.[20] Use

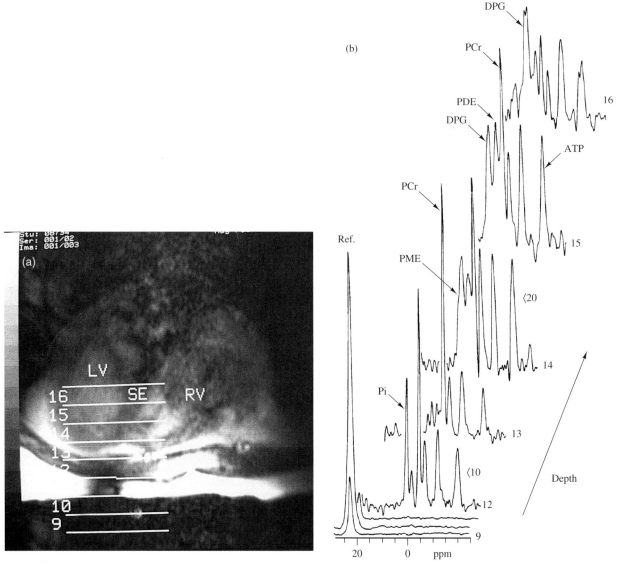

Figure 1   (a) Typical axial ^{1}H surface coil image of the chest and heart of a normal volunteer, annotated to show the location of eight 1 cm thick coronal slices localized in the spectroscopy exam. (b) Cardiac-gated one-dimensional CSI surface coil ^{31}P NMR spectra from the slices show a reference peak from a vial embedded in the coil (slices 9, 10), chest muscle (slices 12, 13), and myocardium (slices 14–16). Scan time for the spectra was about 12 min, in a clinical 1.5 T scanner. (Reproduced by permission of the Radiological Society of North America from Bottomley et al.[20])

of single voxel methods such as DRESS or ISIS requires accurate selection of a suspected region of abnormality a priori: incorrect voxel selection for a stress test, for example, could result in a missed finding for reversible ischemia.

^{1}H NMR spectroscopy studies of myocardial CR in patients to date have been performed using the single voxel point-resolved surface coil spectroscopy method (PRESS)[21] and the stimulated echo method (STEAM)[22] with water suppression. Each method localizes a spin-echo or stimulated echo NMR signal to a voxel defined in all three dimensions by the use of three spatially selective NMR pulses (90°–180°–180° for PRESS; 90°–90°–90° for STEAM).

2.2 Coils and NOE

A typical experimental cardiac ^{31}P NMR surface coil probe is pictured in Figure 2.[23] When the sample is the dominant source of the noise detected by the surface coil, the coil diameter should be chosen roughly equal to the depth of the tissue of interest. This leads to heart ^{31}P detectors with typical diameters of 6–12 cm for the anterior wall to mid-wall. Preferably, the signal is excited with a separate, larger rf coil to minimize spatially dependent distortions due to B_1 inhomogeneity.[23] Additional ^{1}H MRI coils ensure correct ^{31}P detection coil placement, which is critical because of the limited range of sensitivity of the surface coil, and possible avoidable contamination from nearby tissue, especially chest muscle.

Sensitivity improvements of up to about $\sqrt{2}$ compared with the best positioned surface coil, and the need for careful positioning is relaxed if phased-array surface detection coils are used.[24] This technology is in use in MRI,[25] and has been implemented in ^{31}P studies of normal volunteers,[24] but not yet in patient studies.

Other factors offering potential sensitivity gains are the use of a prone, as opposed to supine, patient orientation, to bring the heart closer to the detection coil(s);[23] synchronization of NMR acquisitions to the cardiac cycle (see *Cardiac Gating Practice*); and ^{1}H NOE.[26] At 1.5 T, NOEs of $\eta = 0.61 \pm 0.25$ for PCr, $\eta = 0.6 \pm 0.3$ for γ-ATP, and $\eta = 0.3 \pm 0.2$ for β-ATP have been reported for the human heart.[26] Because of the differences in η for PCr and ATP, the NOE may distort the PCr/ATP ratio by up to 20%, necessitating a correction for quantitative intersite comparisons.[26]

3 QUANTIFICATION

3.1 Myocardial PCr/ATP Ratio

If the PCr/ATP ratio is the most important metabolic index that ^{31}P NMR can reliably detect in the heart, it is important to establish its normal range for inter- and intralaboratory comparisons. However, published values from healthy volunteers range from 0.9 ± 0.3[22] to 2.1 ± 0.4.[28] Why the difference? There are two main causes. First, data are distorted to varying extents by partial saturation effects that result from the use of pulse sequence repetition periods, *TR*, that are comparable to or less than the T_1 values of PCr and ATP. Differences in the T_1 values of the two moieties produce differential distortion of PCr/ATP.[29] The 95% confidence intervals for published T_1 values are given in Table 1.[17,19,26,30–35] With a ratio of the T_1 values of PCr and ATP of about 1.9, the distortion or satur-

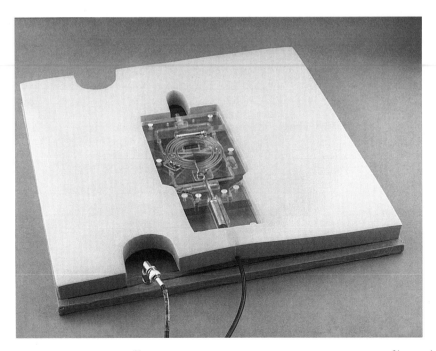

Figure 2 The cardiac ^{31}P NMR surface coil probe used to acquire the localized ^{31}P and ^{1}H NMR data in Figure 1.[23] The patient lies prone on the padded coil set (left), which is positioned on the patient table in the center of the whole body NMR magnet. The coil set comprises a square base with a 0.40 × 0.40 m ^{31}P transmit coil (lower left), and a smaller probe containing a 0.065 m mean diameter spiral ^{31}P detector coil, and a 0.08 × 0.13 m figure-8 ^{1}H receiver coil, used for imaging and shimming (right). Black strips (right) are Velcro, permitting the smaller probe to be positioned at will. Passive diode circuitry minimizes coil interactions[23]

For References see p. 1193

Table 1 Phosphorus-31 NMR Measurements of Phosphate Metabolites in Normal Volunteers

Parameter	Value	Reference
T_1(PCr)	4.37 ± 0.48[a]	30
T_1(γ-ATP)	2.52 ± 0.45[a]	30
T_1(β-ATP)	2.28 ± 0.54[a]	30
[PCr]	11 ± 3 μmol (g wet wt)$^{-1}$	19
	12 ± 4.3 μmol (g wet wt)$^{-1}$	31
	10 ± 2 μmol (g wet wt)$^{-1}$	32
[ATP]	6.9 ± 1.6 μmol (g wet wt)$^{-1}$	19
	7.7 ± 3.0 μmol (g wet wt)$^{-1}$	31
	5.8 ± 1.6 μmol (g wet wt)$^{-1}$	32
P_i/PCr	<0.25	17
	0.14 ± 0.06	33
pH	7.15 ± 0.2	17
	7.15 ± 0.03	33
CK forward rate	0.5 ± 0.2 s^{-1}	34
CK forward flux	6 ± 3 μmol (g wet wt s)$^{-1}$	34
NOE of PCr at 1.5 T	0.61 ± 0.25	26
	0.6 ± 0.1	35
NOE of ATP at 1.5 T	0.6 ± 0.3 (γ-ATP)	26
	0.3 ± 0.2 (β-ATP)	26
	0.4 ± 0.2 (γ- and β-ATP)	35

CK, creatine kinase; γ-ATP and β-ATP, γ- and β- phosphates of ATP.
[a] 95% confidence intervals.

ation factor, F, by which the PCr/ATP ratio measured in the short TR experiment must be multiplied by to yield the true myocardial PCr/ATP ratio, will vary between 1 and ~2, depending on TR and the NMR pulse flip angles.[29] Thus, measurements that are corrected for partial saturation can differ from those that have not, by up to a factor of about two.

Correction of experimental PCr/ATP values for partial saturation can be done by two methods. First, F can be calculated using known myocardial T_1 values and a measurement of the experimental NMR pulse flip angle (e.g. with equation (2) of Bottomley et al.).[29] Due to time constraints, T_1 values are usually either measured on a separate group of individuals, or else published values are adopted. Second, F may be measured directly from the ratio of spectra acquired under partially saturated conditions and fully relaxed conditions. The unlocalized, uniformly excited signal detected by a surface coil on the chest is sufficiently sensitive to permit the measurement of saturation factors in about 6 min, which can easily be incorporated into the study protocol.[5,29] This use of an unlocalized saturation factor depends on the assumption that T_1(PCr)/T_1(ATP) is essentially the same in the chest and heart tissue that contribute to the unlocalized spectrum.[29] The assumption is consistent with the available data,[30] and errors resulting from differences in the ratio can be minimized by using small or Ernst flip angles for excitation.

The second, lesser source of scatter in the reported myocardial PCr/ATP values is that of blood which contains ATP but no PCr. This reduces the apparent PCr/ATP ratio in voxels that intersect the ventricles.[36] The PCr/ATP ratio can be corrected for blood ATP contamination by measuring the amount of blood using the blood DPG signal present in the spectrum as a marker. DPG has a characteristic doublet at 5.4 and 6.3

ppm, near P_i[17] and other phosphomonoesters (PMs). The amount of ATP added by blood is estimated from

$$\text{corrected ATP} = \text{contaminated ATP} - 0.5 \times$$
$$([\text{ATP}]/[\text{DPG}]) \times \text{total DPG signal} \quad (3)$$

where $[\text{ATP}]/[\text{DPG}] \approx 0.30$,[33,37–39] and remembering that DPG has two phosphates.

Blood ATP corrections typically increase PCr/ATP values by a relatively small amount, $13 \pm 6\%$.[20,33,36,39,40] Potential problems with the correction may arise from differences between the T_1 of ATP and DPG for short TR protocols (possibly offset, however, by the effect of inflowing unsaturated blood), variations in the blood ATP/DPG ratio with localization method and patients,[41] and possible contamination of DPG by P_i or PMEs, which would tend to cause overestimation of the amount of blood contamination.

Considering only those human studies in which both saturation and blood corrections were performed[20,31,33,36,40,42–48] or in which blood contamination was ruled out,[3] a consensus is emerging on the normal myocardial PCr/ATP value, as evident in Table 2. The mean value is 1.81 ± 0.14, omitting the highest and lowest values. This should represent the best current estimate for the true PCr/ATP ratio in the normal human heart, since tissue ATP and PCr appear to be 100% NMR visible.[49,50]

3.2 PCr and ATP Concentrations

Tissue [PCr] and [ATP] can be determined by combining the fully corrected PCr/ATP ratio from a voxel, with a measurement of the NMR signal from a concentration reference and a measurement of the volume of myocardium present in the voxel. The reference may lie outside of the subject, which requires that corrections be made for differences in field strength and sensitivity of the NMR coils at the two locations.[16] Alternatively, a reference located at the same position as the heart may be observed in a separate experiment in which differences in coil loading are taken into account.[31] The

Table 2 Some Corrected Normal Human Myocardial PCr/ATP Ratios

Value	Field (T)	Reference
1.80 ± 0.21	1.5	36
1.93 ± 0.21	1.5	20
1.8 ± 0.1	4.0	3
1.95 ± 0.45	1.5	40
1.65 ± 0.26	1.5	33
1.85 ± 0.28	1.5	42
1.8 ± 1	1.5	31
1.71 ± 0.13	1.5	39
1.98 ± 0.07	1.5	43
2.02 ± 0.11	1.5	44
1.42 ± 0.18	1.5	45
1.61 ± 0.18	1.5	46
2.46 ± 0.53	1.5	47
1.6 ± 0.3	2.0	48

tissue volumes must again be estimated by MRI, which may introduce weighting errors when the coil sensitivity is nonuniform across the localized volume.

A quantification method that does not require tissue volumetry employs measurements of the ^1H signal from water, s_W, as a concentration reference, using the identical localization sequence.[32] The water signal is recorded with the same ^{31}P coil used to measure the PCr and ATP metabolite signals, s_P; as a result, both ^{31}P and ^1H spectra are acquired with essentially the same B_1 profile. This is possible because of the enormous concentration advantage of the water signal compared with the metabolites. The method requires calibration with a phosphate reference to determine the ratio of the ^{31}P signal per phosphate to the ^1H signal per proton, C_{PH}, as well as knowledge of the myocardial tissue water content [W], which appears to be relatively constant.[4] The metabolite concentration in a voxel is calculated from:

$$[P] = \frac{2s_P[W]C_{PH}}{s_W} \times \left(\frac{F_P E_P}{F_W E_W}\right) \qquad (4)$$

where $F_P F_W$ are the saturation factors and E the factors accounting for signal decay during TE (e.g. $E = \exp\{TE/T_2^*\}$ where T_2^* is the transverse signal decay time) with subscripts W for water, P for the metabolite (subscript P). The factor of 2 accounts for the two protons on water. The value of C_{PH} is determined in a separate experiment using a standard solution of phosphate.[32] The tissue water signal can be corrected for contamination from water in ventricular blood and in pericardial fat via a ^{31}P DPG measurement and a measurement of the fat resonance in the ^1H spectrum, respectively. The contaminating water signal from ventricular blood is:

$$S(\text{blood}) = \frac{S(DPG)W_B C_{PH}}{[DPG]} \times \left(\frac{F_{DPG}E_{DPG}}{E_B}\right) \qquad (5)$$

The contaminating water signal from fat is $LS(CH_2)$, where $S(CH_2)$ is the lipid ^1H signal, and L is the water content of fat, which is about 15%.[32] The corrected water signal to use in equation (4) is:

$$S_W = (\text{measured water signal}) - S(\text{blood}) - LS(CH_2). \qquad (6)$$

Concentrations reported to date for the human heart are in fair agreement and are summarized in Table 1.

3.3 Total Myocardial Creatine

The detection of the N-methyl ^1H resonance of CR at 3.0 ppm provides a means of quantifying myocardial CR.[4] Unlike PCr/ATP ratios, CR cannot be reliably measured as a ratio from a single ^1H acquisition relative either to water, which is normally suppressed, or to fat resonances, which are often variable because of contamination from pericardial fat. A solution is to acquire two consecutive ^1H spectra under the same conditions, with and without water suppression. The CR peak in the water-suppressed spectrum is measured relative to the water peak in the unsuppressed spectrum. Because ^1H sensitivity is high and because the ^1H T_1 values are shorter than for ^{31}P, the experiment can essentially be done fully relaxed. The CR/water ratios measured with the same STEAM or PRESS timing parameters can be used for patient comparative studies.[4] The

normal CR/water ratio is about 1.3×10^{-3}. Such ratios can be translated into absolute concentration measurements using the known myocardial tissue water content and correcting for the signal loss caused by the STEAM or PRESS echo delay, represented by the E factors as in equation (4):[4]

$$[CR] = \frac{2S_{CR}[W]}{3S_W} \times \left(\frac{F_{CR}E_{CR}}{F_W E_W}\right) \qquad (7)$$

with the extra factor of 3 for the N-methyl protons.

The initial ^1H NMR estimate for normal human heart [CR] is 28 ± 6 μmol (g wet wt)$^{-1}$, which is consistent with canine and human necropsy studies.[4] Concentrations reported to date for the human heart are in agreement, and summarized in Table 1.

3.4 P_i, pH, and Creatine Kinase Reaction Kinetics

Intracellular pH can be measured from the chemical shift of the P_i resonance, which varies from 3.7 to 5.2 ppm relative to PCr over a pH range of 6.0–7.3.[51] In the normal heart, P_i is small and difficult to resolve unambiguously from blood DPG. It is probably better described by inequalities[17] (Table 1). Even with ^1H decoupling, which may reduce the linewidths of the DPG signals and permit a less ambiguous determination of any neighboring P_i, P_i still appears undetectable in about half of the normal subjects studied.[33] When detectable, the normal myocardial pH is about 7.15 (Table 1). It is also possible that P_i is partially NMR invisible.[49,50] It is unlikely, however, that P_i and pH measurements are contaminated by P_i in blood,[52] since P_i is imperceptible in ^{31}P spectra from human blood[39,41] and the $[P_i]$ in whole human blood is listed as only 0.08 mM.[53]

The flux of PCr through the creatine kinase reaction [see equation (1)] in the heart can be measured using the saturation transfer NMR experiment.[34] The experiment involves acquisition of ^{31}P spectra while the γ-phosphate resonance of ATP is being saturated, for example, by application of a long rf pulse tuned to its resonant frequency. When ATP is saturated, the PCr signal declines as PCr is converted to ATP via the forward reaction, in the absence of any refreshment by unsaturated phosphates produced via the reverse reaction. The ratio of PCr signals measured with and without saturation is directly proportional to the first-order rate constant, in units of the T_1 of PCr in the presence of the saturating radiation. Studies in a 4 T whole body instrument indicate that about half of the PCr is turned over per second (Table 1), consistent with data from animals.[34]

4 PATIENT STUDIES

Some data from human heart ^{31}P NMR studies of patients are summarized in Table 3.[5,17,20,28,31,33,36,39,40,42-44,47,48,54,55]

4.1 Cardiomyopathy

Evidence anticipating reductions in myocardial high energy phosphates in human cardiomyopathy and hypertrophic disease comes from ^{31}P NMR studies of animal models,[56-61] and biochemical analyses of patient biopsies taken at surgery.[7,8]

For References see p. 1193

Table 3 Summary of Some Findings for Myocardial PCr/ATP in Patients

Patients	PCr/ATP Controls	Comments	Reference
Cardiomyopathy studies showing significant changes (see text)			
1.3 ± 0.3*	2.1 ± 0.4	MD, BB, CA patients	28
1.4 ± 0.4*	2.1 ± 0.4	HCM patients	28
1.3 ± 0.3***	1.7 ± 0.3	NYHA class 0–II HCM patients	•
1.1 ± 0.4**	1.7 ± 0.3	HCM patients	39
1.1 ± 0.3**	1.6 ± 0.2^a	NYHA class II–III LVH patients	54
1.5 ± 0.31*	1.8 ± 0.2	NYHA class II–IV DCM patients	36
1.4 ± 0.5***	1.9 ± 0.4^a	NYHA class $\geqslant$ III DCM patients	40
1.3 ± 0.5	1.94 ± 0.11	DCM with NYHA class II–III, high mortality	43
1.5 ± 0.1*	2.0 ± 0.1	Aortic valve disease, NYHA class III	44
1.3 ± 0.3**	1.6 ± 0.3	Patients with severe mitral valve disease	48
2 ± 0.4***	2.4 ± 0.5	HCM patients	47
Heart transplant patients			
1.6 ± 0.5**	1.9 ± 0.2	All transplant patients	20
1.6 ± 0.5***	1.9 ± 0.4^a	Patients with rejection (myocyte necrosis)	20
Myocardial infarction			
1.7 ± 0.4 (normal)	1.6 ± 0.4	P_i/ATP elevated 5–9 days-post-MI	17
normal		[PCr]^b, [ATP]^d reduced in old MI	55
1.8 ± 0.5	2.0 ± 0.5	Patients studied 0.5–24 months post-MI	40
1.2 ± 0.3**	1.85 ± 0.28	Patients with fixed ^{201}Tl defects	42
0.9 ± 0.4***	1.8 ± 1.0	Fixed ^{201}Tl defects; [PCr], [ATP] reduced***	31
Ischemia			
0.9 ± 0.2*	1.5 ± 0.3^a	During isometric exercise stress testing	5
1.0 ± 0.3*	1.6 ± 0.2^a	Isometric exercise, reversible ^{201}Tl defects	42

Values are mean ±SD. Data from abstracts are omitted. MD, muscular dystrophy; BB, cardiac beri-beri; CA, cardiac amyloidosis; MI, myocardial infarction. Controls were normal, or LVH patients not in heart failure (indicated by ^{a}),[54] patients with dilated cardiomyopathy NYHA class <III failure,[46] or patients with no rejection.[62]
Statistical analysis versus controls or at rest:[5] *$P < 0.001$; **$P < 0.01$; ***$P < 0.05$.

Human in vivo ^{31}P NMR studies have demonstrated statistically significant reductions in anterior myocardial PCr/ATP ratios in a group of patients with cardiomyopathies due to specific disease (muscular dystrophy, beri-beri, amyloidosis);[28] in patients with HCM;[28,33,39,47] in patients with valve disease who were undergoing treatment for heart failure;[44,48,54] and in patients with DCM who were in heart failure [New York Heart Association (NYHA) clinical classification for heart failure $\geqslant$ II].[36,40,43] One report found a significant negative correlation between the NYHA classification for failure, and the myocardial PCr/ATP ratio.[40] Also, in patients whose NYHA classification was improved by drug therapy, myocardial PCr/ATP values recovered,[40] while patients with significantly reduced PCr/ATP ratios had a poorer survival rate compared with patients with normal PCr/ATP ratios, suggesting that the ratio may be of prognostic value.[43]

A number of other studies have reported no statistically significant changes in myocardial PCr/ATP in LVH,[27,28] or DCM,[27,28,33,62,63] although possible correlations between reduced myocardial PCr/ATP ratio and the NYHA heart failure classification[40] were not explicitly ruled out by these studies. The findings are reviewed in detail elsewhere.[10] Except for the NYHA classification, PCr/ATP ratios have generally correlated weakly with etiology and functional indices of disease severity such as the left ventricular ejection fraction or fractional short-

ening although such links may be proved with ongoing studies.[48]

The factors responsible for the differences in the significant findings for PCr/ATP ratios in cardiomyopathies are probably (i) variations in the range and severity of heart failure in patient study populations, and (ii) variations in statistical sensitivity. Regarding the former, it appears that averaging results from patients in mild and severe failure, or from a group of patients in predominantly mild failure, can mask significant PCr/ATP changes,[40] and that abnormalities might only be seen in subsets of patients in more advanced stages of heart failure.[54] With respect to the second factor, it should be noted that in all studies of HCM, DCM, and LVH so far, the mean myocardial PCr/ATP ratios were lower by 2–54% relative to the corresponding normal control groups with varying statistical significance.[10]

The ^{31}P NMR data linking PCr/ATP changes to heart failure would indicate that abnormal myocardial PCr/ATP ratios may first become detectable in DCM and LVH between NYHA classes II and III,[40,54] which is the point at which physical activity is limited by fatigue, and accompanying symptoms such as palpitation, or dyspnea. This is consistent with studies that conclude that the ability of the cell to sustain adequate levels of ATP is compromised only in the more advanced stages of failure.[64] One hypothesis is that the reduced energy reserve, as demonstrated by the reduction in myocardial PCr/ATP ratios seen by ^{31}P NMR, and by the decreases in myocardial creatine and creatine kinase activity evident in surgical biopsies, may limit the ability of the heart to do work, and lead to contractile dysfunction.[9] On the other hand, in at least two studies of HCM,[15,33] the PCr/ATP ratio was reduced in the absence of a link to heart failure, so it is possible that there are differences between HCM, and DCM and LVH.

In the clinic, ^{31}P NMR might help identify patients in heart failure where physical activity or diagnosis is complicated by other conditions, including age[54] and lung disease. In patients with valve disease, the detection of metabolic abnormalities might find use in surgical planning to minimize permanent injury and maximize benefit.[65] Further work is needed to define where, specifically, the benefits lie.

4.2   Heart Transplant Patients

The idea that changes in myocardial metabolite ratios might predict histological rejection in human heart transplants stems from animal ^{31}P NMR studies, usually of nonimmunosuppressed allografts, which showed metabolic changes prior to the occurrence of histological evidence for acute rejection in the first week or so posttransplantation.[66–70] In the management of heart transplant patients, the standard criterion for assessing the existence of significant allograft rejection of severity sufficient to warrant augmentation of immunosuppressive therapy is histological evidence for myocyte necrosis in endomyocardial biopsies acquired during regular cardiac catheterization procedures. While some earlier conference reports on ^{31}P NMR studies in transplant patients dating from 1988 show reduced myocardial PCr/ATP and PCr/P$_i$ ratios in transplanted hearts, the success with which ^{31}P NMR can reliably predict the outcome of histological evaluations is mixed.

The first published paper on 19 ^{31}P NMR examinations of patients studied up to 5.5 years posttransplantation did find significantly lower resting anterior myocardial PCr/ATP ratios relative to normal controls, consistent with the animal studies.[10,20] However, the results showed agreement between ^{31}P NMR abnormalities and histological evidence for necrosis in only about 60–70% of examinations,[20] suggesting that ^{31}P NMR is not a precise predictor of significant histological rejection in many transplant patients.

To date there is no firm evidence linking the finding of significant reductions in the myocardial PCr/ATP (and possibly PCr/P$_i$) ratio in transplanted hearts to other factors such as hypertrophy, or CAD involving the major vessels. However, one conference report of 13 transplant recipients on 71 occasions found reduced PCr/ATP ratios in the first few weeks after transplantation, suggesting that lower creatine levels or injury or edema following surgery may be possible causes.[71] The observation that PCr/ATP ratios may not precisely predict histological rejection is likely to reflect fundamental differences between histological and metabolic indices. In particular, myocyte necrosis would not cause altered PCr/ATP ratios because dead cells can contribute no high energy phosphates, whereas necrosis is important in histological evaluation. If the PCr/ATP reduction is associated with rejection, it may reflect an earlier phase of the process, which would not be inconsistent with the observed PCr/ATP reduction in the early weeks postsurgery because rejection episodes tend to be more frequent during this period. A much more frequent schedule of NMR examinations and biopsies would be needed to test this hypothesis. Nevertheless, the occurrence of reductions in the high energy phosphate metabolism of a majority of transplant patients is a concern meriting further study.

4.3   Coronary Artery Disease

4.3.1   Myocardial Infarction

Published papers to date show no significant alterations in resting myocardial PCr/ATP ratios in myocardial infarction.[17,40,46,55] Two reports do show significantly reduced resting PCr/ATP in patients with infarction and irreversible ^{201}Tl defects by radionuclide imaging.[31,42] It is unclear whether other factors such as heart failure and/or cardiomopathy may play a role in these patients and could explain the different findings.[10] In recent anterior myocardial infarction, significant elevations in P$_i$ levels can be detected within the first week or so after onset,[17] which is consistent with canine ^{31}P NMR studies showing elevated P$_i$ levels persisting for several days postcoronary occlusion, as a byproduct of PCr and ATP consumption.[72] Biochemical analyses of animal hearts also show that essentially all the ATP and PCr is depleted within the first few hours of ischemic injury that results in cell death.[6] Thus again, as dead cells can contribute no high energy phosphates, it is likely that the normal resting PCr/ATP ratios derive from a mixture of metabolically normal tissue, scar tissue and, possibly, jeopardized myocardium adjacent to, or interspersed with, the infarction. Accordingly the infarction itself may best be characterized by the absence of any contribution to the spectrum.

Evidence for the possible use of ^{31}P NMR for assessing myocardial viability, is provided by a study of 29 patients with recent myocardial infarction.[46] This revealed no changes in PCr/ATP ratios in patients with myocardial stunning after reperfusion, suggesting that relative levels of high-energy phosphates are not depleted in stunned human myocardium.

There is now ^{31}P NMR evidence that the metabolite concentrations are reduced in infarction itself.[31,55] This includes the observation of a significant negative correlation between ATP levels and the size of perfusion deficits in the heart, as quantified by thallium ^{201}Tl radionuclide imaging,[55] and reductions in [PCr] and [ATP] in patients with fixed ^{201}Tl defects compared with those with reversible defects.[31] Concentration measurements made by ^{31}P NMR must be treated cautiously, however, since the tissue volume present in the voxel may be reduced by any wall thinning that occurs after infarction. This may partially offset the signal loss due to infarction, depending on how tissue volume is accounted for in the concentration measurements. Nevertheless, the observations[31,55] are consistent with a model for infarction wherein the myocardial PCr and ATP levels are reduced in the heart spectrum in proportion to the volume of infarcted tissue that intersects the voxel.[10]

Reductions of about 60% in myocardial [CR] can also be detected with water-referenced ^{1}H NMR in patients with myocardial infarction.[4] The results are consistent with animal studies and the concept of metabolic depletion in infarction. Because the resolution and sensitivity are better with ^{1}H NMR than with ^{31}P NMR (and because the CR resonance has three protons compared with the single phosphate group of PCr), [1]H

For References see p. 1193

NMR could provide a useful metabolic means for distinguishing healthy from infarcted nonviable myocardium.

4.3.2 Myocardial Ischemia

In patients with ischemic heart disease involving severe stenosis of the anterior vessels the resting anterior myocardial PCr/ATP ratio is normal,[40] or nearly so.[5] To observe metabolic change corresponding to reversible ischemia, safe and effective stress protocols that can be performed in the NMR magnet for the duration of a localized ^{31}P NMR examination are necessary. Three types of stress have been tried: aerobic exercise involving the lifting of weights with the legs;[73] an isometric exercise, performed, for example, with a handgrip dynamometer;[5,42] and stress induced by pharmaceutical agents such as dobutamine.[45,63] The increase in cardiac work-load indexed by the heart rate–blood pressure product in these protocols is about 70% for an aerobic leg exercise lifting 5 kg weights;[73] about 30–35% with the isometric hand-grip exercise at 30% of the subjects maximum force;[5] and up to 130% (2.3-fold) with dobutamine infusion.[45,63] While producing a lesser increase in cardiac work, the isometric exercise protocol can produce reflex coronary vasoconstriction in stenosed vessels. It also minimizes motion problems during NMR acquisition compared with aerobic exercise, and, in addition, produces vasoconstriction in patients with critical coronary disease.[5] It is also well tolerated by CAD patients, and can be immediately terminated should complications arise.

Phosphorus-31 NMR exercise stress testing of healthy volunteers who are free of significant CAD appears to elicit no significant alterations in anterior myocardial PCr/ATP ratios, with a 30–70% increase in rate–pressure product.[5,73] Dobutamine stress testing at 230% rate–pressure product produced no significant change in seven normal controls, although a recent study of 20 healthy volunteers elicited a statistically significant 14% reduction in myocardial PCr/ATP at a 300% rate–pressure product.[45]

There are two papers reporting ^{31}P findings from isometric hand-grip stress testing of patients with documented anterior wall ischemia.[5,42] In the first hand-grip exercise, the PCr/ATP ratio decreased by 37% in a group of 16 patients with severe anterior stenoses (Figure 3).[5] After exercise, metabolite ratios recovered to near-normal, pre-exercise values. Eleven healthy control subjects, and nine patients with nonischemic heart disease (cardiomyopathy or valve disease) exhibited no PCr/ATP changes during the same exercise, suggesting that the stress-induced changes in PCr/ATP are specific to CAD. Five CAD patients underwent repeat ^{31}P NMR stress-testing after successful revascularization therapy.[5] Prior to therapy, stress provoked a 33% decrease in PCr/ATP in these patients, as in the larger group. Exercise stress testing performed posttherapy produced no change, indicating that the metabolic abnormality resolves with successful clinical outcome.

The second report found a similar 40% decrease in the PCr/ATP ratio in patients with reversible anterior wall ischemia that was confirmed by exercise ^{201}Tl radionuclide imaging.[42] Twelve patients with fixed ^{201}Tl defects indicative of myocardial infarction, as well as normal controls, exhibited no exercise induced PCr/ATP changes. This, and the observation that dobutamine stress testing induced no significant reduction in myocardial PCr/ATP, on average, in patients with DCM,[63]

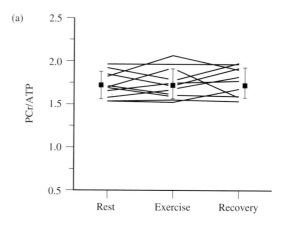

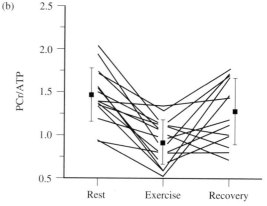

Figure 3 Transient changes in the anterior myocardial PCr/ATP ratio in (a) control subjects free of CAD, and (b) patients with CAD involving the anterior wall, in response to continuous isometric handgrip exercise at 30% of the subjects maximum force.[5] Error bars show means ±SD. (Reproduced by permission of *New England Journal of Medicine* from Weiss et al.[5])

is further evidence that the stress-induced changes may be specific for ischemia.

Present ^{31}P NMR studies are limited to the anterior wall. Whether ^{31}P NMR stress testing might play a role in the evaluation of ischemia in the clinic may depend in part on extending ^{31}P NMR to other regions of the heart, which will necessitate further improvements in sensitivity such as those afforded by NOE and phased array detection coils. Comparative studies of sensitivity relative to existing clinical modalities would then be needed.

4.3.3 Cardiac Model

In summary, the ^{31}P studies of myocardial infarction and reversible ischemia suggest a model in which the cardiac spectrum from a voxel comprises of up to four components: (i) myocardium either with an essentially normal resting PCr/ATP ratio or, possibly, a reduced resting PCr/ATP ratio due to complications arising from chronic infarction such as heart failure or cardiomyopathy; (ii) jeopardized myocardium whose PCr/ATP ratio decreases with stress testing; (iii) scar tissue with lower ATP and PCr concentrations; and (iv) infarcted myocar-

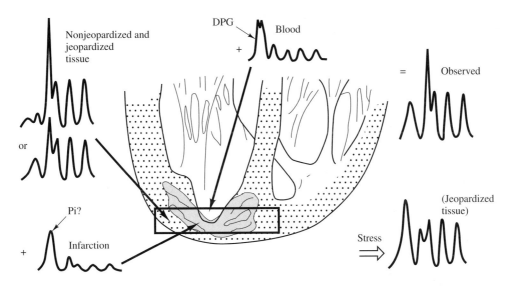

Figure 4 Model for the observed cardiac spectrum as the integral of four potentially distinguishable components: (i) myocardium either with an essentially normal resting PCr/ATP or, possibly, a reduced resting PCr/ATP ratio associated with heart failure; (ii) jeopardized myocardium, possibly only distinguishable from normal or failing heart by stress testing; (iii) a mixture of infarction with no PCr or ATP but perhaps some residual P_i shortly after infarction, and scar tissue, presumably with lower ATP and PCr concentrations; and (iv) contaminating signal from blood[10]

dium with no PCr, ATP, or CR but possibly some residual P_i persisting several days after infarction. Appropriate choice of the ^{31}P or ^1H NMR protocols may resolve these components: measurements of the metabolite concentrations at rest may index the fraction of scar tissue and infarction present in a selected volume, while jeopardized myocardium may be indexed by stress testing (Figure 4).[10]

5 RELATED ARTICLES

Cardiac Gating Practice; Cardiovascular NMR to Study Function; Chemical Shift Imaging; Heart: Clinical Applications of MRI; Magnetization Transfer between Water and Macromolecules in Proton MRI; Proton Decoupling During In Vivo Whole Body Phosphorus MRS; Proton Decoupling in Whole Body Carbon-13 MRS; Rotating Frame Methods for Spectroscopic Localization; Selective Excitation in MRI and MR Spectroscopy; Single Voxel Whole Body Phosphorus MRS.

6 REFERENCES

1. P. A. Bottomley, *Science*, 1985, **229**, 769.
2. P. A. Bottomley, *Radiol.* 1989, **170**, 1.
3. R. S. Menon, K. Hendrich, X. Hu, and K. Ugurbil, *Magn. Reson. Med.*, 1992, **26**, 368.
4. P. A. Bottomley and R. G. Weiss, *Lancet*, 1998, **351**, 714.
5. R. G. Weiss, P. A. Bottomley, C. J. Hardy, and G. Gerstenblith, *N. Engl. J. Med.*, 1990, **323**, 1593.
6. R. B. Jennings and K. A. Reimer, *Am. J. Pathol.*, 1981, **102**, 241.
7. J. L. Swain, R. L. Sabina, R. B. Peyton, R. N. Jones, A. S. Wechsler, and E. W. Holmes, *Proc. Natl. Acad. Sci. USA*, 1982, **79**, 655.
8. J. S. Ingwall, M. F. Kramer, M. A. Fifer, B. H. Lorell, R. Shemin, W. Grossman, and P. D. Allen, *N. Engl. J. Med.*, 1985, **313**, 1050.
9. J. S. Ingwall, *Circulation*, 1993, **87**(Suppl. VII), VII-58.
10. P. A. Bottomley, *Radiology*, 1994, **191**, 593.
11. P. A. Bottomley, C. J. Hardy, P. B. Roemer, and O. M. Mueller, *Magn. Reson. Med.*, 1989, **12**, 348.
12. J. B. Ra, S. K. Hilal, C. H. Oh, and J. K. Mun, *Magn. Reson. Med.*, 1988, **7**, 11.
13. T. B. Parish, D. S. Fieno, S. W. Fitzgerald, and R. M. Judd, *Magn. Reson. Med.*, 1997, **38**, 653.
14. R. J. Kim, J. O. A. C. Lima, E. L. Chen, S. B. Reeder, F. J. Klocke, E. A. Zerhouni, and R. M. Judd, *Circulation*, 1997, **95**, 1877.
15. J. R. Whitman, B. Chance, H. Bode, J. Maris, J. Haselgrove, R. Kelley, B. J. Clark, and A. H. Harken, *J. Am. Coll. Cardiol.*, 1985, **5**, 745.
16. B. Rajagopalan, M. J. Blackledge, W. J. McKenna, N. Bolas, and G. K. Radda, *Ann. N.Y. Acad. Sci.*, 1987, **508**, 321.
17. P. A. Bottomley, R. J. Herfkens, L. S. Smith, and T. M. Bashore, *Radiology*, 1987, **165**, 703.
18. S. Schaefer, J. Gober, M. Valenza, G. S. Karczmar, G. B. Matson, S. A. Camacho, E. H. Botvinick, B. Massie, and M. W. Weiner, *J. Am. Coll. Cardiol.*, 1988, **12**, 1449.
19. P. A. Bottomley, C. J. Hardy, and P. B. Roemer, *Magn. Reson. Med.*, 1990, **14**, 425.
20. P. A. Bottomley, R. G. Weiss, C. J. Hardy, and W. A. Baumgartner, *Radiology*, 1991, **181**, 67.
21. P. A. Bottomley, US patent, 4 480 228, Oct. 30, 1984.
22. J. Frahm, H. Bruhn, M. L. Gyngell, K. D. Merbold, W. Hanicke, and R. Sauter, *Magn. Reson. Med.*, 1989, **9**, 79.
23. P. A. Bottomley and C. J. Hardy, *Philos. Trans. R. Soc. London, Ser. A*, 1990, **333**, 531.

For References see p. 1193

24. C. J. Hardy, P. A. Bottomley, K. W. Rohling, and P. B. Roemer, *Magn. Reson. Med.*, 1992, **28**, 54.
25. P. B. Roemer, W. A. Edelstein, C. E. Hayes, S. P. Souza, and O. M. Mueller, *Magn. Reson. Med.*, 1990, **16**, 192.
26. P. A. Bottomley and C. J. Hardy, *Magn. Reson. Med.*, 1992, **24**, 384.
27. S. Schaefer, J. R. Gober, G. G. Schwartz, D. B. Twieg, M. W. Weiner, and B. Massie, *Am. J. Cardiol.* 1990, **65**, 1154.
28. Y. Masuda, Y. Tateno, H. Ikehira, T. Hashimoto, F. Shishido, M. Sekiya, Y. Imazeki, H. Imai, S. Watanabe, and Y. Inagaki, *Jap. Circ. J.*, 1992, **56**, 620.
29. P. A. Bottomley, C. J. Hardy, and R. G. Weiss, *J. Magn. Reson.*, 1991, **95**, 341.
30. P. A. Bottomley and R. Ouwerkerk, *Magn. Reson. Med.*, 1994, **32**, 137.
31. T. Yabe, K. Mitsunami, T. Inubushi, and M. Kinoshita, *Circulation*, 1995, **92**, 15.
32. P. A. Bottomley, E. A. Atalar, and R. G. Weiss, *Magn. Reson. Med.*, 1996, **35**, 664.
33. A. de Roos, J. Doornbos, P. R. Luyten, L. J. M. P. Oosterwaal, E. E. van der Wall, and J. A. den Hollander, *J. Magn. Reson. Imaging*, 1992, **2**, 711.
34. P. A. Bottomley and C. J. Hardy, *J. Magn. Reson.*, 1992, **99**, 443.
35. H. Kolem, R. Sauter, M. Friedrich, M. Schneider, and K. Wicklow, in 'Cardiovascular Applications of Magnetic Resonance', ed. G. M. Pohost, Futura, Mt. Kisco, NY, 1993, p. 417.
36. C. J. Hardy, R. G. Weiss, P. A. Bottomley, and G. Gerstenblith, *Am. Heart J.*, 1991, **122**, 795.
37. S. Minakami, C. Suzuki, T. Saito, and H. Yoshikawa, *J. Biochem.*, 1965, **58**, 543.
38. E. Beutler, in 'Hematology', 3rd edn., eds. W. J. Williams, E. Beutler, A. J. Erslev, and M. A. Lichtman, McGraw-Hill, New York, 1983, p. 283.
39. H. Sakuma, K. Takeda, T. Tagami, T. Nakagawa, S. Okamoto, T. Konishi, and T. Nakano, *Am. Heart J.*, 1993, **125**, 1323.
40. S. Neubauer, T. Krahe, R. Schindler, M. Horn, H. Hillenbrand, C. Entzeroth, H. Mader, E. P. Kromer, G. A. J. Riegger, K. Lackner, and G. Ertl, *Circulation*, 1992, **86**, 1810.
41. M. Horn, S. Neubauer, M. Bomhard, M. Kadgien, K. Schnackerz, and G. Ertl, *Magma*, 1993, **1**, 55–60.
42. T. Yabe, K. Mitsunami, M. Okada, S. Morikawa, T. Inubushi, and M. Kinoshita, *Circulation*, 1994, **89**, 1709.
43. S. Neubauer, M. Horn, M. Cramer, K. Harre, J. B. Newell, W. Peters, T. Pabst, G. Ertl, D. Hahn, J. S. Ingwall, and K. Kochsiek, *Circulation*, 1997, **96**, 2190.
44. S. Neubauer, M. Horn, T. Pabst, K. Harre, H. Stromer, G. Bertsch, J. Sandstede, G. Ertl, D. Hahn, and K. Kochsiek, *J. Invest. Med.*, 1997, **45**, 453.
45. H. J. Lamb, H. P. Beyerbacht, R. Ouwerkerk, J. Doornbos, B. M. Pluim, E. E. van der Wall, A. van der Laarse, and A. de Roos, *Circulation*, 1997, **96**, 2969.
46. R. Kalil-Filho, C. P. de Albuquerque, R. G. Weiss, A. Mocelin, G. Bellotti, G. Cerri, and F. Pileggi, *J. Am. Coll. Cardiol.*, 1997, **130**, 1228.
47. W. Jung, L. Sieverding, J. Breuer, T. Hoess, S. Widmaier, O. Schmidt, M. Bunse, F. van Erckelens, J. Apitz, O. Lutz, and G. J. Dietze, *Circulation*, 1998, **97**, 2536.
48. M. A. Conway, P. A. Bottomley, R. Ouwerkerk, G. K. Radda, and B. Rajagopalan, *Circulation*, 1998, **97**, 1716.
49. S. M. Humphrey and P. B. Garlick, *Am. J. Physiol.*, 1991, **260**, H6.
50. P. B. Garlick and R. M. Townsend, *Am. J. Physiol.*, 1992, **263**, H497.
51. D. G. Gadian, 'Nuclear Magnetic Resonance and its Applications to Living Systems', Oxford University Press, Oxford, 1982, p. 30.
52. K. M. Brindle, B. Rajagopalan, N. M. Bolas, and G. K. Radda, *J. Magn. Reson.*, 1987, **74**, 356.
53. D. L. Altman and D. A. Dittmer, *FASEB*, 1961, **21**, 29.
54. M. A. Conway, J. Allis, R. Ouwerkerk, T. Niioka, B. Rajagopalan, and G. K. Radda GK, *Lancet*, 1991, **338**, 973.
55. K. Mitsunami, M. Okada, T. Inoue, M. Hachisuka, M. Kinoshita, and T. Inubishi, *Jap. Circ. J.*, 1992, **56**, 614.
56. W. Markiewicz, S. Wu, W. W. Parmley, C. B. Higgins, R. Sievers, T. L. James, J. Wikman-Coffelt, and G. Jasmin, *Circ. Res.*, 1986, **59**, 597.
57. S. A. Camacho, J. Wikman-Coffelt, S. T. Wu, T. A. Watters, E. H. Botvinick, R. Sievers, T. L. James, G. Jasmin, and W. W. Parmley, *Circulation*, 1988, **77**, 712.
58. S. Wu, R. White, J. Wikman-Coffelt, R. Sievers, M. Wendland, J. Garrett, C. B. Higgins, T. James, and W. W. Parmley, *Circulation*, 1987, **75**, 1058.
59. K. Nicolay, W. P. Aue, J. Seelig, C. J. A. van Echteld, T. J. C. Ruigrok, and B. de Kruijff, *Biochim. Biophys. Acta*, 1987, **929**, 5.
60. S. J. Kopp, L. M. Klevay, and J. M. Feliksik, *Am. J. Physiol.*, 1983, **245**, H855.
61. N. Afzal, P. K. Ganguly, K. S. Dhalla, G. N. Pierce, P. K. Singal, and N. S. Dhalla, *Diabetes*, 1988, **37**, 936.
62. W. Auffermann, W. M. Chew, C. L. Wolfe, N. J. Tavares, W. W. Parmley, R. C. Semelka, T. Donnelly, K. Chatterjee, and C. B. Higgins, *Radiology*, 1991, **179**, 253.
63. S. Schaefer, G. G. Schwartz, S. K. Steinman, D. J. Meyerhoff, B. M. Massie, and W. M. Weiner, *Magn. Reson. Med.*, 1992, **25**, 260.
64. S. M. Krause, *Heart Failure*, 1988, 267.
65. Editorial, *Lancet*, 1991, **338**, 981.
66. R. C. Canby, W. T. Evanochko, L. V. Barrett, J. K. Kirklin, D. C. McGiffen, T. T. Sakai, M. E. Brown, R. E. Foster, R. C. Reeves, and G. M. Pohost, *J. Am. Coll. Cardiol.*, 1987, **9**, 1067.
67. C. E. Haug, J. L. Shapiro, L. Chan, and R. Weil, *Transplantation*, 1987, **44**, 175.
68. C. D. Fraser, V. P. Chacko, W. E. Jacobus, R. L. Soulen, G. M. Hutchins, B. A. Reitz, and W. A. Baumgartner, *Transplantation*, 1988, **46**, 346.
69. C. D. Fraser, V. P. Chacko, W. E. Jacobus, P. Mueller, R. L. Soulen, G. M. Hutchins, B. A. Reitz, and W. A. Baumgartner, *J. Heart Transplant.*, 1990, **9**, 197.
70. C. D. Fraser, V. P. Chacko, W. E. Jacobus, G. M. Hutchins, J. Glickson, B. A. Reitz, and W. A. Baumgartner, *Transplantation*, 1989, **48**, 1068.
71. J. O. van Dobbenburgh, N. de Jonge, C. Klopping, J. R. Lahpor, S. R. Woolley, and C. J. A. van Echteld, *Proc. XIIth Ann Mtg. Soc. Magn. Reson. Med.*, New York, 1993, **3**, 1093.
72. P. A. Bottomley, L. S. Smith, S. Brazzamano, L. W. Hedlund, R. W. Redington, and R. J. Herfkens, *Magn. Reson. Med.*, 1987, **5**, 129.
73. M. A. Conway, J. D. Bristow, M. J. Blackledge, B. Rajagopalan, and G. K. Radda, *Br. Heart J.*, 1991, **65**, 25.

Acknowledgements

I would like to thank R. G. Weiss, M. Conway, and B. Rajagopalan for many helpful discussions.

Biographical Sketch

Paul A. Bottomley. *b* 1953. B.Sc. Hon., 1974, Ph.D. 1978, University of Nottingham, UK, with E. Raymond Andrew and Waldo Hinshaw. Research Associate at Johns Hopkins Medical Institutions, 1978–1980. Physicist, G.E. Research and Development Center, 1980–1994. Presently Russell H. Morgan Professor and Director, Division of MR Research, Dept of Radiology, Johns Hopkins University, Baltimore, U.S.A. Approx. 120 publications, 130 conference reports, 27 patents. Gold Medal, Society of Magnetic Resonance in Medicine, 1989; G.E.

Coolidge Medal and Fellow, 1990. Research specialties: in vivo NMR imaging, localized spectroscopy, tissue relaxation times, MRI, human cardiac NMR spectroscopy.

Lung and Mediastinum MRI

Robert J. Herfkens

Stanford University School of Medicine, CA, USA

The utilization of MRI techniques for visualization of lung and mediastinal abnormalities has been greeted with great enthusiasm due to the excellent contrast resolution of MRI. However, the complex motions associated with those respiratory and cardiac events have posed significant challenges to the utilization of MRI for visualizing pulmonary and mediastinal abnormalities. The utilization of motion compensation techniques have greatly improved the image quality of MRI, and its intrinsic qualities have indeed provided unique diagnostic information about thoracic disease.

In order to compensate for the complex motions associated with the heart motion and blood flow, methods for coordinating the acquisition of the images in concert with the cardiac cycle have been developed.[1] Gating or triggering of the acquisition to the cardiac cycle has been obtained through a number of methods; primarily, electrocardiographic (ECG) gating methods have been employed.[2,3] In addition, methods of plethysmographic or pulse oximeter methods have been utilized, but these are somewhat less precise in coordinating the acquisition with the cardiac cycle. The ECG signal triggers the magnetic resonance (MR) scanner as providing a consistent point in the cardiac cycle for initiating multislice or cine imaging. This, however, provides a limitation in imaging of the chest, as the R–R interval determines the repetition time of the scan. This provides a significant limitation to the flexibility of presenting imaging contrast. For an average heart rate of 60 beats per minute, the minimum repetition time would be 1000 ms. Depending upon the patient's physiologic status, the heart rate may vary quite greatly, thus having a significant effect on the overall image contrast, due to varying *TR* times.

Respiratory motion, which is also periodic, causes significant artifacts in MRI. These have a strong impact on chest images. The simplest solution to respiratory artifacts originally was to provide multiple signal averages, which blur the respiratory motion and, in general, improve the signal-to-noise ratio. These increases in imaging time, associated with respiratory averaging, are somewhat inappropriate on modern systems. Techniques that coordinate the phase encoding of the image acquisition with the respiratory cycle, so-called 'respiratory ordered phase encoding' or 'respiratory compensation', can sig-nificantly reduce the amount of respiratory artifacts without unduly increasing the imaging time. Recent fast imaging techniques with fast repetition times as short as 10 ms have allowed breath-held images of the chest. Although these images may introduce some blurring of the cardiac silhouette, improved definition of mediastinal and diaphragmatic structures have been realized.

Basic spin echo techniques gated to the cardiac cycle will provide relatively unique contrast within the mediastinum. The basic contrast associated with relative T_1-weighted imaging shows fat as a relatively high signal intensity, with normal mediastinal structures, such as the esophagus and heart, with intermediate signal intensity.[4] The intrinsic flow sensitivity spin echo images show a relative decrease in signal intensity of the vascular structures. In gated spin echo imaging, however, the signal intensity of major vessels may vary, depending upon when the specific image is obtained in the cardiac cycle. Those images obtained during systole have a relatively strong signal void, and those obtained during diastole may have significant intravascular signal, because of the relatively slow flow. Utilization of flow sensitive sequences, such as gradient-recalled imaging techniques, have improved the ability to differentiate intravascular slow flow signal from soft-tissue signals. Gradient-recalled images show an intrinsic sensitivity to flowing blood due to flow related enhancement, thus demonstrating an increase in signal intensity associated with flow. By utilizing gating with gradient-recalled sequences, phase-encoding signals can be advanced with each cardiac cycle, thus providing a cine image throughout the cardiac cycle, generating sensitivity to flow. In addition, phase contrast techniques have been applied to the cine technique, allowing the demonstration of velocity changes throughout the cardiac cycle (Figure 1). These phase contrast techniques have been used to characterize flow in major vascular structures and may be important in characterizing pulmonary artery flow abnormalities[5] (Figure 2).

When using spin echo sequences, the repetition time is determined by the R–R interval. In order to obtain T_2-weighted images in the chest, techniques which would allow for the gating of the acquisition to multiple heart beats, which is every third or every fourth heart beat, allow for the acquisition of T_2-weighted data. In general, the utilization of T_1-weighted data has been superior for anatomic definition. The intrinsic contrast, as previously noted, where fat is bright and muscle and esophagus are dark, leave any other intermediate signal intensities as being pathologic. The relatively high contrast between flowing blood in spin echo images and other soft tissues also allows for excellent contrast to characterize other pathologic entities related to vascular involvement or improved definition of the spatial relationship associated with the relative signal void of vascular structures (Figure 3).

The ability of MR to generate any plane has been a significant advantage over conventional techniques, such as computerized tomography (CT). The use of coronal planes has significantly enhanced the visualization of abnormalities and the apices of the lungs, and has improved the visualization of lesions associated with the diaphragm. The ability to utilize oblique planes has been extremely helpful in characterizing pathologies associated with the aorta or other structures.[6]

For References see p. 1203

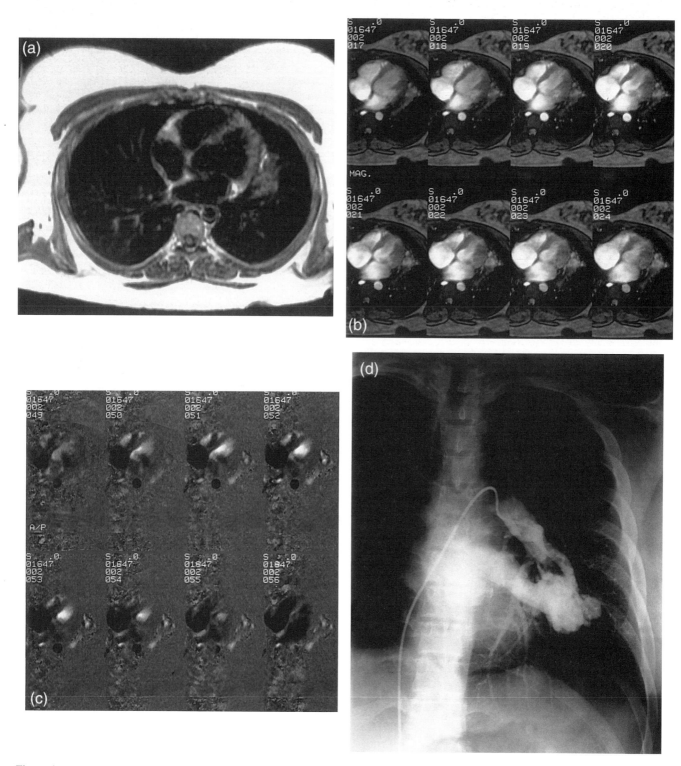

Figure 1   (a) A gated T_1-weighted spin echo image showing prominent soft tissue density along the lateral wall of the heart. This mass in a 32-year-old female was identified on chest X-ray. (b) Eight frames of a cine gradient-recalled imaging sequence from diastole through systole. The mass noted in (a) is shown with signal variations suggesting flow. (c) A cine phase contrast imaging sequence encoded in the superior to inferior direction corresponding to the image shown in (b). Note the significant change in signal intensity during systolic contraction in both the superior and inferior directions in the region of the mass, suggesting that this mass is a pulmonary arterial–venous malformation. (d) A pulmonary angiogram confirming the presence of a pulmonary arterial–venous malformation

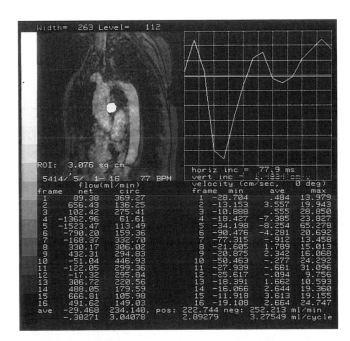

Figure 2 Flow analysis of the right pulmonary artery flow pattern in a patient after lung transplant. This is the native lung and shows a pulsatile flow profile going from positive to negative throughout the cardiac cycle, showing a low net total flow to the lung. Such pulmonary flow analysis may be helpful in determining vascular compliance of the lung, as well as in relative ratios in blood flow in postoperative patients

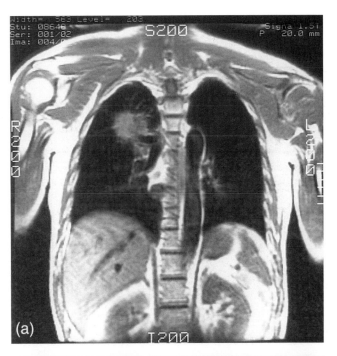

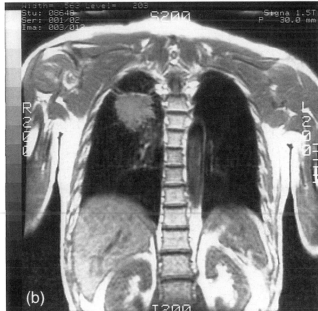

Figure 3 (a) A coronal-gated spin echo image showing the intimate relationship of a bronchiogenic carcinoma with the right upper lobe bronchus. (b) A gated spin echo image posterior to (a), showing a larger portion of the mass and possible association with the chest wall

Another major element which degrades images within the thorax is the susceptibility associated with the multiple air–water interfaces in the pulmonary parenchyma. This marked change in susceptibility has resulted in a very poor delineation of other abnormalities within the lungs themselves. A number of recent techniques have provided significant improvement in the visualization of pulmonary abnormalities; however, they still appear to be significantly less sensitive than CT for detecting pulmonary pathologies.[7–9] The development of fast spin echo imaging techniques has recently been shown to increase the sensitivity to pulmonary parenchyma abnormalities, such as metastasis. However, CT is significantly superior. Projection–reconstruction techniques can utilize extremely short echo times (as short as 100 μs), and thus allow imaging of the pulmonary parenchyma or lung water associated with the pulmonary parenchyma[10] (Figure 4). These techniques remain somewhat experimental but will allow for the demonstration of pulmonary parenchymal signals, which are less sensitive to the susceptibility generated within the lung. These signals represent excellent maps of lung water and show exceptional promise in improving MR's ability for characterizing subtle intrapulmonary abnormalities.

The utilization of paramagnetic contrast agents has also advanced the abilities of thoracic MRI to characterize a number of lesions.[11] The T_1 shortening effects of paramagnetic materials have provided the ability to enhance the differentiation

of atelectatic from existing tumors as well as characterizing nonperfused cystic structures.

MRI has been valuable in characterizing a number of signal intensities related to tissues. The specific signal characteristics of hemorrhage have allowed the identification of short T_1 changes associated with methemoglobin in order to characterize hemorrhagic processes from pulmonary parenchymal diseases.

For References see p. 1203

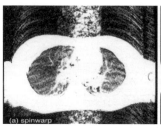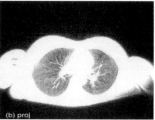

Figure 4 Comparison of images obtained with (a) conventional T_1-weighted gated (1 R–R) multislice spinwarp ($TR \approx 750$ ms, $TE = 15$ ms, 256×2 NEX, scan time 7.6 min, using respiratory compensation and spatial presaturation) and (b) multislice Hadamard projection reconstruction ($TR = 200$ ms, $TE = 0.25$ ms, 512 views $\times$ 8 NEX, scan time 4 min). The images have been windowed to show the lung parenchyma, which causes the flow artifacts in (a), to appear abnormally intense. Note that the projection–reconstruction acquisition has intrinsically lower respiratory and flow artifacts, even without the compensation methods employed in (a), and despite the reduced scan time. The lack of respiratory motion artifacts with the projection–reconstruction sequence is critical to the successful visualization of lung parenchyma

Pulmonary hemorrhage appears as a relatively short T_1 of bright signal intensity on T_1-weighted images. Hemorrhage into the mediastinum or associated with aortic dissections can be characterized due to the signal intensity changes associated with hemoglobin as it changes from oxyhemoglobin to deoxyhemoglobin to methemoglobin.

The mediastinum contains relatively simple contrast relationships for MRI. The esophagus and heart remain intermediate in signal intensity, contrasted to the relatively high signal intensity on T_1-weighted images of thoracic fat. The normal thymus appears as a signal of intermediate intensity; it generally decreases in size with age, and increases in signal intensity with fatty replacement. Of note is the fact that the thymus may rebound in adulthood due to some physiologic stress, thus providing an intermediate anterior mediastinal signal intensity. This, in general, can be differentiated from most pathologic processes on the basis of morphology. Although MRI provides excellent tissue characterization, the ability of MRI in differentiating between benign and malignant changes within the thymus itself is somewhat limited. CT and MRI both provide an excellent technique for evaluating anatomic changes associated with thymic disorders.[12–14]

MRI provides an excellent method for identifying and characterizing abnormalities in the pulmonary hyla. The sensitivity to flow associated with spin echo imaging and the increase in signal intensity from gradient-recalled images provide an excellent format for identifying soft tissue abnormalities in the hyla. MRI appears to be significantly more sensitive than CT for the detection of these abnormalities; however, the characterization of these may be somewhat difficult.[15–17] The pathologic nature of hilar and mediastinal lymph nodes remains largely based on the size of the individual lesions. MRI provides an excellent format for visualizing lymphadenopathy. The ability to image in the coronal plane has increased the sensitivity for the detection of aortapulmonary window adenopathy. The major

advantage of MRI over CT is that the intrinsic contrast relationships allow identification of the lymph nodes and their separation from major vascular structures without the need to administer an iodinated contrast agent. Characterization of pathologies by T_1- and T_2-weighted images, however, has been somewhat disappointing. Both inflammatory and neoplastic adenopathy appear to have prolonged T_2 times. The primary mechanism for distinguishing between benign and malignant adenopathy remains that of lymph node morphology and size.[18,19] Staging of bronchiogenic carcinoma with MRI has provided an excellent means of detecting mediastinal lymph nodes on an anatomic basis. Characterization of lymphomas has also benefited significantly from the detection of mediastinal lymphadenopathy, due to the excellent contrast characteristics of MRI. The relative T_2 prolongation associated with lymphomas has been identified, but its use remains limited due to the lack of ability to characterize changes other than those based on size (Figure 5). The evaluation of patients with recurrent lymphoma has been suggested to be improved by the ability of MRI to characterize increased water content of nodes. In most treated lymphomas, there is a generalized decrease in size and signal intensity related to therapy. The recurrence of lymphoma can be characterized by either an increase in size or an increase in signal intensity. The increase in signal intensity in T_2-weighted images has been suggested to be grounds for suspecting recurrence, and individual patients with changes in signal intensity or size should be assessed for recurrent disease.[20]

The evaluation of bronchiogenic carcinoma is important in determining therapy and prognosis. MRI and CT both provide excellent anatomic images of the primary tumor as well as the ability to characterize the presence of lymphadenopathy. A specific problem for which MRI may provide improved specificity is in the associated changes in atelectasis associated with proximal bronchiogenic lesions.[21,22] MRI has been suggested to be an excellent method for differentiation of obstructed atelectasis from the primary mass. The utilization of paramagnetic contrast material appears to provide significantly improved information over CT, and aids in differentiating mass from atelectasis[23,24] (Figure 6).

Mediastinal cysts can often be confused with neoplastic lesions. The specific characteristics of very long T_2 and very long T_1 associated with mediastinal cysts enable these lesions to be characterized by MRI. The relative homogeneity associated with these benign congenital cysts and their anatomic location provides specific clues as to their etiology. Mediastinal cysts can be divided into duplication cysts, neurenteric cysts, and pleuroparenchymal cysts. In general, they are specifically characterized by their long T_1 and long T_2 and relatively sharp margins. The additional use of an intravenous paramagnetic contrast material will characterize these lesions as being avascular.

Esophageal carcinoma represents a continuing diagnostic dilemma. The presence of the primary mass can easily be identified by the relative increase in soft tissue density associated with the esophagus. MRI provides an excellent method for the staging of esophageal cancer and characterizing the association with vascular structures. The obliteration of fat planes about the aorta and other adjacent vascular structures, such as the left atrium, can be extremely helpful in planning therapy. These changes, however, do not appear to have significantly

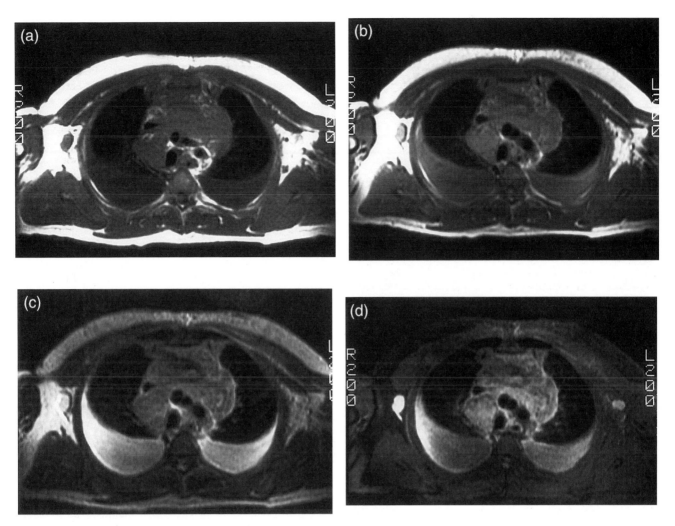

Figure 5 (a) A gated T_1-weighted spin echo image through the upper mediastinum. (b) A spin density gated image through the upper mediastinum. (c) A T_2-weighted gated image through the upper mediastinum. (d) A fat-suppressed, fast spin echo image through the upper mediastinum. Note the bulky adenopathy in the anterior mediastinum and the right paratracheal space. In addition, note the large lymph node in the right axilla. There are bilateral pleural effusions present, which give significantly variable signal intensities, as do all these images, to the relatively long T_1 and T_2 of the effusion, which is affected strongly by flow secondary to cardiac and respiratory motion. Note the difference in the degree of conspicuousness of the axillary node in the fat-suppressed image (d) and the significant difference in the signal characteristics of the mediastinal adenopathy. The fast spin echo image associated with both fat suppression and magnetization transfer effects will have significantly different contrast in this mediastinal adenopathy. The ability to characterize adenopathy, due to signal changes, remains a complex issue

impacted on the ultimate outcome of esophageal cancer (Figure 7).

Approximately 10% of parathyroid glands are ectopic. Two thirds of these appear to arise in the anterior mediastinum. In the presence of hyperparathyroidism unresponsive to resection of the cervical components of the thyroid glands, the mediastinum becomes the primary source. The behavior of hyperplastic parathyroid glands on MRI is characterized by a markedly prolonged T_2 relaxation time. This makes the identification of T_1- and T_2-weighted images extremely important. It should be noted that the identification of a markedly prolonged T_2 signal in the mediastinum should be separated from relatively slow flow in thoracic veins by the utilization of some form of gradient-recalled imaging sequence.

The excellent contrast characteristics of MRI provide an excellent means of staging of thoracic neoplasms. The sensitivity to direct invasion of vascular structures by obliteration of associated short T_1 fat planes provides an excellent method (Figure 8). The ability of MRI to characterize changes along the thoracic wall provides a unique opportunity to demonstrate thoracic wall invasion. The short T_1 associated with subpleural fat and the relatively long T_1 of thoracic neoplasms give MRI great sensitivity in detecting thoracic wall invasion (Figure 9). These changes in thoracic wall invasion observed in both lymphoma and bronchiogenic carcinomas may significantly change therapy.[25] Either identification of the lesion as being nonresectable or the provision of information about thoracic wall invasions may significantly alter the radiation therapy planning.

For References see p. 1203

Similarly, the ability to provide coronal images of Pancoast's tumors may provide important information about resectability and therapy.[26,27]

Pleural effusions are easily identified in MRI scans. The signal intensities were originally thought to provide important information about transudative versus exudative behavior of pleural effusions. However, the strong effect of flow within the pleural effusions provides the dominant mechanism for signal changes in spin echo imaging. Free flowing pleural effusion may even appear as a signal void on spin echo images, and

can easily be identified as areas of increased signal intensity and gradient-recalled images due to their flow related enhancement.[28]

Pulmonary artery abnormalities, such as the presence of pulmonary emboli, have been identified as areas of increased signal intensity on spin echo images. MRI represents an excellent method for characterizing the presence of central pulmonary emboli. In general, chronic pulmonary emboli have intermediate signal intensity on both T_1- and T_2-weighted images. However, the presence of acute pulmonary emboli

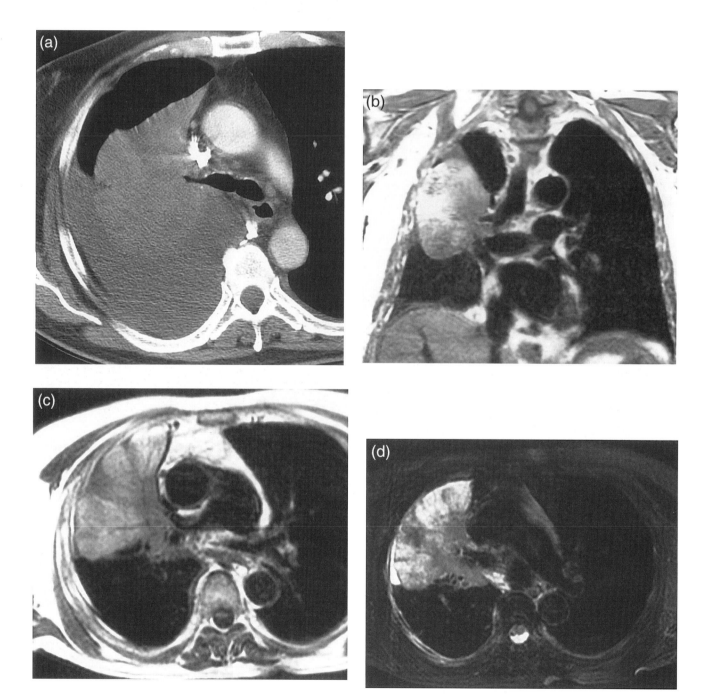

Figure 6   (a)–(d)

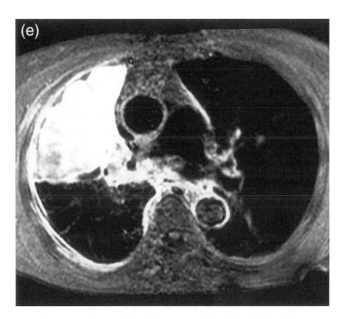

Figure 6 (a) A CT scan obtained from a patient presenting with a right-sided mass and large pleural effusion. (b) A gated coronal image through the same patient as in (a) following a thoracentesis. This image still shows a rather prominent mass, with occlusion of the right upper lobe bronchus. The question of primary mass versus atelectasis is often raised. (c) A gated transverse spin echo image through the mass. Note the relatively high signal intensity of this 'T_1-weighted image' in the periphery of this mass, the finding which is described in atelectatic lung. (d) A fat-saturated fast spin echo image showing marked heterogeneity in the mass but relatively clear delineation between the peripheral components and more central low intensity mass. (e) A T_1-weighted fat-saturated gated fast spin echo image through the same region as in (d) showing marked enhancement of the mass, as well as the adjacent pleura on this side

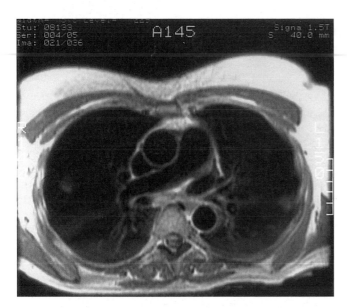

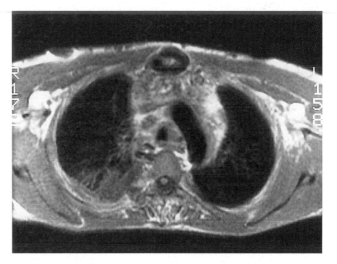

Figure 7 A gated transverse spin echo image through the midthorax shows a subcarinal mass associated with relatively subtle esophageal carcinoma. There is a fat plane between the aorta and mediastinal soft tissue, suggesting no definite invasion. Of more importance are the soft tissue densities in each lung field. These relatively hazy, ill-defined nodules are characteristic of metastatic disease on MRI. CT remains the primary method of screening for pulmonary metastases

Figure 8 A gated spin echo image through the upper mediastinum shows prominent anterior mediastinal adenopathy as well as a prominent para-aortic node. On the right-hand side there is pulmonary parenchymal consolidation and fibrotic changes secondary to prior radiation therapy. Shoddy mediastinal adenopathy is present. Of note is the absence of a significant signal void in the superior vena cava, due to vena caval occlusion

For References see p. 1203

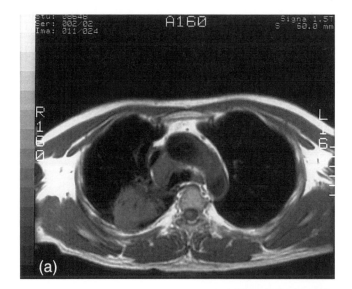

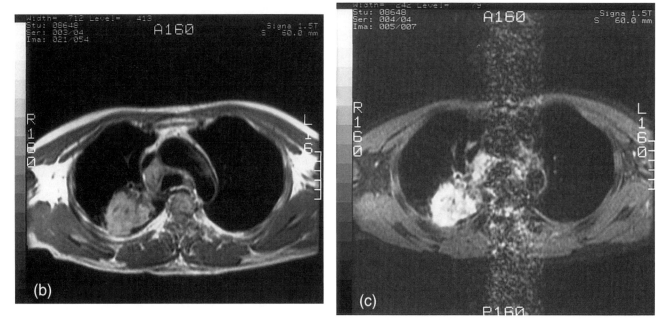

Figure 9 (a) A gated spin echo image through a bronchiogenic carcinoma of the right upper lobe. (b) A gated spin-density-weighted image through the carcinoma. (c) A STIR image obtained through the upper mediastinum. Note the excellent delineation on all images of the primary mass in relation to the pulmonary parenchyma. Paratracheal adenopathy is clearly visible on the spin echo images, but is somewhat obscured on the STIR images. The STIR image does, however, provide much better detail of the mass and its intimate association with the chest wall, suggesting chest wall invasion. In addition, the significant motion artifacts present on the STIR image are due to the relative difficulty of gating inversion–recovery images. The STIR images themselves are useful in investigating chest wall invasion but, due to the relative temporal inefficiency, may not be appropriate for all patients

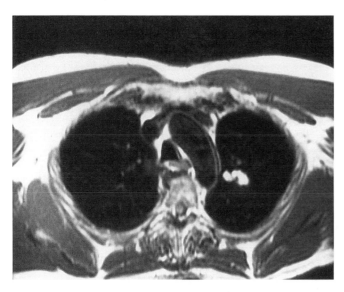

Figure 10 A gated T_1-weighted transverse image through the aortic arch, showing a very high signal intensity area on the left upper lung corresponding to a subacute pulmonary embolus

may be characterized by the presence of an extremely high intensity signal on T_2- and T_1-weighted images due to the presence of methemoglobin in the thrombi (Figure 10). The flow sensitivity of MRI has been utilized to characterize pulmonary arteriovenous malformations. The relatively intermediate signal intensity seen on T_1-weighted images may be present due to relatively slow flow. The sensitivity of gradient-recalled images or phase contrast methods has allowed the characterization of these pulmonary parenchyma abnormalities by demonstrating the flow within them (Figure 1).

In summary, the excellent contrast characteristics of MRI, associated with its sensitivity to flow, make it an excellent method for identifying and characterizing pulmonary and mediastinal pathologies. The relatively limited spatial resolution within the pulmonary parenchyma is a significant limitation; however, the superb contrast sensitivities have ensured the place of MRI in the identification and characterization of thoracic pathologies.

1 RELATED ARTICLES

Breast MRI; Heart: Clinical Applications of MRI; Whole Body Magnetic Resonance Angiography.

2 REFERENCES

1. L. Axel, R. M. Summers, H. Y. Kressel, and C. Charles, *Radiology*, 1986, **160**, 795.
2. L. Hedlund, R. Herfkens, J. Deitz, R. Blinder, R. Nassar, E. Coleman, E. Effman, and C. Putman, *Proc. IVth Ann Mtg. Soc. Magn. Reson. Med.*, London, 1985, **1**, 583.
3. C. B. Higgins, E. H. Botvinick, P. Lanzer, R. J. Herfkens, M. J. Lipton, L. E. Goorn, and L. Kaufman, *Cardiol. Clin.*, 1983, **1**, 527.
4. G. Gamus and D. Sostman, *Am. Rev. Respir. Dis.*, 1989, **139**, 254.
5. N. J. Pelc, R. J. Herfkens, A. Shimakawa, and D. R. Enzmann, *Magn. Reson. Q.*, 1991, **7**, 229.
6. P. Y. Poon, M. J. Bronskill, R. M. Henkelman, D. F. Redeout, M. S. Sholman, G. L. Weisbrod, M. I. Steinhardt, M. J. Dunlap, A. R. J. Ginsberg, and R. Feld, *Radiology*, 1987, **196**, 651.
7. S. L. Primack, J. R. Mayo, T. E. Hartman, R. R. Miller, and N. L. Müller, *J. Comput. Assist. Tomogr.*, 1994, **18**, 233.
8. N. L. Müller, J. R. Mayo, and C. V. Zwirewich, *Am. J. Roentgenol.*, 1992, **158**, 1205.
9. E. H. Moore, W. R. Webb, N. L. Müller, and R. Sollitto, *Am. J. Roentgenol.*, 1986, **146**, 1123.
10. C. J. Bergin, J. M. Pauly, and A. Macovski, *Radiology*, 1991, **179**, 777.
11. C. J. Herold, J. E. Kuhlman, and E. A. Zerhouni, *Radiology*, 1991, **178**, 715.
12. G. M. Glazer, M. B. Orringer, T. L. Chenevert, J. A. Borrello, M. W. Penner, L. E. Quint, K. C. Li, and A. M. Arsers, *Radiology*, 1988, **168**, 429.
13. G. M. Glazer, *Chest*, 1989, **96** (Suppl. 1), 44S.
14. W. R. Webb, M. Sarin, E. A. Zerhouni, R. T. Heelan, G. M. Glazer, and C. Gatsonis, *J. Comput. Assist. Tomogr.*, 1993, **17**, 841.
15. W. R. Webb, C. Gatsonis, E. A. Zerhouni, et al., *Radiology*, 1991, **178**, 705.
16. W. B. Gefter, *Semin. Roentgenol.*, 1990, **25**, 73.
17. O. Musset, P. Grenier, M. F. Carette, G. Frija, M. P. Hauuy, H. T. Desbleds, P. Giraird, J. M. Bigot, and D. Lallemand, *Radiology*, 1986, **160**, 607.
18. G. M. Glazer, B. H. Gross, A. M. Aisen, L. E. Quint, I. R. Francis, and M. B. Orringer, *Am. J. Roentgenol.*, 1985, **145**, 245.
19. H. S. Glazer, R. G. Levitt, J. K. T. Lee, S. Emami, S. Gronemeyer, and W. A. Murphy, *Am. J. Roentgenol.*, 1984, **143**, 729.
20. R. S. Nyman, S. M. Rehn, B. L. Glinelius, M. E. Hagberg, A. L. Memmingtson, and C. J. Sundstrom, *Radiology*, 1989, **170**, 435.
21. J. A. Verschakelen, P. Demaerel, J. Coolen, M. Demeols, G. Marshal, and A. L. Baert, *Am. J. Roentgenol.*, 1989, **152**, 965.
22. C. J. Herold, J. E. Kuhlman, and E. A. Zerhouni, *Radiology*, 1991, **178**, 715.
23. P. M. Bourgouin, T. C. McLoud, J. F. Fitzgibbon, E. J. Mark, J. A. Shepard, E. M. Moore, E. Rummeny, and T. J. Brady, *J. Thorac. Imag.*, 1991, **6**, 22.
24. J. Tobler, R. C. Levitt, H. S. Glazer, J. Moran, E. Crouch, and R. G. Evens, *Invest. Radiol.*, 1987, **22**, 538.
25. R. T. Heelan, B. E. Demas, J. F. Caravelli, N. Martini, M. S. Bains, P. M. McCormack, M. Burt, D. M. Panicek, and A. Mitzner, *Radiology*, 1989, **170**, 637.
26. D. R. Pennes, G. M. Glazer, K. J. Wimbish, B. H. Gross, R. W. Long, and M. B. Orringer, *Am. J. Roentgenol.*, 1985, **144**, 507.
27. A. M. Haggar, J. L. Pearlberg, J. W. Froelich, D. O. Hearshen, G. H. Beure, J. W. Lewis, Jr., G. W. Schkuder, C. Wood, and P. Gniewck, *Am. J. Roentgenol.*, 1987, **148**, 1075.
28. P. Vock, L. W. Hedlund, R. J. Herfkens, E. L. Effmann, M. A. Brown, and C. E. Putman, *Invest. Radiol.*, 1987, **22**, 382.

Biographical Sketch

Robert J. Herfkens. *b* 1949. M.D., 1974, Loyola University, Stritch School of Medicine. Associate Professor and Director of MRI, Stanford University School of Medicine, 1989. Principal investigator on research grants sponsored by Government and private industry. Approx. 100 scientific papers, presented at over 100 national meetings; several textbooks. Main research interest: the development of applications and new techniques directed towards body imaging, most specifically those related to cardiovascular diseases.

For References see p. 1203

Breast MRI

Sylvia H. Heywang-Köbrunner
University of Halle, Germany

Hans Oellinger
Universitäts-Klinikum Rudolf-Virchow, Freie Universität Berlin, Germany

1 INTRODUCTION

On the basis of the high soft tissue contrast of MRI and based on suspected differences of T_1 and T_2 values between benign and malignant breast tissues, MRI was initially considered to be a particularly promising tool for breast imaging. Unfortunately, the expected tissue characterization by MRI could not be demonstrated despite intense further research; signal intensities on various pulse sequences, and in vivo calculated T_1 and T_2 values, varied significantly in both benign and malignant tissues. Both MR parameters and signal intensities seemed to correlate well with the water/content, cells or fibrosis within the corresponding tissues, but did not allow a reliable distinction between benign and malignant tissues. This has proven true until today, even after the application of more sophisticated evaluation such as computer-aided evaluation methods.[1-5] For this reason, plain MRI (= MRI without contrast agent) does not play a role in the detection and diagnosis of breast malignancy nowadays. The disappointing results mentioned above, however, initiated work concerning the use of contrast agents for MR of the breast and the very encouraging results caused its further exploration.[6-8] Today contrast-enhanced MRI of the breast is developing as an important additional tool for breast diagnostics, providing new information which is different from that of conventional imaging and is thus most valuable in certain problem areas of conventional breast diagnostics.[9-28]

In one special area, however, both tomographic imaging and the high soft tissue contrast of plain MRI has proved helpful; this special area concerns the evaluation of the integrity of silicone implants.[29-33] This chapter will provide an overview of these two major applications of breast MRI; plain MRI for the evaluation of implant integrity and contrast-enhanced (CE) MRI of the breast.

2 PLAIN MRI — DETECTION AND DIAGNOSIS OF IMPLANT FAILURE

Based on the latest knowledge concerning potential hazards caused by leaking or ruptured implants and free silicone, the detection of possible implant failure has become an important new demand. The capabilities of conventional methods such as mammography and ultrasound for detecting extra- and particularly intracapsular leakage are very limited. Hence the use of MRI in this field has been evaluated.

At present, various combinations of pulse sequences are still being tested. These use differences of T_1, of T_2, and of the reasonance frequency of silicon compared with the surrounding fat, fibrosis, or glandular tissue.[29-33] In general, a combination of at least three pulse sequences is recommended. These should, for example, include a T_1-weighted pulse sequence with and without fat suppression, and one with silicone or water suppression as well as a fast high-resolution T_2-weighted pulse sequence. The suppression of fat, water, or silicone can be obtained by spectrally selective presaturation pulses or by signal nulling with IR sequences. In order to cut potential ruptures orthogonally, these pulse sequences are usually applied in different planes (coronal, transverse, and/or sagittal). For the detection of discrete ruptures, at least one of the pulse sequences should allow slice thicknesses below 2–3 mm. For the other pulse sequences, a slice thickness below 5 mm appears optimal.

The implant itself is generally covered by a fine shell, which in intact implants can usually not be recognized since it is directly attached to the capsule of fibrous tissue which surrounds the implant. When the shell of the implant is ruptured, silicone can leak. Depending on whether the leakage of silicone is confined to the area surrounded by the fibrous capsule, or whether the leak is beyond the fibrous capsule (about 5% of the ruptures), the rupture is called intra– or extracapsular.

The following signs indicating a rupture have so far been described (Figure 1):

1. Linguine sign: The shell is ruptured, silicone remains predominantly within the fibrous capsule, and the collapsed shell floats within the silicone. The tomographic slices show wavy dark lines within the silicone, which represent the collapsed implant shell floating within the silicone. The wavy lines are called 'linguine'.

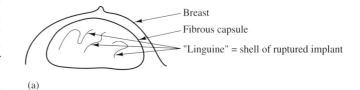

(a)

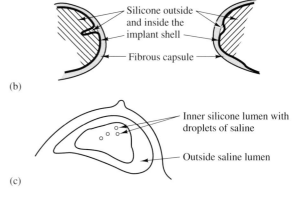

(b)

(c)

Figure 1 Morphological MR signs of implant rupture: (a) 'linguine sign'; (b) 'keyhole sign' or 'reverse c-sign' (right), and (c) 'droplet sign' in double lumen implant

For list of General Abbreviations see end-papers

2. When silicone enters between the shell and the fibrous capsule, silicone becomes visible outside the shell. Depending on the shape of the shell in this area, this sign is called 'c-sign' or 'keyhole sign'.

3. In double lumen implants, saline may enter the silicone lumen of the implant. Since saline and silicone gel do not mix, small droplets become visible; this sign is called the 'droplet sign' or the 'salad oil' sign.

4. Finally small droplets of silicone can be visualized outside the capsule. This sign indicates extracapsular rupture.

To date, with careful examination techniques using several pulse sequences and thin slices, very good sensitivities (above 80%) and specificities (above 90%) have been reported.[29–33]

3 CONTRAST-ENHANCED MRI OF THE BREAST

Based on the unsatisfying results of plain MRI mentioned above, CE MRI of the female breast has been performed and investigated by us since 1985.[6–8,16,17,21,22,28] Other workers have followed and contributed significantly.[9–15,18–20,23–27] Today, contrast-enhanced MRI is gaining increasing importance as an additional tool for breast diagnostics.

3.1 Principles of Contrast-Enhanced MRI of the Breast

The complete breast is examined with thin and contiguous slices before and after the i.v. application of Gd–DTPA. By comparison of the corresponding pre- and postcontrast slices, tissue enhancement can be detected or excluded.

The fact that the large majority, possibly all, invasive breast malignancies do enhance is supported by the following fundamental data:[34–40] (i) an increased number and size of vessels has also been encountered histologically in and around breast malignancies, as studies with special vascular staining show; (ii) according to in vivo animal experiments, vessel sprouting seems to be a prerequisite for tumor growth beyond a size of about 2–3 mm; (iii) tumor growth seems to depend on both increased vascularity and increased vascular permeability; (iv) both increased vascularity and vascular permeability are induced by socalled angiogenesis and permeability factors, numbers of which have already been identified; (v) cells from both invasive and numerous in situ carcinomas were able to induce vascular sprouting when transplanted on an animal cornea, while cells from normal breast tissue could not induce such changes; (vi) it has also been reported that cells from several benign entities were equally well able to produce identical or similar angiogenesis and permeability factors leading to similar vascular changes as in malignancy. Benign changes with similar capabilities included some proliferative dysplasias, healing wounds, and benign tumors.

The above-mentioned observations support may of the MR findings. They may also serve to explain both the reported excellent sensitivity and the limited specificity of CE MRI.

3.2 Techniques

For prognostic reasons, it is generally agreed that breast imaging techniques should be able to image carcinomas of 5 mm and above reliably. Furthermore, it is also desirable to detect in situ carcinomas which usually grow within the ducts where they may extend within areas of very variable size. The ducts themselves, however, rarely exceed 1–3 mm in diameter and may be surrounded by uninvolved nonenhancing tissue. Hence with a tomographic imaging modality such as MRI it is important to minimize slice thickness so as to improve resolution and contrast.

In order to achieve sufficient signal-to-noise ratio, a dedicated single or double breast coil has to be used. Cardiac motion artifacts also need to be eliminated. At present, the best way to achieve the latter is by switching phase and frequency encoding gradients, so that the artifacts cross the axillas instead of the breast. An alternative is the use of coronal slices, where most of the artifacts neither cross the breast nor the axillas. (The use of local presaturation is not useful, since generally larger areas are extinguished than would be disturbed by artifacts. In our experience flow compensation increases echo time without sufficient effect on artifact elimination.)

In order to obtain sufficiently small slices without gaps, 3D imaging is strongly recommended. In this way it is possible, for example, to image the breast with 32 transverse slices of 4 mm slice thickness within slightly more than 1 min. By using a rectangular field of view and coronal slices, either the imaging time can be further reduced by a factor of two, or 64 slices of 2-mm slice thickness are possible within the same imaging time. Thus with present 3D techniques both high temporal resolution or high spatial resolution are possible.

However, when choosing between high spatial and high temporal resolution it should be remembered that with a 5-mm slice thickness in the worst case a 5-mm lesion may only be half contained within two neighboring slices. This means that only half of the signal increase can be measured in each neighboring slice. Hence 2-mm or thinner slices will probably allow better results, particularly with regard to the detection of intraductal enhancement. The optimum temporal resolution is still controversial.[9–17]

As far as the choice of pulse sequences is concerned, the highest sensitivity to Gd–DTPA and hence the best contrast of enhancing lesions can be achieved with 3D gradient echo sequences such as FLASH or spoiled GRASS or with some fat saturation techniques like RODEO or spoiled GRASS with fat saturation.[23–25] Due to their low sensitivity to Gd–DTPA, SE sequences can no longer be recommended.

Since enhancement might be confused with a fat lobule on the postcontrast image alone, elimination of the fat signal has been suggested as a means of improving the detectability of enhancing lesions. Elimination of the fat signal can be achieved in two different ways.

1. On T_1-weighted fat suppression sequences, ideally only enhancing tissues display high signal intensity while dysplasia and suppressed fat exhibit low or no signal intensity. The disadvantage of fat suppression techniques is that they are more sensitive to magnetic field inhomogeneities. Within such inhomogeneities, which may be caused by the patient or the breast coil, enhancement might be underestimated or missed. Thus, greater technical prerequisites need to be fulfilled, i.e. more experience and very careful quality control are necessary.

2. Since fat does not enhance, like all other nonenhancing tissues it is ideally eliminated on subtraction images. Software has now been developed which allows routine subtraction of

For References see p. 1209

all corresponding pre- and postcontrast slices (2×64 slices) within about 1 min. The major disadvantage which has been ascribed to this technique relates to problems which occur when the patient moves. In our experience, this problem occurs with less than 5–10% of all patients with currently available very short imaging times. Motion and its inherent problems are at least readily recognized with this technique

Three-dimensional MIP reconstruction is possible for both fat suppressed and subtracted images. It has been suggested as a means of improving the distinction of enhancing vessels from enhancing ducts.[27]

As far as optimum dosage is concerned, to date only one dose comparison study has been published.[28] In this study, we demonstrated that significantly better results were obtained with FLASH-3D using the higher dosage of 0.16 mmol of Gd–DTPA per kg as compared with the widely used lower dosage of 0.1 mmol of Gd–DTPA per kg. In fact, three out of 54 carcinomatous foci would have been overlooked with the lower dosage. Histologically, these three foci were all smaller than the applied slice thickness of 4 mm.

3.3   Results

The results of CE MRI of the breast are based worldwide on the examination of over 5000 patients.[6–28] Despite still remaining differences in technique and image interpretation, the following results have been confirmed: (a) the great majority of invasive carcinomas are enhanced (see Figure 2); (b) no significant enhancement is usually seen in normal breast tissue, in nonproliferative dysplasia, sclerosed benign tumors, old scarring, and cysts; and (c) variable enhancement, which might cause false positive calls, has been encountered in some proliferative dysplasias, in nonsclerosed benign tumors, and in inflammatory changes.

In over 700 biopsy-proven tissues which we have examined with FLASH-3D, about 75% of the dysplasias enhanced only a little and it was possible to define a threshold so that all carcinomas enhanced more than by this threshold.[16,17]

We also evaluated the shape of enhancement and roughly estimated the speed of enhancement in these 700 tissues. In a similar manner to mammography, we found that an irregular outline of an enhancing lesion was indeed highly suggestive

of malignancy, but neither diffuse nor well-circumscribed enhancement could reliably exclude malignancy (for example, 16% of the diffusely enhancing tissues and 21% of our postoperatively examined well circumscribed lesions, in fact, turned out to be malignant). When the speed of enhancement was estimated from two postcontrast studies (first measurement: 1–5 min post injection; second measurement 6–11 min post injection), fast enhancement was highly suggestive of malignancy but unfortunately delayed enhancement could not rule out malignancy reliably. This result has been confirmed by another separate dynamic study in 79 patients, where selected slices were imaged every 30 s after i.v. injection of Gd–DTPA.[7,16,17]

As to the potential of dynamic contrast studies, controversial results are still being discussed. While Kaiser[9,10] reports a sensitivity and specificity of 97% with dynamic contrast-enhanced MRI, other authors have not been able to reproduce these results, reporting both fast enhancing benign tissues and (what is much more important) slowly enhancing malignancies.[12–17] In our experience the use of a slow enhancement speed as a criterion of benignity is comparable to increasing the threshold above which enhancement is considered to be possibly malignant, i.e., specificity can be improved at the cost of sensitivity.

Since the development of very fast pulse sequences such as Turbo-FLASH or echo planar imaging, it has become possible to image the first pass of a contrast agent within a tissue. Initial results with Turbo-FLASH have been promising.[26] The optimum use of this additional information needs to be carefully determined, in view of its effect on both sensitivity and specificity. When the results of various authors are compared[6–28] (Table 1), excellent sensitivity values overall have been reported for CE MRI. Indeed, the sensitivity exceeds that of conventional imaging. Differences can be explained by different techniques (sometimes thick slices with gaps, the choice of pulse sequence, the dosage of Gd–DTPA), by different guidelines for interpretation, and by different patient selection. On comparison with conventional imaging alone, the sensitivity was however, improved in general. This diagnostic gain was, however, most prominent within mammographically dense and/or distorted tissue, while in fatty breasts and in cases with microcalcifications little or no diagnostic gain was achieved.

Table 1   Sensitivities and Specificities of Enhanced MRI, as Reported in the Literature

Author	Year	Sensitivity of MRI (%)	Specificity of MRI (%)	Sensitivity of Conv.[a] (%)	Specificity of Conv.[a] (%)
Allgayer (1)[c]	1991	90	40	ni	ni
Fischer (2)	1992	95	89	ni	ni
Gilles (3)	1993	93.3	63.1	ni	ni
Harms (3)	1993	94	37	ni	ni
Heywang-Köbrunner (1)	1993	98.6[b]	65.0	60	45
Heywang-Köbrunner (3)	1993	99.5[b]	28.0	82	18
Kaiser (1)	1992	97[b]	97	ni	ni
Lewis-Jones (1)	1991	100	91	100	60
Oellinger (4)	1993	88	80	83	75

[a]Conv. = Sensitivity/Specificity of conventional imaging.
[b]Sensitivity/Specificity based on MR and conventional imaging.
[c](1), problem cases; (2) retrospective study; (3) preoperative patients; (4) evaluation of multicentricity.
ni = not indicated.

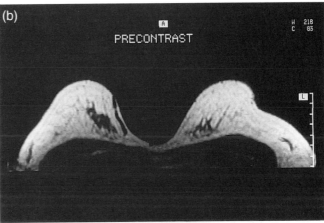

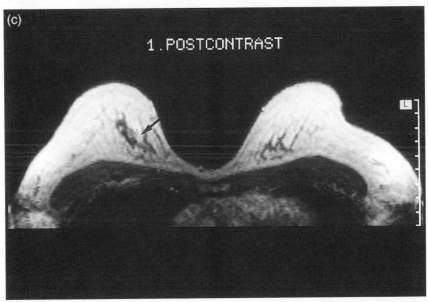

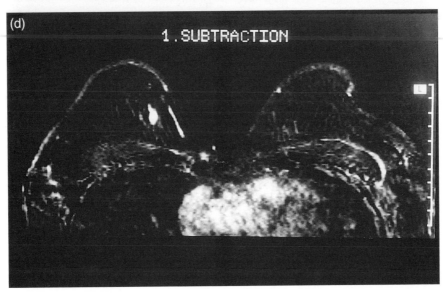

Figure 2   Mammography versus MRI in cases of very dense breasts. (a) Mammography of a 43 year old patient, cranio-caudal view. Very dense right breast, no radiological evidence of malignancy. (b) MR image before injection of the contrast agent; anatomical region of the upper quadrants. Basically fatty tissue and small areas of parenchyma in the upper inner quadrants of both sides. (c) MR image after application of the contrast agent gadolinium–DTPA with moderately strong signal intensity rise in the parenchyma of the inner quadrant of the patient's right breast (see arrow). Same position as in the precontrast MR image shown in (b). (d) The use of the subtraction method (postcontrast minus precontrast) allows the postcontrast enhancement of carcinomas to be seen much better. This method is especially useful in the case of small carcinomas. The histology revealed an invasive ductal carcinoma

For References see p. 1209

With respect to microcalcifications and in situ carcinomas, the following may be stated. Mostly due to patient preselection, experience concerning the capabilities of MRI in this field is still limited to date. Some authors reported misses. Although our in situ carcinomas did enhance, some did so diffusely and many slowly. Thus, detection of in situ carcinomas and their differentiation from proliferative dysplasias is certainly also a question of threshold and interpretation. Those 50–60% of in situ carcinomas which contain microcalcifications are already excellently visualized by mammography. Even though improvement of the mammographic specificity in this field is still desirable, we do not recommend MRI in this case for the following reasons:

1. It is not known as yet whether all in situ carcinomas are sufficiently vascularized to be detectable by contrast-enhanced MRI.
2. Further research must show whether in situ carcinomas can be included in the enhancement thresholds of MRI without risking too many false positive calls. In our own experience, numerous false positive calls occurred particularly in cases with mammographic microcalcifications.
3. Due to the additional possibility of partial volume effects, excellent thin slice techniques will be necessary together with extremely careful application of any threshold.

Thus, we do not recommend reliance on MRI in this field at present and we do not recommend application of MRI as a means of answering these questions. Instead diagnostic decisions should be based on careful analysis of the mammogram and possibly other diagnostic tests such as transcutaneous biopsy.

As far as noncalcified in situ carcinomas are concerned, it is important to mention that seven out of our 21 in situ carcinomas were detected by MRI alone. The in situ carcinomas detected by MRI frequently became apparent as a focal area of enhancement and sometimes enhancement with a ductlike shape.

While the sensitivity results of CE MRI have been very satisfying, improvement of specificity has been a major concern (Table 1). Overall the following possibilities are being discussed:

(i) All available information including that of shape and speed of enhancement should be used for the final diagnosis. Because of the existence of both diffusely and slowly enhancing carcinomas, it is necessary to warn against overinterpretation of shape or speed of enhancement. Optimization of interpretation guidelines in this field (using optimum techniques) will therefore be an important task for the near future.

(ii) In addition, all available information including that of mammography and clinical examination should be used for the final diagnosis. MRI is not a stand-alone method!

(iii) In order to avoid unnecessary biopsy of MR-detected small benign enhancing lesions, MR follow-up of benign-appearing lesions (predominantly well-circumscribed round or oval lesions, which are likely to be small fibroadenomas) is recommended. MR-guided stereotaxic transcutaneaous biopsy should be considered for indeterminate benign-appearing lesions. First experiences with such devices are presently being acquired.

(iv) Finally unnecessary costs should be avoided and optimum diagnostic gain attempted by using MRI exclusively for those indications where definite advantages compared with conventional imaging can be expected.

On the basis of our own experiences over 2000 contrast-enhanced MR studies of the breast, we do not recommend contrast-enhanced MRI for the following indications: (a) evaluation of microcalcifications (see above), and (b) evaluation of those cases where diffuse enhancement is to be expected in any event, since in these cases malignancy can never be excluded in principle. Such cases include clinically obvious inflammatory changes, patients with secretory disease, and patients with known enhancing dysplasia from previous studies.

As an additional method however, CE MRI has been most helpful in diagnostic problems within mammographically dense or distorted tissue. Most interesting indications in these breasts include the search for primary tumors (not detected by conventional imaging) and the exclusion of multicentricity or contralateral involvement in patients with small tumors and very dense tissue (see Figure 2).[15–17,24] Contrast-enhanced MRI is particularly useful for patients with severe scarring after surgery, after radiation therapy, and after silicone implant.[16,20–22] It must, however, be added that MRI should not be used within the first 3–6 months after surgery or within the first 12 months after radiotherapy (RT), since post-therapeutic diffuse enhancement usually impairs evaluation during that period.

3.4 Conclusions and Outlook

The use of a contrast agent is obligatory for MR detection and diagnosis of malignancy. Contrast-enhanced MRI is gaining increasing importance as an additional tool for breast diagnostics. However, because of the increasing interest in the method, optimization and standardization of interpretation is an urgent task for the near future.

In order to develop optimized criteria for interpretation, an international multicenter study is presently being started. The study is supported by Siemens Corporation, Schering, and Berlex Corporations, and includes only centers with a high reputation in both breast imaging and MRI based on histologically proven data only. Both the optimization of interpretation guidelines, and in a second prospective trial the determination of wellfounded and generally recognized accuracy data, are the major goals of this study. These data, as well as further research by other groups, will help to define the optimum role which MRI should play in integrated diagnostic work-up.

MRI is thus gaining its place in breast diagnostics. It must be our concern to use MRI optimally only for those indications where definite diagnostic gain can be achieved for the patient without risking false negative calls and without causing an unacceptable number of false positive calls, which might cause unnecessary concern, costs for additional clarification, or even surgery. Further development and improvements of the technique are still proceeding.

4 RELATED ARTICLES

Cardiac Gating Practice; Echo-Planar Imaging; Gadolinium Chelate Contrast Agents in MRI: Clinical Applications; Gadolinium Chelates: Chemistry, Safety, and Behavior; Whole Body Magnetic Resonance: Fast Low-Angle Acquisition Methods.

5 REFERENCES

1. W. A. Murphy and J. K. Gohagan, 'Magnetic Resonance Imaging', ed. D. D. Stark and W. G. Bradley, Jr., C. V. Mosby, St. Louis, MO, 1987, pp. 861–886.
2. J. I. Wiener, A. C. Chako, C. W. Merten, S. Gross, E. L. Coffey, and H. L. Stein, *Radiology*, 1986, **160**, 299.
3. F. S. Alcorn, D. A. Turner, J. W. Clark, et al., *Radiographics*, 1985, **5**, 631.
4. S. H. Heywang, R. Bassermann, G. Fenzl, W. Nathrath, D. Hahn, R. Beck, I. Krischke, and W. Eiermann, *Eur. J. Radiol.*, 1987, **3**, 175.
5. J. K. Gohagan, A. E. Tome, E. Spitznagel, et al., *Radiology*, 1994.
6. S. H. Heywang, D. Hahn, H. Schmid, I. Krischke, W. Eiermann, R. Bassermann, and J. Lissner, *J. Comput. Assist. Tomogr.*, 1986, **10**, 199.
7. S. H. Heywang, T. Hilbertz, E. Pruss, A. Wolf, W. Permanetter, W. Eiermann, and J. Lissner, *Digitale Bilddiagn.*, 1988, 7.
8. S. H. Heywang, A. Wolf, E. Pruss, T. Hilbertz, W. Eiermann, and W. Permanetter, *Radiology*, 1989, **171**, 95.
9. W. A. Kaiser and E. Zeitler, *Radiology*, 1989, **170**, 681.
10. W. A. Kaiser, *Diagn. Imaging Int.*, 1992, 44.
11. J. P. Stack, A. M. Redmond, M. B. Codd, P. A. Derran, and J. T. Ennis, *Radiology*, 1990, **174**, 491.
12. F. W. Flickinger, J. D. Allison, R. Sherry, and J. C. Wright, *Proc. XIth Ann Mtg. Soc. Magn. Reson. Med.*, Berlin, 1992, p. 955.
13. M. D. Schnall, S. Orel, and L. Muenz, *Proc. XIth Ann Mtg. Soc. Magn. Reson. Med.*, Berlin, 1992, p. 120.
14. U. Fischer, D. von Heyden, R. Vosshenrich, I. Vieweg, and E. Grabbe, *RoeFo*, 1993, **158**, 287.
15. H. Oellinger, S. Heins, B. Sander, et al., *Eur. Radiol.*, 1993, **3**, 223.
16. S. H. Heywang-Köbrunner, 'Contrast-enhanced MRI of the Breast', Karger, Basel/Munchen, 1990.
17. S. H. Heywang-Köbrunner, *Electromedica*, 1991, **61**, 43.
18. B. Allgayer, P. Lukas, W. Loos, and K. Mühlbauer, *Roentgenpraxis*, 1991, **44**, 368.
19. D. Rubens, S. Totterman, A. K. Chacko, K. Kothari, W. Logan-Young, J. Szumowski, J. H. Simon, and E. Zacharich, *AJR*, 1991, **157**, 267.
20. H. G. Lewis-Jones, G. H. Whitehouse, and S. J. Leinster, *Clin. Radiol.*, 1991, **43**, 197.
21. S. H. Heywang-Köbrunner, A. Schlegel, R. Beck, T. Wendt, W. Kellner, B. Lammetzsch, M. Untch, and W. B. Nathrath, *J. Comput. Assist. Tomogr.*, 1993, **17**, 891.
22. S. H. Heywang-Köbrunner, T. Hilbertz, R. Beck et al., *J. Comput. Assist. Tomogr.*, 1990, **14**, 348.
23. W. B. Pierce, S. E. Harms, D. P. Flamig, R. H. Giffey, W. P. Evans, and J. G. Hugins, *Radiology*, 1991, **181**, 757.
24. S. E. Harms, D. P. Flamig, K. L. Hesley, M. D. Meiches, R. A. Jensen, W. P. Evans, D. A. Savina, and R. V. Wells, *Radiology*, 1993, **187**, 493.
25. W. Whitney, R. J. Herfkens, J. Silverman, D. Ikeda, J. Brumbaugh, S. Jeffrey, L. Esserman, J. Frederickson, C. Meyer, and G. Glover, *Proc. XIIth Ann Mtg. Soc. Magn. Reson. Med.*, New York, 1993, p. 856.
26. C. Boetes, R. D. Mus, J. O. Barentsz, J. H. Hendriks, R. Holland, and J. H. Ruijs, *Radiology*, 1993, **189**, 301.
27. H. Oellinger, B. Sander, U. Quednau, J. Hadijuana, H. Schoenegg, and F. Felix, *JMRI*, 1993, **3**, 109.
28. S. H. Heywang-Köbrunner, J. Haustein, C. Pohl, W. M. Bauer, W. Eirmann, W. Permanetter, *Radiology*, 1994, **191**, 639.
29. R. F. Brem, C. M. C. Tempany, and E. A. Zerhouni, *J. Comput. Assist. Tomogr.*, 1992 **16**, 157.
30. D. P. Gorczyca, S. Sinha, and C. Y. Ahn, *Radiology*, 1992, **185**, 407.
31. L. I. Everson, H. Parantainen, A. E. Stillman, T. Dethi, M. C. Toshager, and B. Cunningham, *Radiology*, 1993, **189**, 301.
32. W. A. Berg, N. D. Anderson, B. W. Chang, E. A. Zerhouni, and T. E. Kuhlman, *Proc. XIIth Ann Mtg. Soc. Magn. Reson. Med.*, New York, 1993, p. 123.
33. S. Mukundan, W. T. Dixon, R. C. Nelson, D. C. Monticciolo, and J. Bostwick, *Radiology*, 1993, **189**, 301.
34. A. Ottinetti and A. Sapino, *Breast Cancer Res. Treat.*, 1988, **11**, 241.
35. C. H. Blood and B. R. Zetter, *Biochim. Biophys. Acta*, 1990, **1032**, 89,
36. H. M. Jensen, J. Chen, M. R. De Vault, and A. E. Lewis, *Science*, 1982, **218**, 293.
37. N. Weidner, J. P. Semple, W. R. Welch, and J. Folkman, *N. Engl. J. Med.*, 1991, **324**, 1.
38. M. A. Gimbrone, S. B. Leapman, R. S. Cotran, and J. Folkman, *J. Exp. Med.*, 1972, **136**, 261.
39. I. F. Tannock, *Br. Cancer Res.*, 1968, **22**, 258.
40. D. T. Connolly, D. M. Heuvelman, R. Nelson, J. V. Olander, B. L. Eppley, J. J. Delfino, N. R. Siegel, R. M. Leimgruber, and J. Feder, *J. Clin. Invest.*, 1989, **84**, 1470.

Biographical Sketches

S. H. Heywang-Köbrunner. b 1956. Dr.med., 1982 University of Munich. Fellowship in breast imaging in the US (Georgetown University, Emory University, Betty Ford Breast Diagnosis Center, Washington DC) 1982–83. Radiology residency with special training in CT and MRI, University of Munich (Prof. Dr. J. Lissner); staff member and assistant professor, University of Munich, 1990–92, assistant professor, University of Leipzig, 1993, associate professor, University of Halle, 1994–present. Approx. 150 publications. Research specialties: since 1985, investigation of CE MRI of the breast clinical MRI, breast imaging.

H. Oellinger. b 1948. Dr.med., 1989, Free University of Berlin. M.S.M.E. New Jersey Institute of Technology, Fulbright Fellow, 1974–76; site engineer for conventional and nuclear power plants (Siemens, Erlangen, Germany) 1976–82; Radiology Department (Strahlenklinik und Poliklinik, FUB, Prof. Dr. Dr h.c.. R. Felix) with special training in CT and MRI, 1989–present. Research interests: clinical MRI, breast and pelvic imaging.

Liver, Pancreas, Spleen, and Kidney MRI

David Stark and Ashley Davidoff
University of Massachusetts, MA, USA

1 INTRODUCTION

The upper abdomen presents unique diagnostic challenges. Numerous organs with diverse physiological functions are in

For References see p. 1216

close proximity with each other, abut the cardiopulmonary system through a thin diaphragm, and have open contact with the dependent organs of the pelvis. Thus, tumors, inflammatory disease, and other pathology of the upper abdomen may present clinically with symptoms attributable to an organ involved only secondarily, or they may appear to arise from the thorax or pelvis, leading the clinician away from the primary problem. Imaging of the abdomen is confounded by its large anatomic area, the mobility of the organs and their fluids, and the presence of tissues of varying imaging characteristics.

This article briefly introduces the advantages and limitations of magnetic resonance imaging (MRI) methods as applied to the evaluation of abdominal cancer. After the technical introduction, which provides an understanding of the philosophy of probing the abdomen for known or unsuspected disease, the remainder of the discussion reviews the MRI characteristics of the major upper abdominal neoplasms.

Virtually all diagnostic medicine, and certainly diagnostic imaging, is guided by three hierarchical objectives:

1. detection of disease;
2. characterization of disease; and
3. staging of disease.

Detection itself is the most important challenge, as it begins with the patients themselves deciding whether a symptom is sufficiently abnormal to warrant a visit to the doctor. After several clinical steps, imaging is most commonly used to determine whether an anatomic area (head, neck, chest, abdomen, pelvis, or musculoskeletal system) or physiologic organ system (e.g. hepatobiliary system) is normal or abnormal. This simple, binary decision is critical in selecting patients for more intensive investigation at considerable expense and no small risk of iatrogenic complications.

Characterization of disease commences once an abnormality is found. In the abdomen, nearly all patients develop benign masses if they live long enough (e.g. renal cyst or nodular hyperplasia of the prostate). Where imaging for other reasons results in detection of these benign masses, imaging must also solve the diagnostic dilemma it created and, at low cost, identify these lesions as benign and of no clinical significance.

Staging is essential to determining prognosis and therapy once treatable disease has been characterized. In many ways, staging is a subset of detection as it is the process by which multiple lesions are discriminated from a single abnormality, and the extent of disease within an organ or spread to adjacent areas is measured. Since staging often involves monitoring of treatment, it presents unique problems because the background normal structures may be altered by surgery, radiation, or progression of the disease itself. As society raises ethical and economic questions concerning the merit of treating various conditions, imaging becomes more important as a noninvasive low-cost method of determining which treatments are appropriate and effective.

2 IMAGING TECHNIQUES

MRI is similar to computerized tomography (CT) in providing a full field of view of the abdomen, and a permanent record of the entire examination, to a comparable level of anatomic detail (spatial resolution). MRI has the further advantage of allowing selection of sagittal, coronal, or oblique planes of

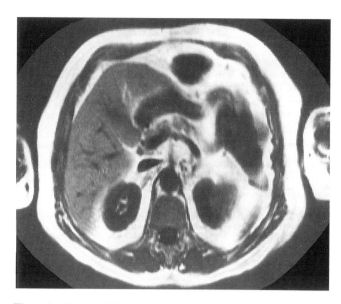

Figure 1 Normal abdomen. T_1-weighted spin echo image at the level of the pancreatic head shows a normal common bile duct, gall bladder, hepatic artery, and confluence of the splenic and portal veins. The absence of a motion artifact and the low signal intensity of kidneys relative to that of the liver are clues that this is a short *TE* sequence

section. However, in the abdomen virtually all imaging is performed in the transverse plane (see Figure 1).

MRI in the 1990s, like CT in the 1970s, is getting faster, with the expectation that speed will solve the problem of motion. Debate rages concerning the optimal methods for obtaining T_1 or T_2 contrast and the best signal-to-noise ratio (S/N) per unit time. These technical details are addressed elsewhere in this volume.

3 IMAGE QUALITY

Specific magnetic resonance (MR) pulse sequences and their user-selectable timing parameters are of utmost importance for lesion–liver contrast and image quality. MR image quality can be quantified by calculating the lesion–liver contrast-to-noise ratio (CNR) and the liver S/N.[1] Motion artifacts contribute most to (systematic) noise in abdominal imaging. Vascular ghost artifacts arising from the aorta and inferior vena cava may obscure focal hepatic abnormalities, especially in the left lobe. Changing the phase-encoding direction results in artifacts projected outside the region of interest.[2,3] Signal averaging or implementation of presaturation pulses are highly effective methods of motion artifact reduction in the upper abdomen.

The CNR is an effective parameter for quantitating pulse sequence performance with respect to lesion detection. A useful rule is that spleen–liver CNR will, on average, match cancer–liver CNR. This rule holds for all pulse sequences, on all machines, and at all field strengths, since, on average, cancer and spleen have the same proton density, T_1 and T_2.

Thus, the spleen can serve as a reference signal intensity, and the spleen–liver CNR will be predictive of pulse sequence performance for liver cancer detection.

4 PULSE SEQUENCE PERFORMANCE

Theoretical calculations suggest that the optimal pulse sequence for hepatic tumor detection varies with field strength. However, clinical results have not confirmed any significant field strength differences. Performance variations due to gradients, other hardware, and software appear to be more substantial. The dominant factor appears to be the level of commitment of the practicing radiologist who implements and maintains quality imaging protocols tailored to the specific machine and its generation (upgrade level). In an area such as the abdomen, of which relatively few MRI examinations are conducted, the necessary expertise and attention to detail is too often lacking.

Properly performed T_1- or T_2-weighted sequences yield comparable CNR values and hence efficacy at liver cancer detection. Cancer or lymphoma of the spleen is poorly detected by any method, since cancer and spleen have the same proton density, T_1, and T_2, on average. T_1-weighted sequences generally offer greater S/N per unit time than do T_2-weighted sequences. This fact, combined with excellent fat–tissue T_1 contrast, leads to the choice of short TE T_1-weighted sequences for imaging the pancreas, kidneys, adrenal glands, and the remainder of the retroperitoneum and mesentery.

While T_1-weighted images offer superior anatomic resolution and are generally superior for detection of abdominal pathology, it is well established that T_2-weighted sequences are preferred for tissue characterization, especially in the liver and adrenal glands. Less is known about the kidney, where both benign (solid or cystic) and malignant lesions are also common.

4.1 Contrast Media

Parenchymal tissue contrast is enhanced by intravenous or arterial administration of magnetopharmaceuticals. Commonly gadolinium diethylenetriaminetetraacetic acid, Gd-DTPA, (Magnevist) or the newer nonionic contrast agents, such as Gd-DTPA-BMA (Omniscan) or GdDO3A (ProHance) are given by the intravenous route. These agents are designed for rapid excretion and do not significantly interact with body physiology. Binding to blood proteins or cells is negligible. Excretion is by passive glomerular filtration. Thus, these agents, introduced to the blood stream, will penetrate capillaries in the body and have an extracellular distribution, although in the brain they are limited to the vascular space by the blood–brain barrier. It is the molecular weight of these agents (less than 1000 Da) that determines their diffusion into the extracellular space of the abdominal organs.

Intravenous contrast agents are concentrated in the renal tubules, distinguishing functioning renal tissue from all but the most vascular kidney tumors. Unfortunately, in the liver, spleen, and pancreas, the blood supply, capillary permeability, and capacity of the extracellular space of tumors is quite variable. The kinetics and degree of enhancement of the tumors is often described in general terms such as 'hypovascular' or 'hypervascular'. Unfortunately, enormous biological variation exists, and tumors of a common cell type (e.g. colon adenocarcinoma) show wide variations in enhancement characteristics.

Efforts to enhance selectively liver parenchyma and improve the contrast between normal liver and cancer include rapid ('dynamic-bolus') intravenous infusion to exploit the dual (portal venous and hepatic arterial) vascular supply of normal liver as opposed to cancers, which are typically supplied by arteries and may have fewer vessels than normal liver. Dissatisfaction with intravenous techniques has stimulated attempts to administer contrast agents at angiography via selective cannulation of the hepatic artery or via the superior mesenteric artery, looking at the portal venous phase of contrast circulation. This latter approach may improve tumor–liver image contrast, and some authors have claimed improved sensitivity for detecting liver lesions. However, due to variations in the portal venous perfusion of the liver, nonneoplastic perfusion inhomogeneities create numerous false-positive diagnoses of cancer, lowering specificity and limiting the value of this technique for planning therapy.

For more than a decade, investigators have attempted to adapt cholegraphic agents, taken up by hepatocytes and excreted into the biliary system, for use in the detection of liver cancer.[4] Similarly, colloidal or particulate magnetopharmaceuticals have been used to target the normal hepatic Kupffer cells in an attempt to contrast liver against cancer, which does not show phagocytic uptake of magnetic particles. Unfortunately, attempts to target particulate agents in bulk to normal liver have been complicated by toxic side effects. Newer superparamagnetic iron oxide agents under development may show improved safety profiles.[5]

5 THE LIVER

5.1 Metastases

The liver is the most common organ in the body to which cancer spreads from primary disease elsewhere. Gastrointestinal neoplasms, including those in the pancreas, nearly always involve the liver before metastasizing to other organs. In addition, breast, lung, renal, and ovarian and other common malignancies spread hematogenously, lymphatically or transperitoneally to the liver. Indeed, liver failure is one of the most common causes of death in cancer patients, and liver metastases serve as a basis for determining prognosis and monitoring therapy.

Serologic studies of liver function, including enzymes leaked into the blood stream when the liver is injured [serum glutamic oxaloacetate transaminase, (SGOT), serum glutamic pyruvic transaminase (SGPT), and alkaline phosphatase], have such poor sensitivity and specificity that they cannot be justified for use as an independent diagnostic test. When metastatic cancer to the liver is suspected or a reasonable risk exists, imaging must be performed. Unfortunately, imaging also has its limitations and pitfalls.

Metastatic spread of cancer is thought to begin as a single cell or a small cluster of cells, well below the resolution limit of imaging. Indeed, imaging cannot reliably detect liver lesions smaller than 5 mm in size. As a metastasis grows exponentially it is therefore submillimeter in size for the vast majority of its 'life cycle' before it enlarges to the point of causing symptoms and ultimately killing the patient. Therefore, it is obvious that the majority of metastases present in an individual, or speaking epidemiologically, the majority of metastases present in a population, are below the threshold size for detection by any means.

For References see p. 1216

5.2 Imaging: Detection

Liver cancer detection does offer clinically useful information because most patients with metastatic cancer have multiple lesions, some of which are sufficiently large (>1 cm diameter) to be detected reliably. The accuracy of imaging for detecting liver metastases is the subject of endless disputes and revision of the radiological literature. The major problem with the literature is the lack of suitable 'gold standards'. It is simply impossible to collect a large series of patients for whom autopsy and histologic confirmation of imaging findings are available. Thus, radiologists frequently resort to comparative studies where one modality is compared to another, and a third technique (such as surgery) or a third imaging modality is used as the arbiter of 'truth'. Such comparative studies consistently overestimate the accuracy of detection of liver lesions.[6-9]

It is now generally accepted that MRI is the best test, followed by CT, ultrasonography, and then angiography. CT angioportography (CTAP) techniques are equal to or slightly better than MRI; however, the angiographic CT methods are generally not available and are nearly 10 times more expensive. Therefore, when the clinician is able to focus the diagnostic question of the detection of liver metastases, MRI is now the preferred technique. Unfortunately, clinicians often have mixed objectives and request inspection of the adrenal glands, retroperitoneum, and other abdominal organs. As a practical matter, in such unfocused clinical situations, ultrasonography or CT are preferred, despite their inaccuracy, as widely available, cheap, and, arguably, acceptable (i.e. 'cost-effective') compared to MRI.

In summary, for patients at risk of metastatic cancer it is an economic and public policy issue whether the best test (MRI) is made available. Given the cost of misdirected cancer therapy in patients with undiagnosed metastases (in societies where cancer treatment is available), it is likely that diagnostic imaging pays for itself. Unfortunately, neither MRI nor CT is available to many patients. For a variety of nonscientific reasons, doctors often accept ultrasonography as the primary screening test, notwithstanding the hidden costs of false-negative and false-positive examinations.

The dominant worldwide use of ultrasonography for liver imaging may in fact be cost-effective, as this rapid and inexpensive method does reliably identify the majority of patients having metastatic cancer. Since the treatments available have such a poor outcome, is it any wonder that misdiagnosis and inappropriate therapy combined do not noticeably alter this miserable outcome? Despite government pressure to sacrifice health and life that is not measurable or statistically cost-effective, medical ethics dictates that in those cases having a solitary or questionable liver lesion at ultrasonography, CT should be made available. Setting aside economic issues, state-of-the-art oncologic practice requires use of either CT or MRI to screen for liver metastases in the abdomen.

5.3 Lesion Characterization

MRI is more likely than CT to demonstrate internal nodular structure, rings and hemorrhage. In particular, T_2-weighted MR images are useful as they show edema at the border of active lesions which may be manifested as rings or as geographic zones of increased signal intensity. T_1-weighted images are less

Table 1 Relaxation Times for Liver, Hemangioma, Metastases, and Hepatocellular Carcinoma at 0.6 T[a]

Tissue	T_1	T_2
Normal liver	499 ± 140	48 ± 11
Hemangioma	1010 ± 497	143 ± 51
Metastases	691 ± 100	71 ± 21
Hepatocellular carcinoma	569 ± 133	87 ± 17

[a]Values are the mean $\pm$ SD calculated from four or more spin echo measurements.

sensitive to edema; as a result, metastases are virtually always the same size or larger on T_2-weighted images.

5.4 Staging

Staging of hepatic neoplasms principally involves identifying the lobe and segment of disease. Patients with larger tumor burdens and receiving chemotherapy can be monitored by scan-to-scan comparison of tumor diameters or estimates of tumor volume. Some idea about the extent of differentiation that is possible is given by Table 1.

Surgical approaches to hepatic metastases depend primarily on confidence that all of the tumors have been identified. If there is reason to believe that a solitary metastasis exists or that a cohort of a few similar sized lesions represent the only disease in the liver, then the question of resection depends upon the functional hepatic reserve and the technical ability to remove the lesions. Lesions invading an adjacent structure, such as the diaphragm or inferior vena cava, or lesions in difficult areas, such as the porta hepatis, make surgery challenging and risky. Undoubtedly, the greatest risk to resective therapy is undiagnosed residual disease.

6 HEPATOCELLULAR CARCINOMA

Also known as 'hepatoma', primary malignancy of hepatocytes [i.e. hepatocellular carcinoma (HCC)] comprises 1% of all cancers in the USA. Worldwide, variation in nutritional factors and the incidence of viral hepatitis[10,11] account for major differences in the incidence and nature of HCC. It is far more common in Japan, the rest of Asia, and sub-Saharan Africa in association with chronic viral hepatitis. Articles can be found in the literature citing a wide range of sensitivity, specificity, and overall accuracy for ultrasonography, CT and MRI.[13,14,15] It is evident that HCC is more difficult to detect with any of the imaging methods than is metastatic cancer. The principal reason for this difficulty is that HCC most often occurs against a background of chronic hepatitis, fatty liver and cirrhosis.[11-14]

Fatty change within malignant hepatocytes serves to decrease tumor–liver contrast on conventional MR images, although the presence of fat can be exploited to increase contrast if chemical shift selective techniques are used.

Ongoing regeneration and areas of nodular hyperplasia serve to mask or mimic hepatoma. Bands of scar tissue in the liver have increased water content, and the long T_1/T_2 signal intensity characteristics are very similar to hepatoma.

6.1   Lesion Characterization

Characterization of hepatoma and discrimination from metastatic cancer or benign hepatic masses is rarely possible with ultrasonography, CT, or scintigraphy. MRI, however, has the ability to detect three features characteristic of hepatoma. Each of these features is present in approximately 30% of hepatomas, and therefore one or more features can be found in a majority of cases.

First, a capsule of compressed liver or scar tissue may create a sharp boundary with adjacent liver tissue. The capsule has long T_1/long T_2 signal intensity characteristics, and is usually best seen on T_1-weighted images. The capsule must be distinguished from low signal intensity blood vessels on T_1-weighted spin echo images. Hepatic adenoma may also show peripheral capsules indistinguishable from those of hepatoma. However, hepatic adenoma is an unusual tumor with different demographics. Tumor and capsules can be distinguished from the rings of metastases, as the latter are better seen on T_2-weighted images and are not well seen at all on T_1-weighted images. Images from a hepatic adenoma study are shown in Figure 2.

Second, hepatomas accumulate intracellular triglyceride. Fatty accumulation can be detected using a variety of chemical shift methods such as phase contrast frequency-selective imaging. Although fatty accumulation may occur in injured hepatocytes in other conditions, including adenoma, focal nodular hyperplasia, hepatitis, and dietary disturbances, it is usually possible to distinguish focal fatty infiltration from fat within a mass, as the latter displaces normal blood vessels, while focal fat infiltration usually does not.

Third, hepatoma has a propensity to grow into hepatic and portal veins, and MRI is able to demonstrate this morphology to advantage. Certainly ultrasonography also has this capability; however, the ability of ultrasonography to distinguish the solid tumor from adjacent liver makes it more difficult to follow the tumor into vessels.[17]

7   BENIGN DISEASE: CAVERNOUS HEMANGIOMA

Hemangioma is a vascular malformation characterized by a cavernous collection of blood spaces (Figure 3). Blood flow is very slow and virtually undetectable, except for the occasional peripheral site of venous entry. Hemangioma is exceedingly common in both sexes and all races around the world. Approximately 15–20% of the adult population has such a lesion.[8]

7.1   Lesion Characterization

MRI has been remarkably successful in identifying the fluid content of cavernous hemangiomas by its long T_2 relaxation time (Table 1).[15,16] Long TE spin echo images, which reduce the signal intensity of solid tissue relative to fluid, show cavernous hemangiomas, cerebrospinal fluid, bile, and gastric contents as extremely bright structures. Properly performed, with TE in excess of 120 ms and successful suppression of motion artifact, cavernous hemangiomas are homogeneous and

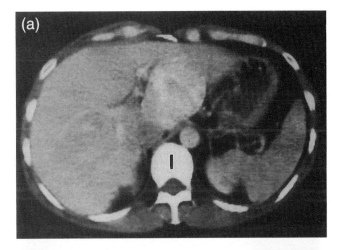

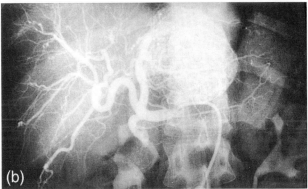

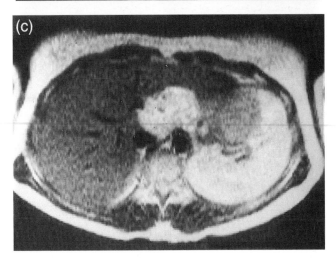

Figure 2 Hepatic adenoma. (a) A CT scan enhanced with bolus infusion of iodinated contrast media shows a large enhancing lesion in the left hepatic lobe. This finding is nonspecific and resembles cancer or any other solid neoplasm. (b) A catheter angiogram suggests a benign solid tumor, most likely hepatic adenoma. (c) MRI (SE 1500/100) shows a heterogenous mass that is isointense with the spleen, consistent with any solid neoplasm; surgery confirmed the diagnosis

sharply circumscribed on the MR image. Tumors, on the other hand, are ill-defined, heterogeneous, and lower in signal intensity due to their gelatinous (solid) make-up.

Contrast-enhanced blood-pool studies provide an alternative method of identifying hemangiomas and other vascular lesions.

For References see p. 1216

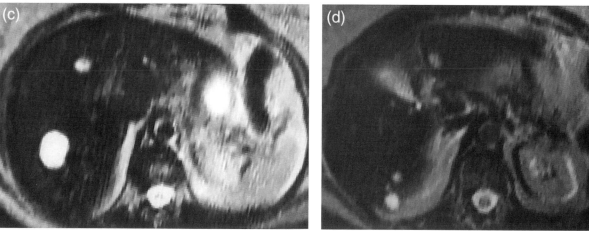

Figure 3 Cavernous hemangioma of the liver. (a) A CT scan shows a 3 cm lesion in the posterior segment, right hepatic lobe, with a hypodense appearance identical to cancer. (b) MRI (SE 300/15) shows the lesion to have a low intensity, similar to cancer (very slightly darker than the spleen). Note a second, 1 cm lesion in the anterior segment not seen on the CT scan. (c) MRI (SE 2400/180) shows the lesions to be hyperintense relative to spleen, and nearly isointense with cerebrospinal fluid; this is consistent with blood in cavernous hemangiomas. (d) Lesions as small as 3 mm diameter can be characterized by MRI

CT uses iodine based diagnostic pharmaceuticals; scintigraphy uses technetium-labeled red blood cells, and MRI uses the paramagnetic agent Gd-DTPA. In addition, iron oxide particles and other magnetic contrast media have been shown to be effective blood-pool markers. The concept behind these studies depends principally on the kinetics and distribution of the vascular agents. Cavernous hemangiomas tend to enhance uniformly over 5–30 min and may retain contrast longer than the circulating blood pool. Solid tumors show more rapid wash-in and wash-out of the contrast agent and usually show less peak enhancement than do cavernous hemangiomas. As these findings depend upon the cardiac output, circulation time, timing of contrast administration, and the specifics of the perfusion pattern of the vascular malformation, contrast-enhanced studies are less predictable than plain T_2-weighted MRI. Nevertheless, MRI can be used in both methods.

Controlled studies have shown that MRI can discriminate cavernous hemangioma from solid neoplasm with an overall accuracy of 90%. If, in addition to this T_2-weighted technique, a contrast study is done, the accuracy increases further. With MRI one can perform both an unenhanced and enhanced scan within a single 1-h examination.

8   THE PANCREAS

Adenocarcinoma is a malignant mucinous neoplasm of the pancreas that arises from the ductal epithelium. It is the fourth most common cause of death from cancer in the USA and accounts for 3% of all cancers. Adenocarcinoma represents 75% of nonendocrine pancreatic malignancies. The disease usually affects people over the age of 70 years, with a male/female ratio of 1.5:1.0. Proposed etiologic factors include cigarette smoking and coffee consumption, alcohol consumption and dietary intake of fat, protein and a high number of calories.[7,18,19]

Disease states of the pancreas that are reportedly associated with the development of pancreatic carcinoma include diabetes mellitus and familial pancreatitis. The etiologic association with chronic pancreatitis is not clear.

Because the pancreas is not encased by a capsule, early spread into the surrounding retroperitoneal fat is common. Local structures then become involved, including the portal vein, superior mesenteric vein, splenic vein, gastroduodenal artery, bile duct and duodenum. Local lymph-node involvement and hematogenous spread to the liver via the portal vein are also seen.

8.1 Imaging: Detection

Lesions in the head of the pancreas are generally visualized by both ultrasonography and CT. The former is limited by surrounding bowel gas or obesity in about 30% of patients. Maneuvers such as altering respiration, filling the stomach with water, placing the patient in the upright position and repeat scanning are helpful. Lesions in the tail of the pancreas are difficult to detect by ultrasonography, but visualization can be improved by using the spleen as a window. With CT, tail and body lesions are sometimes obscured by a partial volume artifact from a bowel not filled with contrast. Intraoperative ultrasound scanning is very sensitive to small and impalpable tumors, particularly for small, functioning tumors such as insulinomas.

MRI has little or no role at the present time due to inferior anatomic resolution and no better contrast than ultrasonography or CT.

8.2 Characterization

Ultrasonography, CT, and MRI are equally nonspecific in distinguishing pancreatic cancer from benign masses. Pancreatitis is the most important differential diagnosis. In most cases, radiologically guided fine-needle aspiration biopsy (FNAB) is necessary for diagnosis.

9 THE SPLEEN

The spleen rarely causes clincial symptoms and is most often imaged as a bystander to a study directed at another abdominal organ. Trauma and suspected rupture is the most frequent indication for imaging the spleen itself. While MRI would be preferred, CT is usually selected because of its availability and resistance to motion artifacts.

Oncologic disease within the spleen is infrequently identified, even though more than half of patients with widespread malignancy may have splenic involvement at autopsy. Melanoma probably accounts for most of the metastases found in the spleen, although various studies have described the origin of metastases, in decreasing order of frequency, as lung, prostate, colon, stomach, melanoma, ovary, and pancreas.

9.1 Lymphoma

Imaging of lymphoma has traditionally been insensitive to splenic involvement. Splenic size is unreliable, as one-third of patients with splenomegaly do not have lymphomatous involvement, and one-third of patients without splenomegaly have such an involvement. In patients with aggressive disease, such as the poorly differentiated nodular type, involvement becomes probable, but not necessarily visible.[20]

Ultrasonography, CT, and scintigraphy have unsatisfactory accuracy rates, ranging between 54% and 75%. MRI is beginning to show some promise. When the water content of lymphoma is high, such as in the large lesions of histiocytic lymphoma, they become detectable on T_2-weighted sequences. MRI using superparamagnetic iron oxide is showing promise in demonstrating splenic involvement, but these research findings need further clinical verification.[21]

10 THE KIDNEY

Renal disease, like diseases of the liver and pancreas, can be divided between diffuse and focal processes. Most diffuse disease is inflammatory or metabolic in nature.

Solid renal masses are cancers until proven otherwise, usually at the expense of nephrectomy given the habit of urologists to avoid referral of patients for FNAB, a less expensive alternative practice by radiologists. Since the aging population at risk of cancer has a high prevalence of benign simple cysts, it has been a challenge to identify benign lesions noninvasively, in order to save good kidneys from the knife. Unfortunately, features demonstrable by ultrasonography, CT, and MRI add little to the low pretest probability of malignancy. The few cancers picked up as cystic structures with nodular irregularities or septations by ultrasonography, calcifications or density changes on CT, or hemorrhage on MRI are outnumbered by similar findings in nonmalignant cysts falsely diagnosed as positive for malignancy.

While the identification of renal cysts by MRI should be as straightforward as identification of hepatic hemangioma, comparable research has not been performed to establish quantitative or qualitative criteria. Furthermore, renal cell carcinoma is more difficult than other cancers to distinguish from cysts because of its hypervascularity, tendency for necrosis, and the resultant long T_1/T_2, approaching that of cysts.

MRI studies in the kidney have largely consisted of anatomic descriptions recapitulating the findings known from CT, albeit less clearly. The sole application of MRI to the diagnosis and management of renal disease is inspection of the draining veins for extension of tumor. For all other questions, ultrasonography and CT are preferred. As a result, very little renal MRI is practiced.

11 SUMMARY

Abdominal imaging is historically anatomic, based on the delineation of high contrast fatty tissue boundaries. Ultrasonography and CT are widely available, familiar to clinicians, competitive in diagnostic quality, and cheap. MRI of the abdomen is best established in the liver for detection and differential diagnosis of focal masses.

For References see p. 1216

12 RELATED ARTICLES

In Vivo Hepatic MRS of Humans; Kidney, Prostate, Testicle, and Uterus of Subjects Studied by MRS; Lung and Mediastinum MRI; Male Pelvis Studies Using MRI; MRI of the Female Pelvis.

13 REFERENCES

1. R. E. Hendrick, T. R. Nelson, and W. R. Hendee, *Magn. Reson. Imag.*, 1984, **2**, 193.
2. G. M. Bydder, J. M. Pennock, R. E. Steiner, S. Khenia, J. A. Payne, and I. R. Young, *Magn. Reson. Imag.*, 1985, **3**, 251.
3. D. R. Bailes, D. J. Gilderdale, G. M. Bydder, A. G. Collins, and D. M. Firmin, *J. Comput. Assist. Tomogr.*, 1985, **9**, 835.
4. Y. M. Tsang, M. Chen, and G. Elizondo, in 'Sixth Annual Meeting, Society for Magnetic Resonance Imaging, Boston, MA, 1988', 6 (S1), 124.
5. D. D. Stark, R. Weissleder, G. Elizondo, P. F. Hahn, S. Sains, L. E. Todel, G. J. Wittenberg, and J. T. Ferrucci, *Radiology*, 1988, **168**, 297.
6. D. D. Stark, J. Wittenberg, R. J. Butch, and J. T. Ferrucci, *Radiology* 1987, **165**, 399.
7. D. G. Mitchell and D. D. Stark, 'Hepatobiliary MRI', Mosby, St Louis, 1992.
8. K. Okuda, K. G. Ishak, 'Neoplasms of the Liver', Springer, Berlin, 1987.
9. J. P. Heiken, P. J. Weyman, J. K. Lee, D. M. Balfe, D. Picus, E. M. Brunt, and M. W. Flyte, *Radiology*, 1989, **171**, 47.
10. P. F. Hahn, D. D. Stark, S. Saini, E. Rummeny, G. Elizonilo, R. Weissleder, J. Wittenberg, and J. T. Ferrucci, *Am. J. Roentgenol.*, 1990, **154**, 287.
11. M. Ebara, M. Ohto, Y. Watanabe, K. Kimura, H. Saisho, Y. Tsuchiya, K. Ohuda, N. Arimuzu, F. Konilo, and H. Ikcheva, *Radiology*, 1986, **159**, 371.
12. P. R. Ros, B. J. Murphy, J. E. Buck, G. Olmedilla, and Z. Goodman, *Gastrointest. Radiol.*, 1990, **15**, 233.
13. H. Yoshida, Y. Itai, K. Ohtomo, T. Kohubo, M. Minami, and N. Yashiro, *Radiology*, 1989, **171**, 339.
14. E. Rummeny, R. Weissleder, D. D. Stark, S. Saini, C. C. Compton, W. Bennett, P. F. Hohn, J. Wittenburg, R. A. Malt, and J. T. Ferruci, *Am. J. Roentgenol.*, 1989, **152**, 63.
15. Y. Itai, K. Ohtomo, S. Furui, T. Yamauchi, M. Minomi, and N. Yashiro, *Am. J. Roentgenol.*, 1985, **145**, 1195.
16. D. D. Stark, R. C. Felder, J. Wittenberg, S. Saini, R. J. Butch, M. E. White, R. R. Edelman, P. R. Mueller, J. F. Simeone, and A. M. Cohen, *Am. J. Roentgenol.*, 1985, **145**, 213.
17. D. D. Stark, P. F. Hahn, C. Trey, M. E. Clouse, and J. T. Ferrucci, *Am. J. Roentgenol.*, 1986, **146**, 1141.
18. A. L. Cubilla and P. J. Fitzgerald, *Clin. Bull.*, 1978, **8**, 143.
19. R. E. Schultz and N. J. Finkler, *Mt. Sinai J. Med.*, 1980, **47**, 622.
20. M. Federle and A. A. Moss, *CRC Crit. Rev. Diagn. Imaging*, 1983, **10**, 1.
21. R. Weissleder, G. Elizondo, D. D. Stark, P. F. Hahn, J. Margel, J. F. Gonzalez, F. Saini, W. E. Todel, and J. T. Farruci, *Am. J. Roentgenol.*, 1989, **152**, 175.

Biographical Sketch

David D. Stark. *b* 1952. B.A., 1974, M.D., 1978, Harvard University, USA. Postdoctoral work, University of California, San Francisco. Successively, instructor, assistant professor and associate professor of radiology, Harvard Medical School, Massachusetts General Hospital. Currently Professor of Radiology, University of Massachusetts Medical Center, USA. Approx. 200 publications. Current research specialties: contrast media, clinical applications of MRI, quantitation of tissue iron.

For list of General Abbreviations see end-papers

Abdominal MRA

Paolo Pavone, Andrea Laghi, Carlo Catalano, Valeria Panebianco, Francesco Fraioli, Isabella Baeli, and Roberto Passariello

University of Rome, 'La Sapienza', Italy

1 INTRODUCTION

The study of abdominal vessels by MRA is a new and interesting perspective available in a routine clinical examination. The main problem which has limited such applications of MRA is the sensitivity of this method to motion artifacts due to breathing, cardiac motion, peristalsis, and pulsatile blood flow. However, the improvement in technical equipment has made it possible to reduce these problems and to obtain an excellent evaluation of abdominal vessels.[1,2]

To date, both the time-of-flight (TOF) and phase contrast (PC) techniques are used and each has its own advantages and disadvantages.[3-5] The TOF method has a faster acquisition time and a higher signal-to-noise ratio (SNR), but it is very sensitive to saturation effects; as a consequence it is only possible to work with small volumes, particularly when a 3D technique is employed.[6]

Recently, a new method of study using a dynamic intravenous injection of gadolinium-DTPA in conjunction with a 3D TOF acquisition technique has been developed.[7] The use of gadolinium shortens the blood relaxation time and overcomes in-flow saturation; hence it is possible to image arteries with a low flow velocity and large volumes. These vessels can therefore be imaged in any direction without saturation problem, i.e. using a coronal plane for the aorta and so reducing the acquisition time. However, the overlapping of venous structures occurs and presaturation pulses are not effective.

The phase contrast technique can be used over large fields of view because of its lack of sensitivity to the spin saturation effect. Other advantages are the complete cancellation of stationary tissues and the possibility of quantifying flow velocity. By comparison, PC techniques require longer acquisition times and are very sensitive to motion artifacts, especially if such motion is not constant during image acquisition.

In the evaluation of abdominal vessels, contrast-enhanced MRA is currently considered the only technique able to provide anatomical images of all major vessels, allowing an optimal image quality, with improved diagnosis.[6] Using this technique, all the datasets needed can be acquired during a breath-hold, provided that enhanced gradient systems are available to reduce repetition and echo time dramatically.

This overcomes the old problem with the conventional TOF or PC methods in which long acquisition times and the complexity of flow effects resulted in a myriad of artifacts, rendering a considerable fraction of examinations unreadable.

Furthermore, arterial blood is clearly visualized because it contains the paramagnetic contrast agents. The presence of paramagnetic agents such as a gadolinium chelate rapidly reduces the T_1 relaxation time of blood from 1400 ms to 200–400 ms depending on the dose and the rate of injection.

Obviously the visualization of blood depends on the presence of contrast in the arterial system while the 3D imaging data are being collected; this makes the technique very sensitive to the timing of the intravenous contrast administration relative to data acquisition.[7]

Great efforts have been made in recent years to improve further the image quality of enhanced MRA. These have been aimed at the improvement of the pulse sequences on the one hand and the contrast agents used on the other.

Improvement of the pulse sequence for enhanced MRA must involve one single and important issue: minimizing dephasing phenomena, thus reducing the signal loss related to phase dispersion of flowing spins, mostly evident in stenotic (jet flow) and tortuous vessels (turbulent flow). These effects are, in fact, still evident even when enhanced techniques are used, despite the fact that these newer sequences visualize directly the effect of the contrast agent (T_1 shortening) rather than the flow phenomena. Reduction of phase dispersion is achieved by using either optimized gradient rephasing pulses or, in a more effective way, by reducing the echo time of the pulse sequence. With current commercial equipment, a minimum echo time of 1.5 ms can be used, and the improvement achieved in moving to this from the 3 ms available on older equipment is really dramatic, with more consistent evidence about small vessels and better definition of the true diameter of every vessel imaged by MRA. Moving to shorter echo time has only one important requisite: that the gradient characteristics of the equipment are optimized. Current equipment must have a gradient strength of at least 20 mT m^{-1} and a rise time of 400–800 μs.[8] Specialized equipments for use in the cardiovascular system are being produced that have a gradient strength of 40 mT m^{-1}, although they are currently being offered only for research purposes.

Another way to improve the image quality in enhanced MRA is to use more effective contrast agents. Conventional gadolinium chelates have basic pharmacokinetics, with intravascular–extracellular diffusion. In practice, these agents pass rapidly through the capillary membrane and very rapidly (within 5 min) reach pseudoequilibium between the vascular and the interstitial concentration. There are two ways to improve the characteristics of the contrast agent for enhanced MRA: by using a contrast agent with the same pharmacokinetic characteristics but having an increased relaxivity (the capacity to influence the local magnetism and, therefore, the T_1 relaxation time of blood), or by using contrast agents with prolonged blood half-life, having a lower capacity to diffuse through the capillary membrane.

The first approach is used with a new gadolinium chelate Gd-Bopta (Multihance, Bracco). This contrast agent was initially proposed only for hepatobiliary use, because of its consistent biliary excretion, but has now also been applied for general purposes, because of its good characteristics. In particular, in the vascular system, this contrast agent is able to provide a very high enhancement of the vascular structure because its relaxivity is almost twice as great as that of other gadolinium chelates in human plasma (R1 9.7 versus 4.9 for Gd-DTPA). Early studies have demonstrated that evaluation of abdominal vessels by enhanced MRA is improved by this contrast agent. The second approach is to use contrast agents that are not able to cross the capillary membrane rapidly. The first class of such contrast agents are the macromolecular compounds, where a long molecular chain with large molecular weight is bonded to gadolinium. Schering has proposed a gadolinium chelate covalently bound to a polymer composed of polylysin (Gadomer 17). This compound has a molecular weight of 30–35 kDa, with high relaxivity, good acute tolerance, low viscosity, and low osmolarity. In early studies, this contrast agent has been able to improve the enhancement of vessels and to provide prolonged evidence of the vascular system in enhanced MRA. The second class of contrast agent that is not able to cross the capillary membrane rapidly is represented by one single molecule: MS 325, first synthesized by EPIX and currently produced by Mallinkrodt. This contrast agent is rapidly bound to human serum albumin after being injected into the vascular system, with a strong but short-lasting link. In fact, after a few hours all the contrast agent injected is free from this link and is excreted through the kidneys. The result is that this agent provides a very high enhancement of the vascular system that is prolonged in time. The advantage of using intravascular agents is that the whole vascular system can be imaged more safely, with images being acquired in different anatomic areas, without the need for giving repeated injections of contrast agent at each anatomic level. In conclusion, there is no doubt that contrast-enhanced MRA is the technique of choice for abdominal MRA. Hardware improvements are to be expected and the use of newer contrast agents will definitely contribute to increasing the image quality and clinical value of this technique.

2 CLINICAL APPLICATIONS

2.1 Abdominal Aorta

The main clinical application of aortic MRA is the study of aortic aneurysms in terms of the extent and size of the pathology as well as the involvement of collateral vessels, identification of either an aortic stenosis or a thrombus in an atherosclerotic vessel and evaluation of either the patency or the complications of a graft.[9]

In the study of aortic aneurysms MRA is complementary to conventional spin echo images. In fact, a complete evaluation of the diameter of the lesion is possible only after a study performed using spin echo sequences. However, MRA has a major role in defining the size of the true lumen, in depicting the neck of the aneurysm, and a potential role in the quantification of the blood flow. Figure 1 illustrates the utility of MRA in the evaluation of an aortic aneurysm. By reconstructing the images with maximum intensity projection (MIP), it is

For References see p. 1222

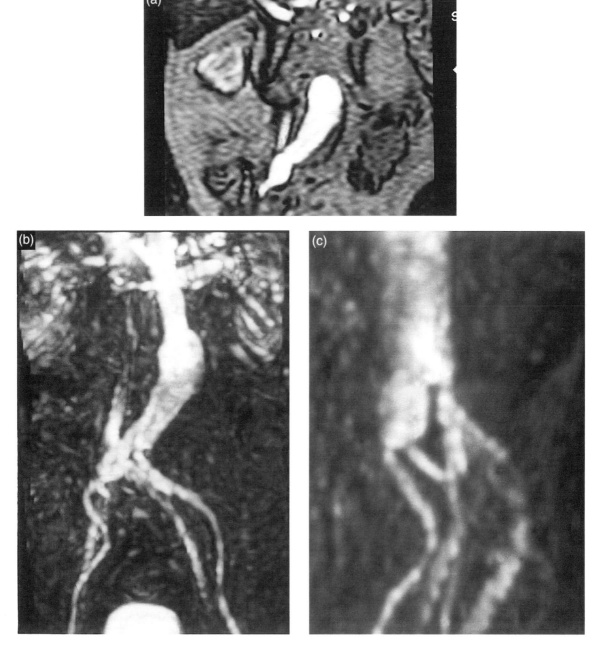

Figure 1 Abdominal aortic aneurysm; evaluation with the 3D TOF technique after gadolinium injection. (a) Single coronal slice depicting the true lumen of the aneurysm; the image also allows the thrombolytic part to be recognized. (b) MIP reconstruction of the same aneurysm; good visualization of the abdominal aorta from the renal arteries to the iliac vessels is obtained. (c) Magnification of the iliac bifurcation with evidence of involvement of the right common iliac artery and a tight stenosis at the origin of the left common iliac artery

possible to obtain a clear definition of the relationship between the aneurysm and the other vessels arising from the aorta. One study[10] showed a sensitivity of 67% and a specificity of 96% for MRA in the identification of proximal stenosis of the renal artery involved by the aneurysm (Figure 2), and a sensitivity of 67% and a specificity of 100% for stenosis of the celiac trunk and of the superior mesenteric artery.

Enhanced MRA is particularly useful in the evaluation of abdominal aortic aneurysms. In fact, the presence of blood tur-

bulence has limited to a large extent the use of nonenhanced techniques, with areas of flow void related to spin dephasing. With enhanced MRA all these problems are overcome, and the use of a coronal acquisition makes it possible to image the whole aorta and iliaco-femular district in one single field of view (FOV of 50 cm can easily be used). The absence of excretion toxicity from gadolinium compounds also permits a complete study in patients with suspected renal arterial stenosis or involvement. Currently, noninvasive 3D imaging studies

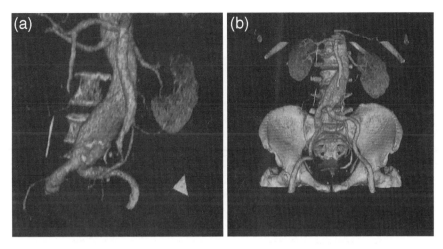

Figure 2 Volume-rendering reconstruction clearly visualizes the celiac trunk, mesenteric artery and both renal arteries in a patient with aneurysm

(either CT or MRI) are recommended for the evaluation of abdominal aortic aneurysms, avoiding the use of invasive catheter angiography. In particular, the aim is to determine the treatment that is to be performed in each patient; endovascular placement of aortic stent grafts can be developed as the treatment of abdominal aortic aneurysms, and all the data needed to evaluate if the graft can be placed can be collected.

A study has compared the results of MRA and CTA (CT X-ray angiography) for evaluating the information needed for graft placement.[11] Both techniques agreed with catheter angiography (as gold standard) in the measurement of the abdominal aorta at the level of its largest diameter and at the level of the renal artery ostium. Another measurement that was obtained consistently with good agreement with both methods was the definition of the distance between the renal arterial ostium and the origin of the aneurysm. For tube graft placement, a distal nonaneurysmal segment of at least 15 mm is mandatory. This distance was also measured equally well by CTA and MRA (Figures 3 and 4).

Another important issue is the evaluation of accessory renal arteries. In initial studies, MRA had generally performed very poorly, with sensitivities lying between 28 and 92%. More recent studies have reported an accuracy of 100%. The importance of breath-hold MRA is shown in all these studies. In general, 12% of patients with abdominal aortic aneurysms present with accessory renal arteries, and information about them is crucial for proper treatment planning. CTA and MRA now also offer similar results in this field.

Other information that is provided by MRA concerns the patency of the mesenteric artery: in some cases the inferior mesenteric artery may be patent and this can cause revascularization of the aneurysmal lumen through reverse flow. The status of the iliac vessels has also to be evaluated. The possibility of stenosis has to be considered, since the endovascular stent graft is of fixed expanded diameter once introduced (18 F requires a minimum vessel diameter of 6 mm). Involvement of the iliac arteries through aneurysmal dilatation does not prevent the placement of a stent graft, but size and location should be considered carefully in deciding on the type of graft to be used and the best approach to its location.

Material arising from thromboses adhering to the walls can result in underestimation of aortic diameter and misinterpretation of the proximal extent of an abdominal aortic aneurysm. Evaluation of images acquired in the axial and coronal planes with gradient echo sequences straight after the MRA sequences (when contrast agent is still enhancing the vascular lumen) permits a detailed evaluation of the thrombus.

Some advantages for MRA do exist compared with CTA: the use of safer contrast agents, which can be used as well in patients with initial renal insufficiency; the use of nonionizing radiation; the possibility of acquiring images directly in the coronal plane, with a more consistent longitudinal (or z axis) resolution; and faster postprocessing analysis (as a 3D MRA dataset is composed of far fewer images than those provided by CTA). However, MRA also has some drawbacks, including the usual contraindications for MR in general (such as

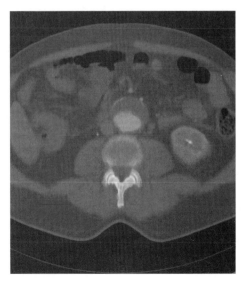

Figure 3 CT axial image with the aneurysm lumen surrounded by thrombolytic material

For References see p. 1222

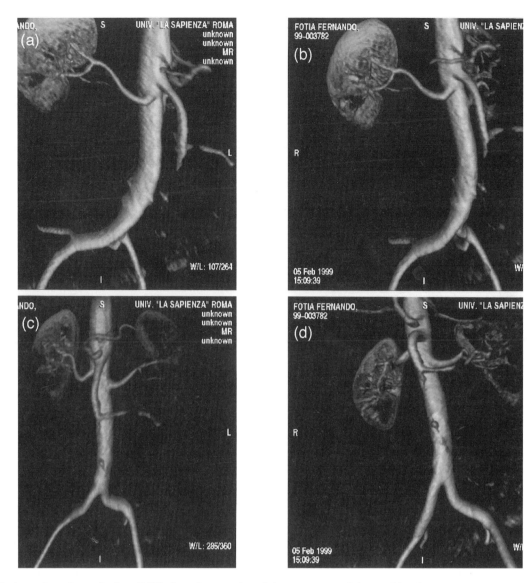

Figure 4   Maximum intensity projections (MIP) show the extension of the aneurysm and the renal involvement

implanted pacemakers). Further, inaccuracy in the detection of small accessory renal arteries must be considered, although this problem may be overcome by continuing technical improvements that allow imaging with better spatial resolution and visualization of very small vessels.

Another field of application for MRA is in the evaluation of leaks after the placement of endovascular stent grafts.[12] Follow-up of endovascular grafts is required in order to evaluate the success of the treatment and the development of complications. In the early phase after placement of an endovascular graft, it is important to ensure that the procedure has led to technical success and that there is no leakage of blood between the graft and the native lumen of the aneurysm, which is better known as endoleak. The two major complications observed in the first series reported by Moore[13] were stent fractures (23%) and endoleaks (44%); 23% of the endoleaks had spontaneous closure. Stent fractures have been corrected fol-

lowing a modification in the stent hook design by the manufacturer. However, the endoleaks remain a significant problem and may be associated with an increase in aneurysmal diameter at 1 year.[14]

In the long-term follow-up of aneurysms treated with a stent graft, it is necessary to monitor the caliber of the lumen of the aneurysm at regular intervals. In fact there have been several case reports of ruptured aneurysms after stent-graft repair. A noninvasive technique is required to monitor all these potential complications, and catheter angiography should be limited to the role of guidance of inventional procedures needed to correct such complications. Ultrasound may be able to provide consistent information with high accuracy in defining the presence of leaks or in evaluating the caliber and patency of the aorta and of the iliac vessels.[15] The value of CT has also been advocated;[16] with current helical acquisition it is possible to reconstruct angiographic projection images. CT is considered to be the required investigation for the EUROSTAR registry.[17]

With MRA, both types of endoleak can be easily detected by the evaluation of areas of increased signal intensity detected after contrast agent injection. Moreover, 3D images provided by MRA are able to define both the patency and morphology of the stent graft and its relative position compared with the origin of the renal arteries.

2.2 Renal Arteries

Several types of examination have been used during the past few years for the evaluation of renovascular hypertension, but many have been abandoned because of their invasivity. Although digital subtraction angiography (DSA) and conventional angiography have become increasingly important because of the success of percutaneous procedures, they still remain too invasive as screening method procedures.

It is now possible to propose MRA of the renal arteries as an accurate, noninvasive diagnostic procedure in the evaluation of renal artery stenosis, especially of the ostium of the renal vessel.[18] An example of MRA of the renal arteries is shown in Figure 5. Recent data show the high accuracy of MRA, which is similar in this case to conventional angiography in detecting both the normal and accessory renal arteries as well as the anomalous vessels. According to the literature, MRA has a sensitivity varying between 89% and 94% and a specificity of between 95% and 97%.[19,20]

The evaluation of renal arteries has been performed principally with DSA, ultrasound, and conventional CT during the past few years.

Different etiologies are implicated in renovascular disease and, in most cases, the renal arteries are affected as part of the entire vascular system. Because of the continuous advances in percutaneous endoluminal therapies, the best possible imaging method is required in the detection of renovascular diseases.

According to the literature, MRA allows the visualization of all the main arteries as well as accessory arteries. In the detection of stenosis, the MRA sensitivity and specificity amounted to 93% and 90%, respectively, with a negative predictive value of 96%.[21,22]

In contrast to other noninvasive techniques such as scintigraphy or ultrasound, contrast-enhanced 3D MRA allows visualization of accessory renal arteries with a high degree of accuracy. The visualization of accessory arteries is especially important in the evaluation of renal transplant donors. Imaging must be completely accurate before surgical revascularization or nephrectomy in living-related renal donors can be contemplated.

As far as atherosclerotic lesions are concerned, the diagnostic performance of 3D MRA is still limited, particularly when all forms of stenosis are considered; in fact, as seen in the literature, there have been a considerable number of false-positive and false-negative findings.[23] Therefore, the gold standard diagnostic procedure for detecting arterial stenosis still remains angiography, though this is an invasive examination that exposes the patient to ionizing radiation as well as iodinated contrast material, which may cause allergic reactions and/or nephrotoxicity. Therefore, according to the literature,[24] contrast-enhanced 3D MRA is, as yet, only a promising approach to obtaining contrast arteriography but is comparable to arteriography in the assessment of renal artery morphology and pathologic states. It does not involve the risk of arterial catheterization, ionizing radiation, or nephrotoxic contrast material.

2.3 MR Venography (MRV)

Studies of abdominal veins are an interesting application of MRA. It is possible to evaluate the normal and pathologic anatomy of the portal system, including the splenic vein, the superior mesenteric vein (SMV), and the eventual collateral vessels.[25] The relatively slow and homogeneous venous blood flow allows high-quality images to be obtained in almost all patients. The large FOV permits the imaging of the whole spleno-portal axis and the collateral vessels at the same time. This option is not available in any other imaging technique such as DSA, duplex scan, or color Doppler ultrasonography.

The main technical difficulties arising from respiratory and peristaltic motion, and to the different flow direction and velocity in abdominal vessels, have now been overcome by new image acquisition procedures. The breath-hold coronal TOF technique is the preferred method with which to image the abdominal veins. An image can be acquired in about 8 s and in a few minutes all the abdominal vessels can be displayed. It is, therefore, possible to obtain a selected evaluation of the portal vein or any of the other vessels by acquiring the images in the plane corresponding to the vessel to be studied. Because of saturation effects, all the vessels not in that plane will not be depicted.

To obtain a more selective evaluation of the venous system, it is possible to use more sophisticated techniques such as bolus tracking and selective presaturation. Those two options allow the clinician to define accurately the flow direction in a vessel and to distinguish a thrombus from flowing blood. Furthermore, they permit the selective suppression of the signal coming from specific arterial vessels or veins. After acquisition of the data, a postprocessing MIP technique can be used to reconstruct the images. Thus, it is possible to rotate the vessels

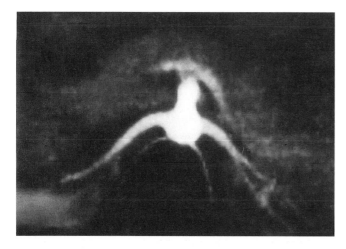

Figure 5 Axial MIP reconstruction of renal arteries; good evaluation of the distal part of the vessels is also obtained

For References see p. 1222

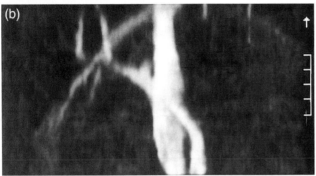

Figure 6 2D TOF MR venography. (a) MIP reconstruction shows the inferior vena cava (IVC) with the hepatic vessels and the portal vein. (b) Good evaluation of the portal vein and its bifurcation. The IVC and one hepatic vessel are also depicted

and to show them on the monitor in the most useful projection and it is also possible to recognize strange anatomic variants in this way.

Portal hypertension represents one of the major fields of application of MR venography (MRV) in the abdomen (Figure 6). Because of the large FOV, MRV allows the whole splenoportal axis and its collateral vessels to be shown at the same time in almost all patients. It is possible to evaluate the diameter of the portal vein and, with adequate techniques, to obtain quantitative information about the blood flow. Compared with color Doppler ultrasonography, which can actually be considered to be the method of choice for the screening and follow-up of patients with portal hypertension, MRV is not operator-dependent; it is not influenced by the presence of ascites, which is very common in these patients, or by colic meteorism and patient habitus, all of which disturb a sonographic examination.

Another recent clinical indication for MRV is the evaluation of complex venous disease in cases of liver transplantation. MRV is very accurate and has a good correlation with DSA and duplex ultrasound. Moreover, MRV displays images in different orientations: thus the reconstructed images can be easily interpreted by the surgeon and complex structures can be demonstrated better than by any alternative imaging method.

Following liver transplantation, complications may arise as a result of anastomosis failure or stenosis. In this case the examination technique is ultrasonography, which has an important limitation in depicting the anastomosis. MRV can also show the condition through the paramagnetic effects of the ma-

terial in the anastomosis, which creates a lack of signal. Additionally, anastomosis on the hepatic artery can be depicted by MRA in a high percentage of patients.

Finally, another recent field of application for MRV is the study of the patency of both surgical and percutaneous shunts of the ortho-systemic structure (TIPS). MRV makes it possible to obtain a noninvasive postsurgical follow-up. The anastomosis can be identified, flow direction established, and both size and patency of the shunts evaluated. After TIPS procedures the position of the stent can also be accurately determined.

There are a few pitfalls and limitations in the use of MRV in the portal system. At a venous confluence, such as the junction of superior mesenteric and splenic veins, a signal void may occur simulating a thrombus. In order not to misunderstand this artifact, gradient echo images can be acquired in different orientations through the questionable area. Changing the slice orientation can also be useful sometimes so as to avoid the in-plane saturation effect. Usually, this is not as common for the portal system as it is for the aorta or the inferior vena cava, but a verticalization of the portal axis together with slow flow, as in a cirrhotic patient, may impair signal acquisition.

In conclusion, MRV of the portal system is extremely accurate and complete for the evaluation of these vessels. The large field of view, the insensitivity to patient habitus and the graphic presentation format make it very useful for preoperative assessment in patients with portal hypertension.

3 RELATED ARTICLES

Liver, Pancreas, Spleen, and Kidney MRI; Peripheral Vasculature MRA; Phase Contrast MRA; Time-of-Flight Method of MRA.

4 REFERENCES

1. R. R. Edelman, H. P. Mattle, D. J. Atkinson, and H. M. Hoogewoud, *Am. J. Roentgenol.*, 1990, **154**, 937.
2. H. Bosmans, G. Marchal, P. van Hecke, and P. Vanhoenacker, *Clin. Imag.*, 1992, **16**, 152.
3. D. G. Nishimura, *Magn. Reson. Med.*, 1990, **14**, 194.
4. C. L. Dumoulin, S. P. Souza, M. F. Walker, and W. Wagle, *Magn. Reson. Med.*, 1989, **9**, 139.
5. C. L. Dumoulin, E. K. Yucel, P. Vock, S. P. Souza, F. Terrier, F. L. Steinberg, and H. Weymuller, *J. Comput. Assist. Tomogr.*, 1990, **14**, 779.
6. J. S. Lewin, G. Laub, and R. Hausmann, *Radiology*, 1991, **179**, 261.
7. J. L. Creasy, R. R. Price, T. Presbrey, D. Goins, C. L. Partain, and R. M. Kessler, *Radiology*, 1990, **175**, 280.
8. A. N. Shetty, K. G. Bis, T. G. Vrachliotis, M. Kirsch, A. Shirkhoda, and R. Ellwood, *JMRI*, 1998, **8**, 603.
9. R. Passariello, P. Pavone, C. Catalano, A. Laghi, L. Marsili, and M. Di Girolamo, *Ann. Chir. Gynaecol.*, 1993, **82**, 87.
10. J. A. Kaufman, E. K. Yucel, A. C. Waltman, S. C. Geller, C. A. Athanosoulis, and M. R. Prince, *Radiology*, 1993, **189(P)**, 174.
11. A. Siegfried, M. D. Thurnher, R. Dorffner, P. Polterauer, and J. Lammer, *Radiology*, 1997, **205**, 341.
12. L. Engellau, E. M. Larsson, U. Albrechtsson, T. Jonung, E. Ribbe, J. Thorne, Z. Zdanowski, and L. Norgren, *Eur. J. Vasc. Endovasc. Surg.*, 1998, **15**, 212.

13. W. S. Moore, *J. Endovasc. Surg.*, 1997, **4**, 182.

14. J. S. Matsumura, *J. Vasc. Surg.*, 1998, **4**, 606.

15. J. Golzarian, L. Dussaussois, H. T. Abada, P. A. Gevenois, D. van Gansbeke, J. Ferreira, and J. Struyven, *Am. J. Roentgenol.*, 1998, **171**, 329.

16. D. T. Sato, C. D. Goff, R. T. Gregory, K. D. Robinson, K. A. Carter, B. R. Herts, H. B. Vilsack, R. G. Gayle, F. N. Parent III, R. J. de Masi, and G. H. Maier, *J. Vasc. Surg.*, 1998, **28**, 657.

17. P. L. Harris, *J. Endovasc. Surg.*, 1997, **4**, 72.

18. M. Strotzer, C. M. Pruell, A. Geissler, S. M. Kohler, B. R. Kraemer, and J. Gmeinwieser, *Radiology*, 1993, **189P**, 189.

19. T. M. Grist, T. W. Kennell, I. A. Sproat, E. M. Flath, J. C. McDermott, and M. M. Majkiwcyz, *Radiology*, 1993, **189P**, 190.

20. T. F. Hany, D. A. Leung, T. Pfammatter, and J. F. Debatin, *Invest. Radiol.*, 1998, **33**, 653.

21. S. Miller, F. Schick, S. H. Duda, T. Nagele, U. Hahn, F. Teufl, M. Muller-Schimpfle, C. M. Erley, J. M. Albes, and C. D. Claussen, *Magn. Reson. Imag.*, 1998, **16**, 1005.

22. D. Kim, R. R. Edelman, K. C. Kent, D. J. Porter, and J. J. Skillman, *Radiology*, 1990, **174**, 727.

23. M. R. Prince, *JMRI*, 1998, **8**, 511.

24. D. B. Stafford-Johnson, C. A. Lerner, M. R. Prince, S. N. Kazanjian, D. L. Narasimham, A. B. Leichtman, and K. J. Cho. *Magn. Reson. Imag.*, 1997, **15**, 13.

25. H. B. Gehl, K. Bohndorf, K. C. Klose, and R. W. Gunther, *J. Comput. Assist. Tomogr.*, 1990, **14**, 619.

Biographical Sketches

P. Pavone, *b* 1957. M.D., 1981, Rome. Studies on CT and vascular and interventional radiology, 1981–87. Researcher on MRI at the University of L'Aquila, Italy, 1988–93; researcher at University of Rome 'La Sapienza', 1993–present. Approx. 300 publications. Current research specialties; experimental and clinical studies of the abdominal organs and vascular system by MRI.

A. Laghi. *b* 1967. M.D., 1992, Rome. Resident, University of Rome 'La Sapienza', 1992–present. Approx. 60 publications. Current research specialties: experimental and clinical studies of the abdominal organs and vascular system by MRI.

C. Catalano. *b* 1965. M.D., 1990, Rome. Residency, University of L'Aquila, 1990–1994. Fellow, University of Rome 'La Sapienza', 1994–present. Approx. 120 publications. Current research specialties: experimental and clinical studies of the abdominal organs and vascular system by MRI.

V. Panebianco. *b* 1967. M.D., 1994, Rome. Resident, University of Rome 'La Sapienza', 1993–present. Approx. 30 publications. Current research specialties: clinical studies of the abdominal organs by MRI.

F. Fraioli. *b* 1973. M.D., 1998, Rome. Residency University of Rome 'La Sapienza', 1993–present. Approx. 10 publications. Current research specialties: experimental and clinical studies of the abdominal and vascular system by MRI.

I. Baeli. *b* 1974. M.D. 1999, Rome. Residency University of Rome 'La Sapienza', 1993–present. Approx. 10 publications. Current research specialties: experimental and clinical studies of the abdominal and vascular system by MRI.

R. Passariello. *b* 1942. M.D. 1965, Rome. Residency University Padua, 1965–1967. Full Professor of Radiology, University of L'Aquila, 1986–1991. Full Professor of Radiology and Chairman of the 2nd Department of Radiology, University of Rome 'La Sapienza', 1991–present. Approx. 500 publications. Organization of more than 65 Congresses. 1994–present Vice-President of the European Congress of Radiology '95 and '97. Chairman of the European Seminars on Diagnostic and Interventional Radiology, 1990–present. Member of the Executive Board of the NICER Programme. 1990–present.

Tissue Behavior Measurements Using Phosphorus-31 NMR

Simon D. Taylor-Robinson

Hammersmith Hospital, London, UK

and

Claude D. Marcus

Hôpital Robert Debré, Reims, France

1 INTRODUCTION

Phosphorus-31 MRS provides a noninvasive method of assessment of mobile phosphorus-containing compounds. A typical in vivo ^{31}P MR spectrum contains seven resonances (Figure 1). Phospholipid cell membrane precursors, adenosine monophosphate (AMP) and glycolytic intermediates (sugar phosphates) contribute to the phosphomonoester (PME) peak. Phospholipid cell membrane degradation products and endoplasmic reticulum contribute to the phosphodiester (PDE) peak. Information on tissue bioenergetics can be obtained from inorganic phosphate (Pi), phosphocreatine (PCr) and the three nucleoside triphosphate resonances. A measurement of intracellular pH (pHi) can be calculated from the chemical shift of the Pi peak.

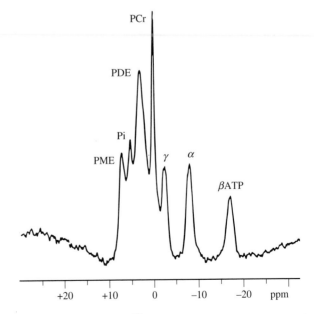

Figure 1 An unlocalized ^{31}P MR spectrum from the head of a healthy adult volunteer. There are seven resonances. PME, phosphomonoester; Pi, inorganic phosphate; PDE, phosphodiester; PCr, phosphocreatine; ATP, adenosine triphosphate

For References see p. 1228

Pathological processes involving hypoxia and ischemia are particularly amenable to ^{31}P MRS assessment using absolute or relative quantitation of the PCr and Pi peaks. The PCr/Pi ratio has been the most commonly used marker of tissue energy reserve under these conditions. Phosphorus-31 MRS may also provide an indication of the viability of isolated donor organs prior to transplantation.

In this article, we discuss the role of in vivo ^{31}P MRS in the examination of tissue bioenergetics in human studies.

2   ^{31}P MRS AND CELLULAR ENERGY STATUS

The energy metabolism of each cell is dependent on the synthesis and utilization of compounds which contain high-energy phosphate bonds such as ATP and PCr.[1] ATP is present in all cells, but PCr is limited to those tissues containing creatine and the enzyme creatine kinase (CK), such as skeletal muscle and brain. ATP has a pivotal role in cellular bioenergetics. The demand for ATP is usually reasonably constant under normal resting conditions. The hydrolysis of ATP to ADP (adenosine 5'-diphosphate) and Pi releases potential energy from high-energy phosphate bonds for all activities involved in maintaining intracellular homeostasis and the specialized functions which may be unique to each cell type. Phosphocreatine acts as an energy reservoir in tissues such as muscle and brain. The enzyme CK splits PCr to provide an energy source for ATP resynthesis (Figure 2). Oxidative phosphorylation, which involves an electron transport chain in the mitochondrial membrane, is the process which provides most of the ATP for each cell under conditions of adequate oxygen supply. In a range of situations the requirements for ATP cannot be met by oxidative phosphorylation in the mitochondria. For example, oxidative phosphorylation is impaired in hypoxia and may be inadequate in normal exercising muscle. The shortfall in ATP production is met by glycolysis in the cytoplasm. Lactic acid may accumulate as a consequence of this process. Phosphocreatine is utilized under these conditions and as PCr falls, Pi increases, but any reduction in ATP is minimized because of the buffering effect of CK. Only when the PCr pool is completely consumed do the tissue ATP levels fall appreciably, leading to a rise in both ADP and Pi.

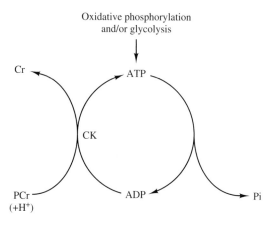

Figure 2   The interrelationship between ATP and PCr. Cr, creatine; CK, creatine kinase; ADP, adenosine diphosphate

For list of General Abbreviations see end-papers

Mitochondrial oxidative metabolism may also be affected by changes in pHi,[2] but the rate of ATP biosynthesis remains constant under normal conditions. The concentration of ADP is a major factor in the rate of ATP production.[2,3] The majority of ADP is not directly detectable using NMR methods because of tissue binding. The ability of a particular tissue to synthesize ATP may be calculated from the phosphorylation potential, given by equation (1)

$$[ATP]/[ADP][Pi] \tag{1}$$

A further indication of cellular energy status is given by equation (2):

$$[PCr]/[Cr][Pi] \tag{2}$$

Absolute or relative concentrations of tissue PCr, Pi, and ATP can be obtained using ^{31}P MRS. The rate of ATP biosynthesis, the phosphorylation potential, ADP concentrations, and the kinetics of CK may be calculated from ^{31}P MRS measurements.[2,3] The PCr/Pi and PCr/ATP ratios are commonly used MR indices of cellular energy status or 'bioenergetic reserve', as they reflect equations (1) and (2).

The oxidative phosphorylation pathway can be assessed using the PCr/Pi ratio. Glycolytic intermediates contribute to the PME peak and therefore an indirect assessment of glycolysis may be obtained from quantitation of this resonance. This is of particular importance in assessment of glycolytic disorders in muscle and in dynamic studies of liver metabolism.

3   ^{31}P MRS AND INTRACELLULAR pH

Phosphorus-31 MRS can be used to measure intracellular pH from the chemical shift of the Pi peak with reference to the PCr resonance in tissues such as brain and muscle where PCr is present. In tissues such as liver where PCr is absent the reference used is αATP. The MR signal from Pi is thought not to represent the total intracellular levels of Pi. It is unclear why the remainder is not detected, but it may be bound in the mitochondria. It is not known whether there are pH differences between the cytoplasm and the mitochondria. Split Pi resonances representing intracellular compartmentation have not been observed in human ^{31}P MRS in vivo.

Different body tissues are more susceptible to ischemia and hypoxia than others. In normal exercising muscle, anaerobic glycolysis may take place with the accumulation of lactic acid as a normal sequence of events. The large PCr reservoir in muscle ensures an energy source for these anaerobic reactions. The accumulation of lactate in the brain is more likely to be of pathological consequence because cerebral function is particularly sensitive to hypoxic insults. The measurement of intracellular acidosis may be used as a marker of hypoxia–ischemia and cellular dysfunction in conditions such as stroke or birth asphyxia. Under these circumstances, the PCr/Pi ratio is reduced. Lactate accumulation leads to a reduction in pHi and a change in the chemical shift of Pi. Intracellular pH may change with time: for example, an intracellular alkalosis can develop in ischemic brain tissue of stroke patients over an extended time period.[4] The underlying mechanisms behind such pH changes are not known. However, the MR measurement of pHi may be used to discriminate between diseased and healthy tissue in combination with indices of bioenergetic reserve such as PCr/Pi.

4 ^{31}P MRS AND PHOSPHOLIPID METABOLISM

The PME and PDE peaks are multicomponent. Phosphoethanolamine (PE) and phosphocholine are cell membrane precursors and contribute to the PME peak with signal from glycolytic intermediates and AMP. Glycerophosphorylethanolamine (GPE) and glycerophosphorylcholine (GPC), which are cell membrane degradation products, contribute to the PDE resonance. Phosphoenolpyruvate and endoplasmic reticulum form other contributory factors. The relative contributions of these compounds to the PME and PDE peaks may change with disease. In situations where there is rapid cell turnover, the PME resonance may be elevated due to an increase in PE and phosphocholine. Similarly, under conditions of rapid cell death, after tumor embolization or chemotherapy, for example, an increased contribution of GPE and GPC to the PDE peak may be expected. Phosphorus–proton decoupling can be used to resolve further these resonances in vivo.

5 ^{31}P MRS AND TRANSPLANT ORGAN VIABILITY

Organ transplantation is a steadily expanding surgical field. Increased patient survival rates have been achieved because of improved operative techniques, anti-rejection chemotherapy, and methods for harvesting and preserving donor organs.

The success of any transplant procedure is dependent on the quality of the donor graft and this is a reflection of organ storage methods. The use of more physiological preservation fluids has led to increased storage times, allowing donor organs to be transported between transplant centers. Despite these advances, tissue damage, caused by cold preservation, is an important factor in patient morbidity and mortality. Phosphorus MRS provides a noninvasive assessment of the viability of the isolated donor organ prior to transplantion. The standard indices of bioenergy reserve such as PCr/ATP and PCr/Pi ratios may be appropriate markers in heart transplantation, but the ^{31}P MR spectra from healthy liver and kidney contain no appreciable PCr. Therefore, other indices such as the PME/Pi ratio have to be employed. Adenosine 5′-triphosphate begins to degenerate to ADP and Pi immediately each organ has been harvested. With time, ADP further degenerates to AMP which contributes to the PME peak. The PME/Pi ratio may reflect the ability of the isolated organ to rephosphorylate AMP to ATP on transplantation. Specific transplant studies are considered later in this chapter.

6   CLINICAL APPLICATIONS

The development of whole body magnets has allowed clinical ^{31}P MRS studies to be undertaken in patients with a variety of pathological processes, facilitating comparisons with healthy volunteers and offering the possibility of disease monitoring in response to treatment. MR measurements of bioenergetic reserve such as the PCr/Pi ratio have been proposed as predictors of outcome in hypoxic and/or ischemic conditions such as birth asphyxia. A review of the role of ^{31}P MRS in some of the major disease processes follows.

6.1 ^{31}P MRS and the Adult Brain

6.1.1   Stroke

Stroke is the most common adult neurological condition and is a prominent cause of mortality in developed countries. Early diagnosis of ischemia facilitates more appropriate treatment, and delay may result in an irreversible loss of neuronal function. Ischemia and the consequent tissue hypoxia result in a depletion in PCr, ATP levels being maintained initially. The acute stages of stroke are characterized by a decreased PCr/Pi ratio in the ^{31}P MR spectrum and an intracellular acidosis.[5] These changes may be detectable before MRI changes become evident.[5] A combination of ^{31}P MRS and MRI may aid early diagnosis, and help to monitor the brain's response to treatment.

Persistent cerebral ischemia results in irreversible cell damage and neuronal death. A reduction in phosphorus signal has been seen in patients with chronic stroke, consistent with a reduction in viable cells in the area of infarction.[4] The pHi has been noted to change with time, resulting in a rebound intracellular alkalosis which may persist.[4] The reasons underlying this change remain unclear.

6.1.2   Transient Ischemia

A transient ischemic attack is defined as a reversible neurological deficit that lasts for 24 h or less. The diagnosis is therefore retrospective and treatment is aimed at preventing recurrence. One study of two patients suggested that the total phosphorus signal was reduced in the absence of MRI changes.[4] This may be due to ischemia, but remains difficult to explain.

6.1.3   Epilepsy

Temporal lobe epilepsy may be unresponsive to standard antiepileptic therapy and a small number of patients require surgical resection of the epileptogenic focus. The definition of the pathological area needs careful preoperative planning. MRI studies have been used to obtain hippocampal volumes[6] but the role of ^{31}P MRS has been limited. A reduced PME, probably as a result of underlying hippocampal sclerosis, has been noted in some studies. An increased Pi and an unexplained intracellular alkalosis have been noted in most studies,[4,7] all of which have involved relatively small numbers of patients.

6.1.4   Alzheimer's Disease

The results of ^{31}P MRS studies have been disappointing. Bottomley and colleagues[8] found no changes in either metabolite ratios or absolute concentrations of metabolites.

6.1.5   Multiple Sclerosis

This condition is common in temperate climates and is characterized by episodes of focal neurological deficit which relapse and remit over a period of many years. The classic histological lesions are plaques of demyelination. MRI has

For References see p. 1228

revolutionized the diagnosis of this condition, but the results of ^{31}P MRS studies are less clear-cut. In one study[9] there was a decrease in the PCr/ATP ratio in some patients with active disease, whereas in another study[10] there was an increase in the PCr/ATP ratio with disease activity.

6.1.6 Other Cerebral Conditions

Decreased PCr/Pi has been observed in the cerebral ^{31}P MR spectra from patients with migraine and mitochondrial cytopathies. This suggests an alteration in bioenergetic reserve.

Chronic hepatic encephalopathy is defined as the neuropsychiatric impairment observed in patients with cirrhosis of the liver. The results of some ^{31}P MRS studies have suggested altered brain energy metabolism in these patients (see also *Systemically Induced Encephalopathies: Newer Clinical Applications of MRS*).

6.2 ^{31}P MRS and Pediatric Brain Studies

The normal ^{31}P MR spectrum from a healthy neonatal brain is significantly different from adult spectra. The PME signal is much larger in the neonate and this varies with gestational age. Azzopardi and colleagues[11] found that the PME is smaller and the PDE larger in the healthy full-term infant than the healthy preterm infant, related most probably to the changes in membrane lipids with myelin formation. The PCr/Pi ratio increases with gestational age, indicating an increased phosphorylation potential with brain development. The measured intracellular pH appears not to vary with the gestational age of healthy newborn infants.

6.2.1 Hypoxic–Ischemic Encephalopathy

Birth asphyxia of the newborn infant has been extensively investigated using ^{31}P MRS. This condition is almost unique because there are well-defined MR indices of prognosis. Reduced PCr/Pi ratio has been correlated with outcome.[12] Spectra obtained in the initial 24 h after birth may be normal, but a fall in PCr and a rise in Pi (reduced PCr/Pi ratio) may develop over the ensuing hours and days (Figure 3). This reflects defective oxidative phosphorylation as a result of birth trauma. Intracellular pH tends to rise in a delayed response to the hypoxic–ischemic insult and there may also be a reduction in ATP levels. The metabolite ratios tend to return to normal within 2 weeks in neonates who recover. The reduction in PCr/Pi ratio is proportional to the degree of subsequent neurodevelopmental impairment and to reduced cranial growth in the first year of life.[12] In the severest cases, where the neonates subsequently die, the PCr and ATP may be almost undetectable. The Pi often rises out of proportion to the reduction in PCr. This delayed or secondary response to ischemic injury is poorly understood, but may be partly due to the toxic effects of neurotransmitters such as glutamate, which may induce mitochondrial membrane disruption through the generation of free radicals.[13] Phosphorus-31 MR spectroscopy may be utilized to monitor the effectiveness of treatment designed to prevent this secondary energy failure. The development of suitable therapeutic regimens is an area of current and future research. However, the PCr/Pi ratio is already being used as a predictor of patient outcome and may give the pediatrician insight into planning future management decisions.

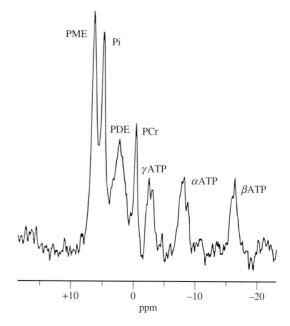

Figure 3 An unlocalized ^{31}P MR spectrum from the head of a birth asphyxiated neonate. The PCr peak is reduced and the Pi peak is elevated

6.3 Myocardial Metabolism

Heart muscle is well supplied with oxygen and mitochondria. It relies on glycolysis to a much smaller extent than skeletal muscle, which has a higher concentration of glycolytic enzyme and fewer mitochondria. The regulation of cardiac metabolism and its relation to mechanical function has been widely studied in animals and humans[14,15] (see also *Cardiovascular NMR to Study Function* and *NMR Spectroscopy of the Human Heart*).

6.3.1 Measurements in Normal Myocardium

Some of the major problems encountered in cardiac studies are related to signal contamination of the myocardium from the chest wall and from blood circulating through the cardiac chambers. The chest wall contains about four times more PCr than the normal myocardium. Signal from blood usually includes appreciable amounts of ATP and PDE, and an intense 2,3-diphosphoglycerate (2,3-DPG) resonance, which may obscure the Pi and PME resonances. Under such circumstances the pHi and the PCr/Pi ratio cannot be measured.[16]

Inorganic phosphate and pHi measurements still remain difficult in humans. Localization techniques may be used to minimize signal contamination from chest wall or blood. Correction for residual blood contamination and saturation effects may also be made.

The PCr/ATP ratio is the most frequently used index of bioenergy reserve in cardiac studies. Results from human and animal studies are comparable.[17] This ratio remains relatively constant during the cardiac cycle and in exercise. It is also highly reproducible. The PDE/ATP ratio is difficult to measure accurately because of interference from the 2,3-DPG signal. The available data show considerable biological variation in

were significant ($p < 0.05$). There was no significant difference between the absolute values for cirrhotics and hepatitics.

In a study of patients with alcohol liver disease, including five patients with cirrhosis, Angus et al.[42] found no change in any of the metabolite ratios in the patients with cirrhosis. There was no change in Pi/ATP, suggesting no evidence of impaired cellular energetics. The hepatic pH was also similar in patients with cirrhosis and in controls. This patient group had previously been found to have established cirrhosis, and had been abstinent from alcohol for at least 6 months.

Absolute concentrations were measured using one-dimensional chemical shift imaging (CSI) techniques in five healthy volunteers and five patients with alcoholic cirrhosis.[29] Absolute metabolite levels were calculated with reference to an external standard. Metabolite ratios were not altered in cirrhosis compared with controls, although absolute concentrations of all hepatic metabolites tended to be lower. However, only the reduction in βATP, of 31%, was significant. Histological evidence suggested that the reduction in ATP levels reflected fewer functioning liver cells per volume of liver, since functional cells had been destroyed and replaced by fibrosis. The amount of ATP per liver parenchymal cell was thought to be unchanged. The authors suggested that metabolite ratios were of limited diagnostic value in the assessment of alcoholic cirrhosis, if they were unsupported by quantitative analysis.

Fourteen patients with liver cirrhosis of differing severity were examined using one-dimensional CSI methods.[43] Patients were divided into two groups according to the severity of their liver disease using Child's classification and the aminopyrine breath (AB) test. The PME/total phosphorus ratio was significantly higher in patients with mild and severe cirrhosis. There was no change in the PME T_1 value in cirrhotics. There was a significant negative linear correlation of PME with percentage dose of $^{14}CO_2$ excreted over 2 h in the AB test. The pH values were significantly elevated in mild cirrhosis (pH 7.45), but not in severe cirrhosis (pH 7.36), compared with controls (pH 7.29). This was the first paper highlighting the clinical potential of hepatic ^{31}P MRS as a noninvasive means of assessing the severity of liver cirrhosis.

In a study of a group of 86 patients with histologically proven cirrhosis of varying etiology and functional grade (Menon et al.[45]), patients with liver disease showed a significantly higher median PME/ATP ($p < 0.0001$), PME/PDE ($p < 0.0001$), PME signal height ratio (SHR) ($p < 0.0001$), and Pi SHR ($p < 0.02$), and a lower median PDE/ATP ($p < 0.001$) and PDE SHR ($p < 0.001$) (the SHR was obtained by dividing the peak height at TR 5 s by that at TR 0.5 s to yield a T_1-related SHR). The magnitude of these changes significantly and progressively increased with worsening functional state (Figure 5).

Etiological differences were noted in patients with compensated cirrhosis. Spectra from patient with cirrhosis secondary to viral hepatitis showed a significantly higher PME/ATP, Pi/ATP, and PME/PDE, and those with cirrhosis secondary to primary sclerosing cholangitis showed a significantly lower Pi/ATP than did other etiological groups.[45] However, spectral appearances did not vary with etiology of cirrhosis in decompensated patients. In vitro ^{31}P MRS of perchloric extracts of biopsy samples of liver tissue obtained from 10 patients with cirrhosis at transplant hepatectomy showed increases in levels of PE and PC, and a reduction in levels of GPE and GPC.

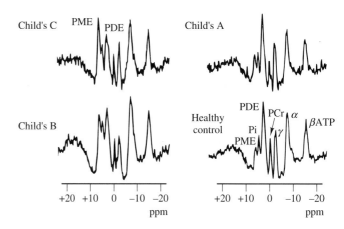

Figure 5 Hepatic ^{31}P MR spectra collected using a two-dimensional CSI sequence with TR T_p-5 s, pulse angle 45° from a healthy volunteer, and patients with Child's A, B and C functional grades of alcoholic cirrhosis

These changes suggest regenerative activity in cirrhotic livers. The reduction in soluble PDE in the aqueous extracts did not quantitatively account for the reduced PDE resonance seen in vivo, and the changes seen in vivo may therefore be partly due to reduction in contributions from hepatic endoplasmic reticulum, resulting from replacement of hepatocytes by fibrous tissue.

In order to determine the feasibility and utility of dynamic hepatic ^{31}P MRS, Dufour et al. studied six healthy subjects and nine patients with nonalcoholic cirrhosis after an intravenous infusion of a fructose load (250 mg kg^{-1}).[44] The basal spectra between the healthy and cirrhotic subjects had comparable metabolite areas, except that the contribution of PDE was significantly smaller in cirrhotic patients than in healthy subjects ($33 \pm 5\%$ versus $38 \pm 5\%$; $p < 0.05$). In both groups, the relative PME peak increased and then returned to basal values, Pi decreased then rapidly increased and overshot basal values and slowly returned to basal values, and βATP decreased and slowly returned to basal values. However, in the cirrhotic group the magnitude of these changes were reduced; PME increased to $9 \pm 5\%$ versus $20 \pm 8\%$ in controls, Pi decreased by $5 \pm 4\%$ versus $11 \pm 3\%$ of total area ($p < 0.005$). These measurements correlated with the severity of the impairment of liver function measured by galactose-elimination capacity.

5.5 Alcoholic Hepatitis

As with studies on alcoholic cirrhosis the hepatic MRS findings varied according to the MRS technique used, the degree of quantitation achieved and the clinical state of the patient. Meyerhoff et al.[41] found that in 10 patients absolute concentrations of metabolites were decreased by 25–44%, but there was no change in metabolite ratios. Also, intracellular pH was more alkaline in the patient group compared with controls.

Angus et al.[42] measured the metabolite ratios to be abnormal in 16 patients with hepatitis, none of whom had consumed alcohol within the previous 72 h. PME/Pi and PME/ATP were elevated, and the PME level showed a significant positive correlation with the severity of alcoholic hepatitis, assessed by histology. Pi/ATP and pH were similar to control values. These

For References see p. 1240

data were similar to findings from four other patients previously studied by the same group.[28]

5.6   Viral Hepatitis

A number of patients with viral hepatitis have been studied, generally as a subgroup of a study of a heterogeneous patient group. Elevated PME/ATP has been reported in five patients with acute viral hepatitis,[28] three patients with viral hepatitis B,[41] and one patient with non-A non-B hepatitis.[7] Oberhaensli et al. found that the high phosphomonoesters returned to normal as liver function became normal, suggesting they were probably associated with liver regeneration 1–2 weeks after the onset of jaundice when the investigations were performed.[28]

5.7   Chronic Hepatitis

Serial changes in phosphorus metabolites were monitored after fructose infusion in five patients with chronic hepatitis.[46] Following 0.5 g kg^{-1} fructose the increase in PME at 15–20 min (151 ± 49% of the preadministration value) was significantly less than in healthy volunteers ($p < 0.05$). Also, the rebound of Pi at 35–40 min (126 ± 42%) was significantly less than that in healthy volunteers ($p < 0.05$). These findings were interpreted as suggesting that reduced fructose utilization is caused by impaired fructose transport into liver cells, thus indicating that this is a promising method in the functional evaluation of certain diffuse liver diseases.

5.8   Glycogen Storage Disease

Over 10 forms of glycogen storage disease resulting from inherited enzyme defects have been characterized. The different diseases are characterized by a storage of glycogen in abnormal quantities and/or synthesis of glycogen with an abnormal structure.

Beckmann et al.[47] obtained hepatic ^{13}C MRS spectra from one patient with type IIIA glycogen storage disease, which is characterized by an increased glycogen concentration of abnormal structure in liver and muscle, due to inactivation of the amylo-1,6-glucosidase debrancher enzyme. They found glycogen levels, measured from the C-1 resonance, to be increased by a factor of 2–3 compared with well-trained athletic normals.

A hepatic ^{31}P MRS study of two patients with glucose 6-phosphatase deficiency (glycogen storage disease type 1A) showed that, after an overnight fast, PME was increased and Pi was low–normal (Pi/PME was markedly reduced).[48] The increase in PME was attributed to accumulation of sugar phosphates (mainly glycolytic intermediates), on the basis of chemical shift measurements. After 1 g kg^{-1} oral glucose, hepatic sugar phosphates decreased by 40–64% and reached normal levels, whereas Pi increased by 40–130%. Liver Pi levels remained elevated in both patients 30 min after ingestion of glucose. Liver PME and Pi levels did not change in four control subjects after a glucose load. These high levels of PME were interpreted as suggesting that the activity of residual glucose 6-phosphatase may be enhanced, thus increasing hepatic glucose production and reducing the degree of hypoglycemia during fasting. The finding of large fluctuation in hepatic Pi levels may be directly related to the increase in uric acid production typically seen in these patients.

5.9   Galactose Intolerance

Galactosemia is an autosomal recessive disorder caused by a deficiency of the enzyme UDPglucose:α-D-galactose-1-phosphate uridyltransferase (EC 1.7.7.12). Liver damage is thought to be caused by accumulation of metabolites of galactose, although the exact mechanism is unclear. In a study of two galactosemic patients,[49] an oral load of 20 mg kg^{-1} produced significant changes in the hepatic ^{31}P MRS spectrum of one of the patients. The peak at 5.2 ppm increased on two occasions to about twice its original size 60 min after galactose administration. In vitro animal studies showed this increase was largely due to galactose 1-phosphate.

5.10   Miscellaneous

Hypothyroidism is known to affect nearly every organ and organ system in the body. However, in seven patients with severe hypothyroidism there were no differences in the hepatic ^{31}P MRS spectra compared with controls, either before or after treatment with thyroid hormone substitution therapy.[50]

In a study of seven patients with iron overload[28] (ferritin levels 600 to >3000 mg mL^{-1}), the hepatic ^{31}P MRS resonances were all broadened, resulting in part from susceptibility effects of Fe^{3+}, which is stored as ferritin and hemosiderin in liver lysosomes. The line broadening correlated with the degree of iron overload.

6   FOCAL LIVER DISEASE

The majority of applications of NMR to focal liver disease have involved tumors, and these have recently been reviewed by Negendank.[51] Ideally, the tumor spectrum should be compared with the normal spectrum of cells from which the neoplasm arises, which is possible for primary liver cancers, but not for secondary liver tumors. A further problem arises because the normal liver spectrum has prominent PME and PDE signals, and as these are signals that are abnormal in tumors it can be difficult to define the tumor spectrum.

The ^{31}P MRS characteristics of hepatic tumors in adults are summarized in Table 1 and Figure 6. A number of technical points must be noted: in the majority of studies peak intensities were reported as relative ratios of areas, and molar quantitation was only attempted in two studies; most spectra were reported to be contaminated with signals from background tissue, but few authors provided estimates of the extent of contamination; and data were partially saturated, so that the effect of any difference in T_1 was not taken into account when comparing cancer and normal hepatic spectra. Despite these comments, a number of points can be made. The most obvious and most consistent abnormality is an elevation of PME. The change in PDE is less consistent, with some groups reporting an elevation and others reporting a reduction. Changes in the Pi signal showed no obvious pattern. The pH in tumors was either similar or slightly higher than that of normal tissue. Furthermore there were no obvious differences between different types of tumor, either between primary and secondary tumors or between secondary tumors from different sites. It is interesting that there appears to be no obvious compromise of energy

Table 1 ^{31}P MRS Spectral Parameters for Patients with Hepatic Malignancy ($n \geqslant 2$)

Diagnosis	n	MRS findings[a]
Primary malignancy		
Hepatocellular carcinoma[52]	2	PME/ATP ↑
Hepatocellular carcinoma[53]	7	PME/ATP ↑, if tumor occupied more than 50% of VOI
Hepatocellular carcinoma[54]	3	PME/Pi ↑, PME/ATP ↑
Hepatocellular carcinoma[55]	2	PME ↑
Various[56]	3	PME/PDE ↑
Secondary malignancy		
Lymphoma[57]	22	PME/ATP ↑, PME/Pi ↑
Adenocarcinoma[56]	11	PME/PDE ↑
Carcinoid[56]	14	PME/PDE ↑
Various[52]	3	PME/ATP ↑, ATP ↓, Pi ↓
Various[28,59]	3	PME ↑
Breast carcinoma[54]	2	PME/ATP ↑, PME/Pi ↑, PDE/Pi ↑, pH ↑
Various[58]	19	PME/ATP ↑, PDE/ATP ↑, Pi/ATP ↑
Colorectal carcinoma[55]	10	PME ↑
Various[53]	30	PME/ATP ↑, if tumor occupied more than 50% of VOI

[a]VOI, volume of interest.

metabolism, and effects of ischemia or hypoxia do not dominate the tumor spectrum.

Thus, in distinguishing neoplasms from normal tissue, only PME can be considered a diagnostic discriminant. Since proton decoupling during acquisition of the ^{31}P MRS data has not been performed in human tumor studies, there is no direct information on individual components in vivo. However, high-resolution MRS of perchloric acid extracts of tissue obtained at the time of surgery show that phosphoethanolamine and phosphocholine were elevated in four hepatocellular carcinomas and four secondary tumors.[1,56]

It is hoped that MRS will provide a marker of treatment response, particularly in the early stages of therapy. For example, hepatic embolization combined with intraarterial administration of cytostatic changes (chemoembolization) may be used to treat primary and metastatic cancers to the liver. As an acute response to chemoembolization in three patients, ATP, PME, and/or PDE concentrations diminished, whereas Pi concentrations increased or stayed relatively constant.[52] Long-term follow-up after chemoembolization showed elevated PME/ATP and increased ATP concentrations in the absence of changes by conventional imaging techniques.[52] In another study of 10 patients with liver metastases from colorectal carcinoma and two patients with hepatocellular carcinoma a marked increase in Pi and a decrease in ATP were observed during the first few hours after local chemotherapy of chemoembolization and later, PME increased and PDE decreased.[55] In a study of two patients with carcinoid syndrome, successful arterial embolization was accompanied by a decrease in PME/PDE and an increase in Pi/ATP, whereas in a patient in whom the tumor blood supply was not effectively interrupted there was little change in metabolite ratios.[7]

In a study of 22 patients with lymphoma, 11 were re-examined after chemotherapy treatment.[57] Before treatment, PME/ATP and PME/Pi were elevated and approximately related to clinical stage, although there were some notable discrepancies. Liver phosphomonoester levels decreased following chemother-apy in all but one of the patients who had abnormally high ratios in their initial spectra; this decrease was seen as early as 1 day and as late as 2 weeks after commencing treatment. In patients with initially normal PME levels, treatment did not produce a fall in PMEs. These findings suggest that detection of falling PME levels indicates that the drugs are reaching the target cells and affecting tumor cell metabolism, which may be of considerable clinical importance.

A study of two children with neuroblastoma[60] examined on a number of occasions from the ages of 1–9 months, showed the hepatic ^{31}P MRS spectrum from the region of pathology to be abnormal with elevated PME levels in the initial spectroscopic examinations. Further treatment showed a reduction in PME levels, and the spectra finally became similar to the normal subject (aged 3 months).

A small number of other focal liver lesions have been studied.[54,58] The metabolite ratios from two patients with cavernous hemangiomas were apparently normal, although the S/N was reduced.[54] The in vitro high-resolution MRS spectra from a multiloculated cyst showed elevated PME and reduced PDE, and this was similar to data from extracts of tumor tissue.[1]

7 DRUGS AND THE LIVER

The liver is the major site of drug metabolism. Lipid-soluble drugs are converted to more water-soluble forms by a group of hepatic mixed-function enzymes, including cytochrome P450. These processes facilitate excretion of the drugs in urine or bile. Drug metabolism firstly involves oxidation, reduction, or demethylation of the drug, followed by conjugation of the derivatives produced with glucrudine, sulfate, or glutathione. These conjugates are excreted in the urine and bile as they cannot be reabsorbed by renal tubular or bile ductular cells. A number of factors affect drug metabolism, including the microsomal enzyme system (which will influence the speed of

For References see p. 1240

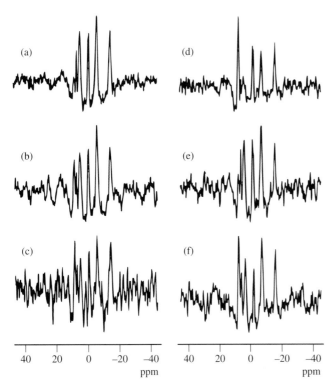

40   20   0   −20   −40   40   20   0   −20   −40
ppm   ppm

Figure 6 Representative ^{31}P hepatic MRS spectra obtained from a healthy adult volunteer and patients with hepatic malignancies of varying histology using CSI techniques, *TR*-1 s, pulse angle 45°. (a) Spectrum from a healthy 22-year-old female acquired using two-dimensional CSI; nominal planar resolution, 30 mm; total number of data collections, 256. (b) Spectrum from a 61-year-old man with squamous cell carcinoma obtained from a voxel containing tumor acquired using four-dimensional CSI; nominal resolution, 30 mm × 30 mm × 30 mm; total number of data collections 2048. (c) Spectrum from a 59-year-old man with carcinoid liver metastases obtained from a voxel containing tumor acquired using four-dimensional CSI; nominal resolution, 20 mm × 20 mm × 20 mm; total number of data collections 2048. (d) Spectrum from a 56-year-old man with carcinoid liver metastases obtained from a voxel containing tumor acquired using four-dimensional CSI; nominal resolution, 30 mm × 30 mm × 30 mm; total number of data collections 2048. (e) Spectrum from a 79-year-old man with carcinoid liver metastases obtained from a voxel containing tumor acquired using four-dimensional CSI; nominal resolution, 30 mm × 30 mm × 30 mm; total number of data collections 2048. (f) Spectrum from a 53-year-old man with hepatocellular carcinoma obtained from a voxel containing tumor acquired using four-dimensional CSI; nominal resolution, 30 mm × 30 mm × 30 mm; total number of data collections 2048. (Reproduced by permission of Heydon & Son from I. J. Cox, J. D. Bell, C. J. Peden, R. A. Iles, C. S. Foster, P. Watanapa, and R. C. N. Williamson, *NMR Biomed.*, 1992, **5**, 114)

metabolism), the route of administration, the liver blood flow, and competitive inhibition.

The majority of studies involving alcohol abuse have reported the spectral characteristics of resultant liver damage, rather than the effects of alcohol per se.[28,29,41–45] However, it may be important to distinguish between the two effects, particularly as the metabolic consequences of an alcohol load may persist for a few hours to a few weeks and may depend on the degree of liver injury.

In a study of three healthy volunteers, acute alcohol ingestion (0.5–1.0 g kg^{-1} alcohol) was associated with a transient but significant elevation in PME/ATP.[61] In 14 patients with minimal liver injury, active drinking was associated with elevation in mean PME/ATP ($p = 0.12$) and in mean PDE/ATP ($p < 0.0001$); abstinence from alcohol was associated with a prompt (3–7 day) reduction in PME/ATP and with a reduction in PDE/ATP over a longer timescale (3–28 days). In eight patients with alcoholic cirrhosis, active drinking was associated with elevation in mean PME/ATP ($p < 0.05$) and mean PDE/ATP ($p = 0.4$); abstinence from alcohol had no effect on the mean PME/ATP, although the mean PDE/ATP fell to within or below the reference range. The reversible elevation in PME/ATP in healthy volunteers given an alcohol load and in chronic alcohol abusers while actively drinking was interpreted as reflecting changes in hepatic redox potential as a result of obligatory alcohol metabolism. The irreversible elevation in PME/ATP, observed in patients with cirrhosis, probably reflects changes associated with hepatocyte regeneration. The reversible elevation in PDE/ATP observed in chronic alcohol abusers while actively drinking most likely reflects induction of hepatocyte endoplasmic reticulum.

Self-poisoning with acetaminophen accounts for many emergency hospital admissions. Overdoses cause cell damage through increased oxidation of the drug as the conjugation pathways are saturated. Early prediction of outcome in patients with acetaminophen overdose is difficult; for example, aspartate aminotransferase levels peak late after the ingestion of the drug and correlate poorly with prognosis. While the prothrombin time gives some indication of synthetic ability, methods for directly assessing cell viability are sought. Dixon and colleagues[62] studied 18 patients with acetaminophen poisoning and found that the concentrations of all metabolites fell in parallel with a decrease in the synthetic ability of the liver. ATP and PDE fell to about 20% of their normal concentrations in severely affected patients. The reduction in PDE was interpreted as a reduction in endoplasmic reticulum.

The direct observation of 5FU and its metabolites in the human liver was first reported in 1987.[63] Signals from 5FU and one of its catabolites, α-fluoro-β-alanine (FBAL), were observed in the human liver in patients undergoing cancer chemotherapy (Figure 7). The pharmacokinetics of 5FU in the tumors of 11 patients with carcinoma at various sites were studied, including three patients with tumor located in the liver.[64] A long liver tumor pool of 5FU was observed in six of 11 tumors, including one of three patients with carcinoma in the liver. The halflife of this 'trapped' pool was 0.33–1.3 h, considerably longer than the halflife of 5FU in blood (5–15 min). Neither the anabolites or catabolites of 5FU were detected by ^{19}F MRS. Patient response to chemotherapy appeared to correlate with the extent of trapping of free 5FU in tumors.

Semmler et al.[65] reported hepatic ^{19}F MRS data from eight patients with liver tumors, receiving intra-arterial 5FU. Contributions from liver and tumor could not be distinguished in the spectrum. The time constants for the kinetics of 5FU ranged from 8 to 75 min, whereas the time constants for FBAL were either approximately 15 or 50 min. A broad peak comprising nucleoside and nucleotide anabolites was detected in one patient.

More recently, Findlay et al.[66] studied hepatic ^{19}F MRS in 26 colorectal cancer patients treated with a continuous low

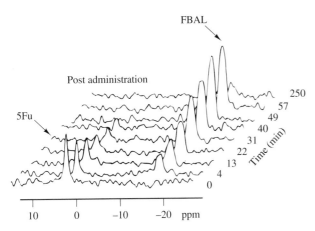

Figure 7 Serial ^{19}F spectra acquired with a surface coil over the liver of a patient (66 year old male) in a 1.5 T MRI system, operating at 59.8 MHz for ^{19}F. Each spectrum is the result of 128 FIDs using a 250 μs, 90° pulse with 512 complex points collected over a period of 8.5 min. Actual spectral width was 29 to −54 ppm. A shifted sine-bell squared apodization was used for S/N enhancement. (Reproduced by permission of Pergamon Journals Ltd. from W. Wolf, M. J. Albright, M. S. Silver, H. Weber, U. Reichardt, and R. Sauer, *Magn. Reson. Imag.*, 1987, **5**, 165)

dose intravenous infusion of 5FU, until the point of refractory disease, at which time interferon-α was added with the objective of modulating 5FU activity. In patients observed by MRS during the first 8 weeks of 5FU treatment, those with a visible 5FU signal were likely to respond to treatment ($p = 0.017$). At the time of interferon-α addition, MRS showed that seven patients developed new or increased 5FU signals, and four patients showed a signal from the active metabolites of 5FU. The patients who exhibited a new or increased 5FU signal were more likely to show further response to interferon-α ($p = 0.007$).

The in vivo pharmacokinetics of a model drug, fleroxacin, was studied in healthy human subjects.[67] After oral administration a signal was detected in the hepatic ^{19}F MRS spectrum, and also muscle, and monitored over a period of 24 h. Pharmacokinetic data for the liver were obtained, combining MRS results with high-performance liquid chromatography (HPLC) analysis of plasma: $t_{max} = 1.4$ h; $C_{max} = 53$ μmol L^{-1}; $t_{1/2} = 4.4$ h (fast phase) and 10.8 h (slow phase).

8 FUTURE APPLICATIONS

Hepatic MRS can provide a direct measure of hepatic function, for example from the baseline spectrum or following a metabolic challenge. Since the indirect clinical and laboratory measures of hepatic function may be subject to extrahepatic influences, direct noninvasive measurement is of importance. A range of studies have been made, as described above, and these can be extended, for example, specifically to study gluconeogenesis in diabetics and tumor-bearing patients using ^{13}C or ^{31}P MRS combined with an alanine stress test. In addition, the interpretation that metabolite changes in end-stage liver disease can provide an index of hepatocyte regeneration, rather than

reflecting production of fibrous tissue or inflammatory cell activity, suggests that hepatic ^{31}P MRS may provide a specific marker which could be used to guide management decisions and as a prognostic assessment in patients with liver disease. Further studies correlating functional grades and etiology to absolute metabolite levels are appropriate.

The clinical management of a range of pathologies relies on the interpretation of histological results from liver biopsy. Since liver biopsy has a defined risk, a noninvasive technique of providing the same information would be of value, particularly when serial liver biopsies are required, for example in patients undergoing liver transplantation. In such patients hepatic MRS may be used to provide a guide to optimal timing of transplantation, to indicate good or poor liver metabolic status for noninvasive assessment of donor livers prior to transplantation[68] and also to provide a useful early marker of rejection after transplantation.

For patients with hepatic malignancy, ^{31}P MRS may provide an early marker of response to treatment, and ^{19}F MRS can be used to measure the pharmacokinetics of specific anticancer drugs and their metabolites. Other areas of study include the metabolism of the liver in patients with tumor at distant sites. Hepatic MRS may provide insights into the mechanisms responsible for cachexia.

Many of the clinical and biochemical applications of hepatic MRS to date have relied on the interpretation of changes in metabolite ratios, rather than absolute metabolite concentrations. However, in a number of pathologies, metabolite ratios alone do not adequately reflect the underlying metabolic changes, particularly if there is an alteration in ATP levels. Recourse to absolute metabolite concentrations is therefore required. This may be difficult to achieve within the confines of a clinical examination because the low S/N of hepatic spectra means it is time consuming to measure the NMR characteristics of individual resonances. Improvements in sensitivity and/or spectral resolution may be achieved using proton decoupling techniques, higher field strengths, and more efficient rf coil arrangements. However, detailed interpretation of clinical data will continue to rely on results from animal models. A wide range of metabolite resonances can be detected using ^{1}H MRS,[1] and, if the technical problems in acquiring hepatic ^{1}H MRS spectra can be overcome, hepatic ^{1}H MRS will provide additional and new information.

In conclusion, hepatic MRS can provide a noninvasive measure of hepatic function, which has a range of clinical and biochemical applications in diagnosis, prognosis, and assessment of treatment efficacy in liver disease.

9 RELATED ARTICLES

Animal Methods in MRS; Applications of ^{19}F-NMR to Oncology; Cells and Cell Systems MRS; Chemical Shift Imaging; High-Field Whole Body Systems; pH Measurement In Vivo in Whole Body Systems; Proton Decoupling During In Vivo Whole Body Phosphorus MRS; Quantitation in In Vivo MRS; Single Voxel Whole Body Phosphorus MRS; Spatial Localization Techniques for Human MRS; Spectroscopic Studies of Animal Tumor Models; Susceptibility Effects in Whole Body Experiments; Tissue and Cell Extracts MRS; Tissue

For References see p. 1240

Behavior Measurements Using Phosphorus-31 NMR; Whole Body Studies: Impact of MRS.

10   REFERENCES

1. J. D. Bell, I. J. Cox, J. Sargentoni, C. J. Peden, D. K. Menon, C. S. Foster, P. Watanapa, R. A. Iles, and J. Urenjak, *Biochim. Biophys. Acta*, 1993, **1225**, 71.
2. R. A. Iles, A. N. Stevens, and J. R. Griffiths, *Prog. NMR Spectrosc.*, 1982, **15**, 49.
3. R. D. Cohen, R. M. Henderson, R. A. Iles, J. P. Monson, and J. A. Smith, 'Metabolic Acidosis', Ciba Foundation Symposium 84, Pitman, London, 1982, p. 20.
4. R. D. Oberhaensli, G. J. Galloway, D. J. Tayor, P. J. Bore, and G. K. Radda, *Br. J. Radiol.*, 1986, **59**, 695.
5. P. R. Luyten, G. Bruntink, F. M. Sloff, J. W. A. H. Vermeulen, J. I. van der Heijden, J. A. den Hollander, and A. Heerschap, *NMR Biomed.*, 1989, **4**, 177.
6. D. J. Meyerhoff, G. S. Karczmar, G. B. Matson, M. D. Boska, and M. W. Weiner, *NMR Biomed.*, 1990, **3**, 17.
7. I. J. Cox, D. K. Menon, J. Sargentoni, D. J. Bryant, A. G. Collins, G. A. Coutts, R. A. Iles, J. D. Bell, I. S. Benjamin, S. Gilbey, H. J. F. Hodgson, and M. Y. Morgan, *J. Hepatol.*, 1992, **14**, 265.
8. E. J. Murphy, B. Rajagopalan, K. M. Brindle, and G. K. Radda, *Magn. Reson. Med.*, 1989, **12**, 282.
9. T. E. Bates, S. R. Williams, and D. G. Gadian, *Magn. Reson. Med.*, 1989, **12**, 145.
10. R. K. Gupta and R. D. Moore, *J. Biol. Chem.*, 1980, **255**, 3987.
11. S. M. Cohen, *J. Biol. Chem.*, 1983, **258**, 14 294.
12. R. A. Iles, I. J. Cox, J. D. Bell, L. M. Dubowitz, F. Cowan, and D. J. Bryant, *NMR Biomed.*, 1990, **3**, 90.
13. M. Barany, D. G. Spigos, E. Mok, P. N. Venkatasubramanian, A. C. Wilbur, and B. G. Langer, *Magn. Reson. Imaging*, 1987, **5**, 393.
14. P. Canioni, J. Alger, and R. G. Shulman, *Biochemistry*, 1983, **22**, 4974.
15. T. Jue, J. A. B. Lohman, R. J. Ordidge, and R. G. Shulman, *Magn. Reson. Med.*, 1987, **5**, 377.
16. T. Jue, D. L. Rothman, B. A. Tavitian, and R. G. Shulman, *Proc. Natl. Acad. Sci. USA*, 1989, **86**, 1439.
17. D. L. Rothman, I. Magnusson, L. D. Katz, R. G. Shulman, and G. I. Shulman, *Science*, 1991, **254**, 573.
18. M. Ishihara, H. Ikehira, S. Nishikawa, T. Hashimoto, K. Yamada, J. Shishido, T. Ogino, K. Cho, S. Kobayashi, M. Kawana, T. Matumoto, T. A. Iinuma, N. Arimizu, and Y. Tateno, *Am. J. Physiol. Imaging*, 1992, **7**, 32.
19. N. Beckmann, R. Fried, I. Turkalj, J. Seelig, U. Keller, and G. Stalder, *Magn. Reson. Med.*, 1993, **29**, 583.
20. J. Alger, K. Behar, D. L. Rothman, and R. G. Shulman, *J. Magn. Reson.*, 1984, **56**, 334.
21. A. Heerschap, P. R. Luyten, J. I. van der Heyden, L. J. M. P. Oosterwaal, and J. A. den Hollander, *NMR Biomed.*, 1989, **2**, 124.
22. M. Saner, G. McKinnon, and P. Boesiger, *Magn. Reson. Med.*, 1992, **28**, 65.
23. J. L. Evelhoch, *Invest. New Drugs*, 1989, **7**, 5.
24. M. D. Boska, B. Hubesch, D. J. Meyerhoff, D. B. Tweig, G. S. Karczmar, G. B. Matson, and M. W. Weiner, *Magn. Reson. Med.*, 1990, **13**, 228.
25. H. Bomsdorf, T. Helzel, D. Kunz, P. Roschmann, O. Tschendel, and J. Wieland, *NMR Biomed.*, 1988, **1**, 151.
26. H. Barfuss, H. Fischer, D. Hentschel, R. Ladebeck, A. Oppelt, R. Wittig, W. Duerr, and R. Oppelt, *NMR Biomed.*, 1990, **3**, 31.
27. W. P. Aue, *Rev. Magn. Reson. Med.*, 1986, **1**, 21.
28. R. Oberhaensli, B. Rajagopalan, G. J. Galloway, D. J. Taylor, and G. K. Radda, *Gut*, 1990, **31**, 463.
29. V. Rajanayagam, R. R. Lee, Z. Ackerman, W. G. Bradley, and B. D. Ross, *J. Magn. Reson. Imaging*, 1992, **2**, 183.
30. I. R. Young, I. J. Cox, G. A. Coutts, and G. M. Bydder, *NMR Biomed.*, 1989, **2**, 329.
31. I. J. Cox, D. J. Bryant, B. D. Ross, I. R. Young, D. G. Gadian, G. M. Bydder, S. R. Williams, A. L. Busza, and T. E. Bates, *Magn. Reson. Med.*, 1987, **5**, 186.
32. I. J. Cox, G. A. Coutts, D. G. Gadian, P. Ghosh, J. Sargentoni, and I. R. Young, *Magn. Reson. Med.*, 1991, **17**, 53.
33. M. J. Blackledge, R. D. Oberhaensli, P. Styles, and G. K. Radda, *J. Magn. Reson.*, 1987, **71**, 331.
34. S. D. Buchhal, W. J. Thoma, J. S. Taylor, S. J. Nelson, and T. R. Brown, *NMR Biomed.*, 1989, **2**, 298.
35. P. C. Dagnelie, D. K. Menon, I. J. Cox, J. D. Bell, J. Sargentoni, G. A. Coutts, J. Urenjak, and R. A. Iles, *Clin. Sci.*, 1992, **83**, 183.
36. F. Terrier, P. Vock, J. Cotting, R. Ladebeck, J. Reichen, and D. Hentschel, *Radiology*, 1989, **171**, 557.
37. C. Segebarth, A. R. Grivegnee, R. Longo, P. R. Luyten, and J. A. den Hollander, *Biochemie*, 1991, **73**, 105.
38. I. Magnusson, D. L. Rothman, L. D. Katz, R. G. Shulman, and G. I. Shulman, *J. Clin. Invest.*, 1992, **90**, 1323.
39. R. D. Oberhaensli, B. Rajagopalan, D. J. Taylor, G. K. Radda, J. E. Collins, J. V. Leonard, H. Schwarz, and N. Herschkowitz, *Lancet*, 1987, **ii**, 931.
40. J. E. Seegmiller, R. M. Dixon, G. J. Kemp, P. W. Angus, T. E. McAlindon, P. Dieppe, B. Rajagopalan, and G. K. Radda, *Proc. Natl. Acad. Sci. USA*, 1990, **87**, 8326.
41. D. J. Meyerhoff, M. D. Boska, A. M. Thomas, and M. W. Weiner, *Radiology*, 1989, **173**, 393.
42. P. W. Angus, R. M. Dixon, B. Rajagopalan, N. G. Ryley, K. J. Simpson, T. J. Peters, D. P. Jewell, and G. K. Radda, *Clin. Sci.*, 1990, **78**, 33.
43. T. Munakata, R. D. Griffiths, P. A. Martin, S. A. Jenkins, R. Shields, and R. H. T. Edwards, *NMR Biomed.*, 1993, **6**, 168.
44. J.-F. Dufour, C. Stoupis, F. Lazeyras, P. Vock, F. Terrier, and J. Reichen, *Hepatology*, 1992, **15**, 835.
45. D. K. Menon, J. Sargentoni, S. D. Taylor-Robinson, J. D. Bell, I. J. Cox, D. J. Bryant, G. A. Coutts, K. Rolles, A. K. Burroughs, and M. Y. Morgan, *Hepatology*, 1995, **21**, 417.
46. H. Sakuma, K. Itabashi, K. Takeda, T. Hirano, Y. Kinosada, T. Nakagawa, M. Yamada, and T. Nakano, *J. Magn. Reson. Imaging*, 1991, **1**, 701.
47. N. Beckmann, J. Seelig, and H. Wick, *Magn. Reson. Med.*, 1990, **16**, 150.
48. R. D. Oberhaensli, B. Rajagopalan, D. J. Taylor, G. K. Radda, J. E. Collins, and J. V. Leonard, *Pediatr. Res.*, 1988, **23**, 375.
49. B. Kalderon, R. M. Dixon, B. Rajagopalan, P. W. Angus, R. D. Oberhaensli, J. E. Collins, J. V. Leonard, and G. K. Radda, *Pediatr. Res.*, 1992, **32**, 39.
50. K. D. Hagspiel, C. von Weymarn, G. McKinnon, R. Haldemann, B. Marincek, and G. K. von Schulthess, *J. Magn. Reson. Imaging*, 1992, **2**, 527.
51. W. Negendank, *NMR Biomed.*, 1992, **5**, 303.
52. D. J. Meyerhoff, G. S. Karczmar, F. Valone, A. Venook, G. B. Matson, and M. W. Weiner, *Invest. Radiol.*, 1992, **27**, 456.
53. I. R. Francis, T. L. Chenevert, B. Gubin, L. Collomb, W. Ensminger, S. Walker-Andrews, and G. M. Glazer, *Radiology*, 1991, **180**, 341.
54. G. M. Glazer, S. R. Smith, T. L. Chenevert, P. A. Martin, A. N. Stevens, and R. H. Edwards, *NMR Biomed.*, 1989, **1**, 184.
55. A. Schilling, B. Gewiese, G. Berger, J. Boese-Landgraf, F. Fobbe, D. Stiller, U. Gallkowski, and K. J. Wolf, *Radiology*, 1992, **182**, 887.
56. I. J. Cox, J. D. Bell, C. J. Peden, R. A. Iles, C. S. Foster, P. Watanapa, and R. C. N. Williamson, *NMR Biomed.*, 1992, **5**, 114.

For list of General Abbreviations see end-papers

57. R. M. Dixon, P. W. Angus, B. Rajagopalan, and G. K. Radda, *Br. J. Cancer*, 1991, **63**, 953.
58. G. Brinkmann, and U. H. Melchert, *Magn. Reson. Imaging*, 1992, **10**, 949.
59. R. D. Oberhaensli, D. Hilton-Jones, P. J. Bore, L. J. Hands, R. P. Rampling, and G. K. Radda, *Science*, 1986, **2**, 8.
60. J. M. Maris, A. E. Evans, A. C. Mclaughlin, G. J. D'Angio, L. Bolinger, H. Manos, and B. Chance, *New Engl. J. Med.*, 1985, **312**, 1500.
61. D. K. Menon, M. Harris, J. Sargentoni, S. Taylor-Robinson, I. J. Cox, and M. Y. Morgan, *Gastroenterology*, 1995, **108**, 776.
62. R. M. Dixon, P. W. Angus, B. Rajagopalan, and G. K. Radda, *Hepatology*, 1992, **16**, 943.
63. W. Wolf, M. J. Albright, M. S. Silver, H. Weber, U. Reichardt, and R. Sauer, *Magn. Reson. Imaging*, 1987, **5**, 165.
64. C. A. Presant, W. Wolf, M. J. Albright, K. L. Servis, R. Ring, D. Atkinson, R. L. Ong, C. Wiseman, M. King, D. Blayney, P. Kennedy, A. El-Tahtawy, M. Singh, and J. Shani, *J. Clin Oncol*, 1990, **8**, 1868.
65. W. Semmler, P. Bachert-Baumann, F. Guckel, F. Ermark, P. Schlag, W. J. Lorenz, and G. van Kaick, *Radiology*, 1990, **174**, 141.
66. M. P. Findlay, M. O. Leach, D. Cunningham, D. J. Collins, G. S. Payne, J. Glaholm, J. L. Mansi, and V. R. McCready, *Ann. Oncol.*, 1993, **4**, 597.
67. P. Jynge, T. Skjetne, I. Gribbestad, C. H. Kleinbloesem, H. F. Hoogkamer, O. Antonsen, J., Krane, O. E. Bakoy, K. M. Furuheim, and O. G. Nilsen, *Clin. Parmacol. Ther.*, 1990, **48**, 481.
68. R. F. Wolf, R. L. Kamman, E. L. Mooyaart, E. B. Haagsma, R. P. Bleichrodt, and M. J. Slooff, *Transplantation*, 1993, **55**, 949.

Biographical Sketch

Isobel Jane Cox. *b* 1959. B.A. (Nat. Sci.), 1981, University of Cambridge, UK; M.Sc., 1982, Ph.D., 1984 (supervisors Peter S. Belton and Robin K. Harris), University of East Anglia, UK. Introduced to clinical MRS on joining I.R. Young's team at GEC Hirst Research Centre. Lecturer in Diagnostic Radiology, RPMS, 1986–present. Approx. 50 publications. Current research speciality: development and applications of liver and brain MRS.

Thermal Therapies in the Body Monitored by MRI

Margaret A. Hall-Craggs, S. Smart, and A. Gillams
The Middlesex Hospital, UCLH, London, UK

1 INTRODUCTION

The purpose of minimally invasive therapy (MIT) is to deliver effective treatment with the minimum disruption and damage to normal surrounding tissue. Consequently the morbidity of the procedure may be reduced compared with an open operation. Measures of this may be such factors as reduced scarring and deformity (as in the breast), more rapid postprocedural recovery, and reduced hospital stay (as in intra-abdominal procedures). In some cases, therapy may be used where there is no other treatment option, as in tumor metastases to multiple lobes in the liver.

Treatment using MIT is fundamentally different in benign and malignant disease. In benign disease, total eradication of the target tissue may not be necessary and size reduction may be adequate. This is the case with breast fibroadenomata and also in uterine fibroid disease where amelioration of menorrhagia or infertility is the treatment aim. With malignant disease, surgical principles preside; in general, tumor resection/destruction with an adequate margin is the treatment aim. Further MIT must not impose additional risks such as increasing the rate of local recurrence or distant disease spread, or reducing disease survival.

A host of MIT techniques have been developed since the mid-1970s; these include the thermal ablative techniques to be discussed in this chapter. Other are percutaneous excision, photodynamic therapy, interstitial radiotherapy/brachytherapy, chemical ablations, laparoscopic surgery, lithotripsy, and intraluminal/intravascular procedures such as stenting, embolization, dilatation, etc.

2 METHODS OF THERMAL ABLATION

There are four principal methods of thermal ablation. Three of these involve heating of the sample—interstitial laser photocoagulation (ILP), radiofrequency (rf)-induced coagulation (RFC) and focused ultrasound (FUS)—while the fourth, cryoblation, uses cooling.

2.1 Interstitial Laser Photocoagulation

For ILP, fiber optic cables that emit light only from their tips are placed into the target region. These are coupled to a laser source and the light is turned on for the duration of the treatment. Theoretically the frequency of the light has to match an absorption band of some (unspecified) chromophore within the target region so the light energy can be converted into heat. There have been a number of approaches to improving the coupling of light to the tissue and these include the use of a focusing lens at the fiber tip and diffuser tip fibers. However it appears that laser energy deposition is primarily a thermal process. Pre-charring the optical fiber causes the fiber to behave as a point heat source (paradoxically reducing light transmission) and increases the size of the laser burn.[1] The use of low laser powers (<5W) allows the slow and controlled development of lesions over extended periods (300–1000s).

In our practice, we have used the semiconductor diode laser as it offers a number of advantages. It is small, compact, and easily transportable; it does not require a cooling system and operates from a conventional phase electrical source.

For References see p. 1248

2.2 Radiofrequency

RFC uses rf alternating current source applied through needles tipped with rf electrodes. Frictional heating is induced via ionic agitation arising from the impedance of the tissue to alternating currents. The rf source typically operates at up to 500 kHz and will often interfere with the MR acquisition at all field strengths. Such interference with the MR measurement can be reduced or eliminated with an appropriate choice of electronic filters, or by gating the rf source off during MR signal reception. This latter approach can easily be performed by simple modification of pulse sequence code. To complete the electric circuit for the applied AC, grounding pads must be applied to the patient. Care must be taken with these to prevent the possibility of local skin burns at the site of the pads. RFC can form lesions faster than ILP; however, because the deposition of the heat depends upon the electrical characteristics of the tissue, the formation of the lesion may be inhomogeneous, especially in regions of tissue boundaries.

2.3 Focused Ultrasound

FUS uses an array of ultrasound sources, typically operating at 1.5 MHz, the energy of which converges at a focus within the body. In this region, a significant amount of acoustic energy is converted into heat. The volume of the hyperthermia is very well defined for FUS; as a result, a sharp boundary may be observed between normothermic regions and those bound for necrosis. The duration of the treatment depends upon the ultrasound parameters but is typically in the region of tens of minutes. Lesions as small as 1 mm^3 can be created; consequently, high-resolution imaging techniques, such as MRI, are necessary for their accurate detection and monitoring.

2.4 Cryotherapy

Cryosurgery will typically involve the use of a needle that is continuously cooled by liquid nitrogen and which is insulated along the length of its shaft except for at the tip. The tip of the needle acts as a heat sink for the tissue and so acts as a point source for cooling. Thermal conduction increases the volume of cooled tissue. There are a variety of mechanisms responsible for the subsequent necrosis, and these depend upon the organ undergoing the cryosurgery. As for FUS, there is a well-defined boundary between normal and treated tissues.

3 ROLE OF IMAGING IN THERMAL ABLATION

The principal roles of imaging in ablative treatments are to define the extent of disease, to guide instruments, to monitor treatment during the therapeutic procedure, and to show the effect of therapy. In practice, a combination of imaging modalities are used to achieve these aims, including computed tomography (CT), ultrasound, and MR. However, there are a number of specific properties that have made MR the subject of much research interest. The exploitation of temperature-sensitive sequences can potentially be used to map areas of tissue damage during an ablative procedure, and this may facilitate treatment modification at the time of delivery to improve efficacy. Other advantages include multiplanar imaging, which

facilitates the guidance and accurate placement of tools. The lack of ionizing radiation is of importance to both the patient and the operator with lengthy therapeutic procedures.

4 MAGNET AND EQUIPMENT CONSIDERATIONS

Interventional procedures have been described both in conventional tubular high-field systems and in lower-field magnets designed to have more open access. In open systems, the tissue of interest can be located within the 'active volume' of the magnet for MRI, and the surgeon may have access to this volume from several sides. The use of the more conventional tubular high-field magnets is hindered by limited bore diameter, which limits the size of interventional devices, and by the restricted access owing to the magnet length, which prevents realtime MR-monitored adjustments of interventional tool position. On open access systems, rapidly and dynamically acquired MRI scans can assist with the positioning of interventional devices. Fast MRI may also allow near realtime visualization of physical changes within the tissue during interventional procedures. Typical procedures include MR-guided biopsies, tissue excisions, and thermal ablations. In this article we shall describe aspects of thermal ablation treatment in the human body as monitored by MRI.

All interventional devices located within the region of treatment must be made of MR-compatible materials. Such tools are available and are usually made of alloys or ceramics that present a low magnetic susceptibility difference to the body. However residual susceptibility artifacts, dependent upon the applied magnetic field strength and on the orientation of B_0 relative to the interventional device, persist and may compromise image quality in their locale, especially when using gradient echo methods. The materials used in the manufacture of these MR-compatible devices are often soft, too flexible, and are difficult to tool to sharp and cutting tips. They are also expensive.

For ILP therapy it is possible to use MR-compatible needles to position the optical fibers and then to withdraw the needles leaving the fiber in place; this eliminates the susceptibility problem. For RFC and cryotherapy, the needles must remain in place; consequently susceptibility artifacts are unavoidable. This is not necessarily a severe problem for FUS as the sonic transducers are positioned around the body not within the target region of tissue.

5 MR METHODOLOGY

Fast imaging methods are required for interventional therapies to facilitate tool guidance and to monitor therapy. Long scan times are also precluded by gross patient motion arising from respiration or peristalsis. The use of navigator echoes may allow the correction of some but not all of these motions. Prospective ordering of the phase-encode steps (e.g., respiratory ordered phase encoding, ROPE) can reduce the degree of motion artifact but requires good calibration and is sensitive to any respiratory irregularity. The use of breath-holding, especially when controlled under general anesthesia, allows imaging free of motion artifacts for scan times of up to about 30 s.

Open-access scanners tend to operate at lower field strengths than those used for routine clinical imaging and this reduces the possible signal-to-noise ratio (SNR). The rf and gradient hardware of the open-access scanner also tends to be of lower specification than that of a high-field research-grade scanner. Some fast imaging protocols, e.g., echo planar imaging, may, therefore, not be available. Further the inherently low SNR realistically achievable with low-field scanners and the persistence of susceptibility artifacts, arising from interventional tools, may make echo planar imaging impractical for some of these applications.

Rapid image reconstruction and display is necessary for realtime monitoring. This may be achieved more quickly by by-passing the scanner and using custom built hardware and software. Methods such as keyhole imaging and local-look scanning have also been used to enhance image update rates.

5.1 Sources of Contrast for Monitoring Thermal Ablations

Successful monitoring of hyperthermic ablations requires a temperature-dependent source of contrast for the heated region. It is important that any contrast seen is a direct consequence of the heating effect rather than resulting from a subsequent effect such as coagulation. Most successful studies demonstrating realtime monitoring of the effects of temperature in thermal ablations have, to date, been performed in model systems and generally using high-field systems. Currently favored mechanisms of temperature-dependent contrasts are T_1 based, where a focal change in the signal amplitude is seen when running under conditions of partial saturation, and proton resonance frequency, where changes in the signal phase with temperature are seen. The proton resonance frequency method is essentially a chemical shift effect and so is most successful at high field. Molecular diffusion has also been proposed as a method for introducing temperature-dependent contrast. However, this method has not been used widely as it is extremely sensitive to bulk patient motion, and because of the potential for anisotropic diffusion to introduce orientation-dependent contrast.

For cryogenic ablations, rapid and accurate monitoring of the region of treatment is possible. As the tissue water solidifies, its T_2 time is greatly reduced; as a result, no echo signal can be measured using conventional MRI hardware. The border of frozen and unfrozen tissues is extremely well defined and for situations where multiple cooling probes are used, any pockets of normothermy can be readily identified allowing repositioning of probes to encompass a given volume of interest. However, direct and unequivocal MR thermometry is precluded, at all field strengths, by the absence of signal from those regions that are frozen.

6 MR-MONITORED THERMAL ABLATION PROCEDURES

6.1 Interstitial Laser Photocoagulation in Breast Disease

Most work has been performed using the semiconductor diode laser (805 nm) and Nd:Yag (1064 nm) pulse lasers.[2–5] Studies have shown that a power of 2–2.5 W applied for about 500 s will generate lesions of up to 10 mm. Lesion size and reproducibility can be improved by pre-charring the fiber. Larger burns (of up to 40 mm) can be generated by using multiple fibers or pull-back techniques. In a dose escalation study, Akimov has shown that at higher power (over 3 W), tumor vaporization can lead to tissue explosions.[4]

In most studies, patients are treated with local anesthesia and sedation alone; general anesthesia is unnecessary.

6.1.1 Benign Breast Tumors

In our institution we have treated just under 50 benign fibroadenomas of the breast with ILP. Diagnosis is made before treatment by triple assessment and either cytology (C2) or core biopsy. Lesions of up to 4 cm diameter have been treated with between one and four fibers at 2–2.5 W (1000–4000 J). Follow-up for at least 1 year by clinical and ultrasound examination is available in 14 of these patients; in each case the mass has become impalpable by 1 year. In 13 patients, the tumor could not be seen on ultrasound by 12 months' follow-up and in the remainder a 9 mm remnant persisted.[6]

MR has a very limited role to play in the monitoring of ILP therapy to benign tumors. Fibroadenomas have variable enhancement characteristics; some enhance very intensely and others not at all. Consequently contrast-enhanced MR is not of consistent value. Those lesions that do enhance before treatment show areas of devascularization after treatment, but this is not of any clear clinical value. Ultrasound is a quick method of localizing needles within lesions and the treatment parameters are relatively standard. Unlike malignant disease, the treatment of margins is not critical; therefore, accurate monitoring of therapy is not essential.

6.1.2 Breast Cancer

All studies have reported that ILP can cause cell necrosis in breast cancers.[3–5] Treatment of small relatively spherical tumors can be complete. Treatment of larger lesions is more challenging as modeling the treatment area to conform with irregular large masses and ensuring treated margins is much more difficult.

Most studies have examined ILP where treatment has been followed by early surgical excision of the tumor and there are very few long-term follow-up data. There are small numbers reported where ILP has been the sole method of treatment. Akimov treated seven women with ILP alone. Two of these died of other causes and without tumor recurrence; a third had disease-free survival of over 5 years. In our own recent experience in a cohort of selected elderly women who are undergoing ILP and delayed excision at 3 months, we have been able to show complete tumor ablation sustained over the 3-month period in small numbers.

Complete tumor ablation and treatment of an adequate margin is essential in the management of breast cancer. The surgical literature has shown that incomplete tumor excision is a major risk factor for local tumor recurrence. Yet it is the margins that are most difficult to treat with ILP because, when the fiber is placed in the center of a lesion, the periphery does not reach as high a temperature as the center of the tumor. In one patient treated with ILP alone, Akimov has shown peripheral tumor recurrence. We have found that when there is inadequate tumor ablation, viable cells are invariably found around the periphery of the mass. Unless margins can be reliably and con-

For References see p. 1248

sistently treated, ILP cannot be accepted as a reasonable form of primary therapy in breast cancer.

Contrast-enhanced MR imaging has an essential role to play before therapy to map accurately the extent of the tumor and thus facilitate appropriate case selection. Treatment changes can be seen at high field during ILP using T_1-weighting sequences. Areas of low signal develop around the fiber tip within a minute of treatment starting and these areas broadly correlate with the histologic extent of tumor necrosis.[3,5] Consequently, MR does appear to have the potential to monitor treatment in close to realtime. Ideally, areas of signal change should be matched to the tumor. This is not easy in practice as the majority of tumors are only visible on contrast-enhanced scans. Repeated use of contrast causes enhancement of normal breast parenchyma and the loss of contrast between normal and abnormal breast tissue.

Following treatment, delayed contrast-enhanced MR can be used to show the effects of treatment. Areas of devascularization (nonenhancement) occur within the treated area. Occasionally rim enhancement may occur around an adequately treated lesion as a result of peripheral inflammation.

The only significant complications of ILP reported are collateral burns to skin and muscle. The former can be avoided by irrigating the skin with chilled saline during treatment. Burns to the pectoralis muscle can be avoided by accurate placement of the fiber tip.

6.2   Radiofrequency, Cryotherapy, and Focused Ultrasound in Breast Disease

Experience using these thermal treatments in breast disease is very limited. Preliminary reports of the use of rf in vivo[7] and in vitro[8] have shown that histologically confirmed areas of cell necrosis occur. MR monitoring of rf therapy during treatment is limited by the need to have the metallic probe in position during the procedure. There are limited reports of the use of FUS in breast tumors.[9,10] Prolonged treatment times reduce patient acceptance of the procedure, and patient movement compromises treatment. Cryotherapy has been shown to produce tissue necrosis in breast cancers[11] but no significant clinical studies have been reported.

6.3   Bone and Soft Tissue Tumors

There are limited reports of the application of MIT to bone and soft tissue tumors. ILP and rf are being used routinely in the ablation of benign osteoid osteomas. As the tumors are best seen on CT and treatment parameters are standard, MRI has no role.

There has been some preliminary experience with the treatment of bone metastases with MIT, but it is difficult to justify these experimental therapies when local radiotherapy is effective, widely available, and well tolerated.

We have limited experience treating soft tissue metastases from soft tissue and bone sarcomas with PDT, rf, and ILP. In these patients (with chemo- and radio-insensitive tumors who have all undergone multiple previous operations), we have had variable results ranging from complete tumor ablation in patients with small volume disease to no effect in patients with massive recurrence.

MR has been useful in these patients to stage local disease and for showing post-treatment changes. For complex tumors, needle placement is facilitated by MR as frequently the tissue/tumor contrast is not sufficient to enable accurate tumor localization with CT guidance.

6.4   Fibroid Disease

ILP and rf have been used to treat uterine fibroids (leiomyomata) percutaneously, endoluminally, and at laparoscopy. The advantage of the last approach is that the peritoneum, which is very susceptible to thermal damage, can be directly visualized. The main role described for MR is in the monitoring of treatment response: fibroids frequently exhibit an iceberg effect and the extent of local damage cannot be accurately assessed by direct visualization. There is no published experience of per-procedural MR guidance to our knowledge.

6.5   MR-Guided Ablation of Liver Tumors

The four principal techniques used to achieve percutaneous hepatic tumor ablation are ILP, RFC, cryotherapy, and percutaneous ethanol injection.

A number of centers including our own are now routinely using rf and ILP to ablate hepatic metastases. In our own center most procedures are performed under local anesthesia and conscious sedation, with approximately 15–20% requiring general anesthesia. The 0.2 T Siemens Viva open-interventional MR system provides good access for either a subcostal or intercostal approach to the right lobe of the liver, or an anterior oblique approach to left lobe lesions with the patient in the supine position. A body flex coil provides adequate coverage of the area of interest and good access for needle positioning. Initial breath-hold T_i-weighted gradient echo images (repetition time (TR); echo time (TE) 152 ms, 9 ms; flip angle 70°) scans are performed to locate the tumor. If tumor visualization is not adequate on baseline images, then liver-specific contrast agents are given (Figure 1). Either super-paramagnetic iron oxide particles (Ferumoxide, 15 mmol kg^{-1}, Guerbet) or T_1-shortening manganese-based agents (Mangafodipir trisodium Teslascan, 0.5 ml kg^{-1}, Nycomed Amersham) will provide a long period of enhancement, facilitating tumor visualization. Needle placement can be performed entirely using MR guidance or by using a combination of ultrasound and MR. Generally we have favored the use of ultrasound for speed of needle placement, reserving MR for tumors not easily visualized by ultrasound and for confirmation or finessing of the final needle position. If ultrasound guidance is used, the patient is positioned within the coil on the MR table, and then the table is undocked and moved across the 5 G line, approximately 2 m from the magnet, to the ultrasound machine.

6.5.1   Needle Visualization

Needle positioning is confirmed using multiplanar, often double-oblique T_1 or T_2 gradient echo images (Figure 1). As needle visualization is reduced when it is aligned with the B_0 field, a direct vertical puncture is best avoided. Other mechanisms for increasing needle visualization include swapping the phase- and frequency-encoding directions. Greater magnetic susceptibility and, therefore, better needle positioning can be

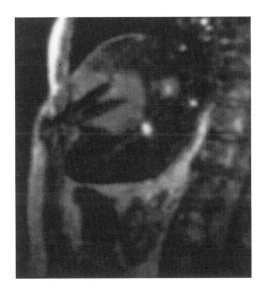

Figure 1   Coronal section through the liver following Endorem during interstitial laser photocoagulation therapy to a colorectal metastasis. The large tumor is seen as a relatively bright mass in the low signal liver. Two 19 gauge MR-compatible needles are shown positioned within the mass

achieved with the frequency encode perpendicular to the needle direction. Other techniques include increasing the number of averages of using breath-hold imaging. In practice breath-hold imaging is very difficult to achieve in sedated patients and if general anesthesia is being performed will require the addition of muscle paralysis and intubation.

For laser therapy, up to eight bare-tip fibers are used inserted through 19 G hollow, MR-compatible needles (Cook, Europe). For RFC, MR-compatible single rf electrodes are inserted directly. Where electrode clusters are required, guide needles are used as MR-compatible electrode arrays are still in development.

6.5.2   Monitoring the Treatment

We use a T_1-weighted fast inversion–recovery sequence (*TR*, 1068 ms; *TE* 48 ms; time to inversion, 80 ms; number of signals averaged, 1 in 59 s; collimation, 10 mm; matrix, 182×256) to monitor the changes produced by thermal ablation (Figure 2). On this sequence, untreated tumor is seen as high signal intensity against lower signal intensity liver. Over time, an area of reduced signal intensity develops at the treatment site and becomes progressively lower in signal intensity and larger in size. Around the area of low signal intensity, a high signal intensity ring of edema is also visualized. Although the area of low signal intensity correlates with the final area of thermal ablation on contrast-enhanced CT at 18 h, in general the MR image underestimates the final extent of necrosis. We perform monitoring scans at 10-min intervals throughout the treatment.

Treatment-sensitive sequences could theoretically provide more accurate monitoring information but are difficult to implement in the liver in vivo. Minimal signal intensity changes requiring image registration and subtraction are inaccurate. The shape of the liver is not constant but changes on respiration. A 13% error in liver registration in the craniocaudal direction has

been reported.[12] Further thermal injury itself produces a change in the liver shape and contour. Motion artifact becomes increasingly problematic during treatment when the patient is sedated and unable to breath-hold. Therefore, although ex vivo and animal experimentation suggest that both T_1 contrast and phase frequency shift will provide accurate temperature information[13–15] both Kettenbach and Vogl report difficulties with susceptibility and motion in vivo.[16,17] Vogl uses phase frequency shift at 1.5 T with the relatively simple indicator of a 50% reduction in signal to terminate treatment; and this is considered sufficient for tumor ablation by this group.[17]

The use of rf suffers from the additional problem of interference with MR image acquisition; as a result, monitoring sequences are performed at breaks in the treatment cycle. Steiner et al. found good correlation between phase frequency shift imaging at 0.5 T immediately at treatment end and final ablation volume.[15] However, liver cooling is rapid (of the order of 50 s) and, therefore, there is only limited opportunity to acquire the necessary information.

For cryotherapy, MR-compatible cryoprobes have been developed. The iceball is well visualized but tends to underestimate the final lesion size.[18,19]

Percutaneous ethanol injection, unlike the other ablation techniques, is reserved for the treatment of hepatocellular carcinoma. Although ex vivo experiments have suggested the use of water-suppressed T_2 fast-spin echo images for monitoring the

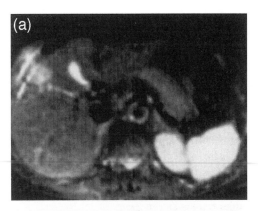

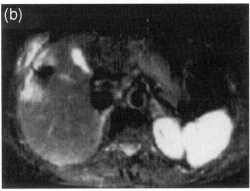

Figure 2   Transverse short tau inversion recovery images through the liver before (a) and after (b) interstitial laser photocoagulation (ILP) to a hepatic metastasis, showing treatment effect. The high-signal peripheral metastasis shows a low-signal focus within the center of the tumor following 8-fiber ILP to the mass. These changes confirmed a treatment effect but underestimated the changes seen on follow-up contrast-enhanced computed tomography

For References see p. 1248

actual injection and animal experiments have suggested that ethanol produces low signal on all sequences, in practice a range of signal intensity changes are produced and MR assessment relies on the absence enhancement during arterial phase imaging to indicate complete ablation.[20–23]

6.5.3 Long-term Follow-up

Most groups prefer CT for long-term follow-up because of cost and availability.

6.5.4 Conclusions

MR guidance offers improved multiplanar imaging compared with other imaging techniques; although quantitative temperature measurement is not yet a reality, MR does provide some monitoring information.

6.6 MR-Guided Ablation in the Prostate

Failure of traditional treatments and the associated morbidity has driven a multifaceted search for alternative therapies for prostate treatment. There are now a number of image-guided treatment options under evaluation for localized prostate cancer, including cryotherapy, brachytherapy, and photodynamic therapy. Laser therapy has been used in benign prostatic hypertrophy. Fluoroscopy, CT and transrectal ultrasound have all been used to guide prostatic interventions. Of these, CT and fluoroscopy provide no information about the internal architecture of the prostate and are unidimensional; transrectal ultrasound offers significant detail of the internal prostatic architecture but does not show the needle trajectory from perineum to prostate. In addition, the presence of the probe in the rectum holds the rectal wall against the prostate gland. MR offers multiplanar imaging, any plane, any obliquity, good prostate detail, and images of the full needle trajectory and of neighboring structures without distortion of the anatomy.

6.6.1 Photodynamic Therapy in the Prostate

We have used interventional MR to guide needle placement for photodynamic therapy in the prostate. Necrosis is produced by the interaction of light of a particular wavelength with a photosensitizing agent. The exact mechanism is unknown but a popular theory is that short-lived oxygen radicals are produced and these cause cell death. Necrosis does not occur immediately and the oxygen radicals are MR occult; therefore, MR does not have a monitoring role during the treatment. However, MR is excellent in guiding needle placement (Figures 3 and 4). As many as eight, 19 G hollow needles are introduced transperineally into the peripheral zone of the prostate gland. Multiplanar T_1-weighted gradient echo images are obtained to confirm needle position and spacing. Laser fibers are introduced through the needles and the tissue is illuminated with red light. Approximately a 16 mm sphere of necrosis is produced around each laser fiber. Treatment is targeted to areas containing tumor on mapping biopsy and away from the urethra, neurovascular bundles, and external sphincter. Imaging assessment of the extent of necrosis is performed using gadolinium-enhanced T_1-weighted sequences at high field (1.5 T). Necrosis is appreciated as an area of nonenhancement at the treatment site (Figure 5). In comparison, CT demonstrates lim-

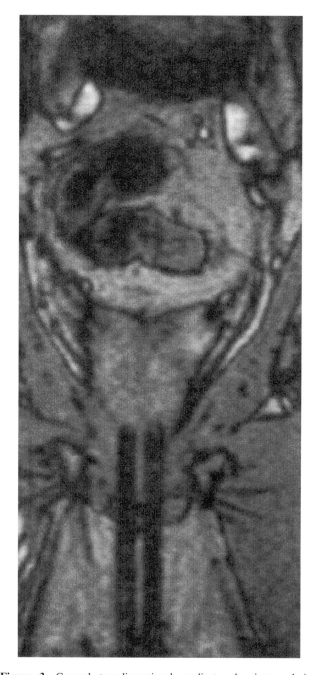

Figure 3 Coronal two-dimensional gradient echo image during photodynamic therapy to the prostate. Two 19 G MR-compatible needles are inserted within the gland and MR is used to confirm the position of the needles

ited tissue contrast and both gray scale and Doppler ultrasound do not show any specific features.

6.6.2 Brachytherapy

MR guidance of seed insertion has been reported in nine patients at 0.5 T.[24] MR allowed confirmation of seed position in three planes. Both T_1-weighted fast gradient echo and proton density fast spin echo sequences have been used satisfactorily.[25,26]

For list of General Abbreviations see end-papers

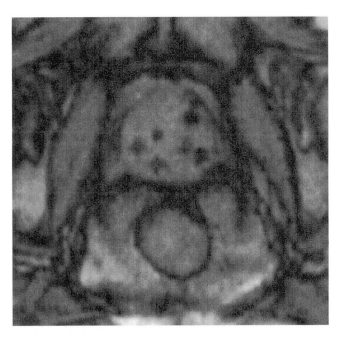

Figure 4 Transverse two-dimensional gradient echo image through the prostate during multifiber photodynamic therapy to the gland. Multiple needles are positioned within the tumor in posterior and left-sided lobes of the gland

6.6.3   Cryotherapy

During cryotherapy, the iceball can be well visualized, although actual temperature measurement is not available. MR has also been used in the follow-up of cryotherapy-induced changes. Contrast-enhanced T_1-weighted images show necrosis as areas of nonenhancement. Expected changes on follow-up include a reduction in prostatic volume of 30–52%, loss of zonal architecture in 81–100%, and soft tissue thickening of the rectal wall (47%) or around the capsule and neurovascular bundles in the majority (89%).[27–29]

6.6.4   Laser-induced Thermotherapy in Benign Prostatic Hypertrophy

MR has been used for monitoring of thermal ablation in benign prostatic hypertrophy and for follow-up assessment. On-line monitoring is feasible using phase-frequency shift[30] providing an accurate prediction of final lesion size in ten of 13 patients in one study.[31] Follow-up MR showed lack of gadolinium enhancement in necrotic areas and volume reduction.[32]

6.6.5   Conclusion

MR guidance and follow-up is preferable to other techniques in prostate ablation; for cryotherapy and thermal techniques there is the added advantage of treatment monitoring.

7   NAVIGATIONAL DEVICE

As indicated above, accurate placement of tools is essential for thermal therapies to ensure that lesions are accurately loca-

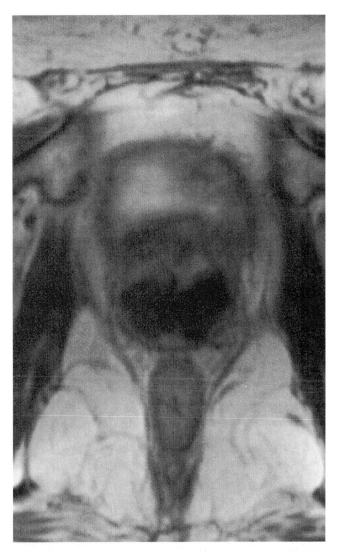

Figure 5 Transverse contrast-enhanced T_1-weighted image through the prostate 5 days after photodynamic therapy (PDT). A lobulated region of nonenhancement is seen within the gland that corresponds to the region of devascularization after PDT. Similar changes are seen after cryotherapy

lized and damage to collateral tissue is avoided. Although freehand placement is usually adequate where lesions are superficial and large, the same is not true for deep and small lesions. In these cases, tool placement may be facilitated by navigational devices. There are a number of systems being developed currently. We are investigating the use of a passive navigation device for interactive definition of the slice selection plane. The device can potentially increase the speed with which needles are positioned as it is possible to define the image slice automatically, during needle insertion, according to the orientation of the needle itself. The biopsy needle within its holder is attached to the passive navigation probe (Figure 6). The reference point for image slice selection can be set relative to the tip of the needle. The stereoscopic camera detects the position and rotation of the probe so as to define the slice selection axis. In situations where multiple needles require

For References see p. 1248

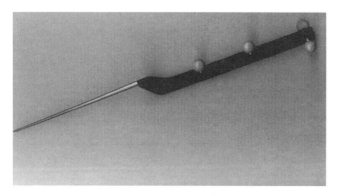

Figure 6 Navigational pointing device. A number of highly reflective balls are mounted on the handle of the pointing device, the light being detected by the stereoscopic camera. Tools may be mounted onto the device to facilitate accurate placement within the target

accurate positioning, the device reduces any ambiguity in identifying needles and so increases the safety of the procedure.

8 SUMMARY

Temperature-sensitive MRI can be achieved for the monitoring of thermal treatments in the body within a realistic scan time. Low-field open-access scanners offer good surgical access to the patient. However, the image quality is poor compared with that expected from high-field closed-bore magnet systems. Several of the temperature-dependent mechanisms of contrast are field strength dependent; consequently, for these, high-field magnets should be preferable for the monitoring of thermal interventions in the body. The development of shorter and wider bore high-field magnets should also assist in the promotion and acceptance of MRI monitoring of thermal interventions in the body, head, and extremities.

9 RELATED ARTICLES

MR-Guided Biopsy, Aspiration, and Cyst Drainage; Temperature Measurement Using In Vivo NMR.

10 REFERENCES

1. S. A. Harries, Z. Amin, M. E. F. Smith, W. R. Lees, J. Cooke, M. G. Cook, J. H. Scurr, M. W. Kissin, and S. G. Bown, *Br. J. Surg.*, 1994, **81**, 1617.
2. H. Mumtaz, M. A. Hall-Craggs, G. Buonnacorssi, M. Kissin, and S. Bown, *Radiology*, 1996, **201**(P), 177.
3. H. Mumtaz, M. A. Hall-Craggs, A. Wotherspoon, M. Paley, G. Buonaccorsi, Z. Amin, I. Wilkinson, M. Kissin, T. Davidson, I. Taylor, and S. Bown, *Radiology*, 1996, **200**, 651.
4. A. B. Akimov, V. E. Seregin, K. V. Rusanov, E. G. Tyurina, T. A. Glushko, V. P. Nevzorov, O. J. Nevzorova, and E. V. Akimova, *Lasers Surg. Med.*, 1998, **22**, 257.
5. S. E. Harms, H. Mumtaz, V. S. Klimberg, C. L. Cowan, S. Korourian, and W. B. Hyslop, *Radiology*, 1998, **209**(P), 468.
6. L. M. Lai, M. A. Hall-Craggs, H. Mumtaz, P. M. Ripley, M. Kissin, T. I. Davidson, C. Saunders, I. Taylor, and S. G. Bown, *Breast*, 1999, **8**, 89.
7. R. L. Birdwell, S. S. Jeffrey, E. L. Kermit, D. M. Ikeda, and K. W. Nowels, *Radiology*, 1998, **209**(P), 198.
8. T. Boehm, I. Hilger, W. Mueller, J. R. Reichenbach, M. Fleck, and W. A. Kaiser, *Radiology*, 1998, **209**(P), 198.
9. H. E. Cline, J. F. Schenck, R. D. Watkins, K. Hynynen, and F. A. Jolesz, *Magn. Reson. Med.*, 1993, **30**, 98.
10. J. W. Hand, C. C. Vernon, and M. V. Prior, *Int. J. Hypertherm.*, 1992, **8**, 587.
11. R. W. Rand, R. P. Rand, F. Eggerding, L. DenBesten, and W. King, *Surg. Gynecol. Obstet.*, 1987, **165**, 392.
12. D. L. Wilson, A. Carrillo, and L. Zheng, *JMRI*, 1998, **8**, 77.
13. T. Kahn, T. Harth, B. Schwabe, H. J. Schwarzmaier, and U. Modder, *Roto. Fortschr. Geb. Rontgenstr. Neuen. Bildgeb. Verfahr.*, 1997, **167**, 187.
14. J. Olsrud, R. Wirestam, S. Brockstedt, A. M. Nilsson, K. G. Tranberg, F. Stahlberg, and B. R. Persson, *Phys. Med. Biol.*, 1998, **43**, 2597.
15. P. Steiner, R. Botnar, B. Dubno, G. G. Zimmermann, G. S. Gazelle, and J. F. Debatin, *Radiology*, 1998, **206**, 803.
16. J. Kettenbach, S. G. Silverman, N. Hata, K. Kuroda, P. Saiviroonporn, G. P. Zientara, P. R. Morrison, S. G., Hushek, P. M. Black, R. Kikinis, and F. A. Jolesz, *JMRI*, 1998, **8**, 933.
17. T. J. Vogl, M. G. Mack, A. Roggan, R. Straub, K. C. Eichler, P. K. Muller, V. Knappe, and R. Felix, *Radiology*, 1998, **209**, 381.
18. J. Tacke, G. Adam, and R. Speetzen, *Magn. Reson. Med.*, 1998, **39**, 354.
19. H. P. Klotz, R. Flury, A. Schonenberger, J. F. Debatin, G. Uhlschmid, and F. Largiader, *Comput. Aided Surg.*, 1997, **6**, 340.
20. D. S. Lu, S. Sinha, J. Lucas, K. Farahani, R. Lufkin and K. Lewin, *JMRI*, 1997, **7**, 303.
21. H. Shinmoto, R. V. Mulkern, K. Oshio, S. G. Silverman, V. M. Colucci, and F. A. Jolesz, *J. Comput. Assist. Tomogr.*, 1997, **21**, 82.
22. T. Fujita, L. Honjo, K. Ito, K. Takano, S. Koike, H. Okazaki, T. Matsumoto, and N. Matsunaga, *J. Comput. Assist. Tomogr.*, 1998, **22**, 379.
23. C. D. Becker, M. Grossholz, G. Mentha, A. Roth, E. Giostra, P. A. Schneider, and F. Terrier, *Cardiovasc. Intervent. Radiol.*, 1997, **20**, 204.
24. A. V. D'Amico, R. Cormack, C. M. Tempany, S. Kumar, G. Topulos, H. M. Kooy, and C. N. Coleman, *Int. J. Radiat. Oncol. Biol. Phys.*, 1998, **42**, 507.
25. D. F. Dubois, B. R. Prestidge, L. A. Hotchkiss, W. S. Bice, Jr, and J. J. Prete, *Int. J. Radiat. Oncol. Biol. Phys.*, 1997, **39**, 1037.
26. M. A. Moerlaand, H. K. Wijrdeman, R. Beersma, C. J. Bakker, and J. J. Battermann, *Int. J. Radiat. Oncol. Biol. Phys.*, 1997, **37**, 927.
27. C. L. Kalbhen, H. Hrciack, K. Shinohara, M. Chen, F. Parivar, J. Kurhanewicz, D. B. Vigneron, and P. R. Carroll, *Radiology*, 1996, **198**, 807.
28. H. Howighorst, J. Dorsam, M. V. Knopp, W. Jugowski, S. O. Schoenberg, M. Wiesel, M. Essig, and G. Van-Kaick, *Roto. Fortschr. Geb. Rontgenstr. Neuen. Bildgeb. Verfahr.*, 1998, **168**, 44.
29. A. D. Vellet, J. Saliken, B. Donnelly, E. Raber, R. F. McLaughlin, D. Wiseman, and N. H. Ali-Ridha, *Radiology*, 1997, **203**, 653.
30. U. G. Mueller Lisse, A. F. Heuck, M. Thoma, R. Muschter, P. Schneede, E. Weninger, S. Faber, A. Hofstetter, and M. F. Reiser, *JMRI*, 1998, **8**, 31.

31. R. A. Boni, T. Sulser, W. Jochum, B. Romanowski, J. F. Debatin, and G. P. Krestin, *Radiology*, 1997, **202**, 232.
32. U. G. Mueller Lisse, A. Heuck, P. Schneede, R. Muschter, J. Scheidler, A. G. Hofstetter, and M. F. Reiser, *J. Comput. Assist. Tomogr.*, 1996, **20**, 273.

Pediatric Body MRI

Rosalind B. Dietrich and Gerald M. Roth

University of California, Irvine, Orange, CA, USA

1 INTRODUCTION

Although initially slow to become established, magnetic resonance imaging (MRI) now plays a vital role in the diagnostic imaging evaluation of many pediatric diseases and disorders. The ability to produce multiplanar images in a noninvasive manner without the use of ionizing radiation and with only minimal patient preparation has made MRI a useful complement to other cross-sectional modalities such as ultrasound and computerized tomography (CT). The most frequent indications for an MRI study in the pediatric population include the evaluation of congenital anomalies and the characterization of tumors and other mass lesions.

2 CONGENITAL ANOMALIES

MRI plays an important role in evaluating both complex congenital anomalies, where ultrasound is inadequate, and simple congenital anomalies, where sonography is not possible, is incomplete or is suboptimal. Congenital anomalies are well visualized with MRI due to its provision of excellent anatomical detail and its multiplanar capabilities. The majority of lesions are adequately assessed using multiplanar T_1- and T_2-weighted spin echo imaging. Occasionally, additional sequences are required, especially if the presence of fat or flowing blood is to be determined.

The ability of MRI to differentiate vessels containing flowing blood from other mediastinal structures without the use of bolus injection of contrast material makes it an ideal choice for evaluating anomalies of the great vessels.[1,2] Preoperative mapping may alleviate the need for angiography in selected patients. In addition, in such lesions as the double aortic arch and the right aortic arch with an aberrant left subclavian artery and a left ligamentum arteriosum, MRI can evaluate for the presence of esophageal and/or tracheal compression that may be associated with these lesions. In the entity known as the 'pulmonary sling', where the left pulmonary artery arises from the right pulmonary artery and passes between the trachea and the esophagus, the presence of tracheal compression may be similarly assessed. Other, more rare, symptomatic abnormalities, as well as the asymptomatic great vessel anomalies can also be evaluated. In patients with a right aortic arch, MRI can determine the presence or absence of mirror image branching and can demonstrate any associated cardiac anomalies. Cardiac gated T_1-weighted images in the axial and coronal planes can often provide sufficient information for these diagnoses to be made, although sometimes images parallel to the course of the aorta are very useful.

The most common congenital pediatric abnormalities involving the lung include congenital lobar emphysema, cystic adenomatoid malformation, and sequestration. The first two entities can be evaluated using MRI, although plain film radiography and/or CT examination are usually sufficient. MRI, however, can be very useful in cases of sequestration, as it may be able to demonstrate the anomalous feeding vessels and draining veins.[3] The corresponding lung tissue may appear solid or aerated on MRI, depending on its connection, if any, with the bronchial tree.

Abnormalities of the diaphragm, such as diaphragmatic hernias and eventrations, can also be evaluated by MRI. Coronal and/or sagittal T_1-weighted images are useful in determining the presence or absence of a portion of the diaphragm, as well as identifying which, if any, of the normal abdominal components are in the chest.[4] In patients with anterior abdominal wall defects, such as gastroschisis and omphalocele, MRI can be used to determine which abdominal organs have protruded through the defect and define their position (Figure 1).

In the abdomen, congenital anomalies of the hepatobiliary system such as choledochal cysts, biliary atresia, and polycystic liver disease can all be evaluated using MRI, but are more commonly diagnosed using ultrasound. MRI plays a more important role in the visualization of abnormal vascular anatomy in such anomalies as the Budd–Chiari malformation and discontinuity of the inferior vena cava. In the Budd–Chiari malformation, MRI may identify obliteration of the hepatic veins and may help determine if a congenital web or a thrombus is responsible for the condition.

In the retroperitoneum, abnormalities in renal position and development are readily seen with MRI. Renal ectopia, whether high (intrathoracic) or low (pelvic) can be differentiated from renal agenesis more easily with MRI than with ultrasound, since visualization of the abdominopelvic cavity by MRI is not limited by the presence of bowel gas or bone, as it is with ultrasound. Fusion anomalies, such as crossed fused ectopia and horseshoe kidney are also well demonstrated by MRI.[5–7] Crossed fused ectopia, like most congenital renal anomalies, is best imaged using coronal T_1-weighted images, and is diagnosed when an abnormally positioned fused kidney is visualized on one side and there is no identifiable renal tissue on the contralateral side. Horseshoe kidneys, on the other hand, are easier to diagnose on axial T_1-weighted images, as the fusion of the lower poles is often more apparent in this plane. In children with renal agenesis, the ipsilateral adrenal gland develops a discoid rather than a chevron-shaped appearance, and can be seen as an elongated linear structure on coronal magnetic resonance (MR) images.[4]

Although all of the renal cystic diseases can be demonstrated using MRI, ultrasound remains the primary imaging modality for this class of diseases. MRI is most useful in clari-

For References see p. 1255

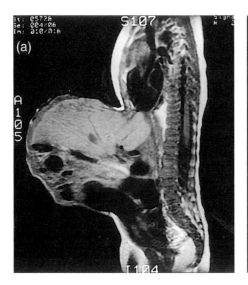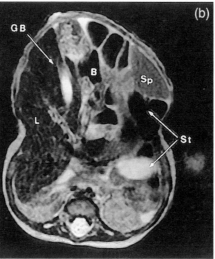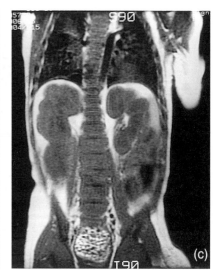

Figure 1 Omphalocele. (a) Sagittal view (SE 350/11): multiple loops of bowel and a large volume of the liver have herniated into the omphalocele sac. (b) Axial view (FSE 4000/102Ef): liver (L), gall bladder (GB), bowel (B), spleen (Sp) and stomach (St) are identified in the omphalocele sac. (c) Coronal view (SE 350/12): both kidneys are abnormally superior in position, lying just below the diaphragm

fying confusing cases and in demonstrating complications such as hemorrhage. Simple renal cysts appear as homogeneous, well-defined masses on the MR image and are of low signal intensity on T_1-weighted sequences and of very high signal intensity on T_2-weighted sequences. The cyst wall and the fluid in the cyst may be indistinguishable. If hemorrhage into the cyst occurs, or if the cyst has a high protein concentration for another reason, inhomogeneous high signal intensity can be demonstrated on both T_1- and T_2-weighted sequences. Despite these guidelines, there is a wide range of variability in the appearance of hemorrhagic cysts, and in some cases it is not always possible to distinguish a hemorrhagic cyst from an infected cyst or even from a neoplasm.

MRI is an excellent modality for the evaluation of congenital anomalies of the genital tract. In disorders of sexual differentiation, MRI can clearly show the absence of, or abnormal location of, pelvic organs in children with ambiguous genitalia or genetic abnormalities.[7-9] Imaging in both the axial and sagittal planes, and with both T_1- and T_2-weighted sequences is often necessary in these patients with Turner syndrome, testicular feminization, hermaphrodism, or one of the various forms of pseudohermaphrodism.

MRI can also evaluate the spectrum of Müllerian duct anomalies, ranging from uterus didelphys (two uteri, two cervices, and two vaginas) through bicornuate uterus (two uteri, single vagina, and cervix) to uterus septus (single uterus, cervix, and vagina with a uterine septum).[10-12] The distinction between the last two entities is crucial if surgical intervention is planned. MRI can often make this distinction in postpubertal girls. On T_2-weighted images in a plane axial to the body of the uterus, the bicornuate uterus will demonstrate a medium signal intensity strip of myometrium separating the two low signal intensity junctional zones that the septate uterus will not. MRI can also evaluate the urinary anomalies, such as renal agenesis or renal malposition, which are often associated with Müllerian duct anomalies.

Renal anomalies can also occur in patients with the Mayer–Rokitansky–Kuster–Hauser syndrome.[13] In this syndrome, which is characterized by aplasia of both the uterus and upper vagina, the spectrum of renal anomalies includes unilateral agenesis, unilateral or bilateral ectopia, horseshoe kidney, malrotation, and collecting system abnormalities. Vertebral anomalies have also been reported. Because MRI can be used to evaluate the vagina noninvasively, it is the modality of choice for patients with this disorder, as well as for patients with isolated vaginal atresia.

In patients with hematometrocolpos, T_1- and T_2-weighted images can demonstrate a high signal intensity collection within a markedly dilated upper vagina and uterine cavity[14-16] (Figure 2). These dilated structures may compress the adjacent bladder or rectum. The fluid collection may extend into the fallopian tube, leading to a dilated tortuous fallopian tube visualized extending into the abdomen.

MRI has also proved useful in identifying undescended testes in male children prior to surgery. Ultrasound is performed initially and, if it is unsuccessful, MRI is then performed.[17,18] The undescended testis is usually in the inguinal canal, but if it is in the abdomen or pelvis, the testis can be adjacent to the lateral bladder wall, the psoas muscle, the iliac vessels or in the retroperitoneum or the superficial inguinal pouch. On T_1-weighted images, the testis is of medium signal intensity and can frequently be identified if it is surrounded by fat, which is of high signal intensity. On T_2-weighted images, the central portion of the testis is of high signal intensity with a surrounding rim of medium signal intensity, and therefore can be easily distinguished from muscles and lymphadenopathy.

Anorectal anomalies, such as imperforate anus and ectopic anus, are also well visualized using MRI. Both these disorders are due to failure of descent of the hindgut. In imperforate anus, the rectum terminates in a blind pouch, whereas in ectopic anus, the more common of the two entities, there is a

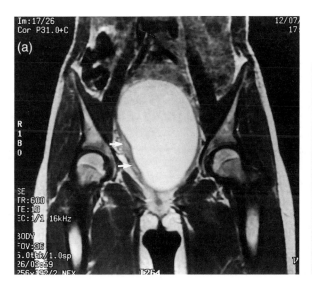
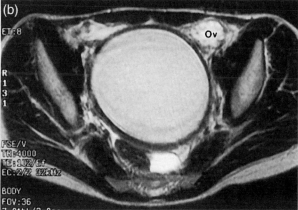

Figure 2 Hematometros involving one side of a uterus didelphys. (a) Coronal view (SE 600/18): high signal intensity blood products are identified in a markedly distended uterine cavity; a second uterine cavity (arrows) with an endometrial stripe is identified compressed on the right. (b) Axial view (FSE 4000/102Ef): the distended left uterine cavity occupies most of the volume of the pelvis, displacing the left ovary (Ov) anteriorly

fistulous connection between the pouch and another structure, such as the perineum, vestibule, vagina, urethra, bladder, or cloaca. Plain film radiography, contrast examination, and ultrasonography have all been helpful in the preoperative evaluation of these patients. MRI, however, can noninvasively give a multiplanar view of the hindgut, puborectalis sling and adjacent structures, defining the anatomy to better advantage.[19–22] In this regard, axial and coronal T_1-weighted images can demonstrate the anatomy of the puborectalis sling, determine whether or not it is hypoplastic, and define its relationship to the hindgut.

3 TUMORS AND OTHER MASS LESIONS

Due to its multiplanar capability, its ability to distinguish vessels with flowing blood from other structures, without the need for bolus injection of contrast material, and its superior contrast resolution compared with that of CT, MRI is an extremely useful tool in the evaluation of mass lesions.[4] Once a lesion has been discovered on a chest radiograph or on an abdominal or pelvic ultrasound, MRI can help to characterize the lesion, identify the organ of origin, define its extent, clarify its relationship to adjacent vessels, and evaluate for distant metastasis. MRI can classify a lesion as either cystic, solid, or mixed. Although some solid lesions such as teratomas, lipomas, and hemorrhagic lesions may demonstrate characteristic MR appearances, the majority of solid lesions are impossible to differentiate using signal intensity alone. Most such solid lesions are of medium signal intensity on T_1-weighted images and of high signal intensity on T_2-weighted images. However, when the signal characteristics of a lesion are combined with information about its organ of origin, its relationship to adjacent vessels and/or its sites of metastasis, a definitive diagnosis can often be made. MRI can be used before treatment, to plan a surgical approach or a radiation port, as well as after treat-

ment, to assess for residual or recurrent tumor after surgery, chemotherapy and/or radiation therapy.

Gadolinium chelate contrast agents may be useful in characterizing vascular masses and in evaluating renal and perirenal lesions.[23] In these instances, T_1-weighted images obtained dynamically during, and/or immediately after, intravenous contrast administration may more clearly define the borders of the lesion, characterize its internal architecture, accentuate surrounding adenopathy, and/or uniquely define the histology of the lesion.

In the thorax, MRI is particularly useful in the evaluation of masses arising in the posterior mediastinum, as it may noninvasively demonstrate the presence or absence of intraspinal extension.[24,25] These posterior mediastinal masses are most frequently of neurogenic origin and include such entities as neuroblastoma, ganglioneuroma, and neurofibroma. They are all of medium signal intensity on T_1-weighted images and of high signal intensity on T_2-weighted images. Coronal and axial images are useful in demonstrating intraspinal extension, if any, and in clarifying the relationship of the lesion to adjacent vasculature.

MRI can also demonstrate lesions in the anterior and middle compartments of the mediastinum, and may even have a slight edge over CT in selected cases.[26] For example, coronal and/or sagittal images may help define the extent of neck masses that secondarily involve the mediastinum or lung apices. These lesions include cystic hygroma, hemangioma, and fibromatosis. Fat selective sequences may also be helpful in distinguishing mediastinal teratomas from other mass lesions (Figure 3). Additionally, as with posterior compartment masses, MRI may help clarify the relationship of the lesion to adjacent vessels and other structures, due to its multiplanar capability, its ability to distinguish vessels with flowing blood from other structures, without the need for bolus injection of contrast material, and its superior contrast resolution.

In the abdomen and pelvis, MRI is an excellent modality with which to differentiate masses arising from the liver, kid-

For References see p. 1255

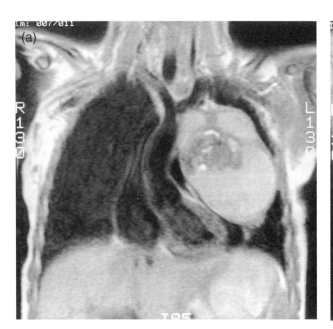
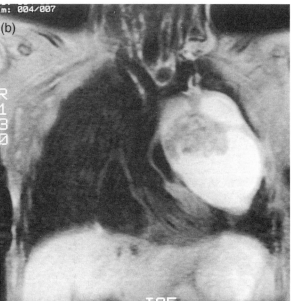

Figure 3 Mediastinal teratoma. (a) Coronal view (SE 625/11): a mediastinal mass is identified with locules of differing signal intensity; a few small foci of bright signal intensity are noted within the mass. (b) Coronal view (SE 625/11, chemical lipid presaturation): using a lipid-selective presaturation pulse, the formerly bright areas in the lesion become dark, verifying that they represent fat; the presence of fat in the lesion histologically characterizes it as a teratoma

ney, adrenal gland, and paraspinal regions. Some of the more common lesions arising from these areas warrant discussion.

In children with hemangioendotheliomas, MRI can often demonstrate the mass in the liver and may be able to map the feeding and draining vessels. An abrupt caliber decrease in the abdominal aorta distal to the origin of the celiac axis may also be observed.[4] Serial imaging after gadolinium chelate administration can often uniquely identify the hepatic lesion as a hemangioendothelioma due to its distinctive temporal pattern of peripheral to central enhancement.

The most common malignant tumors of the liver, hepatoblastomas and hepatocellular carcinomas, can also be evaluated using MRI. Both lesions demonstrate variable signal intensity on both T_1- and T_2-weighted images (Figure 4). Because MRI can usually identify the portal and hepatic veins, it can assess for vascular invasion by the tumor and

can also define the extent of the tumor with respect to the segmental anatomy of the liver, thus aiding in determining resectability of the lesion.

Wilms' tumor is the most common solid abdominal mass and the most common primary renal neoplasm in children, with a peak incidence between the ages of 1 and 3 years. An increased incidence of Wilms' tumor is noted in children with aniridia, hemihypertrophy, neurofibromatosis, genitourinary abnormalities, and the Beckwith–Wiedemann syndrome. The tumor presents most often as a unilateral abdominal mass, although the lesion is bilateral in 5–10% of cases. The initial diagnosis is usually made by ultrasound, which demonstrates a solid renal mass. MRI, however, may be superior to both ultrasound and CT in diagnosing and defining the extent of Wilms' tumor.[27,28] In this area of the body, coronal T_1-weighted images can often define the organ of origin of a

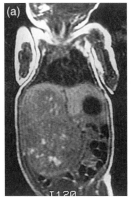
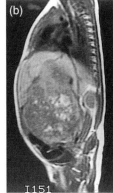
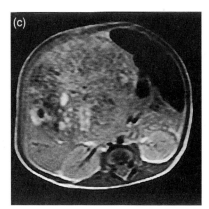
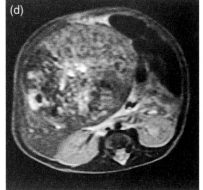

Figure 4 Hepatoblastoma. (a) Coronal and (b) sagittal (SE 300/20) views: a large mass, hetereogeneous in signal intensity, projects inferiorly from the liver and invades the portal vein; the speckled high signal intensity areas probably represent hemorrhage. (c, d) Axial views (SE 2500/30,80): although markedly heterogeneous, the lesion is predominantly low in signal intensity

For list of General Abbreviations see end-papers

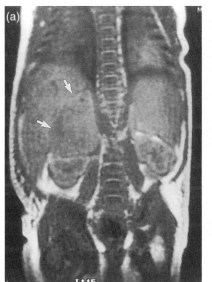
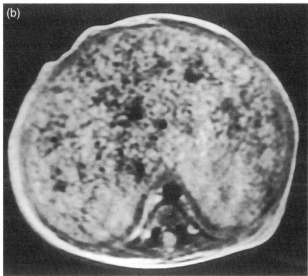

Figure 5 Neuroblastoma. (a) Coronal view (SE 300/16): an enlarged right adrenal gland (arrows) displaces the right kidney inferiorly. (b) Axial view (FSE 2500/17Ef): the liver is also enlarged and filled with metastases, leading to a salt and pepper heterogeneity to its signal intensity

lesion extremely well, separating renal lesions from those arising in the adjacent liver or adrenal gland. With a combination of coronal and axial planes, and T_1- and T_2-weighting, a study can define the full extent of the lesion, its possible invasion of adjacent organs or vessels, as well as assess for the presence of associated lymphadenopathy. In selected cases, intravenous administration of a gadolinium chelate and/or use of MR angiographic techniques may be necessary.[23]

Neuroblastoma is the most common extracranial solid malignant tumor in children. Neuroblastoma and its more differentiated forms, ganglioneuroblastoma and ganglioneuroma, arise from primitive sympathetic neuroblasts and therefore may arise from the adrenal gland (most commonly) or from anywhere else along the sympathetic chain from the nasopharynx to the presacral region. Children with neuroblastoma may present with a palpable abdominal mass or with symptoms referable to metastases, such as bone pain. The staging of neuroblastoma is based on both local extent as well as on the presence or absence of metastases. Clinically, however, it is more important to determine resectability of the lesion, as surgery remains the treatment of choice. On MRI, neuroblastoma is usually of medium signal intensity on T_1-weighted images and of high signal intensity on T_2-weighted images. It is less well defined than Wilms' tumor and if it arises from the adrenal gland it can displace the kidney inferiorly and/or laterally (Figure 5). As with Wilms' tumor, MRI can play a role in defining the full extent of a lesion, identifying its organ of origin, assessing for possible invasion of adjacent organs or vessels, and demonstrating the presence or absence of metastases.[29–31] Evidence of tumor extending into the spinal canal or encasing the retroperitoneal vessels can classify the neuroblastoma as unresectable (Figure 6).

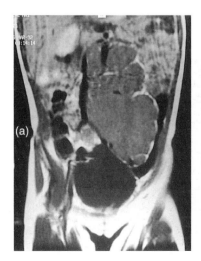
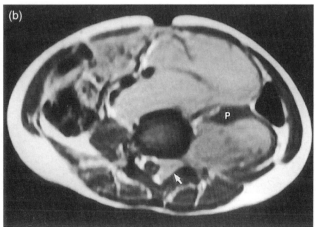

Figure 6 Neuroblastoma. (a) Coronal view (SE 616/15): a large multilobulated medium signal intensity mass surrounds the aorta, displacing it and the inferior vena cava to the right and anteriorly. (b) Axial view (SE 2128/15): a portion of the mass is seen extending into the bony spinal canal (arrow); the left psoas muscle (P) is elevated and displaced to the left by this high signal intensity tumor

For References see p. 1255

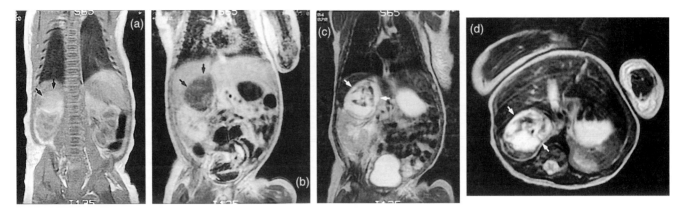

Figure 7   Neonatal adrenal hemorrhage. (a, b) Coronal views (SE 450/11): an enlarged right adrenal gland (arrows) is identified (a) that does not demonstrate enhancement after gadolinium administration (b). (c) Coronal view (FSE 4000/102Ef) and (d) axial view (FSE 3500/102Ef): blood products in varying stages of oxidation are identified in the enlarged right adrenal gland (arrows)

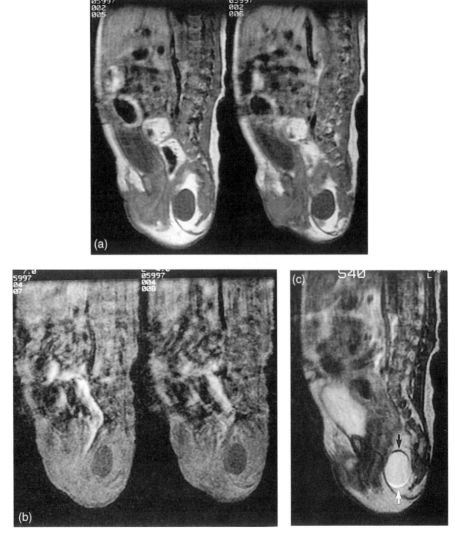

Figure 8   Presacral teratoma. (a) Sagittal view (SE 800/23): a presacral mass is identified with components that are hypointense, isointense, and hyperintense to muscle. (b) Sagittal view (SE 800/16, chemical lipid presaturation): using a lipid-selective presaturation pulse, the formerly hyperintense component is suppressed to the same degree as the adjacent subcutaneous fat; the presence of fat in the lesion histologically characterizes it as a teratoma. (c) Sagittal view (FSE 3000/102Ef): the presence of chemical shift spatial misregistration can also be used to verify the presence of fat in this lesion (arrows); when this artifact is present, lipid-selective presaturation pulses are not necessary to characterize the lesion

For list of General Abbreviations see end-papers

Neonatal adrenal hemorrhage is another entity that may present as an asymptomatic abdominal mass. Differentiation of neonatal adrenal hemorrhage from neuroblastoma is crucial, and can usually be done using ultrasound or MRI. On ultrasonography, neonatal adrenal hemorrhage is usually anechoic and avascular, in contradistinction to neuroblastoma, which is usually echogenic and vascular. In some instances, adrenal hemorrhage does not appear cystic on ultrasonography and MRI may be extremely useful in helping to make this distinction. On MR images neonatal adrenal hemorrhage behaves similarly to hematomas elsewhere in the body, with a changing MR appearance as the hematoma evolves (Figure 7). More specifically, on T_1-weighted images acute hematomas are isointense to muscle, whereas subacute hematomas are hyperintense to muscle. Neuroblastoma usually demonstrates homogeneous medium signal intensity on T_1-weighted images. On T_2-weighted images, neonatal adrenal hemorrhage may be of high signal intensity or of low signal intensity, depending on the chemical state of the blood products. In rare cases of cystic neuroblastoma with hemorrhage, however, the distinction may be difficult. In these cases, evaluation of the liver for the presence of metastasis (low signal intensity on T_1-weighted images and high signal intensity on T_2-weighted images) may help make the diagnosis of neuroblastoma.

Rhabdomyosarcoma is the most common pediatric soft tissue sarcoma and can occur in almost any primary site except the brain. In children, it is most frequently found in the pelvis and genitourinary tract or in the head and neck. In the genitourinary tract, the tumor most frequently arises in the bladder; other common sites are the urethra, prostate, and vagina. Bladder lesions most commonly arise from the submucosa of the trigone or the bladder base and then infiltrate the bladder wall and adjacent structures including the urethra, prostate, vagina, and uterus. Because of the propensity for local invasion, it can be difficult to determine if the primary site was the bladder or the prostate in males or the bladder or the vagina in females. The multiplanar capabilities of MRI make it well suited for demonstrating bladder wall thickening, demarcating the inferior, lateral and posterior extent of the tumor, and detecting distant metastases.[6,24]

Sacrococcygeal teratomas are the most common tumors of the caudal region in children. These lesions are derived from all three germinal cell layers and have a characteristic MR appearance. Specifically, on T_1- and T_2-weighted images a large presacral mass is identified that contains rounded, well-defined areas of different signal intensity. Often one of these locules will contain fat, which can be demonstrated using a lipid-selective presaturation pulse or by simply looking for chemical shift misregistration. (Figure 8). Since these lesions displace the rectum anteriorly, as do all lesions in the presacral space, they can be distinguished from lesions arising in the pelvic cavity. The differential diagnosis for presacral masses in infants also includes anterior meningocele, rectal duplication, neuroblastoma, lymphoma, and lipoma.

4 SUMMARY

During the relatively short time that MRI has been applied to the evaluation of pediatric diseases and disorders, it has proven to be a useful tool in the diagnostic evaluation of a variety of entities, many of which we have touched upon in this article. More detailed information about some of these diseases and disorders can be found in the references given at the end of this article, and it is the authors' hope that the reader will learn from these references and, perhaps in time, add to them.

5 RELATED ARTICLES

Abdominal MRA; Brain MRS of Infants and Children; Lung and Mediastinum MRI; Male Pelvis Studies Using MRI; MRI of the Female Pelvis; Liver, Pancreas, Spleen, and Kidney MRI.

6 REFERENCES

1. B. D. Fletcher and M. D. Jacobstein, *Am. J. Roentgenol.*, 1986, **146**, 941.
2. G. S. Bisset III, J. L. Strife, D. R. Kirks, and W. W. Bailey, *Am. J. Roentgenol.*, 1987, **149**, 251.
3. M. L. Pessar, R. L. Soulen, J. S. Kan, S. Kadir, and E. A. Zerhouni, *Pediatr. Radiol.*, 1988, **18**, 229.
4. R. B. Dietrich, in 'Magnetic Resonance Imaging', 2nd edn, ed. D. D. Stark and W. G. Bradley, Jr, Mosby-Year Book, St Louis, 1992, Vol. 2, Chap. 59.
5. R. B. Dietrich and H. Kangarloo, *Radiology*, 1986, **159**, 215.
6. R. B. Dietrich, in 'Magnetic Resonance Imaging of Children', ed. M. D. Cohen and M. K. Edwards, B. C. Decker, Philadelphia, 1990, Chap. 21.
7. A. Daneman and D. J. Alton, *Radiol. Clin. North Am.*, 1991, **29**, 351.
8. J. Gambino, B. Caldwell, R. B. Dietrich, I. Walot, and H. Kangarloo, *Am. J. Roentgenol.*, 1992, **158**, 383.
9. E. Secaf, H. Hricak, and C. A. Gooding, *Pediatr. Radiol.*, 1994, **24**, 291.
10. H. Hricak and M. J. Popovich, in 'Magnetic Resonance Imaging of the Body', 2nd edn, ed. C. B. Higgins, H. Hricak, and C. A. Helms, Raven, New York, 1992, Chap. 31.
11. M. C. Mintz, D. I. Thickman, D. Gussman, and H. Y. Kressel, *Am. J. Roentgenol.*, 1987, **148**, 287.
12. B. M. Carrington, H. Hricak, R. N. Nuruddin, E. Secaf, R. K. Laros Jr, and E. C. Hill, *Radiology*, 1990, **176**, 715.
13. H. Hricak, Y. C. F. Chang, and S. Thurnher, *Radiology*, 1988, **169**, 169.
14. K. Togashi, K. Nishimura, K. Itoh, I. Fujisawa, Y. Nakano, K. Torizuka, H. Ozasa, and M. Oshima, *Radiology*, 1987, **162**, 675.
15. C. Hugosson, H. Jorulf, and Y. Bakri, *Pediatr. Radiol.*, 1991, **21**, 281.
16. R. B. Dietrich and H. Kangarloo, *Radiology*, 1987, **163**, 367.
17. P. J. Fritzsche, H. Hricak, B. A. Kogan, M. L. Winkler, and E. A. Tanagho, *Radiology*, 1987, **164**, 169.
18. A. H. Troughton, J. Waring, and A. Longstaff, *Clin. Radiol.*, 1990, **41**, 178.
19. S. J. Pomeranz, N. Altman, J. J. Sheldon, T. A. Tobias, K. P. Soila, L. J. Jakus, and M. Viamonte, *Magn. Reson. Imag.*, 1986, **4**, 69.
20. Y. Sato, K. C. Pringle, R. A. Bergman, W. T. C. Yuh, W. L. Smitt, R. T. Soper, and E. A. Franken, *Radiology*, 1988, **168**, 157.
21. K. McHugh, N. E. Dudley, and P. Tarr, *Pediatr. Radiol.*, 1998, **25**, 33.
22. A. Vade, H. Reyes, and A. Wilbur, *Pediatr. Radiol.*, 1989, **19**, 179.
23. D. D. Kidney, R. B. Dietrich, and A. K. Goyal, *Pediatr. Radiol.*, 1998, **28**, 322.

For References see p. 1255

24. R. B. Dietrich and H. Kangarloo, *Am. J. Roentgenol.*, 1986, **146**, 251.

25. M. J. Siegel, G. A. Jamroz, H. S. Glazer, and C. L. Abramson, *J. Comput. Assist. Tomogr.*, 1986, **10**, 593.

26. M. J. Siegel and G. D. Luker, *MRI Clin. North Am.*, 1996, **4**, 599.

27. T. G. Belt, M. D. Cohen, J. A. Smith, D. A. Cory, S. McKenna, and R. Weetman, *Am. J. Roentgenol.*, 1986, **146**, 955.

28. H. Kangarloo, R. B. Dietrich, R. M. Ehrlich, M. I. Boechat, and S. A. Feig, *Urology*, 1986, **28**, 203.

29. B. D. Fletcher, S. Y. Kopiwoda, S. E. Strandjord, A. D. Nelson, and S. P. Pickering, *Radiology*, 1985, **155**, 699.

30. M. D. Cohen, R. M. Weetman, A. J. Provisor, W. McGuire, S. McKenna, B. Case, A. Siddiqui, D. Mirksh, and I. Seo, *Am. J. Roentgenol.*, 1984, **143**, 1241.

31. B. D. Fletcher and S. C. Kaste, *Urol. Radiol.*, 1992, **14**, 263.

Biographical Sketches

Rosalind B. Dietrich. *b* 1953. M.B., Ch.B, 1976, University of Manchester School of Medicine, UK. Internship and radiology residency, Cedars-Sinai Medical Center, Los Angeles, CA, USA, 1979–84. Successively, fellow and Assistant Professor of Pediatric Radiology, University of California, Los Angeles, CA, USA, 1984–1990. Professor of Radiology and Director of MRI, University of California, Irvine, Orange, CA, USA, 1990–1995. Director of Research 1996–present. Approx. 65 publications. Research specialties: applications of MRI in the evaluation of the pediatric brain and body; MRI of brain maturation and white matter diseases.

Gerald M. Roth. *b* 1962. A.B. (biochemical sciences), 1984, Harvard University, USA; M.D., 1988, Columbia University College of Physicians and Surgeons, USA. Internship, Cedars-Sinai Medical Center, Los Angeles, CA, USA. Radiology residency, Hospital of the University of Pennsylvania, Philadelphia, PE, USA, 1989–1993. Successively, MRI Fellow, Faculty, Department of Radiological Sciences, University of California at Irvine, 1993–present. Approx. 5 publications. Research interests: applications of MRI in the chest, abdomen and pelvis; optimization of MR scan protocols.

Male Pelvis Studies Using MRI

Hedvig Hricak

Memorial Sloan-Kettering Cancer Center, New York, NY, USA

and

William T. Okuno

University of California, San Francisco, CA, USA

1 INTRODUCTION

Superb soft tissue contrast resolution and multiplanar imaging capability have placed MR imaging in an eminent

For list of General Abbreviations see end-papers

position in the evaluation of pelvic anatomy and pathology. In the male pelvis, much work has been performed in studying the prostate gland, seminal vesicles, testes, penis, and urethra. While the results of MR imaging in these areas were promising from the very beginning, these applications are still wrapped in controversy, and the indications are not generally accepted by many clinicians. As MR techniques keep evolving and overall image resolution is improved, MRI is gaining recognition and converting many skeptics into ardent advocates.

2 PROSTATE GLAND

2.1 Anatomy

The mature prostate is composed of both glandular and nonglandular tissue. The glandular portion is differentiated into three major zones: peripheral (70%); central (25%); and transition (5%) zones.[1] A small amount of glandular tissue is also present in the periurethral glands. Zonal differentiation is clinically significant because most (68%) prostate carcinomas arise in the peripheral zone whereas benign prostatic hyperplasia (BPH) usually originates in the transition zone.[2] Nonglandular tissue within the prostate includes the urethra and anterior fibromuscular stroma.

The normal prostate gland demonstrates a homogeneous intermediate signal intensity on T_1-weighted images, regardless of patient age or magnetic field strength; T_2-weighted images are necessary to delineate the zonal anatomy of the prostate (Figure 1).[3] The peripheral zone is of high signal intensity, equal to or greater than that of the adjacent periprostatic fat. In contrast, the central zone has a lower signal intensity than the surrounding peripheral zone.[3] These differences in signal characteristics are thought to be related to the presence of

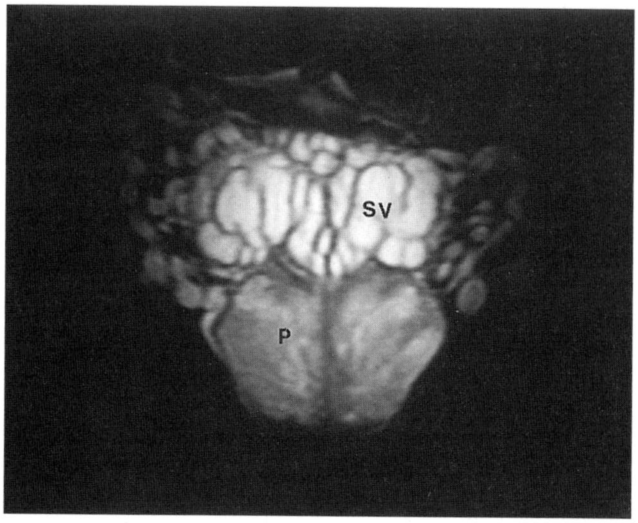

Figure 1 Prostate gland, normal anatomy: T_2-weighted image, endorectal coil, coronal plane of section. The peripheral zone (P) of the prostate gland demonstrates homogeneous high signal intensity. Seminal vesicles (SV) are seen cephalad to the prostate gland. They are convoluted in Nature and have a grape-like appearance

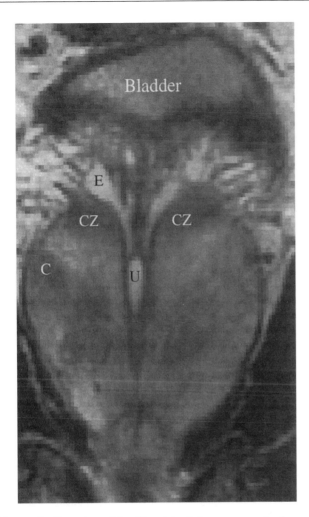

Figure 2 Prostatic utricle: T_2-weighted image, coronal plane of section. The utricle (U) is seen as a midline cyst of high signal intensity extending superior to the verumontanum. In the right peripheral zone, the low signal intensity region is a focus of prostate cancer (C). E, right ejaculatory duct; CZ, central gland

cysts are midline structures that arise at the level of the verumontanum and extend cephalad (Figure 2). Their signal intensity is nonspecific and depends on the nature of their content, ranging from serous to proteinaceous and hemorrhagic fluid.[5]

2.2.2 Inflammatory Disease

MR imaging does not play a role in the diagnosis of prostatic inflammatory disease. When discovered incidentally, acute prostatitis will demonstrate diffuse glandular enlargement and the peripheral zone sometimes exhibits a low signal intensity on the T_2-weighted image.[6] Thus far, there is only limited experience in MR imaging of prostate abscess. If this diagnosis is clinically suspected, transrectal ultrasound (TRUS) or CT should be the modalities of choice.

2.2.3 Benign Prostatic Hyperplasia

Benign prostatic hyperplasia (BPH) is a glandular and/or stromal proliferation in the transition zone and periurethral glands. The prostate gland is enlarged, and often the hyperplastic tissue is separated from the compressed peripheral zone tissue by the surgical pseudocapsule (Figure 3). MR imaging of BPH includes a spectrum of findings, based on the varying histologic composition and depending on the imaging sequence utilized.[7] Although MRI provides excellent morphologic definition of BPH, it is not possible to differentiate benign from malignant disease of the prostate.

The multiplanar imaging capability of MRI allows accurate volumetric measurement of the hyperplastic prostate gland and is especially useful in glands greater than 100 g, where ultrasound is limited in accuracy.[8] Because of its accuracy in volumetric measurement and depiction of zonal anatomy, MR imaging has been used in the follow-up of patients on androgen deprivation therapy in order to monitor the changes in

striated muscle and fewer glandular elements in the central zone.[3] The transition zone also has a lower signal intensity than the peripheral zone. The signal intensities of the central and transition zones are similar at all imaging parameters, and the two can be differentiated only by knowledge of their respective anatomic locations.[3] On Gd chelate-enhanced T_1-weighted images, the zonal anatomy of the gland is depicted but not as consistently as on T_2-weighted images.[4]

2.2 Pathology

2.2.1 Congenital Anomalies

Agenesis and hypoplasia of the prostate are frequently encountered with other congenital anomalies of the genitourinary tract. While MRI can establish the diagnosis, this frequently is an associated finding and not the clinical reason for the study.

Congenital cysts of the prostate are the most common anomalies of this gland. Prostatic utricle and Müllerian duct

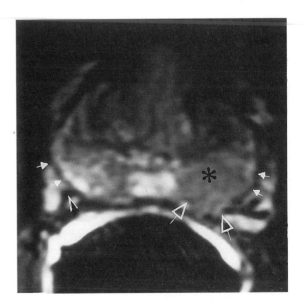

Figure 3 Benign nodular hyperplasia: T_2-weighted image, coronal plane of section. The prostate gland is enlarged. The heterogeneous signal intensity is secondary to benign nodular hyperplasia. A large hyperplastic nodule (*) protrudes into the urinary bladder (B)

For References see p. 1264

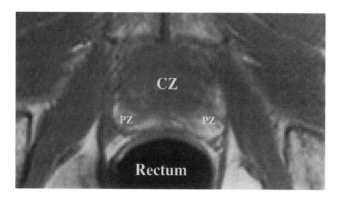

Figure 4 Postbiopsy hemorrhage: T_1-weighted image, transaxial plane of section. In this patient, hemorrhage after a prostate biopsy is seen as increased signal intensity diffusely in the peripheral zone (PZ) and more focally in the right central gland (CZ). Normally on T_1-weighted images, the entire prostate shows homogeneous intermediate signal intensity

gland size and in differential volume reduction in zones of the gland.[9]

2.2.4 Prostate Carcinoma

Cancer of the prostate is the most common cancer in men and ranks as the second most common cause of cancer death in American men.[2,10] MR imaging is not applicable for the detection of prostate carcinoma. The strength of MRI is in the staging of biopsy-proven prostate cancer, although this application continues to be controversial and is a subject of numerous ongoing investigations. On T_2-weighted MR images, prostatic carcinoma (PCa) is most commonly shown as a decreased signal intensity area within the high-signal-intensity normal peripheral zone. The detection of PCa on MRI is applicable only to the tumors located in the peripheral zone. The normal low or heterogeneous signal intensity of the transition and central zones precludes tumor detection.[11] Even in the peripheral zone, tumor detection may be hampered by postbiopsy changes (Figure 4).[11] Depending on the time interval between biopsy and MRI scan, the biopsy changes may cause either under- or overstaging of tumor presence and extent. Recent studies have demonstrated that MRI should not be performed for at least 2 weeks after biopsy.[12] The finding of low signal intensity within the peripheral zone is not specific for prostate cancer.[13] A number of other etiologies such as chronic prostatitis, dystrophic changes, scar, previous trauma, postbiopsy hemorrhage, radiation, and hormonal changes can all cause low signal intensity within the peripheral zone.[13]

Recently, *in vivo* MR spectroscopic imaging of the prostate gland has been developed to increase the specificity of cancer detection in the peripheral zone. This three-dimensional proton spectroscopic technique allows the simultaneous acquisition of multiple spectra from the prostate gland with a tissue voxel resolution down to 0.24 ml.[14,15] Since the metabolic changes of prostate cancer cause an increase in choline and a decrease in citrate levels, the ratio of the area under the choline spectroscopic peak to that under the citrate peak is increased in areas of cancer. The increased choline/citrate

ratio in areas of prostate cancer enables the differentiation of cancer from the other causes of low T_2 signal in the peripheral zone listed above (Figure 5).[16] In addition, the ability to estimate the spatial extent of prostate cancer has important implications for assessing the efficacy of various cancer treatments.[17]

Prostate cancer, particularly the mucinous type, can demonstrate high signal intensity indistinguishable from the surrounding peripheral zone.[18] While prostate cancer detection rates as high as 92% have been reported,[11] the results of large multicenter studies are disappointingly low, with only 60% of lesions greater than 5 mm in any one dimension being detected on MRI scans.[19] Attempts have also been made to measure tumor volume by MRI. As with ultrasound, the results are inaccurate, and smaller lesions are overestimated in 44% of patients and larger lesions underestimated in 55% of patients.[19]

Prostate cancer staging can follow either the TNM or Jewitt classification.[20] The TNM stage 1 or Jewitt stage A are tumors not applicable to MRI detection. Most of those cancers are within the transition zone, the area where MRI can neither depict nor stage the disease. TNM stage 2–Jewitt B disease indicates tumor confined to the prostate gland. The low-signal-intensity tumor is seen within the peripheral zone, and while the lateral margin can bulge, the bulge is smooth in contour.[21,22] With the endorectal coil, direct visualization of the prostate capsule increases the confidence level in the evaluation of TNM stage 2. In TNM stage 3 and Jewitt C disease, the findings of importance are extracapsular extension and seminal vesicle invasion.

MRI findings of extracapsular extension on endorectal coil include (a) bulge of the prostate gland with an irregular margin, (b) contour deformity with step-off or angulated margin, (c) breech of the capsule with direct tumor extension, (d) obliteration of the fat in the rectoprostatic angle, and (e) asymmetry, of the neurovascular bundles (Figure 6).[12,21,22] Seminal vesicle invasion is diagnosed when there is (a) demonstration of contiguous low-signal-intensity tumor extension into and around the seminal vesicles, and/or (b) tumor extension along the ejaculatory duct resulting in nonvisualization of the ejaculatory duct, decreased signal intensity of seminal vesicles, and loss of seminal vesicle wall on T_2-weighted images (Figure 7).

While transaxial planes of section are essential in the evaluation of extracapsular extension, the invasion of the seminal vesicles is facilitated by the evaluation of transaxial and coronal planes of section.[12,23–25] In recently reported studies using the Jewitt classification and endorectal coil, the accuracy for extracapsular extension was 82% and accuracy for seminal vesicle invasion 97%.[12] These reports are in accordance with previously published results by Schnall et al.,[24] although they are higher than the latest publications by Chelsky et al.[25] In the evaluation of lymph node metastases, reports in the literature testify to the superiority of MRI over CT. However, none of the studies has sufficiently high numbers of positive nodes for statistically meaningful analysis. The accuracy of staging of cancer of the prostate depends on the type of coil used and appears to be most accurate when the combination of endorectal and surface multicoil system is used.[11,19–26] Contrast-enhanced images do not contribute to either tumor detection or staging except in rare instances when they help in the detection of seminal vesicle invasion.[4]

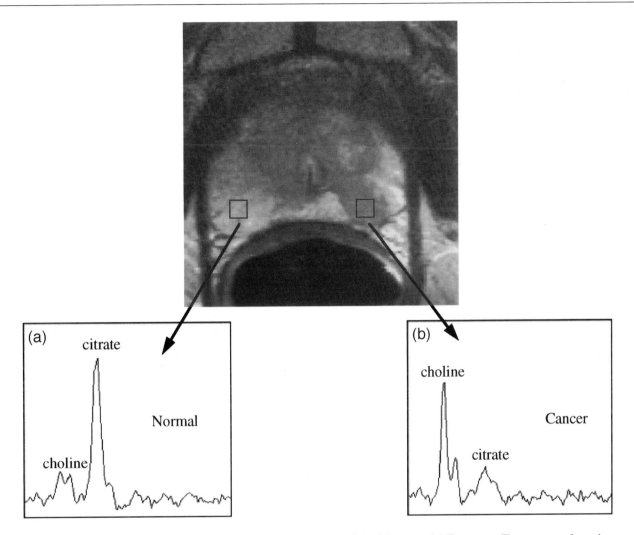

Figure 5 MR spectroscopic imaging of the prostate gland: transaxial T_2-weighted image and MR spectra. The spectrum from the normal peripheral zone (a) shows the normal relationship between the choline and citrate peaks. The spectrum from the prostate cancer (b) shows increased choline and reduced citrate levels

3 SEMINAL VESICLES

3.1 Anatomy

The seminal vesicles are paired, androgen-dependent accessory sex glands. On T_1-weighted images, normal seminal vesicles demonstrate homogeneous medium signal intensity similar to that of the adjacent pelvic muscle. On T_2-weighted images, seminal vesicles demonstrate a grape-like configuration with the high-signal-intensity fluid differentiated from the low signal intensity of the inner convolutions and outer wall (Figure 1).[27] The size and the signal intensity of the seminal vesicles depends on patient's hormonal status and will decrease with patient's age, hormonal replacement therapy, radiation therapy, or severe alcoholism.[27] On contrast-enhanced T_1-weighted images, the internal architecture of the seminal vesicles is depicted with the convolutions and wall demonstrating enhancement while the vesicular fluid remains of low signal intensity.[4,6] Although great variation in the size of normal seminal vesicles can occur in adult men of the same age, there is a tendency for a decrease in size with advancing years.

Seminal vesicles are usually symmetric in size, but asymmetry has been reported in up to 10% of patients.[6]

3.2 Pathology

3.2.1 Congenital Anomalies

Congenital anomalies of the seminal vesicles include the absence of the seminal vesicles and more commonly seminal vesicle cysts. In diagnosing the absence of seminal vesicles, T_1-weighted transaxial images are most useful.[27] In the evaluation of seminal vesicle cysts, MRI provides precise anatomical localization but the signal intensity depends on its fluid composition.[5] Blood is often present within the cyst with its signal intensity depending on the age of the hemorrhage.

3.2.2 Inflammatory Disease

Inflammation of the seminal vesicles is usually associated with epidydimitis and the associated ascending spread of infection through the prostate gland. The MRI appearance varies with the stage of inflammation. Seminal vesicles are often

For References see p. 1264

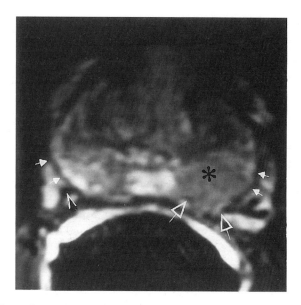

Figure 6 Prostate carcinoma with extracapsular extension on the left: T_2-weighted image, transaxial plane of section. Carcinoma (*) demonstrates low signal intensity. Prostate capsule (white solid arrows). There is a breach of the capsule (open white arrows) with direct cancer invasion to the neurovascular bundle. Normal neurovascular bundle on the right side (black and white arrowhead)

enlarged in the acute stage and small in the chronic phase. Hemorrhage within the seminal vesicles is common in subacute infection.[27] Chronic inflammation results in small atrophic seminal vesicles often of lower than normal signal intensity on T_2-weighted images.[6]

3.2.3 Tumors

The benign tumors of seminal vesicles (leiomyomas predominate) are more common than primary malignant neoplasms

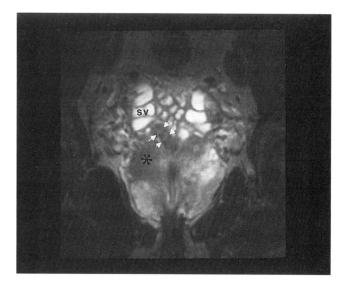

Figure 7 Prostate carcinoma with extracapsular and seminal vesicle extension: T_2-weighted image, coronal plane of section. Prostate cancer (*) is seen at the base of the prostate gland on the right. Irregular outer margin indicates extracapsular invasion. There is also evidence of direct tumor invasion (white arrows) into the right seminal vesicle (SV)

which are usually adenocarcinomas. The vast majority of malignant tumors of the seminal vesicles are secondary usually from prostate carcinoma, but extension from cancer of the bladder or rectum can be seen as well. The differentiation between benign and malignant tumors is usually based on the morphologic findings.[28] Benign tumors appear as smooth, well-marginated masses while malignant invasion of the seminal vesicles usually results in masses of low signal intensity and are irregular in configuration.[28] Loss of normal adjacent tissue planes also suggests secondary involvement of the seminal vesicles (Figure 7). Tumor extension into the seminal vesicles is best seen on T_2-weighted images, in the sagittal and coronal planes.[6,28]

4 PENIS/URETHRA

4.1 Anatomy

The penis is composed of three erectile bodies: the two lateral corpora cavernosa and the ventromedial corpus spongiosum (Figure 8).[29] Each of the three erectile bodies is enveloped by a fibrous sheath—the tunica albuginea. Buck's fascia—a common fibrous sheath—divides the penis into its dorsal compartment by enclosing the two corpora cavernosa and the ventral compartment by enclosing the corpus spongiosum. The male urethra extends from the bladder neck to the fossa navicularis in the glans penis. It is divided into the prostatic, membranous, and penile portions of the urethra. On T_1-weighted images, the three erectile bodies demonstrate medium signal intensity, and they all increase in signal intensity on T_2-weighted images (Figure 8). While the corpus spongiosum demonstrates homogeneous high signal intensity, the normal corpora cavernosa can be either homogeneous or heterogeneous in signal intensity depending on the blood volume within the erectile tissue.[29] The tunica albuginea and Buck's fascia appear as a low-signal-intensity stripe surrounding the corporeal bodies on T_2-weighted images and provide excellent contrast with the high-signal-intensity corpora.[29] The use of contrast enhancement allows demonstration of the penile anatomy on T_1-weighted images.[6]

4.2 Pathology

4.2.1 Congenital Anomalies

Epispadia, hypospadia, or duplication of the penis (penis diphallus) are the three groups of most commonly encountered congenital anomalies of the penis. The ability of MRI to delineate each of the corpora allows precise definition of the type of anomaly and provides depiction of associated anomalies involving the perineum and the remaining genitourinary tract.[29] The clinical indications, however, are rare, the most common being evaluation of penis diphallus.

4.2.2 Inflammatory Disease

Peyronie's disease is an inflammatory condition of unknown etiology characterized by the development of fibrous plaque involving the tunica albuginea and extending into the corpora cavernosa. On T_2-weighted MR images, the lower-signal-intensity plaque can be depicted in contrast to the higher-signal-

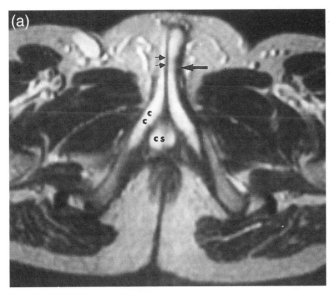 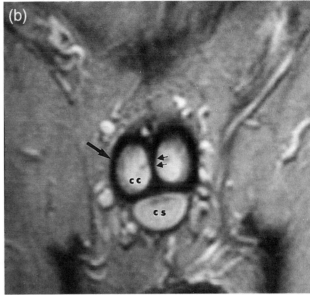

Figure 8 Normal anatomy of the penis. (a) Transaxial and (b) coronal plane of section: T_2-weighted images. Corpora cavernosa (cc), corpus spongiosum (cs), tunica albuginea (short arrows), and Buck's fascia (long black arrow)

intensity corpora cavernosa.[30,31] MRI is not the primary method for diagnosing this condition. MRI, however, can provide information about the location and size of the plaque as well as the degree of cavernosal involvement. Furthermore, MRI can differentiate between the acute and chronic form of Peyronie's disease, especially when contrast media are used.

4.2.3   Trauma

Penile trauma usually results from direct blunt injury to the erect penis causing fracture of the penis or rupture of the corpora cavernosa. On T_2-weighted MR images, the diagnosis of penile fracture is based on the finding of interruption of the normal low-signal-intensity tunica albuginea. Peripenile hematoma is often detected.[29,32]

MRI is valuable in the presurgical evaluation of urethral trauma (grade 3) associated with complex pelvic injury. Separation between the prostatic apex, membranous, and bulbous urethra can occur in superior anteroposterior or lateral direction. Misalignment of greater than 2 cm in the superior direction and greater than 1 cm in the lateral direction necessitates a different surgical approach (perineal versus suprapubic with removal of the symphysis pubis).[33,34] MRI can greatly assist in planning the surgical approach but thin-section T_2-weighted images in all three orthogonal planes are essential for the assessment of this complicated problem.[34]

4.2.4   Tumors

Penile carcinoma accounts for only about 1% of all male cancers, and carcinoma of the male urethra is even more rare.[35] When cancers are located in the glans penis or in the penile shaft, the locoregional tumor extent can usually be determined by physical examination. However, when the tumor involves the root of the penis or bulbomembranous urethra, the clinical evaluation is limited and MRI has a role in local staging of disease.[29] On MRI, penile and urethral carcinomas are usually not

distinguishable as most lesions requiring radiologic assessment are advanced and have spread to involve adjacent structures. The use of T_2-weighted images demonstrates the tumor of different signal intensity (usually lower) as compared with the adjacent normal high-signal-intensity erectile bodies (Figure 9). MRI, however, is not specific for cancer detection and cannot be distinguished from inflammatory disease.[6,29] This is especially true for smaller lesions of the urethra. However, MRI can be of great help in locoregional tumor staging, and in therapy planning—surgery versus radiation therapy.[29] Metastasis to the penis can also be well evaluated by MRI. Metastasis demonstrates diffuse low signal intensity infiltrating the erectile bodies on T_2-weighted images. Differentiation between primary and metastatic lesions is not possible.[29]

5   TESTIS

5.1   Anatomy

On T_1-weighted images, the testis demonstrates intermediate signal intensity, similar to that of the corpora cavernosa or corpus spongiosum and lower than that of subcutaneous fat. On T_2-weighted images, the testis demonstrates high signal intensity in contrast to the lower-signal-intensity tunica albuginea (a dense fibrous connective tissue capsule) covering the testis (Figure 10).[36] The mediastinum testis has signal characteristics similar to tunica albuginea and is seen as a low-signal-intensity stripe invaginating into the high-signal-intensity testicular parenchyma (Figure 10).[36] The parietal and visceral layers of the tunica vaginalis are often separated by a small amount of normal serous fluid. As in a hydrocele, this normal fluid surrounds the testis completely except for the posteromedial bare area. The signal intensity of the epididymis is similar to the testis on

For References see p. 1264

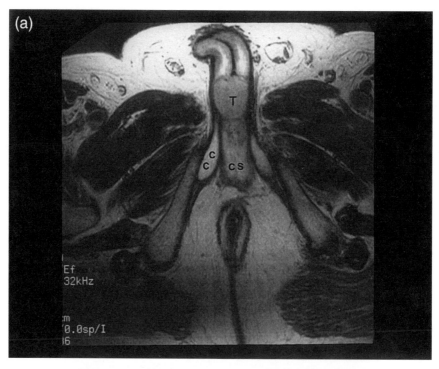

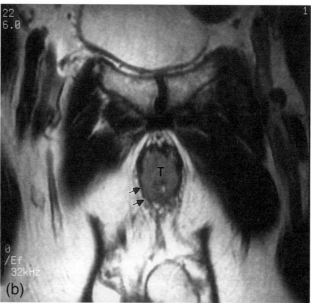

Figure 9 Cancer of the penis. (a) Transaxial and (b) coronal plane of section: T_2-weighted images. Tumor (T) is seen invading both corpora cavernosa (cc), spectum corpora cavernosa as well as corpus spongiosum (cs). In the coronal plane of section, the zonal anatomy of the penis is not visualized, and there is demonstration of tumor extension through the tunica albuginea (small black arrows)

T_1-weighted images and much lower than the testis on T_2-weighted images.[36]

5.2 Pathology

5.2.1 Undescended Testis

An undescended testis is defined as a testis located outside the scrotum. MR diagnosis of undescended testes relies on the finding of an elliptical mass demonstrated along the expected path of testicular descent.[37] However, it has recently been reported that dynamic gadolinium-enhanced MR angiography improves the detection of atrophic undescended testes by showing enhancement of the pampiniform plexus.[38] When the undescended testis is in the inguinal canal, it is usually oval in shape while the intraabdominal testis assumes a more rounded configuration. The large field of view and multiplanar image capability of MRI allows precise identification of the undescended testis as being either high scrotal, intracanalicular, or intraabdominal in location. When the intraabdominal testis is

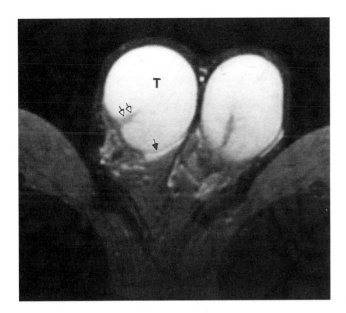

Figure 10 Normal testis, transaxial plane of section: T_2-weighted image. Testis (T) demonstrates homogeneous high signal intensity. Tunica albuginea (solid arrow) is seen as a low-signal-intensity stripe. Mediastinum testis (open arrows)

located close to the internal inguinal ring, it is easily depicted by MRI. However, high abdominal testis is difficult to demonstrate with MRI and CT is the procedure of choice. MR imaging allows differentiation between the undescended testis, gubernaculum, and lymph nodes.[37] Lymph nodes lie outside the expected descent of the testis. The differentiation between the gubernaculum and the testis relies on their respective differences in signal intensity. The gubernaculum characteristically demonstrates very low signal intensity on both T_1- and T_2-weighted images as its predominant histologic composition is fibrous tissue. The signal intensity of the testis varies depending on the degree of atrophy but is never as low as a fibrous cord.

5.2.2 Testicular Tumors

Testicular tumor is the most common malignancy among men of 15 to 34 years of age although it accounts for only 1% of all cancers in males.[35] The majority (90–95%) of testicular tumors are malignant germ cell tumors. The homogeneous high signal intensity of the normal testicular tissue serves as an excellent background for the depiction of intratesticular pathology.[36,39,40] On T_2-weighted images, testicular tumors usually demonstrate a lower signal intensity than the adjacent normal testicular tissue (Figure 11). Although MR imaging has a high sensitivity rate for the detection of testicular tumors, its findings are variable and are not specific for tumor histology. While on T_2-weighted images testicular tumors are mostly hypointense to the testis, they may demonstrate a signal intensity similar or even higher than testicular tissue. Tumors can be homogeneous or heterogeneous in signal pattern. While seminomas are typically homogeneous and hypointense, the nonseminomatous tumors are often heterogeneous and exhibit various signal intensities.[40] However, exceptions to the rule can be encountered.[36] Furthermore, differentiation between primary and secondary testicular tumors is not possible, and the

MR appearance of benign tumors is similar to testicular cancer.[41]

Testicular cysts can usually be differentiated from testicular tumors especially when contrast media are used. They appear as sharper marginated lesions with fluid signal intensity characteristics.[36]

5.2.3 Inflammatory Disease

Acute epididymitis is the most common inflammatory lesion in the scrotum. In acute epididymitis, the epididymis is enlarged and demonstrates heterogeneous and often higher signal intensity than normal. Hemorrhage may complicate acute epididymitis producing signal intensity commensurate with its age. Reactive hydrocele is often present. In chronic epididymitis, the signal intensity of the epididymis is reduced, a finding best appreciated on T_2-weighted images.[39] Associated orchitis, when present, appears as a homogeneous or heterogeneous hypointense lesion within the normal sized or enlarged testis. The signal intensity change is most commonly present in the mediastinum region. Fibrotic thickening of the tunica albuginea can occur following epididymitis, and the thickening may sometimes be difficult to differentiate from small testicular cancer, thus the term 'pseudotumor of the tunica albuginea' has been introduced.[42]

5.2.4 Spermatic Cord Torsion

Torsion is considered a surgical emergency since delaying intervention may result in irreversible damage to the testis. MRI plays no role and is not a primary imaging modality in diagnosing testicular torsion. Reports on MRI findings in acute torsion in humans are sporadic and not well documented. The testis may be normal or enlarged and of heterogeneous signal intensity. With testicular torsion, the spermatic cord can

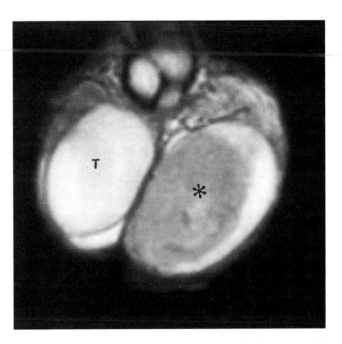

Figure 11 Seminoma of the left testis. Testicular tumor (*) demonstrates low signal intensity as compared with the adjacent remaining higher signal intensity testicular tissue. Normal right testis (T)

For References see p. 1264

enlarge, exhibiting a high signal intensity due to edema. A twisted cord can be seen as multiple low-intensity curvilinear structures rotating in a 'whirlpool' pattern which is best seen in a plane perpendicular to its axis.[43] Torsion of the testicular or epididymal appendices can be identified with MRI by their typical location and the signal characteristics of hemorrhage.[43]

5.2.5 Fluid Collection and Benign Scrotal Masses

Fluid collections whether hydrocele, hematocele, or pyocele all indicate abnormal fluid located between the parietal and visceral layers of the tunica vaginalis. The signal of hydrocele is typical for fluid (low signal intensity on T_1-weighted image and high intensity on T_2-weighted image).[36,39] The signal intensity of hematocele varies with the age of the hemorrhagic fluid, and the proteinaceous component of pyocele usually makes the fluid signal intensity on T_1-weighted images higher than that of hydrocele. Spermatocele is a retention cyst of small tubules which connect the rete testis to the head of epididymis. On MRI scan, the diagnosis of spermatocele can be made by its typical location (usually in the head of the epididymis) and its cystic nature.[6]

In scrotal pathology, MRI serves as a problem-solving modality when ultrasound findings are equivocal, technically inadequate, or there is a discrepancy between the physical examination and ultrasound findings. Recent studies have shown that when ultrasound and physical examination of the scrotum are inconclusive, MRI can improve patient management and reduce treatment costs.[37–46]

6 RELATED ARTICLES

Coils for Insertion into the Human Body; MRI of the Female Pelvis.

7 REFERENCES

1. J. E. McNeal, *Monogr. Urol.*, 1983, **4**, 5.
2. T. A. Stamey, J. E. McNeal, F. S. Freiha, and E. A. Redwine, *J. Urol.*, 1988, **139**, 1235.
3. H. Hricak, G. C. Dooms, J. E. McNeal, A. S. Monk, M. Marotti, A. Avalloro, M. Pelzer, E. C. Proctor, and E. A. Tanagho, *Am. J. Roentgenol.*, 1987, **148**, 51.
4. S. A. Mirowitz, J. J. Brown, and J. P. Heiken, *Radiology*, 1993, **186**, 153.
5. S. Thurnher, H. Hricak, P. R. Carroll, R. S. Pobiel, and R. A. Frilly, *Radiology*, 1988, **167**, 631.
6. H. Hricak, in 'MRI of the Pelvis. A Text Atlas', eds. H. Hricak and B. M. Carrington, Martin Dunitz, London, 1991, p. 313.
7. W. G. Way, Jr., J. J. Brown, J. K. Lee, E. Gutierrez, and G. L. Andriole, *Magn. Reson. Imag.*, 1992, **10**, 341.
8. A. Rahmouni, A. Yang, C. M. Tempany, T. Frenkel, J. Epstein, P. Walsh, P. K. Leichner, C. Ricci, and E. Zerhouni, *J. Comput. Assist. Tomogr.*, 1992, **16**, 935.
9. C. M. Tempany, A. W. Partin, E. A. Zerhouni, S. J. Zinreich, and P. C. Walsh, *Prostate*, 1993, **22**, 39.
10. S. H. Landis, T. Murray, S. Bolden, and P. A. Wingo, *CA Cancer J. Clin.*, 1999, **49**, 8.
11. M. D. Schnall, R. E. Lenkinski, H. M. Pollack, Y. Imai, and H. Y. Kressel, *Radiology*, 1989, **172**, 570.
12. S. F. Quinn, D. A. Franzini, T. A. Demlow, D. R. Rosencrantz, J. Kim, R. M. Hanna, and J. Szumowski, *Radiology*, 1994, **190**, 323.
13. K. Lovett, M. D. Rifkin, P. A. McCue, and H. Choi, *JMRI*, 1992, **2**, 35.
14. J. Kurhanewicz, D. B. Vigneron, S. J. Nelson, H. Hricak, J. M. MacDonald, B. Konety, and P. Narayan, *Urology*, 1995, **45**, 459.
15. J. Kurhanewicz, D. B. Vigneron, H. Hricak, P.Narayan, P. Carroll, and S. J. Nelson, *Radiology*, 1996, **198**, 795.
16. Y. Kaji, J. Kurhanewicz, H. Hricak, D. L. Sokolov, L. R. Huang, S. J. Nelson, and D. B. Vigneron, *Radiology*, 1998, **206**, 785.
17. J. Kurhanewicz, D. B. Vigneron, H. Hricak, F. Parivar, S. J. Nelson, K. Shinohara, and P. Carroll, *Radiology*, 1996, **200**, 489.
18. E. Outwater, M. L. Schiebler, J. E. Tomaszewski, M. D. Schnall, and H. Y. Kressel, *JMRI*, 1992, **2**, 597.
19. M. D. Rifkin, E. A. Zerhouni, C. A. Gatsonis, L. E. Quint, D. M. Paushter, J. I. Epstein, W. Hamper, P. C. Walsh, and B. J. McNeil, *N. Engl. J. Med.*, 1990, **323**, 621.
20. M. Graf, P. Hermanek, R. V. P. Hutter, L. H. Sobin, G. Wagner, and C. Wittekind (eds.), 'TNM Atlas: Illustrated Guide to the TNM/pTNM-Classification of Malignant Tumors, 4th edn', Springer-Verlag, Berlin, 1997.
21. E. K. Outwater, R. O. Petersen, E. S. Siegelman, L. G. Gomella, C. E. Chernesky, and D. G. Mitchell, *Radiology*, 1994, **193**, 333.
22. K. K. Yu, H. Hricak, R. Alagappan, D. M. Chernoff, P. Bacchetti, and C. J. Zaloudek, *Radiology*, 1997, **202**, 697.
23. H. Hricak, G. C. Dooms, R. B. Jeffrey, A. Arallone, D. Jacobs, W. K. Benton, P. Narayan, and E. A. Tanagho, *Radiology*, 1987, **162**, 331.
24. M. D. Schnall, Y. Imai, J. Tomaszewski, H. M. Pollack, R. Lenkinski, and H. Y. Kressel, *Radiology*, 1991, **178**, 797.
25. M. D. Schnall, R. E. Lenkinski, H. M. Pollack, Y. Imai, and H. Y. Kressel, *Radiology*, 1989, **172**, 570.
26. M. J. Chelsky, M. D. Schnall, E. J. Seidmon, and H. M. Pollack, *J. Urol.*, 1993, **150**, 391.
27. E. Secaf, R. N. Nuruddin, H. Hricak, R. D. McClure, and B. Demas, *Am. J. Roentgenol.*, 1991, **156**, 989.
28. R. D. McClure and H. Hricak, *Urology*, 1986, **27**, 91.
29. H. Hricak, M. Marotti, T. J. Gilbert, T. F. Lue, L. H. Wetzel, J. W. McAninch, and E. A. Tanagho, *Radiology*, 1988, **169**, 683.
30. G. Helweg, W. Judmaier, W. Buchberger, K. Wicke, H. Oberhauser, R. Knapp, O. Ennemoser, and D. Zur Nedden, *Am. J. Roentgenol.*, 1992, **158**, 1261.
31. R. Vosshenrich, I. Schroeder-Printzen, W. Weidner, U, Fischer, M. Funke, and R. H. Ringert, *J. Urol.*, 1995, **153**, 1122.
32. M. Fedel, S. Venz, R. Andreessen, F. Sudhoff, and S. A. Loening, *J. Urol.*, 1996, **155**, 1924.
33. C. M. Dixon, H. Hricak, and J. W. McAninch, *J. Urol.*, 1992, **148**, 1162.
34. Y. Narumi, H. Hricak, N. A. Armenakas, C. M. Dixon, and J. W. McAninch, *Radiology*, 1993, **188**, 439.
35. D. M. Parkin, P. Pisani, and J. Ferley, *CA Cancer J. Clin.*, 1999, **49**, 33.
36. S. Thurnher, H. Hricak, P. R. Carroll, R. Pobiel, and R. A. Frilly, *Radiology*, 1988, **167**, 631.
37. P. J. Fritzsche, H. Hricak, B. A. Kogan, M. L. Winkler, and E. A. Tanagho, *Radiology*, 1987, **164**, 169.
38. W. W. Lam, P. K. Tam, V. H. Ai, K. L. Chan, W. Cheng, F. L. Chan, and L. Leong, *J. Pediatr. Surg.*, 1998, **33**, 123.
39. L. L. Baker, P. C. Hajek, T. K. Burkhard, L. Dicepua, H. M. Landa, G. R. Leopold, J. R. Hesselink, and R. F. Mattrey, *Radiology*, 1987, **163**, 93.
40. J. O. Johnson, R. F. Mattrey, and J. Phillipson, *Am. J. Roentgenol.*, 1990, **154**, 539.
41. F. V. Coakley, H. Hricak, and J. C. Presti Jr, *Urol. Clin. North Am.* 1998, **25**, 375.

42. R. B. Poster, B. A. Spirt, A. Tamsen, and B. V. Surya, *Radiology*, 1989, **173**, 561.

43. M. A. Trambert, R. F. Mattrey, D. Levine, and D. P. Merthoty, *Radiology*, 1990, **175**, 53.

44. H. Derouet, H. U. Braedel, G. Brill, K. Hinkeldey, J. Steffens, and M. Ziegler, *Urol. Ausg.*, 1993, **32**, 327.

45. B. M. Cramer, E. A. Schlegel, and J. W. Thueroff, *Radiographics*, 1991, **11**, 9.

46. A. D. Serra, H. Hricak, F. V. Coakley, B. Kim, A. Dudley, A. Morey, B. Tschumper, and P. R. Carroll, *Urology*, 1998, **51**, 1018.

Biographical Sketches

Hedvig Hricak. *b* 1946. M.D., 1970, University of Zagreb, Croatia; Ph.D., 1992, Karolinska Institute, Stockholm, Sweden. University of California, San Francisco, Faculty in Radiology, 1982–99; Faculty in Urology, 1986–present; Faculty in Radiation Oncology, 1991–99. Department of Radiology, Sloan-Kettering Cancer Center, New York, 2000–present. Involved in clinical MR imaging of gynecologic and urologic diseases since the introduction of this technique. Over 150 papers on MR imaging in peer-reviewed journals and co-author of four major textbooks on the subject. Member of numerous editorial boards of prestigious journals in radiology and radiation oncology and serves on national and international committees of prestigious professional organizations.

William T. Okuno. *b* 1967. M.D., 1993, University of Illinois, USA. Radiology Residency, Massachusetts General Hospital, 1994–1998; Abdominal Imaging Fellowship, University of California, San Francisco, 1998–1999; Lutheran General Hospital, Park Ridge, Illinois, 1999–present. Special interests: abdominal imaging.

MRI of the Female Pelvis

Robert C. Smith, Michael J. Varanelli, Leslie M. Scoutt, and Shirley McCarthy

Yale University School of Medicine, New Haven, CT, USA

1 INTRODUCTION

The article will describe the clinical applications and technical considerations of MRI of the female pelvis. Due to its improved tissue contrast capability (compared with computerized tomography and ultrasound), MRI is the technique of choice for depicting normal uterine and cervical anatomy as well as a variety of benign and malignant conditions affecting these structures. MRI is also capable of delineating a number of nongynecological abnormalities of the pelvis.

Recent technical advances have dramatically improved the resolution that can be obtained when imaging the female pelvis. At the same time, these technical improvements have greatly reduced the imaging time. Both of these factors should improve the accuracy of MRI not only for detecting abnormalities but in the staging of gynecological malignancies. Cost–benefit analyses have shown MRI of the female pelvis to be cost-effective.

2 MRI TECHNIQUE

Imaging of the female pelvis requires both T_1- and T_2-weighted images.[1] T_1-weighted images are useful for lymph node detection and to characterize regions which contain blood products and/or fat. T_2-weighted images are necessary to depict the internal anatomy of the gynecological organs, characterize masses and other abnormalities, and determine the origin of a mass as uterine or ovarian.

T_1-weighted images are typically acquired in the axial plane from the level of the aortic bifurcation through the pubic symphysis. The *TR* (repetition time) used is between 500 and 600 ms, and the minimum possible *TE* (echo time) is used (usually 10–15 ms). Superior and inferior saturation pulses are routinely used to diminish the intravascular signal and therefore help decrease pulsation artifacts and distinguish lymph nodes from vascular structures. In some cases, use of a longer *TE* (up to 20 ms) will help diminish intravascular signal. Respiratory compensation is always utilized to diminish ghost artifacts from the high-signal subcutaneous fat.[2]

Other typical parameters would include a field of view (FOV) of 24–30 cm, section thickness of 5 mm, intersection spacing of 2.0–2.5 mm, a frequency matrix size of 256, a phase matrix size of 128 or 192, and one or two signal averages. If there is a large amount of bowel within the pelvis, glucagon can be administered to help diminish artifacts from bowel peristalsis. In the majority of cases, glucagon is not necessary.

T_2-weighted images should be acquired through the uterus and ovaries in at least two orthogonal planes. The main purpose of sagittal images is to depict the uterine and cervical zonal anatomy. Depending upon the orientation of the uterus, the axial or coronal plane can be chosen to image the uterus in a plane perpendicular to its long axis. Both the axial and coronal planes will usually depict the ovaries to good advantage. The T_2-weighted images should be acquired using fast spin echo (FSE) or turbo spin echo (TSE) pulse sequences.[3,4] Using these, high-resolution T_2-weighted images can be acquired through the entire pelvis in less than 5 min.

3 FSE IMAGING

When FSE T_2-weighted images are acquired in a body coil, typical imaging parameters would include a *TR* of 4500–6000 ms, a *TE* of 100–130 ms, a 5 mm section thickness, a 2.0–2.5 mm intersection spacing, a minimum phase matrix size of 192, a FOV of 24–28 cm, and two to four signal averages. If a small FOV is used with a phase matrix size of 256, this usually necessitates use of four signal averages to maintain the signal-to-noise ratio (S/N) at an acceptable level.

In addition to the above parameters, the echo train length (ETL) and the echo spacing (ESP) must also be specified when using FSE sequences. With conventional spin echo sequences, a single phase-encoding step is acquired during each *TR* interval. Even if multiple echoes are measured (as with proton

For References see p. 1274

density and T_2-weighted images), each is acquired with the same phase-encoding step. The multiple echoes are used to generate multiple images (of different TE) at each imaging location. In order to reconstruct an image, the number of phase-encoding steps must equal the size of the phase matrix. It is this fact which makes the imaging time of conventional spin echo sequences so long.

FSE sequences acquire multiple phase-encoding steps during each TR interval. This is achieved by applying multiple 180° pulses after each 90° pulse and separately phase encoding the resulting spin echoes. The ETL is equal to the number of separately phase-encoded echoes acquired following each 90° pulse. The maximum value of the ETL is usually 16 or 32. Recent development of 'single-shot' techniques using a half-Fourier reconstruction allows an ETL of 128 or 192 and acquisition of each image in about 1 s. Once all of the separately phase-encoded echoes have been acquired in one section, data are then acquired in the next section. The spacing between the separate phase-encoded echoes is the ESP. The minimum value of ESP is usually 6–8 ms and depends on the bandwidth. In almost all circumstances the minimum value of ESP should be used.

Regardless of the imaging technique, each phase-encoding step uses a different strength phase-encoding gradient. Echoes acquired with the weak phase-encoding gradients provide most of the signal (and hence contrast) of an image, and echoes acquired with the strong phase-encoding gradients provide most of the spatial resolution of an image. With the fast spin echo technique, multiple separately phase-encoded echoes are used to generate a single image at each location. The overall contrast of such images is determined by the echo times at which the weak phase-encoding steps are acquired.[5–7] If the weak phase-encoding steps are used for the early echoes, the image will appear as either proton density or T_1 weighted (depending on the TR). If the weak phase-encoding steps are used for the late echoes, the image will appear as T_2 weighted.

For this reason, the echo time of an FSE image is usually referred to as an effective TE. When the user selects an effective TE, the phase-encoding steps are assigned to the appropriate echoes of each multiecho train following each 90° pulse. Severe artifacts can result from this technique. Late echoes have diminished signal due to T_2 decay. If the strong phase-encoding gradients are applied to the late echoes (as with proton density- and T_1-weighted FSE images), there may be no useful signal left. Since the strong phase-encoding gradients provide the spatial resolution of an image, without these phase-encoding steps there will be a loss of spatial resolution. This is manifest as image blurring, which can be severe.

Image blurring is minimized by using a shorter ETL (usually 4 or 6) when acquiring proton density- or T_1-weighted FSE images. T_2-weighted FSE images never suffer from this blurring effect since the strong phase-encoding gradients are always applied to the early echoes. Early echoes suffer minimal signal loss due to T_2 decay. Therefore, T_2-weighted FSE images can usually be acquired with the maximum possible value of ETL.

The imaging time of a conventional spin echo sequence is given by $TR \times$ (phase matrix size) $\times$ (number of signal averages). The imaging time of an FSE sequence is this same product divided by ETL. However, when using FSE sequences, the number of sections that can be acquired for a given TR will be diminished. This results from the fact that many more echoes must be measured in each section before moving on to the next section. Depending on the choice of ETL, this will usually necessitate the use of a longer TR with FSE sequences.

The shorter the ETL, the more sections that can be acquired per TR interval. For example, using a TR of 3000 and an ETL of 8 will give approximately the same number of sections as using a TR of 6000 and an ETL of 16. The imaging times will be the same. The advantage of using a longer TR is that it gives a true T_2-weighted image. The trade-off is that use of a shorter ETL will improve the S/N since fewer late echoes are acquired. One must always keep in mind that the maximum value of TE that can be achieved is equal to ETL $\times$ ESP. Therefore, use of a shorter ETL will limit the maximal value of TE.

When imaging the pelvis in a body coil with FSE sequences, one can use either an ETL of 8 (with a TR of 3000–4000 ms) or an ETL of 16 (with a TR of 6000–8000 ms). It is recommended that a 192 or 256 phase matrix is used. There is usually a significant relative image degradation when a 128 phase matrix size is used with FSE images.

4 MULTICOIL IMAGING

When imaging with a single surface coil there is a marked improvement in the S/N, but the FOV that can be achieved is severely limited (roughly of the order of the diameter of the coil). In addition, the relative phase of the signal will depend upon the orientation of the surface coil within the magnet. If multiple surface coils are used together and their signals are combined into a circuit, there can be signal loss due to phase differences between the two signals. This can be minimized by careful orientation of the coils.

A multicoil (also called a phased-array coil) consists of multiple surface coils which act independently in a receive-only mode. Each separate coil of the multicoil inputs its signal into separate receiver channels. The signals from the individual coils are then used to reconstruct an individual image for each coil. These separate images are then recombined into a single composite image. In this case the relative phases of the individual signals is irrelevant, as the signals are only combined after magnitude reconstruction. By using a multicoil one gets the improved S/N of a surface coil, and the FOV that can be achieved is comparable to a body coil.[8–10] The only disadvantage of a multicoil is that four separate receiver channels must be used, which is expensive. All pelvic imaging should now be performed using a multicoil and FSE pulse sequences.

When using a multicoil, the signal from subcutaneous fat immediately adjacent to the coils will be markedly increased and can result in significant ghost artifacts. This can be eliminated by the placement of saturation pulses (within the field of view) through the subcutaneous fat both anteriorly and posteriorly.

5 NORMAL ANATOMY

On T_1-weighted images the uterus has a homogeneous low signal intensity. The appearance is similar to that of the uterus on computed tomography images, as no internal architecture is

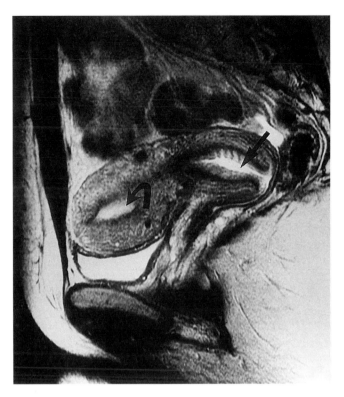

Figure 1 Sagittal FSE multicoil image demonstrating the zonal anatomy of the uterus. The endometrium is the central bright stripe (curved arrow). Note the surrounding low-signal junctional zone and intermediate-signal outer myometrium. Within the cervix, the central high signal represents the canal (straight arrow) and contained mucus. The next layer is the intermediate-intensity signal of the cervical mucosa which is surrounded by the low-intensity signal of the fibrous stroma. Finally, the outer cervical stroma has an intermediate signal intensity and is continuous with the myometrium

Ultrasound studies have been performed in an attempt to determine an approximate upper limit of normal for endometrial stripe thickness in postmenopausal patients.[15,16] Regardless of hormonal replacement therapy, an upper limit of 8 mm has been suggested. No similar large series have as yet been performed with MRI. However, a comparative study of ultrasound and MRI has shown that the MRI measurement of endometrial thickness is almost always smaller than the corresponding ultrasound measurement.[17]

The cervix also has a unique zonal anatomy on T_2-weighted MRI (Figure 1). Four distinct zones of signal intensity have been described:[4,18] a thin inner bright zone which corresponds to the endocervical canal; an adjacent thin zone of intermediate signal intensity corresponding to the cervical mucosa; a thicker zone of very low signal intensity thought to correspond to the predominantly fibrous portion of the wall of the cervix; and an outer zone which is continuous with the myometrium and is isointense with the myometrial signal.

Previous studies have shown little variation in the appearance of the cervix during the course of the menstrual cycle.[13] In addition, there is no apparent difference in the appearance of cervical zonal anatomy between premenopausal and postmenopausal women, or between those using and not using oral contraceptives.

The normal MRI appearance of the vagina is less complex. The vaginal wall typically shows low signal intensity, and the vaginal canal typically appears as a bright stripe of high signal intensity corresponding to vaginal secretions. The vagina is surrounded by the high signal intensity perivaginal venous plexus.

The ovaries appear homogeneously hypointense on T_1-weighted images. Follicles appear as small, round, homogeneous areas of high signal intensity on T_2-weighted images. The ovarian stroma is of low to intermediate signal intensity on T_2-weighted images. With high-resolution imaging, it is almost always possible to identify the ovaries in premenopausal patients, and most of the time in postmenopausal patients.

visible. On T_2-weighted images, a zonal architecture of the uterus is readily identified. The appearance of the zonal anatomy depends upon whether the patient is pre- or postmenopausal and the phase of the menstrual cycle.[11–13]

In virtually all premenopausal patients, three distinct zones of signal intensity can be seen in the uterine corpus (Figure 1).[12] An inner zone of high signal corresponds to the endometrium. An adjacent zone of low signal intensity corresponds to the so-called junctional zone. Histological studies have shown this zone to represent an inner layer of myometrium which has an increased nuclear area.[14] The remainder of the myometrium has intermediate signal intensity but can be quite variable, depending on the hormonal milieu of the patient.

Previous studies have shown that the inner bright zone progressively increases in thickness during the menstrual cycle with a peak thickness occurring late in the secretory phase. The thickness may vary from 4 mm early in the proliferative phase to 13 mm late in the secretory phase. The junctional zone shows no significant change in thickness during the menstrual cycle. In patients taking oral contraceptives (combined estrogen and progestin), endometrial thickness is markedly reduced after several months. In addition, the outer myometrium of these patients shows relatively increased signal intensity compared with those not using oral contraceptives.

6 BENIGN DISEASES OF THE UTERUS AND CERVIX

The most common benign mass of the uterus is the leiomyoma, also commonly referred to as a fibroid. These masses are sharply circumscribed, usually spherical in shape, and typically have low signal intensity on all pulse sequences. When fibroids become large, they can undergo degeneration, which is usually manifest as central increased signal intensity on T_2-weighted images.

MRI is uniquely able to determine the intrauterine location of these masses. Fibroids are usually described as being submucosal, intramural, or subserosal in location. Submucosal fibroids are commonly associated with abnormal menstrual bleeding (Figure 2). Intramural fibroids are not usually associated with abnormal menstrual bleeding but are a common cause of uterine enlargement (Figure 3). Subserosal fibroids can be on a stalk and therefore can undergo torsion. Subserosal fibroids can also be confused with an adnexal mass on physical examination and on other imaging modalities.

When fibroids cause significant clinical symptoms and surgery is contemplated, the surgical approach is dependent upon the precise location of the fibroids. Some submucosal fibroids

For References see p. 1274

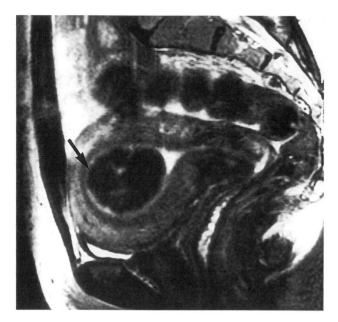

Figure 2 Sagittal FSE multicoil image demonstrating a submucosal fibroid (arrow)

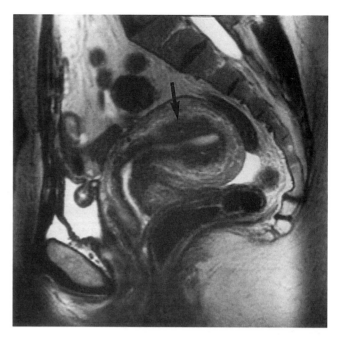

Figure 4 Sagittal FSE multicoil image demonstrating adenomyosis (arrow). Note the diffuse irregular thickening of the junctional zone with tiny areas of hyperintensity characteristic of the disease

can be removed hysteroscopically. Intramural and subserosal fibroids can only be removed using a transabdominal approach. In addition to their location, another important surgical consideration is the size and increased vascularity of these lesions. Hormonal therapy is sometimes used to decrease their size and vascularity prior to possible surgery. MRI can be used to follow precisely the size and vascularity during such therapy.

Adenomyosis is defined as the presence of endometrial tissue within the myometrium. This tissue is usually not functional. The appearance of adenomyosis on MRI scans is rather characteristic.[19] It appears as focal or diffuse ill-defined thickening of the junctional zone (Figure 4). It has low signal

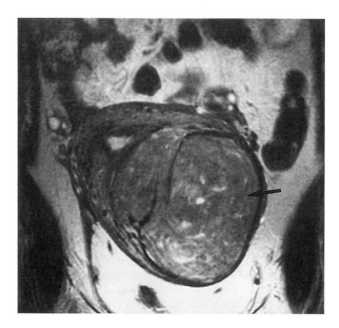

Figure 3 Coronal FSE multicoil image demonstrating a large intramural fibroid (arrow) displacing the endometrial stripe

intensity on T_2-weighted images, being isointense with the junctional zone. In some cases, adenomyosis can appear as small foci of high signal intensity within the myometrium on T_2-weighted images. Adenomyosis can result in uterine enlargement and abnormal menstrual bleeding. It is therefore important to differentiate this condition from fibroids. MRI is particularly useful in making this distinction.

Endometrial hyperplasia is thought to represent a physiological response of the endometrium to unopposed estrogenic stimulation. Some forms of endometrial hyperplasia are known to be precursors of endometrial carcinoma. On MRI, endometrial hyperplasia appears as thickening of the high-signal endometrium on T_2-weighted images (Figure 5).[1] However, MRI cannot distinguish between hyperplasia and early carcinoma.

Endometrial polyps can be sessile or pedunculated masses projecting into the endometrial cavity. They can be associated with abnormal uterine bleeding. However, endometrial carcinoma can sometimes have a polypoid configuration and cannot be reliably distinguished from simple polyps. On T_2-weighted images, polyps may appear as intermediate-signal masses (with respect to the low-signal junctional zone and the high-signal endometrium) but can also be isointense with the endometrium. Gadolinium-enhanced T_1-weighted images may be useful in detecting endometrial polyps. The enhancing polyp can be outlined by nonenhancing fluid within the endometrial cavity.

Pedunculated submucosal fibroids can also appear as polypoid filling defects within the endometrial cavity on MRI. Their signal intensity is usually significantly lower than that of endometrial polyps, which helps in their distinction.

The two most common uterine anomalies that need to be differentiated are the septate uterus and the bicornuate uterus.[20] In the septate uterus, the external contour of the uterus is normal (Figure 6). In the bicornuate uterus (Figure 7), there are

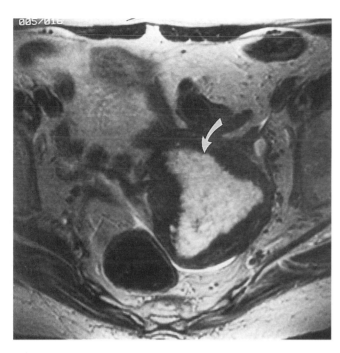

Figure 5 Axial FSE multicoil image demonstrating diffuse thickening of the endometrium (arrow) consistent with endometrial hyperplasia

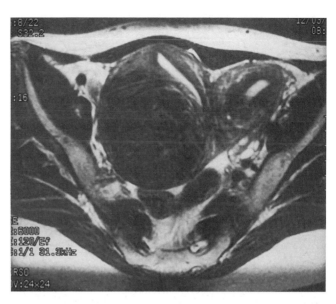

Figure 7 Axial FSE multicoil image of a bicornuate uterus. The contour of the uterine fundus is abnormal, with a large indentation separating the two endometrial cavities. There is a large low-signal intensity leiomyoma in the right horn

two separate horns of the endometrial cavity, and the external contour of the uterus is significantly indented at the fundus (as there is no uterine tissue in the space between the two horns).

By imaging through the fundus along the long axis of the uterus, these two entities can be reliably distinguished. The signal characteristics of the septum may reflect myometrial or fibrous tissue, and hence signal behavior is not useful in classifying anomalies. The septate uterus is more commonly associated with infertility, and can be treated surgically through a hysteroscopic approach.

There are only a few benign lesions of the cervix commonly seen on MRI. Nabothian cysts represent dilated cervical glands

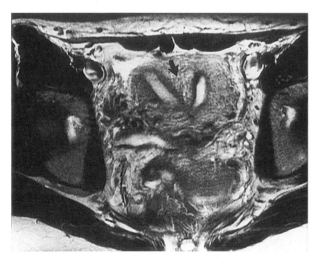

Figure 6 Axial FSE multicoil image of a septate uterus. The contour of the uterine fundus is normal, and a large myometrial septum (arrow) is present

usually seen following inflammation. These are very commonly visualized on MRI as homogeneous, round, sharply circumscribed areas of very high signal intensity on T_2-weighted images. Rarely, fibroids can be seen originating from the cervix. These are otherwise identical to those originating within the uterus.

7 MALIGNANT DISEASE OF THE UTERUS

Endometrial carcinoma is the most common invasive malignancy of the female genital tract. Unlike cervical carcinoma, there is not a well documented progression from precursor lesions to invasive carcinoma. Known risk factors for the development of endometrial carcinoma include obesity, nulliparity, late menopause, diabetes mellitus, hypertension, polycystic ovarian syndrome, estrogen-producing tumors, and unopposed exogenous estrogen supplementation.

MRI has been shown to be useful in evaluating patients with known endometrial carcinoma. The vast majority of patients with endometrial carcinoma (up to 75%) will have stage I disease at the time of diagnosis. Tumor grade and depth of myometrial invasion are the two important prognostic factors which can affect therapy in these patients.[21–23] Tumor grade will be known based upon histological findings. However, depth of myometrial invasion can only be determined preoperatively with the use of imaging studies.

On T_1-weighted images, most endometrial carcinomas are isointense with the uterus unless they contain hemorrhagic areas. On T_2-weighted images, most endometrial carcinomas (large enough to be detected as distinct masses) have a signal intensity intermediate between normal endometrium (higher signal intensity) and normal myometrium (lower signal intensity).

Multiple studies have been performed evaluating the ability of MRI to depict accurately the depth of myometrial inva-

For References see p. 1274

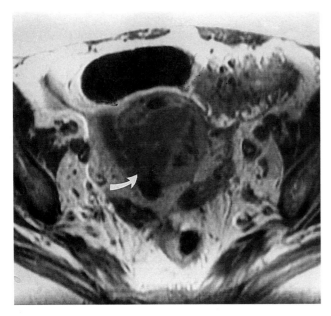

Figure 8 Axial multicoil image acquired following gadolinium administration demonstrates a large tumor extending beyond the uterus (arrow)

sion.[24–31] Using T_2-weighted images and/or gadolinium-enhanced T_1-weighted images, these studies have shown an accuracy of 75–95% in distinguishing superficial from deep myometrial invasion (Figure 8). More importantly, these studies have shown that an intact junctional zone has a 100% negative predictive value (NPV) in excluding myometrial invasion, and that segmental disruption of an otherwise intact junctional zone has a 100% positive predictive value (PPV) in detecting at least superficial myometrial invasion.

Unfortunately, most patients with endometrial carcinoma are postmenopausal, and uterine zonal anatomy may not be as conspicuous as that seen in premenopausal patients. Thus, these data are based on a small number of patients. In addition, patients with large intraluminal polypoid tumors can have significant expansion of the endometrial cavity with resultant distortion of zonal anatomy. Accurate assessment of myometrial invasion can be impossible in such cases.[32]

In patients without visible zonal anatomy, the presence of myometrial invasion must be presumed, based upon the appearance of the endometrial/myometrial interface. If this interface is irregular, invasion is presumed to be present, and if this interface is smooth, invasion is presumed to be absent. However, these findings do not have a high PPV or NPV, and this limits their usefulness.

All patients with clinically suspected stage I endometrial carcinoma must have evaluation of the cervix for possible tumor involvement. Endocervical curettage can be unreliable in this assessment. A few studies indicate an NPV of nearly 100% for MRI detection of cervical involvement.[25,29] The positive data regarding MRI evaluation of cervical involvement in patients with endometrial carcinoma are more limited. This is because only a small number of patients will have cervical involvement, since the vast majority have stage I disease at the time of presentation. In addition, some patients may have dis-

tention of the cervical canal by clot or debris, which can give a false-positive diagnosis of cervical involvement.

Uterine sarcomas are relatively rare tumors accounting for only 3–5% of all uterine cancers.[33] The three most common histological variants are malignant mullerian mixed tumor (MMMT), leiomyosarcoma (LMS), and endometrial stromal sarcoma (ESS). MMMT and LMS each account for about 40% of all uterine sarcomas, while ESS accounts for 10–15%. The staging system for uterine sarcomas is the same as that used for endometrial carcinoma.

The incidence of sarcomatous change in preexisting uterine leiomyomas is reported to be between 0.1 and 0.8%.[33] A small percentage (4%) of these patients will have a history of prior pelvic radiotherapy. The presenting clinical symptoms include vaginal bleeding, pelvic pain, and pelvic mass. The diagnosis should be suspected if rapid uterine growth occurs, especially in a postmenopausal patient.

MMMTs are histologically composed of a mixture of sarcoma and carcinoma. Almost all of these tumors occur after menopause. A history of prior pelvic radiotherapy can be elicited in up to 35% of cases.[33] Postmenopausal bleeding is the most common presentation. The tumor usually grows as a large polypoid mass with areas of necrosis and hemorrhage. It spreads in a manner identical to endometrial carcinoma but tends to be more aggressive with significant myometrial invasion in almost all cases.

Endometrial stromal tumors are rare tumors composed of cells resembling normal endometrial stroma. They can be divided into three types based upon mitotic activity, vascular invasion, and prognosis.[33] The endometrial stromal nodule is a benign lesion confined to the uterus. Endolymphatic stromal myosis infiltrates the myometrium, may extend beyond the uterus, and can metastasize. ESS, the third type of stromal tumor, is differentiated from stromal myosis mainly on the basis of mitotic activity and its much more aggressive course. Stromal tumors usually occur in perimenopausal patients. The MRI appearance of LMS and endometrial stromal tumors is not well known. The MRI findings in a series of seven patients with surgically proven MMMT showed this tumor to have an appearance similar to endometrial carcinoma.[34] The only feature that might suggest the diagnosis is that these tumors are usually very large and show deep myometrial invasion. However, when detected early, these tumors have an appearance identical to early endometrial carcinoma.

Trophoblastic tissue can give rise to a variety of tumors.[35] Complete hydatidiform mole, partial hydatidiform mole, invasive mole, and choriocarcinoma arise from villous trophoblastic tissue, while placental site trophoblastic tumor arises from non-villous trophoblastic tissue.

Complete hydatidiform mole has been shown to be the result of a purely paternal conceptus. This genetic abnormality results in trophoblastic differentiation and proliferation without development of an embryo. Partial hydatidiform mole is seen almost exclusively in triploid conceptuses, with two paternal chromosomal complements and one maternal chromosomal complement. Often a malformed fetus is found in association with a partial mole. The partial mole itself usually arises from only a portion (or part) of the placenta. Clinical presentation of these entities can be abnormal uterine enlargement, vaginal bleeding, elevated human chorionic gonadotropin levels, or the absence of fetal heart sounds.

Invasive moles are thought to develop in previously existing complete moles. They are most commonly seen in the early months following evacuation of a complete hydatidiform mole. The hallmark of this entity is myometrial invasion. Choriocarcinoma is a malignant neoplasm which most commonly occurs after a molar pregnancy (sometimes remotely) but can be seen after a normal pregnancy, abortion, ectopic pregnancy, and possibly de novo.

Placental site trophoblastic tumor arises from the nonvillous trophoblast which infiltrates the placental site in normal pregnancy. It is considered an atypical form of choriocarcinoma. The importance of recognizing this tumor type pathologically lies in its less aggressive behavior compared with choriocarcinoma. This behavior makes this tumor amenable to surgery but often resistant to chemotherapy.

There are few data regarding the role of MRI in the evaluation and management of patients with trophoblastic tumors.[36,37] MRI is usually performed in patients with a known diagnosis of persistent mole following therapy or patients developing invasive mole or choriocarcinoma. A few studies have shown that prior to therapy most tumors show heterogeneous signal intensity on T_2-weighted images and distort or obliterate the normal zonal anatomy. These tumors only rarely appear as endometrial masses. Many tumors show markedly increased vascularity as evidenced by visualization of tortuous, dilated vessels within the tumor and/or the adjacent myometrium.

Previous studies have also shown that in patients responding to chemotherapy there was a progressive decrease in uterine size, tumor size, tumor vascularity, and a progressive improvement in visualization of normal zonal anatomy. All patients completely responding to therapy will show normal uterine size, normal zonal anatomy and no evidence of tumor following completion of chemotherapy.

8 MALIGNANT DISEASE OF THE CERVIX

Invasive cervical cancer is thought to develop over time from noninvasive precursor lesions.[38] These precursor lesions are referred to as cervical intraepithelial neoplasia (CIN). CIN is divided pathologically into three grades: CIN 1 (minor dysplasia), CIN 2 (moderate dysplasia), and CIN 3 (severe dysplasia). CIN 3 is synonymous with carcinoma-in-situ. Available evidence indicates that up to 40% of CIN 3 lesions and a lesser proportion of CIN 1 and CIN 2 lesions would progress to invasive cancer if untreated. Imaging studies play no role in the detection of cervical carcinoma or its precursor lesions.

On T_1-weighted images, cervical carcinoma is of intermediate signal intensity and usually isointense with the uterine corpus and cervix. On T_2-weighted images, cervical carcinoma is usually of intermediate signal intensity relative to the low-signal fibrous cervical stroma and high-signal cervical and endometrial canals. It is the sharp contrast with the fibrous stroma that makes T_2-weighted imaging so crucial in depicting the tumor and its depth of invasion.

Most patients with CIN or microinvasion will have a normal appearance on MRI.[39] However, some patients with early invasive disease can also have a normal appearance on MRI. Prior studies have shown that the detection of a macroscopic lesion on MRI had a 100% PPV in determining the presence of at least invasive disease. A normal appearance on MRI has a less than 100% NPV in excluding the presence of invasive disease. Thus, a normal MRI requires further investigation to exclude early invasive disease. It is important to note that there are few studies in the literature evaluating this point, and few studies have been performed with high-resolution imaging.

The therapy of cervical carcinoma is dependent upon the stage of the disease at the time of diagnosis. The most crucial determination is the presence or absence of tumor invasion into the parametrium. Patients without parametrial invasion will usually be treated surgically. The presence of parametrial invasion precludes surgical therapy.

A number of MRI studies have shown that the finding of a completely intact ring of low-signal intensity cervical stroma has a 100% NPV in excluding parametrial invasion (Figure 9).[39–42] Unfortunately, focal areas of disruption of the stromal ring or full-thickness involvement of the fibrous stroma by tumor have a significantly lower than 100% PPV in determining parametrial invasion

The ability of MR imaging to accurately determine vaginal, pelvic sidewall, bladder, and rectal involvement in patients with cervical carcinoma has been evaluated in only a limited number of studies. The NPV of MR in excluding involvement of these structures is probably close to 100%. The PPV is difficult to assess because of the small number of positive cases.

There is little if any role for gadolinium-enhanced MR imaging in the staging of cervical carcinoma.[24,27,43] Several studies have indicated that gadolinium-enhanced images consistently overestimate the depth of cervical invasion. In addition, there is overestimation of involvement of the parametrium, bladder, and vagina with gadolinium-enhanced images.

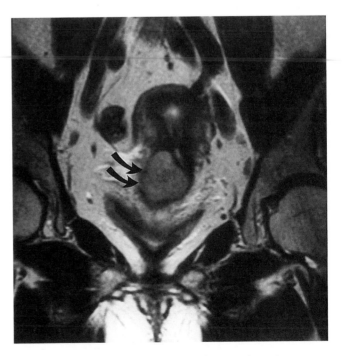

Figure 9 Coronal FSE multicoil image demonstrating a large tumor in the right cervix. The thin rim of intact fibrous stroma (curved arrows) indicates that parametrial invasion is absent

For References see p. 1274

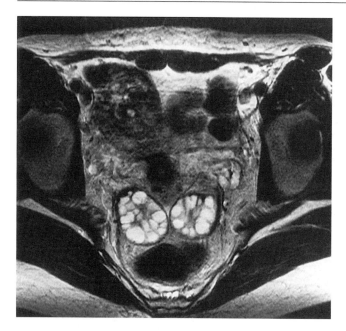

Figure 10 Axial FSE multicoil image demonstrating multiple follicles within the ovaries. The appearance suggests polycystic ovarian disease

9 BENIGN DISEASE OF THE OVARIES

In premenopausal patients the normal ovaries almost always contain multiple small follicles (Figure 10). These appear as round, unilocular, sharply circumscribed high-signal areas on T_2-weighted images. They are usually less than 1.0 cm across but can attain a size of 2.5 cm and still be considered normal follicles. A corpus luteum cyst or follicular cyst can have an identical appearance to normal follicles other than their slightly larger size. They can sometimes contain hemorrhage, which will be seen as a high signal on T_1-weighted images. Theca lutein cysts can be seen in association with excessive levels of hCG, generally associated with trophoblastic disease. These cysts are usually multilocular, bilateral, and very large.

It is well known that even postmenopausal patients may have simple cysts within their ovaries.[44] Studies in the ultrasound literature have shown that if a cyst is simple, less than 3.0 cm in size, and shows no abnormal flow on Doppler evaluation, it is most probably benign. There are no similar studies in the MRI literature.

An infrequently seen benign disorder of the ovaries is polycystic ovarian disease. This is thought to result from unopposed estrogenic stimulation and chronic anovulation. This disorder is commonly associated with obesity, hirsutism, and oligomenorrhea. On MRI, the ovaries in these patients usually appear mildly enlarged, and contain multiple small cysts in the periphery with an abundant central stroma.

A common cystic lesion of the adnexal region is the paratubal cyst. These cysts develop from wolffian duct remnants within the broad ligament, separate from the ovary. These are almost always benign in nature. Their appearance is indistinguishable from simple ovarian cysts, other than their extraovarian origin. It is this latter point that is key to their diagnosis. Although there is a paucity of published data, paratubal cysts usually displace an otherwise normal appearing ovary and therefore appear as a separate structure.[45]

Another cystic adnexal lesion is the Gartner's duct cyst.[45] This develops in a remnant of the müllerian duct. Their signal characteristics are identical to other cystic lesions. They can be diagnosed by their paravaginal location (separate from the ovaries) and usually tubular configuration. They are usually well demonstrated on high-resolution MRI. These patients commonly present with dyspareunia.

Benign epithelial tumors of the ovary include serous and mucinous cystadenomas (Figure 11). The appearance of these tumors at MR imaging is variable, and can be indistinguishable from their malignant counterparts. Benign cystadenomas can appear as simple cystic intraovarian masses. The presence of internal septations, solid components, or papillary projections makes malignancy more likely. Intravenous gadolinium can sometimes be helpful in detecting solid components, papillary projections or even septations.[46]

Ovarian fibromas, fibrothecomas, and Brenner tumors are solid ovarian masses that can contain fibrous tissue, and, therefore, may show diffuse or focal areas of low signal intensity on all pulse sequences. Fibromas are typically extremely low signal intensity, while fibrothecomas typically contain multiple high-signal foci. One must be careful to distinguish these masses from pedunculated subserosal fibroids. The only way to do so definitively is to identify them as intraovarian.

Two ovarian lesions which can occasionally be difficult to distinguish are the ovarian dermoid and an ovarian endometrioma.[47] The dermoid tumor is the most common form of benign teratoma, and can be definitively characterized by its fat content. Endometriosis is a disease characterized by rests of normal functional endometrium in abnormal locations. The most common location of endometriosis is the ovary. This can occur as a large mass, referred to as an endometrioma, which can be characterized by its blood content.

On MRI, a dermoid and an endometrioma can have identical signal characteristics on both T_1- and T_2-weighted

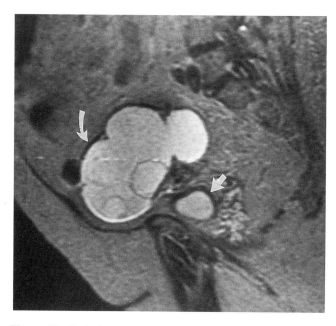

Figure 11 Sagittal conventional spin echo image of a mucinous cystadenoma (curved arrow). Note the thin septations and lack of solid elements. The signal behavior of the mass is identical to urine in the bladder (straight arrow) on all sequences

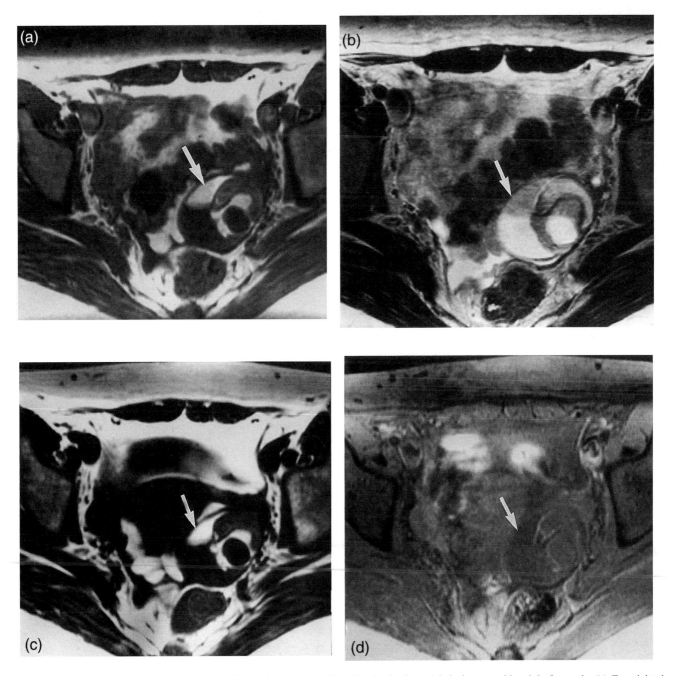

Figure 12 Complex dermoid containing both fluid and fat elements. Note that the fat (arrow) is isointense with pelvic fat on the (a) T_1-weighted and (b) T_2-weighted images. On the (c) water suppression and (d) fat suppression images, the fat (arrows) suppresses on the latter sequence

images. The most definitive way to distinguish these lesions is by using sequences which selectively suppress the signal from fat- or water-containing structures.[47] The fat in dermoids is of high signal on T_1-weighted images obtained with and without water suppression, and is of low signal on T_1-weighted images obtained with fat suppression (Figure 12). Endometriomas appear as a high signal on T_1-weighted images obtained with and without fat suppression, and appear as a low signal on T_1-weighted images obtained with water suppression.

Unfortunately, endometriomas are indistinguishable from other hemorrhagic cysts. Multiple such cysts make the diagnosis of endometriosis more likely (Figure 13). Sometimes

hemosiderin can be visualized within and on the surface of the ovaries. Hemosiderin appears as tiny foci of low signal, most conspicuous on T_2-weighted images. MRI is insensitive in detecting small implants of endometriosis involving bowel, the bladder, or other peritoneal surfaces.

10 MALIGNANT DISEASE OF THE OVARIES

The most common malignant neoplasms of the ovary are of epithelial origin, particularly serous and mucinous cystadenocarcinomas (Figure 14). These tumors often appear primarily

For References see p. 1274

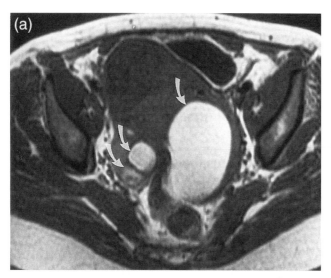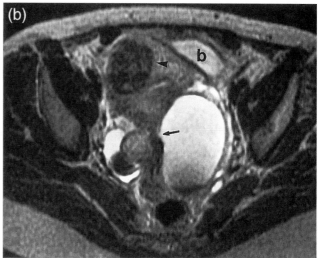

Figure 13 Axial conventional spin echo images demonstrating endometriosis. Note on (a) the intermediate weighted sequence, multiple hyperintense lesions (arrows) are seen. On (b) the T_2-weighted image, some of these lesions become hypointense whereas others remain hyperintense consistent with blood of various ages. Also note tethering of the rectum (arrow) secondary to adhesions, an intramural fibroid (arrowhead) and the bladder (b) pushed anteriorly

cystic in nature. On MRI, the cystic portions will appear as very high-signal areas on T_2-weighted images. They also, however, commonly contain internal septations and solid elements. The solid components are of lower signal on the T_2-weighted images, and often show enhancement following intravenous administration of gadolinium.[46] However, these tumors can sometimes appear indistinguishable from simple cysts. In general, even when purely cystic in appearance, these masses will be larger in size than other simple benign cysts. A variety of other malignant neoplasms that are primarily solid or mixed can involve the ovaries.

Ovarian malignancies usually spread from the surface of the ovaries into the peritoneal cavity. Most commonly, metastases involve the omentum, mesentery, and peritoneal surface of solid viscera. Due to artifacts from the bowel, MRI is not as sensitive as computerized tomography in detecting these metastases. Intravenous gadolinium is essential if one attempts to detect these metastases on MRI.

11 RELATED ARTICLES

Male Pelvis Studies Using MRI; Multi Echo Acquisition Techniques Using Inverting Radiofrequency Pulses in MRI; Whole Body Machines: NMR Phased Array Coil Systems.

12 REFERENCES

1. R. C. Smith and S. M. McCarthy, *Radiol. Clin. N. Am.*, 1994, **32**, 109.
2. M. L. Wood and R. M. Henkelman, *Med. Phys.*, 1986, **13**, 794.
3. R. C. Smith, C. Reinhold, R. C. Lange, T. R. McCauley, R. Kier, and S. McCarthy, *Radiology*, 1992, **184**, 665.
4. R. C. Smith, C. Reinhold, T. R. McCauley, R. C. Lange, R. T. Constable, R. Kier, and S. M. McCarthy, *Radiology*, 1992, **184**, 671.
5. J. Hennig, A. Nauerth, and H. Friedburg, *Magn. Reson. Med.*, 1986, **3**, 823.
6. J. Hennig and H. Friedburg, *Magn. Reson. Imag.*, 1988, **6**, 391.
7. R. V. Mulkern, P. S. Melki, P. Jakab, N. Higuchi, and F. A. Jolesz, *Med. Phys.*, 1991, **18**, 1032.
8. C. E. Hayes, and P. B. Roemer, *Magn. Reson. Med.*, 1990, **16**, 181.
9. C. E. Hayes, N. Hattes, and P. B. Roemer, *Magn. Reson. Med.*, 1991, **18**, 309.
10. P. B. Roemer, W. A. Edelstein, C. E. Hayes, S. P. Souza, and O. M. Mueller, *Magn. Reson. Med.*, 1990, **16**, 192.
11. C. L. Janus, H. P. Wiczyk, and N. Laufer, *Magn. Reson. Imag.*, 1988, **6**, 669.

Figure 14 Axial FSE multicoil image of a clear cell adenocarcinoma of the ovary (arrows). Note both solid and cystic elements indicative of a malignancy

12. A. R. Lupetin, in 'Magnetic Resonance Imaging', ed. D. D. Stark and W. G. Bradley, C. V. Mosby, St Louis, 1988, p. 1270.

13. S. McCarthy, C. Tauber, and J. Gore, *Radiology*, 1986, **160**, 119.

14. L. M. Scoutt, S. D. Flynn, D. J. Luthringer, T. R. McCauley, and S. M. McCarthy, *Radiology*, 1991, **179**, 403.

15. M. C. Lin, B. B. Gosink, S. I. Wolf, M. R. Feldesman, C. A. Stuenkel, P. S. Braly, and D. H. Pretorius, *Radiology*, 1991, **180**, 427.

16. T. J. Dubinsky, H. R. Parvey and N. Maklad, *Am. J. Roentgenol.*, 1997, **169**, 145.

17. D. G. Mitchell, L. Schonholz, P. L. Hilpert, R. G. Pennell, L. Blum, and M. D. Rifkin, *Radiology*, 1990, **174**, 827.

18. T. R. McCauley, L. M. Scoutt, and S. B. Flynn, *JMRI*, 1991, **1**, 319.

19. D. G. Mitchell, *Radiol. Clin. N. Am.*, 1992, **30**(4), 777.

20. J. S. Pellerito, S. M. McCarthy, M. B. Doyle, M. G. Glickman, and A. H. Decherney, *Radiology*, 1992, **183**, 795.

21. W. T. Creasman, C. P. Morrow, B. N. Bundy, M. D. Homesley, J. E. Graham, and P. B. Heller, *Cancer*, 1987, **60**, 2035.

22. W. T. Creasman and J. C. Weed in 'Gynecologic Oncology', 2nd edn, ed. M. Coppleson, Churchill Livingstone, London, 1992, p. 780.

23. M. H. Lutz, P. B. Underwood, A. Kreutner, and M. C. Miller, *Gynecol. Oncol.*, 1978, **6**, 83.

24. Y. Hirano, K. Kubo, Y. Hirai, S. Okada, K. Yamada, S. Sawano, T. Yamashita, and Y. Hiramatsu, *RadioGraphics*, 1992, **12**, 243.

25. H. Hricak, J. L. Stern, M. R. Fisher, L. G. Shapeero, M. L. Winkler, and C. G. Lacey, *Radiology*, 1987, **162**, 297.

26. H. Hricak, L. V. Rubinstein, G. M. Gherman, and N. Karstaedt, *Radiology*, 1991, **179**, 829.

27. H. Hricak, B. Hamm, R. C. Semelka, C. E. Cann, T. Nauert, E. Secaf, J. L. Stern, and K. J. Wolf, *Radiology*, 1991, **181**, 95.

28. H. H. Lien, V. Blomlie, C. Trope, J. Kaern, and V. M. Abeler, *Am. J. Roentgenol.*, 1991, **157**, 1221.

29. H. V. Posniak, M. C. Olson, C. M. Dudiak, M. J. Castelli, J. Dolan, R. A. Wisniewski, J. H. Isaacs, S. K. Sharma, and V. Bychkov, *RadioGraphics*, 1990, **10**, 15.

30. S. Sironi, G. Taccagni, P. Garancini, C. Belloni, and A. Del-Maschio, *Am. J. Roentgenol.*, 1992, **158**, 565.

31. S. Sironi, E. Colombo, G. Villa, G. Taccagni, C. Belloni, P. Garancini, and A. DelMaschio, *Radiology*, 1992, **185**, 207.

32. L. M. Scoutt, S. M. McCarthy, S. D. Flynn, R. C. Lange, F. Long, R. C. Smith, S. K. Chambers, E. I. Kohorn, P. Schwartz and J. T. Chambers, *Radiology*, 1995, **194**, 567.

33. J. R. Lurain and M. S. Piver in 'Gynecologic Oncology', 2nd edn, ed. M. Coppleson, Churchill Livingstone, London, 1992, p. 827.

34. L. G. Shapeero and H. Hricak, *Am. J. Roentgenol.*, 1989, **153**, 317.

35. F. J. Paradinas, in 'Gynecologic Oncology', 2nd edn, ed. M. Coppleson, Churchill Livingstone, London, 1992, p. 1013.

36. J. W. Barton, S. M. McCarthy, E. I. Kohorn, L. M. Scoutt, and R. C. Lange, *Radiology*, 1993, **186**, 163.

37. H. Hricak, B. E. Demas, C. A. Braga, M. R. Fisher, and M. L. Winkler, *Radiology*, 1986, **161**, 11.

38. M. Coppleson, K. H. Atkinson, and F. C. Dalrymple, in 'Gynecologic Oncology', 2nd edn, ed. M. Coppleson, Churchill Livingstone, London, 1992, p. 571.

39. K. Togashi, K. Nishimura, T. Sagoh, S. Minami, S. Noma, I. Fujisawa, Y. Nakano, J. Konishi, H. Ozasa, I. Konishi, and T. Mori, *Radiology*, 1989, **171**, 245.

40. H. Hricak, C. G. Lacey, L. G. Sandles, Y. C. F. Chang, M. L. Winkler, and J. L. Stern, *Radiology*, 1988, **166**, 623.

41. S. H. Kim, B. I. Choi, H. P. Lee, S. B. Kang, Y. M. Choi, M. C. Han, and C. W. Kim, *Radiology*, 1990, **175**, 45.

42. S. Sironi, C. Belloni, G. L. Taccagni, and A. DelMaschio, *Am. J. Roentgenol.*, 1991, **156**, 753.

43. Y. Yamashita, M. Takahashi, T. Sawada, K. Miyazaki, and H. Okamura, *Radiology*, 1992, **182**, 643.

44. D. Levine, B. B. Gosink, S. I. Wolf, M. R. Feldesman, and D. H. Pretorius, *Radiology*, 1992, **184**, 653.

45. R. Kier, *Am. J. Roentgenol.*, 1992, **158**, 1265.

46. S. K. Stevens, H. Hricak, and J. L. Stern, *Radiology*, 1991, **181**, 481.

47. R. Kier, R. C. Smith, and S. M. McCarthy, *Am. J. Roentgenol.*, 1992, **158**, 321.

Biographical Sketches

Robert Smith. b 1960. B.A. (Mathematics), 1981, Johns Hopkins University, M.D., 1985, Yale University. Radiology Resident and MR Fellow, Yale University, 1986–91. Assistant Professor, Yale University, 1991–96. Associate Professor, and Director of MRI, Yale University, 1996–97. Research interests: fast spin echo MRI, clinical applications of multicoil arrays, basic MRI physics.

Michael J. Varanelli. b 1970. B.Sc. (Biology), Bucknell University, 1992. M.D., University of Pittsburgh, 1996. Radiology Resident, Yale University, 1997–present.

Leslie M. Scoutt, b 1952. B.A. (Biology), Wesleyan University, 1974, M.D., University of Rochester, 1978. Radiology Resident, Beth Israel Hospital, Boston, MA, 1982–85. Cross-Sectional Imaging Fellow, Yale University, 1985–87. Assistant Professor, Yale University, 1987–94. Associate Professor, Yale University, 1994–present. Research interests: MRI and ultrasound of the female pelvis, breast and vascular ultrasound.

Shirley McCarthy. b 1949. B.A. (Biological Sciences), 1971, State University of New York, Albany Ph.D. (Mammalian Physiology), 1975, Cornell University, M.D., 1975, Yale University School of Medicine. Associate Professor, Yale University, 1989–present. Chief of MRI, Yale University, 1987–present. Research interests: gynecological imaging, cost-effective analyses of MRI use.

Kidney, Prostate, Testicle, and Uterus of Subjects Studied by MRS

Michael W. Weiner

University of California, San Francisco, CA, USA

1 KIDNEY

1.1 Proton MRS of Kidney

Shah et al.[1] reported a technique for using the stimulated echo amplitude mode (STEAM) sequence for ^{1}H MRS of the human kidney. The results demonstrated the presence of trimethylamines (TMAs). A prominent peak observed at 5.8 ppm

For References see p. 1283

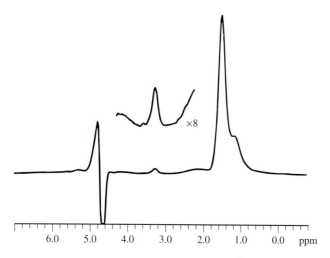

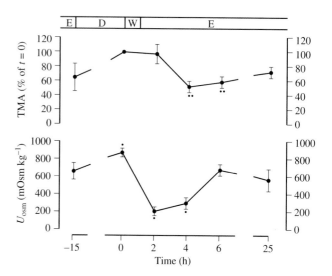

Figure 1 Water suppressed, volume localized ^{1}H spectrum from kidney. Water suppression consisted of a Silver–Hoult phase swept adiabatic fast passage pulse for selective water inversion followed by an inversion–recovery time of 0.8 second to allow the water Z magnetization to null. This sequence was delivered at the start of the volume localized stimulated echo sequence. $TE = 68$ ms, $TM = 44$ ms, $TR = 3$ s. Number of scans = 128. Filtering was 5 Hz of exponential line broadening. The resonances are water (4.75 ppm), lipids (0.9–1.4 ppm), and TMA (3.25 ppm)—which is shown expanded ($\times 8$) above the main spectrum. (Reproduced by permission of the National Academy of Sciences from Avison et al.[2])

Figure 2 Time-course of changes in medullary TMA levels and U_{osm} in the four volunteers studied. TMA levels are expressed as the TMA area (%) at the point of maximal dehydration (i.e. $t = 0$). Error bars are $\pm$ SEM. $^*p < 0.05$ versus $t = 0$ value for the TMA time-course. $^{**}p < 0.05$ versus $t = -15$ for the U_{osm} time-course. E, euvolemic; D, dehydration; W, water load. (Reproduced by permission of the National Academy of Sciences from Avison et al.[2])

in medullary TMAs (Figure 2). Water loading caused a water diuresis and a significant reduction in medullary TMAs within 4 hours. These results are consistent with the view that TMAs may play an osmoregulatory role in the medulla of the normal human kidney.

was from urea. Avison et al.[2] used volume localized ^{1}H MRS to detect and measure changes in medullary trimethylamines in the human kidney (Figure 1). Proton magnetic resonance spectra were obtained from the human renal medulla using a stimulated echo localization sequence.

In addition to residual water and lipid, TMAs were identified at 3.25 ppm. In normal volunteers, overnight dehydration led to a significant increase of urine osmolality, and an increase

1.2   Phosphorus-31 MRS of Kidney

Jue et al.[3] demonstrated that ^{31}P MRS signals can be obtained from the normal human kidney. Matson et al.[4] also

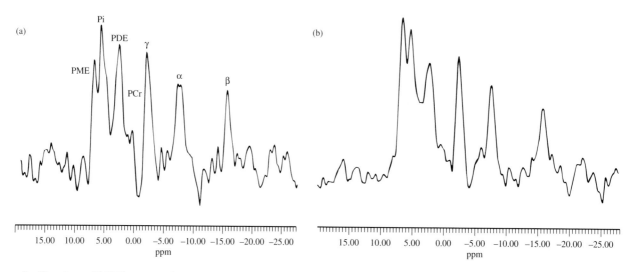

Figure 3 Phosphorus-31 ISIS spectra of (a) the healthy normal kidney and (b) of the well functioning kidney transplant. Acquisition parameters: $TR = 2.0$ seconds, acquisition time = 1 hour, $90°$ pulse set for the region of interest, distance of the center of the volume of interest (VOI) from the 14 cm surface coil = 70 mm for normal kidney and 42 mm for transplanted kidney, size of the VOI is $25 \times 45 \times 55$ mm = 62 ml for normal kidney and $25 \times 50 \times 54$ mm = 68 ml for transplanted kidney. (Reproduced by permission of Springer-Verlag from Boska et al.[5])

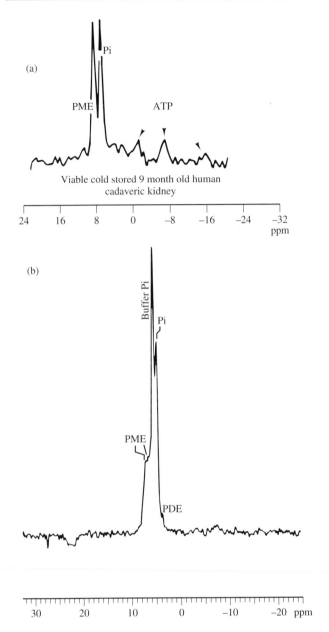

Figure 4 Ex situ in vivo magnetic resonance spectrum of (a) a viable pediatric and (b) an adult kidney. PME, Pi, PDS (phosphodiester), and ATP (adenosine triphosphate) peaks are defined. (Reproduced by permission of Williams & Wilkins from Bretan et al.[6])

demonstrate the feasibility of obtaining ^{31}P magnetic resonance spectra from human kidneys.

Bretan et al.[6,7] reported their clinical experience with pretransplant assessment of renal viability using ^{31}P MRS to study 40 renal transplant recipient patients (Figure 4). The purpose of their study was to develop and investigate the use of MRS in the clinical transplant setting by correlation of pretransplant MRS parameters with subsequent renal function. Kidneys were studied during simple hypothermic storage within their sterile containers using an external ^{31}P MRS surface coil. Mean storage times were about 38 hours. Cold storage times did not correlate with subsequent clinical renal function. However, selected ^{31}P MRS data did. ATP was present in 11 kidneys and was associated with the best subsequent renal function. Only 36% of these patients required dialysis, compared with 71 patients without detectable ATP who required a posttransplant dialysis. Kidneys with ATP had the highest PME/Pi (Pi, inorganic phosphate) ratios; in general, the higher the PME/Pi ratio, the better the renal function posttransplant. The intracellular pH did not correlate with subsequent renal function. The authors suggested that MRS provided a better correlation with renal function after transplantation than existing methods.

Grist et al.[8] used ^{31}P MRS to investigate the effects of rejection on renal transplants (Figures 5 and 6). The PDE/PME (PDE, phosphodiester) and Pi/ATP ratios in the transplants with rejection differed significantly from the corresponding metabolite ratios of patients without rejection. A PDE/PME ratio exceeding 0.8 had a sensitivity of 100% and a specificity of 86% for predicting rejection. A Pi/ATP ratio greater that 0.6 had a sensitivity of 72% and a specificity of 86% for predicting rejection. The authors concluded that ^{31}P MRS may be useful as a noninvasive method for evaluating renal metabolism during episodes of transplant rejection.

demonstrated the feasibility of obtaining ^{31}P magnetic resonance spectra from human kidneys using the image selected in vivo spectroscopy (ISIS) localization technique. Boska et al.[5] obtained spatially localized ^{31}P magnetic resonance spectra from healthy normal human kidneys and from well functioning renal allographs (Figure 3). Little or no phosphocreatine (PCr) in all spectra verified the absence of muscle contamination and was consistent with proper volume localization. The PME/ATP ratio (PME, phosphomonoester) was slightly elevated in transplanted kidneys (1.1) compared with normal healthy kidneys (0.8). Despite the practical problems produced by organ depth, respiratory movement, and tissue heterogeneity, these results

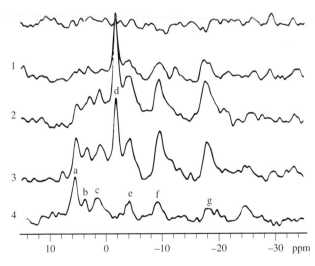

Figure 5 Contiguous ^{31}P magnetic resonance spectra obtained from slices of kidney. Note the presence of significant PCr at the surface, consistent with muscle tissue. The deepest slice shows a large PME peak and little PCr, consistent with renal tissue. Peaks from left to right correspond to PMEs (a), Pi (b), PDEs (c), PCr (d) and γ- (e), α- (f), and β- (g) phosphates of ATP. (Reproduced by permission of Williams & Wilkins from Grist et al.[8])

For References see p. 1283

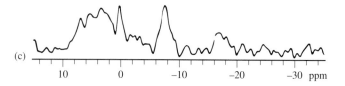

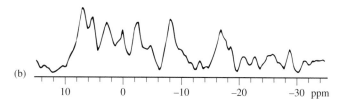

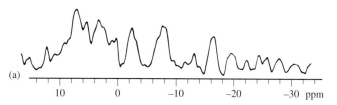

Figure 6 Phosphorus-31 magnetic resonance spectra from a patient with (a) a normally functioning renal allograft, (b) a patient with cyclosporine nephrotoxicity, and (c) a patient with moderate cellular rejection. Note the increase in Pi and PDE in the patient with rejection compared with the normal control subject or the patient with cyclosporine toxicity. PCr is a contaminant from extrarenal tissue. (Reproduced by permission of Williams & Wilkins from Grist et al.[8])

2 PROSTATE

2.1 Proton MRS of Prostate

Normal prostate has a very high concentration of citrate, a unique feature of this tissue related to the function of prostate cells to secrete citrate into the semen. Most of the MRS of prostate has focused on investigating changes of citrate associated with benign prostatic hypertrophy (BPH) and prostatic carcinoma. Schick et al.[9] investigated the signal characteristics of citrate at low field strengths at 1.5 T using spatially selected spectroscopy and theoretical methods. In vivo localized spectroscopy of small volume elements (2 ml) using a double spin echo method within the prostate gland provided citrate signals. Volume selected proton spectra with different echo times were recorded.

Thomas et al.[10] performed ^{1}H MRS of normal and malignant human prostates in vivo (Figures 7 and 8). The results demonstrated that water-suppressed ^{1}H MRS spectra could be obtained from the prostate. Normal healthy prostate showed a large resonance from citrate. The spectrum from a malignant human prostate showed a much lower level of citrate. The results also suggested that the low concentration of citrate might be useful for the identification of prostate cancer.

Schnall et al.[11] performed localized ^{1}H MRS of the human prostate in vivo using an endorectal surface coil. High levels of citrate were observed in all regions of normal prostate and benign prostatic hypertrophy. The citrate levels in regions con-

taining tumor were variable. The presence of high citrate levels in one case of prostate cancer was confirmed from extracts.

Schiebler et al.[12] reported high-resolution ^{1}H MRS of human prostate perchloric acid extracts (Figure 9). The citrate peak area was higher in benign prostatic hyperplasia than in adenocarcinoma. However, citrate peak areas from the normal peripheral zones were not significantly different from those found in adenocarcinomas. A sharp peak at 2.05 ppm that was seen in four out of thirteen adenocarcinoma samples and only one out of thirteen in the BPH samples was assigned to N-acetylneuraminic acid. Fowler et al.[13] also obtained ^{1}H NMR spectra from perchloric extracts of tissue samples from human prostate. Statistically significant differences between the normals, the benign prostatic hypertrophy, and the cancer groups occurred for metabolite ratios of creatine, citrate, and phosphorylcholine. None of the ratios correlated with the Gleason grade of the cancer samples. Different sections of large tumors often yielded substantially different ratios.

Yacoe et al.[14] reported in vitro ^{1}H MRS of normal and abnormal prostate cells (Figure 10).

Proton MRS was used to determine if cell strains derived from prostatic cancers could be distinguished from normal prostate. Prostatic cancer cells had lower concentrations of citrate compared with normal prostate epithelium, but the differences were small and not statistically significant. However, both the cancer and normal prostate cells were washed, which may have removed diffusable citrate.

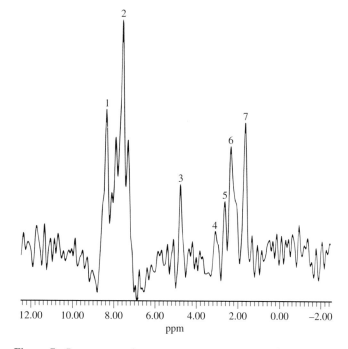

Figure 7 Proton magnetic resonance spectrum prostate in a healthy 26-year-old subject with TE = 40 ms. The total acquisition time was 1.5 minutes. Resonance assignments: 1, 2, water and MDPA from the external standard, situated in the plane of the surface coil; 3, residual water; 4, spermine/creatine/PCr; 5, 6, the methylene protons of citrate; 7, methylene protons of spermine and lipids. (Reproduced by permission of the Society for Magnetic Resonance Imaging from Thomas et al.[10])

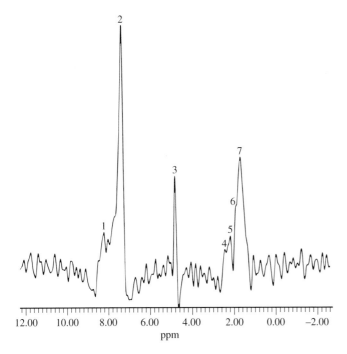

Figure 8 Proton magnetic resonance spectrum of a malignant human prostate with binomial water suppression. Number of scans = 16; repetition time, 5 seconds; *TE* = 80 ms; total acquisition time = 80 seconds. The residual water and the resonances from the external standard were identified at 4.8 and 7.8 ppm, respectively. A 16-step phase cycling was used. Resonance assignments: 1, 2, external standard; 3, residual water; 4, 5, C-2 and C-5 protons of the citrate molecule; 6, 7, methylene protons of spermine and lipids. (Reproduced by permission of the Society for Magnetic Resonance Imaging from Thomas et al.[10])

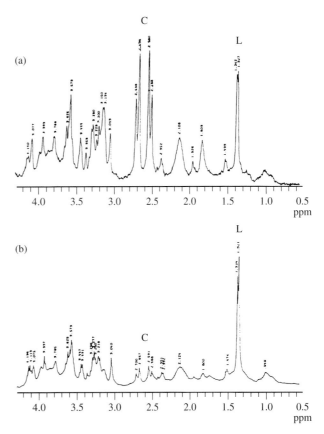

Figure 9 (a) In vitro ^{1}H NMR spectra at 360 MHz, demonstrating a normal pattern with a large citrate peak seen in a sample of benign prostatic hyperplasia with a glandular predominance (C, citrate; L, lactate). Standard not shown. (b) In vitro ^{1}H NMR spectra at 360 MHz for adenocarcinoma showing the expected low citrate peak. Standard not shown. (Reproduced by permission of Williams & Wilkins from Schiebler et al.[12])

Kurhanewicz et al.[15] performed ^{1}H MRS and enzymatic assays of human prostate adenocarcinomas and prostate DU145 xenographs grown in nude mice. The results showed that citrate concentrations in primary human adenocarcinomas were significantly lower than those observed for normal benign hyperplastic prostatic tissue. There was a 10-fold reduction of citrate associated with DU145 xenographs compared with primary prostate cancer. These findings support the hypothesis that citrate concentrations are low in prostate cancer.

In the past several years Kurhanewicz, Vigneron and their colleagues have published a series of reports using ^{1}H MRSI of the prostate in a clinical setting. They have used an endorectal coil[16] and a high spatial resolution technique.[17] This approach has been used to study the effects of hormone ablation[18] and cryosurgery,[19] and for the detection of local recurrence.[20,21]

2.2 Phosphorus-31 MRS of Prostate

Kurhanewicz et al.[22] reported the use of a ^{31}P MRS transrectal probe for studies of the human prostate (Figures 11–13). The preliminary results indicated that transrectal ^{31}P MRS may characterize ^{31}P metabolites in normal prostates, benign prostatic hyperplasia, and malignant prostates. The preliminary

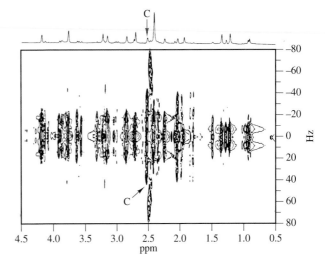

Figure 10 Two-dimensional *J* resolved spectrum of an HClO$_4$ extract of a normal peripheral zone epithelial cell strain, obtained at 500 MHz. The chemical shift is resolved in the horizontal axis and the *J* coupling constant of complex peaks is resolved in the vertical axis. The complex peaks assigned to citrate (C) are pointed out. A one-dimensional projection in the chemical shift dimension is plotted at the top. (Reproduced by permission of Williams & Wilkins from Yacoe et al.[14])

For References see p. 1283

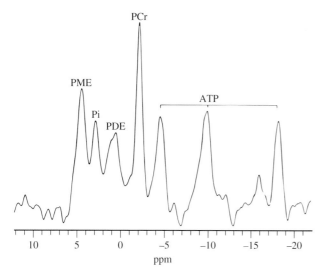

Figure 11 Phosphorus-31 magnetic resonance spectrum of a normal prostate from a normal subject (26 years old), TR = 20 seconds; NS = 100. (Reproduced by permission of the Society for Magnetic Resonance Imaging from Thomas et al.[23])

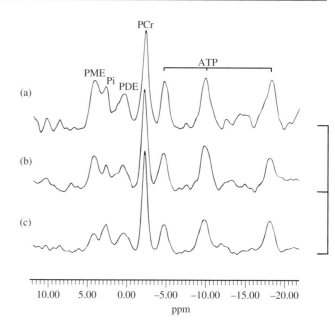

Figure 13 Phosphorus-31 magnetic resonance spectra from (a) upper, (b) middle, and (c) lower regions of the prostate. (Reproduced by permission of the Society for Magnetic Resonance Imaging from Thomas et al.[23])

results suggested that malignant prostates are characterized by significantly decreased levels of PCr and increased levels of PME compared to healthy prostates. Thomas et al. evaluated some of the problems encountered with transrectal ^{31}P MRS of human prostate to determine the optimal conditions for these studies.[23] The authors investigated the reproducibility of ^{31}P MRS, regional differences of ^{31}P metabolites, the T_1 relaxation times, and metabolic alterations associated with disease. The

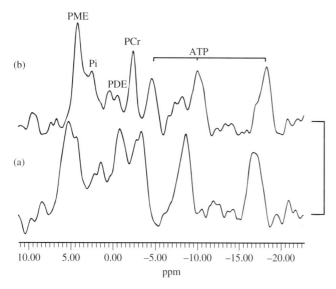

Figure 12 Phosphorus-31 magnetic resonance spectra from human prostates in patients with (a) BPH and (b) prostatic cancer. (Reproduced by permission of the Society for Magnetic Resonance Imaging from Thomas et al.[23])

PME/ATP ratio was highest in the upper region and lowest in the lower region. Similarly, the PDE/ATP ratio was highest in the upper region and lowest in the lower region. In contrast, the PCr/ATP was lowest in the upper region but was increased in the lower region.

The PME/ATP ratio of normal subjects (0.09 ± 0.1) was increased in the patients with BPH (1.5 ± 0.1) and significantly increased in patients with cancer (1.7 ± 0.2). The PCr/ATP ratio in normal subjects (1.5 ± 0.2) was not significantly reduced in BPH, but it was reduced to 0.9 in prostatic cancer. The PME/PCr ratio in normal subjects was 0.7, was not significantly increased in BPH to 1.4 but was significantly in prostate cancer to 2.2.

Hering and Muller reported ^{31}P MRS and ^{1}H MRI of the human prostate with a transrectal probe.[24] Fourteen patients were evaluated with ^{1}H MRI and seven patients with ^{31}P MRS. The PME/ATP ratio was higher in patients with prostatic cancer.

Narayan et al. investigated the ability of ^{31}P MRS to characterize normal human prostate as well as prostate with benign and prosthetic hypertrophy and malignant neoplasms (Figure 14).[25] Normal prostate had PCr/ATP, PME/ATP, and PME/PCr ratios of 1.2, 1.1, and 0.9, respectively. Malignant prostates had PCr/ATP ratios that were lower than normal prostates. Malignant prostates had PME/ATP ratios that were higher than normal prostates.

Using the PME/PCr ratio it was possible to differentiate metabolically malignant prostates from normal prostates with no overlap of individual ratios.

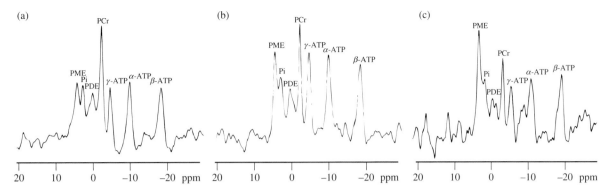

Figure 14 In vivo ^{31}P magnetic resonance prostate spectra. (a) Normal volunteer, (b) patient with BPH, and (c) patient with prostatic cancer. (Reproduced by permission of Williams & Wilkins from Narayan et al.[25])

2.3 Carbon-13 MRS of Prostate

Halliday et al. obtained high-quality, high-resolution proton decoupled natural abundance ^{13}C NMR spectra from various human tumors, including prostate (Figure 15).[26]

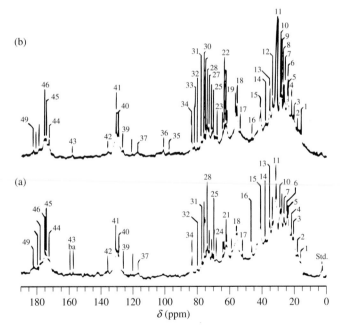

Figure 15 Natural abundance proton decoupled 100.6 MHz ^{13}C NMR spectrum of (a) human prostate with benign prostatic hypertrophy. This proton decoupled spectrum was taken from 2.24 g of tissue, at a temperature of 310 K, with a total of 17 271 scans. (b) A spectrum of poorly differentiated adenocarcinoma of human prostate from the same individual as (a), taken from 3.51 g of tissue with the same parameters, except for the use of 11 241 scans in total. It should be noted that peak 11 is truncated. Assignments for the numbered resonances are given elsewhere.[26] These spectra are representative of seven hyperplastic prostates, three of which contained adenocarcinoma, plus a fourth sample of adenocarcinoma, with the following exception: lactate was increased relative to the control tissue only in the depicted case and resonances from acidic mucins were not present in all of the tumors. (Reproduced by permission of Williams & Wilkins from Halliday et al.[26])

When prostatic adenocarcinoma was compared with adjacent hyperplastic tissue, the tumors were found to contain larger amounts of triacylglycerols, smaller amounts of citrate, and acid mucins. The citrate-to-lipid ratio appeared to differentiate malignant from nonmalignant prostates.[26] Halliday et al. obtained ^{13}C NMR spectra from prostate tumor cell lines.[27] The results showed that the amount of taurine was increased, and tyrosine was decreased in androgen sensitive rat prostatic tumors in comparison to androgen responsive malignant or normal tissue. The authors concluded that the amount of these amino acids discriminates androgen-sensitive from insensitive rat prostatic tissues. Sillerud et al. followed this up with an in vivo ^{13}C NMR study of human prostate (Figure 16).[28]

High levels of citrate were measured in the human prostate in vivo as well as the tissue samples of human and rat prostate in vitro.

3 TESTICLE

3.1 Phosphorus-31 MRS of Testicle

Chew et al. investigated the clinical feasibility of ^{31}P MRS to assess the metabolic integrity of the human testicle (Figures 17 and 18).[29] The PME/ATP ratio was greatly reduced in the abnormal testicle; furthermore, the PME/PDE ratio was also reduced in patients with primary testicular failure.

In patients with azoospermia, there were significant differences in the same peak area ratios between patients with primary testicular failure and those with chronic tubular obstruction. Bretan et al. compared ^{31}P MRS of human testicle with conventional semen analysis.[30] The glycerophosphorylcholine (GPC)/total phosphate ratio in azoospermic men after vasectomy significantly differed from the control GPC/total phosphate ratio, which appropriately reflected complete vasal occlusion. The results suggested that a significant portion of seminal GPC is derived from epidymal secretion and that ^{31}P MRS is useful for monitoring GPC/total phosphate levels when assessing epidymal function in male infertility.

For References see p. 1283

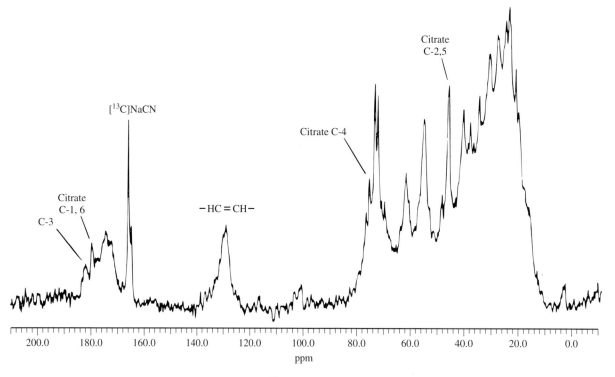

Figure 16 Proton decoupled, natural abundance 100.614 MHz ^{13}C NMR in vitro spectrum from 3.88 g of human benign hypertrophic prostate tissue sample. This spectrum is the average of 13 600 scans at a temperature of 283 K. The signal at 165 ppm is from 15 μl of a 0.15 M [^{13}C]sodium cyanide standard. (Reproduced by permission of Williams & Wilkins from Sillerud et al.[28])

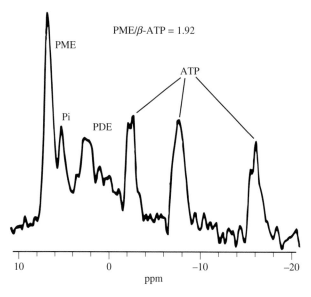

Figure 17 Characteristic ^{31}P magnetic resonance spectrum from the in vivo normal human testicle. Present are the three peaks due to ATP, small signals from PDE and Pi, and a large contribution from the PME peak. This spectrum was acquired with 400 signals averaged in 13.5 minutes. (Reproduced by permission of the Radiological Society of North America from Chew et al.[29])

4 UTERUS

High-resolution ^{1}H MRS was used as an adjunct to conventional and histological diagnosis of cervical neoplasia.[31]

Cervical biopsy specimens were examined with ^{1}H MRS and the results compared with histology. A high-resolution lipid spectrum was observed in 39 of 40 invasive carcinomas whereas 119 preinvasive samples showed little or no lipids but were characterized by a strong unresolved peak between 3.8

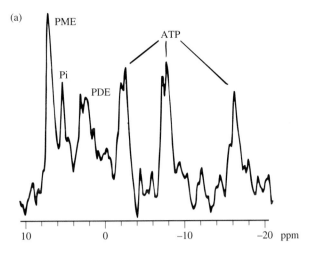

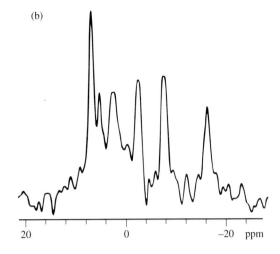

Figure 18 (a) Phosphorus-31 magnetic resonance spectrum from an azoospermic testicle due to primary testicular failure. It is characterized by the same peaks found in the normal spectrum, but the PME/β-ATP peak area ratio is substantially lower (1.32 in this case). (b) Phosphorus-31 magnetic resonance spectrum from an azoospermic testicle due to chronic ductal obstruction. It is characterized by the same peaks found in the normal spectrum, but the PME/β-ATP peak area ratio is lower than normal (1.51 in this case). (Reproduced by permission of the Radiological Society of North America from Chew et al.[29])

and 4.2 ppm. Peak ratios of the methylene/methyl and the unresolved/methylene resonances allowed accurate distinction between invasive and preinvasive malignancy.

5   RELATED ARTICLES

In Vivo Hepatic MRS of Humans; NMR Spectroscopy of the Human Heart; Proton Decoupling During In Vivo Whole Body Phosphorus MRS; Proton Decoupling in Whole Body Carbon-13 MRS; Quantitation in In Vivo MRS; Whole Body Studies: Impact of MRS.

6   REFERENCES

1. N. J. Shah, T. A. Carpenter, I. D. Wilkinson, L. D. Hall, A. K. Dixon, C. E. L. Freer, K. Prosser, and D. B. Evans, *Magn. Reson. Med.*, 1991, **20**, 292.
2. M. J. Avison, D. L. Rothman, T. W. Nixon, W. S. Long, and N. J. Siegel, *Proc. Natl. Acad. Sci. USA*, 1991, **88**, 6053.
3. T. Jue, D. L. Rothman, J. A. B. Lohman, E. W. Hughes, C. C. Hanstock, and R. G. Shulman, *Proc. Natl. Acad. Sci. USA*, 1988, **85**, 971.
4. G. B. Matson, D. B. Twieg, G. S. Karczmar, T. J. Lawry, J. R. Gober, M. Valenza, M. D. Boska, and M. W. Weiner, *Radiology*, 1988, **169**, 541.
5. M. D. Boska, D. J. Meyerhoff, D. B. Twieg, G. S. Karczmar, G. B. Matson, and M. W. Weiner, *Kidney Int.*, 1990, **38**, 294.
6. P. N. Bretan, N. Baldwin, A. C. Novick, A. Majors, K. Easley, T. Ng, N. Stowe, P. Rehm, S. B. Streem, and D. R. Steinmuller, *Transplantation*, 1989, **48**, 48.
7. P. N. Bretan, N. Baldwin, A. C. Novick, A. Majors, K. Easley, T. C. Ng, N. Stowe, S. Streem, D. Steinmuller, P. Rehm, and R. Go, *Transplant. Proc.*, 1989, **21**, 1266.
8. T. M. Grist, H. C. Charles, and H. D. Sostman, *Am. J. Roentgenol.*, 1991, **156**, 105.
9. F. Schick, H. Bongers, S. Kurz, W. I. Jung, M. Pfeffer, and O. Lutz, *Magn. Reson. Med.*, 1993, **29**, 38.
10. M. A. Thomas, P. Narayan, J. Kurhanewicz, P. Jajodia, and M. W. Weiner, *J. Magn. Reson.*, 1990, **87**, 610.
11. M. D. Schnall, R. Lenkinski, B. Milestone, and H. Y. Kressel, *Proc. 9th Ann Mtg. Soc. Magn. Reson. Med.*, New York, 1990, p. 288.
12. M. L. Schiebler, K. K. Miyamoto, M. White, S. J. Maygarden, and J. L. Mohler, *Magn. Reson. Med.*, 1993, **29**, 285.
13. A. H. Fowler, A. A. Pappas, J. C. Holder, A. E. Finkbeiner, G. V. Dalrymple, M. S. Mullins, J. R. Sprigg, and R. A. Komoroski, *Magn. Reson. Med.*, 1992, **25**, 140.
14. M. E. Yacoe, G. Sommer, and D. Peehl, *Magn. Reson. Med.*, 1991, **19**, 429.
15. J. Kurhanewicz, R. Dahiya, J. M. Macdonald, L. H. Chang, T. L. James, and P. Narayan, *Magn. Reson. Med.*, 1993, **29**, 149.
16. H. Hricak, S. White, D. Vigneron, J. Kurhanewicz, A. Kosco, D. Levin, J, Weiss, P. Narayan, and P. Carroll, *Radiology*, 1994, **193**, 703.
17. J. Kurhanewicz, D. Vigneron, H. Hricak, P. Carroll, P. Narayan, and S. Nelson, *Radiology*, 1996, **198**, 795.
18. H. Chen, H. Hricak, C. L. Kalbhen, J. Kurhanewicz, D. Vigneron, J. Weiss, and P. Carroll, *Am. J. Roentgenol.*, **166**, 1157.
19. J. Kurhanewicz, H. Hricak, D. B. Vigneron, S. Nelson, F. Parivar, K. Shinohara, and P. R. Carroll, *Radiology*, 1996, **200**, 489.
20. F. Parivar, H. Hricak, J. Kurhanewicz, K. Shinohara, D. B. Vigneron, S. J. Nelson, and P. R. Carroll, *Urology*, 1996, **48**, 594.
21. F. Parivar and J. Kurhanewicz, *Curr. Opin. Urol.*, 1998, **8**, 83.
22. J. Kurhanewicz, A. Thomas, P. Jajodia, M. W. Weiner, T. L. James, D. B. Vigneron, and P. Narayan, *Magn. Reson. Med.*, 1991, **22**, 404.
23. M. A. Thomas, P. Narayan, J. Kurhanewicz, P. Jajodia, T. L. James, and M. W. Weiner, *J. Magn. Reson.*, 1992, **99**, 377.
24. F. Hering and S. Muller, *Urolog. Res.*, 1991, **19**, 349.
25. P. Narayan, P. Jajodia, J. Kurhanewicz, A. Thomas, J. MacDonald, B. Hubesch, M. Hedgcock, C. M. Anderson, T. L. James, E. A. Tanagho, and M. Weiner, *J. Urol.*, 1991, **146**, 66.
26. K. R. Halliday, C. Fenoglio-Preiser, and L. O. Sillerud, *Magn. Reson. Med.*, 1988, **7**, 384.
27. K. R. Halliday, L. O. Sillerud, and D. Mickey, *Proc. 11th Ann Mtg. Soc. Magn. Reson. Med.*, Berlin, 1992, p. 492.
28. L. O. Sillerud, K. R. Halliday, R. H. Griffey, C. Fenoglio-Preiser, and S. Sheppard, *Magn. Reson. Med.*, 1988, **8**, 224.

29. W. M. Chew, H. Hricak, R. D. McClure, and M. F. Wendland, *Radiology*, 1990, **177**, 743.
30. P. N. Bretan, D. B. Vigneron, R. D. McClure, H. Hricak, R. A. Tom, M. Moseley, E. A. Tanagho, and T. L. James, *Urology*, 1989, **33**, 116.
31. E. J. Delikatny, P. Russell, J. C. Hunter, R. Hancock, K. H. Atkinson, C. van Haaften-Day, and C. E. Mountford, *Radiology*, 1993, **188**, 791.

Acknowledgements

Supported by NIH grant R01AG10897 and the DVA Medical Research Service.

Biographical Sketch

Michael W. Weiner. *b* 1940. B.S., 1961, Johns Hopkins. M.D., 1965, SUNY Upstate Medical Center. Intern and Resident, Mount Sinai Hospital 1965–67. Resident and Fellow in Metabolism, Yale University, 1967–70. Fellow in Biochemistry Institute for Enzyme Research, University of Wisconsin, 1970–72. Faculty at University of Wisconsin, 1971–74, Stanford University, 1974–80, University of California San Francisco, 1980–present. Director, Magnetic Resonance Department, Veterans Affairs Medical Center. Professor of Medicine and Radiology, University of California San Francisco. Research interests include application of MRI to investigation of human metabolism and diagnosis of disease.

Design and Use of Internal Receiver Coils for MRI

Nandita M. deSouza
Imperial College School of Medicine, London, UK

David J. Gilderdale
Engineering Consultant, UK

and

Glyn A. Coutts
Marconi Medical Systems, UK

1 INTRODUCTION

The motivation for investigating the use of internal coils in MRI and MRS is based on the fact that the closer an MR signal detector is placed to its target the better the signal-to-noise ratio (SNR) of the resultant data. Abragam described the significance of the 'filling factor' of an object in a coil[1] and Hoult and Richards developed the theory of the SNR of the MR experiment round this concept.[2] SNR is maximized if the coil

size is matched to that of the region of interest; coils located very close to the target to be studied can achieve dramatic improvements in SNR relative to conventional external enveloping coils. Internal access for such small local coils may be gained via natural orifices or using other minimally invasive techniques.

In the first part of this article the design and construction of the current range of internal coils used for human MRI is reviewed. Discussion of types that are in development and which may become of clinical relevance in the next few years follows, including descriptions of their clinical use where relevant. The second section describes the main clinical applications to date. Development has been gathering pace during the 1990s, and it is expected that within the next few years the variety of devices will increase substantially.

1.1 Historical Note

The first implanted or inserted coils to be described were intended for spectroscopic investigations.[3,4] These were followed by initial studies of devices for MRI.[5,6] The first clinical studies using an inflatable rectal probe (Medrad Inc., Pittsburgh, USA) were published in 1989.[7] Various studies were undertaken with this probe (e.g., that of Schnall *et al.*[8]), but it was not until 1992 that other designs began to appear. The first of these was a coil designed for the imaging of the cervix;[9] this was followed by probes for the anus[10,11] and an alternative design for the rectum.[12] Other coils were used in conjunction with endoscopes, principally for gastroscopy,[13,14] and smaller versions were designed for such applications as imaging the inner ear[15], and the female urethra.[16]

All these devices are designed for insertion through natural orifices to acquire imaging (or in some instances, spectroscopic) data from adjacent regions of interest. Another class of detector acts as markers. An early design used a small coil to mark the tip of a catheter.[17] Later versions of this concept were designed to generate signals along significant lengths of conductor, mimicking a guidewire. These included stub aerials[18,19] and extended coil windings.[20] Of these, the aerial version gives a wider field of view and may eventually prove useful for local imaging.[19] We have used the same approach to mark the location of a naso-gastric tube (unpublished data), and it is likely that other clinical applications will develop over the next few years. A specialist application is in the development of marker (fiducial) technology, which provides a spatial reference point. Fiducials can be passive (as developed primarily for frameless stereotaxy[21]) or active, in which rf circuitry is used to enhance or characterize signals. The latter typically use tiny vials of an MR-visible material enclosed within the windings of a very small coil.[22] Fiducial markers may also be integrated into internal coils or biopsy devices in order to track their trajectories.[23]

2 DESIGN CONSIDERATIONS

2.1 General Features

Insertable coils may be designed for use in one of three ways: (i) as an enveloping coil with the principal target for study being tissue within the coil (the cervical coil described here uses this strategy); (ii) as an 'inside-out' coil in which a winding design normally used in MRI to produce images from the volume enclosed within it is used to obtain data from outside it (e.g., the endoanal coil); (iii) as a coil used in a manner analogous to that of a conventional surface coil (e.g., the endorectal coil).

Insertable coils can be used as receivers only (either single or in arrays) or as combined transmit/receivers. In general, the coils are directly coupled to the input receiver via a cable. However, in certain specific applications (e.g., endoscopic) the coil may be inductively coupled to the external receiver.

Coil size is limited by the size of the orifice through which it is inserted and by other structures en route to its intended location. The useful field-of-view of the coil is a function of its dimensions. Typically, for a saddle coil in which the winding width is a quarter of its length, sensitivity drops to about one third of that at the coil surface at a distance equal to the diameter of the former on which it is wound. While it is arguable as to what is acceptable from a SNR point of view, a rule of thumb is that the useful field-of-view of a coil is approximately equal to twice the winding width.

Patient safety is a key requirement. In most instances, electrical connections have to be made to the outside world and there are potential risks from conductors behaving as aerials, or from the formation of conducting loops. Insulation of conductors and biocompatibility of constructional materials are, therefore, important. A further complication is the sterilization of devices. Unless the devices are cheap enough to be discarded, consideration of the requirements for cleaning and sterilization for re-use is necessary. In order to avoid motion artefact degrading imaging quality, the coil also needs to be immobilized relative to both the machine measurement frame and the tissue surrounding it during data acquisition.

2.2   Transmit/Receive Operation

In typical MRI studies where spatial encoding is used with the coil in receive mode, voxels are smaller than the coil dimensions and coherence of signal phase is preserved within each voxel. The SNR is much lower in spectroscopic data acquisitions than in imaging studies, and the voxel size is correspondingly larger. The filling factor gains in SNR associated with MRI do not then apply. This is because the rapid changes in flux direction relative to the small coil result in partial signal cancellation if the whole volume round the coil is excited uniformly by an external transmitter (Figure 1) and there is a subsequent reduction in the net SNR in the large voxel studied. A transmit/receive coil avoids this signal loss because the phase and magnitude of the transmit field mirrors the flux pattern in the receive mode.

Internal transmit/receive coils may also be used for imaging in cases where the volume of excitation must be restricted. This has proved useful in the case of the twisted pair guidewire tracking device and in the cervical coil where rf irradiation to a pregnant uterus may be undesirable.

2.3   Tuning, Matching, and Decoupling

Solenoidal magnets commonly operate between proton resonant frequencies of 21 and 65 MHz, while magnets with transaxial fields currently have proton resonant frequencies in

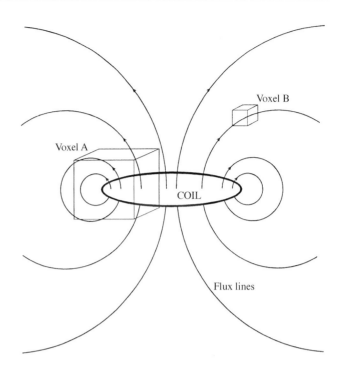

Figure 1   The effect on signal of the voxel size relative to coil size. In a voxel large enough to envelop flux in opposing directions (A), spins in regions of opposing flux cancel each other and reduce the net signal. In a small voxel (B), the flux is unidirectional and all the spins contribute to the net signal. If, however, the coil is used to transmit as well as receive, the flux reversal applies in both transmit and receive modes even in a large voxel and all the spins contribute to the net signal

the range of 8.1–25 MHz. Since the windings are constrained to be similar at all fields for any one class of coil, and are of low inductance, all rely on substantial tuning capacitors mounted close to the imaging field-of-view for good performance. The requirement for small components with minimal ferromagnetic content is an important consideration particularly for capacitors.

Figure 2 shows the tuning, matching, and decoupling circuitry for the endorectal array described by Gilderdale et al.[24] This circuitry is designed for receive-only operation and includes the components needed for active decoupling of rf transmitter pulses. The component values, and circuit detail, are those for operation at 0.5 T (21.3 MHz) in a solenoidal magnet. Active decoupling is used to ensure that the coupling between the receiver coil and the transmit field is eliminated and is independent of the time-dependent rf B_1 fields, levels, though it does require that suitable control lines are available from the machine being used.

2.4   Direct or Inductive Coupling

The elimination of cable connections to internal coils is very attractive in some circumstances; for example, if there is reason to implant a detector (in the same way as a pacemaker) and leave it in position while the patient continues with other activities between repetitive scans. The approach is also potentially valuable in circumstances where there may be technical

For References see p. 1295

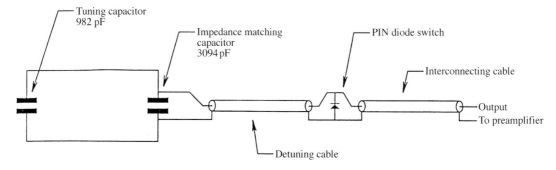

Figure 2   Circuitry associated with the endorectal array

difficulties in making connections, where there are safety issues if direct coupling is used, and where cabling becomes an embarrassment (either because of the number or the extent of the cables).[4,22,25]

Normally the external and internal coils are loosely coupled, since the latter are typically at a distance from the former. The external coils are several times larger in dimension than the internal coils; reciprocity means that the system is generally applicable to reception of signals as well as transmission. Figure 3 is an illustration of the technique used with a small fiducial coil, used either as a position marker or to track motion. The small coil can act as a flux amplifier which enables a 90° flip angle to be achieved within the marker sample while the magnetization of the surrounding tissue experiences negligible disturbance. This allows motion tracking to be interleaved with image data acquisition without the latter being compromised.

2.5   Orientation of Coils Relative to the Main Field

Certain coil designs (e.g., enveloping solenoid used for imaging the uterine cervix) give near optimal performance because anatomic features allow them to be orientated perpendicularly to the fixed magnetic flux density field B_0. However, in situations where the orientation is not fixed (e.g., in endoscopic applications) significant improvements in SNR of internal coils

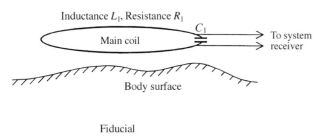

Figure 3   Coupling of a small inserted or implanted coil and a large external one. The shaded volume is a small vial of an appropriate NMR signal-generating material around which is wound a small tuned coil of inductance L_2 and resistance R_2, which is tuned to the system resonance frequency by a capacity C_2. The large coil, which is connected to the scanner, has respective tuning parameters L_1, R_1, and C_1. Its quality factor (Q, reflecting the efficiency of the system), is given by $\omega L_1/R_1$, where ω is frequency

and implanted devices can be achieved with a quadrature design. At least one of the two quadrature windings must be orthogonal to the main field and so capable of detecting excited signals. Three orthogonal windings, though increasingly hard to implement in a small unit, ensure signal detection at all times.

Because of anatomic considerations for coil insertion, the formers are often cylindrical. This implies that windings are either short solenoids or saddles. Variants of these designs were the basis of most external receiver coils employed in early MRI systems using cryogenic magnets.[26]

2.6   Artefacts

The sensitivity profile of a small internal coil results in very high signal adjacent to the coil surface and obscures image detail in this region. In general, a correction algorithm needs to be applied to the images to optimize their interpretation. In some applications it is possible to use the known sensitivity profile of the receiver coil to generate the correction.

As with conventional external coils, loss of signal occurs wherever the flux lines are parallel to the B_0 field. If this happens within the imaging region of interest it can cause 'blind spot' artefacts. This is of particular relevance in endoscopic applications and can be counteracted by the use of quadrature designs (cf. Section 2.5).

The most pernicious artefact encountered with internal coils is caused by motion, in which the coils move relative to the tissues of interest. Even when the coil is apparently firmly situated within the body, any patient motion can re-orientate the coil relative to the target source of signals. In endoscopic applications where the coil is located at, or ahead of, the tip of a flexible endoscope, the coil is liable to uncontrollable small movements. Reduction of motion artefact can be accomplished using fast imaging techniques; however, these not only suffer from a lack of robustness but also the small fields-of-view and thin slices used with internal coils make enormous demands on the machine gradients. Other well-established methods for motion artefact control use navigator echoes and phase-encode reordering, but again with the restricted field-of-view from the internal receiver, anatomic landmarks are difficult to identify clearly. Currently we are investigating the use of fiducial markers attached to the imaging coil former to generate reference signals that permit correction of the incoming data.[27]

Motion artefact can also be controlled by providing a firm anchor in the form of a clamp for the coils wherever possible. The unit shown in Figure 4 is suitable for clamping coils for

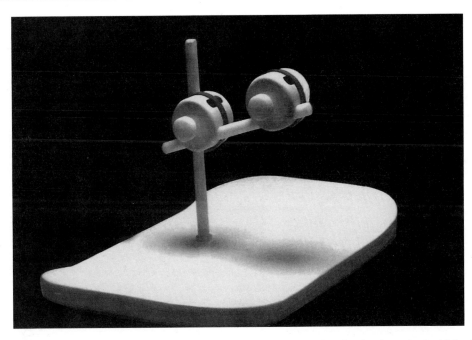

Figure 4 A clamp used to immobilize coils. For pelvic applications, the baseplate is placed under the patient's thighs. The double bosshead provides sufficient adjustment to cope with variations for the different coils and for individual patient requirements

use in the pelvis. The whole assembly is made from Delrin, with the baseplate being underneath the patient's thighs and so being held in position by their weight. The double bosshead arrangement allows sufficient flexibility in orientation for a range of coils, and the whole unit is simple to clean.

3 IMAGING TECHNIQUE

Imaging techniques at 0.5 T (Asset) or at 1.0 T (HPQ Vista, Picker International, Highland Heights, OH) include T_1-weighted spin echo [repetition time (*TR*) range 720–820 ms; echo time (*TE*) 20 ms], T_2-weighted spin echo (*TR* 2500 ms; *TE* 80 ms), and fast spin echo (*TR* 4500 ms; *TE* 96 ms) sequences. Transverse and sagittal 3.5 mm contiguous slices obtained with a 192×256 matrix, two to four signal averages, and a 10–12 cm field-of-view, provide optimal resolution while maintaining the SNR. In addition, a short inversion time (107 ms at 0.5 T) inversion recovery sequence (STIR) (*TR* 2500 ms; *TE* 30 ms) provides robust fat suppression and enables delineation of lesions with long T_2 times, for example fluid collections and tumors.

4 CONSTRUCTION AND APPLICATION OF TYPICAL COILS

Coil formers are made of an acetyl homopolymer (Delrin), as this is mechanically easy to handle and machine, has good rf properties, and is a nontoxic, biocompatible material that meets sterilization requirements. Delrin is water resistant and retains its shape despite repeated soaking in noxious sterilization fluids. The coils are wound from solid copper wire and are coated with an epoxy resin that is water resistant, biocompatible, and provides a seal for the electronics. In our experience, a typical coil has lasted for several hundred patient examinations without requiring renewal. The following sections describe the range of coils that we have developed and indicate what may be possible as capabilities expand.

4.1 Cervical Coils

Intravaginal enveloping cervical coils have been designed for use at 0.5, 1.0, and 1.5 T to operate in receive mode. A range of coil diameters has been investigated, 37 mm being a good compromise between accommodation of the cervix and ease of insertion. The coil is mounted on a cylindrical Delrin former attached to a 20 cm long Delrin handle (Figure 5a).

The anatomy of the uterine cervix provides a rare opportunity to exploit the superior SNR performance of a solenoid coil in a conventional MR scanner where B_0 is parallel to the long axis of the patient. Although the use of a coil in the vagina to image the cervix has been described by others,[28] the coils used had a simple loop geometry. The use of an enveloping cervical ring provides the best resolution of the cervix and the adjacent parametrium; with good positioning, such a coil provides a uniform signal from the structure it encircles. The coil is easy to position following digital examination, particularly in the anteverted uterus. However, even in retroverted uteri, a speculum examination is unnecessary. Although artefacts may obscure a few millimetres of the surrounding vaginal vault because of the proximity of the coil, these artefacts cause no problems in diagnostic accuracy. Also, the coil assists with interpretation by distending the vaginal vault and separating the vaginal wall and cervix.

4.1.1 Normal Anatomy

Both T_1- and T_2-weighted images show the normal cervix to consist of two distinct stromal zones and the mucosa surrounding the central canal.[29] The mucosa has a relatively high signal

For References see p. 1295

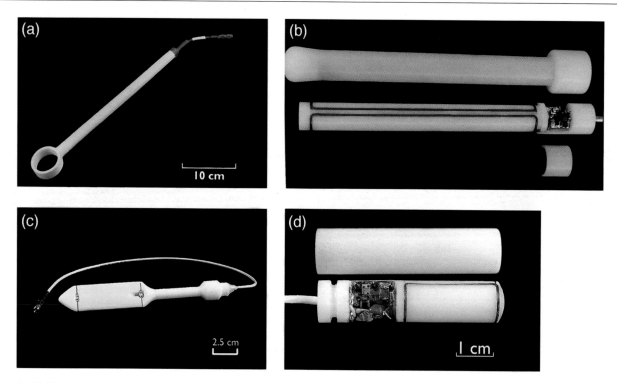

Figure 5 Coils for internal use. (a) Cervical coil; (b) cylindrical anal coil; (c) solid endorectal coil; (d) coil for use with a gastroscope or colonoscope

on both T_1- and T_2-weighted images. Mucosal detail can be appreciated with this technique (pixel size 0.6 mm) (Figure 6). It shows a smooth and regular outline in the nulliparous cervix and a more irregular and indented outline in the parous cervix.

In addition, dilated glands filled with secretions are sometimes seen in the latter.

On administration of 0.1 mmol kg^{-1} body weight gadopentate dimeglumine, brisk mucosal enhancement begins at 30 s

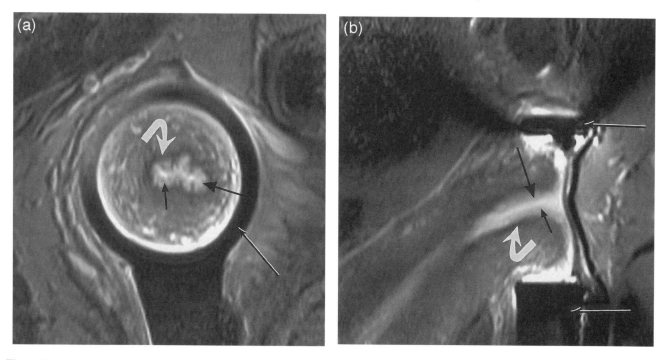

Figure 6 Images of the normal cervix using T_2-weighted spin echo (repetition time 2500 ms; echo time 80 ms) images through the cervix in (a) transverse and (b) sagittal planes. The coil is seen as an outer ring of signal void (black and white arrows). The mucosal layer is irregular (small straight arrow). There is an inner zone of low signal intensity (curved arrow) that corresponds to a highly cellular layer and an outer intermediate signal intensity zone (large arrow) that is less cellular

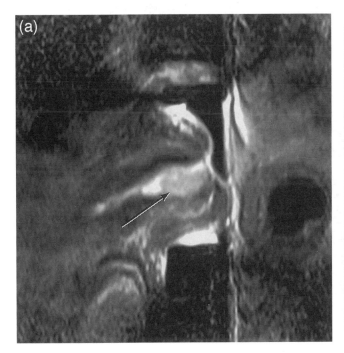

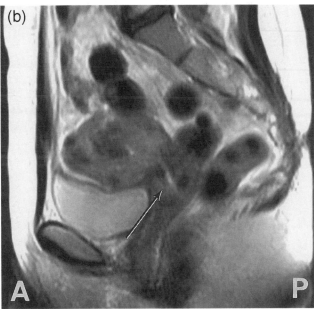

Figure 7 Images of Stage Ib squamous carcinoma of the cervix. Sagittal T_2-weighted spin echo (repetition time 2500 ms; echo time 80 ms) images through the cervix using (a) a cervical coil and (b) a phased array coil. The small tumor with intermediate signal intensity (arrows) is seen in both instances. However the margins are clearly defined in (a) and are difficult to differentiate in (b). There is no extension to parametrium or vagina. This was confirmed on the histology of the radical hysterectomy specimen. A, anterior; P, posterior

after injection and peaks at 120 s. The fibromuscular stroma enhances more slowly, with the outer ring showing more prominent enhancement, making zonal differentiation maximum at 90 s.

4.1.2 Cervical Cancer

There is good agreement between the high-resolution MRI findings obtained with an endovaginal coil and the pathological findings in patients with clinical evidence of stage I cervical cancer.[30] With standard body coil images, it is often difficult to assess the integrity of the stromal ring (Figure 7). The cervical coil makes such assessments easier, which has been confirmed by comparison with histology.[30]

Parametrial invasion may be recognized when there is loss of the low-signal-intensity stripe of normal stroma of the cervix together with diffuse or localized abnormal signal intensity within the parametrial region. Parametrial tumor spread may be overestimated because of peritumoral inflammatory tissue, which is well known as a cause of tumor overstaging.[30] With carcinoma of the cervix, intravenous contrast agents are of limited value,[31] although dynamic contrast enhancement has been reported to be valuable in the accurate assessment of cervical invasion by tumor.[32] We used dynamic contrast-enhanced imaging in three slice positions over a 5 min period, though in no case did these images provide information not available on T_2-weighted images.

High-resolution images of the cervix are particularly important in defining tumors that are less than 1 cm³ in volume because the definition of their extent has a significant impact on clinical decision making.[33] High-resolution images also may

guide the approach to the histologic dissection of the resected tumor and thus not only optimize the surgical approach but also facilitate longer-term management.

4.2 Anal and Urethral Coils

Anal and urethral coils are simple saddle windings when implemented for solenoidal magnets. A variety of coil sizes have been developed for use at different field strengths and for different applications. Cylindrical endoanal coils have diameters ranging from 9 to 12 mm, and lengths of 75–100 mm: the largest version is shown in Figure 5b. Variants of the basic design include arrays and the integration of a pair of fiducial coils.

A wide variety of pathology in adults and children has been imaged using these coils, including infection, trauma, tumors and congenital malformations of the anal sphincter.[33–38] The superior image resolution allows identification of small fistulae-in-ano, assesses muscle defects, and defines the extent of tumors.

4.2.1 Normal Anatomy

From medial to lateral, several layers are recognized. A layer of mucosa with high signal is surrounded by a region of submucosa and subepithelial muscle (musculus submucosae ani) with low signal. The abundance of collagen and elastic tissue probably accounts for the short T_2 time of this region (Figure 8). The internal sphincter is distinguished by its high signal intensity on all sequences and often appears crescentic,

For References see p. 1295

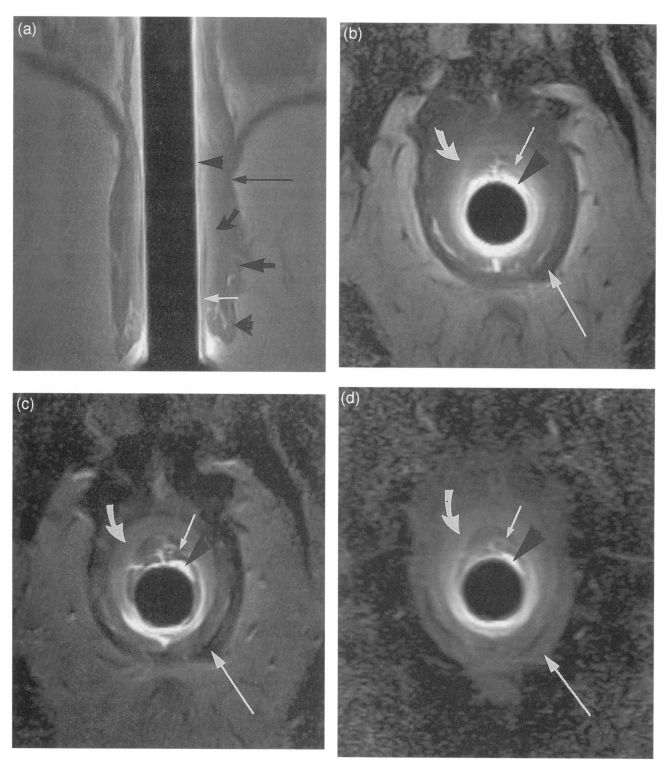

Figure 8   Normal anal anatomy in a 44-year-old female. Spin echo T_1-weighted [repetition time (*TR*) 720 ms; echo time (*TE*) 20 ms] images in (a) coronal and (b) transverse planes; (c) spin echo transverse T_2-weighted image (*TR* 2500 ms; *TE* 80 ms); (d) STIR (short inversion time inversion recovery; inversion time 107 ms; *TR* 2500 ms; *TE* 30 ms) images at a mid-sphincteric level in two planes. The coil is seen as a central signal void. The mucosa (black and white head) gives a very high signal on all sequences. The submucosa (small white arrow) is low signal. The internal sphincter is a homogeneous band (curved arrow) and has a particularly high signal in (c) and (d). The external sphincter is relatively low signal on all sequences. In (a) it is seen in its three components, subcutaneous (lower broad arrowhead), superficial (short black arrows), and deep (long black arrow)

being thicker anteriorly. The characteristic high signal appearance of the internal sphincter is not merely a result of its closeness to the coil; persistence of this high signal pattern on correction with a signal intensity profile from phantom data suggests that it is intrinsic to the smooth muscle of the internal sphincter. Its presence in vitro also suggests that it is not a consequence of the normal resting tone of the internal sphincter.[38] The intersphincteric region has a high signal on T_1- and T_2-weighted spin echo images: this is nulled on the STIR images, which is consistent with the presence of fat. The conjoined longitudinal muscle and external sphincter have a low signal intensity (Figure 8). The oblique coronal plane is particularly useful in demonstrating the three components of the external sphincter (subcutaneous, superficial, and deep) (Figure 8) and their relation to the levator ani.

Like the model described by Oh and Kark,[39] the female anal sphincter is shorter anteriorly than posteriorly. Inferiorly, the anterior margin of the superficial and deep external sphincter are continuous with and inseparable from the perineal body (Figure 8b). The transverse perineal muscle bridges the inferior part of the external sphincter anteriorly (Figure 8c). Superiorly, perineal glands directly abut the internal sphincter anteriorly and laterally, separating it from the muscle of the vaginal wall.

4.2.2 Perianal Sepsis

The change in T_1 and T_2 times associated with infection produces high soft tissue contrast and enables abscesses and fistulous tracks to be demonstrated. Although previous studies using a conventional body coil have detailed the efficacy of MRI in demonstrating fistulae[40] and changes in Crohn's disease,[41] the excellent image quality provided by an internal coil results in further improvement. With more traditional anal endosonographic techniques,[42] tissue contrast is less good, the technique is more operator dependent, and it is no more accurate than digital examination under general anesthesia.[43] With endoanal imaging, in simple as well as in complex fistulae, it is relatively easy to identify a collection and define its extent (Figure 9b) as well as to follow the primary tracks in multiple planes to find the site of the internal fistula opening.

The position and course of the track in relation to the anal verge and the levator ani can be documented from the coronal and sagittal images. The resolution of the technique (pixel size 0.6 mm×0.6 mm) ensures that even small collections of the order of a few millimeters can be identified. Surgical management critically depends on the detection of these small collections, the site of the primary track, and the position of the internal opening.

4.2.3 Fecal Incontinence from Surgical Trauma

Perineal surgery may inadvertently cause a breach of the external or internal sphincter. The external sphincter is particularly at risk during procedures such as hemorrhoidectomy, vulvectomy, and colposuspension. The internal sphincter is particularly vulnerable in lateral sphincterotomies for chronic anal fissure when complete division of the internal sphincter may result in fecal leakage.

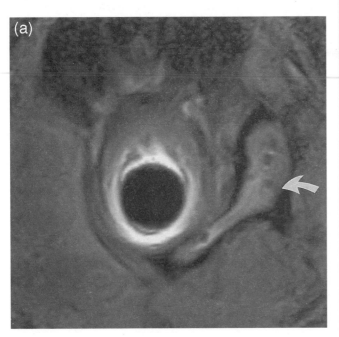
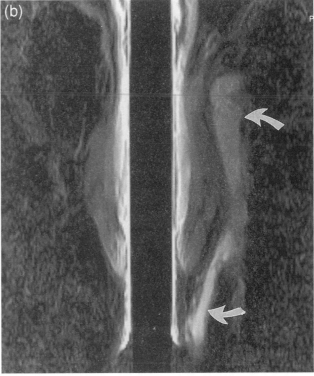

Figure 9 Images of the anal sphincter in anal sepsis using a 12 mm, endoanal coil; (a) transverse T_2-weighted spin echo [repetition time (*TR*) 2500 ms; echo time (*TE*) 80 ms]; (b) coronal short inversion time inversion recovery (STIR) (inversion time 107 ms; *TR* 2500 ms; *TE* 30 ms). The endoanal coil appears as a central signal void. An extrasphincteric collection is seen at 3 o'clock in (a) (arrow); however, there is no extension above the levators in (b) (arrows)

For References see p. 1295

4.2.4 Fecal Incontinence from Obstetric Trauma

Early-onset fecal incontinence Fecal soiling immediately following childbirth is a dramatic symptom and, because intact sphincter musculature is essential to maintain continence, this suggests that part of the external sphincter has been ruptured. Defects in both internal and/or external sphincter are usually strikingly apparent on endoanal MRI. Tears involving the internal sphincter are immediately recognized because of the profound loss of the high signal intensity normally seen in this homogeneous muscle ring. In patients with rectovaginal fistulae, long T_2 tracks are seen to course anteriorly between anus and vagina.

Although endoanal sonography can provide anatomic information, particularly of internal sphincter damage in obstetric trauma,[44] it poorly defines the echogenic external sphincter against the echogenic perianal fat. Endoanal MRI not only shows internal sphincter damage but also accurately delineates how much of the external sphincter has been disrupted.

Late-onset fecal incontinence In patients presenting with fecal incontinence starting several years after childbirth, a different pattern of sphincter abnormality is seen. The internal sphincter is relatively normal in thickness, but marked atrophy of the superficial and deep components of the external sphincter is evident (Figure 10). The subepithelial muscle is thicker in the fecally incontinent group compared with that in age-matched normal controls,[38] probably because of compensatory hypertrophy.

4.2.5 Fecal Incontinence in Systemic Conditions

Patients with scleroderma and fecal incontinence show a striking abnormality of the anterior sphincter musculature. There is buckling forward of the anterior rectal wall and anal sphincter musculature, with descent of rectal air and feces below the levators into the upper half of the anal canal.[45] There is also a significant reduction in internal anal sphincter bulk compared with age-matched fecally incontinent subjects who do not have scleroderma.[45] Patients with scleroderma but without fecal incontinence sometimes show these features.

4.2.6 Tumor

In patients with low carcinoma of the rectum, the extent of tumor invasion of the sphincter can be delineated with endoanal MRI.[36] Irregular lobulated masses of high signal intensity may be seen (Figure 11) on T_2-weighted sequences, which allow visualization of tumor invasion and breakthrough of the muscular coat of the anorectum. If lack of anal invasion is confirmed on endoanal MRI patients may be offered an anterior resection.

The multiplanar facility of MRI allows clear delineation of tumor extent and invasion particularly in relation to the levator floor. The exact distance from each of the components of the external sphincter and the extent of internal sphincter involvement can be accurately asserted prior to surgical planning.

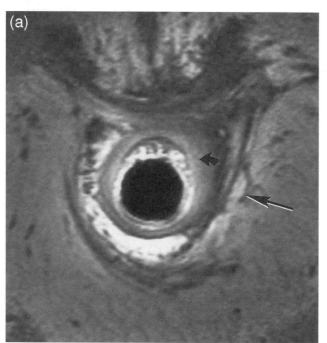

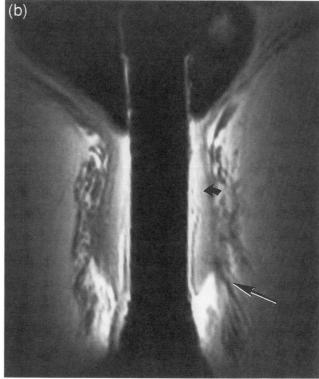

Figure 10 Sphincter atrophy in a patient with long-standing fecal incontinence following childbirth. Spin echo images (repetition time 720 ms; echo time 20 ms) T_1-weighted in (a) transverse and (b) coronal planes. Following traumatic childbirth several years earlier, there is focal atrophy of the internal sphincter (small arrow) in this patient and more global atrophy of the external sphincter (large arrow)

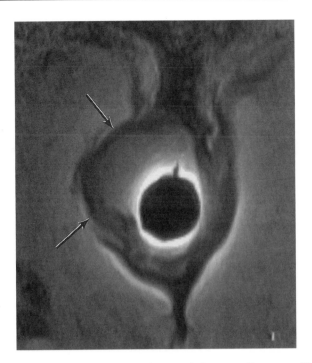

Figure 11 Squamous cell carcinoma of the anus. Transverse T_1-weighted spin echo image (repetition time 660 ms; echo time 20 ms) showing an intermediate signal lump (arrows) in the right anterior quadrant between 9 and 12 o'clock involving the subcutaneous external sphincter

4.2.7 Congenital Anorectal Anomalies

Smaller diameter (7 and 9 mm) coils are used for defining sphincter abnormalities in children with anorectal anomalies. The sensitive region of the various diameters of coil is chosen to suit the dimensions of the anal sphincter in the various age groups. The technique delineates the thickness and position of the muscle components as well as the presence of fistulae with neighboring structures (Figure 12).

The use of an anal coil causes minor discomfort but is acceptable and well tolerated in unsedated children of more than 8 years of age and appears to be better tolerated than digital examination. Children aged less than 8 years normally require sedation for an MRI scan even with an external coil.

4.3 Endorectal Coils

The balloon type of endorectal coil was the first internal coil to be developed.[5] We developed a similar version of our own to operate at 0.5 T as coils for that field strength were not available commercially. Subsequently, a smaller rigid device was developed which both operators and patients preferred,[12] and this device is described here.

This coil (Figure 5c) is essentially a surface coil designed to be placed in the rectum posterior to the prostate gland, and consists of an 8.5 cm × 2.5 cm semicylindrical former. The flat upper surface is used as the base for the 6.5 cm × 2.5 cm winding. This solid, one-piece surface coil is placed in the rectum. The tip of the coil is tapered and it is easy to insert and position following standard digital rectal examination in the lateral decubitus position.[12] Our solid one-piece endorectal coil is

easy to handle and set-up, is re-usable, and does not require external tuning and matching. Also, unlike the commercially available coil, there are no artefacts from susceptibility effects of the air in the balloon, which degrade the prostatic images. The solid design also allows incorporation of a needle guide, and with further refinements will enable MR-guided transrectal biopsy.

4.3.1 Normal Anatomy

The zonal anatomy of the prostate may be identified clearly on the high-resolution T_2-weighted images obtained using the endorectal coil. The outer homogeneous zone of high signal intensity (peripheral zone) forms the bulk of the gland in young males. The inner zone (central zone), which histologically comprises the 'transitional' lobe of the prostate, increases in size with age and displaces the peripheral zone laterally.

4.3.2 Benign Prostatic Hypertrophy

In benign prostatic hypertrophy, the enlarged central zone of the prostate forms a nodular, inhomogeneous adenoma that often includes cystic nodules and calcification. With increase in size it commonly causes compression of the prostatic urethra and leads to bladder outflow obstruction. This normally requires intervention with conventional surgery or minimally invasive thermal ablative techniques. In the latter instances, intraoperative image guidance is becoming increasingly popular. We have demonstrated that intraoperative changes seen on

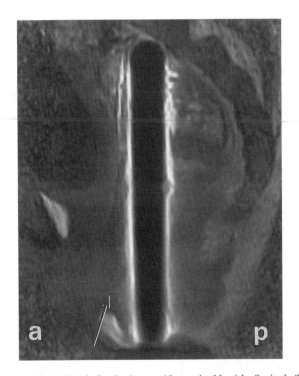

Figure 12 Perineal fistula in an 18-month-old girl. Sagittal T_1-weighted spin echo image (repetition time 720 ms; echo time 20 ms) through the mid-sphincter (a, anterior, p, posterior). The 7 mm coil is seen as a central signal void. There is generally good sphincter bulk. A high signal track is seen to extend anteriorly (arrow). It communicates with an opening in the posterior vaginal fourchette that was seen to discharge feces

For References see p. 1295

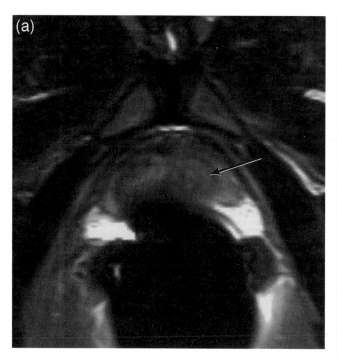
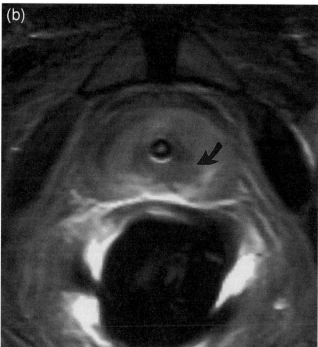

Figure 13 Appearances after endoscopic laser prostatectomy for benign hypertrophy. Transverse T_2-weighted spin echo images (repetition time 2500 ms; echo time 80 ms) before (a) and immediate after (b) a four-quadrant ablation of the prostate using a noncontact laser. (a) The small central adenoma (arrow) was causing significant symptoms. (b) There is overall swelling of the gland with loss of the normal zonal differentiation. A periurethral ring of low signal is seen (curved arrow) that is likely to represent coagulative necrosis

endorectal imaging may be used to predict more long-term outcome;[46] the development of a periurethral band of low signal at the time of surgery correlated with a significant reduction in symptoms at 3-month follow-up (Figure 13).

4.3.3 Prostate Cancer

Endorectal imaging is routinely used in the detection of prostate cancer (Figure 14). Malignant nodules are usually recognized as low-signal foci within the peripheral lobes: however, they are sometimes iso- or even hyperintense compared with background and, therefore, are difficult to detect. Also the multifocal nature of this disease means that small malignant foci may go unrecognized. Recently, dynamic contrast-enhanced techniques have been used to map out the full site and extent of malignant foci,[47] and this approach may be adopted to monitor the response to medical therapy.

4.4 Endoscopic Coils

A gastroscope and colonoscope coil design is shown in Figure 5d for comparison (see *Gastroscopy and Colonoscopy*). This uses a Delrin former with a saddle winding and connecting cable. The coil unit is located ahead of an MR-compatible gastroscope or colonoscope and the cable is passed through the 3 mm channel usually allocated for biopsy devices.

4.5 Miscellaneous Coils

Use of small coils in the external auditory meatus, within blood vessels, and for dental applications have been investi-

gated. The problems with such coils include achieving a useful field-of-view from something with a diameter small enough to be located around or within the structure being studied.

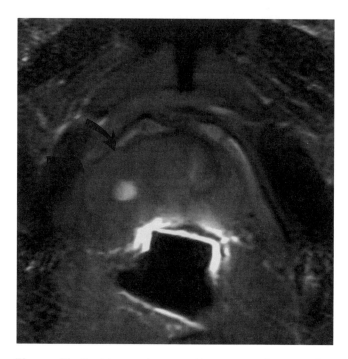

Figure 14 Prostate carcinoma with extracapsular extension. Transverse T_2-weighted spin echo image (repetition time 2500 ms; echo time 80 ms) through the mid-prostate. A large mass is seen on the right that extends into the periprostatic tissues (curved arrows)

5 CONCLUDING REMARKS

The use of internal coils allows high-resolution images of the part under study. Improved performance is achieved by superior coupling to the tissue of interest while minimizing coupling to irrelevant regions. A solid design with external immobilization results in re-usable coils without susceptibility artefacts.

6 RELATED ARTICLES

Coils for Insertion into the Human Body; Gastroscopy and Colonoscopy; Male Pelvis Studies Using MRI; MRI of the Female Pelvis.

7 REFERENCES

1. A. Abragam, 'The Principles of Nuclear Magnetism', Clarendon Press, Oxford, 1961, p. 71
2. D. I. Hoult, and R. E. Richards, *J. Magn. Reson.*, 1976, **24**, 71.
3. H. L. Kantor, R. W. Briggs, and R. S. Balaban, *Circ. Res.*, 1984, **55**, 261.
4. A. L. Benabid, M. Decorps, C. Remy, J. P. Albrand, M. Retenauer, and P. Blondet, *Proc. IVth Annu. Mtg. Soc. Magn. Reson. Med.*, London, 1985, p. 441.
5. M. D. Schnall, C. Barlow, V. H. Subramanian, and J. S. Leigh Jr, *J. Magn. Reson.*, 1986, **68**, 161.
6. R. R. Harman, P. C. Butsen, A. S. Hall, I. R. Young, and G. M. Bydder, *Magn. Reson. Med.*, 1988, **6**, 49.
7. M. D. Schnall, R. E. Lenkinsi, H. M. Pollack, Y. Inai, and H. Y. Kressel, *Radiology*, 1989, **172**, 570.
8. M. D. Schnall, T. Connick, C. E. Hayes, R. E. Lenkinski, and H. Y. Kressel, *JMRI*, 1992, **2**, 229.
9. C. J. Baudouin, W. P. Soutter, D. J. Gilderdale, and G. A. Coutts, *Magn. Reson. Med.*, 1992, **24**, 196.
10. N. M. deSouza, W. A. Kmiot, R. Puni, A. S. Hall, C. I. Bartram, and G. M. Bydder, *Gut*, 1995, **37**, 284.
11. S. M. Hussain, J. Stoker, J. W. Kuiper, W. R. Schouten, J. C. den Hollander, and J. C. Lameris,, *Radiology*, 1995, **197**, 671.
12. N. M. deSouza, D. J. Gilderdale, R. Puni, G. A. Coutts, and I. R. Young, *JMRI*, 1996, **6**, 801.
13. N. M. deSouza, A. H. Gibbons, G. A. Coutts, A. S. Hall, R. Puni, J. Calam, and I. R. Young, *Minim. Invas. Ther.*, 1996, **4**, 23.
14. A. D. Williams, G. A. Coutts, A. S. Hall, W. A. Kmiot, D. J. Gilderdale, M. Dinneen, I. R. Young, and N. M. deSouza, *Proc. VIth Annu. Mtg. (Int.) Soc. Magn. Reson. Med.*, Sydney, 1998, p. 472.
15. D. J. Gilderdale, D. J. Bryant, G. M. Bydder, and I. R. Young, *JMRI*, 1992, **2**(P), 153.
16. D. J. Gilderdale, N. M. deSouza, G. A. Coutts, and I. R. Young, *Proc. IVth Annu. Mtg. (Int.) Soc. Magn. Reson. Med.*, New York, 1996, p. 1437.
17. C. L. Dumoulin, S. P. Souza and R. D. Darrow, *Magn. Reson. Med.*, 1993, **29**, 411.
18. G. C. McKinnon, J. F. Debatin, D. A. Leung, S. Wildermuth, D. J. Holtz, and G. K. von Schulthess, *Proc. IInd Annu. Mtg. Soc. Magn. Reson. Med.*, San Francisco, 1994, p. 429.
19. O. Ocali, and E. Atalar, *Magn. Reson. Med.*, 1997, **31**, 112.
20. M. Burl, G. A. Coutts, D. J. Herlihy, R. Hill-Cottingham, J. F. Eastham, J. V. Hajnal, and I. R. Young, *Magn. Reson. Med.*, 1999, **41**, 636.
21. G. H. Barnett, D. W. Kormos, W. P. Steiner, and J. Weisenberger, *J. Neurosurg.*, 1992, **78**, 510.
22. M. Burl, G. A. Coutts, and I. R. Young, *Magn. Reson. Med.*, 1996, **36**, 491.
23. G. A. Coutts, D. J. Gilderdale, K. M. Chui, L. Kasuboski, and N. M. deSouza, *Magn. Reson. Med.*, 1998, **40**, 908.
24. D. J. Gilderdale, G. A. Coutts, and N. M. deSouza, *Proc. Vth Sci. Mtg. (Int.) Soc. Magn. Reson. Med.*, Vancouver, 1997, p. 1509.
25. R. R. Harman, I. R. Young, G. M. Bydder, P. C. Butsen, and D. Spencer, *Proc. Vth Annu. Mtg. Soc. Magn. Reson. Med.*, Montreal, 1986, p. 57.
26. D. I. Hoult, and P. C. Lauterbur, *J. Magn. Reson.*, 1976, **24**, 71.
27. G. A. Coutts, D. J. Gilderdale, A. D. Williams, and N. M. deSouza, *Proc. VIth Annu. Mtg. (Int.) Soc. Magn. Reson. Med.*, Sydney, 1998, p. 724.
28. B. N. Milestone, M. D. Schnall, R. E. Lenkinski, and H. Y. Kressel, *Radiology*, 1991, **180**, 91.
29. N. M. deSouza, I. C. Hawley, J. E. Schwieso, D. J. Gilderdale, and W. P. Soutter, *Am. J. Roentgenol.*, 1994, **163**, 607.
30. N. M. deSouza, D. J. Scoones, T. Krausz, D. J. Gilderdale, and W. P. Soutter, *Am. J. Roentgenol.*, 1996, **166**, 553.
31. S. Thurner, *Am. J. Roentgenol.*, 1992, **159**, 1243.
32. Y. Yamashita, M. Takahashi, T. Sawada, K. Miyazaki, and H. Okamura, *Radiology*, 1992, **182**, 643.
33. N. M. deSouza, G. A. J. McIndoe, C. Hughes, W. P. Mason, M. K. Chui, T. Krausz, and W. P. Soutter, *Br. J. Obstet. Gynaecol.*, 1998, **105**, 500.
34. N. M. deSouza, W. A. Kmiot, R. Puni, A. S. Hall, C. I. Bartram, and G. M. Bydder, *Gut*, 1995, **37**, 284.
35. N. M. deSouza, R. Puni, D. J. Gilderdale, and G. M. Bydder, *Magn. Reson. Q.*, 1995, **11**, 45.
36. N. M. deSouza, A. S. Hall, R. Puni, D. J. Gilderdale, I. R. Young, and W. A. Kmiot, *Dis. Colon Rectum*, 1996, **39**, 926.
37. N. M. deSouza, D. J. Gilderdale, D. K. MacIver, and H. C. Ward, *Am. J. Roetgenol.*, 1997, **169**, 201.
38. N. M. deSouza, R. Puni, A. Zbar, D. J. Gilderdale, G. A. Coutts, and T. Krausz, *Am. J. Roentgenol.*, 1996, **167**, 1465.
39. C. Oh, and A. E. Kark, *Br. J. Surg.*, 1972, **50**, 717.
40. P. H. Luniss, P. G. Barker, A. H. N. Sultan, *et al.*, *Dis. Colon Rectum*, 1994, **37**, 708.
41. J. J. Tjandra, and G. R. J. Sissons, *Aust. N. Z. J. Surg.*, 1994, **64**, 470.
42. P. J. Law, R. W. Talbot, C. I. Bartram, and J. M. A. Northover, *Br. J. Surg.*, 1989, **76**, 752.
43. S. Choen, S. Burnett, C. I. Bartram, and R. J. Nicholls, *Br. J. Surg.*, 1991, **78**, 445.
44. S. J. D. Burnett, C. Spence-Jones, C. T. M. Speakman, M. A. Kamm, C. N. Hudson, and C. I. Bartram, *Br. J. Radiol.*, 1991, **64**, 225.
45. N. M. deSouza, A. D. Williams, H. J. Wilson, D. J. Gilderdale, G. A. Coutts, and C. M. Black, *Radiology*, 1998, **208**, 529.
46. N. M. deSouza, R. Flynn, A. G. Coutts, D. J. Gilderdale, A. S. Hall, R. Puni, M. Chui, D. N. F. Harris, and E. A. Kiely, *Am. J. Roentgenol.*, 1995, **164**, 1429.
47. A. R. Padhani, J. E. Husband, and G. J. M. Parker, *Diagn. Imag. Eur.*, 1996, **June**, 20.

Acknowledgements

We should like to acknowledge the support of Picker International in the development of the probes, of the Wellcome Trust who provided the 0.5 T machine resource on which much of the work was done, and of the Medical Research Council who supported the work during the earlier part of the program. We would like to thank our clinical collaborators Mr E. Kiely, Dr J. Waxman, Mr W. Kmiot, Mr G. A. McIndoe, and Mr W. P. Soutter. We are also grateful to Mrs M. C. Crisp for her secretarial assistance.

For References see p. 1295

Biographical Sketches

Nandita M. de Souza. *b* 1957. B.S.c. Hons (Physiology) 1978, MBBS, 1983, University of Newcastle upon Tyne, UK, M.R.C.P. 1986, F.R.C.P. 1991, M.D. 1996. House Surgeon and Physician and Senior House Physician, Newcastle upon Tyne 1983–85. Registrar in internal medicine 1986–87. Registrar and Senior Registrar in Diagnostic Radiology, Guy's Hospital, London, 1987–92. Senior Research Fellow in MR Imaging, Royal Postgraduate Medical School, Hammersmith Hospital, 1993–97. Senior Lecturer 1997–present. Research Interests: internal receiver coils, interventional MRI.

David J. Gilderdale. *b* 1947. B.S.c. telecommunications, Ph.D. computer simulation as an aid to lighting design. Research Assistant, Plymouth Polytechnic, 1974–78. Electronic Engineer, Central Research Laboratories, Thorn-EMI Ltd, 1978–81. Senior Scientist, GEC Hirst Research Centre, 1981–84. Senior Lecturer, Department of Electrical Engineering, Plymouth Polytechnic, UK, 1981–88. Engineering Consultant, 1988–present.

Glyn A. Coutts. *b* 1959. B.A. (Physics) University of Oxford, UK, 1980, D.Phil. University of Oxford 1985. Senior Research Associate at Picker Research UK, 1987–99. Marconi Medical Systems, UK, 1999–present. Approx. 30 publications in the field of MRI and MRS. Current interest: interventional MRI.

Gastroscopy and Colonoscopy

Nandita M. deSouza

Imperial College School of Medicine, London, UK

Glyn A. Coutts and David J. Larkman

Marconi Medical Systems, UK

and

David J. Gilderdale

Engineering Consultant, UK

1 INTRODUCTION

Since the 1960s, endoscopy has revolutionized the diagnosis and treatment of both upper and lower gastrointestinal (GI) disease and has become the gold standard in management. However, detection of disease is limited to visualization of the surface mucosa. The extent of invasion beneath the surface may be shown with biopsy, but lesions with no superficial component or those with infiltration remote from the biopsy site may be missed.[1] Ultrasound probes have been used in conjunction with endoscopy to address this problem, and transducers that can be passed down flexible gastroscopes[2] or

bronchoscopes[3] have been used in order to locate lesions and demonstrate their vascularity. However, low soft-tissue contrast limits the application of this technique.

MRI provides a valuable method for visualizing soft tissue; however, with conventional external whole-body or phased array receiver coils, image resolution of small areas of interest is generally low. Dedicated surface receiver coils can greatly improve image resolution of adjacent structures, and intracavitary coils have been used in the rectum and vagina for imaging the adjacent prostate and cervix.[4,5] It is possible to use coils of this type in conjunction with a nonferromagnetic endoscope for MRI of lesions of the upper and lower GI tract.

2 MR-COMPATIBLE ENDOSCOPE DESIGN

Ferromagnetic materials are normally used in the construction of endoscopes. These not only produce an unacceptable degree of artefact but may be subject to movement forces when placed within the magnetic field. It is now possible to produce endoscopes from MR-compatible components. We have developed a MR-compatible gastroscope and colonoscope that contains no ferromagnetic materials (Endoscan Ltd, Burgess Hill, UK). The general handling attributes are similar to those of conventional instruments. An alloy of chrome is used as a substitute for the normal stainless steel guide wires. The conventional steel shroud, which provides integral strength, is replaced by copper braid. Each of the substitute materials is tested to ensure there is no significant MR artefact. The whole assembly is encased in a heat-shrink sleeve that seals the instrument in an acceptable fashion.

The tip of the instrument is associated with the MR receiver coil and must be modified accordingly. In early versions of the endoscope, the coil was separated from the endoscope but used in conjunction with it. The coil could, therefore, be advanced away from the tip itself so that artefacts associated with components in the tip were not critical. The tip endpiece in this instrument is configured from an amalgam for the viewing lens, irrigation, and biopsy channels. Although polymers would be preferable, forming an impervious seal to the optics proved impossible and alloys of gold-plated copper were used instead. A newer version of the endoscope uses a coil design integral to the tip itself, which means that the tip must be artefact free and the tip design has to be appropriately modified to accommodate the coil circuitry.

The finished instrument incorporates conventional steering controls, an eyepiece, a light guide with an illumination source (Olympus CLE-3 with 150 W dichroic bulb), and a biopsy channel 2.8 mm in internal diameter (Figure 1). The length of the upper GI endoscope from eyepiece to tip is 1 m with a diameter of 12.5 mm, which equals that of a therapeutic gastroscope. The colonoscope is 1.7 m in length and 14 mm in diameter.

3 RECEIVER COIL DESIGN

The simplest design uses a solid coil that is separate from the endoscope itself but can be used in conjunction with it. The outer diameter of a rigid coil cannot be larger than that of the endoscope itself. Such a coil may be advanced some dis-

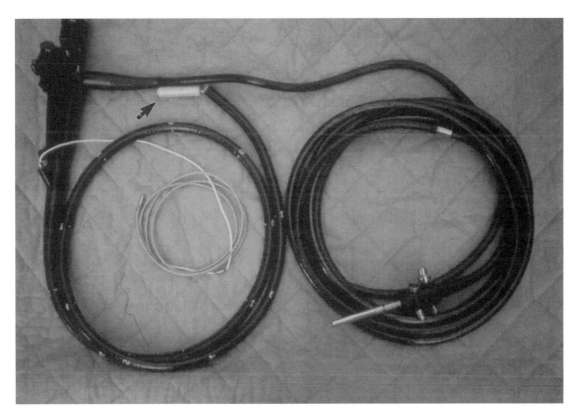

Figure 1 MR-compatible endoscope and receiver coil (arrow). The appearance and handling of the endoscope are identical to a standard instrument. The receiver coil is the same diameter as the endoscope itself

tance from the endoscopic viewing lens or pulled back up to it. It is maintained in the latter position during intubation of the pharynx. Advantages of its rigidity are not only ease of maintenance but also that it does not require tuning on every occasion. We have used a single turn saddle geometry receiver coil 25 mm in length and 10 mm in diameter wound on an acetyl homopolymer (Delrin) former. The conductor is of 18 swg varnish insulated copper and the on-board electronics contain no ferro- or paramagnetic materials. The receiver coil is connected to the external electronics by a specially constructed 50 ohm coaxial cable 2.1 mm in external diameter in a low-friction polytetrafluoroethylene (PTFE) jacket purpose built with MR-compatible components. The cable is accommodated within the biopsy channel of the endoscope.

The addition of an immobilization device to the coil, such as an inflatable balloon or suction cup, would greatly improve image quality. Inflatable balloon devices are currently in use with gastroscopic[2] and bronchoscopic[3] transducers. However, inflatable designs require individual tuning and matching and are subject to distortion while the endoscope is being maneuvered.

Using the receiver soil separately from the tip itself has several disadvantages: it reduces maneuverability of the instrument, compromises optical field-of-view, and decreases suction (because its cable is fed through the biopsy channel). The incorporation of the receiver coil into the wall of the endoscope itself improves maneuverability and optical viewing. However, it is more challenging to construct and has important implications for housing and repair to the electronics. In particular, immersion of the electronics during sterilization of the endo-

scope must be considered. In addition, the tip of the endoscope must be completely MR compatible in order to avoid unacceptable artefact on the images.

In our design, the coil incorporates an electronic switch that renders the coil electrically invisible during the transmit phase of the MR sequence. Computer controlled switching of rf is used to achieve this. Diodes are selected for their optimal rf switching characteristics while avoiding the ferromagnetic materials that are routinely found in conventional diodes. The small inductors used to form part of the rf switch have minimal detrimental effect on coil performance. A further circuit is provided to transform the impedance of the coil to match that of the transmission cable, thereby minimizing signal degradation within the cable. The capacitors used for both coil tuning and impedance transformation are selected for their high quality factor (Q), while avoiding the ferromagnetic nickel barrier terminations that are often preferred for capacitors in a mass-production environment. The board required to accommodate the additional coil control circuitry is typically 11.5 mm×10 mm and is placed immediately behind the coil.

The signal-to-noise ratio (SNR) figures for coil performance are often misleading. When compared with the standard pelvic phased array coil, the gastroscope coil performs significantly better at distances up to 2 cm from the coil surface. A clinically acceptable field-of-view for diagnostic purposes is obtained up to 2.5 cm from the coil surface. However, coil orientation relative to the fixed magnetic flux density field, B_0, also has a significant effect on the local field-of-view. For a single coil, the regions in which the flux is parallel to B_0 produce no useful signal. Employing two or more coils each with

For References see p. 1303

a different orientation such that the blind spot of one is compensated for by the other may overcome this problem. Such coil arrays are crucial in imaging a tortuous structure such as the gut, where coil orientation is extremely variable.

4   SCANNER DESIGN

Modern short-bore magnet designs can be usefully adapted to perform MR endoscopy. Our experience has been with a short-bore design Picker Asset, where the distance from the magnet face to center is 60 cm and an operator can reach at arm's length into the center of the bore (Figure 2). Traditional MR scanner designs in which the patient is at a distance down the long magnet bore do not provide sufficient patient access and are unsuitable. The newer 'interventional' MR scanners (C-arm, double doughnut) specifically address the issues of patient accessibility, and their open-plan features are designed to cater for endoscopic procedures.

4.1   Pulse Sequences

Imaging strategies for use with the gastroscopic receiver coil seek a balance between two major considerations. On the

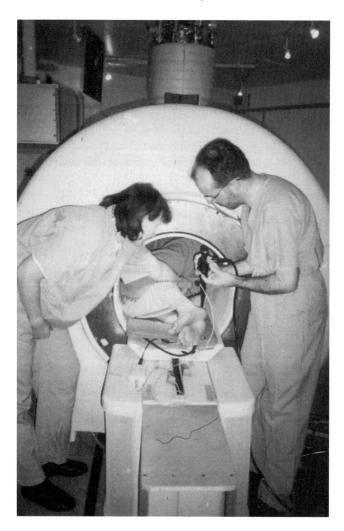

Figure 2   MR endoscopic procedure being performed

one hand the coil gives a large gain in sensitivity over a limited region that can be exploited to reduce patient scan times as well as to provide high-resolution images of immediately adjacent tissue. In particular, image wrap around is not an issue when choosing small fields-of-view. On the other hand, it is important to avoid image degradation associated with motion. The exact sources of motion artefact depend on the positioning of the coil. Major contributions arise from the pulsatile motion of the heart and great vessels in the thorax and from respiratory motion, as well as from patient movement if scan times are too long. If no immobilization is provided for the receiver coil, the sensitivity profile is such that smearing of the very high signal from regions immediately adjacent to the coil occurs and can swamp signal from regions of interest a few centimeters from the coil.

There are many well-known strategies for dealing with motion artefact in MR imaging. In the upper GI tract, cardiac gating is useful. Typically a standard gated spin echo sequence will have a repetition time (TR) of 700–800 ms, so T_1-weighted images, which provide useful anatomic landmarks, may be obtained. These sequences do not have a high gradient demand; consequently, with normal imaging gradients (10 mT m^{-1} with a rise time of 1 ms), 3 mm slice thicknesses and 10 cm fields-of-view can be easily achieved. A small reduction in the achievable resolution and the number of slices that can be scanned (although still more than enough to cover the extent of the coil) allows the use of gated fast-spin echo sequences, which significantly reduce the scan times. With longer pseudo-echo times, these sequences give more T_2-weighted images.

Although gating greatly improves image quality, images collected in this way are still subject to extensive motion arte-fact. In these sedated patient examinations, breath-hold techniques are impractical while the requirement for high resolution means that single-shot techniques that have proved useful for body imaging, such as HASTE (half-Fourier acquisition single-shot turbo spin echo),[6] require gradient performances well above those that are generally available. Single-slice, fast-scanning techniques may be used to further reduce artefact by 'freezing' motion. An alternative technique is to use gated fast gradient echo sequences with segmented k-space acquisition.[7,8] At a TR of 15 ms, 32 lines or more of data can be collected per R–R interval and scan times of a few seconds are possible. Collection of the full image data can be divided over a few heartbeats. Optimization of the SNR depends on further refinements of the sequences, such as varying the flip angle through each cycle of data acquisition.[9] Preparation pulses can be used before the data collection in each cardiac cycle in order to achieve the required image contrast. In combination with this, centrally ordered phase encoding seeks to collect views from the center of k-space early in each cycle of data collection.

In the colon, cardiac motion is not an issue; however residual bowel air reduces contact between the receiver coil and the bowel wall and facilitates movement of the coil within a distended lumen. Modified gradient optimized gradient echo sequences with a TR of 8–15 ms and TE of 3.7–6.1 ms can achieve a 1–25 s acquisition. In addition, versions of this sequence may be used with an inversion prepulse of 300–700 ms to maximize contrast between bowel wall layers.

Finally navigator echoes for line-by-line quantification of motion, together with postcorrection or phase encode re-ordering techniques, have been used in, for instance, cardiac imaging.[10,11] For the application here, the limited field-of-view of the imaging coil means that navigator echoes are impractical. We have investigated exploiting similar techniques by adding an additional small receiver coil (internal diameter ~3 mm) within the tip of the endoscope, attached to a separate receiver channel and containing a small MR-visible fiducial marker (~1 mm diameter).[12] Motion of the endoscope tip can then be monitored by application of four gradient projections,[13] acquired in 60 ms, before each line of data acquisition, from which the current position of the fiducial can be measured. In situations where fast scan times are not necessary for patient tolerance, this method allows full exploitation of the coil sensitivity for high-resolution imaging within the gradient performance of clinical scanners, as well as permitting the full range of diagnostic image contrast.

5 EX VIVO IMAGING

From inner to outer, four layers are seen in normal bowel (Figure 3). First, a layer of high signal intensity on both T_1 and T_2-weighted images that represents the mucosa. Second, there is a layer of low signal intensity on T_1-weighted imaging that represents the muscularis mucosa. This layer is intermediate in signal intensity on T_2-weighted imaging and, therefore, indis-

tinguishable from the adjacent outer layer. Third, a layer of high signal intensity on T_1-weighted imaging is seen that has two counterparts on the T_2-weighted images: an inner layer of intermediate signal intensity and an outer rim of lower signal intensity. These correspond to the submucosa. Finally, a homogeneous thick outer layer has low signal intensity on both T_1- and T_2-weighted imaging; this represents the muscularis propria.

6 IN VIVO STUDIES

6.1 Upper Gastrointestinal Tract

The procedure initially follows standard endoscopy practices. Following local anesthesia to the back of the pharynx, the patient's esophagus is intubated with the coil in advance of the tip of the endoscope. After visual inspection of the upper GI tract, the coil is placed adjacent to a lesion or region of interest under direct vision and the patient positioned with the region to be imaged in the center of the magnetic field.[14]

Use of a short-bore scanner (0.5 T, 60 cm from magnet face to center) allows access to the patient's mouth. Imaging strategies described in Section 4 are then used to study normal anatomy (Figure 4), the extent of tumors (Figure 5), varices

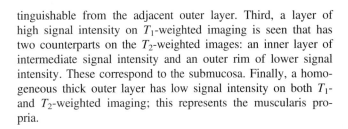

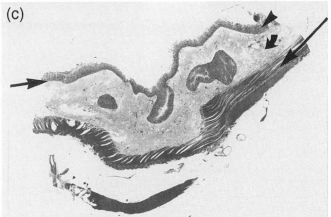

Figure 3 The normal colon ex vivo showing normal colonic layers. (a) Transverse spin echo T_1-weighted image [repetition time (*TR*) 355 ms; echo time (*TE*) 20 ms]; (b) T_2-weighted image (*TR* 2500 ms; *TE* 80 ms); (c) histologic section. The innermost ring of high signal intensity represents mucosal epithelium (arrowhead). The next thin layer of low signal intensity represents the muscularis mucosa (short arrow). The submucosa is seen as a band of high signal intensity in (a) and as an inner ring of intermediate signal intensity and an outer ring of low signal intensity in (b) (curved arrow). The muscularis propria is a thick band of low signal intensity in both (a) and (b) (long arrows). This correlates with the appearance on histologic section

For References see p. 1303

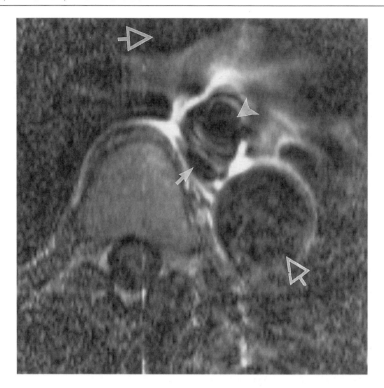

Figure 4 The normal mid-esophagus. Spin echo T_1-weighted transverse image (repetition time 660 ms; echo time 20 ms) showing normal esophagus (arrowhead), ascending and descending aorta (open arrows), and azygos vein (short arrow)

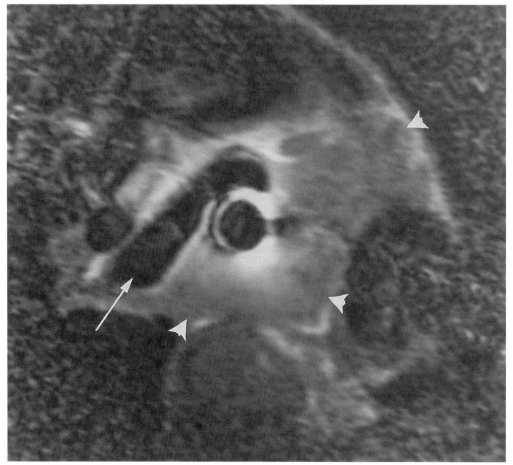

Figure 5 Carcinoma of the esophagus. Spin echo T_1-weighted transverse image (repetition time 660 ms; echo time 20 ms) through the mid-esophagus showing a large mass of intermediate signal intensity in the mediastinum (arrowheads) surrounding the esophagus. A pulmonary vessel is seen passing through it (arrow)

For list of General Abbreviations see end-papers

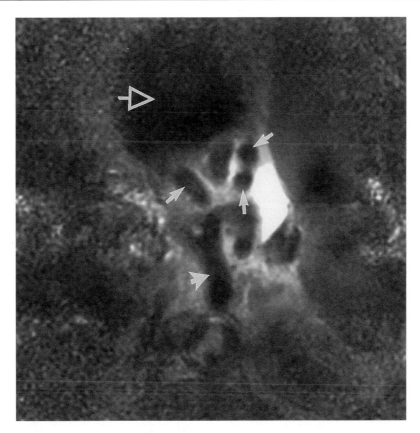

Figure 6 Gastric varices. Electrocardiograph-gated spin echo T_1-weighted transverse image (repetition time 660 ms; echo time 20 ms) at the level of the gastric fundus (open arrow) and coeliac axis (arrowhead). Numerous signal voids (arrows) at the splenic hilum represent recurrence of varices in the left gastric territory

not detectable on visual inspection (Figure 6), and post-surgical appearances at the gastroesophageal junction.

6.2 Lower Gastrointestinal Tract

Patient preparation is identical to that used for a standard colonoscopy. During visual inspection of the colon, note is taken of areas where imaging may be of interest. On completion of inspection, the receiver coil is placed adjacent to these areas and images obtained. As in the ex vivo studies, three layers of bowel wall can be identified: mucosa, submucosa, and muscularis propria (Figure 7).

7 SAFETY ISSUES

7.1 Technical Factors

An alternating magnetic field gives rise to an electric field in a plane normal to the magnetic field. This statement of Faraday's law may be made more concisely as

$$\oint \boldsymbol{E} \mathrm{d}l = \frac{-\mathrm{d}}{\mathrm{d}t} \int \boldsymbol{B} \mathrm{d}A$$

where $\boldsymbol{E}$ is the induced electric field strength along a path of length dl, t is time, and $\boldsymbol{B}$ is the inducing magnetic flux density over an area dA, all variables being vector qualities.

If we take the relatively simple case of a closed circuit path of radius r in free space in a plane normal to the magnetic flux, then $\oint \boldsymbol{E} \mathrm{d}l = \boldsymbol{E}.2\pi r$. The corresponding rate of flux

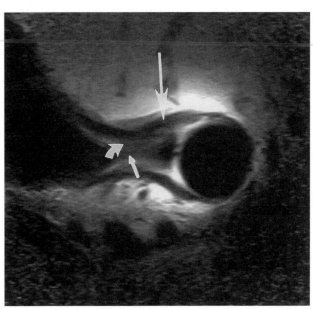

Figure 7 Normal colon. In vivo spin echo T_1-weighted transverse image (repetition time 660 ms; echo time 20 ms) through the sigmoid colon showing a trilaminar wall structure of mucosa (intermediate signal intensity; short arrow), submucosa (high signal intensity; curved arrow), and muscularis propria (low signal intensity; long arrow)

For References see p. 1303

change is:

$$-\int \frac{\mathrm{d}\boldsymbol{B}}{\mathrm{d}t}\mathrm{d}A$$

which will have magnitude

$$\left|\frac{\mathrm{d}\boldsymbol{B}}{\mathrm{d}t}\right|\pi r^2.$$

Then

$$\boldsymbol{E}.2\pi r = \frac{\mathrm{d}\boldsymbol{B}}{\mathrm{d}t}\pi r^2$$

and

$$\boldsymbol{E} = \frac{\mathrm{d}\boldsymbol{B}}{\mathrm{d}t}\frac{r}{2}$$

Consequently, the induced electric field increases linearly with the enclosed loop radius.

If part of this circular path is replaced by a conducting medium, a copper wire for example, the resulting electric field within the wire will be zero and the electric field around this loop will be concentrated into the lower conductivity portion of the loop outside the wire. Since, for a given diameter, the total loop electromotive force ($\boldsymbol{E}.2\pi r$) is fixed, increasing the length of the wire portion of the loop to create a smaller gap will increase the electric field in the gap proportionately. Should this gap be now filled with a nonmetallic medium capable of supporting electric currents (e.g., human tissue), this relatively high electric field could give rise to currents sufficiently high to produce significant tissue heating and a potential safety hazard.

Such a situation can exist within a MR scanner if suitable precautions are not taken. The $\boldsymbol{B}_1$ magnetic field, alternating at the Larmor frequency, typically irradiates the whole body. Under normal operating conditions, the induced electric fields and their associated currents are small enough that the resultant tissue heating is well within safety guidelines. However, the introduction of a conductor, for example an electrocardiograph lead positioned so as to form a large loop, can result in a large electric field between the ends of the loop. If this high-field region is adjacent to the body surface, the local electric current can produce tissue heating that far exceeds the safety guidelines, resulting, in extreme cases, in localized tissue burning. This conducting-loop problem could equally apply to a colonoscope, for example as it performs loops within the abdomen. This hazard is minimized by providing a sufficiently thick layer of insulating material of high dielectric strength between conductive material within the colonoscope and the external tissue, so ensuring that electric fields are confined largely within these layers and that their levels are kept within reasonable limits.

The coil is coated in an epoxy-based adhesive commonly used in medical devices. It is biologically inert and nontoxic, meeting US pharmacopeia biocompatibility standards, thus avoiding allergic reactions and long-term toxic effects. In addition, it is chemical and water resistant, which allows

For list of General Abbreviations see end-papers

sterilization of the coil by soaking it in a solution of glutaraldehyde.

7.2 Clinical Factors

Upper GI endoscopy carries a significant morbidity and mortality.[15] In a large prospective audit across 36 hospitals to include over 14 000 examinations, the death rate was 1 in 2000 and the morbidity rate 1 in 200; cardiorespiratory problems were most prominent and there was a strong relationship between the lack of monitoring and the use of high-dose benzodiazepines and adverse outcomes. A lack of recovery area and poor staffing exacerbates this problem. No complications have been experienced in our unit so far, but as the procedure is considerably longer than a standard upper GI endoscopy and often requires an additional dose of sedative, an increase in the morbidity and mortality must be anticipated. Also although the patient is accessible, this is less than achieved in a standard procedure and all monitoring requires increased vigilance. Continuous oxygen and close observation of the sedated patient during the entire period of imaging must be maintained.

Performing endoscopy in the MR environment, therefore, requires additional considerations. The MR suite must be equipped with full resuscitation facilities nearby. This includes oxygen available to the patient in the scanner and a MR-compatible trolley to transfer a patient in an emergency to a space outside the scanner room where normal resuscitation procedure may be carried out. Finally, all staff must be trained in metal safety and in resuscitation practice peculiar to the MR environment.

8 FUTURE APPLICATIONS

MR endoscopy will provide a useful means of staging neoplastic masses in the upper and lower GI tract. It may also be used as a way of monitoring or guiding therapeutic interventions administered endoscopically. For example, injection or banding esophageal varices may benefit from being performed with simultaneous cross-sectional images so that the completeness and efficacy of the procedure can be monitored in realtime. Laser treatments for the palliative recanalization of esophageal carcinoma may also be monitored in this way, thus reducing the risks of perforation or inadequate treatment and optimizing treatment outcome.

9 CONCLUDING REMARKS

MR imaging using a surface coil placed within the GI tract during an endoscopic procedure enables visualization of mural and extramural lesion extension and may provide a useful adjunct to diagnostic upper and lower GI endoscopy. Formal evaluation of this technique and comparison with ultrasound, computed tomography, and surgical findings is warranted.

10 RELATED ARTICLES

Coils for Insertion into the Human Body; Design and Use of Internal Receiver Coils for MRI.

11 REFERENCES

1. T. W. Rice, G. A. Boyce, and M. V. Sivak, *J. Thorac. Cardiovasc. Surg.*, 1991, **101**, 536.
2. M. D. Rifkin, S. J. Gordon, and B. B. Goldberg, *Radiology*, 1984, **151**, 175.
3. B. B. Goldberg, R. M. Steiner, J. B. Liu, D. A. Merton, G. Articolo, J. R. Cohn, J. Gottlieb, B. L. McCombe, and P. W. Spirn, *Radiology*, 1994, **190**, 233.
4. B. N. Milestone, M. D. Schnall, R. E. Lenkinski, and H. Y. Kressel, *Radiology*, 1991, **180**, 91.
5. N. M. deSouza, I. C. Hawley, J. E. Schwieso, D. J. Gilderdale, and W. P. Soutter, *Am. J. Roentgenol.*, 1994, **163**, 607.
6. T. Miyazaki, Y. Yamashita, T. Tsuchigame, H. Yamamoto, J. Urata, and M. Takahashi, *Am. J. Roentgenol.*, 1996, **166**, 1297.
7. A. Haase, J. Frahm, and K. D. Matthaei, *J. Magn. Reson.*, 1986, **67**, 258.
8. A. Haase, *Magn. Reson. Med.*, 1990, **12**, 77.
9. M. K. Stehling, *Magn. Reson. Imag.*, 1992, **10**, 165.
10. Y. Wang, R. C. Grimm, J. P. Felmlee, S. J. Riederer, and R. L. Ehman, *Magn. Reson. Med.*, 1996, **36**, 117.
11. A. M. Taylor, P. Jhooti, D. N. Firmin, and D. J. Pennell, *JMRI*, 1999, **9**, 395.
12. G. A. Coutts, D. J. Gilderdale, K. M. Chui, L. Kasuboski, and N. M. deSouza, *Magn. Reson. Med.*, 1998, **40**, 908.
13. C. L. Dumoulin, S. P. Souza, and R. D. Darrow, *Magn. Reson. Med.*, 1993, **29**, 411.
14. N. M. deSouza, A. H. Gibbons, G. A. Coutts, A. S. Hall, R. Puni, J. Calam, and I. R. Young, *Minim. Invas. Ther.*, 1995, **4**, 277.
15. A. M. Thompson, K. G. Park, F. Kerr, and A. Munro, *Br. J. Surg.*, 1992, **79**, 1046.

Acknowledgements

We gratefully acknowledge the help we received in undertaking the work described here from Picker International Inc., Cleveland, OH, and Endoscan Ltd, Burgess Hill, UK.

The work was supported by MedLINK grant P37.

Biographical Sketches

Nandita M. deSouza. *b* 1957. B.Sc. Hons (Physiology) 1978, MBBS, 1983, University of Newcastle upon Tyne, UK, M.R.C.P. 1986, F.R.C.P. 1991, M.D. 1996. House Surgeon and Physician and Senior House Physician, Newcastle upon Tyne 1983–85. Registrar in internal medicine 1986–87. Registrar and Senior Registrar in Diagnostic Radiology, Guy's Hospital, London 1987–92. Senior Research Fellow in MR Imaging, Royal Postgraduate Medical School, Hammersmith Hospital, 1993–97. Senior Lecturer 1997–present. Research Interests: internal receiver coils, interventional MRI.

Glyn A. Coutts. *b* 1959. B.A. (Physics) University of Oxford, UK, 1980; D.Phil, University of Oxford 1985. Senior Research Associate at Picker Research UK, 1987–99. Marconi Medical Systems, UK, 1999–present. Approximately 30 publications in the field of MRI and MRS. Current interest: interventional MRI.

David J. Larkman. *b* 1969. 1993, Ph.D. Physics, Manchester University, 1997, BSc (Hons) Physics, Liverpool John Moores University, Research Scientist at the Robert Steiner MRI Unit, Hammersmith Hospital, London, UK, 1997–99. Marconi Medical Systems, UK, 1999–present. Publications as conference and journal papers on small-coil MR, fast imaging using multicoil techniques, and artefact control in MR images.

David J. Gilderdale. *b* 1947. B.Sc. telecommunications, Ph.D. computer simulation as an aid to lighting design. Research Assistant, Plymouth Polytechnic, 1974–78. Electronic Engineer, Central Research Laboratories, Thorn-EMI Ltd, 1978–81. Senior Scientist, GEC Hirst Research Centre, 1981–84. Senior Lecturer, Department of Electrical Engineering, Plymouth Polytechnic, 1981–88. Engineering Consultant, 1988–present.

For References see p. 1303

Part Fourteen
MR of the Musculo-Skeletal System

Imaging and Spectroscopy of Muscle

Chris Boesch and Roland Kreis
University of Bern, Switzerland

1 INTRODUCTION

Examinations of extremities and skeletal muscle have been among the first applications of NMR in the human body. At the advent of in vivo NMR, magnets were available with sufficiently high field strength for magnetic resonance spectroscopy (MRS) but their limited bore size allowed only small animals or human extremities to be examined. The development of surface coils (see *Spatial Localization Techniques for Human MRS*) allowed localized acquisition of NMR signals—a technique that was mainly used for ^{31}P MRS and to a lesser extent for ^{13}C MRS. For a while, classical ^{31}P MRS experiments (see *Peripheral Muscle Metabolism Studied by MRS*, *Tissue Behavior Measurements Using Phosphorus-31 NMR*) were the major application of MRS and led to a substantial contribution to knowledge in muscle physiology. In the following years, the development of whole-body magnets introduced NMR applications to clinical medicine (see *Whole Body Studies: Impact of MRS*). MRI of extremities, muscle tissue, and joints became a routine diagnostic tool (see *ESR Probes as Field Detectors in MRI*, *MRI of Musculoskeletal Neoplasms*, *Skeletal Muscle Evaluated by MRI*). The combination of MRI and MRS in whole-body magnets led not only to improved anatomical localization defined by gradients but also to better description of the sensitive volume by MRI in the same magnet. The wider bore of these systems allowed for examinations of larger muscle groups, for example those of the thigh. Gradient-based volume selection and improved localization sequences particularly promoted the development of ^{1}H MRS methods.

The number of NMR applications in perfused muscles and animals is huge but will not be covered in this chapter, which focuses on human skeletal muscle. At present, publications on ^{31}P MRS in small-bore and whole-body systems still represent a considerable portion of the physiological applications of MR (see also *Peripheral Muscle Metabolism Studied by MRS*). An increasing number of new applications, however, combine and use other MR techniques and other nuclei. Some of these applications will be described in this chapter.

2 STRUCTURAL ORDER AND COMPARTMENTATION IN MUSCLE TISSUE

Skeletal muscle is a biological structure that is highly organized at different spatial levels. (a) Skeletal muscles are composed of various types of muscle fiber with inherently different composition and metabolism. (b) Bulk fat along fasciae around muscles forms macroscopic plates that run almost parallel with the axes of extremities. (c) Muscle fibers may extend parallel to the muscle ('fusiform' arrangement) or with a certain 'pennation' angle between muscle and fibers ('unipennate' with one major direction or 'bipennate' if two different types of fiber orientation exist that are attached to one tendon). (d) Myofibrils form the well-known striated structures (sarcomeres) including actin and myosin molecules, which are spatially organized on the microscopic level. (e) Membranes and cell organelles such as mitochondria separate a specific metabolite from the same metabolite in the cytoplasm, leading to different pools and transport processes.

Structural order and compartmentation in biological tissue can influence MR signals such that measurable parameters or artifacts are generated, depending on the way the experiment and observer are using or neglecting that information. Several effects, which will be discussed below, can be attributed to structure or compartmentation.

1. the relaxation times of muscle, tendons, and cartilage, depend on fiber orientation and type;[1–3]

2. the muscle- or subject-specific metabolite content, energy kinetics, and recruitment pattern based on distributions of fiber types;[3–7]

3. the isotropic and ordered compartments, as distinguished by [23]Na double quantum filtered (DQF) MR spectra;[8,9]

4. susceptibility effects, which are different for extramyocellular (bulk) (EMCL) fat and for intramyocellular lipids (spherical droplets) (IMCL);[10,11]

5. magnetization transfer (MT) of the invisible fixed pool of molecules to the moving and visible pool;[1]

6. restricted diffusion measurable by specialized MRI and MRS sequences;[12–16]

7. the dipolar coupling effects of creatine (Cr) and/or phosphocreatine (PCr) resonances owing to anisotropic motional averaging;[17,18]

8. differences in the NMR visibility of Cr/PCr and other metabolites as a result of compartmentation and/or limited motional freedom.[19–21]

MR is a unique tool with which to investigate order and compartmentation in biological tissues. Even if these are not the major target of a specific examination, their effects should always be kept in mind when interpreting images and spectra.

3 USE OF MRI FOR HUMAN MUSCLE PHYSIOLOGY

3.1 MRI for Volume Localization

The combination of MRI and MRS improves the anatomical localization of spectra considerably. MRS examinations in small-bore systems often have to rely on palpation and approximate positioning of the surface coil. With appropriate imaging sequences, it was possible to show that palpation incorrectly identified flexor muscle margins by more than 15 mm in 50% of attempts.[22] In principle, MRI makes it possible to place the region of interest with an accuracy of about 1 mm. Even if images are acquired prior to MRS, it may not be feasible to place the sensitive volume of a surface coil with the same precision. However, the subsequent use of gradient-based localization schemes and a proper fixation of the examined extremity allow for a volume selection within adequate

For References see p. 1315

accuracy. Since the fiber orientation, composition, and activation are muscle specific and can vary even within one muscle,[23] combined MRI/MRS systems are, therefore, necessary to obtain MR parameters from one specific muscle type, as shown in several examples below.

3.2   Activity-dependent MRI

The amount and distribution of water in skeletal muscle is changed by muscular activity.[2,23–27] Subsequent variations in relaxation times make MR signal intensity highly sensitive to changes in the water distribution, oxygenation, pH, and other factors. Fleckenstein et al.[28] suggested imaging parameters with which it was possible to detect and document muscular activity (see also *Skeletal Muscle Evaluated by MRI*). This technique can be used to study intersubject variations of normal anatomy and muscular activation. Ergometers to be used in an MR system have to be redesigned to fit into the restricted space of the patient bore. The technique described can help to document the activity imposed by such an ergometer. While it is clear that activity-dependent MRI is a valuable and practical tool to investigate muscular activity, it is not so clear which changes in MR parameters lead to this effect. It seems that water inflow and pH changes contribute only partially to the observed increase in T_2 relaxation.[25,27]

3.3   Muscle Volume Determination using MRI

Muscle mass is a variable parameter and is dependent on training, ageing, temporary immobilization, disease, and other factors. Knowing muscle mass is, therefore, crucial for an estimation of training effects and also for calculations of total body content of metabolites and components of the muscular cell. Many localized observations of physiological parameters, such as biopsies, need additional data on the absolute volume of the muscle, since changes in concentration could be a result of increased muscle mass (i.e. simple dilution with unchanged total content) or an effective increase in total content.

It is obvious that MRI with its excellent soft tissue contrast is predestined to be used for in vivo morphometric measurements.[4,29–33] However, all approaches have specific advantages and disadvantages, for example three-dimensional versus multislice techniques, nonlinearity of the spatial representation, different ways of image analysis, and partial volume effects.

The image acquisition can be done either by three-dimensional sequences or using sequential two-dimensional data (i.e. multislice techniques) (Figure 1). As long as discrete slices are evaluated, the so-called Cavalieri principle needs to be respected.[31] It states that the starting point of a series of slices should be arbitrarily set such that every position has equal probability. This avoids systematic over- or underestimation of a volume through specific sampling of the structure.

In principle, images can be measured in one acquisition as long as the muscle does not exceed the homogeneous volume of the magnet, which is typically about 50 cm in diameter. However, the linearity of the gradients within the homogeneous volume and subsequently the accuracy of the mapping is significantly reduced in the outer portions of the volume because of the nonlinearity of the gradients. Therefore, imaging of

larger muscles may be more appropriate in multiple steps. The imperfect spatial accuracy of MR systems is usually improved by retrospective image-correction algorithms, which use the known spatial characteristics of the gradients. This procedure is acceptable for diagnostic images where the relative positions are much more important than the absolute dimensions. For quantitative measurements, however, a continuous monitoring of these effects is necessary, especially if variations over longer time periods are investigated, such as training effects on muscular volume.

The cross-sectional area and actual volume can be determined by several methods:[29,31–33] simple threshold, voxel counting within operator-defined borders, elaborated tissue segmentation algorithms, or point counting. Simple threshold methods need a homogeneous rf sensitivity of the coil since substantial inhomogeneities in signal intensity will lead to wrong assignment of voxels. Computed tissue segmentation, often based on different MRI scans, uses several steps to distinguish between tissues, in many cases with human interaction. Image intensity threshold can be followed by boundary tracing, edge detection filters, morphological erosion, seed-growing algorithms,[29,30] etc. Point counting[31] is an interactive method very popular in morphometry, for example to analyze electron micrographs. Grids are placed over the region of interest (Figure 1) and crosses that hit the structure of interest are counted. The probability for a hit is proportional to the area covered by the structure. If large enough numbers are counted (typically over 100 hits per structure), variations resulting from the counting procedure are negligible and the number of hits represents the desired area almost perfectly. Point counting is a very robust method but is very time consuming and, to a certain extent, operator dependent.

All digitization methods are somewhat susceptible to partial volume effects: the misinterpretation of voxels that contain more than one type of tissue. If a T_1-weighted MR sequence is used that leads to a strong signal of fat, voxels that are only half filled with adipose tissue will produce sufficient signal to be counted as fat. Since this happens in all voxels at the border of fat and muscle tissue, fat content will be systematically overestimated. Optimized imaging sequences and appropriate windowing of the signal intensity may help to reduce this problem.

3.4   Use of MRI for Determination of Muscle Fiber Orientation and Composition

In addition to assessment of gross morphology, histological characterization of muscle tissue by means of MR is desirable. So far, fiber orientations in muscle and the 'pennation angle' between muscle fibers and axis of the muscle have been measured by ultrasound. It has been proposed that MR images would show sufficient morphological detail to see striations generated by fat, which runs parallel and between the muscle fascicles, and that this would allow for an assessment of the fascicle pennation.[4,33] One interesting finding of such a three-dimensional analysis was a considerable variation of pennation angles within one muscle;[4] for example a range of 5 to 50° was observed in the vastus medialis. This noninvasively obtained information helps to evaluate muscle mechanics. Further suggestions have been made to use the effect of

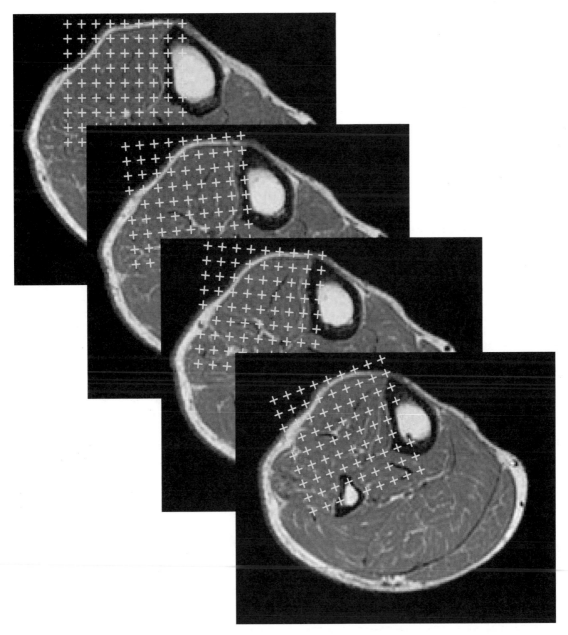

Figure 1 Selected axial slices of the lower leg with an illustration of morphometry based on the point counting method. A grid is placed arbitrarily over the anatomical structure to be measured—here the tibialis anterior muscle—and points lying within the structure of interest are counted. The number of counted points is proportional to the volume of the structure when a series of slices is evaluated. The error introduced by the placement of the grid and the counting becomes negligible if more than 100 points are counted overall

anisotropic diffusion[1,14,16] within heart or skeletal muscle tissue to visualize fiber orientation in MR images.[13,15]

So far, fiber composition has mainly been assessed by ^{31}P MRS using differences in PCr content and pH (see *Peripheral Muscle Metabolism Studied by MRS*). The method requires relatively large sensitive volumes and, therefore, it is inherently susceptible to partial volume effects: a mixture of signals from different muscles that are partially activated and resting. Activity-dependent MR images and gradient-based volume selection may improve these observations in future.

The determination of fiber composition by MRI is based on differences in relaxation times of water.[3] Using the effect of gadolinium diethylenetriamine pentaacetic acid (DTPA) on MR

images of rabbit muscle, larger extracellular space has been observed in slow-twitch (red) muscle (see *Skeletal Muscle Evaluated by MRI*).[7]

4 USE OF MRS FOR HUMAN MUSCLE PHYSIOLOGY

4.1 The Use of Different Nuclei to Observe Muscular Metabolism

The high-energy phosphates visible in the ^{31}P MR spectrum and the pH titration of the inorganic phosphate (Pi) relative to

For References see p. 1315

PCr made ^{31}P MRS the favorite tool for biochemists and physiologists interested in muscular metabolism (see *Peripheral Muscle Metabolism Studied by MRS*).

The application of ^{13}C MRS for examinations of muscular metabolism attracted the attention of physiologists because glycogen is 100% visible in the ^{13}C MR spectrum despite the very large molecular weight of this molecule. However, the technical requirements for ^{1}H-decoupled ^{13}C MRS are considerably higher than those usually provided in a standard clinical scanner.

Until recently, there had only been some exploratory studies using ^{1}H MRS. This is particularly remarkable since ^{1}H is the most frequently used nucleus in brain examinations. Originally, the reasons for this obvious disinterest were of a technical nature (water suppression, eddy currents for short echo times), but later the sparse attention to ^{1}H MRS of muscle may have resulted from a general feeling that its information content would be rather low. It will be shown below that this impression was wrong.

Other nuclei such as [23]Na have only rarely been used to study muscle metabolism.[8,9] However, we expect that studies of order on the molecular level will benefit from the potential of less popular nuclei in the future.

Some very elegant examinations of muscular metabolism use combinations of different nuclei, for example ^{13}C with ^{31}P MRS:[34–37] the ^{13}C studies follow the depletion and recovery of glycogen while ^{31}P is used to monitor phosphorylation. Combinations of different nuclei and the fusion of MRS and MRI in the same session will be vital approaches in the future development of MR of human muscle metabolism and will prove the enormous versatility of MR.

4.2 Classical ^{31}P MRS of Human Muscle

Many of the classical studies of human muscle using ^{31}P MRS are covered in other monographs in these volumes and these provide a wide range of references to work in this area (see *Peripheral Muscle Metabolism Studied by MRS, Phosphorus-31 Magnetization Transfer Studies In Vivo, Proton Decoupling During In Vivo Whole Body Phosphorus MRS, Single Voxel Whole Body Phosphorus MRS, Tissue Behavior Measurements Using Phosphorus-31 NMR, Whole Body Studies: Impact of MRS, pH Measurement In Vivo in Whole Body Systems*). The first report on a muscular disease documented by ^{31}P MRS—McArdle disease—promoted the application of MRS in vivo. However, widespread distribution of ^{31}P MRS in clinical routine and the use of this modality in all-day diagnostics did not occur as expected and it remained an excellent, but somewhat specialized, tool for research in physiology and pathology.

Beside the well-known observation of high-energy phosphates by ^{31}P MRS and the determination of intracellular pH, MT experiments have also examined the creatine kinase reaction (see also *Phosphorus-31 Magnetization Transfer Studies In Vivo*). The very recent development of genetically manipulated mice (e.g. creatine kinase knock-out mice) may help in the understanding of many aspects of ^{31}P MRS.

Proton decoupling during acquisition of phosphorus spectra (see *Proton Decoupling During In Vivo Whole Body Phosphorus MRS*) allows signals such as those from glycerophosphorylcholine and glycerophosphorylethanolamine

to be distinguished, which may help to improve understanding of membrane metabolism.

Diffusion-weighted ^{31}P MRS of PCr and Pi—together with MT techniques—may provide more information on the creatine kinase reaction and the role and nature of Cr/PCr as energy shuttle, reservoir, or buffer in the myocyte.

4.3 Application of ^{13}C MRS in Human Muscle

The main application of ^{13}C MRS in human muscle has been the observation of glycogen depletion and repletion, with and without labeling of blood glucose at the C1 position by ^{13}C, and in some experiments controlled by clamp techniques.[38–40] Specific labeling of other compounds and subsequent isotopomer analysis has been very successful in whole animals and isolated organs but has hardly been used in humans. Despite the fact that ^{13}C MRS is extremely useful and elegant, only a few groups worldwide are using this method in studies of human metabolism. This can be explained by the technical requirements (broadband acquisition, ^{1}H decoupling) and by the low signal-to-noise ratio (SNR), which would benefit from field strengths higher than those available in routine scanners. Studies with substrates labeled with ^{13}C are expensive and clamp studies require additional experience. The increasing number of MR systems at field strengths of 3–4.7 T will hopefully overcome the technical limitations and prove the value of ^{13}C MRS studies.

Four main factors characterize ^{13}C MRS: (a) the low gyromagnetic ratio, which is about 25% of that of ^{1}H; (b) the low natural abundance of ^{13}C (most carbons are NMR invisible in the form of ^{12}C); (c) the large chemical shift dispersion of about 250 ppm compared with about 10 ppm for ^{1}H; and (d) the direct chemical bonds to ^{1}H of most ^{13}C atoms. All these factors have methodological consequences. The low gyromagnetic ratio leads not only to a lower resonance frequency but also to an inherently low sensitivity of ^{13}C, which is about 0.016 compared with [1]H. The natural abundance of 1.1% for ^{13}C further reduces signal intensity. However, the low natural abundance does not only have negative consequences; it also allows for the use of ^{13}C-labeled substrates as tracers, which is obviously not feasible for ^{1}H or [31]P. The large chemical shift dispersion per se is advantageous since the separation of different resonances increases. This makes an unspoiled observation of a specific metabolite much easier than in the much more crowded ^{1}H MR spectrum. However, the large chemical shift dispersion leads to a spatial misregistration that may be intolerable when popular gradient-based volume localization methods are used directly. The fact that gradients have not been widely used for localization of ^{13}C MR spectra is also a consequence of the very short T_2 of glycogen; the signals would have decayed at echo times that are typically used in point-resolved spectroscopy (PRESS) or stimulated echo acquisition (STEAM) sequences. (For additional information on localization sequences see *Spatial Localization Techniques for Human MRS*). The use of image-selected in vivo spectroscopy (ISIS) overcame the problem of fast transverse relaxation since the magnetization is kept in the longitudinal direction during the localization procedure.[41] Localization of ^{13}C nuclei can also be accomplished indirectly via the coupled ^{1}H nuclei and subsequent polarization transfer.[42–46] Because of very short [1]H T_2, these methods are not useful for localization of glycogen. Most

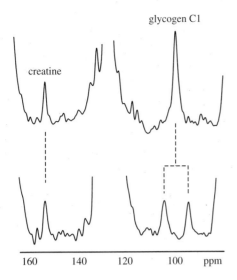

Figure 2 Carbon-13 MR spectrum of the thigh at 1.5 T (GE SIGNA, Milwaukee, WI), without (lower trace) and with (upper trace) continuous wave decoupling of the protons (decoupler S.M.I.S., UK). The decoupling and NOE build up of the doublet of glycogen C1 leads to an improved signal-to-noise ratio that allows a reduction of measurement time (lower trace 10 000 scans, upper trace 4000 scans)

applications, therefore, apply surface coils for localized transmission and reception of the signals, using pulse-and-acquire sequences. Nevertheless, volume localization of ^{13}C MR spectra is often a critical issue for several reasons: (a) adjacent muscle groups may have differences in metabolite levels but usually contribute to an overall signal; (b) huge signals from subcutaneous fat may mask the tiny resonances from metabolites (e.g., signal contributions from muscular glycogen in the abdominal wall may not be separable from liver glycogen); (c) sensitivity profiles of surface coils are complicated and make absolute quantification much more difficult than methods with a rectangular shape of the selected region. Some methods use bottles filled with known solutions that replace the human body to calibrate the metabolite levels;[47] others use Cr as an internal standard.[48] The chemical bonds between carbon atoms and adjacent ^{1}H lead to substantial heteronuclear spin–spin coupling and to splitting of the resonance lines, which further reduce the amplitudes. With ^{1}H decoupling by irradiation on a second rf channel, it is possible to cancel the coupling effect, restoring the full resonance amplitude (Figure 2). In addition, irradiation on the ^{1}H frequency prior to acquisition leads to the so-called nuclear Overhauser enhancement (NOE), which results in an additional 50–90% in signal intensity[49] under in vivo conditions (Figure 2).

Glycogen is a large molecule with a molecular weight of 10^7 to 10^9. Since macromolecules are usually not visible by MRS because of their very short T_2 relaxation, it was surprising that glycogen turned out to be 100% visible.[50,51] It seems that the internal molecular mobility leads to effective decoupling from fast relaxation processes. The observation of glycogen in human muscle (and liver) is meanwhile one of the 'workhorse' applications of in vivo MRS. NOE effect and short repetition times guarantee a reasonable SNR within approximately 10 min even at 1.5 T. Higher field strengths, however, are desirable to increase SNR and/or temporal resolution.

Studies of noninsulin-dependent diabetes mellitus (NIDDM) patients[36] showed a slower glycogen repletion compared with healthy volunteers. Combination with ^{31}P MRS was then used to identify the limiting step in glycogen synthesis more closely (see below). Patients suffering from glycogen storage diseases[48,52] were found to have higher levels of glycogen in muscle and liver. Studies on normal subjects[37,43,47,53,54] revealed important physiological information on glycogen as an energy reserve for the human muscle, for example changing with different diets, with different exercise intensity, and as a function of the insulin-dependent and independent control mechanisms.

Studies of lipid metabolism using ^{13}C MRS would be very attractive since the degree of saturation and chain length could be evaluated.[55–57] However, the difficulties in separating signals from subcutaneous adipose tissue and muscular lipids (IMCL, see below) restrict this method to studies of bulk fat so far.

Since ^{13}C has a very low natural abundance, substrates can be enriched at specific positions in the molecule. This can either be used to distinguish between 'old' and 'new' fractions of a metabolite pool, e.g., to identify glycogen that is newly built in the muscle, or it can be used to distinguish between several potential biochemical pathways. One of the most popular applications is labeling of glucose at the C1 position and the follow-up by ^{13}C MRS during its incorporation into muscular glycogen in clamp studies.[36,50]

Most ^{13}C labeling experiments with subsequent isotopomer analysis (i.e., analysis of the coupling patterns of mixtures of the same compound with different degrees of labeling), have been done in animals, perfused organs, or plasma samples.[58–63] An example in skeletal muscle is the observation of [1-^{13}C]-glucose incorporation into intramuscular [1-^{13}C]-glycogen, [3-^{13}C]-lactate, and [3-^{13}C]-alanine.[59] Carbon-13 MRS requires much higher enrichment of the ^{13}C isotope than does mass spectrometry; it does, however, have the advantage of organ selectivity (e.g., compared with analysis of breathing air) and chemical specificity.

4.4 Application of ^{1}H MRS in Human Muscle

4.4.1 General Features

Proton MR spectra of skeletal muscle were believed to be swamped by the large signal from fat, which would render all other metabolites invisible. Recent work has shown that the spectrum is indeed very complex, but it is rich in information and reflects several aspects of the physiology of exercise.

A typical ^{1}H MR spectrum of human muscle tissue acquired by a PRESS sequence (see *Spatial Localization Techniques for Human MRS*) is shown in Figure 3. Based on high-resolution spectra of muscle extracts and in vivo rat and frog muscle, the peaks visible in spectra of human muscle in vivo have been assigned early on[5,64–68] to the following compounds: lipids (methyl protons 0.9 ppm, aliphatic chain methylene protons 1.3 ppm, aliphatic protons near double bonds and carboxyl group at 1.7–2.5 ppm and protons of unsaturated carbons at 5.4 ppm); Cr/PCr (methyl at 3.03 ppm (Cr3), methylene at 3.93 ppm (Cr2)); trimethylammonium group-containing metabolites (TMA), such as choline, phosphocholine, glycerophosphocholine, and carnitine at 3.2 ppm; a line that has been tentatively assigned to taurine at approximately 3.4 ppm, depending on the orientation of the muscle; and two his-

For References see p. 1315

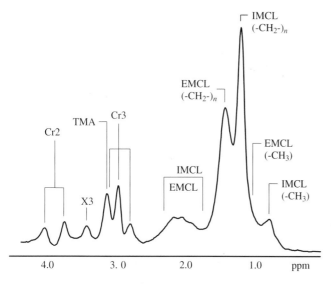

Figure 3 PRESS-localized ¹H MR spectrum of the human tibialis anterior muscle. The orientation of the fibers in this muscle parallel to the magnetic field leads to the separation of the signals from intra- (IMCL) and extramyocellular lipids (EMCL), and the splitting of the creatine (methylene Cr2, methyl Cr3) resonances. Trimethylammonium (TMA) covers one part of the Cr3 triplet and the coupling partner of resonance X3, which can be tentatively assigned to taurine. Acetylcarnitine at 2.13 ppm is not visible at rest; while lactate at 1.3 ppm is only visible with editing techniques and during workload/ischemia

tidine protons of the dipeptide carnosine at 7.0 and 8.0 ppm (anserine in some species). A peak at the unusual position of 78 ppm (Figure 4), which is invisible under the experimental conditions used for Figure 3, could be detected and assigned to deoxymyoglobin.[69,70] Lactate at 1.3 ppm can also be observed only under specific experimental conditions.[6,21,68,71–73]

As seen above, skeletal muscle is highly organized at different spatial levels and many aspects of order influence ¹H MR spectra. These aspects are summarized in a previous paragraph and for the different metabolites separately below.

4.4.2 Lactate

The signal of the methyl group of lactate is overlaid by the much larger signals from lipids in human muscle. However, using the hetero- or homonuclear spin–spin interaction and editing techniques by inversion and decoupling,[68] zero- or double-quantum filters[6,21,71–73] are able to observe the lactate signal without the huge overlapping resonances. Promoted by the physiological importance of this metabolite, observations of lactate have been among the first applications of ¹H MRS in human muscle.

4.4.3 pH

A second important parameter for the description of muscular metabolism is the intracellular pH, usually determined by

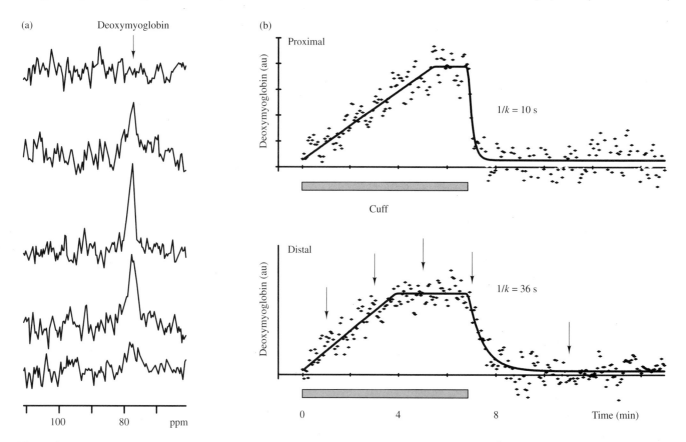

Figure 4 (a) Spectra showing the resonance of deoxymyoglobin from the lower leg in a patient with peripheral arterial disease. (b) Signal development during ischemia and recovery proximal and distal to the lesion (au, arbitary units). Arrows indicate time points when the spectra shown on the left were acquired. The time resolution in the graphs is 3.5 s, one spectrum represents a 16 s acquisition (30 Hz apodization). The time constant for recovery is 10 s proximal and 36 s distal from the lesion (unpublished data; clinical support by Dr I. Baumgartner is greatly appreciated)

the frequency difference between Pi and PCr in the ^{31}P MR spectrum. Proton MRS provides the same information, since the aromatic signals of histidine in the carnosine molecule are titrating and can, therefore, be used to determine intracellular pH. A comparison between titration observed in ^{31}P and ^{1}H MR spectra[68] showed the accuracy of ^{1}H MRS. Because of the fairly low concentrations of carnosine, the ^{1}H MRS measurement may be less sensitive than ^{31}P MRS during exercise. However, ^{1}H MRS has the advantage of identical sensitivity independent of exercise scheme, while the undershoot of Pi after exercise and the low Pi content in resting muscle can make pH measurements with ^{31}P MRS impossible in some instances.

4.4.4 Myoglobin

Proton ^{1}H MRS can also be used during exercise or ischemia to estimate the degree of intracellular oxygenation by measuring deoxymyoglobin resonances. The signals from oxymyoglobin are in the normal frequency range of an ^{1}H MR spectrum and, therefore, are overlapped by many other resonances at 1.5 T. Only at 7 T, the resonance of the γ-methyl group of Val-E11 at -2.8 ppm can be observed separately under oxygenated conditions.[74] However, while this resonance disappears with deoxygenation, the F8 proximal histidyl N-δ proton shifts to 78 ppm,[69,70,75] and becomes visible also at moderate field strength. This histidine resonance has dramatically shortened T_2 relaxation times and is, therefore, very broad, but the chemical shift separation from the large resonances between 0 and 10 ppm is sufficient to detect it reliably. Figure 4 shows the increase of deoxymyoglobin during ischemia in human skeletal muscle, observed at 1.5 T.

4.4.5 Acetylcarnitine

After heavy workload, a peak at 2.13 ppm can be detected in localized spectra of human muscle,[76] while spectra from resting muscle do not feature a sharp singlet at this position in general. Based on its chemical shift, its singlet nature, its approximate concentration, and physiological behavior, the peak was tentatively assigned to the acetyl group of acetylcarnitine.[76] Carnitine has long been known for its vital role in the transportation of fatty acids into mitochondria for β-oxidation. More recently, it was recognized that carnitine is equally important as a buffering system for a potential surplus of acetyl groups.[77] The export of acetyl groups in the form of acetylcarnitine out of the mitochondria helps to keep the acetyl CoA/coenzyme A ratio balanced such that the nonacetylated form of the co-factor can play its roles in the TCA cycle and in pyruvate dehydrogenation.

4.4.6 Creatine

Considering the concentration and chemical shift differences of Cr and PCr in muscle tissue, one could expect that the resulting signals would be well above the noise level and would represent the sum of both, i.e., total Cr, since these species would not be separable at 1.5 T. A detailed analysis of Cr in ^{1}H MR spectra, however, showed several unexpected features[17–20,78] such as an orientation-dependent dipolar splitting for several compounds that were not assigned at the very beginning.[17] In a subsequent Cr-loading study,[78] one of these metabolites has been identified as Cr and/or PCr. The methyl-

ene group of Cr/PCr appears as a prominent dipolar doublet and the methyl group as a less well-defined triplet (see Figure 5). This finding was unexpected since in vivo MRS had been seen as spectroscopy of metabolite solutions where dipolar coupling is averaged out and dipolar effects are restricted to relaxation processes. Using more elaborate methods such as one-dimensional zero- and double-quantum filtering, two-dimensional J-resolved spectroscopy, two-dimensional constant time COSY, and longitudinal order separation spectroscopy,[18] it was possible to prove unequivocally that peaks of Cr methylene result from a pair of dipolar coupled protons.[78] Most other resonance peaks, including the methyl group of Cr, were also found to be—at least partially—dependent on the angle between muscle fibers and magnetic field. The form of the orientation dependence and the approximate size of the coupling have been confirmed at a higher field in rat muscle.[19] One can now speculate about the mechanisms behind the observed effects. Irrespective of the chemical nature of the metabolites involved, there are three main explanations to account for incomplete temporal averaging of dipolar couplings. First of all, the observed molecules may be hindered from isotropic tumbling by being constrained into small elongated spaces between the actin/myosin chains. Second, they might be temporarily bound to macromolecules that themselves are strictly ordered within the muscle cells. Finally, dipolar-coupled peaks might originate from large molecules that are permanently bound to ordered structures in muscle but have enough side-chain mobility to be observable and partially average dipolar couplings. The most likely explanation of the three is that PCr and/or free Cr together with their hydration spheres are large enough to be hindered from isotropic tumbling in the elongated

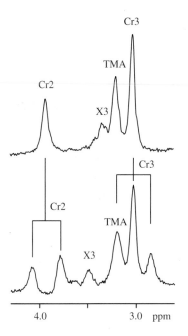

Figure 5 Spectra from the same voxel in the tibialis anterior muscle show the effect of orientation upon creatine Cr2 (methylene) and Cr3 (methyl). The lower spectrum has been acquired with the leg parallel to the magnetic field, the upper spectrum at the magic angle (54° between muscle and field). At the magic angle, splitting owing to dipolar coupling vanishes. TMA, trimethylammonium; X3 tentatively assigned to taurine

For References see p. 1315

spaces between actin/myosin chains. Because of charge distributions on these molecules, specific or unspecific coupling to the actin/myosin complexes may exacerbate the ordering. It has been postulated previously that some of this space is inaccessible to PCr. Since Cr and PCr are involved in the creatine kinase equilibrium and may serve not only as a short-term energy storage but also for some forms of energy transportation, a motional restriction as observed in these experiments would be of crucial importance since the overall reaction could be limited in vivo by transport processes.

Additional experiments lead to further questions about the exact nature of the Cr resonances: (a) in a postmortem study of rat muscle, the doublet of Cr methylene was shown to disappear from the MR spectrum on a time scale similar to the disappearance of PCr;[19] and (b) in human skeletal muscle, Cr/PCr during and after heavy exercise showed decrease and recovery similar to PCr in the ^{31}P MR spectrum.[20] It is surprising that depletion and recovery appear to be related to the PCr content of human muscle, since the Cr signals in ^{1}H MRS have hitherto always been associated with total Cr, not PCr. This finding is, however, in agreement with the signal behavior observed in rat muscle postmortem.[19] A straightforward but still questionable explanation for that observation would be that only the protons from PCr are detected by ^{1}H MRS while free Cr is invisible. The reduced visibility of Cr may or may not have similar roots to the incomplete dipolar averaging, MT effects, restricted diffusion, anisotropy of relaxation times, and the questioned visibility of lactate. The calculation of ADP concentrations based on ATP, free Cr, pH, and the creatine kinase reaction constant would be questionable if reduced visibility of Cr was an indication for reduced availability of this substrate. Therefore, findings on reduced visibility of creatine would have a serious effect on the modeled phosphorylation kinetics derived from ^{31}P MRS.

Diffusion-weighted spectroscopy may be another tool to identify restricted motion of specific metabolites. An excellent overview of diffusion effects in human muscle can be found in Nicolay et al.[12] This study shows restricted motion of PCr compared with water in rat hindleg muscle.

4.4.7 Intramyocellular Lipids

IMCL are stored in droplets in the cytoplasm of muscle cells and are a form of stored energy readily accessed during long-term exercise. When Schick et al.[11] compared the lipid resonances in calf muscle and fat tissue, they observed two compartments of triglycerides with a resonance frequency shift of 0.2 ppm. They assigned the resonance at 1.5 ppm to lipids in fat cells and hypothesized that the resonance at 1.3 ppm could be attributed to lipids located inside muscle cells (i.e., in their cytoplasm) experiencing different bulk susceptibility (see Figure 3). This assignment was verified by Boesch et al.[10] who demonstrated that IMCL signals scale linearly with voxel size, water and Cr signals, while EMCL signals do not. Furthermore the resonance frequency shift of EMCL and the orientation dependence of their spectral pattern could be attributed to the spatial arrangement of these lipids (plate-like structures for EMCL versus spheroid droplets for IMCL). Experimental data agree very well with theoretical estimations of susceptibility effects. Inter- and intraindividual reproducibility studies indicate that the error of the method is about 6% and that IMCL levels differ significantly between identical muscles in different

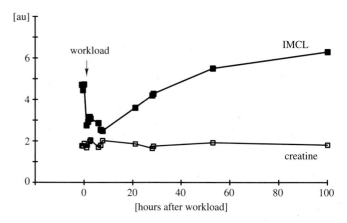

Figure 6 Depletion and repletion of intramyocellular lipids (IMCL) in tibialis anterior muscle after strenuous exercise (3 h bicycle training). In this example, the recovery phase of IMCL can be characterized by an exponential fit with a time constant of about 40 h. Further experiments have shown that recovery is strongly dependent on the diet. Creatine levels stay constant within the experimental accuracy (au, arbitary units; graph adapted from Boesch et al.[10])

subjects, different muscles in the same subject, as well as intraindividually in the same muscle when measured at 1 week intervals. The accuracy of IMCL determination by ^{1}H MRS is sufficient to follow IMCL depletion and recovery after exercise in single individuals (Figure 6). Dietary modulation of IMCL levels has been investigated by biopsy studies, with the well-known advantages and disadvantages of this invasive method. A consequence of the invasiveness is the scarcity of results and follow-up measurements. With ^{1}H MRS it is now feasible to follow IMCL depletion and repletion more or less continuously and in different muscle groups.

A correlation of increased IMCL levels and insulin sensitivity has been shown by ^{1}H MRS in diabetic patients.[79,80] Since it is not trivial to separate EMCL and IMCL, the comparison of measurements in obese patients with normal weight volunteers could be prone to systematic errors. Comparing cohorts of patients with volunteers, weight and muscular fat infiltration should be matched between the groups to avoid systematic differences between the resulting EMCL contaminations of the spectra.

5   CONCLUSIONS

MRI and MRS of muscular physiology has occurred in distinct separate centers with different equipment for a considerable time. A combination of the two complementary modalities is now developing while classical pulse and acquire ^{31}P MRS has become less prominent. Multinuclear MRS in whole-body systems using combinations of MRI and MRS with elaborate MR techniques such as diffusion weighting, editing, and MT seem to lead the way to future MR studies of muscle physiology. Less popular nuclei such as [23]Na have the potential to elucidate order in biological tissues further. Decoupling of ^{31}P and ^{13}C spectra, and the use of whole-body magnets at higher field, have already promoted the application of MRS in humans. This use will increase as soon as a larger number of versatile, high-field MR systems is installed.

Recent results obtained with localized ^{1}H MRS in skeletal muscle have shown that it is a very attractive tool to study muscle physiology both at rest and during exercise. It now appears that with ^{1}H MRS alone one can, in principle, determine pH, oxygenation, lactate levels, substrate use (IMCL, potentially also glycogen in addition to ^{13}C MRS), acetyl group buffering, and possibly PCr levels. It has also become thoroughly evident that the molecular basics of muscle tissue leads to most interesting NMR findings that can by no means be modeled by a single isotropic compartment consisting of an aqueous solution. Incompletely averaged dipolar couplings, reduced visibility of Cr, MT effects on Cr, the postulated compartmentation of lactate, and the susceptibility induced separation between IMCL and EMCL are all reflections of the complex and partly oriented architecture of muscle tissues, where (partial) compartmentation, chemical exchange, varying susceptibility, structural anisotropy, and protein interactions are essential ingredients.

6 RELATED ARTICLES

ESR Probes as Field Detectors in MRI; MRI of Musculoskeletal Neoplasms; Peripheral Muscle Metabolism Studied by MRS; pH Measurement In Vivo in Whole Body Systems; Phosphorus-31 Magnetization Transfer Studies In Vivo; Proton Decoupling During In Vivo Whole Body Phosphorus MRS; Proton Decoupling in Whole Body Carbon-13 MRS; Quantitation in In Vivo MRS; Single Voxel Localized Proton NMR Spectroscopy of Human Brain In Vivo; Single Voxel Whole Body Phosphorus MRS; Skeletal Muscle Evaluated by MRI; Spatial Localization Techniques for Human MRS; Tissue Behavior Measurements Using Phosphorus-31 NMR; Whole Body Studies: Impact of MRS.

7 REFERENCES

1. R. M. Henkelman, G. J. Stanisz, J. K. Kim, and M. J. Bronskill, *Magn. Reson. Med.*, 1994, **32**, 592.
2. W. C. Cole, A. D. LeBlanc, and S. G. Jhingran, *Magn. Reson. Med.*, 1993, **29**, 19.
3. J. M. Bonny, M. Zanca, O. Boespflug-Tanguy, V. Dedieu, S. Joandel, and J. P. Renou, *Magn. Reson. Imag.*, 1998, **16**, 167.
4. S. H. Scott, C. M. Engstrom, and G. E. Loeb, *J. Anat.*, 1993, **182**, 249.
5. J. Hu, M. R. Willcott, and G. J. Moore, *J. Magn. Reson.*, 1997, **126**, 187.
6. J. A. Kmiecik, C. D. Gregory, Z. P. Liang, D. E. Hrad, P. C. Lauterbur, and M. J. Dawson, *Magn. Reson. Med.*, 1997, **37**, 840.
7. I. K. Adzamli, F. A. Jolesz, A. R. Bleier, R. V. Mulkern, and T. Sandor, *Magn. Reson. Med.*, 1989, **11**, 172.
8. R. Reddy, L. Bolinger, M. Shinnar, E. Noyszewski, and J. S. Leigh, *Magn. Reson. Med.*, 1995, **33**, 134.
9. T. Kushnir, T. Knubovets, Y. Itzhak, U. Eliav, M. Sadeh, L. Rapoport, E. Kott, and G. Navon, *Magn. Reson. Med.*, 1997, **37**, 192.
10. C. Boesch, H. Slotboom, H. Hoppeler, and R. Kreis, *Magn. Reson. Med.*, 1997, **37**, 484.
11. F. Schick, B. Eismann, W. I. Jung, H. Bongers, M. Bunse, and O. Lutz, *Magn. Reson. Med.*, 1993, **29**, 158.
12. K. Nicolay, A van der Toorn, and R. M. Dijkhuizen, *NMR Biomed.*, 1995, **8**, 365.
13. A. van Doorn, P. H. Bovendeerd, K. Nicolay, M. R. Drost, and J. D. Janssen, *Eur. J. Morphol.*, 1996, **34**, 5.
14. P. J. Basser, *NMR Biomed.*, 1995, **8**, 333.
15. T. G. Reese, R. M. Weisskoff, R. N. Smith, B. R. Rosen, R. E. Dinsmore, and V. J. Wedeen, *Magn. Reson. Med.*, 1995, **34**, 786.
16. E. W. Hsu and S. Mori, *Magn. Reson. Med.*, 1995, **34**, 194.
17. R. Kreis and C. Boesch, *J. Magn. Reson. Series B*, 1994, **104**, 189.
18. R. Kreis and C. Boesch, *J. Magn. Reson. Series B*, 1996, **113**, 103.
19. V. Ntziachristos, R. Kreis, C. Boesch, and B. Quistorff, *Magn. Reson. Med.*, 1997, **38**, 33.
20. R. Kreis, B. Jung, J. Slotboom, J. Felblinger, and C. Boesch, *J. Magn. Reson.*, 1999, **137**, 350.
21. L. Jouvensal, P. G. Carlier, and G. Bloch, *Magn. Reson. Med.*, 1997, **38**, 706.
22. J. L. Fleckenstein, L. A. Bertocci, R. L. Nunnally, R. W. Parkey, and R. M. Peshock, *Am. J. Roentgenol.*, 1989, **153**, 693.
23. E. R. Weidman, H. C. Charles, R. Negro-Vilar, M. J. Sullivan, and J. R. MacFall, *Invest. Radiol.*, 1991, **26**, 309.
24. T. Ogino, H. Ikehira, N. Arimizu, H. Moriya, K. Wakimoto, S. Nishikawa, H. Shiratsuchi, H. Kato, F. Shishido, and Y. Tateno, *Ann. Nucl. Med.*, 1994, **8**, 219.
25. L. L. Ploutz-Snyder, S. Nyren, T. G. Cooper, E. J. Potchen, and R. A. Meyer, *Magn. Reson. Med.*, 1997, **37**, 676.
26. G. Yue, A. L. Alexander, D. H. Laidlaw, A. F. Gmitro, E. C. Unger, and R. M. Enoka, *J. Appl. Physiol.*, 1994, **77**, 84.
27. H. A. Cheng, R. A. Roberts, J. P. Letellier, A. Caprihan, M. Icenogle, and L. J. Haseler, *J. Appl. Physiol.*, 1995, **79**, 1370.
28. J. L. Fleckenstein, R. C. Canby, R. W. Parkey, and R. M. Peshock, *Am. J. Roentgenol.*, 1988, **151**, 231.
29. M. A. Elliott, G. A. Walter, H. Gulish, A. S. Sadi, D. D. Lawson, W. Jaffe, E. K. Insko, J. S. Leigh, and K. Vandenborne, *MAGMA*, 1997, **5**, 93.
30. N. Mitsiopoulos, R. N. Baumgartner, S. B. Heymsfield, W. Lyons, D. Gallagher, and R. Ross, *J. Appl. Physiol.*, 1998, **85**, 115.
31. N. Roberts, L. M. Cruz-Orive, N. M. K. Reid, D. A. Brodie, M. Bourne, and R. H. T. Edwards, *J. Microsc.*, 1993, **171**, 239.
32. T. Fukunaga, R. R. Roy, F. G. Shellock, J. A. Hodgson, M. K. Day, P. L. Lee, H. Kwong-Fu, and V. R. Edgerton, *J. Orthop. Res.*, 1992, **10**, 926.
33. M. V. Narici, L. Landoni, and A. E. Minetti, *Eur. J. Appl. Physiol.*, 1992, **65**, 438.
34. K. F. Petersen, R. Hendler, T. Price, G. Perseghin, D. L. Rothman, N. Held, J. M. Amatruda, and G. I. Shulman, *Diabetes*, 1998, **47**, 381.
35. D. L. Rothman, R. G. Shulman, and G. I. Shulman, *J. Clin. Invest.*, 1992, **89**, 1069.
36. G. I. Shulman, D. L. Rothman, T. Jue, P. Stein, R. A. DeFronzo, and R. G. Shulman, *N. Engl. J. Med.*, 1990, **332**, 223.
37. R. Roussel, P. G. Carlier, J. J. Robert, G. Velho, and G. Bloch, *Proc. Natl. Acad. Sci. USA*, 1998, **95**, 1313.
38. N. Beckman, in 'Carbon-13 NMR spectroscopy of biological systems', ed. N. Beckmann, Academic Press, New York, 1995, p. 269.
39. J. Seelig and A. P. Burlina, *Clin. Chim. Acta*, 1992, **206**, 125.
40. R. Badar-Goffer and H. Bachelard, *Essays Biochem.*, 1991, **26**, 105.
41. R. Gruetter, E. J. Novotny, S. D. Boulware, D. L. Rothman, G. F. Mason, G. I. Shulman, R. G. Shulman, and W. V. Tamborlane, *Proc. Natl. Acad. Sci. USA*, 1992, **89**, 1109.
42. R. Gruetter, G. Adriany, H. Merkle, and P. M. Andersen, *Magn. Reson. Med.*, 1996, **36**, 659.
43. M. Saner, G. McKinnon, and P. Boesiger, *Magn. Reson. Med.*, 1992, **28**, 65.

For References see p. 1315

44. R. Kreis, J. Slotboom, J. Felblinger, and C. Boesch, *Proc. Vth Ann Mtg. (Int.) Soc. Magn. Reson. Med.*, Vancouver, 1997, p. 1438.
45. H. Watanabe, Y. Ishihara, K. Okamoto, K. Oshio, T. Kanamatsu, and Y. Tsukada, *J. Magn. Reson.*, 1998, **134**, 214.
46. A. J. van den Bergh, H. J. van den Boogert, and A. Heerschap, *J. Magn. Reson.*, 1998, **135**, 93.
47. R. Taylor, T. B. Price, L. D. Katz, R. G. Shulman, and G. I. Shulman, *Am. J. Physiol.*, 1993, **265**, E224.
48. P. Jehenson, D. Duboc, G. Bloch, M. Fardeau, and A. Syrota, *Neuromuscul. Disord.*, 1991, **1**, 99.
49. G. Ende and P. Bachert, *Magn. Reson. Med.*, 1993, **30**, 415.
50. R. Roussel, P. G. Carlier, C. Wary, G. Velho, and G. Bloch, *Magn. Reson. Med.*, 1997, **37**, 821.
51. R. Taylor, T. B. Price, D. L. Rothman, R. G. Shulman, and G. I. Shulman, *Magn. Reson. Med.*, 1992, **27**, 13.
52. N. Beckman, J. Seelig, and H. Wick, *Magn. Reson. Med.*, 1990, **16**, 150.
53. A. van den Bergh, S. Houtman, A. Heerschap, N. J. Rehrer, H. J. van den Boogert, B. Oeseburg, and M. T. E. Hopman, *J. Appl. Physiol.*, 1996, **81**, 1495.
54. T. B. Price D. L. Rothman, R. Taylor M. J. Avison, G. I. Shulman, and R. G. Shulman, *J. Appl. Physiol.*, 1994, **76**, 104.
55. C. Wary, G. Bloch, P. Jehenson, and P. G. Carlier, *Anticancer Res.*, 1996, **16**, 1479.
56. C. T. Moonen, R. J. Dimand, and K. L. Cox, *Magn. Reson. Med.*, 1988, **6**, 140.
57. E. L. Thomas, S. C. Cunnane, and J. D. Bell, *NMR Biomed.*, 1998, **11**, 290.
58. J. G. Jones, M. A Solomon, A. D. Sherry, F. M. H. Jeffrey, and C. R. Malloy, *Am. J. Physiol.*, 1998, **275**, E843.
59. B. M. Jucker, A. J. Rennings, G. W. Cline, K. F. Petersen, and G. I. Shulman, *Am. J. Physiol.*, 1997, **273**, E139.
60. E. D. Lewandowski, C. Doumen, L. T. White, K. F. LaNoue, L. A. Damico, and X. Yu, *Magn. Reson. Med.*, 1996, **35**, 149.
61. A. D. Sherry, P. Zhao, A. Wiethoff, and C. R. Malloy, *Magn. Reson. Med.*, 1994, **31**, 374.
62. L. A. Bertocci, J. G. Jones, C. R. Malloy, R. G. Victor, and G. D. Thomas, *J. Appl. Physiol.*, 1997, **83**, 32.
63. H. Bachelard, *Dev. Neurosci.*, 1998, **20**, 277.
64. S. R. Williams, D. G. Gadian, E. Proctor, D. B. Sprague, D. F. Talbot, I. R. Young, and F. F. Brown, *J. Magn. Reson.*, 1985, **63**, 406.
65. P. A. Narayana, J. D. Hazle, E. F. Jackson, L. K. Fotedar, and M. V. Kulkarni, *Magn. Reson. Imag.*, 1988, **6**, 481.
66. M. Barany and P. N. Venkatasubramanian, *NMR Biomed.*, 1989, **2**, 7.
67. H. Bruhn, J. Frahm, M. L. Gyngell, K. D. Merboldt, W. Haenicke, and R. Sauter, *Magn. Reson. Med.*, 1991, **17**, 82.
68. J. W. Pan, J. R. Hamm, H. P. Hetherington, D. L. Rothman, and R. G. Shulman, *Magn. Reson. Med.*, 1991, **20**, 57.
69. Z. Wang, E. A. Noyszewski, and J. S. Leigh, *Magn. Reson. Med.*, 1990, **14**, 562.
70. C. Brillault-Salvat, E. Giacomini, L. Jouvensal, C. Wary, G. Bloch, and P. G. Carlier, *NMR Biomed.*, 1997, **10**, 315.
71. J. E. van Dijk, D. K. Bosman, R. A. F. M. Chamuleau, and W. M. M. J. Bovee, *Magn. Reson. Med.*, 1991, **22**, 493.
72. R. E. Hurd and D. Freeman, *NMR Biomed.*, 1991, **4**, 73.
73. G. Bloch, L. Jouvensal, and P. G. Carlier, *Magn. Reson. Med.*, 1995, **34**, 353.
74. U. Kreutzer, D. S. Wang, and T. Jue, *Proc. Natl. Acad. Sci. USA*, 1992, **89**, 4731.
75. E. A. Noyszewski, E. L. Chen, R. Reddy, Z. Wang, and J. S. Leigh, *Magn. Reson. Med.*, 1997, **38**, 788.
76. R. Kreis, B. Jung, S. Rotman, J. Slotboom, J. Felblinger, and C. Boesch, *Proc. Vth Ann Mtg. (Int.) Soc. Magn. Reson. Med.*, Vancouver, 1997, p. 162.
77. E. P. Brass and W. R. Hiatt, *Life Sci.*, 1994, **54**, 1383.
78. R. Kreis, M. Koster, M. Kamber, H. Hoppeler, and C. Boesch, *Magn. Reson. Med.*, 1997, **37**, 159.
79. J. Machann, F. Schick, S. Jacob, O. Lutz, H. U. Häring, and C. D. Claussen, *MAGMA*, 1998, p. 220.
80. D. T. Stein, L. S. Szczepaniak, R. L. Dobbins, P. Snell, and J. D. McGarry, *Proc. VIth Ann Mtg. Soc. (Int.) Magn. Reson. Med.*, Sydney, 1998, p. 388.

Biographical Sketch

Chris Boesch. *b* 1951. M.S. (Physics diploma) ETH Zurich (Swiss Federal Institute of Technology) Switzerland, 1976; Ph.D. ETH Zurich, 1979 (High-resolution NMR studies of polypeptide conformation, supervisor K. Wüthrich); Studies in Medicine, 1977–86; M.D. University Zurich 1986. Research assistant at the University Children's Hospital in Zurich, Switzerland: Installation of a 2.35T/40 cm MR system in a clinical setting, patient-monitoring systems for pediatric and intensive care patients, MRI and MRS studies of brain development, 1985–90; Professor and Director of MR Spectroscopy and Methodology at the University of Bern, Switzerland, 1991–present. Current research specialties: in vivo spectroscopy (^{1}H, ^{31}P, and ^{13}C) of brain, muscle, and liver; methodology of in vivo NMR: patient monitoring.

Roland Kreis. *b* 1958. M.S. (diploma in chemistry) ETH (Federal Institute of Technology) Zurich, Switzerland, 1983; Ph.D. ETH Zurich, 1989 (Zero-field NMR, supervisor R. R. Ernst). Boswell Fellow at Caltech and Huntington Medical Research Institutes, Pasadena, CA, USA, 1989–91; Assistant Professor at Department for Clinical Research (MR Spectroscopy and Methodology) University of Bern, Switzerland, 1991–present. Research interests: quantitative clinical spectroscopy using methods for optimized data acquisition and pro-Breakcessing while studying cerebral, musculoskeletal and cardiac (patho) physiology.

Peripheral Vasculature MRA

Cathy Maldjian
Mount Sinai Medical Center, New York, NY, USA

and

Mitchell D. Schnall
University of Pennsylvania Medical Center, Philadelphia, PA, USA

1 INTRODUCTION

While conventional angiography has been the 'gold standard' in the evaluation of peripheral vascular disease, recent technical advances have made magnetic resonance angiography

(MRA) a favorable alternative in many instances. Multiplanar imaging capabilities and noninvasive acquisition of data provide distinct advantages over contrast angiography. Overnight hospital stays, and potential morbidity from contrast injection and from obtaining arterial access may be avoided with MRA. Furthermore, distal runoff may be obscured by inadequate contrast filling in up to 70% of cases.[1–4] This has mainly been attributed to contrast dilution in intramuscular collaterals before reaching the reconstituted vessels of interest. MRA allows cross-sectional imaging, which facilitates the evaluation of the vessel lumen for stenotic lesions. Impaired detection of stenotic lesions due to interference from overlying bone is not a problem, although it may be in contrast angiography. Quantitation of flow, in particular by cine-phase contrast, is also possible with MRA. Three-dimensional data sets may be obtained by MRA. Gravimetric layering of contrast, which may prevent filling in anterior branches of slow-flowing vessels, is obviated with MRA. Pulsatility artifacts tend not be a problem in studying collateral reconstitution past a high level stenotic lesion, since these vessels often adopt a monophasic rather than a triphasic wave pattern. However, with the advent of digital subtraction angiography, a higher level of technical proficiency and accuracy can be achieved. Inplane resolution is superior with cut-film angiography compared with MRA. Skeletal bony anatomy, sequential demonstration of retrograde flow, metal artifacts, and total length of vascular anatomy, may be better appreciated with contrast angiography. Also, contraindications to MR, such as pacemakers, defibrillators, and metallic prosthetic valves, prohibit its use in these instances.

Various artifacts of MRA will be discussed later in further detail. Bearing all this in mind, MRA may play an integral part in the presurgical evaluation of lower-extremity ischemia by providing the surgeon with the arterial anatomy in a noninvasive manner. It may afford decreased intraoperative times. Stenosis, occlusions, and native target outflow vessels favorable for bypass procedures may be identified noninvasively. In addition, reconstituted vessels may potentially be seen to better advantage.

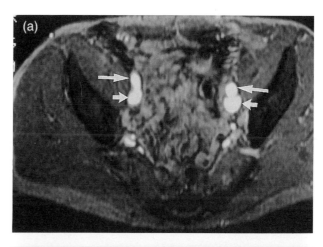

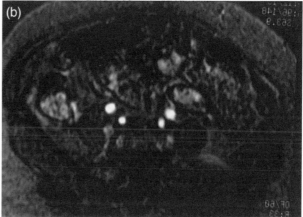

Figure 1 Single slice from a 2D TOF angiogram taken: (a) without saturation bands demonstrating both arteries (larger arrows) and veins (smaller arrows); (b) with inferior saturation. The venous structures have been saturated out and only arterial structures remain

2 TECHNIQUES

Currently, the predominant technique for imaging the peripheral vasculature consists of two-dimensional (2D) time-of-flight (TOF). Single sections are imaged perpendicular to the plane of flow. Short *TR* is utilized to saturate stationary spins, which imparts high intensity to flowing blood where spins are unsaturated. This flow-related enhancement allows visualization of the peripheral vasculature [Figure 1(a)]. Further selection of the arterial or venous system can be accomplished by utilizing a saturation band [Figure 1(b)].

A saturation band positioned distally to the imaging section would provide an arterial study. This relationship would be reversed in the head and neck, where venous blood generally travels inferiorly rather than superiorly and arterial blood courses superiorly rather than inferiorly. During arterial diastole, there is a component of flow reversal which allows arterial blood to enter the presaturation slab during systole and subsequently the imaging section during diastolic reversal of flow if the slab is positioned close enough to the imaging section. This creates undesirable dark, horizontal, stripe artifacts.

For this reason, a saturation gap is often utilized. A saturation gap is a distance placed between the saturation pulse and the imaging plane. The larger the gap, the less likely that retrograde arterial flow from the saturation slab will reach the imaging section. On the other hand, a larger gap would also enable venous blood within the gap to enter the imaging slice without first being saturated. Therefore, some venous blood may also be inadvertently imaged and potentially confused for arterial flow. This phenomenon is called 'venous bleedthrough' (Figure 2). It is therefore evident that the size of the saturation gap is critical and that the optimal saturation gap size may vary depending upon the arterial and venous flow rates at the particular site of interest. In addition to the above techniques, we also utilize flow compensation at our institution.

The 2D axial images obtained are stacked on top of each other to create a three-dimensional (3D) data set. Projection angiograms are created by the MIP algorithm. A ray is projected through this 3D data set, and each pixel in the projection is designated a value based on the maximum intensity encountered within the pixel.

At our institution, we utilize a *TR* of 33–36 ms, a *TE* of 6.2–7.7 ms, and a 60° flip angle. Our 2D TOF imaging proto-

For References see p. 1323

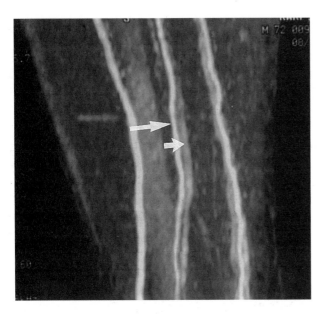

Figure 2 MIP projection obtained from a 2D TOF arteriogram in the region of the calf, demonstrating high-signal arterial vessels (large arrow) and lower signal venous vessels (small arrow). The veins are visible due to the wide saturation gap (2 cm) used in this case

col entails imaging at several stations. The pelvis is imaged with a 32 cm field-of-view (FOV), 2.5 mm thickness, 128×256 matrix, and a body coil. No skip is utilized. Distal thigh, knee, leg, ankle, and foot are imaged with an extremity coil at a 16 cm field of view. Ninety 2 mm images per station are obtained with the knee coil. The pelvis and one lower extremity are imaged in approximately 1.5 h. Station 1 is at the forefoot, where the dorsal and plantar arches form from the dorsalis pedis and posterior tibial arteries [Figure 3(a)]. Upper and lower margins of the coil are marked to ensure that all vessels are examined, which in the General Electric knee coil is up to 20 cm long. In patients with peripheral vascular disease, the flow distally tends to be nonpulsatile, thus a narrower saturation gap is utilized. This also limits venous bleed-through. The arches frequently cannot be seen, since they course horizontally and become saturated. Retrograde flow, sometimes seen in the dorsalis pedis, susceptible to saturation by the inferior saturation slab, may create the impression of stenosis on the MIP.

The second station examines the anterior tibial, posterior tibial, and peroneal arteries [Figure 3(b)]. Calcaneal and perforator branches arise from the peroneal artery, and these course posteriorly and anteriorly, respectively. The calcaneal artery (communicator branch) may function as a collateral to the plantar aspect of the foot, while the perforator branch coursing through the interosseous membrane may function as a dorsal collateral. The posterior tibial artery bifurcates into medial and lateral plantar arteries at the ankle.

The third station ends at the mid-calf [Figure 3(c)]. The posterior tibial artery is most medial and usually anterior to the peroneal artery. The anterior tibial artery is located very superficially at the talus, where the dorsalis pedis forms, and it travels in the tibialis anterior muscle within the anterior compartment. The peroneal artery parallels the fibula. Intramuscular

collaterals may accompany disease at the trifurcation. Venous bleed-through may mimic such arterial collaterals.

The fourth station, at the knee, is where the popliteal artery and the trifurcation are evaluated [Figure 3(d)]. Basis images are critical, especially for assessing stenosis of the anterior tibial artery. This area represents the most variable for the number of vessels seen. Anatomic variation, occlusion, and collaterals all contribute to this phenomenon. The length of the region examined is also variable due to individual height discrepancies.

The fifth station examines the common femoral artery to the superficial femoral artery (SFA)–popliteal artery junction [Figure 3(e)]. The SFA is medial and superficial. The profunda femoral artery is posterior and lateral in location, and terminates at the mid-thigh with associated muscular branches. The SFA has no muscular branches. Hunter's canal is contained in this series. Stenoses at this level are common. Faster arterial inflow and venous outflow at this station are conducive to a shorter TR (33 ms) and a larger saturation gap (1 cm), respectively.

The sixth station (pelvis) covers the distal abdominal aorta to the common femoral arteries [Figure 3(f)]. The aorta normally bifurcates into the common iliac arteries, which further bifurcate into internal and external iliac arteries. The common femoral arteries subsequently give rise to the deep and superficial femoral arteries. The faster flowing veins at this level require an even larger saturation gap (2 cm) at this station.

3   CLINICAL DATA

Clinical data thus far has shown 2D TOF MRA to be a promising technique. Mulligan et al. studied 140 lesions and found a 71% correlation between 2D TOF MRA and conventional angiography of the abdominal aorta and distal-runoff vessels.[5] MRA alone proved inaccurate in six out of 140 cases for planning a surgical approach. As this was one of the early studies in this area, these inaccuracies to some extent may reflect inadequacies of the early techniques. Later studies by other investigators produced more promising results, probably due to technical advancements. Color-duplex ultrasound (US) demonstrated 93% concurrence for infrainguinal disease. While MRA was superior to US in evaluating the iliac regions due to limited visualization with US in that area, image quality with MRA was inconsistent. Workers concluded that both duplex US and MRA were suboptimal for surgical planning. Owen et al. found that most of the discordance between 2D TOF MRA and conventional angiography was due to enhanced detection of runoff vessels on MRA[6] (Figure 4). Interventional planning was altered in 16% of cases (12 patients) due to improved visualization with MRA of stenosis and vessels not seen on conventional angiography.

Carpenter et al. studied 51 patients with peripheral vascular disease.[7] Two-dimensional TOF MRA evaluation demonstrated that in 48% of cases, MRA provided additional information to contrast angiography without loss of other information. In 22% of cases, the MRA data were responsible for altering surgical treatment. Target vessels were observed in 18% of cases on MRA that were undetected on contrast angiography. Identification of these target vessels enabled the surgeons to perform

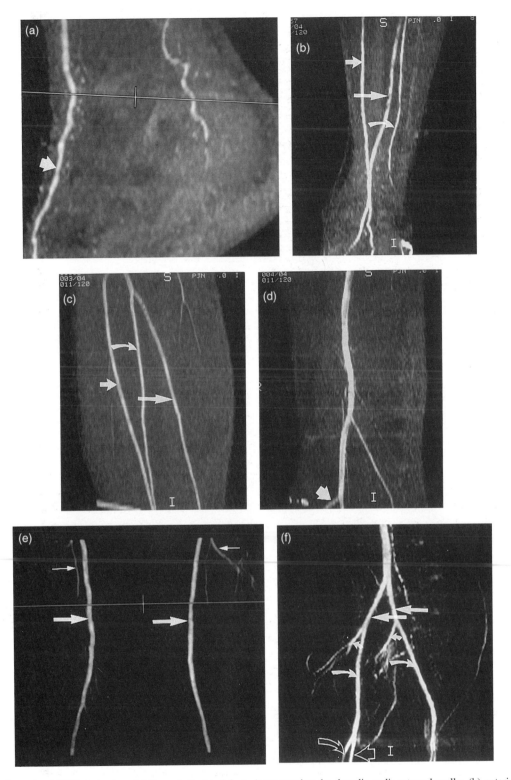

Figure 3 Two-dimensional TOF projection: (a) through the hind foot, demonstrating the dorsalis pedis artery dorsally; (b) anteriorly through the ankle. The anterior tibial (small arrow), peroneal (curved arrow), and posterior tibial arteries (large arrow) are demonstrated; (c) anteriorly in the region of the mid-calf, demonstrating the anterior tibial artery (small arrow) and tibial peroneal trunk bifurcating into posterior tibial (large arrow) and peroneal arteries (curved arrow); (d) through the region of the knee, demonstrating the popliteal artery and the take-off of the anterior tibial artery (arrow); (e) obtained with the body coil, demonstrating the superficial femoral arteries bilaterally from upper thigh through Hunter's canal (large arrows). The distal profunda femoral arteries are also identified (small arrows); (f) through the pelvis, demonstrating the aorta, common iliac arteries (largest arrows), internal (small curved arrows) and external iliac (larger curved arrows) arteries, and common deep (curved open arrow) and superficial (open arrow) femoral arteries. Note the excellent visualization of the femoral artery bifurcation

For References see p. 1323

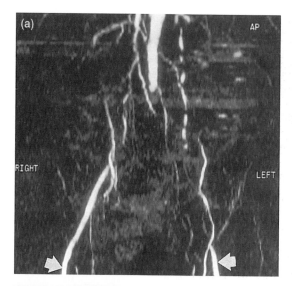

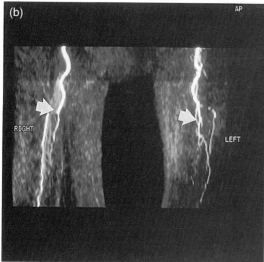

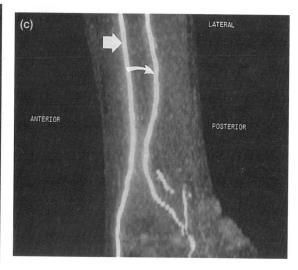

Figure 4   Representative images from the magnetic resonance arteriogram of a patient who presents with lymph-threatening ischemia. (a) Images through the pelvis demonstrating bilateral iliac artery occlusion and reconstruction of the SFA bilaterally (arrows). (b) Images obtained through the upper thighs. Bilateral superficial femoral artery occlusions are demonstrated (arrows). (c) Images through the lower left calf. The presence of an anterior tibial artery crossing the ankle to form the dorsalis pedis (arrow) and a peroneal artery (curved arrow) with a calcaneal branch crossing the ankle to form the plantar arches is demonstrated. On the basis of these MR images, this patient went on to have revascularization procedure without contrast angiogram

limb-salvaging procedures. Similarly, two cases (4%) demonstrated MRA findings of target runoff vessels which were not seen on conventional angiography, where the information was used to formulate a better interventional plan for bypass grafting. These investigators demonstrated superior sensitivity of MRA compared with contrast angiography in evaluating distal runoff vessels.[8] Other workers also found that 2D TOF suppressed veins better than the 3D techniques.[9]

In evaluating for stenosis, Chien et al.[9] suggest black-blood techniques to prevent nonvisualization of turbulent spins from phase incoherence.[5] Bright-blood methods, on the other hand, may overestimate a stenosis due to phase incoherence.

Mulligan et al. evaluated stenosis from MIP, which simulates the unidimensionality of angiographic studies; however, axial-basis images are more accurate for such assessments.[5]

Also, since they utilize the large FOV, the in-plane resolution of the basis images is worsened. This also compromises the resolution of the MIP.

Cambria et al. reported 98% concurrence between 2D TOF MRA and contrast angiography in 178 patients with a 100% concurrence below the inguinal ligament.[10] Intraoperative confirmation was obtained in the seven bypass grafts performed on the basis of the MRA data alone. Hertz et al. also reported a high correlation between 2D TOF MRA and contrast angiography in assessing arterial stenosis in 19 patients[11] (Figure 5). For the popliteal to the dorsalis pedis area, Carpenter et al.[7] showed MRA to be superior to conventional angiography, with a 48% discordance rate between the two techniques.[6] This difference appeared to be more progressively pronounced the further distally one examined. Up to 50% of stenosis may be

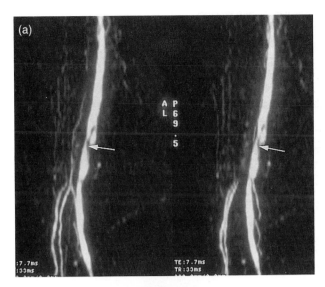

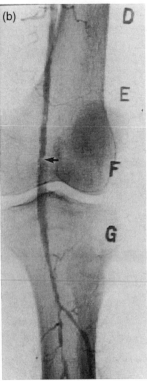

Figure 5 Two-dimensional TOF arteriogram: (a) on a patient who had recently had an angioplasty, demonstrating recurrent stenosis with an intimal flap (arrows); (b) obtained 1 week prior, demonstrating the stenosis (arrow)

overlooked with conventional angiography if the plaque is en face with no available edge for imaging.

We have found 2D TOF MRA to be accurate in assessing peripheral stenosis. Cross-sectional images are particularly useful, in that they may demonstrate more information than contrast angiography. We showed good agreement in blinded interobserver interpretation (91%).[11] Also, lesions distal and proximal to stenosis were seen equally well with MRA, with no appreciable discrepancy between the two.[11] We investigated 41 patients for aorto–iliac disease.[12] Only six cases showed a

discrepancy between MRA and contrast angiography. Two of these proved to be false-positive on MRA, and two proved to be false-negative. The other two underestimated a greater than 50% stenosis seen on angiography. In the pelvis and lower extremity, we obtained a 62% concordance in a 103-patient series, with no false-positives or false-negatives on MRA.[13] Contrast angiography failed to show 162 open segments that were confirmed surgically.

We have found MRA to be reliable in assessing vessel patency. Baum et al. demonstrated a 0.7 correlation coefficient between contrast angiography and 2D TOF MRA for determining the parameters of a stenosis.[14] We compared contrast angiography with MRA results in 60 patients with peripheral vascular disease. ROC-curve analysis of one blinded observer demonstrated that location and patency yielded a high correlation between angiography and MRA at our institution.[15]

4 PHASE CONTRAST (PC)

Phase contrast techniques utilize motion as an imaging tool, whereby flow is detected in all three axes. A technique relying on phase discrepancy, which measures phase shift of moving spins to determine their velocities, was proposed by Bryant et al.[16] Other investigators demonstrated that cardiac gating imparts accuracy to the calculated arterial flow rates.[17] Various phases of arterial flow could also be better segregated and evaluated. Applications of PC in peripheral vascular disease soon followed. A 2D TOF with PC velocity-sensitive technique demonstrated significantly increased velocities at systole in diseased peripheral vessels after angioplasty, whereas normal vessels demonstrated no alteration in flow.[18] A one spatial- and one-dimensional-variant of PC may enhance tracing of the triphasic flow in the popliteal artery.[19]

In our recent experience with two-axis cine-PC angiography for arterial wave patterns in diseased vessels, we have observed monophasicity beyond points of critical stenosis. Simultaneous measurements of velocities at several sites may be implemented with a combined excitation technique to obtain cine-PC data in one dimension, with cardiac gating to generate a second dimension (time).[20] Measurements of arterial compliance are made possible. Addressing arterial compliance is useful in atherosclerotic disease, where compliance is decreased and arterial waveforms are altered. Recently, investigators have demonstrated similar velocity measurements in common iliac arteries with a cine-PC technique with a surgically attached US flow probe on the artery.[21] Velocity-encoded cine MRA was compared with color-coded US in ten healthy subjects above and below the trifurcation and showed good correlation.[22] The classic triphasic pattern was seen. These investigators suggest that velocity-encoded cine MRA may help to evaluate the hemodynamic significance of a stenotic lesion.[23] This information may complement anatomical information obtained from standard 2D TOF MRA. They suggest that a single MRA examination may derive all the information that previously might have been obtained by performing both contrast angiography and duplex US. ECG-triggered 2D PC techniques, where velocity encoding varies with the cardiac cycle, may improve the contrast-to-noise ratio, most notably in small vessels, where up to 260% improved signal has been noted.[24] In summary, PC techniques appear to be very promising.

For References see p. 1323

5 3D MRA

While we do not advocate the use of 3D techniques, we will briefly describe their applications. While in 2D imaging, many slices are imaged in sequence, in 3D imaging a volume is imaged. Acquisition of data from a larger volume, as in 3D TOF, enhances spatial resolution and therefore increases S/N. Imaging times are also decreased. Flow compensation is easier with 3D rather than 2D TOF MRA due to the increased gradient amplitude required for slice selection in 2D TOF. In 3D techniques, resolution along the Y and Z axes is obtained by phase-encoding techniques. Therefore, the slice selection gradient is relatively unburdened. In contradistinction, an extra lobe on the gradient waveform would be necessary to implement flow compensation with 2D techniques, and this would consequently increase the TE. Therefore, shorter TEs are possible with a 3D modality when utilizing flow compensation in comparison with a 2D modality. So, less dephasing of moving spins occurs with 3D flow compensation. Smaller voxel sizes may also be utilized with 3D imaging. The main disadvantages of 3D imaging techniques, which significantly hinder their usefulness, include decreased sensitivity to slow flows and prolonged reconstruction times of a 3D volume-based data set. In the 3D method, slow flow may incur saturation as it traverses further into the saturation slab. By the same reasoning, venous blood would saturate more readily compared with arterial blood. Therefore 3D MRA may be of greater value in imaging fast flowing and small vessels. Its primary benefits have been seen in the area of neurovascular imaging.

6 ARTIFACTS

Clip artifacts and orthopedic prostheses and artifacts related to susceptibility differences at various interfaces may simulate stenosis or occlusion of a vessel (Figure 6). Local magnetic field variations may be large enough to cause phase dispersion, yielding a net signal intensity vector of zero. Frequently, there is an associated high-signal intensity rind near the metal or air due to misregistration in frequency and slice selection directions as a result of magnetic susceptibility. Gradient echo sequences are particularly vulnerable to these effects.

Due to the length of the examination movement may occur in some patients. Displacement may appear as one translocated, or multiple translocated, slices on the MIP (Figure 7). Motion may create the appearance of a vascular lesion, and individual sections must be reviewed. Peristalsis or respiratory motion may confer high signal intensity on bowel material by 'refreshing stationary spins'. Muscle contraction may confer increased signal intensity to muscle in the same fashion as flowing motion in vessels produces flow-related enhancement.

As discussed previously, pulsatile flow may generate dark, horizontal bands on the MIP, as each section is not obtained at identical times in the cardiac cycle. Thus, varying signal intensities are obtained in each section. This banding artifact, or 'Venetian-blind effect,' is more pronounced in vessels with triphasic flow (Figure 8). Cardiac gating and an increased saturation gap are two ways of counteracting these effects. Phase ghosting, also produced by vessel pulsatility, may be counteracted with EKG gating. Apparent absence of retrograde

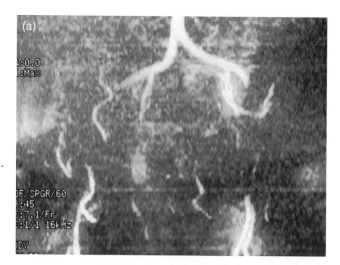

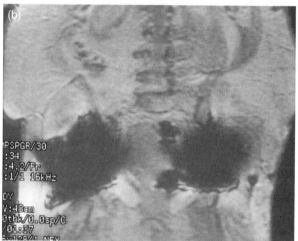

Figure 6 (a) Frontal projection from a 2D TOF arteriogram through the pelvis demonstrating complete absence of iliac arteries distal to the common iliac artery bilaterally. (b) Coronal scout image demonstrating large artifacts due to the presence of hip prostheses bilaterally

arterial flow may occur with the use of a saturation slab which is primarily intended to saturate venous flow. Arterial pulsatility may be confirmed by removing the saturation slab or by acquiring velocity-encoded information.

Turbulence artifacts near a site of stenosis, with its dephasing effects, causes high flow with spin-phase incoherence. However, this is not much of a problem peripherally where flow is less pulsatile. Inspection of axial images for high-grade stenosis would explain the findings.

Potential pitfalls of MIP images occur when a high signal intensity interferes with visualization of a blood vessel, such as from a blood clot. Conversely, a filling defect may become nonapparent if it is surrounded by high signal intensity flowing blood. Also, when section thickness exceeds that of in-plane resolution, normal horizontal flow may appear obstructed on the MIP. In the same vein, pseudobeating may occur with tortuous oblique vessels. This may manifest as a 'stair-step' artifact on the MIP. This may be counteracted by thinner slices, slice overlap, and reacquisition perpendicular to the vessel. Inplane or 'horizontal' flow refers to areas where, due to the

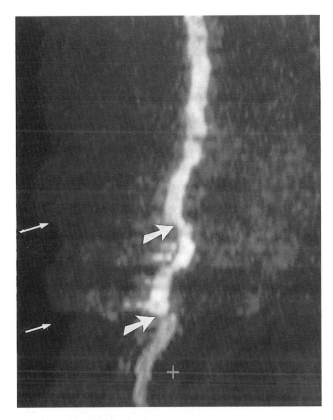

Figure 7 Lateral MIP projection demonstrating the effect of motion on the MR arteriogram. Displacement of segments of the vessel can simulate stenosis (arrows). Motion can be confirmed by observing motion of the skin line (small arrows)

orientation or tortuosity of a vessel, flow occurs in a roughly horizontal rather than vertical plane. Since flow in the horizontal plane would not be detected because of saturation, the vascular anatomy would be obscured or underrepresented as

stenotic. Examples of this include the anterior tibial artery take-off, which is roughly horizontal or perpendicular to the popliteal artery, and tortuous iliac bifurcations.

7 RELATED ARTICLES

Abdominal MRA; Head and Neck Studies Using MRA; Phase Contrast MRA; Time-of-Flight Method of MRA.

8 REFERENCES

1. K. R. Patel, L. Semel, and R. H. Clauss, *J. Vasc. Surg.*, 1988, **7**, 531.
2. J. B. Ricco, W. H. Pearce, J. S. T. Yao, W. R. Flinn, and J. J. Bergan, *Ann. Surg.*, 1983, **198**, 646.
3. R. Scarpato, R. Gembarowicz, S. Farber, T. F. O'Donnell, J. J. Kelly, A. D. Callow, and R. A. Deterling, *Arch. Surg.*, 1981, **116**, 1053.
4. D. P. Flanigan, L. R. Williams, T. Keifer, J. J. Schuler, and A. J. Behrend, *Surgery*, 1982, **92**, 627.
5. S. A. Mulligan, T. Matsuda, P. Lanzer, G. M. Gross, W. D. Routh, F. S. Keller, D. B. Koslin, L. L. Berland, M. D. Fields, and M. Doyle, *Radiology*, 1991, **178**, 695.
6. R. S. Owen, R. A. Baum, J. P. Carpenter, G. A. Holland, and C. Cope, *Radiology*, 1993, **187**, 627.
7. J. P. Carpenter, R. S. Owen, R. A. Baum, C. Cope, C. F. Barker, H. D. Berkowitz, M. A. Golden, and L. J. Perloff, *J. Vasc. Surg.*, 1992, **16**, 807.
8. R. S. Owen, M. Sheline, J. Listerud, and H. Y. Kressel, *Proc. Xth Ann. Mtg. Soc. Magn. Reson. Med., San Francisco, 1991*, p. 138.
9. D. Chien, A. Goldmann, and R. R. Edelman, *Proc. XIth Ann. Mtg. Soc. Magn. Reson. Med., New York, 1992*, p. 3111.
10. R. P. Cambria, E. K. Yucel, D. C. Brewster, G. L'Italien, J. P. Gertler, G. M. Lamuragha, J. A. Kaufman, A. C. Waltman, and W. M. Abbott, *J. Vasc. Surg.*, 1993, **17**, 1050.
11. S. M. Hertz, R. A. Baum, R. S. Owen, G. A. Holland, D. R. Logan, and J. P. Carpenter, *Am. J. Surg.*, 1993, **166**, 112.
12. R. A. Baum, G. A. Holland, D. R. Logan, J. P. Carpenter, and C. Cope, *J. Vasc. Intervent. Radiol.*, 1993, **4**, 59.
13. R. A. Baum, G. A. Holland, D. R. Logan, J. P. Carpenter, and C. Cope, *J. Vasc. Intervent. Radiol.*, 1993, **4**, 15.
14. R. A. Baum, G. A. Holland, D. R. Logan, S. Hertz, J. P. Carpenter, R. S. Owen, and C. Cope, *J. Vasc. Intervent. Radiol.*, 1993, **4**, 60.
15. M. L. Schiebler, J. Listerud, G. A. Holland, R. Owen, R. Baum, and H. Kressel, *Invest. Radiol.*, 1992, **27**, S90.
16. D. J. Bryant, J. A. Payne, D. N. Firman, and D. B. Longmore, *J. Comput. Assist. Tomogr.*, 1984, **8**, 588.
17. G. L. Naylor, D. N. Firman, and D. B. Longmore, *J. Comput. Assist. Tomogr.*, 1986, **10**, 715.
18. M. Koch, S. E. Maier, I. Baumgartner, K. D. Hagspiel, C. Von Weymarn, P. Boesinger, A. Bollinger, and G. K. Von Schultess, *Proc. Xth Ann. Mtg. Soc. Magn. Reson. Med., San Francisco, 1991*, p. 137.
19. V. Dousset, F. W. Wehrli, A. Louie, and J. Listerud, *Radiology*, 1991, **179**, 437.
20. C. L. Dumoulin, D. J. Doorly, and C. G. Caro, *Magn. Reson. Med.*, 1993, **29**, 44.
21. L. R. Pelc, N. J. Pelc, S. C. Rayhill, L. J. Castro, G. H. Glover, R. J. Herfkens, D. C. Miller, and R. B. Jeffrey, *Radiology*, 1992, **185**, 809.
22. G. R. Caputo, T. Masui, G. A. Gooding, J. M. Chang, and C. Higgins, *Radiology*, 1992, **182**, 387.

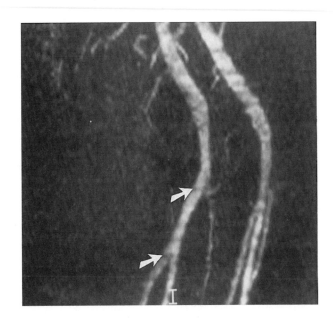

Figure 8 Oblique magnetic resonance arteriogram projection through the pelvis demonstrating high and low signal stripes (arrows) representing banding from pulsatile flow

23. G. R. Caputo, C. B. Higgins, *Invest. Radiol.*, 1992, **27**, S97–S102.
24. J. S. Swan, D. M. Weber, T. M. Grist, M. M. Wojtowycz, F. R. Kovosec, and C. A. Mistretta, *Radiology*, 1992, **184**, 813.

Biographical Sketches

Cathy Maldjian. *b* 1965. B.A. Columbia University, New York, 1986; M.D. University of Medicine and Dentistry of New Jersey, 1990. Diagnostic Radiology Residency, Mount Sinai Medical Center (New York), 1995. Magnetic Resonance Imaging Fellowship at the Hospital of the University of Pennsylvania following residency. Approx. 10 publications. Research interests include clinical applications of MRI.

Mitchell D. Schnall. B.A., physics, University of Pennsylvania, 1982, M.D., Ph.D., University of Pennsylvania, 1986. Residency in Radiology at the Hospital of the University of Pennsylvania, 1991. Assistant Professor of Radiology at the University of Pennsylvania 1991–1994. Currently Associate Professor of Radiology and Chief of the MRI Section.

Skeletal Muscle Evaluated by MRI

James L. Fleckenstein

University of Texas Southwestern Medical Center, Dallas, TX, USA

1 INTRODUCTION

Clinical evaluation of skeletal muscle has long been hampered by difficulty in assessing the morphology and functional integrity of skeletal muscles. Proton magnetic resonance imaging (MRI) represents a major advance in the diagnosis and management of patients with muscle disease by probing beyond the relatively bland surface of skin to identify focal muscle structural lesions, to determine their extent, and to characterize their composition, and guide invasive procedures and monitor therapies. The purpose of this chapter is to review advances made by MRI in understanding the quality of muscle in health and disease.

2 MR TECHNIQUES

MRI is used to examine patients' muscular anatomy noninvasively and determine which muscles are abnormal in size and shape. MRI further characterizes the quality of muscle by discriminating between the mesenchymal alterations of muscle fat and edema, depending on the pulse sequences used.[1–4] Fat

is detected on T_1-weighted images (short *TR*, short *TE*) by high signal intensity due to its short T_1 time constant. Because fat has a long T_2 time constant, it also manifests high signal intensity on T_2-weighted (long *TR*, long *TE*) sequences.

Increased tissue water leads to increased spin density and elevated T_1 and T_2 relaxation times; hence, muscle edema is detectable using MRI.[1,2] The long T_1 can be manifested by decreased signal intensity on heavily T_1-weighted spin echo images. However, in cases in which the sequence is also sensitive to changes in proton density and T_2, the change in T_1 is frequently not sufficient to result in a net change in signal intensity.[2]

Using conventional T_1-weighted and T_2-weighted spin echo sequences, difficulty is sometimes encountered in differentiating intramuscular fat from edema, especially when they coexist. This is because both edema and fat are hyperintense to muscle on long *TR*/long *TE* sequences. This accounts in part for why fat-suppression techniques improve detection of muscle edema.

Multiple fat suppression sequences have been developed that improve detection of muscle edema.[1–3] One of these employs an inversion pulse that nulls signal from tissue having a T_1 time equal to that of fat [short inversion time inversion recovery (STIR)]. STIR has the additional advantage of producing heightened lesion conspicuity due to additive effects to signal intensity caused by edema-associated increases in lesion spin density and T_1 and T_2 times.[3] Other sequences employ frequency selective pulses to null the fat signal. These techniques generally require a longer *TE* to achieve the same degree of lesion conspicuity as STIR sequences at the same *TR*. Important from a financial perspective, a variety of fat-suppression sequences can be incorporated into fast scan techniques, so that high sensitivity to muscle edema can be realized with short scan times. This makes muscle imaging an economically viable, as well as informative, application of MRI.

3 NORMAL MUSCLE ANATOMY AND PHYSIOLOGY

Like computerized tomography and ultrasound, MRI aids physiological studies of muscle size by providing quantitative morphological data regarding the mass of muscle performing work. While cross-sectional area (CSA) has most frequently been used for this purpose, the safety and high spatial resolution of MRI allow for accurate determinations of muscle volume, eliminating errors inherent in modeling volume estimates from single CSA measurements. Incorporation of the muscle fiber pennation angle into estimates of size has further defined morphological determinants of muscle strength.[5] The ability of MRI to accurately define muscle fascicle architecture also enhances MRI assessment of muscle morphology.[6] A variety of these techniques has been used to monitor changes in muscle volume during training[7] and detraining.[8]

Compositional alterations of muscle that can potentially be quantitated with MRI include fiber type, water content and compartmentation, and fat content.[9–25] MRI of muscle fiber type has been studied in both animals and humans, although with conflicting results. In rat soleus muscle, which is nearly all type I (slow twitch, oxidative) fibers, T_1 and T_2 relaxation

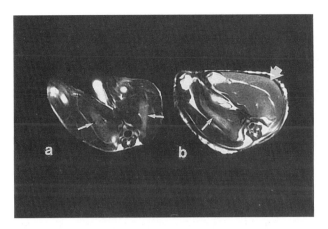

Figure 1 Effects of fiber type heterogeneity and ischemic muscle contraction visible by MRI. Coronal 2000/60 images of rabbit thighs before (a) and after (b) ischemic muscle stimulation. Note on the prestimulation image that a single muscle (semitendinosus, arrows) has a higher signal intensity than nearby muscles. This is a normal finding in this breed of rabbit and reflects a high proportion of oxidative fibers. After stimulation of the left sciatic nerve, signal intensity increases markedly within the left lateral thigh muscles (arrowhead). Because aortic ligation and death preceded the stimulation, blood flow could not have contributed to signal intensity changes in the stimulated muscles, supporting data in humans that blood flow is not critical in mediating transient MRI changes of signal intensity due to exercise. (Reproduced by permission of the Radiological Society of N. America from J. L. Fleckenstein, R. G. Haller, L. A. Bertocci, R. W. Parkey, and R. M. Peshock, *Radiology*, 1992, **183**, 25)

times are longer than in gastrocnemius muscle, which has a greater proportion of type II (fast twitch, glycolytic) fibers. This difference correlates with a greater extracellular water content in the soleus muscle.[9] Muscles having such markedly disparate fiber type proportions can easily be differentiated from other muscles on MRI (Figure 1).[2] Although human muscle has much less variability in fiber type proportion than does animal muscle, a study in humans indicated that the percentage of type II fibers positively correlated with T_2 times.[10] This result is the opposite of what would be predicted from the animal data and so additional studies are needed to determine the accuracy and validity of this MRI application.

MRI measurement of muscle water content can be applied not only to studies of fiber typing, but also to changes that occur as a result of muscular contraction,[11–22] diuresis,[11] and denervation atrophy.[23,24] As an example, the effects of exercise on the MRI appearance of exercise muscle will be examined in detail. During low-intensity muscular contractions, increases of muscle water primarily occur in the extracellular space; during maximal intensity exercise small increases also occur in the intracellular water space.[26] These changes in muscle water content/compartmentation are associated with transient increases in muscle spin density and T_2 times, while T_1 times are relatively less affected.[12] Interestingly, postexercise hyperemia is neither required[12,13] nor sufficient[14] for the effect to be observed (Figure 1). Animal studies suggest that increases in muscle extracellular water content underlie an increase in the number of proton spins having long T_2 decay times.[9,11] This has led some investigators to propose that an increase in this water space is the primary determinant of the changes visible

on MRI.[12,13,15] A role for changes in intracellular water content has also been proposed[16] and debated.[12] These MRI-visible alterations in muscle water are speculated to result from increased muscle osmolality due to accumulation of lactate and other ions which cause osmotic shifts of water between intracellular and extracellular water compartments. This conclusion is supported by absence of the normal postexercise muscle T_2 variations in patients in whom muscle lactate accumulation is absent due to defective muscle phosphorylase (glycogenosis V, McArdle's disease)[14] and by a high correlation between the magnitude of T_2 change and fall in pH in healthy volunteers.[17]

Magnetization transfer contrast techniques, exploiting selective saturation of a pool of 'bound' water protons, were applied in two studies to improve the understanding of water compartmentation during MRI of exercise in humans.[15,18] However, the studies provided conflicting results and conclusions. Another area of apparent controversy regards changes in the T_2 time as a function of work performed. While some studies suggested that T_2 increases in direct proportion to work intensity, the linearity of this relationship has been questioned.[19] Taking these data together it is likely that T_2 varies linearly with work in some regimens but not in others. Recognition that a limit exists in the magnitude of T_2 response (~30%)[19] suggests the existence of a limit to the amount of water that muscle can imbibe from the vasculature during exertion and/or to the magnitude of changes of muscle water compartmentation/binding that can occur during exercise.

Although the precise mechanism(s) involved in exercise-induced changes in muscle relaxation times during exercise remains unknown, the changes in image contrast between strongly recruited muscle and less active muscle have been exploited in a number of interesting practical applications: diagnosis of disorders of muscle energy metabolism;[14] chronic exertional compartment syndromes;[20] identification of muscle recruitment patterns relevant to MR spectroscopy studies of exercise (Figure 2);[21,22] assessment of manufacturers' claims of

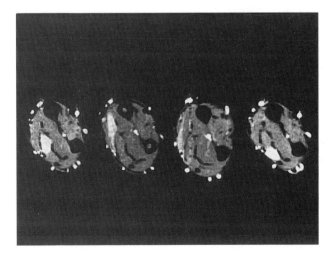

Figure 2 Finger-specific components of the flexor digitorum superficialis. Four discrete parts of this muscle can be demonstrated by selective exercise of individual fingers. From left to right are the index, long, ring, and small components of the flexor digitorum superficialis at the mid-forearm. The bones are the radius and ulna. (Reproduced by permission of Raven Press from J. L. Fleckenstein et al.[2])

For References see p. 1329

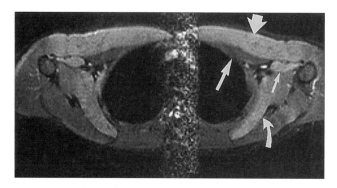

Figure 3 MRI of 'fly' exercise. Using a commercially available exercise device this healthy subject performed arm abduction against resistance within 1 min of scan acquisition. Note that the pectoralis major (arrowhead), subscapularis (curved arrow), and coracobrachialis (small arrow) are strongly stressed while the pectoralis minor is relatively unstressed (arrow). Therefore, the device does not stress the pectoralis group homogeneously, despite claims of the manufacturer to the contrary

muscle recruitment patterns using commercially available exercise equipment (Figure 3).

4 MRI OF MUSCLE PATHOLOGY

MRI has been applied to the evaluation of a broad range of neuromuscular and orthopedic muscle disorders. One of the critical issues that faces a clinician evaluating a patient with neuromuscular disease is determining whether the disease is primarily neurogenic or muscular in origin. This is of particular importance in pediatrics because patients with spinal muscular atrophies (SMA) may present with similar clinical findings to patients with muscular dystrophies, particularly of the Duchenne type (DMD). Early attempts to differentiate SMA from DMD employed ultrasound of the extremities and reported that in SMA the overall volume of muscle was decreased, compared with that of subcutaneous fat.[27,28] As an additional diagnostic clue, MRI has disclosed selective sparing of specific muscles in DMD, particularly of the gracilis, semimembranosus, and sartorius.[25,29] A more recent study sought to distinguish SMA of the Kugelberg–Welander type (KW) from DMD.[30] It was reported that in KW, muscle deterioration tended to be more diffuse than in DMD and the previous finding of generalized muscle atrophy in SMA was corroborated. A tendency toward relatively selective involvement of type II muscles in DMD was also observed. While more work needs to be performed to assess the capability of MRI to distinguish between neurogenic and primary muscle diseases, results to date indicate that MRI may be helpful in this distinction.

Muscle dysfunction that results from peripheral neuropathy has also been evaluated with MRI.[24] MRI was found able to detect edema-like changes in muscles affected by traumatic or compressive peripheral nerve lesions (Figure 4). These changes were anticipated based on results from an animal study in which proton relaxation times were shown to be prolonged in denervated muscle due to fiber atrophy and a resultant increase in the extracellular water space.[23] Like electromyography, MRI was limited in its ability to detect muscle abnormalities in the first few weeks of denervation. On the other hand, denervation

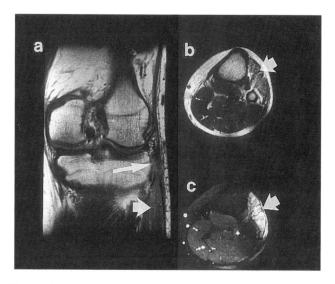

Figure 4 MRI of subacute leg muscle denervation: edema-like change. Lateral collateral ligament injury (arrow, 500/40) (a) and subsequent scar formation resulted in compression of common peroneal nerve (not shown). Note that signal intensity of denervated anterior leg muscles mimics edema, being normal on coronal T_1-weighted image (arrowhead) (a), and increased on axial T_2-weighted image (arrowhead, 2000/60) (b) and STIR (arrowhead, 1500/30/100) (c)

was readily visible on edema-sensitive sequences when denervation had occurred prior to 1 month before the MRI scan. While edema-like change dominated the appearance of muscle in the first year of denervation, fatty change of the muscle was observed in more long-standing denervation. Muscle hypertrophy is a relatively rare result of denervation, and while the finding may be prominent in patients with a remote history of poliomyelitis, it may also be observed relatively early after insult to peripheral nerves.[2]

Unlike disease of nerve and muscle, neuromuscular junction dysfunction, such as seen in myasthenia gravis, has yet to be reported to have an abnormal appearance on MRI, or any other imaging modality. On the other hand, primary myopathies, including dystrophies, idiopathic inflammatory myopathies, metabolic myopathies, and congenital myopathies, display various muscle abnormalities on MRI.[25,29,31,32] These findings, including edema-like change and fatty infiltration of muscles, tend to be nonspecific in terms of distribution. For example, selective sparing of the sartorius and gracilis is a feature not only of DMD (Figure 5), but also of polymyositis (Figure 6), congenital myopathies,[29] and metabolic myopathies (Figure 7). Selective involvement of the same muscles is a feature of some mitochondrial myopathies[2] but may be seen in centronuclear myopathy. The character of the imaging abnormality is also nonspecific, in that edema-like change on MRI may be seen in denervation, necrosis, and inflammation.[2] While not pathognomonic for specific disease processes, the MRI abnormalities are useful in directing invasive procedures, such as biopsy.[31,32] The objective nature of imaging abnormalities can also be used to monitor response to therapy, which is otherwise limited by the subjective aspects of the patients' sense of well being and by limitations in assessing muscular strength.

In the field of orthopedics, muscle injuries are extremely common and include muscle strains and contusions, delayed

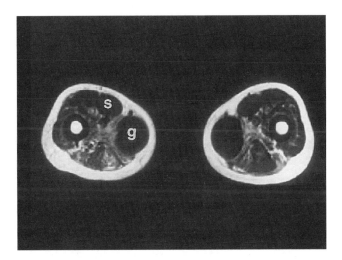

Figure 5 Muscle atrophy and hypertrophy in muscular dystrophy. Symmetric, proximal diminution in muscle volume is evident on a 500/30 sequence. Note sparing of the gracilis (g) and sartorius (s). (Reproduced by permission of Raven Press from J. L. Fleckenstein et al.[2])

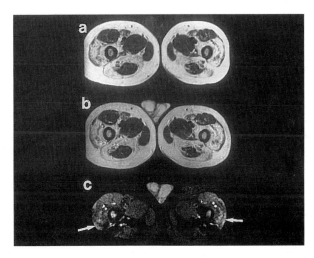

Figure 7 Glycogenosis: T1W (500/30) (a) and T2W (2000/60) (b) images of the thighs in a patient with phosphofructokinase deficiency demonstrate marked replacement of most thigh muscles by high signal intensity fat. STIR (c) suppresses signal intensity from fat, while areas of high signal intensity identify regions of coexistent muscle edema in the vastus lateralis (arrows). Such edematous areas should be avoided during biopsy when glycogenoses are considered because muscle necrosis may produce a small amount of fetal myophosphorylase in patients who usually lack the adult form of that enzyme (McArdle's disease, Glycogenosis V)

onset muscle soreness, and chronic overuse syndromes.[33–37] Associated injuries and sequelae of injuries are important determinants of the prognosis of muscle injuries. Because MRI is sensitive to both muscle trauma and associated abnormalities, it has been aggressively applied to these orthopedic issues.

Muscle strain is defined as an indirect injury to muscle caused by excessive stretch. Although various clinical schemes of grading severity of muscle strains have been advanced, it is acknowledged that clinical evaluation of muscle strains is diffi-

cult, even more so than injuries of tendons or bones. MRI aids in assessing integrity of muscles, myotendinous junctions, fascia, and the tendoosseous unit. MRI identifies edema within and/or around injured muscles, depending on the stage of healing.[33,34]

The myotendinous junction is frequently the point of rupture and the extent of associated fascial or tendinous tear can be addressed by MRI (Figures 8–10). When a fascial or tendinous tear is small (Figure 8), the injury can safely be managed con-

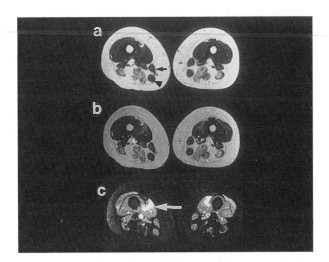

Figure 6 Chronic polymyositis. The extensive high signal intensity throughout the thigh muscles on 500/30 (a), and 2000/60 (b), images implies fatty change. STIR suppresses signal intensity from fat, while high signal intensity identifies regions of coexistent muscle edema (arrow) (c). Note that muscle edema is easy to identify only with fat suppression. Such edematous areas are of particular interest during biopsy when inflammatory myopathy is suspected. Note sparing of multiple muscles, including sartorius (arrow) (a) and gracilis (arrowhead) (a)

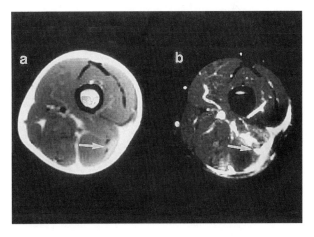

Figure 8 Important MRI findings in muscle injury: partial tendon tear. Associated injuries to fasciae and tendons are important to quantitate since small fibrous tears may require no operative intervention. Note partial biceps femoris tendon tear (arrow, 600/20) (a) and STIR (b), and the superior delineation of perifascial fluid using STIR. The lesion was conservatively managed. (Reproduced by permission of Raven Press from J. L. Fleckenstein et al.[2])

For References see p. 1329

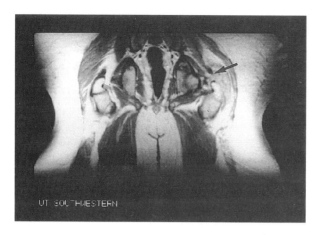

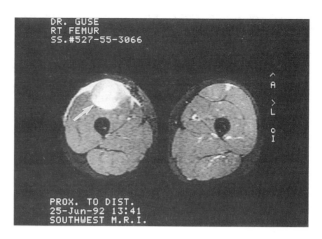

Figure 9 Important MRI findings in muscle injury: complete avulsion. When muscle tearing is complete, myotendinous avulsion occurs. Complete myotendinous avulsion of a juvenile football player's gluteus medius is easily appreciated on the coronal spin density-weighted image (arrow, 2000/30). The patient was treated by immobilization in a body cast for 12 weeks. (Reproduced by permission of Raven Press from J. L. Fleckenstein et al.[2])

Figure 11 Important MRI findings in muscle injury: large volume edema. Factors that suggest a longer convalescent period after injury include a large volume of muscle edema, such as in an axial STIR image of an acutely strained rectus femoris muscle (arrow). (Courtesy of S. C. Schultz, Fort Worth, TX)

servatively. However, when a rupture is complete, or nearly complete (Figures 9 and 10), early surgery may be indicated; a lapse of even a short time may cause an inferior functional result, due to muscle fibrosis and retraction. Fluid collections also frequently accompany strains. These can themselves be a cause of swelling and weakness in the absence of fascial tear. The use of MRI to distinguish focal hematomas from swollen, edematous muscles can guide clinical management; the former may benefit from drainage while the latter are often treated with wrapping procedures for compression and support of the injured area.[4]

Since recurrent muscle strains can devastate elite athletes, it is noteworthy that MRI in muscle injuries can provide information regarding the prognosis of muscle strain. Two studies have reported on MRI findings that are associated with poor outcome. These studies indicated that the occurrence of focal fluid collections (Figure 10), relatively large volume of abnormality (Figure 11) and fibrosis (Figure 12), correlate with either recurrence of muscle strain, delayed convalescent interval, or both.[35,36] It is interesting that MRI alterations of strained muscles typically persist for longer than any other clinical evidence of injury.[33,34] One could speculate that reinstitution of exercise in this setting might be harmful to the healing muscle, but this has not yet been studied. A more im-

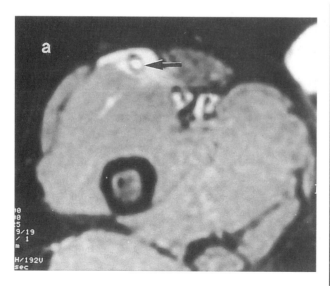

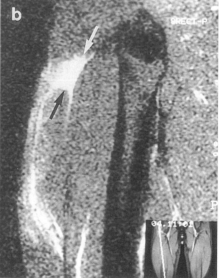

Figure 10 Important MRI findings in muscle injury: fluid collection. Rupture and retraction of the posterior head of the rectus femoris in an elite kicker with recurrent muscle strain reveals a 'ganglion-like' fluid collection within a portion of the rectus femoris on axial STIR image (arrow) (a), corresponding to fluid collecting between the retracted posterior head of the rectus femoris and its origin (arrows, sagittal STIR) (b). This injury was treated by resection of the avulsed muscle head

For list of General Abbreviations see end-papers

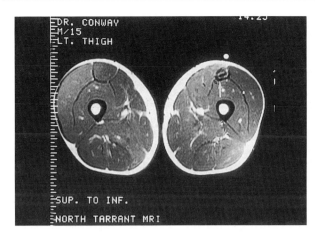

Figure 12 Important MRI findings in muscle injury: scar. MRI-visible scar formation in myotendinous tears is associated with recurrence of strains. Scar is characterized by excessive deposition of signal poor tissue and muscle atrophy (arrow). (Courtesy of S. C. Schultz, Fort Worth, TX)

mediately practical implication of the delayed disappearance of edema from muscle after injurious exercise is that one may detect evidence of previous muscle injury on MRI after the patient forgets the inciting event. This suggests a potential source of serendipitously observed MRI abnormalities. On the other hand, this radiographic finding may be the only clue to the true origin of the patient's musculoskeletal complaints.[4]

Other sequelae of muscle injury detectable by MRI include muscle calcification and ossification. Calcific myonecrosis is a delayed complication of muscle injury in which progressive calcification of an injured muscle is associated with slow development of a mass lesion of the extremities. This condition is a rare complication but its radiology is well described, allowing for a confident preoperative diagnosis.[37] Muscle ossification (myositis ossificans) is also detectable by MRI but its appearance is nonspecific early in its development, mimicking neoplastic disease. As it matures, ossified muscle may be identified by peripheral low signal intensity, corresponding to the outer zone of ossification (see *MRI of Musculoskeletal Neoplasms*).

5 CONCLUSION

MRI circumvents traditional obstacles in the clinical evaluation of muscle physiology and pathology. MRI detection of muscle size, fiber composition and orientation, and water shifts are promising research arenas in the field of exercise physiology. The high sensitivity of MRI in detecting muscle edema and fat allows improved delineation of the distribution and composition of neuromuscular and orthopedic disorders. This sensitivity can be used to substantiate muscle as the source of musculoskeletal pain, weakness, or stiffness in a broad range of patients and on the basis of positive, objective findings rather than by exclusion.

6 RELATED ARTICLES

Inversion–Recovery Pulse Sequence in MRI; Magnetization Transfer Contrast: Clinical Applications; MRI of Musculoskeletal Neoplasms; Peripheral Joint Magnetic Resonance Imaging; Peripheral Muscle Metabolism Studied by MRS.

7 REFERENCES

1. J. L. Fleckenstein, B. T. Archer, B. A. Barker, J. T. Vaughn, R. W. Parkey, and R. M. Peshock, *Radiology*, 1991, **179**, 499.
2. J. L. Fleckenstein, P. T. Weatherall, L. A. Bertocci, M. Ezaki, R. G. Haller, R. Greenlee, W. W. Bryan, and R. M. Peshock, *Magn. Reson. Q.*, 1991, **7**, 79.
3. A. J. Dwyer, J. A. Frank, V. J. Sank, J. W. Reinig, A. M. Hickey, and J. L. Doppman, *Radiology*, 1988, **168**, 827.
4. J. L. Fleckenstein and F. G. Shellock, *Top. Magn. Reson. Imaging*, 1991, **3**, 50.
5. T. Fukunaga, R. R. Roy, F. G. Shellock, J. A. Hodgson, M. K. Day, P. L. Lee, H. Kwong-Fu, and V. R. Edgerton, *J. Orthop. Res.*, 1992, **10**, 928.
6. S. H. Scott, C. M. Engstrom, and G. E. Loeb, *J. Anat.*, 1993, **182**, 249.
7. W. J. Roman, J. L. Fleckenstein, J. Stray-Gundersen, S. E. Alway, R. Peshock, and W. J. Gonyea, *J. Appl. Physiol.*, 1993, **74**, 750.
8. T. Fukunaga, K. Day, J. H. Mink, and V. R. Edgerton, *Med. Sci. Sports Med.*, 1991, **23**, 5110.
9. J. F. Polak, F. A. Jolesz, and D. F. Adams, *Invest. Radiol.*, 1988, **23**, 107.
10. S. Kuno, S. Katsuta, T. Inouye, I. Anno, K. Matsumoto, and M. Akisada, *Radiology*, 1988, **169**, 567.
11. E. Le Rumeur, J. de Certaines, P. Toulouse, and P. Rochcongar, *Mag. Reson. Imaging*, 1987, **5**, 267.
12. B. Archer, J. L. Fleckenstein, L. A. Bertocci, R. G. Haller, B. Barker, R. W. Parkey, and R. M. Peshock, *J. Magn. Reson. Imaging*, 1992, **2**, 407.
13. J. L. Fleckenstein, R. C. Canby, R. W. Parkey, and R. M. Peshock, *Am. J. Roentgenol.*, 1988, **151**, 231.
14. J. L. Fleckenstein, R. G. Haller, S. F. Lewis, B. T. Archer, B. R. Banker, J. Payne, R. W. Parkey, and R. M. Peshock, *J. Appl. Physiol.*, 1991, **71**, 961.
15. X. P. Zhu, S. Zhao, and I. Isherwood, *Br. J. Radiol.*, 1992, **65**, 39.
16. F. G. Shellock, T. Fukunaga, J. H. Mink, and V. R. Edgerton, *Am. J. Roentgenol.*, 1991, **156**, 765.
17. E. R. Weidman, H. C. Charles, R. Negro-Vilar, M. J. Sullivan, and J. R. MacFall, *Invest. Radiol.*, 1991, **26**, 309.
18. K. T. Mattila, M. E. Komu, S. K. Koskinen, and P. T. Niemi, *Acta Radiol.*, 1993, **34**, 559.
19. J. L. Fleckenstein, D. Watumull, D. D. McIntire, L. A. Bertocci, D. P. Chason, and R. M. Peshock, *J. Appl. Physiol.*, 1993, **74**, 2855.
20. A. Amendola, C. H. Rorabeck, D. Vellett, W. Vezina, B. Rutt, and L. Nott, *Am. J. Sports Med.*, 1990, **18**, 29.
21. J. L. Fleckenstein, L. A. Bertocci, R. L. Nunnally, R. W. Parkey, and R. M. Peshock, *Am. J. Roentgenol.*, 1989, **153**, 693.
22. J. L. Fleckenstein, D. Watumull, L. A. Bertocci, R. W. Parkey, and R. M. Peshock, *J. Appl. Physiol.*, 1992, **72**, 1974.
23. J. F. Polak, F. A. Jolesz, and D. F. Adams, *Invest. Radiol.*, 1988, **23**, 365.
24. J. L. Fleckenstein, D. Watumull, K. E. Conner, M. Ezaki, R. G. Greenlee, Jr., W. W. Bryan, D. P. Chason, R. W. Parkey, R. M. Peshock, and P. D. Purdy, *Radiology*, 1993, **187**, 213.
25. W. A. Murphy, W. G. Totty, and J. E. Carroll, *Am. J. Roentgenol.*, 1986, **146**, 565.

For References see p. 1329

26. G. Sjogaard, R. P. Adams, and B. Saltin, *Am. J. Physiol.*, 1985, **248**, R190.

27. J. Z. Heckmatt, N. Pier, and V. Dubowitz, *J. Clin. Ultrasound*, 1988, **16**, 171.

28. A. Lamminen, J. Jaaskelainen, J. Rapola, and I. Suramo, *J. Ultrasound Med.*, 1988, **7**, 505.

29. A. E. Lamminen, *Br. J. Radiol.*, 1990, **63**, 946.

30. D. Suput, A. Zupan, A. Sepe, and F. Demsar, *Acta Neurol. Scand.*, 1993, **87**, 118.

31. A. Pitt, J. L. Fleckenstein, R. G. Greenlee, D. K. Burns, W. W. Bryan, and R. G. Haller, *Mag. Res. Imaging*, 1993, **11**, 1093.

32. J. H. Park, S. J. Gibbs, R. R. Price, C. L. Partain, and A. E. James, Jr., *Radiology*, 1991, **179**, 343.

33. J. L. Fleckenstein, P. T. Weatherall, R. W. Parkey, J. A. Payne, and R. M. Peshock, *Radiology*, 1989, **172**, 793.

34. F. G. Shellock, T. Fukunaga, J. H. Mink, and V. R. Edgerton, *Am. J. Roentgenol.*, 1991, **156**, 765.

35. S. J. Pomeranz and R. S. Heidt, Jr., *Radiology*, 1993, **189**, 897.

36. A. Greco, M. T. McNamara, R. M. B. Escher, G. Trifilio, and J. Parienti, *J. Comput. Assist. Tomogr.*, 1991, **15**, 994.

37. D. L. Janzen, D. G. Connell, and B. J. Vaisler, *Am. J. Roentgenol.*, 1993, **160**, 1072.

Biographical Sketch

James L. Fleckenstein. *b* 1957. B.S. 1979, M.D. 1984, University of Washington. Medical internship, 1984–1985, University of Texas. Radiology residency 1985–1989, University of Texas. Assistant instructor, Radiology, University of Texas, 1989–1990. Director of Neuro MRI, Division of Neuroradiology, and Associate Professor of Radiology at University of Texas Southwestern Medical Center, 1991–present. Approx. 200 publications. Research specialties: MRI of muscle in health and disease.

Peripheral Joint Magnetic Resonance Imaging

Paul S. Hsieh

Kaiser Permanente Medical Center, San Diego, CA, USA

and

John V. Crues III

RadNet Management, Inc., Los Angeles, CA, USA

1 INTRODUCTION

MRI is an important tool in the radiological evaluation of peripheral joint disease. Previously available noninvasive imaging techniques include X-ray and ultrasound imaging. X-ray techniques (plain radiographs and computerized tomography) are able to image cortical bone but are not sensitive in imaging

medullary bone disease or intra-articular soft tissues.[1,2] Ultrasound can evaluate soft tissue structures but is limited in its ability to visualize internal structures in joints and detect some soft tissue diseases.[3] Nuclear medicine bone scintigraphy has also been used to detect articular pathology, but is not commonly used because of its lack of specificity and poor spatial resolution.[4,5] MRI is capable of demonstrating the anatomy of the various components of peripheral joints in exquisite detail, including muscles, tendons, ligaments, nerves, blood vessels, fat, and osseous structures.[6,7] In addition to its high spatial resolution, MRI displays excellent contrast between musculo-skeletal soft tissues because differences in chemical structures lead to different T_1 and T_2 values. These differences in relaxation times can be exploited to generate images with striking contrast between normal tissues and between normal and pathologic tissues. Intra-articular injection of contrast material (arthrography) allows X-ray techniques to visualize the surfaces of intra-articular structures, but X-ray arthrography is both invasive and insensitive to numerous pathologic conditions that do not manifest with surface irregularities.[8] MR arthrography is now widely used for the evaluation of specific articular pathology because it combines the sensitivity of arthrography for surface abnormalities with high soft-tissue contrast.[9–11] For these reasons, many consider MRI to be the modality of choice for noninvasive imaging of peripheral joint disease.

2 BASIC TECHNIQUES

Most MRI of peripheral joints is performed with two-dimensional spin echo pulse sequences.[6] This is a basic (90–180°) sequence, with the time between successive 90° pulses denoted *TR* and that between 90° pulse and the signal acquisition denoted *TE*. A slice selection gradient is applied during each rf pulse to limit the excitation to the desired anatomic plane of interest. Perpendicular gradients are also used to provide frequency and phase encoding. The basic pulse sequence is repeated 128, 192, or 256 times, each time with a different phase-encoding gradient. The signals or echoes from each acquisition are collected and processed using a two-dimensional Fourier transform to generate the final medical image. Many varieties of the basic scheme are now in clinical use.[12]

By adjusting the *TR* and *TE*, the signal can be manipulated to have lesser or greater degrees of T_1 or T_2 contrast. Hence, an image acquired with a sequence with a short *TR* (~600 ms) and short *TE* (~15 ms) is referred to as a T_1-weighted image. Similarly, if the *TR* is long (~2000–4000 ms) and the *TE* is long (~80 ms), this is referred to as a T_2-weighted image. If the *TR* is long but the *TE* is short, then many refer to this as a proton density-weighted image.

Most structures in peripheral joints have fairly characteristic T_1 and T_2 values that make them easy to distinguish on the spin echo images. For instance, fat has a short T_1, so it has bright signal intensity on T_1-weighted images. Ligaments, tendons, and bone cortex have a paucity of mobile protons and therefore demonstrate low signal intensity on all pulse sequences. Muscle tissue also has a fairly long T_1 and therefore looks dark on a T_1-weighted image, although not as dark as tendon or ligament. Bone marrow can have intermediate to bright intensity on T_1-weighted images depending on the fat content (nonhematopoietic or 'yellow' marrow has more fat than hematopoietic or 'red' marrow). Fluid within joints has a long T_1 and a long T_2; hence it will appear dark on T_1-weighted images and bright on T_2-weighted images.

Most pathologic tissues have increased edema compared with their normal counterparts. This increased water content causes prolongation of their T_1 and T_2 values. Hence, on T_1-weighted images, abnormal tissues tend to look dark, whereas on T_2-weighted images, these tissues generally look abnormally bright. Most lesions are fairly obvious on T_2-weighted images. However, the presence of bright signal on a T_2-weighted image is a nonspecific finding: many disease processes (i.e., tumor, infection, trauma, etc.) can cause this appearance and other clues (such as lesion morphology and location) must be sought in order to make a more specific diagnosis.

In particular, abnormalities within ligamentous, tendinous, and other fibrocartilaginous structures (such as the menisci in the knee and the labra in the shoulder) manifest themselves as foci of abnormally increased signal within a usually homogeneously dark structure.[4,13] Normally, there are few mobile water molecules within these structures capable of generating any significant signal. The water that does exist within them is bound to the large macromolecules such as cartilage and is incapable of free translation and rotation.[14] The protons within these bound water molecules have very short T_2 values and therefore do not generate any detectable signal. However, if the collagen matrix undergoes degeneration with microscopic tears, then small amounts of water can be trapped in the interstices. These water molecules are more mobile and therefore have slightly longer T_2 values, which can be detected on short TE images (i.e., T_1-weighted and proton density-weighted images). The T_2 values are still too short to generate signal on longer TE images (i.e. true T_2-weighted images). There may also be a secondary T_1 shortening effect that might contribute signal on T_1-weighted images. With degenerative disruption of macromolecules, water protons may be exposed to protons deep in the macromolecules. This close proximity between the two sets of protons allows for some magnetization exchange to occur, with resulting shortening of the T_1 time.[14] The hallmark of this type of pathologic process is increased signal on the proton density-weighted images but *not* on the T_2-weighted images.[13]

In the presence of more significant trauma, macroscopic amounts of free fluid can be present. Clinical examples would include complete rotator cuff tears and large meniscal tears with development of intrameniscal cysts.[4,13] In these settings, the abnormal area would show increased signal on *both* the proton density- and T_2-weighted images (see Figures 1 and 2).

Another commonly used pulse sequence is the STIR or *s*hort *t*au *i*nversion *r*ecovery sequence. This pulse sequence consists of ($180°-\tau-90°-180°$), where τ is the delay time between the initial $180°$ inversion pulse and the $90°$ pulse. In a STIR sequence, τ is set to the null point of fat, i.e. the length of time it takes for the fat to recover to a point of zero longitudinal magnetization after an initial inversion. At a typical field strength of 1.5 T, the appropriate value of τ is approximately 160 ms.

In a STIR sequence, the signal intensity of the tissue is directly related to its T_1 and T_2 values. Tissues with short T_1 and T_2 values such as fat show little or no signal, whereas tissues with long T_1 and T_2 values have bright signal. Because most pathologic tissues have prolonged T_1 and T_2 relaxation times they are bright on STIR images even more so than on

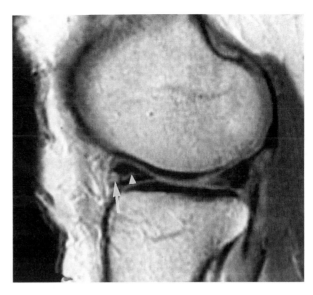

Figure 1 A sagittal proton density-weighted image of a knee (*TR* 2000, *TE* 25) demonstrating tear within the lateral meniscus. These are visualized as increased signal (white arrowhead) within the normal low signal of the menisci. A larger area of bulk fluid is seen anteriorly (white arrow) representing an intrameniscal cyst

T_2-weighted images. This is especially valuable when imaging at mid and low magnetic fields (0.2–0.5 *T*). However, as with a T_2-weighted image, bright signal is not specific for any one disease process and other information (such as morphology and location of the lesion) must be considered to make a more specific diagnosis.

Another set of pulse sequences commonly used in musculoskeletal MRI is the RARE sequence,[15] which is also known as *f*ast *s*pin *e*cho (FSE) or *t*urbo *s*pin *e*cho.[16] In this pulse sequence, an initial $90°$ pulse is followed by a rapid succession

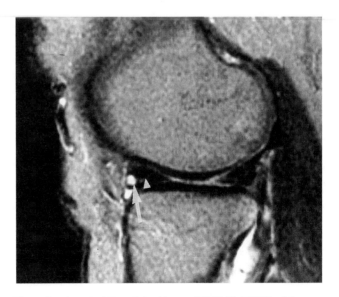

Figure 2 A sagittal T_2-weighted image (*TR* 2000, *TE* 80) of the same knee as in Figure 1 clearly showing the bulk fluid (white arrow) of the intrameniscal cyst. The T_2 of the bulk fluid is prolonged, making this fluid visible. However, the meniscal tears (white arrowhead) are considerably less apparent because their T_2 values are not as prolonged

For References see p. 1336

of 180° pulses to generate a series of echoes known as an echo train. Typically 4, 8, or 16 echoes are generated per excitation, each preceded by a different phase-encoding gradient. Hence, several lines of k-space are acquired per excitation, resulting in a considerable saving in scanner time. However, the TE values are not uniform within the image: some of the echoes making up the image will have shorter TEs, whereas others will have longer TEs. This can result in a variety of image artifacts including blurring and loss of resolution and changes in image contrast.[17] Also, because some pathological processes (such as meniscal tears in knees) demonstrate increased signal on short TE sequences but not on long TE sequences, it is possible that the mix of long and short TEs in a FSE image might make it less sensitive to subtle lesions than a standard spin echo image.[18] For this reason, at our institution the following compromise is used: for the key anatomic plane of a joint (which is the sagittal plane for knees and the oblique coronal plane for shoulders), a slower standard spin echo pulse sequence is used to generate proton density and T_2-weighted images. For the other anatomic planes, FSE is used. In this way, a reasonable balance between scanning time and diagnostic accuracy is maintained. However, as interpreters get more experience with the vagaries of these sequences on individual scanners, more sequences are being converted to the faster techniques. Spectral fat saturation is often advocated with the use of FSE imaging in the musculoskeletal system to eliminate increased signal from fat on FSE T_2-weighted images.[19]

Gradient echo pulse sequences are only infrequently used in MRI of peripheral joints. They are fairly sensitive for the same pathological processes that are detected by short TE spin echo sequences. However, the soft tissue contrast is worse with gradient echo sequences than with spin echo sequences. This is because the contrast in gradient echo sequences is dependent on differences in T_2^* rather than in T_2.[20] In clinical MR imaging, most of the T_2^* effect is caused by field inhomogeneities and other factors not dependent on the biomedical make-up of the tissue. Hence, the signal from the tissues undergoes T_2 decay before the differences in the true tissue T_2s can manifest themselves. For this reason, most soft tissues have very similar appearances and signal intensities on gradient echo images, regardless of the tissue type or the degree of involvement by pathology. Nevertheless, some investigators use gradient echo sequences because the short TRs allow 3-dimensional Fourier transform acquisitions in a reasonable time period for isotropic high-resolution imaging. This may be valuable in imaging the labrum and articular cartilage.[13,21]

In certain settings, intravenous or intraarticular gadolinium may prove useful for diagnosis. Gadolinium is a paramagnetic metal which can be bound to an organic chelating agent such as diethylenetriaminepentaacetic acid (DTPA). In the doses used in clinical practice, the main effect of intravenous gadolinium is to shorten the T_1 relaxation times of the perfused tissues. This causes them to appear brighter on T_1-weighted images. Some investigators have found this useful in evaluating possible recurrences of soft-tissue neoplasms following surgical resection.[22,23] Intravenous gadolinium can also be useful in determining if a soft-tissue mass with long T_1 and long T_2 is cystic or solid. A cystic lesion would not show any signal enhancement following contrast administration whereas a solid lesion would. Similarly, in the setting of a soft-tissue infection, it can be very difficult to tell the difference between an area of

phlegmonous and inflamed solid tissue and a drainable fluid collection. On a postgadolinium image, the phlegmon should enhance, whereas the fluid collection should not enhance.[24,25]

Gadolinium can also be injected into a joint space in the form of a dilute mixture of saline and gadolinium-DTPA.[9,10] This forms a positive contrast on T_1-weighted images that distends the joint capsule and helps delineate adjacent structures. This type of MR arthrography has proven particularly useful in the shoulder, where subtle abnormalities in the cartilaginous labra can be better appreciated following contrast injection.[26] However, this converts the MR study from a noninvasive procedure to an invasive procedure, with resultant discomfort and potential risk to the patient. (Some radiologists also inject saline without gadolinium, which produces a similar arthrographic effect on T_2-weighted images).[26] MR knee anthrography is helpful in some patients who have had previous meniscal repair.[27]

3 DISEASE ENTITIES

3.1 Trauma

MR has proven to be a valuable tool in the evaluation of acute trauma to peripheral joints as well as in evaluation of chronic trauma and degeneration. For instance, within the knee joint (the most commonly imaged peripheral joint), MR can be used to detect partial and complete tears of key anatomic structures including the medial and lateral collateral ligaments, the anterior and posterior cruciate ligaments, and the medial and lateral menisci.[6] Injuries to these structures can be detected by identifying abnormally increased signal within these normally low-signal structures. Other important radiographic signs include alterations in the contour and morphology of these structures, and the presence of edema in the surrounding soft tissues (manifested as high signal intensity on T_2-weighted images). These principles apply to evaluation of ligamentous and tendinous, and muscular structures at all joints, including the rotator cuff of the shoulder and the achilles tendon in the ankle (see Figures 1–3).[13,28,29]

MRI has also proven to be sensitive in the detection of radiographically occult fractures.[28,30] Often these fractures produce a characteristic linear pattern of bone marrow edema, which can be visualized as a region of low signal intensity within the marrow on T_1-weighted images and high signal intensity within the marrow on T_2-weighted images. Detection of these fractures can have important therapeutic implications, so MRI should be strongly considered in the setting of a patient with clinical findings suspicious for fracture with normal radiographs. Many investigators feel that MR is at least as sensitive for detection of such fractures as the other major technique, bone scinitigraphy, and is more specific (see Figure 4).[31,32]

Traumatic lesions of the articular cartilage and subchondral bone can also be detected with MRI. In an acute setting, these are known as osteochondral fractures, whereas in the setting of chronic repetitive microtrauma, these lesions are referred to as osteochondritis dissecans. Pertinent MR findings which may be seen include disruption or thinning of the articular cartilage, alterations of the contour of low-signal subchondral bony plate (which may be accompanied by abnormal high signal within

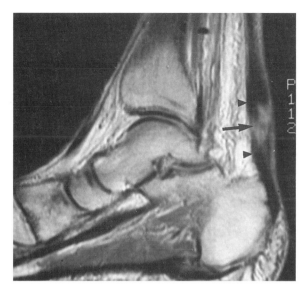

Figure 3   A sagittal proton density-weighted image (*TR* 2200, *TE* 20) of the ankle showing a tear of the achilles tendon. The tear is the area of increased signal (arrow) within the normally dark substance of the tendon (arrowheads). Notice the focal thickening of the tendon at the site of the tear

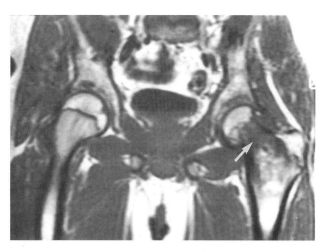

Figure 5   A coronal T_1-weighted image (*TR* 800, *TE* 20) of the hips showing marrow edema caused by osteomyelitis in the left femoral neck (white arrow). The signal intensity is similar to that of the edema caused by the fracture in Figure 4, but the abnormal area is more extensive and ill-defined. This is more compatible with an infectious process. Edema is a nonspecific finding; clinical history and morphologic clues are often necessary to distinguish between various disease entities

the normal low signal of cortical bone), and edema in the adjacent bone marrow.[33]

3.2   Infection

Another common use of MRI is in the evaluation of suspected infections of peripheral joints.[24,34] MR can detect the presence of abnormal fluid within a joint space, but cannot determine if the fluid is sterile or infected (i.e., if the patient has a bland joint effusion or a septic arthritis). MR is also useful in the evaluation of osteomyelitis. The principal finding in osteomyelitis is bone marrow edema (manifested by the usual low signal intensity in the marrow on the T_1-weighted images and high signal intensity on the T_2-weighted image). Osteo-

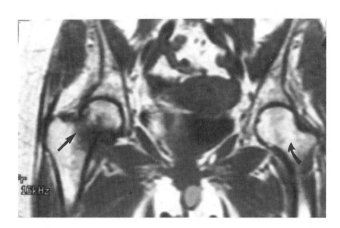

Figure 4   A coronal T_1-weighted image (*TR* 600, *TE* 15) of the hips showing a fracture of the right femoral neck (straight arrow). The abnormal low signal in the marrow is caused by edema. For comparison, the marrow signal in the left femoral neck is normal (curved arrow)

myelitis is also often accompanied by inflammatory changes in the adjacent soft tissues. If an area of soft tissue edema is identified, then MR with intravenous gadolinium administration can help to determine if there is a component of drainable fluid within the inflamed phlegmonous region.[24,25] This is extremely valuable in the evaluation of the diabetic foot.[34] As discussed above, a fluid collection should not enhance, whereas the phlegmonous component should (see Figure 5).

3.3   Arthritis

The role of MRI in evaluation of arthritis patients is fairly limited. MR can detect the findings seen in osteoarthritis, including the cartilage loss, subchondral sclerosis and cyst formation, and osteophytes.[35] In rheumatoid arthritis, MR may be helpful in delineating the exact extent of the inflamed pannus tissue. On standard MR images, the pannus can look similar to joint fluid, but after the administration of intravenous gadolinium, the pannus should enhance intensely. MR is also sensitive for early detection of erosions, which may also be helpful in patients with rheumatoid arthritis or other erosive arthritides.[36]

3.4   Ischemic Disease

MR is sensitive in the evaluation of ischemic bone disease including avascular necrosis and bone infarcts.[37,38] The devascularized portion of bone becomes edematous initially. Later, as fibrovascular tissue replaces granulation tissue the T_2 relaxation time shortens (Figure 6).

Some investigators have found that dynamic gadolinium enhancement studies are useful in assessing the prognosis of involved bone.[39,40] For instance, with avascular necrosis of the femoral head and scaphoid fractures, patients with lesions which enhanced following intravenous gadolinium adminis-

For References see p. 1336

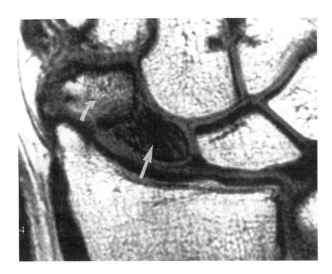

Figure 6 A coronal T_1-weighted image (*TR* 500, *TE* 21) of the wrist showing avascular necrosis of the proximal pole of the scaphoid bone (straight arrow), with abnormally low signal intensity within the marrow. The distal pole of the scaphoid bone has relatively normal signal intensity within the marrow (curved arrow)

tration may have an improved prognosis over patients whose lesions did not enhance. Further work remains to be done to see if this is true in other areas of the body.

3.5 Neoplasms

MR has proven valuable in staging of osseous and soft tissue neoplasms.[22,41–45] With MRI, one can evaluate the anatomic extent of the tumor and the integrity of adjacent neurovascular bundles. MR is helpful in determining whether a tumor of soft tissue origin involves adjacent bone and if a tumor of bony origin involves adjacent soft tissue structures. The extent of the bone marrow involvement can also be evaluated with MRI. Most neoplasms have prolonged T_1 and T_2 values, resulting in the typical dark appearance on T_1-weighted images and bright on T_2-weighted images. The contrast between the tumor and bone marrow or subcutaneous fat is most marked on T_1-weighted or STIR images because fat has such a short T_1 in contrast with the long T_1 of the tumor. On the other hand, T_2-weighted images are better at demonstrating the difference between the tumor and normal muscle or normal neurovascular structures.

These tumors are often surrounded by a halo of edema, which also has similar signal characteristics to the main tumor. In this case, the area of signal abnormality on the images is larger than the size of the actual tumor. In this setting, evaluation with intravenous contrast and fat-saturated T_1-weighted images may be valuable.[46,47]

MRI is not very specific for most tumor cell types. The only major exceptions are lipomas, which are nearly entirely composed of fat.[48] These can be identified by the characteristic signal intensity equal to normal fat on all pulse sequences, as well as by secondary signs such as chemical shift artifact. If a fat suppression pulse sequence is used (where rf energy is applied to the fat peak prior to image acquisition), or a STIR series is acquired, the signal from lipomas should disappear. If

a tumor demonstrates all of these signal characteristics, and does not contain any significant amounts of nonfatty tissue, then the diagnosis of benign lipoma can be confidently made.

Most soft tissue tumors have a nonspecific appearance, and the main role of MR is to evaluate for anatomic extent. This is useful in planning biopsies, surgical resections, and/or radiation therapy. MRI is also helpful in the evaluation of suspected recurrence following resection. In this setting, some investigators have found that intravenous gadolinium is helpful in distinguishing recurrent tumor from normal postoperative reaction.[29,49,50] Examples of bone and soft tissue tumors are shown in Figures 7–10.

4 FUTURE DIRECTIONS

There are several technical advances that may have promising applications in musculoskeletal MRI. One of these is MR microscopy. With current clinical MRI, the spatial resolution is on the order of 0.5–1.0 mm per pixel. With specialized gradients and other MR microscopy techniques, this can be improved by a factor of approximately 10. If this is done, then it becomes possible to evaluate structures such as articular cartilage in much greater detail. Instead of appearing as a thin stripe of signal, 2 or 3 pixels thick, the articular cartilage will be a broad band, 20 or 30 pixels thick. (See Figure 11). It may then be possible to evaluate subtle pathology within the various sublayers of articular cartilage.[51–53] Any technique that can

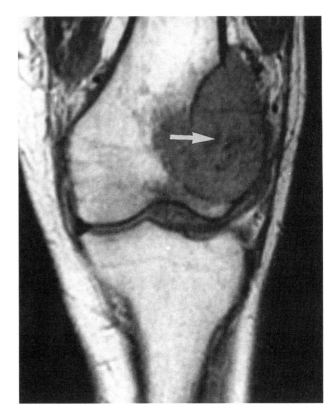

Figure 7 A coronal T_1-weighted image (*TR* 800, *TE* 20) showing a large destructive mass in the lateral femoral condyle (arrow), caused by metastatic renal cell carcinoma

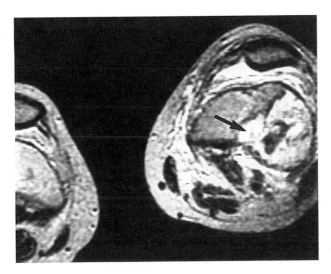

Figure 8 A transaxial T_2-weighted image (*TR* 2117, *TE* 80) of the same lesion shown in Figure 7, showing increased signal intensity within the mass (arrow) caused by T_2 prolongation

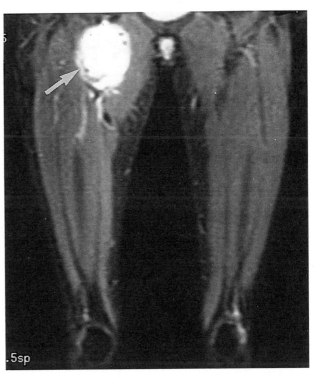

Figure 10 A coronal STIR image (*TR* 2200, *TE* 35, τ 160) showing the same lesion as that shown in Figure 9 (arrow). The mass is much more conspicuous on the STIR image

determine the presence or absence of early changes in rheumatological disease is also helpful, particularly for evaluating the efficacy of various treatments.

Other potentially important advances are the various diffusion and perfusion imaging techniques. The musculoskeletal system is well suited for application of these techniques because these body parts can be kept fairly immobile during

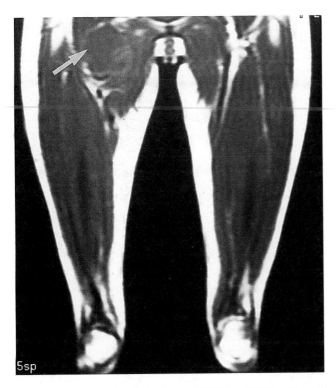

Figure 9 A coronal T_1-weighted image (*TR* 600, *TE* 20) of the right thigh showing a round mass in the medial soft tissues (white arrow). The mass (a malignant fibrous histiocytoma) is difficult to identify because it has similar signal intensity to normal muscle on this sequence

scanning, which is a requirement for successful use of these techniques. Although this has not yet been proven, it is possible that some pathological processes may manifest themselves as alterations in either local perfusion or alterations in the local diffusion coefficients before one sees evidence of gross edema and prolongation of T_1 and T_2 relaxation times.[54]

Magnetization transfer contrast is another technique that may prove useful in musculoskeletal application, particularly in articular cartilage. Some investigators have demonstrated a large magetization transfer effect due to the high degree of interaction between water and collagen macromolecules in articular cartilage. Specifically, it has been shown that the majority of the magnetization transfer effect is caused by interactions between water and the collagen matrix, and not between water and the proteoglycan component of cartilage.[55] This finding may prove useful in evaluation of subtle cartilaginous pathology, perhaps in conjuction with MR microscopy.

MR spectroscopy has existed for many years. Some work has been done using spectroscopy to evaluate metabolic diseases of muscles, but none of these techniques is in routine clinical use as yet. In recent years several manufacturers have developed smaller, relatively inexpensive scanners designed to scan extremity joints. These scanners are often 50 to 80% less expensive than traditional whole-body scanners and can be installed in standard clinic examining rooms (e.g. Esaote, Genoa, Italy). Though these scanners typically operate at low magnetic fields (0.2 T) and produce relatively noisy images, the price, convenience, and comfort is highly attractive to

For References see p. 1336

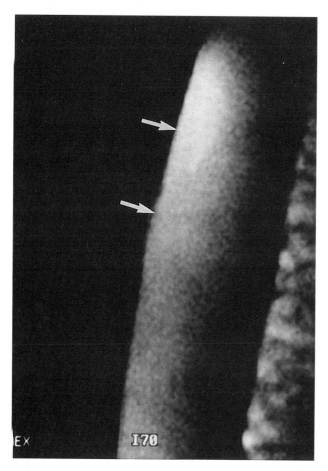

Figure 11 A T_1-weighted MR microscopy image showing the articular cartilage as a thick band of intermediate signal intensity (arrows)

many patients. These devices are currently the fastest growing segment of the MR market.

5 CONCLUSIONS

In summary, MRI is a powerful tool in noninvasive evaluation of abnormalities of peripheral joints. The combination of high spatial resolution and sensitivity to local alterations in water content and T_1 and T_2 relaxation times makes it ideal for demonstrating pathological processes. MR has a role in the evaluation of many disease processes including trauma, infection, vascular compromise, and neoplasm. Additional techniques that may prove clinically useful in the future include MR microscopy, diffusion and perfusion imaging, magnetization transfer contrast, MR spectroscopy, and speciality scanners.

6 RELATED ARTICLES

Gadolinium Chelate Contrast Agents in MRI: Clinical Applications; Gadolinium Chelates: Chemistry, Safety, and Behavior; MRI of Musculoskeletal Neoplasms; Skeletal Muscle Evaluated by MRI.

7 REFERENCES

1. J. Tehranzadeh, W. Mnaymneh, C. Ghavam, G. Morillo, and B. J. Murphy, *J. Comput. Assist. Tomogr.*, 1989, **13**, 466.
2. W. P. Chan, P. Lang, M. P. Stevens, K. Sack, S. Majumdar, D. W. Stoller, C. Basch, and H. K. Genant, *Am. J. Roentgenol.*, 1991, **157**, 799.
3. P. Dragorot and C. Claussen, *ROFO*, 1980, **133**, 185.
4. J. V. Crues and D. W. Stoller, in 'Magnetic Resonance Imaging of the Knee', 2nd edn., eds., J. Mink, M. Reicher, J. V. Crues, and A. Deutsch, Raven Press, New York, 1993, pp. 91–140.
5. I. P. C. Murray, J. Dixon, and L. Kohan, *Clin. Nucl. Med.*, 1990, **15**, 828.
6. J. H. Mink, M. A. Reicher, J. V. I. Crues, and A. L. Deutsch, 'Magnetic Resonance Imaging of the Knee', 2nd edn. eds. J. Mink, M. Reicher, J. V. Crues, and A. Deutsch, Raven Press, New York, 1993, p. 474.
7. D. W. Stoller, 'Magnetic Resonance Imaging in Orthopedic and Sports Medicine', Lippincott, Philadelphia, PA, 1993, 1127.
8. D. Resnick and M. Niwayama, eds. 'Diagnosis of Bone and Joint Disorders', 2nd edn. W. B. Saunders, Philadelphia, PA, 1988, Vol. 2, 4174+.
9. B. Flannigan, S. Kursunoglu-Brahme, S. Snyder, R. Karzel, W. Del Pizzo and D. Resnick, *Am. J. Roentgenol.*, 1990, **155**, 829.
10. J. Hodler, S. Kursunoglu-Brahme, S. J. Snyder, V. Cervilla, R. P. Karzel, M. E. Schweitzer, B. D. Flannigan, and D. Resnick, *Radiology*, 1992, **182**, 4316.
11. J. W. Helgason, V. P. Chandnani, and J. S. Yu, *Am. J. Roentgenol.*, 1997, **168**, 1473.
12. J. V. I. Crues, in 'MRI of the Knee', eds. J. H. Mink et al., Raven Press, New York, 1993, p. 1.
13. J. V. Crues and R. Ryu, in 'Magnetic Resonance Imaging', eds. D. Stark and W. G. Bradley, Mosby, St. Louis, MO, 1992, Vol. 2, pp. 2424–2458.
14. S. Koenig, R. D. Brown III, and R. Ugolini, *Magn. Reson. Med.*, 1993, **29**, 77.
15. J. Hennig and H. Friedburg, *Magn. Reson. Imaging*, 1988, **6**, 391.
16. J. Listerud, S. Einstein, E. Outwater, H. Y. Kressel, *Magn. Reson. Q.*, 1992, **8**, 199.
17. R. T. Constable and J. C. Gore, *Magn. Reson. Med.*, 1992, **28**, 9.
18. D. A. Rubin, J. B. Kneeland, J. Listerud, S. J. Underberg-Davis, and M. K. Dalinka, *Am. J. Roentgenol.*, 1994, **162**, 1131.
19. J. A. Carrino, T. R. McCauley, L. D. Katz, R. C. Smith, and R. C. Lange, *Radiology*, 1997, **202**, 533.
20. E. M. Haacke and J. A. Tkach, *Am. J. Roentgenol.*, 1990, **155**, 951.
21. M. P. Recht, G. A. Piraino, J. P. Schils, and G. H. Belhobek, *Radiology*, 1996, **198**, 209.
22. M. J. Geirnaerdt, J. L. Bloem, F. Eulderink, P. C. Hogendoorn, and A. H. Taminiau, *Radiology*, 1993, **186**, 813.
23. K. Herrlin et al., *Acta Radiol.*, 1990, **31**, 233.
24. B. C. Dangman, F. A. Hoffer, F. F. Rand, and E. J. O'Rourke, *Radiology*, 1992, **182**, 743.
25. M. F. Reiser and M. Naegele, *J. Magn. Reson. Imaging*, 1993, **3**, 307.
26. P. F. J. Tirman, A. E. Stauffer, J. V. Crues III, R. M. Turner, W. M. Nottage, W. E. Schobert, B. D. Rubin, D. L. Janzen, and R. C. Linares, *Arthroscopy*, 1993, **9**, 550.
27. A. L. Deutsch, J. H. Mink, J. M. Fox, M. J. Friedman, and S. M. Howell, *Magn. Res. Q.*, 1992, **8**, 23.
28. A. L. Deutsch, J. H. Mink, and R. Kerr, 'MRI of the Foot and Ankle'. New York, Raven Press, 1992, p. 378.

29. J. H. Mink, in 'MRI of the Foot and Ankle', eds. A. L. Deutsch, J. H. Mink, and R. Kerr, Raven Press, New York, 1992.

30. T. C. Lynch, J. V. Crues III, F. W. Morgan, W. E. Sheehan, L. P. Harter, and R. Ryu, *Radiology*, 1989, **171**, 761.

31. G. A. Bogost, E. K. Lizerbram, and J. V. Crues III, *Radiology*, 1995, **197**, 263.

32. A. Vellet, P. Marks, P. Fowler, and T. Munro, *Radiology*, 1991, **178**, 271.

33. G. M. Blum, P. F. J. Tirman, and J. V. Crues III, in 'MRI of the Knee', 2nd. edn., eds. J. H. Mink, M. A. Reicher, J. V. Crues III, and A. L. Deutsch, Raven Press, New York, 1993, pp. 295–332.

34. A. Wang, D. Weinstein, L. Greenfield, L. Chiu, R. Chambers, C. Stewart, G. Hung, F. Diaz, and T. Ellis, *Magn. Reson. Imaging*, 1990, **8**, 675.

35. C. P. Sabiston, M. E. Adams, and D. K. Li, *J. Orthop. Res.*, 1987, **5**, 164.

36. M. O. Senac, Jr., D. Deutsch, B. H. Bernstein, P. Stanley, J. V. Crues III, D. W. Stoller, and J. Mink, *Am. J. Roentgenol.*, 1988, **150**, 873.

37. H. J. Mankin, *N. Engl. J. Med.*, 1992, **326**, 1473.

38. D. G. Mitchell, M. E. Steinberg, M. K. Dalinka, V. M. Rao, M. Fallon, and H. Y. Kressel, *Clin. Orthop.*, 1989, **244**, 60.

39. H. Tsukamoto, Y. S. Kang, L. C. Jones, M. Cova, C. J. Herold, E. McVeigh, D. S. Hungerford, and E. A. Zerhouni, *Invest. Radiol.*, 1992, **4**, 275.

40. M. Cova, Y. S. Kang, H. Tsukamoto, L. C. Jones, E. McVeigh, B. L. Neff, C. J. Herold, J. Scott, D. S. Hungerford, and E. A. Zerhouni, *Radiology*, 1991, **179**, 535.

41. M. J. A. Geirnaerdt, J. Hermans, J. L. Bloem, H. M. Kroon, T. L. Pope, A. H. M. Taminiau, and P. C. W. Hogendoorn, *Am. J. Roentgenol.*, 1997, **169**, 1097.

42. J. S. Jelinek, M. J. Kransdorf, B. M. Shmookler, A. J. Aboulafia, and M. M. Malawer, *Radiology*, 1993, **186**, 455.

43. J. S. Jelinek, M. J. Kransdorf, B. M. Shmookler, A. A. Aboulafia, and M. M. Malawer, *Am. J. Roentgenol.*, 1994, **162** 919.

44. J. S. Jelinek, M. D. Murphey, M. J. Kransdorf, B. M. Shmookler, M. M. Malawer, and R. C. Hur, *Radiology*, 1996, **201**, 837.

45. D. G. Varma, A. G. Ayala, S. Q. Guo, L. A. Moulopoulos, E. E. Kim, and C. Charnsangavej, *J. Comput. Assist. Tomogr.*, 1993, **17**, 414.

46. S. L. Hanna, B. D. Fletcher, D. M. Parham, and M. F. Bugg, *J. Magn. Reson. Imaging*, 1991, **1**, 441.

47. P. Lang, G. Honda, T. Roberts, *et al.*, *Radiology*, 1995, **197**, 83.

48. T. H. Berquist, R. L. Ehman, B. F. King, C. G. Hodgman, and D. M. Iistrup, *Am. J. Roentgenol.*, 1990, **155**, 1251.

49. R. Erlemann, J. Sciuk, A. Bosse, J. Ritter, C. R. Kusnierz-Glaz, P. E. Peters, and P. Wuisman, *Radiology*, 1990, **175**, 791.

50. R. Erlemann, M. Reiser, P. Peters, P. Vasallo, B. Nommensen, C. R. Kusnierz-Glaz, J. Ritter, and A. Roessner, *Radiology*, 1989, **171**, 767.

51. J. Rubenstein, M. Recht, D. G. Disler, J. Kim, and R. M. Henkelman, *Radiology*, 1997, **204**, 15.

52. J. D. Rubenstein, J. G. Li, S. Majumdar, and R. M. Henkelman, *Am. J. Roentgenol.*, 1997, **169**, 1089.

53. K. B. Lehner, H. P. Rechl, J. K. Gmeinwieser, A. F. Heuck, H. P. Lukas, and H. P. Kohl, *Radiology*, 1989, **170**, 495.

54. D. Le Bihan, *Radiology*, 1998, **207**, 305.

55. D. K. Kim, T. L. Ceckler, V. C. Hascall, A. Calabro, and R. S. Balaban, *Magn. Reson. Med.*, 1993, **29**, 211.

Biographical Sketches

Paul S. Hsieh. *b* 1962. B.S. (Mathematics), MIT, 1984; M.D. University of Michigan, 1989; residency in Diagnostic Radiology, Mallinckrodt Institute of Radiology, 1993; MRI training with John V. Crues, 1994; Faculty in Musculoskeletal Radiology at the Mallinckrodt Insti-

tute of Radiology 1994–1997. Currently staff radiologist, Kaiser Permanente, San Diego, CA. Research interests: musculoskeletal MRI.

John V. Crues, III. *b* 1949. A.B., 1972, Harvard; M.S., 1975, physics, with Charles Slichter, University of Illinois; M.D., 1979 Harvard Medical School, residency in Internal Medicine 1982 and Radiology 1985 at Cedars-Sinai Medical Center. Currently Director of Magnetic Resonance at Cedars-Sinai Medical Center. Former President of the International Society for Magnetic Resonance in Medicine. Currently Medical Director of RadNet Management, Inc. and President, ProNet Imaging in Los Angeles, CA. Approx. 200 publications. Current research specialty: musculoskeletal MRI, picture archiving and communication systems.

Peripheral Muscle Metabolism Studied by MRS

Peter A. Martin, Henry Gibson, and Richard H. T. Edwards

Magnetic Resonance Research Centre and Muscle Research Centre, The University of Liverpool, Liverpool, UK

1 INTRODUCTION

Muscle is a unique biological machine providing the main method of generating motive force, work, and power: it constitutes some 40% of total body cell mass in normal man. It is easily accessible for study and capable of great versatility in performance from a delicate touch to a powerful punch or from running a sprint to a marathon. Muscle metabolic rates can rapidly increase by over 100-fold. The body has nearly 600 muscles divided into three groups, skeletal (striated), visceral (non-striated), and cardiac; cardiac muscle is dealt with elsewhere (see *NMR Spectroscopy of the Human Heart*). Here we are only concerned with the muscles of the limbs, i.e., 'peripheral' skeletal muscle, which was one of the first tissues to be studied by magnetic resonance spectroscopy (MRS). Diseases of muscle (myopathies) are rare and often chronic and disabling, largely due to wasting of muscle tissue, which may be progressively replaced by fat or fibrous tissue, such as in the inherited X-linked muscular dystrophies. This creates a problem in trying to monitor muscle biochemistry because of serious partial volume effects, i.e., any volume of muscle under study contains an abnormally high proportion of fat or fibrous tissue as is readily seen on muscle biopsy. Quite different are the specific cellular enzyme and membrane defects of muscle, where this fat replacement is absent or less evident. They rep-

For References see p. 1346

resent 'Nature's experiments' where the impairment is due to interference with some essential metabolic pathway. They constitute an interesting group of disorders which are particularly amenable to study by in vivo MRS. Magnetic resonance imaging (MRI) has been confined largely to the study of anatomy and gross pathology (see *Skeletal Muscle Evaluated by MRI*), whereas MRS has mostly been confined to metabolic studies of phosphorus (^{31}P)-containing compounds, although some work has been done using hydrogen (^{1}H)[1,2] and carbon-13 (^{13}C).[3] Fortunately, the phosphorus-containing compounds visible by MRS—phosphocreatine (PCr), adenosine triphosphate (ATP) and inorganic phosphate (Pi)—are of great interest since they are involved in the energy metabolism of the muscle cells enabling MRS to be used to study muscle energetics noninvasively.

Table 1 Summary of broad characteristics of main fiber types in human skeletal muscle

Characteristic	Slow twitch, Type I	Fast twitch, Type II
Contraction time	Long	Short
Relaxation time	Long	Short
Myosin ATPase activity	Low	High
Fatigability	Low	High
Phosphocreatine content	Low	High
Oxidative enzyme activity	High	Low
Capillary density	High	Low
Mitochondria	Numerous	Few
Glycogen content	No difference	No difference
Fat content	High	Low
Myoglobin content	High	Low

2 MUSCLE STRUCTURE AND FUNCTION

Human muscle comprises at least two populations of fibers with different functional and metabolic characteristics. In health, the distribution of fibers in particular muscles reflects the physiological function of the muscle, whether for postural control (e.g. soleus) or for rapid movement (e.g. biceps). Muscles are organized in bundles (fascicles) and form into groups with a particular gross structure characteristic of that muscle (e.g. pennation angle, which can be determined by MRI[4]).

Muscles work together as 'agonists' and 'antagonists' to achieve a particular force or movement. The protein chemistry and fine structural features responsible for force generation in muscle are beyond the resolution of whole body MR systems, and are, thus, not covered here.

The muscle fibers (cells) have an outer membrane, the sarcolemma, filled with a fluid, sarcoplasm, containing the endoplasmic reticulum, mitochondria, and the many peripherally located nuclei. The myofibrils, which constitute a large component of the cell, consist of a matrix of interdigitating actin and myosin protein chains. Muscle performance is achieved by the action of actin and myosin chains sliding over each other, shortening the length of the sarcomere and thus the muscle. Muscles contract in response to a nerve impulse and each muscle cell must therefore have a neuronal connection. However a single neurone will have from a few to several hundred branches, each connecting with muscle cells that all contract together. This group forms a motor unit and the recruitment of an increasing number of motor units is one of the main factors governing the control and precision of any force generated.

Initiation of a contraction depends on a nerve signal being transmitted to the muscle via a chemical messenger acetylcholine. This causes an 'action potential' to be produced in the form of a region of electrical depolarization on the sarcolemmal membrane. The action potential rapidly travels the length of the muscle cell releasing calcium from the lateral cisternae of the endoplasmic reticulum. The calcium release activates actomyosin ATPase and triggers myofibrillar cross bridge interaction and thus force generation. A single action potential produces a single twitch involving the *entire* muscle fiber. Gradation of strength depends on changing the number of muscle fibers that are active. For a prolonged contraction of greater force, multiple stimuli are used, and if the frequency of stimulation is high enough the individual twitches fuse together to produce a smoothly sustained *tetanic* contraction, which lasts longer and is several-fold stronger than a single twitch. The two basic types of skeletal muscle fibers contract at different speeds; the 'red', slow-twitch (or type I) fibers are best suited to prolonged aerobic exercise, whereas the 'white' fast-twitch (or type II) fibers are best adapted to high intensity exercise that is largely anaerobic; intermediate fiber types also occur, and their prevalence depends on the activity history of the muscle (see Table 1). The importance of the recognition of two main fiber type populations in human muscles is that regional metabolic inhomogeneity can develop within the muscle as a consequence of particular forms of muscular activity in which individual populations of muscle become fatigued to different extents, resulting in their discrete appearance in the spectra. [e.g. split inorganic phosphate (Pi) peak]. Another consideration which can confuse this interpretation is that the sensitive volume of tissue studied may include more than one agonist muscle group, of which only one may be active.

3 MUSCLE AS A BIOLOGICAL MACHINE

When considering muscle function it is important to recognize the relationships between energy, force, work, and power. In the human body energy, measured in joules (J), is available for muscle metabolism via either the long- or the short-term energy supply processes. The maximum force, measured in newtons (N), that a muscle can generate (its strength), depends on the cross-sectional area of active muscle fibers. Comparison of working muscles requires that they are doing the same amount of work, i.e. force × distance, or generating the same power output (force × distance/time) which is measured in watts (W). The work done by the muscle will equal the energy expended only if the muscle is 100% efficient. Most muscular activity is about 20% efficient, the remainder of the energy being dissipated as heat. The maximum power output of the muscle depends on its maximum force *and* its velocity of shortening. During isometric exercise, when the muscle contracts without altering its length, external power output is zero;

however the force–time integral can be used as an indicator of muscular activity.[5]

4 PRACTICAL ASPECTS OF ERGOMETRY

Of vital importance in the study of any particular muscle is the need to be able to measure accurately and objectively its force, work, and power output. In the body, muscles in vivo form part of a complex machine, with their diverse attachments to bones and the agonist/antagonist arrangement by which they work together or against each other respectively. Sensible force measurement depends on the careful design of the ergometers used. The 'man–machine interface' thus becomes of paramount importance to provide an objective work standard against which NMR measurements may be correlated. Exercise can be dynamic such as running or walking or isometric (static) as in holding various body postures or a heavy weight. Although much human activity consists of dynamic exercise the constraints of space within the magnet system mean that isometric exercise is more easily studied. Although many ergometers are available, their use in conjunction with MR systems is limited, not least because human muscle studies by MRS carried out in the 1.5–2 T horizontal bore, whole body magnet systems, necessitate that the subject should lie down in the magnet. MR systems have magnets that are designed to have good homogeneity, but they also produce strong fringe fields. Commercially available ergometers are made of ferromagnetic materials or involve electrical motors or dynamos, which means that they will not work properly in the close proximity of strong magnetic fields of the MR system; such ergometers will in turn destroy the homogeneity of the magnetic field.

In the future, we may expect both these problems to be resolved as new magnet designs are becoming available that will enable the subject to stand up within the magnet system and carry out exercise routines such as running or cycling, which are more in keeping with normal human activity. Secondly, improved magnetic shielding is reducing the stray field of the magnet systems to such an extent that conventional ergometers can be used within 1 m or so of the MR magnet. However it is unlikely that the need for specially designed nonmagnetic ergometers will disappear; indeed the opportunities opened up by this new field will require new and more sophisticated designs.

Figure 1 shows an example of an apparatus that has been used for the study of isometric exercise in the quadriceps muscle during both voluntary and electrically stimulated exercise studies.[6] The subject lies supine with the knees bent and resting at an angle of 120° over a polystyrene foam block. The ankle is restrained by a strap which is attached to a force transducer while a wide belt across the waist prevents subjects from using their back and abdominal muscles to aid in the exercise. Stimulation electrodes strapped on with crepe bandages are applied both superior and inferior of the quadriceps to allow electrical stimulation of the muscle. The electrodes are connected via screened leads and a radiofrequency filter to a commercial stimulator and consist of 2 cm² of copper over a conducting gel pad, 5 cm × 10 cm, cut from a commercial defibrillator pad. The force transducer consists of an aluminum bar and strain gauge connected to a preamplifier, the output of which passes out of the scan room via a radiofrequency filter.

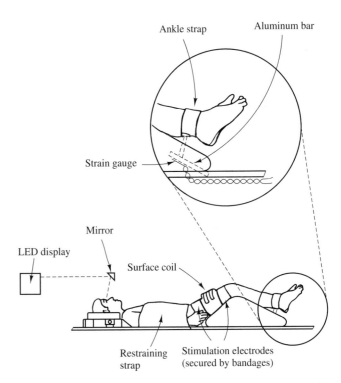

Figure 1 Apparatus for the study of the quadriceps muscles

Visual and audio feedback of the measured force are provided to the subject from a LED bar graph and tone generator to enable the subject to make voluntary contractions to prescribed target forces. Red, yellow and green LEDs under control from the spectrometer's computer are used to tell the subject when to exercise and to provide warning of forthcoming electrical stimulation. The limb may be made ischemic (blood supply stopped) by simply inflating a thigh sphygmomanometer cuff to 100 mmHg above systolic blood pressure, thereby providing a closed system trapping metabolites and preventing oxidative recovery processes. Measurement of myographic activity is also possible without deterioration of the magnetic homogeneity.[7]

5 MRS TECHNIQUES FOR MUSCLE STUDIES

The spectacular improvement in MRI has not been matched by a comparable improvement in in vivo MRS over the 15 or so years since the introduction of the first small-bore in vivo spectroscopy systems. Clinical applications of spectroscopy have been slow to appear, and it is still not a routine diagnostic tool. An illustration of the range and quality of spectra that can be obtained is shown in Figure 2. This shows a set of spectra from the forearm of a normal boy and a dystrophic boy with Duchenne muscular dystrophy.[8] The ^1H spectrum for the dystrophic forearm shows a decrease in the water peak and an increase in the fat peak. The ^{13}C spectrum shows a lot more peaks, but at the resolution achievable with the 1.5–2.3 T of most in vivo systems, this is difficult to interpret with many overlapping peaks. The peaks due to CH_3 and CH_2 groups are clearly visible, however, and the dystrophic forearm shows a general increase in the ^{13}C signal, which is most obvious in fat

For References see p. 1346

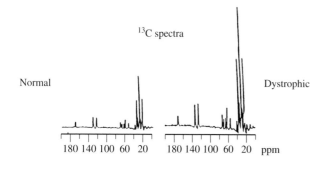

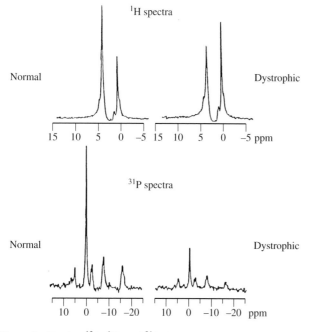

Figure 2 Resting ^{13}C, ^{1}H, and ^{31}P NMR spectra of the forearm. The spectra on the right are from a 9-year-old with Duchenne's dystrophy, and on the left are from an age-matched control with similar skin thickness

signals in the CH$_2$ region. The phosphorus spectrum shows a large peak due to PCr and three peaks due to ATP, as well as a small peak due to Pi. The dystrophic ^{31}P spectrum shows a decrease in the total signal of phosphorus metabolites as the fat displaces the muscle. The utility of MRS for muscle studies depends primarily on the quality of the spectra that can be obtained. Quantitative analysis requires as good a signal-to-noise ratio (S/N) as possible for all the peaks of interest. Qualitative studies can tolerate a poorer S/N but here extra care needs to be taken in the interpretation of results. Many factors interact to influence the spectral quality, e.g. the field strength and homogeneity of the magnet, the size and location of the muscle to be studied, the magnetic sensitivity of the nucleus, e.g. ^{31}P, ^{13}C, ^{1}H, the type of radiofrequency coil to be used, the MR localization technique, and the type of muscular activity to be studied.

The first localization technique used in muscle studies was 'surface coil localization'.[9] This is suitable for all nuclei and relies entirely on the localized but inherently nonuniform response of a simple loop antenna (the surface coil) to restrict the region from which signals would be obtained (see **Surface**

Coil NMR: Detection with Inhomogeneous Radiofrequency Field Antennas); when both transmitter and receiver surface coils are used they have a sensitive volume which forms a hemisphere penetrating to a depth of roughly one coil radius. This was later supplemented by the use of magnetic field localization techniques of which the most widely used for muscle studies has been the topical magnetic resonance (TMR) method.[10] This used a static, high order (Z4) gradient to spoil deliberately the magnet homogeneity over all but the chosen small diameter, central region of the magnet; any signals coming from this uniform central volume would be sharp peaks, whereas any signals coming from outside this volume would appear as a broad hump in the baseline, which could be removed from the final spectra via a convolution difference technique.

Topical magnetic resonance had the big advantage that good localization (superior to that of a surface coil alone) can be obtained from a single acquisition, making it suitable for studies requiring time resolution of the order of 1 or 2 s. It had the disadvantage of relying on static magnetic field gradients, which could not be switched rapidly on and off, and which only produced a uniform field at the center of the magnet, requiring the muscle of interest to be located there. This put physical and anatomical constraints on what could be studied. Thus, the TMR method rapidly fell out of favor once magnets with switched gradients became available holding the promise of combining imaging and spectroscopy in studies and pulsed gradient localization techniques, allowing localization of any volume within the (larger) homogenous region in the center of the magnet. An alternative localization strategy that has been used by some groups is to use one-dimensional, 'Rotating Frame' chemical shift imaging;[11] this has not been widely adopted, however, largely because of the T_2 distortion to which it is susceptible. This shows itself primarily as a decrease in the area of the β-ATP peak which is a particular disadvantage since this peak is often used as an internal reference for quantitation.

Of the many gradient localization techniques that have been introduced over the last 10 years, few of them have found practical application in muscle studies. This is because most muscle studies have used ^{31}P spectroscopy. Carbon-13 spectroscopy has been done at natural abundance,[12,13] but generally the most useful results are obtained with the introduction of ^{13}C labeled compounds, but with greatly increased cost. Water-suppressed ^{1}H spectroscopy has not yet been widely applied to muscle and for the most part unsuppressed ^{1}H spectra are only of limited value. Compared with protons, phosphorus nuclei tend to have short T_2 relaxation times,[14] which makes them unsuitable for any localization technique that relies on long echo time, spin, or stimulated echoes, e.g. Pixel-RESolved Spectrocopy (PRESS) and the STimulated Echo Acquisition Method (STEAM). One simple localization technique which has found some application in muscle spectroscopy is Depth-REsolved Surface coil Spectroscopy (DRESS). This consists of a conventional slice selection followed by collection of an FID signal, though the loss of some signal, particularly from β-ATP, is a disadvantage, and baseline distortions caused by the long delays between excitation and data acquisition make subsequent processing and any attempts at accurate quantitation difficult.[15] The best true volume localization technique for ^{31}P spectroscopy is the Image-Selected In vivo Spectroscopy tech-

Table 2 ^{31}P relaxation times in human skeletal muscle

	Field (T)	α-ATP	β-ATP	γ-ATP	PCr	Pi	Ref.
T_1(s)	1.5	3.603	4.310	4.755	5.517	4.017	16
T_2(ms)	1.5	22.2^a	8.1^a	16.1^a	424.3	204.7	14

^{a}These measured values are probably lower than the true in vivo values due to the effects of J coupling which were not taken into account in this study[17]

nique (ISIS). This has the advantage that it does not use spin echoes and thus is capable of giving good, undistorted spectra suitable for quantitation, but ISIS suffers from the serious disadvantage that it is a differencing technique requiring a minimum of eight signals for full volume localization. This makes it highly susceptible to motion artifacts and thus completely unsuitable for exercise studies. The quality of localization is affected by T_1 effects and thus, for muscle studies, it should ideally be run at repetition times (*TR*) of 15 s, making it unsuitable for short-duration time course studies.

Because of the problems associated with volume localization techniques most muscle studies still use surface coil localization alone. This presents some difficulties for quantitation since the nonuniform response of the coil results in a variation of the flip angle throughout the sensitive volume causing variation in signal strength due to T_1 and T_2 effects (see Table 2). This has been ameliorated a little with the introduction of specially designed rf pulses designed to give uniform flip angles over a larger proportion of the coil's sensitive volume (see **Surface Coil NMR: Detection with Inhomogeneous Radiofrequency Field Antennas**).

In vivo spectroscopy is currently only capable of 'seeing' narrow lines produced from long T_2 metabolites, i.e. those in solution in the cytoplasm. These metabolites become MR invisible or only give broad lines, which appear as a hump in the baseline, when in a viscous medium or when bound to the mitochondrial matrix. This is an advantage in that only the cytoplasmic metabolites are important in assessing the muscle energy metabolism during exercise.

In vivo quantification has always been difficult. As a result, many muscle studies report their results in terms of ratios of peak areas. An approximation to quantitation can then be made if appropriate assumptions are made, e.g., the total MRS visible phosphorus concentration or the ATP concentration. This enables estimates to be made of the ^{31}P-containing metabolite concentrations. A common reference method is by comparison with β-ATP concentration. Improved quantitation can be obtained if some form of external or internal standard is used. Internal standards have to be inherent in the muscle under study and thus one method is to collect a ^{1}H spectrum and to use the water peak as a reference. One problem with this is that the water content of muscle varies, especially during exercise. Assumptions about the water concentration cannot therefore be used. To overcome this, several water spectra at different *TR*s must be obtained and the true H_2O concentration calculated, but again this is not suitable for exercise studies. An acceptable method is to use an external phantom, often attached to the coil as a reference, and the ^{31}P concentration in vivo is determined by comparison with that from a series of different concentration phosphate solutions external to the limb studied or suitably substituted for the limb after the study.

6   METABOLIC PATHWAYS

Most MRS studies of muscle concern the biochemical pathways (see Table 3) for the supply and utilization of energy. The first reaction to be considered is the hydrolysis of ATP to ADP, catalyzed by myosin ATPase, which produces the

Table 3   Energy sources for muscular activity

Short-term (anaerobic) energy sources

ATP hydrolysis:

$$\text{Adenosine triphosphate (ATP)} + H_2O \xrightarrow{\text{myosin-ATPase}} \text{adenosine diphosphate (ADP)} + \text{inorganic phosphate (Pi)} + \text{energy}$$

Creatine kinase reaction:

$$\text{Phosphocreatine} + \text{ADP} + 0.9\ H^+ \xrightarrow{\text{creatine kinase}} \text{creatine} + \text{ATP}$$

Anaerobic glycolysis:

$$\text{Glycogen (glycosyl unit)} + 3\ \text{Pi} + 3\ \text{ADP} \longrightarrow 2\ \text{lactate} + 3\ \text{ATP}$$

Long-term (aerobic) sources

Oxidative phosphorylation:

$$\text{Glycogen/Glucose/Free fatty acids/Free amino acids} \longrightarrow \text{NADH}$$

$$\text{NADH} + 1.5\ H^+ + 0.5 O_2 + 3\text{ADP} + 3\text{Pi} \longrightarrow H_2O + \text{NAD}^+ + 3\text{ATP}$$

For References see p. 1346

energy-driving muscular contraction. Normal muscle only contains a small amount of cytosolic ATP and so can only sustain contractile activity for a very short period of time before fresh supplies of ATP must be made available. These supplies are obtained not from stores of ATP itself but via a reservoir of energy in the form of PCr which is used in the cytosol by the creatine kinase reaction to recycle ADP rapidly back to ATP again. The resulting creatine (Cr) is transported to the mitochondria of the cell where the reverse reaction occurs catalyzed by creatine kinase and Cr is converted back to PCr at the expense of mitochondrial ATP; this is the oxidative phosphorylation reaction or mitochondrial respiration.

Under aerobic conditions, i.e. when the O_2 supply from the blood is maintained, mitochondrial respiration supplies most of the muscles' energy requirements in the form of ATP. This results in production of one-tenth of a mole of H^+, from carbonic acid, per mole of ATP. When the energy demand exceeds the mitochondrial capacity or when anaerobic (ischemic) conditions exist, only PCr and glycogenolysis can supply the ATP for contraction. When anaerobic metabolism takes place, large amounts of lactate are therefore produced from glycogenolysis, giving two-thirds of a mole of H^+ per mole of ATP leading to a large decrease in intracellular pH.[18] This fall in pH may have profound effects on pH-sensitive metabolic processes (see e.g., Table 1). Phosphocreatine hydrolysis consumes protons whereas PCr synthesis produces protons. Fortunately it is easy to measure pH by ^{31}P MRS, and after taking into account the effects of buffering and proton efflux, i.e. loss of protons from the cell, the rate of proton production can be calculated. Thus the effect of pH on ATP fluxes can be allowed for. Furthermore, it is believed that pH affects cross bridge kinetics[19] leading to a reduction in force generation, i.e. fatigue.

7   THE STUDY OF MUSCLE USING ^{31}P MRS

7.1   Estimation of pH

Inorganic phosphate (Pi) in vivo has a pK_a of about 6.75 at physiological pH and exists in an equilibrium between two forms $H_2PO_4^-$ and HPO_4^{2-}. These give rise to two separate phosphate resonances, 2.3 ppm apart, which undergo rapid chemical exchange (10^9–10^{10} s^{-1}), resulting in a spectrum containing a single resonance whose frequency depends on the relative amounts of the two moieties. Since the equilibrium between the two forms of Pi is pH-dependent, the chemical shift of the Pi peak can be used as an indicator of pH by measuring its chemical shift either relative to an external standard such as methylenediphosphonate or relative to internal standards such as the pH-insensitive resonances due to the phosphorus PCr peak or the proton water signal. In the normal, resting, human forearm muscle of Figure 2, the chemical shift of Pi from PCr is 5.00 ppm corresponding to a pH of 7.15, whereas during exercise the chemical shift of Pi falls to 4.66 ppm corresponding to a pH of 6.88, due to the presence of lactate. The pH may be calculated from the equation.

$$pH = 6.75 + \log[(\delta - 3.27)/(5.69 - \delta)] \qquad (1)$$

where δ is the chemical shift difference, in ppm, between the Pi and PCr peaks.

7.2   The Buffering Capacity of Muscle

A knowledge of the buffering capacity is necessary for determination of ATP fluxes since the H^+ ions are involved in the equilibria (see Table 3). During the initial part of aerobic exercise the assumption can be made that the proton efflux can be neglected, enabling the glycogenolytic rate to be estimated in the same way as for ischemic exercise. This assumption falls down when the buffering capacity has been exhausted and the pH starts to fall. The cytosolic buffering capacity of skeletal muscle depends on Pi (pK = 6.57), bicarbonate (pK = 6.1), and other buffers, largely imidazole groups in histidine residues. It has been shown[20] that in a closed system where the total CO_2 is constant (e.g. during ischemia) the buffer capacity β is given by:

For Pi:

$$\beta = 2.3 \, [\text{Pi}]/\{[1 + 10^{(\text{pH} - 6.75)}] \, [1 + 10^{(6.75 - \text{pH})}]\} \qquad (2)$$

where β is measured in slykes (i.e. mmol L^{-1} per pH unit).

For bicarbonate:

$$\beta = 2.3 \, S p\text{CO}_2 \, 10^{(\text{pH} - 6.1)}/\{[1 + 10^{(\text{pH} - 6.1)}] \, [1 + 10^{(6.1 - \text{pH})}]\} \qquad (3)$$

where S is the solubility of CO_2. Taking $p\text{CO}_2$ as 5 kPa and S as 0.3 mmol L^{-1} kPa^{-1} then at its resting pH in closed muscle β is less than 5 slykes.

For other buffers: $\beta = 20$–30 slykes; inferred by analysis of ^{31}P MRS data and from measurements in muscle homogenates.[20]

7.3   Muscle at Rest

Many studies have been performed on resting muscle but these have frequently been hampered by the wide range of physiological states of the muscle under study, largely due to the intrinsic variations in the level of training in the populations under study. One of the first and most important findings from studies of resting muscle has been the observation that the concentration of PCr is consistently higher when measured by ^{31}P MRS than when measured by freeze-clamped needle biopsy.[8] This is almost certainly due to the rapid hydrolysis of PCr to Pi during the freeze-clamping process and this interpretation can be supported by the observation that the total phosphorus concentration [PCr + Pi] obtained by both methods is approximately the same. No such discrepancies between the techniques have been observed with ATP, possibly due to equilibration of the creatine kinase reaction during the extraction procedures. Most quantitative ^{31}P MRS studies on resting muscle have relied not on true quantitation, but on the assumption that the ATP levels are reasonably constant. This assumption requires care as there is evidence that changes in the resting levels of metabolites depend on the prior history of the muscle. If the muscle has been involved in exercise involving lengthening contractions,[21] the PCr/Pi ratio is significantly reduced up to 1 h after exercise; the reduction continues with the ratio reaching a minimum at 1 day postexercise and remaining low for between 3 and 10 days postexercise. Similarly, abnormal spectra have been reported for up to 2–3 days after short-term exercise, even when no muscle fiber damage is thought to have occurred. Much evidence is now accumulating to show that type I and type II fibers have different PCr/Pi ratios, with type II fibers having elevated PCr and ATP compared with type I. The resting ratios and changes discussed above may be related to the different fiber-type ratios found in normal subjects.[22]

7.4 Exercise and Fatigue

While much information about muscle energy status can be obtained at rest, it is during exercise that the most dramatic changes are seen owing to the high metabolic exchanges associated with muscular activity. Particular interest lies in the study of individuals with defects of metabolism, which can give information about metabolic processes not normally accessible. This is discussed further in Section 8. Metabolic requirements may be determined at various workloads and metabolic processes may be related to physiological changes in function. Muscle fatigue, the decline in force or power output with prolonged activity, has received much attention in this respect, but is complicated by the type of muscular activity undertaken and the many cellular physiological factors that appear to be interrelated with the chemical changes taking place. For this reason, and the importance of maintaining and improving muscle performance in sports and disease, the various known mechanisms contributing to fatigue are described below.

The central contribution to fatigue can be measured by comparing force output between alternate periods of voluntary contractions with that produced from electrical stimulation of a peripheral nerve. If the voluntary force output falls more than the stimulated force output the difference is due to fatigue of central motor control mechanisms rather than of the muscle itself.[23]

Except in the rare neuromuscular disease myasthenia gravis, fatigue due to failure of the neuromuscular transmission is rare. Much of the research of the last half century in this field has been directed toward gaining an understanding of the extent to which impaired energy supply or electromechanical coupling failure is the dominant problem in a particular form of muscular activity.[24,25] Fatigue of the muscle itself can be classified according to the response to different frequencies of electrical stimulation. Fatigue that is produced more with high-frequency electrical stimulation (HFF) occurs in myasthenia gravis. Low-frequency fatigue (LFF), in which there is selective impairment of force generation at low frequency, i.e., reduced 20:50 Hz force ratio, can occur for a long time after ischemia[26] or an eccentric muscular contraction[27] when force generation at high frequency has recovered.

From the energetics point of view HFF and LFF can be distinguished by the fact (from needle biopsy and MR studies early in recovery from severe ischemic exercise) that with HFF, recovery of force may occur before full recovery of PCr/β-ATP (or Pi/PCr), due to rapid recovery of membrane excitation (compound muscle action potential; CMAP). With LFF, force is still reduced 1 h after exercise when CMAP and PCr/β-ATP have recovered.[28]

7.4.1 ATP Turnover in Ischemic Exercise

In ischemic (anaerobic) exercise (i.e. when the blood supply to the muscle has been occluded), ATP is produced from two sources, the hydrolysis of PCr catalyzed by creatine kinase (= PCr depletion) and glycogenolysis, the breakdown of glycogen to lactic acid.[29] Knowledge of [PCr] and pH can be used to estimate the rate of ATP synthesis. Since decay of the PCr peak can be measured directly, the rate of ATP synthesis from the hydrolysis of PCr (D) can also be measured directly:

$$D = -\delta[\text{PCr}]/\delta t \qquad (4)$$

Hydrolytic ATP synthesis also results in a net proton consumption at the rate:

$$\text{net proton consumption} = D/[1 + 10^{(\text{pH} - 6.75)}] \qquad (5)$$

The production of ATP via glycogenolysis (L) can be estimated from the effect of lactic acid production on pH. Since during ischemic exercise protons cannot escape from the system, glycogenolysis (from glycosyl units) produces 2 mol of lactic acid thereby releasing 3 mol of ATP.[29] Thus:

$$L = (3/2)\{D/[1 + 10^{(\text{pH} - 6.75)}] - \beta \delta \text{pH}/\delta t\} \qquad (6)$$

where β is the cytosolic buffering capacity of the muscle (derived as above). Since the total rate of ATP synthesis (F) is:

$$F = D + L \qquad (7)$$

the total turnover of ATP is:

$$F = D + (3/2)\{D/[1 + 10^{(\text{pH} - 6.75)}] - \beta \delta \text{pH}/\delta t\} \qquad (8)$$

7.4.2 ATP Turnover in Aerobic Exercise

In aerobic (oxidative) exercise ATP is produced from PCr and the oxidative synthesis of ATP in the mitochondria. As before, the rate of depletion of PCr and thus the rate of hydrolytic ATP production is readily accessible by MRS. However the presence of oxidative ATP synthesis and the fact that the system is no longer closed, i.e. there is a net efflux of protons from the cells, complicates the assessment of total ATP production. One strategy has been to compare aerobic with ischemic exercise, making use of the power output relationships derived from graded ischemic exercise to provide the 'missing data' from aerobic exercise studies at the same power output.

At the start of aerobic exercise, proton efflux can be neglected and the glycogenolytic rate estimated as for ischemic exercise. At the end of ischemic exercise, when the pH is falling, estimates of glycogenolytic ATP synthesis rates must be corrected for the large proton efflux from the system. This can be done by assuming the proton efflux rate has a linear relationship to pH and [PCr] and thus can be inferred from the initial phase of recovery from exercise.

7.5 Recovery

The study of postexercise recovery provides an opportunity of studying the kinetics of recovery of muscle metabolites, and in particular it can be used to provide information on the pH-dependence of proton efflux, which can also be applied as a correction during the analysis of aerobic exercise.[30] After the

For References see p. 1346

completion of exercise, the accumulated ADP continues to stimulate mitochondrial respiration to resynthesize ATP; as a consequence, the cytoplasmic PCr pool is replenished through the creatine kinase equilibrium. The PCr replenishment during recovery from exercise has been shown to depend entirely on mitochondrial respiration by the absence of any metabolic recovery when the muscle is made ischemic after exercise by rapid inflation of a sphygmomanometer cuff.[31] Thus, we can consider the rate of PCr resynthesis to be a good index of mitochondrial function.[32]

The recovery of PCr after exercise appears to follow an exponential time course and experimentally is usually treated as such although it has been shown to be biphasic,[33] showing an initial rapid rise followed by a much slower rate of recovery back to resting level. The rate of PCr resynthesis is dependent on the extent of intracellular acidosis, which in turn depends on the work rate during exercise. The final rate of PCr recovery is dependent on the rate of pH recovery and a direct linear relationship has been shown between the value of intracellular pH at the end of exercise and the rate of PCr recovery.[34]

At the end of exercise, intracellular pH continues to fall until a minimum which, depending on the work rate during the exercise, occurs approximately 1 min after the end of the exercise period. The higher the work rate the later pH recovery begins (see Figure 3). The mechanisms that control intracellular pH are not well known though it is likely that active transport mechanisms are involved.[35]

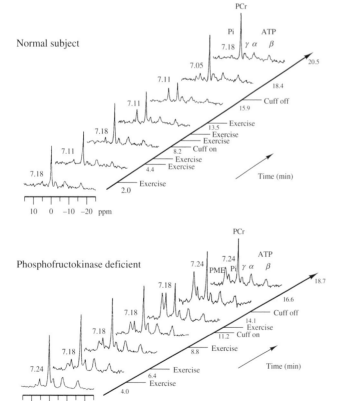

Figure 3 A comparative ^{31}P NMR exercise study of a patient with phosphofructokinase deficiency and a normal subject

During recovery [Pi] generally mirrors [PCr]. The Pi accumulated during exercise is transported into the mitochondria where it is used in the phosphorylation of ADP. Like PCr, Pi recovery is biphasic.[36] After the end of exercise and during the period in which pH is still falling Pi recovery is fast. Subsequently, as the pH recovers, Pi recovery slows down, and [Pi] decreases to undetectable levels for several minutes before it again reappears and recovers to resting levels. During this time a temporary decrease in the total [PCr + Pi] occurs, which otherwise remains constant throughout exercise. When exhaustive exercise has been carried out, such as to show a reduction of [ATP], e.g. strenuous aerobic exercise, then all recovery processes are substantially impaired.[37]

8   STUDIES OF MUSCLE DISEASE

Much of what is known about muscle energy metabolism in disease has come from needle biopsy studies. MRS affords a noninvasive approach to the study of metabolism in disease, but the diagnostic value of MRS in muscle is still yet to be shown. The advantage of MRS here is in continually providing information of metabolic levels during exercise (whereas repeated biopsy samples were previously required), although resting spectra can provide valuable metabolic information in certain myopathies such as in dystrophy[38] and muscle injury.[21] The question of whether metabolic abnormalities are secondary to other disease processes is becoming of increasing interest clinically in view of the consequences of disease with respect to muscle pain, performance, and fatigue. Some examples are shown in Table 4. A further application has been to study the effects of therapy, for example, in altering muscle energetics, where enzyme defects have limited the energy supply for contraction, such as glucose infusion in McArdle's disease (myophosphorylase deficiency in which glycolysis is impaired),[55] drug trials in Duchenne dystrophy,[38] and the consequences of insulin infusion on phosphate metabolism.[56]

The early applications of MRS in disease, in particular to patients with clear metabolic defects, provided a novel means to study the normal physiological mechanisms limiting exercise. Such patients can be considered to represent 'Nature's experiments' allowing examination of metabolic processes under conditions that would normally not be possible in human muscle. McArdle's disease[57] represents an example where a defect in metabolism may manifest itself as a failure of membrane excitation, highlighting the importance of glycolysis for the maintenance of membrane excitability. No lactate is produced in these individuals and hence no acidic shift in the Pi peak occurs, making it possible to measure initial rates of PCr resynthesis and K_m for ADP control of oxidative phosphorylation.[51] Figure 3 illustrates the metabolic changes seen during exercise in a patient with phosphofructokinase deficiency,[58] which is arguably a more significant metabolic defect than McArdle's disease, in that oxidation of blood-borne glucose is not possible. Accumulation of fructose 6-phosphate traps Pi resulting in a relatively small rise in Pi, and again no change in pH occurs. In these patients fatigue occurs rapidly, but this cannot be attributed to Pi or H$^+$.

Another group of the metabolic myopathies is represented by the mitochondrial abnormalities. Those patients affected

Table 4 Diseases studied by MRS of skeletal muscle (after Barbiroli[39])

Disease	Nucleus	Cause or defect	Conclusions from MRS	Ref.
Congenital heart disease	^{31}P	Impaired O$_2$ cyanosis	Resting pH and Pi elevated; abnormal PCr depletion and acidification during exercise; prolonged recovery times	40
Duchenne/Becker dystrophy	^{31}P, ^{1}H	Inherited lack of dystrophin	Progressive replacement of muscle tissue by fat; high resting Pi/PCr ratio and slightly alkaline pH; PDE peak increases with age; altered energy metabolism in carriers	41–43
Mitochondrial myopathy	^{31}P	Various enzyme defects	Abnormal transfer function: slow recovery of PCr after exercise	44,45
Encephalomyopathy	^{31}P	Mitochondrial enzyme defects	Abnormal transfer function and slow recovery of PCr and pH; abnormal brain energy metabolism	46,47
Glycogenosis	^{31}P	Phosphofructokinase deficiency	Limited acidosis, abnormal build up of PME peak during exercise	48,49
		Phosphoglycerate mutase	Limited acidosis and abnormally raised PME peak during exercise	50
McArdle's disease	^{31}P	Myophosphorylase deficiency	Lack of acidosis during sustained aerobic or ischemic exercise	51,48
Sickle cell anemia	^{31}P	Muscle ischemia	Reduced total ^{31}P compared with normals	52
Malignant hyperthermia	^{31}P	Anesthetic-induced hyperthermia	High resting Pi/PCr ratio; slow post-exercise recovery of PCr/Pi ratio	53
Peripheral vascular disease	^{31}P	Relative ischemia	Acid shift and a fall in PCr/Pi with exercise	54

show abnormalities in oxidative phosphorylation and consequently a high Pi and low PCr,[59] and slow PCr resynthesis.[60,61] A number of these patients have been studied with various defects in mitochondrial function, with slowed PCr and ADP recovery, although the most reliable indication comes from the resting spectrum.

Abnormalities in MRS parameters in Duchenne muscular dystrophy have been discussed earlier (see Section 5). The question of whether there is a reduction in energy state of the dystrophic tissue has been difficult to answer owing to the increasing proportion of muscle tissue replaced with fat and connective tissues as the disease progresses,[38] and highlights an inherent problem in acquiring a signal from an inhomogenous tissue. Attempts to correct the acquired phosphorus signal by dilution effects from noncontractile tissue have been made from biopsy samples and indeed several studies have shown a reduced PCr/ATP ratio in dystrophic muscle. Moreover, a reduction in PCr/Pi ratio suggests impairment of mitochondrial function[62] which is likely to be secondary to the dystrophic process, probably owing to disuse of the muscle, rather than contributing to the damage process, as greater changes are seen in patients with enzyme defects in whom his-

tological evidence of damage is not apparent. Indeed, ^{31}P MRS has been reported to be sufficiently sensitive to show abnormalities in metabolism in patients with minimal or no muscle weakness.[39]

The most marked changes in muscle metabolism demonstrated in disease are probably secondary in origin, particularly where impairment of blood flow is evident. Examples include peripheral vascular disease,[54] sickle cell anemia,[52] and congenital heart disease.[40] It is likely that many more disease states where some degree of fatigue or muscle impairment occurs may also demonstrate abnormalities in metabolism of muscle.

Although MRS has found application in the study of muscle disease for purely scientific purposes, it has yet to find any great application as a routine diagnostic test for muscle disease. This is primarily due to the specialized nature of the MRS examination and the great expense involved when many less expensive procedures are available, e.g. needle biopsy, followed by chemical and genetic analysis. Furthermore, these other procedures are capable of providing information on a greater range of metabolites or gene products than is currently possible by MRS. There is little doubt however that MRS will continue to provide a useful tool for the scientific elucidation

For References see p. 1346

of muscle biochemistry in the future, particularly when combined with the use of labeled compounds such as ^{13}C-labeled glucose as tracers.

9 RELATED ARTICLES

Animal Methods in MRS; Body Fat Metabolism: Observation by MR Imaging and Spectroscopy; MRI in Clinical Medicine; MRI of Musculoskeletal Neoplasms; NMR Spectroscopy of the Human Heart; Proton Decoupling in Whole Body Carbon-13 MRS; Quantitation in In Vivo MRS; Skeletal Muscle Evaluated by MRI; Surface Coil NMR: Detection with Inhomogeneous Radiofrequency Field Antennas; Tissue and Cell Extracts MRS; Tissue Behavior Measurements Using Phosphorus-31 NMR; Whole Body Studies: Impact of MRS.

10 REFERENCES

1. J. W. Pan, J. R. Hamm, D. L. Rothman, and R. G. Shulman, *Proc. Natl. Acad. Sci. U.S.A.*, 1988, **85**, 7836.
2. J. W. Pan, J. R. Hamm, H. P. Hetherington, D. L. Rothman, and R. G. Shulman, *Magn. Reson. Med.*, 1991, **20**, 57.
3. R. Taylor, T. B. Price, D. L. Rothman, R. G. Shulman, and G. I. Shulman, *Magn. Reson. Med.*, 1993, **27**, 13.
4. M. V. Narici, L. Landoni, and A. E. Minetti, *Eur. J. Appl. Physiol.*, 1992, **65**, 438.
5. M. Boska, *NMR Biomed.*, 1991, **4**, 173.
6. P. A. Martin, H. Gibson, S. Hughes, and R. H. T. Edwards, *Proc. Xth Ann Mtg. Soc. Magn. Reson. Med.*, San Francisco, 1991, p. 547.
7. P. Vestergaard-Poulsen, C. Thomsen, T. Sinkjaer, M. Stubgaard, A. Rosenfalck, and O. Henriksen, *Electroencephalogr. Clin. Neurophysiol.*, 1992, **85**, 402.
8. R. H. T. Edwards, M. J. Dawson, D. R. Wilkie, R. E. Gordon, and D. Shaw, *Lancet*, 1982, 725.
9. D. I. Hoult, S. J. W. Busby, D. G. Gadian, G. K. Radda, R. E. Richards, and R. J. Seeley, *Nature (London)*, 1974, **252**, 285.
10. R. E. Gordon, P. E. Hanley, and D. Shaw, *Prog. Nucl. Magn. Reson. Spectrosc.*, 1981, **15**, 1.
11. J. F. Dunn, G. K. Kemp, and G. K. Radda, *NMR Biomed.*, 1992, **5**, 154.
12. T. B. Price, D. L. Rothman, M. J. Avison, P. Buonamico, and R. G. Shulman, *J. Appl. Physiol.*, 1991, **70**, 1836.
13. R. Taylor, T. B. Price, L. D. Katz, R. G. Shulman, and G. I. Shulman, *Am. J. Physiol.*, 1993, **265**, 224.
14. C. Thomsen, K. E. Jensen, and O. Henriksen, *Magn. Reson. Imag.*, 1989, **7**, 557.
15. P. A. Bottomley, T. B. Foster, and R. D. Darrow, *JMRI*, 1984, **59**, 338.
16. C. Thomsen, K. E. Jensen, and O. Henriksen, *Magn. Reson. Imag.*, 1989, **7**, 231.
17. W. I. Jung, K. Straubinger, M. Bunse, S. Widmaier, F. Schick, K. Kuper, G. Dietze, and O. Lutz, *Proc. XIIth Ann Mtg. Soc. Magn. Reson. Med.*, New York, 1993, **30**, 138.
18. K. Sahlin, *Acta Physiol. Scand., Suppl.*, 1978, **455**, 1.
19. A. Fabiato and F. Fabiato, *J. Physiol. London*, 1978, **276**, 233.
20. G. K. Kemp, D. J. Taylor, P. Styles, and G. K. Radda, *NMR Biomed.*, 1993, **6**, 73.
21. K. McCully, Z. Argov, B. P. Boden, R. L. Brown, W. J. Bank, and B. Chance, *Muscle Nerve*, 1988, **11**, 212.
22. M. J. Kushmerick, T. M. Moerland, and R. W. Wiseman, *Proc. Natl. Acad. Sci. U.S.A.*, 1992, **89**, 7521.
23. B. Bigland-Ritchie, D. A. Jones, G. P. Hosking, and R. H. T. Edwards, *Clin. Sci.*, 1978, **54**, 609.
24. 'Human Muscle Fatigue: Physiological Mechanisms', eds. R. Porter and J. Whelan, Ciba Foundation Symposium 82, Pitman Medical, London, 1981.
25. 'Neuromuscular Fatigue', eds. A. J. Sargeant and D. Kernell, R. Neth. Acad. Arts & Sci., Amsterdam, 1993.
26. R. H. T. Edwards, D. K. Hill, D. A. Jones, and P. A. Merton, *J. Physiol. London*, 1977, **272**, 769.
27. D. J. Newham, K. R. Mills, B. M. Quigley, and R. H. T. Edwards, *Clin. Sci.*, 1983, **64**, 55.
28. R. G. Miller, D. Giannini, H. S. Milner-Brown, R. B. Layzer, A. P. Koretsky, D. Hooper, and M. W. Weiner, *Muscle Nerve*, 1987, **10**, 810.
29. G. K. Kemp, C. H. Thompson, P. R. Barnes, and G. K. Radda, *Proc. Ist Ann Mtg. Int. Soc. Magn. Reson. Med.*, Dallas, 1994, **31**, 248.
30. G. K. Kemp, D. J. Taylor, and G. K. Radda, *NMR Biomed.*, 1993, **6**, 66.
31. D. J. Taylor, P. J. Bore, P. Styles, D. G. Gadian, and G. K. Radda, *Mol. Biol. Med.*, 1983, **1**, 77.
32. G. K. Kemp, D. J. Taylor, C. H. Thompson, P. Styles, L. J. Hands, B. Rajagopalan, and G. K. Radda, *NMR Biomed.*, 1993, **6**, 302.
33. D. L. Arnold, P. M. Matthews, and G. K. Radda, *Proc. IIIrd Ann Mtg. Soc. Magn. Reson. Med.*, New York, 1984, **1**, 307.
34. D. Bendahan, S. Confort-Gouny, G. Kozak-Reiss, and P. J. Cozzone, *FEBS Lett.*, 1990, **272**, 155.
35. I. H. Nadshus. *Biochem. J.*, 1988, **250**, 1.
36. S. Iotti, R. Funicello, P. Zaniol, and B. Barbiroli, *Biochem. Biophys. Res. Commun.*, 1991, **176**, 1204.
37. D. J. Taylor, P. Styles P. M. Matthews, D. L. Arnold, D. G. Gadian, P. J. Bore, and G. K. Radda, *Proc. Vth Ann Mtg. Soc. Magn. Reson. Med.*, Montreal, 1986, **3**, 44.
38. R. D. Griffiths, E. B. Cady, R. H. T. Edwards, and D. R. Wilkie, *Muscle Nerve*, 1985, **8**, 760.
39. B. Barbiroli, *Magn. Reson. Spectrosc. Biol. Med.*, 1992, **20**, 369.
40. I. Adatia, G. J. Kemp, D. J. Taylor, G. K. Radda, B. Rajagopalan, and S. G. Haworth, *Clin. Sci.*, 1993, **85**, 105.
41. R. J. Newman, P. J. Bore, L. Chan, D. L. Gadian, P. Styles, D. Taylor, and G. K. Radda, *Br. Med. J.*, 1982, **284**, 1072.
42. D. Younkin, P. Berman, J. Sladky, C. Chee, W. Bank, and B. Chance, *Neurology*, 1987, **37**, 165.
43. B. Barbiroli, R. Funicello, A. Ferlini, P. Montagna, and P. Zaniol, *Muscle Nerve*, 1992, **15**, 344.
44. D. L. Arnold, D. J. Taylor, and G. K. Radda, *Ann. Neurol.*, 1985, **18**, 189.
45. Z. Argov, W. J. Bank, J. Maris, P. Peterson, and B. Chance, *Neurology*, 1987, **37**, 257.
46. D. J. Hayes, D. Hilton-Jones, D. L. Arnold, G. Galloway, P. Styles, J. Duncan, and G. K. Radda, *J. Neurol. Sci.*, 1985, **71**, 105.
47. B. Barbiroli, P. Montagna, P. Cortelli, P. Martinelli, T. Sacquegna, P. Zaniol, and E. Lugaresi, *Cephalalgia*, 1990, **10**, 263.
48. D. Duboc, P. Jehenson, S. Tran Dinh, C. Marsac, A. Syrota, and M. Fardeu, *Neurology*, 1987, **37**, 663.
49. Z. Argov, W. J. Bank, J. Maris, J. S. Leigh, Jr., and B. Chance, *Ann. Neurol.*, 1987, **22**, 46.
50. Z. Argov, W. J. Bank, B. Boden, Y. I. Ro, and B. Chance, *Arch. Neurol.*, 1987, **44**, 614.
51. G. K. Radda, *Biochem. Soc. Trans.*, 1986, **14**, 517.
52. S. L. Norris, J. R. Gober, L. J. Haywood, J. Halls, W. Boswell, P. Colletti, and M. Terk, *Magn. Reson. Imag.*, 1993, **11**, 119.
53. J. Olgin, H. Rosenberg, G. Allen, R. Seestedt, and B. Chance, *Anesth. Analg. (Cleveland)*, 1991, **72**, 36.
54. M. A. Zatina, H. D. Berkowitz, G. M. Gross, J. M. Maris, and B. Chance, *J. Vasc. Surg.*, 1986, **3**, 411.

55. S. F. Lewis, R. G. Haller, J. D. Cook, and R. L. Nunnally, *J. Appl. Physiol.*, 1985, **59**, 1991.

56. D. J. Taylor, S. W. Coppack, T. A. D. Cadoux-Hudson, G. J. Kemp, G. K. Radda, K. N. Frayn, and L. L. Ng, *Clin. Sci.*, 1991, **81**, 123.

57. B. McArdle, *Clin. Sci.*, 1951, **10**, 13.

58. R. H. T. Edwards, *Muscle Nerve.*, 1984, **7**, 599.

59. D. G. Gadian, G. K. Radda, B. D. Ross, J. Hockaday, P. Bore, D. Taylor, and P. Styles, *Lancet ii*, 1981, 774.

60. G. K. Radda, P. J. Bore, D. G. Gadian, B. D. Ross, P. Styles, D. J. Taylor, and J. Morgan-Hughes, *Nature, (London)*, 1982, **295**, 608.

61. R. H. T. Edwards, R. D. Griffiths, and E. B. Cady, *Clin. Physiol.*, 1985, **5**, 93.

62. B. Chance, S. Eleff, J. S. Leigh, Jr., D. Sokolow, and A. Sapega, *Proc. Natl. Acad. Sci. U.S.A.*, 1981, **78**, 6714.

Biographical Sketches

Peter A. Martin. *b* 1954. B.Sc. (Loughborough), Ph.D. (Dunelm). Applications scientist at Oxford Research Systems Limited, 1980–85. Operations manager of Liverpool University's Magnetic Resonance Research Centre 1986–present. Research interests include the applications of neural networks to spectral analysis, proton spectroscopy of the brain, and the quantitation of muscle physiology via MRI and MRS.

Henry Gibson. *b* 1961. B.Sc. (Liverpool), M.Sc. (King's College, London), Ph.D. (Liverpool). Honorary nonclinical lecturer in medicine, Research interests; quantification of muscle physiology via MRI and MRS and ultrasonography of muscle.

Richard H. T. Edwards. *b* 1939. B.Sc., M.B., B.S., Ph.D. (London), F.R.C.P., Codirector of the Jerry Lewis Muscle Research Centre, Royal Postgraduate Medical School. Honorary Consultant Respiratory Physician, Hammersmith Hospital; Professor of Human Metabolism, Hospital Medical School and Head of Department of Medicine, University College, London; Professor and Head of Department of Medicine, University of Liverpool, 1984–present. Director, Magnetic Resonance Research Centre, 1986–present.

MRI of Musculoskeletal Neoplasms

Johan L. Bloem

Leiden University Medical Center, The Netherlands

1 INTRODUCTION

Staging of disease is the single most important reason for performing MRI in patients with musculoskeletal tumors. Other indications for MRI in these patients are detection, specific diagnosis, chemotherapy monitoring, and detection of recur-rence. The last two indications are becoming increasingly important. After a short discussion about technique, this article deals with clinical applications of MRI in patients with primary musculoskeletal tumors, tumor-like lesions and metastases.

2 TECHNIQUE

When a bone lesion is potentially malignant, imaging studies (to allow local staging) and biopsy are needed. Imaging studies are performed prior to histologic biopsy, because they can be used to plan the biopsy procedure. Furthermore, biopsy induces reactive changes (edema, hemorrhage), which interfere with accuracy of staging. Cytologic biopsy is a minimally invasive procedure that can be useful in soft tissue tumors, prior to MRI, to differentiate between broad categories such as lymphoma or sarcoma metastasis. Conventional radiographs have to be present when MRI is planned and executed. This should be a golden rule and will avoid both major disasters and silly mistakes.

High- or low-frequency (low or high field) magnetic resonance (MR) systems can be used for imaging musculoskeletal tumors. It is possible to stage bone tumors accurately at 0.5 T. Whenever possible, dedicated coils should be used in order to increase spatial resolution. It is important to use both T_1- and T_2-weighted sequences.

T_1-weighted sequences are used for intramedullary staging and T_2-weighted sequences are used for defining soft tissue extension and cortical involvement. A large number of pulse sequences, such as STIR (short τ inversion–recovery), gradient echo (GE) imaging, fast or turbo spin echo (TSE), chemical shift imaging, magnetization transfer contrast added to various pulse sequences, etc., are available. For imaging of musculoskeletal tumors, conventional spin echo or TSE sequences are best used to obtain T_1-weighted images. For T_2-weighting conventional spin echo sequences, or preferably TSE sequences, are used. Contrast between tumor and normal tissue, especially fat containing tissue, is greatly enhanced by combining TSE with fat selective presaturation. It is important to make sure that enough T_2 weighting is achieved. At 0.5 T, a TSE sequence with a *TR* of 3000–4000 ms and an effective *TE* of 150 ms has proven to be as good as the conventional T_2-weighted spin echo sequences in our practice. When fat suppression is used, the TE should be shortened. Contrast between tumor and muscle on T_2-weighted SE or TSE images is superior to that on native T_2^*-proton density weighted, or enhanced GE images. STIR–TSE sequences are very sensitive, but the specificity and low signal-to-noise ratio are low. When T_1 or T_2-weighted images are mentioned in this article, I refer to both conventional spin echo and TSE sequences.

Routinely, multiple imaging planes are used: the transverse plane (T_2-weighted) for soft tissue extension, and sagittal or coronal planes (T_1-weighted) for intraosseous extension (Figure 1). The sagittal and coronal planes can be angulated to be parallel to the long bones. Additional planes are used when necessary, for instance the transverse plane for intraosseous staging to rule out partial volume effects of cortical bone which may occur in the sagittal and coronal plane in thin tubular bones in children. Slice thickness varies between 2 and 10 mm and the number of excitations will be 1–4, depending on various parameters such as slice thickness, *TR*, *TE*, type of coil,

For References see p. 1355

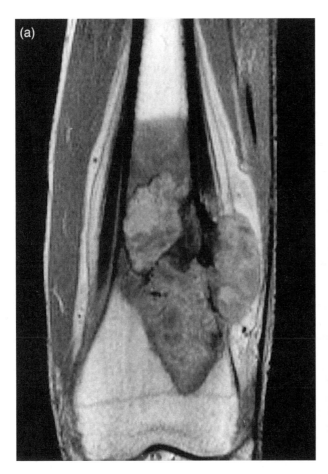

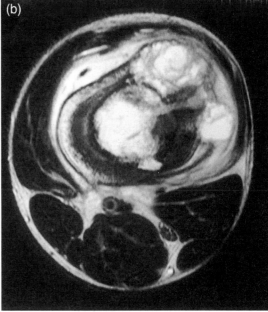

Figure 1 Osteosarcoma in an 18-year-old male imaged with a 0.5 T MRI system. (a) Coronal T_1-weighted (600/20) MR image displays the intraosseous tumor margins very well. Medial soft tissue extension and periosteal reaction are also depicted. The tumor has an inhomogeneous low to intermediate signal intensity. (b) Axial T_2-weighted TSE (3570/150) MR image. Contrast between the high signal intensity of viable tumor and the low signal intensity of muscle is excellent. However, both tumor and fat have a similar high signal intensity. Ossified periosteal reaction and ossified tumor display a signal void

and field strength used. Echo train length will vary between 3 and 8.

Gadolinium enhanced studies are performed following intravenous injection of 0.1–0.2 mmol Gd chelate/kg body weight with a T_1 spin echo or TSE technique and dynamic magnetization prepared GE imaging. The T_1-weighted sequences are combined with fat selective presaturation. It is not advisable to combine STIR sequences with gadolinium, since suppression of signal from enhancing tissue is possible. Out-of-phase GE images may further enhance contrast as compared with in-phase GE images.

Gadolinium can be used in patients with poor renal function, and can be removed from the body with dialysis.

3 MRI CHARACTERISTICS

Bone marrow and fat exhibit a high signal intensity (white) on T_1-weighted sequences, whereas cortex, fibrous cartilage, and ligaments are seen as signal void (black) structures. Hya-

line cartilage and muscle have an intermediate signal intensity (gray). Relative signal intensities of muscle, fat, and bone marrow decrease somewhat on T_2-weighted spin echo sequences. This is in contrast to the increase of signal intensity of structures containing a large amount of water, such as hyaline cartilage. Fat displays a high signal intensity on TSE images.

Distribution of areas with abnormal signal intensity is important, not only for differentiating tumors and tumor-like lesions, but also in identifying normal variants. The low to intermediate signal intensity areas of red bone marrow in the axial skeleton and femur diaphysis in adults and also in the periphery of the skeleton in adolescents and children on T_1-weighted images may be confusing.[1] On STIR sequences, the high signal intensity of red marrow is easily differentiated from the absence of signal from the nulled yellow marrow. Low signal intensity areas within bone marrow on GE images are due to susceptibility effects secondary to the presence of bone trabeculae and cortex, and iron in red bone marrow.

Like most pathologic tissues, musculoskeletal tumors usually have prolonged T_1 and T_2 relaxation times (Figure 1). Therefore, osteolytic tumors have a relatively high signal intensity on T_2-weighted images and a relatively low signal intensity on T_1-weighted images. Because of their low spin density and short T_2 relaxation time, osteosclerotic components or calcifications within the tumor have a low signal intensity on both T_1- and T_2-weighted images. Small calcifications detected by plain radiographs or computerized tomography (CT) often cannot be detected by MRI. Liquefaction or lacunae caused by necrosis within a tumor can often be identified because the T_1 and T_2 relaxation times are even longer than those of viable tumor.

Cortical bone and periosteal reaction can be evaluated not only with CT, but also with MRI. Normal cortical bone is represented as a signal void area, whereas intracortical lesions almost invariably show areas of increased signal intensity relative to cortex, on T_1-weighted, gadolinium enhanced and T_2-weighted images. Likewise, mineralized periosteal reaction is seen as a solid or layered signal void area, whereas the nonmineralized cellular cambium layer of the periosteum has a high signal intensity on T_2-weighted images.

Only in selected cases does MRI allow a specific diagnosis to be made. The inhomogeneity of the tumor consisting of viable tumor with different histologic components, necrosis, hemorrhage, and reactive changes make it impossible to differentiate histologic types by relaxation times. MRI increases specificity only when unusual features such as high signal intensity on T_1-weighted images, low signal intensity on T_2-weighted images, or specific morphology or enhancement patterns are observed. Examples of lesions with a high signal intensity on T_1-weighted sequences are fat and subacute (one week to several months old) hematoma. The signal intensity of hematoma depends on the sequential degradation of hemoglobin, cell lysis, and field strength of the MR system.[2] The high signal intensity representing methemoglobin is initially seen in the periphery, and subsequently extends towards the center of the hematoma. Interstitial hemorrhage is more diffuse than hematoma and is accompanied by edema resulting in aspecific prolonged T_1 and T_2 relaxation times, irrespective of age.

Although hemorrhage and hematoma may occur in any tumor or hemophiliac pseudotumor, telangiectatic osteosarcoma and aneurysmal bone cyst are the lesions to consider when hemorrhage is found.[3] Telangiectatic osteosarcoma will display malignant features such as indistinct margins and inhomogeneity, whereas aneurysmal bone cyst often contains multiple compartments and displays well-defined margins. Layering or a fluid level, as can be demonstrated with MRI, is more often seen in aneurysmal bone cyst than in telangiectatic osteosarcoma. These fluid levels may also be encountered in chondroblastoma, giant cell tumor, especially when a secondary aneurysmal bone cyst is present, and occasionally in other tumors as well.[3] Another reason for hematoma is surgical intervention. Since hematoma and edema may be quite extensive following histological needle or open biopsy, imaging studies are preferably performed prior to invasive procedures.

High signal intensity may also be encountered in lipoma. On T_2-weighted images the signal intensity will remain identical to that of subcutaneous fat. Liposarcomas may contain areas consisting of well-differentiated fat; these areas have the same appearance on MRI as benign lipomas. Liposarcomas, however, invariably contain large areas (myxoid liposarcoma) or some areas (lipoblastic liposarcoma) of poorly differentiated

fat and mesenchymal tissue, which have the same MRI characteristics as other high-grade malignancies.

The signal intensity of hemangioma and arteriovenous malformations is variable. The signal intensity may be high on T_1-weighted sequences due to adipose tissue within the lesion and due to slow flow through dilated sinuses and vessels (Figure 2). Because of slow flow, the serpiginous vascular channels typically display a high signal intensity on T_2-weighted images. Signal void represents high flow in vascular channels. Fat predominates in asymptomatic vertebral hemangiomas. Vascularity is often more pronounced in symptomatic hemangiomas.[4]

A low signal intensity on T_2-weighted sequences, secondary to a low spin density and/or short T_2 relaxation time, may indicate the presence of calcification, osteoid, collagen, fibrosis, hemosiderin, or bone cement. Low signal intensity areas on T_2-weighted images may be present in pigmented villonodular synovitis, giant cell tumor of tendon sheath, elastofibroma, fibromatoses, fibrous dysplasia, neuroma (Morton neuroma), etc.[3] The signal intensity reflects the amount of collagen and fibrous tissue present. Cellular areas will be seen as high signal intensity areas on T_2-weighted images, whereas the paucicellular areas will exhibit a low signal intensity on all pulse sequences.

Low signal intensity masses located within ligaments (Achilles tendon) representing xanthomas may be encountered in patients suffering from familial hypercholesterolemia. This is caused by deposition of cholesterol in combination with dense collagen fibers.

The flow void phenomenon, representing rapid flow in vascular channels, such as found in arteriovenous malformations, may also be a cause for low signal intensity on T_2-weighted images. This is in contrast to the characteristics of cavernous or capillary hemangioma.

The morphology, distribution, and localization of disorders can also be used in MRI to characterize a lesion. Indistinct margins, a soft tissue lesion with size larger than 3 cm, and heterogeneity indicate malignancy, whereas well-defined margins and homogeneity are more indicative of a benign lesion.[3,5] Of course there are exceptions to these general statements. Aggressive fibromatosis is a benign lesion characterized by infiltrative growth and a very high recurrence rate. Other benign lesions such as active eosinophilic granuloma may also present with indistinct margins, just as malignant lesions may present with well-defined margins.

Edema, characterized by an increased signal intensity on T_2-weighted images, a decreased signal intensity on T_1-weighted images, and indistinct margins, is a poor indicator of malignancy. Extensive soft tissue and bone marrow edema is frequently encountered in osteoid osteoma, osteoblastoma, and chondroblastoma. Edema in these benign entities is often more pronounced than in malignant tumors. Peritumoral high signal intensity in the soft tissue of patients with primary malignant tumors may represent only edema or a reactive zone also containing tumor. Edema is found in and around many lesions, including eosinophilic granuloma, fracture, bone bruise, necrosis, and transient osteoporosis.[3]

Neurofibroma is easily recognized on both CT and MRI when it presents as the classic dumb-bell tumor of the vertebral canal. Plexiform neurofibromatosis extends along neural bundles in a lobulated fashion, and is thus also easily recognized. When located in the retroperitoneum or pelvis, these plexiform

For References see p. 1355

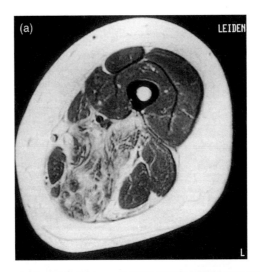

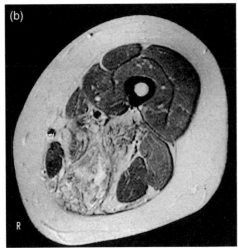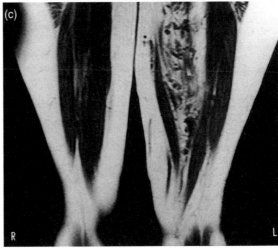

Figure 2 Hemangioma in the adductor compartment of the left thigh imaged with a 1.5 T MRI system. (a) The high signal intensity and morphology on this axial T_1-weighted (600/20) spin echo image are typical of hemangioma. A soft tissue sarcoma would occupy more space. (b) After intravenous administration of 0.1 mmol Gd-DTPA/kg body weight, diffuse enhancement resulting in an increase in signal intensity is appreciated. This is consistent with the presence of well-vascularized cellular tissue. (c) On the coronal T_1-weighted (600/20) spin echo image, the same features are visualized. The signal void of rapid flow within the hemangioma is shown to best advantage

neurofibromas may be quite extensive. The signal intensity of neurofibroma is determined by its prolonged T_1 and T_2 relaxation times, low to intermediate signal intensity on T_1-weighted images, and a very high signal intensity on T_2-weighted images. When present, collagen and fibrous tissue decrease the signal intensity (usually centrally).

The distribution of areas with abnormal signal intensity is important, not only for differentiating between tumors and tumor-like lesions, but also for identifying normal variants. The low to intermediate signal intensity areas of red bone marrow in the axial skeleton and femur diaphysis in adults and also in the periphery of the skeleton in adolescents and children on T_1-weighted images may be confusing.

When reviewing an MR examination, clinical data and radiographs are of paramount importance. Many tumor-like lesions, such as those due to Paget's disease, pseudoaneurysms, polymyositis, osteomyelitis, and stress fractures, can be identified in the proper clinical setting.

Osteomyelitis will, in contrast to Ewing's sarcoma, usually display only reactive changes in the surrounding soft tissue. However, a soft tissue abscess may be encountered.

Stress fractures present as irregular or band-like (perpendicular to cortical bone) areas of low signal intensity on T_1-weighted images. The high signal intensity seen on T_2-weighted or STIR sequences represents edema.

4   GADOLINIUM ENHANCED MRI

The pharmokinetics of gadolinium chelates are similar to iodinated contrast agents. This paramagnetic contrast agent acts indirectly by facilitating T_1 and T_2 relaxation processes through an alteration of the local magnetic environment. Shortening of both relaxation times is a function of the concentration of the gadolinium complex. Initially, an increase in the gadolinium complex concentration results in a predominantly shortened T_1 relaxation time and an ensuing increased signal intensity on T_1-weighted images (Figure 2). With an increasing concentration of the gadolinium complex, the signal intensity drops because of the dominant effect on the T_2^* relaxation time (dephasing). With the concentrations used in clinical practice (0.1–0.2 mmol/kg body weight for intravenous administration), the shortening of T_1 relaxation time supervenes. Only in areas where a high concentration is reached, such as in the urinary bladder or early in a dynamic sequence during bolus administration in an artery, can a signal void occur.

Well perfused, often cellular, tumor components are able to accumulate a high concentration of gadolinium, which results, through shortening of the T_1 relaxation time, in a high signal intensity on short *TR/TE* (600/20) images, as opposed to some tumor components, such as sclerosis, old fibrosis and liquefaction, which are not well perfused and therefore do not show a high signal intensity on the postcontrast images. In general, the presence or absence of late enhancement is not a very good indicator of malignancy. Still, gadolinium may increase tissue characterization.

Although the intravenous administration of Gd-DTPA (DTPA, diethylenetriaminepentaacetic acid) may assist in differentiating viable tumor from liquefaction and edema, neovascularity in necrotic areas may also be enhanced. Differentiation between viable and necrotic tumor is, as a consequence, not always possible. Dynamic MRI may assist in differentiating tumor from reactive tissue, since viable tumor usually enhances faster than reactive tissue. Dynamic sequences, obtained during gadolinium complex administration, with a temporal resolution of at least 3 s can be used to help to differentiate benign from malignant soft tissue tumors. Peripheral enhancement starting within 6 s after arterial enhancement without further increase in enhancement suggests malignancy.[6]

Static short *TR/TE* spin echo images may suggest a specific diagnosis in selected cases. Well-differentiated cartilaginous tumors containing large paucicellular cartilage fields or cartilage fields with mucoid degeneration demonstrate no or only slight enhancement. Only the margin and cellular septae in the periphery and within the tumor enhance. Because of the lobulated gross anatomy of enchondroma and low grade (grade I and II) chondrosarcoma, the postcontrast MRI scans of these tumors exhibit a septal-like or curvilineair (serpengineous) enhancement pattern[7] (Figure 3). Mature enchondromas do not enhance. Enchondromas and osteochondromas may, however, enhance in the adult population 10 s or later after arrival of contrast agent in the artery. Central or peripheral chondrosarcoma grade 1 typically enhances within 10 s of arterial enhancement. Benign cartilaginous lesions in children or in association with sursae (mechanical stress) may also enhance early. Only high-grade cartilaginous tumors such as chondrosarcoma grade III and mesenchymal chondrosarcoma exhibit the aspecific homogeneous or inhomogeneous enhancement displayed by most tumors.

Other cartilage containing tumors such as chondroid osteosarcoma may also show areas with curvilineair enhancement.

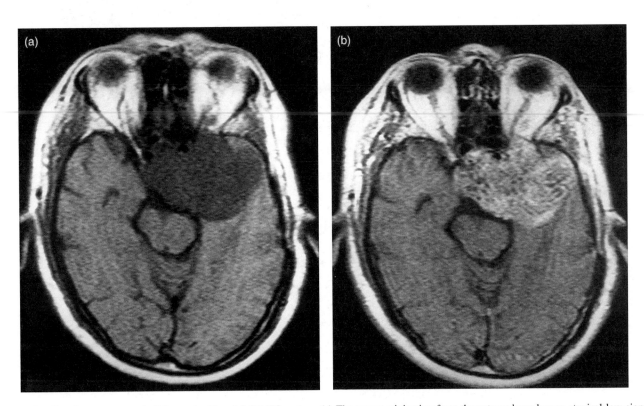

Figure 3 Chondrosarcoma grade II imaged with a 1.5 T MRI system. (a) The tumor originating from the petrous bone has an atypical low signal intensity on this axial T_1-weighted (550/20) spin echo image. (b) The same sequence is repeated 5 min after intravenous administration of 0.1 mmol Gd-DTPA. A serpentiginous enhancement pattern consistent with the presence of fibrovascular septations of well-differentiated chondrosarcoma is visualized. Compression on the temporal lobe and the brainstem is present

For References see p. 1355

Usually other criteria such as morphology will assist in differentiating these lesions. A chondroid osteosarcoma is less homogeneous than a well-differentiated cartilaginous tumor and often contains large areas with osteoid, which can be recognized.

The curvilinear enhancement of well-differentiated cartilaginous tumors must be differentiated from thin peripheral enhancement of the pseudocapsule of lesions such as eosinophilic granuloma and solitary bone cyst.

5   LOCAL TUMOR STAGING

Adequate, wide or radical surgery has been a prerequisite for proper treatment of patients with primary malignant musculoskeletal tumors. The choice of therapy depends on a number of factors such as the grade of malignancy, the response to neoadjuvant therapy, local tumor extension, and the presence or absence of regional or distant metastases. The surgical staging system developed by Enneking takes all these factors, with the exception of the response to neoadjuvant therapy, into account.

The biologic aggressiveness, indicated by the grade, is the key factor in the selection of the surgical margin required to achieve control. Four different surgical procedures, which imply four different margins, are recognized: intralesional, marginal, wide, and radical. The exact anatomic location of tumor is the key factor in selecting how a required tumor-free margin can be accomplished.

5.1   Bone Marrow Involvement

For extremities, a plane through the long axis of a long bone is chosen, i.e. sagittal or coronal. MRI is superior to technetium bone scanning because tumor related hyperemic osteoporosis is not depicted on MRI and is therefore not a source of false-positive readings. MRI has an almost perfect correlation ($r = 0.99$) with pathologic/morphologic examination; CT has a less substantial correlation ($r = 0.93$); ^{99m}Tc methylene diphosphonate (MDP) scintigraphy has a weak correlation ($r = 0.69$).[3,8]

The osseous tumor margin may be obscured by the presence of intraosseous edema. Often a double margin can be visualized. The margin nearest to the center of the tumor represents the true tumor margin, whereas the outer margin represents the margin of the edematous reactive zone towards normal bone marrow. Differentiation can be facilitated by using gadolinium. Tumor and the edematous reactive zones display different enhancement patterns: a well-vascularized tumor will enhance more rapidly than the reactive zone, but usually (for instance in osteosarcoma) the intraosseous reactive zone will enhance more than the tumor itself on images taken 5 min after contrast injection. Furthermore, enhancement of the edematous reactive zone is, as a rule, more homogeneous than that of tumor. The intraosseous reactive zone is usually not a clinical problem, because it disappears after one or two cycles of chemotherapy. The true tumor margin is then easily identified.

Caution is needed in the diagnosis of skip metastases. Skip metastases, as may be encountered in patients with osteosarcoma and Ewings's sarcoma, are detected with bone scintigraphy unless the size is below the detection threshold.

Skip lesions of less than 5 mm may thus pose a diagnostic problem. Imaging of the entire bone on T_1-weighted images by using a large field of view may be of help.

5.2   Cortical Involvement

Destruction of cortical bone is not a diagnostic problem since it can be evaluated on plain radiographs and, if necessary, with tomography or CT. MR images are also able to visualize the status of cortex. Invasion of cortex by tumor is best shown on T_2-weighted images as a disruption of the cortical line and replacement of cortex by the high signal intensity of tumor. The sensitivity and specificity of MRI (92% and 99% respectively) were not found to be significantly higher than the sensitivity and specificity (91% and 98% respectively) of CT.[3,8] MRI is superior to CT in visualizing sclerotic osteosarcomas, because the signal intensity of osteosclerotic tumor is still slightly higher than the low signal intensity of normal cortex. The high density of osteosclerotic tumor and cortex may be indistinguishable on CT.

5.3   Involvement of Muscular Compartments

MRI is significantly superior (sensitivity 97%, specificity 99%) to CT in identifying muscle compartments containing tumor.[3,8] The superior performance of MRI is based on the display of tumor relative to muscular compartments with, compared to CT, superior contrast in the ideal imaging plane. The presence of peritumoral edema is, as a rule, easily identified because of its slightly different signal intensity compared to that of tumor, and especially because of the fading margins of edema as opposed to the distinct margin of the (pseudo)capsule of the tumor. However, accurate delineation of the tumor–edema interface can be rather difficult. Differences in signal intensity on multiple echoes and enhancement following administration of gadolinium may be used to differentiate tumor from edema.

5.4   Vascular Involvement

Vessels are more often displaced than encased by tumor. The relationship between tumor and neurovascular bundle is easily evaluated on T_2-weighted images because normal flow in a vessel results in low or absent signal intensity, the so-called 'flow void phenomenon.' The lumen of the vessel may have a higher signal intensity due to paradoxal enhancement, even-echo rephasing, or slow flow caused by compression. CT (sensitivity 36%, specificity 94%) and MRI (sensitivity 92%, specificity 98%) provide more information than angiography (sensitivity 75%, specificity 71%) because large vessels are, especially with MRI, well visualized in relation to the tumor.[3,8]

5.5   Joint Involvement

CT (sensitivity 94%, specificity 90%) and MRI (sensitivity 95%, specificity 98%) are both able to demonstrate joint involvement with high accuracy.[3,8] Joint involvement is sometimes more accurately demonstrated on MR images than on CT, because the articular surfaces may be parallel to the transverse CT plane. Cartilage is an effective barrier that is not easily crossed by tumor. Osteosarcoma and giant cell tumor are the tumors that are able to cross cartilage. When assessing possible joint involvement, the number of false-positive readings is much higher than the number of false-negative readings. When in doubt, the joint usually is not affected. Joint effusion with or without hemorrhage, often but not always a secondary sign of a contaminated joint, is easily identified on CT and MRI.

6   RECURRENCE

Detection of recurrent or residual tumor following initial treatment is a challenging problem for diagnostic imaging. A large recurrent tumor mass may be detected at clinical examination, with radiographs or by CT. Small recurrent tumors are difficult to define in relation to posttherapy changes

caused by surgery, radiation therapy, and chemotherapy. The matter is further complicated when fixation devices have been used in reconstructive surgical procedures. These devices cause susceptibility artifacts on MRI. Even when no hardware is used, susceptibility artifacts, caused by metallic particles left behind after instrumentation, can degrade the MR images.

Despite these drawbacks, MRI can be used to detect recurrence (Figure 4). A recurrence is very likely (sensitivity 96–100%) when, following surgery, with or without chemotherapy, a mass is seen on MRI which is characterized by a high signal intensity of Gd-DTPA enhanced or T_2-weighted images.[9] Cystic masses without tumor have a high signal intensity on T_2-weighted images, but do not enhance after intravenous administration of Gd-DTPA. A tumor recurrence is very unlikely when, following surgery, no enhancement or a low signal intensity on T_2-weighted images is found. However, knowledge of the signal intensity of the primary tumor prior to therapy is crucial, especially when the primary tumor was characterized by a low signal intensity on T_2-weighted images.

When an equivocal lesion is detected, follow-up studies, biopsy or other imaging studies such as radiographs or ultra-

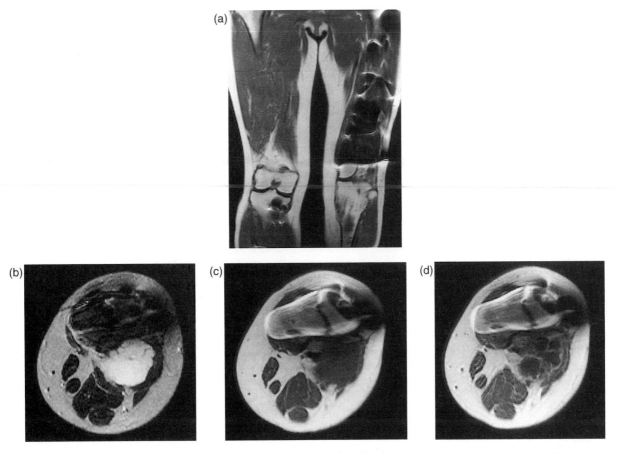

Figure 4 MRI images obtained with a 1.5 T imager of a patient with enchondromatosis and recurrent chondrosarcoma after resection and reconstructive surgery. (a) Coronal large field of view, T_1-weighted (600/20) spin echo images show cartilaginous tumors in both proximal tibias. Marked susceptibility artifacts, secondary to metallic hardware left by the orthopedic surgeon, are seen in the left femur and proximal right tibia. (b) Axial T_2-weighted (2000/100) spin echo images show a lobulated soft tissue recurrence posterior to the artifacts within the femur. (c) The soft tissue mass has an atypical low signal intensity on this T_1-weighted (600/20) image. (d) After injection of Gd-DTPA the serpentiginous enhancement pattern indicative of recurrent well-differentiated chondrosarcoma is easily appreciated, despite the marked susceptibility artifacts

For References see p. 1355

sound may be helpful. Radiation therapy adds to the confusion because it may induce inflammatory, reactive changes that may be indistinguishable from tumor recurrence. These reactive changes are characterized by a high signal intensity on static Gd-DTPA enhanced and T_2-weighted images, and may persist for more than a year. However, reactive lesions usually confine themselves to the space available and, unlike tumor recurrence, do not present themselves as space occupying expansile masses. Dynamic MRI can be used to differentiate reactive tissue from viable tumor. Viable tumor typically enhances 6 s after arrival of gadolinium complex in the artery. When detection of early recurrence is of vital importance, or when there is a high risk of recurrence, a baseline study obtained 3–6 months after surgery may be of help.

7   MONITORING CHEMOTHERAPY

Preoperative (neoadjuvant) chemotherapy of bone sarcomas has increased the feasibility of limb salvage procedures. As a consequence of chemotherapy, a soft tissue tumor mass may shrink and, in combination with encapsulation of the residual extramedullary tumor mass, improve surgical conditions (down staging). However, identification of good and poor respondents is a challenging and controversial exercise.[10] Assessment of viable and necrotic tumor at histology is the gold standard. Spontaneous necrosis of up to 50% is not uncommon in high-grade malignancies. Therefore a histologic good response is defined as the presence of 10% or less viable tumor tissue. A change of tumor volume and signal intensity in osteosarcomas and Ewing sarcoma may correlate with response to chemotherapy. Despite major limitations, several qualitative and quantitative MR parameters contribute to the differentiation of good and poor respondents during and after preoperative chemotherapy.

Increase in tumor volume without hemorrhage and increase of signal intensity on T_2-weighted images in patients with osteosarcoma, even after the first cycle of chemotherapy, indicates poor response.[11] At present there are no reliable criteria for identifying good respondents on native MRI.

Reduction of tumor volume in Ewing sarcoma is characteristically seen in all patients with Ewing sarcoma after successful, or unsuccessful, chemotherapy.[10] A 75% decrease in tumor volume, or complete absence of a residual soft tissue mass after chemotherapy, are consistent with good response. Resolution of the soft tissue mass in Ewing sarcoma is frequently accompanied by reactive subperiosteal bone formation, which may result in the development of an inhomogeneous cuff of tissue encircling the original cortex. Progressive ossification of the periosteal mass may reflect healing of these tumors, but the presence of minimal residual disease in this peripheral area cannot be excluded with any imaging modality.

A well-defined rim of low signal intensity forming a margin for the extramedullary tumor compartment represents a fibrous pseudocapsule continuous with the periosteum; however, there is no correlation with the percentage of tumor necrosis.[10,11] Although this qualitative sign cannot be used as a differentiating criterion between good and poor respondents,

the improved tumor demarcation can facilitate the surgical resection.[10,11]

Following chemotherapy, the amount of viable residual tumor can be assessed by dynamic gadolinium-enhanced images. Viable tumor enhances within 6 s after arterial enhancement. Non-viable tumor components enhance later or not at all.[12]

8   BONE METASTASES

Bone metastases primarily occur in the red bone marrow of the axial skeleton. We therefore focus here on the vertebral column. Conventional radiographs have a very low yield in depicting skeletal metastases when these are still located in the bone marrow. The origin of metastatic deposits is within the bone marrow of the vertebral body. Metastatic tumor may fill the marrow and leave cortical bone intact. Detection of metastatic disease can occur at a much earlier stage when destruction of cortical bone is present. The contours of collapsed vertebral bodies can suggest the nature of the underlying disease, but conventional radiography alone is unreliable in differentiating between benign and malignant causes of vertebral body collapse.

MRI can be performed using several pulse sequences; spin echo, TSE, STIR, GE, and out-of-phase chemical shift imaging. Each pulse sequence has its own imaging features and will exhibit normal bone marrow and pathology in a characteristic manner. The signal intensities reflect the histology of the tissue. Typically, metastases display a low signal intensity on T_1-weighted sequences and a high signal intensity on T_2-weighted sequences. There are exceptions, such as in sclerotic metastases which exhibit a low signal intensity on all pulse sequences, and lipoblastic metastases which have a relatively high signal intensity on T_1-weighted images.

In the STIR pulse sequence, yellow marrow signal is nulled, such that bone marrow appears black. T_1 and T_2 values other than those for fat are additive. Lytic metastases, because of their high water content, produce an increase in both T_1 and T_2 relaxation times. This will enhance the contrast between lytic metastases and bone marrow. Often lytic metastases are better depicted on STIR images than on T_1-weighted spin echo images. Early edema (10–14 days) after radiation therapy can typically be seen earlier on STIR images than on T_1- or T_2-weighted spin echo images.

Contrast between normal bone marrow and tumor can also be increased by using the difference in resonance frequency between aliphatic and water protons. Out-of-phase images will increase image contrast in the case of lytic bone marrow metastases. Fat suppression with presaturation pulses may further increase contrast on T_2-weighted sequences. Fat presaturation may be used in combination with opposed-phase imaging.

Gadolinium will further increase sensitivity when used in combination with fat suppression. It may also increase the specificity of MRI, as metastatic deposits in vertebral bodies will exhibit diffuse enhancement of signal intensity on T_1-weighted images, whereas osteoporotic collapsed vertebral bodies will show band-like enhancement. Some authors, however, have found that the use of Gd-DTPA can be disappointing in the differential diagnosis of malignant and benign tissues.

T_1- and T_2-weighted (STIR or fat suppression) images will constitute a sufficient routine imaging protocol. Additional imaging in orthogonal planes will render valuable information in the case of soft tissue involvement. Gadolinium should be reserved for selected cases such as differential diagnosis between malignant and benign collapsed vertebrae and where there is suspected leptomeningeal tumor spread.

MRI is more sensitive in the detection of vertebral metastases than is bone scintigraphy.[13] This is not surprising since MRI visualizes bone marrow directly rather than depending on secondary signs such as new bone formation.

Although the MR characteristics of metastatic disease do not always allow reliable differentiation between benign and malignant disease, MRI is still more specific than bone scintigraphy in distinguishing between benign and malignant disease. For instance, by means of MR examination it is usually possible to make a reliable distinction between degenerative bone disease in the vertebral column and malignant infiltration. Morphologic characteristics of the lesion and adjacent disk must be taken into account. Loss of height of the intervertebral disk can be seen in degenerative disease and in diskitis, whereas the shape and height of the disk is usually preserved in metastatic diseases. In diffuse bone marrow lesions, the intervertebral disk may show a high signal intensity relative to the decreased signal intensity of abnormal bone marrow on T_1-weighted SE images.

The morphology of vertebral bodies and the signal intensity changes of bone marrow, particularly in relation to the vertebral endplates, may assist in differentiating between benign and malignant compression fractures. In old osteoporotic compression fractures the signal intensity is typically normal (fat), whereas in malignant compression fractures an abnormal signal intensity due to replacement of bone marrow is seen. In the (sub)acute phase, an inhomogeneous increase in signal intensity on STIR or T_2-weighted images, or a sharply delineated isointense vertical band of preserved normal (fatty) bone marrow along the dorsal aspect of the compressed body, indicates a benign fracture. The abnormal signal intensity in benign fractures may have the shape of a horizontal band. Fractures at multiple levels with preservation of normal bone marrow, vertebral body fragmentation, and disk rupture also indicate benign disease. Signs in favor of malignancy are: homogeneous signal intensity changes; convex anterior, and especially posterior, contour; cortical destruction; multiple levels with abnormal signal intensity, but without fracture; and abnormal signal intensity in posterior elements. Paraspinal masses are more conspicuous in malignant disease, but can be seen in both traumatic and malignant cases. Care must be taken in interpreting MR examinations in recently collapsed vertebral bodies as they can show signal intensities indistinguishable from metastatic disease.

Currently, bone scintigraphy remains the screening procedure of choice because it is readily available, and it allows imaging of the entire skeleton in a time effective way. However, secondary to increased sensitivity, specificity, and faster pulse sequences, the role of MRI is increasing. MRI can visualize metastatic disease when bone scintigraphy is falsely negative. Thus MRI is currently indicated when a strong clinical suspicion is combined with a negative bone scan.

MRI may also elucidate the true nature of hot spots in the vertebral column in cancer patients, as it can often assist in making a distinction between malignant and benign disease. MRI can also be helpful in the guidance of biopsies.

In order to rule out compressive myelopathy or to establish soft tissue extension of tumor tissue, multiplanar MRI offers unique imaging features. The use of MRI is helpful in determining the local extent of metastatic disease when planning palliative surgery or radiation therapy.

9 RELATED ARTICLES

Imaging of Trabecular Bone; Skeletal Muscle Evaluated by MRI.

10 REFERENCES

1. S. G. Moore and K. L. Dawson, *Radiology*, 1990, **175**, 219.
2. J. M. Gomori and R. I. Grossman, *RadioGraphics*, 1988, **8**, 427.
3. J. L. Bloem, H. C. Holscher, and A. H. M. Taminiau, in 'MRI and CT of the Musculoskeletal System', ed. J. L. Bloem and D. J. Sartoris, Williams & Wilkins, Baltimore, 1992, Chap. 15.
4. J. D. Laredo, E. Assouline, F. Gelbert, M. Wybier, J. J. Merland, and J. M. Tubiana, *Radiology*, 1990, **177**, 467.
5. M. J. Kransdorf, J. S. Jelinek, and R. P. Moser, *Radiol. Clin. North Am.*, 1993, **31**, 359.
6. H. D. van der Woude, K. L. Verstraete, P. C. W. Hogendoorn, A. H. M. Taminiau, J. Hermans, and J. L. Bloem, *Radiology*, 1998, **208**, 821.
7. M. J. A. Geirnaerdt, J. L. Bloem, F. Eulderink, P. C. W. Hogendoorn, and A. H. M. Taminiau, *Radiology*, 1993, **186**, 813.
8. J. L. Bloem, A. H. M. Taminiau, F. Eulderink, J. Hermans, and E. K. J. Pauwels, *Radiology*, 1988, **169**, 805.
9. D. Vanel, L. G. Shapeero, T. de Baere, R. Gilles, A. Tardivon, J. Genin, and J. M. Guinebretiere, *Radiology*, 1994, **190**, 263.
10. H. D. van der Woude, J. L. Bloem, and P. C. W. Hogendoorn, *Skeletal Radiol.*, 1998, **27**, 145.
11. H. C. Holscher, J. L. Bloem, D. Vanel, J. Hermans, M. A. Nooy, A. H. Taminian, and M. Henry-Anar, *Radiology*, 1992, **182**, 839.
12. H. D. van der Woude, J. L. Bloem, K. L. Verstraete, A. H. M. Taminiau, M. A. Nooy, and P. C. W. Hogendoorn, *Radiology*, 1995, **165**, 593.
13. P. R. Algra and J. L. Bloem, in 'MRI and CT of the Musculoskeletal Systems', ed. J. L. Bloem and D. J. Sartoris, Williams & Wilkins, Baltimore, 1992, Chap. 16.

Acknowledgements

The contributions of M. Geirnaerdt, H. C. Holscher, H. J. van der Woude, A. H. M. Taminiau, P. Hogendoorn, F. Eulderink, M. A. Nooy, and H. M. Kroon, are gratefully acknowledged.

Biographical Sketch

Johan L. (Hans) Bloem. *b* 1954. M.D., 1979, Ph.D., 1988, Leiden University, The Netherlands. Visiting professor, Charles Gairdner Hospital, Perth; Thomas Jefferson University, Philadelphia. NMR program at Leiden University, 1983–present. Currently, Professor and Chairman

For References see p. 1355

of Radiology at Leiden University. Approx. 100 publications; editor of *MRI and CT of the Musculoskeletal System*. Research interest: MRI of the musculoskeletal system.

Imaging of Trabecular Bone

Felix W. Wehrli

University of Pennsylvania Medical School, Philadelphia, PA, USA

1 INTRODUCTION

Bone is a composite material consisting of an inorganic phase—calcium apatite, $Ca_{10}(PO_4)_6(OH)_2$, corresponding to about 65% of total volume—and an organic phase—essentially collagen—accounting for most of the remaining 35%. From an architectural point of view, bone can be subdivided into cortical and trabecular, the latter providing most of the strength of the axial skeleton (e.g., the vertebral column) and the portions of the appendicular skeleton near the joints. Trabecular bone is made up of a three-dimensional network of struts and plates, the trabeculae, which are on the order of $100–150\,\mu m$ in width and spaced $300–1000\,\mu m$ apart.

Like engineering materials, trabecular bone derives its mechanical strength from its inherent elastic properties, its volume density, and its structural arrangement. Bone is constantly renewed through a process called 'bone remodeling', a term referring to a dynamic equilibrium between bone formation and bone resorption, controlled by two essential types of cells: the osteoblasts—bone-forming cells—and the osteoclasts—bone-resorbing cells. During bone formation, osteoblasts eventually become imbedded in bone, turning into osteocytes, which presumably act as piezoelectric sensors transmitting signals to the osteoblasts to induce bone formation. Since the seminal work of Wolff,[1] it has been known that bone grows in response to the forces to which it is subjected (see, for example, Roesler[2]). Therefore, weightlessness and physical inactivity are well-known factors inducing bone loss.

The most common pathologic process leading to bone loss is osteoporosis.[3] Among the various etiologies, postmenopausal osteoporosis, which results from increased osteoclast activity, is the most frequent form of the disease, afflicting a substantial fraction of the elderly female population and, increasingly, the male population. The most common clinical manifestations are fractures of the hip and vertebrae. If detected early, calcium supplements and estrogen replacement are effective forms of therapeutic intervention. Further, the development of drugs inhibiting osteoclastic activity is in progress.

Bone mineral density is the most widely invoked criterion for fracture risk assessment, typically measured by dual energy X-ray absorptiometry (DEXA), which is based on the measurement of the attenuation coefficient in a quantitative radiographic procedure, or by quantitative computed tomography (QCT). Whereas both methods measure bone mineral density with sufficient precision, neither provides information on the properties or structural arrangement of the bone. NMR, however, has the potential to probe structure as well as chemical composition of bone, both relevant to biomechanical competence.

2 DIRECT DETECTION OF BONE MINERAL BY ^{31}P NMR

The difficulties of detecting phosphorus in the solid state in vivo are considerable but do not seem insurmountable. The problems are symptomatic of high-resolution NMR in the solid state in general: a combination of long T_1 (on the order of minutes) and short T_2 (on the order of $100\,\mu s$), as well as additional line broadening by anisotropic chemical shift and dipolar coupling.

Brown et al. first demonstrated the feasibility of quantitative analysis by solid state ^{31}P spectroscopy in human limbs as a means of measuring bone mineral density noninvasively.[4] Imaging adds an additional level of complexity, since the short lifetime of the signal demands short gradient duration, a requirement that can only be reconciled with gradients of high amplitude and large slew rates, which are both difficult to achieve in large sample volumes. Ackerman et al. produced one-dimensional spin echo images at 7.4 T with echo times on the order of 1 ms and flip-back 180° rf pulses as a means to restore the longitudinal magnetization, inverted by the phase reversal pulse.[5] The same group of workers reported two-dimensional images of chicken bone at 6 T by means of a combination of back-projection and ^{1}H–^{31}P cross polarization for sensitivity enhancement, with echo times as short as $200\,\mu s$.[5] In the cross-polarization technique, the ^{31}P magnetization is derived from that of dipolar-coupled protons, which have shorter T_1 and thus permit shorter pulse sequence recycling times. Making use of the dependence of the cross-polarization rate on the proton–phosphorus dipolar coupling, the same workers showed that different phosphate species can be distinguished, demonstrating the presence of a minor HPO_4^{2-} species in immature chicken bone.[7] Very recently, Wu et al. have measured three-dimensional mineral density of hydroxyapatite phantoms and specimens of bone ex vivo.[8] While this work is at an early stage, it clearly has unique potential for nondestructive assaying of the chemical composition and its age-related changes of bone parameters that might in part explain the increased fragility of bone in older individuals.

3 IMPLICATIONS OF BONE DIAMAGNETISM ON NMR LINE BROADENING

Another approach toward assessing the properties of trabecular bone—specifically, as its architectural arrangement is concerned—exploits the diamagnetic properties of bone mineral. By virtue of the higher atomic number of its elemental composition (i.e., calcium and phosphorus), mineralized bone is more diamagnetic than marrow constituents in the trabecular marrow cavities, which consist mainly of water and lipids (i.e., oxygen, carbon, and hydrogen).

Note that in this article, the following definitions and notation will be used for the contribution from magnetic field inhomogeneity to the total dephasing rate $R_2^* \equiv 1/T_2^*$, irrespective of the notation in the original literature: $1/T_2' \equiv R_2' \approx \frac{1}{2}\gamma \Delta H_z = 1/T_2^* - 1/T_2$, with ΔH_z representing the full width at half maximum of the magnetic field histogram in the sampling volume such as the imaging voxel and R_2' the effective transverse relaxation rate.

It is well known that near the interface of two materials of different magnetic susceptibility, and depending on the geometry of the interface, the magnetic field is inhomogeneous. Among the first to investigate these effects systematically were Glasel and Lee, who studied deuteron relaxation of beads of different size and susceptibility, suspended in ^2H_2O.[9] Specifically, they showed that the linewidth $1/\pi T_2^*$ scaled with $\Delta\chi$, the difference in volume susceptibility between the beads and deuterium oxide. The transverse relaxation rate $1/T_2$ was found to increase linearly with reciprocal bead size, an observation that could be reconciled with diffusion in the induced magnetic field gradients. Similar phenomena were reported by Davis et al., who measured proton NMR linewidths at 5.9 T in powdered bone suspended in various solvents and found T_2^* to decrease with decreasing grain size and thus increased surface-to-volume ratio.[10] Rosenthal et al. measured R_2' in specimens from human vertebrae following marrow removal and immersion in water at 0.6 T, reporting a value of $6.1\,s^{-1}$, more than an order of magnitude smaller than that found for powdered bone at 5.9 T.[11] Wehrli et al. first reported a gradient-echo-based method for measuring the line width in vivo in humans in vertebral trabecular bone marrow and found R_2^* in the vertebrae to be increased by a factor of two to three relative to those in the intervertebral discs.[12]

One of the earliest studied magnetically heterogeneous systems in biology is lung tissue, where the local magnetic field distortions are caused by air in the alveoli, and some of the concepts described here have parallels in imaging pulmonary parenchyma.[13] Transverse relaxation enhancement from diffusion in intrinsic microscopic gradients has received increased attention in conjunction with the blood oxygenation level-dependent (BOLD) contrast phenomenon, resulting from physiologic variations of deoxyhemoglobin in capillary vessels during functional activation.[14,15] However, since trabeculae are considerably larger than the venules of the capillary bed (100–200 μm versus 10–20 μm), and diffusion of the protons in the marrow spaces is small (on the order of $10^{-5}\,cm^2\,s^{-1}$), diffusion-induced shortening of T_2 is expected to be negligible. Consequently, the effect of the susceptibility-induced inhomogeneous field is essentially line broadening.

Suppose there is a distribution $\Delta B(x,y,z)$ across the sample volume, such as an imaging voxel of dimension $\Delta x\,\Delta y\,\Delta z$; then the transverse magnetization $M_{xy}(t)$ can be written as

$$M_{x,y}(t) = \frac{M_0}{\Delta x\,\Delta y\,\Delta z} e^{-t/T_2}$$
$$\times \int_{\Delta x}\int_{\Delta y}\int_{\Delta z} e^{i\gamma\Delta B(x,y,z)t}\,dx\,dy\,dz \quad (1)$$

For a Lorentzian field distribution, the integral in Equation (1) can be described as an additional damping term, yielding

$$M_{x,y}(t) = M_0 e^{-t/T_2} e^{-t/T_2'} \quad (2)$$

T_2' is, therefore, the time constant for inhomogeneity-induced spin dephasing. While the line broadening, in general, of course, is not Lorentzian, it will be seen subsequently that the assumption of a single exponential time constant is often a valid approximation.

3.1 Susceptibility of Bone

Although there is plenty of evidence for the susceptibility hypothesis, a quantitative determination of the volume magnetic susceptibility of bone has not been reported until recently. In preliminary experiments the author determined the susceptibility of bone using a susceptibility matching technique.[16] For this purpose, powdered bone from bovine femoral head was suspended in a cylindrical sample tube with a coaxial capillary containing water and serving as a reference, both aligned along the axis of a superconducting magnet. Potassium ferricyanide, $K_4Fe(CN)_6$, which is highly diamagnetic, was then added incrementally to the suspension. This operation resulted in a decrease in linewidth and an increase in the bulk magnetic susceptibility (BMS) shift. For concentric cylinders, the BMS shift is given as[17]

$$\Delta\nu = \frac{\gamma B_0}{2\pi}\frac{|\chi_d - \chi_w|}{3} \quad (3)$$

with χ_w and χ_d being the volume susceptibilities of water and ferricyanide solution, respectively. Whereas the critical concentration at which the solution matched the susceptibility of bone could not be attained owing to limited solubility of $K_4Fe(CN)_6$, the matching concentration was determined by extrapolation of the line broadening–concentration curve, from which a BMS shift of -0.95 ± 0.13 ppm was obtained (corresponding to $\chi_d - \chi_w = -2.85$ ppm). Hence, bone is considerably less diamagnetic than calcium hydroxyapatite.

Based on these earlier experimental approaches, Hopkins and Wehrli conducted a more rigorous study to determine the absolute susceptibility of bone, with potassium chloride as the diamagnetic additive.[18] Specifically, they showed that the susceptibility of the suspension, χ_{susp}, is related to the line broadening $\Gamma-\Gamma_0$ [difference between the line width in the presence (subscript p) and absence (subscript 0) of bone powder], as follows:

$$\chi_{susp} = \chi_p - \frac{1-f_p}{f_p k H_0}(\Gamma - \Gamma_0) \quad (4)$$

where f_p represents the volume fraction of the bone powder and k is an empirical proportionality constant, which is a function of particle size, shape, orientation, diffusion, and so forth. Therefore, extrapolation of the straight line defined by equation (4) to $\Gamma-\Gamma_0=0$ directly provides the susceptibility of the powder. The concentration-dependent decrease in line width and increase in BMS shift is shown in Figure 1. The susceptibility of bovine rib bone was found to be -11.3 ($\pm0.25)\times10^{-6}$ (S.I.), indicating bone to be about 2.3 ppm more diamagnetic than water, which is somewhat less than the earlier experiments suggested.

For References see p. 1367

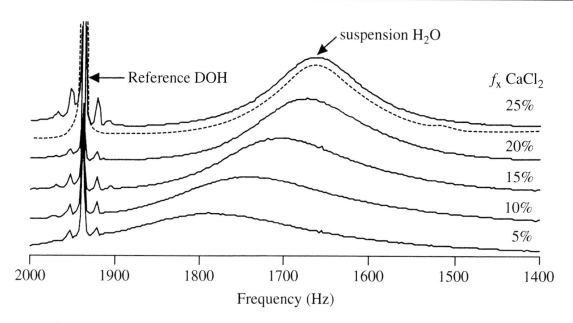

Figure 1 Spectra of ^{1}H at 400 MHz from suspensions of bovine rib bone in $CaCl_2$ solutions of increasing volume fraction of salt, ranging from 5 to 25% (bottom to top). The dotted line spectrum offset from the top spectrum is the best fit Lorentzian line used to estimate the spectral parameters. Small symmetric resonances about the DOH reference line are spinning sidebands. (With permission from Hopkins and Wehrli[18])

3.2 Theoretical Considerations and Computer Modeling

Consider two adjoining materials of susceptibilities χ_m and χ_b. The induced magnetic surface charge density σ at some location on the interface between the two materials is then given as

$$\sigma = \Delta\chi \boldsymbol{H}_0 \cdot \boldsymbol{n} \qquad (5)$$

where $\boldsymbol{n}$ is the unit vector normal to the interface and $\Delta\chi = \chi_m - \chi_b$. The additional field $\boldsymbol{H}_i(\boldsymbol{r})$ resulting from the magnetic charges at the phase boundary can be estimated from the Coulomb integral[19]

$$\boldsymbol{H}_i(\boldsymbol{r}) = \int \sigma(\boldsymbol{r}') \frac{\boldsymbol{r} - \boldsymbol{r}'}{|\boldsymbol{r} - \boldsymbol{r}'|^3} \, \mathrm{d}\boldsymbol{S}' \qquad (6)$$

where the integration is over the surface S of the interface, with $\boldsymbol{r}'$ and $\boldsymbol{r}$ representing the locations of the source and field points, respectively.

The induced field is thus proportional to the difference in susceptibility of the two adjoining materials, the strength of the applied field, and the inverse square of the distance between source and field location. For an array of trabeculae, the field should be highly inhomogeneous, and a relationship is expected to exist between the magnetic field distribution within the volume of interest and the number density, thickness, and orientation of trabeculae.

Ford et al.[20] developed a three-dimensional model that resembles strut-like trabecular bone as found in the vertebrae[21] to predict the line broadening behavior of protons in the marrow spaces, as a means to investigate the structural dependence of R_2'. The model consists of a tetragonal lattice of interconnected parallelepipeds ('struts') of square cross-section, differing in susceptibility from the medium by $\Delta\chi$. When oriented so that the two parallel faces of each transverse strut are normal to the

direction of the applied field $\boldsymbol{H}_0$ then, according to Equation (5), these two faces will have uniform charge density $\sigma = \pm\Delta\chi \boldsymbol{H}_0$. The other faces of the transverse (and longitudinal) struts will have $\sigma = 0$, and so do not contribute to $\boldsymbol{H}_i(\boldsymbol{r})$. If the field is oriented arbitrarily relative to the lattice, all faces will be uniformly polarized, with a charge density $\sigma = \Delta\chi \boldsymbol{H}_0 \cos \alpha$ where α is the angle between $\boldsymbol{H}_0$ and the unit vector normal to any given polarized surface. An analytical expression exists for the induced field given by the integral of Equation (6) for a rectangular lamina; consequently, the total field at any one point in space can be calculated as a sum of contributions from all charge-bearing faces. In this manner, a histogram of the field for the unit cell of this lattice was obtained by randomly placing field points within the unit cell, and R_2' was calculated by fitting the Fourier transform of the field histogram to a decaying exponential. The model predicts nearly exponential decay within the experimentally practical range of echo times (about 10–50 ms). Further, R_2' is predicted to increase with both the number density of transverse struts and their thickness. Since the latter two quantities scale with material density, this finding appears unremarkable, since it would imply that R_2' merely measures bone mineral density. However, if both strut thickness and number density are varied in an opposite manner (so as to keep the material density constant), the model indicates that R_2' will increase as strut thickness decreases and number density increases. These predictions, which have been confirmed in analogous physical models,[22] underscore the importance of the distribution of the material, and suggest that different etiologies of bone loss (e.g., trabecular thinning as opposed to loss of trabecular elements) might be distinguishable.

Extending the numerical approaches described above, Hwang and Wehrli computed the magnetic field distribution in trabecular bone of human and bovine origin on the basis of a surface model derived from isotropic high-resolution three-dimensional NMR images.[23] Surfaces were modeled with triangular elements of constant magnetic surface charge density, which allows the induced field to be computed from the charged surfaces. The method was applied to computing histograms of the induced fields in specimens of trabecular bone. The width of the induced field distributions was found to be narrowest when the polarizing field was parallel to the preferred orientation of the trabeculae, confirming previous experimental findings,[24] which provides further support for the anisotropic nature of the effect. Figure 2 shows a histogram of the induced magnetic field, derived from three-dimensional MR

micrograms of bovine tibia with the magnetic field oriented along two orthogonal directions, together with a projection image of the specimen.

The dependence of the induced field, expressed in terms of the reversible contribution to R_2', was also investigated by Yablonskiy and Haacke, who derived an analytical expression for R_2' in a model of trabecular bone consisting of a distribution of mutually orthogonal cylindrical columns and struts.[25] They showed that beyond a critical time within which signal decay is Gaussian, signal evolution can be described by a single exponential time constant:

$$R_2' \propto \Delta\chi H_0 \left[\varsigma_h + \left(\varsigma_v - \frac{1}{2}\varsigma_h \right) \sin^2 \vartheta \right] \qquad (7)$$

where $\Delta\chi$ is the susceptibility difference between bone and marrow, ς_h and ς_v are the densities of the struts and columns, and ϑ is the angle between the magnetic field $\boldsymbol{B}_0$ and the columns. Yablonskiy et al. demonstrated the angular dependence of R_2' in a simple trabecular bone phantom composed of parallel polyethylene filaments and found their experimental findings to agree well with their theory.[26] Similar findings were reported by Selby et al. in microphantoms consisting of cylindrical Pyrex rods.[27] Previously, Chung et al. had demonstrated the anisotropic behavior of R_2^*, in vivo in the distal radius where trabeculae are highly ordered, following the anatomic axis.[24] This observation was confirmed by Yablonskiy et al., who found R_2' in the radius to be twice as large with the axis of the wrist perpendicular compared with parallel to the direction of the field.[26]

4 RELATIONSHIP BETWEEN TRABECULAR ARCHITECTURE, LINE BROADENING, AND MECHANICAL COMPETENCE

Trabecular bone is well known to be anisotropic, with the orientation of the trabeculae following the major stress lines. In the vertebrae, for example, the preferred orientation of the trabeculae is along the body axis, in response to the compressive forces acting in this direction. The role of the horizontal trabeculae is to act as cross ties preventing failure by buckling. It has also been shown that, during aging, horizontal trabeculae are lost preferentially.[28]

If the static magnetic field is applied parallel to the inferior–superior axis of the vertebrae, the horizontal trabeculae (i.e., those orthogonal to the field) are polarized and, therefore, are expected to be the principal cause of the susceptibility-induced line broadening. Chung et al. measured the mean spacing of horizontal trabeculae (i.e., the reciprocal of horizontal trabecular number density) in cadaver specimens of human lumbar trabecular bone after bone marrow removal and suspension of the bone in water, using NMR microscopy and digital image processing.[29] They found a positive correlation between water proton R_2', measured at 1.5 T, and mean number density of the horizontal trabeculae ($r = 0.74$, $p < 0.0001$).

The critical role of the horizontal trabeculae in the vertebrae in conferring compressive strength is illustrated with the correlation between R_2' measured with the polarizing field parallel to the body axis and Young's modulus of elasticity (stiffness) for compressive loading. Figure 3(a) shows the relationship

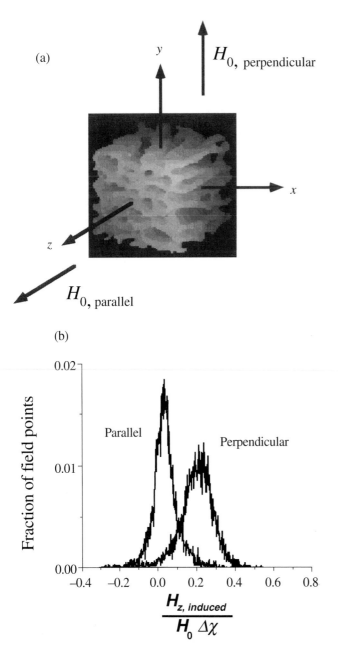

Figure 2 Induced magnetic field in trabecular bone from bovine tibia. (a) Shaded three-dimensional surface display derived from three-dimensional NMR micrograms. (b) Histogram of the induced magnetic field resulting from the bone's diamagnetism with respect to the marrow constituents

For References see p. 1367

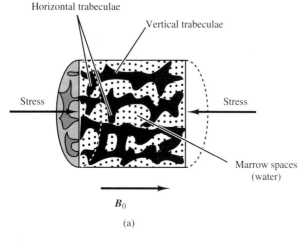

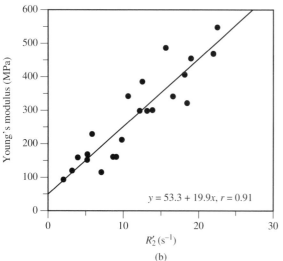

Figure 3 (a) Cross-section through a cylindrical trabecular bone specimen (schematic) used for R_2', structural, and stress analysis. The cylinder axis is parallel to the anatomic inferior–superior axis, aligned with the external field polarizing predominantly horizontal trabeculae, which cause line broadening of the proton resonances in the marrow spaces. Compressive loading is applied along the cylinder axis. (b) Young's modulus of elasticity obtained from compression tests in 22 cylindrical specimens from the lumbar vertebral bodies of 16 human subjects aged 24–86 years, plotted as a function of R_2' for the water protons in the intertrabecular spaces ($r = 0.91$, $p < 0.0001$). (Modified from Chung et al.[29])

between anatomic axis, trabecular orientation, the orientation of the applied magnetic field, and the direction of compressive loading. A strong association between stiffness and R_2' exists over a wide range of values ($r = 0.90$), corresponding to trabecular bone of very different morphologic composition Figure 3(b). From these data, it is inferred that a global measurement of R_2' in trabecular bone is able to predict the compressive strength of this highly complicated structure. Subsequently, Jergas et al. evaluated the ability of R_2^* to predict the elastic modulus for uniaxial loading in specimens of the human proximal tibia.[30] They found correlations ranging from 0.87 to 0.95 in specimens in which the marrow was removed, but

much weaker associations in another set of specimens where the measurements were conducted with the bone marrow intact.

5 IN VIVO QUANTITATIVE NMR OF TRABECULAR BONE

5.1 Measurement and Data Analysis

Bone marrow has cellular (hematopoietic) and fatty components, with the relative fractions varying widely, depending on anatomical site and age. The major chemical constituents of the two types of marrow are water and fatty acid triglycerides. This chemical heterogeneity of bone marrow complicates in vivo measurement of T_2^*. A linewidth measurement by means of image-guided localized spectroscopy has the advantage of providing T_2^* for each spectral component.[31,32] Schick, in an excellent review on bone marrow NMR in vivo, described the potential of localized spectroscopy as a means to probe osteoporosis in the calcaneus, showing dramatic reductions in the linewidth of the CH_2 lipid resonance in response to trabecular bone loss.[32]

Image-based (nonspectrally resolved) techniques, typically conducted by means of gradient echo[33] or asymmetric spin echo techniques,[34] have the advantage of providing information at multiple skeletal locations rapidly, allowing generation of maps of R_2^*[35] or R_2'.[36] By collecting an array of images with incrementally stepped time for inhomogeneity dephasing (gradient echo delay or echo offset), the pixel amplitudes can be fitted to some model for signal decay. These methods are less sensitive to magnetic field inhomogeneity arising from effects unrelated to susceptibility-induced gradients, since the field across an imaging voxel of a few cubic millimeters is, in general, quite homogeneous.

The presence of multiple chemically shifted constituents causes an amplitude modulation that has the characteristics of an interferogram. The latter can be expressed as the modulus of the vector sum of the individual phase-modulated spectral components:[33,37]

$$I(t) = \left| \sum_{i=1}^{n} \boldsymbol{I}_i \right| = \left[\sum_{i=1}^{n} \sum_{j=1}^{n} I_{0i} e^{-t/T_{2i}^*} I_{0j} e^{-t/T_{2j}^*} \cos(\Delta\omega_{ij} t) \right]^{1/2} \quad (8)$$

where I_{0i} is the initial amplitude of the ith chemically shifted constituent, $\Delta\omega_{ij}$ is the chemical shift difference in $\mathrm{rad\,s^{-1}}$ between nuclei i and j, t is the dephasing time (e.g., the echo time TE in a gradient echo), and the summation is over all spectral components n. Typically, the most abundant spectral components are those of the CH_2 protons of fatty acids and of water, which are separated by chemical shift (δ) = 3.35 ppm. It has been shown that $I(t)$ can be fitted to a two-component interferogram with T_2^* and fat and water signal amplitudes as adjustable parameters and assuming $T_{2,\,\mathrm{fat}}^* \approx T_{2,\,\mathrm{water}}^*$.[33]

Multiparameter curve fits are hampered by the difficulty of locating the global minimum, and require a relatively large number of images. One alternative is to suppress one spectral component,[35] which sacrifices some of the signal-to-noise ratio and is relatively sensitive to global magnetic field inhomogeneity, demanding that the field across the sample volume vary less than the chemical shift. Another approach to suppress the modulation is to sample the interferogram at the modulation frequency, ideally in such a manner that the two components are in phase with one another.[38] This condition is satisfied for sampling at multiples of the modulation period $T = 2\pi/\gamma\delta H_0$, which is 4.65 ms at 1.5 T field strength.

Another approach consists of deconvolving the amplitude-modulated signal with a reference signal.[39] If the marrow composition is known (e.g., all fatty, such as in most of the appendicular skeleton), the reference signal could be derived from subcutaneous fat or marrow in the diaphysis, locations where no line broadening occurs. In the time domain, deconvolution can be achieved by simple division of the marrow signal at the location of interest (trabecular bone marrow) by the reference signal, resulting in a demodulated signal that fits to a decaying exponential of rate constant R_2'. Difficulties of this approach lie in the uncertainties of the marrow composition and large-scale static field inhomogeneities, which may differ at the region of interest and the reference region.

Rather than computing T_2^* by fitting the mean signal from a two-dimensional array of pixels (often called the 'region of interest'), it is desirable to perform this calculation pixel by pixel for the generation of T_2^* maps. This, however, requires that misregistration be minimal between acquisitions, which may, for example, be achieved with a multiple echo pulse sequence that collects gradient echoes of the same polarity at multiples of the chemical shift modulation period. The precision achievable in vivo with this technique is illustrated with the data obtained from five separate scans in the same test subject in Figure 4. The data also show that chemical shift modulation is completely suppressed and that the decay is exponential to a high degree (Figure 4b). Excellent precision was

also reported by Funke et al., who used a 16-echo gradient echo pulse sequence to give a coefficient of variation of 2.5% from 11 successive scans, which included repositioning following each scan.[40]

Since variations in R_2 (for example from variations in marrow composition) could mask effects resulting from changes in trabecular architecture or density, a direct measurement of R_2' would be preferable. The asymmetric echo technique[34,35] measures R_2' but is inherently inefficient as it requires separate image acquisitions for each time increment. Recently, Ma and Wehrli reported a new multislice pulse sequence capable of measuring both R_2' and R_2 in a single scan.[36] The method, termed GESFIDE (gradient echo sampling of FID and echo), is based on sampling the descending and ascending portion of a Hahn echo with a train of gradient echoes. The transient signal before and after the phase-reversal rf pulse decays with rate constants $R_2^* = R_2 + R_2'$ and $R_2^- = R_2 - R_2'$, respectively. If, further, the time interval between successive gradient echoes is set equal to the fat–water chemical shift modulation period (see above), R_2^* and R_2^- can be determined by curve fitting and R_2 and R_2' determined algebraically. Salient features of the method are its insensitivity to rf pulse imperfections, as well as its high precision and efficiency. Figure 5 shows the evolution of the GESFIDE signal in a region where $R_2' > R_2$, along with a computed R_2' parameter image.

5.2 Dependence of R_2^* and R_2' on Image Voxel Size and Field Strength

For a perturber that is smaller than the imaging voxel but much larger than the molecular scale (also referred to as 'meso-

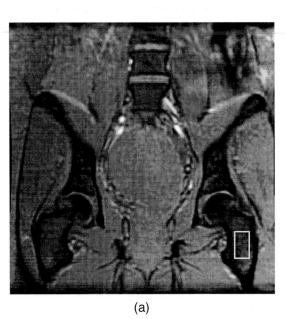

(a)

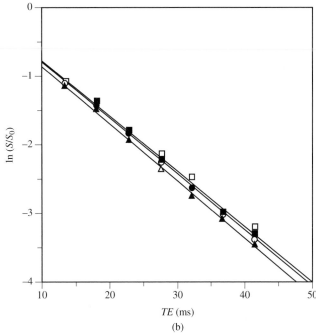

(b)

Figure 4 (a) First of a series of eight coronal gradient echo images for simultaneous measurement of T_2^* in the hip and lumbar spine, obtained by collecting 128 data samples every 4.65 ms, from a single gradient echo train so as effectively to demodulate the signal (see text for details). (b) Plot of signal measured in the trochanter [see region indicated in (a)] versus echo time, obtained from five successive scans. Solid lines are linear least-square fits, affording $T_2^* = 12.55 \pm 0.38$ ms

For References see p. 1367

(a)   (b)

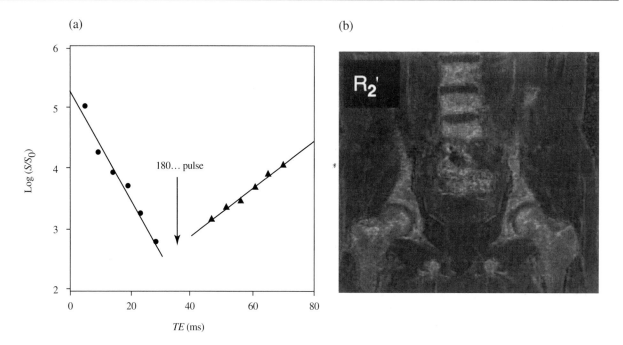

Figure 5 The GESFIDE method. (a) Evolution of the gradient echo signal from the region of interest in the greater trochanter before and after the phase-reversal rf pulse (S, signal; points obtained from five successive scans). Prior to the phase-reversal pulse, the signal evolves with a time constant $R_2^* = R_2 + R_2'$, subsequently with $R_2^- = R_2 + R_2'$. The change in the sign of the slope indicates that $R_2' \gg R_2$. (b) Computed R_2' map. Note the high intensity for structures pertaining to trabecular bone, consistent with enhanced R_2' as a result of susceptibility-induced line broadening. (Modified from Ma and Wehrli[36])

scopic' scale[41]) the susceptibility-induced line broadening should be independent of image voxel size. However, if the voxel size decreases below the typical range of gradients induced by the field perturbing trabeculae, then the likelihood of this voxel falling in a region sufficiently removed from the field gradients induced by trabeculae increases. As a consequence, the mean T_2' is expected to increase with decreasing pixel size. Majumdar and Genant studied this effect in trabecular bone of various densities.[35] They found that the T_2' histogram becomes wider and more asymmetric with decreasing pixel size. However, if the voxel size is large relative to the range of the gradients, the value of T_2' becomes independent of voxel size, which is the case for pixels on the order of 1.5–2 mm. The distribution of T_2' as a function of image resolution is shown in Figure 6.

The induced magnetic field is proportional to the polarizing static field. In the absence of diffusion (static dephasing regime) one would, therefore, expect a linear relationship between trabecular bone marrow R_2' and field strength. Parizel et al. reported the field strength dependence of R_2^* from measurement of the gradient echo signal decay at 1.5, 1.0, and 0.2 T in vertebral bone marrow in a single volunteer[42] and found R_2^* to increase with field strength. More recently, Song et al. performed a detailed study of the field strength dependence of R_2 and R_2' in the calcaneus at 1.5 and 4 T by means of the GESFIDE technique.[43] They found R_2' to scale nearly proportionally with field strength while R_2 was almost field-strength invariant. The data, which show that at 4 T R_2' accounts for nearly 90% of the total relaxation rate ($R_2^* = R_2 + R_2'$), are summarized in Table 1.

5.3 Clinical Studies in Patients with Osteoporosis

The most likely association involving R_2^* is with bone marrow density (BMD). In fact, the theory developed by Yablonskiy and Haacke predicts a linear relationship between R_2' and the volume fraction of the perturber.[25] A possible re-

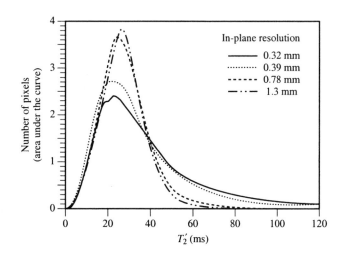

Figure 6 Smoothed T_2' histograms obtained from T_2' maps computed from axial images in the epiphysis of the distal femur, a site of dense trabeculation. As image resolution increases (smaller pixel size), the histogram broadens and becomes skewed. Means are found to vary between 41.7 ms at 0.32 mm pixels size and 25.7 ms at 1.25 mm pixel size. (Modified, with permission, from Majumdar and Genant[35])

Table 1 Field dependence of cancellous bone marrow relaxation rates

Field (T)	Relaxation rate ±SD (s⁻¹)[a]	
	R_2	R_2'
1.5	16.4 ± 1.6	65.4 ± 11.2
4.0	19.2 ± 3.4	178.4 ± 37.6

[a]SD, standard deviation from values for five subjects. R_2' (at 4 T)/R_2' (at 1.5 T) = 2.73. Source: Song et al.[43]

lationship between these parameters was suggested by the observed T_2^* shortening in the distal femur from the diaphysis (lowest trabeculation) toward the metaphysis and epiphysis of the bone (highest trabeculation).[31] Majumdar et al. determined R_2' in intact specimens of human vertebral trabecular bone after bone marrow removal and suspending the bone in saline.[44] The measurements afforded a linear correlation between R_2' and BMD (in mg cm⁻³), obtained from QCT, with a slope of 0.20 ± 0.02 s⁻¹ mg⁻¹ cm³ ($r = 0.92$). The same group extended these studies in vivo in normal volunteers to the distal radius and proximal tibia.[35] Confirming earlier work, they found both BMD and R_2' to depend on the anatomical site of measurement. In excellent agreement with in vitro data, they measured 0.20 ± 0.01 s⁻¹ mg⁻¹ cm³ ($r = 0.88$) for the combined data from both anatomic sites.

Since inception of the T_2^* method, several groups have evaluated its potential for assessing osteoporosis and have compared it with other modalities, notably DEXA and QCT.[33,40,45–47]

An early pilot study on a small group of patients with clinically established osteoporosis of the spine ($n = 12$) and an equal number of control subjects showed the former to have significantly prolonged bone marrow T_2^* in lumbar vertebra L5.[33] A larger follow-up study was designed to explore whether image-based measurements of T_2^* could provide an index of the integrity of trabecular bone as a possible criterion for predicting fracture risk. The value of R_2^* was measured in 146 non-black women at 1.5 T field strength by means of five successive gradient echoes, spaced 4.6 ms apart to minimize fat–water chemical shift modulation. Data were fitted to an exponential model and data sets with $p > 0.05$ for the Pearson correlation rejected. The control population ($n = 77$, mean age 46.6 ± 14.9 years) consisted of women with mean spinal BMD >0.9 g cm⁻² (DEXA) or >90 mg cm⁻³ (QCT), and no vertebral deformities. The patient population ($n = 59$; mean age 59.7 ± 10.2 years) was made up of women with osteoporosis of the spine, exhibiting at least one radiographic deformity of the thoracic vertebrae and/or BMD below the cutoff for controls. The extent of deformity was determined as a mean deformity index, DI_{av}. The value of R_2^* was significantly lower in osteoporosis for all L vertebrae ($p < 0.001$), except for L1. The best discriminator was the average of L3–L5 (R_{2av}^*) for which means and standard errors obtained were 64.8 ± 1.2 s⁻¹ and 53.4 ± 1.2 s⁻¹ in controls and osteoporotics, respectively ($p < 0.0001$). Both R_2^* and BMD correlated with DI_{av}, the correlation with R_2^* being slightly stronger ($r = 0.40$; $p < 0.0005$, versus $r = 0.36$; $p < 0.001$). Finally, R_{2av}^* was significantly correlated with mean BMD ($r = 0.54$; $p < 0.0001$, slope = 31.4 s⁻¹ g cm⁻², $p < 0.0001$).

Overall, these findings corroborate the results of the prior study in that subjects with osteoporosis have lower R_2^* values of their vertebral marrow and show that MR may have the potential to distinguish patients with fractures from those without this condition. Nevertheless, the results fell short of demonstrating MRI's superiority over bone densitometry.

A more recent study from the same laboratory, undertaken with more advanced methodology, involving measurements of R_2' (rather than R_2^*) at multiple skeletal sites, indicates the complementary nature of this parameter to BMD.[48] In this work, the rate constants R_2' and R_2 and the marrow fat fraction were measured in the lumbar vertebrae and proximal femur by the GESFIDE method.[36] Sixteen gradient echoes were collected, eight each before and after the phase-reversal pulse, at TE 4.6, 6.9, 9.2, 11.5, 13.8, 18.4, 23.0, and 27.6 ms at 57.0, 61.6, 66.2, 70.8, 73.1, 75.4, 77.7, and 80 ms, the latter being an rf echo. Fat and water signal amplitudes were computed from echoes 1–3 by three-point Dixon processing,[49] and from these, the volume fractions of the two constituents. The value of R_2' was moderately (positively) correlated with BMD at all sites ($r = 0.46–0.69$; $p < 0.0001$), albeit with different slopes, indicative of the different trabecular orientation relative to the static field at the various anatomical locations. For example, in the femoral neck, the slope was twice that found for the average R_2' from lumbar vertebrae 2–4 (38.0 s⁻¹ g cm⁻² versus 19.2 s⁻¹ g cm⁻²). The value of R_2' classified the subjects well at all sites, but the strength of the discrimination was greater at the femoral sites, a finding that agrees with a small patient study by Machann et al.,[50] and suggests that skeletal sites rich in yellow marrow (such as the calcaneus) are better predictors of osteoporosis than measurements in the vertebrae. This observation can be understood when the different susceptibilities of fat (i.e., yellow marrow) and water (i.e., hematopoietic marrow) are considered. It is well known that fat is less diamagnetic than water (see, for example, Hopkins and Wehrli[18]) thus resulting in a greater absolute susceptibility difference $\Delta\chi$ between bone and bone marrow, which increases the sensitivity of R_2' to variations in bone volume fraction [Equation (7)]. In fact, when the spine R_2' values are normalized by correcting for varying bone marrow composition, achieved by computing R_2' as it would be observed for an all yellow marrow composition, the slope of the R_2' versus BMD correlation increased from 19.2 to 28.4 s⁻¹ g cm⁻².

Finally, the value of R_2' in this study was found to be predictive of vertebral fracture status. The latter was defined by measuring vertebral deformities in the thoracic and lumbar spine in terms of standard deviations from normalcy and by selecting a threshold beyond which a fracture was considered present. When only a single parameter was included in the logistic regression, DEXA BMD was found to predict fracture status better than R_2', with Ward's triangle exhibiting the strongest correlation ($r^2 = 0.48$). Further, the femoral sites were more predictive than the lumbar spine, a finding that applies to both R_2' and BMD. However, combining MR and BMD significantly improved prediction. The strongest association was found for the combination of R_2' measured in the greater trochanter and BMD at the Ward's triangle, affording a 30% increase in the strength of the correlation relative to BMD alone ($r^2 = 0.62$). By contrast, the combination of multiple BMD sites did not yield stronger associations. These data are promising in that they underscore the complementary nature of

For References see p. 1367

R_2' and the potential role of structure in affecting the trabecular bone's mechanical competence.

6 HIGH-RESOLUTION IMAGING OF TRABECULAR BONE STRUCTURE

6.1 In Vitro Cancellous Bone NMR Microscopy

Whereas the measurement of the induced magnetic field inhomogeneity provides structural information indirectly, high-resolution MRI at microscopic dimensions has the potential for nondestructive mapping of three-dimensional trabecular morphology, as an alternative to conventional microscopy from sections[51] or tomographic X-ray microscopy.[52–54] Bone is well suited for imaging by NMR since it appears with background intensity and, therefore, provides excellent contrast with marrow, which has high signal intensity. Ideally, the image voxel size is smaller than a typical structural element (i.e., a trabecula), in which case partial volume blurring is minimal and the resulting histogram is bimodal, allowing segmentation (generation of a binary image) by setting the intensity threshold midway between the marrow peak and background. Clearly, the voxel size needed to satisfy this requirement depends on the trabecular width, which is species dependent. Bovine trabeculae are 200–300 μm thick, those in humans between 100 and 200 μm, whereas the trabecular thickness in rats is only 50–70 μm, consequently demanding considerably higher resolution.

Most of the work reported so far has been conducted on small-bore microimaging systems at 300 or 400 MHz proton resonance in human cadaver or biopsy specimens[55–58] or animals, at resolutions ranging from 25 to 120 μm. Ex vivo, the quality of the images is optimized by substituting the marrow with gadolinium-doped water as a means to optimize the signal-to-noise ratio. The background gradients from the susceptibility-induced fields can cause artifacts in the form of signal loss from intravoxel phase dispersion [Equation (1)]. Since these phase losses are recoverable—assuming diffusion to be negligible—spin echo detection is advantageous. Alternatively, the dephasing time should be minimal; this can, for example, be achieved with projection–reconstruction techniques, as applied in microimaging of lung parenchyma.[59]

At marginal resolution the accuracy of the 3D structures derived from NMR images can be improved with such techniques as subvoxel tissue classification.[60] Based on a statistical model for the noise and partial volume averaging, the number of subvoxels containing marrow can be determined and the most likely spatial arrangement ascertained using probabilistic arguments for the interaction between adjacent subvoxels. Chung et al. developed algorithms for the measurement of the structural parameters of interest (bone volume fraction, mean trabecular thickness, and mean trabecular plate separation[55] based on two-dimensional images. These techniques, however, are merely the digital imaging adaptations of stereology, the methodology practiced by the histomorphometrist who performs measurements on anatomic sections which are then extrapolated to the third dimension.[61] However, trabecular bone is inherently three-dimensional and anisotropic. Figure 7 shows three orthogonal images from bovine trabecular bone illustrating the bone's anisotropic structure, along with a surface projection image enhanced by subvoxel classification.[60]

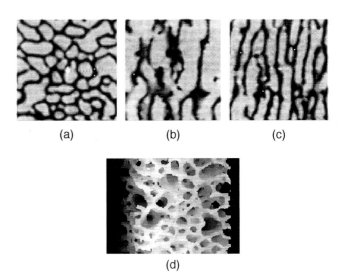

(a)   (b)   (c)

(d)

Figure 7 Images at microscopic dimension showing trabecular structure: (a)–(c) orthogonal views obtained from a 64^3 array of 3D spin echo array of images acquired at 9.4 T on a specimen of trabecular bone from bovine tibia (114 μm $\times$ 114 μm $\times$ 139 μm voxel size); (a) transverse and (b), (c) longitudinal sections. Note the preferential orientation of the trabeculae along the inferior–superior direction. (d) Shaded surface display of the same array of data, resolution-enhanced by means of a subvoxel tissue classification technique using Bayesian segmentation[60]

6.2 Relationship between Architecture and Strength

Knowledge of the three-dimensional architecture of cancellous bone allows one of the key questions to be addressed, i.e., whether structure is predictive of the bone's strength. While still controversial, it is known that apparent density (essentially bone volume fraction, i.e., the amount of bone present per unit volume) predicts anywhere from 40–80% of Young's modulus (and thus ultimate strength); the remainder is generally attributed to architecture.[62]

To investigate whether architecture is predictive of the modulus of elasticity for compressive loading, Hwang et al. acquired three-dimensional images at 78 μm isotropic resolution of cadaver specimens cored from the ultradistal radius, after the cores were tested nondestructively in compression along the bone's anatomic axis. Rather than segmenting the images, bone volume fraction (BVF) images were generated by fitting the histogram to a two-peaked noiseless histogram convolved with Rician noise, using maximum likelihood methods. In this manner the true BVF can be found for each pixel as the probability of finding bone at that location. This idea was then extended to two-point probabilities, for example the probability of finding bone at two neighboring locations x_i and x_{i+n} along a row of voxels, $P(x_i, x_{i+n}) = \text{BVF}(x_I)\text{BVF}(x_{i+n})$. Averaging the two-point probabilities over all locations yields the spatial auto-correlation function (AFC). Because of the quasiregular nature of the trabecular lattice, the AFC has a maximum at the average spacing between trabeculae along that direction. This concept can be extended and other parameters characterizing the trabecular network defined. One useful parameter is tubularity, which quantifies how 'tubular' the bone is along its

	Male 76 years	Male 80 years	Male 53 years
Bone volume fraction	0.16	0.14	0.12
Young's modulus (MPa)	316	676	543

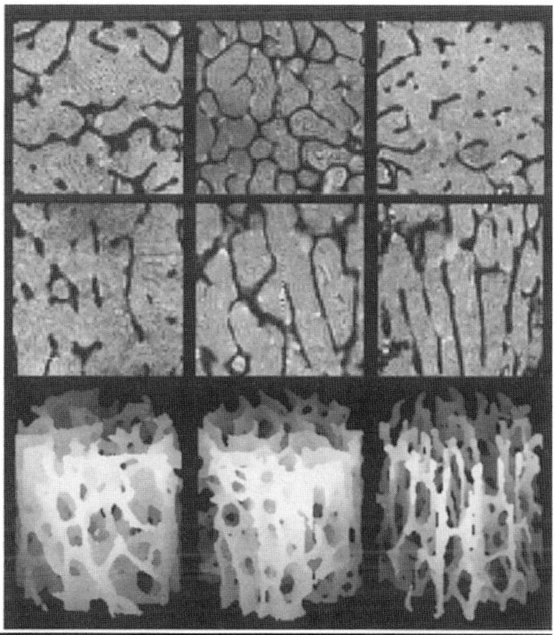

Figure 8 NMR images of three samples of trabecular bone illustrating its structural variability. Note that the sample from the 76-year-old man has the lowest modulus in spite of having the highest bone volume fraction. However, the bone of the 80- and 50-year-old donors is considerably more tubular, which explains their greater Young's modulus.

anatomic axis, which is the direction along the radius (z). Tubularity is given as $AFC_z(1)/AFC(0)$, where $AFC_z(1)$ is the average of the product of BVF in corresponding voxels of successive slices. The significance of this parameter in determining cancellous bone strength is illustrated with Figure 8 showing cross-sectional and projection images of three radius specimens of varying BVF. It is noteworthy that the specimen with the highest BVF is only half as strong as the two other specimens that are both more tubular. A detailed analysis on specimens from 23 cadavers of widely varying BVF, tubularity, and longitudinal spacing (spacing between trabeculae orthogonal to the direction of loading) jointly best predicted Young's modulus, accounting for over 90% of its variance (as opposed to BVF, which explained only 50%). Clearly, a perfectly tubular structure could not fail by loading in the direction of these trabeculae. However, as the deviation from such an ideal geometry decreases, the transverse trabeculae, acting as cross-ties, become increasingly important.

6.3 In Vivo Microimaging

In vivo imaging of the microarchitecture of cancellous bone at voxel sizes sufficient to resolve individual trabeculae is considerably more challenging than measurements on small specimens in vitro. The main difficulty is to achieve sufficient signal to noise ratio in scan times tolerated by the patient, typi-

cally on the order of 10 min. A second problem is subject movement, which, even on a minute scale on the order of a pixel (100–200 μm), can cause blurring and artifacts, precluding derivation of accurate structural parameters. Restraining the portion of the body being scanned is usually not sufficient to prevent motional blurring. However, the incorporation of navigator echoes into the pulse sequence proves to be effective.[63] Additional echoes, which are not phase-encoded, generate projections of the object from which the displacement can be measured and a commensurate phase correction applied to the k-space data. The effect of alternate navigator echoes incorporated into a three-dimensional spin echo sequence was found to eliminate motion degradation effectively (Figure 9).[43]

The third problem is the considerably lower resolution achievable in vivo, which, while sufficient to visualize the trabecular architecture, demands more sophisticated methods for image restoration and analysis. In spite of these difficulties, a growing body of literature has accumulated during the past few years, highlights the potential of in vivo micro-MRI to probe bone structure.[64–68]

The image voxel sizes reported for cancellous bone imaging in humans range from 2×10^{-5} mm^3 in the distal phalanx of the middle finger[69] to 2×10^{-2} mm^3 in the calcaneus.[68] At these resolutions, the histogram is monomodal (instead of bimodal as it is at higher resolution) and, therefore, does not provide simple criteria for segmenting the images into bone

For References see p. 1367

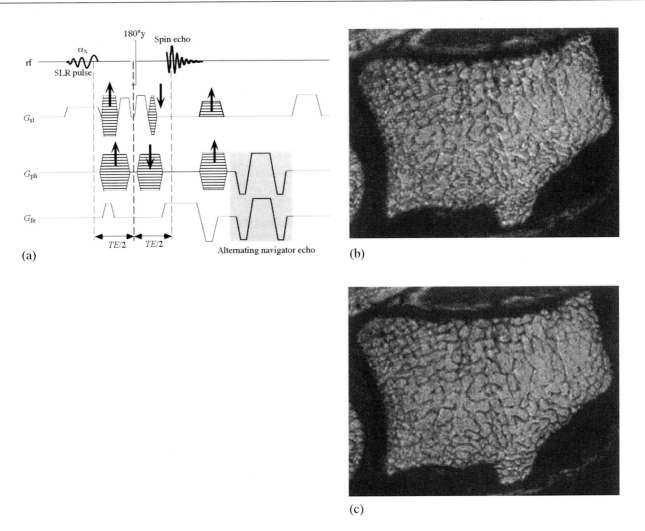

Figure 9 (a) FLASE (fast large-angle spin echo) sequence with navigator echoes alternating between frequency and phase-encoding axes (137 $\mu m^3 \times 137$ $\mu m^3 \times 350$ μm^3 voxel size). The zeroeth moment of all three gradients are nulled before the navigators are acquired. Also, to maintain steady state, the navigator gradients are refocused after collecting motion data. (b) Image of the distal radius from a 56-year-old female at this voxel size, before motion correction. (c) Image after motion correction. (With permission from Song et al.[43])

and bone marrow. The limitations imposed by marginal resolution to segment the images and derive structural parameters by image processing has been addressed in various ways. Majumdar et al. set the threshold at half the peak maximum (observed for cortical bone) in the grayscale-inverted histogram.[65] In recent work on the calcaneus, Link et al. compared structural measures in postmenopausal subjects who had suffered hip fractures with those in age-matched controls.[68] They found that both apparent trabecular bone volume and apparent trabecular separation were stronger predictors of hip fracture than BMD of the proximal femur. Gordon et al. applied an adaptive threshold to eliminate errors from intensity variations caused by the inhomogeneous reception profile of the surface coil, by thresholding the image against the low-pass filtered version of itself, followed by a second threshold at 50% of the maximum.[66] From the threshold and subsequent skeleton images they determined hole size and a connectivity index and found these parameters to be correlated with subject age (positively and negatively).

Whereas the above methods cannot retrieve the true BVF, they, nevertheless, can provide parameters that can effectively characterize the trabecular network and provide clinically relevant information. True BVF images have been obtained in vivo in the distal radius by means of a histogram deconvolution technique.[70] In brief, a model histogram is convolved with Rician noise (the type of noise encountered in magnitude NMR images[71]) and the result compared with the observed histogram. The resulting error then serves as input to improve the noiseless histogram, and the process is repeated until the error falls below a predetermined level. Finally, the actual noiseless image is computed using both intensity and connectivity arguments. Wehrli et al. obtained BVF images in this manner in the distal radius of a small cohort of patients with and without vertebral fractures.[67] The parameters that proved to be successful in predicting the elastic modulus in vitro were also evaluated to explore whether architecture is associated with the degree of vertebral deformities, measured in terms of a fracture index, D_{fract}. Whereas none of the individual structural measures was found to correlate with the D_{fract}, a highly significant relationship ($R = 0.78$; $p = 0.0016$) was found between D_{fract} and a function of tubularity and longitudinal spacing. These data further emphasize the role of architecture in determining trabecular bone's resistance to fracture, and highlight the prospects of in vitro micro-MRI as a noninvasive modality to assess trabecular microstructure.

7 RELATED ARTICLES

Lung and Mediastinum: A Discussion of the Relevant NMR Physics; Susceptibility and Diffusion Effects in NMR Microscopy; Susceptibility Effects in Whole Body Experiments.

8 REFERENCES

1. J. Wolff, 'Das Gesetz der Transformation der Knochen', A. Hirschwald, Berlin, 1892.
2. H. Roesler, *J. Biomech.*, 1987, **20**, 1025.
3. B. Riggs and L. Melton, 'Osteoporosis', Raven Press, New York, 1988.
4. C. E. Brown, J. R. Allaway, K. L. Brown, and H. Battocletti, *Clin. Chem.*, 1987, **33**, 227.
5. J. L. Ackerman, D. P. Raleigh, and M. J. Glimcher, *Magn. Reson. Med.*, 1992, **25**, 1.
6. J. Moore, L. Garrido, and J. Ackerman, *Magn. Reson. Med.*, 1995, **33**, 293.
7. Y. Wu, M. Glimcher, C. Rey, and J. Ackerman, *J. Mol. Biol.*, 1994, **244**, 423.
8. Y. Wu, J. L. Ackerman, D. A. Chesler, J. Li, R. M. Neer, J. Wang, and M. J. Glimcher, *Calcif. Tissue Int.*, 1998, **62**, 512.
9. J. Glasel and K. Lee, *J. Am. Chem. Soc.*, 1974, **96**, 970.
10. C. A. Davis, H. K. Genant, and J. S. Dunham, *Invest. Radiol.*, 1986, **21**, 472.
11. H. Rosenthal, K. R. Thulborn, D. I. Rosenthal, S. H. Kim, and B. R. Rosen, *Invest. Radiol.*, 1990, **25**, 173.
12. F. W. Wehrli, J. C. Ford, D. A. Gusnard, and J. Listerud, *Proc. VIIIth Annu. Mtg Soc. Magn. Reson. Med.*, Amsterdam, 1989, **1**, 217.
13. D. C. Ailion, T. A. Case, D. Blatter, A. H. Morris, A. Cutillo, C. H. Durney, and S. A. Johnson, *Bull. Magn. Reson.*, 1984, **6**, 130.
14. K. K. Kwong, J. W. Belliveau, D. A. Chesler, I. Goldberg, R. Weisskoff, B. Poncelet, D. N. Kennedy, B. Hoppel, M. Cohen, R. Turner, H. Cheng, T. Brady, and B. Rosen, *Proc. Natl. Acad. Sci. USA*, 1992, **89**, 5675.
15. S. Ogawa, D. W. Tank, R. Menon, J. M. Ellerman, S. G. Kim, H. Merkle, and K. Ugurbil, *Proc. Natl. Acad. Sci. USA*, 1992, **89**, 5951.
16. P. C. Lauterbur, B. V. Kaufman, and M. K. Crawford, in 'Biomolecular Structure and Function', ed. P. F. Agris. Academic Press, New York, 1978, p. 343.
17. S. Chu, Y. Xu, J. Balschi, and C. Springer, *Magn. Reson. Med.*, 1990, **13**, 239.
18. J. A. Hopkins and F. W. Wehrli, *Magn. Reson. Med.*, 1997, **37**, 494.
19. A. Morrish, 'The Physical Principles of Magnetism', Robert E. Krieger, Malabar, FL, 1983.
20. J. C. Ford, F. W. Wehrli, and H. W. Chung, *Magn. Reson. Med.*, 1993, **30**, 373.
21. L. J. Gibson, *J. Biomech.*, 1985, **18**, 317.
22. K. Engelke, S. Majumdar, and H. K. Genant, *Magn. Reson. Med.*, 1994, **31**, 380.
23. S. Hwang and F. Wehrli, *J. Magn. Reson. B*, 1995, **109**, 126.
24. H. Chung and F. W. Wehrli, *Proc. XIIth Annu. Mtg Soc. Magn. Reson. Med.*, New York, 1993, **1**, 138.
25. D. A. Yablonskiy and E. M. Haacke, *Magn. Reson. Med.*, 1994, **32**, 749.
26. D. A. Yablonskiy, W. R. Reinus, E. M. Haacke, and H. Stark, *Magn. Reson. Med.*, 1996, **37**, 214.
27. K. Selby, S. Majumdar, D. C. Newitt, and H. K. Genant, *J. Magn. Reson. Imaging*, 1996, **6**, 549.
28. L. Mosekilde, *Bone*, 1990, **11**, 67.
29. H. Chung, F. W. Wehrli, J. L. Williams, and S. D. Kugelmass, *Proc. Natl. Acad. Sci. USA*, 1993, **90**, 10250.
30. M. Jergas, S. Majumdar, J. Keyak, I. Lee, D. Newitt, S. Grampp, H. Skinner, and H. Genant, *J. Comput. Assist. Tomogr.*, 1995, **19**, 472.
31. J. C. Ford and F. W. Wehrli, *Magn. Reson. Med.*, 1991, **17**, 543.
32. F. Schick, *Prog. Nucl. Magn. Reson. Spect.*, 1996, **29**, 169.
33. F. W. Wehrli, J. C. Ford, M. Attie, H. Y. Kressel, and F. S. Kaplan, *Radiology*, 1991, **179**, 615.
34. G. L. Wismer, R. B. Buxton, B. R. Rosen, C. Fisel, R. Oot, T. Brady, and K. Davis, *J. Comput. Assist. Tomogr.*, 1988, **12**, 259.
35. S. Majumdar and H. K. Genant, *JMRI*, 1992, **2**, 209.
36. J. Ma and F. W. Wehrli, *J. Magn. Reson. B*, 1996, **111**, 61.
37. F. W. Wehrli, T. G. Perkins, A. Shimakawa, and F. Roberts, *Magn. Reson. Imag.*, 1987, **5**, 157.
38. J. C. Ford and F. W. Wehrli, *JMRI*, 1992, **2**(P), 103.
39. F. W. Wehrli, J. Ma, J. A. Hopkins, and H. K. Song, *J. Magn. Reson.*, 1998, **131**, 61.
40. M. Funke, H. Bruhn, R. Vosshenrich, O. Rudolph, and E. Grabbe, *Fortschr. Geb. Rontgen. (Neuen Bildgeb Verfahr)*, 1994, **161**, 58.
41. D. A. Yablonskiy, *Magn. Reson. Med.*, 1997, **39**, 417.
42. P. M. Parizel, B. van Riet, B. A. van Hasselt, J. W. Van Goethem, L. van den Hauwe, H. A. Dijkstra, P. J. van Wiechem, and A. M. De Schepper, *J. Comput. Assist. Tomogr.*, 1995, **19**, 465.
43. H. K. Song, F. W. Wehrli, and J. Ma, *JMRI*, 1997, **7**, 382.
44. S. Majumdar, D. Thomasson, A. Shimakawa, and H. K. Genant, *Magn. Reson. Med.*, 1991, **22**, 111.
45. H. Sugimoto, T. Kimura, and T. Ohsawa, *Invest. Radiol.*, 1993, **28**, 208.
46. F. W. Wehrli, J. C. Ford, and J. G. Haddad, *Radiology*, 1995, **196**, 631.
47. S. Grampp, S. Majumdar, M. Jergas, D. Newitt, P. Lang, and H. K. Genant, *Radiology*, 1996, **198**, 213.
48. F. W. Wehrli, J. A. Hopkins, H. K. Song, S. N. Hwang, J. D. Haddad, and P. J. Snyder, *Proc. VIth Annu. Mtg (Int.) Soc. Magn. Reson. Med.*, Sydney, 1998, p. 456.
49. G. H. Glover and E. Schneider, *Magn. Reson. Med.*, 1991, **18**, 371.
50. J. Machann, F. Schick, D. Seitz, et al, *Proc. IVth Annu. Mtg (Int.) Soc. Magn. Reson. Med.*, New York, 1996, **2**, 1093.
51. A. Odgaard, K. Andersen, F. Melsen, and H. J. Gundersen, *J. Microsc.*, 1990, **159**, 335.
52. L. A. Feldkamp. S. A. Goldstein, A. M. Parfitt, G. Jesion, and M. Kleerekoper, *J. Bone Miner. Res.*, 1989, **4**, 3.
53. J. H. Kinney, N. E. Lane, and D. L. Haupt, *J. Bone Miner. Res.*, 1995, **10**, 264.
54. P. Rüegsegger, B. Koller, and R. Muller, *Calcif. Tissue Int.*, 1996, **58**, 24.
55. H. W. Chung, F. W. Wehrli, J. L. Williams, S. D. Kugelmass, and S. L. Wehrli, *Proc. Natl. Acad. Sci. USA*, 1995, **10**, 803.
56. H. W. Chung, F. W. Wehrli, J. L. Williams, and S. L. Wehrli, *J. Bone Miner. Res.*, 1995, **10**, 1452.
57. S. N. Hwang, F. W. Wehrli, and J. L. Williams, *Med. Phys.*, 1997, **24**, 1255.

For References see p. 1367

58. M. Wessels, R. P. Mason, P. P. Antich, J. E. Zerwekh, and C. Y. Pak, *Med. Phys.*, 1997, **24**, 1409.

59. S. L. Gewalt, G. H. Glover, L. W. Hedlund, G. P. Cofer, J. R. MacFall, and G. A. Johnson, *Magn. Reson. Med.*, 1993, **29**, 99.

60. Z. Wu, H. Chung, and F. W. Wehrli, *Magn. Reson. Med.*, 1994, **31**, 302.

61. A. M. Parfitt, in 'Bone Histomorphometry: Techniques and Interpretation', ed. R. R. Recker, CRC Press, Boca Raton, FL, 1981, p. 53.

62. M. J. Ciarelli, S. A. Goldstein, J. L. Kuhn, D. D. Cody, and M. B. Brown, *J. Orthop. Res.*, 1991, **9**, 674.

63. R. L. Ehman and J. P. Felmlee, *Radiology*, 1989, **173**, 255.

64. S. Majumdar, D. Newitt, M. Jergas, A. Gies, E. Chiu, D. Osman, J. Keltner, J. Keyak, and H. Genant, *Bone*, 1995, **17**, 417.

65. S. Majumdar, H. K. Genant, S. Grampp, D. C. Newitt, V.-H. Truong, J. C. Lin, and A. Mathur, *J. Bone Miner. Res.*, 1997, **12**, 111.

66. C. L. Gordon, C. E. Webber, N. Christoforou, and C. Nahmias, *Med. Phys.*, 1997, **24**, 585.

67. F. W. Wehrli, S. N. Hwang, J. Ma, H. K. Song, J. C. Ford, and J. G. Haddad, *Radiology*, 1998, **207**, 833; erratum in *Radiology*, 1998, **206**, 347.

68. T. M. Link, S. Majumdar, P. Augat, J. C. Lin, D. Newitt, Y. Lu, N. E. Lane, and H. K. Genant, *J. Bone Miner. Res.*, 1998, **13**, 1175.

69. H. Jara, F. W. Wehrli, H. Chung, and J. C. Ford, *Magn. Reson. Med.*, 1993, **29**, 528.

70. S. N. Hwang and F. W. Wehrli, *Int. J. Imaging Syst. Technol.*, 1999, **10**, 186.

71. H. Gubdjartsson and S. Patz, *Magn. Reson. Med.*, 1995, **34**, 910.

Biographical Sketch

Felix W. Wehrli. *b* 1941. M.S., 1967, Ph.D., 1970, chemistry, Swiss Federal Institute of Technology, Switzerland. NMR application scientist, Varian AG, 1970–79; Executive Vice President, Bruker Instruments, Billerica, 1979–82; NMR Application Manager, General Electric Medical Systems, Milwaukee, 1982–88. Currently Professor of Radiologic Science and Biophysics, University of Pennsylvania Medical School. Editor-in-Chief of *Magnetic Resonance in Medicine* 1991–present. Over 100 publications. Current research specialty: NMR imaging of connective tissues, biomaterials, specifically trabecular bone.

Subject Index

Note: Figures and Tables are indicated (in this index) by *italic page numbers*; abbreviations are those listed in end-papers (of this book)